Probability and its Applications

Published in association with the Applied Probability Trust

Editors: S. Asmussen, J. Gani, P. Jagers, T.G. Kurtz

Probability and its Applications

Azencott et al.: Series of Irregular Observations. Forecasting and Model Building. 1986

Bass: Diffusions and Elliptic Operators. 1997

Bass: Probabilistic Techniques in Analysis. 1995

Berglund/Gentz: Noise-Induced Phenomena in Slow-Fast Dynamical Systems: A Sample-Paths Approach. 2006

Biagini/Hu/Øksendal/Zhang: Stochastic Calculus for Fractional Brownian Motion and Applications. 2008

Chen: Eigenvalues, Inequalities and Ergodic Theory. 2005

Costa/Fragoso/Marques: Discrete-Time Markov Jump Linear Systems. 2005

Daley/Vere-Jones: An Introduction to the Theory of Point Processes I: Elementary Theory and Methods. 2nd ed. 2003, corr. 2nd printing 2005

Daley/Vere-Jones: An Introduction to the Theory of Point Processes II: General Theory and Structure. 2nd ed. 2008

de la Peña/Gine: Decoupling: From Dependence to Independence, Randomly Stopped Processes, U-Statistics and Processes, Martingales and Beyond. 1999

de la Peña/Lai/Shao: Self-Normalized Processes. 2009

Del Moral: Feynman-Kac Formulae. Genealogical and Interacting Particle Systems with Applications. 2004

Durrett: Probability Models for DNA Sequence Evolution. 2002, 2nd ed. 2008

Ethier: The Doctrine of Chances. Probabilistic Aspects of Gambling. 2010

Feng: The Poisson–Dirichlet Distribution and Related Topics. 2010

Galambos/Simonelli: Bonferroni-Type Inequalities with Equations. 1996

Gani (ed.): The Craft of Probabilistic Modelling. A Collection of Personal Accounts. 1986

Gut: Stopped Random Walks. Limit Theorems and Applications. 1987

Guyon: Random Fields on a Network. Modeling, Statistics and Applications. 1995

Kallenberg: Foundations of Modern Probability. 1997, 2nd ed. 2002

Kallenberg: Probabilistic Symmetries and Invariance Principles. 2005

Last/Brandt: Marked Point Processes on the Real Line. 1995

Molchanov: Theory of Random Sets. 2005

Nualart: The Malliavin Calculus and Related Topics, 1995, 2nd ed. 2006

Rachev/Rueschendorf: Mass Transportation Problems. Volume I: Theory and Volume II: Applications. 1998

Resnick: Extreme Values, Regular Variation and Point Processes. 1987

Schmidli: Stochastic Control in Insurance. 2008

Schneider/Weil: Stochastic and Integral Geometry. 2008

Shedler: Regeneration and Networks of Queues. 1986

Silvestrov: Limit Theorems for Randomly Stopped Stochastic Processes. 2004

Thorisson: Coupling, Stationarity and Regeneration. 2000

Stewart N. Ethier

The Doctrine of Chances

Probabilistic Aspects of Gambling

 Springer

Stewart N. Ethier
Department of Mathematics
University of Utah
155 South 1400 East
Salt Lake City UT 84112-0090
USA
ethier@math.utah.edu

Series Editors:

Søren Asmussen
Department of Mathematical Sciences
Aarhus University
Ny Munkegade
8000 Aarhus C
Denmark
asmus@imf.au.dk

Peter Jagers
Mathematical Statistics
Chalmers University of Technology
and Göteborg (Gothenburg) University
412 96 Göteborg
Sweden
jagers@chalmers.se

Joe Gani
Centre for Mathematics and its Applications
Mathematical Sciences Institute
Australian National University
Canberra, ACT 0200
Australia
gani@maths.anu.edu.au

Thomas G. Kurtz
Department of Mathematics
University of Wisconsin - Madison
480 Lincoln Drive
Madison, WI 53706-1388
USA
kurtz@math.wisc.edu

ISSN 1431-7028
ISBN 978-3-540-78782-2   e-ISBN 978-3-540-78783-9
DOI 10.1007/978-3-540-78783-9
Springer Heidelberg Dordrecht London New York

Library of Congress Control Number: 2010927487

Mathematics Subject Classification (2010): 60-02, 91A60, 60G40, 60C05

Cover design: WMXDesign

Printed on acid-free paper

Springer is part of Springer Science+Business Media (www.springer.com)

Preface

I have found many thousands more readers than I ever looked for. I have no right to say to these, You shall not find fault with my art, or fall asleep over my pages; but I ask you to believe that this person writing strives to tell the truth. If there is not that, there is nothing.

William Makepeace Thackeray, *The History of Pendennis*

This is a monograph/textbook on the probabilistic aspects of gambling, intended for those already familiar with probability at the post-calculus, pre-measure-theory level.

Gambling motivated much of the early development of probability theory (David 1962).[1] Indeed, some of the earliest works on probability include Girolamo Cardano's [1501–1576] *Liber de Ludo Aleae* (*The Book on Games of Chance*, written c. 1565, published 1663), Christiaan Huygens's [1629–1695] "De ratiociniis in ludo aleae" ("On reckoning in games of chance," 1657), Jacob Bernoulli's [1654–1705] *Ars Conjectandi* (*The Art of Conjecturing*, written c. 1690, published 1713), Pierre Rémond de Montmort's [1678–1719] *Essay d'analyse sur les jeux de hasard* (*Analytical Essay on Games of Chance*, 1708, 1713), and Abraham De Moivre's [1667–1754] *The Doctrine of Chances* (1718, 1738, 1756). Gambling also had a major influence on 20th-century probability theory, as it provided the motivation for the concept of a martingale.

Thus, gambling has contributed to probability theory. Conversely, probability theory has contributed much to gambling, from the gambler's ruin formula of Blaise Pascal [1623–1662] to the optimality of bold play due to Lester E. Dubins [1920–2010] and Leonard J. Savage [1917–1971]; from the solution of le her due to Charles Waldegrave to the solution of chemin de fer due to John G. Kemeny [1926–1992] and J. Laurie Snell [1925–]; from the duration-of-play formula of Joseph-Louis Lagrange [1736–1813] to the optimal proportional betting strategy of John L. Kelly, Jr. [1923–1965]; and from

---

[1] See Maistrov (1974, Chapter 1, Section 2) for a different point of view.

the first evaluation of the banker's advantage at trente et quarante due to Siméon-Denis Poisson [1781–1840] to the first published card-counting system at twenty-one due to Edward O. Thorp [1932–]. Topics such as these are the principal focus of this book.

Is gambling a subject worthy of academic study? Let us quote an authority from the 18th century on this question. In the preface to *The Doctrine of Chances*, De Moivre (1718, p. iii) wrote,

> Another use to be made of this Doctrine of Chances is, that it may serve in Conjunction with the other parts of the Mathematicks, as a fit introduction to the Art of Reasoning; it being known by experience that nothing can contribute more to the attaining of that Art, than the consideration of a long Train of Consequences, rightly deduced from undoubted Principles, of which this Book affords many Examples.

We also quote a 20th-century authority on the same question. In *Le jeu, la chance et le hasard*, Louis Bachelier [1870–1946] (1914, p. 6) wrote,[2]

> It is almost always gambling that enables one to form a fairly clear idea of a manifestation of chance; it is gambling that gave birth to the calculus of probability; it is to gambling that this calculus owes its first faltering utterances and its most recent developments; it is gambling that allows us to conceive of this calculus in the most general way; it is, therefore, gambling that one must strive to understand, but one should understand it in a philosophic sense, free from all vulgar ideas.

Certainly, there are other applications of probability theory on which courses of study could be based, and some of them (e.g., actuarial science, financial engineering) may offer better career prospects than does gambling! But gambling is one of the only applications in which the probabilistic models are often *exactly* correct.[3] This is due to the fundamental simplicity of the nature of the randomness in games of chance. This simplicity translates into an elegance that few other applications enjoy.

The book consists of two parts. Part I ("Theory") begins with a review of probability, then turns to several probability topics that are often not covered in a first course (conditional expectation, martingales, and Markov chains), then briefly considers game theory, and finally concludes with various gambling topics (house advantage, gambler's ruin, betting systems, bold play, optimal proportional play, and card theory). Part II ("Applications") discusses a variety of casino games, including six games in which successive coups are independent (slot machines, roulette, keno, craps, house-banked poker, and video poker) and four games with dependence among coups (faro, baccarat, trente et quarante, and twenty-one). Within each group, chapters are ordered according to difficulty but are largely independent of one another and can be read in any order. We conclude with a discussion of poker, which is in a class by itself.

---

[2] Translation from Dubins and Savage (1976).

[3] Here, and throughout the book (perhaps with the exception of Section 13.2), we model the ideal, or *benchmark*, game, the game as it is intended to be played by the manufacturer of the dice, cards, wheels, machines, etc.

The only contemporary book with comparable content and prerequisites is Richard A. Epstein's [1927–] *The Theory of Gambling and Statistical Logic* (1967, 1977, 2009). Epstein's book is fun to read but is not entirely suitable as a textbook: It is a compendium of results, often without derivations, and there are few problems or exercises to reinforce the reader's understanding. Our aim was not only to supply the missing material but to provide more-self-contained and more-comprehensive coverage of the principal topics. We have tried to do this without sacrificing the "fun to read" factor.

Although there is enough material here for a two-semester course, the book could be used for a one-semester course, either by covering some subset of the chapters thoroughly (perhaps assigning other chapters as individual projects) or by covering every chapter less than thoroughly. In an NSF-sponsored Research Experience for Undergraduates (REU) summer program at the University of Utah in 2005, we adopted the latter approach using a preliminary draft of the book. Fred M. Hoppe, in a course titled "Probability and Games of Chance" at McMaster University in spring 2009, adopted the former approach, covering Chapters 1, 2, 17, 3, 15, and 6 in that order.

The book is not intended solely for American and Canadian readers. Money is measured in units, not dollars, and European games, such as chemin de fer and trente et quarante, are studied. This is appropriate, inasmuch as France is not only the birthplace of probability theory but also that of roulette, faro, baccarat, trente et quarante, and twenty-one.

With few exceptions, all random variables in the book are discrete.[4] This allows us to provide a mathematically rigorous treatment, while avoiding the need for measure theory except for occasional references to the Appendix. Each chapter contains a collection of problems that range from straightforward to challenging. Some require computing. Answers, but not solutions, will be provided at the author's web page (http://www.math.utah.edu/~ethier/). While we have not hesitated to use computing in the text (in fact, it is a necessity in studying such topics as video poker, twenty-one, and Texas hold'em), we have avoided the use of computer simulation, which seems to us outside the spirit of the subject. Each chapter also contains a set of historical notes, in which credit is assigned wherever possible and to the best of our knowledge. This has necessitated a lengthy bibliography. In many cases we simply do not know who originated a particular idea, so a lack of attribution should not be interpreted as a claim of originality. We frequently refer to the generic gambler, bettor, player, dealer, etc. with the personal pronoun "he," which has the old-fashioned interpretation of, but is much less awkward than, "he or she."

A year or two ago (2008) was the tercentenary of the publication of the first edition of Montmort's *Analytical Essay on Games of Chance*, which can

---

[4] The only exceptions are nondiscrete limits of sequences of discrete random variables. These may occur, for example, in the martingale convergence theorem.

be regarded as the first published full-length book on probability theory.[5] As Todhunter (1865, Article 136) said of Montmort,

> In 1708 he published his work on Chances, where with the courage of Columbus he revealed a new world to mathematicians.

A decade later De Moivre published his equally groundbreaking work, *The Doctrine of Chances*. Either title would be suitable for the present book; we have chosen the latter because it sounds a little less intimidating.

Acknowledgments: I am grateful to Nelson H. F. Beebe for technical advice and assistance and to Davar Khoshnevisan for valuable discussions. Portions of the book were read by Patrik Andersson, R. Michael Canjar, Anthony Curtis, Anirban DasGupta, Persi Diaconis, Marc Estafanous, Robert C. Hannum, Fred M. Hoppe, Robert Muir, Don Schlesinger, and Edward O. Thorp, as well as by several anonymous reviewers for Springer and AMS. I thank them for their input. A fellowship from the Center for Gaming Research at the University of Nevada, Las Vegas, allowed me to spend a month in the Special Collections room of the UNLV Lied Library, and the assistance of the staff is much appreciated.

I would also like to thank several others who helped in various ways during the preparation of this book. These include David R. Bellhouse, István Berkes, Mr. Cacarulo, Renzo Cavalieri, Bob Dancer, Régis Deloche, Edward G. Dunne, Marshall Fey, Carlos Gamez, Susan E. Gevaert, James Grosjean, Norm Hellmers, Thomas M. Kavanagh, David A. Levin, Basil Nestor, Marina Reizakis, Michael W. Shackleford, Larry Shepp, Arnold Snyder, George Stamos, and Zenfighter.

Finally, I am especially grateful to my wife, Kyoko, for her patience throughout this lengthy project.

Dedication: The book is dedicated to the memory of gambling historian Russell T. Barnhart [1926–2003] and twenty-one theorist Peter A. Griffin [1937–1998], whom I met in 1984 and 1981, respectively. Their correspondence about gambling matters over the years fills several thick folders (neither used e-mail), and their influence on the book is substantial.

Salt Lake City, December 2009 Stewart N. Ethier

---

[5] Cardano's *Liber de Ludo Aleae* comprises only 15 (dense) pages of his *Opera omnia* and Huygens's "De ratiociniis in ludo aleae" comprises only 18 pages of van Schooten's *Exercitationum Mathematicarum*.

Contents

List of Notation

symbol	meaning	page of first use
♠	end of proof or end of example[1]	6
:=	equals by definition (=: is also used)	3
≡	is identically equal to, or is congruent to	59
≈	is approximately equal to	9
~	is asymptotic to ($a_n \sim b_n$ if $\lim_{n\to\infty} a_n/b_n = 1$)	48
$\mathbf{N}$	the set of positive integers	58
$\mathbf{Z}_+$	the set of nonnegative integers	91
$\mathbf{Z}$	the set of integers	58
$\mathbf{Q}$	the set of rational numbers	108
$\mathbf{R}$	the set of real numbers	21
$|x|$	absolute value of real x; modulus of complex x	6
x^+	nonnegative part of the real number x ($:= \max(x,0)$)	22
x^-	nonpositive part of the real number x ($:= -\min(x,0)$)	64
$\boldsymbol{x} \cdot \boldsymbol{y}$	inner product of $\boldsymbol{x}, \boldsymbol{y} \in \boldsymbol{R}^d$	34
$|\boldsymbol{x}|$	Euclidean norm of vector $\boldsymbol{x} \in \boldsymbol{R}^d$ ($:= (\boldsymbol{x} \cdot \boldsymbol{x})^{1/2}$)	177
$\boldsymbol{A}^{\mathsf{T}}$	transpose of matrix (or vector) $\boldsymbol{A}$	176
$0.4\overline{92}$	$0.4929292\cdots$ (repeating decimal expansion)	19
$\text{sgn}(x)$	sign of x ($:= 1, 0, -1$ if $x > 0$, $= 0$, < 0)	170
$|A|$	cardinality of (number of elements of) the finite set A	3
$x \in A$	x is an element of A	3
$A \subset B$	A is a (not necessarily proper) subset of B	3
$B \supset A$	equivalent to $A \subset B$	13

[1]In discussions of card games, the symbol ♠ signifies a spade.

symbol	meaning	page of first use
$A \cup B$	union of A and B	11
$A \cap B$	intersection of A and B	11
A^c	complement of A	11
$A - B$	set-theoretic difference $(:= A \cap B^c)$	11
$A \times B$	cartesian product of A and B	15
A^n	n-fold cartesian product $A \times \cdots \times A$	25
1_A	indicator r.v. of event A or indicator function of set A	33
$x \vee y$	$\max(x, y)$	56
$x \wedge y$	$\min(x, y)$	6
$\ln x$	natural (base e) logarithm of x	30
$\log_2 x$	base-2 logarithm of x	48
$\lfloor x \rfloor$	the greatest integer less than or equal to x	39
$\lceil x \rceil$	the least integer greater than or equal to x	30
$\varnothing$	empty set	12
$(n)_k$	$:= n(n-1) \cdots (n-k+1)$ if $k \geq 1$, $(n)_0 := 1$	4
$n!$	n factorial $(:= (n)_n = n(n-1) \cdots 1$ if $n \geq 1$, $0! := 1)$	4
$\binom{n}{k}$	binomial coefficient n choose k $(:= (n)_k / k!)$	4
$\binom{n}{n_1, \ldots, n_r}$	multinomial coefficient $(:= n!/(n_1! \cdots n_r!))$	5
$\delta_{i,j}$	Kronecker delta $(:= 1$ if $i = j$, $:= 0$ otherwise$)$	20
(a, b)	open interval $\{x : a < x < b\}$	22
$[a, b)$	half-open interval $\{x : a \leq x < b\}$	27
$(a, b]$	half-open interval $\{x : a < x \leq b\}$	140
$[a, b]$	closed interval $\{x : a \leq x \leq b\}$	21
$\xrightarrow{d}$	converges in distribution to	42
$N(0, 1)$	standard-normal distribution	42
$\Phi(x)$	standard-normal cumulative distribution function	42
$\phi(x)$	standard-normal density function $(:= \Phi'(x))$	63
a.s.	almost surely	42
i.o.	infinitely often	43
i.i.d.	independent and identically distributed	43
g.c.d.	greatest common divisor	137

Positive, negative, increasing, and decreasing are used only in the strict sense. When the weak sense is intended, we use nonnegative, nonpositive, nondecreasing, and nonincreasing, respectively.

Part I
Theory

Chapter 1
Review of Probability

Mr. Arthur Pendennis did not win much money in these transactions with Mr.
Bloundell, or indeed gain good of any kind except a knowledge of the odds at hazard,
which he might have learned out of books.

William Makepeace Thackeray, *The History of Pendennis*

The reader is assumed to be familiar with basic probability, and here we
provide the definitions and theorems, without proofs, for easy reference. We
restrict our attention to discrete random variables but not necessarily to
discrete sample spaces. A number of examples are worked out in detail, and
problems are provided for those who need additional review.

1.1 Combinatorics and Probability

The set Ω (omega) of all possible outcomes of a random experiment is called
the *sample space*. Let us first consider the case in which Ω is finite. Let $n \geq 2$,
let $\Omega = \{o_1, o_2, \ldots, o_n\}$, let $p_1, p_2, \ldots, p_n$ be positive numbers that sum to 1,
and assign probability p_i to outcome o_i for $i = 1, 2, \ldots, n$. An *event* E is a
subset of Ω, and the *probability* of an event $E \subset \Omega$ is defined to be the sum
of the probabilities of the outcomes in E:

$$\mathrm{P}(E) := \sum_{1 \leq i \leq n:\, o_i \in E} p_i. \tag{1.1}$$

This leads to possibly the oldest result in probability theory:

Theorem 1.1.1. *Under the assumption that all outcomes in a finite sample
space Ω are equally likely, the probability of an event $E \subset \Omega$ is given by*

$$\mathrm{P}(E) = \frac{|E|}{|\Omega|}, \tag{1.2}$$

S.N. Ethier, *The Doctrine of Chances*, Probability and its Applications,
DOI 10.1007/978-3-540-78783-9_1, © Springer-Verlag Berlin Heidelberg 2010

where $|E|$ denotes the cardinality of (or the number of outcomes in) E.

The only difficulty in applying Theorem 1.1.1 is in counting the numbers of outcomes in E and in Ω. This can often be done with the help of *combinatorial analysis*, with which the next seven theorems are concerned.

Theorem 1.1.2. *Consider a task that requires completing r subtasks in order, where $r \geq 2$. Suppose there are n_1 ways to complete the first subtask; no matter which way is chosen, there are n_2 ways to complete the second subtask; no matter which ways are chosen for the first two subtasks, there are n_3 ways to complete the third subtask; ... no matter which ways are chosen for the first $r - 1$ subtasks, there are n_r ways to complete the rth subtask. Then there are $n_1 n_2 \cdots n_r$ ways to complete the task.*

We define

$$n! := n(n - 1) \cdots 2 \cdot 1 \tag{1.3}$$

for each positive integer n. The symbol $n!$ is read "n factorial." It will be convenient to define $0! := 1$.

Theorem 1.1.3. *The number of permutations of n distinct items taken k at a time (i.e., the number of ways to choose k out of n distinct items, taking the order in which the items are chosen into account) is*

$$(n)_k := n(n - 1) \cdots (n - k + 1) = \frac{n!}{(n - k)!}, \tag{1.4}$$

assuming $1 \leq k \leq n$.

Notice that $n! = (n)_n$. Thus, $n!$ is the number of permutations of n distinct items. It will be convenient to define $(n)_0 := 1$ for each nonnegative integer n.

Theorem 1.1.4. *The number of combinations of n distinct items taken k at a time (i.e., the number of ways to choose k out of n distinct items, without regard to the order in which the items are chosen) is*

$$\binom{n}{k} := \frac{(n)_k}{k!} = \frac{n!}{k! \, (n - k)!}, \tag{1.5}$$

assuming $0 \leq k \leq n$.

The symbol $\binom{n}{k}$ is read "n choose k" and is called a *binomial coefficient*. It is useful to be aware that

$$\binom{n}{0} = \binom{n}{n} = 1, \qquad \binom{n}{1} = \binom{n}{n-1} = n, \qquad \binom{n}{k} = \binom{n}{n-k}. \tag{1.6}$$

Another useful identity is the one that generates Pascal's triangle, namely

$$\binom{n + 1}{k} = \binom{n}{k - 1} + \binom{n}{k} \tag{1.7}$$

Table 1.1 The first seven rows of Pascal's triangle. Row $n+1$ contains the $n+1$ binomial coefficients $\binom{n}{0}, \ldots, \binom{n}{n}$.

$$
\begin{array}{ccccccccccccc}
&&&&&& 1 &&&&&& \\
&&&&& 1 && 1 &&&&& \\
&&&& 1 && 2 && 1 &&&& \\
&&& 1 && 3 && 3 && 1 &&& \\
&& 1 && 4 && 6 && 4 && 1 && \\
& 1 && 5 && 10 && 10 && 5 && 1 & \\
1 && 6 && 15 && 20 && 15 && 6 && 1
\end{array}
$$

for $1 \le k \le n$. See Table 1.1 for Pascal's triangle. The reason that the quantities $\binom{n}{k}$ are called binomial coefficients is that they appear as coefficients in the *binomial theorem*:

Theorem 1.1.5. *For all real a and b and positive integers n,*

$$
(a+b)^n = \sum_{k=0}^{n} \binom{n}{k} a^k b^{n-k}, \tag{1.8}
$$

where $0^0 := 1$.

Theorem 1.1.6. *The number of ways in which to partition a set of n distinct items into r specified subsets, the first having $n_1 \ge 0$ elements, the second having $n_2 \ge 0$ elements, $\ldots$, the rth having $n_r \ge 0$ elements, where $n_1 + n_2 + \cdots + n_r = n$, is*

$$
\binom{n}{n_1, n_2, \ldots, n_r} := \frac{n!}{n_1! \, n_2! \cdots n_r!}. \tag{1.9}
$$

This is also the number of permutations of n items of r distinct types, with n_1 of the first type, n_2 of the second type, $\ldots$, n_r of the rth type.

The quantities $\binom{n}{n_1, \ldots, n_r}$ are called *multinomial coefficients* because they appear as coefficients in the *multinomial theorem*:

Theorem 1.1.7. *For all real $a_1, a_2, \ldots, a_r$ and positive integers n,*

$$
(a_1 + \cdots + a_r)^n = \sum_{n_1 \ge 0, \ldots, n_r \ge 0: \, n_1 + \cdots + n_r = n} \binom{n}{n_1, \ldots, n_r} a_1^{n_1} \cdots a_r^{n_r}, \tag{1.10}
$$

where $0^0 := 1$.

Of course, the special case of the multinomial theorem in which $r = 2$ coincides with the binomial theorem, and multinomial coefficients with $r = 2$

are usually written as binomial coefficients. For example, $\binom{52}{5}$ is preferred to $\binom{52}{5,47}$.

Theorem 1.1.8. *The number of terms in the sum in (1.10) is $\binom{n+r-1}{r-1}$ or, equivalently, $\binom{n+r-1}{n}$. This is also the number of ways to distribute n indistinguishable balls into r specified urns.*

Of course, the number of ways to distribute n *distinguishable* balls into r specified urns is r^n by Theorem 1.1.2.

Example 1.1.9. *Two-dice totals* When rolling a pair of indistinguishable dice (e.g., two red dice), the number of distinguishable outcomes is $\binom{2+6-1}{2} = 21$ by Theorem 1.1.8, but these outcomes are not equally likely. On the other hand, when rolling a pair of distinguishable dice (e.g., one red die and one green die), the number of distinguishable outcomes is $6 \cdot 6 = 36$ by Theorem 1.1.2, and these outcomes *are* equally likely. We list them in Table 1.2, together with the dice totals and their probabilities.

Table 1.2 The results of tossing two distinguishable dice.

outcomes	total	probability
$(1,1)$	2	$1/36$
$(1,2),(2,1)$	3	$2/36$
$(1,3),(2,2),(3,1)$	4	$3/36$
$(1,4),(2,3),(3,2),(4,1)$	5	$4/36$
$(1,5),(2,4),(3,3),(4,2),(5,1)$	6	$5/36$
$(1,6),(2,5),(3,4),(4,3),(5,2),(6,1)$	7	$6/36$
$(2,6),(3,5),(4,4),(5,3),(6,2)$	8	$5/36$
$(3,6),(4,5),(5,4),(6,3)$	9	$4/36$
$(4,6),(5,5),(6,4)$	10	$3/36$
$(5,6),(6,5)$	11	$2/36$
$(6,6)$	12	$1/36$

We conclude that the probability π_j of rolling a total of $j \in \{2,3,4,\ldots,12\}$ is given by

$$\pi_j = \frac{(j-1) \wedge (13-j)}{36} = \frac{6 - |j-7|}{36}, \qquad (1.11)$$

a formula that will be cited frequently in the sequel. Clearly, the probabilities (1.11) are not affected by the colors of the dice, so (1.11) is equally valid for a pair of indistinguishable dice.   ♠

Example 1.1.10. *Poker hands.* Poker is played with a standard 52-card deck. By such a deck we mean that each card is described by its *denomination*, namely 2, 3, 4, 5, 6, 7, 8, 9, 10, J (jack), Q (queen), K (king), or A (ace), and its *suit*, namely ♣ (club), ♢ (diamond), ♡ (heart), or ♠ (spade). Sometimes we will denote denomination 10 by T to avoid two-digit numbers. A poker hand consists of five cards. A hand is said to be *in sequence* if its denominations consist (after rearrangement if necessary) of 5-4-3-2-A or 6-5-4-3-2 or ... or A-K-Q-J-T. Notice that the ace plays a special role, in that it can appear as either the low card or the high card in a hand that is in sequence. A *straight flush* contains five cards in sequence and of the same suit, a *flush* contains five cards of the same suit but not in sequence, and a *straight* contains five cards in sequence but not of the same suit. A *royal flush* is an ace-high straight flush.

To describe the other types of hands, we let d_0, d_1, d_2, d_3, d_4 denote, respectively, the numbers of denominations in a hand represented $0, 1, 2, 3, 4$ times, and we note that

$$d_0 + d_1 + d_2 + d_3 + d_4 = 13, \qquad d_1 + 2d_2 + 3d_3 + 4d_4 = 5. \qquad (1.12)$$

We define $\boldsymbol{d} := (d_0, d_1, d_2, d_3, d_4)$ to be the *denomination multiplicity vector* of the hand. A hand is ranked *four of a kind* if $\boldsymbol{d} = (11, 1, 0, 0, 1)$, *full house* if $\boldsymbol{d} = (11, 0, 1, 1, 0)$, *three of a kind* if $\boldsymbol{d} = (10, 2, 0, 1, 0)$, *two pair* if $\boldsymbol{d} = (10, 1, 2, 0, 0)$, *one pair* if $\boldsymbol{d} = (9, 3, 1, 0, 0)$, and *no pair* if $\boldsymbol{d} = (8, 5, 0, 0, 0)$ and if the five cards are neither in sequence nor of the same suit. For example, the hand consisting of A♠-A♣-8♠-8♣-9♢ has two denominations (A and 8) represented twice and one denomination (9) represented once; the remaining 10 denominations are not represented. Thus, $d_2 = 2$, $d_1 = 1$, and $d_0 = 10$, and we see that this hand is ranked two pair.

The probability that a randomly dealt five-card poker hand has denomination multiplicity vector $\boldsymbol{d}$ (equal to $(11, 1, 0, 0, 1)$, $(11, 0, 1, 1, 0)$, $(10, 2, 0, 1, 0)$, $(10, 1, 2, 0, 0)$, $(9, 3, 1, 0, 0)$, or $(8, 5, 0, 0, 0)$) is given by

$$\frac{\binom{13}{d_0, d_1, d_2, d_3, d_4} \prod_{i=1}^{4} \binom{4}{i}^{d_i}}{\binom{52}{5}}. \qquad (1.13)$$

The multinomial coefficient is the number of ways to choose the hand's denominations, while the product of powers of binomial coefficients is the number of ways to choose the suits for the chosen denominations. (We have omitted the $i = 0$ term in the product because it is unnecessary. The $i = 4$ term is also unnecessary, but we have included it for clarity.) See Table 1.3 for the numerators of these probabilities. By a separate argument, the case $\boldsymbol{d} = (8, 5, 0, 0, 0)$ must be broken down into the four ranks straight flush, flush, straight, and no pair. ♠

In Example 1.1.9 we counted outcomes by *enumeration*, that is, by creating a list. In Example 1.1.10 we counted outcomes with the help of combinatorial analysis. Certainly, the latter approach is more elegant. However, the former approach is sometimes the only viable method. The next example illustrates this point.

Table 1.3 The five-card poker-hand frequencies. For each expression that is the product of two factors, the first is the number of ways to choose the hand's denominations, and the second is the number of ways to choose the suits for the chosen denominations.

rank	number of ways	
straight flush*	$\binom{10}{1}\binom{4}{1}$	40
four of a kind	$\binom{13}{11,1,0,0,1}\left[\binom{4}{1}\binom{4}{4}\right]$	624
full house	$\binom{13}{11,0,1,1,0}\left[\binom{4}{2}\binom{4}{3}\right]$	3,744
flush	$\left[\binom{13}{5}-\binom{10}{1}\right]\binom{4}{1}$	5,108
straight	$\binom{10}{1}\left[\binom{4}{1}^5-\binom{4}{1}\right]$	10,200
three of a kind	$\binom{13}{10,2,0,1,0}\left[\binom{4}{1}^2\binom{4}{3}\right]$	54,912
two pair	$\binom{13}{10,1,2,0,0}\left[\binom{4}{1}\binom{4}{2}^2\right]$	123,552
one pair	$\binom{13}{9,3,1,0,0}\left[\binom{4}{1}^3\binom{4}{2}\right]$	1,098,240
no pair	$\left[\binom{13}{5}-\binom{10}{1}\right]\left[\binom{4}{1}^5-\binom{4}{1}\right]$	1,302,540
sum	$\binom{52}{5}$	2,598,960

*including royal flush

Example 1.1.11. *Twenty-one-dealer sequences.* For the purposes of this example, we need to know only a few of the rules of twenty-one, or blackjack. We assume that the game is dealt from a single standard 52-card deck. Aces have value 1 or 11 as specified below, court cards (J, Q, K) have value 10, and every other card has value equal to its nominal value. Suits are irrelevant. The dealer receives two cards initially (one face up) and additional cards one at a time as needed to achieve a total of 17 or greater. The first ace has value 11 unless that would result in a total greater than 21, in which case it has value 1. Every subsequent ace has value 1. A total that includes an ace valued as 11 is called a *soft total*; every other total is called a *hard total*. For example, if the dealer is dealt $(A, 5)$, he then has a soft total of 16 and requires another card. If his third card is a 6, he then has a hard total of 12 and requires another card. If his fourth card is a 7, he then has a hard total of 19, which is his final total.

Let us define a *twenty-one-dealer sequence* to be a finite sequence $a_1, \ldots, a_k$ of positive integers, none of which exceeds 10, and at most four of which are equal to 1, at most four of which are equal to 2, and so on, such that k is the smallest integer $j \geq 2$ for which

$$a_1 + \cdots + a_j \geq 17 \tag{1.14}$$

or

$$1 \in \{a_1, \ldots, a_j\} \quad \text{and} \quad 7 \leq a_1 + \cdots + a_j \leq 11. \tag{1.15}$$

Observe that (1.14) signifies a hard total of $a_1 + \cdots + a_j$ and that, since 1s play the role of aces, (1.15) signifies a soft total of $a_1 + \cdots + a_j + 10$. Clearly, the order of the terms is crucial: $8, 8, 10$ is a twenty-one-dealer sequence but $10, 8, 8$ and $8, 10, 8$ are not. In general, if $a_1, \ldots, a_k$ is a twenty-one-dealer sequence, then its length k satisfies $2 \leq k \leq 10$.

How many twenty-one-dealer sequences are there? We do not know how to answer this question using combinatorial analysis. Therefore, we resort to the crude but effective method of enumerating all such sequences. By ordering them in reverse-lexicographical order, we ensure that no sequence is overlooked. The list is displayed in Table 1.4, and we see that the answer to our question is 48,532.

Although the twenty-one-dealer sequences are obviously not equally likely, we can nevertheless apply Theorem 1.1.1 to find the probability of each such sequence. Letting

$$k_j := |\{1 \leq i \leq k : a_i = j\}|, \qquad j = 1, 2, \ldots, 10, \tag{1.16}$$

we find that the probability of the twenty-one-dealer sequence $a_1, a_2, \ldots, a_k$ is

$$\frac{(4)_{k_1}(4)_{k_2} \cdots (4)_{k_9}(16)_{k_{10}}}{(52)_k}. \tag{1.17}$$

Here the random experiment consists merely of dealing out k cards in succession.

We now regard these 48,532 sequences as the outcomes of a random experiment and use Table 1.4 and (1.1) to find the probabilities of the various possible dealer final totals. The totals of interest are 17, 18, 19, 20, and 21, with 22–26 collectively describing a dealer *bust*. Further, a two-card 21 (a *natural*) should be distinguished from a 21 comprising three or more cards.

For example,

$$P(\text{dealer has two-card 21}) = P(10, 1) + P(1, 10)$$

$$= 2\frac{(4)_1(16)_1}{(52)_2} = \frac{32}{663} \approx 0.048265460. \tag{1.18}$$

The remaining cases require the use of a computer, and results are displayed in Table 1.5. ♠

We have limited our attention so far to finite sample spaces, but this is far too restrictive. We could extend (1.1) to countably infinite sample spaces, but even that is too restrictive. (Consider, for example, the random

Table 1.4 A partial list of the 48,532 twenty-one-dealer sequences, in reverse lexicographical order. (Rules: single deck, dealer stands on soft 17.)

seq. no.	sequence	total	probability
1	$10, 10$	20	$(16)_2/(52)_2$
2	$10, 9$	19	$(4)_1(16)_1/(52)_2$
3	$10, 8$	18	$(4)_1(16)_1/(52)_2$
4	$10, 7$	17	$(4)_1(16)_1/(52)_2$
5	$10, 6, 10$	26	$(4)_1(16)_2/(52)_3$
6	$10, 6, 9$	25	$(4)_1(4)_1(16)_1/(52)_3$
$\vdots$			
286	$10, 2, 1, 1, 1, 1, 3$	19	$(4)_4(4)_1(4)_1(16)_1/(52)_7$
287	$10, 2, 1, 1, 1, 1, 2$	18	$(4)_4(4)_2(16)_1/(52)_7$
288	$10, 1$	21	$(4)_1(16)_1/(52)_2$
289	$9, 10$	19	$(4)_1(16)_1/(52)_2$
290	$9, 9$	18	$(4)_2/(52)_2$
$\vdots$			
15,110	$4, 2, 2, 2, 2, 1, 1, 1, 1, 4$	20	$(4)_4(4)_4(4)_2/(52)_{10}$
15,111	$4, 2, 2, 2, 2, 1, 1, 1, 1, 3$	19	$(4)_4(4)_4(4)_1(4)_1/(52)_{10}$
15,112	$4, 2, 2, 2, 1$	21	$(4)_1(4)_3(4)_1/(52)_5$
15,113	$4, 2, 2, 1$	19	$(4)_1(4)_2(4)_1/(52)_4$
15,114	$4, 2, 1$	17	$(4)_1(4)_1(4)_1/(52)_3$
15,115	$4, 1, 10, 10$	25	$(4)_1(4)_1(16)_2/(52)_4$
15,116	$4, 1, 10, 9$	24	$(4)_1(4)_1(4)_1(16)_1/(52)_4$
$\vdots$			
42,532	$2, 1, 1, 1, 1, 6, 2, 2, 3$	19	$(4)_4(4)_3(4)_1(4)_1/(52)_9$
42,533	$2, 1, 1, 1, 1, 6, 2, 2, 2$	18	$(4)_4(4)_4(4)_1/(52)_9$
42,534	$2, 1, 1, 1, 1, 5$	21	$(4)_4(4)_1(4)_1/(52)_6$
42,535	$2, 1, 1, 1, 1, 4$	20	$(4)_4(4)_1(4)_1/(52)_6$
42,536	$2, 1, 1, 1, 1, 3$	19	$(4)_4(4)_1(4)_1/(52)_6$
42,537	$2, 1, 1, 1, 1, 2$	18	$(4)_4(4)_2/(52)_6$
42,538	$1, 10$	21	$(4)_1(16)_1/(52)_2$
42,539	$1, 9$	20	$(4)_1(4)_1/(52)_2$
$\vdots$			
48,527	$1, 1, 1, 1, 2, 6, 2, 2, 3$	19	$(4)_4(4)_3(4)_1(4)_1/(52)_9$
48,528	$1, 1, 1, 1, 2, 6, 2, 2, 2$	18	$(4)_4(4)_4(4)_1/(52)_9$
48,529	$1, 1, 1, 1, 2, 5$	21	$(4)_4(4)_1(4)_1/(52)_6$
48,530	$1, 1, 1, 1, 2, 4$	20	$(4)_4(4)_1(4)_1/(52)_6$
48,531	$1, 1, 1, 1, 2, 3$	19	$(4)_4(4)_1(4)_1/(52)_6$
48,532	$1, 1, 1, 1, 2, 2$	18	$(4)_4(4)_2/(52)_6$

experiment consisting of an infinite sequence of coin tosses.) Therefore, we take an axiomatic approach in what follows. We begin by introducing the required definitions.

Table 1.5 The probabilities of the twenty-one dealer's various final totals, rounded to nine decimal places. (Rules: single deck, dealer stands on soft 17.)

dealer total	no. of sequences	probability
17	5,134	.145 829 659
18	5,243	.138 063 176
19	5,433	.134 820 214
20	5,455	.175 806 476
21^3	5,433	.073 629 613
21^2	2	.048 265 460
bust	21,832	.283 585 403
sum	48,532	1.000 000 000

[3]three or more cards   [2]two cards (natural)

We define the *union* of two events E and F by

$$E \cup F := \{o \in \Omega : o \in E \text{ or } o \in F \text{ (or both)}\} \tag{1.19}$$

and the *intersection* by

$$E \cap F := \{o \in \Omega : o \in E \text{ and } o \in F\}. \tag{1.20}$$

The *complement* of E is

$$E^c := \{o \in \Omega : o \notin E\}. \tag{1.21}$$

We will also occasionally use

$$F - E := F \cap E^c. \tag{1.22}$$

We can extend the binary operations, union and intersection, to finite or countably infinite collections of events. Given events $E_1, E_2, \ldots$, their *union* is given by

$$E_1 \cup E_2 \cup \cdots = \bigcup_{i=1}^{\infty} E_i := \{o \in \Omega : o \in E_i \text{ for some } i\}, \tag{1.23}$$

and their *intersection* is given by

$$E_1 \cap E_2 \cap \cdots = \bigcap_{i=1}^{\infty} E_i := \{ o \in \Omega : o \in E_i \text{ for every } i \}. \qquad (1.24)$$

Incidentally, unions, intersections, and complements apply also to arbitrary sets (not just events) and will occasionally be used in that way. The operations (1.21), (1.23), and (1.24) are related by *De Morgan's laws*:

$$\left(\bigcup_{i=1}^{\infty} E_i \right)^c = \bigcap_{i=1}^{\infty} E_i^c, \qquad \left(\bigcap_{i=1}^{\infty} E_i \right)^c = \bigcup_{i=1}^{\infty} E_i^c. \qquad (1.25)$$

Given events $E_1, E_2, \ldots$, we say that they are *mutually exclusive* (or *pairwise disjoint*) if no two of them can occur simultaneously, that is, if $E_i \cap E_j = \varnothing$ for all $i \neq j$. We can now state the four *axioms of probability*:

Axiom 1.1.12. *The collection of events contains the sample space Ω and is closed under complementation and under countable unions.*

Axiom 1.1.13. $P(E) \geq 0$ *for every event E.*

Axiom 1.1.14. *If $E_1, E_2, \ldots$ are mutually exclusive events, then*

$$P\left(\bigcup_{i=1}^{\infty} E_i \right) = \sum_{i=1}^{\infty} P(E_i). \qquad (1.26)$$

Axiom 1.1.15. $P(\Omega) = 1$.

When Ω is finite or countably infinite, it is possible to define the collection of events to be the collection of all subsets of Ω. But if Ω is uncountable, such a definition leads to complications, so instead we simply adopt Axiom 1.1.12. By that axiom and De Morgan's laws, the collection of events is also closed under countable intersections. Axiom 1.1.14 is called *countable additivity*. If we take $E_1 = E_2 = \cdots = \varnothing$ in Axiom 1.1.14 and use Axiom 1.1.15, we find that $P(\varnothing) = 0$. It follows that countable additivity implies *finite additivity*: If $n \geq 2$ and $E_1, E_2, \ldots, E_n$ are mutually exclusive events, then

$$P\left(\bigcup_{i=1}^{n} E_i \right) = \sum_{i=1}^{n} P(E_i). \qquad (1.27)$$

Notice that our definition (1.1), with every subset $E \subset \Omega$ being an event, satisfies the axioms.

These axioms allow us to establish several useful theorems, the first of which is concerned with the monotonicity of probability.

Theorem 1.1.16. *If E and F are events with $E \subset F$, then $P(E) \leq P(F)$, and in fact $P(F - E) = P(F) - P(E)$.*

Corollary 1.1.17. $P(E^c) = 1 - P(E)$ *for every event E.*

This simple corollary is called the *complementation law*. The next result is known as the *inclusion-exclusion law*. It generalizes the familiar formula

$$P(E_1 \cup E_2) = P(E_1) + P(E_2) - P(E_1 \cap E_2). \tag{1.28}$$

Theorem 1.1.18. *Given events* $E_1, E_2, \ldots, E_n$, *define*

$$S_1 := \sum_{i=1}^{n} P(E_i), \qquad S_2 := \sum\sum_{1 \leq i < j \leq n} P(E_i \cap E_j), \tag{1.29}$$

and so on. More generally, for $1 \leq m \leq n$, *define*

$$S_m := \sum \cdots \sum_{1 \leq i_1 < i_2 < \cdots < i_m \leq n} P(E_{i_1} \cap E_{i_2} \cap \cdots \cap E_{i_m}). \tag{1.30}$$

Then

$$P(E_1 \cup E_2 \cup \cdots \cup E_n) = \sum_{m=1}^{n} (-1)^{m-1} S_m. \tag{1.31}$$

The next result contains several inequalities related to the inclusion-exclusion law. The first is often called *Boole's inequality*.

Theorem 1.1.19. *Under the assumptions and notation of Theorem 1.1.18,*

$$P(E_1 \cup E_2 \cup \cdots \cup E_n) \leq S_1, \tag{1.32}$$

$$P(E_1 \cup E_2 \cup \cdots \cup E_n) \geq S_1 - S_2, \tag{1.33}$$

$$P(E_1 \cup E_2 \cup \cdots \cup E_n) \leq S_1 - S_2 + S_3, \tag{1.34}$$

$$P(E_1 \cup E_2 \cup \cdots \cup E_n) \geq S_1 - S_2 + S_3 - S_4, \tag{1.35}$$

and so on.

The last theorem in this section is the first that requires countable additivity.

Theorem 1.1.20. (a) *Given a sequence of events satisfying* $E_1 \subset E_2 \subset \cdots$,

$$P\left(\bigcup_{i=1}^{\infty} E_i\right) = \lim_{i \to \infty} P(E_i). \tag{1.36}$$

(b) *Given a sequence of events satisfying* $E_1 \supset E_2 \supset \cdots$,

$$P\left(\bigcap_{i=1}^{\infty} E_i\right) = \lim_{i \to \infty} P(E_i). \tag{1.37}$$

Finally, we generalize the first inequality in Theorem 1.1.19. The result is called *countable subadditivity*.

Corollary 1.1.21. *Given an arbitrary sequence of events* $E_1, E_2, \ldots$,

$$P\left(\bigcup_{i=1}^{\infty} E_i\right) \le \sum_{i=1}^{\infty} P(E_i).\tag{1.38}$$

Example 1.1.22. *Méré's problem.* In 1654, the Chevalier de Méré raised the question of whether the probability of at least one six in four tosses of a single die is equal to the probability of at least one double six in 24 tosses of a pair of dice. We can easily evaluate both probabilities using the complementation law and Theorems 1.1.1 and 1.1.2. The probability of at least one six in four tosses of a single die is

$$1 - P(\text{no sixes in four tosses of a single die})$$

$$= 1 - \frac{5^4}{6^4} = \frac{671}{1{,}296} \approx 0.517747,\tag{1.39}$$

while the probability of at least one double six in 24 tosses of a pair of dice is

$$1 - P(\text{no double sixes in 24 tosses of a pair of dice})$$

$$= 1 - \frac{(35)^{24}}{(36)^{24}} \approx 0.491404.\tag{1.40}$$

(The second probability is the ratio of two 38-digit integers, but it does not seem useful to display them.) Méré had predicted the nonequality of the two probabilities based on empirical evidence.   ♠

Example 1.1.23. *Rencontre.* The game of rencontre ("encounter" or "coincidence" in French) has been studied by Montmort, De Moivre, Laplace, Euler, and others. There are several versions of this game, but the one described by Montmort, or actually his simplification of it, is as follows. Consider a deck of n distinct cards, which for convenience we will assume are labeled $1, 2, \ldots, n$. For specificity, we also label the positions of the cards in the deck as follows: With the cards face down, the top card in the deck is in position 1, the second card is in position 2, and so on. The cards are well shuffled and cut, and then dealt out one by one. The dealer is said to win if, for some $j \in \{1, 2, \ldots, n\}$, the card labeled j is in position j. What is the probability P_n that the dealer wins?

For $j = 1, 2, \ldots, n$, let E_j be the event that the card labeled j is in position j. The problem is to evaluate $P_n := P(E_1 \cup E_2 \cup \cdots \cup E_n)$. We use the inclusion-exclusion law. If $1 \le m \le n$ and $1 \le i_1 < i_2 < \cdots < i_m \le n$, then

$$P(E_{i_1} \cap E_{i_2} \cap \cdots \cap E_{i_m}) = \frac{(n-m)!}{n!},\tag{1.41}$$

and hence, for $m = 1, 2, \ldots, n$, S_m of (1.30) is given by

$$S_m = \binom{n}{m} \frac{(n-m)!}{n!} = \frac{1}{m!}.\tag{1.42}$$

We conclude from Theorem 1.1.18 that

$$P_n = \mathrm{P}(E_1 \cup E_2 \cup \cdots \cup E_n) = \sum_{m=1}^{n} \frac{(-1)^{m-1}}{m!} = 1 - \sum_{m=0}^{n} \frac{(-1)^m}{m!}. \qquad (1.43)$$

Notice that P_n converges rapidly to $1 - e^{-1} \approx 0.632120559$ as $n \to \infty$. ♠

Probabilities are frequently stated in terms of odds, and it is occasionally necessary to convert from one to the other. Given an event E, to say that the *odds against* E are β to α (or the *odds in favor of* E are α to β) simply means that $\mathrm{P}(E) = \alpha/(\alpha + \beta)$. Here α and β are positive numbers (typically, but not necessarily, integers). Notice that the odds factors α and β can be scaled arbitrarily, that is, both can be multiplied by the same positive number without effect. For example, to say that the odds against E are β to α is equivalent to saying that they are β/α to 1.

The odds just defined are often referred to as the *true odds*, to distinguish them from the payoff odds. Suppose that an event E offers *payoff odds* of β to α (briefly, E pays β to α), and that the bettor stakes 1 unit on E. If E occurs, he wins β/α units, otherwise he loses 1 unit. In the case of a win, the casino returns his stake of 1 unit together with his profit of β/α units, for a total of $(\alpha + \beta)/\alpha$ units. In particular, the payoff odds of β *to* α are sometimes stated as $\alpha + \beta$ *for* α. Here α and β are positive numbers (typically, but not necessarily, integers). Again, notice that the odds factors α and β can be scaled arbitrarily. For example, to say that an event pays β to α is equivalent to saying that it pays β/α to 1. If an event pays 1 to 1, it is said to pay *even money*.

Consider, for example, a single number (zero, say) on an unbiased 37-number roulette wheel. The probability that zero will occur at the next coup is $1/37$, so the odds against zero occurring are 36 to 1. However, the payoff odds for the occurrence of zero are only 35 to 1, which can also be stated as 36 *for* 1.

1.2 Independence and Conditional Probability

Consider two random experiments that are unrelated to each other, and assume that Theorem 1.1.1 on p. 3 applies to both. Let Ω_1 and Ω_2 be the two sample spaces, and let $E_1 \subset \Omega_1$ and $F_2 \subset \Omega_2$ be events. If we define

$$\Omega := \Omega_1 \times \Omega_2, \quad E := E_1 \times \Omega_2, \quad F := \Omega_1 \times F_2, \qquad (1.44)$$

then Ω is the sample space for the joint random experiment, to which Theorem 1.1.1 on p. 3 still applies, E is the event that E_1 occurs in the first random experiment, and F is the event that F_2 occurs in the second random experiment. Furthermore,

$$P(E \cap F) = P(E_1 \times F_2) = \frac{|E_1 \times F_2|}{|\Omega_1 \times \Omega_2|}$$

$$= \frac{|E_1 \times \Omega_2|}{|\Omega_1 \times \Omega_2|} \frac{|\Omega_1 \times F_2|}{|\Omega_1 \times \Omega_2|} = \frac{|E|}{|\Omega|} \frac{|F|}{|\Omega|} = P(E)P(F). \quad (1.45)$$

Although this is not the most general situation under which events E and F are unrelated, we use (1.45) to motivate the next definition.

In general, events E and F are said to be *independent* if $P(E \cap F) = P(E)P(F)$. More generally, events $E_1, E_2, \ldots, E_n$ $(n \geq 2)$ are said to be *independent* if

$$P(E_{i_1} \cap E_{i_2} \cap \cdots \cap E_{i_m}) = P(E_{i_1})P(E_{i_2}) \cdots P(E_{i_m}) \quad (1.46)$$

whenever $2 \leq m \leq n$ and $1 \leq i_1 < i_2 < \cdots < i_m \leq n$. Finally, a countably infinite collection of events $E_1, E_2, \ldots$ is said to be *independent* if $E_1, E_2, \ldots, E_n$ are independent for every $n \geq 2$.

Example 1.2.1. *Outcome 1 before outcome 2 in repeated independent trials.* Given a random experiment that has exactly three possible outcomes, referred to as outcomes 1, 2, and 3, with probabilities $p_1 > 0$, $p_2 > 0$, and $p_3 > 0$ $(p_1 + p_2 + p_3 = 1)$, consider a sequence of independent trials, at each of which the given random experiment is performed. What is the probability that outcome 1 occurs at least once before the first occurrence of outcome 2? For $n = 1, 2, \ldots$, let E_n be the event that outcome 1 occurs for the first time at trial n and prior to the first occurrence of outcome 2. Then $E_1, E_2, \ldots$ are mutually exclusive, so by Axiom 1.1.14 on p. 12,

$$P\left(\bigcup_{n=1}^{\infty} E_n\right) = \sum_{n=1}^{\infty} P(E_n) = \sum_{n=1}^{\infty} p_3^{n-1} p_1 = \frac{p_1}{1 - p_3} = \frac{p_1}{p_1 + p_2}. \quad (1.47)$$

To justify the second equality, we can write $E_n = F_1 \cap \cdots \cap F_{n-1} \cap G_n$, where F_j $(1 \leq j \leq n - 1)$ is the event that outcome 3 occurs at trial j and G_n is the event that outcome 1 occurs at trial n. Then $F_1, \ldots, F_{n-1}, G_n$ are independent events by the assumed independence of the trials, and therefore $P(E_n) = p_3^{n-1} p_1$.

This result is more useful than it may at first appear. In particular, the sample space for the original experiment may have more than three possible outcomes, and the roles of outcomes 1, 2, and 3 above may be played by three mutually exclusive events whose union is Ω and whose probabilities are p_1, p_2, and p_3.

For example, in repeated rolls of a pair of dice, the probability of rolling a total of 6 at least once before the first occurrence of a total of 7 is $\pi_6/(\pi_6 + \pi_7) = 5/11$, where we are using (1.11) on p. 6.   ♠

The *conditional probability* of an event E, given the occurrence of an event D, is defined by

$$P(E \mid D) := \frac{P(D \cap E)}{P(D)}, \tag{1.48}$$

provided that $P(D) > 0$. Notice that, if D and E are independent, then $P(E \mid D) = P(E)$, that is, the conditional probability of an event E, given an independent event D, is equal to the (unconditional) probability of E.

We rarely use definition (1.48) to evaluate conditional probabilities. Instead, we can usually evaluate them as unconditional probabilities. Two examples should suffice to explain the idea.

Example 1.2.2. *Pass-line bet at craps.* The game of craps is played by rolling a pair of dice repeatedly. Except for some less important wagers, only the total of the two dice matters, and the probabilities of the various totals are given by (1.11) on p. 6. The principal bet at craps is called the *pass-line bet* and is initiated prior to the initial roll of the dice, which is called the *come-out roll*. The bet is won if the shooter rolls 7 or 11 (a *natural*) on the come-out roll. It is lost if the shooter rolls 2, 3, or 12 (a *craps number*) on the come-out roll. The only other possibility is that the shooter rolls a number belonging to

$$\mathscr{P} := \{4, 5, 6, 8, 9, 10\} \tag{1.49}$$

on the come-out roll, which establishes that number as the shooter's *point*. He continues to roll the dice until he either wins by repeating his point or loses by rolling 7. A win pays even money.

Let us introduce some events. For $j = 2, 3, 4, \ldots, 12$, we let D_j be the event that j is rolled on the come-out roll. We let E be the event that the pass-line bet is won, and, for each $j \in \mathscr{P}$, we let E_j be the event that, beginning with the second roll of the dice, j appears before 7. Then

$$P(E \mid D_j) = P(E_j \mid D_j) = P(E_j) = \frac{\pi_j}{\pi_j + \pi_7}, \qquad j \in \mathscr{P}. \tag{1.50}$$

Here the first equality uses the fact that, given that point j is established on the come-out roll, events E and E_j are equivalent (i.e., $D_j \cap E = D_j \cap E_j$). The second equality is a consequence of the independence of D_j and E_j; this independence is due to the fact that D_j depends on only the result of the come-out roll, while E_j depends on only the results of subsequent rolls. Finally, we use Example 1.2.1 (together with the notation (1.11) on p. 6) for the third equality.

We continue with this example in Example 1.2.8 below. ♠

Example 1.2.3. *Drawing to a four-card flush.* Consider the game of five-card draw poker (or video poker). Given that a player is dealt four cards of one suit and a fifth card of another, what is the conditional probability of completing the flush (or straight flush) with a one-card draw? (Here the card of the odd suit is replaced by a card drawn from the residual 47-card deck.) We let D be the event that the player is dealt four cards of one suit and a fifth card of another, and we let E be the event that he completes the flush (or straight flush). We could certainly use the definition (1.48) to evaluate $P(E \mid D)$. We would simply need to evaluate the ratio of

$$P(D \cap E) = \binom{4}{1} \frac{\binom{13}{4,1,8}\binom{39}{1,0,38}}{\binom{52}{5,1,46}} \quad \text{to} \quad P(D) = \binom{4}{1} \frac{\binom{13}{4}\binom{39}{1}}{\binom{52}{5}}. \tag{1.51}$$

But a much simpler approach is to regard the event D as having occurred and to choose a new sample space accordingly. Specifically, the random experiment consists of drawing a single card from the residual 47-card deck, in which nine cards of the four-card-flush suit remain. All 47 cards being equally likely, we conclude that $P(E \mid D) = 9/47$. ♠

The next result is often called the *multiplication law* for conditional probabilities.

Theorem 1.2.4. *Let $E_1, E_2, \ldots, E_n$ be events with $P(E_1 \cap \cdots \cap E_{n-1}) > 0$.*
Then

$$P(E_1 \cap \cdots \cap E_n) = P(E_1) \prod_{m=2}^{n} P(E_m \mid E_1 \cap \cdots \cap E_{m-1}). \tag{1.52}$$

Here is a generalization of Example 1.2.1.

Example 1.2.5. *Each of the outcomes $1, 2, \ldots, n$ before outcome $n + 1$ in repeated independent trials.* Given a random experiment that has exactly $n+2$ possible outcomes, referred to as outcomes $1, 2, \ldots, n + 2$, with probabilities $p_1 > 0$, $p_2 > 0$, $\ldots$, $p_{n+1} > 0$, $p_{n+2} \geq 0$ $(p_1 + p_2 + \cdots + p_{n+2} = 1)$, consider a sequence of independent trials, at each of which the given random experiment is performed. What is the probability of the event E that each of the outcomes $1, 2, \ldots, n$ occurs at least once before the first occurrence of outcome $n + 1$?

There are at least two possible approaches. The first is to specify the order in which the outcomes $1, 2, \ldots, n$ first occur and apply Theorem 1.2.4 and the result of Example 1.2.1. We find that

$$P(E) = \sum_{\sigma \in S_n} \prod_{m=1}^{n} \frac{p_{\sigma(m)}}{p_{\sigma(m)} + \cdots + p_{\sigma(n)} + p_{n+1}}$$

$$= p_1 \cdots p_n \sum_{\sigma \in S_n} \prod_{m=1}^{n} \frac{1}{p_{\sigma(m)} + \cdots + p_{\sigma(n)} + p_{n+1}}, \tag{1.53}$$

where S_n is the group of permutations of $\{1, 2, \ldots, n\}$. The sum has $n!$ terms.

The second is based on the inclusion-exclusion law (Theorem 1.1.18 on p. 13) and the result of Example 1.2.1. For $i = 1, 2, \ldots, n$, define E_i to be the event that outcome $n + 1$ occurs at least once before the first occurrence of outcome i. Then

$$P(E) = P(E_1^c \cap \cdots \cap E_n^c) = 1 - P(E_1 \cup \cdots \cup E_n)$$

$$= 1 - \sum_{m=1}^{n} (-1)^{m-1} \sum_{1 \le i_1 < \cdots < i_m \le n} \cdots \sum \frac{p_{n+1}}{p_{i_1} + \cdots + p_{i_m} + p_{n+1}}$$

$$= \sum_{A \subset \{1,2,\ldots,n\}} (-1)^{|A|} \frac{p_{n+1}}{\sum_{i \in A} p_i + p_{n+1}}. \tag{1.54}$$

The sum has only 2^n terms.

For example, in repeated rolls of a pair of dice, the probability of rolling each of the totals 4, 5, 6, 8, 9, and 10 at least once before the first appearance of a total of 7 can be found from (1.54) with $n = 6$ and $(p_1, p_2, \ldots, p_7) = (\pi_4, \pi_5, \pi_6, \pi_8, \pi_9, \pi_{10}, \pi_7)$. Here we are using the notation (1.11) on p. 6, and we find that the probability in question is approximately 0.062168159. ♠

We will refer to the next result as the *conditioning law*.

Theorem 1.2.6. *Given a finite or countably infinite index set I, let $\{D_i,\ i \in I\}$ be a collection of mutually exclusive events whose union is an event of probability 1, and assume that $P(D_i) > 0$ for each $i \in I$. If E is an event, then*

$$P(E) = \sum_{i \in I} P(D_i) P(E \mid D_i). \tag{1.55}$$

The next result is known as *Bayes's law*.

Theorem 1.2.7. *Assume, in addition to the hypotheses of Theorem 1.2.6, that $P(E) > 0$. Then*

$$P(D_j \mid E) = \frac{P(D_j) P(E \mid D_j)}{P(E)} = \frac{P(D_j) P(E \mid D_j)}{\sum_{i \in I} P(D_i) P(E \mid D_i)} \tag{1.56}$$

for each $j \in I$.

Example 1.2.8. *Pass-line bet at craps, continued.* We continue with Example 1.2.2. Recall that, for $j = 2, 3, 4, \ldots, 12$, we let D_j be the event that j is rolled on the come-out roll, and we let E be the event that the pass-line bet is won. We can find the probability of E by conditioning on the result of the come-out roll. By Theorem 1.2.6 and Example 1.2.2,

$$P(E) = \sum_{j=2}^{12} P(D_j) P(E \mid D_j)$$

$$= \pi_7 + \pi_{11} + \sum_{j \in \mathscr{P}} \pi_j \frac{\pi_j}{\pi_j + \pi_7} = \frac{244}{495} = 0.4\overline{92}. \tag{1.57}$$

We will elaborate on this example in Chapter 15 and elsewhere. ♠

Example 1.2.9. *Application of Bayes's law to twenty-one.* Suppose that a twenty-one player has been dealt a 10-valued card and a 6, while the dealer's upcard is a 10-valued card. (Assume a single-deck game.) The player is asked whether he wants a third card. What are the probabilities of the player's various final totals (at least those of 21 or less), assuming that he decides to take one, and only one, card? For $j = 1, 2, 3, 4, 5$, let D_j be the event that the player's final total is $16 + j$. The solution appears to be straightforward:

$$P(D_j) = \frac{4}{49}, \quad j = 1, 2, 3, 4, 5. \tag{1.58}$$

However, this is not quite correct because there is a rule that we have not yet explained: A dealer natural (two-card 21) beats a player 21 if the latter comprises three or more cards. Therefore, if the dealer's upcard is a ten-valued card, he checks to see if his downcard is an ace before proceeding. In particular, if the player is allowed to hit his hard 16, he knows that the dealer does not have a natural. This additional information changes the probabilities slightly.

Let E be the event that the dealer has a natural (i.e., his downcard is an ace). Then it is $P(D_j \mid E^c)$ that we seek ($j = 1, 2, 3, 4, 5$), not simply $P(D_j)$. For this we can apply Bayes's law. Indeed,

$$\begin{aligned} P(D_j \mid E^c) &= \frac{P(D_j)P(E^c \mid D_j)}{P(E^c)} = \frac{P(D_j)[1 - P(E \mid D_j)]}{1 - P(E)} \\ &= \frac{(4/49)(1 - (4 - \delta_{j,1})/48)}{1 - 4/49} \\ &= \begin{cases} 4/48 & \text{if } j = 1, \\ (44/48)(4/45) & \text{if } j \in \{2, 3, 4, 5\}, \end{cases} \end{aligned} \tag{1.59}$$

where $\delta_{j,1}$ is the Kronecker delta ($= 1$ if $j = 1$; $= 0$ otherwise). Intuitively, knowledge that the dealer does not have an ace as his downcard means that there are still four aces left for the player among the 48 cards available (the 52-card deck less two 10-valued cards, one 6, and the dealer's non-ace downcard, whatever it may be). This gives (1.59) for $j = 1$. If $j \in \{2, 3, 4, 5\}$, the complementary probability that the player does not draw an ace is $44/48$. But now there are only 45 cards available (the 52-card deck less two 10-valued cards, one 6, and four aces), of which four are of denomination j. (The fact that one of those 45 is the dealer's downcard does not affect the probabilities since we have no other information about it.)

The probabilities (1.58), which are applicable before the player knows that the dealer does not have a natural, are called *prior probabilities*. The probabilities (1.59), which are applicable after the player knows that the dealer does not have a natural, are called *posterior probabilities*. ♠

We conclude this section with a useful observation concerning conditional probabilities. Fix an event D with $P(D) > 0$. We define the probability Q by

$$Q(E) := P(E \mid D), \quad E \text{ an event in } \Omega. \tag{1.60}$$

Then it is easy to check that the fact that P satisfies Axioms 1.1.12–1.1.15 on pp. 12–12 implies the same is true for Q. Thus, all results derived from the axioms apply to Q. For example, Q satisfies the complementation law, a fact we just used in (1.59). When applying the results of this section to Q, it suffices to notice that

$$Q(F \mid E) = \frac{Q(E \cap F)}{Q(E)} = \frac{P(E \cap F \mid D)}{P(E \mid D)}$$

$$= \frac{P(D \cap E \cap F)}{P(D \cap E)} = P(F \mid D \cap E). \tag{1.61}$$

For example, given a finite or countably infinite index set J, let $\{E_j, \ j \in J\}$ be a collection of mutually exclusive events whose union is an event of probability 1, and assume that $P(D \cap E_j) > 0$ for each $j \in J$. If F is an event, then

$$P(F \mid D) = \sum_{j \in J} P(E_j \mid D) P(F \mid D \cap E_j). \tag{1.62}$$

1.3 Random Variables and Random Vectors

Given a sample space Ω, a *discrete random variable* X is a real-valued function on Ω that assumes at most countably many values and for which $\{X = x\} := \{o \in \Omega : X(o) = x\}$ is an event for every $x \in \mathbf{R}$. The *distribution* of X is the function $p : \mathbf{R} \mapsto [0, 1]$ defined by

$$p(x) := P(X = x), \qquad x \in \mathbf{R}. \tag{1.63}$$

The values of p must sum to 1. If $p(x) = 1$ for some $x \in \mathbf{R}$, we say that the distribution of X (or X itself) is *degenerate*. Although we will not have much occasion to use it, we note that the *cumulative distribution function* of X is the function $F : \mathbf{R} \mapsto [0, 1]$ given by $F(x) := P(X \leq x)$.

Occasionally, we allow extended-real-valued discrete random variables, in which case $\mathbf{R}$ above is replaced by $[-\infty, \infty]$.

We begin by considering several special distributions. Given a finite set $K \subset \mathbf{R}$, the *discrete uniform distribution* on K is the distribution of the random variable X that assumes each of the values in K with equal probability, that is,

$$P(X = k) = \frac{1}{|K|}, \qquad k \in K. \tag{1.64}$$

For example, the result of the toss of a single die is described by a random variable having the discrete uniform distribution on $\{1, 2, 3, 4, 5, 6\}$.

The *hypergeometric distribution* with parameters n, N, and K is the distribution of the number of black balls in a sample of size n taken without replacement from an urn containing N balls, of which K are black and $N - K$ are white. If X is hypergeometric(n, N, K), then

$$P(X = k) = \frac{\binom{K}{k}\binom{N-K}{n-k}}{\binom{N}{n}}, \qquad k = (K + n - N)^+, \ldots, K \wedge n; \qquad (1.65)$$

here $x^+ := \max(x, 0)$ and $x \wedge y := \min(x, y)$.

A *Bernoulli trial* is a random experiment with only two possible outcomes, which may be called *success* and *failure*. Several distributions are defined in terms of a sequence of independent Bernoulli trials, each trial having the same success probability $p \in (0, 1)$.

The *binomial distribution* with parameters n and p is the distribution of the number of successes in n independent Bernoulli trials, each with success probability p. If X is binomial(n, p), then

$$P(X = k) = \binom{n}{k} p^k (1 - p)^{n-k}, \qquad k = 0, 1, \ldots, n. \qquad (1.66)$$

The *geometric distribution* with parameter p is the distribution of the number of trials needed to achieve the first success in a sequence of independent Bernoulli trials, each with success probability p. If X is geometric(p), then

$$P(X = k) = (1 - p)^{k-1} p, \qquad k = 1, 2, \ldots. \qquad (1.67)$$

It will be convenient to define the *shifted geometric distribution* with parameter p as the distribution of the number of failures prior to the first success. If X is shifted geometric(p), then

$$P(X = k) = (1 - p)^k p, \qquad k = 0, 1, \ldots, \qquad (1.68)$$

and $X + 1$ is geometric(p).

The *negative binomial distribution* with parameters n and p is the distribution of the number of trials needed to achieve the nth success in a sequence of independent Bernoulli trials, each with success probability p. If X is negative binomial(n, p), then

$$P(X = k) = \binom{k-1}{n-1} p^n (1 - p)^{k-n}, \qquad k = n, n + 1, \ldots. \qquad (1.69)$$

It will be convenient to define the *shifted negative binomial distribution* with parameters n and p as the distribution of the number of failures prior to the nth success. If X is shifted negative binomial(n, p), then

$$P(X = k) = \binom{k+n-1}{n-1} p^n (1 - p)^k, \qquad k = 0, 1, \ldots, \qquad (1.70)$$

and $X + n$ is negative binomial(n, p). (To verify that these probabilities sum to 1, we can use Theorem A.2.10 on p. 749.)

The last special distribution we mention here is the *Poisson distribution* with parameter $\lambda > 0$. If X is Poisson(λ), then

$$P(X = k) = \frac{e^{-\lambda} \lambda^k}{k!}, \qquad k = 0, 1, \ldots. \tag{1.71}$$

Although the Poisson distribution does not have a direct interpretation in terms of a sequence of Bernoulli trials, it is often used as an approximation of the binomial$(n, \lambda/n)$ distribution for large n, since

$$\lim_{n \to \infty} \binom{n}{k} (\lambda/n)^k (1 - \lambda/n)^{n-k} = \frac{e^{-\lambda} \lambda^k}{k!}, \qquad k = 0, 1, \ldots. \tag{1.72}$$

Example 1.3.1. *Keno 10-spot ticket.* The game of keno is played as follows. The keno player chooses 10 distinct numbers from the set $\{1, 2, \ldots, 80\}$, not necessarily at random, marks them on his keno ticket, and pays an entry fee. The casino then randomly draws 20 numbers from the set $\{1, 2, \ldots, 80\}$ without replacement. If we denote by X the number of the player's 10 numbers that are among the casino's 20 numbers, the player then receives a payoff depending on X and the size of his entry fee. The relevant probabilities are given by (1.65) with $n = 20$, $N = 80$, and $K = 10$. We conclude that

$$P(X = k) = \frac{\binom{10}{k}\binom{80-10}{20-k}}{\binom{80}{20}}, \qquad k = 0, 1, \ldots, 10. \tag{1.73}$$

By virtue of Problem 1.21 on p. 56, we can rewrite (1.73) as

$$P(X = k) = \frac{\binom{20}{k}\binom{80-20}{10-k}}{\binom{80}{10}}, \qquad k = 0, 1, \ldots, 10. \tag{1.74}$$

In fact, (1.74) is preferable to (1.73) because the denominator (and therefore the numerator) is smaller. For more on keno, see Chapter 14. ♠

Example 1.3.2. *John Smith's problem.* In 1693 John Smith asked Samuel Pepys, who in turn asked Isaac Newton, for the answer to the following question: Which is more likely, at least one six in 6 rolls of a die, at least two sixes in 12 rolls of a die, or at least three sixes in 18 rolls of a die? More generally, let E_n be the event of obtaining at least n sixes in $6n$ rolls of a die. Then

$$P(E_n) = \sum_{k=n}^{6n} \binom{6n}{k} \left(\frac{1}{6}\right)^k \left(\frac{5}{6}\right)^{6n-k} = 1 - \sum_{k=0}^{n-1} \binom{6n}{k} \left(\frac{1}{6}\right)^k \left(\frac{5}{6}\right)^{6n-k}, \tag{1.75}$$

and we find that

$$P(E_1) \approx 0.665102, \quad P(E_2) \approx 0.618667, \quad P(E_3) \approx 0.597346. \qquad (1.76)$$

Not surprisingly, Newton was able to answer the question correctly. ♠

Example 1.3.3. *Méré's problem, revisited.* Recalling Méré's problem (Example 1.1.22 on p. 14), we can use the binomial distribution to express the probabilities in a more natural form. The probability of at least one six in four tosses of a single die is

$$1 - P(\text{no sixes in four tosses of a single die})$$

$$= 1 - \binom{4}{0}\left(\frac{1}{6}\right)^0\left(\frac{5}{6}\right)^4 = \frac{671}{1{,}296} \approx 0.517747, \qquad (1.77)$$

while the probability of at least one double six in 24 tosses of a pair of dice is

$$1 - P(\text{no double sixes in 24 tosses of a pair of dice})$$

$$= 1 - \binom{24}{0}\left(\frac{1}{36}\right)^0\left(\frac{35}{36}\right)^{24} \approx 0.491404. \qquad (1.78)$$

These are the same results we derived in Example 1.1.22 on p. 14 without the benefit of the concept of independence.

We can also use the geometric distribution. Let X be the number of rolls of a single die needed to produce a six, and let Y be the number of rolls of a pair of dice needed to produce a double six. Then X is geometric($1/6$), and Y is geometric($1/36$). Consequently, the probability of at least one six in four tosses of a single die is

$$P(X \le 4) = 1 - P(X \ge 5) = 1 - \sum_{k=5}^{\infty}\left(\frac{5}{6}\right)^{k-1}\left(\frac{1}{6}\right)$$

$$= 1 - (5/6)^4 = 671/1{,}296 \approx 0.517747, \qquad (1.79)$$

while the probability of at least one double six in 24 tosses of a pair of dice is

$$P(Y \le 24) = 1 - P(Y \ge 25) = 1 - \sum_{k=25}^{\infty}\left(\frac{35}{36}\right)^{k-1}\left(\frac{1}{36}\right)$$

$$= 1 - (35/36)^{24} \approx 0.491404. \qquad (1.80)$$

Often there are several ways to approach a problem. ♠

We postpone an example of the negative binomial distribution to Chapter 2 (Example 2.1.1 on p. 76).

Example 1.3.4. *Poisson approximation.* Consider a video-poker player who plays 1,000 hands of five-card draw per day for 365 consecutive days. What is the probability that he is dealt a pat royal flush exactly once; at least once? We have $n = 365{,}000$ independent Bernoulli trials, each with success probability $p = 4/2{,}598{,}960 = 1/649{,}740$. Let $\lambda = np$. The exact binomial and approximate Poisson probabilities of exactly one success are

$$\binom{n}{1} p(1-p)^{n-1} \approx 0.320319294 \quad \text{and} \quad e^{-\lambda}\lambda \approx 0.320318939. \qquad (1.81)$$

For at least one success, they are

$$1 - \binom{n}{0}(1-p)^n \approx 0.429797431 \quad \text{and} \quad 1 - e^{-\lambda} \approx 0.429797186. \qquad (1.82)$$

We see that the approximation is quite good, but because the exact probability is not much more difficult to compute than the approximation, the result (1.72) is primarily of theoretical, rather than of practical, interest.   ♠

Suppose that the discrete random variables $X_1, X_2, \ldots, X_r$ are defined on the same sample space and therefore refer to the same random experiment. We then say that they are *jointly distributed* and call $\boldsymbol{X} := (X_1, X_2, \ldots, X_r)$ a *discrete random vector*. (Typically, $r \geq 2$, but it will be convenient to allow $r = 1$.) The *joint distribution* of $X_1, X_2, \ldots, X_r$ (or the distribution of $\boldsymbol{X}$) is the function $p : \mathbf{R}^r \mapsto [0, 1]$ defined by

$$p(\boldsymbol{x}) := \mathrm{P}(\boldsymbol{X} = \boldsymbol{x}) = \mathrm{P}((X_1, \ldots, X_r) = (x_1, \ldots, x_r)). \qquad (1.83)$$

If $r = 2$, the probabilities of the joint distribution are often displayed in a two-dimensional array. The row sums determine the *marginal distribution* of X_1, while the column sums determine the *marginal distribution* of X_2. This is a consequence of

$$\mathrm{P}(X_1 = x_1) = \sum_{x_2 \in \mathbf{R}} \mathrm{P}((X_1, X_2) = (x_1, x_2)) \qquad (1.84)$$

and

$$\mathrm{P}(X_2 = x_2) = \sum_{x_1 \in \mathbf{R}} \mathrm{P}((X_1, X_2) = (x_1, x_2)). \qquad (1.85)$$

If $r \geq 2$, we define the *marginal distribution* of X_i $(1 \leq i \leq r)$ in terms of the joint distribution of $X_1, X_2, \ldots, X_r$ by

$$\mathrm{P}(X_i = x_i) = \sum_{(x_1, \ldots, x_{i-1}, x_{i+1}, \ldots, x_r) \in \mathbf{R}^{r-1}} \mathrm{P}((X_1, \ldots, X_r) = (x_1, \ldots, x_r)).$$
$$(1.86)$$

We say that $X_1, \ldots, X_r$ are *independent* if their joint distribution is the product of their marginal distributions, that is, if

$$\mathrm{P}(\boldsymbol{X} = \boldsymbol{x}) = \mathrm{P}(X_1 = x_1) \cdots \mathrm{P}(X_r = x_r) \qquad (1.87)$$

for all $\boldsymbol{x} \in \mathbf{R}^r$. Equivalently, $X_1, \ldots, X_r$ are independent if the events $\{X_1 = x_1\}, \ldots, \{X_r = x_r\}$ are independent for all $x_1, \ldots, x_r \in \mathbf{R}$.

More generally, if $(\boldsymbol{X}_1, \ldots, \boldsymbol{X}_r)$ is a discrete random vector whose components are themselves discrete random vectors, then $\boldsymbol{X}_1, \ldots, \boldsymbol{X}_r$ are said to be *independent* if the events $\{\boldsymbol{X}_1 = \boldsymbol{x}_1\}, \ldots, \{\boldsymbol{X}_r = \boldsymbol{x}_r\}$ are independent for all $\boldsymbol{x}_1, \ldots \boldsymbol{x}_r$ of the appropriate dimensions.

Finally, a sequence of discrete random variables (or random vectors) $X_1, X_2, \ldots$ is said to be *jointly distributed* if $X_1, X_2, \ldots$ are defined on the same sample space. The sequence is said to be *independent* if $X_1, X_2, \ldots, X_n$ are independent for each $n \geq 2$.

The distribution of a random variable is called a *univariate distribution*, whereas the distribution of a random vector of two or more dimensions is called a *multivariate distribution*. The hypergeometric and binomial distributions are univariate distributions that have multivariate analogues, namely the *multivariate hypergeometric distribution* and the *multinomial distribution*.

First, consider an urn containing N balls, of which N_1 are of type 1, ..., N_r are of type r. A sample of size n is taken without replacement, and we define X_1 to be the number of type-1 balls in the sample, ..., X_r to be the number of type-r balls in the sample. Then $(X_1, X_2, \ldots, X_r)$ has the multivariate hypergeometric$(n, N, N_1, \ldots, N_r)$ distribution, and

$$\mathrm{P}(\boldsymbol{X} = \boldsymbol{n}) = \frac{\binom{N_1}{n_1}\binom{N_2}{n_2} \cdots \binom{N_r}{n_r}}{\binom{N}{n}}, \qquad (1.88)$$

provided $\boldsymbol{n} = (n_1, \ldots, n_r)$ has nonnegative integer components that sum to n and $n_i \leq N_i$ for $i = 1, \ldots, r$. Notice that, for $i = 1, \ldots, r$, the marginal distribution of X_i is hypergeometric(n, N, N_i).

Second, consider n independent trials, each of which has r possible outcomes labeled $1, 2, \ldots, r$ with probabilities $p_1, p_2, \ldots, p_r$. Define X_1 to be the number of type-1 outcomes, ..., X_r to be the number of type-r outcomes. Then $(X_1, X_2, \ldots, X_r)$ has the multinomial$(n, p_1, \ldots, p_r)$ distribution, and

$$\mathrm{P}(\boldsymbol{X} = \boldsymbol{n}) = \binom{n}{n_1, \ldots, n_r} p_1^{n_1} \cdots p_r^{n_r}, \qquad (1.89)$$

provided $\boldsymbol{n} = (n_1, \ldots, n_r)$ has nonnegative integer components that sum to n. Notice that, for $i = 1, \ldots, r$, the marginal distribution of X_i is binomial(n, p_i).

Example 1.3.5. *Dead-man's hand.* The poker hand consisting of two pair, aces and eights, is called the *dead-man's hand* (see the chapter notes). The probability of being dealt the dead-man's hand in five-card poker is given by the multivariate hypergeometric distribution:

$$\frac{\binom{4}{2}\binom{4}{2}\binom{44}{1}}{\binom{52}{5}} = \frac{33}{54,145} \approx 0.000609475. \qquad (1.90)$$

On the other hand, if we were to make the unrealistic assumption that cards are dealt *with* replacement (sometimes expressed by assuming an *infinite deck*), the relevant probability would be given by the multinomial distribution:

$$\binom{5}{2,2,1}\left(\frac{4}{52}\right)^2\left(\frac{4}{52}\right)^2\left(\frac{44}{52}\right)^1 = \frac{330}{371{,}293} \approx 0.000888786. \qquad (1.91)$$

The reason that (1.91) is larger than (1.90) is that the replacement of the first ace and the first eight makes the drawing of the second ace and the second eight more likely.   ♠

1.4 Expectation and Variance

Next we define the *expectation* (or *expected value* or *mean*) of a discrete random variable X. Roughly speaking, it is simply the weighted average of the values of X with weights equal to the probabilities with which those values are assumed. For a more precise formulation, we proceed in two steps.

First, if X is a discrete random variable with values in $[0, \infty]$, we define its *expectation* to be

$$E[X] := \sum_{x \in [0,\infty]} x P(X = x), \qquad (1.92)$$

where we adopt the convention that

$$\infty \cdot p := \begin{cases} \infty & \text{if } p > 0, \\ 0 & \text{if } p = 0, \end{cases} \qquad (1.93)$$

The sum in (1.92) has at most countably many nonzero terms. It is possible that $E[X] = \infty$, either because $P(X = \infty) > 0$ or because the series in (1.92) diverges.

Second, if X is a discrete random variable with values in $\mathbf{R}$ and $E[|X|] < \infty$, we say that X has *finite expectation*. In this case it is easy to check that

$$E[|X|] = \sum_{x \in [0,\infty)} x P(|X| = x) = \sum_{x \in \mathbf{R}} |x| P(X = x), \qquad (1.94)$$

and we define the *expectation* of X to be

$$E[X] := \sum_{x \in \mathbf{R}} x P(X = x). \qquad (1.95)$$

Again, the sum in (1.95) has at most countably many nonzero terms, and by
(1.94) it is either a finite sum or an absolutely convergent series.

Example 1.4.1. *Expected payout for the Fey "Liberty Bell" slot machine.*
Let us consider one of the first three-reel slot machines, the "Liberty Bell,"
designed and built by Charles Fey in San Francisco in 1899. (See Section 12.4
for more information.) The player inserts a coin into the slot and pulls the
handle. This causes three reels, on each of which 10 symbols are marked, to
rotate independently. When they come to rest, one symbol on each reel is
visible, and the three resulting symbols determine the number of coins paid
out to the player in accordance with the pay table (see Table 1.6), which is
printed on the outside of the machine for all to see.

Table 1.6 Pay table for the Fey "Liberty Bell" slot machine. The payout is
the number of coins paid out from a one-coin bet.

reel 1	reel 2	reel 3	payout
bell	bell	bell	20
♡	♡	♡	16
◇	◇	◇	12
♠	♠	♠	8
Ω	Ω	★	4
Ω	Ω	not ★	2

To determine the expected number of coins paid out we need additional
information, namely the reel-strip labels displayed in Table 1.7. Actually, the
symbol-inventory table, also shown in Table 1.7, contains sufficient information. For example, Table 1.6 shows that the result[1] Ω-Ω-★ pays out 4 coins,
and Table 1.7 shows that there are five Ω symbols on reel 1, five Ω symbols
on reel 2, and two ★ symbols on reel 3. It follows that this event can occur
in $5 \cdot 5 \cdot 2 = 50$ ways, and since there are $(10)^3 = 1{,}000$ possible outcomes,
which may be assumed equally likely, the probability of the event Ω-Ω-★ is
$50/1{,}000$. The other five positive payouts can be analyzed in the same way
(see Table 1.8).

Letting R denote the number of coins paid out from a one-coin bet, we
evaluate the expected value of R as

$$\mathrm{E}[R] = 20\left(\frac{2}{1{,}000}\right) + 16\left(\frac{1}{1{,}000}\right) + 12\left(\frac{3}{1{,}000}\right)$$
$$+ 8\left(\frac{8}{1{,}000}\right) + 4\left(\frac{50}{1{,}000}\right) + 2\left(\frac{200}{1{,}000}\right)$$
$$= \frac{756}{1{,}000} = 0.756. \tag{1.96}$$

[1] Here Ω represents a horseshoe, not to be confused with the sample space Ω.

Table 1.7 Reel-strip labels and symbol-inventory table from the Fey "Liberty Bell" slot machine.

stop no.	reel 1	reel 2	reel 3
1	bell	bell	bell
2	Ω	Ω	◇
3	♠	♠	★
4	Ω	Ω	♠
5	◇	◇	bell
6	Ω	Ω	◇
7	♠	♠	♡
8	Ω	Ω	★
9	♡	♡	♠
10	Ω	Ω	◇

symbol	reel 1	reel 2	reel 3
bell	1	1	2
♡	1	1	1
◇	1	1	3
♠	2	2	2
Ω	5	5	0
★	0	0	2
total	10	10	10

Table 1.8 Payout analysis for the Fey "Liberty Bell" slot machine.

reel 1	reel 2	reel 3	payout	number of ways		product
bell	bell	bell	20	$1 \cdot 1 \cdot 2 =$	2	40
♡	♡	♡	16	$1 \cdot 1 \cdot 1 =$	1	16
◇	◇	◇	12	$1 \cdot 1 \cdot 3 =$	3	36
♠	♠	♠	8	$2 \cdot 2 \cdot 2 =$	8	64
Ω	Ω	★	4	$5 \cdot 5 \cdot 2 =$	50	200
Ω	Ω	not ★	2	$5 \cdot 5 \cdot 8 =$	200	400
total					264	756

Notice that we have omitted the term corresponding to $R = 0$ because it does not contribute. For completeness, Table 1.8 shows that $P(R = 0) = 1 - P(R \geq 1) = 1 - 264/1{,}000 = 736/1{,}000 = 0.736$.   ♠

If X denotes the gambler's profit from a wager, we say that the wager is *subfair* if $E[X] < 0$, *fair* if $E[X] = 0$, and *superfair* if $E[X] > 0$. Slot machines are always subfair, with the possible exception of machines offering progressive jackpots.

An alternative way to measure the central value of a random variable is in terms of its median. We say that m is a *median* of a discrete random variable X if

$$P(X \leq m) \geq \frac{1}{2} \quad \text{and} \quad P(X \geq m) \geq \frac{1}{2}. \tag{1.97}$$

Occasionally, this definition does not uniquely determine m, and the set of m for which (1.97) holds is an interval. This is the case, for example, if X is discrete uniform on $\{1, 2, 3, 4, 5, 6\}$. In such cases the definition of the median is ambiguous and requires clarification.

A different expectation formula is available for nonnegative, integer-valued random variables.

Theorem 1.4.2. *Let X be a nonnegative integer-valued random variable. Then*

$$E[X] = \sum_{n=1}^{\infty} P(X \geq n). \tag{1.98}$$

Example 1.4.3. *Mean and median of the geometric distribution.* Let X be geometric(p). Then, for each $n \geq 1$,

$$P(X \geq n) = \sum_{k=n}^{\infty} (1-p)^{k-1} p = (1-p)^{n-1}. \tag{1.99}$$

It follows that (1.97) holds with m being a positive integer if and only if $(1-p)^{m-1} \geq \frac{1}{2} \geq (1-p)^m$, or if and only if $(\ln 2)/|\ln(1-p)| \leq m \leq 1 + (\ln 2)/|\ln(1-p)|$. We conclude that, unless $(\ln 2)/|\ln(1-p)|$ is a positive integer,

$$\text{median}(X) = \left\lceil \frac{\ln 2}{|\ln(1-p)|} \right\rceil = \left\lceil \frac{\ln 2}{p + \frac{1}{2}p^2 + \frac{1}{3}p^3 + \cdots} \right\rceil \leq \left\lceil \frac{\ln 2}{p} \right\rceil. \tag{1.100}$$

On the other hand, by Theorem 1.4.2 and (1.99),

$$E[X] = \sum_{n=1}^{\infty} P(X \geq n) = \sum_{n=1}^{\infty} (1-p)^{n-1} = \frac{1}{p}. \tag{1.101}$$

The latter result is the more important one.

For example, let Y be the number of rolls of a pair of dice needed to produce a double six. Recalling (1.80) on p. 24 and noting that $P(Y \leq$

$25) = 1 - (35/36)^{25} \approx 0.505532$, we find that median$(Y) = 25$. Since $(\ln 2)/|\ln(35/36)| \approx 24.605$ and $36 \ln 2 \approx 24.953$, the same conclusion follows from the formula in (1.100), and equality holds in the inequality there. Of course, $\mathrm{E}[Y] = 36$.   ♠

Next we need a generalization of (1.94).

Theorem 1.4.4. *Let g be a real-valued function on the range of the discrete random variable X such that either $g(X) \geq 0$ or $g(X)$ has finite expectation. Then*

$$\mathrm{E}[g(X)] = \sum_{x \in \mathbf{R}} g(x)\mathrm{P}(X = x). \tag{1.102}$$

Example 1.4.5. *St. Petersburg paradox.* The *St. Petersburg game* is a hypothetical game that dates back to the early 18th century. A player tosses a fair coin until heads turns up. If the random variable N denotes the number of tosses required, the player is paid $X := 2^{N-1}$ units. What is a fair entry fee?

Notice that N is geometric$(\frac{1}{2})$. Consequently, by Theorem 1.4.4,

$$\mathrm{E}[X] = \mathrm{E}[2^{N-1}] = \sum_{n=1}^{\infty} 2^{n-1} \left(\frac{1}{2}\right)^n = \sum_{n=1}^{\infty} \frac{1}{2} = \infty. \tag{1.103}$$

This suggests that the game is favorable to the player no matter how large the entry fee is. Yet few players would be willing to pay more than a few units to play. That is the "paradox." See Example 1.5.11 on p. 48 and Problem 1.34 on p. 59 for two possible resolutions.   ♠

Let n be a positive integer. If X is a discrete random variable with $\mathrm{E}\big[|X|^n\big] < \infty$, we say that X has *finite nth moment* and define its *nth moment* to be $\mathrm{E}[X^n]$. The *variance* and the *standard deviation* of a discrete random variable X with finite second moment are defined by

$$\mathrm{Var}(X) := \mathrm{E}[(X - \mathrm{E}[X])^2] \quad \text{and} \quad \mathrm{SD}(X) := \sqrt{\mathrm{Var}(X)}. \tag{1.104}$$

The variance (and hence the standard deviation) of X is a measure of how spread out the values of X are. The standard deviation has the advantage of being measured in the same units as X itself is. We say that a discrete random variable has *finite variance* if it has finite second moment. The following result says that the variance of a discrete random variable can be evaluated as the second moment minus the square of the mean.

Theorem 1.4.6. *If the discrete random variable X has finite variance, then*

$$\mathrm{Var}(X) = \mathrm{E}[X^2] - (\mathrm{E}[X])^2. \tag{1.105}$$

Example 1.4.7. *Variance of payout for the Fey "Liberty Bell" slot machine.* Letting R denote the number of coins paid out from a one-coin bet, we know $\mathrm{E}[R]$ from Example 1.4.1. It follows that the variance of R is

$$\mathrm{Var}(R) = \mathrm{E}[R^2] - (\mathrm{E}[R])^2$$

$$= (20)^2 \left(\frac{2}{1{,}000}\right) + (16)^2 \left(\frac{1}{1{,}000}\right) + (12)^2 \left(\frac{3}{1{,}000}\right)$$

$$+ (8)^2 \left(\frac{8}{1{,}000}\right) + (4)^2 \left(\frac{50}{1{,}000}\right) + (2)^2 \left(\frac{200}{1{,}000}\right) - \left(\frac{756}{1{,}000}\right)^2$$

$$= \frac{3{,}028{,}464}{1{,}000{,}000} = 3.028464. \tag{1.106}$$

Due to the lack of a substantial jackpot, this is a rather low variance relative to those of modern slot machines. ♠

A different second-moment formula is available for nonnegative, integer-valued random variables.

Theorem 1.4.8. *Let X be a nonnegative integer-valued random variable. Then*

$$\mathrm{E}[X^2] = \sum_{n=1}^{\infty} (2n-1)\mathrm{P}(X \geq n). \tag{1.107}$$

Example 1.4.9. *Variance of the geometric distribution.* By Theorem 1.4.8, (1.99), and (1.101), if X is geometric(p), then $\mathrm{E}[X] = 1/p$ and

$$\mathrm{E}[X^2] = \sum_{n=1}^{\infty} (2n-1)(1-p)^{n-1} = \frac{2}{p} \sum_{n=1}^{\infty} n(1-p)^{n-1}p - \sum_{n=1}^{\infty} (1-p)^{n-1}$$

$$= \frac{2}{p} \mathrm{E}[X] - \mathrm{E}[X] = \frac{2-p}{p^2}. \tag{1.108}$$

Consequently, by Theorem 1.4.6, $\mathrm{Var}(X) = \mathrm{E}[X^2] - (\mathrm{E}[X])^2 = (1-p)/p^2$. ♠

Next we generalize Theorem 1.4.4.

Theorem 1.4.10. *Let g be a real-valued function on the range of the discrete random vector $\boldsymbol{X}$ such that either $g(\boldsymbol{X}) \geq 0$ or $g(\boldsymbol{X})$ has finite expectation. Then*

$$\mathrm{E}[g(\boldsymbol{X})] = \sum_{\boldsymbol{x}} g(\boldsymbol{x})\mathrm{P}(\boldsymbol{X} = \boldsymbol{x}). \tag{1.109}$$

The next theorem expresses the linearity of expectation.

Theorem 1.4.11. *If X is a discrete random variable with finite expectation and a is a constant, then aX has finite expectation and $\mathrm{E}[aX] = a\mathrm{E}[X]$. If $n \geq 2$ and each of the jointly distributed discrete random variables $X_1, \ldots, X_n$ has finite expectation, then so does $X_1 + \cdots + X_n$ and*

$$\mathrm{E}[X_1 + \cdots + X_n] = \mathrm{E}[X_1] + \cdots + \mathrm{E}[X_n]. \tag{1.110}$$

The result (1.110) is particularly useful when the random variables assume only two values, 0 and 1. We define the *indicator random variable* 1_E of the event E by

$$1_E := \begin{cases} 1 & \text{on } E, \\ 0 & \text{on } E^c. \end{cases} \tag{1.111}$$

(In particular, 1_E is defined on all of Ω, $\{1_E = 1\} = E$, and $\{1_E = 0\} = E^c$, so 1_E is a discrete random variable.)

Example 1.4.12. *Means of binomial, hypergeometric, and negative binomial distributions.* Let X be binomial(n, p). We interpret X as the number of successes in n independent Bernoulli trials, each with success probability p. If, for $i = 1, \ldots, n$, we let E_i be the event that a success occurs at trial i, then

$$X = 1_{E_1} + \cdots + 1_{E_n}, \tag{1.112}$$

so

$$E[X] = P(E_1) + \cdots + P(E_n) = p + \cdots + p = np. \tag{1.113}$$

Let X be hypergeometric(n, N, K). We interpret X as the number of black balls in a sample of size n taken without replacement from an urn containing N balls, of which K are black and $N - K$ are white. If, for $i = 1, 2, \ldots, n$, we let E_i be the event that the ith ball drawn is black (here we assume without loss of generality that balls are drawn sequentially), then (1.112) holds and therefore so does (1.113) with $p := K/N$, since

$$P(E_i) = \frac{K(N-1)_{n-1}}{(N)_n} = \frac{K}{N} \tag{1.114}$$

for $i = 1, \ldots, n$.

Let X be negative binomial(n, p). We interpret X as the number of trials needed to achieve the nth success in a sequence of independent Bernoulli trials, each with success probability p. If we let X_1 be the number of trials needed to achieve the first success, X_2 the number of *additional* trials needed to achieve the second success, and so on, then $X_1, X_2, \ldots$ are independent geometric(p) random variables and $X = X_1 + \cdots + X_n$, hence

$$E[X] = E[X_1] + \cdots + E[X_n] = \frac{1}{p} + \cdots + \frac{1}{p} = \frac{n}{p} \tag{1.115}$$

by Example 1.4.3.   ♠

Example 1.4.13. *Rebated loss.* Consider a game of independent coups in which a winning bet pays a to 1 and occurs with probability p. Otherwise the bet is lost. We assume that the game is subfair, so that $ap + (-1)(1 - p) = (a + 1)p - 1 < 0$. Suppose that the casino offers to rebate a proportion r of a player's loss (if any) if he agrees to bet one unit at each of n coups. How large must n be to ensure that the casino's expected profit is positive?

Let X be the number of player wins in n coups, so that X is binomial(n, p). Then the player's profit before applying the rebate is $aX + (-1)(n - X) = (a+1)X - n$, so his profit after applying the rebate is $h_r((a+1)X - n)$, where

$$h_r(y) := (1-r)y + ry^+, \tag{1.116}$$

where $y^+ := \max(y, 0)$. Consequently, we need to evaluate

$$\begin{aligned}
\mathrm{E}[h_r((a+1)X - n)] &= (1-r)\mathrm{E}[(a+1)X - n] + r\mathrm{E}[((a+1)X - n)^+] \\
&= (1-r)((a+1)p - 1)n \tag{1.117} \\
&\quad + r\sum_{k=0}^{n}((a+1)k - n)^+ \cdot \binom{n}{k}p^k(1-p)^{n-k}.
\end{aligned}$$

Here we have used the binomial mean from Example 1.4.12. This sum can be evaluated numerically by noticing that

$$\binom{n}{k+1}p^{k+1}(1-p)^{n-(k+1)} \bigg/ \binom{n}{k}p^k(1-p)^{n-k} = \frac{(n-k)p}{(k+1)(1-p)}. \tag{1.118}$$

Numerical values are displayed in Table 1.9.   ♠

We take this opportunity to introduce, for later use, the *expectation of a random vector*. Let $\boldsymbol{X} = (X_1, \ldots, X_r)$ be a random vector each of whose components has finite expectation. We then define

$$\mathrm{E}[\boldsymbol{X}] := (\mathrm{E}[X_1], \ldots, \mathrm{E}[X_r]). \tag{1.119}$$

To see how this might be useful, consider a game with d betting opportunities, let X_i be the gambler's profit per unit bet on the ith betting opportunity, and let b_i be the (nonrandom) amount bet on the ith betting opportunity. Then the gambler's expected profit is

$$\mathrm{E}[b_1 X_1 + \cdots + b_d X_d] = b_1\mathrm{E}[X_1] + \cdots + b_d\mathrm{E}[X_d], \tag{1.120}$$

which with $\boldsymbol{b} := (b_1, \ldots, b_d)$ can be expressed more concisely as

$$\mathrm{E}[\boldsymbol{b} \cdot \boldsymbol{X}] = \boldsymbol{b} \cdot \mathrm{E}[\boldsymbol{X}]. \tag{1.121}$$

Theorem 1.4.14. *If $X_1, X_2, \ldots, X_n$ are independent discrete random variables, each with finite expectation, then $X_1 X_2 \cdots X_n$ has finite expectation and*

$$\mathrm{E}[X_1 X_2 \cdots X_n] = \mathrm{E}[X_1]\,\mathrm{E}[X_2] \cdots \mathrm{E}[X_n]. \tag{1.122}$$

The next result is called the *Cauchy–Schwarz inequality*.

Theorem 1.4.15. *If X and Y are jointly distributed discrete random variables with finite second moments, then XY has finite expectation and*

$$(\mathrm{E}[XY])^2 \le \mathrm{E}[X^2]\mathrm{E}[Y^2]. \tag{1.123}$$

Moreover, equality holds in (1.123) if and only if either $\mathrm{P}(X = 0) = 1$ or there exists a constant a such that $\mathrm{P}(Y = aX) = 1$.

Table 1.9 Player's expected profit from n coups at three games (pass-line bet at craps, even-money bet at 38-number roulette, single-number bet at 38-number roulette) when the casino rebates half of his loss. The player's optimal number of coups is shown in the fifth line, and the minimum number of coups needed to ensure that the casino's expected profit is positive is shown in the third line from the bottom.

$p = 244/495,\ a = 1$		$p = 18/38,\ a = 1$		$p = 1/38,\ a = 35$	
n	expectation	n	expectation	n	expectation
89	.959 742	3	.257 618	230	8.778 139
90	.949 512	4	.219 182	231	8.786 200
91	.960 062	5	.273 977	232	8.791 992
92	.949 837	6	.235 800	233	8.795 512
93	.960 161	7	.275 100	234	8.796 758
94	.949 939	8	.237 137	235	8.795 728
95	.960 045	9	.266 779	236	8.792 422
96	.949 828	10	.229 003	237	8.786 839
97	.959 722	11	.251 903	238	8.778 980
$\vdots$	$\vdots$	$\vdots$	$\vdots$	$\vdots$	$\vdots$
379	.012 111	25	.053 852	927	.068 214
380	.002 272	26	.017 138	928	.039 835
381	.002 278	27	.017 797	929	.010 353
382	−.007 559	28	−.018 813	930	−.020 235
383	−.007 578	29	−.019 485	931	−.051 928
384	−.017 413	30	−.055 996	932	−.084 730

Given two jointly distributed discrete random variables X and Y, each with finite variance, we define the *covariance* between X and Y by

$$\operatorname{Cov}(X, Y) := \operatorname{E}[(X - \operatorname{E}[X])(Y - \operatorname{E}[Y])]. \tag{1.124}$$

If the variances are also nonzero, then we define the *correlation* between X and Y by

$$\operatorname{Corr}(X, Y) := \frac{\operatorname{Cov}(X, Y)}{\operatorname{SD}(X)\operatorname{SD}(Y)}. \tag{1.125}$$

By the Cauchy–Schwarz inequality, $-1 \le \operatorname{Corr}(X, Y) \le 1$.

Theorem 1.4.16. *Given two jointly distributed discrete random variables X and Y, each with finite variance,*

$$\operatorname{Cov}(X, Y) = \operatorname{E}[XY] - \operatorname{E}[X]\operatorname{E}[Y]. \tag{1.126}$$

In particular, if X and Y are independent, each with finite variance, then by Theorem 1.4.14 they are *uncorrelated*, that is, $\text{Cov}(X, Y) = 0$.

Theorem 1.4.17. *If X is a discrete random variable with finite variance and a is a constant, then aX has finite variance and $\text{Var}(aX) = a^2\text{Var}(X)$. If $X_1, X_2, \ldots, X_n$ are jointly distributed discrete random variables, each with finite variance, then $X_1 + \cdots + X_n$ has finite variance and*

$$\text{Var}\left(\sum_{i=1}^{n} X_i\right) = \sum_{i=1}^{n} \text{Var}(X_i) + 2 \sum_{1 \le i < j \le n} \sum \text{Cov}(X_i, X_j). \qquad (1.127)$$

More generally, if X and Y are jointly distributed discrete random variables, each with finite variance, and if a and b are constants, then $\text{Cov}(aX, bY) = ab\,\text{Cov}(X, Y)$. If $X_1, X_2, \ldots, X_n, Y_1, Y_2, \ldots, Y_m$ are jointly distributed discrete random variables, each with finite variance, then

$$\text{Cov}\left(\sum_{i=1}^{n} X_i, \sum_{j=1}^{m} Y_j\right) = \sum_{i=1}^{n}\sum_{j=1}^{m} \text{Cov}(X_i, Y_j). \qquad (1.128)$$

Example 1.4.18. *Variances of binomial, hypergeometric, and negative binomial distributions.* Let X be binomial(n, p). Using the notation of Example 1.4.12,

$$\begin{aligned} \text{Var}(X) &= \text{Var}(1_{E_1}) + \cdots + \text{Var}(1_{E_n}) \\ &= p(1 - p) + \cdots + p(1 - p) = np(1 - p). \end{aligned} \qquad (1.129)$$

Let X be hypergeometric(n, N, K). Using the notation of Example 1.4.12,

$$\begin{aligned} \text{Var}(X) &= \sum_{i=1}^{n} \text{Var}(1_{E_i}) + 2 \sum_{1 \le i < j \le n} \sum \text{Cov}(1_{E_i}, 1_{E_j}) \\ &= n\frac{K}{N}\left(1 - \frac{K}{N}\right) + 2\binom{n}{2}\left(\frac{K(K-1)}{N(N-1)} - \left(\frac{K}{N}\right)^2\right) \\ &= n\frac{K}{N}\left(1 - \frac{K}{N}\right)\frac{N-n}{N-1}. \end{aligned} \qquad (1.130)$$

With $p := K/N$ as in Example 1.4.12, the variance in (1.130) coincides with the variance in (1.129) except for the factor $(N - n)/(N - 1)$, which is called the *finite-population correction factor*.

Let X be negative binomial(n, p). Using the notation of Example 1.4.12,

$$\text{Var}(X) = \sum_{i=1}^{n} \text{Var}(X_i) = \sum_{i=1}^{n} \frac{1-p}{p^2} = \frac{n(1-p)}{p^2} \qquad (1.131)$$

by Example 1.4.9.   ♠

Example 1.4.19. *Hedging bets.* We say that one *hedges* a bet if, by making one or more additional bets, the variance of the composite bet is less than the variance of the original bet. Let us consider an example of this.

Let X denote the player's profit per unit bet from a pass-line bet, and let Y denote his profit per unit bet from an *any-craps bet*. The former was discussed in Examples 1.2.2 on p. 17 and 1.2.8 on p. 19, and the latter is a one-roll bet that a 2, 3, or 12 will appear. A win pays 7 to 1. The joint distribution of X and Y is given by

$$P((X,Y) = (i,j)) = \begin{cases} 244/495 & \text{if } (i,j) = (1,-1), \\ 55/495 & \text{if } (i,j) = (-1,7), \\ 196/495 & \text{if } (i,j) = (-1,-1), \\ 0 & \text{otherwise.} \end{cases} \qquad (1.132)$$

One can check that $\text{Var}(Y) = 512/81$ and $\text{Cov}(X,Y) = -3{,}904/4{,}455$. Let us now assume a bet of one unit on the pass line and b units on any craps. Then the variance of the composite bet satisfies

$$\text{Var}(X + bY) = \text{Var}(X) + 2b\,\text{Cov}(X,Y) + b^2\text{Var}(Y) < \text{Var}(X) \qquad (1.133)$$

if and only if

$$0 < b < 2\,\frac{|\text{Cov}(X,Y)|}{\text{Var}(Y)} = \frac{61}{220} \approx 0.277273. \qquad (1.134)$$

Typically, one chooses b to be $1/7$ in this hedge, so that a win from the any craps bet cancels the loss from the pass-line bet. We hasten to add that, as a general rule, one should not hedge one's bets in a subfair game. ♠

Let $I \subset \mathbf{R}$ be an interval. A function $f : I \mapsto \mathbf{R}$ is said to be *convex* (resp., *concave*) if

$$f(\lambda x + (1 - \lambda)y) \le \lambda f(x) + (1 - \lambda)f(y) \qquad (\text{resp.,} \ \ge) \qquad (1.135)$$

whenever $x, y \in I$, $x < y$, and $0 < \lambda < 1$. It is said to be *strictly convex* (resp., *strictly concave*) if the same statement holds with strict inequality in (1.135). Since f is convex if and only if $-f$ is concave, we will formulate our results in terms of convexity.

Lemma 1.4.20. *Assume that $f : I \mapsto \mathbf{R}$ is twice continuously differentiable. If $f'' \ge 0$ on I, then f is convex. If $f'' > 0$ on I, then f is strictly convex.*

The next theorem is known as *Jensen's inequality.* We continue to assume that $I \subset \mathbf{R}$ is an interval.

Theorem 1.4.21. *Let $f : I \mapsto \mathbf{R}$ be convex, and let X be a discrete random variable with values in I. If both X and $f(X)$ have finite expectation, then*

$$f(\mathrm{E}[X]) \le \mathrm{E}[f(X)]. \tag{1.136}$$

If, in addition, f is strictly convex, then equality holds in (1.136) if and only if X is degenerate.

Example 1.4.22. *Arithmetic-geometric mean inequality.* Let X be a discrete random variable with values in $I := (0, \infty)$ such that both X and $\ln(X)$ have finite expectation. Since $f(x) := \ln(x)$ is strictly concave on I, Jensen's inequality tells us that $\ln(\mathrm{E}[X]) \ge \mathrm{E}[\ln(X)]$, hence

$$e^{\mathrm{E}[\ln(X)]} \le \mathrm{E}[X]. \tag{1.137}$$

Moreover, strict inequality holds unless X is degenerate. The left side of (1.137) is called the *geometric mean* of X and the right side may be called the *arithmetic mean* of X, so the inequality is known as the *arithmetic-geometric mean inequality*. It follows that

$$(a_1 \cdots a_n)^{1/n} \le \frac{a_1 + \cdots + a_n}{n} \tag{1.138}$$

for all positive numbers $a_1, \ldots, a_n$, where $n \ge 2$, with strict inequality unless $a_1 = \cdots = a_n$. ♠

The *probability generating function* $f(z)$ of a nonnegative integer-valued random variable X is given by

$$f(z) := \mathrm{E}[z^X] = \sum_{n=0}^{\infty} \mathrm{P}(X = n)z^n, \tag{1.139}$$

and is defined for all complex z for which the series converges. An important consequence of Theorem 1.4.14 is that the probability generating function of a sum of independent nonnegative integer-valued random variables is the product of the individual probability generating functions. Here is an application of this property.

Example 1.4.23. *Distribution of the n-dice totals.* Let $U_1, U_2, \ldots, U_n$ be independent discrete uniform on $\{1, 2, 3, 4, 5, 6\}$, representing the results of n distinguishable (without loss of generality) dice. Then $U_1 + \cdots + U_n$ has probability generating function

$$\mathrm{E}[z^{U_1 + \cdots + U_n}] = (\mathrm{E}[z^{U_1}])^n = \left(\frac{z + z^2 + z^3 + z^4 + z^5 + z^6}{6} \right)^n$$

$$= \left(\frac{z(1 - z^6)}{6(1 - z)} \right)^n = 6^{-n} z^n (1 - z^6)^n (1 - z)^{-n} \tag{1.140}$$

if $z \ne 1$. We can apply the binomial theorem to the factor $(1 - z^6)^n$, and for the last factor we can use (by Theorem A.2.10 on p. 749)

$$(1-z)^{-n} = \sum_{k=0}^{\infty} \binom{k+n-1}{n-1} z^k, \qquad |z| < 1. \tag{1.141}$$

We conclude that

$$E[z^{U_1+\cdots+U_n}] = 6^{-n} z^n \sum_{j=0}^{n} \binom{n}{j} (-1)^j z^{6j} \sum_{k=0}^{\infty} \binom{k+n-1}{n-1} z^k$$

$$= \sum_{s=n}^{6n} \left[\frac{1}{6^n} \sum_{j=0}^{\lfloor (s-n)/6 \rfloor} (-1)^j \binom{n}{j} \binom{s-6j-1}{n-1} \right] z^s \tag{1.142}$$

whenever $|z| < 1$, where the second equality follows by noting that whenever $n + 6j + k = s$ and $k \geq 0$, we have $k + n = s - 6j$ and $j \leq (s-n)/6$. In particular,

$$P(U_1 + \cdots + U_n = s) = \frac{1}{6^n} \sum_{j=0}^{\lfloor (s-n)/6 \rfloor} (-1)^j \binom{n}{j} \binom{s-6j-1}{n-1} \tag{1.143}$$

for $s = n, n+1, \ldots, 6n$. ♠

Example 1.4.24. *Probability that exactly j of n events occur.* Given an integer $n \geq 2$, let $E_1, \ldots, E_n$ be arbitrary events, and define

$$N := \sum_{i=1}^{n} 1_{E_i} \tag{1.144}$$

to be the number of the events $E_1, \ldots, E_n$ that occur. We claim that

$$P(N = j) = \sum_{k=j}^{n} (-1)^{k-j} \binom{k}{j} S_k \tag{1.145}$$

for $j = 0, 1, \ldots, n$, where

$$S_0 := 1, \qquad S_1 := \sum_{i=1}^{n} P(E_i), \qquad S_2 := \sum_{1 \leq i < j \leq n} P(E_i \cap E_j), \tag{1.146}$$

and so on. More generally, for $1 \leq m \leq n$, S_m is given by (1.30) on p. 13. Since $P(E_1 \cup E_2 \cup \cdots \cup E_n) = P(N \geq 1) = 1 - P(N = 0)$, this result can be regarded as a generalization of the inclusion-exclusion law.

To prove (1.145), the probability generating function of N can be written

$$\sum_{j=0}^{n} P(N = j) z^j = E[z^N]$$

$$= \mathrm{E}\left[\prod_{i=1}^{n}(1_{E_i^c} + z\, 1_{E_i})\right]$$

$$= \mathrm{E}\left[\prod_{i=1}^{n}\{1 + (z-1)1_{E_i}\}\right]$$

$$= 1 + (z-1)S_1 + (z-1)^2 S_2 + \cdots + (z-1)^n S_n$$

$$= \sum_{k=0}^{n}(z-1)^k S_k$$

$$= \sum_{k=0}^{n}\sum_{j=0}^{k}\binom{k}{j}z^j(-1)^{k-j}S_k$$

$$= \sum_{j=0}^{n}\left[\sum_{k=j}^{n}(-1)^{k-j}\binom{k}{j}S_k\right]z^j, \qquad (1.147)$$

and comparing coefficients of z^j, we have (1.145). ♠

Finally, we show that generating functions can be useful even when the underlying sequence of coefficients is not necessarily a probability distribution.

Example 1.4.25. *Success runs.* Fix a positive integer r, and let $p \in (0,1)$ and $q := 1 - p$. For each $n \geq 0$ we let P_n denote the probability that, in a sequence of n independent Bernoulli trials, each with success probability p, there is a run of at least r consecutive successes. Observe that $P_0 = P_1 = \cdots = P_{r-1} = 0$, $P_r = p^r$, and

$$P_{n+1} = P_n + (1 - P_{n-r})qp^r, \qquad n \geq r. \qquad (1.148)$$

Indeed, a success run of length r or more occurs in $n+1$ trials if it occurs in n trials, or if it occurs in $n+1$ but not n trials. In the latter case, the first $n-r$ trials cannot have a success run of length r or more, trial $n-r+1$ must result in failure, and the last r trials must result in r consecutive successes. Letting $Q_n := 1 - P_n$ for each $n \geq 0$, we have $Q_0 = Q_1 = \cdots = Q_{r-1} = 1$, $Q_r = 1 - p^r$, and

$$Q_{n+1} = Q_n - qp^r Q_{n-r}, \qquad n \geq r. \qquad (1.149)$$

Introducing the generating function

$$f(z) := \sum_{n=0}^{\infty} Q_n z^n, \qquad |z| < 1, \qquad (1.150)$$

we find that

$$(1 - z + qp^r z^{r+1})f(z)$$

$$= Q_0 + (Q_1 - Q_0)z + \cdots + (Q_{r-1} - Q_{r-2})z^{r-1}$$

$$+ (Q_r - Q_{r-1})z^r + \sum_{n=r}^{\infty}(Q_{n+1} - Q_n + qp^r Q_{n-r})z^{n+1}$$

$$= 1 - p^r z^r, \tag{1.151}$$

where the first equality uses (1.150) and the second uses (1.149) and the values of $Q_0, Q_1, \ldots, Q_r$, and hence

$$f(z) = \frac{1 - p^r z^r}{1 - z(1 - qp^r z^r)}$$

$$= (1 - p^r z^r)\sum_{m=0}^{\infty} z^m (1 - qp^r z^r)^m$$

$$= (1 - p^r z^r)\sum_{m=0}^{\infty}\sum_{k=0}^{m}(-1)^k \binom{m}{k}(qp^r)^k z^{m+kr}$$

$$= (1 - p^r z^r)\sum_{n=0}^{\infty} A(n,r)z^n, \tag{1.152}$$

where

$$A(n,r) := \sum_{k=0}^{\lfloor n/(r+1)\rfloor}(-1)^k \binom{n - kr}{k}(qp^r)^k. \tag{1.153}$$

It follows that

$$Q_n = [\text{coefficient of } z^n \text{ in } f(z)] = A(n,r) - p^r A(n-r,r) \tag{1.154}$$

for all $n \geq r$. We conclude that

$$P_n = 1 - A(n,r) + p^r A(n-r,r), \qquad n \geq r. \tag{1.155}$$

If n is much larger than r, then the sum (1.153) contains large binomial coefficients with alternating signs, and high-precision arithmetic may be required for its numerical evaluation.

Let us consider a special case. We take $r = 4$, $n = 21$, and $p = \frac{1}{2}$. Since

$$A(21,4) = \binom{21}{0}2^{-0} - \binom{17}{1}2^{-5} + \binom{13}{2}2^{-10} - \binom{9}{3}2^{-15} + \binom{5}{4}2^{-20} \tag{1.156}$$

and

$$A(17,4) = \binom{17}{0}2^{-0} - \binom{13}{1}2^{-5} + \binom{9}{2}2^{-10} - \binom{5}{3}2^{-15}, \tag{1.157}$$

we have

$$P_{21} = 1 - A(21,4) + 2^{-4}A(17,4) = \frac{521{,}063}{1{,}048{,}576} \approx 0.496924, \qquad (1.158)$$

which can be interpreted as the probability of at least four consecutive heads in 21 fair coin tosses.   ♠

1.5 Law of Large Numbers and Central Limit Theorem

This section contains several limit theorems, so we begin by defining three types of limits of a sequence of random variables. If $X_1, X_2, \ldots$ are jointly distributed discrete random variables and are jointly distributed with the random variable X, we say that X_n converges to X *with probability 1* or *almost surely* and write $X_n \to X$ a.s. if

$$P\left(\lim_{n \to \infty} X_n = X \right) = 1. \qquad (1.159)$$

If $Y_1, Y_2, \ldots$ are discrete random variables not necessarily jointly distributed and c is a constant, we say that Y_n converges *in probability* to c and write $Y_n \to c$ in probability if

$$\lim_{n \to \infty} P(|Y_n - c| < \varepsilon) = 1 \qquad (1.160)$$

for every $\varepsilon > 0$. If $Z_1, Z_2, \ldots$ are discrete random variables not necessarily jointly distributed, we say that Z_n converges *in distribution* to the standard-normal distribution and write $Z_n \to N(0,1)$ in distribution or

$$Z_n \xrightarrow{d} N(0,1) \qquad (1.161)$$

if

$$\lim_{n \to \infty} P(Z_n \leq z) = \Phi(z) := \frac{1}{\sqrt{2\pi}} \int_{-\infty}^{z} e^{-x^2/2}\, dx \qquad (1.162)$$

for every real number z. The function Φ is called the *standard-normal cumulative distribution function*, and its approximate values can be either looked up in a table or evaluated in a spreadsheet.

In fact it will be useful to generalize the notion of almost sure convergence by defining an event E to hold *almost surely* (or *a.s.*) if $P(E) = 1$. For example, $X = Y$ a.s. if $P(X = Y) = 1$.

We begin with two inequalities, first *Markov's inequality* and then *Chebyshev's inequality*.

Lemma 1.5.1. *If X is a nonnegative discrete random variable with finite mean, then*

$$P(X \geq \varepsilon) \leq \frac{E[X]}{\varepsilon} \qquad (1.163)$$

for every $\varepsilon > 0$.

Lemma 1.5.2. *Let X be a discrete random variable with finite variance. Then*

$$P(|X - E[X]| \geq \varepsilon) \leq \frac{\mathrm{Var}(X)}{\varepsilon^2} \qquad (1.164)$$

for every $\varepsilon > 0$.

We can now state the *weak law of large numbers*.

Theorem 1.5.3. *Let $X_1, X_2, \ldots$ be a sequence of independent and identically distributed (i.i.d.) discrete random variables with finite mean μ, and put $S_n := X_1 + \cdots + X_n$ for each $n \geq 1$. Then $S_n/n \to \mu$ in probability.*

If we assume, in addition to the hypotheses of Theorem 1.5.3, that X_1 has finite variance σ^2, then Chebyshev's inequality (Lemma 1.5.2) implies that

$$P\left(\left|\frac{S_n}{n} - \mu\right| \geq \varepsilon\right) \leq \frac{\mathrm{Var}(S_n/n)}{\varepsilon^2} = \frac{\sigma^2}{n\varepsilon^2}, \qquad (1.165)$$

and the theorem follows.

Let $E_1, E_2, \ldots$ be a sequence of events. The event $\{E_n$ infinitely often$\}$ is defined by

$$\{E_n \text{ i.o.}\} := \left\{\sum_{n=1}^{\infty} 1_{E_n} = \infty\right\} = \bigcap_{n=1}^{\infty} \bigcup_{m=n}^{\infty} E_m. \qquad (1.166)$$

The next result is known as the *Borel–Cantelli lemma*.

Lemma 1.5.4. *If $E_1, E_2, \ldots$ is a sequence of events for which $\sum_{n=1}^{\infty} P(E_n) < \infty$, then $P(E_n \text{ i.o.}) = 0$.*

We now state the *strong law of large numbers*.

Theorem 1.5.5. *Let $X_1, X_2, \ldots$ be a sequence of i.i.d. discrete random variables with finite mean μ, and put $S_n := X_1 + \cdots + X_n$ for each $n \geq 1$. Then $S_n/n \to \mu$ a.s.*

If we assume, in addition to the hypotheses of Theorem 1.5.5, that X_1 has finite fourth moment, then we can give an elementary proof of the strong law. It is enough to treat the case $\mu = 0$, and then apply that case to the sequence $X_1 - \mu, X_2 - \mu, \ldots$. First, by the multinomial theorem (Theorem 1.1.7 on p. 5) and the i.i.d. mean-zero property,

$$E[S_n^4] = E\left[\sum_{m_1 \geq 0, \ldots, m_n \geq 0: \, m_1 + \cdots + m_n = 4} \cdots \sum \binom{4}{m_1, \ldots, m_n} X_1^{m_1} \cdots X_n^{m_n}\right]$$

$$= \binom{n}{1}\binom{4}{4} E[X_1^4] + \binom{n}{1, 1, n-2}\binom{4}{3} E[X_1^3 X_2]$$

$$+ \binom{n}{2}\binom{4}{2}\mathrm{E}[X_1^2 X_2^2] + \binom{n}{1,2,n-3}\binom{4}{2,1,1}\mathrm{E}[X_1^2 X_2 X_3]$$

$$+ \binom{n}{4}\binom{4}{1,1,1,1}\mathrm{E}[X_1 X_2 X_3 X_4].$$

$$= n\mathrm{E}[X_1^4] + 3n(n-1)(\mathrm{E}[X_1^2])^2, \qquad n \geq 4. \tag{1.167}$$

Second, by Markov's inequality (Lemma 1.5.1) and (1.167), there is a constant K depending only on the distribution of X_1 such that, for every $\varepsilon > 0$,

$$\mathrm{P}(|S_n/n| \geq \varepsilon) = \mathrm{P}((S_n/n)^4 \geq \varepsilon^4) \leq \frac{\mathrm{E}[S_n^4]}{n^4 \varepsilon^4} \leq \frac{K}{n^2 \varepsilon^4}, \qquad n \geq 4, \tag{1.168}$$

and therefore, by the Borel–Cantelli lemma, $\mathrm{P}(|S_n/n| \geq \varepsilon \text{ i.o.}) = 0$ for every $\varepsilon > 0$. This implies that $\mathrm{P}(\lim_{n\to\infty} S_n/n = 0) = 1$, as required.

Corollary 1.5.6. *Let $X_1, X_2, \ldots$ be a sequence of i.i.d. discrete nonnegative random variables with infinite mean, and put $S_n := X_1 + \cdots + X_n$ for each $n \geq 1$. Then $S_n/n \to \infty$ a.s.*

Example 1.5.7. *3-4-5-times free odds and the concept of house advantage.* We have already introduced the pass-line bet at craps (Examples 1.2.2 on p. 17 and 1.2.8 on p. 19). A free odds bet associated with a pass-line bet becomes available when a point is established. This is in effect a side bet that the original bet will be won, and it pays fair odds. More precisely, the payoff odds are 2 to 1 if the point is 4 or 10, 3 to 2 if the point is 5 or 9, and 6 to 5 if the point is 6 or 8. Because it is a fair bet, its size is limited: If the casino offers 3-4-5-times free odds, the pass-line bettor is permitted to bet no more on free odds than 3 times the amount of his pass-line bet if the point is 4 or 10, 4 times the amount of his pass-line bet if the point is 5 or 9, and 5 times the amount of his pass-line bet if the point is 6 or 8. Let us assume in what follows that he makes a maximal free odds bet.

Notice that the amount bet is random. If we denote by B the amount bet and by X the gambler's profit, we can evaluate the joint distribution of B and X much as in Example 1.2.8 on p. 19. The results are summarized in Table 1.10. (Actually, we do not need the joint distribution for this example, but it may help to clarify the wager in question.)

Let $(B_1, X_1), (B_2, X_2), \ldots$ be a sequence of independent, identically distributed random vectors whose common distribution is that of (B, X), representing the results of independent repetitions of the original one-unit pass-line bet with 3-4-5-times free odds. Then

$$\frac{-(X_1 + \cdots + X_n)}{B_1 + \cdots + B_n} \tag{1.169}$$

represents the ratio of the gambler's cumulative loss after n such wagers to his total amount bet. Dividing both numerator and denominator by n and applying the strong law of large numbers twice, we see that this ratio

Table 1.10 The distribution of (B, X) for a one-unit pass-line bet with 3-4-5-times free odds. (B = amount bet, X = gambler's profit.)

event	B	X	probab. $\times 990$
natural 7 or 11	1	1	220
craps 2, 3, or 12	1	-1	110
point 4 or 10 made	4	7	55
point 4 or 10 missed	4	-4	110
point 5 or 9 made	5	7	88
point 5 or 9 missed	5	-5	132
point 6 or 8 made	6	7	125
point 6 or 8 missed	6	-6	150

converges almost surely to

$$H_0(B, X) := \frac{-E[X]}{E[B]} = \frac{7/495}{34/9} = \frac{7}{1,870}, \tag{1.170}$$

which we define to be the *house advantage* of the wager (B, X). Chapter 6 explores this topic in greater detail. ♠

Example 1.5.8. *Asymptotic behavior of a betting system.* Let Y_0 be a constant random variable, and let $(X_1, Y_1), (X_2, Y_2), \ldots$ be a sequence of i.i.d. discrete random vectors with X_1 and Y_1 having finite expectation. If we define

$$S_n = X_1 + \cdots + X_n \quad \text{and} \quad T_n = Y_0 + \cdots + Y_{n-1}, \tag{1.171}$$

then

$$S_n/n \to \mu := E[X_1] \quad \text{and} \quad T_n/n \to \nu := E[Y_1] \tag{1.172}$$

with probability 1 by the strong law of law numbers. We claim that therefore

$$\frac{1}{n^2} \sum_{k=1}^{n} T_k X_k \to \frac{\nu\mu}{2} \tag{1.173}$$

with probability 1.

To see how this might be useful, let Y_0 be a positive integer, let $X_1, X_2, \ldots$ be i.i.d. with $P(X_1 = 1) = p$ and $P(X_1 = -1) = 1 - p$, where $0 < p < \frac{1}{2}$, and define $Y_n := -X_n$ for $n = 1, 2, \ldots$. Consider a gambler whose initial bet is Y_0 units and who increases his bet by one unit after each loss and decreases it by one unit after each win, at least until his bet size reaches 0, at which time he stops betting. Then the sum in (1.173) represents the gambler's fortune after

n independent coups, each of which is won with probability p, provided he has not stopped betting at that point. Here $\nu\mu/2$ becomes $-(1-2p)^2/2 < 0$, showing that, with positive probability the gambler's fortune tends to $-\infty$ at a quadratic rate. This betting system is essentially the *d'Alembert system*, which is discussed in detail in Example 8.1.5 on p. 289.

To prove that (1.173) holds with probability 1, it will suffice by the strong law of large numbers to demonstrate that (1.173) holds on the event

$$\left\{\frac{S_n}{n} \to \mu, \ \frac{T_n}{n} \to \nu, \ \frac{1}{n}\sum_{k=1}^{n}|X_k| \to \mathrm{E}\big[|X_1|\big]\right\}. \tag{1.174}$$

We begin by writing

$$\frac{1}{n^2}\sum_{k=1}^{n}T_kX_k = \frac{1}{n^2}\sum_{k=1}^{n}\left(\frac{T_k}{k} - \nu\right)kX_k + \frac{\nu}{n^2}\sum_{k=1}^{n}kX_k. \tag{1.175}$$

The first of the two terms on the right is bounded in absolute value by

$$\frac{1}{n}\sum_{k=1}^{n}\left|\frac{T_k}{k} - \nu\right||X_k|, \tag{1.176}$$

and this tends to 0 on the event (1.174). The second term on the right side of (1.175) equals

$$\frac{\nu}{n^2}\sum_{k=1}^{n}\sum_{j=1}^{k}X_k = \frac{\nu}{n^2}\sum_{j=1}^{n}\sum_{k=j}^{n}X_k = \frac{\nu}{n^2}\sum_{j=1}^{n}(S_n - S_{j-1})$$

$$= \frac{\nu S_n}{n} - \frac{\nu}{n}\sum_{l=1}^{n-1}\frac{l}{n}\left(\frac{S_l}{l} - \mu\right) - \frac{\nu\mu}{2}\left(1 - \frac{1}{n}\right), \tag{1.177}$$

where $S_0 := 0$. This converges to the required limit on the event (1.174). ♠

Example 1.5.9. *Kelly system.* Consider a game of chance that pays even money and is advantageous to the gambler. Let X represent the gambler's profit per unit bet, so that $\mathrm{P}(X = 1) = p$ and $\mathrm{P}(X = -1) = 1 - p$, where $\frac{1}{2} < p < 1$.

Now let $X_1, X_2, \ldots$ be i.i.d. with common distribution that of X. Here X_n represents the gambler's profit per unit bet at coup n. A gambler is said to use *proportional play* if he bets a fixed proportion $f \in [0, 1]$ of his current fortune at each coup. In this case, his fortune F_n after n coups satisfies $F_n = F_{n-1}(1 + fX_n)$ for each $n \geq 1$, and therefore

$$F_n = F_0 \prod_{l=1}^{n}(1 + fX_l), \qquad n \geq 1. \tag{1.178}$$

To determine the optimal choice of the betting proportion f, we define

$$r_n(f) := n^{-1} \ln(F_n/F_0), \qquad n \geq 1, \tag{1.179}$$

and

$$\mu(f) := \mathrm{E}[\ln(1 + fX)] = p \ln(1 + f) + (1 - p) \ln(1 - f) \tag{1.180}$$

for all $f \in [0, 1)$. Notice that we can interpret $r_n(f)$ as the average geometric rate of growth of the proportional bettor's fortune over the first n coups. Since $X_1, X_2, \ldots$ are i.i.d., so too are $\ln(1 + fX_1), \ln(1 + fX_2), \ldots$. By the strong law of large numbers,

$$\lim_{n \to \infty} r_n(f) = \lim_{n \to \infty} \frac{1}{n} \sum_{l=1}^{n} \ln(1 + fX_l) = \mu(f) \quad \text{a.s.,} \tag{1.181}$$

so $\mu(f)$ can be interpreted as the long-term geometric rate of growth of the proportional bettor's fortune.

The choice f^* of f that maximizes $\mu(f)$, namely $f^* = 2p - 1$, results in a betting system known as the *Kelly system* or as optimal proportional play. Chapter 10 explores this system in greater detail and generality. ♠

Example 1.5.10. *Elementary renewal theorem.* Let $T_1, T_2, \ldots$ be a sequence of i.i.d. positive-integer-valued random variables with mean $\mu \in [1, \infty]$, and put $R_k := T_1 + \cdots + T_k$ for each $k \geq 1$ and $R_0 := 0$. Define $N_n := \max\{k \geq 0 : R_k \leq n\}$ for each $n \geq 1$. We think of $R_1, R_2, \ldots$ as the sequence of times at which renewals occur, so that $T_1, T_2, \ldots$ is the sequence of times between successive renewals and N_n is the number of renewals by time n. We claim that

$$\lim_{n \to \infty} \frac{N_n}{n} = \frac{1}{\mu} \quad \text{a.s.,} \tag{1.182}$$

where $1/\infty := 0$. To see this, observe that $R_{N_n} \leq n < R_{N_n+1}$ and hence

$$\frac{R_{N_n}}{N_n} \leq \frac{n}{N_n} < \frac{R_{N_n+1}}{N_n + 1} \frac{N_n + 1}{N_n} \tag{1.183}$$

for all $n \geq T_1$. Now $N_1 \leq N_2 \leq \cdots$ and clearly $\lim_{n \to \infty} N_n = \infty$ a.s. Therefore, letting $n \to \infty$ in (1.183), the strong law of large numbers (Theorem 1.5.5) or Corollary 1.5.6 implies (1.182).

Since $N_n \leq n$ for each $n \geq 1$, the bounded convergence theorem (Theorem A.2.1 on p. 748) implies that

$$\lim_{n \to \infty} \frac{\mathrm{E}[N_n]}{n} = \frac{1}{\mu}. \tag{1.184}$$

This is known as the *elementary renewal theorem*. For a stronger result, see Theorem 4.4.1 on p. 146. ♠

Example 1.5.11. *St. Petersburg paradox, revisited.* Example 1.4.5 on p. 31 introduced the St. Petersburg game, which pays out $X := 2^{N-1}$ units to the player, where N is geometric($\frac{1}{2}$), with the result that $E[X] = \infty$. Let $N_1, N_2, \ldots$ be i.i.d. as N, and define $X_n := 2^{N_n - 1}$ and $S_n := X_1 + \cdots + X_n$ for each $n \geq 1$. By Corollary 1.5.6, $\lim_{n\to\infty} S_n/n = \infty$ a.s. It seems clear that the entry fee that one should be willing to pay to play this game should depend in some way on the number of coups one is allowed to play. In fact, it can be shown that $S_n/(\frac{1}{2}n \log_2 n) \to 1$ in probability. Thus, an entry fee of $\frac{1}{2}\log_2 n$ units per coup is fair (in an asymptotic sense) for the player who plays n coups. See Problem 1.52 on p. 62 for the proof.   ♠

The asymptotic behavior of $n!$ is given by *Stirling's formula*:

Theorem 1.5.12. *As $n \to \infty$,*

$$n! \sim \left(\frac{n}{e}\right)^n \sqrt{2\pi n}, \tag{1.185}$$

where $a_n \sim b_n$ means that $a_n/b_n \to 1$.

Our next result is *De Moivre's limit theorem.*

Theorem 1.5.13. *Let $E_1, E_2, \ldots$ be a sequence of independent events, each with probability $p \in (0,1)$, and put $S_n := 1_{E_1} + \cdots + 1_{E_n}$ for each $n \geq 1$. Then*

$$\frac{S_n - np}{\sqrt{np(1-p)}} \xrightarrow{d} N(0,1). \tag{1.186}$$

Notice that S_n here is binomial(n, p), so Theorem 1.5.13 can be proved directly from Stirling's formula (Theorem 1.5.12), though not without difficulty.

Like the binomial distribution (sampling with replacement), the hypergeometric distribution (sampling without replacement) is also asymptotically normal.

Theorem 1.5.14. *Let $S_{n,N,K}$ be hypergeometric(n, N, K). Then*

$$\frac{S_{n,N,K} - n(K/N)}{\sqrt{n(K/N)(1 - K/N)(N-n)/(N-1)}} \xrightarrow{d} N(0,1), \tag{1.187}$$

provided $n, N, K \to \infty$ in such a way that $n/N \to \alpha \in (0,1)$ and $K/N \to p \in (0,1)$.

Here is the *central limit theorem* in the i.i.d. setting.

Theorem 1.5.15. *Let $X_1, X_2, \ldots$ be a sequence of i.i.d. discrete random variables, each with finite mean μ and finite variance $\sigma^2 > 0$, and put $S_n := X_1 + \cdots + X_n$ for each $n \geq 1$. Then*

$$\frac{S_n - n\mu}{\sqrt{n\sigma^2}} \xrightarrow{d} N(0,1). \tag{1.188}$$

If we assume, in addition to the hypotheses of Theorem 1.5.15, that X_1 has finite third moment, then we can sketch an elementary proof. It is enough to treat the case $\mu = 0$ and $\sigma^2 = 1$, and then apply that case to the sequence $(X_1 - \mu)/\sigma, (X_2 - \mu)/\sigma, \ldots$. First, by Problem 1.55(b) on p. 63, it suffices to prove that

$$\lim_{n \to \infty} \mathrm{E}[f(S_n/\sqrt{n})] = \frac{1}{\sqrt{2\pi}} \int_{-\infty}^{\infty} f(z) \, e^{-z^2/2} \, dz \qquad (1.189)$$

for arbitrary bounded continuous $f : \mathbf{R} \mapsto \mathbf{R}$ with three bounded continuous derivatives. Let $Y_1, Y_2, \ldots$ be a sequence of i.i.d. random variables, independent of $X_1, X_2, \ldots$, with $\mathrm{P}(Y_1 = 1) = \mathrm{P}(Y_1 = -1) = \frac{1}{2}$, and put $T_n := Y_1 + \cdots + Y_n$. Define, for $1 \le k \le n$,

$$Z_{n,k} := X_1 + \cdots + X_{k-1} + Y_{k+1} + \cdots + Y_n, \qquad (1.190)$$

and observe that

$$\mathrm{E}[f(S_n/\sqrt{n})] - \mathrm{E}[f(T_n/\sqrt{n})]$$
$$= \sum_{k=1}^{n} \left\{ \mathrm{E}\left[f\left(\frac{Z_{n,k} + X_k}{\sqrt{n}}\right)\right] - \mathrm{E}\left[f\left(\frac{Z_{n,k} + Y_k}{\sqrt{n}}\right)\right] \right\}. \qquad (1.191)$$

But, by a third-order Taylor expansion,

$$\left| \mathrm{E}\left[f\left(\frac{Z_{n,k} + X_k}{\sqrt{n}}\right) - \left\{ f\left(\frac{Z_{n,k}}{\sqrt{n}}\right) + \frac{X_k}{\sqrt{n}} f'\left(\frac{Z_{n,k}}{\sqrt{n}}\right) + \frac{X_k^2}{2n} f''\left(\frac{Z_{n,k}}{\sqrt{n}}\right) \right\} \right] \right|$$
$$\le \frac{\mathrm{E}[|X_1|^3]}{6n^{3/2}} \sup_{x \in \mathbf{R}} |f'''(x)|, \qquad (1.192)$$

and since $\mu = 0$ and $\sigma^2 = 1$, the left side of (1.192) is simply

$$\left| \mathrm{E}\left[f\left(\frac{Z_{n,k} + X_k}{\sqrt{n}}\right)\right] - \mathrm{E}\left[f\left(\frac{Z_{n,k}}{\sqrt{n}}\right)\right] - \frac{1}{2n} \mathrm{E}\left[f''\left(\frac{Z_{n,k}}{\sqrt{n}}\right)\right] \right|. \qquad (1.193)$$

Now the same inequality holds with X_k replaced by Y_k (and $\mathrm{E}[|X_1|^3]$ replaced by 1), so by (1.191), there is a constant K depending only on the distribution of X_1 such that

$$|\mathrm{E}[f(S_n/\sqrt{n})] - \mathrm{E}[f(T_n/\sqrt{n})]| \le \frac{K}{\sqrt{n}} \sup_{x \in \mathbf{R}} |f'''(x)|. \qquad (1.194)$$

Since we know that

$$\lim_{n \to \infty} \mathrm{E}[f(T_n/\sqrt{n})] = \frac{1}{\sqrt{2\pi}} \int_{-\infty}^{\infty} f(z) \, e^{-z^2/2} \, dz \qquad (1.195)$$

by De Moivre's limit theorem (Theorem 1.5.13) and Problem 1.55(a) on p. 63, (1.189) follows.

Of course, Theorem 1.5.13 is a special case of Theorem 1.5.15. In either case we are led to the useful approximation

$$P\left(\frac{S_n - n\mu}{\sqrt{n\sigma^2}} \leq z\right) \approx \Phi(z) \tag{1.196}$$

with $\mu = p$ and $\sigma^2 = p(1-p)$ in the case of Theorem 1.5.13. Typically, the left side of (1.196) is a step function in z, while the right side of (1.196) is of course continuous in z. If the left side is constant on the interval $[a, b)$ but not on any strictly larger interval, then we have some discretion over the choice of z, and the choice $z = (a+b)/2$ seems to give the most reliable approximation. The next example illustrates this *continuity correction*.

Example 1.5.16. *Normal approximation to a binomial.* If an experimenter tosses a pair of dice 500 times, what is the probability that the number of sevens obtained is 74 or fewer? We can use Theorem 1.5.13 with $n = 500$ and $p = \pi_7 = 1/6$. Then

$$P(S_n \leq 74) = P\left(\frac{S_n - np}{\sqrt{np(1-p)}} \leq \frac{74.5 - 500(1/6)}{\sqrt{500(1/6)(5/6)}}\right) \tag{1.197}$$

$$\approx \Phi\left(\frac{74.5 - 500(1/6)}{\sqrt{500(1/6)(5/6)}}\right) = \Phi\left(-\frac{53}{50}\right) \approx 0.144572.$$

The exact probability is

$$P(S_n \leq 74) = \sum_{k=0}^{74} \binom{500}{k}\left(\frac{1}{6}\right)^k \left(\frac{5}{6}\right)^{500-k} \approx 0.144115, \tag{1.198}$$

so with the continuity correction, the approximation is reasonably good. ♠

Example 1.5.17. *Normal approximation to a hypergeometric.* If exactly half of the cards are dealt from a six-deck (312-card) shoe, what is the probability that at least one-third of the undealt cards have denomination 10, J, Q, or K? This is of interest in twenty-one. Notice that at least 52 of the undealt cards are of the specified type if and only if at most 44 of the dealt cards are of the specified type. We can use Theorem 1.5.14 with $n = 156$, $N = 312$, and $K = 96$. Then

$$P(S_{n,N,K} \leq 44) = P\left(\frac{S_{n,N,K} - n(K/N)}{\sqrt{n(K/N)(1 - K/N)(N-n)/(N-1)}}\right.$$

$$\left. \leq \frac{44.5 - 156(4/13)}{\sqrt{156(4/13)(9/13)(156/311)}}\right)$$

$$\approx \Phi\left(\frac{44.5 - 156(4/13)}{\sqrt{156(4/13)(9/13)(156/311)}}\right)$$

$$\approx \Phi(-0.857266) \approx 0.195649, \tag{1.199}$$

where again we have used a continuity correction. The exact probability is

$$P(S_{n,N,K} \leq 44) = \sum_{k=0}^{44} \frac{\binom{96}{k}\binom{216}{156-k}}{\binom{312}{156}} \approx 0.195300, \qquad (1.200)$$

so the approximation is again quite good. ♠

Example 1.5.18. *Fey "Liberty Bell" and the bell-shaped curve.* Let R denote the number of coins paid out from a one-coin bet on the Fey "Liberty Bell" slot machine of Examples 1.4.1 on p. 28 and 1.4.7 on p. 31. We put

$$\mu := \mathrm{E}[R] = 0.756 \quad \text{and} \quad \sigma^2 := \mathrm{Var}(R) = 3.028464. \qquad (1.201)$$

We also let $R_1, R_2, \ldots$ be a sequence of i.i.d. random variables with common distribution that of R, representing the results of a sequence of coups at this machine. Then

$$\overline{R}_n := \frac{R_1 + R_2 + \cdots + R_n}{n} \qquad (1.202)$$

represents the proportion of coins paid out after n coups.

By the weak law of large numbers,

$$\lim_{n \to \infty} P(|\overline{R}_n - \mu| < \varepsilon) = 1, \qquad \varepsilon > 0, \qquad (1.203)$$

and by the central limit theorem,

$$\lim_{n \to \infty} P\left(|\overline{R}_n - \mu| \leq z_{1-\alpha/2}\frac{\sigma}{\sqrt{n}}\right) = 1 - \alpha \qquad (1.204)$$

for $0 < \alpha < 1$, where $z_{1-\alpha/2} := \Phi^{-1}(1 - \alpha/2)$ is a standard-normal *quantile*. Both results tell us that the proportion $\overline{R}_n$ of coins paid out after n coups converges to μ in some sense, but because the central limit theorem uses the variance and not just the mean, it is able to quantify the rate at which this convergence occurs. ♠

Example 1.5.19. *Proportional play and the central limit theorem.* Under the assumptions and notation of Example 1.5.9, we define

$$\sigma^2(f) := \mathrm{Var}(\ln(1 + fX_1)) > 0. \qquad (1.205)$$

By the central limit theorem,

$$Z_n := \frac{\sqrt{n}}{\sigma(f)}\left(\frac{1}{n}\ln\left(\frac{F_n}{F_0}\right) - \mu(f)\right) \xrightarrow{d} N(0,1). \qquad (1.206)$$

Hence

$$F_n = F_0 e^{\mu(f)n + \sigma(f)\sqrt{n}Z_n}. \qquad (1.207)$$

This helps to clarify the asymptotic behavior of the proportional bettor's
fortune.   ♠

1.6 Problems

1.1. *Twenty-one subsets.* Find the number of distinguishable subsets of a
52-card twenty-one deck. (If the cards were distinguishable, as in poker for
example, the answer would be 2^{52}. But in twenty-one, suits are disregarded
and denominations 10, J, Q, and K are indistinguishable.)

1.2. *Baccarat coups.* Baccarat is played with eight standard 52-card decks
mixed together. As in twenty-one, suits are disregarded and denominations
10, J, Q, and K are indistinguishable. Two two-card hands are dealt (one to
player and one to banker), and to each hand a third card may or may not be
drawn. Thus, each coup requires at least four cards and at most six cards. An
upper bound on the number of distinguishable coups that has been suggested
is

$$\binom{6+10-1}{6}\binom{6}{2,2,1,1} = (5{,}005)(180) = 900{,}900, \tag{1.208}$$

where the first factor represents the number of distinguishable six-card subsets and the second factor represents the number of ways of dividing six cards
into two two-card hands and two third cards.

(a) Noticing that this bound counts some coups more than once, give a
smaller upper bound that avoids such duplication.

(b) After reading the rules of baccarat (Section 19.1), find the exact number of distinguishable coups.

1.3. *A combinatorial identity.* Prove the following combinatorial identity:
For integers $n \geq k \geq 1$,

$$\binom{n}{k} = \sum_{l=k}^{n} \binom{l-1}{k-1}. \tag{1.209}$$

Hint: Solution 1: Let $\mathscr{S} := \{A \subset \{1, \ldots, n\} : |A| = k\}$, and observe that

$$\mathscr{S} = \bigcup_{l=k}^{n} \{A \in \mathscr{S} : \max A = l\}. \tag{1.210}$$

Solution 2: Use (1.7) on p. 4 and induction.

1.4. *Three-dice totals.* Show, without using the concept of independence,
that the probability p_j of rolling a total of $j \in \{3, 4, 5, \ldots, 18\}$ with three
dice is given by

$$p_j = \frac{\binom{j-1}{2} \wedge (28 - |2j - 21|) \wedge \binom{20-j}{2}}{216}. \qquad (1.211)$$

Hint: By Theorem 1.1.8 on p. 6, there are $\binom{3+6-1}{3} = 56$ distinguishable outcomes if the dice are indistinguishable. Each of these outcomes appears $\binom{3}{3} = 1$, $\binom{3}{2} = 3$, or $\binom{3}{1,1,1} = 6$ times in the list of $6^3 = 216$ equally likely outcomes when the dice are distinguishable.

1.5. *Finer classification of no-pair poker hands.* By Table 1.3 on p. 8, 1,302,540 of the 2,598,960 poker hands are classified as no pair. Determine the numbers of such hands that can be described as ace high; king high; ... ; and 7 high. (Note that none can be described as 6 high because 6-5-4-3-2 is a straight.)

1.6. *Number of distinct poker hands.* Let us define an equivalence relation on the set of the $\binom{52}{5}$ poker hands. Two hands are *equivalent* if neither is ranked higher than the other. It is not necessary that both hands can appear in a showdown. For example, a full house comprising three aces and two kings outranks a full house comprising three aces and two queens, even if it is not possible for both hands to appear in a showdown. (Actually, it *is* possible in Texas hold'em!)

Let us be more precise about the ranking of hands within the nine categories of Table 1.3 on p. 8. (Hands within different categories are ranked as the categories are ranked.) Straight flushes are ranked according to the highest denomination (which is 5 in the case of 5-4-3-2-A). Four-of-a-kind hands are ranked first according to the denomination appearing four times and second according to the denomination of the fifth card. Full houses are ranked first according to the denomination appearing three times and second according to the denomination appearing twice. Flushes are ranked first according to the highest denomination, second according to the next-highest denomination, and so on. Straights are ranked according to the highest denomination (which is 5 in the case of 5-4-3-2-A). Three-of-a-kind hands are ranked first according to the denomination appearing three times, second according to the higher of the two remaining denominations, and third according to the denomination of the fifth card. Two pair are ranked first according to the denomination of the higher pair, second according to the denomination of the lower pair, and third according to the denomination of the fifth card. One-pair hands are ranked first according to the denomination of the pair, second according to the highest of the three unpaired denominations, third according to the second-highest of the three unpaired denominations, and fourth according to the denomination of the fifth card. No-pair hands are ranked first according to the highest denomination, second according to the next-highest denomination, and so on.

(a) Find the number of equivalence classes.

(b) More generally, for each of the nine categories, find the number of equivalence classes and the size of each equivalence class.

(c) Which equivalence class (or classes) determines the median hand?

1.7. *Round-the-corner straights in poker.* Suppose, in Example 1.1.10 on p. 6, we had defined a hand to be *in sequence* if its denominations consisted (after rearrangement if necessary) of 5-4-3-2-A or 6-5-4-3-2 or ... or A-K-Q-J-T or 2-A-K-Q-J or 3-2-A-K-Q or 4-3-2-A-K. Rederive the five-card poker-hand frequencies in Table 1.3 on p. 8, showing that P(one pair or better) $= \frac{1}{2}$.

1.8. *Five-card poker-hand frequencies when seven cards are dealt.* In several versions of poker (e.g., seven-card stud, Texas hold'em), players are dealt seven cards with which to make the best five-card poker hand. Table 1.11 shows the frequencies of each of the nine categories. Derive combinatorial formulas for these numbers.

Hint: Notice that there are 11 denomination multiplicity vectors $\boldsymbol{d}$ and that a generalization of (1.13) on p. 7 applies. However, in four of the 11 cases $\boldsymbol{d}$ does not determine the category of the hand and a finer classification is needed.

Table 1.11 The five-card poker-hand frequencies when seven cards are dealt.

rank	no. of ways	rank	no. of ways
straight flush*	41,584	three of a kind	6,461,620
four of a kind	224,848	two pair	31,433,400
full house	3,473,184	one pair	58,627,800
flush	4,047,644	no pair	23,294,460
straight	6,180,020		
sum			133,784,560

*including royal flush

1.9. *Median five-card poker hand when seven cards are dealt.* Find the median hand when, as in Problem 1.8, seven cards are dealt with which to make the best five-card poker hand.

1.10. *Poker-dice hands.* Poker dice can be played by rolling five standard dice simultaneously. There are no suits in poker dice. We define the *multiplicity vector* $\boldsymbol{d} := (d_0, d_1, \ldots, d_5)$ of the outcome by letting $d_0, d_1, \ldots, d_5$ be the numbers of die labels $(1, 2, \ldots, 6)$ appearing $0, 1, \ldots, 5$ times, respectively. Let us say that the outcome is ranked *five of a kind* if $\boldsymbol{d} = (5, 0, 0, 0, 0, 1)$, *four of a kind* if $\boldsymbol{d} = (4, 1, 0, 0, 1, 0)$, *full house* if $\boldsymbol{d} = (4, 0, 1, 1, 0, 0)$, *three of a kind* if $\boldsymbol{d} = (3, 2, 0, 1, 0, 0)$, *two pair* if $\boldsymbol{d} = (3, 1, 2, 0, 0, 0)$, *one pair* if $\boldsymbol{d} = (2, 3, 1, 0, 0, 0)$, and *no pair* if $\boldsymbol{d} = (1, 5, 0, 0, 0, 0)$. There are no straights, flushes, or straight flushes. Even though the dice are typically indistinguishable, it is convenient to regard them as distinguishable, in which case there are $6^5 = 7,776$ equally likely outcomes. For each $\boldsymbol{d}$ listed above, find the number of outcomes with multiplicity vector $\boldsymbol{d}$.

1.11. *Length distribution of twenty-one-dealer sequences.* Extend Table 1.4 on p. 10, evaluating the probabilities of the twenty-one-dealer's various hand lengths (i.e., number of cards). Also find the number of sequences of each length.

1.12. *Rencontre with d denominations and s suits.* In Example 1.1.23 on p. 14, we assumed a deck of n distinct cards labeled $1, 2, \ldots, n$. Here we assume a deck with d denominations, labeled $1, 2, \ldots, d$, and s suits, hence ds distinct cards. The dealer is said to win if, for some $j \in \{1, 2, \ldots, d\}$, the card in position j has denomination j. Find the probability that the dealer wins. Evaluate for $d = 13$ and $s = 4$.

1.13. *Maturity of chances.* In an 1870 publication (see the chapter notes), we find the following statement:

> In a game of chance, the oftener the same combination has occurred in succession, the nearer we are to the certainty that it will not recur at the next cast or turn up. This is the most elementary of the theories on probabilities; it is termed the *maturity of the chances.*

Justify or criticize this theory.

1.14. *Outcome 1 before outcome 2 by conditioning.* Derive the result of Example 1.2.1 on p. 16 using the conditioning law, conditioning on the result of the first trial.

1.15. *A dice problem.* Find the probability that the each of the totals 2, 3, 4, 5, 6, 8, 9, 10, 11, and 12 appears before the first 7 when rolling a pair of dice repeatedly.

1.16. *Another dice problem.* Find the probability that the totals 6 and 8 both appear before two 7s appear when rolling a pair of dice repeatedly. Do the same for the event that 5 and 9 both appear before two 7s.

1.17. *Still another dice problem.* Find the probability that the total 7 will appear twice *consecutively* before the first appearance of the total 12 when rolling a pair of dice repeatedly.

Hint: Condition on the result of the first roll.

1.18. *A fair craps game.*

(a) Consider a craps game played with fair six-sided dice, but instead of the labels $1, 2, 3, 4, 5, 6$ appearing on each die, the labels $1, 1, 4, 4, 6, 6$ appear on one die and the labels $1, 2, 3, 5, 5, 6$ appear on the other. Show that the probability of winning a pass-line bet is $\frac{1}{2}$.

(b) Find all other relabelings of the dice that have this property and such that the number of distinct labels on one die plus the number of distinct labels on the other die is no less than eight. (For each die, the labels are assumed to satisfy $1 \leq i_1 \leq i_2 \leq \cdots \leq i_6 \leq 6$.)

1.19. *Three-door Monty.* A contestant on the television game show *Let's Make a Deal* must choose one of three doors. Behind one of them is a new car, and behind the other two are goats. He chooses Door 1. The host, Monty Hall, then opens Door 2 to reveal a goat and offers the contestant the opportunity to switch his choice to Door 3. Should he switch?

As stated, the problem is not well posed. Let us make the following assumptions: The car is placed behind one of the three doors at random. The host always opens a door to reveal a goat (not just when the contestant selects the door hiding the car), and when the host has a choice of doors to open, he makes his choice at random. Finally, the host always offers the contestant the opportunity to switch (again, not just when the contestant selects the door hiding the car).

1.20. *Bayes's law and craps.* Recall the rules of the pass-line bet at craps (Examples 1.2.2 on p. 17 and 1.2.8 on p. 19). On the come-out roll the probabilities of the various totals are given by (1.11) on p. 6. Regard these as the prior probabilities. (a) Find the posterior probabilities, given that the pass-line bet is won. (b) Find the posterior probabilities, given that the pass-line bet is lost.

1.21. *A property of the hypergeometric distribution.* Let n, N, and K be positive integers with $n \vee K \leq N$. Prove that the hypergeometric(n, N, K) distribution coincides with the hypergeometric(K, N, n) distribution. More explicitly, show that

$$\frac{\binom{K}{k}\binom{N-K}{n-k}}{\binom{N}{n}} = \frac{\binom{n}{k}\binom{N-n}{K-k}}{\binom{N}{K}}, \qquad k = (n + K - N)^+, \ldots, n \wedge K. \qquad (1.212)$$

In other words, the parameters n and K can be interchanged without affecting the distribution. See Example 2.2.6 on p. 82 for a more probabilistic approach.

1.22. *A natural at twenty-one.* In twenty-one a *natural* is a two-card total of 21 (i.e., an ace and a ten-valued card, in either order). In a two-handed game (player vs. dealer) of single-deck twenty-one in which no cards have yet been seen, find the probability that either the player or the dealer (or both) is dealt a natural. Use (1.28) on p. 13.

1.23. *Fermat's card problem.* Consider a 40-card deck with four each of ten denominations (e.g., one could eliminate the 8s, 9s, and 10s from the 52-card deck). What is the probability that four cards, drawn at random without replacement, are of different suits?

1.24. *Montmort's poker-hand formula.* Consider a deck of cards labeled by the ordered pairs (i, j), where $i = 1, 2, \ldots, d$ and $j = 1, 2, \ldots, s$. In effect, there are d denominations and s suits, hence ds cards. If a hand comprises n such cards, we define the *denomination multiplicity vector* $\boldsymbol{d} := (d_0, d_1, \ldots, d_n)$ of the hand by letting $d_0, d_1, \ldots, d_n$ be the numbers of denominations $(1, 2, \ldots, d)$ appearing $0, 1, \ldots, n$ times, respectively. Note that

$$d_0 + d_1 + \cdots + d_n = d, \qquad d_1 + 2d_2 + \cdots + nd_n = n. \qquad (1.213)$$

Show that the probability that a randomly chosen hand (sampled without replacement) has denomination multiplicity vector $\boldsymbol{d}$ is

$$\binom{d}{d_0, d_1, \ldots, d_n} \binom{s}{1}^{d_1} \binom{s}{2}^{d_2} \cdots \binom{s}{s}^{d_s} \bigg/ \binom{ds}{n}, \qquad (1.214)$$

thereby generalizing (1.13) on p. 7.

1.25. *Montmort's dice formula.* Consider a die having d equally likely faces, labeled $1, 2, \ldots, d$. If we roll n such dice, we define the *multiplicity vector* $\boldsymbol{d} := (d_0, d_1, \ldots, d_n)$ of the outcome by letting $d_0, d_1, \ldots, d_n$ be the numbers of die labels $(1, 2, \ldots, d)$ appearing $0, 1, \ldots, n$ times, respectively. Note that

$$d_0 + d_1 + \cdots + d_n = d, \qquad d_1 + 2d_2 + \cdots + nd_n = n. \qquad (1.215)$$

Show that the probability that the outcome has multiplicity vector $\boldsymbol{d}$ is

$$\frac{1}{d^n} \binom{d}{d_0, d_1, \ldots, d_n} \frac{n!}{(1!)^{d_1} (2!)^{d_2} \cdots (n!)^{d_n}}. \qquad (1.216)$$

1.26. *Relationship between Montmort's formulas.* Show directly that (1.214) converges to (1.216) as $s \to \infty$. Explain probabilistically why we would expect this to be so.

1.27. *Sic bo.* The game of sic bo is played by rolling three dice. If the player bets on number $j \in \{1, 2, 3, 4, 5, 6\}$, he is paid k to 1 if number j appears k times ($k = 1, 2, 3$), and he loses his bet if number j does not appear. Let X be the number of times that number j appears, and let Y be the player's profit from a one-unit bet. Observe that X is binomial$(3, 1/6)$ and $Y = X - 1_{\{X=0\}}$. Use this observation to find $E[Y]$.

1.28. *Cinq et neuf.* The game of cinq et neuf ("five and nine" in French) has a structure similar to that of craps. The player rolls a pair of dice. He wins if he rolls a total of 3 or 11 or a pair (i.e., $(1,1), (2,2), \ldots, (6,6)$). He loses if he rolls a total of 5 or 9. If the player rolls a nonpair total of 4, 6, 8, or 10 or a total of 7, that number becomes his *point*. He continues to roll the dice until he wins by repeating his point (including pairs) or loses by rolling a total of 5 or 9. A win pays even money. Find the probability of a win for the player and his expected profit per unit bet.

1.29. *Two-up.* The game of two-up is played by tossing a pair of fair coins, repeatedly if necessary. Each toss results in *heads* (two heads), *tails* (two tails), or *odds* (one of each). A bet can be made that the spinner (coin tosser) will obtain three heads before a single tails and before five consecutive odds.

A win pays 7.5 to 1. Notice that at most 15 tosses are required to resolve this bet. (If it is unresolved after 14 tosses, then the 14 tosses must consist of oooohoooohoooo, and the next toss is decisive.)

(a) Find the expected profit from a one-unit bet.

(b) Find the expected number of tosses needed to resolve the bet.

Hint: Let $p_n(h, c)$ be the probability of accumulating h heads with the last c tosses being odds, after n tosses without resolution, where $0 \leq h \leq 2$, $0 \leq c \leq 4$, and $n \geq 1$. Express $p_{n+1}(h, c)$ in terms of the matrix $(p_n(i, j))_{0 \leq i \leq 2,\, 0 \leq j \leq 4}$. After generating these probabilities recursively, derive (a) and (b) from them.

1.30. *Mean and variance of the Poisson distribution.* Find the mean and variance of a Poisson random variable with parameter $\lambda > 0$.

1.31. *Discrete Dirichlet distribution.* Fix an integer $r \geq 2$ and define

$$\Delta := \left\{ \boldsymbol{p} = (p_1, \ldots, p_r) : p_1 \geq 0, \ldots, p_r \geq 0, \sum_{i=1}^{r} p_i = 1 \right\}. \qquad (1.217)$$

Given $N \in \mathbf{N}$ and $\boldsymbol{m} = (m_1, \ldots, m_r) \in \{2, 3, \ldots\}^r$, define $\Delta_N := \{ \boldsymbol{p} \in \Delta : N\boldsymbol{p} \in \mathbf{Z}^r \}$ and let $\boldsymbol{X}_N = (X_{N,1}, \ldots, X_{N,r})$ be a Δ_N-valued random vector with distribution

$$\mathrm{P}(\boldsymbol{X}_N = \boldsymbol{p}) = C_N(\boldsymbol{m})^{-1} p_1^{m_1 - 1} \cdots p_r^{m_r - 1}, \qquad \boldsymbol{p} \in \Delta_N, \qquad (1.218)$$

where

$$C_N(\boldsymbol{m}) := \sum_{\boldsymbol{p} \in \Delta_N} p_1^{m_1 - 1} \cdots p_r^{m_r - 1}. \qquad (1.219)$$

Then $\boldsymbol{X}_N$ is said to have the *discrete Dirichlet distribution* with parameters N and $m_1, \ldots, m_r$.

(a) Define $\Delta^* := \{ (p_1, \ldots, p_{r-1}) : (p_1, \ldots, p_r) \in \Delta \}$ and

$$I(\boldsymbol{m}) := \int_{\Delta^*} p_1^{m_1 - 1} \cdots p_r^{m_r - 1} \, dp_1 \cdots dp_{r-1}, \qquad (1.220)$$

where $p_r := 1 - p_1 - \cdots - p_{r-1}$. Show that $C_N(\boldsymbol{m}) \sim N^{r-1} I(\boldsymbol{m})$ as $N \to \infty$.

(b) Write $I(\boldsymbol{m})$ as an iterated integral and use integration by parts to show that

$$I(\boldsymbol{m}) = \frac{(m_1 - 1)! \cdots (m_r - 1)!}{(m_1 + \cdots + m_r - 1)!}. \qquad (1.221)$$

(c) Use parts (a) and (b) to prove that $\lim_{N \to \infty} \mathrm{E}[X_{N,i}] = m_i/(m_1 + \cdots + m_r)$ for $i = 1, \ldots, r$.

This problem will be needed in Section 13.2.

1.32. *Expectation of a discrete nonnegative random variable.* Let X be a discrete nonnegative random variable.

(a) Show that

$$E[X] = \int_0^\infty P(X \geq t)\,dt. \tag{1.222}$$

(b) Show also that this generalizes Theorem 1.4.2 on p. 30.

Hint: (a) In $\sum_x x P(X = x)$, write the factor x as $\int_0^\infty 1_{[0,x]}(t)\,dt$, and then interchange summation and integration. (For justification, see Theorem A.2.7 on p. 749.)

1.33. *Monotonicity of expectation.* Let X and Y be jointly distributed discrete random variables, not necessarily having finite expectation, such that $0 \leq X \leq Y$. Show that $E[X] \leq E[Y]$. Be careful to avoid the argument that $E[Y] - E[X] = E[Y - X] \geq 0$ because the left side of the equation may be of the form $\infty - \infty$. This generalizes the first part of Theorem 1.1.16 on p. 12.

1.34. *Expected utility and the St. Petersburg paradox.* Let $u : (0, \infty) \mapsto \mathbf{R}$ be the player's "utility function" at the St. Petersburg game. This means that $u(x)$ is the "utility" of x units to the player. (See Section 5.3 for more information.) Let N be geometric($\frac{1}{2}$) and $X := 2^{N-1}$. We saw in Example 1.4.5 on p. 31 that $E[u(X)] = \infty$ if $u(x) \equiv x$. Find $E[u(X)]$ if $u(x) \equiv \sqrt{x}$ and if $u(x) \equiv \log_2 x$.

1.35. *mth moment of a discrete nonnegative random variable.* Let X be a discrete nonnegative random variable.

(a) Show that

$$E[X^m] = \int_0^\infty m t^{m-1} P(X \geq t)\,dt, \qquad m \geq 1. \tag{1.223}$$

(b) Show also that this generalizes Theorem 1.4.8 on p. 32.

Hint: (a) Apply Problem 1.32.

1.36. *Mean and median of the number of rolls needed to achieve all results.*

(a) Find the mean and median of the number of rolls of a single die needed to achieve all six results.

(b) Find the mean and median of the number of rolls of a pair of dice needed to achieve all 11 totals.

Hint: (a) Let $p_n(j)$ be the probability that n dice show j distinct faces $(1 \leq j \leq 6 \wedge n)$, and show that these probabilities are recursive in n. Then use them to find the median. For the mean, write the random variable in question as the sum of independent geometric random variables.

(b) Let X be the number of tosses required. Use the inclusion-exclusion law (Theorem 1.1.18 on p. 13) to evaluate $P(X \leq n)$ and hence the median. Use these probabilities and Theorem 1.4.2 on p. 30 to find the mean.

1.37. *Laplace's problem on the French Lottery.* In the French Lottery, each draw consists of randomly selecting five of the numbers $1, 2, 3, \ldots, 90$. (Within each draw, numbers are selected without replacement; between draws, numbers are replaced.) Find the mean and median of the number of draws necessary for each of the 90 numbers to be selected at least once.

1.38. *A correlation inequality.* Let X be a discrete random variable with values in an interval I, and suppose that f and g are nondecreasing functions on I such that $f(X)$ and $g(X)$ have finite variance. Show that

$$\text{Cov}(f(X), g(X)) \geq 0. \tag{1.224}$$

Hint: Let X_1 and X_2 be i.i.d. as X, and note that $(f(X_1) - f(X_2))(g(X_1) - g(X_2)) \geq 0$.

1.39. *Mean and variance of the total of three dice.* Find the mean and variance of the total of three dice, first using (1.211) and then using Theorems 1.4.11 on p. 32 and 1.4.17 on p. 36. Which approach is easier?

1.40. *Distribution of the two- and three-dice totals.* Show that the formula for the distribution of the n-dice totals derived in Example 1.4.23 on p. 38 reduces when $n = 2$ to (1.11) on p. 6 and when $n = 3$ to (1.211).

1.41. *Distribution of the n-dice totals by recursion.* Let $U_1, U_2, \ldots$ be a sequence of i.i.d. discrete uniform random variables on $\{1, 2, 3, 4, 5, 6\}$. For each positive integer n, let $S_n := U_1 + \cdots + U_n$, and argue that

$$P(S_n = s) = \frac{1}{6} \sum_{u=1}^{6} P(S_{n-1} = s - u) \tag{1.225}$$

for $s = n, n+1, \ldots, 6n$ and $n \geq 2$. Use this to find the distributions of S_2, S_3, S_4, S_5, and S_6. Give exact results, not decimal approximations.

1.42. *Jacob Bernoulli's game.* In this game a die is rolled, and if the result is $n \in \{1, 2, 3, 4, 5, 6\}$, then n dice are rolled. If the total of the n dice is greater than 12, the player wins. If it is less than 12, he loses. If it is equal to 12, a push is declared. A win pays even money. Find the probabilities of a win, a loss, and a push as well as the player's expected profit per unit bet.

1.43. *Distribution of the n-dice totals by inclusion-exclusion.* Let n be a positive integer, and assume that the 6^n outcomes in $\Omega := \{1, 2, \ldots, 6\}^n$ are equally likely. Given a positive integer s satisfying $n \leq s \leq 6n$, evaluate the probability of

$$F := \{(i_1, \ldots, i_n) \in \Omega : i_1 + \cdots + i_n = s\}, \tag{1.226}$$

using the inclusion-exclusion law (Theorem 1.1.18 on p. 13), obtaining the result of Example 1.4.23 on p. 38 by a different method. More specifically, let

$$E := \{(i_1, \ldots, i_n) \in \mathbf{N}^n : i_1 + \cdots + i_n = s\}, \tag{1.227}$$

$$E_k := \{(i_1, \ldots, i_n) \in \mathbf{N}^n : i_k \geq 7, \ i_1 + \cdots + i_n = s\} \tag{1.228}$$

for $k = 1, 2, \ldots, n$, and observe that

$$F = E - \bigcup_{k=1}^{n} E_k. \tag{1.229}$$

Use Theorem 1.1.8 on p. 6 to argue that $|E| = \binom{s-1}{n-1}$, and then show that

$$P(F) = \frac{1}{6^n} \sum_{j=0}^{\lfloor (s-n)/6 \rfloor} (-1)^j \binom{n}{j} \binom{s-6j-1}{n-1}. \tag{1.230}$$

1.44. *Cumulative distribution of the n-dice totals.* Use (1.143) on p. 39 and Problem 1.3 to show that

$$P(U_1 + \cdots + U_n \leq s) = \frac{1}{6^n} \sum_{j=0}^{\lfloor (s-n)/6 \rfloor} (-1)^j \binom{n}{j} \binom{s-6j}{n} \tag{1.231}$$

for $s = n, n+1, \ldots, 6n$.

1.45. *Sicherman dice.* Sicherman dice are fair six-sided dice, but instead of the numbers $1, 2, 3, 4, 5, 6$ appearing on each die, the numbers $1, 2, 2, 3, 3, 4$ appear on one die and the numbers $1, 3, 4, 5, 6, 8$ appear on the other.

(a) Show that the distribution of the sum of these dice is nevertheless given by (1.11) on p. 6.

(b) Are there any other pairs of fair dice with this property, with each face labeled by a positive integer? (The restriction to positive integers rules out subtracting 1 from each face of one die and adding 1 to each face of the other die.)

Hint: (b) Use probability generating functions and the fact that the polynomial $\sum_{k=1}^{6} z^k$ factors uniquely as $z(1 + z)(1 + z + z^2)(1 - z + z^2)$.

1.46. *Nonuniformity of dice totals.* Consider dice whose six faces are labeled $1, 2, 3, 4, 5, 6$ but are not necessarily equally likely.

(a) Show that there does not exist a die with the property that, when two identical copies of it are tossed, the 11 possible totals, 2–12, are equally likely.

(b) More generally, show that there do not exist two dice with the property that, when the two dice are tossed, the 11 possible totals, 2–12, are equally likely. Are probability generating functions useful here?

1.47. *Probability that at least j of n events occur.* Use the result of Example 1.4.24 on p. 39 to show that, if $n \geq 2$, $E_1, \ldots, E_n$ are arbitrary events, and N is defined by (1.144) on p. 39, then

$$P(N \geq j) = \sum_{k=j}^{n} (-1)^{k-j} \binom{k-1}{j-1} S_k \tag{1.232}$$

for $j = 1, \ldots, n$, where $S_1, \ldots, S_n$ are as in (1.146) on p. 39.

1.48. *Distribution of the number of coincidences in rencontre.* Use the result of Example 1.4.24 on p. 39 to find the distribution of the number of coincidences at rencontre (Example 1.1.23 on p. 14). Show that, as $n \to \infty$, this distribution converges to the Poisson distribution with parameter 1.

1.49. *Proof of Borel–Cantelli lemma.* Use Theorem 1.1.20(b) on p. 13 to prove the Borel–Cantelli lemma (Lemma 1.5.4 on p. 43).

1.50. *Second Borel–Cantelli lemma.* If $E_1, E_2, \dots$ is a sequence of independent events for which $\sum_{n=1}^{\infty} P(E_n) = \infty$, show that then $P(E_n \text{ i.o.}) = 1$. This is a partial converse of the Borel–Cantelli lemma (Lemma 1.5.4 on p. 43).

1.51. *A fair game that leads to ruin.* Let $X_1, X_2, \dots$ be independent with

$$P(X_n = n^2 - 1) = n^{-2} = 1 - P(X_n = -1), \qquad n \geq 1. \tag{1.233}$$

Notice that $E[X_n] = 0$ for all $n \geq 1$, but use the Borel–Cantelli lemma (Lemma 1.5.4 on p. 43) to show that $S_n := X_1 + \cdots + X_n$ satisfies $S_n/n \to -1$ a.s.

1.52. *A weak law of large numbers for the St. Petersburg paradox.* Recall the St. Petersburg game (Example 1.4.5 on p. 31). Let $N_1, N_2, \dots$ be i.i.d. geometric($\frac{1}{2}$), and let $X_n := 2^{N_n - 1}$ and $S_n := X_1 + \cdots + X_n$ for each $n \geq 1$. Show that $S_n/(\frac{1}{2} n \log_2 n) \to 1$ in probability.

Hint: Let $X_j^{(n)} := X_j 1_{\{X_j \leq \frac{1}{2} n \log_2 n\}}$ and $\bar{X}_j^{(n)} := X_j 1_{\{X_j > \frac{1}{2} n \log_2 n\}}$ for $j = 1, \dots, n$. Then, for each $\varepsilon > 0$,

$$P\left(\left| \frac{S_n}{\frac{1}{2} n \log_2 n} - 1 \right| \geq \varepsilon \right)$$

$$\leq P\left(\left| \frac{X_1^{(n)} + \cdots + X_n^{(n)}}{\frac{1}{2} n \log_2 n} - 1 \right| \geq \varepsilon \right) + P(\bar{X}_1^{(n)} + \cdots + \bar{X}_n^{(n)} \neq 0)$$

$$\leq P\left(\left| \frac{X_1^{(n)} + \cdots + X_n^{(n)} - E[X_1^{(n)} + \cdots + X_n^{(n)}]}{\frac{1}{2} n \log_2 n} \right| \geq \frac{\varepsilon}{2} \right)$$

$$+ nP(X_1 > \tfrac{1}{2} n \log_2 n) \tag{1.234}$$

for all n sufficiently large. Now apply Chebyshev's inequality to the first probability on the right.

1.53. *On the fairness of Feller's solution of the St. Petersburg paradox.* Problem 1.52 can be interpreted as saying that an entry fee of $\frac{1}{2} \log_2 n$ units per coup is fair (in an asymptotic sense) for the player who plays n coups (see Example 1.5.11 on p. 48). Nevertheless, show that $\limsup_{n \to \infty} S_n/(\frac{1}{2} n \log_2 n) = \infty$ a.s.

Hint: Note that $S_n \geq X_n$ for all $n \geq 1$. For each $K > 1$, apply the second Borel–Cantelli lemma (Problem 1.50) to the sequence of events $E_n := \{X_n/(\frac{1}{2} n \log_2 n) \geq K\}$ ($n \geq 2$).

1.54. *Bounds on the tail of the standard normal.* Define $\Phi(z)$ by (1.162) on p. 42 and $\phi(z) := \Phi'(z)$ for all $z \in \mathbf{R}$. Show that

$$(z^{-1} - z^{-3})\phi(z) < 1 - \Phi(z) < z^{-1}\phi(z), \qquad z > 0. \tag{1.235}$$

This tells us the asymptotic behavior of $1 - \Phi(z)$ as $z \to \infty$.

Hint: Show that minus the derivatives of the three expressions above satisfy similar inequalities, then integrate.

1.55. *Convergence in distribution to a standard normal.* Let $Z_1, Z_2, \ldots$ be discrete random variables not necessarily defined on the same sample space.

(a) Show that, if

$$\lim_{n \to \infty} \mathrm{P}(Z_n \leq z) = \Phi(z) \tag{1.236}$$

for every real number z, then

$$\lim_{n \to \infty} \mathrm{E}[f(Z_n)] = \frac{1}{\sqrt{2\pi}} \int_{-\infty}^{\infty} f(z) e^{-z^2/2} \, dz \tag{1.237}$$

for every bounded continuous $f : \mathbf{R} \mapsto \mathbf{R}$.

(b) Conversely, show that, if (1.237) holds for every bounded continuous $f : \mathbf{R} \mapsto \mathbf{R}$ with three bounded continuous derivatives, then (1.236) holds for every real number z.

1.56. *Convergence in distribution of a ratio.* For each $n \geq 1$, let X_n and Y_n be jointly distributed discrete random variables. Suppose that $X_n \to N(0,1)$ in distribution and $Y_n \to 1$ in probability. Show that $X_n/Y_n \to N(0,1)$ in distribution.

Hint: Let $0 < \varepsilon < 1$. Show first that, if $z \geq 0$, then

$$\begin{aligned}
\mathrm{P}(X_n \leq (1-\varepsilon)z) - \mathrm{P}(|Y_n - 1| \geq \varepsilon) \\
\leq \mathrm{P}(X_n/Y_n \leq z) \leq \mathrm{P}(X_n \leq (1+\varepsilon)z) + \mathrm{P}(|Y_n - 1| \geq \varepsilon).
\end{aligned} \tag{1.238}$$

1.57. *John Smith's problem, revisited.* For each $m \geq 1$, let S_m be the number of sixes in m tosses of a die, and, for each $n \geq 1$, put $E_n := \{S_{6n} \geq n\}$ as in Example 1.3.2 on p. 23.

(a) Show that $\lim_{n \to \infty} \mathrm{P}(E_n) = \frac{1}{2}$.

(b) Show that, as suggested by (1.76) on p. 24, $\mathrm{P}(E_n)$ is decreasing in n.

Hint: (b) Justify the following steps:

$$\begin{aligned}
\mathrm{P}(E_{n+1}) &= \mathrm{P}(S_{6n} + (S_{6(n+1)} - S_{6n}) \geq n+1) \\
&= \mathrm{P}(S_{6n} \geq n) + \sum_{j=0}^{6} [\mathrm{P}(S_{6n} \geq n+1-j) - \mathrm{P}(S_{6n} \geq n)]\mathrm{P}(S_6 = j) \\
&< \mathrm{P}(S_{6n} \geq n) + \mathrm{P}(S_{6n} = n) \sum_{j=0}^{6} (j-1)\mathrm{P}(S_6 = j)
\end{aligned}$$

$$= \mathrm{P}(E_n). \tag{1.239}$$

1.58. *Stirling's formula and the central limit theorem.* Let $X_1, X_2, \ldots$ be i.i.d. Poisson(1) random variables, and put $S_n := X_1 + \cdots + X_n$ for each $n \geq 1$. Prove Stirling's formula (Theorem 1.5.12 on p. 48) by completing the following steps.

(a) Show that S_n is Poisson(n) and therefore that

$$\mathrm{E}\left[\left(\frac{S_n - n}{\sqrt{n}}\right)^-\right] = \frac{(n/e)^n \sqrt{n}}{n!}, \qquad n \geq 1. \tag{1.240}$$

(b) Use Chebyshev's inequality (Lemma 1.5.2 on p. 43) to show that

$$\mathrm{P}\left(\left(\frac{S_n - n}{\sqrt{n}}\right)^- \geq x\right) \leq \mathrm{P}\left(\left|\frac{S_n - n}{\sqrt{n}}\right| \geq x\right) \leq \frac{1}{(x \vee 1)^2} \tag{1.241}$$

for each $x > 0$ and $n \geq 1$.

(c) Use Problem 1.32, the central limit theorem (Theorem 1.5.15 on p. 48), and the dominated convergence theorem (Theorem A.2.8 on p. 749) to show that

$$\mathrm{E}\left[\left(\frac{S_n - n}{\sqrt{n}}\right)^-\right] \to \int_0^\infty (1 - \Phi(x))\, dx = \frac{1}{\sqrt{2\pi}}. \tag{1.242}$$

1.7 Notes

The history of probability theory has been studied by Todhunter (1865), David (1962), Maistrov (1974), Hacking (1975), Stigler (1986), Edwards (1987), and Hald (1990, 1998), among others. Here we emphasize the history of the topics treated in this chapter.

The first book on probability theory was written by Girolamo Cardano [1501–1576] and titled *Liber de Ludo Aleae* (*The Book on Games of Chance*, written c. 1565, published 1663; see Ore 1953). Theorem 1.1.1 (equally likely outcomes) can be attributed to Cardano (Ore 1953, p. 148). As Hald (1990, p. 39) pointed out,

> Cardano's most advanced result is the "multiplication rule" for finding the odds for and against the recurrence of an event every time in a given number of repetitions of a game.

He also obtained the distributions of the two- and three-dice totals. Thus, the frequently stated assertion that the study of probability theory was initiated in 1654 by the correspondence between Blaise Pascal [1623–1662] and Pierre de Fermat [1601–1665] on the problem of points (Example 2.1.1) should not be interpreted to mean that nothing was known about probability before the middle of the 17th century. As Ore (1953, p. 143–144) observed, the undervaluing of Cardano's contributions is likely due to the difficulty in understanding his book. For example, Todhunter (1865, Article 4) wrote that the work "is so badly printed as to be scarcely intelligible." Nevertheless, Ore (1953) studied the book carefully and explained most of the material that puzzled Todhunter.

Combinatorial analysis is much older than probability theory. Edwards (1987) provided an excellent history of the subject, and we summarize the relevant facts. Page citations in this paragraph are from Edwards (1987). The earliest results came from India. Theorem 1.1.4 (combinations) dates back to 850 (p. 27), when the Jain mathematician Mahavira [c. 800–c. 870] wrote *Ganita Sara Sangraha*, though the special case of $\binom{n}{2}$ dates back to Anicius Manlius Severinus Boethius [c. 480–524] in about 510 (pp. 20–21). Theorem 1.1.6 (multinomial coefficients as numbers of permutations) first appeared in 1150 (p. 29) in the Hindu mathematician Bhaskara's [1114–1185] *Lilavati*. Theorem 1.1.3 (permutations) was known to Bhaskara but is probably older (p. 29), and certainly the "$n!$ rule" is much older, the case $n = 6$ having occurred in a work of the Jains in India in about 300 B.C. (p. 24). Theorems 1.1.3 and 1.1.4 first appeared in the West in 1321 in the work of Levi ben Gerson [1288–1344] (pp. 34–35), who lived in France, while Theorem 1.1.6 first appeared in the West in 1636 in the work of Marin Mersenne [1588–1648] (pp. 46, 113). The binomial theorem (Theorem 1.1.5) is of Persian origin (p. 52). It appeared in 1429 in Jamshid al-Kashi's [1380–1429] *Key to Arithmetic*, but is likely much older. Cardano is responsible for its first appearance in the West in 1570 (p. 52). The multinomial theorem (Theorem 1.1.7) was discovered by Gottfried Wilhelm von Leibniz [1646–1716] in 1695 (p. 113). Theorem 1.1.8 can be credited to Nicolo Fontana Tartaglia [1500–1557] in 1523 (in the form of the number of distinguishable outcomes of n indistinguishable dice, p. 37) and more explicitly to Jacob Bernoulli [1654–1705] (1713; 2006, p. 245). Pascal's (1665) triangle, incidentally, predates Pascal by many centuries (p. 32), but attaching his name to it acknowledges his fundamental work *Traité du triangle arithmétique* (written 1654, published 1665). See Edwards (1987) for the primary sources for this paragraph.

The frequencies of the three-dice totals were derived in a Latin poem titled "De Vetula," written in the mid-13th century (Bellhouse 2000), perhaps by Richard de Fournival [1201–c. 1260]. This is the earliest known probability calculation. Galileo Galilei [1564–1642] derived these same frequencies in a paper perhaps written between 1613 and 1623 (David 1962, p. 64 and Appendix 2), though this was some 50 years after Cardano. Undoubtedly, the frequencies of the two-dice totals (Example 1.1.9) are just as old, though we do not know of any pre-Cardano derivations.

The five-card poker probabilities (Example 1.1.10) date back to 1879 or earlier, as explained in Section 22.5. The formula (1.13), which includes five of the nine probabilities, is a special case of a result of Pierre Rémond de Montmort [1678–1719] (1708, Proposition 14, p. 97). Of course Montmort did not discuss this particular application because poker had not yet been invented. Surprisingly, all modern published derivations of the poker probabilities that we have seen evaluate the nine probabilities one by one, failing to take advantage of Montmort's formula.[2] These include Scarne (1949, Chapter 25), Thorp (1966c, pp. 130–132), Feller (1968, p. 487), Haigh (1999, pp. 283–284), Alspach (2000b), Grimmett and Stirzaker (2001, Problem 1.8.33), Hannum and Cabot (2005, pp. 167–168), Packel (2006, pp. 56–57), and Epstein (2009, pp. 248–249). Example 1.1.11 (twenty-one-dealer sequences) is essentially from Griffin (1999, p. 158).

De Morgan's laws are due to Augustus De Morgan [1806–1871] (1847). Axioms 1.1.12–1.1.15 are equivalent to the six axioms of Andrei N. Kolmogorov [1903–1987] (1933) and to the modern definition of a probability measure, with the collection of events being what is called a σ-algebra. Our axiomatic treatment allows us to largely avoid the subject of measure theory. Perhaps the first author to use countable additivity in a problem was Jacob Bernoulli (1690) (Hald 1990, p. 185). The inclusion-exclusion law was used by Montmort (1713, Proposition 16, p. 46) but never formalized by him. A general statement was first provided by Abraham De Moivre [1667–1754] (1718, Problem 25); see below. Boole's inequality is named for George Boole [1815–1864].

Example 1.1.22 was posed by Antoine Gombaud, Chevalier de Méré [1607–1684], as related in a letter from Pascal to Fermat dated July 29, 1654 (David 1962, pp. 235–236):[3]

> I have not time to send you the proof of a difficulty which greatly puzzled M. de Méré, for he is very able, but he is not a geometrician (this, as you know, is a great defect) and he does not even understand that a mathematical line can be divided *ad infinitum* and he believes that it is made up of a finite number of points, and I have never been able to rid him of this idea. If you could do that, you would make him perfect.
>
> He told me that he had found a fallacy in the theory of numbers, for this reason:
>
> If one undertakes to get a six with one die, the advantage in getting it in 4 throws is as 671 is to 625.
>
> If one undertakes to throw 2 sixes with two dice, there is a disadvantage in undertaking it in 24 throws.
>
> And nevertheless 24 is to 36 (which is the number of pairings of the faces of two dice) as 4 is to 6 (which is the number of faces of one die).
>
> This is what made him so indignant and which made him say to one and all that the propositions were not consistent and that Arithmetic was self-contradictory: but you will very easily see that what I say is correct, understanding the principles as you do.

Example 1.1.23 (rencontre) was first addressed by Montmort (1708, p. 54–64), who assumed 13 cards and therefore called the game *treize*. De Moivre (1718, Problems 24 and 25) found a slightly different solution to rencontre, which as suggested in his preface he was quite pleased with:

---

[2] Uspensky (1937, p. 26) recognized the applicability of the formula in this context.

[3] Translation by Maxine Merrington.

In the 24th and 25th Problems, I explain a new sort of Algebra, whereby some Questions relating to Combinations are solved by so easy a Process, that their solution is made in some measure an immediate consequence of the Method of Notation. I will not pretend to say that this new Algebra is absolutely necessary to the Solving of those Questions which I make to depend on it, since it appears by Mr. *De Monmort's* Book, that both he and Mr. *Nicholas Bernoully* have solved, by another Method, many of the cases therein proposed: But I hope I shall not be thought guilty of too much Confidence, if I assure the Reader, that the Method I have followed has a degree of Simplicity, not to say of Generality, which will hardly be attained by any other Steps than by those I have taken.

This new sort of algebra led to De Moivre's inclusion-exclusion law, under the *exchangeability* condition that, for $m = 1, \ldots, n$,

$$P(E_{i_1} \cap \cdots \cap E_{i_m}) = P(E_1 \cap \cdots \cap E_m), \quad 1 \le i_1 < \cdots < i_m \le n. \quad (1.243)$$

Leonhard Euler [1707–1783] (1753), Pierre-Simon Laplace [1749–1827] (1820, pp. 236–245), and Eugène Catalan [1814–1894] (1837), among others, also studied the game of rencontre.

The notion of independence has long been underappreciated by gamblers; in fact, the belief that a series of failures makes a success more likely is often referred to as the *maturity of chances*, the *gambler's fallacy*, or the *Monte Carlo fallacy*. Edgar Allan Poe [1809–1849] tried to explain this in "The Mystery of Marie Rogêt" (1842–1843) but confused the truth with the fallacy (Poe 1984, p. 554):

> Nothing, for example, is more difficult than to convince the merely general reader that the fact of sixes having been thrown twice in succession by a player at dice, is sufficient cause for betting the largest odds that sixes will not be thrown in the third attempt. A suggestion to this effect is usually rejected by the intellect at once. It does not appear that the two throws which have been completed, and which lie now absolutely in the Past, can have influence upon the throw which exists only in the Future. The chance for throwing sixes seems to be precisely as it was at any ordinary time—that is to say, subject only to the influence of the various other throws which may be made by the dice. And this is a reflection which appears so exceedingly obvious that attempts to controvert it are received more frequently with a derisive smile than with anything like respectful attention.

Example 1.2.1 (outcome 1 before outcome 2) was certainly known to De Moivre (1738, Problem 42) but was likely known even earlier. Examples 1.2.2 and 1.2.8 (craps) were known to John Arbuthnot [1667–1735] (1692, p. 90), Montmort (1708, p. 114), and De Moivre (1718, Problem 47) because, although craps did not exist until the 19th century, its ancestor hazard required an identical calculation. See Section 15.4 for additional information. The multiplication law for conditional probabilities (Theorem 1.2.4) was known to De Moivre (1738, pp. 7–8).

Example 1.2.5 (outcomes $1, \ldots, n$ before outcome $n+1$) is due to De Moivre (1738, Problem 42), who found the solution (1.53). The dice problem in the example is also due to De Moivre, who noticed that, since $\pi_4 = \pi_{10}$, $\pi_5 = \pi_9$, and $\pi_6 = \pi_8$, only $\binom{6}{2,2,2} = 90$ terms would be needed. Nevertheless, he resorted to an approximation that yielded the answer $1{,}024/15{,}015 \approx 0.068198$. A second "still more exact" approximation gave about 0.065860. Scarne (1961, pp. 664–665) posed the same dice problem and solved it using De Moivre's first approximation, apparently without realizing that it was only an approximation. Thorp (1964), correcting Scarne, provided the exact answer (≈ 0.062168), together with the formula (1.54). See Grosjean (2009, pp. 480–484) for a recursive approach.

The conditioning law (Theorem 1.2.6, often called the law of total probability) was used without comment by Montmort and De Moivre. It was formalized in the derivation of Bayes's law, which is named for Thomas Bayes [1702–1761] (1764). He found the posterior distribution of the binomial parameter, assuming a continuous uniform prior distribution. The discrete formulation of Bayes's law (Theorem 1.2.7) is apparently due to Laplace (1774), who referred to it by the descriptive term, "the probability of causes," at least in the case in which the "causes" are equally likely. However, Stigler (1983) has provided evidence, albeit inconclusive, that the originator of Bayes's law was Nicholas Saunderson [1682–1739]. Example 1.2.9 (Bayes's law and twenty-one) was motivated by Griffin (1999, Chapter 12).

There is an amusing story on how the term "random variable" was chosen. In the late 1940s, William Feller [1906–1970] was working on his book *An Introduction to Probability Theory and Its Applications* (1968; first published in 1950), while Joseph L. Doob [1910–2004] was working on his book *Stochastic Processes* (1953). As Doob recalled (Snell 1997),

> While writing my book I had an argument with Feller. He asserted that everyone said "random variable" and I asserted that everyone said "chance variable." We obviously had to use the same name in our books, so we decided the issue by a stochastic procedure. That is, we tossed for it and he won.

The hypergeometric distribution appeared first in the fourth problem of Christiaan Huygens [1629–1695] (1657), and its general form was given by Jacob Bernoulli (1713; 2006, Part 3, Problem 6) and De Moivre (1712). The binomial distribution with $p = \frac{1}{2}$ was known to Pascal in 1654, while the general case is due to Johann Bernoulli [1667–1748] in 1710 (Montmort 1713, p. 295), though apparently known to Jacob Bernoulli as early as 1689 (Edwards 1987, p. 119). The negative binomial distribution is due to Montmort (1713, p. 245), where it appears in connection with the problem of points (Example 2.1.1). The Poisson distribution is named for Siméon-Denis Poisson [1781–1840] (1837), with some anticipation by De Moivre (1712, Problems 5–7).

For the history of Example 1.3.1 (keno), see Section 14.4. As for Example 1.3.2, John Smith posed the problem to Samuel Pepys [1633–1703]. Pepys in turn wrote to Isaac Newton [1642–1727] on November 22, 1693, as follows (David 1962, p. 125):

> The Question:
> A has 6 dyes in a box with which he is to fling a 6;

B has in another box 12 dyes with which he is to fling 2 sixes;
C has in another box 18 dyes with which he is to fling 3 sixes;
Q.—whether B and C have not as easy a chance as A at even luck?

Newton responded on November 26 by pointing out that the problem was ambiguous, suggesting that "a 6" should be "at least one 6," that "2 sixes" should be "at least 2 sixes," and so on. Pepys replied on December 9 that this was indeed the correct interpretation, and Newton submitted his solution on December 16.

The multivariate hypergeometric distribution is due to Montmort (1708, Proposition 13, p. 94), although a special case appears as Huygens's (1657) third problem. The multinomial distribution with equal probabilities is due to Montmort (1708, pp. 136–140), with the general case due to De Moivre (1712). The dead-man's hand is the hand James Butler "Wild Bill" Hickok [1837–1876] is said to have held on August 2, 1876, when he was shot in the back of the head by Jack "Crooked Nose" McCall [1852/1853–1877] at Nuttall and Mann's No. 10 Saloon in Deadwood, Dakota Territory (DeArment 1982, p. 336). Most agree that Hickok held two black aces and two black 8s, but there is disagreement about the fifth card. See Wilson (2008, Chapter 1) for a comprehensive discussion.

The concept of expectation is often ascribed to Huygens (1657) but, as pointed out by Edwards (1982), it had already been used by Pascal in 1654 in his solution of the problem of points. Tables 1.6–1.8 are from Fey (2002, p. 242), except for the reel-strip labels, which were obtained by opening up one of these machines, courtesy of Marshall Fey. The median of the geometric distribution was found by De Moivre (1712), and the mean by Nicolaus Bernoulli [1687–1759] (Montmort 1713, Problem 4, p. 402). Todhunter (1865, Article 231) expressed this mean as an infinite series, but his formula is incorrect.

A version of the St. Petersburg paradox was proposed by Nicolaus Bernoulli in a letter to Montmort (1713, Problem 5, p. 402). It was discussed at length by Daniel Bernoulli [1700–1782] (1738) in a paper read at the St. Petersburg Academy of Sciences and was therefore subsequently named after that city. The term "standard deviation" was introduced by Karl Pearson [1857–1936] in 1893 (Stigler 1986, p. 328), whereas the term "variance" was first used by Ronald A. Fisher [1890–1962] (1918).

The linearity of expectation is due to De Moivre (1738, p. 8):

> If I have several Expectations upon several Sums, it is very evident that my Expectation upon the whole is the Sum of the Expectations I have upon the particulars.

Example 1.4.13 (rebates on loss) is due to Griffin (1991, Chapter 7). See Grosjean and Chen (2007) and Grosjean (2009, Chapter 34) for the more difficult case in which the number of coups is not specified in advance. For a discussion of actual casino practices concerning rebated losses, see Castleman (2004, Chapter 7). The Cauchy–Schwarz inequality is named for Augustin-Louis Cauchy [1789–1857] (1821) and Hermann A. Schwarz [1843–

1921] (1885). The former found it for sums and the latter for integrals. It is sometimes referred to as the Cauchy–Bunyakovsky inequality because Viktor Y. Bunyakovsky's [1804–1889] (1859) contribution predates Schwarz's. The concept of correlation is due to Francis Galton [1822–1911] (1888), who originally called it "co-relation." The term "covariance" first appeared around 1930 in the work of Ronald A. Fisher. Jensen's inequality is due to Johan Ludwig William Valdemar Jensen [1859–1925] (1906). In Example 1.4.19 "hedging" is defined a bit differently from its traditional gambling usage. According to Maurer (1950), to hedge is to "cover a wager with a compensating bet to avoid a loss or to break even."

The formula for the distribution of dice totals (Example 1.4.23 and Problem 1.43) is due to De Moivre (stated 1712; proved 1738, pp. 35–39) and Montmort (1713, pp. 46–50). De Moivre introduced probability generating functions for his proof, while Montmort used an inclusion-exclusion argument (Problem 1.43). See Henny (1973) and Hald (1990) for expositions of Montmort's argument. The recursive formula of Problem 1.41 is due to Jacob Bernoulli (1713; 2006, pp. 152–153) and Montmort (1713, p. 51). Todhunter (1865, p. 624) referred to (1.143) as De Moivre's formula, but clearly Montmort deserves equal credit.

Example 1.4.24 (generalization of inclusion-exclusion law) is due to Camille Jordan [1838–1922] (1867), though the result had previously been proved by De Moivre (1718) under the exchangeability condition (1.243). See Takács (1967) for a more complete history. Example 1.4.25 (success runs) is due to De Moivre (1738, Problem 88), who did not provide a proof or an explicit formula. This was done by Thomas Simpson [1710–1761] (1740, Problem 24). Our derivation follows Todhunter (1865, Article 325) and, more explicitly, Uspensky (1937, Chapter 5, Section 3). The special case $r = 4$, $n = 21$, and $p = \frac{1}{2}$ is from De Moivre (1738, pp. 245–247).

Markov's inequality is named for Andrei A. Markov [1856–1922]. Chebyshev's inequality was derived by Pafnuty L. Chebyshev [1821–1894] (1867). It had previously been found by Irénée-Jules Bienaymé [1796–1878] (1853). See Seneta (1998) for further discussion.

The weak law of large numbers was first established by Jacob Bernoulli no later than 1690 (published 1713; 2006, p. 337) under the assumptions of Theorem 1.5.13 (De Moivre's limit theorem). See Hald (1990) for the precise formulation and proof by Jacob Bernoulli, a sharpened version due to Nicolaus Bernoulli (Montmort 1713, pp. 388–394), and subsequent improvements. The term "law of large numbers" was coined by Poisson (1835). The simple proof of the weak law sketched in Section 1.5 is due to Chebyshev (1867). The proof of the i.i.d. weak law of large numbers assuming only finite expectation (Theorem 1.5.3) was first obtained by Aleksandr Y. Khinchin [1894–1959] (1929).

The Borel–Cantelli lemma was proved by Émile Borel [1871–1956] (1909) assuming independence of the events and by Felix Hausdorff [1868–1942] (1914) and Francesco Paolo Cantelli [1875–1966] (1917) in general. The strong law of large numbers was proved essentially under the assumptions of Theorem 1.5.13 by Borel (1909). Cantelli (1917) proved the strong law under a fourth-moment assumption, much as in the argument sketched in Section 1.5. The general version (Theorem 1.5.5) is due to Kolmogorov (1930). Example 1.5.8 (asymptotic behavior of a betting system) is due to Borovkov (1997). Example 1.5.9 is essentially a result of Kelly (1956); see Section 10.5 for additional details.

Example 1.5.11 (including Problem 1.52) on the St. Petersburg paradox is due to Feller (1945). The idea of an entry fee that depends on the number of coups goes back to Joseph Bertrand [1822–1900] (1888, Section 51). For a good survey of the history of the St. Petersburg paradox, see Samuelson (1977). It would be inaccurate to say the Feller's result resolved the paradox, and much work has been done since; see, for example, Martin-Löf (1985) and Csörgő and Simons (1996).

Stirling's formula (Theorem 1.5.12) is due to James Stirling [1692–1770] (1730) and De Moivre (1730), but it was the former who first found the constant $\sqrt{2\pi}$ that appears in the formula. De Moivre's limit theorem (Theorem 1.5.13) was proved by De Moivre (1733) in the case $p = \frac{1}{2}$ and simply stated in the general case. The theorem is often referred to as the De Moivre–Laplace limit theorem (e.g., Feller 1968), with Laplace sharing the credit because he made De Moivre's argument more rigorous and was the first to show that $\lim_{z\to\infty} \Phi(z) = 1$.

Like the law of large numbers, the central limit theorem was a collaborative effort of a number of mathematicians over a period of more than a century. The first formulation is due to Laplace in 1810–1812 (see Laplace 1820). He was followed by Poisson in 1824, 1829, and 1837 (see Poisson 1837), Friedrich Wilhelm Bessel [1784–1846] in 1838, and Cauchy in 1853. At this point the theorem was rigorously established only for bounded random variables. Then came the work of the Russian school. Chebyshev, Markov, and Aleksandr M. Lyapunov [1857–1918] (1900) made fundamental contributions. The version of the central limit theorem we have stated as Theorem 1.5.15 is due to Jarl Waldemar Lindeberg [1876–1932] (1922), and the proof sketched in Section 1.5 is Lindeberg's (1922) but is based on an idea of Lyapunov (1900). The term "central limit theorem" first appeared in a paper of Georg Pólya [1887–1985] (1920). Here "central" refers to the importance of the theorem, not to the centering of S_n by its mean. Example 1.5.16 (normal approximation to a binomial) was motivated by an experiment described by Wong (2005, Chapter 8).

Problem 1.1 is due to Thorp (2000a, p. 126). Problem 1.2 was motivated by Griffin (1999, pp. 216–217), who gave the bound (1.208). Problem 1.4 dates back to the 13th century, as explained above. Problem 1.5 is well known; see for example Scarne (1961, p. 582). Problem 1.6 was posed by Thorp (1966b) and solved by Thorp (1966c, p. 132) and Marsh (1967). Marsh complained that the problem was "ill-defined," but his objections are without merit. Problem 1.7 (round-the-corner straights) is due to Takahasi and Futatsuya

(1984), who also investigated whether n-denomination, k-suit, r-card poker
has this property for any (n, k, r) with $r \geq 3$ other than $(13, 4, 5)$. The poker-
hand frequencies when seven cards are dealt (Problem 1.8) are relatively
recent, it appears. The first to publish the correct numbers was Hill (1997),
as far as we know. A combinatorial derivation, such as the one given by
Alspach (2000c), is preferable. Problem 1.9 is also from Hill (1997). Problem
1.10 is well known; see for example Scarne (1961, p. 305).

Problem 1.12 (rencontre) was addressed by Montmort (1713, p. 324) and
Nicolaus Bernoulli [1695–1726]. Bernoulli gave the exact result [actually $(1 +
P)/2$] as a ratio of 18-digit integers; see also De Moivre (1718, Problem 26).
The quotation in Problem 1.13 comes from Steinmetz (1870, p. 255), who in
turn quoted from Robert-Houdin's *L'art de gagner à tous les jeux*; Robert-
Houdin got his information from a successful gamester named Raymond.
Problem 1.15 was solved by Scarne (1961, pp. 664–665) using De Moivre's
(1738, Problem 42) approximation, but the result was off by nearly a factor
of 3; Thorp (1964) provided the correct solution. See also Grosjean (2009,
pp. 480–484). Problem 1.16 was solved incorrectly by Scarne (1961, p. 668).
(It was also misstated there but subsequently stated correctly in Scarne 1974,
p. 827.) It was correctly solved by Sklansky (2001, pp. 50–54). Problem 1.17
was discussed by Griffin (1991, Chapter 6). Problem 1.18(a) (fair craps) is
based on an idea of Edward A. Fredenburg, which was patented in 1998 (U.S.
patent 5,746,428).

Problem 1.19 is the famous Monty Hall problem, due to Selvin (1975a,b)
and named for the actual host of the U.S. television game show, which
aired from 1963 to 1991 (with gaps). The problem became widely known
when Marilyn vos Savant, said to have an IQ of 228 (eight standard deviations above the mean), posed it in her widely distributed newspaper column
(vos Savant 1990a,b, 1991). Many readers, including some mathematicians,
criticized vos Savant for the answer she provided when in fact it was correct. Even Paul Erdős [1913–1996], the most prolific mathematician of the
20th century, got it wrong (Hoffman 1998, Chapter 6). See Tierney (1991),
Morgan, Chaganty, Dahiya, and Doviak (1991), Snell and Vanderbei (1995),
Puza, Pitt, and O'Neill (2005), and Rosenhouse (2009) for further discussion
of the problem.

Problem 1.21 (property of hypergeometric) is due to De Moivre (1718,
Problem 20). Problem 1.23 was posed by Fermat and was one of Huygens's
(1657) five problems. David (1962, p. 119) surprisingly remarked,

> Huygens gives this chance as 1000 to 8139 and so does Todhunter, but I make it
> 1000 to 9139.

There is no inconsistency here: Huygens and Todhunter were speaking in
terms of odds (8,139 to 1,000) while David was speaking in terms of probabilities (1,000/9,139). Problem 1.24 generalizes (1.13) and is due to Montmort as noted above (Montmort 1708, Proposition 14, p. 97). Problem
1.25 includes the poker-dice probabilities (Problem 1.10) and is also due to
Montmort (1713, Proposition 15, p. 44). In modern notation it was stated

by Sodano and Yaspan (1978), who described it as a "neglected probability formula," apparently unaware of its origin.

Sic bo (Problem 1.27) is an old game of Chinese origin that has much in common with the game of chuck-a-luck. Problem 1.28 (cinq et neuf) comes from Jacob Bernoulli's *Ars Conjectandi* (1713; 2006, Part 3, Problem 16). The game of two-up (Problem 1.29) has been described as "a form of national sport" in Australia (O'Hara 1988, p. 119). The result of the problem is due to Griffin (2000), though his solution is different from the one we have suggested. The discrete Dirichlet distribution of Problem 1.31 is not the one described by Johnson, Kotz, and Balakrishnan (1997, Section 43.6). Problem 1.34, which provides a possible explanation for the St. Petersburg paradox in terms of expected utility, is due to Gabriel Cramer [1704–1752] (in a 1728 letter to Nicolaus Bernoulli) and to Daniel Bernoulli (1738, 1954). Problem 1.36 is a special case of the famous *coupon collector's problem*; part (a) for the median was posed and solved by De Moivre (1712, Problem 19). The recursive method for part (a) is from Feller (1968, p. 284). The representation in terms of independent geometric random variables is standard. The median in Problem 1.37 (French Lottery) was obtained by Laplace (1820, Volume 2, Chapter 2). Problem 1.38 is a special case of the FKG inequality (Fortuin, Kasteleyn, and Ginibre 1971).

Problem 1.41 goes back to Jacob Bernoulli (1713; 2006, pp. 152–153). This is also the source of Problem 1.42 (Part 3, Problem 14). Problem 1.43 is from Montmort (1713, Proposition 16, p. 46); see Henny (1973) for further discussion. Problem 1.44 is due to Simpson (1740, p. 53). Problem 1.45 is due to Col. George Sicherman of Buffalo, NY, according to Gardner (1978); the solution suggested for part (b) is from Gallian (1994, pp. 265–267). Problem 1.46 comes from Grimmett and Stirzaker (2001, Problems 2.7.12 or 5.12.36). Problem 1.47 is from Feller (1968, p. 109) but is undoubtedly much older. Problem 1.48 (Poisson distribution in rencontre) is a result of Catalan (1837). Problem 1.50 (second Borel–Cantelli lemma) is due to Borel (1909). Problem 1.52 is due to Feller (1945; 1968, Section 10.4), and Problem 1.53 is a result of Yuan Shih Chow [1924–] and Herbert Robbins [1915–2001] (1961). Problem 1.54 is from Feller (1968, pp. 175, 179) but likely goes back to Laplace. Problem 1.55 is a special case of a result of Lévy (1937). Problem 1.56 is a special case of a theorem of Slutsky (1925). The monotonicity in Problem 1.57 was established by Theodore W. Chaundy [1889–1966] and J. E. Bullard (1960), though the argument suggested here is that of Grimmett and Stirzaker (2001, Problem 3.8.5). Problem 1.58 is due to Wong (1977). It should be noted that the proof of the central limit theorem (as we have sketched it) depends on De Moivre's limit theorem, which in turn depends on Stirling's formula, so the argument of Problem 1.58 is slightly circular, though the circularity can be avoided.

Chapter 2
Conditional Expectation

Poor Percy, about whose means and expectations she had in the most natural way
in the world asked information from me, was not perhaps a very eligible admirer for
darling Rosey.

William Makepeace Thackeray, *The Newcomes*

Conditional expectations are expectations with respect to conditional probabilities. The reader may already have some familiarity with this topic, but
here we go into greater detail than is typical in an introductory probability
course. This material is essential for much of what follows.

2.1 Conditioning on an Event

Recall that we define the expectation of a discrete random variable in two
steps. First, if Y is a discrete random variable with values in $[0, \infty]$, we define
its expectation to be

$$E[Y] := \sum_{y \in [0, \infty]} y P(Y = y), \tag{2.1}$$

where

$$\infty \cdot p := \begin{cases} \infty & \text{if } p > 0, \\ 0 & \text{if } p = 0. \end{cases} \tag{2.2}$$

Second, if Y is a discrete random variable with values in $\mathbf{R}$ and $E\big[|Y|\big] < \infty$,
we say that Y has finite expectation and define its expectation to be

$$E[Y] := \sum_{y \in \mathbf{R}} y P(Y = y). \tag{2.3}$$

S.N. Ethier, *The Doctrine of Chances*, Probability and its Applications,
DOI 10.1007/978-3-540-78783-9_2, © Springer-Verlag Berlin Heidelberg 2010

Now suppose that D is an event with $P(D) > 0$. Then

$$Q(E) := P(E \mid D) \tag{2.4}$$

defines a probability Q that, with the same sample space Ω and the same collection of events, satisfies Axioms 1.1.12–1.1.15 on p. 12. In particular, we can use Q to define the notion of conditional expectation. Again, we proceed in two steps.

First, if Y is a discrete random variable with values in $[0, \infty]$, we define the *conditional expectation of Y given D* by

$$E[Y \mid D] := \sum_{y \in [0, \infty]} y P(Y = y \mid D), \tag{2.5}$$

where we again use the convention (2.2). Second, if Y is a discrete random variable with values in $\mathbf{R}$ and $E\big[|Y| \mid D\big] < \infty$, we say that Y has *finite conditional expectation given D* and define the *conditional expectation of Y given D* by

$$E[Y \mid D] := \sum_{y \in \mathbf{R}} y P(Y = y \mid D). \tag{2.6}$$

Perhaps the first use of conditional expectation occurred in the 17th century:

Example 2.1.1. *Problem of points.* This problem, solved by Fermat and Pascal in 1654, is sometimes said to have initiated the study of probability theory. (See the chapter notes for more history.) Two players, A and B, each stake one unit and compete in a sequence of independent Bernoulli trials. At each trial either A is awarded one point (probability p) or B is awarded one point (probability $q := 1 - p$). The first player to accumulate a specified number of points is declared the winner and receives the total stake of two units. If the game is stopped prior to completion with A needing j points for a win and B needing k points for a win, how should the total stake be divided?

Let $D_{j,k}$ be the event that A needs j points for a win and B needs k points for a win. Let E be the event that A wins the game when it is played to its natural conclusion. Then the random variable

$$Y := 2 \cdot 1_E - 1 \tag{2.7}$$

denotes A's profit when the game is completed.

An equitable solution is that the total stake be divided in such a way that A's profit from the division coincide with A's conditional expected profit were the game to be completed, given the current score. In other words, A should receive a proportion $P_{j,k}$ of the total stake, where

$$2P_{j,k} - 1 = E[Y \mid D_{j,k}]. \tag{2.8}$$

Since
$$\mathrm{E}[Y \mid D_{j,k}] = 2\mathrm{P}(E \mid D_{j,k}) - 1, \tag{2.9}$$

we find that
$$P_{j,k} = \mathrm{P}(E \mid D_{j,k}) \tag{2.10}$$

is the required proportion.

Now, given that A needs j points for a win and B needs k points for a win, and assuming that the game is played to completion, A will win if he scores his jth additional point in the next $j+k-1$ trials. Using the negative binomial distribution, this is

$$\mathrm{P}(E \mid D_{j,k}) = \sum_{l=j}^{j+k-1} \binom{l-1}{j-1} p^j q^{l-j}. \tag{2.11}$$

Alternatively, one can imagine $j + k - 1$ additional trials being played, regardless of whether they are all necessary. Then A will win if he wins at least j of these trials, so, using the binomial distribution,

$$\mathrm{P}(E \mid D_{j,k}) = \sum_{l=j}^{j+k-1} \binom{j+k-1}{l} p^l q^{j+k-1-l}. \tag{2.12}$$

For example, if the game is stopped with A needing one point and B needing two points, then A should receive a proportion of the total stake equal to $P_{1,2} = p + pq$ by (2.10) and (2.11) and $P_{1,2} = 2pq + p^2$ by (2.10) and (2.12). The two expressions are algebraically equivalent since $q := 1 - p$.

An alternative approach to this problem consists of letting

$$E_{j,k} := \mathrm{E}[1_E \mid D_{j,k}] = \mathrm{P}(E \mid D_{j,k}) \tag{2.13}$$

denote the conditional expectation, given $D_{j,k}$, of the proportion of the total stake returned to A upon completion of the game. Conditioning on the result of the next trial (Theorem 1.2.6 on p. 19), we get

$$E_{j,k} = pE_{j-1,k} + qE_{j,k-1}, \qquad j,k \geq 1. \tag{2.14}$$

Noting that $E_{0,k} = 1$ and $E_{j,0} = 0$ for all $j,k \geq 1$, we can derive successively

$$E_{1,1} = p, \quad E_{1,2} = p + qp, \quad E_{2,1} = p^2, \quad E_{1,3} = p + qp + q^2 p, \tag{2.15}$$

and so on. The general formula, given by the right side of (2.11) or (2.12), can then be established by induction using (2.14). ♠

Another approach to conditional expectation is given in the following lemma.

Lemma 2.1.2. *Let $D \subset \Omega$ be an event with $\mathrm{P}(D) > 0$, and let Y be a discrete random variable on Ω. If the conditional expectation of Y given D*

exists (i.e., if either (2.5) or (2.6) applies), then

$$E[Y \mid D] = E[Y\, 1_D]/P(D). \tag{2.16}$$

Proof. This is a consequence of the definitions, using

$$\{Y = y\} \cap D = \{Y\, 1_D = y\}, \qquad y \neq 0. \tag{2.17}$$

Note that omission of the $y = 0$ term in (2.5) or (2.6) has no effect. ♠

We can now state a generalization of Theorem 1.2.6 on p. 19, which we will refer to as the *conditioning law for expectations*.

Theorem 2.1.3. *Given a finite or countably infinite index set I, let $\{D_i,\ i \in I\}$ be a collection of mutually exclusive events whose union is an event of probability 1, and assume that $P(D_i) > 0$ for each $i \in I$. If the discrete random variable Y on Ω has finite expectation, then*

$$E[Y] = \sum_{i \in I} P(D_i)E[Y \mid D_i]. \tag{2.18}$$

Remark. Taking Y to be the indicator of the event E in (2.18), we obtain Theorem 1.2.6 on p. 19.

Proof. If we apply Lemma 2.1.2, we have

$$E[Y] = \sum_{i \in I} E[Y\, 1_{D_i}] = \sum_{i \in I} P(D_i) \frac{E[Y\, 1_{D_i}]}{P(D_i)} = \sum_{i \in I} P(D_i)E[Y \mid D_i], \quad (2.19)$$

as required. The justification of the first equality uses Theorem 1.4.11 on p. 32 if I is finite and the dominated convergence theorem (Theorem A.2.5 on p. 748) if I is infinite. ♠

Example 2.1.4. *Expected number of rolls in a pass-line decision at craps.* The pass-line bet at craps was discussed in Examples 1.2.2 on p. 17 and 1.2.8 on p. 19. Let Y be the number of rolls in a pass-line decision. Then the expected value of Y can be found by conditioning on the result of the come-out roll and using the mean of the geometric distribution from Example 1.4.3 on p. 30. We obtain

$$E[Y] = \left(1 - \sum_{j \in \mathscr{P}} \pi_j\right) + \sum_{j \in \mathscr{P}} \pi_j \left(1 + \frac{1}{\pi_j + \pi_7}\right)$$

$$= 1 + \sum_{j \in \mathscr{P}} \pi_j \frac{1}{\pi_j + \pi_7} = \frac{557}{165} = 3.3\overline{75}. \tag{2.20}$$

Notice that we can skip the first step in the calculation by regarding Y as 1 plus the number of rolls beyond the come-out roll. For more about craps, see Chapter 15. ♠

2.2 Conditioning on a Random Vector

The usual way in which the sample space Ω is partitioned in applying Theorem 2.1.3 on p. 78 is according to the values of a random variable or random vector. We treat the more general case of a random vector (a random variable is a one-dimensional random vector). Let the discrete random vector $\boldsymbol{X}$ and the discrete random variable Y be jointly distributed. Then

$$D_i := \{\boldsymbol{X} = \boldsymbol{x}_i\}, \qquad i \in I, \tag{2.21}$$

defines a suitable partition if (a) I is finite or countably infinite, (b) the vectors in $\{\boldsymbol{x}_i,\ i \in I\}$ are distinct, (c) $\mathrm{P}(\boldsymbol{X} = \boldsymbol{x}_i) > 0$ for each $i \in I$, and (d) $\mathrm{P}(\boldsymbol{X} = \boldsymbol{x}_i$ for some $i \in I) = 1$. If Y has finite expectation, then Theorem 2.1.3 on p. 78 becomes

$$\mathrm{E}[Y] = \sum_{i \in I} \mathrm{P}(\boldsymbol{X} = \boldsymbol{x}_i)\mathrm{E}[Y \mid \boldsymbol{X} = \boldsymbol{x}_i]$$

$$= \sum_{\boldsymbol{x}} \mathrm{P}(\boldsymbol{X} = \boldsymbol{x})\mathrm{E}[Y \mid \boldsymbol{X} = \boldsymbol{x}], \tag{2.22}$$

where $\mathrm{E}[Y \mid \boldsymbol{X} = \boldsymbol{x}]$ is defined arbitrarily if $\boldsymbol{x}$ is such that $\mathrm{P}(\boldsymbol{X} = \boldsymbol{x}) = 0$. This suggests defining the *conditional expectation of Y given $\boldsymbol{X}$* as the random variable

$$\mathrm{E}[Y \mid \boldsymbol{X}] := h(\boldsymbol{X}), \quad \text{where} \quad h(\boldsymbol{x}) := \mathrm{E}[Y \mid \boldsymbol{X} = \boldsymbol{x}]. \tag{2.23}$$

In particular, if $\boldsymbol{X}$ assumes a value $\boldsymbol{x}$ with probability 0, then $\mathrm{E}[Y \mid \boldsymbol{X}]$ is defined arbitrarily on the event $\{\boldsymbol{X} = \boldsymbol{x}\}$. However, on an event of probability 1 it is unambiguously defined, and (2.22) becomes

$$\mathrm{E}[Y] = \mathrm{E}[\mathrm{E}[Y \mid \boldsymbol{X}]]. \tag{2.24}$$

Similarly, if E is an event such that the discrete random vector $\boldsymbol{X}$ and 1_E are jointly distributed, we define the *conditional probability of E given $\boldsymbol{X}$* as the random variable

$$\mathrm{P}(E \mid \boldsymbol{X}) := h(\boldsymbol{X}), \quad \text{where} \quad h(\boldsymbol{x}) := \mathrm{P}(E \mid \boldsymbol{X} = \boldsymbol{x}). \tag{2.25}$$

We then have, by analogy with (2.24),

$$P(E) = E[P(E \mid \boldsymbol{X})]. \tag{2.26}$$

Of course, this is just the special case of (2.24) in which $Y = 1_E$. A special case of (2.25), with $\boldsymbol{X}$ and Y as in the preceding paragraph, is

$$P(Y = y \mid \boldsymbol{X}) := h(\boldsymbol{X}), \quad \text{where} \quad h(\boldsymbol{x}) := P(Y = y \mid \boldsymbol{X} = \boldsymbol{x}). \tag{2.27}$$

This is called the *conditional distribution of Y given $\boldsymbol{X}$* (or *given $\boldsymbol{X} = \boldsymbol{x}$*).

Incidentally, if $\boldsymbol{X} = (X_1, \dots, X_n)$, we often write $E[Y \mid \boldsymbol{X}]$ and $P(E \mid \boldsymbol{X})$ as

$$E[Y \mid X_1, \dots, X_n] \quad \text{and} \quad P(E \mid X_1, \dots, X_n). \tag{2.28}$$

Example 2.2.1. *Insurance at twenty-one.* In twenty-one, if the dealer's up-card is an ace, the player may make an *insurance bet*, for an amount up to one half the size of his basic bet. The insurance bet pays 2 to 1 if the dealer has a natural (i.e., if his downcard is a ten-valued card) and is lost otherwise. What is the player's (conditional) expected profit from an insurance bet?

We assume a single-deck game. It will be convenient to relabel the 13 denominations A, 2–9, 10, J, Q, K by the values 1, 2–9, 10, 10, 10, 10. Let us denote the 52 cards by random variables $X_1, X_2, \dots, X_{52}$. Specifically, $(X_1, X_2, \dots, X_{52})$ assumes each of the $52!/[(4!)^9 16!]$ permutations of four 1s, four 2s, ..., four 9s, and 16 10s with equal probability.

Let Y_n denote the player's profit from a one-unit insurance bet made when the first n cards have been seen. More precisely, we define

$$Y_n := 2 \cdot 1_{\{X_{n+1}=10\}} + (-1)1_{\{X_{n+1}\neq 10\}} = 3 \cdot 1_{\{X_{n+1}=10\}} - 1, \tag{2.29}$$

where X_{n+1} is the dealer's downcard. (This formulation has the advantage that Y_n is well defined even if the dealer's upcard is not an ace, though the bet will not be available in that case.) Then, given that the first n cards $X_1, X_2, \dots, X_n$ have been seen, Y_n has conditional expectation

$$\begin{aligned} E[Y_n \mid X_1, X_2, \dots, X_n] &= 3P(X_{n+1} = 10 \mid X_1, \dots, X_n) - 1 \\ &= 3p_n(X_1, \dots, X_n) - 1, \end{aligned} \tag{2.30}$$

where

$$p_n(X_1, \dots, X_n) := \frac{16 - \sum_{i=1}^{n} 1_{\{X_i=10\}}}{52 - n} \tag{2.31}$$

is the proportion of 10-valued cards in the *unseen* deck.

If the player makes an insurance bet after seeing only his own hand (X_1, X_2) and the dealer's ace $(X_3 = 1)$, then his conditional expected profit, on the event $\{X_3 = 1\}$, is

$$\begin{aligned} E[Y_3 \mid X_1, X_2, X_3] &= 3p_3(X_1, X_2, 1) - 1 \\ &= 3\left(\frac{16 - 1_{\{X_1=10\}} - 1_{\{X_2=10\}}}{49} \right) - 1, \end{aligned} \tag{2.32}$$

which is $-1/49$, $-4/49$, or $-7/49$, depending on whether he has 0, 1, or 2 10-valued cards in his hand. ♠

Because conditional expectations are just expectations with respect to conditional probabilities, most of the results of Chapter 1 about expectations carry over to conditional expectations. However, for the sake of completeness, we state a few of these generalizations explicitly. We begin with the *linearity of conditional expectation*.

Theorem 2.2.2. *Let $\boldsymbol{X}$ and Y be discrete and jointly distributed. If Y has finite expectation and a is a constant, then aY has finite expectation and $\mathrm{E}[aY \mid \boldsymbol{X}] = a\mathrm{E}[Y \mid \boldsymbol{X}]$. Let $n \geq 2$ and $\boldsymbol{X}$ and $Y_1, \ldots, Y_n$ be discrete and jointly distributed. If each of $Y_1, \ldots, Y_n$ has finite expectation, then so does $Y_1 + \cdots + Y_n$ and*

$$\mathrm{E}[Y_1 + \cdots + Y_n \mid \boldsymbol{X}] = \mathrm{E}[Y_1 \mid \boldsymbol{X}] + \cdots + \mathrm{E}[Y_n \mid \boldsymbol{X}]. \tag{2.33}$$

Next we turn to the *conditional form of Jensen's inequality*.

Theorem 2.2.3. *Let $I \subset \mathbf{R}$ be an interval, and let $f : I \mapsto \mathbf{R}$ be convex. Let $\boldsymbol{X}$ be a discrete random vector and let Y be a discrete random variable with values in I, jointly distributed with $\boldsymbol{X}$. If each of Y and $f(Y)$ has finite expectation, then*

$$f(\mathrm{E}[Y \mid \boldsymbol{X}]) \leq \mathrm{E}[f(Y) \mid \boldsymbol{X}]. \tag{2.34}$$

If, in addition, f is strictly convex, then equality holds in (2.34) if and only if the conditional distribution of Y given $\boldsymbol{X}$ is degenerate.

The next result concerns conditional expectation and independence and is often useful in evaluating conditional expectations

Theorem 2.2.4. *Let $\boldsymbol{X}$ and $\boldsymbol{Y}$ be independent discrete random vectors, and let f be a real-valued function on the range of $(\boldsymbol{X}, \boldsymbol{Y})$ such that $f(\boldsymbol{X}, \boldsymbol{Y})$ has finite expectation. Then*

$$\mathrm{E}[f(\boldsymbol{X}, \boldsymbol{Y}) \mid \boldsymbol{X}] = g(\boldsymbol{X}), \quad \text{where} \quad g(\boldsymbol{x}) := \mathrm{E}[f(\boldsymbol{x}, \boldsymbol{Y})]. \tag{2.35}$$

Proof. It suffices to notice that

$$\mathrm{E}[f(\boldsymbol{X}, \boldsymbol{Y}) \mid \boldsymbol{X} = \boldsymbol{x}] = \mathrm{E}[f(\boldsymbol{x}, \boldsymbol{Y}) \mid \boldsymbol{X} = \boldsymbol{x}] = g(\boldsymbol{x}), \tag{2.36}$$

where the first equality uses Lemma 2.1.2 on p. 77 and the second uses the assumed independence. ♠

Corollary 2.2.5. *Let $\boldsymbol{X}$ and $\boldsymbol{Y}$ be independent discrete random vectors, and let f be a real-valued function on the range of $\boldsymbol{Y}$ such that $f(\boldsymbol{Y})$ has finite expectation. Then*

$$\mathrm{E}[f(\boldsymbol{Y}) \mid \boldsymbol{X}] = \mathrm{E}[f(\boldsymbol{Y})]. \tag{2.37}$$

Example 2.2.6. *A property of the hypergeometric distribution.* For $1 \leq n \leq N$, define

$$C_{N,n} := \{\boldsymbol{x} = (x_1, \ldots, x_n) \in \{1, 2, \ldots, N\}^n : x_1 < x_2 < \cdots < x_n\}. \quad (2.38)$$

Given positive integers n, N, K with $n \vee K \leq N$, let $\boldsymbol{X}$ and $\boldsymbol{Y}$ be independent random vectors with discrete uniform distributions

$$\mathrm{P}(\boldsymbol{X} = \boldsymbol{x}) = \frac{1}{\binom{N}{n}}, \quad \boldsymbol{x} \in C_{N,n}, \qquad \mathrm{P}(\boldsymbol{Y} = \boldsymbol{y}) = \frac{1}{\binom{N}{K}}, \quad \boldsymbol{y} \in C_{N,K}. \quad (2.39)$$

Define the function $f : C_{N,n} \times C_{N,K} \mapsto \{0, 1, \ldots, n \wedge K\}$ by

$$f(\boldsymbol{x}, \boldsymbol{y}) := |\{x_1, \ldots, x_n\} \cap \{y_1, \ldots, y_K\}|. \quad (2.40)$$

In other words, $f(\boldsymbol{x}, \boldsymbol{y})$ is the number of elements that the set consisting of the components of $\boldsymbol{x}$ and the set consisting of the components of $\boldsymbol{y}$ have in common. Then, for $k = (n + K - N)^+, \ldots, n \wedge K$,

$$g_k(\boldsymbol{y}) := \mathrm{P}(f(\boldsymbol{X}, \boldsymbol{y}) = k) = \frac{\binom{K}{k}\binom{N-K}{n-k}}{\binom{N}{n}}, \qquad \boldsymbol{y} \in C_{N,K}, \quad (2.41)$$

and

$$h_k(\boldsymbol{x}) := \mathrm{P}(f(\boldsymbol{x}, \boldsymbol{Y}) = k) = \frac{\binom{n}{k}\binom{N-n}{K-k}}{\binom{N}{K}}, \qquad \boldsymbol{x} \in C_{N,n}. \quad (2.42)$$

Notice that $g_k(\boldsymbol{y})$ is constant in $\boldsymbol{y}$ and $h_k(\boldsymbol{x})$ is constant in $\boldsymbol{x}$. Therefore, by Theorem 2.2.4 and (2.26),

$$
\begin{aligned}
\frac{\binom{K}{k}\binom{N-K}{n-k}}{\binom{N}{n}} &= \mathrm{E}[g_k(\boldsymbol{Y})] \\
&= \mathrm{E}[\mathrm{P}(f(\boldsymbol{X}, \boldsymbol{Y}) = k \mid \boldsymbol{Y})] \\
&= \mathrm{P}(f(\boldsymbol{X}, \boldsymbol{Y}) = k) \\
&= \mathrm{E}[\mathrm{P}(f(\boldsymbol{X}, \boldsymbol{Y}) = k \mid \boldsymbol{X})] \\
&= \mathrm{E}[h_k(\boldsymbol{X})] = \frac{\binom{n}{k}\binom{N-n}{K-k}}{\binom{N}{K}}
\end{aligned}
\quad (2.43)
$$

for $k = (n + K - N)^+, \ldots, n \wedge K$.

If we apply this result to keno (Example 1.3.1 on p. 23), we find that formulas (1.73) and (1.74) on p. 23 are both correct, provided the house chooses its 20 numbers at random *or* the player chooses his 10 numbers at random *or* both sets of numbers are chosen at random and independently. While the combinatorial proof of Problem 1.21 on p. 56 is quicker, the probabilistic proof given here is more revealing.   ♠

The next result is known as the *factorization property* of conditional expectation.

Theorem 2.2.7. *Let the discrete random vector $\boldsymbol{X}$ and the discrete random variable Y be jointly distributed, and let f be a real-valued function on the range of $\boldsymbol{X}$ such that each of $f(\boldsymbol{X})Y$ and Y has finite expectation. Then*

$$\mathrm{E}[f(\boldsymbol{X})Y \mid \boldsymbol{X}] = f(\boldsymbol{X})\mathrm{E}[Y \mid \boldsymbol{X}]. \tag{2.44}$$

Remark. The theorem says that a multiplicative function of the conditioning random vector can be factored out of the conditional expectation as though it were a constant.

Proof. This follows from

$$\mathrm{E}[f(\boldsymbol{X})Y \mid \boldsymbol{X} = \boldsymbol{x}] = \mathrm{E}[f(\boldsymbol{x})Y \mid \boldsymbol{X} = \boldsymbol{x}] = f(\boldsymbol{x})\mathrm{E}[Y \mid \boldsymbol{X} = \boldsymbol{x}], \tag{2.45}$$

where the first equality uses Lemma 2.1.2 on p. 77.   ♠

Corollary 2.2.8. *Let f be a real-valued function on the range of the discrete random vector $\boldsymbol{X}$ such that $f(\boldsymbol{X})$ has finite expectation. Then*

$$\mathrm{E}[f(\boldsymbol{X}) \mid \boldsymbol{X}] = f(\boldsymbol{X}). \tag{2.46}$$

Example 2.2.9. *Red dog.* The game of red dog was introduced in the 1980s but is rarely found today. It is played with a d-deck shoe. (Usually, $d = 6$.) Denominations are ranked in the usual way (2, 3, 4, 5, 6, 7, 8, 9, 10, J, Q, K, A) and suits do not matter. After the player makes a bet, the dealer deals two cards face up. There are three possibilities:

If the denominations of the two cards are equal, a third card is dealt. If its denomination matches that of the first two cards, the player is paid 11 to 1. Otherwise the result is a push.

If the denominations of the two cards are consecutive, then the result is a push.

If the denominations of the two cards are neither equal nor consecutive, then the dealer announces the *spread*, that is, the number of denominations that lie strictly between the two displayed denominations. The player then has the opportunity to increase his wager by an amount not to exceed that of the original wager. Finally, a third card is dealt, and if its denomination lies strictly between those of the first two cards, then the player wins. He is paid 5 to 1 if the spread is one, 4 to 1 if the spread is two, 2 to 1 if the spread is three, and 1 to 1 if the spread is four or more. Otherwise the player loses the amount of his bet.

It will be convenient to relabel the 13 denominations 2–10, J, Q, K, A as 2–14. We denote the $52d$ cards by random variables $X_1, X_2, \ldots, X_{52d}$. Specifically, $(X_1, X_2, \ldots, X_{52d})$ assumes each of the $(52d)!/[(4d)!]^{13}$ permutations of $4d$ 2s, $4d$ 3s, ..., $4d$ 14s with equal probability. Of course, only X_1, X_2, and X_3 will play a role in our analysis.

We let Y denote the player's profit from his initial one-unit bet. Then we can evaluate the conditional expectation of Y, given the first two cards X_1 and X_2. For example,

$$\mathrm{E}[Y \mid X_1 = X_2] = 11\left(\frac{4d - 2}{52d - 2}\right) + 0\left(1 - \frac{4d - 2}{52d - 2}\right) = \frac{44d - 22}{52d - 2}. \quad (2.47)$$

If the spread is denoted by

$$S := |X_1 - X_2| - 1, \quad (2.48)$$

then $X_1 = X_2$ if and only if $S = -1$, and X_1 and X_2 are consecutive if and only if $S = 0$. In particular,

$$\mathrm{E}[Y \mid S = -1] = \frac{44d - 22}{52d - 2} \quad \text{and} \quad \mathrm{E}[Y \mid S = 0] = 0. \quad (2.49)$$

If we further define the payoff function $a : \{1, 2, \ldots, 11\} \mapsto \{1, 2, 4, 5\}$ by

$$a(1) := 5, \quad a(2) := 4, \quad a(3) := 2, \quad a(s) := 1 \text{ if } s \geq 4, \quad (2.50)$$

then, on the event $\{S \geq 1\}$,

$$\mathrm{E}[Y \mid S] = a(S)\frac{(4d)S}{52d - 2} + (-1)\left(1 - \frac{(4d)S}{52d - 2}\right)$$

$$= [a(S) + 1]\frac{(4d)S}{52d - 2} - 1. \quad (2.51)$$

The values of these conditional expectations of Y are displayed in Table 2.1.

The main point is that, in the situations in which the player has the opportunity to increase his bet size (namely, when $S \geq 1$), his conditional expected profit is positive if and only if $S \geq 7$, regardless of the value of $d \geq 1$. Therefore, a reasonable strategy is to double one's bet if $S \geq 7$ and to leave it unchanged otherwise.

Now the probabilities of the conditioning events are easy to evaluate:

$$P(S = -1) = 13\frac{\binom{4d}{2}\binom{48d}{0}}{\binom{52d}{2}} \quad (2.52)$$

and

$$P(S = s) = (12 - s)\frac{\binom{4d}{1}^2\binom{44d}{0}}{\binom{52d}{2}}, \quad s = 0, 1, \ldots, 11. \quad (2.53)$$

These values are also given in Table 2.1.

Table 2.1 Evaluation of the conditional expectation of Y.

event E	$\binom{52d}{2}\mathrm{P}(E)$	$(52d-2)\mathrm{E}[Y \mid E]$
$\{S=-1\}$	$104d^2 - 26d$	$44d - 22$
$\{S=0\}$	$192d^2$	0
$\{S=1\}$	$176d^2$	$-28d + 2$
$\{S=2\}$	$160d^2$	$-12d + 2$
$\{S=3\}$	$144d^2$	$-16d + 2$
$\{S=4\}$	$128d^2$	$-20d + 2$
$\{S=5\}$	$112d^2$	$-12d + 2$
$\{S=6\}$	$96d^2$	$-4d + 2$
$\{S=7\}$	$80d^2$	$4d + 2$
$\{S=8\}$	$64d^2$	$12d + 2$
$\{S=9\}$	$48d^2$	$20d + 2$
$\{S=10\}$	$32d^2$	$28d + 2$
$\{S=11\}$	$16d^2$	$36d + 2$
total	$\binom{52d}{2}$	

We therefore conclude that the player's expected profit from an initial one-unit bet, assuming the suggested strategy, is given by

$$\mathrm{E}[(1 + 1_{\{S \geq 7\}})Y] = \mathrm{E}[\mathrm{E}[(1 + 1_{\{S \geq 7\}})Y \mid S]] = \mathrm{E}[(1 + 1_{\{S \geq 7\}})\mathrm{E}[Y \mid S]]$$

$$= \sum_{s=-1}^{6} \mathrm{P}(S = s)\mathrm{E}[Y \mid S = s] + 2\sum_{s=7}^{11} \mathrm{P}(S = s)\mathrm{E}[Y \mid S = s]$$

$$= -\frac{4d(456d^2 + 210d - 143)}{\binom{52d}{2}(52d - 2)}, \tag{2.54}$$

where we have used Theorem 2.2.7 in the second equality. When $d = 6$, this is $-420{,}792 / \left[\binom{312}{2}310\right] \approx -0.027978.$ ♠

The next theorem describes the *tower property* of conditional expectations.

Theorem 2.2.10. *Let the discrete random vector $\boldsymbol{X}$ and the discrete random variable Z be jointly distributed with Z having finite expectation. Let $\boldsymbol{f}$ be a vector-valued function on the range of $\boldsymbol{X}$, and put $\boldsymbol{Y} := \boldsymbol{f}(\boldsymbol{X})$. Then*

$$\mathrm{E}[\mathrm{E}[Z \mid \boldsymbol{X}] \mid \boldsymbol{Y}] = \mathrm{E}[Z \mid \boldsymbol{Y}]. \tag{2.55}$$

Proof. It suffices to show that

$$\mathrm{E}[\mathrm{E}[Z \mid \boldsymbol{X}] \mid \boldsymbol{Y} = \boldsymbol{y}] = \mathrm{E}[Z \mid \boldsymbol{Y} = \boldsymbol{y}] \tag{2.56}$$

or, by Lemma 2.1.2 on p. 77, that

$$E[E[Z \mid \boldsymbol{X}] 1_{\{\boldsymbol{f}(\boldsymbol{X})=\boldsymbol{y}\}}] = E[Z 1_{\{\boldsymbol{f}(\boldsymbol{X})=\boldsymbol{y}\}}]. \tag{2.57}$$

Applying Theorem 2.2.7 to the left side of (2.57), we see that (2.57) follows from (2.24).   ♠

Example 2.2.11. *A sampling problem.* Consider an urn containing N balls, of which N_1 are labeled 1, N_2 are labeled 2, and N_3 are labeled 3, with N_1, N_2, N_3 being positive integers that sum to N. Fix a positive integer n less than N. First, n balls are drawn from the urn without replacement, and then a final ball is drawn from the remaining $N - n$ balls. What are the conditional probabilities that ball $n + 1$ is labeled 1, 2, and 3, respectively, given that the sample of n contains i balls labeled 1 $(0 \le i \le n \wedge N_1)$?

Let X be the number of 1s in the sample of n and let Y be the number of 2s. Let Z be the label of ball $n + 1$. Clearly,

$$P(Z = 1 \mid X = i) = \frac{N_1 - i}{N - n}. \tag{2.58}$$

Less obvious is the result that, by Theorem 2.2.10,

$$
\begin{aligned}
P(Z = 2 \mid X = i) &= E[P(Z = 2 \mid X, Y) \mid X = i] \\
&= E\left[\sum_j P(Z = 2 \mid X = i,\ Y = j)\, 1_{\{Y=j\}} \,\middle|\, X = i \right] \\
&= \sum_j P(Z = 2 \mid X = i,\ Y = j) P(Y = j \mid X = i) \\
&= \sum_{j=(n-i-N_3)^+}^{(n-i)\wedge N_2} \frac{N_2 - j}{N - n} \cdot \frac{\binom{N_2}{j}\binom{N_3}{n-i-j}}{\binom{N_2+N_3}{n-i}} \\
&= \frac{N_2 - (n - i)[N_2/(N_2 + N_3)]}{N - n}. \tag{2.59}
\end{aligned}
$$

(Alternatively, we could have cited (1.62) on p. 21.) Here we have used the mean of the hypergeometric distribution from Example 1.4.12 on p. 33. Similarly,

$$P(Z = 3 \mid X = i) = \frac{N_3 - (n - i)[N_3/(N_2 + N_3)]}{N - n}. \tag{2.60}$$

Let us illustrate these results with a numerical example. Given that 2 aces and 15 nonaces have been dealt from a standard 52-card deck, what is the conditional probability that the next card to be dealt is 10, J, Q, or K? We define the type 1 balls to be the aces, the type 2 balls to be the 10s, Js, Qs, and Ks, and the type 3 balls to be the 2s–9s. By (2.59), the answer to our question is

$$\frac{16 - (17 - 2)[16/(16 + 32)]}{52 - 17} = \frac{11}{35}. \tag{2.61}$$

This type of calculation is useful in twenty-one.   ♠

We need one last definition. If the discrete random vector $\boldsymbol{X}$ and the discrete random variable Y are jointly distributed, and if Y has finite variance, then we define the *conditional variance of Y given $\boldsymbol{X}$* by

$$\operatorname{Var}(Y \mid \boldsymbol{X}) := \operatorname{E}[(Y - \operatorname{E}[Y \mid \boldsymbol{X}])^2 \mid \boldsymbol{X}]. \tag{2.62}$$

The first result is the analogue of Theorem 1.4.6 on p. 31.

Theorem 2.2.12. *Under the above assumptions on $\boldsymbol{X}$ and Y,*

$$\operatorname{Var}(Y \mid \boldsymbol{X}) = \operatorname{E}[Y^2 \mid \boldsymbol{X}] - (\operatorname{E}[Y \mid \boldsymbol{X}])^2. \tag{2.63}$$

Proof. We expand the square in (2.62) and apply Theorems 2.2.2 and 2.2.7.
♠

The following result is the *conditioning law for variances*, analogous to (2.24).

Theorem 2.2.13. *Under the above assumptions on $\boldsymbol{X}$ and Y,*

$$\operatorname{Var}(Y) = \operatorname{E}[\operatorname{Var}(Y \mid \boldsymbol{X})] + \operatorname{Var}(\operatorname{E}[Y \mid \boldsymbol{X}]). \tag{2.64}$$

Proof. By Theorem 2.2.12 above and Theorem 1.4.6 on p. 31,

$$\operatorname{E}[\operatorname{Var}(Y \mid \boldsymbol{X})] = \operatorname{E}[Y^2] - \operatorname{E}[(\operatorname{E}[Y \mid \boldsymbol{X}])^2] \tag{2.65}$$

and

$$\operatorname{Var}(\operatorname{E}[Y \mid \boldsymbol{X}]) = \operatorname{E}[(\operatorname{E}[Y \mid \boldsymbol{X}])^2] - (\operatorname{E}[Y])^2. \tag{2.66}$$

The result follows from Theorem 1.4.6 on p. 31 by adding these two equations.
♠

We conclude this section by extending Example 2.1.4 on p. 78.

Example 2.2.14. *Variance of the number of rolls in a pass-line decision at craps.* The pass-line bet at craps was discussed in Examples 1.2.2 on p. 17, 1.2.8 on p. 19, and 2.1.4 on p. 78.

Let Y be the number of rolls in a pass-line decision, and let X be the result of the come-out roll. Specifically, X has distribution (1.11) on p. 6. With

$$\mathscr{S} := \{2, 3, 4, \ldots, 12\} \quad \text{and} \quad \mathscr{P} := \{4, 5, 6, 8, 9, 10\}, \tag{2.67}$$

we have

$$\operatorname{E}[Y \mid X] = \begin{cases} 1 & \text{if } X \in \mathscr{S} - \mathscr{P}, \\ 1 + 1/(\pi_X + \pi_7) & \text{if } X \in \mathscr{P}, \end{cases} \tag{2.68}$$

and

$$\operatorname{Var}(Y \mid X) = \begin{cases} 0 & \text{if } X \in \mathscr{S} - \mathscr{P}, \\ (1 - \pi_X - \pi_7)/(\pi_X + \pi_7)^2 & \text{if } X \in \mathscr{P}, \end{cases} \tag{2.69}$$

where we have used the mean and variance of the geometric distribution from Examples 1.4.3 on p. 30 and 1.4.9 on p. 32. By Theorem 2.2.13,

$$
\begin{aligned}
\mathrm{Var}(Y) &= \sum_{j \in \mathscr{P}} \pi_j \frac{1 - \pi_j - \pi_7}{(\pi_j + \pi_7)^2} + \sum_{j \in \mathscr{P}} \pi_j \frac{1}{(\pi_j + \pi_7)^2} - \left(\sum_{j \in \mathscr{P}} \pi_j \frac{1}{\pi_j + \pi_7} \right)^2 \\
&= \frac{245{,}672}{27{,}225} \approx 9.023765,
\end{aligned} \tag{2.70}
$$

where we subtracted 1 from $\mathrm{E}[Y \mid X]$ before evaluating its variance using Theorem 1.4.6 on p. 31.   ♠

2.3 Problems

2.1. *Problem of points for three players.*

(a) Three players of equal skill, A, B, and C, play a game that randomly awards one point to one of the players at each round, and successive rounds are independent. They agree to continue until some player has won a specified number of points. If the game is discontinued with A lacking one point and B and C each lacking two, how should the stakes be divided?

(b) Generalize part (a). Specifically, assume that A, B, and C have probabilities p, q, and r of winning a round and lack i, j, and k points when the game is discontinued.

2.2. *Conditioning on outcome 1 before outcome 2.* Recall the assumptions of Example 1.2.1 on p. 16. (A random experiment has exactly three possible outcomes, referred to as outcomes 1, 2, and 3, with probabilities $p_1 > 0$, $p_2 > 0$, and $p_3 > 0$, where $p_1 + p_2 + p_3 = 1$. We consider a sequence of independent trials, at each of which the specified random experiment is performed.) For $i = 1, 2$, let N_i be the number of trials needed for outcome i to occur, and put $N := N_1 \wedge N_2$.

(a) Show that N is independent of $1_{\{N_1 < N_2\}}$.

(b) Evaluate $\mathrm{E}[N_1 \mid N_1 < N_2]$.

(c) Roll a pair of dice until a total of 6 or 7 appears. Given that 6 appears before 7, what is the (conditional) expected number of rolls?

2.3. *Mean and variance of the geometric distribution via conditioning.* By conditioning on the result of the first trial, find the mean of the geometric(p) distribution, where $0 < p < 1$. Use a similar argument to find the variance.

2.4. *Fermat's game.* A and B compete against each other with a pair of dice. A wins if he rolls a total of 6, while B wins if he rolls a total of 7. A has the first roll, B the following two, A the following two, B the following two, and so on.

(a) Find the probability that A wins.

(b) Find the expected number of rolls required to resolve the game.

2.5. *Mean duration of a cinq-et-neuf decision.* Find the expected number of tosses of a pair of dice necessary to complete a game of cinq et neuf (Problem 1.28 on p. 57).

2.6. *Conditional expected number of rolls in a pass-line decision.* Find the conditional expected number of rolls in a pass-line decision (Example 1.2.2 on p. 17), given that the pass-line bet is won; ... given that the pass-line bet is lost.

Hint: By Problem 2.2, $E[N\,1_{\{N_1<N_2\}}] = E[N]\,P(N_1 < N_2) = p_1/(p_2+p_2)^2$.

2.7. *Expected length of the shooter's hand at craps.* In craps a shooter continues to roll the dice until he loses a pass-line decision (Example 1.2.2 on p. 17) by rolling 7. The sequence of rolls from the initial come out to the seven out is called the *shooter's hand*. To put it another way, the shooter rolls the dice for N pass-line decisions, where N has the geometric distribution with parameter

$$q := \sum_{j \in \mathscr{P}} \pi_j \frac{\pi_7}{\pi_j + \pi_7} = \frac{196}{495}. \tag{2.71}$$

Now from Example 2.1.4 on p. 78 we know that the expected number of rolls in a single pass-line decision is $\mu = 557/165$. Letting L denote the length of the shooter's hand (i.e., the number of rolls), evaluate $E[L]$ by conditioning on whether $N = 1$ to show that

$$E[L] = \mu + (1 - q)E[L]. \tag{2.72}$$

Assume for now that $E[L] < \infty$. (This will be confirmed in Example 3.2.6 on p. 103.)

2.8. *Expected number of occurrences of outcome 1 before the first occurrence of outcome 2.*

(a) Under the assumptions of Problem 2.2, find the expected number of times outcome 1 occurs before the first occurrence of outcome 2.

(b) When rolling a pair of dice, find the expected number of times the total 4 (resp., 5; 6; 8; 9; 10) appears before the first total of 7.

2.9. *Expected number of Bernoulli trials until m consecutive successes.* Let $X_1, X_2, \ldots$ be a sequence of i.i.d. random variables with $P(X_1 = 1) = p$ and $P(X_1 = 0) = 1 - p$, where $0 < p < 1$. For each $m \geq 1$, define

$$N_m := \min\{n \geq m : (X_{n-m+1}, \ldots, X_n) = (1, 1, \ldots, 1)\} \tag{2.73}$$

and show that

$$E[N_m] = \frac{1}{p} + \frac{1}{p^2} + \cdots + \frac{1}{p^m}. \tag{2.74}$$

Hint: Use induction, evaluating $E[N_m]$ by conditioning on N_{m-1}.
See Problem 3.10 on p. 113 for a generalization.

2.10. *A Poisson number of Bernoulli trials.* Fix $\lambda > 0$ and $0 < p < 1$, and let N be Poisson(λ). Assume that the conditional distribution of Y, given that $N = n$, is binomial(n, p) for $n = 0, 1, \ldots$.
  (a) Find the mean and variance of Y by conditioning on N.
  (b) Find the distribution of Y by conditioning on N.
  Hint: (b) First find the probability generating function of N, take its kth derivative, and evaluate at $1 - p$.

2.11. *Conditional expectation given a sum.* Fix $n \geq 2$, let $X_1, \ldots, X_n$ be n i.i.d. discrete random variables with finite expectation, and put $S_n := X_1 + \cdots + X_n$. Show that $\mathrm{E}[X_1 \mid S_n] = S_n/n$.

2.12. *Probability that each of k possible outcomes occurs at least once in n trials.* This problem essentially generalizes Problem 1.36 on p. 59. Consider n independent trials, each of which has k possible outcomes $1, \ldots, k$ with probabilities $p_1 > 0, \ldots, p_k > 0$, where $p_1 + \cdots + p_k = 1$. What is the probability that each of the k outcomes occurs at least once?
  (a) First, obtain a solution using the inclusion-exclusion law (Theorem 1.1.18 on p. 13).
  (b) Second, derive a recursive solution by conditioning. More explicitly, for $2 \leq j \leq k$ and $m \geq j$, let $A_{m,j}$ be the event that each of the outcomes $1, \ldots, j$ (with probabilities $p_1/P_j, \ldots, p_j/P_j$, where $P_j := p_1 + \cdots + p_j$) occurs at least once in m trials. Show that

$$P(A_{m,j}) = \sum_{l=1}^{m-(j-1)} \binom{m}{l} (p_j/P_j)^l (1 - p_j/P_j)^{m-l} P(A_{m-l,j-1}), \qquad (2.75)$$

and explain how to use this to obtain $P(A_{n,k})$.
  (c) Both approaches, (a) and (b), simplify when $p_1 = \cdots = p_k = 1/k$. In the case of (b), however, it may be easier to start over, fixing k and defining $p_m(j)$ to be the probability of j distinct results in m trials, $1 \leq j \leq m \wedge k$. Argue that these probabilities are recursive in m, and use this to obtain $p_n(k)$.
  (d) Find the median number of coups of an unbiased 38-number roulette wheel needed for each of the 38 numbers to appear at least once. Which of the three above approaches is computationally simplest?

2.13. *Conditional multinomial distribution.* Let the random vector $(\boldsymbol{X}, \boldsymbol{Y})$ have the multinomial($n, (\boldsymbol{p}, \boldsymbol{q})$) distribution. (Here $\boldsymbol{X}$ and $\boldsymbol{p}$ have the same dimension, while $\boldsymbol{Y}$ and $\boldsymbol{q}$ have the same dimension. The sum of the components of $\boldsymbol{p}$ plus the sum of those of $\boldsymbol{q}$ is 1.) Show that the conditional distribution of $\boldsymbol{Y}$, given $\boldsymbol{X} = \boldsymbol{k}$, is also multinomial, and identify the parameters. Is a similar property true of the multivariate hypergeometric distribution?

2.14. *Pairwise independence and the hypergeometric distribution.* With $\boldsymbol{X}$ and $\boldsymbol{Y}$ independent as in Example 2.2.6 on p. 82 and with f as in (2.40) on

p. 82, argue that $\boldsymbol{X}$, $\boldsymbol{Y}$, and $Z := f(\boldsymbol{X}, \boldsymbol{Y})$ are pairwise independent but not independent (unless $n = N = K$).

2.15. *Conditional expectation as a projection.* Let the discrete random vector $\boldsymbol{X}$ and the discrete random variable Y be jointly distributed, and assume that Y has finite second moment. Show that the bounded function h on the range of $\boldsymbol{X}$ that minimizes $\mathrm{E}[(Y - h(\boldsymbol{X}))^2]$ is given by $h(\boldsymbol{x}) = \mathrm{E}[Y \mid \boldsymbol{X} = \boldsymbol{x}]$.

2.16. *Conditional independence.* Let $\boldsymbol{X}$, $\boldsymbol{Y}$, and $\boldsymbol{Z}$ be jointly distributed discrete random vectors. We say that $\boldsymbol{Y}$ and $\boldsymbol{Z}$ are *conditionally independent given $\boldsymbol{X}$* if

$$P(\boldsymbol{Y} = \boldsymbol{y}, \ \boldsymbol{Z} = \boldsymbol{z} \mid \boldsymbol{X}) = P(\boldsymbol{Y} = \boldsymbol{y} \mid \boldsymbol{X})P(\boldsymbol{Z} = \boldsymbol{z} \mid \boldsymbol{X}) \qquad (2.76)$$

for all $\boldsymbol{y}$ and $\boldsymbol{z}$. (If $\boldsymbol{X}$ is independent of $(\boldsymbol{Y}, \boldsymbol{Z})$, then this is equivalent to the independence of $\boldsymbol{Y}$ and $\boldsymbol{Z}$.) Show that $\boldsymbol{Y}$ and $\boldsymbol{Z}$ are conditionally independent given $\boldsymbol{X}$ if and only if

$$P(\boldsymbol{Z} = \boldsymbol{z} \mid \boldsymbol{X}, \boldsymbol{Y}) = P(\boldsymbol{Z} = \boldsymbol{z} \mid \boldsymbol{X}) \qquad (2.77)$$

for all $\boldsymbol{z}$.

2.17. *Variance of a sum of conditionally i.i.d. random variables.* Fix $n \geq 2$ and let $X, Y_1, \ldots, Y_n$ be jointly distributed discrete random variables. By analogy with Problem 2.16, we say that $Y_1, \ldots, Y_n$ are *conditionally i.i.d. given X* if

$$P(Y_1 = y_1, \ \ldots, \ Y_n = y_n \mid X) = P(Y_1 = y_1 \mid X) \cdots P(Y_n = y_n \mid X) \quad (2.78)$$

for all $y_1, \ldots, y_n$. Show that, if $Y_1, \ldots, Y_n$ are conditionally i.i.d. given X, then

$$\mathrm{Var}(Y_1 + \cdots + Y_n) = n^2 \mathrm{Var}(Y_1) - n(n-1)\mathrm{E}[\mathrm{Var}(Y_1 \mid X)]. \qquad (2.79)$$

2.18. *Conditional expectation and stopping times.* Let $X_0, X_1, \ldots$ be a sequence of discrete random variables with values in some countable set S, and let N be a random variable with values in $\mathbf{Z}_+ \cup \{\infty\}$ such that

$$1_{\{N \leq n\}} = f_n(X_0, \ldots, X_n), \qquad n \geq 0, \qquad (2.80)$$

for suitable functions $f_n : S^{n+1} \mapsto \{0, 1\}$. (Such an N is called a *stopping time*.) If $g : S \mapsto \mathbf{R}$ is a bounded function, show that

$$\mathrm{E}[g(X_{(n+1) \wedge N}) \mid X_0, \ldots, X_n] = \mathrm{E}[g(X_{n+1}) \mid X_0, \ldots, X_n] \quad \text{on } \{N \geq n+1\}, \qquad (2.81)$$

and

$$\mathrm{E}[g(X_{(n+1) \wedge N}) \mid X_0, \ldots, X_n] = g(X_N) \quad \text{on } \{N \leq n\}. \qquad (2.82)$$

Hint: For (2.81) use Theorem 2.2.7 on p. 83 with $f := 1_{\{N \geq n+1\}}$.

2.4 Notes

Conditional expectation is as old as expectation. Early authors did not see the distinction as important.

The problem of points (Example 2.1.1) was posed by Luca Pacioli [1445–1517] in *Summa de arithmetica, geometria, proportioni e proportionalita* (1494). According to David (1962, p. 37), he wrote,

> A and B are playing a fair game of *balla*. They agree to continue until one has won six rounds. The game actually stops when A has won five and B three. How should the stakes be divided?

Pacioli answered 5 : 3 rather than the correct 7 : 1 (see (2.15)). Subsequent unsuccessful solutions were given by Girolamo Cardano [1501–1576] (1539), Nicolo Fontana Tartaglia [1500–1557] (1556), Giovanni Francesco Peverone [1509–1559] in 1558, and Lorenzo Forestani in 1603. Cardano, however, was the first to realize that the answer should depend on only the numbers of points lacked by A and B.

Actually, Pacioli did not originate the problem of points, and as a result of research by Toti Rigatelli (1985) (see Schneider 1988), we know that the problem was correctly solved around 1400 by an unknown Florentine mathematician, at least in the special case in which A lacks one point and B lacks three points.

Some 250 years later, the problem of points was communicated to Blaise Pascal [1623–1662] by Antoine Gombaud, the Chevalier de Méré [1607–1684], and solved in correspondence with Pierre de Fermat [1601–1665] in the summer of 1654. As Siméon-Denis Poisson [1781–1840] (1837, p. 1) put it,

> A problem relative to games of chance, proposed to an austere Jansenist by a man of the world, was the origin of the calculus of probabilities.

Pascal and Fermat considered only the case $p = \frac{1}{2}$. First, Fermat proposed the solution (2.12), or, more precisely, he proposed enumerating the 2^{j+k-1} equally likely outcomes, determining for each whether it was a win for A or for B. The formula (2.12) first appeared explicitly in a tract appended to Pascal's *Traité du triangle arithmétique* (written 1654, published 1665). Pascal responded to Fermat (David 1962, p. 231),[1]

> Your method is very sound and is the one which first came to my mind in this research; but because the labour of the combination is excessive, I have found a short cut and indeed another method which is much quicker and neater [...]

Pascal's "quicker and neater" method was essentially (2.14), which he explained to Fermat as follows (David 1962, pp. 231–232):[1]

> Here, more or less, is what I do to show the fair value of each game, when two opponents play, for example in three games, and each person has staked 32 pistoles.

---

[1] Translation by Maxine Merrington

Let us say that the first man had won twice and the other once; now they play another game, in which the conditions are that, if the first wins, he takes all the stakes, that is 64 pistoles, if the other wins it, then they have each won two games, and therefore, if they wish to stop playing, they must each take back their own stake, that is, 32 pistoles each.

Then consider, Sir, if the first man wins, he gets 64 pistoles, if he loses he gets 32. Thus if they do not wish to risk this last game, but wish to separate without playing it, the first man must say: "I am certain to get 32 pistoles, even if I lose I still get them; but as for the other 32, perhaps I will get them, perhaps you will get them, the chances are equal. Let us then divide these 32 pistoles in half and give one half to me as well as my 32 which are mine for sure." He will then have 48 pistoles and the other 16.

Let us suppose now that the first had won two games and the other had won none, and they begin to play a new game. The conditions of this new game are such that if the first man wins it, he takes all the money, 64 pistoles; if the other wins it they are in the same position as in the preceding case, when the first man had won *two* games and the other, *one*.

Now, we have already shown in this case, that 48 pistoles are due to the one who has two games: thus if they do not wish to play that new game, he must say: "If I win it I will have all the stakes, that is 64; if I lose it, 48 will legitimately be mine; then give me the 48 which I have in any case, even if I lose, and let us share the other 16 in half, since there is as good a chance for you to win them as for me." Thus he will have 48 and 8, which is 56 pistoles.

Let us suppose, finally, that the first had won one game and the other none. You see, Sir, that if they begin a new game, the conditions of it are such that, if the first man wins it he will have *two* games to *none*, and thus by the preceding case, 56 belong to him; if he loses it, they each have one game, then 32 pistoles belong to him. So he must say: "If you do not wish to play, give me 32 pistoles which are mine in any case, let us take half each of the remainder taken from 56. From 56 set aside 32, leaving 24, then divide 24 in half, you take 12 and give me 12 which with 32 makes 44."

A recent discussion of the problem of points, written for nonmathematicians, argued that "Fermat's solution is simply much better." (Devlin 2008, pp. 62–63.) Moreover,

> Pascal's solution to the problem of points was far more complicated than Fermat's and required some sophisticated—and daunting!—algebra.

We disagree that Fermat's solution is "simply much better" and that Pascal's solution is "far more complicated." In fact, the latter assertion is belied by Pascal's own words, quoted above in translation. Fermat's solution required a clever trick, namely allowing the game to continue beyond its natural conclusion, to evaluate the relevant probabilities. Pascal's solution was based on conditioning and recursion, a technique that has been used time and time again over the past 350 years and is still used today. Undoubtedly, Pascal's solution has had more impact.

Johann Bernoulli [1667–1748] obtained (2.12) for arbitrary p, and communicated it to Pierre Rémond de Montmort [1678–1719] in 1710 (see Montmort 1713, p. 295). Actually, the result was first published by Abraham De Moivre [1667–1754] (1712). The solution (2.11) involving the negative binomial distribution is due to Montmort (1713, p. 245). See Takács (1994) for a survey article on the problem of points, and see Llorente (2007) for a recent application to gambling.

Example 2.1.4 (mean duration of craps decision) is due to Brown (1919).

Regarding conditional expectation as a random variable is relatively recent. Andrei N. Kolmogorov [1903–1987] (1933) defined conditional expectation for arbitrary (not just discrete) random variables, and the theorems of Section 2.2 were certainly known to him, though are perhaps older in the discrete setting.

Example 2.2.6 (interchangeability of the parameters of the hypergeometric distribution) was known to De Moivre (1718, Problem 20), as noted in Section 1.7. The derivation here is essentially from Davidson and Johnson (1993). The game of red dog (Example 2.2.9) was introduced by Terrance W. Oliver in the early 1980s and was successful for a short period, particularly in Atlantic City. Because it was based on the old game of acey-deucey, it was not patented. Example 2.2.11 (a sampling problem) is due to Marc Estafanous (personal communication, 2005). The result of Example 2.2.14 (variance of duration of craps decision) was first found by Smith (1968) using a different method.

Problem 2.1, in the special case stated, was first solved by Fermat in his correspondence with Pascal in 1654 (David 1962, pp. 92–94; Edwards 1987, pp. 144–147). Pascal at first misunderstood Fermat's solution, apparently overlooking the fact that the order in which the points are won matters in the case of three players but not in the case of two players. He did arrive at the correct answer, however, using his recursive method. Generalizations in terms of the multinomial distribution were found by Montmort (1708; 1713, pp. 242, 253, 371) and De Moivre (1712; 1718, Problem 8). A generalization of (2.11) was given by De Moivre (1730; 1738, Problems 6 and 69).

Problem 2.2 is from Ross (2006, p. 279), who observed that the conditional distribution of $1_{\{N_1 < N_2\}}$ given N clearly does not depend on N, whereas it is not obvious that the conditional distribution of N given $1_{\{N_1 < N_2\}}$ does not depend on $1_{\{N_1 < N_2\}}$. Problem 2.3 (mean and variance of geometric by conditioning) dates back at least as far as Louis Bachelier [1870–1946] (1912, p. 12). Problem 2.4(a) (Fermat's game) was first posed by Fermat in a letter to Christiaan Huygens [1629–1695] in June 1656. It was solved by Huygens in a letter to Carcavi on July 6, 1656 (Hald 1990, p. 74). Problem 2.6 is from Ross (2006, p. 370). See Section 15.4 for the history of Problem 2.7. As for Problem 2.12, which like Problem 1.36 is a version of the coupon collector's problem, (a) is from De Moivre (1712), (b) is from Ross (2003, pp. 133–134), (c) is from Feller (1968, p. 284), and (d) is from Griffin (1991, p. 33). Problem 2.14 was motivated by a result of Perlman and Wichura (1975).

Chapter 3
Martingales

You have not played as yet? Do not do so; above all avoid a martingale, if you do.

William Makepeace Thackeray, *The Newcomes*

A martingale is a sequence of discrete random variables indexed by a time parameter with the property that the conditional expectation of a future term given the past and present terms is the present term. It can be thought of as a stochastic model for the gambler's fortune in a fair game. In Section 3.1 we formalize and generalize this definition and motivate it with examples. In Section 3.2 we prove the optional stopping theorem for martingales, which will play an important role in what follows. In Section 3.3 we prove the martingale convergence theorem.

3.1 Definitions and Examples

We begin with the definitions. Let $\{X_n\}_{n\geq 0}$ be a sequence of jointly distributed discrete random variables (or random vectors). This sequence is called a *filtration* and is our basic source of randomness. Often the filtration is indexed by the positive integers rather than the nonnegative ones; in such cases X_0 is understood to be a constant for the purpose of the definitions.

We say that a sequence of discrete random variables $\{M_n\}_{n\geq 0}$ is a *martingale with respect to* $\{X_n\}_{n\geq 0}$ if there exists a sequence $\{f_n\}_{n\geq 0}$ of nonrandom functions such that

$$M_n = f_n(X_0, \ldots, X_n), \qquad E\big[|M_n|\big] < \infty, \tag{3.1}$$

and

$$E[M_{n+1} \mid X_0, \ldots, X_n] = M_n \tag{3.2}$$

for all $n \geq 0$.

S.N. Ethier, *The Doctrine of Chances*, Probability and its Applications, DOI 10.1007/978-3-540-78783-9_3, © Springer-Verlag Berlin Heidelberg 2010

Notice that (3.2) implies $\mathrm{E}[M_{n+1}] = \mathrm{E}[M_n]$, so the sequence of expectations $\mathrm{E}[M_0], \mathrm{E}[M_1], \ldots$ is constant. One can think of a martingale as a stochastic model for the gambler's fortune in a fair game. Under this interpretation, X_0 is constant, X_n $(n \geq 1)$ represents the result of the nth coup, M_0 is the gambler's initial fortune, and M_n $(n \geq 1)$ represents his fortune after n coups. Equation (3.1) says that the gambler's fortune following the nth coup depends solely on the results of the first n coups (there is no other source of randomness) and has finite expectation, while (3.2) expresses the fairness requirement: The gambler's conditional expected profit from coup $n + 1$ is 0.

It will be convenient to define also at this time the notions of supermartingale and submartingale. If $\{M_n\}_{n \geq 0}$ satisfies the definition of a martingale with respect to $\{X_n\}_{n \geq 0}$ but with the = sign in (3.2) replaced by $\leq$ (resp., $\geq$), we say that $\{M_n\}_{n \geq 0}$ is a *supermartingale* (resp., *submartingale*) *with respect to* $\{X_n\}_{n \geq 0}$.

The sequence of expectations $\mathrm{E}[M_0], \mathrm{E}[M_1], \ldots$ is nonincreasing for a supermartingale and nondecreasing for a submartingale. Just as a martingale models the gambler's fortune in a fair game, a supermartingale (resp., submartingale) models the gambler's fortune in a subfair (resp., superfair) game. (This suggests that the prefixes super- and sub- were not chosen correctly! There is, however, a good reason for this choice; see Problem 4.6 on p. 152.)

If $\{M_n\}_{n \geq 0}$ is a martingale (resp., supermartingale, submartingale) with respect to $\{X_n\}_{n \geq 0}$, then, by Theorem 2.2.10 on p. 85, $\{M_n\}_{n \geq 0}$ is a martingale (resp., supermartingale, submartingale) with respect to $\{M_n\}_{n \geq 0}$, and we express this by saying simply that $\{M_n\}_{n \geq 0}$ is a martingale (resp., supermartingale, submartingale). This terminology is useful when the filtration does not play an important role, as for example in the case of the martingale convergence theorem.

Example 3.1.1. *A gambler's ruin model.* Suppose $p > 0$, $q > 0$, $r \geq 0$, and $p + q + r = 1$. Let $\{X_n\}_{n \geq 1}$ be i.i.d. with $\mathrm{P}(X_1 = 1) = p$, $\mathrm{P}(X_1 = -1) = q$, and $\mathrm{P}(X_1 = 0) = r$. Let us define M_0 to be a nonnegative constant and

$$M_n := M_0 + X_1 + \cdots + X_n, \qquad n \geq 1. \tag{3.3}$$

Then (3.1) holds and $M_{n+1} = M_n + X_{n+1}$, so, by Theorem 2.2.2 on p. 81, Corollary 2.2.8 on p. 83, and Corollary 2.2.5 on p. 81,

$$\begin{aligned}
\mathrm{E}[M_{n+1} \mid X_1, \ldots, X_n] &= M_n + \mathrm{E}[X_{n+1} \mid X_1, \ldots, X_n] \\
&= M_n + \mathrm{E}[X_{n+1}] \\
&= M_n + p - q, \qquad n \geq 0,
\end{aligned} \tag{3.4}$$

and $\{M_n\}_{n \geq 0}$ is a martingale (resp., supermartingale, submartingale) with respect to $\{X_n\}_{n \geq 1}$ if $p = q$ (resp., $p \leq q$, $p \geq q$). This is an example in which X_0 does not play a role, which is why we condition on $X_1, \ldots, X_n$ instead of $X_0, \ldots, X_n$. Furthermore, (3.4) shows that, regardless of the values of p, q, and r,

$$M_n - n(p - q), \qquad n \geq 0, \tag{3.5}$$

is a martingale with respect to $\{X_n\}_{n \geq 1}$.

There are other martingales with respect to $\{X_n\}_{n \geq 1}$. Consider $M_0 := 1$ and

$$M_n := (q/p)^{X_1 + \cdots + X_n}, \qquad n \geq 1. \tag{3.6}$$

Then $M_{n+1} = M_n (q/p)^{X_{n+1}}$, so by Theorem 2.2.7 on p. 83 and Corollary 2.2.5 on p. 81,

$$\begin{aligned}
E[M_{n+1} \mid X_1, \ldots, X_n] &= M_n E[(q/p)^{X_{n+1}} \mid X_1, \ldots, X_n] \\
&= M_n E[(q/p)^{X_{n+1}}] \\
&= M_n[(q/p)p + (q/p)^{-1}q + r] \\
&= M_n, \qquad n \geq 0,
\end{aligned} \tag{3.7}$$

and $\{M_n\}_{n \geq 0}$ is a martingale with respect to $\{X_n\}_{n \geq 1}$. Suppose that $p = q$, define $S_0 := 0$ and $S_n := X_1 + \cdots + X_n$ for each $n \geq 1$, and consider

$$M_n := S_n^2 - n(1 - r), \qquad n \geq 0. \tag{3.8}$$

Then $M_{n+1} = (S_n + X_{n+1})^2 - (n+1)(1-r) = M_n + 2S_n X_{n+1} + X_{n+1}^2 - (1-r)$, so by Theorem 2.2.2 on p. 81, Corollary 2.2.8 on p. 83, Theorem 2.2.7 on p. 83, and Corollary 2.2.5 on p. 81,

$$\begin{aligned}
E[M_{n+1} \mid X_1, \ldots, X_n] &= M_n + E[2S_n X_{n+1} + X_{n+1}^2 - (1 - r) \mid X_1, \ldots, X_n] \\
&= M_n + 2S_n E[X_{n+1}] + E[X_{n+1}^2] - (1 - r) \\
&= M_n,
\end{aligned} \tag{3.9}$$

and again $\{M_n\}_{n \geq 0}$ is a martingale with respect to $\{X_n\}_{n \geq 1}$. $\spadesuit$

We will not cite the theorems of Chapter 2 quite so carefully in what follows because they should by now be second nature.

We next consider the effect of varying bet sizes.

Example 3.1.2. *Even-money bets with varying bet sizes.* Consider the same game as in Example 3.1.1, but suppose now that the gambler varies his bets according to some system. Let B_n be the bet size at coup n, and assume that there exists a sequence $\{b_n\}_{n \geq 1}$ of nonrandom functions satisfying

$$B_1 = b_1 \geq 0 \ \text{(a constant)}, \qquad B_n = b_n(X_1, \ldots, X_{n-1}) \geq 0, \quad n \geq 2. \tag{3.10}$$

The point is that, when the gambler places his bet at the nth coup, the result X_n of that coup is still unknown to him, but the results of all previous coups are known. Notice that $E[B_n] < \infty$ for each $n \geq 1$ because B_n assumes only finitely many values (in fact, at most 3^{n-1} values) by (3.10). Consider

$$M_n := M_0 + \sum_{l=1}^{n} B_l X_l, \qquad n \geq 0. \tag{3.11}$$

Since $M_{n+1} = M_n + B_{n+1} X_{n+1}$, we have

$$\begin{aligned}
\mathrm{E}[M_{n+1} \mid X_1, \ldots, X_n] &= M_n + \mathrm{E}[B_{n+1} X_{n+1} \mid X_1, \ldots, X_n] \\
&= M_n + B_{n+1} \mathrm{E}[X_{n+1} \mid X_1, \ldots, X_n] \\
&= M_n + B_{n+1} \mathrm{E}[X_{n+1}] \\
&= M_n + B_{n+1}(p - q), \tag{3.12}
\end{aligned}$$

where the second equality uses Theorem 2.2.7 on p. 83, and we conclude that $\{M_n\}_{n \geq 0}$ is a martingale (resp., supermartingale, submartingale) with respect to $\{X_n\}_{n \geq 1}$ if $p = q$ (resp., $p \leq q$, $p \geq q$). ♠

We need one more definition at this point. As in the definition of a martingale, we assume a given filtration $\{X_n\}_{n \geq 0}$ representing the basic source of randomness. We say that N is a *stopping time with respect to* $\{X_n\}_{n \geq 0}$ if N is a random variable with values in $\mathbf{Z}_+ \cup \{\infty\}$ such that there exists a sequence $\{g_n\}_{n \geq 0}$ of nonrandom functions with the property that

$$1_{\{N \leq n\}} = g_n(X_0, X_1, \ldots, X_n), \qquad n \geq 0. \tag{3.13}$$

In other words, the occurrence of the event $\{N \leq n\}$ is completely determined by the random variables $X_0, \ldots, X_n$. If we think of N as the coup at which the gambler makes his final bet, then $\{N \leq n\}$ is the event that the gambler makes no bets after the nth coup. Such an event can depend only on $X_0, \ldots, X_n$. It cannot depend on X_{n+1}, for this would require prescience on the part of the gambler.

Example 3.1.3. *Even-money bets with varying bet sizes and random stopping.* Let $\{X_n\}_{n \geq 1}$ and $\{B_n\}_{n \geq 1}$ be as in Example 3.1.2, and let N be a stopping time with respect to $\{X_n\}_{n \geq 1}$. Then

$$M_n := M_0 + \sum_{l=1}^{n \wedge N} B_l X_l = M_0 + \sum_{l=1}^{n} 1_{\{N \geq l\}} B_l X_l, \qquad n \geq 0, \tag{3.14}$$

defines a martingale (resp., supermartingale, submartingale) with respect to $\{X_n\}_{n \geq 1}$ if $p = q$ (resp., $p \leq q$, $p \geq q$). In fact, this follows from Example 3.1.2 because, by (3.13) with $X_0 := 0$,

$$1_{\{N \geq l\}} = 1 - 1_{\{N \leq l-1\}} = 1 - g_{l-1}(0, X_1, \ldots, X_{l-1}), \tag{3.15}$$

as required by (3.10). ♠

Example 3.1.4. *Martingale system.* Here we assume $0 < p \leq \frac{1}{2}$ and $q := 1 - p$, so that pushes are impossible (or are regarded as inconclusive). In the *martingale system*, the gambler doubles his bet size after each loss, and stops betting after the first win. (Of course, he can restart the system after the first win, but here we ignore that possibility.) In terms of the notation of the preceding example,

$$B_n = 2B_{n-1} 1_{\{X_{n-1}=-1\}}, \qquad n \geq 2, \tag{3.16}$$

and therefore, if money is measured in units of the initial bet size,

$$B_1 = 1, \qquad B_n = 2^{n-1} 1_{\{X_1=\cdots=X_{n-1}=-1\}}, \quad n \geq 2. \tag{3.17}$$

See Table 8.1 on p. 277 for clarification.

Now let N be the number of coups needed to achieve the first win, that is, $N := \min\{n \geq 1 : X_n = 1\}$. Then (3.14) becomes

$$M_n = \begin{cases} M_0 - (1 + 2 + \cdots + 2^{n-1}) = M_0 - (2^n - 1) & \text{if } n \leq N - 1, \\ M_0 - (1 + 2 + \cdots + 2^{N-2}) + 2^{N-1} = M_0 + 1 & \text{if } n \geq N. \end{cases}$$
$$\tag{3.18}$$

This equation tells us that, if the gambler makes n bets and fails to win any of them, he will have lost $2^n - 1$ units. If he continues betting until he finally wins a bet, he will have won one unit. Noting that N has the geometric(p) distribution, we see that $P(N < \infty) = 1$ and

$$P(M_N = M_0 + 1) = 1, \tag{3.19}$$

which would seem to make this an infallible system. The difficulty is that, after a sufficiently long sequence of losses, the system will call for a bet that either exceeds what is left of the gambler's resources or exceeds the house maximum betting limit. In either case, the system must be aborted.   ♠

3.2 Optional Stopping Theorem

The optional stopping theorem gives conditions that ensure that, if $\{M_n\}_{n\geq 0}$ is a supermartingale and N is a stopping time, both with respect to $\{X_n\}_{n\geq 0}$, then $E[M_N] \leq E[M_0]$. This tells us that no betting system can turn a sequence of subfair (or fair) wagers into a superfair one, a principle known as the conservation of fairness. We begin with a lemma that is of some interest in its own right.

Lemma 3.2.1. *Let $\{X_n\}_{n\geq 0}$ be a filtration, let N be a stopping time with respect to $\{X_n\}_{n\geq 0}$, and let $\{M_n\}_{n\geq 0}$ be a sequence of discrete random variables with the property that there exists a sequence $\{f_n\}_{n\geq 0}$ of nonrandom functions such that*

$$M_n = f_n(X_0, \ldots, X_n), \qquad E\big[|M_n|\big] < \infty, \tag{3.20}$$

and

$$\mathrm{E}[M_{n+1} \mid X_0, \ldots, X_n] \leq M_n \quad \text{on} \quad \{N \geq n+1\} \qquad (3.21)$$

for all $n \geq 0$. *Then* $\{M_{n \wedge N}\}_{n \geq 0}$ *is a supermartingale with respect to* $\{X_n\}_{n \geq 0}$.

In particular, if $\{M_n\}_{n \geq 0}$ *is a supermartingale and* N *is a stopping time, both with respect to* $\{X_n\}_{n \geq 0}$, *then* $\{M_{n \wedge N}\}_{n \geq 0}$ *is a supermartingale with respect to* $\{X_n\}_{n \geq 0}$.

Proof. It suffices to observe that

$$
\begin{aligned}
1_{\{N \geq n+1\}} & \mathrm{E}[M_{(n+1) \wedge N} \mid X_0, \ldots, X_n] \\
&= \mathrm{E}[1_{\{N \geq n+1\}} M_{(n+1) \wedge N} \mid X_0, \ldots, X_n] \\
&= \mathrm{E}[1_{\{N \geq n+1\}} M_{n+1} \mid X_0, \ldots, X_n] \\
&= 1_{\{N \geq n+1\}} \mathrm{E}[M_{n+1} \mid X_0, \ldots, X_n] \\
&\leq 1_{\{N \geq n+1\}} M_n \\
&= 1_{\{N \geq n+1\}} M_{n \wedge N}, \qquad n \geq 0,
\end{aligned} \qquad (3.22)
$$

and

$$
\begin{aligned}
1_{\{N \leq n\}} & \mathrm{E}[M_{(n+1) \wedge N} \mid X_0, \ldots, X_n] \\
&= \mathrm{E}[1_{\{N \leq n\}} M_{(n+1) \wedge N} \mid X_0, \ldots, X_n] \\
&= \mathrm{E}[1_{\{N \leq n\}} M_{n \wedge N} \mid X_0, \ldots, X_n] \\
&= 1_{\{N \leq n\}} M_{n \wedge N}, \qquad n \geq 0.
\end{aligned} \qquad (3.23)
$$

Here we have used Theorem 2.2.7 on p. 83 three times and Corollary 2.2.8 on p. 83 once. Summing these two results gives the desired conclusion.

The proof is related to Problem 2.18 on p. 91. ♠

We can now state the *optional stopping theorem*.

Theorem 3.2.2. *Let* $\{X_n\}_{n \geq 0}$ *be a filtration, and let* $\{M_n\}_{n \geq 0}$ *be a supermartingale and* N *be a stopping time, both with respect to* $\{X_n\}_{n \geq 0}$. *Assume that one of the following hypotheses holds:*

(a) $\mathrm{P}(N < \infty) = 1$ *and* $M_n \geq 0$ *for all* $n \geq 0$.

(b) $\mathrm{P}(N < \infty) = 1$, M_N *has finite expectation, and*

$$\lim_{n \to \infty} \mathrm{E}\big[|M_n| 1_{\{N \geq n+1\}}\big] = 0. \qquad (3.24)$$

(c) $\mathrm{E}[N] < \infty$ *and there exists a constant* C *such that*

$$\mathrm{E}\big[|M_{n+1} - M_n| \mid X_0, \ldots, X_n\big] \leq C \text{ on } \{N \geq n+1\} \qquad (3.25)$$

for all $n \geq 0$.

Then M_N *has finite expectation and*

$$E[M_N] \leq E[M_0].\qquad(3.26)$$

Alternatively, let $\{X_n\}_{n\geq 0}$ *be a filtration, and let* $\{M_n\}_{n\geq 0}$ *be a sub-martingale (resp., martingale) and* N *be a stopping time, both with respect to* $\{X_n\}_{n\geq 0}$. *Under hypothesis* (b) *or* (c), M_N *has finite expectation and* (3.26) *holds with the inequality reversed (resp., replaced by an equality).*

Proof. First,

$$E\big[|M_{n\wedge N}|\big] \leq E\Big[\max_{0\leq l\leq n}|M_l|\Big] \leq E\Big[\sum_{l=0}^{n}|M_l|\Big] < \infty.\qquad(3.27)$$

By Lemma 3.2.1, $E[M_{(n+1)\wedge N}] \leq E[M_{n\wedge N}]$ for all $n \geq 0$, and this implies that

$$E[M_{n\wedge N}] \leq E[M_0], \qquad n \geq 0.\qquad(3.28)$$

Under hypothesis (a) the theorem follows from (3.28) by Fatou's lemma (Lemma A.2.2 on p. 748). Under hypothesis (b),

$$\begin{aligned}
\big|E[M_N] - E[M_{n\wedge N}]\big| &\leq E\big[|M_N - M_{n\wedge N}|\big] \\
&\leq E\big[|M_N|\,1_{\{N\geq n+1\}}\big] + E\big[|M_n|\,1_{\{N\geq n+1\}}\big],
\end{aligned}\qquad(3.29)$$

and this tends to 0 as $n \to \infty$ by the dominated convergence theorem (Theorem A.2.5 on p. 748) and (3.24).

Under hypothesis (c) the dominated convergence theorem (Theorem A.2.5 on p. 748) applies because, with empty sums equal to 0,

$$|M_N - M_{n\wedge N}| \leq \sum_{k=n}^{N-1}|M_{k+1} - M_k| \leq \sum_{k=0}^{\infty}|M_{k+1} - M_k|\,1_{\{N\geq k+1\}}\qquad(3.30)$$

and

$$\begin{aligned}
E\Big[\sum_{k=0}^{\infty}&|M_{k+1} - M_k|\,1_{\{N\geq k+1\}}\Big] \\
&= \sum_{k=0}^{\infty}E\big[E\big[|M_{k+1} - M_k|\,1_{\{N\geq k+1\}} \mid X_0,\dots,X_k\big]\big] \\
&= \sum_{k=0}^{\infty}E\big[E\big[|M_{k+1} - M_k| \mid X_0,\dots,X_k\big]\,1_{\{N\geq k+1\}}\big] \\
&\leq C\sum_{k=0}^{\infty}P(N \geq k + 1) \\
&= CE[N] < \infty,
\end{aligned}\qquad(3.31)$$

where the first equality uses Corollary A.2.4 on p. 748, and the last equality uses Theorem 1.4.2 on p. 30, and the proof is complete. ♠

Example 3.2.3. *Even-money bets with varying bet sizes and random stopping, continued.* Let us assume the conditions of Example 3.1.3 on p. 98, together with $p - q \leq 0$ (the game is subfair or fair). In particular, $\{X_n\}_{n \geq 1}$ is as in Example 3.1.1 on p. 96, $\{M_n\}_{n \geq 0}$ is as in (3.11) on p. 98, and N is a stopping time with respect to $\{X_n\}_{n \geq 1}$ satisfying $P(N < \infty) = 1$. Then the following conclusions are consequences of the optional stopping theorem (Theorem 3.2.2).

(a) If $M_n \geq 0$ for all $n \geq 0$, then M_N has finite expectation and

$$\mathrm{E}[M_N] \leq M_0. \tag{3.32}$$

(b) If M_N has finite expectation and (3.24) holds, then (3.32) holds.

Conclusion (3.32) says that, in a subfair (or fair) game with even-money payoffs and independent coups, the gambler's expected profit cannot be positive, regardless of the betting system and stopping rule used, provided only that the conditions of (a) or (b) are satisfied. (a) requires that the gambler's fortune never become negative. This can be assured if he never bets more than he has (i.e., $B_n \leq M_{n-1}$ for all $n \geq 1$). (b) eliminates this nonnegativity condition, replacing it with another condition. ♠

Example 3.2.4. *Martingale system, continued.* Let us consider the martingale system played by a gambler with unlimited credit in a casino with no house limit. We further assume that $p \leq \frac{1}{2}$ and $q := 1 - p$. Let N be the number of coups needed to achieve a win. By (3.19) on p. 99, conclusion (3.32) fails, and therefore both of our hypotheses must fail. As for hypothesis (a), the nonnegativity assumption fails because no matter the value of M_0, the gambler's fortune M_n will become negative after sufficiently many consecutive losses. As for hypothesis (b), M_N has finite expectation, but (3.24) fails. Indeed,

$$\mathrm{E}\big[|M_n| \mathbf{1}_{\{N \geq n+1\}}\big] = |M_0 - (2^n - 1)|q^n \not\to 0 \tag{3.33}$$

since $q \geq \frac{1}{2}$.

It is interesting to determine the expected size of the largest bet. The largest bet is the one made at the coup at which the first win occurs, hence it is 2^{N-1} units. Its expected value is

$$\mathrm{E}[2^{N-1}] = \sum_{n=1}^{\infty} 2^{n-1} q^{n-1} p = \infty, \tag{3.34}$$

again since $q \geq \frac{1}{2}$. When $p = \frac{1}{2}$, we knew this already because the distribution of the size of the largest bet in the martingale system is the same as the distribution of the amount paid out in the St. Petersburg game (Example 1.4.5 on p. 31).

Of course, our assumptions of unlimited credit and no house limit are both unrealistic. In the more realistic situation in which the gambler cannot bet more than he has, and in any case cannot bet more than a specified house limit, hypotheses (a) and (b) apply. See Example 8.1.1 on p. 276 for further details.   ♠

An immediate consequence of the optional stopping theorem is a result known as *Wald's identity*:

Corollary 3.2.5. *Let* $(X_1, Y_1), (X_2, Y_2), \ldots$ *be a sequence of i.i.d. discrete random vectors with* X_1 *having finite expectation, and put* $S_0 := 0$ *and* $S_n := X_1 + \cdots + X_n$ *for each* $n \geq 1$. *Let* N *be a stopping time with respect to* $\{(X_n, Y_n)\}_{n \geq 1}$ *having finite expectation. Then* S_N *has finite expectation and*

$$E[S_N] = E[N]E[X_1]. \tag{3.35}$$

Remark. The sequence $Y_1, Y_2, \ldots$ does not play an essential role in this result and can be omitted. Example 3.2.6 explains why it is included.

Proof. If we define

$$M_n := S_n - nE[X_1], \qquad n \geq 0, \tag{3.36}$$

then $\{M_n\}_{n \geq 0}$ is a martingale with respect to $\{(X_n, Y_n)\}_{n \geq 1}$, and the conclusion follows from the optional stopping theorem (Theorem 3.2.2) under hypothesis (c). Indeed, note that

$$\begin{aligned}
E\big[|M_{n+1} - M_n| \mid (X_1, Y_1), \ldots, (X_n, Y_n)\big] \\
= E\big[|X_{n+1} - E[X_1]| \mid (X_1, Y_1), \ldots, (X_n, Y_n)\big] \\
= E\big[|X_{n+1} - E[X_1]|\big] \\
\leq 2E\big[|X_1|\big], \qquad n \geq 0,
\end{aligned} \tag{3.37}$$

so (3.25) is satisfied, and the conclusion $E[M_N] = E[M_0] = 0$ implies (3.35).
                                                                            ♠

Example 3.2.6. *Expected length of the shooter's hand at craps.* Let $T_1, T_2, \ldots$ be i.i.d. with common distribution (1.11) on p. 6, representing the results of a sequence of dice rolls. Using the notation

$$\mathscr{S} := \{2, 3, 4, \ldots, 12\} \quad \text{and} \quad \mathscr{P} := \{4, 5, 6, 8, 9, 10\}, \tag{3.38}$$

define the sequence of i.i.d. random vectors $(X_1, Y_1), (X_2, Y_2), \ldots$ in terms of $T_1, T_2, \ldots$ by letting

$$X_1 := \begin{cases} 1 & \text{if } T_1 \in \mathscr{S} - \mathscr{P}, \\ \min\{k \geq 2 : T_k = T_1 \text{ or } T_k = 7\} & \text{if } T_1 \in \mathscr{P}, \end{cases} \tag{3.39}$$

$$Y_1 := \begin{cases} 1 & \text{if } T_1 \in \{7, 11\} \text{ or both } T_1 \in \mathscr{P} \text{ and } T_{X_1} = T_1, \\ -1 & \text{otherwise;} \end{cases} \tag{3.40}$$

X_2 and Y_2 are defined similarly in terms of $T_{X_1+1}, T_{X_1+2}, \ldots$; X_3 and Y_3 are defined similarly in terms of $T_{X_1+X_2+1}, T_{X_1+X_2+2}, \ldots$; and so on. The proof that the resulting sequence $(X_1, Y_1), (X_2, Y_2), \ldots$ is i.i.d. is left to the reader (Problem 3.2 on p. 111).

Recalling Examples 1.2.2 on p. 17, 1.2.8 on p. 19, and 2.1.4 on p. 78, we see that X_n is the number of rolls in the nth pass-line decision, while Y_n is the profit from a one-unit bet on the nth pass-line decision. According to the rules of craps, a shooter continues to roll the dice until he *sevens out*, that is, until he loses a pass-line decision by rolling 7. In other words, he rolls the dice for N pass-line decisions, where

$$N := \min\{n \geq 1 : X_n \geq 2 \text{ and } Y_n = -1\}. \tag{3.41}$$

Consequently, the length L of the shooter's hand (i.e., the number of rolls) is given by $L := S_N = X_1 + \cdots + X_N$. Now, N has a geometric distribution with parameter

$$q := \sum_{j \in \mathscr{P}} \pi_j \frac{\pi_7}{\pi_j + \pi_7} = \frac{196}{495} \tag{3.42}$$

and therefore with mean q^{-1} (Example 1.4.3 on p. 30). Using $\mu := \mathrm{E}[X_1] = 557/165$ by Example 2.1.4 on p. 78, we have, by Wald's identity (Corollary 3.2.5)

$$\mathrm{E}[L] = \mathrm{E}[N]\mathrm{E}[X_1] = q^{-1}\mu = \frac{1{,}671}{196} \approx 8.525510. \tag{3.43}$$

Compare with Problem 2.7 on p. 89.

Notice that N is a stopping time with respect to $(X_1, Y_1), (X_2, Y_2), \ldots$, but not with respect to $X_1, X_2, \ldots$. This is why we introduced the sequence $Y_1, Y_2, \ldots$ into the statement of Corollary 3.2.5. There is an alternative form of Wald's identity (Problem 3.6 on p. 112) that requires only that $1_{\{N \leq n\}}$ be independent of $X_{n+1}, X_{n+2}, \ldots$ (but not necessarily a function of $X_0, \ldots, X_n$) for each $n \geq 0$, and this would also apply to the present example. ♠

We would like to extend Theorem 3.2.2 to what might be called non-adapted supermartingales. Let $\{M_n\}_{n \geq 0}$ be a sequence of jointly distributed discrete random variables (and jointly distributed with $\{X_n\}_{n \geq 0}$). We say that $\{M_n\}_{n \geq 0}$ is a *nonadapted martingale with respect to* $\{X_n\}_{n \geq 0}$ if

$$\mathrm{E}\big[|M_n|\big] < \infty \tag{3.44}$$

and

$$\mathrm{E}[M_{n+1} \mid X_0, \ldots, X_n] = \mathrm{E}[M_n \mid X_0, \ldots, X_n] \qquad (3.45)$$

for all $n \geq 0$.

If $\{M_n\}_{n\geq0}$ satisfies the definition of nonadapted martingale with respect to $\{X_n\}_{n\geq0}$ but with the $=$ sign in (3.45) replaced by $\leq$ (resp., $\geq$), we say that $\{M_n\}_{n\geq0}$ is a *nonadapted supermartingale* (resp., *nonadapted submartingale*) *with respect to* $\{X_n\}_{n\geq0}$. Lemma 3.2.1 and Theorem 3.2.2 have direct analogues in this context.

Lemma 3.2.7. *Let $\{X_n\}_{n\geq0}$ be a filtration, let N be a stopping time with respect to $\{X_n\}_{n\geq0}$, and let $\{M_n\}_{n\geq0}$ be a sequence of jointly distributed discrete random variables (and jointly distributed with $\{X_n\}_{n\geq0}$) with the property that*

$$\mathrm{E}\big[|M_n|\big] < \infty, \qquad (3.46)$$

and

$$\mathrm{E}[M_{n+1} \mid X_0, \ldots, X_n] \leq \mathrm{E}[M_n \mid X_0, \ldots, X_n] \quad \text{on} \quad \{N \geq n+1\} \qquad (3.47)$$

for all $n \geq 0$. Then $\{M_{n\wedge N}\}_{n\geq0}$ is a nonadapted supermartingale with respect to $\{X_n\}_{n\geq0}$.

In particular, if $\{M_n\}_{n\geq0}$ is a nonadapted supermartingale and N is a stopping time, both with respect to $\{X_n\}_{n\geq0}$, then $\{M_{n\wedge N}\}_{n\geq0}$ is a nonadapted supermartingale with respect to $\{X_n\}_{n\geq0}$.

Proof. The proof is virtually identical to that of Lemma 3.2.1.   ♠

We can now state the *optional stopping theorem* for nonadapted martingales.

Theorem 3.2.8. *Let $\{X_n\}_{n\geq0}$ be a filtration, and let $\{M_n\}_{n\geq0}$ be a nonadapted supermartingale and N be a stopping time, both with respect to $\{X_n\}_{n\geq0}$. Assume that either (a) $\mathrm{P}(N < \infty) = 1$ and $M_n \geq 0$ for all $n \geq 0$, or (c) $\mathrm{E}[N] < \infty$ and there exists a constant C such that (3.25) holds. Then M_N has finite expectation and (3.26) holds.*

Alternatively, let $\{X_n\}_{n\geq0}$ be a filtration, and let $\{M_n\}_{n\geq0}$ be a nonadapted submartingale (resp., nonadapted martingale) and N be a stopping time, both with respect to $\{X_n\}_{n\geq0}$. Under hypothesis (c), M_N has finite expectation and (3.26) holds with the inequality reversed (resp., replaced by an equality).

Proof. The proof is virtually identical to that of Theorem 3.2.2.   ♠

Example 3.2.9. *Nonadapted supermartingales and craps.* Consider the craps system that calls for a one-unit pass-line bet on each come-out roll, together with a one-unit come bet on each point roll. A *come bet* is mathematically identical to a pass-line bet, except that it is initiated while a point has been established by the shooter but not yet resolved. It is natural to describe the underlying randomness by a sequence of i.i.d. dice totals $T_1, T_2, \ldots$

having common distribution (1.11) on p. 6. With $\mathscr{S}$ and $\mathscr{P}$ as in (3.38), the pass-line or come bet initiated prior to the nth roll of the dice is resolved following roll N_n, where

$$N_n := \begin{cases} n & \text{if } T_n \in \mathscr{S} - \mathscr{P}, \\ \min\{k \geq n+1 : T_k = T_n \text{ or } T_k = 7\} & \text{if } T_n \in \mathscr{P}. \end{cases} \tag{3.48}$$

The bettor's profit from a one-unit pass-line or come bet initiated prior to the nth roll of the dice is described by

$$X_n = 1_{\{T_n \in \{7,11\}\}} - 1_{\{T_n \in \{2,3,12\}\}}$$
$$+ 1_{\{T_n \in \mathscr{P}\}}\left(1_{\{T_{N_n} = T_n\}} - 1_{\{T_{N_n} = 7\}}\right). \tag{3.49}$$

Notice that the sequence $X_1, X_2, \ldots$ is identically distributed with common mean $\mathrm{E}[X_1] = -7/495$. However, it is not an independent sequence because, if $1 \leq m < n$, then X_m and X_n both depend on $T_n, T_{n+1}, \ldots$. Nevertheless, letting

$$M_0 := 0, \qquad M_n := \sum_{l=1}^{n} X_l, \quad n \geq 1, \tag{3.50}$$

we find that

$$\mathrm{E}[M_{n+1} - M_n \mid T_1, \ldots, T_n] = \mathrm{E}[X_{n+1} \mid T_1, \ldots, T_n]$$
$$= \mathrm{E}[X_{n+1}] = -7/495 \tag{3.51}$$

for all $n \geq 0$, hence $\{M_n\}_{n \geq 0}$ is a nonadapted supermartingale with respect to $\{T_n\}_{n \geq 1}$. We can obtain additional nonadapted supermartingales with respect to $\{T_n\}_{n \geq 1}$ by introducing betting systems as in Example 3.1.2 on p. 97. ♠

3.3 Martingale Convergence Theorem

We begin with the *upcrossing inequality*. Given a supermartingale $\{M_n\}_{n \geq 0}$, a finite interval (a, b), and a positive integer N, we define the (random) number $U_N(a, b)$ of *upcrossings* of (a, b) by $\{M_n\}_{n \geq 0}$ by time N to be the largest nonnegative integer n such that there exist integers

$$0 \leq k_1 < l_1 < k_2 < l_2 < \cdots < k_n < l_n \leq N \tag{3.52}$$

with

$$M_{k_i} \leq a \quad \text{and} \quad M_{l_i} \geq b, \qquad i = 1, 2, \ldots, n. \tag{3.53}$$

Lemma 3.3.1. *Under the above assumptions and notation,*

$$\mathrm{E}[U_N(a,b)] \leq \frac{\mathrm{E}[(M_N - a)^-]}{b - a}. \tag{3.54}$$

Proof. Think of $M_n - M_{n-1}$ as representing a gambler's profit per unit bet at the nth coup. If $B_1, B_2, \ldots$ represent his bet sizes, then

$$F_n := \sum_{l=1}^{n} B_l(M_l - M_{l-1}) \tag{3.55}$$

represents his cumulative profit after n coups, and if B_n is a bounded, non-random function of $M_0, \ldots, M_{n-1}$ for each $n \geq 1$, then $\{F_n\}_{n \geq 0}$ is a super-martingale. Now let us specify that $B_1 := 1_{\{M_0 \leq a\}}$ and

$$B_n := 1_{\{B_{n-1}=1,\, M_{n-1}<b\}} + 1_{\{B_{n-1}=0,\, M_{n-1} \leq a\}}, \quad n \geq 2. \tag{3.56}$$

In words, we start betting one unit as soon as $M \leq a$ and continue as long as $M < b$. Once $M \geq b$ we stop betting until $M \leq a$ again, restarting our betting at one unit. This process repeats ad infinitum. We claim that

$$F_N \geq (b - a)U_N(a,b) - (M_N - a)^-. \tag{3.57}$$

This follows from the fact that F increases by at least $b - a$ during each upcrossing, and if betting has begun again by time N without completing an upcrossing, the worst-case scenario would be an additional loss of $(M_N - a)^-$ units. Since $\{F_n\}_{n \geq 0}$ is a supermartingale,

$$0 = \mathrm{E}[F_0] \geq \mathrm{E}[F_N] \geq (b - a)\mathrm{E}[U_N(a,b)] - \mathrm{E}[(M_N - a)^-], \tag{3.58}$$

and the desired result follows.   ♠

We can now state the *martingale convergence theorem*.

Theorem 3.3.2. *Let $\{M_n\}_{n \geq 0}$ be a supermartingale with the property that $\sup_{n \geq 0} \mathrm{E}[|M_n|] < \infty$. Then the random variable $M_\infty := \lim_{n \to \infty} M_n$ exists a.s., and*

$$\mathrm{E}[|M_\infty|] \leq \sup_{n \geq 0} \mathrm{E}[|M_n|] < \infty. \tag{3.59}$$

If also $M_n \geq 0$ for all $n \geq 0$, then $\mathrm{E}[M_\infty] \leq \mathrm{E}[M_0]$.

Remark. Even though each of the random variables M_n is discrete, M_∞ need not be. In that case, see Section A.2 (Appendix) for the definitions of $\mathrm{E}[|M_\infty|]$ and $\mathrm{E}[M_\infty]$.

Proof. Having defined $U_N(a,b)$ above in terms of $\{M_n\}_{n \geq 0}$ for each pair of real numbers a and b with $a < b$ and each positive integer N, let us further define

$$U_\infty(a,b) := \sup_{N \geq 1} U_N(a,b) = \lim_{N \to \infty} U_N(a,b) \tag{3.60}$$

and
$$E_{a,b} := \{U_\infty(a,b) = \infty\}. \qquad (3.61)$$

Noting that $\mathrm{E}[(M_N-a)^-] \le \mathrm{E}[|M_N-a|] \le |a|+\sup_{n\ge 0}\mathrm{E}[|M_n|]$, Lemma 3.3.1 and Fatou's lemma (Lemma A.2.2 on p. 748) imply that

$$\mathrm{E}[U_\infty(a,b)] \le \liminf_{N\to\infty} \mathrm{E}[U_N(a,b)] \le \frac{|a| + \sup_{n\ge 0}\mathrm{E}[|M_n|]}{b-a} < \infty, \qquad (3.62)$$

and therefore $\mathrm{P}(E_{a,b}) = 0$.

We observe that

$$\begin{aligned}
E &:= \left\{ \lim_{n\to\infty} M_n \text{ does not exist in } [-\infty,\infty] \right\} \\
&= \left\{ \liminf_{n\to\infty} M_n < \limsup_{n\to\infty} M_n \right\} \\
&= \bigcup_{a,b\in\mathbf{Q}:\ a<b} \left\{ \liminf_{n\to\infty} M_n < a < b < \limsup_{n\to\infty} M_n \right\} \\
&\subset \bigcup_{a,b\in\mathbf{Q}:\ a<b} E_{a,b}, \qquad\qquad\qquad\qquad (3.63)
\end{aligned}$$

hence $\mathrm{P}(E) = 0$ by countable subadditivity (Corollary 1.1.21 on p. 13). Thus, M_∞ exists in $[-\infty,\infty]$ with probability 1. Finally, Fatou's lemma (Lemma A.2.2 on p. 748) gives

$$\mathrm{E}[|M_\infty|] \le \liminf_{n\to\infty} \mathrm{E}[|M_n|] \le \sup_{n\ge 0} \mathrm{E}[|M_n|] < \infty. \qquad (3.64)$$

Thus, M_∞ exists in $\mathbf{R}$ with probability 1.

The last conclusion of the theorem follows by noting that, if $M_n \ge 0$ for each $n \ge 1$, then $\sup_{n\ge 0}\mathrm{E}[|M_n|] = \mathrm{E}[M_0]$. ♠

Example 3.3.3. *Kelly system, continued.* Recall the Kelly system from Example 1.5.9 on p. 46. Let X represent the bettor's profit per unit bet, so that $\mathrm{P}(X=1) = p$ and $\mathrm{P}(X=-1) = 1-p$, where $\frac{1}{2} < p < 1$. Now let $X_1, X_2, \ldots$ be i.i.d. with common distribution that of X. The proportional bettor's fortune F_n after n coups is given by

$$F_n = F_0 \prod_{l=1}^{n}(1 + fX_l), \qquad n \ge 1, \qquad (3.65)$$

where $F_0 > 0$ and $f \in (0,1)$ denote his initial fortune and betting proportion. The Kelly optimal bettor chooses the betting proportion $f^* = 2p - 1$. Let us see how the Kelly bettor's fortune compares with that of a generic bettor for the same sequence of wins and losses.

Let f_n° be the generic bettor's betting proportion at the nth coup, and assume that

$$f_1^\circ = f_1, \qquad f_n^\circ = f_n(X_1, \dots, X_{n-1}), \quad n \geq 2, \tag{3.66}$$

where f_1 is a constant in $[0, 1)$, and f_n is a nonrandom $[0, 1)$-valued function of $n - 1$ variables. We define the Kelly bettor's fortune after n coups as well as the generic bettor's fortune after n coups as

$$F_n^* := F_0 \prod_{l=1}^{n}(1 + f^* X_l), \qquad F_n^\circ := F_0 \prod_{l=1}^{n}(1 + f_l^\circ X_l). \tag{3.67}$$

We claim that $\{F_n^\circ / F_n^*\}_{n \geq 0}$ is a martingale with respect to $\{X_n\}_{n \geq 1}$. Indeed,

$$\mathrm{E}\left[\frac{F_n^\circ}{F_n^*} \middle| X_1, \dots, X_{n-1}\right] = \frac{F_{n-1}^\circ}{F_{n-1}^*}\mathrm{E}\left[\frac{1 + f_n^\circ X_n}{1 + f^* X_n} \middle| X_1, \dots, X_{n-1}\right] = \frac{F_{n-1}^\circ}{F_{n-1}^*} \tag{3.68}$$

since

$$\begin{aligned}
&\mathrm{E}\left[\frac{1 + f_n^\circ X_n}{1 + f^* X_n} \middle| X_1, \dots, X_{n-1}\right] \\
&= \mathrm{E}\left[\frac{1}{1 + f^* X_n} \middle| X_1, \dots, X_{n-1}\right] + f_n^\circ \mathrm{E}\left[\frac{X_n}{1 + f^* X_n} \middle| X_1, \dots, X_{n-1}\right] \\
&= \mathrm{E}\left[\frac{1}{1 + f^* X_n}\right] + f_n^\circ \mathrm{E}\left[\frac{X_n}{1 + f^* X_n}\right] \\
&= \mathrm{E}\left[\frac{1}{1 + f^* X_n}\right] \\
&= \frac{p}{1 + (2p - 1)} + \frac{1 - p}{1 - (2p - 1)} \\
&= 1, \tag{3.69}
\end{aligned}$$

where the third equality uses the fact that f^* maximizes $\mu(f) := \mathrm{E}[\ln(1 + f X_n)]$ and therefore $\mu'(f^*) = 0$.

Consequently, by the martingale convergence theorem,

$$\lim_{n \to \infty} \frac{F_n^\circ}{F_n^*} \text{ exists in } [0, \infty) \text{ a.s., and } \mathrm{E}\left[\lim_{n \to \infty} \frac{F_n^\circ}{F_n^*}\right] \leq 1. \tag{3.70}$$

This is an optimality property of the Kelly system. It says that, on average, the Kelly bettor performs at least as well as the generic bettor. See Section 10.3 for additional results of this nature. ♠

Theorem 3.3.4. *Let $\{M_n\}_{n \geq 0}$ be a martingale with $M_0 = 0$ and uniformly bounded increments. More precisely, we assume there exists a (nonrandom) constant L such that $|M_n - M_{n-1}| \leq L$ for each $n \geq 1$. Letting*

$$C := \left\{ \lim_{n \to \infty} M_n \text{ exists in } \mathbf{R} \right\} \tag{3.71}$$

and

$$D := \Big\{ \liminf_{n \to \infty} M_n = -\infty \text{ and } \limsup_{n \to \infty} M_n = \infty \Big\}, \tag{3.72}$$

we have $P(C \cup D) = 1$.

Remark. The events C and D stand for "converges" and "diverges."

Proof. Given a positive integer K, let $N_K := \min\{n \geq 1 : M_n > K\}$. Then $\{M_{n \wedge N_K}\}_{n \geq 0}$ is a martingale bounded above by $K+L$, and therefore $\{K+L-M_{n \wedge N_K}\}_{n \geq 0}$ is a nonnegative martingale. Thus, the martingale convergence theorem applies, and we find that $\lim_{n \to \infty} M_n$ exists in $\mathbf{R}$ a.s. on the event $\{N_K = \infty\}$. Since K was arbitrary, $\lim_{n \to \infty} M_n$ exists in $\mathbf{R}$ a.s. on the event

$$\bigcup_{K=1}^{\infty} \{N_K = \infty\} = \Big\{ \limsup_{n \to \infty} M_n < \infty \Big\}. \tag{3.73}$$

Applying this to the martingale $\{-M_n\}_{n \geq 0}$, we find that $\lim_{n \to \infty} M_n$ exists in $\mathbf{R}$ a.s. on the event $\{\liminf_{n \to \infty} M_n > -\infty\}$ as well. These two conclusions are equivalent to the conclusion of the theorem. ♠

Recall the Borel–Cantelli lemma (Lemma 1.5.4 on p. 43). There is a partial converse to that result, often called the second Borel–Cantelli lemma, stated in Problem 1.50 on p. 62. Here we prove Lévy's generalization of the second Borel–Cantelli lemma, which we will call the *conditional Borel–Cantelli lemma*.

Theorem 3.3.5. *Let $\{X_n\}_{n \geq 1}$ be a filtration, and let $E_1, E_2, \ldots$ be a sequence of events such that there exists a sequence $\{f_n\}_{n \geq 1}$ of nonrandom functions for which $1_{E_n} = f_n(X_1, \ldots, X_n)$ for each $n \geq 1$. Define*

$$E := \{E_n \text{ i.o.}\}, \qquad F := \Big\{ \sum_{n=1}^{\infty} P(E_n \mid X_1, \ldots, X_{n-1}) = \infty \Big\}. \tag{3.74}$$

Then $1_E = 1_F$ a.s.

Remark. If $E_1, E_2, \ldots$ are independent, we can take $X_n = 1_{E_n}$ for each $n \geq 1$ to get the second Borel–Cantelli lemma.

Proof. Define the martingale $\{M_n\}_{n \geq 0}$ with respect to $\{X_n\}_{n \geq 1}$ by $M_0 := 0$ and

$$M_n := \sum_{l=1}^{n} \{1_{E_l} - P(E_l \mid X_1, \ldots, X_{l-1})\}, \qquad n \geq 1, \tag{3.75}$$

and note that Theorem 3.3.4 applies. On the event C of that theorem,

$$\sum_{n=1}^{\infty} 1_{E_n} = \infty \text{ if and only if } \sum_{n=1}^{\infty} P(E_n \mid X_1, \ldots, X_{n-1}) = \infty. \tag{3.76}$$

On the event D of that theorem,

$$\sum_{n=1}^{\infty} 1_{E_n} = \infty \quad \text{and} \quad \sum_{n=1}^{\infty} P(E_n \mid X_1, \ldots, X_{n-1}) = \infty. \qquad (3.77)$$

By Theorem 3.3.4, $P(C \cup D) = 1$ and the conclusion follows. ♠

3.4 Problems

3.1. *Convex functions of martingales.*
   (a) If $\{M_n\}_{n\geq 0}$ is a martingale with respect to $\{X_n\}_{n\geq 0}$ and with values in an interval I, if $f : I \mapsto \mathbf{R}$ is convex, and if $f(M_n)$ has finite expectation for each $n \geq 0$, show that $\{f(M_n)\}_{n\geq 0}$ is a submartingale with respect to $\{X_n\}_{n\geq 0}$.
   (b) If $\{M_n\}_{n\geq 0}$ is a submartingale with respect to $\{X_n\}_{n\geq 0}$ and with values in an interval I, if $f : I \mapsto \mathbf{R}$ is convex and nondecreasing, and if $f(M_n)$ has finite expectation for each $n \geq 0$, show that $\{f(M_n)\}_{n\geq 0}$ is a submartingale with respect to $\{X_n\}_{n\geq 0}$.

3.2. *Justification of craps analysis.* Show that the sequence of random vectors $(X_1, Y_1), (X_2, Y_2), \ldots$ defined in Example 3.2.6 on p. 103 is i.i.d.

3.3. *Quitting when ahead in a fair game.* Let $X_1, X_2, \ldots$ be i.i.d. with common distribution $P(X_1 = 1) = P(X_1 = -1) = \frac{1}{2}$, and put $S_0 := 0$ and $S_n := X_1 + \cdots + X_n$ for each $n \geq 1$. Consider the martingale $\{S_n\}_{n\geq 0}$ and the stopping time $N := \min\{n \geq 1 : S_n = 1\}$, both with respect to $\{X_n\}_{n\geq 1}$, and show that $E[N] = \infty$, for otherwise Wald's identity would apply and lead to a contradiction.

3.4. *Distribution of number of coups until ahead in a fair game.*
   (a) With the notation of Problem 3.3, let $g(u) := E[u^{X_1}] = \frac{1}{2}(u^{-1} + u)$ for all $u > 1$, and show that the optional stopping theorem (Theorem 3.2.2) applies to the martingale

$$M_n := u^{S_n} g(u)^{-n}, \qquad n \geq 0, \qquad (3.78)$$

where $u > 1$, and the stopping time N.
   (b) Given $v \in (0, 1)$, use the quadratic formula to solve $g(u) = 1/v$ for u, and then use part (a) to show that

$$E[v^N] = \frac{1 - \sqrt{1 - v^2}}{v}, \qquad 0 < v < 1. \qquad (3.79)$$

   (c) By expanding $1 - \sqrt{1 - v^2}$ is a Taylor series about 0, use part (b) to show that

$$P(N = 2m + 1) = \left[\frac{1}{m+1}\binom{2m}{m}\right]\frac{1}{2^{2m+1}}, \qquad m \geq 0. \tag{3.80}$$

The quantities within brackets in (3.80) are called the *Catalan numbers*.

3.5. *Expected total amount bet in an "infallible" system.*

(a) Show that condition (b) of the optional stopping theorem (Theorem 3.2.2) on p. 100 can be replaced by the following: $|M_{n \wedge N}| \leq Y$ for all $n \geq 0$, where Y is a random variable with finite expectation.

(b) Consider the betting system of Example 3.1.3 on p. 98. If $p \leq q$ and $P(M_N > M_0) = 1$ as in the martingale system (Example 3.1.4 on p. 98), show that the expected total amount bet is infinite, i.e., $E[B_1 + \cdots + B_N] = \infty$.

3.6. *Alternative form of Wald's identity.* Let $X_1, X_2, \ldots$ be a sequence of i.i.d. discrete random variables with X_1 having finite expectation, and put $S_0 := 0$ and $S_n := X_1 + \cdots + X_n$ for each $n \geq 1$. Let the random variable $N : \Omega \mapsto \mathbf{Z}_+ \cup \{\infty\}$ have the property that $1_{\{N \leq n\}}$ is independent of $X_{n+1}, X_{n+2}, \ldots$ for each $n \geq 0$, and assume that $E[N] < \infty$. Show that S_N has finite expectation and

$$E[S_N] = E[N]E[X_1]. \tag{3.81}$$

This proves Wald's identity (Corollary 3.2.5 on p. 103) under slightly different assumptions on N.

Hint: Write $S_N = \sum_{n=1}^{\infty} 1_{\{N \geq n\}} X_n$ and use Theorem A.2.6 on p. 748.

3.7. *Jacob Bernoulli's game and Wald's identity.* In Jacob Bernoulli's game (Problem 1.42 on p. 60), a fair die is rolled, and if the result is $N \in \{1, 2, 3, 4, 5, 6\}$, then N dice are rolled. If the total of the N dice is greater than 12, the player wins. If it is less than 12, he loses. If it is equal to 12, a push is declared. A win pays even money. Use Wald's identity to find the expected total of the N dice.

3.8. *A three-player ruin problem.* Initially, three players have a, b, and c units of capital, respectively, where a, b, and c are positive integers. Games are independent and each game consists of choosing two players at random and transferring one unit from the first-chosen to the second-chosen player. Once a player is ruined, he is ineligible for further play. Let S_1 be the number of games required for one player to be ruined, and let S_2 be the number of games required for two players to be ruined. Find $E[S_1]$ and $E[S_2]$.

Hint: Let (X_n, Y_n, Z_n) be the numbers of units possessed by the three players after the nth game. Show that $M_n := X_n Y_n Z_n + (n/3)(a + b + c)$ and $N_n := X_n Y_n + X_n Z_n + Y_n Z_n + n$ are martingales with respect to $\{(X_n, Y_n, Z_n)\}_{n \geq 1}$, at least up to times S_1 and S_2, respectively.

3.9. *Alternative condition for the optional stopping theorem.* Show that condition (c) of Theorem 3.2.2 on p. 100 can be weakened to the following: $P(N < \infty) = 1$ and there exist constants $C_1, C_2, \ldots$ such that

$$E\big[|M_{n+1} - M_n| \,\big|\, X_0, \ldots, X_n\big] \leq C_{n+1} \text{ on } \{N \geq n + 1\} \tag{3.82}$$

and

$$\mathrm{E}\left[\sum_{n=1}^{N} C_n\right] < \infty. \qquad (3.83)$$

3.10. *Expected number of Bernoulli trials until a specified pattern appears.*
Let $X_1, X_2, \ldots$ be a sequence of i.i.d. random variables with $\mathrm{P}(X_1 = 0) = p_0$
and $\mathrm{P}(X_1 = 1) = p_1 := 1 - p_0$, where $0 < p_0 < 1$. Let $(u_1, \ldots, u_m)$ be a
specified pattern of 0s and 1s of length $m \geq 1$, and define

$$N := \min\{n \geq m : (X_{n-m+1}, \ldots, X_n) = (u_1, \ldots, u_m)\}. \qquad (3.84)$$

The problem is to find $\mathrm{E}[N]$, the expected number of trials until the pattern
$(u_1, \ldots, u_m)$ first occurs.

(a) Consider a sequence a bettors indexed by the positive integers. Each
bettor has an initial fortune of one unit. Bettor j is inactive until coup j, at
which time he bets one unit that $X_j = u_1$ and is paid fair odds ($p_{u_1}^{-1} - 1$ to
1). If he loses the bet, he is ruined and makes no further bets. If he wins and
$m = 1$, he stops betting. If he wins and $m \geq 2$, he bets his entire fortune ($p_{u_1}^{-1}$
units) at coup $j+1$ that $X_{j+1} = u_2$ and is paid fair odds ($p_{u_2}^{-1} - 1$ to 1). If he
loses the bet, he is ruined and makes no further bets. If he wins and $m = 2$,
he stops betting. If he wins and $m \geq 3$, he bets his entire fortune ($p_{u_1}^{-1} p_{u_2}^{-1}$
units) at coup $j+2$ that $X_{j+2} = u_3$ and is paid fair odds ($p_{u_3}^{-1} - 1$ to 1). This
process continues for at most m coups. For each $n \geq 1$, let $M_{n,j}$ denote the
profit of bettor j following the nth coup (so $M_{0,j} = \cdots = M_{j-1,j} = 0$). Find
an explicit formula for $M_{n,j}$ in terms of $X_1, X_2, \ldots$ and show that $\{M_{n,j}\}_{n\geq 0}$
is a martingale with respect to $\{X_n\}_{n\geq 1}$.

(b) Define $M_n := M_{n,1} + M_{n,2} + \cdots$ for each $n \geq 0$ (the sum is finite),
and notice that $\{M_n\}_{n\geq 0}$ is a martingale with respect to $\{X_n\}_{n\geq 1}$. Apply
the optional stopping theorem (with stopping time N) to show that

$$\mathrm{E}[N] = \sum_{1\leq i\leq m:\, (u_{m-i+1},\ldots,u_m)=(u_1,\ldots,u_i)} (p_{u_1} \cdots p_{u_i})^{-1}. \qquad (3.85)$$

(c) Generalize the above result to the case in which there are k possible
outcomes at each trial, not just two.

3.11. *Probability that pattern 1 appears before pattern 2.* Under the assumptions of Problem 3.10, let $(v_1, \ldots, v_l)$ be a second pattern, and assume that
neither pattern is embedded in the other. Find the probability that pattern
$(u_1, \ldots, u_m)$ appears before pattern $(v_1, \ldots, v_l)$.

Hint: Let $N(u_1, \ldots, u_m)$ be the stopping time and $\{M_n(u_1, \ldots, u_m)\}_{n\geq 0}$
be the martingale of Problem 3.10. Stop the martingale

$$M_n(u_1, \ldots, u_m) - M_n(v_1, \ldots, v_l), \qquad n \geq 0, \qquad (3.86)$$

at time $N(u_1, \ldots, u_m) \wedge N(v_1, \ldots, v_l)$.

3.12. *A partial-product martingale.* Let $X_1, X_2, \ldots$ be a sequence of i.i.d. nonnegative random variables with $E[X_1] = 1$. Show that, if $M_0 := 1$ and $M_n := X_1 X_2 \cdots X_n$ for each $n \geq 1$, then $\{M_n\}_{n \geq 0}$ is a nonnegative martingale with respect to $\{X_n\}_{n \geq 1}$. By the martingale convergence theorem, we know that $M_\infty := \lim_{n \to \infty} M_n$ exists in $[0, \infty)$ a.s. and $E[M_\infty] \leq 1$. Use the strong law of large numbers and Jensen's inequality to show that, unless X_1 is degenerate (i.e., $P(X_1 = 1) = 1$), we have $P(M_\infty = 0) = 1$.

3.13. *Bachelier's martingale.* Consider a game that, at each coup, pays α to β with probability p, where $0 < p < 1$; otherwise the amount of the bet is lost. A gambler with initial fortune β bets his entire fortune at each coup until he loses. Under what conditions on α, β, and p is the gambler's sequence of fortunes a martingale? ... a supermartingale? ... a submartingale?

3.14. *De Moivre's martingale.* Let $X_1, X_2, \ldots$ be a sequence of i.i.d. integer-valued random variables, and let λ_0 be a root of the equation $E[\lambda^{X_1}] = 1$. (If λ_0 is complex, it suffices to notice that expectations, conditional expectations, and martingales extend easily to complex-valued discrete random variables.) Let $S_n := X_1 + \cdots + X_n$ and $M_n := \lambda_0^{S_n}$ for each $n \geq 0$, where $S_0 := 0$.
(a) Show that $\{M_n\}_{n \geq 0}$ is a martingale with respect to $\{X_n\}_{n \geq 1}$.
(b) In De Moivre's example, $P(X_1 = 1) = p$ and $P(X_1 = -1) = 1 - p$, where $0 < p < 1$. In this case, show that $\lim_{n \to \infty} M_n$ exists in $\mathbf{R}$ a.s.
(c) Find another example to show that $\lim_{n \to \infty} M_n$ may fail to exist, even in $[-\infty, \infty]$, a.s.

3.15. *Borel's martingale.* Consider the generalization of the martingale system (Example 3.1.4 on p. 98) in which (3.16) on p. 99 is replaced by

$$B_n = (1 - \alpha X_{n-1}) B_{n-1}, \qquad n \geq 2, \qquad (3.87)$$

where $X_1, X_2, \ldots$ are i.i.d. with $P(X_1 = 1) = p \in (0, 1)$ and $P(X_1 = -1) = 1 - p$, as in the example. Here $0 < \alpha \leq 1$, and the case $\alpha = 1$ corresponds to the martingale system. Assume that $B_1 = 1$.
(a) Show that the analogue of (3.18) on p. 99 is

$$M_n = M_0 + \alpha^{-1} \left[1 - \prod_{l=1}^{n} (1 - \alpha X_l) \right], \qquad n \geq 1. \qquad (3.88)$$

While the gambler's initial fortune M_0 is a nonnegative constant, we allow M_n to become negative.
(b) For what values of α (perhaps depending on p) does $\lim_{n \to \infty} M_n$ exist a.s.?
(c) Assuming that $p = \frac{1}{2}$, show that $\{M_n\}_{n \geq 0}$ is a martingale with respect to $\{X_n\}_{n \geq 1}$. Does the martingale convergence theorem apply?

3.16. *Doob's martingale.* Let $Y, X_0, X_1, X_2, \ldots$ be jointly distributed discrete random variables, with Y having finite expectation. Show that

$$M_n := \mathrm{E}[Y \mid X_0, X_1, \ldots, X_n], \qquad n \geq 0, \tag{3.89}$$

defines a martingale $\{M_n\}_{n\geq 0}$ with respect to $\{X_n\}_{n\geq 0}$. Does the martingale convergence theorem apply?

3.17. *Insurance at twenty-one.* Recall the notation of Example 2.2.1 on p. 80. Specifically, $(X_1, X_2, \ldots, X_{52})$ assumes each of the $52!/[(4!)^9 16!]$ permutations of four 1s, four 2s, $\ldots$, four 9s, and 16 10s with equal probability, and

$$Y_n := 3 \cdot 1_{\{X_{n+1}=10\}} - 1, \qquad 0 \leq n \leq 51. \tag{3.90}$$

In terms of these sequences, we define $M_0 := \mathrm{E}[Y_0]$ and

$$M_n := \mathrm{E}[Y_n \mid X_1, \ldots, X_n], \qquad 1 \leq n \leq 51. \tag{3.91}$$

Show that $\{M_n\}_{0\leq n\leq 51}$ is a martingale with respect to $\{X_n\}_{1\leq n\leq 51}$.

3.18. *A generalization of Jacob Bernoulli's game and the martingale convergence theorem.* In this game a fair die is rolled, and if the result is Z_1, then Z_1 dice are rolled. If the total of the Z_1 dice is Z_2, then Z_2 dice are rolled. If the total of the Z_2 dice is Z_3, then Z_3 dice are rolled, and so on. Let $Z_0 := 1$. Find a positive constant μ such that $\{Z_n/\mu^n\}_{n\geq 0}$ is a martingale. What does the martingale convergence theorem tell us about this?

3.5 Notes

Paul Lévy [1886–1971] (1937, pp. 233–234) can be credited with the concept of a martingale, having introduced what we now call martingale difference sequences. Jean-André Ville [1910–1989] (1939, p. 99) was the first to state the modern definition, and the concept was fully developed by Joseph L. Doob [1910–2004] (1940, 1949, 1953). The first use of the term "martingale" as a stochastic process was by Ville in 1939. Doob called them "chance variables with the property $\mathscr{E}$" in 1940, but used "martingale" in his next publication on the subject (Doob 1949). He later recalled (Snell 1997) that the name "property $\mathscr{E}$" was chosen because Lévy had used "property $\mathscr{C}$" and $\mathscr{E}$ was the next available letter ($\mathscr{D}$ would have been considered immodest).

Why was the term "martingale" chosen? Since the 18th century the betting system in which one doubles one's stake after each loss (Example 3.1.4) has been called the martingale system. According to Russell T. Barnhart (personal communication, 1997),

> [T]he French often use *martingale* to mean just any kind of progression or gambling system.

This assertion is not supported by *Trésor de la langue française* (Imbs 1974), where the usual definition is given, but it is supported by Albigny's (1902)

Les martingales modernes: Nouveau recueil du jouer de roulette et de trente-et-quarante contenant les meilleurs systèmes. The book is a compilation of 23 betting systems, or "modern martingales," offered to gamblers who might find the traditional martingale too dangerous. Further support is provided by a book that Ville may have been familiar with, Marcel Boll's [1886–1958] (1936, pp. 83–84) *La chance et les jeux de hasard*:

> We will call a *martingale* a set of rules that determines the manner in which one stakes the capital on hand so as to have such and such a chance of increasing one's capital in such and such a proportion. The counterpart lies, of course, in the probability that one has of losing one's entire bankroll, of being ruined, or, as is said, of being busted, or "cleaned out."

Louis Bachelier [1870–1946] (1912, p. 38) used the term "martingale" for the betting system in which one doubles one's stake after each *win*, and noted that such a martingale cannot make a game advantageous or disadvantageous if at each coup the game is fair. Ville (1939), and Richard von Mises [1883–1953] (1928) before him, were concerned with betting systems at fair games. With betting systems (albeit typically at subfair games) already known as martingales (in France), Ville's choice of "martingale" for the stochastic process that describes the gambler's cumulative fortune when using a betting system at a fair game was a natural and elegant one.

Why were betting systems known as martingales long before Ville? See Section 8.4 for a possible explanation.

As for the correspondence between subfair games and supermartingales, Thomas M. Cover (Golomb, Berlekamp, Cover, Gallager, Massey, and Viterbi 2002) wrote,

> I hasten to add that Doob has recanted this remark [concerning Shannon's (1948) paper on information theory] many times, saying that it and his naming of super martingales (processes that go down instead of up) are his two big regrets.

Incidentally, the power of martingale theory was not immediately recognized. In a review of Doob's (1949) second paper on martingales, the martingale concept was described as an "unpromising looking object" (M. 1950).

Long before the appearance of the optional stopping theorem, the English physicist Lord Raleigh [Strutt, John] [1842–1919] (1877) demonstrated a remarkably clear understanding of its consequences:

> Two players, A and B, toss for pennies. A has the option of continuing or stopping the game at any moment as it suits him. Has he, in consequence of this option, any advantage over B? [...]
>
> In order to examine the matter more closely, let us suppose that A has originally 1000 pennies, and that he proposes to continue the game until he has won 10, and then to leave off. Under these circumstances, it is clear that in no case can B lose more than 10, whereas A, if unlucky, may lose his whole stock before he has an opportunity of carrying off B's. The case is in fact exactly the same as if B had originally only 10 pennies, and the agreement were to continue the game until either A or B was ruined. The problem thus presented was solved long ago (see Todhunter's *History of Probabilities*, p. 62); and the result, as might have been expected, is that

the odds are exactly 100 : 1 that B will be ruined. But it does not follow from this that the arrangement is in any degree advantageous for A; for, if A loses, he loses a sum one hundred times as great as that which he gains from B in the other (and more probable) contingency.

He went on to argue that

> [T]he question is entirely altered by the introduction of indefinite credit. There is no object, of course, in insisting on perpetual payments, and a credit may properly be allowed to the extent of the actual resources of the parties; but the case is very different when insolvency is permitted. In order to make a comparison, let us suppose, in our previous example, that A has no fortune of his own but is allowed a credit of 1000. If he wins 10 from B without first losing 1000 himself, he retires a victor, and his actual poverty is not exposed. But how does the matter stand if the luck is against him, and he comes to the end of his credit before securing his prize? When called upon to pay at the termination of the transaction, he has no means of doing so, and thus B is defrauded of his 1000, which in the long run would otherwise compensate him for the more frequent losses of 10. The advantage which A possesses depends entirely, as it seems to me, on the credit which is allowed him, but to which he is not justly entitled, and is of exactly the same nature as that enjoyed by any man of straw, who is nevertheless allowed to trade.

The concept of a stopping time, the optional stopping theorem (Theorem 3.2.2), the upcrossing inequality (Lemma 3.3.1), and the martingale convergence theorem (Theorem 3.3.2) are due to Doob (1940). Our proof of the upcrossing inequality is from Williams (1991, Chapter 11). Wald's identity (Corollary 3.2.5) is named for Abraham Wald [1902–1950], who used it in his development of sequential analysis (Wald 1945). See Section 15.4 for the history of Example 3.2.6. Theorem 3.2.8 and Example 3.2.9 are from Ethier (1998). Example 3.3.3 is due to Finkelstein and Whitley (1981). Theorem 3.3.4 is essentially from Doob (1953, p. 320), while Theorem 3.3.5 is due to Lévy (1937, p. 249).

Problem 3.5 was motivated by Problem 12.9.15 of Grimmett and Stirzaker (2001). The source of the game of Problem 3.7 is Jacob Bernoulli [1654–1705] (1713; 2006, Part 3, Problem 14). Problem 3.8 was posed by Engel (1993) and the martingale solution is due to Stirzaker (1994). The result of Problem 3.10 is apparently due to William Feller [1906–1970] (1968, p. 328), who used renewal theory. The approach via martingales is due to Li (1980). See Ross (2003, Section 3.6.4) for another approach. Problem 3.11 is due to Penney (1969) and may therefore be called *Penney ante*, and our hint follows Li (1980). Problem 3.13 is due to Bachelier (1912, pp. 37–38). For more information on De Moivre's martingale (Problem 3.14), see Section 7.5. Problem 3.15 is due to Émile Borel [1871–1956] (1949). Problem 3.17 is generalized in Section 11.3. Problem 3.18 is a special case of a well-known theorem in branching processes first proved by Joseph L. Doob. It can be shown that W is positive with probability 1, has mean 1, variance $1/3$, and is absolutely continuous with a continuous density (Athreya and Ney 1972, pp. 9, 34, 36).

Chapter 4
Markov Chains

The past is broken away. The morrow is before us.

William Makepeace Thackeray, *The History of Pendennis*

A Markov chain is a sequence of discrete random variables indexed by a time parameter with the property that the conditional probability of a future event, given the present state and the past history, does not depend on the past history. Section 4.1 presents several examples, and Section 4.2 introduces the notions of transience and recurrence and shows how to evaluate absorption probabilities. Section 4.3 is concerned with the asymptotic behavior of the n-step transition probabilities of irreducible, aperiodic Markov chains, and Section 4.4 proves the discrete renewal theorem in the aperiodic setting.

4.1 Definitions and Examples

Let S be a countable (finite or countably infinite) set. An *S-valued random variable* X is a function from a sample space Ω into S for which $\{X = x\}$ is an event for every $x \in S$. Here S need not be a subset of $\mathbf{R}$ (or of $[0, \infty]$ or of $\mathbf{R}^d$), so this extends the notion of a discrete random variable (or vector). The concepts of distribution, jointly distributed random variables, and so on, extend in the obvious way. The expectation of X, however, is not meaningful unless $S \subset \mathbf{R}$ (or $S \subset [0, \infty]$ or $S \subset \mathbf{R}^d$). On the other hand, the conditioning random variables in a conditional expectation may be S-valued, and all of the results about conditional expectation generalize without difficulty.

A matrix $\boldsymbol{P} = (P(i,j))_{i,j \in S}$ with rows and columns indexed by S is called a *one-step transition matrix* if $P(i,j) \geq 0$ for all $i,j \in S$ and

S.N. Ethier, *The Doctrine of Chances*, Probability and its Applications, DOI 10.1007/978-3-540-78783-9_4, © Springer-Verlag Berlin Heidelberg 2010

$$\sum_{j \in S} P(i,j) = 1, \qquad i \in S. \tag{4.1}$$

In particular, the row sums of a one-step transition matrix are equal to 1. We call $P(i,j)$, the entry in row i and column j of the matrix $\boldsymbol{P}$, a *one-step transition probability*.

We say that $\{X_n\}_{n \geq 0}$ is a *Markov chain* in a countable state space S with one-step transition matrix $\boldsymbol{P}$ if $X_0, X_1, \ldots$ is a sequence of jointly distributed S-valued random variables with the property that

$$\mathrm{P}(X_{n+1} = j \mid X_0, \ldots, X_n) = \mathrm{P}(X_{n+1} = j \mid X_n) = P(X_n, j) \tag{4.2}$$

for all $n \geq 0$ and $j \in S$. We think of the sequence $X_0, X_1, \ldots$ as being indexed by time, and if we regard time n as the present, the first equation in (4.2), known as the *Markov property*, says that the conditional distribution of the state of the process one time step into the future, given its present state as well as its past history, depends only on its present state. The second equation in (4.2) tells us that

$$\mathrm{P}(X_{n+1} = j \mid X_n = i) = P(i,j) \tag{4.3}$$

does not depend on n. This property is called *time homogeneity*.

The distribution of X_0 is called the *initial distribution* $\boldsymbol{\pi}$ and is given by

$$\pi(i) := \mathrm{P}(X_0 = i), \qquad i \in S. \tag{4.4}$$

A Markov chain can be described by specifying its state space, its initial distribution, and its one-step transition matrix. Usually the initial distribution plays a secondary role.

Example 4.1.1. *Simple random walk in* $\mathbf{Z}$. Let $X_1, X_2, \ldots$ be i.i.d. with $\mathrm{P}(X_1 = 1) = p$ and $\mathrm{P}(X_1 = -1) = 1 - p$, where $0 < p < 1$, and put $S_0 := 0$ and $S_n := X_1 + \cdots + X_n$ for each $n \geq 1$. Then

$$\begin{aligned}
\mathrm{P}(S_{n+1} = j \mid S_0, S_1, \ldots, S_n) &= \mathrm{P}(S_n + X_{n+1} = j \mid S_0, S_1, \ldots, S_n) \\
&= P(S_n, j), \tag{4.5}
\end{aligned}$$

where $P(i, i+1) := p$, $P(i, i-1) := 1 - p$, and $P(i,j) := 0$ if $|j - i| \neq 1$. Here we are using Theorem 2.2.4 on p. 81. Therefore, $\{S_n\}_{n \geq 0}$ is a Markov chain in $\mathbf{Z}$, and it can be used to model the cumulative profit of a gambler who bets one unit at each coup in a sequence of independent coups, assuming that each bet is won with probability p (and paid even money) and lost otherwise. ♠

Example 4.1.2. *Oscar system.* The Oscar betting system is a more interesting system than that of the preceding example. The first bet is one unit. After a loss, the bet size remains the same. After a win, the bet size is increased by one unit, but with this possible exception: The gambler never bets more than necessary to achieve a cumulative profit of one unit. Once the one-unit cumulative profit is achieved, the gambler can quit or start anew. See Table 8.4 on p. 287 for clarification.

The system can be modeled by a Markov chain. Let $X_1, X_2, \ldots$ be i.i.d. with

$$P(X_1 = 1) = p \quad \text{and} \quad P(X_1 = -1) = 1 - p, \quad (4.6)$$

where $0 < p < 1$, with X_n representing the gambler's profit per unit bet at the nth coup. Let Y_n denote the number of units the gambler needs to reach his one-unit goal following the nth coup, and let Z_n denote the gambler's bet size at coup $n + 1$. Then

$$Y_{n+1} = Y_n - Z_n X_{n+1} \quad (4.7)$$
$$Z_{n+1} = [Z_n + 1_{\{X_{n+1}=1\}}] \wedge Y_{n+1} \quad (4.8)$$

if $(Y_n, Z_n) \neq (0, 0)$. Furthermore, the assumptions require that $(Y_0, Z_0) = (1, 1)$. We will assume that the gambler starts anew after achieving his goal, and therefore we put $(Y_{n+1}, Z_{n+1}) = (1, 1)$ if $(Y_n, Z_n) = (0, 0)$. (In effect, we are assuming that the gambler, after achieving his goal, ignores the next coup and then restarts.)

It follows easily that $\{(Y_n, Z_n)\}_{n \geq 0}$ is a Markov chain in the state space

$$S := \{(0,0)\} \cup \{(i, j) \in \mathbf{N} \times \mathbf{N} : i \geq j\} \quad (4.9)$$

with initial state $(Y_0, Z_0) = (1, 1)$ and one-step transition probabilities

$$P((i,j),(k,l)) := \begin{cases} p & \text{if } (k,l) = (i - j, (j + 1) \wedge (i - j)), \\ 1 - p & \text{if } (k,l) = (i + j, j), \\ 0 & \text{otherwise,} \end{cases} \quad (4.10)$$

for all $i \geq j \geq 1$, and

$$P((0,0),(k,l)) := \begin{cases} 1 & \text{if } (k,l) = (1,1), \\ 0 & \text{otherwise.} \end{cases} \quad (4.11)$$

Notice that one can write down the one-step transition matrix directly from the description of the model and the interpretations of the random variables, without having to rely on (4.7) and (4.8). We will return to this example in Section 4.2. ♠

Example 4.1.3. *Labouchere system.* The Labouchere betting system is even more complicated than the Oscar system. Here the gambler's bet size at each coup is determined by a list of positive integers kept on his score sheet, which is updated after each coup. Let us denote such a list by

$$\boldsymbol{l} = (l_1, \ldots, l_j), \quad (4.12)$$

where $j \geq 0$ denotes its length. (A list of length 0 is empty, denoted by $\boldsymbol{l} = \varnothing$.)

Given a list (4.12), the gambler's bet size at the next coup is

$$b(\boldsymbol{l}) := \begin{cases} 0 & \text{if } j = 0, \\ l_1 & \text{if } j = 1, \\ l_1 + l_j & \text{if } j \geq 2, \end{cases} \qquad (4.13)$$

that is, the sum of the extreme terms on the list. (This is just the sum of the first and last terms, except when the list has only one term or is empty.) Following the resolution of this bet, the list (4.12) is updated as follows: After a win, it is replaced by

$$\boldsymbol{l}[1] := \begin{cases} \varnothing & \text{if } j \in \{0, 1, 2\}, \\ (l_2, \ldots, l_{j-1}) & \text{if } j \geq 3, \end{cases} \qquad (4.14)$$

that is, the extreme terms are canceled. After a loss, it is replaced by

$$\boldsymbol{l}[-1] := \begin{cases} \varnothing & \text{if } j = 0, \\ (l_1, \ldots, l_j, b(\boldsymbol{l})) & \text{if } j \geq 1, \end{cases} \qquad (4.15)$$

that is, the amount just lost is appended to the list, if nonempty, as a new last term. Once the list becomes empty, it remains so, and no further bets are made. See Table 8.3 on p. 284 for clarification.

Again, the system can be modeled by a Markov chain. Let $X_1, X_2, \ldots$ be i.i.d. with $\mathrm{P}(X_1 = 1) = p$ and $\mathrm{P}(X_1 = -1) = 1 - p$, where $0 < p < 1$, with X_n representing the gambler's profit per unit bet at the nth coup. Let $\boldsymbol{L}_n$ denote the list as updated after the nth coup. The initial state is $\boldsymbol{L}_0 = \boldsymbol{l}^0$, where $\boldsymbol{l}^0$ is the initial list (e.g., $\boldsymbol{l}^0 = (1, 2, 3, 4)$ is a popular choice), and

$$\boldsymbol{L}_n := \boldsymbol{L}_{n-1}[X_n], \qquad n \geq 1, \qquad (4.16)$$

where the meaning of the bracket is as in (4.14) or (4.15). Let B_n denote the amount bet at the nth coup, so that

$$B_n := b(\boldsymbol{L}_{n-1}), \qquad n \geq 1. \qquad (4.17)$$

Finally, let

$$Y_n := \text{length of the list } \boldsymbol{L}_n, \qquad n \geq 0. \qquad (4.18)$$

We claim that $\{\boldsymbol{L}_n\}_{n \geq 0}$ and $\{Y_n\}_{n \geq 0}$ are Markov chains but $\{B_n\}_{n \geq 1}$ is not. Indeed, $\{\boldsymbol{L}_n\}_{n \geq 0}$ is a Markov chain in the countable state space of all lists with one-step transition probabilities

$$P(\boldsymbol{l}, \boldsymbol{m}) := \begin{cases} p & \text{if } \boldsymbol{m} = \boldsymbol{l}[1], \\ 1 - p & \text{if } \boldsymbol{m} = \boldsymbol{l}[-1], \qquad \boldsymbol{l} \neq \varnothing, \\ 0 & \text{otherwise,} \end{cases} \qquad (4.19)$$

and

$$P(\varnothing, \boldsymbol{m}) := \begin{cases} 1 & \text{if } \boldsymbol{m} = \varnothing, \\ 0 & \text{otherwise.} \end{cases} \qquad (4.20)$$

while $\{Y_n\}_{n \geq 0}$ is a Markov chain in $\mathbf{Z}_+$ with one-step transition probabilities

$$Q(i, j) := \begin{cases} p & \text{if } j = (i - 2)^+, \\ 1 - p & \text{if } j = i + 1, \qquad i \geq 1, \\ 0 & \text{otherwise,} \end{cases} \qquad (4.21)$$

and

$$Q(0, j) := \begin{cases} 1 & \text{if } j = 0, \\ 0 & \text{otherwise.} \end{cases} \qquad (4.22)$$

We express (4.20) (resp., (4.22)) by saying that state $\varnothing$ (resp., state 0) is *absorbing*. We leave it to the reader to show that the Markov property fails for $\{B_n\}_{n \geq 1}$ (Problem 4.2 on p. 150). ♠

Example 4.1.4. *A 64-state Markov chain in craps.* Consider a sequence of i.i.d. random variables $T_1, T_2, \ldots$ corresponding to a sequence of dice rolls. The common distribution is given by (1.11) on p. 6. We recall that the set of *point numbers* was defined as

$$\mathscr{P} := \{4, 5, 6, 8, 9, 10\}. \qquad (4.23)$$

For each $n \geq 0$ let X_n be the subset of $\mathscr{P}$ consisting of those point numbers that have occurred at least once by the nth roll and since the last 7 (or since the first roll if no 7 has occurred). Of course, $X_0 := \varnothing$. More precisely,

$$X_n = \{j \in \mathscr{P} : T_m = j \text{ for some } m \in \{1, 2, \ldots, n\}$$
$$\text{for which } 7 \notin \{T_{m+1}, \ldots, T_n\}\}. \qquad (4.24)$$

Then $\{X_n\}_{n \geq 0}$ is a Markov chain in the state space S comprising all subsets of $\mathscr{P}$, and its one-step transition matrix $\boldsymbol{P}$ is given by

$$P(A, B) := \begin{cases} \pi_j & \text{if } j \in \mathscr{P} - A \text{ and } B = A \cup \{j\}, \\ \pi_7 & \text{if } B = \varnothing, \\ 1 - \pi_7 - \sum_{i \in \mathscr{P} - A} \pi_i & \text{if } B = A, \end{cases}$$
$$\qquad (4.25)$$

if $A \neq \varnothing$, and

$$P(\varnothing, B) := \begin{cases} \pi_j & \text{if } j \in \mathscr{P} \text{ and } B = \{j\}, \\ 1 - \sum_{i \in \mathscr{P}} \pi_i & \text{if } B = \varnothing. \end{cases} \qquad (4.26)$$

We will return to this example in Section 4.3.   ♠

Given a Markov chain $\{X_n\}_{n \geq 0}$ in the state space S with one-step transition matrix $\boldsymbol{P}$, we observe that, for every $m \geq 1$, $i_0, i_1, \ldots, i_m \in S$, and $n \geq 0$,

$$\begin{aligned} P(X_{n+1} &= i_1, \ldots, X_{n+m} = i_m \mid X_n = i_0) \\ &= P(X_{n+1} = i_1 \mid X_n = i_0)P(X_{n+2} = i_2 \mid X_n = i_0, X_{n+1} = i_1) \\ &\quad \cdots P(X_{n+m} = i_m \mid X_n = i_0, X_{n+1} = i_1, \ldots, X_{n+m-1} = i_{m-1}) \\ &= P(i_0, i_1)P(i_1, i_2) \cdots P(i_{m-1}, i_m), \end{aligned} \qquad (4.27)$$

provided that $P(X_n = i_0) > 0$. (If any of the other conditional probabilities are undefined, they can be defined arbitrarily.) It follows that

$$\begin{aligned} P(X_{n+m} &= i_m \mid X_n = i_0) \\ &= \sum_{i_1, \ldots, i_{m-1} \in S} \cdots \sum P(X_{n+1} = i_1, \ldots, X_{n+m-1} = i_{m-1}, X_{n+m} = i_m \mid X_n = i_0) \\ &= \sum_{i_1, \ldots, i_{m-1} \in S} \cdots \sum P(i_0, i_1)P(i_1, i_2) \cdots P(i_{m-1}, i_m), \end{aligned} \qquad (4.28)$$

so we define the *m-step transition matrix* $\boldsymbol{P}^m$ of the Markov chain by

$$P^m(i, j) := \sum_{i_1, \ldots, i_{m-1} \in S} \cdots \sum P(i, i_1)P(i_1, i_2) \cdots P(i_{m-1}, j). \qquad (4.29)$$

Notice that the superscript m can be interpreted as an exponent, this is, the m-step transition matrix is the mth power of the one-step transition matrix. This is valid both when S is finite and when S is countably infinite. It is easy to check that this allows us to generalize (4.2), obtaining

$$P(X_{n+m} = j \mid X_0, \ldots, X_n) = P(X_{n+m} = j \mid X_n) = P^m(X_n, j) \qquad (4.30)$$

for all $n \geq 0$, $m \geq 1$, and $j \in S$.

Example 4.1.5. *A Markov chain related to faro.* As cards are dealt from a standard 52-card deck and exposed one by one, the composition of the unseen deck evolves according to a Markov chain. If n cards have been dealt ($n = 0, 1, \ldots, 52$), we let

$$\boldsymbol{D}_n := (d_0, d_1, d_2, d_3, d_4), \qquad (4.31)$$

where d_0, d_1, d_2, d_3, d_4 are the numbers of denominations represented $0, 1, 2, 3, 4$ times, respectively, in the unseen deck of size $52 - n$. Thus, $\{\boldsymbol{D}_n\}_{n \geq 0}$ is a Markov chain in the state space

$$S := \{\boldsymbol{d} = (d_0, d_1, d_2, d_3, d_4) \in \mathbf{Z}_+^5 : d_0 + d_1 + d_2 + d_3 + d_4 = 13\}, \quad (4.32)$$

which has $\binom{13+5-1}{5-1} = 2{,}380$ states (Theorem 1.1.8 on p. 6). Its initial state is $\boldsymbol{D}_0 = (0, 0, 0, 0, 13)$ and its one-step transition matrix $\boldsymbol{P}$ is given by

$$P(\boldsymbol{d}, \boldsymbol{d} - \boldsymbol{e}_j + \boldsymbol{e}_{j-1}) := \frac{jd_j}{\mu(\boldsymbol{d})}, \qquad j = 1, 2, 3, 4, \quad (4.33)$$

for all $\boldsymbol{d} \in S - \{(13, 0, 0, 0, 0)\}$, where $\boldsymbol{e}_0 := (1, 0, 0, 0, 0)$, $\boldsymbol{e}_1 := (0, 1, 0, 0, 0)$, and so on, and the function $\mu : S \mapsto \{0, 1, \ldots, 52\}$, defined by

$$\mu(\boldsymbol{d}) := d_1 + 2d_2 + 3d_3 + 4d_4, \quad (4.34)$$

is the number of cards remaining. Finally, state $(13, 0, 0, 0, 0)$ is absorbing. Notice that $P(\mu(\boldsymbol{D}_n) = (52 - n)^+) = 1$ for all $n \geq 0$.

Now in the game of faro (see Chapter 18 for details), the first card is exposed. The game proceeds by exposing two cards at each *turn*, or hand, until only one card remains unseen. Thus, only the subsequence $\{\boldsymbol{D}_{2n+1}\}_{n \geq 0}$ is of interest. But this subsequence is itself a Markov chain. Its initial state is $\boldsymbol{D}_1 = (0, 0, 0, 1, 12)$, and its one-step transition matrix is $\boldsymbol{P}^2$, the two-step transition matrix for the original chain. This Markov chain is explored further in Section 18.2. ♠

Example 4.1.6. *Median length of the shooter's hand at craps.* In Example 3.2.6 on p. 103 we found the expected length of the shooter's hand at craps. Here we find the median length with the help of a Markov chain. The state space for the chain is

$$S := \{\varnothing\} \cup \mathscr{P} \cup \{\Delta\}, \quad (4.35)$$

where $\mathscr{P} := \{4, 5, 6, 8, 9, 10\}$, with the following interpretation: The chain is in state $\varnothing$ if the shooter is coming out. It is in state $j \in \mathscr{P}$ if the point j has been established on a come-out roll but not yet resolved. Finally, it is in the absorbing state Δ if the shooter has sevened out. It is clear that the one-step transition matrix $\boldsymbol{P}$ satisfies

$$P(\varnothing, j) = \begin{cases} \pi_2 + \pi_3 + \pi_7 + \pi_{11} + \pi_{12} & \text{if } j = \varnothing, \\ \pi_j & \text{if } j \in \mathscr{P}, \\ 0 & \text{if } j = \Delta, \end{cases} \quad (4.36)$$

$$P(i, j) = \begin{cases} \pi_i & \text{if } j = \varnothing, \\ 1 - \pi_i - \pi_7 & \text{if } j = i, \\ 0 & \text{if } j \in \mathscr{P} - \{i\}, \\ \pi_7 & \text{if } j = \Delta, \end{cases} \quad i \in \mathscr{P}, \quad (4.37)$$

and

$$P(\Delta, j) = \begin{cases} 1 & \text{if } j = \Delta, \\ 0 & \text{otherwise.} \end{cases} \tag{4.38}$$

We can use (1.11) on p. 6 to evaluate the one-step transition matrix $\boldsymbol{P}$ as

$$36\boldsymbol{P} = \begin{array}{c} \\ \varnothing \\ 4 \\ 5 \\ 6 \\ 8 \\ 9 \\ 10 \\ \Delta \end{array} \begin{pmatrix} \varnothing & 4 & 5 & 6 & 8 & 9 & 10 & \Delta \\ 12 & 3 & 4 & 5 & 5 & 4 & 3 & 0 \\ 3 & 27 & 0 & 0 & 0 & 0 & 0 & 6 \\ 4 & 0 & 26 & 0 & 0 & 0 & 0 & 6 \\ 5 & 0 & 0 & 25 & 0 & 0 & 0 & 6 \\ 5 & 0 & 0 & 0 & 25 & 0 & 0 & 6 \\ 4 & 0 & 0 & 0 & 0 & 26 & 0 & 6 \\ 3 & 0 & 0 & 0 & 0 & 0 & 27 & 6 \\ 0 & 0 & 0 & 0 & 0 & 0 & 0 & 36 \end{pmatrix}. \tag{4.39}$$

Because $\pi_j = \pi_{14-j}$ for $j = 4, 5, 6$, it is possible to reduce the number of states to five: $\varnothing$, 4-10, 5-9, 6-8, and Δ. However, we prefer not to make this simplification here.

Now notice that $P(L \le n) = P^n(\varnothing, \Delta)$ for each $n \ge 1$ because a shooter sevens out in n rolls or less if and only if the Markov chain, starting in state $\varnothing$, visits state Δ at the nth step (not necessarily for the first time), using the fact that Δ is absorbing. In particular,

$$P(L \le 5) = P^5(\varnothing, \Delta) = \frac{1{,}067{,}911}{2{,}519{,}424} < \frac{1}{2}, \tag{4.40}$$

$$P(L \le 6) = P^6(\varnothing, \Delta) = \frac{1{,}266{,}739}{2{,}519{,}424} > \frac{1}{2}, \tag{4.41}$$

hence median$(L) = 6$. Thus, the median-to-mean ratio, namely

$$\frac{\text{median}(L)}{E[L]} = \frac{6}{1{,}671/196} = \frac{392}{557} \approx 0.704 \tag{4.42}$$

(see Example 3.2.6 on p. 103), is similar to that for the geometric distribution:

$$\frac{\text{median}(\text{geometric}(p))}{E[\text{geometric}(p)]} \approx \ln 2 \approx 0.693, \tag{4.43}$$

at least for small p (Example 1.4.3 on p. 30). ♠

Example 4.1.7. *Shuffling.* Consider a deck of N distinct cards, which for convenience we will assume are labeled $1, 2, \ldots, N$. For specificity, we also label the positions of the cards in the deck as follows: With the cards face down, the top card in the deck is in position 1, the second card is in position 2, and so on. In particular, when the cards are dealt, the card in position 1 is dealt first.

We define a *permutation* π of $(1, 2, \ldots, N)$ to be a one-to-one mapping $\pi : \{1, 2, \ldots, N\} \mapsto \{1, 2, \ldots, N\}$. The *symmetric group* S_N is the group of all permutations of $(1, 2, \ldots, N)$, with group operation given by composition of mappings, that is, $\pi_1 \pi_2 := \pi_2 \circ \pi_1$, where $(\pi_2 \circ \pi_1)(i) := \pi_2(\pi_1(i))$. The number of permutations in S_N is of course $N!$.

Think of $\pi(i) = j$ as meaning that the card in position i is moved to position j under the permutation. A *shuffle* can be defined to be a random permutation Π, that is, an S_N-valued random variable. A shuffle can be used to define the one-step transition matrix $\boldsymbol{P}$ of a Markov chain in S_N by means of the formula

$$P(\pi_1, \pi_2) = \mathrm{P}(\pi_1 \Pi = \pi_2). \tag{4.44}$$

In general, a one-step transition matrix $\boldsymbol{P}$ is said to be *doubly stochastic* if column sums (in addition to row sums) are equal to 1. The one-step transition matrix $\boldsymbol{P}$ of (4.44) is doubly stochastic (as is that of Example 4.1.1) because

$$\sum_{\pi_1 \in S_N} P(\pi_1, \pi_2) = \sum_{\pi_1 \in S_N} \mathrm{P}(\pi_1 \Pi = \pi_2) = \sum_{\pi_1 \in S_N} \mathrm{P}(\pi_2 \Pi^{-1} = \pi_1) = 1. \tag{4.45}$$

We will describe a simple but unrealistic model of shuffling in Section 4.3 and will study a more realistic model in Section 11.1. ♠

4.2 Transience and Recurrence

Given $i \in S$, let us introduce the notation $\mathrm{P}_i(\cdot) := \mathrm{P}(\cdot \mid X_0 = i)$, with the understanding that the initial distribution is such that $\mathrm{P}(X_0 = i) > 0$. By (4.27) on p. 124 with $n = 0$,

$$\mathrm{P}_i(X_1 = i_1, \ldots, X_m = i_m) = P(i, i_1) P(i_1, i_2) \cdots P(i_{m-1}, i_m) \tag{4.46}$$

for all $i_1, \ldots, i_m \in S$. It follows from this that

$$\mathrm{P}_i(X_1 = i_1, \ldots, X_m = i_m, X_{m+1} = j_1, \ldots, X_{m+n} = j_n)$$
$$= \mathrm{P}_i(X_1 = i_1, \ldots, X_m = i_m) \mathrm{P}_{i_m}(X_1 = j_1, \ldots, X_n = j_n) \tag{4.47}$$

for all $i_1, \ldots, i_m, j_1, \ldots, j_n \in S$. Given subsets $A_1, \ldots, A_m, B_1, \ldots, B_n \subset S$ and $j \in S$ with $A_m = \{j\}$, we can sum (4.47) to obtain

$$\mathrm{P}_i(X_1 \in A_1, \ldots, X_m \in A_m, X_{m+1} \in B_1, \ldots, X_{m+n} \in B_n)$$
$$= \mathrm{P}_i(X_1 \in A_1, \ldots, X_m \in A_m) \mathrm{P}_j(X_1 \in B_1, \ldots, X_n \in B_n). \tag{4.48}$$

Given $j \in S$, let us introduce the notation T_j for the *first hitting time* of state j (or first return time if starting in state j) and N_j for the number of visits to state j (excluding visits at time 0). More precisely,

$$T_j := \min\{n \geq 1 : X_n = j\} \quad \text{and} \quad N_j := \sum_{n=1}^{\infty} 1_{\{X_n = j\}}, \tag{4.49}$$

where $\min \varnothing = \infty$. If also $i \in S$, we define

$$f_{ij} := P_i(T_j < \infty) = P_i(N_j \geq 1). \tag{4.50}$$

This is the probability that the Markov chain, starting in state i, ever visits state j (or ever returns to state i if $j = i$). We define state j to be *transient* if $f_{jj} < 1$ and to be *recurrent* if $f_{jj} = 1$.

Given $i, j \in S$ and $m \geq 1$, we deduce from (4.48) that

$$\begin{aligned}
P_i(N_j \geq m) &= \sum_{n_1, \ldots, n_m \geq 1} \cdots \sum P_i(X_1 \neq j, \ldots, X_{n_1-1} \neq j, X_{n_1} = j, \\
&\qquad\qquad X_{n_1+1} \neq j, \ldots, X_{n_1+n_2-1} \neq j, X_{n_1+n_2} = j, \\
&\qquad\qquad\qquad \vdots \\
&\qquad\qquad X_{n_1+\cdots+n_{m-1}+1} \neq j, \ldots, X_{n_1+\cdots+n_m-1} \neq j, \\
&\qquad\qquad\qquad\qquad\qquad\qquad\qquad\qquad X_{n_1+\cdots+n_m} = j) \\
&= \sum_{n_1 \geq 1} P_i(X_1 \neq j, \ldots, X_{n_1-1} \neq j, X_{n_1} = j) \\
&\quad \cdot \sum_{n_2 \geq 1} P_j(X_1 \neq j, \ldots, X_{n_2-1} \neq j, X_{n_2} = j) \\
&\qquad\qquad \vdots \\
&\quad \cdot \sum_{n_m \geq 1} P_j(X_1 \neq j, \ldots, X_{n_m-1} \neq j, X_{n_m} = j) \\
&= f_{ij}(f_{jj})^{m-1}. \tag{4.51}
\end{aligned}$$

Letting $m \to \infty$ and applying Theorem 1.1.20(b) on p. 13, we find that

$$P_i(N_j = \infty) = \begin{cases} 0 & \text{if } j \text{ is transient,} \\ f_{ij} & \text{if } j \text{ is recurrent.} \end{cases} \tag{4.52}$$

Another useful calculation, by Theorem 1.4.2 on p. 30 and (4.51), is

$$\begin{aligned}
E_i[N_j] &= \sum_{m=1}^{\infty} P_i(N_j \geq m) = \sum_{m=1}^{\infty} f_{ij}(f_{jj})^{m-1} \\
&= \begin{cases} f_{ij}/(1 - f_{jj}) < \infty & \text{if } j \text{ is transient,} \\ \infty & \text{if } j \text{ is recurrent and } f_{ij} > 0. \end{cases} \tag{4.53}
\end{aligned}$$

Theorem 4.2.1. *For a Markov chain in S with one-step transition matrix $\boldsymbol{P}$, state $j \in S$ is*

$$\text{transient if } \sum_{n=1}^{\infty} P^n(j,j) < \infty, \qquad (4.54)$$

$$\text{recurrent if } \sum_{n=1}^{\infty} P^n(j,j) = \infty. \qquad (4.55)$$

Proof. Given $i, j \in S$, we have, by Corollary A.2.4 on p. 748,

$$\mathrm{E}_i[N_j] = \mathrm{E}_i\left[\sum_{n=1}^{\infty} 1_{\{X_n=j\}}\right] = \sum_{n=1}^{\infty} \mathrm{P}_i(X_n = j) = \sum_{n=1}^{\infty} P^n(i,j), \qquad (4.56)$$

so the result is immediate from (4.53) and (4.56) with $i = j$. ♠

Lemma 4.2.2. *Let $i, j \in S$ be distinct. If state i is recurrent and $f_{ij} > 0$, then state j is also recurrent and $f_{ji} = 1$.*

Proof. Since $f_{ij} > 0$, there exists $r \in \mathbf{N}$ such that $P^r(i,j) > 0$, and we can assume that r is chosen minimally. Hence there exists $i_1, \dots, i_{r-1} \in S$ such that $P(i,i_1)P(i_1,i_2)\cdots P(i_{r-1},j) > 0$. Moreover, $i_1, \dots, i_{r-1} \neq i$, for otherwise r could be chosen smaller. It follows that $f_{ji} = 1$, for if f_{ji} were less than 1, we would have

$$\mathrm{P}_i(X_n \neq i \text{ for all } n \geq 1) \geq P(i,i_1)P(i_1,i_2)\cdots P(i_{r-1},j)(1 - f_{ji}) > 0, \qquad (4.57)$$

contradicting the recurrence of i. In particular, there exists $s \in \mathbf{N}$ such that $P^s(j,i) > 0$, so

$$P^{s+n+r}(j,j) \geq P^s(j,i)P^n(i,i)P^r(i,j), \qquad n \geq 1. \qquad (4.58)$$

Summing this inequality over all $n \geq 1$ and using Theorem 4.2.1, we find that the recurrence of i implies the recurrence of j. ♠

Let us define a Markov chain in S with one-step transition matrix $\boldsymbol{P}$ to be *irreducible* if $f_{ij} > 0$ for all $i, j \in S$. Lemma 4.2.2 tells us that if a Markov chain in S with one-step transition matrix $\boldsymbol{P}$ is irreducible, then either all states in S are transient or all are recurrent. This allows us to refer to an irreducible Markov chain as either *transient* or *recurrent*.

Theorem 4.2.3. *Let $\{X_n\}_{n\geq 0}$ be an irreducible Markov chain in S with one-step transition matrix $\boldsymbol{P}$. If it is transient, then*

$$\mathrm{P}_i\left(\bigcup_{j \in S}\{X_n = j \text{ i.o.}\}\right) = 0, \qquad i \in S, \qquad (4.59)$$

and $\sum_{n=1}^{\infty} P^n(i,j) < \infty$ for all $i, j \in S$. If it is recurrent, then

$$P_i\left(\bigcap_{j \in S} \{X_n = j \text{ i.o.}\}\right) = 1, \qquad i \in S, \tag{4.60}$$

and $\sum_{n=1}^{\infty} P^n(i,j) = \infty$ for all $i, j \in S$.

Remark. In particular, an irreducible transient chain visits no state infinitely often, while an irreducible recurrent chain visits every state infinitely often.

Proof. The conclusions concerning the infinite series follow immediately from (4.53) and (4.56). The conclusions (4.59) and (4.60) follow from (4.52) and Lemma 4.2.2. ♠

Example 4.2.4. *Simple symmetric random walk in* $\mathbf{Z}^d$. Define $e_1, \ldots, e_d \in \mathbf{Z}^d$ by letting $e_1 := (1, 0, \ldots, 0)$, $e_2 := (0, 1, 0, \ldots, 0)$, and so on. Let us consider the Markov chain in $\mathbf{Z}^d$ with one-step transition probabilities of the form

$$P(\boldsymbol{j}, \boldsymbol{k}) := \begin{cases} (2d)^{-1} & \text{if } \boldsymbol{k} = \boldsymbol{j} \pm e_i \text{ for some } i \in \{1, \ldots, d\}, \\ 0 & \text{otherwise.} \end{cases} \tag{4.61}$$

The resulting Markov chain is called a *simple symmetric random walk in* $\mathbf{Z}^d$. It is clearly irreducible. Is it transient or recurrent?

By Theorem 4.2.1 it suffices to determine the convergence or divergence of the series whose nth term is

$$P^{2n}(\mathbf{0}, \mathbf{0}) = \sum_{(n_1, \ldots, n_d) \in \mathbf{Z}_+^d : n_1 + \cdots + n_d = n} \binom{2n}{n_1, n_1, n_2, n_2, \ldots, n_d, n_d} \left(\frac{1}{2d}\right)^{2n}$$

$$= \binom{2n}{n} (2d)^{-2n} \sum_{(n_1, n_2, \ldots, n_d) \in \mathbf{Z}_+^d : n_1 + \cdots + n_d = n} \binom{n}{n_1, \ldots, n_d}^2. \tag{4.62}$$

In particular, if $d = 1$, then, by Stirling's formula (Theorem 1.5.12 on p. 48),

$$P^{2n}(0,0) = \binom{2n}{n} 2^{-2n} \sim \frac{1}{\sqrt{\pi n}} \text{ as } n \to \infty, \tag{4.63}$$

so the simple symmetric random walk in one dimension is recurrent. If $d = 2$, then

$$P^{2n}((0,0),(0,0)) = \binom{2n}{n} 4^{-2n} \sum_{n_1=0}^{n} \binom{n}{n_1}\binom{n}{n-n_1}$$

$$= \left[\binom{2n}{n} 2^{-2n}\right]^2 \sim \frac{1}{\pi n} \text{ as } n \to \infty, \tag{4.64}$$

so the simple symmetric random walk in two dimensions is also recurrent.

We claim that the simple symmetric random walk in dimension $d \geq 3$ is transient. Letting

$$M_d(n) := \max_{(n_1,\ldots,n_d)\in\mathbf{Z}_+^d :\, n_1+\cdots+n_d=n} \binom{n}{n_1,\ldots,n_d},\qquad (4.65)$$

(4.62) and the multinomial theorem (Theorem 1.1.7 on p. 5) tell us that

$$P^{2n}(\mathbf{0},\mathbf{0}) \leq \binom{2n}{n} 2^{-2n} d^{-n} M_d(n) =: U_d(n).\qquad (4.66)$$

We will show that the series whose nth term is $U_d(n)$ converges.

Suppose that $\mathbf{n}^\circ = (n_1^\circ, \ldots, n_d^\circ)$ achieves the maximum in (4.65). Then

$$\binom{n}{\mathbf{n}^\circ} \geq \binom{n}{\mathbf{n}^\circ + \mathbf{e}_i - \mathbf{e}_j}\qquad (4.67)$$

for $i, j = 1, \ldots, d$. In particular, $n_i^\circ + 1 \geq n_j^\circ$ for $i, j = 1, \ldots, d$, or

$$|n_i^\circ - n_j^\circ| \leq 1, \qquad i, j = 1, \ldots, d.\qquad (4.68)$$

This implies that

$$M_d(md + r) = \binom{md+r}{m,\ldots,m,m+1,\ldots,m+1},\qquad (4.69)$$

where m is a positive integer, $r \in \{0, 1, \ldots, d-1\}$, and $m+1$ appears in the multinomial coefficient exactly r times. A straightforward calculation shows that

$$\lim_{m\to\infty} \frac{d^{-(md+r)} M_d(md+r)}{d^{-md} M_d(md)} = 1,\qquad (4.70)$$

and it follows that $\lim_{m\to\infty} U_d(md+r)/U_d(md) = 1$ for $r = 0, 1, \ldots, d-1$. Therefore, it suffices to analyze the asymptotic behavior of $U_d(n)$ as $n \to \infty$ through positive integer multiples of d. In this case, Stirling's formula implies that

$$U_d(n) = \binom{2n}{n} 2^{-2n} d^{-n} \binom{n}{n/d,\ldots,n/d} \sim c_d\, n^{-d/2},\qquad (4.71)$$

where $c_d := 2^{-(d-1)/2}(d/\pi)^{d/2}$, and we conclude that

$$\sum_{n=1}^\infty P^{2n}(\mathbf{0},\mathbf{0}) \leq \sum_{n=1}^\infty U_d(n) < \infty, \qquad d \geq 3,\qquad (4.72)$$

as required.   ♠

Example 4.2.5. *Oscar system, continued.* The Markov chain defined in Example 4.1.2 on p. 120 is clearly irreducible. We claim that it is transient if $p < \frac{1}{2}$ and recurrent if $p \geq \frac{1}{2}$.

We first treat the case $0 < p < \frac{1}{2}$. Our strategy is to show that there exists a positive integer m such that

$$P_{(m,1)}(N = \infty) > 0, \tag{4.73}$$

where $N := \min\{n \geq 1 : (Y_n, Z_n) = (0,0)\}$. This would then imply that $f_{(m,1),(0,0)} < 1$, and therefore transience would hold (Lemma 4.2.2). Let $m \in \mathbf{N}$ be arbitrary. Given the sequence of i.i.d. random variables $X_1, X_2, \ldots$ satisfying (4.6) on p. 121, where $0 < p < \frac{1}{2}$, we can define $\{(Y_n, Z_n)\}_{n \geq 0}$ in S via $(Y_0, Z_0) = (m, 1)$ and

$$Y_{n+1} = Y_n - Z_n X_{n+1} \tag{4.74}$$
$$Z_{n+1} = [Z_n + 1_{\{X_{n+1}=1\}}] \wedge Y_{n+1} \tag{4.75}$$

for $0 \leq n < N$. (Recall (4.7) and (4.8) on p. 121.) Let us also define $\{(Y_n^\circ, Z_n^\circ)\}_{n \geq 0}$ in $\mathbf{Z} \times \mathbf{N}$ via $(Y_0^\circ, Z_0^\circ) = (m, 1)$ and

$$Y_{n+1}^\circ = Y_n^\circ - Z_n^\circ X_{n+1} \tag{4.76}$$
$$Z_{n+1}^\circ = Z_n^\circ + 1_{\{X_{n+1}=1\}} \tag{4.77}$$

for all $n \geq 0$. Clearly,

$$P((Y_n^\circ, Z_n^\circ) = (Y_n, Z_n) \text{ for } 0 \leq n < N^\circ) = 1, \tag{4.78}$$

where $N^\circ := \min\{n \geq 1 : (Y_n^\circ, Z_n^\circ) = (0,0) \text{ or } Y_n^\circ < Z_n^\circ\}$. Thus, it will suffice to show that $P(N^\circ = \infty) > 0$ for m sufficiently large.

The advantage of $\{(Y_n^\circ, Z_n^\circ)\}_{n \geq 0}$ over $\{(Y_n, Z_n)\}_{n \geq 0}$ is that its asymptotic behavior is more easily determined. Indeed,

$$Y_n^\circ - m = -\sum_{l=1}^{n}\left(1 + \sum_{k=1}^{l-1} 1_{\{X_k=1\}}\right) X_l \tag{4.79}$$

$$Z_n^\circ - 1 = \sum_{k=1}^{n} 1_{\{X_k=1\}} \tag{4.80}$$

for all $n \geq 0$. By Example 1.5.8 on p. 45, $n^{-2}(Y_n^\circ - m) \to p(1 - 2p)/2$ a.s. and $n^{-1}(Z_n^\circ - 1) \to p$ a.s. This implies that

$$\inf_{n \geq 1}(Y_n^\circ - m - (Z_n^\circ - 1)) > -\infty \quad \text{a.s.}, \tag{4.81}$$

where it is important to observe that the distribution of the expression on the left side of the inequality in (4.81) does not depend on m (see (4.79) and (4.80)). It follows that there exists a positive integer m large enough that

$$P(Y_n^\circ - m - (Z_n^\circ - 1) \geq -m + 2 \text{ for all } n \geq 1) > 0, \qquad (4.82)$$

and this is equivalent to $P(Y_n^\circ - Z_n^\circ \geq 1 \text{ for all } n \geq 1) > 0$. We conclude that $P(N^\circ = \infty) > 0$, as required. This proves the transience assertion.

We next treat the case $\frac{1}{2} \leq p < 1$. Our strategy is to show that

$$P_{(1,1)}(N < \infty) = 1, \qquad (4.83)$$

where $N := \min\{n \geq 1 : (Y_n, Z_n) = (0,0)\}$. This would then imply that $f_{(1,1),(1,1)} = 1$, and therefore recurrence would hold. Given the sequence of i.i.d. random variables $X_1, X_2, \ldots$ satisfying (4.6) on p. 121, where $\frac{1}{2} \leq p < 1$, we can define $\{(Y_n, Z_n), \ 0 \leq n \leq N\}$ in S via $(Y_0, Z_0) = (1,1)$ and (4.74) and (4.75) for $0 \leq n < N$. Then

$$E_{(1,1)}[Y_{n+1} \mid X_1, \ldots, X_n] = Y_n - Z_n(2p - 1) \leq Y_n \qquad (4.84)$$

on $\{N \geq n + 1\}$. By Lemma 3.2.1 on p. 99, $\{Y_{n \wedge N}\}_{n \geq 0}$ is a nonnegative supermartingale with respect to $\{X_n\}_{n \geq 1}$. By the martingale convergence theorem (Theorem 3.3.2 on p. 107), $Y := \lim_{n \to \infty} Y_{n \wedge N}$ exists a.s., and $E_{(1,1)}[Y] \leq E_{(1,1)}[Y_0] = 1$. In particular, we have $P_{(1,1)}(Y < \infty) = 1$, and since $|Y_{n+1} - Y_n| \geq 1$ whenever $0 \leq n < N$, it must be the case that (4.83) holds. (The point is that an infinite sequence $\{y_n\}$ cannot converge to a finite limit if $|y_{n+1} - y_n| \geq 1$ for all n.)   ♠

We conclude this section by showing how to evaluate absorption probabilities. Given a Markov chain $\{X_n\}_{n \geq 0}$ in S and a nonempty subset $A \subset S$, we define the *first visiting time* of A by

$$V_A := \min\{n \geq 0 : X_n \in A\}, \qquad (4.85)$$

where $\min \varnothing := \infty$. Notice that V_A is a stopping time with respect to $\{X_n\}_{n \geq 0}$ (see Section 3.1). We begin with an easy lemma.

Lemma 4.2.6. *Given a Markov chain $\{X_n\}_{n \geq 0}$ in S and a nonempty subset $A \subset S$ such that A^c is finite and $P_i(V_A < \infty) > 0$ for every $i \in A^c$, there exists a positive integer M and $\varepsilon \in (0, 1)$ such that*

$$P_i(V_A > nM) < (1 - \varepsilon)^n, \qquad n \geq 1, \qquad (4.86)$$

for each $i \in S$. In particular, $E_i[V_A] < \infty$ for each $i \in S$.

Proof. Since A^c is finite, there exists $\varepsilon \in (0, 1)$ such that $P_i(V_A < \infty) > \varepsilon$ for all $i \in A^c$, and therefore there exists a positive integer M such that $P_i(V_A \leq M) > \varepsilon$ for all $i \in A^c$. Using (4.48) we find that

$$
\begin{aligned}
P_i(V_A > 2M) &= P_i(X_1 \notin A, \ldots, X_M \notin A, X_{M+1} \notin A, \ldots, X_{2M} \notin A) \\
&= \sum_{j \in A^c} P_i(X_1 \notin A, \ldots, X_{M-1} \notin A, X_M = j,
\end{aligned}
$$

$$X_{M+1} \notin A, \ldots, X_{2M} \notin A)$$

$$= \sum_{j \in A^c} P_i(V_A > M, X_M = j) P_j(V_A > M)$$

$$\leq P_i(V_A > M) \max_{j \in A^c} P_j(V_A > M)$$

$$< (1 - \varepsilon)^2 \tag{4.87}$$

for all $i \in A^c$, hence for all $i \in S$. A similar argument gives (4.86). ♠

We first consider absorption probabilities.

Theorem 4.2.7. *Let $\boldsymbol{P}$ be the one-step transition matrix for a Markov chain $\{X_n\}_{n \geq 0}$ in S, let A and B be nonempty mutually exclusive (i.e., $A \cap B = \varnothing$) subsets of S such that $(A \cup B)^c$ is finite, and assume that*

$$P_i(V_A \wedge V_B < \infty) > 0, \qquad i \in (A \cup B)^c. \tag{4.88}$$

Let $g : S \mapsto [0, \infty)$ be a function satisfying $g = 1$ on A, $g = 0$ on B, and

$$g(i) = \sum_{j \in S} g(j) P(i, j), \qquad i \in (A \cup B)^c. \tag{4.89}$$

Then g is given by

$$g(i) = P_i(V_A < V_B), \qquad i \in S. \tag{4.90}$$

Proof. Define $N := V_A \wedge V_B = V_{A \cup B}$, and assume that $X_0 = i \in (A \cup B)^c$. By Lemma 3.2.1 on p. 99, $\{g(X_{n \wedge N})\}_{n \geq 0}$ is a martingale with respect to $\{X_n\}_{n \geq 0}$ because

$$\begin{aligned}
E_i[g(X_{n+1}) \mid X_0, \ldots, X_n] &= \sum_{j \in S} g(j) P_i(X_{n+1} = j \mid X_0, \ldots, X_n) \\
&= \sum_{j \in S} g(j) P(X_n, j) \\
&= g(X_n) \quad \text{on} \quad \{N \geq n+1\} \tag{4.91}
\end{aligned}$$

for all $n \geq 0$. By Lemma 4.2.6, $P_i(N < \infty) = 1$, and g is bounded because $(A \cup B)^c$ is finite. Therefore, the optional stopping theorem (Theorem 3.2.2 on p. 100) implies that

$$g(i) = E_i[g(X_0)] = E_i[g(X_N)] = P_i(V_A < V_B). \tag{4.92}$$

The same must therefore also hold for all $i \in S$. ♠

The next result deals with mean absorption times, among other things.

Theorem 4.2.8. *Let $\boldsymbol{P}$ be the one-step transition matrix for a Markov chain $\{X_n\}_{n \geq 0}$ in S, let A be a nonempty subset of S such that A^c is finite, and* *assume that*

$$P_i(V_A < \infty) > 0, \qquad i \in A^c. \tag{4.93}$$

Let $g : S \mapsto [0, \infty)$ and $h : A^c \mapsto [0, \infty)$ be functions satisfying $g = 0$ on A and

$$g(i) = \sum_{j \in S} g(j) P(i, j) + h(i), \qquad i \in A^c. \tag{4.94}$$

Then g is given by

$$g(i) = \mathrm{E}_i \left[\sum_{n=0}^{V_A - 1} h(X_n) \right], \qquad i \in A^c. \tag{4.95}$$

Proof. Define $N := V_A$, and assume that $X_0 = i \in A^c$. By Lemma 3.2.1 on p. 99,

$$g(X_{n \wedge N}) + \sum_{l=0}^{n \wedge N - 1} h(X_l), \qquad n \geq 0, \tag{4.96}$$

is a martingale with respect to $\{X_n\}_{n \geq 0}$ because

$$\mathrm{E}_i \left[g(X_{n+1}) + \sum_{l=0}^{n} h(X_l) \,\bigg|\, X_0, \ldots, X_n \right]$$

$$= \sum_{j \in S} g(j) P(X_n, j) + h(X_n) + \sum_{l=0}^{n-1} h(X_l)$$

$$= g(X_n) + \sum_{l=0}^{n-1} h(X_l) \quad \text{on} \quad \{N \geq n+1\} \tag{4.97}$$

for all $n \geq 0$. By Lemma 4.2.6, $\mathrm{E}_i[N] < \infty$, and g and h are bounded because A^c is finite. Therefore, the optional stopping theorem (Theorem 3.2.2 on p. 100) implies that

$$g(i) = \mathrm{E}_i[g(X_0)] = \mathrm{E}_i \left[g(X_N) + \sum_{l=0}^{N-1} h(X_l) \right] = \mathrm{E}_i \left[\sum_{l=0}^{N-1} h(X_l) \right], \tag{4.98}$$

as required.   ♠

Example 4.2.9. *Markovian betting systems.* There is a large class of betting systems that are intended for games with even-money payoffs. Let $X_1, X_2, \ldots$ be a sequence of i.i.d. random variables with common distribution (4.6) on p. 121, where $0 < p < 1$, with X_n representing the gambler's profit per unit bet at coup n. (For simplicity, we exclude the possibility of pushes.) A betting system consists of a sequence of bet sizes, $B_1, B_2, \ldots$, with B_n representing the amount bet at coup n and depending only on the results of the previous coups, $X_1, \ldots, X_{n-1}$. More precisely,

$$B_1 = b_1 \geq 0, \qquad B_n = b_n(X_1, \ldots, X_{n-1}) \geq 0, \quad n \geq 2, \qquad (4.99)$$

where b_1 is a constant and b_n is a nonrandom function of $n-1$ variables for each $n \geq 2$.

Most such betting systems are Markovian, that is, there exists a Markov chain $\{Y_n\}_{n \geq 0}$ in a countable state space S and a function $b : S \mapsto [0, \infty)$ such that $B_{n+1} = b(Y_n)$ for all $n \geq 0$. We assume moreover that, given $j \in S$, there exist $j[1] \in S$ and $j[-1] \in S$ such that

$$Y_{n+1} = Y_n[X_{n+1}], \qquad n \geq 0. \qquad (4.100)$$

Once the initial state is specified, the sequence $X_1, X_2, \ldots$ determines the process completely.

Let A represent the set of states (or the single state) in S corresponding to the gambler achieving his goal using this betting system. Let B be the set of states in S corresponding to the gambler failing to achieve his goal and aborting the system. For example, if M is the house betting limit, we might define $B := \{j \in S : b(j) > M\}$. However, we do not require this and assume only that A and B are mutually exclusive with $(A \cup B)^c$ finite. Then, letting $N := V_A \wedge V_B$, we have $\mathrm{E}_i[N] < \infty$ for each $i \in S$. In fact, consider the system of linear equations

$$R(j) = pR(j[1]) + qR(j[-1]) + Q(j), \qquad j \in (A \cup B)^c, \qquad (4.101)$$

where the function $Q : S \mapsto [0, \infty)$ is specified. With $Q \equiv 0$, this system is of the form (4.89), so

$$R(i) = \mathrm{P}_i(V_A < V_B), \qquad i \in S. \qquad (4.102)$$

More generally, it is of the form (4.94) with A playing the role of $A \cup B$, and therefore

$$R(i) = \mathrm{E}_i\left[\sum_{n=0}^{N-1} Q(Y_n) \right], \qquad i \in (A \cup B)^c. \qquad (4.103)$$

For example, with $Q \equiv 1$, we have $R(i) = \mathrm{E}_i[N]$, so $R(i)$ is the expected number of coups starting in state i. With $Q \equiv b$, $R(i)$ is the expected amount bet over the course of the N coups. With $Q \equiv (1-2p)b$, $R(i)$ is the expected amount lost over the course of the N coups.

Section 8.1 has several examples to illustrate this result.   ♠

4.3 Asymptotic Behavior

Let π be a probability distribution on S satisfying

$$\pi(j) = \sum_{i \in S} \pi(i) P(i,j), \qquad j \in S. \tag{4.104}$$

Regarding π as a row vector, this condition is equivalent to $\pi = \pi P$. Iterating, we have

$$\pi = \pi P = \pi P^2 = \cdots = \pi P^n, \qquad n \geq 1. \tag{4.105}$$

In particular, if $\{X_n\}_{n \geq 0}$ is a Markov chain in S with one-step transition matrix P and if X_0 has distribution π, then X_n has distribution π for each $n \geq 1$. For this reason, a distribution π satisfying (4.104) is called a *stationary distribution* for the Markov chain.

We need one more definition to state the main result of this section. The *period* $d(i)$ of state $i \in S$ is defined to be

$$d(i) := \text{g.c.d.} D(i), \qquad D(i) := \{n \in \mathbf{N} : P^n(i,i) > 0\}, \tag{4.106}$$

where g.c.d. stands for greatest common divisor. We first notice that every state of an irreducible Markov chain has the same period.

Lemma 4.3.1. *If $i, j \in S$ are such that $f_{ij} > 0$ and $f_{ji} > 0$, then $d(i) = d(j)$.*

Proof. Let $r, s \in \mathbf{N}$ be such that $P^r(i,j) > 0$ and $P^s(j,i) > 0$. Then

$$P^{r+n+s}(i,i) \geq P^r(i,j) P^n(j,j) P^s(j,i), \qquad n \geq 0, \tag{4.107}$$

where $P^0(j,j) := 1$. This implies that $r + s \in D(i)$, hence $d(i)$ divides $r + s$. Also, if $n \in D(j)$, then $r + n + s \in D(i)$, hence $d(i)$ divides $r + n + s$ and therefore $d(i)$ divides n. Thus, $d(i)$ is a common divisor of $D(j)$, hence $d(i) \leq d(j)$. Reversing the roles of i and j gives the desired result. ♠

This allows us to speak of the *period* of an irreducible Markov chain. If the period is 1, we call the chain *aperiodic*. We can now describe the asymptotic behavior of the n-step transition probabilities of an irreducible aperiodic Markov chain.

Theorem 4.3.2. *If an irreducible aperiodic Markov chain in S with one-step transition matrix P has a stationary distribution π, then it is recurrent and*

$$\lim_{n \to \infty} P^n(i,j) = \pi(j), \qquad i, j \in S. \tag{4.108}$$

Furthermore, $\pi(i) > 0$ for all $i \in S$.

Proof. The recurrence is immediate, for if transience held, then Theorem 4.2.3 on p. 129 would tell us that $\lim_{n \to \infty} P^n(i,j) = 0$ for all $i, j \in S$. But then the identity

$$\pi(j) = \sum_{i \in S} \pi(i) P^n(i,j), \qquad n \geq 1, \; j \in S, \tag{4.109}$$

would imply, by the dominated convergence theorem (Theorem A.2.5 on p. 748), that $\pi(j) = 0$ for all $j \in S$, contradicting the assumption that π is a stationary distribution.

Next, we define the *coupled* Markov chain in $S \times S$ with one-step transition probabilities

$$Q((i,j),(k,l)) := P(i,k)P(j,l), \qquad i,j,k,l \in S. \qquad (4.110)$$

We claim that the assumptions that $\boldsymbol{P}$ is irreducible, aperiodic, and has a stationary distribution imply the same properties for $\boldsymbol{Q}$.

First, given $i \in S$, the set $\{n \in \mathbf{N} : P^n(i,i) > 0\}$ has greatest common divisor 1 and is closed under addition (since $P^{n+m}(i,i) \geq P^n(i,i)P^m(i,i)$). By a simple result from number theory (Theorem A.1.2 on p. 745), there exists $n(i) \in \mathbf{N}$ such that $P^n(i,i) > 0$ for all $n \geq n(i)$. Second, given $i,j \in S$, choose $r(i,j) \in \mathbf{N}$ such that $P^{r(i,j)}(i,j) > 0$. Then $P^n(i,j) \geq P^{n-r(i,j)}(i,i)P^{r(i,j)}(i,j) > 0$ for all $n \geq n(i) + r(i,j)$. Third, given $i,j,k,l \in S$, $Q^n((i,j),(k,l)) = P^n(i,k)P^n(j,l) > 0$ for all $n \geq [n(i) + r(i,k)] \vee [n(j) + r(j,l)]$, and it follows that Q is irreducible and aperiodic.

Next, we observe that $\rho(i,j) := \pi(i)\pi(j)$ defines a stationary distribution $\boldsymbol{\rho}$ for Q. By the first part of the proof, we conclude that the Markov chain with one-step transition matrix $\boldsymbol{Q}$ is recurrent.

Now let $\{(X_n, Y_n)\}_{n \geq 0}$ be a Markov chain in $S \times S$ with one-step transition matrix $\boldsymbol{Q}$ and with initial distribution to be specified later. We define

$$T := \min\{n \geq 1 : X_n = Y_n\} = \min_{i \in S} T_{(i,i)}. \qquad (4.111)$$

where $T_{(i,i)} := \min\{n \geq 1 : (X_n, Y_n) = (i,i)\}$, and we notice that $\mathrm{P}(T < \infty) = 1$ since $\mathrm{P}(T_{(i,i)} < \infty) = 1$ for each $i \in S$. Then

$$\begin{aligned}
\mathrm{P}(X_n = j \mid T = m, \ X_m = Y_m = i) \\
= P^{n-m}(i,j) = \mathrm{P}(Y_n = j \mid T = m, \ X_m = Y_m = i) \qquad (4.112)
\end{aligned}$$

for $1 \leq m \leq n$ and $i,j \in S$, hence

$$\begin{aligned}
\mathrm{P}(X_n = j, \ T = m, \ X_m = Y_m = i) \\
= \mathrm{P}(Y_n = j, \ T = m, \ X_m = Y_m = i). \qquad (4.113)
\end{aligned}$$

Summing over $i \in S$ and $m \in \{1, \ldots, n\}$, we find that

$$\mathrm{P}(X_n = j, \ T \leq n) = \mathrm{P}(Y_n = j, \ T \leq n) \qquad (4.114)$$

for all $j \in S$ and $n \geq 1$. It follows that

$$\begin{aligned}
\mathrm{P}(X_n = j) &= \mathrm{P}(X_n = j, \ T \leq n) + \mathrm{P}(X_n = j, \ T \geq n+1) \\
&= \mathrm{P}(Y_n = j, \ T \leq n) + \mathrm{P}(X_n = j, \ T \geq n+1)
\end{aligned}$$

$$\leq P(Y_n = j) + P(T \geq n+1) \tag{4.115}$$

and similarly $P(Y_n = j) \leq P(X_n = j) + P(T \geq n+1)$, and we conclude that

$$|P(X_n = j) - P(Y_n = j)| \leq P(T \geq n+1) \tag{4.116}$$

for all $j \in S$ and $n \geq 1$.

If we now choose the initial distribution so that

$$P(X_0 = i_1) = 1 \quad \text{and} \quad P(Y_0 = i_2) = 1, \tag{4.117}$$

we find from (4.116) that

$$|P^n(i_1, j) - P^n(i_2, j)| \leq P(T \geq n+1) \tag{4.118}$$

for all $i_1, i_2, j \in S$ and $n \geq 1$, that is, the effect of the initial state disappears in the limit. We will need this result later.

For our present purposes, if we choose the initial distribution so that

$$P(X_0 = i) = 1, \qquad P(Y_0 = l) = \pi(l), \quad l \in S, \tag{4.119}$$

we find from (4.116) that

$$|P^n(i, j) - \pi(j)| \leq P(T \geq n+1) \tag{4.120}$$

for all $i, j \in S$ and $n \geq 1$, and (4.108) follows. Finally, by (4.107) and (4.108), if $i, j \in S$, then $\pi(i) > 0$ if $\pi(j) > 0$, so since $\pi(j) > 0$ for some $j \in S$, we have $\pi(i) > 0$ for all $i \in S$. ♠

The next result is complementary to Theorem 4.3.2.

Theorem 4.3.3. *If an irreducible aperiodic Markov chain in S with one-step transition matrix $\boldsymbol{P}$ has no stationary distribution, then*

$$\lim_{n \to \infty} P^n(i, j) = 0, \qquad i, j \in S. \tag{4.121}$$

Proof. We can define the coupled chain as in the proof of Theorem 4.3.2. If it is transient, then, by Theorem 4.2.3 on p. 129,

$$\sum_{n=1}^{\infty} [P^n(i, j)]^2 = \sum_{n=1}^{\infty} Q^n((i, i), (j, j)) < \infty, \tag{4.122}$$

from which (4.121) follows. If the coupled chain is recurrent, then the proof of Theorem 4.3.2 (which because of the recurrence just assumed does not require existence of a stationary distribution) implies that (4.118) must hold for all $i_1, i_2, j \in S$ and $n \geq 1$.

We prove (4.121) in this case by contradiction. Suppose that (4.121) fails. Then there exist $j_1 \in S$ and a subsequence of the positive integers $n_{1,m}$ such that $P^{n_{1,m}}(i, j_1)$ converges to a nonzero limit for every $i \in S$. We denote this limit by $\rho(j_1) > 0$; it is independent of i by (4.118). Letting $j_1, j_2, \ldots$ be an enumeration of S, there exists a subsequence $n_{2,m}$ of $n_{1,m}$ such that $P^{n_{2,m}}(i, j_2) \to \rho(j_2) \geq 0$. Then there exists a subsequence $n_{3,m}$ of $n_{2,m}$ such that $P^{n_{3,m}}(i, j_3) \to \rho(j_3) \geq 0$. Continuing in this way, we find that "diagonal subsequence" $P^{n_{m,m}}(i, j) \to \rho(j) \geq 0$ for all $i, j \in S$, and $\rho(j_1) > 0$. By Fatou's lemma (Lemma A.2.2 on p. 748),

$$\sum_{j \in S} \rho(j) \leq \liminf_{m \to \infty} \sum_{j \in S} P^{n_{m,m}}(i, j) = 1 \qquad (4.123)$$

and

$$\begin{aligned}
\sum_{i \in S} \rho(i) P(i, j) &= \sum_{i \in S} \lim_{m \to \infty} P^{n_{m,m}}(k, i) P(i, j) \\
&\leq \liminf_{m \to \infty} \sum_{i \in S} P^{n_{m,m}}(k, i) P(i, j) \\
&= \liminf_{m \to \infty} \sum_{i \in S} P(k, i) P^{n_{m,m}}(i, j) \\
&= \sum_{i \in S} P(k, i) \rho(j) \\
&= \rho(j), \qquad j, k \in S, \qquad (4.124)
\end{aligned}$$

where the next-to-last equality uses the dominated convergence theorem (Theorem A.2.5 on p. 748). Now strict inequality in (4.124) can be ruled out because both sides sum to $\sum_{j \in S} \rho(j)$. Therefore, $\boldsymbol{\rho}$, normalized by $\sum_{j \in S} \rho(j) \in (0, 1]$, is a stationary distribution for $\boldsymbol{P}$, contradicting our assumption that no stationary distribution exists. ♠

Corollary 4.3.4. *An irreducible aperiodic Markov chain in a finite state space S has a stationary distribution.*

Proof. Let $\boldsymbol{P}$ be the one-step transition matrix, and suppose that a stationary distribution does not exist. Then Theorem 4.3.3 applies, so

$$1 = \lim_{n \to \infty} \sum_{j \in S} P^n(i, j) = \sum_{j \in S} \lim_{n \to \infty} P^n(i, j) = 0, \qquad (4.125)$$

a contradiction. Here the second equality uses the finiteness of S. ♠

Theorem 4.3.5. *Let $\{X_n\}_{n \geq 0}$ be an irreducible aperiodic recurrent Markov chain in S with one-step transition matrix $\boldsymbol{P}$. Then one of the following conclusions holds:*

(a) $E_i[T_i] < \infty$ for all $i \in S$, and $\boldsymbol{P}$ has a unique stationary distribution $\boldsymbol{\pi}$ given by

$$\pi(i) = \frac{1}{\mathrm{E}_i[T_i]}, \qquad i \in S. \tag{4.126}$$

(b) $\mathrm{E}_i[T_i] = \infty$ *for all* $i \in S$, *and* $\boldsymbol{P}$ *has no stationary distribution.*

Remark. If (a) holds, then the chain is said to be *positive recurrent*, and the conclusions of Theorem 4.3.2 hold. If (b) holds, then the chain is said to be *null recurrent*, and the conclusion of Theorem 4.3.3 hold.

Proof. Fix $i \in S$. Recall the first return time to state i, namely $T_i := \min\{n \geq 1 : X_n = i\}$. We generalize this by defining $T_i^1 := T_i$ and

$$T_i^k := \min\{n \geq 1 : X_{T_i^1 + \cdots + T_i^{k-1} + n} = i\}, \qquad k = 2, 3, \ldots, \tag{4.127}$$

so that T_i^k represents the time between the $(k-1)$th and kth returns to state i. We claim that, given $X_0 = i$, $T_i^1, T_i^2, \ldots$ is a sequence of i.i.d. random variables with common distribution that of T_i. A proof can be based on (4.48) on p. 127. For example,

$$\begin{aligned}
\mathrm{P}_i(T_i^1 = m, \ T_i^2 = n) &= \mathrm{P}_i(X_1 \neq i, \ldots, X_{m-1} \neq i, \ X_m = i, \\
&\qquad X_{m+1} \neq i, \ldots, X_{m+n-1} \neq i, \ X_{m+n} = i) \\
&= \mathrm{P}_i(X_1 \neq i, \ldots, X_{m-1} \neq i, \ X_m = i) \\
&\qquad \mathrm{P}_i(X_1 \neq i, \ldots, X_{n-1} \neq i, \ X_n = i) \\
&= \mathrm{P}_i(T_i = m) \, \mathrm{P}_i(T_i = n) \\
&= \mathrm{P}_i(T_i^1 = m) \, \mathrm{P}_i(T_i^2 = n), \qquad m, n \in \mathbf{N}. \tag{4.128}
\end{aligned}$$

Now let us define

$$N_n(i) := \sum_{l=1}^{n} 1_{\{X_l = i\}} = \max\{k \geq 0 : T_i^1 + \cdots + T_i^k \leq n\}, \quad n \geq 1. \tag{4.129}$$

Then, by the elementary renewal theorem of Example 1.5.10 on p. 47,

$$\lim_{n \to \infty} \frac{1}{n} \sum_{l=1}^{n} P^l(i, i) = \lim_{n \to \infty} \frac{\mathrm{E}_i[N_n(i)]}{n} = \frac{1}{\mathrm{E}_i[T_i]}, \tag{4.130}$$

where $1/\infty := 0$. If there is a stationary distribution $\boldsymbol{\pi}$, then Theorem 4.3.2 is applicable, so $P^n(i, i) \to \pi(i)$. Since the Cesàro averages of a convergent sequence converge to the same limit, (4.130) implies that $\mathrm{E}_i[T_i] < \infty$ and $\boldsymbol{\pi}$ is given by (4.126). If there is no stationary distribution, then Theorem 4.3.3 is applicable, so $P^n(i, i) \to 0$. By the same argument, (4.130) implies that $\mathrm{E}_i[T_i] = \infty$. ♠

Example 4.3.6. *A 64-state Markov chain in craps, continued.* The Markov chain of Example 4.1.4 on p. 123 is finite, irreducible, and aperiodic. Consequently, it has a unique stationary distribution $\boldsymbol{\pi}$ and $\lim_{n \to \infty} P^n(A, B) =$

$\pi(B)$ for all $A, B \subset \mathscr{P}$. We could find $\boldsymbol{\pi}$ by solving a system of 64 linear equations in 64 variables, but there is an easier way.

It will be useful to know the following result. If X is a random variable that assumes values in the collection of subsets of $\mathscr{P}$, then, given $B \subset \mathscr{P}$,

$$P(X = B) = \sum_{A:\ A \subset B} (-1)^{|B-A|} P(X \subset A). \qquad (4.131)$$

This is sometimes referred to as the *Möbius inversion formula*, but it is a simple consequence of the inclusion-exclusion law (Theorem 1.1.18 on p. 13). To see this, observe that

$$P(X = B) = P(X \subset B) - P\left(\bigcup_{b \in B} \{X \subset B - \{b\}\} \right), \qquad (4.132)$$

from which (4.131) follows.

Direct evaluation shows that

$$P_\varnothing(X_n \subset A) = \sum_{k=1}^{n} \pi_7 \left(1 - \pi_7 - \sum_{i \in \mathscr{P}-A} \pi_i \right)^{n-k}$$

$$+ \left(1 - \pi_7 - \sum_{i \in \mathscr{P}-A} \pi_i \right)^{n}, \qquad (4.133)$$

with the kth term in the sum representing the probability of 7 at the kth roll and no 7s or point numbers outside of A since. The last term represents the probability of no 7s or point numbers outside of A in n rolls. (Here π_j is from (1.11) on p. 6 and should not be confused with the stationary distribution $\boldsymbol{\pi}$.) Taking the limit as $n \to \infty$, we find that

$$\lim_{n \to \infty} P_\varnothing(X_n \subset A) = \frac{\pi_7}{\pi_7 + \sum_{i \in \mathscr{P}-A} \pi_i}, \qquad (4.134)$$

and therefore, by (4.131),

$$\pi(B) = \lim_{n \to \infty} P_\varnothing(X_n = B) = \sum_{A:\ A \subset B} (-1)^{|B-A|} \frac{\pi_7}{\pi_7 + \sum_{i \in \mathscr{P}-A} \pi_i}. \qquad (4.135)$$

Notice that $\pi(\mathscr{P})$ can be simplified slightly to

$$\pi(\mathscr{P}) = \sum_{A \subset \mathscr{P}} (-1)^{|A|} \frac{\pi_7}{\pi_7 + \sum_{i \in A} \pi_i} \approx 0.062168159, \qquad (4.136)$$

a number we saw in Example 1.2.5 on p. 18. This is not a coincidence but is due to the fact that the probability that each of the six point numbers has occurred since the most recent 7 is equal to the probability that each of the six point numbers will occur before the next 7. Mathematically, a doubly-infinite sequence of dice rolls $(\ldots, T_{-1}, T_0, T_1, \ldots)$ is *reversible*, that is, $(\ldots, T_{-1}, T_0, T_1, \ldots)$ and $(\ldots, T_1, T_0, T_{-1}, \ldots)$ have the same distribution. ♠

Example 4.3.7. *Martingale system and a Markov chain.* In Example 3.1.4 on p. 98 we introduced the martingale system. Let us consider a player who adopts this system repeatedly, restarting after each win. As before we assume that the player has unlimited resources and that there is no maximum betting limit. Then consider the Markov chain $\{X_n\}_{n \geq 0}$ in $\mathbf{Z}_+$ with initial state $X_0 = 0$ and one-step transition probabilities

$$P(i, j) := \begin{cases} p & \text{if } j = 0, \\ q & \text{if } j = i + 1, \\ 0 & \text{otherwise,} \end{cases} \tag{4.137}$$

where $0 < p < 1$ and $q := 1 - p$. The state of the chain can be interpreted as the number of losses since the last win, or as the base-2 logarithm of the amount bet at the next coup.

This Markov chain has a particularly simple structure. It is clearly irreducible and aperiodic. Observe that

$$P_0(T_0 = n) = q^{n-1} p, \qquad n \geq 1, \tag{4.138}$$

that is, the distribution of the first return time to state 0 is geometric(p), hence

$$E_0[T_0] = 1/p \tag{4.139}$$

and the chain is positive recurrent. Its unique stationary distribution $\boldsymbol{\pi}$ satisfies $\pi(0) = p$ by Theorem 4.3.5 and

$$\pi(j) = q\pi(j - 1), \qquad j \geq 1. \tag{4.140}$$

Therefore,

$$\pi(j) = q\pi(j - 1) = \cdots = q^j \pi(0) = q^j p, \qquad j \geq 0, \tag{4.141}$$

that is, the unique stationary distribution is shifted geometric(p). ♠

Example 4.3.8. *Basic example.* Here we generalize Example 4.3.7, replacing the constant success probability p by a sequence of success probabilities $p_0, p_1, p_2, \ldots$, where $0 < p_i < 1$. Of interest is the Markov chain $\{X_n\}_{n \geq 0}$ in $\mathbf{Z}_+$ with one-step transition probabilities

$$P(i, j) := \begin{cases} p_i & \text{if } j = 0, \\ q_i & \text{if } j = i + 1, \\ 0 & \text{otherwise,} \end{cases} \tag{4.142}$$

where $q_i := 1 - p_i$. Our assumptions ensure that the chain is irreducible and aperiodic. Arguing as in (4.138), we have

$$P_0(T_0 \geq n + 1) = \begin{cases} 1 & \text{if } n = 0, \\ q_0 q_1 \cdots q_{n-1} & \text{if } n \geq 1, \end{cases} \tag{4.143}$$

and the chain is recurrent if and only if

$$0 = P_0(T_0 = \infty) = \lim_{n \to \infty} P_0(T_0 \geq n), \tag{4.144}$$

that is, if and only if

$$\lim_{n \to \infty} q_0 q_1 \cdots q_n = 0. \tag{4.145}$$

It is positive recurrent if and only if

$$\sum_{n=1}^{\infty} P_0(T_0 \geq n) = E_0[T_0] < \infty, \tag{4.146}$$

that is, if and only if

$$S := \sum_{n=0}^{\infty} q_0 q_1 \cdots q_n < \infty. \tag{4.147}$$

In the positive recurrent case, the unique stationary distribution π must satisfy

$$\pi(j) = \pi(j-1) q_{j-1}, \qquad j \geq 1, \tag{4.148}$$

and therefore

$$\pi(j) = \pi(0) q_0 q_1 \cdots q_{j-1}, \qquad j \geq 1, \tag{4.149}$$

where $\pi(0) = (1 + S)^{-1}$.

Therefore, in this simple example, all three types of behavior, transience, null recurrence, and positive recurrence, are possible with a suitable choice of the parameters. ♠

Example 4.3.9. *Top-to-random shuffle.* First, let P be a doubly stochastic one-step transition matrix for an irreducible aperiodic Markov chain in a finite state space S. Then, by Theorem 4.3.2,

$$\lim_{n \to \infty} P^n(i, j) = \frac{1}{|S|}, \qquad i, j \in S. \tag{4.150}$$

Indeed, the double stochasticity of P implies that the discrete uniform distribution on S is stationary.

This result can be applied to the card shuffling model of Example 4.1.7 on p. 126, provided the irreducibility and aperiodicity conditions are verified. An example of a "shuffle" for which irreducibility fails would be the *uniform cut*, in which

$$(\Pi(1), \ldots, \Pi(N)) = (K+1, K+2, \ldots, N, 1, 2, \ldots, K), \qquad (4.151)$$

where K has the discrete uniform distribution on $\{0, 1, \ldots, N-1\}$. To see this, notice that the cyclic order of the cards is preserved by the random cut.

The conclusion (4.150), by itself, is not very useful. We would like to know how many shuffles are needed to achieve a satisfactory level of randomness.

An example of a shuffle for which we can answer this question is the *top-to-random* shuffle. In this shuffle, the top card is inserted at random in one of the N positions among the remaining $N-1$ cards. Thus,

$$(\Pi(1), \ldots, \Pi(N)) = (K, 1, 2, \ldots, K-1, K+1, \ldots, N), \qquad (4.152)$$

where K has the discrete uniform distribution on $\{1, 2, \ldots, N\}$. Let T_1 be the number of shuffles needed for a card to be inserted below the original bottom card, let T_2 be the number of additional shuffles needed for a second card to be inserted below the original bottom card, and so on. After $T_1 + \cdots + T_k$ shuffles, all $k!$ arrangements of the k cards below the original bottom card are equally likely. It follows that, after $T_1 + \cdots + T_{N-1} + 1$ shuffles, all $N!$ arrangements of the N cards are equally likely, and perfect randomness is achieved. Now $T_1, T_2, \ldots$ are independent, and T_1 is geometric($1/N$), T_2 is geometric($2/N$), and so on. In particular, by Example 1.4.3 on p. 30,

$$\mathrm{E}\left[\sum_{i=1}^{N-1} T_i + 1\right] = N\left(1 + \frac{1}{2} + \cdots + \frac{1}{N-1} + \frac{1}{N}\right) \sim N \ln N \qquad (4.153)$$

as $N \to \infty$. In fact, by Chebyshev's inequality (Lemma 1.5.2 on p. 43) and the variance of the geometric distribution (Example 1.4.9 on p. 32), we have

$$\frac{1}{N \ln N}\left(\sum_{i=1}^{N-1} T_i + 1\right) \to 1 \qquad (4.154)$$

in probability as $N \to \infty$. Thus, approximately $N \ln N$ top-to-random shuffles are needed to thoroughly mix a deck of N distinct cards. (A better approximation to the expectation is $N(\gamma + \ln N)$, where γ is *Euler's constant*: $\gamma \approx 0.577215665$.) ♠

4.4 Renewal Theorem

We recall the assumptions of Example 1.5.10 on p. 47. Specifically, $T_1, T_2, \ldots$ is a sequence of i.i.d. positive-integer-valued random variables with mean $\mu \in [1, \infty]$, $R_k := T_1 + \cdots + T_k$ for each $k \geq 1$, $R_0 := 0$, $N_n := \max\{k \geq 0 : R_k \leq n\}$ for each $n \geq 1$, and $N_0 := 0$. We think of $R_1, R_2, \ldots$ as the sequence of times at which renewals occur, so that $T_1, T_2, \ldots$ is the sequence of times between successive renewals and N_n is the number of renewals by time n. The elementary renewal theorem (Example 1.5.10 on p. 47) tells us that

$$\lim_{n\to\infty} \frac{\mathrm{E}[N_n]}{n} = \frac{1}{\mu}, \tag{4.155}$$

where $1/\infty := 0$. Here we prove the *renewal theorem*, a stronger conclusion under slightly stronger hypotheses.

Theorem 4.4.1. *Under the above assumptions, let* $p_i := \mathrm{P}(T_1 = i)$ *for each* $i \geq 1$, *and assume that*

$$\mathrm{g.c.d.}\{i \geq 1 : p_i > 0\} = 1. \tag{4.156}$$

Then

$$\lim_{n\to\infty} \mathrm{P}(R_k = n \text{ for some } k \geq 0) = \frac{1}{\mu}. \tag{4.157}$$

Remark. Noting that

$$\frac{\mathrm{E}[N_n]}{n} = \frac{1}{n} \sum_{m=1}^{n} \mathrm{P}(R_k = m \text{ for some } k \geq 0), \tag{4.158}$$

we see that (4.157) implies (4.155) because the Cesàro averages of a convergent sequence converge to the same limit. Thus, the renewal theorem is more delicate than the elementary renewal theorem.

Proof. We define the *age process* $X_0, X_1, \ldots$ by

$$X_n := n - R_{N_n}, \qquad n \geq 0. \tag{4.159}$$

It is easy to check that the age process is a Markov chain in $\mathbf{Z}_+$ with $X_0 = 0$ and one-step transition matrix $\boldsymbol{P}$ given by

$$P(i, 0) = \frac{p_{i+1}}{p_{i+1} + p_{i+2} + \cdots} = 1 - P(i, i+1), \qquad i \geq 0. \tag{4.160}$$

When a renewal occurs, the age process reverts to 0; otherwise it is incremented by one time unit. If $K := \sup\{i \geq 1 : p_i > 0\}$, then the Markov chain is irreducible in the state space $S := \{0, 1, \ldots, K-1\}$ if $K < \infty$ and in the state space $S := \mathbf{Z}_+$ if $K = \infty$. Let us define T_0 to be the first return time to state 0 and observe that

$$\mathrm{P}_0(T_0 = i) = P(0, 1)P(1, 2) \cdots P(i-2, i-1)P(i-1, 0) = p_i \tag{4.161}$$

for all $i \geq 1$, so that T_0 has the same distribution as T_1. In particular, (4.156) ensures aperiodicity of the Markov chain, and since $\mu = \mathrm{E}_0[T_0]$, the chain is positive recurrent if $\mu < \infty$ and null recurrent if $\mu = \infty$.

Assume for now that $\mu < \infty$, and let us find the unique stationary distribution $\boldsymbol{\pi}$ for the Markov chain. It must satisfy

$$\pi(i+1) = \pi(i)P(i, i+1), \qquad i \geq 0, \tag{4.162}$$

or

$$\frac{\pi(i+1)}{p_{i+2} + p_{i+3} + \cdots} = \frac{\pi(i)}{p_{i+1} + p_{i+2} + \cdots} = \cdots = \pi(0) \tag{4.163}$$

for all $i \geq 0$, from which we conclude that

$$\pi(i) = \pi(0)(p_{i+1} + p_{i+2} + \cdots) = \frac{p_{i+1} + p_{i+2} + \cdots}{\mu}, \qquad i \geq 0, \tag{4.164}$$

where we used Theorem 4.3.5 on p. 140 to conclude that $\pi(0) = 1/\mu$. In particular, by Theorem 4.3.2 on p. 137,

$$\lim_{n \to \infty} P(R_k = n \text{ for some } k \geq 0) = \lim_{n \to \infty} P_0(X_n = 0) = \pi(0) = \frac{1}{\mu}. \tag{4.165}$$

Finally, if $\mu = \infty$, then Theorem 4.3.3 on p. 139 tells us that

$$\lim_{n \to \infty} P(R_k = n \text{ for some } k \geq 0) = \lim_{n \to \infty} P_0(X_n = 0) = 0, \tag{4.166}$$

proving (4.157).   ♠

We next define the *residual-lifetime process* $Y_0, Y_1, \ldots$ by

$$Y_n := R_{N_n+1} - n, \qquad n \geq 0. \tag{4.167}$$

Further, in terms of the age process (4.159) and the residual-lifetime process (4.167), we define the *total-lifetime process* $Z_0, Z_1, \ldots$ by

$$Z_n := X_n + Y_n = R_{N_n+1} - R_{N_n}, \qquad n \geq 0. \tag{4.168}$$

Our next result derives the asymptotic distributions of the three processes.

Theorem 4.4.2. *In addition to the assumptions of Theorem 4.4.1, assume that $\mu < \infty$. Then*

$$\lim_{n \to \infty} P(X_n = i) = \frac{p_{i+1} + p_{i+2} + \cdots}{\mu}, \qquad i \geq 0, \tag{4.169}$$

$$\lim_{n \to \infty} P(Y_n = i) = \frac{p_i + p_{i+1} + \cdots}{\mu}, \qquad i \geq 1, \tag{4.170}$$

$$\lim_{n \to \infty} P(Z_n = i) = \frac{i p_i}{\mu}, \qquad i \geq 1. \tag{4.171}$$

Proof. Let $\boldsymbol{\pi}$ denote the stationary distribution (4.164). By the proof of Theorem 4.4.1 and by Theorem 4.3.2 on p. 137, $\lim_{n \to \infty} P(X_n = i) = \pi(i)$ for each $i \geq 0$, and (4.169) follows. Consequently,

$$
\lim_{n \to \infty} P(Y_n = j) = \lim_{n \to \infty} \sum_{i=0}^{\infty} P(X_n = i) P(Y_n = j \mid X_n = i)
$$

$$
= \lim_{n \to \infty} \sum_{i=0}^{\infty} P(X_n = i) \frac{p_{i+j}}{p_{i+1} + p_{i+2} + \cdots}
$$

$$
= \sum_{i=0}^{\infty} \pi(i) \frac{p_{i+j}}{p_{i+1} + p_{i+2} + \cdots}
$$

$$
= \sum_{i=0}^{\infty} \frac{p_{i+1} + p_{i+2} + \cdots}{\mu} \frac{p_{i+j}}{p_{i+1} + p_{i+2} + \cdots}
$$

$$
= \frac{p_j + p_{j+1} + \cdots}{\mu}, \qquad j \geq 1. \tag{4.172}
$$

Here the interchange of limit and sum is justified as follows:

$$
\sum_{i=0}^{\infty} |P(X_n = i) - \pi(i)| = 2 \sum_{i=0}^{\infty} (\pi(i) - P(X_n = i))^+ \to 0. \tag{4.173}
$$

The equality uses the fact that we are considering the difference of two probability distributions, and the limit follows from the dominated convergence theorem (Theorem A.2.5 on p. 748).

Finally,

$$
\lim_{n \to \infty} P(Z_n = j) = \lim_{n \to \infty} \sum_{i=0}^{j-1} P(X_n = i) P(Z_n = j \mid X_n = i)
$$

$$
= \lim_{n \to \infty} \sum_{i=0}^{j-1} P(X_n = i) \frac{p_j}{p_{i+1} + p_{i+2} + \cdots}
$$

$$
= \sum_{i=0}^{j-1} \pi(i) \frac{p_j}{p_{i+1} + p_{i+2} + \cdots}
$$

$$
= \sum_{i=0}^{j-1} \frac{p_{i+1} + p_{i+2} + \cdots}{\mu} \frac{p_j}{p_{i+1} + p_{i+2} + \cdots}
$$

$$
= \frac{j p_j}{\mu}, \qquad j \geq 1, \tag{4.174}
$$

as required.   ♠

Example 4.4.3. *Craps and renewal theory.* Craps is a good source of applications of renewal theory. In Examples 3.2.6 on p. 103 and 4.1.6 on p. 125 we found the mean and median of the length L of the shooter's hand. Specifically, $\mu := E[L] = 1{,}671/196$ and median(L) = 6. We let L play the role of T_1 of this section, so the renewal times are the rolls at which the shooter sevens out. The elementary renewal theorem tells us that the asymptotic expected proportion of rolls at which a seven out occurs is $1/\mu = 196/1{,}671 \approx 0.117295$. With $p_i := \mathrm{P}(L = i)$ for each $i \geq 1$, we have $p_2 > 0$ and $p_3 > 0$, so assumption (4.156) holds. Therefore, the renewal theorem (Theorem 4.4.1) tells us that the probability of a seven out at the nth roll converges to the same limit as $n \to \infty$.

Theorem 4.4.2 yields a conclusion that may, at first glance, seem surprising. Using the method of Example 4.1.6 on p. 125, we can compute

$$\frac{1}{\mu} \sum_{i=1}^{11} i p_i < \frac{1}{2} \quad \text{and} \quad \frac{1}{\mu} \sum_{i=1}^{12} i p_i > \frac{1}{2}. \tag{4.175}$$

It follows from (4.175) that the median length of the shooter's hand *that contains the nth roll* converges to 12 as $n \to \infty$, twice its ordinary value! The explanation is quite simple. By specifying that the hand contain the nth roll, we are choosing a hand not at random but in way that is biased in favor of longer hands. The strategy that calls for betting only on the hand that contains the nth roll for some prespecified large n is unfortunately not a viable one, for it would require prescience. ♠

Example 4.4.4. *Generalized trente et quarante and the renewal theorem.* For the purposes of this example, we need not describe the complete set of rules of the card game trente et quarante ("thirty and forty" in French). It is played with six standard 52-card decks mixed together in a shoe, but here we will assume an infinite-deck shoe (more precisely, sampling with replacement). The denominations A, 2–9, 10, J, Q, K have values 1, 2–9, 10, 10, 10, 10, respectively, and suits are irrelevant for the purpose of this example. Two rows of cards are dealt. In the first row, called Black, cards are dealt until the total value is 31 or greater. In the second row, called Red, the process is repeated. Thus, each row has associated with it a total between 31 and 40 inclusive.

Here we consider a generalization of trente et quarante, which we call $10n$ et $10(n+1)$. If $n = 3$ it coincides with trente et quarante as just described. For other values of $n \geq 1$, the rules are analogous to those of trente et quarante, except that the minimum of each row's total is $10n + 1$ instead of 31. As already noted, we assume sampling with replacement. We denote by $P_n(i)$ the probability of a row total of i $(10n + 1 \leq i \leq 10(n + 1))$. We claim that

$$\lim_{n \to \infty} P_n(10n + j) = \frac{14 - j}{85}, \qquad 1 \leq j \leq 10. \tag{4.176}$$

To see this, let $T_1, T_2, \ldots$ denote the values of the successive cards dealt, let $R_k := T_1 + \cdots + T_k$ be the cumulative total after k cards have been dealt $(k \geq 1)$, let $R_0 := 0$, and let $N_n := \max\{k \geq 0 : R_k \leq n\}$ for each $n \geq 1$. We apply the results of this section with

$$p_i := \frac{1 + 3\delta_{i,10}}{13}, \qquad 1 \le i \le 10, \tag{4.177}$$

and therefore

$$\mu = \sum_{i=1}^{10} i p_i = \frac{85}{13}. \tag{4.178}$$

By the renewal theorem (Theorem 4.4.1) we have

$$\lim_{n \to \infty} P(R_k = 10n + j \text{ for some } k \ge 0) = \frac{1}{\mu} = \frac{13}{85}, \quad 1 \le j \le 10, \tag{4.179}$$

but this is not the probability of interest (except when $j = 1$). Instead, we want the asymptotic distribution of the residual lifetime (4.167):

$$
\begin{aligned}
\lim_{n \to \infty} P_n(10n + j) &= \lim_{n \to \infty} P(R_k = 10n + j \text{ and } T_k \ge j \text{ for some } k \ge 1) \\
&= \lim_{n \to \infty} P(R_{N_{10n}+1} = 10n + j) \\
&= \lim_{n \to \infty} P(Y_{10n} = j) \\
&= \frac{p_j + \cdots + p_{10}}{\mu} \\
&= \frac{14 - j}{85}, \qquad 1 \le j \le 10. \tag{4.180}
\end{aligned}
$$

This formula was used in the late 18th century for the distribution of trente-et-quarante totals (see Section 20.4). Of course, it is incorrect for trente et quarante, even assuming sampling with replacement, but it is correct asymptotically for the game of $10n$ et $10(n + 1)$.   ♠

4.5 Problems

4.1. *Markov chains and repeated rolls of a fair die.* A fair die is rolled repeatedly. Determine which of the following are Markov chains, and for those that are, provide their state spaces and one-step transition matrices. Here $n = 0, 1, 2, \ldots$.
   (a) The largest number M_n in the first n rolls.
   (b) The number N_n of sixes in the first n rolls.
   (c) After the nth roll, the (nonnegative) number A_n of rolls since the last six (with $A_n := n$ if no sixes have appeared).
   (d) After the nth roll, the (positive) number B_n of rolls until the next six.
   (e) $C_n := A_n + B_n$.

4.2. *Non-Markovian aspect of Labouchere bet sizes.* Show that, for the initial list $l^0 := (1, 2, 3, 4)$, the sequence of bet sizes, $B_1, B_2, \ldots$, in the Labouchere system (Example 4.1.3 on p. 121) fails to satisfy the Markov property, that is, the first equation in (4.2) on p. 120 fails. (The fact that $X_0, X_1, \ldots$ in (4.2) on p. 120 is indexed by $\mathbf{Z}_+$ and $B_1, B_2, \ldots$ in (4.17) on p. 122 is indexed by $\mathbf{N}$ is insignificant.)

4.3. *Two-state Markov chains.* Consider the two-state Markov chain in $S := \{0, 1\}$ with one-step transition matrix

$$\boldsymbol{P} := \begin{pmatrix} 1-p & p \\ q & 1-q \end{pmatrix}, \tag{4.181}$$

where $0 < p < 1$ and $0 < q < 1$. Find a formula for $\boldsymbol{P}^n$, valid for each $n \geq 1$. Use the formula to determine $\lim_{n\to\infty} P^n(i, j)$ for all $i, j \in S$.

Hint: Let λ_0 and λ_1 be the eigenvalues of $\boldsymbol{P}$, let $\boldsymbol{\Lambda}$ be the 2×2 diagonal matrix with λ_0 and λ_1 on the diagonal, and find a nonsingular 2×2 matrix $\boldsymbol{B}$ such that $\boldsymbol{P} = \boldsymbol{B}\boldsymbol{\Lambda}\boldsymbol{B}^{-1}$. (The columns of $\boldsymbol{B}$ are right eigenvectors of $\boldsymbol{P}$.) Then $\boldsymbol{P}^n = \boldsymbol{B}\boldsymbol{\Lambda}^n\boldsymbol{B}^{-1}$.

4.4. *Success runs and a Markov chain.* Formulate the result of Example 1.4.25 on p. 40 as a statement about a suitably defined Markov chain. Use this formulation to recalculate the probability of at least four consecutive heads in 21 fair coin tosses.

4.5. *Martingales from Markov chains.* Let $\{X_n\}_{n\geq0}$ be a Markov chain in a countable state space S with one-step transition matrix $\boldsymbol{P}$, and define the operator P on the space of bounded real-valued functions on S by

$$(Pf)(i) := \sum_{j\in S} f(j)P(i, j), \qquad i \in S. \tag{4.182}$$

(a) Given a bounded real-valued function f on S, show that

$$M_n := f(X_n) - f(X_0) - \sum_{l=0}^{n-1}(Pf - f)(X_l) \tag{4.183}$$

is a mean-zero martingale with respect to $\{X_n\}_{n\geq0}$.

(b) Show also that

$$L_n := (M_n)^2 - \sum_{l=0}^{n-1}[P(f^2) - (Pf)^2](X_l) \tag{4.184}$$

is a mean-zero martingale with respect to $\{X_n\}_{n\geq0}$.

(c) Given a positive bounded function f on S that is bounded away from 0, show that

$$K_n := \frac{f(X_n)}{f(X_0)} \Big/ \prod_{l=0}^{n-1}\left(\frac{Pf}{f}\right)(X_l) \tag{4.185}$$

is a mean-one martingale with respect to $\{X_n\}_{n\geq 0}$.

4.6. *Super- and subharmonic functions.* Under the assumptions and notation of Problem 4.5, a bounded real-valued function f on S is said to be *superharmonic* if $f \geq Pf$ on S, *subharmonic* if $f \leq Pf$ on S, and *harmonic* if $f = Pf$ on S. Show that $\{f(X_n)\}_{n\geq 0}$ is a supermartingale if f is superharmonic, a submartingale if f is subharmonic, and a martingale if f is harmonic, in each case with respect to $\{X_n\}_{n\geq 0}$.

4.7. *Relationship between one- and two-dimensional simple symmetric random walks.* In Example 4.2.4 on p. 130 we found that the probability of returning to the origin $(0,0)$ in $2n$ steps for the two-dimensional simple symmetric random walk is the square of the corresponding probability for the one-dimensional simple symmetric random walk, that is,

$$P^{2n}((0,0),(0,0)) = [P^{2n}(0,0)]^2, \qquad n \geq 1. \tag{4.186}$$

Give a proof of this result that does not require evaluation of the $2n$-step transition probabilities.

4.8. *A criterion for recurrence.* Consider an irreducible Markov chain in a countably infinite state space S with one-step transition matrix $\boldsymbol{P}$. Show that it is recurrent if there exist a function $f : S \mapsto [0, \infty)$ and a finite set $A \subset S$ such that $\{i \in S : f(i) \leq M\}$ is finite for every positive integer M and

$$\sum_{j \in S} f(j)P(i,j) - f(i) \leq 0, \qquad i \in A^c. \tag{4.187}$$

Hint: Show that $Y_n := f(X_{n \wedge V_A})$ $(n \geq 0)$ is a supermartingale with respect to $\{X_n\}_{n\geq 0}$, hence $Y_\infty := \lim_{n\to\infty} Y_n$ exists a.s., and $\mathrm{E}_i[Y_\infty] \leq \mathrm{E}_i[Y_0] = f(i)$. Assume transience and argue that $\mathrm{P}_i(V_A < \infty) = 1$ for each $i \in S$. But then $\mathrm{P}_i(X_n \in A \text{ i.o.}) = 1$ for all $i \in S$, hence $\mathrm{P}_i(X_n = j \text{ i.o.}) = 1$ for some $i \in S$ and $j \in A$, contradicting Theorem 4.2.3 on p. 129.

4.9. *A criterion for transience.* Consider an irreducible Markov chain in a countably infinite state space S with one-step transition matrix $\boldsymbol{P}$. Show that it is transient if there exist a function $f : S \mapsto [0, \infty)$ and a finite set $A \subset S$ such that f is positive on A, $\{i \in S : f(i) \geq 1/M\}$ is finite for every positive integer M, and (4.187) holds.

Hint: Show that $Y_n := f(X_{n \wedge V_A})$ $(n \geq 0)$ is a supermartingale with respect to $\{X_n\}_{n\geq 0}$. Assume recurrence and conclude that $\min_{j \in A} f(j) \leq \mathrm{E}_i[f(X_{V_A})] \leq f(i)$ for all $i \in S$, a contradiction for some $i \in S$.

4.10. *Bold play and a Markov chain.* **Bold play** consists of betting, at each coup, either one's entire fortune or just enough to achieve one's goal in the event of a win, whichever amount is smaller. Assume a game of independent coups, each of which is won (and paid even money) with probability p and lost with probability $q := 1 - p$. Assuming a goal of m units (m a positive integer), the bold bettor's fortune is a Markov chain in $S := \{0, 1, \ldots, m\}$ with absorption at 0 and at m.

(a) Determine the one-step transition matrix $\boldsymbol{P}$.

(b) Apply Theorem 4.2.7 on p. 134 to find a system of linear equations uniquely satisfied by $Q(i)$, the probability, starting in state i, of eventually reaching state m.

(c) Apply Theorem 4.2.8 on p. 134 to find a system of linear equations uniquely satisfied by $R(i)$, the expected number of coups, starting in state i, at which a nonzero bet is made.

4.11. *Foster's criterion for positive recurrence.* Consider an irreducible aperiodic Markov chain in a countably infinite state space S with one-step transition matrix $\boldsymbol{P}$. Show that it is positive recurrent if there exist a function $f : S \mapsto [0, \infty)$, a positive number ε, and a finite set $A \subset S$ such that

$$\sum_{j \in S} f(j)P(i,j) - f(i) \leq -\varepsilon, \qquad i \in A^c, \tag{4.188}$$

and

$$\sum_{j \in S} f(j)P(i,j) < \infty, \qquad i \in A. \tag{4.189}$$

(The aperiodicity is not needed but is included because we have defined positive recurrence only in the irreducible aperiodic setting.)

Hint: Show that $Y_n := f(X_{n \wedge V_A}) + (n \wedge V_A)\varepsilon$ $(n \geq 0)$ is a supermartingale with respect to $\{X_n\}_{n \geq 0}$. This leads to $\mathrm{E}_i[V_A] \leq f(i)/\varepsilon$ for all $i \in S$. Letting $T_A := \min\{n \geq 1 : X_n \in A\}$ be the first return time to A, we have

$$\mathrm{E}_i[T_A] = \sum_{j \in A} P(i,j) + \sum_{j \notin A} P(i,j)(1 + \mathrm{E}_j[V_A])$$

$$\leq 1 + \sum_{j \notin A} P(i,j)\frac{f(j)}{\varepsilon} < \infty \tag{4.190}$$

for each $i \in A$ by (4.189), which suffices since A is finite.

4.12. *A random walk in the set of nonnegative integers.* Consider the random walk in $\mathbf{Z}_+$ with one-step transition probabilities

$$P(i,j) := \begin{cases} p & \text{if } j = (i-1)^+, \\ q & \text{if } j = i+1, \qquad i \in \mathbf{Z}_+, \\ 0 & \text{otherwise,} \end{cases} \tag{4.191}$$

where $0 < p < 1$ and $q := 1 - p$. Notice that it is irreducible and aperiodic. Show that it is transient if $p < \frac{1}{2}$, null recurrent if $p = \frac{1}{2}$, and positive recurrent if $p > \frac{1}{2}$. In the latter case determine its unique stationary distribution.

Hint: If there is a stationary distribution, consider its probability generating function.

4.13. *A Markov chain related to the Labouchere system.* Consider the Markov chain in $\mathbf{Z}_+$ with one-step transition probabilities

$$P(i,j) := \begin{cases} p & \text{if } j = (i-2)^+, \\ q & \text{if } j = i+1, \qquad i \geq 1, \\ 0 & \text{otherwise,} \end{cases} \tag{4.192}$$

where $0 < p < 1$ and $q := 1 - p$, and

$$P(0,j) := \begin{cases} 1 & \text{if } j = i_0, \\ 0 & \text{otherwise,} \end{cases} \tag{4.193}$$

where i_0 is a fixed positive integer. This is an irreducible version of $\{Y_n\}_{n \geq 0}$ in Example 4.1.3 on p. 121. (In effect, after reaching state 0, it restarts at state i_0.) Show that it is aperiodic. Show further that it is transient if $p < \frac{1}{3}$, null recurrent if $p = \frac{1}{3}$, and positive recurrent if $p > \frac{1}{3}$.

4.14. *Oscar system and positive recurrence.* In Example 4.2.5 on p. 132 we showed that the Markov chain associated with the Oscar system (Example 4.1.2 on p. 120) is transient for $0 < p < \frac{1}{2}$ and recurrent for $\frac{1}{2} \leq p < 1$. Show that it is positive recurrent for $\frac{1}{2} \leq p < 1$.

Hint for the case $p = \frac{1}{2}$: Apply Foster's criterion (Problem 4.11) with $A := \{(0,0)\}$. Since the left side of (4.188) has the form

$$\Delta f(i,j) := \sum_{(k,l) \in S} P((i,j),(k,l)) f(k,l) - f(i,j)$$

$$= \begin{cases} \Delta_1 f(i,j) + \Delta_2 f(i,j) & \text{if } (i,j) \neq (0,0), \\ f(1,1) - f(0,0) & \text{if } (i,j) = (0,0), \end{cases} \tag{4.194}$$

where

$$\Delta_1 f(i,j) := \tfrac{1}{2} f(i-j, j \wedge (i-j)) + \tfrac{1}{2} f(i+j, j) - f(i,j) \tag{4.195}$$

and

$$\Delta_2 f(i,j) := \tfrac{1}{2} f(i-j, (j+1) \wedge (i-j)) - \tfrac{1}{2} f(i-j, j \wedge (i-j)), \tag{4.196}$$

we should seek a nonnegative function f on S that is concave in the first variable and decreasing in the second variable. Try $f := f_1 + \gamma f_2$, where $f_1(i,j) := i^\alpha$ and

$$
f_2(i,j) := \begin{cases} 0 & \text{if } (i/\beta)^\beta < j \le i, \\ (i - \beta j^{1/\beta})^2 j^{-2} & \text{if } i^\beta < j \le (i/\beta)^\beta \wedge i, \\ i^{2(1-\beta)} - \beta(2-\beta)j^{2(1-\beta)/\beta} & \text{if } 1 \le j \le i^\beta, \\ 0 & \text{if } (i,j) = (0,0), \end{cases} \tag{4.197}
$$

for suitable $\alpha, \beta, \gamma \in (0,1)$. (The third line in the formula for $f_2(i,j)$ is the important one; the second line just smooths out the function.)

4.15. *Geometric rate of convergence for finite Markov chains.* Show that an irreducible aperiodic Markov chain in a finite state space S with one-step transition matrix $\boldsymbol{P}$ and stationary distribution $\boldsymbol{\pi}$ satisfies

$$
|P^n(i,j) - \pi(j)| \le C\rho^n, \qquad i,j \in S,\ n \ge 1, \tag{4.198}
$$

for some constants $C \ge 1$ and $0 < \rho < 1$.

Hint: Use (4.120) on p. 139. Apply Lemma 4.2.6 on p. 133 to the Markov chain $\{(X_n, Y_n)\}_{n\ge 0}$ of the proof of Theorem 4.3.2 on p. 137 and to the set $A := \{(j,j) : j \in S\}$.

4.16. *Parrondo's paradox.* Let $0 < \rho < 1$ and define

$$
p := \frac{1}{2} - \varepsilon, \qquad p_0 := \frac{\rho^2}{1+\rho^2} - \varepsilon, \qquad p_1 := \frac{1}{1+\rho} - \varepsilon, \tag{4.199}
$$

where $\varepsilon > 0$ is sufficiently small that $p_0 > 0$. Consider two games, A and B. In game A the player tosses a p-coin (i.e., p is the probability of heads). In game B, the player tosses a p_0-coin if his current capital is divisible by 3 and a p_1-coin otherwise. In either game he wins one unit with heads and loses one unit with tails. For simplicity, assume that initial capital is 0. *Parrondo's paradox* is that game A and game B are losing games, regardless of ε, whereas the randomly mixed game C, in which a fair coin is tossed to determine whether game A or game B is played, is a winning game for ε sufficiently small. Here "losing" and "winning" are determined by the sign of the long-term expected cumulative profit per game played. The purpose of this problem is to establish these claims.

(a) Let $\{X_n\}_{n\ge 0}$ be the Markov chain in $S = \{0,1,2\}$ tracking the player's current capital modulo 3 under the assumption that the player always plays game B. Determine the one-step transition matrix $\boldsymbol{P}_B$.

(b) Let $\boldsymbol{\pi} = (\pi_0, \pi_1, \pi_2)$ denote the unique stationary distribution for $\boldsymbol{P}_B$. Show that

$$
\pi_0 = \frac{1 - p_1(1 - p_1)}{2 + p_0 p_1^2 + (1 - p_0)(1 - p_1)^2}. \tag{4.200}
$$

(c) Show that the player's long-term expected cumulative profit per game played, under the assumption that the player always plays game B, is

$$\mu_B(\varepsilon) := \lim_{n \to \infty} \frac{1}{n} \sum_{m=1}^{n} (1,0,0) \boldsymbol{P}_B^{m-1} \begin{pmatrix} 2p_0 - 1 \\ 2p_1 - 1 \\ 2p_1 - 1 \end{pmatrix}$$

$$= \pi_0(2p_0 - 1) + (1 - \pi_0)(2p_1 - 1). \tag{4.201}$$

(d) Repeat parts (a)–(c) when the player always plays game A and when the player always plays game C. Show that, given $0 < \rho < 1$, there exists $\varepsilon_0 > 0$, depending on ρ, such that $\mu_A(\varepsilon) < 0$, $\mu_B(\varepsilon) < 0$, and $\mu_C(\varepsilon) > 0$ for $0 < \epsilon < \varepsilon_0$.

4.17. *A finite periodic Markov chain.* Let $m \geq 2$ be an integer, and consider the Markov chain in $S := \{0, 1, \ldots, m-1\} \times \{0, 1, \ldots, m-1\}$ with transition probabilities

$$P((i,j),(k,l)) \tag{4.202}$$

$$:= \begin{cases} p_i & \text{if } (k,l) = (i+1 \pmod{m}, 0) \text{ and } j \leq m-2, \\ q_i & \text{if } (k,l) = (i+1 \pmod{m}, j+1) \text{ and } j \leq m-2, \\ 1 & \text{if } (k,l) = (i+1 \pmod{m}, 0) \text{ and } j = m-1, \end{cases}$$

where $0 < p_i < 1$ and $q_i := 1 - p_i$ for $i = 0, 1, \ldots, m-1$.

(a) Show that this Markov chain is irreducible with period m.

(b) Find a stationary distribution for this Markov chain. (It can be shown to exist and be unique.)

See Problem 12.12 on p. 453 for an application.

4.18. *Renewal theory for Bernoulli trials.* Given $0 < p < 1$, let $T_1, T_2, \ldots$ be i.i.d. geometric(p). For each $n \geq 0$, find the distribution of the age X_n, the residual lifetime Y_n, and the total lifetime Z_n of Section 4.4. Confirm directly that the conclusions of Theorem 4.4.2 on p. 147 hold in this case.

4.19. *Craps and renewal theory.* Consider the craps player who makes a one-unit bet at each toss, either on the pass line (Example 1.2.2 on p. 17) or on come (Example 3.2.9 on p. 105). Let P_n be the expected number of unresolved pass-line bets following the nth roll. Let C_n be the expected number of unresolved come bets following the nth roll. Use the renewal theorem (Theorem 4.4.1 on p. 146) to find $\lim_{n \to \infty} P_n$ and $\lim_{n \to \infty} C_n$.

4.20. *Renewal sequences.* Let $(p_i)_{i \geq 1}$ be a probability distribution on $\mathbf{N}$ satisfying the aperiodicity condition (4.156) on p. 146, and define $u_0 := 1$ and

$$u_n := \sum_{i=1}^{n} p_i u_{n-i}, \qquad n \geq 1. \tag{4.203}$$

Use the renewal theorem (Theorem 4.4.1 on p. 146) to show that $\lim_{n \to \infty} u_n = 1/\mu$, where μ is the mean of the distribution $(p_i)_{i \geq 1}$ and $1/\infty := 0$.

4.21. *Simplified backgammon.* In a simplified model of backgammon called the *single-checker model*, two players take turns rolling a pair of dice, with the winner being the first to accumulate 167 or more points. At each turn the dice roller is awarded a number of points equal to the dice total or, if the two dice show the same number (*doubles*), twice the dice total.

(a) Using Wald's identity (Corollary 3.2.5 on p. 103) and Theorem 4.4.2 on p. 147, estimate the mean number of dice rolls needed for a player to accumulate 167 or more points. Can the expectation be computed exactly?

(b) Suppose a game is incomplete with the player whose turn it is to roll lacking i points and his opponent lacking j points. What is the probability that the first player (lacking i) will win? Evaluate for $i = j = 167$.

4.6 Notes

Finite Markov chains were introduced in 1906 by Andrei A. Markov [1856–1922] (1906). He proved the existence of stationary distributions. The phrase "les chaînes de Markoff" first appeared in the work of Sergei N. Bernstein [1880–1968] (1926).

Examples 4.1.2 and 4.2.5 (Oscar system) and Example 4.1.3 (Labouchere system) are discussed in greater detail in Chapter 8. Example 4.1.6 is due, to the best of our knowledge, to Peter A. Griffin (personal communication, 1987). The transience proof in Example 4.2.5 is due to Borovkov (1997). Example 4.2.4 (simple symmetric random walk in $\mathbf{Z}^d$) dates back to Georg Pólya [1887–1985] (1921).

Andrei N. Kolmogorov [1903–1987] (1936) introduced Markov chains with countably infinite state spaces, though random walks had already been studied. Theorems 4.3.2 and 4.3.3 are due to Kolmogorov (1936, 1937), though the coupling proof of Theorem 4.3.2 is due to Wolfang Doeblin [1915–1940] (1938). The Möbius inversion formula is named for August Möbius [1790–1868]. Example 4.3.8 was termed the "basic example" by Kemeny, Snell, and Knapp (1966, p. 126). Example 4.3.9 (top-to-random shuffle) is from Aldous and Diaconis (1986).

The (discrete) renewal theorem (Theorem 4.4.1) was first proved by Paul Erdős [1913–1996], William Feller [1906–1970], and Harry Pollard [1919–1985] (1949).

Most of the results in this chapter are standard textbook material. Our treatment benefitted from the texts of Billingsley (1995) and Durrett (2004). For Section 4 we used Lawler (2006, Section 6.3). Some authors (e.g., Feller 1968, Billingsley 1995) use the term "persistent" instead of "recurrent," but the latter term seems to have wider acceptance.

Problem 4.1 is from Grimmett and Stirzaker (2001, Exercise 6.1.2). Problem 4.8 is due to Mertens, Samuel-Cahn, and Zamir (1978). Problems 4.9 and 4.11 are due to Foster (1953a). Problem 4.14 is from Ethier (1996). Parrondo's paradox (Problem 4.16) is due to Juan M. R. Parrondo [1964–]. The parametrization of the probabilities in the problem in terms of ρ is from Ethier and Lee (2009); Parrondo took $\rho := 1/3$. For more information, see Harmer and Abbott (2002). Problem 4.17 is from Ethier and Lee (2010). Problem 4.19 is due to Joseph Kupka [1942–] (personal communication, c. 1986), and Problem 4.21 was motivated by Ross, Benjamin, and Munson (2007).

Chapter 5
Game Theory

It's a game where only 2 plays, and where, in coarse, when there's ony [sic] 3, one looks on.

William Makepeace Thackeray, *The Yellowplush Correspondence*

Game theory is concerned with games of strategy, which may or may not be games of chance. In Section 5.1 we introduce matrix games, with emphasis on the games of le her and chemin de fer, and in Section 5.2 we prove the fundamental theorem concerning such games, the minimax theorem. Few house-banked casino games fit into this framework because typically the dealer lacks the authority to make strategy decisions. Poker, however, does have game-theoretic aspects. Finally, in Section 5.3 we introduce the basic ideas of utility theory. It should be emphasized that this chapter is really only a very brief introduction to what has become a vast subject.

5.1 Matrix Games

Game theory is concerned with games in which each player has nontrivial strategy choices. We will limit the discussion to *two-person games*, in which there are only two players, whom we will refer to as player 1 and player 2. If player 1's profit is player 2's loss, and vice versa, the game is called a *zero-sum game*. The rules of a game specify such things as the order of play, the strategy choices for each player, and the payoffs. If the number of strategy choices for each player is finite, the game is said to be a *finite game*. A finite, two-person, zero-sum game is called a *matrix game*. In this section and the next one we study matrix games.

Games of strategy may or may not be games of chance. However, since games of chance are our primary focus, we will confine our attention to games that involve some element of chance, thereby excluding such classics

S.N. Ethier, *The Doctrine of Chances*, Probability and its Applications, DOI 10.1007/978-3-540-78783-9_5, © Springer-Verlag Berlin Heidelberg 2010

as *scissors-paper-stone*. In most casino games, the dealer plays mechanically, lacking the authority to make strategy decisions. Such games are outside the scope of game theory. Poker is the best-known example of a casino game to which game theory is applicable. Another example, not as well known, is chemin de fer.

If player 1 has m strategies, denoted by $1, 2, \ldots, m$, from which to choose, and player 2 has n strategies, denoted by $1, 2, \ldots, n$, from which to choose, then the game is an $m \times n$ matrix game and is completely described by the *payoff matrix*

$$\boldsymbol{A} = \begin{pmatrix} a_{11} & a_{12} & \cdots & a_{1n} \\ a_{21} & a_{22} & \cdots & a_{2n} \\ \vdots & \vdots & & \vdots \\ a_{m1} & a_{m2} & \cdots & a_{mn} \end{pmatrix}, \tag{5.1}$$

in which the entry a_{ij} in row i and column j represents the expected profit of player 1 (or, equivalently, the expected loss of player 2) when player 1 adopts strategy i and player 2 adopts strategy j.

If player 1 is aware that player 2 intends to use strategy j, then he will choose the strategy corresponding to the largest entry in column j. Similarly, if player 2 is aware that player 1 intends to use strategy i, then he will choose the strategy corresponding to the smallest entry in row i. However, in game theory it is assumed their neither player has knowledge of his opponent's choice.

If there exist $i^* \in \{1, 2, \ldots, m\}$ and $j^* \in \{1, 2, \ldots, n\}$ such that

$$\min_{1 \le j \le n} a_{i^* j} = a_{i^* j^*} = \max_{1 \le i \le m} a_{ij^*}, \tag{5.2}$$

that is, $a_{i^* j^*}$ is the minimum entry in row i^* and the maximum entry in column j^*, then (i^*, j^*) is said to be a *saddle point* of the matrix $\boldsymbol{A}$. Saddle points need not be unique.

If (i^*, j^*) is a saddle point of the matrix $\boldsymbol{A}$, then i^* is an optimal strategy for player 1 and j^* is an optimal strategy for player 2, in the following sense. There exists a number v $(= a_{i^* j^*})$ such that, if player 1 adopts strategy i^*, then, no matter which strategy player 2 chooses, the expected profit of player 1 is at least v, and, if player 2 adopts strategy j^*, then, no matter which strategy player 1 chooses, the expected profit of player 1 is at most v. For this reason v is called the *value* of the game.

Before considering our first example, it will be useful to introduce another definition. If $a_{kj} > a_{ij}$ (resp., $a_{kj} \ge a_{ij}$) for $j = 1, \ldots, n$, we say that strategy k for player 1 *strictly dominates* (resp., *dominates*) strategy i, or that strategy i *is strictly dominated by* (resp., *is dominated by*) strategy k. Notice that, if strategy i is strictly dominated by strategy k, player 1 can disregard strategy i, since strategy k will provide him with a greater expected profit, regardless of the strategy chosen by player 2.

Similarly, if $a_{ij} > a_{il}$ (resp., $a_{ij} \geq a_{il}$) for $i = 1, \ldots, m$, we say that strategy l for player 2 *strictly dominates* (resp., *dominates*) strategy j, or that strategy j *is strictly dominated by* (resp., *is dominated by*) strategy l. Notice that, if strategy j is strictly dominated by strategy l, player 2 can disregard strategy j, since strategy l will provide a smaller expected profit to player 1 and therefore a greater one to him, regardless of the strategy chosen by player 1.

Example 5.1.1. *Competing subsets.* Consider a contrived game we refer to as *competing subsets*. Player 1's strategies correspond to the nonempty subsets $S \subset \{1, 2, \ldots, N\}$, where $N \geq 3$ is an integer, and player 2's strategies correspond to the nonempty subsets $T \subset \{1, 2, \ldots, N\}$. Thus, we have a $(2^N - 1) \times (2^N - 1)$ matrix game. The resolution of the game depends on two independent random variables X and Y, both discrete uniform on $\{1, 2, \ldots, N\}$. Player 1 wins one unit from player 2 if $X \in S$ and $Y \notin T$, or if $X \in S, Y \in T$, and $|S| < |T|$. Player 2 wins one unit from player 1 if $X \notin S$ and $Y \in T$, or if $X \in S, Y \in T$, and $|S| > |T|$. In all other cases, no money changes hands.

For example, if $N = 38$, the randomness can be modeled by the results of spinning two independent 38-number roulette wheels. First, each player, without the other's knowledge, chooses a nonempty subset of the 38 numbers on which to bet. Then both wheels are spun. A player wins one unit from his opponent if one of his numbers occurs on the spin of his wheel but none of his opponent's numbers occurs on the spin of his opponent's wheel, or if one of his numbers occurs on the spin of his wheel, one of his opponent's numbers occurs on the spin of his opponent's wheel, and he bet on fewer numbers than did his opponent.

Let us find the payoff matrix $\boldsymbol{A}$ for this game (with arbitrary N). The entry a_{ST}, corresponding to player 1's expected profit when he uses strategy S and player 2 uses strategy T, is given by

$$a_{ST} = \begin{cases} |S|/N - (1 - |S|/N)(|T|/N) & \text{if } |S| < |T|, \\ (|S|/N)(1 - |T|/N) - (1 - |S|/N)(|T|/N) & \text{if } |S| = |T|, \\ (|S|/N)(1 - |T|/N) - |T|/N & \text{if } |S| > |T|. \end{cases} \quad (5.3)$$

For example, if $|S| < |T|$, then player 1 wins one unit with probability $|S|/N$ and loses one unit with probability $(1 - |S|/N)(|T|/N)$. Notice that $a_{ST} = 0$ if $|S| = |T|$. Furthermore, since a_{ST} depends on S and T only through $|S|$ and $|T|$, we see that only the sizes of S and T matter, and so strategies can be identified with elements of $\{1, 2, \ldots, N\}$. Thus, we have an $N \times N$ matrix game.

Let $\boldsymbol{B}$ be the payoff matrix for the game with this understanding. Then the entry b_{ij}, corresponding to player 1's expected profit when he uses strategy i and player 2 uses strategy j, is given by

$$b_{ij} = \begin{cases} i/N - (1 - i/N)(j/N) & \text{if } i < j, \\ 0 & \text{if } i = j, \\ (i/N)(1 - j/N) - j/N & \text{if } i > j. \end{cases} \qquad (5.4)$$

Notice that $b_{ij} = -b_{ji}$ for $i, j = 1, 2, \ldots, N$. A game whose payoff matrix satisfies this condition is called *symmetric*. The intuitive meaning is that the game is unchanged if players 1 and 2 exchange roles.

It is easy to verify that, for $N = 3$, 4, and 6, the matrix $\boldsymbol{B}$ has a saddle point (see Problem 5.7 on p. 194). Therefore, we specialize to the simplest nontrivial case, namely $N = 5$. It is often convenient to label the rows and columns of the payoff matrix (or a fixed multiple of the payoff matrix) by the strategies of players 1 and 2, respectively. Here the payoff matrix, multiplied by 25, has the form

$$\begin{array}{c} \\ 1 \\ 2 \\ 3 \\ 4 \\ 5 \end{array} \begin{array}{ccccc} 1 & 2 & 3 & 4 & 5 \\ \left(\begin{array}{ccccc} 0 & -3 & -7 & -11 & -15 \\ 3 & 0 & 1 & -2 & -5 \\ 7 & -1 & 0 & 7 & 5 \\ 11 & 2 & -7 & 0 & 15 \\ 15 & 5 & -5 & -15 & 0 \end{array} \right). \end{array} \qquad (5.5)$$

The first observation is that strategy 1 for player 1 is strictly dominated by strategy 2 (or 3) for player 1. Similarly, or by symmetry, strategy 1 for player 2 is strictly dominated by strategy 2 (or 3) for player 2. Therefore, the first row and first column of the payoff matrix can be deleted, as they correspond to strategies that would never be chosen. The resulting payoff matrix, again multiplied by 25, has the form

$$\begin{array}{c} \\ 2 \\ 3 \\ 4 \\ 5 \end{array} \begin{array}{cccc} 2 & 3 & 4 & 5 \\ \left(\begin{array}{cccc} 0 & 1 & -2 & -5 \\ -1 & 0 & 7 & 5 \\ 2 & -7 & 0 & 15 \\ 5 & -5 & -15 & 0 \end{array} \right). \end{array} \qquad (5.6)$$

No further reductions are possible through the use of strict dominance because each row contains the largest element of some column.

In Example 5.2.5 on p. 180 we will find the optimal strategies for players 1 and 2.   ♠

Example 5.1.2. *Le her.* The game of le her ("the gentleman" in 17th-century French) is a two-person game played with a standard 52-card deck, and we will continue to refer to the two players as player 1 and player 2. Cards are ranked from lowest to highest in the order A, 2, 3, $\ldots$, 10, J, Q, K, and suits are ignored. A card is dealt face down to each player, and each player may look only at his own card. The object of the game is to have the higher-ranking card at the end of play. First, player 1, if he is not satisfied with his card, can require that player 2 exchange cards with him. The only exception to this rule occurs when player 2 has a king (K), in which case the exchange is void. Second, player 2, if he is not satisfied with his card, whether it be his original card or a new card obtained in exchange with player 1, can exchange it for the next card in the deck. The only exception to this rule occurs when the next card is a king, in which case the exchange is void. This completes the game, and the winner is the player with the higher-ranked card, with player 2 winning in the case of a tie. The game pays even money.

It will be convenient to define the *ranks* of the cards A, 2, 3, ..., 10, J, Q, K as 1, 2, 3, ..., 10, 11, 12, 13, respectively. Let us denote by X, Y, and Z the ranks of the card dealt to player 1, the card dealt to player 2, and the next card in the deck, respectively. Player 1's strategies correspond to the subsets $S \subset \{1, 2, \ldots, 13\}$. Given such an S, player 1 exchanges his card with that of player 2 if and only if $X \in S$. Player 2's strategies correspond to the subsets $T \subset \{1, 2, \ldots, 13\}$. Given such a T, if player 1 fails to exchange his card with that of player 2, player 2 exchanges his card with the next card in the deck if and only if $Y \in T$. Of course, if player 1 exchanges his card with that of player 2, then player 2's decision is clear: He keeps his new card if $X \geq Y$ and exchanges it for the next card in the deck otherwise. Thus, we have a $2^{13} \times 2^{13}$ matrix game.

It is intuitively clear that the only reasonable strategies are of the form $S_i := \{1, \ldots, i\}$ and $T_j := \{1, \ldots, j\}$ for $i, j = 0, 1, \ldots, 13$ (of course, $S_0 = T_0 := \varnothing$). It can be shown that every other strategy is strictly dominated by at least one of these (see Problem 5.9 on p. 194).

Let B_{ij} denote the event that player 1 wins when player 1 uses strategy S_i and player 2 uses strategy T_j for $i, j = 0, 1, \ldots, 13$. We evaluate $\mathrm{P}(B_{ij})$ by conditioning on $\{X = k, Y = l\}$. There are three cases to consider.

Case 1. $k \leq i$. Here player 1 exchanges his card with that of player 2, provided player 2 does not have a king. The only case in which player 1 can win is $k < l < 13$, which forces player 2 to exchange his new card with the next card in the deck. Player 1 wins if $Z < l$ or if $Z = 13$. Therefore,

$$\mathrm{P}(B_{ij} \mid X = k, \, Y = l) = \mathrm{P}(Z < l \text{ or } Z = 13 \mid X = k, \, Y = l) 1_{\{k < l < 13\}}$$
$$= \left(\frac{4(l-1) - 1}{50} + \frac{4}{50} \right) 1_{\{k < l < 13\}}$$
$$= \frac{4l - 1}{50} 1_{\{k < l < 13\}}. \tag{5.7}$$

Case 2. $k > i$, $l \leq j$. Here player 1 keeps his card, while player 2 exchanges his card with the next card in the deck. Player 1 wins if $Z < k$ or if $Z = 13$ and $k > l$. Therefore,

$$\mathrm{P}(B_{ij} \mid X = k, \, Y = l) = \mathrm{P}(Z < k \mid X = k, \, Y = l)$$
$$+ \mathrm{P}(Z = 13 \mid X = k, \, Y = l) 1_{\{k > l\}}$$

$$= \frac{4(k-1) - 1_{\{k>l\}}}{50} + \frac{4 - \delta_{k,13}}{50} 1_{\{k>l\}}$$
$$= \frac{4(k-1) + (3 - \delta_{k,13}) 1_{\{k>l\}}}{50}. \tag{5.8}$$

Case 3. $k > i$, $l > j$. Here both players keep their cards, so

$$P(B_{ij} \mid X = k, \, Y = l) = 1_{\{k>l\}}. \tag{5.9}$$

It follows that

$$P(B_{ij}) = \sum_{k=1}^{i} \sum_{l=1}^{13} \frac{4}{52} \frac{4}{51} \frac{4l-1}{50} 1_{\{k<l<13\}}$$

$$+ \sum_{k=i+1}^{13} \sum_{l=1}^{j} \frac{4}{52} \frac{4 - \delta_{k,l}}{51} \frac{4(k-1) + (3 - \delta_{k,13}) 1_{\{k>l\}}}{50}$$

$$+ \sum_{k=i+1}^{13} \sum_{l=j+1}^{13} \frac{4}{52} \frac{4}{51} 1_{\{k>l\}}. \tag{5.10}$$

Of course, this formula could be further simplified. For example, the first double sum could be written as a cubic polynomial in i. However, there is no need to do this, since our only concern is with the numerical evaluation of (5.10), and this is most reliably done by computer. The payoff matrix $\boldsymbol{A}$ for this game has (i, j) entry

$$a_{ij} = 2P(B_{ij}) - 1. \tag{5.11}$$

The full matrix, multiplied by $(52)_3/2^3 = 16{,}575$, is displayed in Table 5.1.

The payoff matrix can be reduced considerably using strict dominance. Examining it, we see that strategies 0–4 for player 1 are strictly dominated by strategy 5 for player 1, and that strategies 8–13 for player 1 are strictly dominated by strategy 7 for player 1. Eliminating the strictly dominated rows, we are left with the 3×14 payoff matrix corresponding to the shaded rows in Table 5.1.

Next, within the shaded rows in Table 5.1 strategies 0–6 for player 2 are strictly dominated by strategy 7 for player 2, and strategies 9–13 for player 2 are strictly dominated by strategy 8 (or 7) for player 2. Eliminating the strictly dominated columns, we are left with

$$\begin{array}{c} \\ 5 \\ 6 \\ 7 \end{array} \begin{array}{cc} 7 & 8 \\ \begin{pmatrix} 105 & 221 \\ 393 & 429 \\ 453 & 393 \end{pmatrix} \end{array}. \tag{5.12}$$

Table 5.1 Payoff matrix, multiplied by 16,575, for the game of le her. All rows except the shaded ones are strictly dominated.

	0	1	2	3	4	5	6	7	8	9	10	11	12	13
0	−975	−1,987	−2,815	−3,459	−3,919	−4,195	−4,287	−4,195	−3,919	−3,459	−2,815	−1,987	−975	225
1	213	−799	−1,627	−2,271	−2,731	−3,007	−3,099	−3,007	−2,731	−2,271	−1,627	−799	213	1,413
2	1,173	333	−507	−1,167	−1,643	−1,935	−2,043	−1,967	−1,707	−1,263	−635	177	1,173	2,357
3	1,889	1,205	521	−163	−671	−995	−1,135	−1,091	−863	−451	145	925	1,889	3,041
4	2,345	1,801	1,257	713	169	−203	−391	−395	−215	149	697	1,429	2,345	3,449
5	2,525	2,105	1,685	1,265	845	425	173	105	221	521	1,005	1,673	2,525	3,565
6	2,413	2,101	1,789	1,477	1,165	853	541	393	429	649	1,053	1,641	2,413	3,373
7	1,993	1,773	1,553	1,333	1,113	893	673	453	393	517	825	1,317	1,993	2,857
8	1,249	1,105	961	817	673	529	385	241	97	109	305	685	1,249	2,001
9	165	81	−3	−87	−171	−255	−339	−423	−507	−591	−523	−271	165	789
10	−1,275	−1,315	−1,355	−1,395	−1,435	−1,475	−1,515	−1,555	−1,595	−1,635	−1,675	−1,567	−1,275	−795
11	−3,087	−3,099	−3,111	−3,123	−3,135	−3,147	−3,159	−3,171	−3,183	−3,195	−3,207	−3,219	−3,087	−2,767
12	−5,287	−5,287	−5,287	−5,287	−5,287	−5,287	−5,287	−5,287	−5,287	−5,287	−5,287	−5,287	−5,287	−5,143
13	−7,687	−7,687	−7,687	−7,687	−7,687	−7,687	−7,687	−7,687	−7,687	−7,687	−7,687	−7,687	−7,687	−7,687

Finally, in (5.12) strategy 5 for player 1 is strictly dominated by strategy 6 for player 1, so we end up with the 2×2 payoff matrix, multiplied by 16,575, of

$$
\begin{array}{cc}
 & 7 \quad 8 \\
\begin{array}{c} 6 \\ 7 \end{array} & \begin{pmatrix} 393 & 429 \\ 453 & 393 \end{pmatrix}.
\end{array}
\qquad (5.13)
$$

In Example 5.2.6 on p. 182 we will find the optimal strategies for players 1 and 2.   ♠

The next example is even more complicated, but it can be simplified considerably with the aid of the following lemma.

Lemma 5.1.3. *Let $m \geq 2$ and $n \geq 1$ and consider an $m \times 2^n$ matrix game of the following form. Player 1 has m strategies, labeled $1, 2, \ldots, m$. Player 2 has 2^n strategies, labeled by the subsets $T \subset \{1, 2, \ldots, n\}$. Furthermore, for $i = 1, \ldots, m$, there exist $p_i(0) \geq 0$, $p_i(1) > 0$, $\ldots$, $p_i(n) > 0$ with $p_i(0) + p_i(1) + \cdots + p_i(n) = 1$ together with a real number $a_i(0)$, and for $l = 1, 2, \ldots, n$, there exist $m \times 2$ payoff matrices*

$$
\begin{pmatrix}
a_{11}(l) & a_{12}(l) \\
\vdots & \vdots \\
a_{m1}(l) & a_{m2}(l)
\end{pmatrix}
\qquad (5.14)
$$

such that the $m \times 2^n$ matrix game has payoff matrix with (i, T) entry given by

$$
a_{i,T} := p_i(0)a_i(0) + \sum_{l \in T} p_i(l)a_{i1}(l) + \sum_{l \in T^c} p_i(l)a_{i2}(l)
\qquad (5.15)
$$

for $i \in \{1, 2, \ldots, m\}$ and $T \subset \{1, 2, \ldots, n\}$. Here $T^c := \{1, 2, \ldots, n\} - T$. We define

$$
\begin{aligned}
T_1 &:= \{1 \leq l \leq n : a_{i1}(l) < a_{i2}(l) \text{ for } i = 1, 2, \ldots, m\}, \\
T_2 &:= \{1 \leq l \leq n : a_{i1}(l) > a_{i2}(l) \text{ for } i = 1, 2, \ldots, m\}, \qquad (5.16) \\
T_3 &:= \{1, 2, \ldots, n\} - T_1 - T_2,
\end{aligned}
$$

and put $n_0 := |T_3|$. Then, given $T \subset \{1, 2, \ldots, n\}$, strategy T is strictly dominated unless $T_1 \subset T \subset T_2^c$. Thus, the $m \times 2^n$ matrix game can be reduced to an $m \times 2^{n_0}$ matrix game.

Remark. The game can be thought of as follows. Player 1 chooses a strategy $i \in \{1, 2, \ldots, m\}$. Let Z_i be a random variable with distribution $P(Z_i = l) = p_i(l)$ for $l = 0, 1, \ldots, n$. Given that $Z_i = 0$, the game is over and player 1's conditional expected profit is $a_i(0)$. If $Z_i \in \{1, 2, \ldots, n\}$, then player 2 observes Z_i (but not i) and based on this information chooses a "move" $j \in \{1, 2\}$. Given that $Z_i = l$ and player 2 chooses move 1 (resp., move 2), player 1's conditional expected profit is $a_{i1}(l)$ (resp., $a_{i2}(l)$). Thus, player

2's strategies can be identified with subsets $T \subset \{1, 2, \ldots, n\}$, with player 2 choosing move 1 if $Z_i \in T$ and move 2 if $Z_i \notin T$.

Proof. Suppose that the condition $T_1 \subset T \subset T_2^c$ fails. There are two cases. In case 1, there exists $l_0 \in T_1$ with $l_0 \notin T$. Here define $T' := T \cup \{l_0\}$. In case 2, there exists $l_0 \in T$ with $l_0 \notin T_2^c$ (or $l_0 \in T_2$). Here define $T' := T - \{l_0\}$. Then, for $i = 1, 2, \ldots, m$,

$$
\begin{aligned}
a_{i,T'} &= p_i(0)a_i(0) + \sum_{l \in T'} p_i(l)a_{i1}(l) + \sum_{l \in (T')^c} p_i(l)a_{i2}(l) \\
&= p_i(0)a_i(0) + \sum_{l \in T} p_i(l)a_{i1}(l) + \sum_{l \in T^c} p_i(l)a_{i2}(l) \\
&\quad \pm p_i(l_0)(a_{i1}(l_0) - a_{i2}(l_0)) \\
&< p_i(0)a_i(0) + \sum_{l \in T} p_i(l)a_{i1}(l) + \sum_{l \in T^c} p_i(l)a_{i2}(l) \\
&= a_{i,T},
\end{aligned}
\tag{5.17}
$$

where the $\pm$ sign is a plus sign in case 1 and a minus sign in case 2. This tells us that strategy T for player 2 is strictly dominated by strategy T', as required. ♠

Example 5.1.4. *Chemin de fer.* The game of chemin de fer ("railway" in French) is a variant of baccarat that is still played in Monte Carlo. However, present-day rules are more restrictive than they once were, and it will serve our purposes to consider the game as it was played early in the 20th century. Chemin de fer is a two-person game played with a six-deck shoe comprising six standard 52-card decks, hence 312 cards. We will refer to player 1 and player 2 as player and banker, respectively. Denominations A, 2–9, 10, J, Q, K have values 1, 2–9, 0, 0, 0, 0, respectively. The value of a hand, consisting of two or three cards, is the sum of the values of the cards, modulo 10. In other words, only the final digit of the sum is used to evaluate a hand. For example, $5 + 7 \equiv 2 \pmod{10}$ and $5 + 7 + 9 \equiv 1 \pmod{10}$.

Two cards are dealt face down to player and two to banker, and each may look only at his own hand. The object of the game is to have the higher-valued hand (closer to 9) at the end of play. A two-card hand of value 8 or 9 is a *natural*. If either hand is a natural, the game is over and the higher-valued hand wins. Hands of equal value result in a push (no money changes hands). If neither hand is a natural, player then has the option of drawing a third card. If he exercises this option, his third card is dealt face up. Next, banker has the option of drawing a third card. This completes the game, and the higher-valued hand wins. A win for player pays even money. Again, hands of equal value result in a push.

Since nonplayers can bet on player's hand, player's strategy is restricted. He must draw to a hand valued 4 or less and stand on a hand valued 6 or 7. When his hand has value 5, he is free to draw or stand as he chooses. Banker, on whose hand no one can bet, has no restrictions on his strategy. Bets on player's hand pay even money. (We again emphasize that current rules are more restrictive as regards banker's strategy.)

To keep calculations to a minimum, we will assume that cards are dealt *with replacement*, recognizing that this is only an approximation. The probability that a two-card hand has a value of 0 is

$$\left(\frac{4}{13}\right)^2 + 9\left(\frac{1}{13}\right)^2 = \frac{25}{169}, \tag{5.18}$$

as we see by conditioning on the value of the first card dealt, while the probability that a two-card hand has the value $i \in \{1, 2, \ldots, 9\}$ is

$$\frac{4}{13} \cdot \frac{1}{13} + \frac{1}{13} \cdot \frac{4}{13} + 8\left(\frac{1}{13}\right)^2 = \frac{16}{169}. \tag{5.19}$$

Let X denote the value of player's two-card hand and let Y denote the value of banker's two-card hand. On the event $\{X \leq 7,\ Y \leq 7\}$, let X_3 denote the value of player's third card if he draws, and let $X_3 := \varnothing$ if he stands. Similarly, let Y_3 denote the value of banker's third card if he draws, and let $Y_3 := \varnothing$ if he stands. As the rules specify, player has only two strategies, which we will denote by $S_5 := \{0, 1, 2, 3, 4, 5\}$ (draw to 5) and $S_4 := \{0, 1, 2, 3, 4\}$ (stand on 5). In general, $S \subset \{0, 1, \ldots, 7\}$ denotes the strategy in which player draws if $X \in S$ and stands otherwise. Banker, on the other hand, has a strategy for each subset $T \subset \{0, 1, \ldots, 7\} \times \{0, 1, \ldots, 9, \varnothing\}$. Specifically, suppose that $X \leq 7$ and $Y \leq 7$. Then banker draws if $(Y, X_3) \in T$ and stands otherwise. It follows that chemin de fer is a 2×2^{88} matrix game.

Our first step is to show that Lemma 5.1.3 applies, allowing us to reduce the game to a much more manageable 2×2^4 matrix game. Let us denote by G_{ST} player's profit from a one-unit bet when he adopts strategy S and banker adopts strategy T, so that $a_{ST} := \mathrm{E}[G_{ST}]$ is the (S, T) entry in the payoff matrix. Then

$$\begin{aligned}
a_{ST} &= \mathrm{E}[G_{ST}] \\
&= \mathrm{P}(X \in \{8, 9\},\ X > Y) - \mathrm{P}(Y \in \{8, 9\},\ Y > X) \\
&\quad + \mathrm{E}[G_{ST}\,1_{\{X \leq 7,\ Y \leq 7\}}] \\
&= \mathrm{E}[G_{ST}\,1_{\{X \leq 7,\ Y \leq 7\}}] \tag{5.20} \\
&= \sum_{j=0}^{7}\sum_{k=0}^{9} \mathrm{P}(X \in S,\ Y = j,\ X_3 = k)\mathrm{E}[G_{ST} \mid X \in S,\ Y = j,\ X_3 = k] \\
&\quad + \sum_{j=0}^{7} \mathrm{P}(X \in S^c,\ Y = j,\ X_3 = \varnothing)\mathrm{E}[G_{ST} \mid X \in S^c,\ Y = j,\ X_3 = \varnothing]
\end{aligned}$$

for $S = S_5$ and $S = S_4$, and for $T \subset \{0, 1, \ldots, 7\} \times \{0, 1, \ldots, 9, \varnothing\}$, where $S^c := \{0, 1, \ldots, 7\} - S$.

Let us now define, for $j \in \{0, 1, \ldots, 7\}$ and $k \in \{0, 1, \ldots, 9\}$,

$$a_{S,l}(j, k) := \mathrm{E}[G_{ST} \mid X \in S, \ Y = j, \ X_3 = k]$$
$$a_{S,l}(j, \varnothing) := \mathrm{E}[G_{ST} \mid X \in S^c, \ Y = j, \ X_3 = \varnothing] \tag{5.21}$$

for $S = S_5$ and $S = S_4$; $l = 1$ if (j, k) (or $(j, \varnothing)$) belongs to T; and $l = 2$ if (j, k) (or $(j, \varnothing)$) belongs to $T^c := (\{0, 1, \ldots, 7\} \times \{0, 1, \ldots, 9, \varnothing\}) - T$. Defining also

$$p_S(j, k) := \mathrm{P}(X \in S, \ Y = j, \ X_3 = k),$$
$$p_S(j, \varnothing) := \mathrm{P}(X \in S^c, \ Y = j, \ X_3 = \varnothing), \tag{5.22}$$

we have

$$a_{ST} = \sum_{(j,k) \in T, \ k \neq \varnothing} p_S(j, k) a_{S,1}(j, k) + \sum_{(j,\varnothing) \in T} p_S(j, \varnothing) a_{S,1}(j, \varnothing) \tag{5.23}$$
$$+ \sum_{(j,k) \in T^c, \ k \neq \varnothing} p_S(j, k) a_{S,2}(j, k) + \sum_{(j,\varnothing) \in T^c} p_S(j, \varnothing) a_{S,2}(j, \varnothing),$$

which has the form (5.15) with $m = 2$, $n = 88$, $p_S(0) = \mathrm{P}(X \in \{8, 9\}$ or $Y \in \{8, 9\})$, and $a_S(0) = 0$. It remains to evaluate T_1, T_2, and T_3 of the lemma. For this we need to evaluate $a_{S,l}(j, k)$ and $a_{S,l}(j, \varnothing)$ in (5.21).

Observe that

$$a_{S,l}(j, k)$$
$$= \mathrm{E}[G_{ST} \mid X \in S, \ Y = j, \ X_3 = k]$$
$$= \sum_{i \in S} \frac{\mathrm{P}(X = i, \ Y = j, \ X_3 = k)}{\mathrm{P}(X \in S, \ Y = j, \ X_3 = k)} \mathrm{E}[G_{ST} \mid X = i, \ Y = j, \ X_3 = k]$$
$$= \sum_{i \in S} \mathrm{P}(X = i \mid X \in S) \mathrm{E}[G_{ST} \mid X = i, \ Y = j, \ X_3 = k] \tag{5.24}$$

if $k \neq \varnothing$ and that

$$a_{S,l}(j, \varnothing)$$
$$= \mathrm{E}[G_{ST} \mid X \in S^c, \ Y = j, \ X_3 = \varnothing]$$
$$= \sum_{i \in S^c} \frac{\mathrm{P}(X = i, \ Y = j, \ X_3 = \varnothing)}{\mathrm{P}(X \in S^c, \ Y = j, \ X_3 = \varnothing)} \mathrm{E}[G_{ST} \mid X = i, \ Y = j, \ X_3 = \varnothing]$$
$$= \sum_{i \in S^c} \mathrm{P}(X = i \mid X \in S^c) \mathrm{E}[G_{ST} \mid X = i, \ Y = j, \ X_3 = \varnothing]. \tag{5.25}$$

Let us define the function $M : \mathbf{Z}_+ \mapsto \{0, 1, \ldots, 9\}$ by $M(r) \equiv r \pmod{10}$. Then there are four cases to consider:

Case 1. $i \in S$, $(j, k) \in T$, $k \neq \varnothing$. Here

$$
\begin{aligned}
\mathrm{E}[G_{ST} \mid X = i, \, Y = j, \, X_3 = k] \\
= \mathrm{P}(M(i+k) > M(j+Y_3) \mid X = i, \, Y = j, \, X_3 = k) \\
- \mathrm{P}(M(i+k) < M(j+Y_3) \mid X = i, \, Y = j, \, X_3 = k) \\
= \frac{M(i+k) + 3 \cdot 1_{\{M(i+k)>j\}}}{13} - \frac{9 - M(i+k) + 3 \cdot 1_{\{M(i+k)<j\}}}{13} \\
= \frac{2M(i+k) - 9 + 3 \operatorname{sgn}(M(i+k) - j)}{13}.
\end{aligned}
\tag{5.26}
$$

Case 2. $i \in S$, $(j, k) \in T^c$, $k \neq \varnothing$. Here

$$
\begin{aligned}
\mathrm{E}[G_{ST} \mid X = i, \, Y = j, \, X_3 = k] &= 1_{\{M(i+k)>j\}} - 1_{\{M(i+k)<j\}} \\
&= \operatorname{sgn}(M(i+k) - j).
\end{aligned}
\tag{5.27}
$$

Case 3. $i \in S^c$, $(j, \varnothing) \in T$. Here

$$
\begin{aligned}
\mathrm{E}[G_{ST} \mid X = i, \, Y = j, \, X_3 = \varnothing] \\
= \mathrm{P}(i > M(j+Y_3) \mid X = i, \, Y = j, \, X_3 = \varnothing) \\
- \mathrm{P}(i < M(j+Y_3) \mid X = i, \, Y = j, \, X_3 = \varnothing) \\
= \frac{i + 3 \cdot 1_{\{i>j\}}}{13} - \frac{9 - i + 3 \cdot 1_{\{i<j\}}}{13} \\
= \frac{2i - 9 + 3 \operatorname{sgn}(i - j)}{13}.
\end{aligned}
\tag{5.28}
$$

Case 4. $i \in S^c$, $(j, \varnothing) \in T^c$. Here

$$
\mathrm{E}[G_{ST} \mid X = i, \, Y = j, \, X_3 = \varnothing] = 1_{\{i>j\}} - 1_{\{i<j\}} = \operatorname{sgn}(i - j).
\tag{5.29}
$$

Finally, by (5.18) and (5.19), we have

$$
\begin{aligned}
\mathrm{P}(X = i \mid X \in S) &= (16 + 9\delta_{i,0})/(16|S| + 9), & i &\in S, & \tag{5.30} \\
\mathrm{P}(X = i \mid X \in S^c) &= 1/|S^c|, & i &\in S^c. & \tag{5.31}
\end{aligned}
$$

This suffices to complete the evaluation of (5.24) and (5.25). For example, substituting (5.30) and either (5.26) or (5.27) into (5.24), we find that

$$
\begin{aligned}
a_{S_4,1}(5, 4) &= \sum_{i \in S_4} \mathrm{P}(X = i \mid X \in S_4) \mathrm{E}[G_{S_4 T} \mid X = i, \, Y = 5, \, X_3 = 4] \\
&= \sum_{i=0}^{4} \frac{16 + 9\delta_{i,0}}{89} \frac{2M(i+4) - 9 + 3 \operatorname{sgn}(M(i+4) - 5)}{13} \\
&= \frac{300}{1,157}
\end{aligned}
\tag{5.32}
$$

and

$$a_{S_4,2}(5,4) = \sum_{i \in S_4} \mathrm{P}(X = i \mid X \in S_4)\mathrm{E}[G_{S_4 T} \mid X = i,\ Y = 5,\ X_3 = 4]$$

$$= \sum_{i=0}^{4} \frac{16 + 9\delta_{i,0}}{89}\,\mathrm{sgn}(M(i+4) - 5) = \frac{23}{89} = \frac{299}{1{,}157}. \qquad (5.33)$$

The distinction between the first sums in (5.32) and (5.33), which appear identical, is that $(5,4) \in T$ in (5.32), while $(5,4) \notin T$ in (5.33).

In Table 5.3, we display $(2{,}730)a_{S_5,1}(j,k)$ and $(2{,}730)a_{S_5,2}(j,k)$, and in Table 5.4, we display $(3{,}471)a_{S_4,1}(j,k)$ and $(3{,}471)a_{S_4,2}(j,k)$, in both cases for $j = 0, 1, \ldots, 7$ and $k = 0, 1, \ldots, 9, \varnothing$. This tells us what T_1, T_2, and T_3 are, and the results are summarized in Table 5.2. T_1 (resp., T_2) is the set of pairs (j,k) for which the first (j,k) entry is greater than (resp., is less than) the second in both Table 5.3 and Table 5.4. In particular, $|T_3| = 4$.

Table 5.2 Banker's optimal move in chemin de fer, indicated by D (draw) or S (stand), except in the four cases indicated by $*$ in which it depends on player's strategy.

banker's two-card total	player's third card ($\varnothing$ if player stands)										
	0	1	2	3	4	5	6	7	8	9	$\varnothing$
0–2	D	D	D	D	D	D	D	D	D	D	D
3	D	D	D	D	D	D	D	D	S	$*$	D
4	S	$*$	D	D	D	D	D	D	S	S	D
5	S	S	S	S	$*$	D	D	D	S	S	D
6	S	S	S	S	S	S	D	D	S	S	$*$
7	S	S	S	S	S	S	S	S	S	S	S

Requiring that banker make the optimal move in the 84 cases that do not depend on player's strategy, we have reduced the game to a 2×2^4 matrix game, and our next step is to find the resulting payoff matrix. Banker's 16 remaining strategies are described by whether he draws or stands in the four uncertain cases, namely $(j,k) = (3,9), (4,1), (5,4), (6,\varnothing)$ (in that order). For example, the strategy DSDS corresponds to banker drawing with $(j,k) \in \{(3,9), (5,4)\}$ and standing with $(j,k) \in \{(4,1), (6,\varnothing)\}$.

It suffices to use equations (5.23), etc., together with

Table 5.3 Assuming player draws to 5, first entry is player's conditional expected profit when banker draws, multiplied by 2,730, and second entry is player's conditional expected profit when banker stands, multiplied by 2,730. Value of banker's two-card hand is j, and player's third card is k ($= \varnothing$ if player stands).

j		0	1	2	3	4	5	6	7	8	9	$\varnothing$
0		−450	120	540	960	1,380	1,064	844	624	404	184	1,470
		2,080	2,730	2,730	2,730	2,730	2,314	2,314	2,314	2,314	2,314	2,730
1		−696	−30	540	960	1,380	968	652	432	212	−8	1,470
		1,014	2,080	2,730	2,730	2,730	1,898	1,482	1,482	1,482	1,482	2,730
2		−888	−276	390	960	1,380	968	556	240	20	−200	1,470
		182	1,014	2,080	2,730	2,730	1,898	1,066	650	650	650	2,730
3		−1,080	−468	144	810	1,380	968	556	144	−172	−392	1,470
		−650	182	1,014	2,080	2,730	1,898	1,066	234	−182	−182	2,730
4		−1,272	−660	−48	564	1,230	968	556	144	−268	−584	1,470
		−1,482	−650	182	1,014	2,080	1,898	1,066	234	−598	−1,014	2,730
5		−1,464	−852	−240	372	984	818	556	144	−268	−680	1,470
		−2,314	−1,482	−650	182	1,014	1,248	1,066	234	−598	−1,430	2,730
6		−1,560	−1,044	−432	180	792	572	406	144	−268	−680	1,155
		−2,730	−2,314	−1,482	−650	182	182	416	234	−598	−1,430	1,365
7		−1,560	−1,140	−624	−12	600	380	160	−6	−268	−680	525
		−2,730	−2,730	−2,314	−1,482	−650	−650	−650	−416	−598	−1,430	−1,365

Table 5.4 Assuming player stands on 5, first entry is player's conditional expected profit when banker draws, multiplied by 3,471, and second entry is player's conditional expected profit when banker stands, multiplied by 3,471. Value of banker's two-card hand is j, and player's third card is k ($= \varnothing$ if player stands).

j		0	1	2	3	4	5	6	7	8	9	$\varnothing$
							k					
0		−867	−108	426	960	1,494	2,028	1,458	1,032	606	180	1,602
		2,496	3,471	3,471	3,471	3,471	3,471	2,847	2,847	2,847	2,847	3,471
1		−1,236	−333	426	960	1,494	2,028	1,314	744	318	−108	1,602
		897	2,496	3,471	3,471	3,471	3,471	2,223	1,599	1,599	1,599	3,471
2		−1,524	−702	201	960	1,494	2,028	1,314	600	30	−396	1,602
		−351	897	2,496	3,471	3,471	3,471	2,223	975	351	351	3,471
3		−1,812	−990	−168	735	1,494	2,028	1,314	600	−114	−684	1,602
		−1,599	−351	897	2,496	3,471	3,471	2,223	975	−273	−897	3,471
4		−2,100	−1,278	−456	366	1,269	2,028	1,314	600	−114	−828	1,602
		−2,847	−1,599	−351	897	2,496	3,471	2,223	975	−273	−1,521	3,471
5		−2,244	−1,566	−744	78	900	1,803	1,314	600	−114	−828	1,335
		−3,471	−2,847	−1,599	−351	897	2,496	2,223	975	−273	−1,521	2,314
6		−2,244	−1,710	−1,032	−210	612	1,434	1,089	600	−114	−828	801
		−3,471	−3,471	−2,847	−1,599	−351	897	1,248	975	−273	−1,521	0
7		−2,244	−1,710	−1,176	−498	324	1,146	720	375	−114	−828	267
		−3,471	−3,471	−3,471	−2,847	−1,599	−351	−351	0	−273	−1,521	−2,314

$$
p_S(j,k) = \frac{16|S|+9}{169} \frac{16+9\delta_{j,0}}{169} \frac{1+3\delta_{k,0}}{13},
$$

$$
p_S(j,\varnothing) = \frac{16|S^c|}{169} \frac{16+9\delta_{j,0}}{169},
$$

$$(5.34)$$

to obtain the 2×16 payoff matrix, of which, for typographical reasons, we display the transpose multiplied by $(13)^6/16$ in Table 5.5. Keep in mind that the entries are player's expected profits.

We can reduce this payoff matrix using strict dominance. Specifically, strategies DDSS, SDDS, and SDSS are strictly dominated by strategy DSDS. In addition, strategies DDSD, SDDD, and SDSD are strictly dominated by strategy DSDD. This leaves us with the 2×10 payoff matrix of Table 5.6.

In Example 5.2.7 on p. 183 we will find the optimal strategies for player and banker.   ♠

It would be instructive to include some examples of game theory applied to poker, but we postpone them to Section 22.2.

5.2 Minimax Theorem

Consider an $m \times n$ matrix game with payoff matrix

$$
A = \begin{pmatrix} a_{11} & a_{12} & \cdots & a_{1n} \\ a_{21} & a_{22} & \cdots & a_{2n} \\ \vdots & \vdots & & \vdots \\ a_{m1} & a_{m2} & \cdots & a_{mn} \end{pmatrix},
$$

$$(5.35)$$

so that player 1 has strategies $1, 2, \ldots, m$ and player 2 has strategies $1, 2, \ldots, n$. Each of player 1's m strategies is called a *pure strategy*, as is each of player 2's n strategies.

Let us define

$$
\Delta_m := \left\{ \boldsymbol{p} = (p_1, \ldots, p_m) : p_1 \geq 0, \ldots, p_m \geq 0, \sum_{i=1}^{m} p_i = 1 \right\}.
$$

$$(5.36)$$

A vector $\boldsymbol{p} \in \Delta_m$ describes a *mixed strategy* for player 1. Specifically, if player 1 chooses pure strategy i with probability p_i for $i = 1, \ldots, m$, he is said to use the mixed strategy $\boldsymbol{p}$.

Similarly, a vector $\boldsymbol{q} \in \Delta_n$ describes a mixed strategy for player 2.

If player 1 uses mixed strategy $\boldsymbol{p} \in \Delta_m$ and player 2 independently uses mixed strategy $\boldsymbol{q} \in \Delta_n$, then player 1's expected profit (as well as player 2's expected loss) is

Table 5.5 Transpose of 2×16 payoff matrix, multiplied by $(13)^6/16$, for chemin de fer, obtained by application of Lemma 5.1.3.

	player draws to 5	player stands on 5
DDDD	$-2{,}585$	$-4{,}126$
DDDS	$-4{,}457$	$-3{,}710$
DDSD	$-2{,}586$	$-4{,}111$
DDSS	$-4{,}458$	$-3{,}695$
DSDD	$-2{,}692$	$-4{,}121$
DSDS	$-4{,}564$	$-3{,}705$
DSSD	$-2{,}693$	$-4{,}106$
DSSS	$-4{,}565$	$-3{,}690$
SDDD	$-2{,}656$	$-4{,}021$
SDDS	$-4{,}528$	$-3{,}605$
SDSD	$-2{,}657$	$-4{,}006$
SDSS	$-4{,}529$	$-3{,}590$
SSDD	$-2{,}763$	$-4{,}016$
SSDS	$-4{,}635$	$-3{,}600$
SSSD	$-2{,}764$	$-4{,}001$
SSSS	$-4{,}636$	$-3{,}585$

Table 5.6 Transpose of 2×10 payoff matrix, multiplied by $(13)^6/16$, for chemin de fer, obtained from Table 5.5 by strict dominance.

	player draws to 5	player stands on 5
DDDD	$-4{,}126$	$-2{,}585$
DDDS	$-3{,}710$	$-4{,}457$
DSDD	$-4{,}121$	$-2{,}692$
DSDS	$-3{,}705$	$-4{,}564$
DSSD	$-4{,}106$	$-2{,}693$
DSSS	$-3{,}690$	$-4{,}565$
SSDD	$-4{,}016$	$-2{,}763$
SSDS	$-3{,}600$	$-4{,}635$
SSSD	$-4{,}001$	$-2{,}764$
SSSS	$-3{,}585$	$-4{,}636$

$$\sum_{i=1}^{m}\sum_{j=1}^{n}a_{ij}p_iq_j = \boldsymbol{p}\boldsymbol{A}\boldsymbol{q}^{\mathsf{T}}. \tag{5.37}$$

Note that $\boldsymbol{p}$ and $\boldsymbol{q}$ are row vectors, hence $\boldsymbol{q}^{\mathsf{T}}$ is a column vector.

By adopting the mixed strategy $\boldsymbol{p} \in \Delta_m$, player 1 can ensure an expected profit of at least

$$\min_{\boldsymbol{q}\in\Delta_n} \boldsymbol{p}\boldsymbol{A}\boldsymbol{q}^{\mathsf{T}}, \tag{5.38}$$

and therefore, by choosing $\boldsymbol{p}$ appropriately, he can ensure an expected profit of at least

$$\max_{\boldsymbol{p}\in\Delta_m}\min_{\boldsymbol{q}\in\Delta_n} \boldsymbol{p}\boldsymbol{A}\boldsymbol{q}^{\mathsf{T}}. \tag{5.39}$$

On the other hand, by adopting the mixed strategy $\boldsymbol{q} \in \Delta_n$, player 2 can ensure an expected loss of at most

$$\max_{\boldsymbol{p}\in\Delta_m} \boldsymbol{p}\boldsymbol{A}\boldsymbol{q}^{\mathsf{T}}, \tag{5.40}$$

and therefore, by choosing $\boldsymbol{q}$ appropriately, he can ensure an expected loss of at most

$$\min_{\boldsymbol{q}\in\Delta_n}\max_{\boldsymbol{p}\in\Delta_m} \boldsymbol{p}\boldsymbol{A}\boldsymbol{q}^{\mathsf{T}}. \tag{5.41}$$

It is evident from these remarks that

$$\max_{\boldsymbol{p}\in\Delta_m}\min_{\boldsymbol{q}\in\Delta_n} \boldsymbol{p}\boldsymbol{A}\boldsymbol{q}^{\mathsf{T}} \le \min_{\boldsymbol{q}\in\Delta_n}\max_{\boldsymbol{p}\in\Delta_m} \boldsymbol{p}\boldsymbol{A}\boldsymbol{q}^{\mathsf{T}}, \tag{5.42}$$

and indeed this is easy to verify algebraically.

The fundamental theorem of game theory, better known as the minimax theorem, asserts that equality holds in (5.42). We provide a proof below. First, however, assuming the validity of the minimax theorem, let us denote the common value of the left and right sides of (5.42) by v.

It follows that player 1 can find a mixed strategy $\boldsymbol{p}^* \in \Delta_m$ such that

$$\min_{\boldsymbol{q}\in\Delta_n} \boldsymbol{p}^*\boldsymbol{A}\boldsymbol{q}^{\mathsf{T}} = v, \tag{5.43}$$

and therefore

$$\boldsymbol{p}^*\boldsymbol{A}\boldsymbol{q}^{\mathsf{T}} \ge v, \qquad \boldsymbol{q} \in \Delta_n. \tag{5.44}$$

In words, player 1 can find a mixed strategy $\boldsymbol{p}^* \in \Delta_m$ that ensures an expected profit of at least v, regardless of the mixed strategy $\boldsymbol{q} \in \Delta_n$ used by player 2.

Similarly, player 2 can find a mixed strategy $\boldsymbol{q}^* \in \Delta_n$ such that

$$\max_{\boldsymbol{p}\in\Delta_m} \boldsymbol{p}\boldsymbol{A}(\boldsymbol{q}^*)^{\mathsf{T}} = v, \tag{5.45}$$

and therefore

$$\boldsymbol{p}\boldsymbol{A}(\boldsymbol{q}^*)^\mathsf{T} \le v, \qquad \boldsymbol{p} \in \Delta_m. \tag{5.46}$$

In words, player 2 can find a mixed strategy $\boldsymbol{q}^* \in \Delta_n$ that ensures an expected loss of at most v, regardless of the mixed strategy $\boldsymbol{p} \in \Delta_m$ used by player 1.

A mixed strategy $\boldsymbol{p}^* \in \Delta_m$ satisfying (5.44) is said to be *optimal* for player 1, whereas a mixed strategy $\boldsymbol{q}^* \in \Delta_n$ satisfying (5.46) is said to be *optimal* for player 2. Optimal strategies are not necessarily unique. Since player 1 can ensure an expected profit of at least v by using an optimal strategy $\boldsymbol{p}^* \in \Delta_m$, and since player 2 can ensure an expected loss of at most v by using an optimal strategy $\boldsymbol{q}^* \in \Delta_n$, it follows that, if both players use optimal strategies, then the expected profit to player 1 (as well as the expected loss to player 2) is

$$\boldsymbol{p}^*\boldsymbol{A}(\boldsymbol{q}^*)^\mathsf{T} = v. \tag{5.47}$$

For this reason, v is called the *value* of the game.

To *solve* a game will mean to find an optimal strategy for each player as well as the value of the game. Of course, whenever possible, we will find *all* optimal strategies for each player, but sometimes this is difficult.

Notice that, if there exist $\boldsymbol{p}^* \in \Delta_m$ and $\boldsymbol{q}^* \in \Delta_n$ such that both (5.44) and (5.46) hold for some real number v (not necessarily the value of the game), then v *is* the value of the game and $\boldsymbol{p}^*$ and $\boldsymbol{q}^*$ are optimal mixed strategies. Indeed,

$$
\begin{aligned}
\max_{\boldsymbol{p}\in\Delta_m} \min_{\boldsymbol{q}\in\Delta_n} \boldsymbol{p}\boldsymbol{A}\boldsymbol{q}^\mathsf{T} &\ge \min_{\boldsymbol{q}\in\Delta_n} \boldsymbol{p}^*\boldsymbol{A}\boldsymbol{q}^\mathsf{T} \\
&\ge v \\
&\ge \max_{\boldsymbol{p}\in\Delta_m} \boldsymbol{p}\boldsymbol{A}(\boldsymbol{q}^*)^\mathsf{T} \\
&\ge \min_{\boldsymbol{q}\in\Delta_n} \max_{\boldsymbol{p}\in\Delta_m} \boldsymbol{p}\boldsymbol{A}\boldsymbol{q}^\mathsf{T},
\end{aligned}
\tag{5.48}
$$

so by (5.42) equality must hold throughout.

Our first aim is to prove the minimax theorem, and then we will return to the examples of the preceding section. We begin with two lemmas. The first is a special case of the *supporting hyperplanes lemma* from linear algebra.

Lemma 5.2.1. *Let C be a closed convex set in $\mathbf{R}^m$, and suppose that $\boldsymbol{0} \notin C$. Then there exists $\boldsymbol{b} \in \mathbf{R}^m$ such that $\boldsymbol{b} \cdot \boldsymbol{x} > 0$ for all $\boldsymbol{x} \in C$.*

Remark. If $\boldsymbol{b} \ne \boldsymbol{0}$, then the set of all $\boldsymbol{x} \in \mathbf{R}^m$ such that $\boldsymbol{b} \cdot \boldsymbol{x} = 0$ is a *hyperplane* in $\mathbf{R}^m$ (a line in $\mathbf{R}^2$, a plane in $\mathbf{R}^3$, etc.) that contains the origin. Under the assumptions of the lemma, the conclusion is that there exists a hyperplane in $\mathbf{R}^m$ containing the origin such that C lies entirely on one side of the hyperplane.

Proof. Let $\boldsymbol{b}$ be an element of C that minimizes the distance to the origin. The existence of $\boldsymbol{b}$ is a consequence of the fact that, given $\boldsymbol{y}_0 \in C$, the continuous function $f(\boldsymbol{y}) := |\boldsymbol{y}|$ achieves its minimum on the compact set

$\{ \boldsymbol{y} \in C : |\boldsymbol{y}| \leq |\boldsymbol{y}_0| \}$. Given $\boldsymbol{x} \in C$ and $0 < \lambda < 1$, the convexity of C implies that $\lambda \boldsymbol{x} + (1 - \lambda) \boldsymbol{b} \in C$, and therefore

$$\lambda^2 |\boldsymbol{x} - \boldsymbol{b}|^2 + 2\lambda \boldsymbol{b} \cdot (\boldsymbol{x} - \boldsymbol{b}) + |\boldsymbol{b}|^2 = |\lambda \boldsymbol{x} + (1 - \lambda) \boldsymbol{b}|^2 \geq |\boldsymbol{b}|^2. \tag{5.49}$$

Subtract $|\boldsymbol{b}|^2$ from each side, divide by 2λ, and let $\lambda \to 0$ to get $\boldsymbol{b} \cdot (\boldsymbol{x} - \boldsymbol{b}) \geq 0$, or

$$\boldsymbol{b} \cdot \boldsymbol{x} \geq |\boldsymbol{b}|^2 > 0, \tag{5.50}$$

as required. ♠

The next result is called the *lemma of the alternative*.

Lemma 5.2.2. *Given an $m \times n$ matrix $\boldsymbol{A}$, either there exists $\boldsymbol{q}^* \in \Delta_n$ such that all components of $\boldsymbol{q}^* \boldsymbol{A}^\mathsf{T}$ are nonpositive, or there exists $\boldsymbol{p}^* \in \Delta_m$ such that all components of $\boldsymbol{p}^* \boldsymbol{A}$ are positive.*

Proof. Let $\boldsymbol{a}_1, \ldots, \boldsymbol{a}_n \in \mathbf{R}^m$ be the n rows of $\boldsymbol{A}^\mathsf{T}$, and define $\boldsymbol{e}_1, \ldots, \boldsymbol{e}_m \in \mathbf{R}^m$ by $\boldsymbol{e}_1 := (1, 0, \ldots, 0)$, $\boldsymbol{e}_2 := (0, 1, 0, \ldots, 0)$, and so on. Let $C \subset \mathbf{R}^m$ be the *convex hull* of $\boldsymbol{a}_1, \ldots, \boldsymbol{a}_n, \boldsymbol{e}_1, \ldots, \boldsymbol{e}_m$, that is,

$$C := \{ c_1 \boldsymbol{a}_1 + \cdots + c_n \boldsymbol{a}_n + c_{n+1} \boldsymbol{e}_1 + \cdots + c_{n+m} \boldsymbol{e}_m :$$
$$(c_1, \ldots, c_n, c_{n+1}, \ldots, c_{n+m}) \in \Delta_{n+m} \}. \tag{5.51}$$

If $\boldsymbol{0} \in C$, then there exists $(c_1, \ldots, c_n, c_{n+1}, \ldots, c_{n+m}) \in \Delta_{n+m}$ such that

$$c_1 \boldsymbol{a}_1 + \cdots + c_n \boldsymbol{a}_n = -(c_{n+1} \boldsymbol{e}_1 + \cdots + c_{n+m} \boldsymbol{e}_m). \tag{5.52}$$

It cannot be the case that $c_1 = \cdots = c_n = 0$, for if it were, then $c_{n+1}, \ldots, c_{n+m}$ would also be 0 by the linear independence of $\boldsymbol{e}_1, \ldots, \boldsymbol{e}_m$, and this would contradict the fact that $c_1 + \cdots + c_{n+m} = 1$. Therefore, defining $\boldsymbol{q}^* \in \Delta_n$ by $q_j^* = c_j / (c_1 + \cdots + c_n)$, we have

$$\boldsymbol{q}^* \boldsymbol{A}^\mathsf{T} = q_1^* \boldsymbol{a}_1 + \cdots + q_n^* \boldsymbol{a}_n = (c_1 + \cdots + c_n)^{-1} (c_1 \boldsymbol{a}_1 + \cdots + c_n \boldsymbol{a}_n), \tag{5.53}$$

which has nonpositive components by (5.52).

If $\boldsymbol{0} \notin C$, we can apply Lemma 5.2.1 to deduce the existence of $\boldsymbol{b} \in \mathbf{R}^m$ such that $\boldsymbol{b} \cdot \boldsymbol{x} > 0$ for all $\boldsymbol{x} \in C$. In particular,

$$\boldsymbol{b} \cdot \boldsymbol{a}_1 > 0, \ldots, \boldsymbol{b} \cdot \boldsymbol{a}_n > 0, \; b_1 = \boldsymbol{b} \cdot \boldsymbol{e}_1 > 0, \ldots, b_m = \boldsymbol{b} \cdot \boldsymbol{e}_m > 0. \tag{5.54}$$

Therefore, defining $\boldsymbol{p}^* \in \Delta_m$ by $p_i^* = b_i / (b_1 + \cdots + b_m)$, we have

$$\boldsymbol{p}^* \boldsymbol{A} = (\boldsymbol{p}^* \cdot \boldsymbol{a}_1, \ldots, \boldsymbol{p}^* \cdot \boldsymbol{a}_n) = (b_1 + \cdots + b_m)^{-1} (\boldsymbol{b} \cdot \boldsymbol{a}_1, \ldots, \boldsymbol{b} \cdot \boldsymbol{a}_n), \tag{5.55}$$

which has positive components by (5.54). ♠

We are now ready to prove the *minimax theorem*.

Theorem 5.2.3. *Given an $m \times n$ matrix $\boldsymbol{A}$,*

$$\max_{\boldsymbol{p} \in \Delta_m} \min_{\boldsymbol{q} \in \Delta_n} \boldsymbol{p} \boldsymbol{A} \boldsymbol{q}^{\mathsf{T}} = \min_{\boldsymbol{q} \in \Delta_n} \max_{\boldsymbol{p} \in \Delta_m} \boldsymbol{p} \boldsymbol{A} \boldsymbol{q}^{\mathsf{T}}. \tag{5.56}$$

Proof. Suppose not. Then, by (5.42), there exists a real number α such that

$$\max_{\boldsymbol{p} \in \Delta_m} \min_{\boldsymbol{q} \in \Delta_n} \boldsymbol{p} \boldsymbol{A} \boldsymbol{q}^{\mathsf{T}} < \alpha < \min_{\boldsymbol{q} \in \Delta_n} \max_{\boldsymbol{p} \in \Delta_m} \boldsymbol{p} \boldsymbol{A} \boldsymbol{q}^{\mathsf{T}}. \tag{5.57}$$

Letting $\boldsymbol{1}$ temporarily denote the $m \times n$ matrix of 1s, define $\boldsymbol{B} := \boldsymbol{A} - \alpha \boldsymbol{1}$. We apply Lemma 5.2.2 to the $m \times n$ matrix $\boldsymbol{B}$. There are two possibilities.

The first is that there exists $\boldsymbol{q}^* \in \Delta_n$ such that all components of $\boldsymbol{q}^* \boldsymbol{B}^{\mathsf{T}}$ are nonpositive. It follows that

$$\boldsymbol{p} \boldsymbol{A} (\boldsymbol{q}^*)^{\mathsf{T}} - \alpha = \boldsymbol{p} \boldsymbol{B} (\boldsymbol{q}^*)^{\mathsf{T}} = \boldsymbol{q}^* \boldsymbol{B}^{\mathsf{T}} \boldsymbol{p}^{\mathsf{T}} \leq 0 \tag{5.58}$$

for all $\boldsymbol{p} \in \Delta_m$, or that

$$\min_{\boldsymbol{q} \in \Delta_n} \max_{\boldsymbol{p} \in \Delta_m} \boldsymbol{p} \boldsymbol{A} \boldsymbol{q}^{\mathsf{T}} \leq \max_{\boldsymbol{p} \in \Delta_m} \boldsymbol{p} \boldsymbol{A} (\boldsymbol{q}^*)^{\mathsf{T}} \leq \alpha, \tag{5.59}$$

and this contradicts the second inequality in (5.57).

The second possibility is that there exists $\boldsymbol{p}^* \in \Delta_m$ such that all components of $\boldsymbol{p}^* \boldsymbol{B}$ are positive. It follows that

$$\boldsymbol{p}^* \boldsymbol{A} \boldsymbol{q}^{\mathsf{T}} - \alpha = \boldsymbol{p}^* \boldsymbol{B} \boldsymbol{q}^{\mathsf{T}} > 0 \tag{5.60}$$

for all $\boldsymbol{q} \in \Delta_n$, or that

$$\max_{\boldsymbol{p} \in \Delta_m} \min_{\boldsymbol{q} \in \Delta_n} \boldsymbol{p} \boldsymbol{A} \boldsymbol{q}^{\mathsf{T}} \geq \min_{\boldsymbol{q} \in \Delta_n} \boldsymbol{p}^* \boldsymbol{A} \boldsymbol{q}^{\mathsf{T}} > \alpha, \tag{5.61}$$

and this contradicts the first inequality in (5.57). In either case we have a contradiction, and therefore (5.56) must hold.   ♠

Before turning to examples, it will be useful to isolate some simple consequences of the theorem.

Corollary 5.2.4. *Let $\boldsymbol{A}$ be the payoff matrix for an $m \times n$ matrix game, let $\boldsymbol{p}^* \in \Delta_m$ and $\boldsymbol{q}^* \in \Delta_n$ be optimal mixed strategies for players 1 and 2, respectively, and let v be the value of the game.*

(a) If $p_i^* > 0$, then the ith component of the vector $\boldsymbol{A}(\boldsymbol{q}^*)^{\mathsf{T}}$ is equal to v. Similarly, if $q_j^* > 0$, then the jth component of the vector $\boldsymbol{p}^* \boldsymbol{A}$ is equal to v.

(b) Suppose strategy i for player 1 is strictly dominated by strategy k for player 1 (i.e., $a_{kj} > a_{ij}$ for $j = 1, \ldots, n$). Then $p_i^* = 0$. Similarly, suppose strategy j for player 2 is strictly dominated by strategy l for player 2 (i.e., $a_{ij} > a_{il}$ for $i = 1, \ldots, m$). Then $q_j^* = 0$.

Remark. Part (a) of the corollary says that, if a pure strategy is part of an optimal mixed strategy for player 1, then player 1's expected profit using that pure strategy against any optimal mixed strategy of player 2 is equal to the game's value. Thus, player 1 is indifferent as to which of his viable pure strategies he uses against any of player 2's optimal mixed strategies. When player 2 has a unique optimal mixed strategy, this so-called *indifference principle* can be used to derive it. Similarly, player 2 is indifferent as to which of his viable pure strategies he uses against any of player 1's optimal mixed strategies. When player 1 has a unique optimal mixed strategy, this can be used to derive it.

Part (b) says that a strictly dominated pure strategy is never part of an optimal mixed strategy.

Proof. By virtue of (5.46), all components of the vector $A(q^*)^\mathsf{T}$ are less than or equal to v. By (5.47), $p_i^* = 0$ whenever the ith component of the vector $A(q^*)^\mathsf{T}$ is less than v. This is equivalent to the first assertion in part (a). In particular, if strategy i for player 1 is strictly dominated by strategy k for player 1, then the ith component of the vector $A(q^*)^\mathsf{T}$ is less than the kth component and therefore less than v, and so part (a) applies and yields the first assertion in part (b). The second assertions are proved analogously. ♠

Let us now turn to the examples of the preceding section.

Example 5.2.5. *Solution of competing subsets.* We continue with Example 5.1.1 on p. 161. Our aim here is to find all optimal mixed strategies for the symmetric matrix game with 5×5 payoff matrix (5.5) on p. 162. By Corollary 5.2.4(b), it will suffice to consider the symmetric matrix game with 4×4 payoff matrix (5.6) on p. 162. The symmetry of the game makes the solution quite easy.

We begin by deriving a few properties of symmetric matrix games. Let A be an $m \times m$ (square) matrix satisfying

$$A^\mathsf{T} = -A. \tag{5.62}$$

Then the game has value v and $p^*, q^* \in \Delta_m$ are optimal mixed strategies for players 1 and 2, respectively, if and only if

$$pA(q^*)^\mathsf{T} \leq v \leq p^* A q^\mathsf{T} \tag{5.63}$$

for all $p, q \in \Delta_m$. By (5.62), this is equivalent to

$$qA(p^*)^\mathsf{T} \leq -v \leq q^* A p^\mathsf{T} \tag{5.64}$$

for all $q, p \in \Delta_m$. But the latter conclusion holds if and only the game has value $-v$ and $q^*, p^* \in \Delta_m$ are optimal mixed strategies for players 1 and 2, respectively. It follows that $v = -v$, hence $v = 0$, that is, a symmetric matrix game has value 0. Moreover, we have therefore shown that a mixed strategy $p^* \in \Delta_m$ is optimal for player 1 if and only if it optimal for player 2, a conclusion that is intuitively obvious.

Let $\boldsymbol{p}^* \in \Delta_m$ be an optimal mixed strategy, and let $S := \{i \in \{1, \ldots, m\} : p_i^* > 0\}$. Denote by $\boldsymbol{A}_S$ the principal $|S| \times |S|$ submatrix obtained from $\boldsymbol{A}$ by restricting row and column indices to S. Similarly, let $\boldsymbol{p}_S^*$ be the vector in $\Delta_{|S|}$ obtained from $\boldsymbol{p}^*$ by restricting component indices to S. By Corollary 5.2.4(a),

$$\boldsymbol{A}_S(\boldsymbol{p}_S^*)^\mathsf{T} = \boldsymbol{0}. \tag{5.65}$$

Thus, we can find all optimal mixed strategies by finding, for all nonempty sets $S \subset \{1, \ldots, m\}$, all solutions $\boldsymbol{p}_S^* \in \Delta_{|S|}$ of the system (5.65) having all components positive. Of course, for each such solution, we must check that, when extended to $\boldsymbol{p}^* \in \Delta_m$, it remains an optimal mixed strategy.

As an example, let us consider the symmetric game with 3×3 payoff matrix

$$\boldsymbol{A} = \begin{pmatrix} 0 & a & -b \\ -a & 0 & c \\ b & -c & 0 \end{pmatrix}, \tag{5.66}$$

where $a, b, c > 0$. First, we look for solutions $\boldsymbol{p}^* \in \Delta_3$ of $\boldsymbol{A}(\boldsymbol{p}^*)^\mathsf{T} = \boldsymbol{0}$. The unique such solution is

$$\boldsymbol{p}^* = \left(\frac{c}{a+b+c}, \frac{b}{a+b+c}, \frac{a}{a+b+c} \right). \tag{5.67}$$

Next we consider the principal 2×2 submatrices

$$\begin{pmatrix} 0 & a \\ -a & 0 \end{pmatrix}, \quad \begin{pmatrix} 0 & -b \\ b & 0 \end{pmatrix}, \quad \begin{pmatrix} 0 & c \\ -c & 0 \end{pmatrix}. \tag{5.68}$$

In each case, one strategy is strictly dominated by the other, implying by Corollary 5.2.4(b) that an optimal mixed strategy cannot have both components positive. Finally, none of the three pure strategies, namely $(1, 0, 0)$, $(0, 1, 0)$, and $(0, 0, 1)$, is optimal for $\boldsymbol{A}$, because (5.44) fails, each row of $\boldsymbol{A}$ having a negative entry. We conclude that (5.67) is the unique optimal mixed strategy.

We return to the matrix (5.6) on p. 162, which we denote by $\boldsymbol{A}$ instead of $\boldsymbol{B}$. Note that $\det(\boldsymbol{A}) = 100$, so the only solution of (5.65) with $S = \{2, 3, 4, 5\}$ is the zero solution. In particular, there is no solution with all components positive. Next we consider the principal 3×3 submatrices of $\boldsymbol{A}$, namely

$$\begin{pmatrix} 0 & 1 & -2 \\ -1 & 0 & 7 \\ 2 & -7 & 0 \end{pmatrix}, \begin{pmatrix} 0 & 1 & -5 \\ -1 & 0 & 5 \\ 5 & -5 & 0 \end{pmatrix}, \begin{pmatrix} 0 & -2 & -5 \\ 2 & 0 & 15 \\ 5 & -15 & 0 \end{pmatrix}, \begin{pmatrix} 0 & 7 & 5 \\ -7 & 0 & 15 \\ -5 & -15 & 0 \end{pmatrix}. \tag{5.69}$$

Each of the last two matrices in (5.69) has a strictly dominated strategy, so by Corollary 5.2.4(b) there cannot be a solution of (5.65) with all components positive. The first two matrices in (5.69) are of the form (5.66), and therefore have unique optimal mixed strategies $\left(\frac{7}{10}, \frac{2}{10}, \frac{1}{10} \right)$ and $\left(\frac{5}{11}, \frac{5}{11}, \frac{1}{11} \right)$, respectively. This yields two potential optimal mixed strategies for (5.6) on p. 162, namely $\left(\frac{7}{10}, \frac{2}{10}, \frac{1}{10}, 0\right)$ and $\left(\frac{5}{11}, \frac{5}{11}, 0, \frac{1}{11}\right)$. However, only the last of these two mixed strategies satisfies (5.44). Checking the remaining principal submatrices of A as we did for (5.66), we find that $\left(\frac{5}{11}, \frac{5}{11}, 0, \frac{1}{11}\right)$ is the unique optimal mixed strategy for (5.6) on p. 162, and therefore, by Corollary 5.2.4(b), $\left(0, \frac{5}{11}, \frac{5}{11}, 0, \frac{1}{11}\right)$ is the unique optimal mixed strategy for (5.5) on p. 162, the original payoff matrix. This then completes the solution of the game.   ♠

Example 5.2.6. *Solution of le her.* We continue with Example 5.1.2 on p. 162. We have already reduced the game of le her to a 2×2 matrix game, so we begin by considering a general 2×2 payoff matrix of the form

$$A = \begin{pmatrix} a & b \\ d & c \end{pmatrix}, \tag{5.70}$$

where the unusual ordering of entries was chosen to make the solution easier to remember. For simplicity, we assume that no saddle point exists (otherwise, see Problem 5.3 on p. 193). If $a \geq b$, then $b < c$, for otherwise b would be a saddle point. Hence $d < c$, for otherwise c would be a saddle point. It follows that $a > d$, for otherwise d would be a saddle point. Finally $a > b$, for otherwise a would be a saddle point. Had we started with $a \leq b$, then all inequalities would be reversed. Consequently, there are two possibilities, namely

$$\min(a, c) > \max(b, d) \quad \text{or} \quad \min(b, d) > \max(a, c). \tag{5.71}$$

It is easy to check that no pure strategy is optimal for player 1 or for player 2 (otherwise there would be a saddle point). It follows that the unique optimal strategies have the form $(p^*, 1 - p^*)$ for player 1 and $(q^*, 1 - q^*)$ for player 2, where $0 < p^* < 1$ and $0 < q^* < 1$. By Corollary 5.2.4(a), the value v of the game satisfies

$$(p^*, 1 - p^*) \begin{pmatrix} a & b \\ d & c \end{pmatrix} = (v, v) = \begin{pmatrix} a & b \\ d & c \end{pmatrix} (q^*, 1 - q^*)^{\mathsf{T}}, \tag{5.72}$$

implying that

$$\begin{aligned} ap^* + d(1 - p^*) = bp^* + c(1 - p^*) = v, \\ aq^* + b(1 - q^*) = dq^* + c(1 - q^*) = v. \end{aligned} \tag{5.73}$$

We conclude that

$$p^* = \frac{c - d}{(a - b) + (c - d)}, \qquad q^* = \frac{c - b}{(a - b) + (c - d)}, \tag{5.74}$$

and

$$v = \frac{ac - bd}{(a - b) + (c - d)}. \tag{5.75}$$

Because of (5.71), $a - b$ and $c - d$ are both positive or both negative, hence $0 < p^* < 1$ as we have already noted. The same is true of $a - d$ and $c - b$, hence $0 < q^* < 1$.

Returning to le her, recall that we had reduced the game to the 2×2 matrix game with payoff matrix

$$\begin{array}{c} \quad\quad\quad\quad\quad \text{exchange} \quad \text{exchange} \\ \quad\quad\quad\quad\quad \text{7 or less} \quad\; \text{8 or less} \\ \begin{array}{c} \text{exchange 6 or less} \\ \text{exchange 7 or less} \end{array} \left(\begin{array}{cc} 131/5{,}525 & 143/5{,}525 \\ 151/5{,}525 & 131/5{,}525 \end{array} \right) \end{array} . \qquad (5.76)$$

We conclude from (5.74) and (5.75) that the unique optimal mixed strategy for player 1 is $\left(\frac{5}{8}, \frac{3}{8}\right)$, the unique optimal mixed strategy for player 2 is $\left(\frac{3}{8}, \frac{5}{8}\right)$, and the value of the game is $277/11{,}050 \approx 0.025067873$. Thus, the game is advantageous to player 1, despite the fact that player 2 wins ties. ♠

Example 5.2.7. *Solution of chemin de fer.* We continue with Example 5.1.4 on p. 167. We have reduced the game of chemin de fer to a 2×10 matrix game, so let us consider more generally the $2 \times n$ matrix game with payoff matrix

$$\boldsymbol{A} = \begin{pmatrix} a_{11} & a_{12} & \cdots & a_{1n} \\ a_{21} & a_{22} & \cdots & a_{2n} \end{pmatrix}. \qquad (5.77)$$

Let v be the value of the game, and define $\boldsymbol{e}_1, \ldots, \boldsymbol{e}_n \in \Delta_n$ by $\boldsymbol{e}_1 := (1, 0, \ldots, 0)$, $\boldsymbol{e}_2 := (0, 1, 0, \ldots, 0)$, and so on. Then

$$\begin{aligned} v &= \max_{\boldsymbol{p} \in \Delta_2} \min_{\boldsymbol{q} \in \Delta_n} \boldsymbol{p} \boldsymbol{A} \boldsymbol{q}^{\mathsf{T}} \\ &= \max_{\boldsymbol{p} \in \Delta_2} \min_{1 \leq j \leq n} \boldsymbol{p} \boldsymbol{A} \boldsymbol{e}_j^{\mathsf{T}} \\ &= \max_{0 \leq p \leq 1} \min_{1 \leq j \leq n} \{a_{1j} p + a_{2j}(1 - p)\} \\ &= \max_{0 \leq p \leq 1} \min\{f_1(p), \ldots, f_n(p)\}, \end{aligned} \qquad (5.78)$$

where $f_j : [0, 1] \mapsto \mathbf{R}$ is the first-degree polynomial whose graph connects the points $(0, a_{2j})$ and $(1, a_{1j})$ with a straight line.

Let us suppose that the maximum in (5.78) is uniquely attained at some $p^* \in (0, 1)$, so that in particular

$$v = \min\{f_1(p^*), \ldots, f_n(p^*)\}. \qquad (5.79)$$

Then $(p^*, 1 - p^*)$ is the unique optimal mixed strategy for player 1 (cf. (5.43)). To determine p^* we need only find the value of p that maximizes the continuous piecewise-linear function $p \mapsto \min\{f_1(p), \ldots, f_n(p)\}$ (called the *lower envelope* of the family $f_1, \ldots, f_n$), and this maximum value will be v.

Let us also suppose that the minimum in (5.79) is attained by exactly two of the function values, say $f_{j_1}(p^*)$ and $f_{j_2}(p^*)$, where $1 \leq j_1 < j_2 \leq n$. Let

$q^* \in \Delta_n$ be an optimal mixed strategy for player 2. By Corollary 5.2.4(a), $q_j^* = 0$ unless $j = j_1$ or $j = j_2$. We write $q^* := q_{j_1}^*$, so $1 - q^* = q_{j_2}^*$. Then

$$v = \max_{\boldsymbol{p} \in \Delta_2} \boldsymbol{p} A(\boldsymbol{q}^*)^\mathsf{T} = \max_{0 \le p \le 1} (p, 1 - p) \begin{pmatrix} a_{1j_1} & a_{1j_2} \\ a_{2j_1} & a_{2j_2} \end{pmatrix} (q^*, 1 - q^*)^\mathsf{T}, \qquad (5.80)$$

implying that $(p^*, 1 - p^*)$ and $(q^*, 1 - q^*)$ are optimal mixed strategies for players 1 and 2 of the 2×2 matrix game with payoff matrix

$$\begin{pmatrix} a_{1j_1} & a_{1j_2} \\ a_{2j_1} & a_{2j_2} \end{pmatrix}. \qquad (5.81)$$

Furthermore, our assumptions imply that, if we denote $a_{1j_1}, a_{1j_2}, a_{2j_2}, a_{2j_1}$ by a, b, c, d, respectively, then (5.71) holds, and consequently p^*, q^*, and v are given by (5.74) and (5.75). This then solves the $2 \times n$ matrix game with payoff matrix (5.77), at least under the two assumptions that were made.

We now apply this to the 2×10 matrix game with payoff matrix whose transpose, multiplied by $(13)^6/16$, is given in Table 5.6 on p. 175. The two assumptions are satisfied, with $j_1 = 3$ and $j_2 = 4$, and by the preceding paragraph we need only consider the 2×2 matrix game with payoff matrix, multiplied by $(13)^6/16$, equal to

$$
\begin{array}{cc}
& \text{DSDD} \quad \text{DSDS} \\
\begin{array}{l} \text{player draws to 5} \\ \text{player stands on 5} \end{array} & \begin{pmatrix} -4{,}121 & -3{,}705 \\ -2{,}692 & -4{,}564 \end{pmatrix}.
\end{array} \qquad (5.82)
$$

We find from (5.74) that player's unique optimal mixed strategy (i.e., his draw-stand mixture when holding 5) is $\left(\frac{9}{11}, \frac{2}{11}\right)$, while banker's unique optimal mixed strategy is given by Table 5.7. By (5.75), the value of the game is $v = -679{,}568/[11(13)^6] \approx -0.012799120$. This is player's expected profit (per unit bet) when both player and banker play optimally.

As we have mentioned, the rules of chemin de fer assumed here are those of the early 20th century. Modern rules are much more restrictive. Specifically, player's two pure strategies remain the same, but banker has a decision to make in only two of the 88 strategic situations that can occur. The precise rules are spelled out in Table 5.8. The result is a 2×4 matrix game. Regrettably, the game is not very interesting, as the optimal strategies are pure (see Problem 5.17 on p. 196).

As we will see in Chapter 19, modern baccarat is effectively chemin de fer with no options at all for player or banker. Specifically, player draws to 5 or less, while banker follows the rules of Table 5.2 on p. 171, together with DSDS in the four uncertain cases. It follows from (5.82) that player's expectation is $(-3{,}705)(16)/(13)^6 \approx -0.012281406$. Notice from Table 5.3 on p. 172 that banker's optimal pure strategy is DDDD, which by Table 5.6 on p. 175 would reduce player's expectation to $(-4{,}126)(16)/(13)^6 \approx -0.013676945$. Was this an error on the part of the creators of baccarat?   ♠

Table 5.7 Banker's optimal mixed strategy in chemin de fer, indicated by D (draw) or S (stand).

banker's two-card total	player's third card (∅ if player stands)										
	0	1	2	3	4	5	6	7	8	9	∅
0–2	D	D	D	D	D	D	D	D	D	D	D
3	D	D	D	D	D	D	D	D	S	D	D
4	S	S	D	D	D	D	D	D	S	S	D
5	S	S	S	S	D	D	D	D	S	S	D
6	S	S	S	S	S	S	D	D	S	S	*
7	S	S	S	S	S	S	S	S	S	S	S

*draw-stand mixture is $\left(\frac{859}{2{,}288}, \frac{1{,}429}{2{,}288}\right)$

Table 5.8 Banker's rules in modern chemin de fer, indicated by D (draw), S (stand), or O (optional).

banker's two-card total	player's third card (∅ if player stands)										
	0	1	2	3	4	5	6	7	8	9	∅
0–2	D	D	D	D	D	D	D	D	D	D	D
3	D	D	D	D	D	D	D	D	S	O	D
4	S	S	D	D	D	D	D	D	S	S	D
5	S	S	S	S	O	D	D	D	S	S	D
6	S	S	S	S	S	S	D	D	S	S	S
7	S	S	S	S	S	S	S	S	S	S	S

5.3 Utility Theory

Let X be a discrete random variable assuming finitely many values in an interval $I \subset \mathbf{R}$ and such that

$$P(X = x_i) = p_i, \quad 1 \leq i \leq n, \tag{5.83}$$

where $n \geq 1$, $x_1, \ldots, x_n \in I$ are distinct, and $p_1 > 0, \ldots, p_n > 0$ satisfy $\sum_{i=1}^{n} p_i = 1$. For each $x \in I$ we define the function δ_x on the collection of subsets of I by $\delta_x(B) := 1_B(x)$. Then the probability distribution of X is given by

$$\mu := \sum_{i=1}^{n} p_i \delta_{x_i} \tag{5.84}$$

because

$$\mu(B) = \sum_{i=1}^{n} p_i \delta_{x_i}(B) = \sum_{i=1}^{n} p_i 1_B(x_i) = \sum_{1 \leq i \leq n: \, x_i \in B} p_i = P(X \in B) \tag{5.85}$$

for all subsets $B \subset I$.

Wagers and strategies can be compared in various ways. One is in terms of house advantage, mentioned in Example 1.5.7 on p. 44. Another is in terms of *expected utility*. To each possible value of a gambler's fortune or wealth is associated a numerical value called its utility. Such a *utility function* is subjective and may vary from one gambler to the next. Let u be a utility function on $[0, \infty)$ and consider a gambler with fortune F who bets b units $(0 \leq b \leq F)$ on a set of m of the $36 + z$ numbers of a roulette wheel ($m \in \{1, 2, 3, 4, 6, 12, 18\}$, $z \in \{1, 2\}$). Then his fortune X after the wager is resolved has probability distribution

$$\mu = \left(1 - \frac{m}{36 + z}\right) \delta_{F-b} + \frac{m}{36 + z} \delta_{F+(36/m-1)b}, \tag{5.86}$$

and expected utility

$$E[u(X)] = \left(1 - \frac{m}{36 + z}\right) u(F - b) + \frac{m}{36 + z} u(F + (36/m - 1)b). \tag{5.87}$$

These wagers (F, z fixed; $0 \leq b \leq F$; $m = 1, 2, 3, 4, 6, 12, 18$) have the same house advantage (if $b > 0$) but may have very different expected utilities.

How is the utility function u determined? We take an axiomatic approach, assuming only that the gambler has a preference ordering on the set of probability distributions of discrete random variable assuming finitely many values in the set $[0, \infty)$ of fortunes. For example, given

$$\mu = \frac{20}{38} \delta_{F-1} + \frac{18}{38} \delta_{F+1} \quad \text{and} \quad \nu = \frac{37}{38} \delta_{F-1} + \frac{1}{38} \delta_{F+35}, \tag{5.88}$$

where $F \geq 1$ is specified, we assume that the gambler can decide whether $\mu \succ \nu$ (μ is preferred to ν), $\nu \succ \mu$ (ν is preferred to μ), or $\mu \sim \nu$ (μ is indifferent to ν). Actually, it will be convenient to introduce $\succeq$ ("is preferred to or indifferent to") and to replace $[0, \infty)$ by an arbitrary interval I.

We let $\mathscr{D}$ be the set of all probability distributions of discrete random variables assuming finitely many values in I. Notice that $\mathscr{D}$ is closed under convex combinations, that is,

$$\mu, \nu \in \mathscr{D} \text{ and } p \in [0,1] \quad \text{imply} \quad p\mu + (1-p)\nu \in \mathscr{D}. \tag{5.89}$$

We assume that there is a binary relation $\succeq$ on $\mathscr{D}$ that satisfies the following three axioms.

Axiom 5.3.1. *$\mathscr{D}$ is totally ordered by $\succeq$.*

This means that

$$\text{whenever } \mu, \nu \in \mathscr{D}, \text{ either } \mu \succeq \nu \text{ or } \nu \succeq \mu, \tag{5.90}$$

and

$$\text{if } \lambda, \mu, \nu \in \mathscr{D}, \ \lambda \succeq \mu, \text{ and } \mu \succeq \nu, \text{ then } \lambda \succeq \nu. \tag{5.91}$$

We write $\mu \sim \nu$ if $\mu \succeq \nu$ and $\nu \succeq \mu$. Observe that $\sim$ is an equivalence relation.

Axiom 5.3.2. *If $\lambda, \mu, \nu \in \mathscr{D}$, then the sets*

$$\{p \in [0,1] : \mu \succeq p\lambda + (1-p)\nu\} \quad \text{and} \quad \{p \in [0,1] : p\lambda + (1-p)\nu \succeq \mu\} \tag{5.92}$$

are closed.

This is called the *continuity axiom*. It follows that the set

$$\{p \in [0,1] : \mu \sim p\lambda + (1-p)\nu\} \tag{5.93}$$
$$= \{p \in [0,1] : \mu \succeq p\lambda + (1-p)\nu\} \cap \{p \in [0,1] : p\lambda + (1-p)\nu \succeq \mu\}$$

is also closed.

Axiom 5.3.3. *If $\lambda, \mu, \nu \in \mathscr{D}$, $\lambda \sim \mu$, and $p \in [0,1]$, then*

$$p\lambda + (1-p)\nu \sim p\mu + (1-p)\nu. \tag{5.94}$$

This is called the *independence axiom*, and it has been questioned by some. We begin with several basic lemmas.

Lemma 5.3.4. *If $\lambda, \mu, \nu \in \mathscr{D}$ and $\lambda \succeq \mu \succeq \nu$, then there exists $p \in [0,1]$ such that $\mu \sim p\lambda + (1-p)\nu$.*

Proof. Let

$$A_0 := \{p \in [0,1] : \mu \succeq p\lambda + (1-p)\nu\} \tag{5.95}$$

and

$$A_1 := \{p \in [0,1] : p\lambda + (1-p)\nu \succeq \mu\}. \tag{5.96}$$

Then $0 \in A_0$ and $1 \in A_1$ by assumption, so neither A_0 nor A_1 is empty. By Axiom 5.3.1, $A_0 \cup A_1 = [0,1]$, and by Axiom 5.3.2, both A_0 and A_1 are closed. Since the interval $[0,1]$ is connected, A_0 and A_1 cannot be mutually exclusive, and any p belonging to $A_0 \cap A_1$ satisfies the conclusion of the lemma. ♠

For the next result let us introduce the notation $\mu \succ \nu$ to mean that $\mu \succeq \nu$ but $\mu \not\sim \nu$.

Lemma 5.3.5. *If $\mu \succ \nu$ and $0 < p < 1$, then $\mu \succ p\mu + (1-p)\nu \succ \nu$.*

Proof. Suppose that it is not the case that $\mu \succ p\mu + (1-p)\nu$. Then $p\mu + (1-p)\nu \succeq \mu \succ \nu$, so by Lemma 5.3.4, there exists $q \in [0,1]$ such that $q[p\mu + (1-p)\nu] + (1-q)\nu \sim \mu$, or $qp\mu + (1-qp)\nu \sim \mu$. In particular,

$$A := \{r \in [0,1] : r\mu + (1-r)\nu \sim \mu\} \tag{5.97}$$

is nonempty. A is closed by Axiom 5.3.2 and excludes 0 by assumption, so it contains its infimum $r_0 > 0$. In particular, $r_0\mu + (1-r_0)\nu \sim \mu$, hence by Axiom 5.3.3, $p[r_0\mu + (1-r_0)\nu] + (1-p)\nu \sim p\mu + (1-p)\nu \succeq \mu \succ \nu$. Applying Lemma 5.3.4 once again, there exists $s \in [0,1]$ such that $s\{p[r_0\mu + (1-r_0)\nu] + (1-s)\nu \sim \mu$, or $spr_0\mu + (1-spr_0)\nu \sim \mu$, hence $spr_0 \in A$. But $spr_0 < r_0$, contradicting the definition of r_0. Therefore, $\mu \succ p\mu + (1-p)\nu$.

The assertion that $p\mu + (1-p)\nu \succ \nu$ is proved similarly.   ♠

Lemma 5.3.6. *Assume that $\mu \succ \nu$ and $p, q \in [0,1]$. Then $p\mu + (1-p)\nu \succ q\mu + (1-q)\nu$ if and only if $p > q$.*

Proof. First suppose that $p > q$. By Lemma 5.3.5, $p\mu + (1-p)\nu \succ \nu$, so by the same lemma, $p\mu + (1-p)\nu \succ (q/p)[p\mu + (1-p)\nu] + (1-q/p)\nu = q\mu + (1-q)\nu$. Conversely, suppose that $p \leq q$. By Lemma 5.3.5, $q\mu + (1-q)\nu \succ \nu$, so by the same lemma, $q\mu + (1-q)\nu \succeq (p/q)[q\mu + (1-q)\nu] + (1-p/q)\nu = p\mu + (1-p)\nu$.   ♠

To avoid trivialities, we exclude the possibility in what follows that $\mu \sim \nu$ for all $\mu, \nu \in \mathscr{D}$. Lemma 5.3.6 immediately yields the following extension of Lemma 5.3.4.

Corollary 5.3.7. *If $\lambda, \mu, \nu \in \mathscr{D}$ and $\lambda \succ \mu \succ \nu$, then there exists a unique $p \in (0,1)$ such that $\mu \sim p\lambda + (1-p)\nu$.*

Given $\lambda, \nu \in \mathscr{D}$ with $\lambda \succ \nu$, we define

$$\mathscr{D}_{\lambda\nu} := \{\mu \in \mathscr{D} : \lambda \succeq \mu \succeq \nu\}, \tag{5.98}$$

and we define the function $p_{\lambda\nu} : \mathscr{D}_{\lambda\nu} \mapsto [0,1]$ by

$$\mu \sim p_{\lambda\nu}(\mu)\lambda + (1 - p_{\lambda\nu}(\mu))\nu. \tag{5.99}$$

By Lemma 5.3.6, given $\mu_0, \mu_1 \in \mathscr{D}_{\lambda\nu}$, $p_{\lambda\nu}(\mu_1) > p_{\lambda\nu}(\mu_0)$ if and only if $\mu_1 \succ \mu_0$. Furthermore, the uniqueness in Corollary 5.3.7 implies that $p_{\lambda\nu}$ is linear on $\mathscr{D}_{\lambda\nu}$ in the sense that

$$p_{\lambda\nu}(q\mu_0 + (1-q)\mu_1) = qp_{\lambda\nu}(\mu_0) + (1-q)p_{\lambda\nu}(\mu_1) \tag{5.100}$$

for all $\mu_0, \mu_1 \in \mathscr{D}_{\lambda\nu}$ and $q \in [0,1]$.

Fix $\mu_0, \mu_1 \in \mathscr{D}$ with $\mu_1 \succ \mu_0$ for the remainder of this section. Given $\lambda, \nu \in \mathscr{D}$ with $\lambda \succ \nu$ and $\mu_0, \mu_1 \in \mathscr{D}_{\lambda\nu}$, we define $U_{\lambda\nu}$ on $\mathscr{D}_{\lambda\nu}$ by

$$U_{\lambda\nu}(\mu) := \frac{p_{\lambda\nu}(\mu) - p_{\lambda\nu}(\mu_0)}{p_{\lambda\nu}(\mu_1) - p_{\lambda\nu}(\mu_0)}. \tag{5.101}$$

Notice that $U_{\lambda\nu}(\mu_0) = 0$ and $U_{\lambda\nu}(\mu_1) = 1$.

Lemma 5.3.8. *Suppose $\lambda, \nu, \lambda', \nu' \in \mathscr{D}$ satisfy $\lambda \succ \nu$, $\lambda' \succ \nu'$, and $\mu_0, \mu_1 \in \mathscr{D}_{\lambda\nu} \cap \mathscr{D}_{\lambda'\nu'}$. Let f and g be linear, order-preserving functions on $\mathscr{D}_{\lambda\nu} \cap \mathscr{D}_{\lambda'\nu'}$ with $f(\mu_0) = g(\mu_0)$ and $f(\mu_1) = g(\mu_1)$. Then $f = g$ on $\mathscr{D}_{\lambda\nu} \cap \mathscr{D}_{\lambda'\nu'}$.*

Proof. Let $\mu \in \mathscr{D}_{\lambda\nu} \cap \mathscr{D}_{\lambda'\nu'}$.

If $\mu_1 \succeq \mu \succeq \mu_0$, then there exists $p \in [0,1]$ such that $\mu \sim p\mu_1 + (1-p)\mu_0$, so

$$\begin{aligned}
f(\mu) &= f(p\mu_1 + (1-p)\mu_0) \\
&= pf(\mu_1) + (1-p)f(\mu_0) \\
&= pg(\mu_1) + (1-p)g(\mu_0) \\
&= g(p\mu_1 + (1-p)\mu_0) \\
&= g(\mu). \tag{5.102}
\end{aligned}$$

If $\mu \succeq \mu_1 \succ \mu_0$, then there exists $q \in (0,1]$ such that $\mu_1 \sim q\mu + (1-q)\mu_0$, so

$$\begin{aligned}
qf(\mu) + (1-q)f(\mu_0) &= f(q\mu + (1-q)\mu_0) \\
&= f(\mu_1) \\
&= g(\mu_1) \\
&= g(q\mu + (1-q)\mu_0) \\
&= qg(\mu) + (1-q)g(\mu_0). \tag{5.103}
\end{aligned}$$

Since $f(\mu_0) = g(\mu_0)$ and $q > 0$, we conclude that $f(\mu) = g(\mu)$.

If $\mu_1 \succ \mu_0 \succeq \mu$, a similar argument applies and the proof is complete. ♠

We now define the function U on $\mathscr{D}$ as follows. Given $\mu \in \mathscr{D}$, choose $\lambda, \nu \in \mathscr{D}$ with $\lambda \succ \nu$ such that $\mu, \mu_0, \mu_1 \in \mathscr{D}_{\lambda\nu}$. We then define

$$U(\mu) := U_{\lambda\nu}(\mu). \tag{5.104}$$

Theorem 5.3.9. *The function U on $\mathscr{D}$ defined by (5.104) is well defined, linear, and order-preserving.*

Proof. Given $\mu \in \mathscr{D}$, suppose $\lambda, \nu, \lambda', \nu' \in \mathscr{D}$ satisfy $\lambda \succ \nu$, $\lambda' \succ \nu'$, and $\mu, \mu_0, \mu_1 \in \mathscr{D}_{\lambda\nu} \cap \mathscr{D}_{\lambda'\nu'}$. Then $U_{\lambda\nu}(\mu) = U_{\lambda'\nu'}(\mu)$ by Lemma 5.3.8 because $U_{\lambda\nu}(\mu_0) = 0 = U_{\lambda'\nu'}(\mu_0)$ and $U_{\lambda\nu}(\mu_1) = 1 = U_{\lambda'\nu'}(\mu_1)$. Thus, U is well defined. To show that U is order-preserving and linear, let $\mu, \mu' \in \mathscr{D}$, and choose $\lambda, \nu \in \mathscr{D}$ with $\lambda \succ \nu$ such that $\mu, \mu', \mu_0, \mu_1 \in \mathscr{D}_{\lambda\nu}$. Then $U(\mu) > U(\mu')$ if and only if $\mu \succ \mu'$ because the same property holds for $p_{\lambda\nu}$ and hence $U_{\lambda\nu}$. The linearity is established similarly.   ♠

If we identify the probability distribution δ_x with the real number x, the function U on $\mathscr{D}$ of Theorem 5.3.9 induces a function, denoted by u, on $\mathbf{R}$ given by $u(x) := U(\delta_x)$ and thought of as the *utility* to the gambler of the fortune x. Letting X be a discrete random variable satisfying (5.83), and letting μ be the distribution (5.84), the linearity property of U implies that

$$U(\mu) = U\left(\sum_{i \in I} p_i \delta_{x_i}\right) = \sum_{i \in I} p_i U(\delta_{x_i}) = \sum_{i \in I} p_i u(x_i) = \mathrm{E}[u(X)]. \quad (5.105)$$

This says that $U(\mu)$ is the expected utility of the gambler's fortune X, where X has probability distribution μ. It is for this reason that u is called a *utility function* while U is called an *expected utility function*.

What properties must u satisfy? First, it must be nondecreasing on I. A larger fortune cannot have smaller utility. Second, concavity of u is associated with risk aversion. Let us see why. A gambler is said to be *risk averse* (respectively, *risk seeking*) if, whenever X is a discrete random variable assuming finitely many values in I and having probability distribution μ, we have $\delta_{\mathrm{E}[X]} \succeq \mu$ (resp., $\mu \succeq \delta_{\mathrm{E}[X]}$). Informally, a risk-averse gambler prefers $\mathrm{E}[X]$ to X itself.

Lemma 5.3.10. *A gambler is risk averse if and only if his utility function u is concave. He is risk seeking if and only if u is convex.*

Proof. Risk aversion is equivalent to $U(\delta_{\mathrm{E}[X]}) \geq U(\mu)$ for all discrete random variables X assuming finitely many values in I, or to $u(\mathrm{E}[X]) \geq \mathrm{E}[u(X)]$ for all such X. By Jensen's inequality (Theorem 1.4.21 on p. 37), the latter is equivalent to the concavity of u.

The second assertion is proved similarly.   ♠

Another way to think about these properties of utility functions is in terms of certainty equivalents. If $C_u[X]$ is a number such that

$$u(C_u[X]) = \mathrm{E}[u(X)], \quad (5.106)$$

then we say that $C_u[X]$ is a *certainty equivalent* of X, in that the utility of the certain amount $C_u[X]$ is equal to the expected utility of the random amount X. If u is continuous and increasing on I, then $C_u[X]$ is uniquely determined and given by

$$C_u[X] = u^{-1}(\mathrm{E}[u(X)]). \quad (5.107)$$

For a risk-averse gambler, $C_u[X] \leq \mathrm{E}[X]$, while for a risk-seeking gambler, $C_u[X] \geq \mathrm{E}[X]$.

Example 5.3.11. *Utility functions.*

(a) If we take $I = (0, \infty)$ (instead of $I = [0, \infty)$), then the important logarithmic utility function $u(x) := \ln x$ is well defined. Notice that the certainty equivalent of X (a discrete random variable assuming finitely many positive values) is its *geometric mean*

$$C_u[X] = e^{\mathrm{E}[\ln X]}, \tag{5.108}$$

which by Example 1.4.22 on p. 38 is less than $\mathrm{E}[X]$ unless X is degenerate. The logarithmic utility function plays a minor role in Chapter 10.

(b) Another important utility function (with $I = [0, \infty)$) is the power utility function $u(x) := x^p$ for $p > 0$, which is concave for $0 < p \leq 1$ and convex for $p \geq 1$. Here the certainty equivalent of X (a discrete random variable assuming finitely many nonnegative values) is its L^p *norm* (which is, strictly speaking, not a norm unless $p \geq 1$)

$$C_u[X] = (\mathrm{E}[X^p])^{1/p}. \tag{5.109}$$

The special case $p = 1$ leads to the *objective utility function* $u(x) = x$, for which $C_u[X] = \mathrm{E}[X]$.

(c) We could also consider (with $I = (0, \infty)$ again) $u(x) = -x^{-p}$ for $p > 0$, also known as a power utility function. In the case $p = 1$, the certainty equivalent of X (a discrete random variable assuming finitely many positive values) is its *harmonic mean*

$$C_u[X] = (\mathrm{E}[X^{-1}])^{-1}. \tag{5.110}$$

(d) Utility functions need not be increasing (nondecreasing suffices) and need not be continuous. For example, $u(x) = 1_{[G, \infty)}(x)$ is the utility function of the gambler whose goal is to achieve the fortune $G > 0$; anything less is of no utility, and anything more has no additional utility. Here the concept of certainty equivalent is undefined. Such utility functions play an important role in Chapter 9. ♠

We conclude this section by discussing the local behavior of the certainty equivalent for a risk-averse gambler.

Theorem 5.3.12. *Let u be an increasing, concave, and twice continuously differentiable utility function on the interval I. Assume that the inverse function u^{-1} is also twice continuously differentiable. Let X be a random variable assuming finitely many values and representing the gambler's profit per unit bet. Let b and w denote the (nonrandom) bet size and initial wealth. Assume that $\mathrm{P}(w + bX \in I) = 1$ for $0 \leq b < b_0$. Then*

$$C_u[w + bX] = w + \mathrm{E}[X]b - \alpha(w)\frac{\mathrm{Var}(X)}{2w}b^2 + o(b^2) \tag{5.111}$$

as $b \to 0+$, where $o(b^2)$ denotes an error term that is negligible compared to b^2 in the sense that $o(b^2)/b^2 \to 0$ as $b \to 0+$, and

$$\alpha(w) := -w \frac{u''(w)}{u'(w)} \tag{5.112}$$

is the Arrow–Pratt measure of relative risk aversion.

Remark. (a) Notice that b is typically much smaller than w, so if money is measured in units of w, for example, then b is very small and the conclusion is meaningful.

(b) Examples of utility functions to which the theorem applies include $u(x) = \ln x$, in which case $\alpha \equiv 1$, $u(x) = x^p$, where $0 < p \le 1$, in which case $\alpha \equiv 1 - p$, and $u(x) = -x^{-p}$, where $p > 0$, in which case $\alpha \equiv 1 + p$. In these three cases, $\alpha(w)$ is constant in w.

(c) If $\mathrm{E}[X] > 0$ and $\alpha(w) > 0$, we can maximize the right side of (5.111), ignoring the $o(b^2)$ term, by choosing

$$b = \frac{1}{\alpha(w)} \frac{\mathrm{E}[X]}{\mathrm{Var}(X)} w. \tag{5.113}$$

In the case of $u(x) = \ln x$ and $\alpha \equiv 1$, we obtain the Kelly system (cf. (10.14)), whereas in the case of $u(x) = -x^{-p}$ and $\alpha \equiv 1 + p$, we obtain a fractional Kelly system, the fraction of Kelly being $1/(1+p)$.

Proof. Define $C(b) := C_u[w + bX] - w$ on some interval $[0, b_0]$, so that

$$C(b) = u^{-1}(\mathrm{E}[u(w + bX)]) - w. \tag{5.114}$$

Using calculus, we obtain

$$C'(b) = \frac{\mathrm{E}[u'(w + bX)X]}{u'(u^{-1}(\mathrm{E}[u(w + bX)]))} \tag{5.115}$$

and

$$\begin{aligned} C''(b) = \{ & u'(u^{-1}(\mathrm{E}[u(w + bX)]))\mathrm{E}[u''(w + bX)X^2] \\ & - \mathrm{E}[u'(w + bX)X]u''(u^{-1}(\mathrm{E}[u(w + bX)]))C'(b) \} \\ & \cdot \{ u'(u^{-1}(\mathrm{E}[u(w + bX)])) \}^{-2}. \end{aligned} \tag{5.116}$$

This leads to $C(0) = 0$, $C'(0) = \mathrm{E}[X]$, and $C''(0) = (u''(w)/u'(w))\mathrm{Var}(X)$, and therefore the result follows by a second-order Taylor expansion. ♠

5.4 Problems

5.1. *Elimination of dominated strategies.* By Corollary 5.2.4(b) on p. 179, elimination of a strictly dominated pure strategy does not affect the set of solutions of a game. Prove that a weaker conclusion holds for (not necessarily strictly) dominated strategies: Elimination of a dominated pure strategy does not affect the value of a game.

5.2. *A matching game.* In J. D. Williams's *The Compleat Strategyst* we find the following game.

> "I know a good game," says Alex. "We point fingers at each other; either one finger or two fingers. If we match with one finger, you buy me a Daiquiri. If we match with two fingers, you buy me two Daiquiris. If we don't match I let you off with a payment of a dime. It'll help to pass the time."
>
> Olaf appears quite unmoved. "That sounds like a very dull game—at least in its early stages." His eyes glaze on the ceiling for a moment and his lips flutter briefly; he returns to the conversation with: "Now if you'd care to pay me 42 cents before each game, as partial compensation for all those 55-cent drinks I'll have to buy you, then I'd be happy to pass the time with you."
>
> "Forty-one cents," says Alex.
>
> "All right," sighs Olaf. "You really should pay 42 cents, at least once in every 30 games, but I suppose it won't last that long."

To avoid possible misinterpretation, note that the phrase, "I let you off with a payment of a dime," means that Olaf pays Alex 10 cents.

(a) Solve the game if there is no side payment.

(b) Find the exact side payment from Alex to Olaf required to make the game fair. Explain Olaf's remark in the last paragraph.

5.3. 2×2 *games with saddle points.* Find all solutions of an arbitrary 2×2 matrix game with at least one saddle point. For specificity, write the payoff matrix as in (5.70) on p. 182, and assume that $d \leq a \leq b$, so that $(1,1)$ is a saddle point but not necessarily uniquely. Consider four cases, $d < a < b$, $d < a = b$, $d = a < b$, and $d = a = b$, perhaps with subcases that depend on c.

5.4. *Nonsingular-matrix games.* Let $\boldsymbol{A}$ be the payoff matrix for an $m \times m$ matrix game. Write $\mathbf{1} := (1, 1, \ldots, 1) \in \mathbf{R}^m$, and assume that $\boldsymbol{A}$ is nonsingular and $\mathbf{1}\boldsymbol{A}^{-1}\mathbf{1}^\mathsf{T} \neq 0$. Assuming there exist optimal mixed strategies $\boldsymbol{p}^*, \boldsymbol{q}^* \in \Delta_m$ with $p_i^* > 0$ for $i = 1, \ldots, m$ and $q_j^* > 0$ for $j = 1, \ldots, m$, give formulas for $\boldsymbol{p}^*$, $\boldsymbol{q}^*$, and the value v of the game. Can this last assumption be weakened by assuming only that the components of $\boldsymbol{p}^*$ and $\boldsymbol{q}^*$ in the formulas just found are nonnegative?

5.5. *Montmort's game.* Player 1 has a number of coins in his hand and asks player 2 to guess whether the number is even or odd. If player 2's guess is correct, he wins from player 1 four units when the number is even and one unit when the number is odd. If player 2's guess is incorrect, he wins nothing. Find the optimal strategy for each player and the value of the game.

5.6. *Diagonal-matrix games.* As a special case of Problem 5.4, consider the game with diagonal payoff matrix

$$A = \begin{pmatrix} a_1 & 0 & \cdots & 0 \\ 0 & a_2 & \cdots & 0 \\ \vdots & \vdots & & \vdots \\ 0 & 0 & \cdots & a_m \end{pmatrix}. \tag{5.117}$$

(a) If $a_1, a_2, \ldots, a_m$ are all positive or all negative, solve the game.

(b) If $a_1, a_2, \ldots, a_m$ do not meet the condition in (a), solve the game.

5.7. *Examples with pure optimal strategies.* Consider Example 5.1.1 on p. 161 with $N = 3, 4, 6$. In each case find a saddle point for the payoff matrix and therefore a pure optimal strategy (for both players). Show that, if dominated rows and columns are eliminated one by one, the same strategy remains. Is this optimal strategy unique?

5.8. *Verifying a game's solution* . Consider Example 5.1.1 on p. 161 with $N = 8$. By symmetry, the game's value is 0. Determine whether the strategy (for both players) $\boldsymbol{p}^* = (1/87)(0, 0, 32, 36, 0, 16, 0, 3)$ is optimal. Do the same for the strategy $\boldsymbol{p}^{**} = (1/117)(0, 0, 32, 60, 0, 16, 0, 9)$.

5.9. *Dominance and le her.* Recall Example 5.1.2 on p. 162.

(a) Prove that every pure strategy for player 1 not of the form $S_i := \{1, \ldots, i\}$ $(i = 0, 1, \ldots, 13)$ is strictly dominated by a pure strategy of this form.

(b) Prove also that every pure strategy for player 2 not of the form $T_j := \{1, \ldots, j\}$ $(j = 0, 1, \ldots, 13)$ is dominated by a pure strategy of this form, but not strictly.

Hint: Extend the payoff matrix (5.11) on p. 164 to $a_{S,T}$, where $S, T \subset \{1, 2, \ldots, 13\}$. Here S comprises the denominations that are exchanged by player 1, and T comprises those that are exchanged by player 2 when player 1 does not exchange. Suppose $i \in S$, $i \geq 2$, and $i - 1 \notin S$. With $S' := (S \cup \{i - 1\}) - \{i\}$, show that $a_{S',T} > a_{S,T}$ for all $T \subset \{1, 2, \ldots, 13\}$. Suppose $j \in T$, $j \geq 2$, and $j - 1 \notin T$. With $T' := (T \cup \{j - 1\}) - \{j\}$, show that $a_{S,T'} \leq a_{S,T}$ for all $S \subset \{1, 2, \ldots, 13\}$.

5.10. *Discrete convexity and concavity.* Recall that a twice-continuously differentiable function f on $\mathbf{R}$ is convex (resp., concave) if f' is nondecreasing (resp., nonincreasing). (See Lemma 1.4.20 on p. 37.) By analogy, we say that a finite sequence $a_1, \ldots, a_n$ is *discrete convex* (resp., *discrete concave*) if $a_{i+1} - a_i$ is nondecreasing (resp., nonincreasing) in $i = 1, \ldots, n - 1$.

Let (a_{ij}) be the payoff matrix for an $m \times n$ matrix game, and prove the following. If a_{ij} is discrete concave in i for each $j \in \{1, \ldots, n\}$, then player 1 has an optimal strategy that is a mixture of at most two consecutive pure strategies. If a_{ij} is discrete convex in j for each $i \in \{1, \ldots, m\}$, then player

2 has an optimal strategy that is a mixture of at most two consecutive pure strategies.

Hint: Consider the first assertion. For $j = 1, \ldots, n$ define $f_j(i) := a_{ij}$ on $\{1, \ldots, m\}$ and extend f_j to $[1, m]$ by linear interpolation, so that f_j is concave on $[1, m]$. Let $v := \max_{p \in \Delta_m} \min_{q \in \Delta_n} p A q^{\mathsf{T}}$ be the value of the game, and define $v_0 := \max_{1 \le x \le m} \min_{1 \le j \le n} f_j(x)$. Show that $v \ge v_0$ by definition and that $v \le v_0$ by Jensen's inequality.

5.11. *Le her with an n-deck shoe.* Consider le her played with an n-deck shoe comprising $52n$ cards. Find the value of the game (to player 1) as a function of n. What is the smallest n for which this value is negative?

5.12. *Le her with d denominations and s suits.* Consider an abstract version of le her played with a deck of sd cards in which each of the d denominations $1, 2, \ldots, d$ appears s times. The rules are exactly as in Example 5.1.2 on p. 162 but with the role of the king denomination played by denomination d. Solve the game, and check that the results reduce to those of Examples 5.1.2 on p. 162 and 5.2.6 on p. 182 in the special case $d = 13$ and $s = 4$.

Hint: Use the results of Problem 5.10.

5.13. *Solving a $2 \times n$ game via graphical methods.* Solve the 2×10 game of Table 5.6 on p. 175 by drawing the graphs of $f_1, \ldots, f_{10}$ of (5.78) on p. 183 on the same coordinate system. To achieve sufficient accuracy, use computer graphics.

5.14. *Strict domination by mixtures in chemin de fer.*

(a) Generalize Corollary 5.2.4(b) on p. 179 to the case in which a row (or column) is strictly dominated by a mixture of rows (or columns).

(b) Show that, in the notation of (5.78) on p. 183, if, for some $j \in \{1, 2, \ldots, n\}$, $\min\{f_1(p), \ldots, f_n(p)\} < f_j(p)$ for all $p \in [0, 1]$, then strategy j for player 2 is strictly dominated by a mixture of two of player 2's pure strategies. In the case of Example 5.1.4 on p. 167, use this idea to reduce the 2×10 payoff matrix in Table 5.6 on p. 175 to a 2×5 one.

5.15. *Implementing the optimal strategies at chemin de fer.* We derived the optimal strategies at chemin de fer assuming sampling with replacement.

(a) The only randomization required for player's optimal strategy is when player has 5. Let us assume that player notices the composition of his 5 and draws with $\{0, 5\}$, $\{1, 4\}$, and $\{2, 3\}$, and stands with $\{6, 9\}$. How should he play $\{7, 8\}$ to achieve the optimal draw-stand mixture?

(b) The only randomization required for banker's optimal strategy is when banker has 6 and player stands. Let us assume that banker notices the composition of his 6 in this case and draws with $\{1, 5\}$, $\{2, 4\}$, and $\{7, 9\}$, and stands with $\{0, 6\}$ and $\{3, 3\}$. How should he play $\{8, 8\}$ to achieve the optimal draw-stand mixture?

(c) Is there any advantage to playing in this way?

5.16. *Chemin de fer with less restrictive player rules.* Suppose that player at chemin de fer is required to draw to 3 or less and stand on 6 and 7, but that he may draw or stand on 4 and 5. Suppose further that banker is unrestricted. Assuming sampling with replacement, as in Examples 5.1.4 on p. 167 and 5.2.7 on p. 183, this makes chemin de fer a 4×2^{88} matrix game. Solve the game.

Hint: Use Lemma 5.1.3 on p. 166 and then use strict dominance.

5.17. *Modern chemin de fer.* In the modern version of chemin de fer, player draws to 4 or less, stands on 6 or more, and may draw or stand on 5. Banker's rules are given by Table 5.8 on p. 185. Formulate the game as a 2×4 matrix game, and solve the game.

5.18. *Chemin de fer and sampling without replacement.* In Example 5.1.4 on p. 167, we assumed sampling with replacement to simplify the analysis. Under the more accurate assumption that cards are dealt without replacement from a six-deck shoe comprising 312 cards, notice that chemin de fer is a $2^5 \times 2^{484}$ matrix game. Use strict dominance (especially Lemma 5.1.3 on p. 166) to reduce this game to a more manageable size.

5.19. *Super pan nine.* This game is similar to chemin de fer. We regard it as a two-person game, player vs. banker, although several players vs. one banker is more typical. Super pan nine is dealt from a shoe comprising 8, 10, or 12 nonstandard 36-card decks (denominations 7, 8, 9, and 10 are removed from a standard deck). The remaining denominations A, 2–6, J, Q, K are valued as 1, 2–6, 0, 0, 0, respectively. The value of a hand, consisting of three or four cards, is the sum of the values of the cards, modulo 10.

Three cards are dealt face down to player and three to banker, and each may look only at his own hand. The object of the game is to have the higher-valued hand (closer to 9) at the end of play. First, player has the option of drawing a fourth card. If he exercises this option, his fourth card is dealt face down. Next, banker has the option of drawing a fourth card. This completes the game, and the higher-valued hand wins. A win for player pays even money. Hands of equal value result in a push. Notice that banker makes his decision knowing only whether player has drawn a fourth card, not the value of that card. This makes super pan nine simpler than chemin de fer.

To simplify matters, assume that cards are dealt *with replacement*. Further, assume that player always draws to a hand valued 5 or less and always stands on a hand valued 7 or more. Only with a hand valued 6 does player have a decision to make. Banker's strategy, on the other hand, is assumed unrestricted. Thus, we have a 2×2^{20} matrix game. Solve the game.

5.20. *A symmetric dice game.* Each of two players has an unmarked six-sided fair die and is required to secretly mark each face with a positive integer between 1 and 6 inclusive, subject to the constraints that the numbers chosen sum to 21 and that some number is used more than once (ruling out the standard die). Each player then rolls his die, and if the uppermost faces are unequal, the player with the larger integer wins, receiving one unit from his opponent. Otherwise the game results in a tie and no money changes hands.

Show that this is a symmetric 31×31 matrix game, and find a (nonunique) solution.

5.21. *Allais's paradox.* Let $M = 10^6$ and define

$$\mu_1 := \delta_M, \qquad \mu_2 := \frac{10\delta_{5M} + 89\delta_M + \delta_0}{100}, \qquad (5.118)$$

as well as

$$\mu_3 := \frac{10\delta_{5M} + 90\delta_0}{100}, \qquad \mu_4 := \frac{11\delta_M + 89\delta_0}{100}. \qquad (5.119)$$

It is said that most people prefer μ_1 to μ_2 and μ_3 to μ_4. Show that the preferences $\mu_1 \succ \mu_2$ and $\mu_3 \succ \mu_4$ violate Axioms 5.3.1–5.3.3 on p. 187.

5.5 Notes

The minimax theorem (Theorem 5.2.3) is due to John von Neumann [1903–1957] (1928). The first elementary (nontopological) proof was provided by Jean-André Ville [1910–1989] (1938), and his argument was further simplified by von Neumann and Oskar Morgenstern [1902–1977] in their classic *Theory of Games and Economic Behavior* (1944), said by Poundstone (1992, p. 41) to be "one of the most influential and least-read books of the twentieth century." Our proof coincides with the theirs. For a constructive proof, see Dresher (1961, Section 2.8).

The notions of a dominated strategy, a mixed strategy, and an optimal (or minimax) strategy are due to Émile Borel [1871–1956] in a series of papers from 1921 to 1927 (see, e.g., Borel 1953). Dimand and Dimand (1992) discussed Borel's contributions. Lemma 5.2.1 (separating hyperplanes) "was certainly known to Minkowski as early as 1896," according to Kuhn (2003, p. 46), who cited Hermann Minkowski [1864–1909] (1910, Section 16).

Example 5.1.1 and its solution for $N = 5$ in Example 5.2.5 are due to David Gale [1921–2008] (1960, pp. 232–233), who interpreted it as a five-step silent duel.

> The players advance towards each other in five steps. After each step, a player is allowed to fire or not but he may fire only once during the game, and the probability that a player will hit his opponent if he fires after moving in k steps is $k/5$.

Remarkably, Example 5.1.2 and its solution in Example 5.2.6 predate the minimax theorem by more than two centuries. In 1713 Charles Waldegrave (misidentified as James Waldegrave by Kuhn 1968; see Bellhouse 2007 for the explanation) obtained the explicit minimax solution of le her and sent it to Pierre Rémond de Montmort [1678–1719], who subsequently quoted from Waldegrave's solution in a letter to Nicolaus Bernoulli [1687–1759] dated November 15, 1713. This letter appears in Montmort (1713, pp. 403–412).

(See Kuhn 1968 for an English translation and discussion.) Waldegrave, described by Todhunter (1865, Article 209) as an "English gentleman," acknowledged that a mixed strategy "does not seem to be in the usual rules of play." His result remained unnoticed until Guilbaud [1911–2007] (1961) rediscovered it. This is apparently not well known, for as recently as a few years ago Waldegrave was not credited with solving the game (Gambarelli and Owen 2004). In any case, the famous statistician and geneticist Ronald A. Fisher [1890–1962] (1934b) solved le her independently of Waldegrave, while Dresher (1951) subsequently did the same independently of Waldegrave and Fisher. For an alternative approach to solving le her, see Deloche and Oguer (2007a).

Examples 5.1.4 and 5.2.7 are due to John G. Kemeny [1926–1992] and J. Laurie Snell [1925–] (1957), and Lemma 5.1.3 is a precise formulation of an idea that is implicit in their work. We postpone to Section 19.4 a discussion of the history of chemin de fer, including earlier attempts at a mathematical analysis of the game, as well as later work. It should be noted that it was possible to play chemin de fer optimally as recently as 1963, at Crockford's Club in London. Kendall and Murchland (1964) described "Crockford's Drawing Card," and it is identical to Table 5.8 with one exception: There is a third optional case, namely when banker's two-card total is 6 and player stands. (This, we suggest, is how the game *should* be played.) For an alternative approach to solving chemin de fer, see Deloche and Oguer (2007b).

There are numerous textbooks on game theory. We found the one by Ferguson (2009) particularly useful.

The notion of utility was introduced by Daniel Bernoulli [1700–1782] (1738) as a way of resolving the St. Petersburg paradox (see Example 1.4.5 and Problem 1.34). The axiomatic treatment of the existence of an expected utility function is due to Israel N. Herstein [1923–1988] and John Milnor [1931–] (1953), whose work simplified the original results on this topic due to von Neumann and Morgenstern (1944). The Arrow–Pratt measure of relative risk aversion is due to Kenneth J. Arrow [1921–] (1963) and John W. Pratt (1964), and Theorem 5.3.12 is essentially a result of Pratt. See also Canjar (2007a).

The quote in Problem 5.2 is from Williams (1954, pp. 48–49), which also has a solution of the problem. Problem 5.5 (Montmort's game) was proposed to Nicolaus Bernoulli by Montmort (1713, p. 406); see also Hald (1990, p. 324). Problem 5.9 was solved in an expanded unpublished version of Benjamin and Goldman (1994). Problem 5.10 is due to Howard (1994). Problem 5.12 was solved by Vanniasegaram (2006) and is closely related to work of Benjamin and Goldman (2002). Problem 5.15 was motivated by Downton and Holder (1972). Problem 5.20 was motivated by Finkelstein and Thorp (2007). The Allais paradox is due to Maurice Allais [1911–] (1953). See Machina (1982) for discussion.

Chapter 6
House Advantage

But no man can say I ever took an advantage of him beyond the advantage of the game.

William Makepeace Thackeray, *The History of Henry Esmond*

In Section 6.1 we define the notion of the house advantage of a wager, a numerical index of its unfavorability. The house advantage is the ratio of the gambler's expected loss to his expected amount bet. Actually, there are at least three aspects of the definition that are arguable, the first concerning how pushes are accounted for, the second concerning how bet size is measured in compound wagers, and the third concerning the very meaning of bet size. In Section 6.2 we treat the case of composite wagers, which may or may not be made simultaneously. In Section 6.3 we discuss the volatility of a wager as measured by its variance.

6.1 A Single Wager

We begin by defining the house advantage of a single wager. A wager can be described in terms of a pair (B, X) of jointly distributed random variables, B denoting the amount bet and X denoting the gambler's profit (positive, negative, or zero). The *amount bet* is not necessarily the amount placed on the betting surface or the amount of the entry fee, but rather it is the amount placed at risk. We require

$$ X^- \leq B, \quad \mathrm{E}[B] < \infty, \quad \mathrm{E}[B\,1_{\{X \neq 0\}}] > 0, \quad \mathrm{E}\big[|X|\big] < \infty. \tag{6.1} $$

Recalling that $X^- := -\min(X, 0)$, the first of these inequalities says simply that one cannot lose more than one bets.

S.N. Ethier, *The Doctrine of Chances*, Probability and its Applications,
DOI 10.1007/978-3-540-78783-9_6, © Springer-Verlag Berlin Heidelberg 2010

There are two accepted definitions of the *house advantage* of the wager (B, X). The first is[1]

$$H_0(B, X) := \frac{-E[X]}{E[B]},\tag{6.2}$$

and the second is

$$H(B, X) := \frac{-E[X]}{E[B \, 1_{\{X \neq 0\}}]}.\tag{6.3}$$

In both versions the numerator is the gambler's expected loss. In (6.2) the denominator is the gambler's expected amount bet. In (6.3) the denominator is the gambler's expected amount of *action*. The amount of action is just the amount bet, except in the case of a push (tie). A push provides no action. We will later discuss the distinction between the two formulations.

Thus, the house advantage is a numerical index of the unfavorability of a wager that is invariant under changes in bet size. (Typically, if the bet size is multiplied by a factor of b, so too is the gambler's profit. Keno is an exception.) It allows the gambler to choose, with some justification, one wager over another by virtue of its smaller house advantage.

Let $(B_1, X_1), (B_2, X_2), \ldots$ be a sequence of i.i.d. random vectors whose common distribution is that of (B, X), representing the results of independent repetitions of the original wager. Then

$$\frac{-(X_1 + \cdots + X_n)}{B_1 + \cdots + B_n}\tag{6.4}$$

represents the ratio of the gambler's cumulative loss after n such wagers to his total amount bet, while

$$\frac{-(X_1 + \cdots + X_n)}{B_1 \, 1_{\{X_1 \neq 0\}} + \cdots + B_n \, 1_{\{X_n \neq 0\}}}\tag{6.5}$$

represents the ratio of the gambler's cumulative loss after n such wagers to his total amount of action. Dividing both numerator and denominator by n and applying the strong law of large numbers (Theorem 1.5.5 on p. 43), we see that these ratios converge almost surely to $H_0(B, X)$ and $H(B, X)$. This gives a simple interpretation to the concept of house advantage: $H_0(B, X)$ is the long-term ratio of the gambler's cumulative loss to his total amount bet, while $H(B, X)$ is the long-term ratio of the gambler's cumulative loss to his total amount of action.

Let N be the number of coups needed to achieve a win or a loss, that is, $N := \min\{n \geq 1 : X_n \neq 0\}$, and notice that N is geometrically distributed with success probability $P(X \neq 0)$. An equivalent way to look at $H(B, X)$ is as follows.

---

[1] We will not be doing hypothesis testing in this book (except perhaps in Section 13.2), so there is no chance of confusing H_0 with the null hypothesis.

Theorem 6.1.1. *Under the above assumptions (namely, the random vector* (B, X) *satisfies (6.1) and* $(B_1, X_1), (B_2, X_2), \ldots$ *are i.i.d. as* (B, X)*), the distribution of* (B_N, X_N) *coincides with the conditional distribution of* (B, X) *given* $X \neq 0$. *Moreover,*

$$\mathrm{H}_0(B_1 + \cdots + B_N, X_1 + \cdots + X_N) = \mathrm{H}_0(B, X) \tag{6.6}$$

and

$$\mathrm{H}_0(B_N, X_N) = \mathrm{H}(B, X). \tag{6.7}$$

Remark. Results (6.6) and (6.7) make the distinction between $\mathrm{H}_0(B, X)$ and $\mathrm{H}(B, X)$ more explicit than do (6.2) and (6.3). If a push is regarded as a conclusive outcome of the bet, then $B_1 + \cdots + B_N$ units have been bet by time N and (6.2) is the appropriate definition of house advantage; we describe it as house advantage with *pushes included*. On the other hand, if a push is regarded as merely a delay in the eventual resolution of the bet, then only B_N units have been bet by time N and (6.3) is the appropriate definition of house advantage; we describe it as house advantage with *pushes excluded*.

How should a push be regarded? We believe that the answer depends on the game. Therefore, we consider both formulations in what follows.

Proof. First, notice that for bounded functions $f(b, x)$, we have

$$
\begin{aligned}
\mathrm{E}[f(B_N, X_N)] &= \sum_{n=1}^{\infty} \mathrm{E}[f(B_n, X_n)\, 1_{\{N=n\}}] \\
&= \sum_{n=1}^{\infty} \mathrm{E}[f(B_n, X_n)\, 1_{\{X_1=0,\ldots,X_{n-1}=0,X_n\neq 0\}}] \\
&= \sum_{n=1}^{\infty} \mathrm{E}[1_{\{X_1=0,\ldots,X_{n-1}=0\}}]\mathrm{E}[f(B_n, X_n)\, 1_{\{X_n\neq 0\}}] \\
&= \sum_{n=1}^{\infty} \mathrm{P}(X = 0)^{n-1}\mathrm{E}[f(B, X)\, 1_{\{X\neq 0\}}] \\
&= \frac{\mathrm{E}[f(B, X)\, 1_{\{X\neq 0\}}]}{1 - \mathrm{P}(X = 0)} \\
&= \mathrm{E}[f(B, X) \mid X \neq 0], \tag{6.8}
\end{aligned}
$$

where the first equality uses the fact that $\mathrm{P}(N < \infty) = 1$ and the dominated convergence theorem (Theorem A.2.5 on p. 748), the second uses the definition of N, the third uses independence, the fourth uses the i.i.d. property, the fifth uses the formula for the sum of a geometric series, and the sixth uses the definition of conditional expectation.

Second, (6.6) follows from Wald's identity (Corollary 3.2.5 on p. 103) since

$$\frac{\mathrm{E}[X_1 + \cdots + X_N]}{\mathrm{E}[B_1 + \cdots + B_N]} = \frac{\mathrm{E}[N]\mathrm{E}[X]}{\mathrm{E}[N]\mathrm{E}[B]} = \frac{\mathrm{E}[X]}{\mathrm{E}[B]}, \tag{6.9}$$

while (6.7) uses

$$\frac{\mathrm{E}[X_N]}{\mathrm{E}[B_N]} = \frac{\mathrm{E}[X \mid X \neq 0]}{\mathrm{E}[B \mid X \neq 0]} = \frac{\mathrm{E}[X 1_{\{X \neq 0\}}]}{\mathrm{E}[B 1_{\{X \neq 0\}}]} = \frac{\mathrm{E}[X]}{\mathrm{E}[B 1_{\{X \neq 0\}}]}, \qquad (6.10)$$

a consequence of (6.8).   ♠

We now illustrate the definitions with several examples.

Example 6.1.2. *Wagers with three possible outcomes: Win, loss, or push.*
Let the probabilities of a win, loss, and push be respectively $p > 0$, $q > 0$,
and $r \geq 0$, where $p + q + r = 1$. Suppose that a win pays a to 1, where $a > 0$.
If the bet size is $b > 0$, then $\mathrm{P}(X = ab) = p$, $\mathrm{P}(X = -b) = q$, $\mathrm{P}(X = 0) = r$,
and $\mathrm{P}(B = b) = 1$. We conclude that the house advantage is

$$\mathrm{H}_0(B, X) = -ap + q \qquad \text{or} \qquad \mathrm{H}(B, X) = \frac{-ap + q}{p + q}. \qquad (6.11)$$

Many wagers have this simple structure, including the next three examples.
                                                                                                                ♠

Example 6.1.3. *Roulette: m-number bet.* Let us consider an unbiased wheel
with $z = 1$ or $z = 2$ zeros. A bet on a subset of size m of the set of $36 + z$
numbers is available for $m = 1, 2, 3, 4, 6, 12, 18$ and pays $36/m - 1$ to 1 if
a number in that subset appears, ignoring the "partager" and "en prison"
rules. There is no possibility of a push, so the house advantage is

$$\mathrm{H}_0 = \mathrm{H} = -\left(\frac{36}{m} - 1\right)\frac{m}{36 + z} + 1 - \frac{m}{36 + z}$$

$$= \frac{z}{36 + z} = \begin{cases} 1/37 = 0.\overline{027} & \text{if } z = 1, \\ 1/19 \approx 0.052632 & \text{if } z = 2, \end{cases} \qquad (6.12)$$

regardless of the value of m. Like probabilities, house advantages are often
stated in percentage terms (e.g., $2.7\overline{027}$ percent).   ♠

Example 6.1.4. *Craps: Pass line or don't pass.* The *don't-pass bet* at craps
is an even-money bet that the pass-line bet (Example 1.2.2 on p. 17) will
be lost. However, 12 on the come-out roll is considered a push rather than a
win. As in Example 1.2.8 on p. 19, the probabilities of a win, loss, and push
for the don't-pass bet are $p := 949/1{,}980$, $q := 976/1{,}980$, and $r := 55/1{,}980$,
and the bet pays even money $(a = 1)$. By (6.11), the house advantage is
$\mathrm{H}_0 = 27/1{,}980 = 3/220 = 0.01\overline{36}$ or $\mathrm{H} = 27/1{,}925 = 0.01\overline{402597}$.

The probabilities of a win, loss, and push for the pass-line bet are q, $p + r$,
and 0, respectively, with p, q, and r as in the preceding paragraph. The house
advantage is therefore $\mathrm{H}_0 = \mathrm{H} = -q + p + r = 28/1{,}980 = 7/495 = 0.0\overline{14}$.   ♠

Example 6.1.5. *Craps: Place bets.* A *place bet* on one of the six point numbers (4, 5, 6, 8, 9, and 10) is usually described as a bet that the specified number will appear before 7 appears. It pays 9 to 5 for 4 or 10, 7 to 5 for 5 or 9, and 7 to 6 for 6 or 8. Since the bet can be withdrawn following a nondecision, it is arguably more accurate to describe the bet as a one-roll bet that is won if the specified number appears, is lost if 7 appears, and is pushed otherwise. With the latter formulation, Example 6.1.2 applies. The house advantage for a place bet on 4 (or 10) is

$$H = \frac{-(9/5)\pi_4 + \pi_7}{\pi_4 + \pi_7} = \frac{1}{15} = 0.0\overline{6}. \tag{6.13}$$

Similarly, the house advantage for a place bet on 5 (or 9) is $H = 1/25 = 0.04$, and the house advantage for a place bet on 6 (or 8) is $H = 1/66 = 0.0\overline{15}$. There is nearly universal agreement on these figures. However, were we to use (6.2), the house advantage for a place bet on 4 (or 10) would be

$$H_0 = -(9/5)\pi_4 + \pi_7 = 1/60 = 0.01\overline{6}. \tag{6.14}$$

Similarly, the house advantage for a place bet on 5 (or 9) would be $H_0 = 1/90 = 0.0\overline{1}$, and the house advantage for a place bet on 6 (or 8) would be $H_0 = 1/216 = 0.004\overline{629}$. Those who prefer H_0 as the definition of house advantage either must accept these unconventional figures or are obliged to use the first-stated formulation of the place bets, which implicitly requires such bets to remain in effect until they are won or lost. ♠

There are a number of situations in which the "amount bet" is not what we have defined as the bet size. Here is an example.

Example 6.1.6. *Keno: 20-spot ticket.* The keno player chooses 20 numbers from the set $\{1, 2, \ldots, 80\}$, not necessarily at random, marks them on his keno ticket, and pays a one-unit entry fee. The casino then randomly draws 20 numbers from the set $\{1, 2, \ldots, 80\}$ without replacement. If the player's 20 numbers and the casino's 20 numbers have k numbers in common, where $0 \leq k \leq 20$, then the player receives a payoff of $a(k)$ units (but his entry fee is not returned). The payoffs are $a(0) = 200$, $a(1) = 2$, ..., $a(20) = 50,000$. See Table 6.1 for the complete list. Thus, for example, the payoff odds for no catches are 200 *for* 1. Of particular interest is the fact that the player *always* receives something. These payoffs vary somewhat from one casino to the next, and the latter feature is atypical.

It is easy to see (cf. Example 1.3.1 on p. 23) that

$$P(k) := P(\text{catch } k \text{ of } 20) = \frac{\binom{20}{k}\binom{60}{20-k}}{\binom{80}{20}}, \tag{6.15}$$

so the player's expected loss (in units) can be calculated as

Table 6.1 20-spot keno ticket payoffs (in units) at the MGM Grand (Sept. 2004). Entry fee is one unit (= \$5). Maximum aggregate payout is 50,000 units.

catch	payoff	catch	payoff	catch	payoff
0	200	7	3/5	14	2,500
1	2	8	7/5	15	5,000
2	3/5	9	4	16	10,000
3	1/5	10	10	17	20,000
4	1/5	11	40	18	30,000
5	1/5	12	200	19	40,000
6	1/5	13	1,000	20	50,000

$$1 - \sum_{k=0}^{20} a(k)P(k) \approx 0.260619539. \tag{6.16}$$

To evaluate the house advantage, it remains to divide by the expected amount bet, and here it is important to distinguish between the one-unit entry fee and the amount placed at risk. The latter is only 4/5 units because the player is assured of a payoff of at least 1/5 units. Dividing (6.16) by 4/5, we find that the house advantage for this 20-spot keno ticket is $H_0 = H \approx 0.325774424$.

Here we have not taken into account the maximum aggregate payout. See Section 14.1 for a discussion of this point.   ♠

We turn next to a second point of contention in the definition of house advantage. Frequently, a wager calls for an initial bet and then, conditional on the occurrence of some event, an additional bet becomes available. We will call this a (two-stage) *compound wager*. In evaluating the house advantage under the assumption that the additional wager is accepted when offered, should the bet size be regarded as the initial bet only, or should it be regarded as the sum of the initial bet and the additional bet if any? The two approaches lead to different definitions of house advantage.

To make this distinction more precise, we consider an initial wager (B, X) satisfying (6.1) and suppose that, if event A occurs, an additional wager (B', X'), also satisfying (6.1), is made, where $B' = X' = 0$ on A^c. Then the two formulations of house advantage with pushes included are

$$H_0(B, X + X') = \frac{-E[X + X']}{E[B]} \tag{6.17}$$

and

$$H_0(B + B', X + X') = \frac{-E[X + X']}{E[B + B']}, \tag{6.18}$$

while the two formulations of house advantage with pushes excluded are

$$H(B, X + X') = \frac{-E[X + X']}{E[B \, 1_{\{X+X'\neq 0\}}]} \tag{6.19}$$

and

$$H(B + B', X + X') = \frac{-E[X + X']}{E[(B + B') \, 1_{\{X+X'\neq 0\}}]}. \tag{6.20}$$

Notice that we are extending the definitions of H_0 and H beyond the class of wagers satisfying (6.1) in that we allow (in the case of (6.17) and (6.19)) the amount lost to exceed the amount bet. Further, we need to assume that the denominators in (6.19) and (6.20) are positive.

The interpretations should be clear: $H_0(B + B', X + X')$ is the long-term ratio of the gambler's cumulative loss to his total amount bet, while $H_0(B, X + X')$ is the long-term ratio of the gambler's cumulative loss to the total amount of his initial bets. Similarly, $H(B + B', X + X')$ is the long-term ratio of the gambler's cumulative loss to his total amount of action, while $H(B, X + X')$ is the long-term ratio of the gambler's cumulative loss to the total amount of action from his initial bets.

In our view, (6.18) and (6.20) are the more natural and useful quantities, yet (6.17) and (6.19) seem to be more widely accepted, except at craps.

Example 6.1.7. *Pass-line or don't-pass bet with 3-4-5-times free odds.* An initial pass-line or don't-pass bet (Examples 1.2.2 on p. 17 and 6.1.4), together with a free odds bet (Example 1.5.7 on p. 44), is an example of a compound wager. The initial roll of the dice is called the *come-out roll*, and when a *point* is established, that is, 4, 5, 6, 8, 9, or 10 is rolled, a 3-4-5-*times free odds bet* becomes available. We summarize the needed information about (B, X) and (B', X') in Table 6.2, assuming the initial bet is one unit and a maximal free odds bet is made.

For the pass-line bet with 3-4-5-times free odds, $E[X] = -7/495$, $E[X'] = 0$, and $E[(B+B') \, 1_{\{X+X'\neq 0\}}] = E[B+B'] = 1+3(6/36)+4(8/36)+5(10/36) = 34/9$, so the house advantage is

$$H_0(B, X + X') = H(B, X + X') = \frac{7}{495} \tag{6.21}$$

or

$$H_0(B + B', X + X') = H(B + B', X + X') = \frac{7}{1,870}. \tag{6.22}$$

For the don't-pass bet with 3-4-5-times free odds, $E[X] = -3/220$, $E[X'] = 0$,

$$E[(B + B') \, 1_{\{X+X'\neq 0\}}] = \frac{35}{36} + 6\left(\frac{24}{36}\right) = \frac{179}{36}, \tag{6.23}$$

and $E[B + B'] = 5$, so the house advantage is

Table 6.2 The joint distribution of (B, X) and (B', X') for a one-unit pass-line or don't-pass bet with 3-4-5-times free odds.

event	pass line (B, X)	pass line (B', X')	don't pass (B, X)	don't pass (B', X')	probab. $\times 1{,}980$
natural 7 or 11	$(1, 1)$	$(0, 0)$	$(1, -1)$	$(0, 0)$	440
craps 2 or 3	$(1, -1)$	$(0, 0)$	$(1, 1)$	$(0, 0)$	165
craps 12	$(1, -1)$	$(0, 0)$	$(1, 0)$	$(0, 0)$	55
point 4 or 10 made	$(1, 1)$	$(3, 6)$	$(1, -1)$	$(6, -6)$	110
point 4 or 10 missed	$(1, -1)$	$(3, -3)$	$(1, 1)$	$(6, 3)$	220
point 5 or 9 made	$(1, 1)$	$(4, 6)$	$(1, -1)$	$(6, -6)$	176
point 5 or 9 missed	$(1, -1)$	$(4, -4)$	$(1, 1)$	$(6, 4)$	264
point 6 or 8 made	$(1, 1)$	$(5, 6)$	$(1, -1)$	$(6, -6)$	250
point 6 or 8 missed	$(1, -1)$	$(5, -5)$	$(1, 1)$	$(6, 5)$	300

$$H_0(B, X + X') = \frac{3}{220} \quad \text{or} \quad H(B, X + X') = \frac{27}{1{,}925} \qquad (6.24)$$

or

$$H_0(B + B', X + X') = \frac{3}{1{,}100} \quad \text{or} \quad H(B + B', X + X') = \frac{27}{9{,}845}. \qquad (6.25)$$

The reason we prefer (6.22) to (6.21) is as follows. Suppose gambler A bets one unit on the pass line at each come-out roll and makes no other bets, while gambler B bets $9/34$ units on the pass line at each come-out roll and takes 3-4-5-times free odds when a point is established. Then each gambler has an expected total bet of 1 unit per pass-line decision, but gambler A has an expected loss of $7/495$ units per pass-line decision, while gambler B has an expected loss of only $7/1{,}870$ units per pass-line decision. In particular, gambler B is getting the same expected amount of action (or entertainment value) at lower expected cost. The superiority of gambler B's strategy is reflected in the house advantage if it is defined with respect to the expected total amount bet but not if it is defined with respect to the initial bet.

The same argument applies equally well to the don't-pass bet, so we prefer (6.25) to (6.24) as well. ♠

As the preceding example suggests, strategy can be optimized by minimizing house advantage. To elaborate on this point, we consider an initial wager (B, X) and suppose that an event A can be partitioned as the union of the mutually exclusive events $A_1, \ldots, A_k$, each with positive probability. Let us further assume that, for $j = 1, \ldots, k$, if event A_j occurs, the player has the option of making an additional bet (B_j, X_j), where $B_j = X_j = 0$ on A_j^c. We require that $(B, X), (B_1, X_1), \ldots, (B_k, X_k)$ satisfy (6.1).

Let $C \subset \{1, 2, \ldots, k\}$ and suppose that the gambler makes the additional bet if and only if A_j occurs for some $j \in C$. Then the four definitions of house advantage are as in (6.17)–(6.20) with

$$B' := \sum_{j \in C} B_j \quad \text{and} \quad X' := \sum_{j \in C} X_j. \tag{6.26}$$

What criterion should the player use to decide whether to make the additional bet (B_j, X_j) if A_j occurs? The argument of Example 6.1.7 suggests that one should attempt to minimize

$$H_0\left(B + \sum_{j \in C} B_j, \ X + \sum_{j \in C} X_j\right) \tag{6.27}$$

or

$$H\left(B + \sum_{j \in C} B_j, \ X + \sum_{j \in C} X_j\right) \tag{6.28}$$

by choosing $C \subset \{1, 2, \ldots, k\}$ appropriately. We will focus on (6.27).

Notice that

$$E\left[X + \sum_{j \in C} X_j\right] = E[X] + \sum_{j \in C} P(A_j) E[X_j \mid A_j] \tag{6.29}$$

and

$$E\left[B + \sum_{j \in C} B_j\right] = E[B] + \sum_{j \in C} P(A_j) E[B_j \mid A_j] \tag{6.30}$$

so that we can minimize the numerator of (6.27) by maximizing (6.29) and hence by choosing C to be the set of $j \in \{1, 2, \ldots, k\}$ for which $E[X_j \mid A_j] \geq 0$. The resulting C does not necessarily minimize the ratio (6.27), however. To find the optimal C, it will be convenient to introduce the notation

$$x := E[X], \quad x_j := E[X_j \mid A_j], \quad b := E[B], \quad b_j := E[B_j \mid A_j], \tag{6.31}$$

and $p_j := P(A_j)$. Then the problem is to maximize the ratio

$$\frac{x + \sum_{j \in C} p_j x_j}{b + \sum_{j \in C} p_j b_j} \tag{6.32}$$

by choosing $C \subset \{1, 2, \ldots, k\}$ appropriately. The following theorem shows that this can be done using an iterative procedure.

Theorem 6.1.8. *Given $k \geq 1$, $p_1, \ldots, p_k$ positive, $x, x_1, \ldots, x_k$ real, and $b, b_1, \ldots, b_k$ positive, define*

$$h^* := \max_{C \subset \{1,2,\dots,k\}} \frac{x + \sum_{j \in C} p_j x_j}{b + \sum_{j \in C} p_j b_j}, \tag{6.33}$$

and define the step function

$$G(h) := \frac{x + \sum_{j \in C(h)} p_j x_j}{b + \sum_{j \in C(h)} p_j b_j}, \qquad h \in \mathbf{R}, \tag{6.34}$$

where $C(h) := \{1 \leq i \leq k : x_i/b_i \geq h\}$. Then h^* is the unique fixed point of G. Moreover, with h_0 arbitrary and $h_n := G(h_{n-1})$ for each $n \geq 1$, we have $h_n = h^*$ for all $n \geq k$.

Remark. In other words, starting with $h_0 := 0$, say, we evaluate the terms of the sequence

$$G(0), G(G(0)), G(G(G(0))), \dots \tag{6.35}$$

until they stop changing. The final value is h^* and the optimal C is $C(h^*)$.

Proof. First suppose that h is a fixed point of G. Then, for arbitrary $C \subset \{1, 2, \dots, k\}$,

$$
\begin{aligned}
x + \sum_{j \in C} p_j x_j &- \left(b + \sum_{j \in C} p_j b_j\right) h \\
&= b\left(\frac{x}{b} - h\right) + \sum_{j \in C} p_j b_j \left(\frac{x_j}{b_j} - h\right) \\
&= b\left(\frac{x}{b} - h\right) + \sum_{j \in C \cap C(h)} p_j b_j \left(\frac{x_j}{b_j} - h\right) + \sum_{j \in C - C(h)} p_j b_j \left(\frac{x_j}{b_j} - h\right) \\
&\leq b\left(\frac{x}{b} - h\right) + \sum_{j \in C(h)} p_j b_j \left(\frac{x_j}{b_j} - h\right) \\
&= x + \sum_{j \in C(h)} p_j x_j - \left(b + \sum_{j \in C(h)} p_j b_j\right) h \\
&= 0 \tag{6.36}
\end{aligned}
$$

since $G(h) = h$, and it follows that

$$\frac{x + \sum_{j \in C} p_j x_j}{b + \sum_{j \in C} p_j b_j} \leq h. \tag{6.37}$$

Since C is arbitrary, $h^* \leq h$. But $h = G(h) \leq h^*$, so $h = h^*$.

The case in which G is constant is easy, so we let $f_1 < f_2 < \cdots < f_l$ be the set of discontinuity points of G, where $1 \leq l \leq k$. (This is the set of distinct points in $\{x_1/b_1, \dots, x_k/b_k\}$, with the possible exception of one such point.) We let $I_1 = (-\infty, f_1)$, $I_j := [f_{j-1}, f_j)$ for $j = 2, 3, \dots, l$, and $I_{l+1} := [f_l, \infty)$.

Let us denote the value of G on I_j by g_j for $j = 1, \ldots, l+1$. Given $\alpha, \beta, \gamma, \delta$ with $\beta > 0$ and $\delta > 0$, we note that

$$\frac{\alpha}{\beta} > \frac{\gamma}{\delta} \quad \text{implies} \quad \frac{\alpha}{\beta} > \frac{\alpha + \gamma}{\beta + \delta} > \frac{\gamma}{\delta} \tag{6.38}$$

because $(\alpha + \gamma)/(\beta + \delta)$ is a convex combination of α/β and γ/δ, that is,

$$\frac{\alpha + \gamma}{\beta + \delta} = \frac{\beta}{\beta + \delta} \frac{\alpha}{\beta} + \frac{\delta}{\beta + \delta} \frac{\gamma}{\delta}; \tag{6.39}$$

similar results hold with each $>$ replaced by $<$ or with each $>$ replaced by $=$. In particular,

$$f_j > g_{j+1} \quad \text{implies} \quad f_j > g_j > g_{j+1}; \tag{6.40}$$

similar results hold with each $>$ replaced by $<$ or with each $>$ replaced by $=$. Alternatively, if $(f_j, G(f_j+))$ is below (resp., above) the diagonal, then so is $(f_j, G(f_j))$ and $G(f_j) > G(f_j+)$ (resp., $G(f_j) < G(f_j+)$). If $(f_j, G(f_j+))$ is on the diagonal, then so is $(f_j, G(f_j))$, but this is impossible because f_j is a discontinuity point of G.

This implies that there exists j_0 such that

$$g_1 < g_2 < \cdots < g_{j_0-1} < g_{j_0}, \qquad g_{j_0} > g_{j_0+1} > \cdots > g_{l+1}, \tag{6.41}$$

$f_j < g_j$ for $j = 1, \ldots, j_0 - 1$, and $f_j > g_j$ for $j = j_0, \ldots, l$. (It may clarify matters to sketch the graph of G.) In particular, g_{j_0} is the unique fixed point of G and therefore, by the first part of the proof, equals h^*. Finally, starting with arbitrary h_0, we find that $h_1 = G(h_0) = g_i$ for some $i \in \{1, 2, \ldots, j_0\}$. Continuing, $h_1 \leq h_2 \leq \cdots \leq h^*$, so at most j_0 ($\leq k$) iterations are needed to arrive at h^*. ♠

Suppose we want to minimize (6.28). Noting that

$$\mathrm{E}\left[\left(B + \sum_{j \in C} B_j\right) 1_{N_C}\right] = \mathrm{E}[B \, 1_{N_C}] + \sum_{j \in C} \mathrm{P}(A_j)\mathrm{E}[B_j \, 1_{N_C} \mid A_j] \tag{6.42}$$

$$= \mathrm{E}[B \, 1_{N_C}] + \sum_{j \in C} \mathrm{P}(A_j)\mathrm{E}[B_j \, 1_{\{X+X_j \neq 0\}} \mid A_j],$$

where $N_C := \{X + \sum_{i \in C} X_i \neq 0\}$, we see that Theorem 6.1.8 may not quite apply because the b in (6.32) may depend on C.

We illustrate the theorem with two well-known examples and a contrived one.

Example 6.1.9. *Three Card Poker: Ante-play wager.* Three Card Poker is played with a standard 52-card deck, which is reshuffled between coups.

Here we consider only the principal wager, the so-called ante-play wager (see Section 16.2 for more-complete coverage).

Each player makes an *ante wager* and receives three cards face down. The dealer also receives three cards face down. Each player, after examining his hand, must decide whether to fold or play. If he folds, he loses his ante wager. If he plays, he must make an additional wager, called a *play wager*, equal in size to the ante wager. The dealer then reveals his hand to determine whether it *qualifies* with a rank of queen high or better. If the dealer's hand fails to qualify, the player's ante wager is paid even money and his play wager is pushed. If the dealer's hand qualifies, it is compared with the player's hand. If the player's hand outranks the dealer's hand, both of the player's wagers are paid even money. If the dealer's hand outranks the player's hand, both of the player's wagers are lost. If the player's hand and the dealer's hand are equal in rank, both of the player's wagers are pushed.

The ranking of hands in Three Card Poker is slightly different from that in traditional poker. First, the categories two pair, full house, and four of a kind do not occur. Second, the relative ranks of the categories three of a kind, straight, and flush are reversed, in accordance with their probabilities. Hands in different categories are ranked as the categories are ranked, namely straight flush, three of a kind, straight, flush, one pair, and no pair. Hands in the same categories are ranked with the help of the following rules. Straight flushes are ranked according to the highest denomination (which is 3 in the case of 3-2-A). Three-of-a-kind hands are ranked according to the common denomination. Straights are ranked according to the highest denomination (which is 3 in the case of 3-2-A). Flushes are ranked first according to the highest denomination, second according to the middle denomination, and third according to the lowest denomination. One-pair hands are ranked first according to the denomination of the pair and second according to the denomination of the third card. No-pair hands are ranked first according to the highest denomination, second according to the middle denomination, and third according to the lowest denomination.

There is one remaining detail called the *ante bonus*. Regardless of whether the player plays, regardless of whether the dealer qualifies, and regardless of the outcome of the ante-play wager, if the player's hand is a straight or better, he receives a bonus payoff on his ante wager. This bonus amounts to 1 to 1 for a straight, 4 to 1 for three of a kind, and 5 to 1 for a straight flush.

Three Card Poker is a compound wager, so we can apply Theorem 6.1.8 to determine an optimal strategy. The initial amount bet is the amount of the ante bet, which we take to be one unit. In the notation of Theorem 6.1.8,

$$x = -1 + (1)\frac{\binom{12}{1}\left[\binom{4}{1}^3 - \binom{4}{1}\right]}{\binom{52}{3}} + (4)\frac{\binom{13}{1}\binom{4}{3}}{\binom{52}{3}} + (5)\frac{\binom{12}{1}\binom{4}{1}}{\binom{52}{3}} \qquad (6.43)$$

and $b = 1$. This is the player's expected profit if he folds automatically. Notice that the ante bonus is paid if he has a straight or better.

Let us call two hands *equivalent* if they have the same three denominations and if the corresponding denominations have the same suits after a permutation of $(\clubsuit,\diamondsuit,\heartsuit,\spadesuit)$. This is an equivalence relation in the sense of Theorem A.1.1. Thus, for example, the equivalence class containing $Q\clubsuit\text{-}6\diamondsuit\text{-}3\heartsuit$ has 24 hands, while the one containing $Q\clubsuit\text{-}6\clubsuit\text{-}3\diamondsuit$ has 12 hands and the one containing $Q\clubsuit\text{-}6\clubsuit\text{-}3\clubsuit$ has four hands.

Continuing with the notation of Theorem 6.1.8, there are $k = 1{,}755$ equivalence classes, and to each equivalence class corresponds a probability p_i, a conditional expectation x_i, and a bet size $b_i = 1$. With $h_0 := 0$, we find that

$$h_1 := G(0) = -\frac{686{,}689}{34{,}084{,}400} \approx -0.020147, \qquad (6.44)$$

$$h_2 := G(h_1) = -\frac{59{,}273}{2{,}947{,}840} \approx -0.020107, \qquad (6.45)$$

and $h_3 := G(h_2) = h_2$.

The strategy having house advantage (6.18) equal to the negative of (6.44) is to play with unsuited Q-6-4 or better. This is the accepted optimal strategy in the gambling literature, obtained by making only those play wagers with nonnegative (or, equivalently, positive) conditional expectations. This strategy minimizes house advantage (6.17), which is equal to

$$\frac{13{,}733{,}780}{\binom{52}{3}\binom{49}{3}} = \frac{686{,}689}{20{,}358{,}520} \approx 0.033730. \qquad (6.46)$$

The strategy having house advantage (6.18) equal to the negative of (6.45) is to play with unsuited Q-6-2 or better. With unsuited Q-6-2 and unsuited Q-6-3, the play wagers have negative conditional expectations, but they are less negative than (6.44). Here the player minimizes house advantage (6.18), while house advantage (6.17) with this strategy is equal to

$$\frac{13{,}751{,}336}{\binom{52}{3}\binom{49}{3}} = \frac{1{,}718{,}917}{50{,}896{,}300} \approx 0.033773, \qquad (6.47)$$

which is slightly larger than (6.46).

We could, with additional effort, evaluate the house advantage with pushes excluded, but we do not do so for a very simple reason. After a push in Three Card Poker, one's bet is double what it was initially, so the player is unlikely to regard a push as merely a temporary delay, as he might if the bet were unchanged. Thus, it seems appropriate to include pushes when evaluating the house advantage of the ante-play wager. $\spadesuit$

Example 6.1.10. *Let It Ride: Basic bet.* Let It Ride is played with a standard 52-card deck, which is reshuffled between coups. Here we consider only the basic wager (see Section 16.1 for more-complete coverage).

Each player makes a three-unit bet and receives three cards face down. Two community cards are then dealt face down to the dealer. Each player, after examining his hand, has the option of either reducing his bet size by one unit or letting it ride. The dealer then reveals the first community card, which is each player's fourth card. Each player then has the option of reducing his bet size by one unit (regardless of his previous decision) or letting it ride. The dealer then reveals the second community card, which is each player's fifth and final card. If a player has a pair of 10s or better, he is paid according to the value of his hand. A royal flush pays 1,000 to 1, a straight flush pays 200 to 1, four of a kind pays 50 to 1, a full house pays 11 to 1, a flush pays 8 to 1, a straight pays 5 to 1, three of a kind pays 3 to 1, two pair pays 2 to 1, and a high pair (10s or better) pays 1 to 1. Note that these odds are paid on whatever remains of the player's initial three-unit bet (one, two, or three units). If the player has a pair of 9s or worse, he loses whatever remains of his initial three-unit bet. There is no possibility of a push.

Observe that, although the player begins by placing three units on the betting surface, at that point he is placing only one unit at risk and is making what we will call the *first bet*. After seeing his three-card hand, the player has the option of reclaiming one of his three units. If he lets it ride, he is putting a second unit at risk and is making what we will call the *second bet*. After the first community card is revealed, the player has the option, on the basis of his four-card hand, of reclaiming one of his two or three remaining units. If he lets it ride, he is putting another unit at risk and is making what we will call the *third bet*. Each of the three bets is a one-unit bet.[2]

Let It Ride is a three-stage compound wager rather than a two-stage one, but Theorem 6.1.8 is nevertheless applicable. Recall that our original formulation of a compound wager went like this: We considered an initial wager (B, X) and supposed the existence of an event A that can be partitioned as the union of the mutually exclusive events $A_1, \ldots, A_k$, each with positive probability. We further assumed that, for $j = 1, \ldots, k$, if event A_j occurs, the player has the option of making an additional bet (B_j, X_j), where $B_j = X_j = 0$ on A_j^c. The same setup applies to Let It Ride, except that the events $A_1, \ldots, A_k$ are no longer assumed mutually exclusive. Specifically, $k = k_1 + k_2$ and $A_1, \ldots, A_{k_1}$ partition Ω, as do $A_{k_1+1}, \ldots, A_{k_1+k_2}$, so two of the events can occur simultaneously. This does not affect the applicability of the theorem because it does not require that $p_1 + \cdots + p_k \leq 1$.

The first bet can be shown to have expected profit

$$x = \frac{-968{,}692}{\binom{52}{5}} = -\frac{242{,}173}{649{,}740} \approx -0.372723, \qquad (6.48)$$

and the bet size is $b = 1$.

---

[2] Our terminology departs slightly from casino practice, in which what we have called the first, second, and third bets are placed in circles labeled $, 1, and 2, respectively.

We say that two three-card hands are *equivalent* if they have the same conditional distribution of profit. This is an equivalence relation in the sense of Theorem A.1.1. There are 33 equivalence classes of three-card hands. Let s be the number of straights—disregarding suits—that can be made from the hand, and let h be the number of high cards (10, J, Q, K, A) in the hand. The various equivalence classes include three of a kind, high pair (10s or better), low pair, suited no pair ($0 \leq s \leq 3$, $0 \leq h \leq 3$, $(s, h) \neq (0, 3)$), and unsuited no pair (same conditions on s and h). In each case the probability p_i and the conditional expectation x_i can be evaluated, and the bet size satisfies $b_i = 1$.

We say that two four-card hands are *equivalent* if they have the same conditional distribution of profit. This too is an equivalence relation in the sense of Theorem A.1.1. There are also 33 equivalence classes of four-card hands. With s and h as before, the various equivalence classes include four of a kind, three of a kind, two pair, high pair, low pair, suited no pair ($0 \leq s \leq 2$, $0 \leq h \leq 4$, $(s, h) \neq (0, 4)$), and unsuited no pair (same conditions on s and h). In each case the probability p_i and the conditional expectation x_i can be evaluated, and the bet size satisfies $b_i = 1$.

Therefore, $k = 66$, $p_1 + \cdots + p_{33} = 1$, and $p_{34} + \cdots + p_{66} = 1$. With $h_0 := 0$, we find that

$$G(0+) = -\frac{37{,}963}{1{,}325{,}152} \approx -0.028648, \tag{6.49}$$

$$h_1 := G(0) = -\frac{37{,}963}{1{,}334{,}224} \approx -0.028453, \tag{6.50}$$

$$h_2 := G(h_1) = -\frac{76{,}171}{2{,}678{,}640} \approx -0.028436, \tag{6.51}$$

and $h_3 := G(h_2) = h_2$.

The strategy having house advantage (6.18) equal to the negative of (6.49) consists of a three-card strategy (let it ride with a made payoff or with a suited hand with $s + h \geq 3$) and a four-card strategy (let it ride with a made payoff, a suited hand, or an unsuited no-pair hand with $s = 2$ and $h \geq 1$). This strategy is obtained by letting it ride only with positive conditional expectation, and it minimizes house advantage (6.17), which is equal to

$$\frac{1{,}822{,}224}{\binom{52}{3}(49)_2} = \frac{37{,}963}{1{,}082{,}900} \approx 0.035057. \tag{6.52}$$

The strategy having house advantage (6.18) equal to the negative of (6.50) differs only in the four-card strategy, requiring that the player let it ride in two additional cases: an unsuited no-pair hand with $(s, h) = (1, 4)$ or $(s, h) = (2, 0)$. These latter bets are conditionally fair, so this strategy is obtained by letting it ride only with nonnegative conditional expectation. It also minimizes house advantage (6.17), which is again equal to (6.52).

The strategy having house advantage (6.18) equal to the negative of (6.51) differs from the one just given only in the three-card strategy, requiring that the player let it ride in two additional cases: a suited no-pair hand with $(s, h) = (2, 0)$ and an unsuited no-pair hand with $(s, h) = (3, 3)$. Although these two bets have negative conditional expectations, they are less negative than (6.50). Here the player minimizes house advantage (6.18). With this strategy, house advantage (6.17) is equal to

$$\frac{1{,}828{,}104}{\binom{52}{3}(49)_2} = \frac{76{,}171}{2{,}165{,}800} \approx 0.035170, \tag{6.53}$$

which is slightly larger than (6.52).   ♠

Are we saying that the strategies corresponding to (6.45) and (6.51) are superior to those corresponding to (6.44) and (6.50), contrary to the gambling literature? Not necessarily. Before we explain, it will help to have one more example for comparison purposes, this one a contrived modification of Example 6.1.7.

Example 6.1.11. *Craps: Pass-line bet with 3-4-5-times not-free odds.* Here we assume that, instead of 3-4-5-times free odds, the pass-line bettor is offered 3-4-5-times odds but is charged a one percent commission on the additional amount bet. Thus, in the notation of Theorem 6.1.8, $x = -7/495$, $b = 1$, and

$$p_1 = 6/36, \quad p_2 = 8/36, \quad p_3 = 10/36, \tag{6.54}$$

$$x_1 = -0.03, \quad x_2 = -0.04, \quad x_3 = -0.05, \tag{6.55}$$

$$b_1 = 3.03, \quad b_2 = 4.04, \quad b_3 = 5.05. \tag{6.56}$$

If one wishes to minimize house advantage with respect to initial amount bet, then these not-free odds bets should be avoided, whereas if one wishes to minimize house advantage with respect to total amount bet, then these not-free odds bets, being less disadvantageous than the pass-line bet, should be made.

More precisely, with $h_0 := 0$, we find that

$$h_1 := G(0) = -\frac{7}{495} = -0.0\overline{14}, \tag{6.57}$$

$$h_2 := G(h_1) = -\frac{83}{7{,}535} \approx -0.011015, \tag{6.58}$$

and $h_3 := G(h_2) = h_2$.

The strategy having house advantage (6.18) equal to the negative of (6.57) is to never take the not-free odds. This strategy minimizes house advantage (6.17), which is also equal to (6.57).

The strategy having house advantage (6.18) equal to the negative of (6.58) is to always take the not-free odds. With this strategy, house advantage (6.17) is equal to

$$\frac{83}{1{,}980} = 0.04\overline{19}, \tag{6.59}$$

which is substantially larger than the negative of (6.57).

The distinction can be made more clearly, much as we did in Example 6.1.7. Suppose first that each of gamblers A and B bets one unit on the pass line at each come-out roll, but gambler B takes 3-4-5-times not-free odds when a point is established, and gambler A does not. Then gambler A has an expected loss of 7/495 per unit bet on the pass line, while gambler B has an expected loss of $7/495 + 1/36 = 83/1{,}980$ per unit bet on the pass line. Clearly, gambler A is better off.

On the other hand, suppose gambler A bets one unit on the pass line at each come-out roll and makes no other bets, while gambler B bets 36/137 units on the pass line at each come-out roll and takes 3-4-5-times not-free odds when a point is established. Then each gambler has an expected total amount bet of one unit per pass-line decision, but gambler A has an expected loss of 7/495 units per pass-line decision, while gambler B has an expected loss of 28/7,535 units per pass-line decision, which is smaller. In other words, gambler B is getting the same expected amount of action at lower cost. Clearly, gambler B is better off.

Which of the last two paragraphs is correct? In our view, they are not inconsistent. If the initial amounts bet are equal, gambler A is better off. If the expected total amounts bet are equal, gambler B is better off. Notice that the latter argument may work even if 36/137 is not a permitted bet size, as is likely the case. Indeed, if we replace the amount of gambler B's initial bet by an arbitrary amount b, it suffices that

$$0.262774 \approx 36/137 \leq b < 28/83 \approx 0.337349. \tag{6.60}$$

This ensures both that gambler B's expected total amount bet is at least one unit and that his expected loss is less than 7/495 units because $(7/495)b + [(6/36)(0.03)+(8/36)(0.04)+(10/36)(0.05)]b < 7/495$ if the second inequality in (6.60) holds. ♠

Let us isolate the features that Examples 6.1.9–6.1.11 have in common. Given a compound wager, let (B_0, X_0) be the wager corresponding to the initial bet together with each of the corresponding conditional bets having nonnegative conditional expected profit. Let (B_1, X_1) be the wager corresponding to each of the *additional* conditional bets needed to maximize overall expected profit per expected total amount bet. Suppose that

$$\frac{E[X_0]}{E[B_0]} < \frac{E[X_0 + X_1]}{E[B_0 + B_1]} < 0. \tag{6.61}$$

Then, if there exists a permitted bet size b such that

$$\frac{E[B_0]}{E[B_0 + B_1]} \leq b < \frac{E[X_0]}{E[X_0 + X_1]}, \tag{6.62}$$

we can compare the results of gambler A, who makes an initial bet of one unit and then plays to maximize overall expected profit, and gambler B, who makes an initial bet of b units and then plays to maximize the ratio of overall expected profit to expected total amount bet. Gambler B will wager at least as much as gambler A on average, and gambler B will win more than gambler A on average, that is,

$$bE[B_0 + B_1] \geq E[B_0] \quad \text{and} \quad bE[X_0 + X_1] > E[X_0]. \tag{6.63}$$

It is therefore hard to deny that gambler B has the better strategy of the two.

Now, in Example 6.1.11, (6.62) reduces to (6.60), whereas in Examples 6.1.9 and 6.1.10 the corresponding inequalities become

$$0.996767 \approx \frac{925}{928} \leq b < \frac{3{,}433{,}445}{3{,}437{,}834} \approx 0.998723 \tag{6.64}$$

and

$$0.996195 \approx \frac{166{,}778}{167{,}415} \leq b < \frac{75{,}926}{76{,}171} \approx 0.996784, \tag{6.65}$$

respectively. Unless the initial betting unit is extremely large, there will be no b satisfying (6.64) or (6.65) that corresponds to a permitted bet size. Thus, the argument used to justify the free odds bets (Example 6.1.7) and even the not-free odds bets (Example 6.1.11) at craps does not apply to Three Card Poker (Example 6.1.9) or Let It Ride (Example 6.1.10).

Consequently, to compare the two strategies at Three Card Poker, for example, corresponding to (6.44) and (6.45), we could again consider gamblers A and B, but now each gambler makes an initial bet of one unit. Gambler A plays to maximize overall expected profit and gambler B plays to maximize the ratio of overall expected profit to expected total amount bet. Gambler B loses slightly more on average but he also bets slightly more on average, with the result that the ratio of overall expected loss to expected total amount bet is slightly less for gambler B. Which strategy is superior?

One can make a case for either approach. The gambler wants to choose the strategy with the smallest house advantage. (He accepts the fact that the house will have some advantage as the price he must pay for his entertainment.) The only question is, should the denominator of the house advantage be his initial bet (the one-unit ante) or should it be his expected total amount bet? Those who prefer the former could justify it by arguing that the gambler achieves no additional benefit by making a play wager rather than folding but is merely following the dictates of sound strategy. Those who prefer the latter could justify it by arguing that by making a play wager rather than folding the gambler is putting additional money at risk, and such a bet should be regarded no differently from the initial ante bet.

In games such as Three Card Poker, it is conventional to measure house advantage with respect to initial amount bet, whereas in craps with free odds, it is conventional to measure house advantage with respect to total amount bet. While this appears to be an inconsistency, the above discussion shows that it may be justifiable. The author's preference is to measure house advantage with respect to total amount bet in all cases.

Example 6.1.12. *Laying the odds and conditional advantage.* We take this opportunity to notice a dubious argument that has appeared in the gambling literature. Several authors have stated that one should never lay the odds, that is, one should never make a free odds bet in conjunction with a don't-pass bet. Their justification for this assertion is based on the following observation. Suppose, after betting on don't pass, the point $j \in \{4, 5, 6, 8, 9, 10\}$ is established. Then the player has the advantage over the house because $\pi_7 > \pi_j$. In fact, his conditional expectation per unit bet is

$$\frac{\pi_7}{\pi_j + \pi_7} - \frac{\pi_j}{\pi_j + \pi_7} = \frac{|j - 7|}{12 - |j - 7|} > 0. \tag{6.66}$$

On the other hand, if he lays 3-4-5-times free odds, his conditional expectation per conditional expected number of units bet is

$$\frac{|j - 7|/(12 - |j - 7|)}{1 + 6}. \tag{6.67}$$

Thus, the player lays the odds at the cost of reducing his conditional advantage over the house.

The argument of the preceding paragraph suggests that, in the notation introduced prior to Theorem 6.1.8, if

$$\frac{\mathrm{E}[X + X_j \mid A_j]}{\mathrm{E}[B + B_j \mid A_j]} < \frac{\mathrm{E}[X \mid A_j]}{\mathrm{E}[B \mid A_j]}, \tag{6.68}$$

then the additional bet should be refused. But this is contrary to the criterion just discussed, which recommends a free odds bet on each point because

$$\frac{\mathrm{E}[X] + \sum_{j \in C} \mathrm{P}(A_j)\mathrm{E}[X_j \mid A_j]}{\mathrm{E}[B] + \sum_{j \in C} \mathrm{P}(A_j)\mathrm{E}[B_j \mid A_j]} \tag{6.69}$$

is maximized by $C = \{4, 5, 6, 8, 9, 10\}$. Notice that the criterion (6.68) cannot even be expressed in terms of the parameters in (6.31).

The fallacy here is that one cannot analyze the consequences of an event (as in the criterion (6.68)) in isolation, without considering the consequences of the complementary event. ♠

There are equivalent formulations of our two definitions of house advantage, H_0 and H, that will be useful in the next section. They also help to clarify the relationship between our definitions and one found in the mathematical literature. To state them, we first define the *essential supremum* of a nonnegative discrete random variable Y by

$$\text{ess sup}(Y) := \inf\{M > 0 : P(Y \le M) = 1\}. \tag{6.70}$$

In words, the essential supremum of Y is the supremum of the values of Y, ignoring values that are assumed only with probability 0. An alternative definition of the house advantage of a wager (B, X), which we will call the *cut*, does not depend on B:

$$H_c(X) := \frac{-E[X]}{\text{ess sup}(X^-)}. \tag{6.71}$$

Of course, this is just the ratio of the gambler's expected loss to his amount risked. The cut coincides with our $H_0(X)$ in Examples 6.1.2–6.1.6, but not in Examples 6.1.7–6.1.11. In the latter four cases, the fact that B has a non-degenerate distribution makes (6.71) inapplicable. Instead, we can generalize (6.71) as follows.

Given (B, X) satisfying (6.1), we define the *bet size* by

$$\beta(B, X) := \text{ess sup}(1_{\{B=b\}} X^-) \text{ on } \{B = b\}, \quad b \ge 0, \tag{6.72}$$

and the amount of *action* by

$$\alpha(B, X) := \text{ess sup}(1_{\{B=b\}} X^-) 1_{\{X \ne 0\}} \text{ on } \{B = b\}, \quad b \ge 0. \tag{6.73}$$

Notice that $\beta(B, X) = B$ and $\alpha(B, X) = B 1_{\{X \ne 0\}}$, since by (6.1),

$$\begin{aligned}
\beta(B, X) &= \text{ess sup}(1_{\{B=b\}} X^-) \\
&\le \text{ess sup}(1_{\{B=b\}} B) \\
&\le b = B \text{ on } \{B = b\}, \quad b \ge 0, \tag{6.74}
\end{aligned}$$

and strict inequality cannot hold if $P(B = b) > 0$ because B is defined to be the amount placed at risk. It follows that

$$H_0(B, X) = \frac{-E[X]}{E[\beta(B, X)]} \quad \text{and} \quad H(B, X) = \frac{-E[X]}{E[\alpha(B, X)]}. \tag{6.75}$$

To emphasize that we cannot eliminate B from the definition as we did in (6.71), we consider the following artificial example.

Example 6.1.13. *Two distinct wagers with the same marginal distribution of profit.* The wager

$$(B, X) = \begin{cases} (2, -1) & \text{with probability } 1/37, \\ (2, 2) & \text{with probability } 18/37, \\ (2, -2) & \text{with probability } 18/37, \end{cases} \tag{6.76}$$

corresponds to a two-unit bet on red at a 37-number roulette wheel with partager (half the bet is lost if 0 appears). Clearly, the house advantage is $H_0(B, X) = H(B, X) = 1/74$.

On the other hand, consider the following compound wager, also based on a 37-number roulette wheel. The player bets one unit on red. The wheel is spun, and if 0 appears, he loses the bet. If any other number appears, his bet is unresolved. In this case the player has the opportunity to increase his bet to two units. The wheel is spun again until a nonzero number appears, and if it is red, the player's bet is paid even money; if it is black, the player's bet is lost. Notice the two-stage betting procedure. Assuming that the player makes the additional bet, the wager has the form

$$(B, X) = \begin{cases} (1, -1) & \text{with probability } 1/37, \\ (2, 2) & \text{with probability } 18/37, \\ (2, -2) & \text{with probability } 18/37, \end{cases} \tag{6.77}$$

so the house advantage is $H_0(B, X) = H(B, X) = 1/73$.

Since the marginal distribution of X is the same for (6.76) and (6.77), house advantage cannot be defined solely in terms of the distribution of X. It depends on something else, namely knowledge of the betting procedure. ♠

We conclude this section by recalling that the definition of house advantage implicitly assumes a game of independent coups. In card games characterized by sampling without replacement (see, for example, Chapters 18–21), the independence assumption fails, yet it is nevertheless important to have the concept of house advantage. The definitions (6.2) and (6.3) are of course well defined, but what are their interpretations? For now we simply point out that Theorem 6.1.1 fails in the absence of independence. For example, see Section 18.1.

6.2 Composite Wagers

We now consider several wagers, which may or may not be made simultaneously, and we define the house advantage of the *composite wager*. More precisely, let $(B_1, X_1), \ldots, (B_d, X_d)$ be d jointly distributed (but not necessarily independent) random vectors satisfying (6.1) on p. 199, and let

$$\boldsymbol{B} := (B_1, \ldots, B_d), \qquad \boldsymbol{X} := (X_1, \ldots, X_d). \tag{6.78}$$

As in Section 6.1 the *house advantage* of this composite wager can be defined as either

$$H_0(\boldsymbol{B}, \boldsymbol{X}) := \frac{-E[X_1 + \cdots + X_d]}{E[\beta(\boldsymbol{B}, \boldsymbol{X})]} \tag{6.79}$$

or

$$H(\boldsymbol{B}, \boldsymbol{X}) := \frac{-E[X_1 + \cdots + X_d]}{E[\alpha(\boldsymbol{B}, \boldsymbol{X})]}, \tag{6.80}$$

where $\beta(\boldsymbol{B}, \boldsymbol{X})$ is the amount bet on the composite wager, and $\alpha(\boldsymbol{B}, \boldsymbol{X})$ is the amount of action on the composite wager. The amount bet (resp., the amount of action) on the composite wager is the sum of the amounts bet (resp., the amounts of action) on the d individual wagers, *unless*, owing to cancelation between two or more of the individual wagers, the amount bet (resp., the amount of action) can be reduced to reflect reduced risk. To make this more precise, we generalize (6.72) and (6.73) on p. 218.

First, we define the *bet size* of the composite wager by

$$\beta(\boldsymbol{B}, \boldsymbol{X}) := \text{ess sup}(1_{\{\boldsymbol{B}=\boldsymbol{b}\}}(X_1 + \cdots + X_d)^-) \text{ on } \{\boldsymbol{B} = \boldsymbol{b}\}, \quad \boldsymbol{b} \geq \boldsymbol{0}. \tag{6.81}$$

It is easy to check that $(X_1 + \cdots + X_d)^- \leq X_1^- + \cdots + X_d^-$, and therefore, on the event $\{\boldsymbol{B} = \boldsymbol{b}\}$,

$$\begin{aligned}
\beta(\boldsymbol{B}, \boldsymbol{X}) &\leq \text{ess sup}(1_{\{\boldsymbol{B}=\boldsymbol{b}\}} X_1^-) + \cdots + \text{ess sup}(1_{\{\boldsymbol{B}=\boldsymbol{b}\}} X_d^-) \\
&\leq \text{ess sup}(1_{\{\boldsymbol{B}=\boldsymbol{b}\}} B_1) + \cdots + \text{ess sup}(1_{\{\boldsymbol{B}=\boldsymbol{b}\}} B_d) \\
&\leq b_1 + \cdots + b_d.
\end{aligned} \tag{6.82}$$

This implies that

$$\beta(\boldsymbol{B}, \boldsymbol{X}) \leq B_1 + \cdots + B_d. \tag{6.83}$$

Second, the definition of $\alpha(\boldsymbol{B}, \boldsymbol{X})$, the amount of *action* of the composite wager, is a bit more complicated. Our heuristic formulation above suggests that

$$\alpha(\boldsymbol{B}, \boldsymbol{X}) \leq B_1 1_{\{X_1 \neq 0\}} + \cdots + B_d 1_{\{X_d \neq 0\}}, \tag{6.84}$$

the right side of which can be rewritten as

$$\sum_{I \subset \{1,\ldots,d\}} \left(\sum_{i \in I} B_i \right) 1_{\{X_i \neq 0 \text{ if } i \in I, \, X_i = 0 \text{ if } i \in I^c\}}. \tag{6.85}$$

We define $\alpha(\boldsymbol{B}, \boldsymbol{X})$ as in (6.85) but with $\sum_{i \in I} B_i$ reduced if necessary to reflect reduced risk. More precisely, we define

$$\begin{aligned}
\alpha(\boldsymbol{B}, \boldsymbol{X}) := \sum_{I \subset \{1,\ldots,d\}} \text{ess sup}\left(1_{\{\boldsymbol{B}=\boldsymbol{b}\}} \left(\sum_{i \in I} X_i \right)^- \right) \\
\cdot 1_{\{X_i \neq 0 \text{ if } i \in I, \, X_i = 0 \text{ if } i \in I^c, \, \sum_{i \in I} X_i \neq 0\}} \\
\text{on } \{\boldsymbol{B} = \boldsymbol{b}\}, \quad \boldsymbol{b} \geq \boldsymbol{0}.
\end{aligned} \tag{6.86}$$

Notice that we have included an extra condition in the indicator random variable to account for no action in the case of a push. With this definition, (6.84) holds.

The interpretations of (6.79) and (6.80) are analogous to those of (6.2) and (6.3) on p. 200. Specifically, (6.79) is the ratio of the gambler's expected loss to his expected amount bet, and (6.80) is the ratio of the gambler's expected loss to his expected amount of action.

That strict inequality can hold in (6.83) and (6.84) is illustrated by the next two examples.

Example 6.2.1. *Roulette: 37-across-the-board.* Let us consider the wager consisting of 37 one-unit bets, one on each of the numbers of a 37-number roulette wheel, namely $0, 1, 2, \ldots, 36$. Since the winning number pays 35 to 1, one is assured of winning 35 units and losing 36, for a loss of one unit. The amount bet (i.e., risked) is therefore one unit. In symbols,

$$\beta(\boldsymbol{B}, \boldsymbol{X}) = \alpha(\boldsymbol{B}, \boldsymbol{X}) = 1 < 37 = \sum_{j=0}^{36} B_j = \sum_{j=0}^{36} B_j \, 1_{\{X_j \neq 0\}}. \tag{6.87}$$

Therefore, the house advantage is $H_0 = H = 1$, or 100 percent. We expect that this wager has rarely, if ever, been made. ♠

Example 6.2.2. *Craps: Pass-line and don't-pass bets with 3-4-5-times free odds on the pass line.* Here the idea is that the pass-line and don't-pass bets, each assumed to be for one unit, cancel each other out, whereas the three-, four-, or five-unit free odds bet is fair. Of course, this is not quite correct because of the possibility of a 12 on the come-out roll, which results in a loss to the gambler of one unit. With Y_1 denoting the result of the come-out roll,

$$\begin{aligned}
\beta(\boldsymbol{B}, \boldsymbol{X}) &= 1_{\{Y_1 \in \{2,3,7,11,12\}\}} \\
&\quad + 3 \cdot 1_{\{Y_1 \in \{4,10\}\}} + 4 \cdot 1_{\{Y_1 \in \{5,9\}\}} + 5 \cdot 1_{\{Y_1 \in \{6,8\}\}} \\
&< 1 + 1 + 3 \cdot 1_{\{Y_1 \in \{4,10\}\}} + 4 \cdot 1_{\{Y_1 \in \{5,9\}\}} + 5 \cdot 1_{\{Y_1 \in \{6,8\}\}} \\
&= B_1 + B_2 + B_3
\end{aligned} \tag{6.88}$$

and

$$\begin{aligned}
\alpha(\boldsymbol{B}, \boldsymbol{X}) &= 1_{\{Y_1 = 12\}} + 3 \cdot 1_{\{Y_1 \in \{4,10\}\}} + 4 \cdot 1_{\{Y_1 \in \{5,9\}\}} + 5 \cdot 1_{\{Y_1 \in \{6,8\}\}} \\
&\leq 1 + 1_{\{Y_1 \neq 12\}} + 3 \cdot 1_{\{Y_1 \in \{4,10\}\}} + 4 \cdot 1_{\{Y_1 \in \{5,9\}\}} + 5 \cdot 1_{\{Y_1 \in \{6,8\}\}} \\
&= B_1 \, 1_{\{X_1 \neq 0\}} + B_2 \, 1_{\{X_2 \neq 0\}} + B_3 \, 1_{\{X_3 \neq 0\}}.
\end{aligned} \tag{6.89}$$

The house advantage can be found by noting that the pass-line bet and the don't-pass bet contribute $1/36$ to the gambler's expected loss, whereas the three-, four-, or five-unit free odds bet contributes nothing. On the other hand, the expected amount bet and expected amount of action can be found from the first equalities in (6.88) and (6.89). We obtain

$$H_0 = \frac{1}{112} \quad \text{and} \quad H = \frac{1}{101}. \tag{6.90}$$

One might consider this an example of a hedge, but it fails to satisfy the definition given in Example 1.4.19 on p. 37. ♠

Now let us assume that $E[X_j] \leq 0$ for $j = 1, \ldots, d$. Then we can write

$$H_0(\boldsymbol{B}, \boldsymbol{X}) = \sum_{j=1}^{d} \frac{E[B_j]}{E[\beta(\boldsymbol{B}, \boldsymbol{X})]} H_0(B_j, X_j)$$

$$\geq \sum_{j=1}^{d} \frac{E[B_j]}{E[B_1] + \cdots + E[B_d]} H_0(B_j, X_j) \qquad (6.91)$$

and

$$H(\boldsymbol{B}, \boldsymbol{X}) = \sum_{j=1}^{d} \frac{E[B_j \, 1_{\{X_j \neq 0\}}]}{E[\alpha(\boldsymbol{B}, \boldsymbol{X})]} H(B_j, X_j) \qquad (6.92)$$

$$\geq \sum_{j=1}^{d} \frac{E[B_j \, 1_{\{X_j \neq 0\}}]}{E[B_1 \, 1_{\{X_1 \neq 0\}}] + \cdots + E[B_d \, 1_{\{X_d \neq 0\}}]} H(B_j, X_j).$$

In particular,

$$H_0(\boldsymbol{B}, \boldsymbol{X}) \geq \min_{1 \leq j \leq d} H_0(B_j, X_j) \qquad (6.93)$$

and

$$H(\boldsymbol{B}, \boldsymbol{X}) \geq \min_{1 \leq j \leq d} H(B_j, X_j), \qquad (6.94)$$

that is, the house advantage of the composite wager cannot be less than each of the house advantages of the individual wagers.

Furthermore, it is often the case that equality holds in (6.83) (resp., (6.84)). Then equality holds in (6.91) (resp., (6.92)). If equality holds in (6.91) (resp., (6.92)), then the house advantage of the composite wager is a weighted average of the house advantages of the individual wagers, with weights being proportional to the expected amounts bet (resp., to the expected amounts of action) on the individual wagers. In particular, the house advantage of the composite wager lies between the smallest and the largest of the house advantages of the individual wagers.

The next example describes a typical composite wager.

Example 6.2.3. *Craps: 32-across-the-board.* Here the gambler makes six simultaneous place bets, betting five units on each of 4, 5, 9, and 10, and betting six units on each of 6 and 8. (Recalling from Example 6.1.5 on p. 203 that 4 and 10 pay 9 to 5, 5 and 9 pay 7 to 5, and 6 and 8 pay 7 to 6, we see that the bet sizes were chosen to avoid fractional payoffs.) Observe that equality holds in (6.83) and (6.84) because, in the event of a 7, all six bets are lost. Thus, the composite bet involves no reduction of risk. It follows that equality also holds in (6.91) and (6.92).

Recalling Example 6.1.5 on p. 203, the house advantage of the composite wager is either

$$H_0 = \frac{2[5(1/15) + 5(1/25) + 6(1/66)]}{2[5 + 5 + 6]} = \frac{103}{2,640} \approx 0.039015, \qquad (6.95)$$

a weighted average of $1/15$, $1/25$, and $1/66$, or

$$H = \frac{2[(5/4)(1/15) + (25/18)(1/25) + (11/6)(1/66)]}{2[(5/4) + (25/18) + (11/6)]} = \frac{6}{161} \approx 0.037267, \qquad (6.96)$$

a differently weighted average of $1/15$, $1/25$, and $1/66$. The latter result requires noting that the expected amounts of action for the place bets on 4 (or 10), 5 (or 9), and 6 (or 8) are respectively

$$5(\pi_4 + \pi_7) = \frac{5}{4}, \quad 5(\pi_5 + \pi_7) = \frac{25}{18}, \quad 6(\pi_6 + \pi_7) = \frac{11}{6}. \qquad (6.97)$$

We believe that (6.96) is more appropriate than (6.95) for reasons explained below. ♠

Here it may help to discuss the interpretations of $H_0(\boldsymbol{B}, \boldsymbol{X})$ and $H(\boldsymbol{B}, \boldsymbol{X})$ in more detail. As before, let $(\boldsymbol{B}_1, \boldsymbol{X}_1), (\boldsymbol{B}_2, \boldsymbol{X}_2), \ldots$ be a sequence of i.i.d. random vectors whose common distribution is that of $(\boldsymbol{B}, \boldsymbol{X})$, representing the results of independent repetitions of the original set of d wagers. Write

$$\boldsymbol{B}_n = (B_{n,1}, \ldots, B_{n,d}), \qquad \boldsymbol{X}_n = (X_{n,1}, \ldots, X_{n,d}), \qquad (6.98)$$

so that $B_{n,j}$ is the amount bet on the jth wager at the nth coup, and $X_{n,j}$ is the gambler's profit from the jth wager at the nth coup. Then

$$\frac{-\sum_{l=1}^{n} \sum_{j=1}^{d} X_{l,j}}{\sum_{l=1}^{n} \beta(\boldsymbol{B}_l, \boldsymbol{X}_l)} \qquad (6.99)$$

represents the ratio of the gambler's cumulative loss after n such composite wagers to his total amount bet, while

$$\frac{-\sum_{l=1}^{n} \sum_{j=1}^{d} X_{l,j}}{\sum_{l=1}^{n} \alpha(\boldsymbol{B}_l, \boldsymbol{X}_l)} \qquad (6.100)$$

represents the ratio of the gambler's cumulative loss after n such composite wagers to his total amount of action. Dividing both numerator and denominator by n and applying the strong law of large numbers, we see that these ratios converge almost surely to $H_0(\boldsymbol{B}, \boldsymbol{X})$ and $H(\boldsymbol{B}, \boldsymbol{X})$, respectively.

Example 6.2.4. *Craps: 32-across-the-board, continued.* We continue with Example 6.2.3. Now consider (6.100) in the this context. Because place bets were considered to be one-roll bets in calculating H (see Example 6.1.5 on

p. 203), each coup in (6.100) corresponds to a single roll of the dice. Therefore, implicit in (6.100) is the assumption that winning place bets are replaced immediately by identical bets. This is in fact the way in which many gamblers tend to bet, assuming they do not *press* (i.e., increase) their winning bets. (See the term "California bet" in the chapter notes.)

Alternatively, consider (6.99) in the context of Example 6.2.3. Recall that place bets cannot be regarded as one-roll bets in calculating H_0 if one is to obtain conventional results (see Example 6.1.5 on p. 203). Thus, each coup in (6.99) corresponds to a random number of dice rolls, which must be the number needed to resolve all six place bets. Therefore, implicit in (6.99) is the assumption that winning place bets are not replaced until all six place bets have been won or lost. This is contrary to the way in which most gamblers tend to bet.

Another way to look at this distinction is to compare the weighted averages (6.95) and (6.96). In (6.95) the weights are proportional to the bet sizes only, whereas in (6.96) the weights are proportional to not only the bet sizes but also to the frequencies with which the various place bets are resolved (see (6.97)). The former suggests that the various place bets are made equally often, while the latter suggests that the various place bets are made with frequencies proportional to the frequencies with which they are resolved. As we have already indicated, the latter corresponds to a more typical style of play. Thus, in Example 6.2.3, H seems more appropriate than H_0. ♠

We next illustrate the consistency of the definitions. We rework Example 6.1.7 on p. 205 using (6.91) and (6.92).

Example 6.2.5. *Craps: Pass line or don't pass with 3-4-5-times free odds, revisited.* Here we think of the line bet (pass line or don't pass) and the free odds bet as being two parts of a composite wager. (Although the free odds bet cannot be made alone, let us suppose temporarily that it can be.) Clearly, equality holds in (6.91) and (6.92).

We consider the pass-line bet first. Assuming an initial one-unit bet, the expected amount of action, as well as the expected amount bet, is of course 1. For the 3-4-5-times free odds bet, these two quantities are both equal to $25/9$. The house advantages for the two bets are $H_0 = H = 7/495$ and $H_0 = H = 0$, respectively. Therefore, the house advantage of the pass-line bet with 3-4-5-times free odds is

$$H_0 = H = \frac{(1)(7/495) + (25/9)(0)}{1 + (25/9)} = \frac{7}{1{,}870}, \tag{6.101}$$

in agreement with Example 6.1.7 on p. 205.

Next we consider the don't-pass bet. Assuming an initial one-unit bet, the expected amount bet is 1, while the expected amount of action is $35/36$. For the 3-4-5-times free odds bet, these two quantities are both equal to $6(2/3) = 4$. The house advantages for the two bets are $H_0 = 3/220$ and $H_0 = 0$, respectively; or $H = 27/1{,}925$ and $H = 0$, respectively. Therefore, the house advantage of the don't-pass bet with 3-4-5-times free odds is

$$H_0 = \frac{(1)(3/220) + (4)(0)}{1 + 4} = \frac{3}{1,100} \tag{6.102}$$

or

$$H = \frac{(35/36)(27/1,925) + (4)(0)}{(35/36) + 4} = \frac{27}{9,845}, \tag{6.103}$$

again in agreement with Example 6.1.7 on p. 205. ♠

6.3 Volatility

In Sections 6.1 and 6.2 we used the strong law of large numbers to argue that certain statistics, such as the ratio of the gambler's cumulative loss to his total amount of action, converge with probability one to values, such as the house advantage, that can be computed directly and used to evaluate the unfavorability of the wager. Here we use the central limit theorem to determine how good the resulting approximation is. We will need the following consequence of the central limit theorem.

Theorem 6.3.1. *Let X and Y be jointly distributed discrete random variables with finite second moments and with $P(Y \geq 0) = 1$ together with $P(Y > 0) > 0$. Let $(X_1, Y_1), (X_2, Y_2), \ldots$ be a sequence of i.i.d. random vectors with common distribution that of (X, Y). Then*

$$\frac{\sqrt{n}}{\sigma} \left(\frac{X_1 + \cdots + X_n}{Y_1 + \cdots + Y_n} - \frac{E[X]}{E[Y]} \right) \xrightarrow{d} N(0, 1), \tag{6.104}$$

where

$$\sigma^2 := \frac{E[\{X - (E[X]/E[Y])Y\}^2]}{(E[Y])^2}. \tag{6.105}$$

Proof. Define $S_n := X_1 + \cdots + X_n$ and $T_n := Y_1 + \cdots + Y_n$ for each $n \geq 1$, and put $\mu := E[X]$ and $\nu := E[Y]$. Then

$$\frac{\sqrt{n}}{\sigma} \left(\frac{S_n}{T_n} - \frac{\mu}{\nu} \right) = \frac{\sum_{i=1}^{n} \{X_i - (\mu/\nu)Y_i\}}{\sqrt{n E[\{X - (\mu/\nu)Y\}^2]}} \bigg/ \frac{T_n}{n\nu}, \tag{6.106}$$

and the numerator converges in distribution to the standard normal distribution by the central limit theorem (Theorem 1.5.15 on p. 48), while the denominator converges in probability to 1 by the weak law of large numbers (Theorem 1.5.3 on p. 43). Therefore, the desired conclusion follows from Problem 1.56 on p. 63. ♠

Let (B, X) be a pair of jointly distributed discrete random variables satisfying (6.1) on p. 199 and having finite second moments, with B denoting the amount bet and X denoting the gambler's profit. Let $(B_1, X_1), (B_2, X_2), \ldots$ be a sequence of i.i.d. random vectors whose common distribution is that of (B, X), representing the results of independent repetitions of the original wager. Then

$$\frac{\sqrt{n}}{\sigma_0} \left(\frac{-(X_1 + \cdots + X_n)}{B_1 + \cdots + B_n} - \mathrm{H}_0(B, X) \right) \xrightarrow{d} N(0, 1), \qquad (6.107)$$

where

$$\sigma_0^2 := \frac{\mathrm{E}[\{X + \mathrm{H}_0(B, X)B\}^2]}{(\mathrm{E}[B])^2}. \qquad (6.108)$$

Similarly,

$$\frac{\sqrt{n}}{\sigma} \left(\frac{-(X_1 + \cdots + X_n)}{B_1 1_{\{X_1 \neq 0\}} + \cdots + B_n 1_{\{X_n \neq 0\}}} - \mathrm{H}(B, X) \right) \xrightarrow{d} N(0, 1), \qquad (6.109)$$

where

$$\sigma^2 := \frac{\mathrm{E}[\{X + \mathrm{H}(B, X)B 1_{\{X \neq 0\}}\}^2]}{(\mathrm{E}[B 1_{\{X \neq 0\}}])^2}. \qquad (6.110)$$

Notice that the numerators in (6.108) and (6.110) are second moments of mean-zero random variables. We illustrate these results with two examples.

Example 6.3.2. *Roulette: m-number bet, continued.* We continue with Example 6.1.3 on p. 202. For the roulette bet on m of the $36 + z$ numbers ($m = 1, 2, 3, 4, 6, 12, 18$; $z = 1, 2$), (6.107) and (6.109) are equivalent and become

$$\frac{\sqrt{n}}{\sigma_0} \left(\frac{-(X_1 + \cdots + X_n)}{B_1 + \cdots + B_n} - \mathrm{H}_0 \right) \xrightarrow{d} N(0, 1), \qquad (6.111)$$

where $\mathrm{H}_0 = z/(36 + z)$ (Example 6.1.3 on p. 202) and

$$\sigma_0^2 = \frac{\mathrm{Var}(X)}{(\mathrm{E}[B])^2} = \left(\frac{36}{m} - 1 \right)^2 \frac{m}{36 + z} + \left(1 - \frac{m}{36 + z} \right) - \left(\frac{z}{36 + z} \right)^2$$

$$= \left(\frac{36}{36 + z} \right)^2 \left(\frac{36 + z}{m} - 1 \right). \qquad (6.112)$$

It follows that

$$\lim_{n \to \infty} \mathrm{P}\left(\left| \frac{-(X_1 + \cdots + X_n)}{B_1 + \cdots + B_n} - \mathrm{H}_0 \right| \leq z_{1-\alpha/2} \frac{\sigma_0}{\sqrt{n}} \right) = 1 - \alpha. \qquad (6.113)$$

This tells us how much the ratio of the gambler's cumulative loss in n coups to his total amount bet in n coups can be expected to deviate from the house advantage for large n, at least when bet sizes are constant. Notice that σ_0^2

decreases as m increases, so there is greater volatility from a bet on fewer numbers. ♠

Example 6.3.3. *Craps: Pass line or don't pass with 3-4-5-times free odds, continued.* We continue with Example 6.1.7 on p. 205. For a one-unit pass-line bet with a three-, four-, or five-unit free odds bet, (6.107) applies with $H_0 = 7/1{,}870$ and

$$\sigma_0^2 = \left(\frac{3{,}932}{165} - 2\,\frac{16}{33}\,H_0 + \frac{167}{9}\,H_0^2\right)\left(\frac{34}{9}\right)^{-2} \approx 1.669534. \quad (6.114)$$

Here the volatility is significantly greater than it is for the pass-line bettor who fails to take the odds ($\sigma_0^2 \approx 0.999800$), not surprisingly.

For a one-unit don't-pass bet with a six-unit free odds bet, (6.107) and (6.109) apply with $H_0 = 3/1{,}100$, $H = 27/9{,}845$,

$$\sigma_0^2 = \left(\frac{47{,}789}{1{,}980} + 2\,\frac{487}{660}\,H_0 + 33\,H_0^2\right)5^{-2} \approx 0.965605, \quad (6.115)$$

and

$$\sigma^2 = \left(\frac{47{,}789}{1{,}980} + 2\,\frac{487}{660}\,H + \frac{1{,}187}{36}\,H^2\right)\left(\frac{179}{36}\right)^{-2} \approx 0.976425. \quad (6.116)$$

Here the volatility is slightly less than it is for the don't-pass bettor who fails to lay the odds ($\sigma_0^2 \approx 0.972036$ and $\sigma^2 \approx 1.028369$). ♠

With $(X_1, B_1), (X_2, B_2), \ldots$ as near the beginning of this section, we have considered criteria given by the asymptotic values of

$$\frac{-\sum_{l=1}^{n} X_l}{\sum_{l=1}^{n} B_l} \quad \text{and} \quad \frac{-\sum_{l=1}^{n} X_l}{\sum_{l=1}^{n} B_l\,1_{\{X_l \neq 0\}}}. \quad (6.117)$$

A rather different criterion, which avoids the issue of which denominator in (6.117) is most appropriate, is to maximize the probability of ending a session ahead (or at least even).

Let us assume that the sequence i.i.d. discrete random variables $X_1, X_2, \ldots$ has common distribution that of X, and that X has mean μ and variance σ^2. Put $S_0 := 0$ and $S_n := X_1 + \cdots + X_n$ for each $n \geq 1$. Particularly in situations in which $\mu < 0$, we are suggesting that a reasonable criterion would be to choose the wager that maximizes $P(S_n \geq 0)$, the probability of ending a session of length n ahead (or at least even). For this it has been suggested that we apply a central limit theorem approximation. Specifically,

$$P(S_n \geq 0) = P\left(\frac{S_n - n\mu}{\sqrt{n\sigma^2}} \geq \frac{-n\mu}{\sqrt{n\sigma^2}}\right) \approx 1 - \Phi\left(-\frac{\mu}{\sigma}\sqrt{n}\right). \quad (6.118)$$

Now the right side of (6.118) is decreasing in $-\mu/\sigma$, suggesting the criterion

$$J(X) := \frac{-\mathrm{E}[X]}{\mathrm{SD}(X)}, \tag{6.119}$$

which we will call the *expected loss per standard deviation*.

Although $J(X)$ has some desirable properties ($J(bX) = J(X)$ for all $b > 0$ and $J(-X) = -J(X)$), it is *not* the case that $\mathrm{P}(S_n \geq 0)$ is decreasing in $J(X)$ for small and moderate n, because (6.118) is only an approximation. See Problem 6.17 on p. 234.

In fact, it is not the case that $\mathrm{P}(S_n \geq 0)$ is decreasing in $J(X)$ for large n either, because (6.118) is not asymptotically valid, even after taking logarithms of both sides. But there is a result, namely *Cramér's theorem on large deviations*, that more accurately describes the behavior of $\ln \mathrm{P}(S_n \geq 0)$ as $n \to \infty$. It depends on the notion of the *moment generating function* $M(t)$ of a discrete random variable X, defined by

$$M(t) := \mathrm{E}[e^{tX}], \qquad t \in \mathbf{R}, \tag{6.120}$$

which need not be finite for $t \neq 0$.

Theorem 6.3.4. *Let X be a discrete random variable with finite expectation such that*

$$\mathrm{E}[X] < 0 \qquad \text{and} \qquad \mathrm{P}(X > 0) > 0. \tag{6.121}$$

In fact, assume that the moment generating function $M(t)$ of X is finite in a neighborhood of 0. Let (t_0, t_1) be the maximal open interval on which $M(t)$ is finite, and assume that

$$\rho := \inf_{t \in \mathbf{R}} M(t) = M(\tau) \text{ for some } \tau \in (t_0, t_1). \tag{6.122}$$

Let $X_1, X_2, \ldots$ be i.i.d. as X and put $S_n := X_1 + \cdots + X_n$ for each $n \geq 1$. Then

$$\mathrm{P}(S_n \geq 0) \leq \rho^n, \qquad n \geq 1. \tag{6.123}$$

In fact,

$$\lim_{n \to \infty} \frac{1}{n} \ln \mathrm{P}(S_n \geq 0) = \ln \rho. \tag{6.124}$$

Remark. (a) In particular, given $\varepsilon \in (0, \rho)$, we have $(\rho - \varepsilon)^n \leq \mathrm{P}(S_n \geq 0) \leq \rho^n$, the second inequality holding for all n and the first for all n sufficiently large. This suggests the criterion

$$\rho(X) := \min_t \mathrm{E}[e^{tX}]. \tag{6.125}$$

Unlike (6.119), $\rho(X)$ should be maximized rather than minimized.

(b) The very mild assumption that the infimum of $M(t)$ is achieved in (t_0, t_1) can be eliminated with additional effort.

(c) By Problem 1.54 on p. 63,

$$\lim_{n\to\infty} \frac{1}{n} \ln\left[1 - \Phi\left(-\frac{\mu}{\sigma}\sqrt{n}\right)\right] = -\frac{\mu^2}{2\sigma^2}, \tag{6.126}$$

showing that the approximation (6.118) is inconsistent with (6.124).

Proof. The dominated convergence theorem (Theorem A.2.5 on p. 748) shows that $M'(0) = \mathrm{E}[X] < 0$ and $M''(t) = \mathrm{E}[X^2 e^{tX}] \geq 0$ for $t_0 < t < t_1$, hence M is convex on (t_0, t_1) (Lemma 1.4.20 on p. 37). These facts imply that τ of (6.122) is positive. Conclusion (6.123) is therefore immediate from Markov's inequality (Theorem 1.5.1 on p. 42):

$$\mathrm{P}(S_n \geq 0) = \mathrm{P}(e^{\tau S_n} \geq 1) \leq \mathrm{E}[e^{\tau S_n}] = M(\tau)^n = \rho^n. \tag{6.127}$$

The second conclusion is more delicate. Let us define (on a different sample space) a random variable $\hat{X}$ with distribution

$$\mathrm{P}(\hat{X} = x) = \frac{e^{\tau x}}{\rho}\, \mathrm{P}(X = x), \qquad x \in \mathbf{R}. \tag{6.128}$$

Then $\hat{X}$ has moment generating function $\hat{M}(t) = M(\tau + t)/\rho$, hence mean $\hat{M}'(0) = M'(\tau)/\rho = 0$ and variance $\hat{M}''(0) - (\hat{M}'(0))^2 = M''(\tau)/\rho \in (0, \infty)$. Let $\hat{X}_1, \hat{X}_2, \ldots$ be i.i.d. as $\hat{X}$ and put $\hat{S}_n := \hat{X}_1 + \cdots + \hat{X}_n$ for each $n \geq 1$. Then $\hat{S}_n$ has distribution

$$\mathrm{P}(\hat{S}_n = x) = \frac{e^{\tau x}}{\rho^n}\, \mathrm{P}(S_n = x), \qquad x \in \mathbf{R}, \tag{6.129}$$

as can be seen by induction:

$$\begin{aligned}
\mathrm{P}(\hat{S}_{n+1} = x) &= \sum_{y\in\mathbf{R}} \mathrm{P}(\hat{S}_n = y)\, \mathrm{P}(\hat{X}_{n+1} = x - y) \\
&= \sum_{y\in\mathbf{R}} \frac{e^{\tau y}}{\rho^n}\, \mathrm{P}(S_n = y)\, \frac{e^{\tau(x-y)}}{\rho}\, \mathrm{P}(X_{n+1} = x - y) \\
&= \frac{e^{\tau x}}{\rho^{n+1}} \sum_{y\in\mathbf{R}} \mathrm{P}(S_n = y)\, \mathrm{P}(X_{n+1} = x - y) \\
&= \frac{e^{\tau x}}{\rho^{n+1}}\, \mathrm{P}(S_{n+1} = x). \tag{6.130}
\end{aligned}$$

By (6.129) we have, for each $\varepsilon > 0$,

$$\mathrm{P}(S_n \geq 0) = \sum_{x\geq 0} \frac{\rho^n}{e^{\tau x}}\, \mathrm{P}(\hat{S}_n = x) \geq \frac{\rho^n}{e^{n\varepsilon\tau}} \sum_{0\leq x\leq n\varepsilon} \mathrm{P}(\hat{S}_n = x). \tag{6.131}$$

Now, by the central limit theorem (Theorem 1.5.15 on p. 48), the sum on the right side of (6.131), which equals $\mathrm{P}(\hat{S}_n/\sqrt{n} \in [0, \varepsilon\sqrt{n}])$, converges to $\frac{1}{2}$ as $n \to \infty$, and therefore

$$\liminf_{n \to \infty} \frac{1}{n} \ln P(S_n \geq 0) \geq \ln \rho - \varepsilon\tau. \tag{6.132}$$

Since ε was arbitrary, the desired conclusion follows.   ♠

Next, we reconsider Example 6.1.2 on p. 202.

Example 6.3.5. *Wagers with three possible outcomes, continued.* For the wager with three possible outcomes, win, loss, or push, with respective probabilities $p > 0$, $q > 0$, and $r \geq 0$, where $p + q + r = 1$, and with a win paying a to 1, the expected loss per standard deviation is

$$J = \frac{-ap + q}{\sqrt{a^2 p(1-p) + 2apq + q(1-q)}}. \tag{6.133}$$

On the other hand, the minimum of the moment generating function is

$$\rho = \frac{a+1}{a} q \left(\frac{ap}{q}\right)^{1/(a+1)} + r. \tag{6.134}$$

Many of the wagers at roulette and craps have this simple structure.   ♠

Example 6.3.6. *Roulette: m-number bet, continued.* For the roulette bet on m of the $36 + z$ numbers ($m = 1, 2, 3, 4, 6, 12, 18$; $z = 1, 2$), we have $p = m/(36 + z)$, $q = 1 - p$, $r = 0$, and $a = 36/m - 1$ in Example 6.3.5. It follows that

$$J = \frac{z}{36\sqrt{(36+z)/m - 1}}, \tag{6.135}$$

which is increasing in m for $z = 1$ and for $z = 2$. On the other hand, the minimum of the moment generating function is

$$\rho = \frac{36}{36+z} \left(1 + \frac{z}{36-m}\right)^{(36+m)/36}, \tag{6.136}$$

which is decreasing in m for $z = 1$ and for $z = 2$. Thus, both criteria recommend betting on fewer numbers.   ♠

Example 6.3.7. *Comparing craps wagers.* Let X denote the wager corresponding, in the notation of Example 6.3.5, to $p = 244/495$, $q = 1 - p$, $r = 0$, and $a = 1$ (the pass-line bet at craps). Then

$$J(X) = \frac{7}{495\sqrt{1 - (7/495)^2}} \approx 0.014142828351 \tag{6.137}$$

and

$$\rho(X) = 2\frac{251}{495}\sqrt{\frac{244}{251}} \approx 0.999900005204. \tag{6.138}$$

Let Y denote the wager corresponding to $p = 1/18$, $q = 1 - p$, $r = 0$ and $a = 16$ (the one-roll bet on 11 at craps, assuming more-generous British payoffs). Then

$$J(Y) = 17^{-3/2} \approx 0.014266801473 \qquad (6.139)$$

and

$$\rho(Y) = \frac{17}{16}\frac{17}{18}\left(\frac{16}{17}\right)^{1/17} \approx 0.999900058746. \qquad (6.140)$$

Based on the J criterion (minimize expected loss per standard deviation), we would prefer X to Y. Based on the ρ criterion (maximize the minimum of the moment generating function), we would prefer Y to X. By Theorem 6.3.4, Y is superior in terms of our original criterion, namely maximizing the probability of ending a session of n coups ahead (or even), at least if n is large enough. How large is "large enough"? ♠

6.4 Problems

6.1. *House advantage in terms of odds.* Reconsider Example 6.1.2 on p. 202 without pushes, in which case the outcome of a wager results in a win or a loss, and assume that the true odds against a win are L to W, with W and L positive integers. Suppose also that the payoff odds are w to l, with w and l positive integers. Derive a formula for the house advantage in terms of these four parameters. For example, if the true odds against an event are 6 to 5, and if the payoff odds are 7 to 6, then $L = 6$, $W = 5$, $w = 7$, and $l = 6$. Evaluate the formula in this case.

6.2. *Craps: One-roll bets.* Find the house advantages of the following one-roll bets at craps. Use Problem 6.1 where possible. (a1) Field bet (3, 4, 9, 10, and 11 pay 1 to 1; 2 and 12 pay 2 to 1). (a2) Field bet (3, 4, 9, 10, and 11 pay 1 to 1; 2 pays 2 to 1; and 12 pays 3 to 1). (b) Any craps (2, 3, and 12 pay 7 to 1). (c) Any seven (pays 4 to 1). (d) Craps and 11 (2, 3, and 12 pay 3 to 1; 11 pays 7 to 1). (e1) Two (pays 29 to 1) or 12 (pays 29 to 1). (e2) Two (pays 30 to 1) or 12 (pays 30 to 1). (f1) Three (pays 14 to 1) or 11 (pays 14 to 1). (f2) Three (pays 15 to 1) or 11 (pays 15 to 1). (g1) Horn bet (a composite wager, with one unit on each of 2, 3, 11, and 12; use payoff odds in (e1) and (f1)). (g2) Horn bet (same as (g1), except use payoff odds in (e2) and (f2)).

6.3. *Two-up.* Find the house advantage at the two-up wager described in Problem 1.29 on p. 57.

6.4. *Keno: 16-spot ticket.* Here the keno player choose 16 numbers from the set $\{1, 2, \ldots, 80\}$, not necessarily at random, and the game proceeds otherwise as in Example 6.1.6 on p. 203. The payoffs are as in Table 6.3. Find the house advantage. Note that "free play" is not quite the same thing as a return of the entry fee.

Table 6.3 Sweet-Sixteen keno payoffs (in units) at the Imperial Palace (Jan. 2008). Entry fee is 3 units (one unit is \$1). Maximum aggregate payout is 200,000 units.

catch	payoff	catch	payoff	catch	payoff
0	25	6	2	12	7,500
1	free play	7	free play	13	15,000
2	2	8	10	14	30,000
3	1	9	120	15	50,000
4	1	10	350	16	150,000
5	1	11	1,500		

6.5. *Red dog.* Recall the game of red dog as described in Example 2.2.9 on p. 83, and assume a six-deck shoe.

(a) Find the house advantage (pushes included and excluded).

(b) Does Theorem 6.1.8 on p. 207 suggest an alternative optimal strategy?

6.6. *Craps: Pass-line bet with m-times free odds*

(a) Some casinos offer m-times free odds, regardless of the point. The player makes a pass-line bet, and after a point is established, a free odds bet, for no more than m times the amount of his pass-line bet, becomes available. This is in effect a side bet that the original bet will be won, and it pays fair odds. Find the house advantage of the pass-line bet with m-times free odds.

(b) Here we consider a modification of the compound wager in part (a). Suppose that a pass-line bettor who wishes to make an m-times free odds bet after a point is established must do so at the same time that the pass-line bet is made (namely, prior to the come-out roll). A proportion $1/(1 + m)$ of the bet is paid even money just like a pass-line bet. The remaining proportion $m/(1+m)$ of the bet is pushed if the pass-line bet is resolved on the come-out roll, whereas it is treated as though it were a free odds bet (i.e., it is paid at fair odds) if a point is established. Find the house advantage.

6.7. *Number of iterations required by Theorem 6.1.8.* In Examples 6.1.9–6.1.11 on pp. 209, 211, and 214, only two iterations are needed to reach the fixed point h^* of G of Theorem 6.1.8 on p. 207 starting at $h_0 := 0$. Given a positive integer k, find positive parameters $p_1, \ldots, p_k, b, b_1, \ldots, b_k$, and real parameters $x, x_1, \ldots, x_k$, such that k iterations are needed. More precisely, letting $h_0 := 0$ and $h_n := G(h_{n-1})$ for each $n \geq 1$, we require that $k = \min\{n \geq 0 : G(h_n) = h_n\}$.

6.8. *A composite wager at roulette.* Generalize Example 6.2.1 on p. 221 to the case in which the amounts bet on the individual numbers are arbitrary (possibly zero in some cases).

6.9. *A composite wager at craps.* Suppose one makes three place bets (five units on 5, six units on 6, and six units on 8) and a five-unit field bet (a one-roll bet that pays 1 to 1 for 3, 4, 9, 10, and 11; 2 to 1 for 2; and 3 to 1 for 12) simultaneously. Find the house advantage of the composite wager.

6.10. *Another composite wager at craps.* Suppose one makes a five-unit pass-line bet and, if a point is established, takes the odds for five units if the point is 4, 6, 8, or 10, and for six units if it is 5 or 9. Suppose further that, if a point is established, one makes a place bet on each of the five point numbers that are not established, the individual amounts being those of 32-across-the-board (Example 6.2.3 on p. 222). Find the house advantage of the composite wager. (To avoid ambiguities, assume that place bets are replaced with identical bets if won but are withdrawn if the pass-line bet is won.)

6.11. *32-across-the-board, revisited.* Consider 32-across-the-board, as in Example 6.2.3 on p. 222. Suppose the player replaces winning place bets until 7 finally appears. Use Problem 2.8 on p. 89 to find the expected loss and the expected amount bet, thereby obtaining an alternative derivation of the house advantage.

6.12. *An alternative definition of house advantage.* Given a random variable X representing the gambler's profit and satisfying

$$0 < \mathrm{E}\big[|X|\big] < \infty, \tag{6.141}$$

we define

$$\mathrm{H}^*(X) := \frac{-\mathrm{E}[X]}{\mathrm{E}\big[|X|\big]}. \tag{6.142}$$

We can interpret $\mathrm{H}^*(X)$ as the ratio of the gambler's expected loss to the expected amount of money that changes hands. This alternative definition of the house advantage of the wager X has several properties to recommend it. First, it does not depend on the bet size B. Second, $\mathrm{H}^*(bX) = \mathrm{H}^*(X)$ for all $b > 0$. Third, it does not depend on whether pushes are included or excluded. Fourth, for composite wagers $\boldsymbol{X} = (X_1, \ldots, X_d)$, if we assume that $X_1, \ldots, X_d$ and $X_1 + \cdots + X_d$ satisfy (6.141), then we can simply define

$$\mathrm{H}^*(\boldsymbol{X}) := \mathrm{H}^*(X_1 + \cdots + X_d). \tag{6.143}$$

   (a) Prove the third property of $\mathrm{H}^*(X)$ mentioned above.
   (b) Find the analogue of (6.91) and (6.92) on p. 222.
   (c) Evaluate this alternative house advantage for Examples 6.1.2 on p. 202, 6.1.5 on p. 203, and 6.2.3 on p. 222.
   (d) Find the analogues of (6.107) and (6.108) on p. 226.

6.13. *Monotonicity property of (6.142).* Let X_1 and X_2 satisfy (6.141). Suppose that $\mathrm{P}(X_1 \geq t) \leq \mathrm{P}(X_2 \geq t)$ for all real t, that is, X_1 is *stochastically less than* X_2. Show that $\mathrm{H}^*(X_1) \geq \mathrm{H}^*(X_2)$.

6.14. *A variant of house advantage for composite wagers.* Let $(\boldsymbol{B}_1, \boldsymbol{X}_1), (\boldsymbol{B}_2, \boldsymbol{X}_2), \ldots$ be a sequence of i.i.d. random vectors whose common distribution is that of $(\boldsymbol{B}, \boldsymbol{X})$, and adopt the notation of (6.78) on p. 219 and (6.98) on p. 223. For $j = 1, \ldots, d$, define $N_j := \min\{n \geq 1 : X_{n,j} \neq 0\}$, and define

$$H_1(\boldsymbol{B}, \boldsymbol{X}) := \frac{-E[X_{N_1,1} + \cdots + X_{N_d,d}]}{E[B_{N_1,1} + \cdots + B_{N_d,d}]}. \tag{6.144}$$

Use Theorem 6.1.1 on p. 201 to express $H_1(\boldsymbol{B}, \boldsymbol{X})$ solely in terms of $(\boldsymbol{B}, \boldsymbol{X})$, and deduce the analogue of (6.91) and (6.92) on p. 222. Finally, evaluate H_1 for the composite wager of Example 6.2.3 on p. 222.

6.15. *Another variant of house advantage for composite wagers.* Using the notation of Problem 6.14, define $N := \min\{N_1, \ldots, N_d\}$ and

$$H_2(\boldsymbol{B}, \boldsymbol{X}) := \frac{-E[X_{N,1} + \cdots + X_{N,d}]}{E[B_{N,1} + \cdots + B_{N,d}]}. \tag{6.145}$$

Express $H_2(\boldsymbol{B}, \boldsymbol{X})$ solely in terms of $(\boldsymbol{B}, \boldsymbol{X})$. What is the analogue of (6.91) and (6.92) on p. 222? Finally, evaluate H_2 for the composite wager of Example 6.2.3 on p. 222.

6.16. *A variant of (6.143).* Using the notation of Problem 6.14, define

$$H_1^*(\boldsymbol{X}) := H^*(X_{N_1,1} + \cdots + X_{N_d,d}). \tag{6.146}$$

Obtain the analogue of (6.91) and (6.92) on p. 222, and evaluate H_1^* for the composite wager of Example 6.2.3 on p. 222.

Hint: Recall Example 4.3.6 on p. 141.

6.17. *Monotonicity and the probability of ending a session ahead.* The approximation (6.118) on p. 227 and Theorem 6.3.4 on p. 228, together with Example 6.3.6 on p. 230, suggest that, in the case of the roulette bet on m numbers (out of 38), the probability of ending a session of n coups ahead (or even) should be decreasing in m. By considering $m = 1, 2, 3, 4, 6, 12, 18$, determine for which n $(1 \leq n \leq 100)$ the suggested monotonicity holds.

6.18. *Extending house advantage beyond the i.i.d. case.* Suppose that $(\boldsymbol{B}_1, \boldsymbol{X}_1), (\boldsymbol{B}_2, \boldsymbol{X}_2), \ldots$ is a sequence of jointly distributed, discrete random vectors, each of which satisfies (6.1) on p. 199. Let $1 \leq N_1 < N_2 < \cdots$ be a sequence of stopping times, each with respect to the given sequence, and assume that

$$\begin{aligned}
&((\boldsymbol{B}_1, \boldsymbol{X}_1), \ldots, (\boldsymbol{B}_{N_1}, \boldsymbol{X}_{N_1})), \\
&((\boldsymbol{B}_{N_1+1}, \boldsymbol{X}_{N_1+1}), \ldots, (\boldsymbol{B}_{N_2}, \boldsymbol{X}_{N_2})), \\
&((\boldsymbol{B}_{N_2+1}, \boldsymbol{X}_{N_2+1}), \ldots, (\boldsymbol{B}_{N_3}, \boldsymbol{X}_{N_3})), \\
&\quad\vdots
\end{aligned} \tag{6.147}$$

is a sequence of i.i.d. random vectors of random dimensions. (In other words, the i.i.d. property holds between rows of (6.147), but not necessarily within them.) Argue that

$$\frac{-(X_1 + \cdots + X_{N_n})}{B_1 \, 1_{\{X_1 \neq 0\}} + \cdots + B_{N_n} \, 1_{\{X_{N_n} \neq 0\}}} \tag{6.148}$$

converges a.s. to

$$\mathrm{H} := \frac{-\mathrm{E}[X_1 + \cdots + X_{N_1}]}{\mathrm{E}[B_1 \, 1_{\{X_1 \neq 0\}} + \cdots + B_{N_1} \, 1_{\{X_{N_1} \neq 0\}}]}, \tag{6.149}$$

provided both expectations in (6.149) exist. H can be regarded as a generalization of (6.3) on p. 200. Obtain the analogous generalization of (6.2) on p. 200.

6.19. *Effective house advantage under a dead-chip program.* Many casinos offer incentives to high rollers. One such incentive is a dead-chip program, which works as follows. Instead of buying negotiable chips ("live chips") at par, that is, one live chip per unit of capital invested, the gambler buys nonnegotiable chips ("dead chips") at $1 + \beta$ to 1, that is, $1 + \beta$ dead chips per unit of capital invested, where $\beta > 0$. A bet consisting of one or more dead chips is paid with live chips in the case of a win and is collected by the casino in the case of a loss.

Assume a game as in Example 6.1.2 on p. 202, that is, assume there are only three possible outcomes, win, loss, or push, with respective probabilities $p > 0$, $q > 0$, and $r \geq 0$, where $p + q + r = 1$, and suppose that a win pays a to 1, where $a > 0$. Suppose further that $ap - q < 0$ and assume also independent coups.

(a) The cost of a dead chip is $(1 + \beta)^{-1}$ units. Show that its expected value is ap/q units.

(b) Use Problem 6.18 to show that the effective house advantage H^β of the specified wager, taking the dead-chip bonus β into account, is

$$\mathrm{H}^\beta = \left(\mathrm{H} - \frac{\beta}{1 + \beta} \frac{q}{p + q} \right) \Big/ \left(1 - \frac{p + q}{q} \mathrm{H} \right), \tag{6.150}$$

where H is the ordinary house advantage (pushes excluded) as in Example 6.1.2 on p. 202. By effective house advantage we mean the long-term ratio of the gambler's cumulative loss to his total amount of action, with both numerator and denominator measured in units of live chips.

(c) Show that the effective house advantage is positive if and only if the value of a dead chip is less than its cost. How large must β be for the dead-chip program to be better for the gambler than no program at all? How large can β be before the gambler has an advantage over the house? Answer these last two questions for the even-money bets at 38-number roulette.

Hint: (a) Show that the number of wins prior to the first loss has a shifted geometric distribution with success probability $q/(p+q)$.

(b) In Problem 6.18 let N_n being the coup at which the nth loss occurs. Use part (a) to convert the denominator from units of dead chips to units of live chips.

6.20. *Effective house advantage under a dead-chip program, continued.* The formula (6.150) of Problem 6.19 assumes a game with only three possible outcomes, win, loss, or push, with respective probabilities $p > 0$, $q > 0$, and $r \geq 0$, where $p + q + r = 1$, and payoff odds of a to 1, where $a > 0$.

(a) Generalize the formula to allow the distribution of (B, X) to be arbitrary (discrete, of course).

(b) Apply the formula obtained in part (a) to the game of red dog (Example 2.2.9 on p. 83), assuming a six-deck shoe. How large can β be before the gambler has an advantage over the house?

6.5 Notes

The notion of advantage is nearly as old as the concept of expectation itself. The phrase "l'avantage du banquier" was used by Joseph Sauveur [1653–1716] (1679), Pierre Rémond de Montmort [1678–1719] (1708, p. 4), Leonhard Euler [1707–1783] (1764), and Siméon-Denis Poisson [1781–1840] (1825), among many others. Abraham De Moivre [1667–1754] (1718, p. 155) referred to "the advantage of the setter" at hazard. Other names for house advantage include house edge, house percentage or PC (an acronym for percentage), house take, cagnotte, and vigorish. According to Scarne (1961, pp. 241–242), the term "vigorish" originated around 1907 when John H. Winn banked a craps game in New York City, charging a 5 percent commission on each bet. This commission provided profits to the bank with such strength or vigor that it became known as the vigorish. The *Oxford English Dictionary* (Simpson and Weiner 1989) suggested an alternative etymology: It is probably from Yiddish, an adaptation of the Russian *vyigryš*, meaning gain or winnings.

Thus, the concept of house advantage has been implicit in the gambling literature for more than three centuries. Nevertheless, no single definition is universally accepted. Indeed, as we have noted, there are several points of contention.

The first is the question of whether to include or exclude pushes. Here we have treated both approaches (H_0 and H) because we believe that the answer depends on the game. Scarne (1961, p. 260) and Wilson (1965, pp. 184–185) gave both definitions in the context of the don't-pass bet at craps, but expressed a preference for excluding pushes. Griffin (1991, p. 111) expressed the opposite preference "on the grounds that a player is not compelled to leave his bet out after a tie occurs." Grosjean (2009, p. 469) reached the same conclusion, arguing that excluding pushes "incorrectly treats the Don't bet as a contract bet that must be left on the table after a push [...].” The Griffin/Grosjean approach begs the question of how to handle place bets at craps, which are certainly not contract bets. Shackleford (2005a, p. 35) noted the unconventional house-advantage figure of $1/216$ for the place bet on 6 (Example 6.1.5) but attributed it to “perfectionists.” Wong (2005, p. 164) observed that $1/216$ can be regarded as the house advantage *per roll* for the place bet on 6 after noting that it requires $(\pi_6 + \pi_7)^{-1} = 36/11$ rolls on average to resolve the bet. Hannum and Cabot (2005, pp. 57–58) discussed the issue without explicitly stating a preference. Kilby, Fox, and Lucas (2005, p. 216) argued in favor of including pushes at baccarat because it is more consistent with casino marketing practices. In the game of faro, in which pushes occur more frequently than wins or losses, all of the 18th-century authors (e.g., Montmort 1713, p. 97) effectively excluded pushes by assuming that the bet is left in place until resolution, while Lemmel (1964, p. 111) is the only faro author, to the best of our knowledge, who has included pushes. Thorp (1973) argued in favor of excluding pushes at faro because

> The player tends to think of the resolution of the bet as the basic entity, rather than the turn [i.e., coup] or the other procedural aspects of the game.

Van-Tenac (1847, p. 51) excluded pushes in his discussion of trente et quarante, while Poisson (1825) included them. The editor of a 1976 reprint of Sham (1930) mistakenly characterized Sham's inclusion of pushes as an error, and Shackleford (2005b) mistakenly characterized the exclusion of pushes as an error. Neither approach is wrong, but one or the other may be more useful as a numerical index of the unfavorability of a particular wager.

A second issue, at least in compound wagers, is whether to measure house advantage with respect to the player's initial amount bet or with respect to his total amount bet. (We have emphasized the latter, but we recognize that the former is the accepted approach for games such as twenty-one.) There is general agreement in the gambling literature on this issue but it is inconsistent: In the context of the line bets at craps (pass line and don't pass) with free odds, almost all authors measure house advantage with respect to total amount bet, whereas in the newer compound wagers such as Three Card Poker (Example 6.1.9 and Section 16.2) and Let It Ride (Example 6.1.10 and Section 16.1), there is almost universal agreement that house advantage should be measured with respect to initial amount bet. Hannum and Cabot (2005, pp. 56–57) discussed the matter at some length, arguing that both formulations have correct interpretations. However, the interpretation of house advantage with respect to initial amount bet is a bit awkward. Specifically, it is the long-term ratio of the gambler's cumulative loss *not* to his total amount bet but to the total amount of his initial bets. Ko (1998, pp. 30–31) argued that dividing by expected total amount bet is misleading, citing as evidence the fact that it makes Caribbean Stud Poker (Problem 16.13) appear better than Let It Ride or Casino War (Example 11.2.3), when “it actually is the worst of the three games.” Ko did not explain why it is the worst. Shackleford (2005a) argued in favor of house advantage with respect to initial amount bet on the grounds that it makes it easier for the player to estimate how much he will lose. Nevertheless, he endorsed house advantage with respect to total amount bet for comparing different games (which is, after all, the main purpose of the concept of house advantage), and he introduced the term "element of risk" for this quantity, terminology that has been adopted by Epstein (2009, p. 251).

A third issue, where we have departed from tradition, is our insistence that the amount bet be the amount placed at risk (rather than the amount "in play"). Several examples have convinced us of the desirability of this formulation. One is Example 6.1.6, which is due to Cowles (2003, Chapter 16). He argued that the amount placed at risk, not the amount of the entry fee, should be used in evaluating the house advantage at keno. Specifically, Cowles observed that if one were to add b units to the entry fee and b units to each of the payoffs (including 0), the player's expected loss would remain unchanged, but the house advantage, if calculated with respect to the entry fee rather than with respect to the amount placed at risk, could be made arbitrarily small by choosing b sufficiently large. We find this a persuasive argument. Another example is the game of Let It Ride. Even though the basic wager requires a three-unit bet, it has been recognized by everyone who has written about this game that initially the player is putting only one unit at risk. Because the amount risked is so fundamental to the issue of a wager's desirability, we have incorporated it into our definition of house advantage. This has consequences that some readers may find undesirable, such as the result of Example 6.2.1. There it was argued that betting one unit on each of the 37 numbers of a roulette wheel is a composite wager with a house advantage of 1 (i.e., 100 percent), the point being that such a bettor is certain to lose exactly one unit, and so is risking not 37 units but only a single unit. One could simply accept the various anomalies inherent in the traditional definition, but instead we have decided to modify the definition. See Catlin (2009) for another example that illustrates the distinction.

Section 6.1 is an expanded version of Ethier (2007c).

As noted by Catlin (2000), the issue of initial amount bet vs. total amount bet is closely related to the question of how to determine optimal play at compound wagers. The use of house advantage with respect to initial amount bet leads to the criterion that a conditional wager should be accepted if and only if its conditional expectation is nonnegative (or positive, depending on one's view of fair bets). The use of house advantage with respect to total amount bet leads to what is typically a less restrictive criterion. Theorem 6.1.8 was formulated as a means of evaluating this criterion, and it extends two previous results. Frank Downton [1925–1984] (1969) showed, in a continuous rather than a discrete setting, that if h is a fixed point of G (in our notation), then $h = h^*$. He did not address the issue of the existence of such a fixed point. Donald E. Catlin [1936–] (2000) showed in effect that h^* is a fixed point of

G and proposed the iterative procedure described in the theorem. He did not prove that it converges.

There is some inconsistency in the gambling literature on how to deal with composite wagers. That of Example 6.2.3 is due to Ionescu Tulcea (1981, p. 18), who used the formulation (6.95) (or, equivalently, (6.144)). For essentially the same example, Scarne (1961, p. 298–299) effectively used (6.145), resulting in a house advantage $(1/80 = 0.0125)$ that is inappropriately small, being smaller than each of the individual house advantages. Similarly, Brisman (2004, pp. 139–140) used (6.145) for the example of Problem 6.9. A more plausible solution was found by Catlin (2003).

The argument, discussed in Example 6.1.12, that one should never lay the odds has appeared in Barstow (1979, Chapter 10), Scoblete (1993, pp. 116–118), and other sources. It is clearly flawed logic, but the flaw is rather subtle. Catlin (2000) discussed similar matters without addressing this specific example.

The definition of the cut of a wager X as stated in (6.71) is due to Dubins and Savage (1976, p. 176), and they obtained our (6.93) in this context. An alternative definition of house advantage depending only on the distribution of X is certainly possible and perhaps even desirable. Definition (6.142) was first proposed by Donald Ornstein [1934–] (Dubins and Savage 1976, p. 3), and was suggested also by Scarne (1974, p. 333) as a way of resolving the difficulty he encountered with Example 6.2.3. It was discussed in some detail by Ethier (1982a).

The index $J(X)$ of (6.119) was termed the *gain ratio* by Downton and Holder (1972). It suggests that, when comparing two subfair games with the same expected loss, the one with the larger variance should be preferred. This was Kozek's (1995) rule of thumb. Anderson and Fontenot (1980) pointed out that $P(S_n \geq 0)$ is not decreasing in $J(X)$ for small or moderate n. As noted in Section 6.3, the same assertion is true for large n. An alternative justification of $J(X)$ (actually $J(X)^2$) was given for superfair games by Sileo (1992), and a reparametrization of Sileo's index was introduced and named the *SCORE* (standardized comparison of risk and expectation) by Schlesinger (1999a,b, 2005) with emphasis on applications to the game of twenty-one.

Another index, $I(X) := -2\mathrm{E}[X]/\mathrm{Var}(X)$, was termed the *inequity* by Dubins and Savage (1965, pp. 167, 182), who acknowledged having learned of it from Jess Marcum (see Section 21.5 for more about him). Notice that $I(bX) = (1/b)I(X)$ for $b > 0$.

Theorem 6.3.4 (large deviations) is due to Harald Cramér [1893–1985] (1937). The inequality in (6.123) is due to Herman Chernoff [1923–] (1952). Problem 6.9 was posed by Brisman (2004, pp. 139–140) as noted above. Problem 6.11 was motivated by what used to be called a "California" or a "California bet" (Maurer 1950):

A bet in which the player covers all of the box numbers and takes back the winnings, though not the original bet after each throw.

Problem 6.17 was motivated by observations of Anderson and Fontenot (1980) and Downton (1980a). Problem 6.19 was motivated by Robert C. Hannum [1953–] and Sudhir H. Kale (2004). They defined effective house advantage slightly differently, measuring the gambler's loss in terms of live chips and measuring his action in terms of dead chips. This simplifies (6.150), replacing the denominator by 1, but the use of two different units of measurement raises the question of whether the resulting quantity is dimensionless, as it is ordinarily thought to be. According to Castleman (2004, p. 135), dead chips are used in East Asian casinos.

Chapter 7
Gambler's Ruin

"For God's sake, stop him from gambling, my dear," she said, "or he will ruin himself. [...]"

William Makepeace Thackeray, *Vanity Fair*

This chapter is concerned with finding the probability that, in an independent sequence of identical wagers, the gambler loses L or more units (that is, he is ruined) before he wins W or more units. In Section 7.1, we treat the case of even-money payoffs by deriving an explicit formula. In Section 7.2, we consider more-general integer-valued payoffs. Formulas are available in some special cases, but in general a computational algorithm is needed. Finally, Section 7.3 allows arbitrary payoffs, and here we resort to establishing upper and lower bounds on the probability of interest.

7.1 Even-Money Payoffs

Consider a wager that pays even money. Let p denote the probability of a win for the gambler, q the probability of a loss, and r the probability of a push (tie). Of course, $p + q + r = 1$, and we assume that $p > 0$, $q > 0$, and $r \geq 0$. Let the random variable X represent the gambler's profit from a one-unit wager on this betting opportunity. Then the distribution of X is given by

$$\mathrm{P}(X = 1) = p, \quad \mathrm{P}(X = -1) = q, \quad \mathrm{P}(X = 0) = r. \qquad (7.1)$$

Now let $X_1, X_2, \ldots$ be a sequence of i.i.d. random variables with common distribution that of X, representing the results of independent repetitions of the original one-unit wager. Then

$$S_n := X_1 + X_2 + \cdots + X_n \qquad (7.2)$$

S.N. Ethier, *The Doctrine of Chances*, Probability and its Applications, 241
DOI 10.1007/978-3-540-78783-9_7, © Springer-Verlag Berlin Heidelberg 2010

represents the gambler's cumulative profit after n such wagers; of course, $S_0 = 0$. Given positive integers W and L, we suppose that the gambler stops betting as soon as he wins W units or loses L units. Consequently, his last bet occurs at coup

$$N(-L, W) := \min\{n \geq 1 : S_n = -L \quad \text{or} \quad S_n = W\}, \qquad (7.3)$$

where $\min \varnothing := \infty$. Notice that, on the event $\{N(-L, W) < \infty\}$, which we will show has probability 1, the random variable $S_{N(-L,W)}$, which is the sum of a random number of random variables, represents the gambler's cumulative profit following the resolution of his final bet and is equal to either $-L$ or W. Our first theorem, known as the *gambler's ruin formula*, specifies the probability that the gambler is ruined (i.e., $S_{N(-L,W)} = -L$). Actually, it is slightly more convenient to express the result in terms of the complementary event that the gambler achieves his goal.

Theorem 7.1.1. *Assume that X has distribution (7.1), where $p > 0$, $q > 0$, $r \geq 0$, and $p + q + r = 1$, and let $X_1, X_2, \ldots$ be i.i.d. as X. Then $P(N(-L, W) < \infty) = 1$. Furthermore, if $p \neq q$, then*

$$P(S_{N(-L,W)} = W) = \frac{(q/p)^L - 1}{(q/p)^{L+W} - 1} = 1 - P(S_{N(-L,W)} = -L). \qquad (7.4)$$

If $p = q$, then

$$P(S_{N(-L,W)} = W) = \frac{L}{L + W} = 1 - P(S_{N(-L,W)} = -L). \qquad (7.5)$$

Proof. For simplicity of notation, we write N for $N(-L, W)$ in the proof only. First, put $M := L + W$, define the random variable

$$T := \min\{n \geq 1 : X_{(n-1)M+1} = \cdots = X_{(n-1)M+M} = 1\}, \qquad (7.6)$$

and notice that T has the geometric distribution with success probability p^M. In particular, $E[T] < \infty$. But $N \leq MT$ because if M consecutive wins have occurred, it is impossible that the gambler has avoided both being ruined and achieving his goal. Therefore, $E[N] < \infty$.

By Example 3.1.1 on p. 96, $\{(q/p)^{S_n}\}_{n \geq 0}$ is a martingale with respect to $\{X_n\}_{n \geq 1}$. The optional stopping theorem (Theorem 3.2.2 on p. 100) implies that

$$E[(q/p)^{S_N}] = E[(q/p)^{S_0}] = 1, \qquad (7.7)$$

hence

$$(q/p)^{-L}[1 - P(S_N = W)] + (q/p)^W P(S_N = W) = 1. \qquad (7.8)$$

If $p \neq q$, then (7.4) follows from this.

If $p = q$, then by Wald's identity (Corollary 3.2.5 on p. 103), we find that

$$E[S_N] = E[N]E[X_1] = 0, \qquad (7.9)$$

hence

$$(-L)[1 - P(S_N = W)] + W P(S_N = W) = 0, \qquad (7.10)$$

from which (7.5) follows. ♠

Notice that (7.4) simplifies in the special case in which $L = W$. Indeed,

$$P(S_{N(-W,W)} = W) = \frac{1}{(q/p)^W + 1}, \qquad (7.11)$$

and this holds regardless of whether $p \neq q$ or $p = q$.

Example 7.1.2. *Probability of doubling one's fortune before losing it.* Suppose a gambler has W units and bets one unit per coup on an even-money proposition until he has either doubled his betting capital or lost it. The probability that he doubles it is given by (7.11), with p being the probability of a win and q being the probability of a loss at each coup. Table 7.1 gives numerical results for several choices of W, several choices of p (motivated by 38-number roulette, 37-number roulette, craps, and coin-tossing), and $q := 1 - p$, that is, we assume that $r = 0$ in these examples.

For an alternative interpretation of the results, suppose a gambler wants to double an initial betting capital of 1,000 units. He considers several betting strategies: bet 1,000 units double-or-nothing; bet 100 units per coup; bet 10 units per coup; or bet 1 unit per coup. The probabilities of success correspond to the four rows of Table 7.1. Thus, the gambler's ruin formula of Theorem 7.1.1 can be used to ensure a specified probability of achieving one's goal by choosing the bet size appropriately. This technique is often referred to as *bet sizing.* ♠

Table 7.1 The probability that a gambler will double his fortune of W units before losing it, if on each of a sequence of independent coups, he wins one unit with probability p and loses one unit with probability $q := 1 - p$. Results are from (7.11) and are rounded to six significant digits.

W	$p = \frac{18}{38}$	$p = \frac{18}{37}$	$p = \frac{244}{495}$	$p = \frac{1}{2}$
1	.473 684	.486 486	.492 929	$\frac{1}{2}$
10	.258 533	.368 031	.429 756	$\frac{1}{2}$
100	.000 026 560 7	.004 466 28	.055 804 9	$\frac{1}{2}$
1,000	$.174\ 787 \cdot 10^{-45}$	$.330\ 297 \cdot 10^{-23}$	$.520\ 122 \cdot 10^{-12}$	$\frac{1}{2}$

Other important cases include the limiting cases $L \to \infty$ and $W \to \infty$. Let us define

$$N(-\infty, W) := \min\{n \geq 1 : S_n = W\} \tag{7.12}$$

and

$$N(-L, \infty) := \min\{n \geq 1 : S_n = -L\}, \tag{7.13}$$

where $\min \varnothing := \infty$. Let us further define $S_\infty := \pm\infty$, whichever is appropriate so that $S_{N(-\infty,W)} = -\infty$ if $N(-\infty, W) = \infty$, and $S_{N(-L,\infty)} = \infty$ if $N(-L, \infty) = \infty$. Then we have the following result.

Corollary 7.1.3. *Under the assumptions of Theorem 7.1.1,*

$$P(S_{N(-\infty,W)} = W) = (q/p)^{-W} \wedge 1 \tag{7.14}$$

and

$$P(S_{N(-L,\infty)} = -L) = (q/p)^L \wedge 1. \tag{7.15}$$

Remark. Equation (7.14) gives the probability that a gambler with unlimited resources or credit and no loss limit eventually achieves his goal W. This probability is less than 1 if and only if $p < q$. Equation (7.15) gives the probability that a gambler with initial fortune L with no win limit playing against an opponent with unlimited resources or credit is eventually ruined. This probability is less than 1 if and only if $p > q$.

Proof. We get (7.14) from Theorem 7.1.1 by letting $L \to \infty$. To see that the left side of (7.4) tends to the left side of (7.14), it suffices to notice that

$$\{S_{N(-L,W)} = W\} \subset \{S_{N(-(L+1),W)} = W\}, \tag{7.16}$$

and then apply Theorem 1.1.20(a) on p. 13. The same argument works for (7.15).   ♠

We now turn to the gambler's expected duration of play.

Theorem 7.1.4. *Under the assumptions of Theorem 7.1.1, if $p \neq q$, then*

$$E[N(-L, W)] = \frac{L+W}{q-p}\left(\frac{L}{L+W} - \frac{(q/p)^L - 1}{(q/p)^{L+W} - 1}\right). \tag{7.17}$$

If $p = q$, then

$$E[N(-L, W)] = \frac{LW}{p+q}. \tag{7.18}$$

Proof. Again write N for $N(-L, W)$ in the proof. By Wald's identity (Corollary 3.2.5 on p. 103), we have

$$E[S_N] = E[N]E[X_1] = (p-q)E[N], \tag{7.19}$$

hence, if $p \neq q$,

$$E[N] = (p - q)^{-1}\{(-L)[1 - P(S_N = W)] + WP(S_N = W)\}, \qquad (7.20)$$

and (7.17) follows from this and (7.4).

If $p = q$, then by Example 3.1.1 on p. 96, $\{S_n^2 - (p+q)n\}_{n \geq 0}$ is a martingale with respect to $\{X_n\}_{n \geq 1}$. Applying the optional stopping theorem once again, we find that

$$E[S_N^2] = (p + q)E[N], \qquad (7.21)$$

hence

$$E[N] = (p + q)^{-1}\{(-L)^2[1 - P(S_N = W)] + W^2 P(S_N = W)\}, \qquad (7.22)$$

and (7.18) follows from this and (7.5). ♠

Notice that (7.17) simplifies in the special case in which $L = W$. Indeed,

$$E[N(-W, W)] = \frac{W}{q - p} \frac{(q/p)^W - 1}{(q/p)^W + 1}, \qquad p \neq q, \qquad (7.23)$$

and $E[N(-W, W)] = W^2/(2p)$ if $p = q$.

Example 7.1.5. *Expected number of coups required to double one's fortune or lose it.* Suppose a gambler has W units and bets one unit per coup on an even-money proposition until he has either doubled his betting capital or lost it. The expected number of coups required is given by (7.23), with p being the probability of win and q being the probability of a loss at each coup. Table 7.2 gives numerical results for the same values of W, p, and q considered in Table 7.1 (in particular, we assume that $q := 1 - p$). ♠

Table 7.2 The expected number of coups required for a gambler to double his fortune of W units or lose it, if on each of a sequence of independent coups, he wins one unit with probability p and loses one unit with probability $q := 1 - p$. Results are from (7.23) and are rounded to six significant digits.

W	$p = \frac{18}{38}$	$p = \frac{18}{37}$	$p = \frac{244}{495}$	$p = \frac{1}{2}$
1	1.000 00	1.000 00	1.000 00	1
10	91.757 3	97.656 9	99.345 2	10^2
100	1,899.90	3,666.95	6,282.19	10^4
1,000	19,000.0	37,000.0	70,714.3	10^6

The next result concerns the limiting cases of Theorem 7.1.4.

Corollary 7.1.6. *Under the assumptions of Theorem 7.1.1, if $p > q$, then*

$$\mathrm{E}[N(-\infty, W)] = \frac{W}{p - q}. \tag{7.24}$$

If $p < q$, then

$$\mathrm{E}[N(-L, \infty)] = \frac{L}{q - p}. \tag{7.25}$$

Remark. The results are intuitively clear. Take (7.25), for example. The gambler's expected loss per coup is $q - p$ units, so it should require $L/(q - p)$ coups on average to lose L units.

Proof. For (7.25), we let $W \to \infty$ in (7.17). Since $N(-L, W) \leq N(-L, W+1)$, the result follows from the monotone convergence theorem (Theorem A.2.3 on p. 748). The proof of (7.24) is similar. ♠

It is possible to say a great deal more about $N(-L, W)$. We begin by deriving its probability generating function.

Theorem 7.1.7. *Under the assumptions of Theorem 7.1.1,*

$$\mathrm{E}[z^{N(-L,W)} 1_{\{S_{N(-L,W)}=W\}}] = \frac{\lambda_+(z)^L - \lambda_-(z)^L}{\lambda_+(z)^{L+W} - \lambda_-(z)^{L+W}} \tag{7.26}$$

for $0 < z \leq 1$, where

$$\lambda_\pm(z) := \frac{1 - rz \pm \sqrt{(1 - rz)^2 - 4pqz^2}}{2pz}, \tag{7.27}$$

and

$$\mathrm{E}[z^{N(-L,W)} 1_{\{S_{N(-L,W)}=-L\}}] = \left(\frac{q}{p}\right)^L \frac{\lambda_+(z)^W - \lambda_-(z)^W}{\lambda_+(z)^{L+W} - \lambda_-(z)^{L+W}} \tag{7.28}$$

for $0 < z \leq 1$. In particular, $\mathrm{E}[z^{N(-L,W)}]$ is the sum of (7.26) and (7.28).

Proof. As usual, we write N for $N(-L, W)$ in the proof. For $0 < z \leq 1$, let $\lambda(z)$ satisfy

$$[\lambda(z)p + \lambda(z)^{-1}q + r]z = 1. \tag{7.29}$$

This equation is equivalent to a quadratic equation in $\lambda(z)$ whose roots are given by (7.27). It also ensures that

$$M_n := \lambda(z)^{S_n} z^n, \qquad n \geq 0, \tag{7.30}$$

is a martingale with respect to $\{X_n\}_{n \geq 1}$. Thus, $\mathrm{E}[\lambda(z)^{S_N} z^N] = 1$ by the optional stopping theorem (Theorem 3.2.2 on p. 100), and it follows that

$$\lambda(z)^W \mathrm{E}[z^N 1_{\{S_N=W\}}] + \lambda(z)^{-L} \mathrm{E}[z^N 1_{\{S_N=-L\}}] = 1. \tag{7.31}$$

Substituting $\lambda(z) = \lambda_+(z)$ and $\lambda(z) = \lambda_-(z)$, we have two linear equations in two variables (namely, the two expectations in (7.31)), and noting that $\lambda_+(z)\lambda_-(z) = q/p$, we obtain (7.26) and (7.28).   ♠

Corollary 7.1.8. *Under the assumptions of Theorem 7.1.1, if $p \neq q$, then*

$$
\begin{aligned}
&\mathrm{E}[N(-L, W) \mid S_{N(-L,W)} = W] \\
&\qquad = \frac{L+W}{q-p}\frac{(q/p)^{L+W}+1}{(q/p)^{L+W}-1} - \frac{L}{q-p}\frac{(q/p)^{L}+1}{(q/p)^{L}-1}
\end{aligned}
\tag{7.32}
$$

and

$$
\begin{aligned}
&\mathrm{E}[N(-L, W) \mid S_{N(-L,W)} = -L] \\
&\qquad = \frac{L+W}{q-p}\frac{(q/p)^{L+W}+1}{(q/p)^{L+W}-1} - \frac{W}{q-p}\frac{(q/p)^{W}+1}{(q/p)^{W}-1}.
\end{aligned}
\tag{7.33}
$$

In particular, if $p < q$, then

$$
\mathrm{E}[N(-\infty, W) \mid S_{N(-\infty,W)} = W] = \frac{W}{q-p},
\tag{7.34}
$$

and if $p > q$, then

$$
\mathrm{E}[N(-L, \infty) \mid S_{N(-L,\infty)} = -L] = \frac{L}{p-q}.
\tag{7.35}
$$

Finally, if $p = q$, then

$$
\mathrm{E}[N(-L, W) \mid S_{N(-L,W)} = W] = \frac{W(2L+W)}{3(p+q)}
\tag{7.36}
$$

and

$$
\mathrm{E}[N(-L, W) \mid S_{N(-L,W)} = -L] = \frac{L(L+2W)}{3(p+q)}.
\tag{7.37}
$$

Remark. (a) It is interesting to note that (7.32) and (7.33) are equal if $L = W$, even though $p \neq q$. Also, both formulas are unchanged if the roles of p and q are reversed.

(b) It can be verified that (7.32) and (7.33), together with (7.4), imply (7.17). However, (7.32) and (7.33) cannot be deduced solely from (7.4) and (7.17).

(c) Recall that the results of Corollary 7.1.6 were intuitively clear. Results (7.34) and (7.35), on the other hand, are rather nonintuitive.

Proof. This is a tedious but straightforward calculus problem. First note that $\lambda_+(1) = (q/p) \vee 1$, $\lambda_-(1) = (q/p) \wedge 1$, $\lambda'_+(1) = -\lambda_+(1)/|p-q|$, and $\lambda'_-(1) = \lambda_-(1)/|p-q|$. For (7.32), differentiate (7.26) with respect to z, let $z = 1$, and divide by (7.4). The proof of (7.33) is similar. As in the proof of Corollary 7.1.6, the monotone convergence theorem (Theorem A.2.3 on p. 748), applied to (7.32), leads to (7.34); (7.35) is similar. The proofs of (7.36) and (7.37) are left to the reader. ♠

We recall that we have allowed $r > 0$ in (7.1), simply because the slightly greater generality requires no additional effort. However, it should be noted that the general case ($r \geq 0$) can easily be derived from the special case $r = 0$ (see Problem 7.2 on p. 267).

We turn finally the problem of establishing a time-limited ruin formula. Here it will be convenient to exclude pushes.

Theorem 7.1.9. *Under the assumptions of Theorem 7.1.1, together with the assumption that $r = 0$, or $q = 1 - p$,*

$$P(S_{N(-L,W)} = W, \ N(-L,W) = n) \tag{7.38}$$

$$= 2^n p^{(n+W)/2} q^{(n-W)/2} \frac{1}{M} \sum_{j=1}^{M-1} \sin(\pi j/M) \sin(\pi W j/M) \cos^{n-1}(\pi j/M)$$

and

$$P(S_{N(-L,W)} = -L, \ N(-L,W) = n) \tag{7.39}$$

$$= 2^n p^{(n-L)/2} q^{(n+L)/2} \frac{1}{M} \sum_{j=1}^{M-1} \sin(\pi j/M) \sin(\pi L j/M) \cos^{n-1}(\pi j/M)$$

for each $n \geq 1$, where $M := L + W$. Consequently,

$$P(S_{N(-L,W)} = W, \ N(-L,W) \leq n)$$
$$= P(S_{N(-L,W)} = W) \tag{7.40}$$
$$- \frac{(q/p)^{-W/2}(2\sqrt{pq})^{n+1}}{M} \sum_{j=1}^{M-1} \frac{\sin(\pi j/M) \sin(\pi W j/M) \cos^n(\pi j/M)}{1 - 2\sqrt{pq} \cdot \cos(\pi j/M)}$$

and

$$P(S_{N(-L,W)} = -L, \ N(-L,W) \leq n)$$
$$= P(S_{N(-L,W)} = -L) \tag{7.41}$$
$$- \frac{(q/p)^{L/2}(2\sqrt{pq})^{n+1}}{M} \sum_{j=1}^{M-1} \frac{\sin(\pi j/M) \sin(\pi L j/M) \cos^n(\pi j/M)}{1 - 2\sqrt{pq} \cdot \cos(\pi j/M)}$$

for each $n \geq 1$, where $M := L + W$.

Remark. This result gives the probabilities of achieving the goal and of ruin in exactly n coups as well as in n coups or fewer. The distribution of the duration of play follows. The special case $p = q = \frac{1}{2}$ is included.

Proof. Again, we write N for $N(-L, W)$ in the proof. We begin by evaluating

$$u_n := \mathrm{P}(S_N = W,\ N = n), \tag{7.42}$$

which by (7.26) has generating function

$$f(z) := \sum_{n=0}^{\infty} u_n z^n = \frac{\lambda_+(z)^L - \lambda_-(z)^L}{\lambda_+(z)^M - \lambda_-(z)^M}, \tag{7.43}$$

where $\lambda_+(z)$ and $\lambda_-(z)$ are given by (7.27) with $r = 0$. First note that the denominator in (7.43) has the form

$$\begin{aligned}
\lambda_+&(z)^M - \lambda_-(z)^M \\
&= \left(\frac{1 + \sqrt{1 - 4pqz^2}}{2pz}\right)^M - \left(\frac{1 - \sqrt{1 - 4pqz^2}}{2pz}\right)^M \\
&= (2pz)^{-M} \sum_{k=0}^{M} \binom{M}{k} [1 - (-1)^k](1 - 4pqz^2)^{k/2} \\
&= (2pz)^{-M} \sqrt{1 - 4pqz^2} \sum_{0 \le k \le M,\ k \text{ odd}} \binom{M}{k} 2(1 - 4pqz^2)^{(k-1)/2} \\
&= (2pz)^{-M} \sqrt{1 - 4pqz^2}\ P_M(z), \tag{7.44}
\end{aligned}$$

where $P_M(z)$ is a polynomial of degree $M - 1$ if M is odd, $M - 2$ if M is even. It follows that

$$f(z) = (2pz)^W P_L(z)/P_M(z). \tag{7.45}$$

The degree of the polynomial in the numerator of (7.45) is $M - 1$ if L is odd, $M - 2$ if L is even. Thus, $f(z)$ is the ratio of two polynomials whose degrees differ by at most 1.

The idea of the proof is to expand the rational function $f(z)$ in partial fractions. Thus, we can write

$$f(z) = \frac{A_1}{z_1 - z} + \cdots + \frac{A_{M-1}}{z_{M-1} - z} + B_0 + B_1 z \tag{7.46}$$

if the $M - 1$ (or $M - 2$) roots $z_1, \ldots, z_{M-1}$ of $P_M(z)$ are distinct. To determine the roots and the coefficients, we make a change of variables: Let

$$\theta := \cos^{-1} \frac{1}{2\sqrt{pq}\, z} \in (0, \pi), \tag{7.47}$$

provided z is real with

$$2\sqrt{pq}\, |z| > 1. \tag{7.48}$$

Then $\lambda_+(z)$ and $\lambda_-(z)$ are given by

$$\frac{1 \pm \sqrt{1 - 4pqz^2}}{2pz} = \sqrt{\frac{q}{p}} \left[\frac{1}{2\sqrt{pq}\, z} \pm i \sqrt{1 - \left(\frac{1}{2\sqrt{pq}\, z}\right)^2} \right]$$

$$= \sqrt{q/p}\,(\cos\theta \pm i\sin\theta) = \sqrt{q/p}\; e^{\pm i\theta} \qquad (7.49)$$

(which is which does not matter). It follows from (7.43) that

$$f(z) = \left(\frac{q}{p}\right)^{-W/2} \frac{e^{iL\theta} - e^{-iL\theta}}{e^{iM\theta} - e^{-iM\theta}} = \left(\frac{q}{p}\right)^{-W/2} \frac{\sin(L\theta)}{\sin(M\theta)}. \qquad (7.50)$$

Clearly the roots of the denominator of (7.50) in the interval $(0, \pi)$ are $\theta_j :=$ $\pi j/M$ for $j = 1, \ldots, M - 1$, with corresponding z-values

$$z_j := \frac{1}{2\sqrt{pq}\,\cos\theta_j}, \quad j = 1, \ldots, M - 1 \quad (j \neq M/2 \text{ if } M \text{ is even}). \qquad (7.51)$$

Note that (7.48) is satisfied in a neighborhood of each z_j. We exclude $z_{M/2}$ if M is even because it is not within the domain of our transformation, with the result that the denominator of (7.50) has the correct number of roots. (Actually, the corresponding term in (7.46) will turn out to be zero unless $n = 1$.)

We turn to the determination the coefficients A_j in (7.46). Clearly,

$$A_j = \lim_{z \to z_j} (z_j - z) f(z)$$

$$= -(q/p)^{-W/2} \lim_{z \to z_j} \frac{\sin(L\theta_j)}{\sin(M\theta)/(z - z_j)}$$

$$= -(q/p)^{-W/2} \sin(L\theta_j) \lim_{z \to z_j} \frac{\theta - \theta_j}{\sin(M\theta)} \frac{z - z_j}{\theta - \theta_j}$$

$$= \frac{-(q/p)^{-W/2} \sin(L\theta_j)}{M \cos(M\theta_j)\, \theta'(z_j)}$$

$$= \frac{(q/p)^{-W/2}\,(2\sqrt{pq})}{M} \sin(\theta_j)\sin(W\theta_j)\, z_j^2, \qquad (7.52)$$

where the last equality uses

$$\sin(L\theta_j) = \sin(M\theta_j - W\theta_j)$$
$$= \sin(\pi j)\cos(W\theta_j) - \cos(\pi j)\sin(W\theta_j)$$
$$= -(-1)^j \sin(W\theta_j), \qquad (7.53)$$

$\cos(M\theta_j) = \cos(\pi j) = (-1)^j$, and, by (7.47), $(-\sin\theta)\theta'(z) = -1/(2\sqrt{pq}\, z^2)$. It is unnecessary to determine the coefficients B_0 and B_1 in (7.46).

Now for $|z| < \min_{1 \le j \le M} |z_j|$,

$$f(z) - B_0 - B_1 z = \sum_{j=1}^{M-1} \frac{A_j/z_j}{1 - (z/z_j)} = \sum_{n=0}^{\infty} \sum_{j=1}^{M-1} A_j z^n z_j^{-n-1}. \qquad (7.54)$$

Consequently, for each $n \geq 2$,

$$
\begin{aligned}
&\mathrm{P}(S_N = W, \ N = n) \\
&\quad = \text{coefficient of } z^n \text{ in } f(z) \\
&\quad = \text{coefficient of } z^n \text{ in } f(z) - B_0 - B_1 z \\
&\quad = \sum_{j=1}^{M-1} A_j z_j^{-n-1} \qquad\qquad\qquad\qquad\qquad\qquad (7.55) \\
&\quad = \frac{(q/p)^{-W/2}(2\sqrt{pq})}{M} \sum_{j=1}^{M-1} \sin(\theta_j)\sin(W\theta_j)[2\sqrt{pq}\,\cos(\theta_j)]^{n-1},
\end{aligned}
$$

proving (7.38). Next,

$$\mathrm{P}(S_N = W, \ N \leq n) = \mathrm{P}(S_N = W) - \sum_{l=n+1}^{\infty} \mathrm{P}(S_N = W, \ N = l) \qquad (7.56)$$

for each $n \geq 1$, and we obtain (7.40) from (7.55).

Finally, (7.39) and (7.41) follow from (7.38) and (7.40) by reversing the roles of W and L and of p and q. ♠

We conclude this section by showing that the results of Theorem 7.1.9 simplify considerably if $L = \infty$ or $W = \infty$.

Corollary 7.1.10. *Under the assumptions of Theorem 7.1.9,*

$$\mathrm{P}(N(-\infty, W) = n) = \frac{W}{n}\binom{n}{(n-W)/2} p^{(n+W)/2} q^{(n-W)/2} \qquad (7.57)$$

for $n = W, W+2, W+4, \ldots,$ and

$$\mathrm{P}(N(-L, \infty) = n) = \frac{L}{n}\binom{n}{(n-L)/2} p^{(n-L)/2} q^{(n+L)/2} \qquad (7.58)$$

for $n = L, L+2, L+4, \ldots.$ The probabilities in (7.57) and (7.58) are 0 otherwise.

Proof. Letting $L \to \infty$ in (7.38) (hence $M := L + W \to \infty$), we find that

$$
\begin{aligned}
&\mathrm{P}(N(-\infty, W) = n) \\
&\quad = \mathrm{P}(S_{N(-\infty,W)} = W, \ N(-\infty, W) = n) \qquad\qquad (7.59) \\
&\quad = 2^n p^{(n+W)/2} q^{(n-W)/2} \int_0^1 \sin(\pi x)\sin(\pi W x)\cos^{n-1}(\pi x)\, dx
\end{aligned}
$$

for all $n \geq 1$. Let us denote (7.59) by $P_W(n)$ for each $(W, n) \in \mathbf{N}^2$. Then

$$P_W(n+1) = qP_{W+1}(n) + pP_{W-1}(n), \qquad (W, n) \in \mathbf{N}^2, \qquad (7.60)$$

where $P_0(n) := 0$ for all $n \geq 1$. This can be seen either probabilistically from the left side of (7.59) or analytically from the right side of (7.59) using the identity $\sin(u + v) + \sin(u - v) = 2\sin(u)\cos(v)$. In addition, $P_W(1)$ is given by

$$P_W(1) = \delta_{W,1}, \qquad W \in \mathbf{N}. \qquad (7.61)$$

This too can be derived either probabilistically or analytically. Notice also that (7.60) and (7.61) uniquely determine $P_W(n)$ for all $(W, n) \in \mathbf{N}^2$. To complete the proof of (7.57), it is enough to show that the right side of (7.57) satisfies (7.60) and (7.61); here the binomial coefficient is defined to be 0 if $(n - W)/2$ fails to be a nonnegative integer. This last step is straightforward.

Finally, as in the proof of Theorem 7.1.9, we can derive (7.58) from (7.57) by reversing the roles of W and L and of p and q. ♠

7.2 Integer Payoffs

In the preceding section we assumed that the gambler's profit from the basic wager could be represented by a random variable X with values in $\{-1, 0, 1\}$. In this section we allow X to take on arbitrary integer values. However, we assume the existence of positive integers μ and ν such that

$$P(-\nu \leq X \leq \mu) = 1, \quad P(X = -\nu) > 0, \quad P(X = \mu) > 0. \qquad (7.62)$$

Thus, μ and ν represent the maximum amounts that can be won or lost, respectively, at a single coup. It will be convenient to introduce the notation

$$p_j := P(X = j), \qquad j \in \mathbf{Z}, \qquad (7.63)$$

where $\mathbf{Z}$ denotes the set of all integers, and

$$J := \{j \in \mathbf{Z} : p_j > 0\}. \qquad (7.64)$$

Note that J is finite with $\min J = -\nu$ and $\max J = \mu$.

Now let $X_1, X_2, \ldots$ be a sequence of i.i.d. random variables with common distribution that of X, representing the results of independent repetitions of the original wager. Then

$$S_n := X_1 + X_2 + \cdots + X_n \qquad (7.65)$$

represents the gambler's cumulative profit after n such wagers; of course, $S_0 = 0$.

Given positive integers W and L, we suppose that the gambler stops betting as soon as he wins W or more units or loses L or more units. Consequently, his last bet occurs at coup

$$N(-L, W) := \min\{n \geq 1 : S_n \leq -L \text{ or } S_n \geq W\}. \qquad (7.66)$$

Although no formula exists in general for $P(S_{N(-L,W)} \geq W)$, we can nevertheless evaluate this probability by solving a linear system of difference equations.

Theorem 7.2.1. *Under the above assumptions, suppose that the function $u(i)$, defined for $i = -L - \nu + 1, \ldots, W + \mu - 1$, satisfies the linear system of difference equations*

$$u(i) = \sum_{j \in J} p_j u(i + j), \qquad i = -L + 1, \ldots, W - 1, \qquad (7.67)$$

with boundary conditions

$$u(-L - \nu + 1) = \cdots = u(-L) = 0 \qquad (7.68)$$

and

$$u(W) = \cdots = u(W + \mu - 1) = 1. \qquad (7.69)$$

Then

$$u(0) = P(S_{N(-L,W)} \geq W). \qquad (7.70)$$

More generally,

$$u(i) = P(S_{N(-L-i,W-i)} \geq W - i), \qquad i = -L + 1, \ldots, W - 1. \qquad (7.71)$$

Proof. This is a consequence of Theorem 4.2.7 on p. 134, regarding $\{S_n\}_{n \geq 0}$ as a Markov chain in $\{-L - \nu + 1, \ldots, W + \mu - 1\}$ with absorbing states $-L - \nu + 1, \ldots, -L$ and $W, \ldots, W + \mu - 1$.   ♠

There are at least two ways to use this result to get an explicit formula for the probability in (7.70).

First, we can substitute the boundary conditions in (7.68) and (7.69) into (7.67) to obtain an nonhomogeneous system of $L + W - 1$ linear equations in the $L + W - 1$ variables $u(-L + 1), \ldots, u(W - 1)$. We can then solve for $u(0)$ by Gaussian elimination or Cramer's rule. There is one important case in which this has been successfully done.

Example 7.2.2. *A ruin formula in the case of an m-to-1 payoff.* Let m be a positive integer, and assume that

$$P(X = m) = p, \quad P(X = -1) = q, \quad P(X = 0) = r, \qquad (7.72)$$

where $p > 0$, $q > 0$, $r \geq 0$, and $p + q + r = 1$. Then

$$P(S_{N(-L,W)} \geq W) = 1 - \left(\frac{q}{p+q}\right)^L \frac{a(W)}{a(L+W)}, \tag{7.73}$$

where

$$a(K) := \sum_{j=0}^{\lfloor (K-1)/(m+1) \rfloor} (-1)^j \binom{K-jm-1}{j} \left(\frac{pq^m}{(p+q)^{m+1}}\right)^j. \tag{7.74}$$

To confirm this assertion, we apply Theorem 7.2.1 with

$$u(i) := 1 - \left(\frac{q}{p+q}\right)^{L+i} \frac{a(W-i)}{a(L+W)}, \qquad i = -L+1, \ldots, W-1, \tag{7.75}$$

and with $u(-L) := 0$ and $u(W) = \cdots = u(W+m-1) := 1$. It suffices to verify that

$$u(i) = qu(i-1) + ru(i) + pu(i+m), \qquad i = -L+1, \ldots, W-1. \tag{7.76}$$

The details are straightforward and left to the reader (see Problem 7.9 on p. 268). ♠

Second, we can occasionally get an explicit solution to (7.67) as follows. If λ is a real or complex number such that

$$\sum_{j \in J} p_j \lambda^j = 1, \tag{7.77}$$

then $u(i) := \lambda^i$ is a solution of (7.67). Now (7.77) is equivalent to the polynomial equation of degree $\nu + \mu$,

$$P(\lambda) := \sum_{j \in J} p_j \lambda^{\nu+j} - \lambda^\nu = 0, \tag{7.78}$$

which has $\nu + \mu$ roots.

If these roots, which we denote by $\lambda_0, \lambda_1, \ldots, \lambda_{\nu+\mu-1}$ (with $\lambda_0 := 1$), are distinct, then

$$u(i) := \sum_{k=0}^{\nu+\mu-1} a_k \lambda_k^i \tag{7.79}$$

is the general solution of (7.67). If we can determine the $\nu + \mu$ coefficients $a_0, a_1, \ldots, a_{\nu+\mu-1}$ so that the $\nu + \mu$ boundary conditions (7.68) and (7.69) are satisfied, we will have an explicit solution to (7.67)–(7.69).

More generally, if there are d distinct roots $\lambda_0, \ldots, \lambda_{d-1}$ with $\lambda_0 := 1$ and respective multiplicities $m_0, \ldots, m_{d-1}$ ($m_0 \geq 1, \ldots, m_{d-1} \geq 1$, $m_0 + \cdots + m_{d-1} = \nu + \mu$), then

$$u(i) := \sum_{k=0}^{d-1} \sum_{j=0}^{m_k-1} a_{kj}\, i^j \lambda_k^i \qquad (7.80)$$

is the general solution of (7.67), and the same argument applies.

Example 7.2.3. *A ruin formula in the case of a 2-to-1 payoff.* Assume that

$$\mathrm{P}(X = 2) = p, \quad \mathrm{P}(X = -1) = q, \quad \mathrm{P}(X = 0) = r, \qquad (7.81)$$

where $p > 0$, $q > 0$, $r \geq 0$, and $p + q + r = 1$. Then $\mu = 2$, $\nu = 1$, and

$$\begin{aligned}
P(\lambda) &= p\lambda^3 + r\lambda + q - \lambda \\
&= p\lambda^3 - (p+q)\lambda + q \\
&= (\lambda - 1)(p\lambda^2 + p\lambda - q),
\end{aligned} \qquad (7.82)$$

which has roots $\lambda_0 := 1$ and, using the quadratic formula,

$$\lambda_1 := -\frac{1}{2} + \frac{1}{2}\sqrt{1 + 4q/p}, \qquad \lambda_2 := -\frac{1}{2} - \frac{1}{2}\sqrt{1 + 4q/p}. \qquad (7.83)$$

These three roots are distinct provided $q/p \neq 2$ or, equivalently, $\mathrm{E}[X] \neq 0$, which we now assume. The general solution of (7.67) in this setting is

$$u(i) := a_0 + a_1 \lambda_1^i + a_2 \lambda_2^i. \qquad (7.84)$$

Imposing the boundary conditions $u(-L) = 0$ and $u(W) = u(W + 1) = 1$, we can solve for a_0, a_1, and a_2 to obtain $u(0) = a_0 + a_1 + a_2$ and hence

$$\begin{aligned}
&\mathrm{P}(S_{N(-L,W)} \geq W) \\
&= \frac{(\lambda_1^L - 1)\lambda_2^{L+W}(\lambda_2 - 1) - \lambda_1^{L+W}(\lambda_1 - 1)(\lambda_2^L - 1)}{(\lambda_1^{L+W} - 1)(\lambda_2^{L+W+1} - 1) - (\lambda_1^{L+W+1} - 1)(\lambda_2^{L+W} - 1)}.
\end{aligned} \qquad (7.85)$$

See Problem 7.11 on p. 268 for the case in which $\mathrm{E}[X] = 0$. ♠

The method just described is severely limited by our inability to find all roots of (7.78) except in the simplest cases. Consequently, we now consider a computational method for finding the solution of (7.67)–(7.69).

Since by Theorem 7.2.1 the solution is a probability, we can bound it below and above by

$$u_0^-(i) := 0, \qquad i = -L + 1, \ldots, W - 1, \qquad (7.86)$$

$$u_0^+(i) := 1, \qquad i = -L + 1, \ldots, W - 1. \qquad (7.87)$$

(Here the superscripts do not denote nonpositive and nonnegative parts.) Starting with these initial choices, let us define iteratively

$$u_{n+1}^-(i) = \sum_{j \in J} p_j u_n^-(i + j), \qquad i = -L + 1, \ldots, W - 1, \qquad (7.88)$$

$$u_{n+1}^+(i) = \sum_{j \in J} p_j u_n^+(i+j), \qquad i = -L+1, \dots, W-1, \qquad (7.89)$$

where $u_n^\pm(k) := 0$ if $k = -L-\nu+1, \dots, -L$ and $u_n^\pm(k) := 1$ if $k = W, \dots, W+\mu-1$.

Theorem 7.2.4. *Under the above assumptions,*

$$u_0^-(i) \le u_1^-(i) \le u_2^-(i) \le \cdots \le 1 \qquad (7.90)$$

and

$$u_0^+(i) \ge u_1^+(i) \ge u_2^+(i) \ge \cdots \ge 0 \qquad (7.91)$$

for $i = -L+1, \dots, W-1$. Let $u^-(i)$ and $u^+(i)$ denote the limits of these two sequences. Then

$$u^-(i) = u^+(i) = \mathrm{P}(S_{N(-L-i,W-i)} \ge W-i) \qquad (7.92)$$

for $i = -L+1, \dots, W-1$. In particular,

$$u_n^-(0) \le \mathrm{P}(S_{N(-L,W)} \ge W) \le u_n^+(0), \qquad n \ge 1. \qquad (7.93)$$

Remark. The inequalities (7.93) sandwich the probability of interest between two numbers that are easy to compute. By choosing n large enough, we can determine this probability as accurately as desired.

Proof. Clearly, $u_0^- \le u_1^-$. If $u_{n-1}^- \le u_n^-$, then (7.88) implies that $u_n^- \le u_{n+1}^-$, so (7.90) follows by induction. A similar argument gives (7.91).

Taking limits as $n \to \infty$ in (7.88) and (7.89), we obtain (7.92) from Theorem 7.2.1. Finally, (7.90)–(7.92) imply (7.93). ♠

Example 7.2.5. *Probability that a pass-line bettor taking 3-4-5-times free odds doubles his fortune before losing it.* We consider the one-unit pass-line bettor taking 3-4-5-times free odds (Example 1.5.7 on p. 44). The profit from a single coup is a random variable X with distribution shown in Table 7.3.

What is the probability that the gambler wins 100 or more units before losing 100 or more units? To determine the answer to six significant digits, it is necessary and sufficient to carry out $n = 5{,}065$ iterations. Indeed, we find that

$$u_{5,065}^-(0) \approx 0.468958893, \qquad u_{5,065}^+(0) \approx 0.468959499, \qquad (7.94)$$

and hence by (7.93) that

$$\mathrm{P}(S_{N(-100,100)} \ge 100) \approx 0.468959, \qquad (7.95)$$

rounded to six significant digits.

If the number 100 above were replaced by a significantly larger number such as 100,000, this method may become impractical. See Section 7.3 for an alternative. ♠

Table 7.3 The distribution of profit X from a one-unit pass-line bet with 3-4-5-times free odds.

result	X	probability $\times 990$
natural 7 or 11	1	220
craps 2, 3, or 12	-1	110
point 4, 5, 6, 8, 9, or 10 made	7	268
point 4 or 10 missed	-4	110
point 5 or 9 missed	-5	132
point 6 or 8 missed	-6	150

We turn next to the expected duration of the game.

Theorem 7.2.6. *Under the assumptions of Theorem 7.2.1, suppose that the function $v(i)$, defined for $i = -L - \nu + 1, \ldots, W + \mu - 1$, satisfies the linear system of difference equations*

$$v(i) = 1 + \sum_{j \in J} p_j v(i + j), \qquad i = -L + 1, \ldots, W - 1, \tag{7.96}$$

with boundary conditions

$$v(-L - \nu + 1) = \cdots = v(-L) = 0 \tag{7.97}$$

and

$$v(W) = \cdots = v(W + \mu - 1) = 0. \tag{7.98}$$

Then

$$v(0) = \mathrm{E}[N(-L, W)]. \tag{7.99}$$

More generally,

$$v(i) = \mathrm{E}[N(-L - i, W - i)], \qquad i = -L + 1, \ldots, W - 1. \tag{7.100}$$

Proof. This is a consequence of Theorem 4.2.8 on p. 134, regarding $\{S_n\}_{n \geq 0}$ is a Markov chain in $\{-L - \nu + 1, \ldots, W + \mu - 1\}$ with absorbing states $-L - \nu + 1, \ldots, -L$ and $W, \ldots, W + \mu - 1$. ♠

This result can be used in exactly the same way in which we used Theorem 7.2.1.

Example 7.2.7. *Expected duration in the case of a 2-to-1 payoff.* Let us revisit Example 7.2.3. We assume (7.81), where $p > 0$, $q > 0$, $r \geq 0$, and $p + q + r = 1$. We further require $q/p \neq 2$ or, equivalently, $\mathrm{E}[X] \neq 0$. With λ_1 and λ_2 as in (7.83), the general solution of

$$v(i) = 1 + pv(i + 2) + qv(i - 1) + rv(i) \tag{7.101}$$

is $v = v_g + v_p$, where

$$v_g(i) := b_0 + b_1\lambda_1^i + b_2\lambda_2^i \tag{7.102}$$

is the general solution of the corresponding homogeneous equation as in (7.84), and v_p is a particular solution of the nonhomogeneous equation (7.101). To determine the latter, we use the method of undetermined coefficients. Specifically, we try $v_p(i) := ci$ and find that this provides a solution of (7.101) if $c := (q - 2p)^{-1}$. Imposing the boundary conditions $v(-L) = v(W) = v(W + 1) = 0$, we can solve for b_0, b_1, and b_2 to obtain $v(0) = b_0 + b_1 + b_2$ and hence

$$\mathrm{E}[N(-L, W)] = \frac{1}{q - 2p}\left\{L + \frac{A + B}{C}\right\}, \tag{7.103}$$

where, with $M := L + W$,

$$
\begin{aligned}
A &:= (\lambda_1^L - 1)\{(M + 1)(\lambda_2^M - 1) - M(\lambda_2^{M+1} - 1)\}, \\
B &:= \{M(\lambda_1^{M+1} - 1) - (M + 1)(\lambda_1^M - 1)\}(\lambda_2^L - 1), \\
C &:= (\lambda_1^M - 1)(\lambda_2^{M+1} - 1) - (\lambda_1^{M+1} - 1)(\lambda_2^M - 1).
\end{aligned}
\tag{7.104}
$$

See Problem 7.16 on p. 269 for the case in which $\mathrm{E}[X] = 0$. ♠

As with Example 7.2.3, the method used in Example 7.2.7 has rather limited applicability. However, the computational method based on Theorem 7.2.4 can be extended to the present setting as follows.

Since by Theorem 7.2.6 the solution of (7.96)–(7.98) is nonnegative, we can bound it below by

$$v_0^-(i) = 0, \qquad i = -L + 1, \ldots, W - 1. \tag{7.105}$$

Starting with this initial value, let us define iteratively

$$v_{n+1}^-(i) = 1 + \sum_{j \in J} p_j v_n^-(i + j), \qquad i = -L + 1, \ldots, W - 1, \tag{7.106}$$

where $v_n^-(k) := 0$ if $k = -L - \nu + 1, \ldots, -L$ or if $k = W, \ldots, W + \mu - 1$.

Theorem 7.2.8. *Under the above assumptions,*

$$v_0^-(i) \leq v_1^-(i) \leq v_2^-(i) \leq \cdots \tag{7.107}$$

for $i = -L + 1, \ldots, W - 1$. *Let* $v^-(i)$ *denote the limit of this sequence. Then*

$$v^-(i) = \mathrm{E}[N(-L - i, W - i)] < \infty \tag{7.108}$$

for $i = -L + 1, \ldots, W - 1$. *In particular,*

$$v_n^-(0) \leq v^-(0) = \mathrm{E}[N(-L, W)] < \infty, \qquad n \geq 1. \tag{7.109}$$

Remark. The result (7.109) is less precise than (7.93) because it provides a lower bound but no upper bound. Nevertheless, by iterating until stability is achieved, we can still get useful information.

Proof. The same induction argument used in the proof of Theorem 7.2.4 implies (7.107). Taking limits as $n \to \infty$ in (7.106), we obtain (7.108) from Theorem 7.2.6. Finally, (7.109) is clear.   ♠

Example 7.2.9. *Expected duration for the pass-line bettor taking 3-4-5-times free odds.* We revisit Example 7.2.5. Recall the wager described by Table 7.3. If the gambler plays until he wins 100 or more units or loses 100 or more units, what is the expected number of coups (i.e., pass-line decisions) required? With 10,000 iterations, we find

$$v_{2,000}^-(0) \approx 428.448178,$$
$$v_{4,000}^-(0) \approx 429.858350,$$
$$v_{6,000}^-(0) \approx 429.862846, \tag{7.110}$$
$$v_{8,000}^-(0) \approx 429.862860,$$
$$v_{10,000}^-(0) \approx 429.862860.$$

It is therefore safe to say that

$$\mathrm{E}[N(-100, 100)] \approx 429.863, \tag{7.111}$$

rounded to six significant digits.   ♠

7.3 Arbitrary Payoffs

In Section 7.2 we assumed that the gambler's profit from the basic wager could be represented by a bounded random variable X taking on *integer* values. Here we allow X to take on finitely many *real* values. As before, we let μ and ν be positive real numbers satisfying

$$\mathrm{P}(-\nu \leq X \leq \mu) = 1, \quad \mathrm{P}(X = -\nu) > 0, \quad \mathrm{P}(X = \mu) > 0. \tag{7.112}$$

Again, μ and ν represent the maximum amounts that can be won or lost, respectively, at a single coup.

We define the *probability generating function* of X by

$$f_X(\rho) := \mathrm{E}[\rho^X], \qquad 0 < \rho < \infty. \tag{7.113}$$

A similar definition in Section 1.4 required that X be nonnegative-integer valued. Here we allow X to have finitely many real values, but we require that ρ be positive.

Lemma 7.3.1. *In addition to the above assumptions, suppose that* $\mathrm{E}[X] \neq 0$. *Then there is one and only one* $\rho_0 \in (0,1) \cup (1,\infty)$ *satisfying the equation*

$$f_X(\rho_0) = 1. \tag{7.114}$$

If $\mathrm{E}[X] < 0$, *then* $\rho_0 > 1$. *If* $\mathrm{E}[X] > 0$, *then* $\rho_0 < 1$.

Proof. Let a be the minimum of 1 and the smallest positive value that X assumes. Then X/a assumes no values in $(0,1)$, and therefore $(X/a)(X/a - 1) \geq 0$. It follows that

$$(f_{X/a})''(\lambda) = \mathrm{E}[(X/a)(X/a - 1)\lambda^{X/a-2}] > 0 \tag{7.115}$$

for all $\lambda > 0$, so $f_{X/a}$ is convex (Lemma 1.4.20 on p. 37). Since $f_{X/a}(1) = 1$, $(f_{X/a})'(1) = \mathrm{E}[X]/a$, and, by (7.112), $f_{X/a}(\lambda) > 1$ for sufficiently small $\lambda \in (0,1)$ as well as for sufficiently large $\lambda \in (1,\infty)$, it follows that there exists a unique $\lambda_0 \in (0,1) \cup (1,\infty)$ such that $f_{X/a}(\lambda_0) = 1$. Also, $\lambda_0 > 1$ if $\mathrm{E}[X] < 0$ and $\lambda_0 < 1$ if $\mathrm{E}[X] > 0$. Since

$$f_{X/a}(\lambda) = f_X(\lambda^{1/a}), \tag{7.116}$$

we can deduce the stated conclusions with $\rho_0 := \lambda_0^{1/a}$.   ♠

Now let $X_1, X_2, \ldots$ be a sequence of i.i.d. random variables with common distribution that of X, representing the results of independent repetitions of the original wager. Then

$$S_n := X_1 + X_2 + \cdots + X_n \tag{7.117}$$

represents the gambler's cumulative profit after n such wagers; of course, $S_0 = 0$. Given positive numbers W and L (not necessarily integers), we suppose that the gambler stops betting as soon as he wins at least W units or loses at least L units. Consequently, his last bet occurs at coup

$$N(-L, W) := \min\{n \geq 1 : S_n \leq -L \ \text{ or } \ S_n \geq W\}. \tag{7.118}$$

Here we need to introduce two more parameters. Let $W^* \geq W$ and $L^* \geq L$ be positive numbers satisfying

$$P(S_{N(-L,W)} \in [-L^*, -L] \cup [W, W^*]) = 1. \qquad (7.119)$$

W^* and L^* are the maximum win and loss amounts, taking overshoot into account. They should be chosen as close to W and L, respectively, as possible. To determine their values, let $\varepsilon \in [0, \mu \wedge W]$ and $\delta \in [0, \nu \wedge L]$ satisfy

$$P(S_n \in [-(L - \delta), W - \varepsilon] \text{ whenever } 1 \le n < N(-L, W)) = 1. \qquad (7.120)$$

In words, ε and δ measure how close the gambler's cumulative profit can get to W and $-L$ without actually achieving these values. They should be chosen as large as possible. We can then take

$$L^* = L + \nu - \delta, \qquad W^* = W + \mu - \varepsilon. \qquad (7.121)$$

If W and L are integers and X is integer valued, then (7.120) holds automatically with $\varepsilon = \delta = 1$, though it may also hold for larger values of ε and δ.

Theorem 7.3.2. *Under the above assumptions and notation, if* $E[X] \ne 0$, *then*

$$\frac{\rho_0^L - 1}{\rho_0^{L+W^*} - 1} \le P(S_{N(-L,W)} \ge W) \le \frac{\rho_0^{L^*} - 1}{\rho_0^{L^*+W} - 1}, \qquad (7.122)$$

where ρ_0 is as in Lemma 7.3.1. If $E[X] = 0$, *then*

$$\frac{L}{L + W^*} \le P(S_{N(-L,W)} \ge W) \le \frac{L^*}{L^* + W}. \qquad (7.123)$$

Proof. As usual, we write N for $N(-L, W)$ in the proof only. An argument similar to the one used in the proof of Theorem 7.1.1 on p. 242 shows that $E[N] < \infty$. If $E[X] \ne 0$, then $\{\rho_0^{S_n}\}_{n \ge 0}$ is a martingale (De Moivre's martingale) with respect to $\{X_n\}_{n \ge 1}$ since

$$\begin{aligned}
E[\rho_0^{S_{n+1}} \mid X_1, \ldots, X_n] &= \rho_0^{S_n} E[\rho_0^{X_{n+1}} \mid X_1, \ldots X_n] \\
&= \rho_0^{S_n} E[\rho_0^{X_{n+1}}] \\
&= \rho_0^{S_n}. \qquad (7.124)
\end{aligned}$$

By the optional stopping theorem (Theorem 3.2.2 on p. 100),

$$E[\rho_0^{S_N}] = E[\rho_0^{S_0}] = 1. \qquad (7.125)$$

If $E[X] < 0$, then $\rho_0 > 1$ by Lemma 7.3.1, so with $P := P(S_N \ge W)$, we conclude from (7.125) that

$$\rho_0^{-L^*}(1 - P) + \rho_0^W P \le 1 \le \rho_0^{-L}(1 - P) + \rho_0^{W^*} P. \qquad (7.126)$$

If $E[X] > 0$, then $\rho_0 < 1$ by Lemma 7.3.1, so (7.126) holds with the inequalities reversed. In either case we deduce (7.122).

If $E[X] = 0$, then by Wald's identity (Corollary 3.2.5 on p. 103), we find that

$$E[S_N] = E[X]E[N] = 0, \qquad (7.127)$$

hence with $P := P(S_N \geq W)$, (7.127) implies that

$$(-L^*)(1 - P) + WP \leq 0 \leq (-L)(1 - P) + W^*P. \qquad (7.128)$$

This yields (7.123).   ♠

In the special case (7.1) on p. 241 and with W and L being positive integers, we have $W^* = W$, $L^* = L$, and $\rho_0 = q/p$ if $p \neq q$, hence Theorem 7.3.2 implies Theorem 7.1.1 on p. 242.

An important feature of the bounds of Theorem 7.3.2 is that they are *scale invariant*. This means that, if X, W, L, W^*, and L^* are all multiplied by the same positive constant (in effect, rescaling the basic monetary unit), the bounds remain unchanged, as one would expect (see Problem 7.20 on p. 270).

Before we can apply Theorem 7.3.2, we need a method for evaluating ρ_0 numerically. The following lemma based on *Newton's method* provides a straightforward approach.

Lemma 7.3.3. *Under the assumptions of Lemma 7.3.1 (including* $E[X] \neq 0$*), let a be the minimum of 1 and the smallest positive value that X assumes, and let* λ_0 *be the unique element of* $(0, 1) \cup (1, \infty)$ *satisfying* $f_{X/a}(\lambda_0) = 1$. *If* $E[X] < 0$ *and* $\eta_0 > \lambda_0 > 1$*, or if* $E[X] > 0$ *and* $\eta_0 < \lambda_0 < 1$*, then the sequence* $\{\eta_n\}$*, defined recursively by*

$$\eta_n := \eta_{n-1} - \frac{E[\eta_{n-1}^{X/a}] - 1}{E[(X/a)\eta_{n-1}^{X/a-1}]}, \qquad n \geq 1, \qquad (7.129)$$

converges monotonically to λ_0 *as* $n \to \infty$. *Consequently,* $\{\eta_n^{1/a}\}$ *converges monotonically to* ρ_0 *of Lemma 7.3.1 as* $n \to \infty$.

Proof. As we saw in the proof of Lemma 7.3.1, $f_{X/a}$ is convex. If we apply Newton's method to the equation $f_{X/a}(\lambda) = 1$, the nth iterate is given by

$$\eta_n := \eta_{n-1} - \frac{f_{X/a}(\eta_{n-1}) - 1}{(f_{X/a})'(\eta_{n-1})}, \qquad n \geq 1, \qquad (7.130)$$

which is the same as (7.129). The convexity of $f_{X/a}$ ensures that $\eta_0 > \eta_1 > \eta_2 > \cdots > \lambda_0 > 1$ if $E[X] < 0$ and $\eta_0 < \eta_1 < \eta_2 < \cdots < \lambda_0 < 1$ if $E[X] > 0$, and that the required convergence holds.   ♠

To apply this lemma effectively, a computer will usually be needed. For the reader who does not have easy access to a computer or does not need a high degree of accuracy, we mention in passing a useful but crude approximation, namely,

$$\rho_0 \approx \rho_* := \left(\frac{1 - \mathrm{E}[X]/\sqrt{\mathrm{E}[X^2]}}{1 + \mathrm{E}[X]/\sqrt{\mathrm{E}[X^2]}} \right)^{1/\sqrt{\mathrm{E}[X^2]}} \tag{7.131}$$

when $\mathrm{E}[X] \neq 0$. As for a "justification" of (7.131), consider the random variable Y with

$$\mathrm{P}(Y = y) = p \quad \text{and} \quad \mathrm{P}(Y = -y) = q := 1 - p, \tag{7.132}$$

where $y > 0$ and $p \in (0, \frac{1}{2}) \cup (\frac{1}{2}, 1)$ are chosen so that

$$\mathrm{E}[Y] = \mathrm{E}[X] \quad \text{and} \quad \mathrm{E}[Y^2] = \mathrm{E}[X^2]. \tag{7.133}$$

Specifically, $y(2p - 1) = \mathrm{E}[X]$ and $y^2 = \mathrm{E}[X^2]$, so

$$y = \sqrt{\mathrm{E}[X^2]} \quad \text{and} \quad p = \frac{1}{2} + \frac{\mathrm{E}[X]}{2\sqrt{\mathrm{E}[X^2]}}. \tag{7.134}$$

Now the value of ρ_0 for Y is $(q/p)^{1/y}$, which is (7.131). In view of (7.133), the value of ρ_0 for X should be similar. (Notice that we have not proved anything.)

For a more reliable, but more complicated, approximation to ρ_0 in terms of the moments of X, see Problem 7.19 on p. 269.

Example 7.3.4. *Bounds on the probability that a pass-line bettor taking 3-4-5-times free odds doubles his fortune before losing it.* We consider the one-unit pass-line bettor taking 3-4-5-times free odds, as in Example 7.2.5 on p. 256. The gambler's profit X from a single coup has distribution as in Table 7.3 on p. 257. First, we compute

$$\mathrm{E}[X] = -7/495, \qquad \mathrm{E}[X^2] = 1{,}329/55. \tag{7.135}$$

Approximation (7.131) implies that $\rho_0 \approx 1.001171159$, so we apply Lemma 7.3.3 with $a = 1$ and $\eta_0 = 1.002$. We find that

$$\begin{aligned}
\eta_1 &= 1.001413493075 \\
\eta_2 &= 1.001206096371 \\
\eta_3 &= 1.001171483696 \\
\eta_4 &= 1.001170462536 \\
\eta_5 &= 1.001170461646 \\
\eta_6 &= 1.001170461646
\end{aligned} \tag{7.136}$$

and hence that $\rho_0 \approx 1.001170461646$ after rounding.

Next we apply Theorem 7.3.2 with $W = L$ for several choices of W (all multiples of 100). Note that $\mu = 7$, $\nu = 6$, and $\varepsilon = \delta = 1$ in (7.120). Therefore, $W^* = W + 6$ and $L^* = L + 5$. The results are displayed in Table 7.4. Notice how the accuracy of the bounds seems to increase as W increases (with X fixed). ♠

Table 7.4 Bounds on the probability that a one-unit pass-line bettor taking 3-4-5-times free odds will win W or more units before losing W or more units. Results are from (7.122) and are rounded to six significant digits (lower bounds rounded down, upper bounds rounded up). Maximum possible relative error is $100(\text{upper} - \text{lower})/\text{lower}$.

W	lower bound	upper bound	maximum possible relative error (%)
100	.455 412	.482 245	5.892
1,000	.235 063	.237 370	.981
10,000	.825 411 $\cdot 10^{-5}$	.831 226 $\cdot 10^{-5}$	.704

As in Corollary 7.1.3 on p. 244, we consider next the limiting cases $L \to \infty$ and $W \to \infty$. Let us define

$$N(-\infty, W) := \min\{n \geq 1 : S_n \geq W\} \tag{7.137}$$

and

$$N(-L, \infty) := \min\{n \geq 1 : S_n \leq -L\}. \tag{7.138}$$

Here we take $\min \varnothing = \infty$ and $S_\infty = \pm\infty$, whichever is appropriate so that $S_{N(-\infty,W)} = -\infty$ if $N(-\infty, W) = \infty$, and $S_{N(-L,\infty)} = \infty$ if $N(-L, \infty) = \infty$. Then we have the following result.

Corollary 7.3.5. *Under the assumptions and notation of Theorem 7.3.2,*

$$\rho_0^{-W^*} \wedge 1 \leq P(S_{N(-\infty,W)} \geq W) \leq \rho_0^{-W} \wedge 1 \tag{7.139}$$

and

$$\rho_0^{L^*} \wedge 1 \leq P(S_{N(-L,\infty)} \leq -L) \leq \rho_0^{L} \wedge 1, \tag{7.140}$$

where $\rho_0 := 1$ if $E[X] = 0$.

Remark. It is often the case that $L^* = L$. (A sufficient condition is that X be integer valued with $\nu = 1$ and L be an integer.) If in addition $E[X] > 0$, then (7.140) implies that

$$P(S_{N(-L,\infty)} = -L) = \rho_0^{L}. \tag{7.141}$$

Proof. The argument is analogous to that of Corollary 7.1.3 on p. 244. ♠

Example 7.3.6. *Video poker and the probability of ruin.* In the video poker game "Deuces Wild," the distribution of the payout R from a one-unit bet, assuming a maximum-coin bet and the (minimum-variance) optimal drawing strategy, is given in Table 7.5. See Section 17.2 for complete details. The gambler's profit from such a bet is

$$X := R - 1. \tag{7.142}$$

Table 7.5 The distribution of the payout R from a one-unit bet on the video poker game "Deuces Wild." Assumes maximum-coin bet and the (minimum-variance) optimal drawing strategy.

result	R	probability	probability $\times \binom{52}{5}\binom{47}{5}5/12$
natural royal flush	800	.000 022 083 864	36,683,563
four deuces	200	.000 203 703 199	338,371,902
wild royal flush	25	.001 795 843 261	2,983,079,808
five of a kind	15	.003 201 603 965	5,318,192,488
straight flush	9	.004 119 878 191	6,843,540,140
four of a kind	5	.064 938 165 916	107,868,952,548
full house	3	.021 229 137 790	35,263,774,770
flush *or* straight	2	.073 145 116 685	121,501,539,340
three of a kind	1	.284 544 359 823	472,657,359,726
other	0	.546 800 107 307	908,291,048,815
total		1.000 000 000 000	1,661,102,543,100

We observe first that this is a positive-expectation game:

$$E[X] = \frac{243,403,018}{31,944,279,675} \approx 0.007619612039. \tag{7.143}$$

We therefore find, using (7.141), that the bankroll needed to ensure a ruin probability of at most 0.05, that is, the smallest positive integer L satisfying

$$P(S_{N(-L,\infty)} = -L) \le 0.05 \tag{7.144}$$

is given by

$$L = \left\lceil \frac{\ln 0.05}{\ln \rho_0} \right\rceil. \tag{7.145}$$

From Lemma 7.3.3 we compute $\rho_0 \approx 0.999346832577$, hence 4,585 is the required bankroll. ♠

We turn to the mean duration once again.

Theorem 7.3.7. *Under the assumptions and notation of Theorem 7.3.2, let* $P := P(S_{N(-L,W)} \geq W)$ *and suppose* $P_- \leq P \leq P_+$. *If* $E[X] < 0$, *then*

$$\frac{L + W^*}{|E[X]|}\left(\frac{L}{L + W^*} - P_+\right) \leq E[N(-L, W)] \leq \frac{L^* + W}{|E[X]|}\left(\frac{L^*}{L^* + W} - P_-\right).$$
(7.146)

If $E[X] > 0$, *then*

$$\frac{L^* + W}{E[X]}\left(P_- - \frac{L^*}{L^* + W}\right) \leq E[N(-L, W)] \leq \frac{L + W^*}{E[X]}\left(P_+ - \frac{L}{L + W^*}\right).$$
(7.147)

Proof. The proof is similar to that of Theorem 7.1.4 on p. 244.   ♠

Example 7.3.8. *Bounds on expected duration for the pass-line bettor taking 3-4-5-times free odds.* We consider once again the one-unit pass-line bettor taking 3-4-5-times free odds, as in Example 7.2.5 on p. 256. The gambler's profit X from a single coup has distribution as in Table 7.3 on p. 257. With ρ_0, W, and L as in Example 7.3.4, we apply Theorem 7.3.7 and obtain the results displayed in Table 7.6. Again, the accuracy of the bounds seems to increase as W increases (with X fixed).   ♠

Table 7.6 Bounds on the expected number of coups needed for a one-unit pass-line bettor taking 3-4-5-times free odds to either win W or more units or lose W or more units. Results are from (7.146) and are rounded to six significant digits (lower bounds rounded down, upper bounds rounded up). Maximum possible relative error is 100(upper − lower)/lower.

W	lower bound	upper bound	maximum possible relative error (%)
100	46.5102	823.151	1,670.
1,000	37,042.7	37,740.2	1.883
10,000	707,131.	707,485.	.05001

Finally, we consider the limiting cases.

Corollary 7.3.9. *Under the assumptions and notation of Theorem 7.3.7, if* $E[X] < 0$, *then*

$$\frac{L}{|E[X]|} \leq E[N(-L, \infty)] \leq \frac{L^*}{|E[X]|}.$$
(7.148)

If $E[X] > 0$, *then*

$$\frac{W}{\mathrm{E}[X]} \leq \mathrm{E}[N(-\infty, W)] \leq \frac{W^*}{\mathrm{E}[X]}. \tag{7.149}$$

Remark. Just as for Corollary 7.1.6 on p. 245, the results are intuitively clear.

Proof. The proof is similar to that of Corollary 7.1.6 on p. 245. ♠

7.4 Problems

7.1. *Fermat's likely derivation of the ruin formula in the case* $L = W$. Derive (7.11) on p. 243 by arguing that

$$\frac{\mathrm{P}(S_{N(-W,W)} = -W, \ N(-W, W) = W + 2n)}{\mathrm{P}(S_{N(-W,W)} = W, \ N(-W, W) = W + 2n)} = \frac{q^W}{p^W} \tag{7.150}$$

for $n = 0, 1, 2, \ldots$. To ensure that $\mathrm{P}(N(-W, W) < \infty) = 1$, one can apply Lemma 4.2.6 on p. 133.

7.2. *Ruin formula for games without pushes vs. same for games with pushes.* Recall that $p > 0$, $q > 0$, $r \geq 0$, and $p + q + r = 1$. Show directly that the validity of (7.4) on p. 242 for $r = 0$ implies the validity of (7.4) for $r > 0$.

7.3. *A tridiagonal linear system.* Let $p \in (0, \frac{1}{2}) \cup (\frac{1}{2}, 1)$ and put $q := 1 - p$. Given an integer $M \geq 2$ and a function $A : \{0, 1, \ldots, M\} \mapsto \mathbf{R}$ such that

$$A(j) = pA(j + 1) + qA(j - 1), \qquad j = 1, \ldots M - 1, \tag{7.151}$$

show that

$$A(j) = A(0)\left(1 - \frac{(q/p)^j - 1}{(q/p)^M - 1}\right) + A(M)\frac{(q/p)^j - 1}{(q/p)^M - 1} \tag{7.152}$$

for $j = 1, \ldots, M - 1$. Thus, each of the interior terms $A(1), \ldots, A(M - 1)$ is uniquely determined by the two boundary terms $A(0)$ and $A(M)$.

Hint: Use martingales rather than linear algebra.

7.4. *Variance of duration in even-money case.* Find $\mathrm{Var}(N(-L, W))$ in the case of an even-money proposition, assuming $p \neq q$.

Hint: Let $\mu := \mathrm{E}[X_1]$ and $\sigma^2 := \mathrm{Var}(X_1)$, and show that $M_n := (S_n - n\mu)^2 - n\sigma^2$ is a mean-zero martingale with respect to $\{X_n\}_{n \geq 1}$. Use the optional stopping theorem (Theorem 3.2.2 on p. 100), Theorems 7.1.1 on p. 242 and 7.1.4 on p. 244, and Corollary 7.1.8 on p. 247.

7.5. *Distribution of duration when* $L = W = 2$. Assuming $q := 1 - p$, find the conditional distributions of $N(-2, 2)$ given $S_{N(-2,2)} = 2$ and given $S_{N(-2,2)} = -2$ directly. Confirm the results of Corollary 7.1.8 on p. 247 and Theorem 7.1.9 on p. 248 in this setting. (Note that this case can be regarded as a simple model for a tennis game tied at deuce.)

7.6. *Montmort's formula for the distribution of duration when* $L = W = 3$. Assuming that $p = q = \frac{1}{2}$, verify that

$$P(N(-3,3) \leq 2m+1) = 1 - (3/4)^m, \qquad m \geq 1. \tag{7.153}$$

7.7. *Independence of duration and ruin when* $L = W$.

(a) Use Theorem 7.1.7 on p. 246 to prove, under the assumptions of Theorem 7.1.1 on p. 242, that $N(-L, W)$ is independent of $S_{N(-L,W)}$ when $L = W$.

(b) Use the idea of Problem 7.1 to give an even simpler proof.

7.8. *Ruin formula for 2-to-1 payoff by linear algebra.* Let $L = 2$ and $W = 3$ and solve the system (7.67) on p. 253 in the case (7.81) on p. 255 (after substituting the boundary conditions (7.68) and (7.69) on p. 253) for $u(0)$ by Gaussian elimination or Cramer's rule. Show that the result is consistent with Examples 7.2.2 on p. 253 and 7.2.3 on p. 255.

7.9. *Proof of ruin formula in Example 7.2.2.* Complete the proof of formula (7.73) on p. 254 by verifying (7.67) on p. 253 for the function u defined in Example 7.2.2 on p. 253.

Hint: Note that there are three cases to consider: $i = -L+1$, in which we must use $u(-L) := 0$; $-L+2 \leq i \leq W - m - 1$, in which no boundary conditions are needed; and $W - m \leq i \leq W - 1$, in which we must use $u(W) = \cdots = u(W + m - 1) := 1$. The third case is the easiest, while the first two cases rely on (1.7) on p. 4.

7.10. *Ruin formula for 1-to-2 payoff, unfair case.* Assume that

$$P(X = 1) = p, \quad P(X = -2) = q, \quad P(X = 0) = r, \tag{7.154}$$

where $p > 0$, $q > 0$, $r \geq 0$, and $p + q + r = 1$. Find the gambler's ruin formula analogous to (7.85) on p. 255, assuming $E[X] \neq 0$.

7.11. *Ruin formula for 2-to-1 payoff in fair case.* Extend Example 7.2.3 on p. 255 to the case in which $E[X] = 0$.

Hint: In this case $\lambda_1 := 1$ is a double root of (7.78) on p. 254. Use (7.80) on p. 255.

7.12. *Ruin formula for 3-to-1 payoff.* Assume that

$$P(X = 3) = p, \quad P(X = -1) = q, \quad P(X = 0) = r, \tag{7.155}$$

where $p > 0$, $q > 0$, $r \geq 0$, and $p + q + r = 1$. Find the gambler's ruin formula analogous to (7.85) on p. 255, assuming $E[X] \neq 0$.

Hint: The equation $P(\lambda) := p\lambda^4 - (p+q)\lambda + q = 0$ has roots $\lambda_0 = 1$, $\lambda_1 > 0$, and λ_2 and λ_3 complex conjugates. To see this, divide $P(\lambda)$ by $\lambda - 1$ and apply Cardano's formula (Theorem A.1.5 on p. 746) to the resulting cubic equation.

7.13. *Probabilistic interpretations of* $u_n^-(i)$ *and* $u_n^+(i)$. Show that the functions $u_n^-(i)$ and $u_n^+(i)$ of Theorem 7.2.4 on p. 256 can be represented as

$$u_n^-(i) = P(S_{N(-L-i,W-i)\wedge n} \geq W - i) \qquad (7.156)$$

and

$$u_n^+(i) = P(S_{N(-L-i,W-i)\wedge n} \geq -L - i + 1). \qquad (7.157)$$

7.14. *Ruin probability for pass-line bettor taking 100-times odds.* Here the pass-line bettor is permitted to take free odds for 100 times the amount of his pass-line bet. Find, to six significant digits, the probability that the one-unit pass-line bettor taking 100-times odds wins 500 or more units before losing 500 or more units.

7.15. *Mean duration for 2-to-1 payoff by linear algebra.* Let $L = 2$ and $W = 3$ and solve the system (7.96) on p. 257 in the case (7.81) on p. 255 (after substituting the boundary conditions (7.97) and (7.98) on p. 257) for $v(0)$ by Gaussian elimination or Cramer's rule. Show that the result is consistent with that of Example 7.2.7 on p. 258.

7.16. *Mean duration for 2-to-1 payoff in fair case.* Extend Example 7.2.7 on p. 258 to the case in which $E[X] = 0$.
   Hint: Use (7.80) on p. 255 for v_g (cf. Problem 7.11), and try $v_p(i) := ci^2$.

7.17. *Probabilistic interpretation of* $v_n^-(i)$. Show that the function $v_n^-(i)$ of Theorem 7.2.8 on p. 258 can be represented as

$$v_n^-(i) = E[N(-L - i, W - i) \wedge n]. \qquad (7.158)$$

7.18. *Mean duration for pass-line bettor taking 100-times odds.* As in Problem 7.14, find, to six significant digits, the expected number of coups (i.e., pass-line decisions) needed for the one-unit pass-line bettor taking 100-times odds to win 500 or more units or lose 500 or more units.

7.19. *Moment bounds on* ρ_0. With X as in Lemma 7.3.1 on p. 260, let $m_n := E[X^n]$ denote the nth moment of X. If $m_1 m_3 \leq 3m_2^2/8$ and $m_3 \neq 0$, define

$$\rho_1 := \exp\left\{-3\left(m_2 - \sqrt{(m_2)^2 - (8/3)m_1 m_3}\right)\Big/(2m_3)\right\} \qquad (7.159)$$

and

$$\rho_2 := \exp\left\{-3\left(m_2 + \gamma - \sqrt{(m_2 + \gamma)^2 - (8/3)m_1 m_3}\right)\Big/(2m_3)\right\}, \qquad (7.160)$$

where $\gamma := m_4(\ln \rho_1)^2/6$, and assume that $\rho_2^{-\nu} \vee \rho_2^\mu < 2$. Show that $1 < \rho_2 < \rho_0 < \rho_1$ if $m_1 < 0$, and $\rho_1 < \rho_0 < \rho_2 < 1$ if $m_1 > 0$, where ρ_0 is as in Lemma 7.3.1 on p. 260.
   Hint: Use Taylor's theorem or integration by parts to show that

$$e^x = 1 + x + \frac{1}{2}x^2 + \frac{1}{6}x^3 + \frac{1}{24}x^4 h(x), \tag{7.161}$$

where $h(x) := 4\int_0^1 (1-t)^3 e^{tx}\, dt$. Then apply this to $f_X(\rho) := \mathrm{E}[\rho^X] = \mathrm{E}[e^{(\ln\rho)X}]$. Notice that $\ln\rho_1$ and $\ln\rho_2$ come from the quadratic formula.

7.20. *Scale invariance of ruin bounds.* Show that the bounds in Theorem 7.3.2 on p. 261 are scale invariant, that is, if X, W, L, W^*, and L^* are all multiplied by the same positive constant, the bounds remain unchanged.

7.21. *Accuracy of second-moment approximation in case of 2-to-1 payoff.* With ρ_* as in (7.131) on p. 263, the approximation

$$\mathrm{P}(S_{N(-L,W)} \geq W) \approx \frac{\rho_*^L - 1}{\rho_*^{L+W} - 1} \tag{7.162}$$

has been proposed. To get a sense of how accurate this is, assume (7.81) on p. 255 with $p = 0.33$, $q = 0.67$, and $r = 0$, and compute the relative error in the approximation for $L = 1, 2, \ldots, 99$ and $W = 100 - L$, using Example 7.2.3 on p. 255.

7.22. *Ruin formula for 2-to-1 payoff against adversary with unlimited wealth.* Let L be a positive integer. In the case of (7.81) on p. 255, the probability that the gambler who initially has L units is eventually ruined (assuming an adversary with unlimited wealth or credit and assuming $\mathrm{E}[X] > 0$) can be evaluated in two ways, either letting $W \to \infty$ in (7.85) on p. 255 or using (7.141) on p. 264. Show that the two methods yield the same result.

7.23. *Ruin of a gambler vs. extinction of a branching process.* Let $(p_i)_{i\geq 0}$ be a probability distribution on $\mathbf{Z}_+$ with $p_0 > 0$, only finitely many positive terms, and mean $\mu := p_1 + 2p_2 + 3p_3 + \cdots > 1$. Let $g(\rho)$ be the corresponding probability generating function, that is, $g(\rho) = p_0 + p_1\rho + p_2\rho^2 + \cdots$.

Consider the Galton–Watson branching process $\{Z_n,\ n = 0, 1, 2, \ldots\}$ with initial state $Z_0 = L$ and offspring distribution $(p_i)_{i\geq 0}$. This is a Markov chain in $\mathbf{Z}_+$ with the conditional distribution of Z_{n+1} given $Z_n = k$ being the distribution of the sum of k independent random variables with common distribution $(p_i)_{i\geq 0}$. State 0 is absorbing. (One can think of Z_n as representing the size of generation n. To form generation $n + 1$, each individual in generation n is independently replaced by a random number of offspring with distribution $(p_i)_{i\geq 0}$.) Let ρ_1 be the smallest positive root of the equation $g(\rho) = \rho$. It is well known that the probability of eventual extinction of the branching process (i.e., $\mathrm{P}(Z_n = 0$ for some $n \geq 0)$) is given by ρ_1^L.

Let X be a random variable with values in $\{-1, 0, 1, 2, \ldots\}$ and with distribution $\mathrm{P}(X = i - 1) = p_i$ ($i \geq 0$). Consider a gambler starting with L units and repeatedly playing a game that provides a profit with distribution that of X. Then the probability of ruin, by (7.141) on p. 264, is ρ_0^L with ρ_0 as in Lemma 7.3.1 on p. 260.

Show that $\rho_0 = \rho_1$ and that it is not a coincidence that these two problems have the same answer because they are equivalent.

Hint: Assume that the gambler plays with one-unit chips kept in a stack. He always bets with the top chip and any winnings are placed on the top of the stack. Regard the L initial units as belonging to generation 0, the units won by betting these L units as belonging to generation 1, and so on.

7.24. *Mean duration bounds in fair case.* Generalize Theorem 7.3.7 on p. 266 to the case in which $\mathrm{E}[X] = 0$.

7.5 Notes

Edwards [1935–] (1983) discovered that the original formulation of the gambler's ruin problem, which had always been credited to Christiaan Huygens [1629–1695] (1657), is actually due to Blaise Pascal [1623–1662]. In 1656, two years after he solved the problem of points in his famous correspondence with Pierre de Fermat [1601–1665], Pascal posed the problem to Fermat (in the special case of Theorem 7.1.1 in which $p = 27/216$, $q = 15/216$, $r = 1 - p - q$, and $W = L = 12$), and he considered it so difficult that he doubted whether Fermat could solve it. Both Pascal and Fermat were able to solve the problem. Their correspondence has not survived, but Edwards (1983) speculated that Pascal used a derivation via difference equations while Fermat used a combinatorial argument that required $W = L$ (see Problem 7.1). Huygens (1657) was credited with the first formulation of the problem by Todhunter (1865) only because Huygens's correspondence, which mentions his source, was not published until 1888. Another derivation of the gambler's ruin formula is that of Jan Hudde [1628–1704] in 1665, whose method was used by Jacob Bernoulli [1654–1705] (1713; 2006, Part 1, pp. 189–192). The first published proof of the general formula (7.4) is due to Abraham De Moivre [1667–1754] (1712). The first published proof via difference equations can be credited to Nicolaas Struyck [1687–1769] (1716).

De Moivre's (1712) proof of our Theorem 7.1.1 is clever and bears repeating. (It was reproduced in De Moivre 1718, Problem 9; 1738 and 1756, Problem 7.) Let us refer to the gambler as A and to his opponent as B. We assume that A has L units initially while B has W. Play continues until one of the players has all $L + W$ units. Let us call these units "chips" (De Moivre used the word "counters") and assume that each player stacks his chips and that each coup amounts to a transfer of one chip from the top of one stack to the top of the other. The key idea is to artificially assign values to the chips that make the game fair. Thus, we assign values $q/p, (q/p)^2, \dots, (q/p)^L$ to A's initial stack of chips from bottom to top, and $(q/p)^{L+1}, \dots, (q/p)^{L+W}$ to B's initial stack of chips from top to bottom. At every coup, B risks q/p times as much as A, and since A wins, loses, and pushes each coup with probabilities p, q, and r, respectively, the game is fair. Consequently, at the end of play, the expected (artificial) value of A's profit is

$$P(A\ wins)\left\{\left(\frac{q}{p}\right)^{L+1} + \cdots + \left(\frac{q}{p}\right)^{L+W}\right\}$$

$$- (1 - P(A\ wins))\left\{\frac{q}{p} + \left(\frac{q}{p}\right)^{2} + \cdots + \left(\frac{q}{p}\right)^{L}\right\}, \qquad (7.163)$$

which must be 0. This implies that P(A wins) is given by (7.4) if $p \neq q$ and (7.5) if $p = q$.

A close examination of this derivation reveals that it is essentially the same martingale argument as the one used for the proof of Theorem 7.1.1, except that De Moivre took the optional stopping theorem for granted (and he assumed, rather than proved, that eventually one of the players will have all of the chips). This is remarkable because the concept of a martingale postdates De Moivre's (1712) idea by more than two centuries!

Corollary 7.1.8 and the observation that (7.32) and (7.33) are equal if $L = W$ are due to Stern (1975).

The problem of the duration of play (Theorem 7.1.9), in the words of Todhunter (1865, Article 107) "exercised the highest powers of De Moivre, Lagrange, and Laplace." The first published solution is due to De Moivre (1712). (See also De Moivre 1718, Problem 39; 1738, Problem 63; 1756, Problem 64.) He found, among other results, that

$$P(S_{N(-L,W)} = W,\ N(-L,W) = n) \qquad (7.164)$$

$$= p^{(n+W)/2}q^{(n-W)/2} \sum_{i=-\lfloor(n+W)/(2M)\rfloor}^{\lfloor(n-W)/(2M)\rfloor} \frac{W+2Mi}{n}\binom{n}{(n-W)/2 - Mi}.$$

Nicolaus Bernoulli [1687–1759] (Montmort 1713, pp. 308–314) and Montmort (1713, pp. 268–277) obtained another solution, namely

$$P(S_{N(-L,W)} = W,\ N(-L,W) \leq n) \qquad (7.165)$$

$$= \sum_{k=0}^{\lfloor(n-W)/(2M)\rfloor} p^{kM+W}q^{kM}\left[\sum_{j=0}^{\lfloor(n-W-2kM)/2\rfloor}\binom{n}{j}p^{j}q^{n-W-j-2kM}\right.$$

$$\left. + \sum_{j=0}^{\lfloor(n-W-2kM-1)/2\rfloor}\binom{n}{j}q^{j}p^{n-W-j-2kM}\right]$$

$$- \sum_{k=1}^{\lfloor(n+W)/(2M)\rfloor} p^{kM}q^{kM-W}\left[\sum_{j=0}^{\lfloor(n+W-2kM)/2\rfloor}\binom{n}{j}p^{j}q^{n+W-j-2kM}\right.$$

$$\left. + \sum_{j=0}^{\lfloor(n+W-2kM-1)/2\rfloor}\binom{n}{j}q^{j}p^{n+W-j-2kM}\right].$$

De Moivre (1738, pp. 181–182) remarked,

In the first attempt that I had ever made towards solving the general Problem of the Duration of Play, which was in the Year 1708, I began with the Solution of this LXIVth Problem, well knowing that it might be a Foundation for what I farther wanted, since which time, by a due repetition of it, I solved the main Problem: but as I found afterwards a nearer way to it, I barely published, in my first Essay on those matters, what seemed to me most simple and elegant, still preserving this Problem by me in order to be published when I should think it proper. Now in the year 1713 Mr. *Monmort* printed a Solution of it in a Book by him published upon Chance, in which was also inserted a Solution of the same by Mr. *Nicolas Bernoulli*; and as those two Solutions seemed to me, at first sight, to have some affinity with what I had found before, I considered them with very great attention; but the Solution of Mr. *Nicolas Bernoulli* being very much crouded with Symbols, and the verbal Explication of them too scanty, I own I did not understand it thoroughly, which obliged me to consider Mr. *Monmort*'s Solution with very great attention: I found indeed that he was very plain, but to my great surprize I found him very erroneous; still in my Doctrine of Chances I printed that Solution, but rectified and ascribed it to Mr. *Monmort*, without the least intimation of any alterations made by me; but as I had no thanks for so doing, I resume my right, and now print it as my own: [. . .]

Todhunter (1865, Article 181) noted that De Moivre's criticism of Montmort's solution was "unjustly severe." See Todhunter (1865, Articles 181–184, 302–316) for further discussion the problem. Our formulation is that of Joseph-Louis Lagrange [1736–1813] (1777) and our derivation follows William Feller [1906–1970] (1968). For an alternative proof based on contour integration, see Epstein (1967). However, notice that, for Epstein's formulation to be correct, one must interpret a "play of the game" as a win or a loss, but not a push. For other modern treatments, see Takács (1969), Hald and Johansen (1983), and Hald (1990, Chapters 20 and 23).

Example 7.2.2 is due to Kozek (2002), whose proof was based on Cramer's rule. The proof given here (Problem 7.9) is more straightforward, but the difficulty in such a result is not in proving it but rather in discovering it. Kozek (1995) had earlier given an erroneous formula in this setting, and finding the (subtle) flaw in his argument makes for a good problem. In Example 2.2, we had to solve a system of three linear equations in three variables to derive the formula (7.85). The coefficient matrix was nonsingular in this case, but will it be so in general? Gilliland, Levental, and Xiao (2007) have shown that the answer is affirmative, even when the roots of (7.78) are nondistinct. The iterative method of solving the linear system (7.67)–(7.69) suggested by (7.88) or (7.89) is known in numerical analysis as *Jacobi's iteration method*, named for Carl Jacobi [1804–1851], at least if $p_0 = 0$. Its applicability in this context was pointed out by Griffin (1991, Chapter 9), though the latter's algorithm is not quite of that form.

The idea of bounding the gambler's ruin probability above and below is due to Andrei A. Markov [1856–1922] (1912). Uspensky (1937) treated the case $P(X = -\nu) + P(X = \mu) = 1$ (with μ and ν positive integers), and Feller generalized this in 1950 (see Feller 1968) to the special case of

Theorem 7.3.2 in which X is integer-valued, W and L are integers, $\varepsilon = \delta = 1$ in (7.121). Feller's proof used the theory of linear difference equations. The slightly more general formulation given here is from Ethier and Khoshnevisan (2002). Equation (7.141) is well known and related to the formula for the extinction probability of a supercritical Galton–Watson branching process (Problem 7.23). It was rediscovered in 1999 by Evgeny Sorokin and posted on bjmath.com (Dunbar and B. 1999). See Hoppe (2007) for another application of branching processes to gambling.

Problem 7.6 is from Montmort (1708, p. 184) and Problem 7.7 is due to Samuels (1975). The results of Problem 7.19 are due to Ethier and Khoshnevisan (2002), but the bounds involving ρ_1 were obtained independently by Canjar (2007b). Problem 7.20 is from Ethier and Khoshnevisan (2002). The approximation (7.162) in Problem 7.21 is due to Griffin (1999, Chapter 9, Appendix C, first having appeared in the 1981 edition; 1991, pp. 73–76). The connection with branching processes described in Problem 7.23 is an observation of Munford and Lewis (1981).

Chapter 8
Betting Systems

It is deuced unlucky, to be sure, that he should have won all the little *coups* and lost all the great ones; but there is a plan which the commonest play-man knows, an infallible means of retrieving yourself at play: it is simply doubling your stake. Say, you lose a guinea: you bet two guineas, which if you win, you win a guinea and your original stake: if you lose, you have but to bet four guineas on the third stake, eight on the fourth, sixteen on the fifth, thirty-two on the sixth, and so on. It stands to reason that you cannot lose *always*, and the very first time you win, all your losings are made up to you. There is but one drawback to this infallible process; if you begin at a guinea, double every time you lose, and lose fifteen times, you will have lost exactly sixteen thousand three hundred and eighty-four guineas; a sum which probably exceeds the amount of your yearly income:—mine is considerably under that figure.[1]

William Makepeace Thackeray, "Captain Rook and Mr. Pigeon"

In Section 8.1 we describe and compare six well-known betting systems, namely the martingale, Fibonacci, Labouchere, Oscar, d'Alembert, and Blundell systems. Assuming a house maximum betting limit, each of these systems provides the gambler with a small win with high probability or a large loss with low probability. In Section 8.2 we show, under certain natural assumptions, that none of these systems, nor any other system that one may devise, can turn a sequence of subfair wagers into a superfair one. This principle is known as the conservation of fairness.

8.1 Examples

Here we consider betting systems that are restricted to games with even-money payoffs. Let $X_1, X_2, \ldots$ be a sequence of i.i.d. random variables with

---

[1] 16,384 $(= 2^{15-1})$ is the size of the 15th bet; the total amount lost is 32,767 $(= 2^{15} - 1)$. See Table 8.1.

S.N. Ethier, *The Doctrine of Chances*, Probability and its Applications, DOI 10.1007/978-3-540-78783-9_8, © Springer-Verlag Berlin Heidelberg 2010

common distribution

$$P(X_1 = 1) = p \quad \text{and} \quad P(X_1 = -1) = q, \tag{8.1}$$

where $0 < p < 1$ and $q := 1 - p$, with X_n representing the bettor's profit per unit bet at coup n. (For simplicity, we exclude the possibility of pushes.) A betting system consists of a sequence of bet sizes, $B_1, B_2, \ldots$, with B_n representing the amount bet at coup n and depending only on the results of the previous coups, $X_1, \ldots, X_{n-1}$. More precisely,

$$B_1 = b_1 \geq 0, \qquad B_n = b_n(X_1, \ldots, X_{n-1}) \geq 0, \quad n \geq 2, \tag{8.2}$$

where b_1 is a constant and b_n is a nonrandom function of $n - 1$ variables for each $n \geq 2$. The gambler's fortune F_n after n coups satisfies $F_n = F_{n-1} + B_n X_n$ for each $n \geq 1$, and therefore

$$F_n = F_0 + \sum_{l=1}^{n} B_l X_l, \qquad n \geq 1, \tag{8.3}$$

F_0 being the gambler's initial fortune (a positive constant). In this section we consider six well-known examples of betting systems.

Example 8.1.1. *Martingale system.* In this system, the gambler doubles his bet size after each loss, and stops betting after the first win. (Of course, he can restart the system after the first win, but to simplify the analysis, we ignore this possibility for now.) In terms of our notation,

$$B_n = 2B_{n-1}1_{\{X_{n-1}=-1\}}, \qquad n \geq 2, \tag{8.4}$$

and therefore, if money is measured in units of the initial bet size,

$$B_1 = 1, \qquad B_n = 2^{n-1}1_{\{X_1=\cdots=X_{n-1}=-1\}}, \quad n \geq 2. \tag{8.5}$$

See Table 8.1 for clarification.

Now let N be the number of coups needed to achieve the first win, that is, $N := \min\{n \geq 1 : X_n = 1\}$. Then (8.5) can be simplified to

$$B_n = 2^{n-1}1_{\{n \leq N\}}, \qquad n \geq 1, \tag{8.6}$$

and (8.3) becomes

$$F_n = \begin{cases} F_0 - (1 + 2 + \cdots + 2^{n-1}) = F_0 - (2^n - 1) & \text{if } n \leq N - 1, \\ F_0 - (1 + 2 + \cdots + 2^{N-2}) + 2^{N-1} = F_0 + 1 & \text{if } n \geq N. \end{cases} \tag{8.7}$$

This equation tells us that, if the gambler makes n bets and fails to win any of them, he will have lost $2^n - 1$ units. If he continues betting until he finally wins a bet, he will have achieved a cumulative profit of one unit. In

Table 8.1 An illustration of the martingale system, motivated by the quote of Thackeray on p. 275.

coup	bet size	outcome	cumulative profit
1	$1 = 2^0$	L	$-1 = -(2^1 - 1)$
2	$2 = 2^1$	L	$-3 = -(2^2 - 1)$
3	$4 = 2^2$	L	$-7 = -(2^3 - 1)$
4	$8 = 2^3$	L	$-15 = -(2^4 - 1)$
5	$16 = 2^4$	L	$-31 = -(2^5 - 1)$
6	$32 = 2^5$	L	$-63 = -(2^6 - 1)$
7	$64 = 2^6$	L	$-127 = -(2^7 - 1)$
8	$128 = 2^7$	L	$-255 = -(2^8 - 1)$
9	$256 = 2^8$	L	$-511 = -(2^9 - 1)$
10	$512 = 2^9$	L	$-1{,}023 = -(2^{10} - 1)$
11	$1{,}024 = 2^{10}$	L	$-2{,}047 = -(2^{11} - 1)$
12	$2{,}048 = 2^{11}$	L	$-4{,}095 = -(2^{12} - 1)$
13	$4{,}096 = 2^{12}$	L	$-8{,}191 = -(2^{13} - 1)$
14	$8{,}192 = 2^{13}$	L	$-16{,}383 = -(2^{14} - 1)$
15	$16{,}384 = 2^{14}$	L	$-32{,}767 = -(2^{15} - 1)$
16	$32{,}768 = 2^{15}$	W	$1 = 2^0$

particular,

$$P(F_N = F_0 + 1) = 1, \tag{8.8}$$

which would make this an infallible system indeed, were it not for two complications: First, the gambler cannot bet what he does not have, and second, the house usually imposes a maximum betting limit. Thus, if a bet is called for that exceeds either the gambler's remaining capital or the house betting limit, the system must be aborted.

Given an initial fortune of F_0 units and a house limit of M units (both positive integers), we define

$$m_1 := \lfloor \log_2(F_0 + 1) \rfloor, \quad m_2 := 1 + \lfloor \log_2 M \rfloor, \quad m := m_1 \wedge m_2; \tag{8.9}$$

in particular, m_1 is the integer part of the base-2 logarithm of $F_0 + 1$. For example, if $F_0 = 1{,}000$ and $M = 500$, then $2^8 \leq M < 2^9 \leq F_0 + 1 < 2^{10}$, so $m = m_1 = m_2 = 9$.

Clearly, the gambler achieves his one-unit cumulative profit if and only if the total of all bets that the system calls for, namely $2^N - 1$ units, does not exceed the gambler's initial fortune of F_0 units, *and* the largest bet that the system calls for, namely 2^{N-1} units, does not exceed the house limit of M units. Using the notation (8.9), the probability of this event is

$$P(2^N - 1 \leq F_0, \; 2^{N-1} \leq M) = P(N \leq \log_2(F_0 + 1), \; N \leq 1 + \log_2 M)$$
$$= P(N \leq m_1, \; N \leq m_2)$$
$$= P(N \leq m)$$
$$= 1 - P(X_1 = -1, \ldots, X_m = -1)$$
$$= 1 - q^m. \tag{8.10}$$

Furthermore, on the complementary event $\{N \geq m + 1\}$, only m bets are made and the gambler's cumulative loss is $2^m - 1$ units. We conclude that, over the course of $N \wedge m$ coups, the gambler will achieve a small win (one unit) with high probability $(1 - q^m)$ while incurring a large loss $(2^m - 1$ units) with low probability (q^m). In particular, his expected cumulative profit is given by

$$E[F_{N \wedge m} - F_0] = (1)(1 - q^m) + [-(2^m - 1)]q^m = 1 - (2q)^m, \tag{8.11}$$

and (8.11) is negative, zero, or positive if $p < \frac{1}{2}, p = \frac{1}{2}$, or $p > \frac{1}{2}$, respectively.

Actually, the house limit is a more serious consideration than the player's bankroll, what with the easy availability of credit. In fact, given a house limit of M units, the gambler can ensure that the martingale system will never call for a bet that exceeds his available resources without first exceeding the house limit. Indeed, it suffices to choose an initial fortune F_0 satisfying

$$F_0 \geq 2^{1 + \lfloor \log_2 M \rfloor} - 1. \tag{8.12}$$

This inequality is equivalent to $m_1 \geq m_2$ or $m = m_2$.

Finally, the expected number of coups is given by

$$E[N \wedge m] = p + 2qp + 3q^2 p + \cdots + mq^{m-1}p + mq^m$$
$$= \sum_{l=1}^{m} lq^{l-1}p + mq^m = \sum_{l=1}^{m} \sum_{k=1}^{l} q^{l-1}p + mq^m$$
$$= \sum_{k=1}^{m} \sum_{l=k}^{m} q^{l-1}p + mq^m = \sum_{k=1}^{m} (q^{k-1} - q^m) + mq^m$$
$$= \sum_{k=1}^{m} q^{k-1} = \frac{1 - q^m}{1 - q}, \tag{8.13}$$

and the expected total amount bet is given by

$$E[2^{N \wedge m} - 1]$$
$$= p + 3qp + 7q^2 p + \cdots + (2^m - 1)q^{m-1}p + (2^m - 1)q^m$$
$$= 2p(1 + 2q + \cdots + (2q)^{m-1}) - p(1 + q + \cdots + q^{m-1}) + (2^m - 1)q^m$$
$$= \begin{cases} (1 - (2q)^m)/(1 - 2q) & \text{if } p \neq \frac{1}{2}, \\ m & \text{if } p = \frac{1}{2}. \end{cases} \tag{8.14}$$

Notice that the ratio of gambler's expected cumulative profit (8.11) to his expected total amount bet (8.14) coincides with the expected profit from a single one-unit bet, namely $1-2q$ (or $2p-1$). This is not coincidental; indeed, Section 8.2 shows that all systems have this property.   ♠

Example 8.1.2. *Fibonacci system.* Consider the *Fibonacci sequence* $1, 1, 2,$ $3, 5, 8, 13, 21, 34, 55, 89, 144, 233, 377, \ldots$, in which each term is the sum of the preceding two. Drop the first term, and denote the mth of the remaining terms by f_m. More precisely, $f_1 := 1$, $f_2 := 2$, and

$$f_m := f_{m-2} + f_{m-1}, \qquad m \geq 3. \tag{8.15}$$

In the Fibonacci system, the first bet is one unit (f_1). After a loss of f_m units for some $m \geq 1$, the bet size is increased to f_{m+1} units. After a win of f_m units for some $m \geq 3$, the bet size is decreased to f_{m-2} units. After a win of f_1 or f_2 units, no further bets are made.

Notice that the sequence of bet sizes is Markovian. Indeed, with $X_1, X_2, \ldots$ a sequence of i.i.d. random variables with common distribution (8.1), where $0 < p < 1$ and $q := 1 - p$, we define the random sequence $\{Y_n\}_{n \geq 0}$ by $Y_0 := 1$ and

$$Y_{n+1} := \begin{cases} (Y_n - \frac{3}{2}X_{n+1} - \frac{1}{2})^+ & \text{if } Y_n \geq 1, \\ 0 & \text{if } Y_n = 0. \end{cases} \tag{8.16}$$

Then $\{Y_n\}_{n \geq 0}$ is a Markov chain in $\mathbf{Z}_+$ with initial state $Y_0 := 1$, one-step transition probabilities

$$P(i, j) := \begin{cases} p & \text{if } j = (i-2)^+, \\ q & \text{if } j = i+1, \qquad i \geq 1, \\ 0 & \text{otherwise,} \end{cases} \tag{8.17}$$

and absorbing state 0, and the Fibonacci bettor stakes

$$B_n := f_{Y_{n-1}} \tag{8.18}$$

at the nth coup for each $n \geq 1$, where $f_0 := 0$. See Table 8.2 for clarification. Of course, we have already seen (8.17) in Example 4.1.3 on p. 121.

Let $N := \min\{n \geq 1 : B_{n+1} = 0\} = \min\{n \geq 1 : Y_n = 0\}$ be the number of coups at which a nonzero bet is made. We claim that the gambler's fortune after n coups is given by

$$F_n = F_0 + 2 - f_{Y_n+1}, \qquad 0 \leq n \leq N. \tag{8.19}$$

This can be proved by induction. It is immediate for $n = 0$ since $Y_0 := 1$. If $F_n = F_0 + 2 - f_{Y_n+1}$ for some $0 \leq n \leq N - 1$, then

$$F_{n+1} = F_n + B_{n+1}X_{n+1}$$

Table 8.2 An illustration of the Fibonacci system.

coup	bet size	outcome	cumulative profit
1	$1 = f_1$	L	$-1 = 2 - f_3$
2	$2 = f_2$	L	$-3 = 2 - f_4$
3	$3 = f_3$	L	$-6 = 2 - f_5$
4	$5 = f_4$	L	$-11 = 2 - f_6$
5	$8 = f_5$	W	$-3 = 2 - f_4$
6	$3 = f_3$	L	$-6 = 2 - f_5$
7	$5 = f_4$	L	$-11 = 2 - f_6$
8	$8 = f_5$	L	$-19 = 2 - f_7$
9	$13 = f_6$	L	$-32 = 2 - f_8$
10	$21 = f_7$	W	$-11 = 2 - f_6$
11	$8 = f_5$	W	$-3 = 2 - f_4$
12	$3 = f_3$	L	$-6 = 2 - f_5$
13	$5 = f_4$	W	$-1 = 2 - f_3$
14	$2 = f_2$	W	$1 = 2 - f_1$

$$
\begin{aligned}
&= \begin{cases} F_0 + 2 - f_{Y_n+1} + f_{Y_n} & \text{if } X_{n+1} = 1 \\ F_0 + 2 - f_{Y_n+1} - f_{Y_n} & \text{if } X_{n+1} = -1 \end{cases} \\
&= \begin{cases} F_0 + 2 - f_{Y_n-1} & \text{if } X_{n+1} = 1 \\ F_0 + 2 - f_{Y_n+2} & \text{if } X_{n+1} = -1 \end{cases} \\
&= F_0 + 2 - f_{Y_{n+1}+1},
\end{aligned}
\tag{8.20}
$$

as required, where we have used (8.15) in the third equality. By Example 7.2.3 on p. 255, $E[N] < \infty$ if $p > \frac{1}{3}$, and by Problem 7.11 on p. 268, $P(N < \infty) = 1$ if $p \geq \frac{1}{3}$. Assuming $p \geq \frac{1}{3}$, we deduce from (8.19) that

$$
P(F_N = F_0 + 1) = 1.
\tag{8.21}
$$

Again we have an infallible system, assuming unlimited credit and no house betting limit.

Continuing with this assumption, let us find the distribution of the Fibonacci bettor's largest bet B^*. Here we assume only that $p > \frac{1}{3}$. Let us define the random variable

$$
Y^* := \sup_{n \geq 0} Y_n,
\tag{8.22}
$$

and observe that

$$
B^* = \sup_{n \geq 1} B_n = \sup_{n \geq 1} f_{Y_n-1} = f_{Y^*}.
\tag{8.23}
$$

First, letting

$$\phi_1 := \frac{1}{2} + \frac{1}{2}\sqrt{5} \quad \text{and} \quad \phi_2 := \frac{1}{2} - \frac{1}{2}\sqrt{5} \tag{8.24}$$

be the two roots of the quadratic equation $x^2 - x - 1 = 0$, *Binet's formula*

$$f_m = \frac{\phi_1^{m+1} - \phi_2^{m+1}}{\phi_1 - \phi_2}, \qquad m \geq 1, \tag{8.25}$$

is easily established by induction. In particular, with uninterrupted losses, the Fibonacci bettor's bet size grows geometrically but at a slower rate than the martingale bettor's:

$$\lim_{m \to \infty} \frac{f_{m+1}}{f_m} = \phi_1, \tag{8.26}$$

with ϕ_1 known as the *golden ratio* (≈ 1.618034).

Now we turn to the relevant probabilities. Let T_j denote the hitting time of state j by the Markov chain $\{Y_n\}_{n \geq 0}$. Then, by Example 7.2.3 on p. 255 with $W = 1$ and $L = m$,

$$\begin{aligned}
&P_1(Y^* \geq m + 1) \\
&= 1 - P_1(T_0 < T_{m+1}) \\
&= 1 - \frac{(\lambda_1^m - 1)\lambda_2^{m+1}(\lambda_2 - 1) - \lambda_1^{m+1}(\lambda_1 - 1)(\lambda_2^m - 1)}{(\lambda_1^{m+1} - 1)(\lambda_2^{m+2} - 1) - (\lambda_1^{m+2} - 1)(\lambda_2^{m+1} - 1)} \\
&= \frac{(\lambda_1\lambda_2)^m[\lambda_1\lambda_2(\lambda_2 - \lambda_1) + \lambda_1(\lambda_1 - 1) - \lambda_2(\lambda_2 - 1)]}{(\lambda_1\lambda_2)^m[\lambda_1\lambda_2(\lambda_2 - \lambda_1)] + \lambda_1^m\lambda_1(\lambda_1 - 1) - \lambda_2^m\lambda_2(\lambda_2 - 1)} \\
&\sim \lambda_1^m\left(\frac{\lambda_1\lambda_2(\lambda_2 - \lambda_1) + \lambda_1(\lambda_1 - 1) - \lambda_2(\lambda_2 - 1)}{-\lambda_2(\lambda_2 - 1)}\right) \tag{8.27}
\end{aligned}$$

for each $m \geq 1$, where the subscript on P indicates that we are conditioning on the event $\{Y_0 = 1\}$, and

$$\lambda_1 := -\frac{1}{2} + \frac{1}{2}\sqrt{1 + 4(q/p)} \in (0, 1) \tag{8.28}$$

and

$$\lambda_2 := -\frac{1}{2} - \frac{1}{2}\sqrt{1 + 4(q/p)} \in (-2, -1). \tag{8.29}$$

Consequently,

$$P_1(Y^* = m) = P_1(Y^* \geq m) - P_1(Y^* \geq m + 1) \sim c\lambda_1^m \tag{8.30}$$

for an appropriate positive constant c.

If we assume also that $p \leq \frac{1}{2}$, then $\lambda_1 \geq -\phi_2$, so $\phi_1\lambda_1 \geq -\phi_1\phi_2 = 1$. This is enough to show that

$$E_1[B^*] = E_1[f_{Y^*}] = \sum_{m \geq 1} f_m P_1(Y^* = m) = \infty \tag{8.31}$$

if $\frac{1}{3} < p \leq \frac{1}{2}$ because the mth term in the sum is asymptotic to

$$\frac{\phi_1^{m+1}}{\sqrt{5}} c\lambda_1^m = \frac{c\phi_1}{\sqrt{5}}(\phi_1\lambda_1)^m \geq \frac{c\phi_1}{\sqrt{5}}. \tag{8.32}$$

Finally, we consider the case of a house limit of M units. We let

$$\begin{aligned} N_M &:= \min\{n \geq 1 : B_{n+1} = 0 \text{ or } B_{n+1} > M\} \\ &= \min\{n \geq 1 : Y_n = 0 \text{ or } Y_n > m\}, \end{aligned} \tag{8.33}$$

where m is such that $f_m \leq M < f_{m+1}$. This is the number of coups at which a nonzero bet is made, assuming that the system is aborted once a bet exceeding M units is called for. If

$$F_0 \geq f_{m+2} - 2, \tag{8.34}$$

then

$$\begin{aligned} F_n - B_{n+1} &= F_0 + 2 - f_{Y_n+1} - f_{Y_n} \\ &= F_0 + 2 - f_{Y_n+2} \\ &\geq f_{m+2} - f_{Y_n+2} \geq 0, \qquad 0 \leq n \leq N_M - 1. \end{aligned} \tag{8.35}$$

Thus, assuming (8.34), the Fibonacci bettor can make any bet that is called for, up to the time that the goal is achieved or the system is aborted on account of the house limit.

By the result of Example 7.2.7 on p. 258, $E[N_M]$ is given by (7.103) and (7.104) on p. 258 with $W = 1$, $L = m$, and λ_1 and λ_2 as in (8.28) and (8.29). ♠

Example 8.1.3. *Labouchere system.* In this system, also known as the *cancelation system* and already introduced in Example 4.1.3 on p. 121, the gambler's bet size at each coup is determined by a list of positive integers kept on his score sheet, which is updated after each coup. Let us denote such a list by

$$\boldsymbol{l} = (l_1, \ldots, l_j), \tag{8.36}$$

where $j \geq 0$ denotes its length. (A list of length 0 is empty, denoted by $\boldsymbol{l} = \varnothing$.)

Given a list (8.36), the gambler's bet size at the next coup is

$$b(\boldsymbol{l}) := \begin{cases} 0 & \text{if } j = 0, \\ l_1 & \text{if } j = 1, \\ l_1 + l_j & \text{if } j \geq 2, \end{cases} \tag{8.37}$$

that is, the sum of the extreme terms of the list. (This is just the sum of the first and last terms, except when the list has only one term or is empty.)

Following the resolution of this bet, the list (8.36) is updated as follows: After a win, it is replaced by

$$l[1] := \begin{cases} \varnothing & \text{if } j \in \{0, 1, 2\}, \\ (l_2, \ldots, l_{j-1}) & \text{if } j \geq 3, \end{cases} \tag{8.38}$$

that is, the extreme terms are canceled. After a loss, it is replaced by

$$l[-1] := \begin{cases} \varnothing & \text{if } j = 0, \\ (l_1, \ldots, l_j, b(l)) & \text{if } j \geq 1, \end{cases} \tag{8.39}$$

that is, the amount just lost is appended to the list, if nonempty, as a new last term.

The system is begun with an initial list l^0, a popular choice for which is

$$l^0 = (1, 2, \ldots, j_0) \tag{8.40}$$

for a positive integer j_0. The rules above, together with $X_1, X_2, \ldots$, determine all bet sizes. More precisely, let L_n denote the list as updated after the nth coup. Then $L_0 = l^0$ and

$$L_n := L_{n-1}[X_n], \qquad n \geq 1, \tag{8.41}$$

where the meaning of the brackets is as in (8.38) or (8.39). Let B_n denote the amount bet at the nth coup, so that

$$B_n := b(L_{n-1}), \qquad n \geq 1. \tag{8.42}$$

Once the list becomes empty, it remains so, and no further bets are made. See Table 8.3 for clarification.

Let Y_n be the length of the list after the nth coup. Notice that $\{Y_n\}_{n \geq 0}$ is a Markov chain in $\mathbf{Z}_+$ with initial state 1, transition probabilities (8.17), and absorbing state 0. Using Example 7.2.3 on p. 255 and Problem 7.11 on p. 268, one can show that absorption at 0 occurs with probability 1 if and only if $p \geq \frac{1}{3}$. Let N be the time until absorption, that is, $N = \min\{n \geq 1 : Y_n = 0\}$. Using Example 7.2.7 on p. 258 and Problem 7.16 on p. 269, one can show that $E[N] < \infty$ if and only if $p > \frac{1}{3}$ (see Problem 8.6 on p. 306). In fact, even the distribution of N can be evaluated (Problems 8.7 and 8.8 on p. 307).

Let S_n be the sum of the terms of the list after the nth coup. With F_n as in (8.3), we observe that $F_n + S_n$ does not depend on n. Indeed, if $1 \leq n \leq N$, then

$$F_n - F_{n-1} = B_n X_n \quad \text{and} \quad S_n - S_{n-1} = -B_n X_n, \tag{8.43}$$

implying that $F_n + S_n = F_{n-1} + S_{n-1}$. It follows that

$$F_n = F_0 + S_0 - S_n, \qquad 1 \leq n \leq N. \tag{8.44}$$

Table 8.3 An illustration of the Labouchere system. Here the initial list is $l^0 = (1)$.

coup	list	bet size	outcome	cumulative profit
1	(1)	1	L	-1
2	$(1, 1)$	2	L	-3
3	$(1, 1, 2)$	3	L	-6
4	$(1, 1, 2, 3)$	4	L	-10
5	$(1, 1, 2, 3, 4)$	5	W	-5
6	$(1, 2, 3)$	4	L	-9
7	$(1, 2, 3, 4)$	5	L	-14
8	$(1, 2, 3, 4, 5)$	6	L	-20
9	$(1, 2, 3, 4, 5, 6)$	7	W	-13
10	$(2, 3, 4, 5)$	7	L	-20
11	$(2, 3, 4, 5, 7)$	9	L	-29
12	$(2, 3, 4, 5, 7, 9)$	11	W	-18
13	$(3, 4, 5, 7)$	10	L	-28
14	$(3, 4, 5, 7, 10)$	13	W	-15
15	$(4, 5, 7)$	11	W	-4
16	(5)	5	W	1

In particular, assuming $p \geq \frac{1}{3}$, we have $S_N = 0$ a.s., that is,

$$P(F_N = F_0 + S_0) = 1. \tag{8.45}$$

This equation says that, with probability 1, the gambler wins an amount equal to the sum of the terms of the initial list. Again, this would be an infallible system were it not for the limit on the gambler's resources and the house betting limit.

Let us assume an initial list of the form (8.40). Let M be a positive integer satisfying $M \geq j_0 + 1$ (or $M \geq 1$ if $j_0 = 1$), and assume a house limit of M units. How large must F_0 be to ensure that the gambler can make any bet that is called for, up to the time that the goal is achieved or the system is aborted on account of the house limit? The worst-case scenario is $M - j_0$ (or M if $j_0 = 1$) consecutive losses, which would require an initial fortune of

$$F_0 \geq \begin{cases} \sum_{i=j_0+1}^{M} i = \binom{M+1}{2} - \binom{j_0+1}{2} & \text{if } j_0 \geq 2, \\ \sum_{i=1}^{M} i = \binom{M+1}{2} & \text{if } j_0 = 1. \end{cases} \tag{8.46}$$

This may be unclear because, for example, starting with the initial list $(1, 2, 3, 4)$, a win followed by $n \geq 8$ consecutive losses leaves the gambler with a greater deficit than does $n + 1$ consecutive losses. To justify the claim, notice by induction that the list is increasing from first to last term except possibly for the first two terms, which may be equal. Therefore, the worst-case scenario, by virtue of (8.44), corresponds to the list for which the sum of its terms is maximal and such that the resulting bet size exceeds M for the first time. This is of course $(1, 2, \ldots, M)$ (or $(1, 1, 2, \ldots, M)$ if $j_0 = 1$).

Just as in Example 8.1.1, it would be desirable to evaluate the probability of achieving the goal of S_0 units, as a function of the house limit M (assuming (8.40) and (8.46)). In theory, this can be done using the fact that $\{L_n\}_{n \geq 0}$ is a Markov chain. Let $Q(l^0)$ denote the probability, starting with initial list l^0, that the list becomes empty (i.e., the goal is achieved) before the list reaches a state l with $b(l) \geq M + 1$. Then, using the notation (8.38) and (8.39),

$$Q(l) = pQ(l[1]) + qQ(l[-1]), \qquad 1 \leq b(l) \leq M, \qquad (8.47)$$

where $Q(\varnothing) = 1$ and $Q(l) = 0$ if $b(l) \geq M + 1$. For example, let us consider the simple case in which the initial list is $l^0 = (1)$ and the house limit is $M = 5$. Then (8.47) becomes

$$
\begin{aligned}
Q(1) &= pQ(\varnothing) + qQ(11), \\
Q(2) &= pQ(\varnothing) + qQ(22), \\
Q(11) &= pQ(\varnothing) + qQ(112), \\
Q(12) &= pQ(\varnothing) + qQ(123), \\
Q(22) &= pQ(\varnothing) + qQ(224), \\
Q(23) &= pQ(\varnothing) + qQ(235), \\
Q(112) &= pQ(1) + qQ(1123), \\
Q(123) &= pQ(2) + qQ(1234), \\
Q(1123) &= pQ(12) + qQ(11234), \\
Q(1234) &= pQ(23) + qQ(12345), \\
Q(11234) &= pQ(123) + qQ(112345),
\end{aligned}
\qquad (8.48)
$$

where we have omitted commas between terms of lists (each term has only one digit). (We have also omitted equations, such as $Q(3) = pQ(\varnothing) + qQ(33)$, corresponding to states, such as (3), that cannot be reached from $l^0 = (1)$ without first visiting a state l with $b(l) \geq 6$.) Now $Q(\varnothing) = 1$ and $Q(224) = Q(235) = Q(12345) = Q(112345) = 0$, so we have a closed system of 11 linear equations in 11 variables. Solving for $Q(1)$ we find that

$$Q(1) = \frac{p + pq + p^2 q^3 + 2p^3 q^4 + 4p^3 q^5}{1 - pq^2}. \qquad (8.49)$$

The reader may find it easier to derive (8.49) using a binary tree diagram.

Finally, we want to observe that similar methods can be used to find several expectations of interest using Example 4.2.9 on p. 135. Consider the linear system

$$R(l) = A(l) + pR(l[1]) + qR(l[-1]), \qquad 1 \le b(l) \le M, \qquad (8.50)$$

where $R(\varnothing) = 0$ and $R(l) = 0$ if $b(l) \ge M + 1$. Then $R(l^0)$ is the gambler's expected cumulative profit if $A(l) := b(l)(2p - 1)$, $R(l^0)$ is the expected number of coups if $A(l) := 1$, and $R(l^0)$ is the expected total amount bet if $A(l) := b(l)$. In the special case $l^0 = (1)$ and $M = 5$, we have a closed system of 11 linear equations in 11 variables analogous to (8.48), and the relevant solutions are

$$R(1) = \frac{1 + q + q^2 + 2q^3 + 4pq^4 + 4p^2q^5}{1 - pq^2} \qquad (8.51)$$

for the expected number of coups, and

$$R(1) = \frac{1 + 2q + 3q^2 + 4q^3 + 3pq^3 + 5q^4 + 8pq^4 + 4p^2q^4 + 10pq^5 + 18p^2q^5}{1 - pq^2}$$

$$(8.52)$$

for the expected total amount bet.

For large M the method of the two preceding paragraphs is not practical—there are too many lists needed to assure a closed linear system.   ♠

Example 8.1.4. *Oscar system.* This system was introduced in Example 4.1.2 and discussed further in Example 4.2.5. The first bet is one unit. After a loss, the bet size remains the same. After a win, the bet size is increased by one unit, but with this possible exception: The gambler never bets more than necessary to achieve a cumulative profit of one unit. See Table 8.4 for clarification.

The system can be modeled by a Markov chain. Let Y_n denote the number of units the gambler needs to reach his one-unit goal following the nth coup, and let Z_n denote the gambler's bet size at coup $n + 1$. Then $\{(Y_n, Z_n)\}_{n \ge 0}$ is a Markov chain in the state space

$$S := \{(0,0)\} \cup \{(i,j) \in \mathbf{N} \times \mathbf{N} : i \ge j\} \qquad (8.53)$$

with initial state $(Y_0, Z_0) = (1, 1)$ and transition probabilities

$$P((i,j),(k,l)) := \begin{cases} p & \text{if } (k,l) = (i - j, (j + 1) \wedge (i - j)), \\ q & \text{if } (k,l) = (i + j, j), \\ 0 & \text{otherwise,} \end{cases} \qquad (8.54)$$

for all $i \ge j \ge 1$. We assume, in contrast to Example 4.2.5 on p. 132 that absorption occurs at $(0,0)$.

We now let M be a positive integer, and assume a house limit of M units. Define

$$N_M := \min\{n \ge 0 : (Y_n, Z_n) = (0,0) \text{ or } Z_n \ge M + 1\}. \qquad (8.55)$$

Table 8.4 An illustration of the Oscar system.

coup	bet size	outcome	cumulative profit
1	1	L	-1
2	1	L	-2
3	1	L	-3
4	1	L	-4
5	1	L	-5
6	1	W	-4
7	2	L	-6
8	2	L	-8
9	2	L	-10
10	2	W	-8
11	3	W	-5
12	4	W	-1
13	2	L	-3
14	2	W	-1
15	2	W	1

This is the time after which no further bets are made. We are interested in $Q(1,1)$, the probability of achieving the one-unit goal, where in general

$$Q(i,j) := P_{(i,j)}((Y_{N_M}, Z_{N_M}) = (0,0)) \qquad (8.56)$$

for all $(i,j) \in S$. This is the probability of achieving the one-unit goal starting in state (i,j). The latter probabilities satisfy the linear system

$$Q(i,j) = pQ(i-j, (j+1) \wedge (i-j)) + qQ(i+j, j), \qquad (i,j) \in S, \quad (8.57)$$

where $Q(0,0) = 1$ and $Q(i,j) = 0$ for all $(i,j) \in S$ with $j \geq M+1$.

In fact, there are other cases in which $Q(i,j) = 0$. For $j = 1, \ldots, M$, let

$$I(j) := \left(\sum_{k=j}^{M+1} k \right) - 1, \qquad (8.58)$$

and observe that if $1 \leq j \leq M$ and $i \geq I(j) + 1$, then $(0,0)$ cannot be reached from (i,j) without first visiting $\{(k,l) \in S : l \geq M+1\}$ and hence $Q(i,j) = 0$. Thus, the infinite system (8.57) is really the finite system

$$Q(i,j) = pQ(i-j, (j+1) \wedge (i-j)) + qQ(i+j, j),$$
$$j \leq i \leq I(j), \quad 1 \leq j \leq M, \qquad (8.59)$$

where $Q(0,0) = 1$ and $Q(i,j) = 0$ if $1 \leq j \leq M$ and $i \geq I(j) + 1$. Using the formulas for the sum of the first n integers and the sum of their squares $\left(\sum_{j=1}^{n} j = n(n+1)/2 \text{ and } \sum_{j=1}^{n} j^2 = n(n+1)(2n+1)/6\right)$, we find that the number of equations in the system (8.59) is equal to

$$\sum_{j=1}^{M}(I(j) - j + 1) = \frac{1}{2}\sum_{j=1}^{M}[(M+1)(M+2) - j(j+1)]$$

$$= \frac{M(M+1)(M+2)}{3}. \tag{8.60}$$

For example, if $M = 2$, (8.59) becomes

$$Q(1,1) = p + qQ(2,1)$$
$$Q(2,1) = pQ(1,1) + qQ(3,1) \qquad Q(2,2) = p + qQ(4,2)$$
$$Q(3,1) = pQ(2,2) + qQ(4,1) \qquad Q(3,2) = pQ(1,1) \tag{8.61}$$
$$Q(4,1) = pQ(3,2) + qQ(5,1) \qquad Q(4,2) = pQ(2,2)$$
$$Q(5,1) = pQ(4,2)$$

and the relevant solution is

$$Q(1,1) = (1 - pq - p^2 q^3)^{-1}\left(p + \frac{p^2 q^2 + p^3 q^4}{1 - pq}\right). \tag{8.62}$$

This can be seen by substituting the second column of equations into the first (after solving explicitly for $Q(2,2)$ and $Q(4,2)$), then reducing the new first column by repeated substitution to an equation involving $Q(1,1)$ alone.

In fact, the same argument applies to (8.59) in general. Indeed, it is possible to express the probabilities $Q(j,j), Q(j+1,j), \ldots, Q(I(j),j)$ in terms of $Q(1,1), \ldots, Q(j-1,j-1)$, successively for $j = M, M-1, \ldots, 1$. This leads to a recursive algorithm that is well suited to numerical computation for large M. For example, if $M = 500$, the system has 41,917,000 equations. Nevertheless, a numerical solution is feasible, and when $p = 244/495$, it has been found that $Q(1,1) \approx 0.999740807832$.

How large must F_0 be to ensure that the gambler can make any bet that is called for, up to the time that the goal is achieved or the system is aborted on account of the house limit? At first glance, it might seem that unlimited credit is required, because, starting with an arbitrarily long sequence of losses, the bet size remains at one unit. However, if by aborting the system on account of the house limit we mean not only terminating play when a bet exceeding M units is called for, but also terminating play when the gambler is so far behind that there is no possibility of recovery without eventually having to make a bet exceeding M units, then the answer to the question, by virtue of the discussion above involving (8.58), is $F_0 \geq I(1) = \binom{M+2}{2} - 1$.

Under the same assumptions, we can evaluate several expectations of interest. With the preceding paragraph in mind, we redefine

$$N_M := \min\{n \geq 0 : (Y_n, Z_n) = (0,0) \text{ or } Y_n \geq I(Z_n) + 1\}. \qquad (8.63)$$

This is the first time after which no further bets are made. (We do not include $Y_n \geq M + 1$ as one of the conditions of the definition because one of the two conditions we do include must occur first.) Consider the linear system

$$R(i, j) = A(i, j) + pR(i - j, (j + 1) \wedge (i - j)) + qR(i + j, j),$$
$$j \leq i \leq I(j), \quad 1 \leq j \leq M, \qquad (8.64)$$

where $R(0,0) = 0$ and $R(i, j) = 0$ if $1 \leq j \leq M$ and $i \geq I(j) + 1$. Then

$$R(1, 1) = E[Y_0 - Y_{N_M}] \quad \text{if} \quad A(i, j) := j(2p - 1),$$
$$R(1, 1) = E[N_M] \quad \text{if} \quad A(i, j) := 1, \qquad (8.65)$$
$$R(1, 1) = E\left[\sum_{n=0}^{N_M - 1} Z_n\right] \quad \text{if} \quad A(i, j) := j,$$

representing the gambler's expected cumulative profit, the expected number of coups, and the expected total amount bet, respectively.

For example, if $M = 2$, the relevant solutions are

$$R(1, 1) = (1 - pq - p^2 q^3)^{-1} \qquad (8.66)$$
$$\cdot \left(1 + q + q^2 + q^3 + pq^3 + q^4 + \frac{pq^2(1 + q + q^2 + pq^2)}{1 - pq}\right)$$

for the expected number of coups and

$$R(1, 1) = (1 - pq - p^2 q^3)^{-1} \qquad (8.67)$$
$$\cdot \left(1 + q + q^2 + q^3 + 2pq^3 + q^4 + \frac{2pq^2(1 + q + q^2 + pq^2)}{1 - pq}\right)$$

for the expected total amount bet. For large M, the approach suggested for (8.59) applies here equally well.   ♠

Example 8.1.5. *D'Alembert system.* This system was briefly mentioned in Example 1.5.8 on p. 45. The first bet is one unit. After a loss, the bet size is increased by one unit, and after a win, the bet size is decreased by one unit. The system terminates after an initial win. After an initial loss, the system terminates as soon as the number of wins equals the number of losses or, equivalently, as soon as the system calls for a bet of one unit. See Table 8.5 for clarification.

The idea underlying this system is that a win followed by a loss, *or* a loss followed by a win, leaves the bet size unchanged and increases the gambler's

Table 8.5 An illustration of the d'Alembert system. With an initial loss, betting stops when the number of wins equals the number of losses or, equivalently, when a bet of one unit is called for.

coup	bet size	outcome	cumulative profit
1	1	L	-1
2	2	L	-3
3	3	L	-6
4	4	W	-2
5	3	W	1
6	2	L	-1
7	3	L	-4
8	4	L	-8
9	5	W	-3
10	4	W	1
11	3	W	4
12	2	L	2
13	3	W	5
14	2	W	7

cumulative profit by one unit. Thus, the order of wins and losses is not important (except to the extent that it affects the time at which betting stops), and if the number of wins ever equals the number of losses, the goal is achieved and the gambler's cumulative profit will equal the number of wins.

Before defining the sequence of bet sizes $B_1, B_2, \ldots$, we define a preliminary sequence $B_1^*, B_2^*, \ldots$ by $B_1^* := 1$ and $B_n^* := B_{n-1}^* - X_{n-1}$ for each $n \geq 2$, so that

$$B_n^* = 1 - \sum_{l=1}^{n-1} X_l, \qquad n \geq 1. \tag{8.68}$$

These would be the bet sizes if there were no stopping rule. We let

$$N := \begin{cases} 1 & \text{if } X_1 = 1, \\ \min\{n \geq 2 : B_{n+1}^* = 1\} & \text{if } X_1 = -1, \end{cases} \tag{8.69}$$

where $\min \varnothing := \infty$, and define

$$B_n := B_n^* 1_{\{n \leq N\}}, \qquad n \geq 1. \tag{8.70}$$

These are the correct bet sizes, and N is the number of coups at which bets are made.

We begin by deriving a formula for F_n on the event $\{n \leq N\}$. On this event, letting $S_n := \sum_{l=1}^{n} X_l$, we have

$$F_n = F_0 + \sum_{l=1}^{n} B_l X_l$$

$$= F_0 + \sum_{l=1}^{n} B_l^* X_l$$

$$= F_0 + \sum_{l=1}^{n} \left(1 - \sum_{k=1}^{l-1} X_k\right) X_l$$

$$= F_0 + \sum_{l=1}^{n} X_l - \sum_{1 \leq k < l \leq n} X_k X_l$$

$$= F_0 + \sum_{l=1}^{n} X_l - \frac{1}{2}\left\{\left(\sum_{l=1}^{n} X_l\right)^2 - \sum_{l=1}^{n} X_l^2\right\}$$

$$= F_0 + S_n - \frac{1}{2}(S_n)^2 + \frac{1}{2}n. \tag{8.71}$$

We are interested in F_N on the event $\{N < \infty\}$. Consider two cases. If $N = 1$, then $S_N = X_1 = 1$, whereas if $N \geq 2$, then $B_{N+1}^* = 1$ and therefore $S_N = 0$. Consequently,

$$F_N - F_0 = 1 \vee (N/2) \quad \text{on} \quad \{N < \infty\}, \tag{8.72}$$

as suggested in the second paragraph of this example.

Now let M be a positive integer, and assume a house limit of M units. Then the probability of the event E that the goal is achieved can be obtained from the gambler's ruin formula (Theorem 7.1.1 on p. 242) after conditioning on the result of the first coup. If the first coup is lost and $M \geq 2$, then for E to occur the gambler must win one more coup than he loses before losing $M - 1$ more coups than he wins. Indeed, the former event equalizes wins and losses, whereas the latter event calls for a bet of size $M + 1$. Therefore,

$$P(E) = \begin{cases} p + q((q/p)^{M-1} - 1)/((q/p)^M - 1) & \text{if } p \neq \frac{1}{2}, \\ \frac{1}{2} + \frac{1}{2}(M-1)/M & \text{if } p = \frac{1}{2}, \end{cases} \tag{8.73}$$

which is also valid for $M = 1$. In deriving this formula, we have implicitly assumed that the gambler's initial fortune F_0 is sufficient to permit any bet that the system calls for, up to the time that the goal is achieved or the system is aborted on account of the house limit. What requirement does this impose on F_0? The worst-case scenario is $M - 1$ consecutive losses, which would require an initial fortune of

$$F_0 \geq \sum_{j=1}^{M} j = \binom{M+1}{2}. \qquad (8.74)$$

Let N_M be the number of coups at which bets are made. Arguing as above (but with Theorem 7.1.4 on p. 244 in place of Theorem 7.1.1 on p. 242), we have

$$\mathrm{E}[N_M] = p + q\left\{1 + \frac{M}{q-p}\left(\frac{M-1}{M} - \frac{(q/p)^{M-1}-1}{(q/p)^M - 1}\right)\right\} \qquad (8.75)$$

if $p \neq \frac{1}{2}$, and $\mathrm{E}[N_M] = (M+1)/2$ if $p = \frac{1}{2}$. As for the gambler's expected cumulative profit, (8.71) yields

$$\mathrm{E}[F_{N_M} - F_0] = \mathrm{E}\left[S_{N_M} - \frac{1}{2}(S_{N_M})^2 + \frac{1}{2}N_M\right] \qquad (8.76)$$

$$= p\left(1 - \frac{1}{2}(1)^2\right) + (1 - \mathrm{P}(E))\left(-M - \frac{1}{2}(-M)^2\right) + \frac{1}{2}\mathrm{E}[N_M],$$

and we need only substitute (8.73) and (8.75). Similarly, the gambler's conditional expected cumulative loss, given that the system is aborted, requires (8.71) together with Corollary 7.1.8 on p. 247:

$$-\mathrm{E}[F_{N_M} - F_0 \mid E^c]$$

$$= M + \frac{1}{2}(-M)^2 - \frac{1}{2}\mathrm{E}[N_M \mid E^c] \qquad (8.77)$$

$$= M + \frac{1}{2}M^2 - \frac{1}{2}\left(1 + \frac{M}{q-p}\frac{(q/p)^M + 1}{(q/p)^M - 1} - \frac{1}{(q-p)^2}\right)$$

if $p \neq \frac{1}{2}$, and $-\mathrm{E}[F_{N_M} - F_0 \mid E^c] = M + (M^2-1)/3$ if $p = \frac{1}{2}$.

Finally, in the case of unlimited credit and no house limit, we conclude by letting $M \to \infty$ in (8.73) that $\mathrm{P}(N < \infty) = 1$ if and only $p \geq \frac{1}{2}$, and by letting $M \to \infty$ in (8.75) that $\mathrm{E}[N] < \infty$ if and only if $p > \frac{1}{2}$. ♠

Example 8.1.6. *Blundell system.* This system depends on an even positive-integer parameter K. (In the original formulation, $K = 10$.) The first bet is one unit. The gambler continues to bet one unit until either he achieves a one-unit cumulative profit, in which case he quits, or his deficit reaches K units, in which case his bet size is increased to two units. In general, if his deficit reaches $\binom{m}{2}K$ units ($m \geq 2$) and his bet size is increased to m units, he continues to bet m units until either he recoups all losses, in which case he recommences with a bet of one unit, or his deficit increases by mK units, thereby reaching $\binom{m+1}{2}K$ units, in which case his bet size is increased to $m+1$ units. See Table 8.6 for clarification.

Define $u(W, L)$ to be the probability of winning W units before losing L units when betting one unit at each coup in a sequence of independent

Table 8.6 An illustration of the Blundell system. Here $K = 10$.

coup	bet size	outcome	cumulative profit
1	1	L	−1
2	1	L	−2
⋮	⋮	⋮	⋮
10	1	L	−10
11	2	L	−12
12	2	L	−14
⋮	⋮	⋮	⋮
20	2	L	−30
21	3	L	−33
22	3	W	−30
23	3	W	−27
⋮	⋮	⋮	⋮
31	3	W	−3
32	3	W	0
33	1	L	−1
34	1	W	0
35	1	W	1

coups, each with win probability p and loss probability $q := 1 - p$. Recalling the gambler's ruin formula (Theorem 7.1.1 on p. 242), we see that $u(W, L)$ is given by (7.4) or (7.5) on p. 242. Let M be a positive integer, and assume a house limit of M units. If the system is aborted because it calls for a bet that exceeds the house limit, the gambler's loss will be $\binom{M+1}{2}K$ units. We further assume that $F_0 \geq \binom{M+1}{2}K$.

We claim that the probability P of achieving the one-unit goal satisfies the equation

$$
\begin{aligned}
P = \; & u(1, K) + (1 - u(1, K))u(K/2, K)P \\
& + (1 - u(1, K))(1 - u(K/2, K))u(K, K)P \\
& + (1 - u(1, K))(1 - u(K/2, K))(1 - u(K, K))u(3K/2, K)P \\
& \;\;\vdots \\
& + (1 - u(1, K))\left[\prod_{j=1}^{M-2}(1 - u(jK/2, K))\right]u((M-1)K/2, K)P.
\end{aligned} \tag{8.78}
$$

This can be seen as follows. Let A_1 be the event that the goal is achieved with the bet size never exceeding one unit. For $2 \leq m \leq M$, let A_m be the event that m is the maximum bet size prior to the first recommencement of the system, *and* the goal is achieved. The events $A_1, A_2, \ldots, A_M$ are mutually exclusive and their union is the event that the goal is achieved, so $P = P(A_1) + P(A_2) + \cdots + P(A_M)$, and it suffices to evaluate these probabilities. For A_2 to occur, for example, the gambler must first lose K units before winning one unit while betting one unit; he must then win K units before losing $2K$ units while betting two units; finally, having recommenced the system, he must achieve his one-unit goal. Therefore, by Theorem 1.2.4 on p. 18, $P(A_2) = (1 - u(1, K))u(K/2, K)P$. The other terms are treated similarly, and the claim is established.

It follows from (8.78) that

$$P = \frac{u(1, K)}{1 - (1 - u(1, K)) \sum_{l=1}^{M-1} \left[\prod_{j=1}^{l-1} (1 - u(jK/2, K)) \right] u(lK/2, K)}, \quad (8.79)$$

where empty sums are 0 and empty products are 1.

Let us also define $v(W, L)$ to be the expected number of coups required to win W units or lose L units when betting one unit at each coup in a sequence of independent coups, each with win probability p and loss probability $q := 1 - p$. Recalling Theorem 7.1.4 on p. 244, we see that $v(W, L)$ is given by (7.17) or (7.18) on p. 244.

Consider the equation

$$\begin{aligned}
E = {} & A(1)v(1, K) \\
& + A(2)(1 - u(1, K))v(K/2, K) + (1 - u(1, K))u(K/2, K)E \\
& + A(3)(1 - u(1, K))(1 - u(K/2, K))v(K, K) \\
& \quad + (1 - u(1, K))(1 - u(K/2, K))u(K, K)E \\
& \qquad \vdots \\
& + A(M)(1 - u(1, K)) \left[\prod_{j=1}^{M-2} (1 - u(jK/2, K)) \right] v((M-1)K/2, K) \\
& \quad + (1 - u(1, K)) \left[\prod_{j=1}^{M-2} (1 - u(jK/2, K)) \right] u((M-1)K/2, K)E.
\end{aligned} \qquad (8.80)$$

Then E is the gambler's expected cumulative loss if $A(j) := j(1 - 2p)$, it is the expected number of coups if $A(j) := 1$, and it is the expected total amount bet if $A(j) := j$. The solution of (8.80) is

$$\begin{aligned}
E = {} & \Bigg\{ A(1)v(1, K) \\
& + \sum_{l=2}^{M} A(l)(1 - u(1, K)) \left[\prod_{j=1}^{l-2} (1 - u(jK/2, K)) \right] v((l-1)K/2, K) \Bigg\}
\end{aligned} \qquad (8.81)$$

$$\cdot \left\{ 1 - (1 - u(1, K)) \sum_{l=1}^{M-1} \left[\prod_{j=1}^{l-1} (1 - u(jK/2, K)) \right] u(lK/2, K) \right\}^{-1},$$

where again empty sums are 0 and empty products are 1.

We now address the question of whether the gambler is assured of achieving his goal, assuming unlimited credit and no house limit. Let N be the number of coups needed to achieve the goal, with $N = \infty$ if the goal is never achieved. If we denote (8.79) by P_M, then $P(N < \infty) = \lim_{M \to \infty} P_M$, so the question is equivalent to the question of whether $\lim_{M \to \infty} P_M = 1$. To answer this question, notice that the sum in (8.79) (and (8.81)) has the form

$$\sum_{l=1}^{M-1} \left[\prod_{j=1}^{l-1} (1 - q_j) \right] q_l. \tag{8.82}$$

Since this sum can be interpreted as the probability of at least one success in $M - 1$ independent Bernoulli trials, with respective success probabilities $q_1, q_2, \ldots, q_{M-1}$, it can be rewritten in terms of the complementary probability as

$$1 - \prod_{j=1}^{M-1} (1 - q_j). \tag{8.83}$$

Now the latter will converge to 1 as $M \to \infty$ if and only if $\sum_{j=1}^{\infty} q_j = \infty$ (Theorem A.2.11 on p. 750). With $q_j = u(jK/2, K)$ for each $j \geq 1$, this happens if and only if $p \geq \frac{1}{2}$. In particular, if $p < \frac{1}{2}$, there is again positive probability of failing to achieve the one-unit goal, even allowing unlimited credit and no house limit.   ♠

We now briefly compare the six betting systems discussed in Examples 8.1.1–8.1.6. First assume $\frac{1}{3} \leq p < \frac{1}{2}$, as is typically the case in a gambling casino. If we also assume unlimited credit and no house limit (as is *not* typical), the martingale, Fibonacci, and Labouchere systems are infallible, while the other three systems are not (see Table 8.7).

In the more realistic situation in which there is a house limit of M units, with M a positive integer, each system has a capitalization requirement to ensure that funds will be available to make any bet that the system calls for prior to the goal being achieved or the system being aborted on account of the house limit. For the martingale system, $2M - 1$ units suffice (more precisely, the requirement is (8.12)). For the Fibonacci system, $\lfloor (\phi_1 + 1)M - \frac{3}{2} \rfloor$ units suffice, where ϕ_1 is as in (8.24) (more precisely, the requirement is (8.34)). For each of the other systems, at least $M(M + 1)/2$ units are needed (see Table 8.7).

In Table 8.8 we compare the systems for a specified game ($p = 244/495$) and several choices for the house limit ($M = 5, 25, 100, 500$). We exclude the Labouchere system, which is not amenable to these kinds of calculations. Let us denote the event that the goal is achieved by A, the gambler's loss by

Table 8.7 Comparison of six betting systems. (N is the number of coups needed to complete the system, p is the probability of a win at a given coup, and F_0 is the gambler's initial betting capital.)

betting system	win goal	req.[1] for $N < \infty$ a.s.	req.[1] for $E[N] < \infty$	required F_0 when house limit is M
martingale	1	$p > 0$	$p > 0$	$2^{1+\lfloor \log_2 M \rfloor} - 1$
Fibonacci	1	$p \geq \frac{1}{3}$	$p > \frac{1}{3}$	$f_{\max\{l \geq 1 : f_l \leq M\}+2} - 2$
Labouchere[2]	1	$p \geq \frac{1}{3}$	$p > \frac{1}{3}$	$\binom{M+1}{2}$
Oscar	1	$p \geq \frac{1}{2}$	$p \geq \frac{1}{2}$	$\binom{M+2}{2} - 1$
d'Alembert	$1 \vee (N/2)$	$p \geq \frac{1}{2}$	$p > \frac{1}{2}$	$\binom{M+1}{2}$
Blundell	1	$p \geq \frac{1}{2}$	$p > \frac{1}{2}$	$\binom{M+1}{2}K$

[1]assumes unlimited credit and no house limit   [2]assumes initial list (1)

L, the number of coups by N_M, and the total amount bet by B. The odds against a loss are given by $O := (1 - P(A))^{-1} - 1$ to 1. Table 8.8 evaluates O, $E[L \mid A^c]$, $E[N_M]$, and $E[B]$ in each of the cases mentioned. There is some redundancy here, in that

$$(-1)\frac{O}{O+1} + E[L \mid A^c]\frac{1}{O+1} = E[L] = E[B](1 - 2p). \tag{8.84}$$

The first equality assumes a goal of one unit, so it applies to all except the d'Alembert system. The justification of the second equality, which applies to all six systems, is postponed to the next section. Of course, the gambler wants to maximize O while minimizing $E[L \mid A^c]$, but because the right side of (8.84) is necessarily positive (recall that $p = 244/495$), it must be the case that $E[L \mid A^c] > O$.

If one were to consider only the odds against a loss, the clear choice among the five systems considered here would be the Blundell system for small M and the Oscar system for large M. The odds figure by itself, however, provides an incomplete picture. Of equal importance is the conditional expected loss, given that the system is aborted. The greater the odds against a loss, the greater the amount of capital one must risk to achieve those odds. (This is a rule of thumb to which exceptions exist.) A combination of the two quantities is the gambler's expected loss (see (8.84)), or equivalently his expected total amount bet. Using this criterion, which is perhaps *not* the criterion that most gamblers would select, the martingale system performs best. Indeed, it has certain optimality properties, as we will see in Chapter 9.

Table 8.8 Comparison of five betting systems, assuming a particular game ($p = 244/495$) and a house limit of M units. The system is aborted if it calls for a bet exceeding M units, or if recovery is impossible without making a bet exceeding M units. The condition on F_0 in Table 8.7 is assumed to be satisfied. Results are exact, except that numbers with decimal points are rounded to six significant digits. The Labouchere system is excluded because it is not amenable to exact calculations for large M.

system	$M = 5$	$M = 25$	$M = 100$	$M = 500$
\multicolumn{5}{c}{odds against having to abort the system}			
martingale	6.66998 : 1	28.8303 : 1	115.017 : 1	450.215 : 1
Fibonacci	10.2918 : 1	47.0865 : 1	193.762 : 1	778.133 : 1
Oscar	13.3971 : 1	106.751 : 1	628.798 : 1	3,857.14 : 1
d'Alembert	8.32569 : 1	34.8477 : 1	65.5349 : 1	69.7142 : 1
Blundell[1]	76.6370 : 1	250.118 : 1	264.312 : 1	264.313 : 1
\multicolumn{5}{c}{conditional expected loss, given system is aborted}			
martingale	7	31	127	511
Fibonacci	11	53	231	985
Oscar	14.9195	139.114	1,082.70	12,677.8
d'Alembert	13.0011	233.851	3,619.84	110,321.
Blundell[1]	150	3,250	50,500	1,252,500
\multicolumn{5}{c}{expected number of coups}			
martingale	1.76419	1.96068	2.01120	2.02419
Fibonacci	2.36811	3.00569	3.24728	3.33079
Oscar	4.67097	6.20476	6.88647	7.38865
d'Alembert	3.05655	14.4587	71.4244	465.143
Blundell[1]	28.5345	79.5223	278.035	1,344.16
\multicolumn{5}{c}{expected total amount bet}			
martingale	3.04262	5.14343	7.30407	9.52625
Fibonacci	4.43481	8.69619	13.5203	18.7753
Oscar	7.47751	21.2394	50.9640	161.671
d'Alembert	7.19848	137.118	2,877.57	109,037.
Blundell[1]	66.8215	844.759	13,389.4	333,761.

[1] assumes $K = 10$

It should be mentioned that Table 8.8 compares several betting systems after equalizing the maximum allowable bet. Different results would be obtained if instead we were to equalize initial betting capital and abort the system as soon as betting capital was exhausted. We prefer the present approach because the house betting limit is more rigid than the limit on the gambler's bankroll, due to the easy availability of credit. In the problems we describe other systems of a similar nature (providing the gambler with a small win with high probability or a large loss with low probability), as well as "reversed" versions of the systems considered here, which result in a small loss with high probability and a large win with low probability.

8.2 Conservation of Fairness

The *conservation of fairness* principle says that no betting system can turn a sequence of subfair wagers into a superfair one. We begin by formulating a very simple version of this result, well suited to the examples of the preceding section. We later provide a much more general version of the same result.

As in Section 8.1, let X have distribution $P(X = 1) = p$ and $P(X = -1) = q := 1 - p$, where $0 < p < 1$, and let $X_1, X_2, \ldots$ be a sequence of i.i.d. random variables with common distribution that of X, with X_n representing the gambler's profit per unit bet at coup n. Consider the betting system $B_1, B_2, \ldots$, with B_n representing the amount bet at coup n and depending only on the results of the previous coups, that is,

$$B_1 = b_1 \geq 0, \qquad B_n = b_n(X_1, \ldots, X_{n-1}) \geq 0, \quad n \geq 2, \qquad (8.85)$$

where b_1 is a constant and b_n is a nonrandom function of $n - 1$ variables for each $n \geq 2$. The gambler's fortune F_n after n coups is given by (8.3) on p. 276, that is,

$$F_n = F_0 + \sum_{l=1}^{n} B_l X_l, \qquad n \geq 1, \qquad (8.86)$$

F_0 being his initial fortune (a constant).

Theorem 8.2.1. *Under the above conditions, assume the game is subfair (or at best fair), so $p \leq \frac{1}{2}$. Assume uniformly bounded bet sizes, that is, there exists a constant C such that $0 \leq B_n \leq C$ for all $n \geq 1$. Let N be a stopping time with respect to $\{X_n\}_{n \geq 1}$ such that $P(N < \infty) = 1$.*
   (a) *If (8.86) satisfies $F_n \geq 0$ for all $n \geq 0$, then*

$$E[F_N] \leq F_0. \qquad (8.87)$$

   (b) *Assume that $E[N] < \infty$. Then (8.87) holds; more generally,*

$$E[F_N - F_0] = E\left[\sum_{l=1}^{N} B_l X_l\right] = E\left[\sum_{l=1}^{N} B_l\right](2p - 1). \tag{8.88}$$

Remark. Conclusion (8.87) says that, in a subfair or fair game with even-money payoffs played repeatedly at independent coups with uniformly bounded bet sizes, the gambler's expected cumulative profit cannot be positive, regardless of the betting system used. Part (a) requires that the gambler's fortune never become negative and the stopping time be finite with probability 1. The first of these two conditions can be assured if the gambler never bets more than he has (i.e., $B_n \leq F_{n-1}$ for all $n \geq 1$). Part (b) eliminates this nonnegativity requirement on the gambler's fortune at the expense of requiring that $E[N]$ be finite. It includes the additional conclusion (8.88) that the gambler's expected cumulative profit is equal to the product of his expected total amount bet and his expected profit from a single one-unit bet.

Proof. If $p < \frac{1}{2}$, then $\{F_n\}_{n\geq 0}$ is a supermartingale with respect to $\{X_n\}$, while if $p = \frac{1}{2}$, then it is a martingale. To see this, note that, for $p \leq \frac{1}{2}$,

$$\begin{aligned}
E[F_n \mid X_1, \ldots, X_{n-1}] &= E[F_{n-1} + B_n X_n \mid X_1, \ldots, X_{n-1}] \\
&= F_{n-1} + E[B_n X_n \mid X_1, \ldots, X_{n-1}] \\
&= F_{n-1} + B_n E[X_n \mid X_1, \ldots, X_{n-1}] \\
&= F_{n-1} + B_n E[X_n] \\
&= F_{n-1} + B_n(2p - 1) \\
&\leq F_{n-1} \tag{8.89}
\end{aligned}$$

with equality if $p = \frac{1}{2}$. Part (a) follows from the optional stopping theorem (Theorem 3.2.2 on p. 100, assumption (a)). The same conclusion follows from the hypotheses of part (b) using the same theorem (assumption (c)). Conclusion (8.88) follows by noting that

$$M_n := \sum_{l=1}^{n} B_l[X_l - (2p - 1)] \tag{8.90}$$

is a martingale with respect to $\{X_n\}_{n\geq 1}$, and hence by the optional stopping theorem (Theorem 3.2.2 on p. 100, assumption (c)) we have $E[M_N] = E[M_0] = 0$. ♠

We now apply this result to the examples of the preceding section.

Example 8.2.2. *Six betting systems and the conservation of fairness.* Let us first consider the case of no house limit. We defined a stopping time N that had the same interpretation in each of the six systems: N is the number of coups the gambler needs to achieve his goal. If the goal is never achieved, then $N := \infty$. We have seen that $F_N > F_0$ on $\{N < \infty\}$ in each system. Table 8.7 on p. 296 gives necessary and sufficient conditions for $P(N < \infty) = 1$. Even with these conditions satisfied, conclusion (8.87) of Theorem 8.2.1 fails. What is the explanation?

The main difficulty is that the theorem assumes uniformly bounded bet sizes, and that assumption is not met by any of the six systems. A secondary difficulty is that part (a) of the theorem is not applicable, because, no matter how large F_0 is, there is positive probability that F_n will eventually become negative. This is why we assumed unlimited credit when discussing the case of no house limit. Part (b) requires $E[N] < \infty$, for which necessary and sufficient conditions can be found in Table 8.7 on p. 296. Recalling that $p \leq \frac{1}{2}$ in the theorem, we see that part (b) would not be applicable to the d'Alembert or Blundell systems (or to the Oscar system unless $p = \frac{1}{2}$) even if the uniformly-bounded-bet-size condition could be weakened.

We now turn to the case of a house limit of M units, where M is a positive integer. Let us define the stopping time N_M to be the time after which no further bets are made because the gambler has achieved his goal or the system has been aborted after calling for a bet exceeding the house limit of M units. If the original sequence of bet sizes $B_1, B_2, \ldots$ is replaced by

$$B_n^\circ := B_n 1_{\{n \leq N_M\}}, \tag{8.91}$$

then the sequence $B_1^\circ, B_2^\circ, \ldots$ satisfies (8.85) as well as $0 \leq B_n^\circ \leq M$ for all $n \geq 1$. Part (a) of the theorem applies as long as the initial capitalization requirement is met (see Table 8.7 on p. 296). Part (b) also applies and tells us that (8.88) holds with $N = N_M$, and this finally implies the second equality in (8.84) on p. 296. Here we have used the fact that $E[N_M] < \infty$, which is a consequence of Lemma 4.2.6 on p. 133.

Recall that, in Example 8.1.4 on p. 286, there were two definitions of N_M (namely, (8.55) on p. 286 and (8.63) on p. 289). Theorem 8.2.1 applies in either case, but the second definition (in which the gambler quits as soon as the system is no longer viable) provides more-useful results.   ♠

Before generalizing this result, we want to observe that stronger conclusions are provable under stronger assumptions. We consider two such results. The first says that betting on a random subsequence of coups offers no advantages.

Theorem 8.2.3. *Under the assumptions and notation in the paragraph preceding Theorem 8.2.1, suppose also that $B_n \in \{0,1\}$ for all $n \geq 1$ and that $P(B_n = 1 \text{ i.o.}) = 1$. For each $n \geq 1$, define $N_n = \min\{l \geq 1 : B_1 + \cdots + B_l = n\}$ and $Y_n = X_{N_n}$. Then $Y_1, Y_2, \ldots$ is a sequence i.i.d. random variables with common distribution that of X.*

Remark. Here the gambler decides which coups to bet on based on the results of previous coups. If he bets at all, he bets one unit. We assume that there is no last bet. The nth bet is made at coup N_n, and Y_n is the profit from it. The theorem says that the sequence of results $Y_1, Y_2, \ldots$ is probabilistically identical to the sequence $X_1, X_2, \ldots$ that would have resulted from betting one unit at every coup.

Proof. Here it will be convenient to temporarily introduce the notation $p(1) := p$, $p(-1) := q$. Let $n \geq 1$ be arbitrary, and let $x_1, \ldots, x_n \in \{-1, 1\}$. Then

$$
\begin{aligned}
&\mathrm{P}(Y_1 = x_1, \ldots, Y_n = x_n) \\
&\quad = \mathrm{P}(X_{N_1} = x_1, \ldots, X_{N_n} = x_n) \\
&\quad = \sum_{1 \leq k_1 < \cdots < k_n < \infty} \cdots \sum \mathrm{P}(X_{k_1} = x_1, \ldots, X_{k_n} = x_n, N_1 = k_1, \ldots, N_n = k_n) \\
&\quad = \sum_{1 \leq k_1 < \cdots < k_n < \infty} \cdots \sum \mathrm{P}(X_{k_1} = x_1, \ldots, X_{k_{n-1}} = x_{n-1}, \\
&\qquad\qquad\qquad\qquad\qquad N_1 = k_1, \ldots, N_n = k_n) p(x_n) \\
&\quad = \sum_{1 \leq k_1 < \cdots < k_{n-1} < \infty} \cdots \sum \mathrm{P}(X_{k_1} = x_1, \ldots, X_{k_{n-1}} = x_{n-1}, \\
&\qquad\qquad\qquad\qquad\qquad N_1 = k_1, \ldots, N_{n-1} = k_{n-1}) p(x_n) \\
&\quad = \mathrm{P}(X_{N_1} = x_1, \ldots, X_{N_{n-1}} = x_{n-1}) p(x_n) \\
&\quad = \mathrm{P}(Y_1 = x_1, \ldots, Y_{n-1} = x_{n-1}) p(x_n) \\
&\quad \vdots \\
&\quad = p(x_1) \cdots p(x_n) \\
&\quad = \mathrm{P}(X_1 = x_1, \ldots, X_n = x_n),
\end{aligned} \tag{8.92}
$$

where the third equality uses the fact that

$$
\{N_n = k_n\} = \{B_1 + \cdots + B_{k_n} = n, B_{k_n} = 1\}, \tag{8.93}
$$

which depends only on $X_1, \ldots, X_{k_n - 1}$, and the fourth equality results from summing over all k_n such that $k_n \geq k_{n-1} + 1$. ♠

The next theorem says that the persistent gambler goes broke with probability 1. This assumes of course that $p \leq \frac{1}{2}$.

Theorem 8.2.4. *Under the assumptions and notation in the paragraph preceding Theorem 8.2.1, suppose also that $p \leq \frac{1}{2}$, that*

$$
F_n \geq 0 \text{ for all } n \geq 0, \tag{8.94}
$$

and that for some $\delta > 0$,

$$
B_n \in \{0\} \cup [\delta, \infty), \qquad n \geq 0. \tag{8.95}
$$

Then $\mathrm{P}(B_n \geq \delta \text{ i.o.}) = 0$, that is, with probability 1 there is a last bet. Suppose that the last bet occurs at coup N (which need not be a stopping time). Then $\mathrm{E}[F_N] \leq F_0$.

Remark. Hypothesis (8.94) says that the gambler never bets what he does not have, while hypothesis (8.95) says that there is a minimum allowable nonzero bet. If we assume also that the gambler bets whenever he can (i.e., $B_n \geq \delta$ whenever $F_{n-1} \geq \delta$), it must hold that $N = \min\{n \geq 1 : F_n < \delta\}$ (in which case N *is* a stopping time) and $\mathrm{P}(F_N < \delta) = 1$. This is the sense in which the persistent gambler goes broke with probability 1.

Proof. First, $\{F_n\}_{n \geq 0}$ is a nonnegative supermartingale with respect to $\{X_n\}_{n \geq 1}$, so it converges with probability 1 to a finite limit. It follows that

$$\mathrm{P}(B_n \geq \delta \text{ i.o.}) = \mathrm{P}(|B_n X_n| \geq \delta \text{ i.o.}) = \mathrm{P}(|F_n - F_{n-1}| \geq \delta \text{ i.o.}) = 0. \quad (8.96)$$

With N defined as in the statement of the theorem,

$$\begin{aligned}
\mathrm{E}[F_N] &= \lim_{n \to \infty} \mathrm{E}[F_N 1_{\{N \leq n\}}] \\
&= \lim_{n \to \infty} \mathrm{E}[F_n 1_{\{N \leq n\}}] \leq \lim_{n \to \infty} \mathrm{E}[F_n] \leq F_0, \quad (8.97)
\end{aligned}$$

where the first equality uses (8.94) and the monotone convergence theorem (Theorem A.2.3 on p. 748), the second equality uses the fact $F_n = F_N$ if $n \geq N$, the first inequality uses (8.94), and the second inequality uses the fact that $\mathrm{E}[F_n]$ is nonincreasing in $n \geq 0$.   ♠

We would now like to generalize Theorem 8.2.1 in three directions. First, we want to eliminate the assumption of even-money payoffs, or indeed the assumption that only two results, win or lose, are possible at each coup. Second, we want to weaken the assumption of independent coups. Third, we want to allow multiple betting opportunities at each coup, as is typical in games such as roulette and craps.

Let $Y_1, Y_2, \ldots$ be a sequence of jointly distributed discrete random variables (or random vectors), not necessarily independent or identically distributed, with Y_n describing in some sense the randomness of the nth coup. (For example, in craps Y_n would be the result of the nth roll of the dice.) Suppose that at each coup the gambler has d subfair betting opportunities. Let $X_{n,i}$ be the gambler's profit from a one-unit bet on the ith betting opportunity at the nth coup, and let $B_{n,i}$ be the amount bet on the ith betting opportunity at the nth coup. Put

$$\boldsymbol{X}_n := (X_{n,1}, \ldots, X_{n,d}), \qquad \boldsymbol{B}_n := (B_{n,1}, \ldots, B_{n,d}). \quad (8.98)$$

We assume that

$$\boldsymbol{B}_1 = \boldsymbol{b}_1 \geq \boldsymbol{0}, \qquad \boldsymbol{B}_n = \boldsymbol{b}_n(Y_1, \ldots, Y_{n-1}) \geq \boldsymbol{0}, \quad n \geq 2. \quad (8.99)$$

where $\boldsymbol{b}_1$ is a constant and $\boldsymbol{b}_n$ is a nonrandom function of $n-1$ variables for each $n \geq 2$. (Vector inequalities hold componentwise.) The gambler's fortune F_n after n coups is given by

$$F_n = F_0 + \sum_{l=1}^{n} \boldsymbol{B}_l \cdot \boldsymbol{X}_l, \qquad n \geq 1, \tag{8.100}$$

F_0 being his initial fortune (a positive constant). Here $\boldsymbol{b} \cdot \boldsymbol{x} := \sum_{i=1}^{d} b_i x_i$ is the inner product.

Theorem 8.2.5. *Assume all betting opportunities are subfair (or at best fair), so that*

$$\mathrm{E}[\boldsymbol{X}_n \mid Y_1, \ldots, Y_{n-1}] \leq \boldsymbol{0}, \qquad n \geq 1, \tag{8.101}$$

where conditional expectation given $Y_1, \ldots, Y_{n-1}$ is just unconditional expectation if $n = 1$. Assume uniformly bounded bet sizes, that is, there exists a constant $\boldsymbol{C}$ such that $\boldsymbol{0} \leq \boldsymbol{B}_n \leq \boldsymbol{C}$ for all $n \geq 1$. Let N be a stopping time with respect to $\{Y_n\}_{n \geq 0}$ such that $\mathrm{P}(N < \infty) = 1$.

(a) If (8.100) satisfies $F_n \geq 0$ for all $n \geq 0$, then

$$\mathrm{E}[F_N] \leq F_0. \tag{8.102}$$

(b) Assume that $\mathrm{E}[N] < \infty$. Then (8.102) holds.

(c) Assume that $\mathrm{E}[N] < \infty$. If in addition $\boldsymbol{X}_1, \boldsymbol{X}_2, \ldots$ are identically distributed and $\boldsymbol{X}_n$ is independent of $Y_1, \ldots, Y_{n-1}$ for each $n \geq 2$, then

$$\mathrm{E}[F_N - F_0] = \mathrm{E}\left[\sum_{l=1}^{N} \boldsymbol{B}_l \cdot \boldsymbol{X}_l\right] = \mathrm{E}\left[\sum_{l=1}^{N} \boldsymbol{B}_l\right] \cdot \mathrm{E}[\boldsymbol{X}_1]. \tag{8.103}$$

Proof. First, $\{F_n\}_{n \geq 0}$ is a nonadapted supermartingale because

$$\begin{aligned}
\mathrm{E}[F_n \mid Y_1, \ldots, Y_{n-1}] &= \mathrm{E}[F_{n-1} + \boldsymbol{B}_n \cdot \boldsymbol{X}_n \mid Y_1, \ldots, Y_{n-1}] \\
&= \mathrm{E}[F_{n-1} \mid Y_1, \ldots, Y_{n-1}] + \mathrm{E}[\boldsymbol{B}_n \cdot \boldsymbol{X}_n \mid Y_1, \ldots, Y_{n-1}] \\
&= \mathrm{E}[F_{n-1} \mid Y_1, \ldots, Y_{n-1}] + \boldsymbol{B}_n \cdot \mathrm{E}[\boldsymbol{X}_n \mid Y_1, \ldots, Y_{n-1}] \\
&\leq \mathrm{E}[F_{n-1} \mid Y_1, \ldots, Y_{n-1}]. \tag{8.104}
\end{aligned}$$

Part (a) follows from the nonadapted optional stopping theorem (Theorem 3.2.8, assumption (a)). The same conclusion follows from the hypothesis of part (b) using the same theorem (assumption (c)). Under the hypotheses of part (c), conclusion (8.103) follows by noting that

$$\begin{aligned}
M_n &:= \sum_{l=1}^{n} \boldsymbol{B}_l \cdot (\boldsymbol{X}_l - \mathrm{E}[\boldsymbol{X}_l \mid Y_1, \ldots, Y_{l-1}]) \\
&= \sum_{l=1}^{n} \boldsymbol{B}_l \cdot (\boldsymbol{X}_l - \mathrm{E}[\boldsymbol{X}_1]) \tag{8.105}
\end{aligned}$$

is a nonadapted martingale with respect to $\{Y_n\}$, and hence by Theorem 3.2.8 on p. 105 (assumption (c)) we have $\mathrm{E}[M_N] = \mathrm{E}[M_0] = 0$. ♠

We now consider an example that illustrates the need for the extra generality that Theorem 8.2.5 provides.

Example 8.2.6. *A craps system.* (See Examples 1.2.2 on p. 17, 1.2.8 on p. 19, 1.5.7 on p. 44, and 3.2.9 on p. 105.) Here we assume that the casino offers 3-4-5-times free odds. The system calls for a one-unit pass-line bet on each come-out roll, together with a one-unit come bet on each point roll. Each pass-line and come bet is backed up with a maximal free odds bet. It is a convention that free odds bets on come bets are "off" on come-out rolls. We assume that the gambler stops betting at the conclusion of the *shooter's hand*, that is, the system is completed once the shooter *sevens out*, that is, once he loses a pass-line bet by rolling 7.

To formalize all this, we define

$$\mathscr{S} := \{2, 3, 4, \ldots, 12\} \quad \text{and} \quad \mathscr{P} := \{4, 5, 6, 8, 9, 10\}, \tag{8.106}$$

and we let $T_1, T_2, \ldots$ be a sequence of i.i.d. random variables with common distribution given by (1.11) on p. 6, representing a sequence of dice rolls. We regard the pass-line and come bets as indistinguishable. The pass-line or come bet initiated prior to the nth roll of the dice is resolved following roll N_n, where

$$N_n := \begin{cases} n & \text{if } T_n \in \mathscr{S} - \mathscr{P} \\ \min\{k \geq n+1 : T_k = T_n \text{ or } T_k = 7\} & \text{if } T_n \in \mathscr{P}. \end{cases} \tag{8.107}$$

A one-unit pass-line or come bet is described by

$$X_{n,1} := 1_{\{T_n \in \{7,11\}\}} - 1_{\{T_n \in \{2,3,12\}\}}$$
$$+ 1_{\{T_n \in \mathscr{P}\}}\big(1_{\{T_{N_n} = T_n\}} - 1_{\{T_{N_n} = 7\}}\big), \tag{8.108}$$

and we know that $\mathrm{E}[X_{n,1}] = -7/495$. A free odds bet on point i at the nth roll is described by

$$X_{n,i} := (\pi_7/\pi_i)\, 1_{\{T_n = i\}} - 1_{\{T_n = 7\}}, \qquad i \in \mathscr{P}. \tag{8.109}$$

Because this bet is fair (i.e., $\mathrm{E}[X_{n,i}] = 0$), its use is constrained by the requirement that

$$B_{n,i} \leq (6 - |i - 7|)\, 1_{\cup_{k=1}^{n-1}\{T_k = i,\, T_l \notin \{i,7\} \text{ if } k < l < n\}}, \qquad i \in \mathscr{P}. \tag{8.110}$$

This says that, if the point $i \in \mathscr{P}$ is established by the pass-line or come bet at roll k, and if it is not resolved following roll $n - 1$, then one can make a free odds bet on point i at coup n for no more than $6 - |i - 7|$ units. Notice that at most one event in the union in (8.110) can occur.

In the present system we take $B_{n,1} = 1$ for all $n \geq 1$. We choose $B_{n,i}$ as large as possible consistent with convention, namely,

$$B_{n,i} := (6 - |i - 7|) \, 1_{\cup_{k=1}^{n-1} \{T_k = i, \, T_l \notin \{i,7\} \text{ if } k < l < n\}} \, 1_{\{n \in A^c\}}, \quad i \in \mathscr{P}, \quad (8.111)$$

where A is the (random) set of come-out rolls. Defining $C_1, C_2, \dots$ recursively by $C_1 = 1$ and $C_n = N_{C_{n-1}} + 1$ for each $n \geq 2$, we note that $A = \{C_1, C_2, \dots\}$.

We apply Theorem 8.2.5 with

$$\begin{aligned}
\boldsymbol{X}_n &:= (X_{n,1}, X_{n,4}, X_{n,5}, X_{n,6}, X_{n,8}, X_{n,9}, X_{n,10}), \\
\boldsymbol{B}_n &:= (B_{n,1}, B_{n,4}, B_{n,5}, B_{n,6}, B_{n,8}, B_{n,9}, B_{n,10}),
\end{aligned} \quad (8.112)$$

and $L = \min\{N_{C_n} : n \geq 1, \, C_n < N_{C_n}, \, T_{N_{C_n}} = 7\}$. By (8.103) and Example 3.2.6 on p. 103,

$$\begin{aligned}
\mathrm{E}[F_L - F_0] &= \mathrm{E}\left[\sum_{l=1}^{L} \boldsymbol{B}_l\right] \cdot \mathrm{E}[\boldsymbol{X}_1] = \mathrm{E}\left[\sum_{l=1}^{L} B_{l,1}\right] \mathrm{E}[X_{1,1}] \\
&= \mathrm{E}[N] \mathrm{E}[X_{1,1}] = \frac{1{,}671}{196}\left(-\frac{7}{495}\right) = -\frac{557}{4{,}620}. \quad (8.113)
\end{aligned}$$

The application of Theorem 8.2.5 in this setting is problematic. The point is that F_n does not necessarily represent the gambler's fortune following the nth roll, because bets made at or prior to the nth roll may still be unresolved following the nth roll. This is not an issue with F_L, however, because when the shooter sevens out all outstanding bets are resolved.

It is worth mentioning explicitly that $\boldsymbol{X}_1, \boldsymbol{X}_2, \dots$ are identically distributed but *not* independent, though $\boldsymbol{X}_n$ is independent of $Y_1, \dots, Y_{n-1}$. This is because $\boldsymbol{X}_n = f(Y_n, Y_{n+1}, \dots)$ ($n \geq 1$) for some nonrandom function f. We could not combine the free odds bets with the pass-line and come bets, as we have done elsewhere, because of the convention concerning come-out rolls. Thus, the vector formulation was essential here. ♠

Example 8.2.7. *Betting systems at card games.* The systems of Section 8.1 are often employed at card games, such as trente et quarante and baccarat, to which Theorem 8.2.1 does not apply, owing to its assumption of independent coups. Theorem 8.2.5(a,b), however, may apply, while the independence condition of (c) fails. The key requirement is (8.101), which requires that the available bets be subfair (or at best fair), *regardless of the composition of the remaining deck*. Such a condition is rarely met in twenty-one, for example. ♠

8.3 Problems

8.1. *Repeated application of the martingale system.* Consider repeated application of the martingale system, assuming a game as in Example 8.1.1 on p. 276 and a house limit of M units. Use Problem 6.18 on p. 234 to evaluate the house advantage for this system.

8.2. *Reverse martingale system.* In this system the gambler doubles his bet size after each win, and stops betting after the first loss. In particular, (8.4) on p. 276 is replaced by

$$B_n = 2B_{n-1} 1_{\{X_{n-1}=1\}}, \qquad n \geq 2. \tag{8.114}$$

Use the results of Example 8.1.1 on p. 276 to obtain for the reverse martingale system results analogous to those of the example.

Hint: Unlike in the martingale system, here the size of F_0 is irrelevant (as long as $F_0 \geq 1$).

8.3. *Great martingale system.* This system is identical to the martingale system, except that (8.4) on p. 276 is replaced by

$$B_n = (2B_{n-1} + 1) 1_{\{X_{n-1}=-1\}}, \qquad n \geq 2. \tag{8.115}$$

Thus, the sequence of bet sizes in the absence of a winning coup has the form $1, 3, 7, 15, \ldots$. Recalculate all quantities in Example 8.1.1 on p. 276 for the great martingale system.

8.4. *Alternative analysis of the Fibonacci system.*

(a) Show that a loss followed by a win is equivalent to a win followed by a loss, except when a win terminates the system. Here "equivalent" means that the profit is the same and the bet size at the third coup equals the bet size at the first coup.

(b) Show that two losses and a win, in any order (i.e., LLW or LWL or WLL), result in a profit of 0, except when the system is terminated before the three coups are completed. Show further that in such cases the bet size at the fourth coup equals the bet size at the first coup.

(c) Use the results of (a) and (b) to prove (8.21) on p. 280 without reference to the Markov chain $\{Y_n\}_{n \geq 0}$.

8.5. *Expected total bet in the Fibonacci system.* Consider a Fibonacci bettor with unlimited credit in the absence of a house limit. Assume that $\frac{1}{3} \leq p \leq \frac{1}{2}$, and let N denote the duration of the system and $B_1, B_2, \ldots, B_N$ denote the sequence of bet sizes. Show that $E[B_1 + \cdots + B_N] = \infty$. (This is a weaker but simpler version of the result that $E[B^*] = \infty$, where $B^* := \sup_{n \geq 1} B_n$.)

Hint: If the expectation were finite, then the optional stopping theorem would imply that $E[F_N] \leq F_0$, contradicting (8.21) on p. 280.

8.6. *Mean duration of the Labouchere system.* Consider the Labouchere system with initial list of length $j_0 \geq 1$. Assume unlimited credit, no house limit, and $p > \frac{1}{3}$. Letting N be the number of coups needed to successfully complete the system, evaluate $E[N]$.

Hint: See Example 7.2.7 on p. 258.

8.7. *Distribution of the duration of the Labouchere system.* Consider the Labouchere system with initial list of length $j_0 \geq 1$. Assume unlimited credit, no house limit, and $p \geq \frac{1}{3}$. For each $0 \leq l \leq n$, let $c(n, l)$ be the number of the $\binom{n}{l}$ permutations of l losses and $n - l$ wins for which the Labouchere bettor has not yet completed the system. Also, let $c(0, l) := \delta_{0, l}$.

(a) Let $l_n := \lceil (2n + 1 - j_0)/3 \rceil^+$ and argue that

$$c(n, l) = \begin{cases} c(n - 1, l - 1) + c(n - 1, l) & \text{if } l_n \leq l \leq n, \\ 0 & \text{otherwise,} \end{cases} \tag{8.116}$$

for all $n \geq 1$. Thus, the coefficients $c(n, l)$ can be generated from a modified Pascal triangle.

(b) Letting N be the number of coups needed to successfully complete the system, show that

$$P_{j_0}(N = n) = c(n - 1, l_{n-1})p^{n - l_{n-1} - 1}q^{l_{n-1}} \tag{8.117}$$

if $n \geq \lfloor (j_0 + 1)/2 \rfloor$ and $(n + j_0 - 1)/3 \notin \mathbf{Z}$, and $P_{j_0}(N = n) = 0$ otherwise.

8.8. *Distribution of the duration of the Labouchere system, continued.*

(a) Given positive integers n, m, and k with $n > km$, show that the number of lattice paths, with steps $(1, 0)$ and $(0, 1)$ (east and north), from $(0, 0)$ to (n, m) that lie strictly below the line $y = x/k$ (except at the origin) is

$$\frac{n - km}{n + m}\binom{n + m}{m}. \tag{8.118}$$

(b) Given a positive integer m, show that the number of lattice paths, with steps $(1, 0)$ and $(0, 1)$, from $(0, 0)$ to $(2m, m)$ that never rise above (but may touch) the line $y = x/2$ [respectively, the line $y = (x + 1)/2$] is

$$a_m := \frac{1}{2m + 1}\binom{3m}{m} = \frac{1}{3m + 1}\binom{3m + 1}{m} \tag{8.119}$$

[respectively,

$$b_m := \frac{1}{m + 1}\binom{3m + 1}{m} \ ]. \tag{8.120}$$

(c) Using the notation of the preceding problem and part (b) of this problem, show that for all $m \geq 0$,

$$P_1(N = 3m + 1) = a_m p^{m+1} q^{2m}, \tag{8.121}$$
$$P_1(N = 3m + 2) = b_m p^{m+1} q^{2m+1}, \tag{8.122}$$
$$P_1(N = 3m + 3) = 0. \tag{8.123}$$

(d) Extend the results of part (c) to initial states $1 \leq j_0 \leq 6$ using

$$P_{j_0}(N = n + 1) = pP_{j_0 - 2}(N = n) + qP_{j_0 + 1}(N = n). \tag{8.124}$$

Hint: (a) Use induction on n and Problem 1.3 on p. 52.
(b) Use part (a).

8.9. *Conditional expected deficit after n coups of Labouchere system.* Assume that $p = \frac{1}{2}$, so that the Labouchere bettor's fortune $\{F_{n \wedge N}\}_{n \geq 0}$ is a martingale with respect to $\{X_n\}_{n \geq 1}$, in the notation of Example 8.1.3 on p. 282. Show that $-\mathrm{E}[F_n \mid N \geq n + 1]$ can be expressed in terms of the distribution of N and the sum of the terms of bettor's initial list.

8.10. *Reverse Labouchere system.* This system is identical to the Labouchere system, except that the roles of wins and losses are reversed. (In particular, Table 8.3 on p. 284 illustrates the reverse Labouchere system if Ls and Ws are interchanged in column 4 and if the signs of the cumulative-profit figures are changed in column 5.) Let N denote the duration of the Labouchere system and $\hat{N}$ the duration of the reverse Labouchere system. Argue that

$$\frac{\mathrm{P}_{j_0}(\hat{N} = n)}{\mathrm{P}_{j_0}(N = n)} = (p/q)^{m_n} \tag{8.125}$$

for some integer m_n if $n \geq \lfloor (j_0 + 1)/2 \rfloor$ and $(n + j_0 - 1)/3 \notin \mathbf{Z}$, and that $m_n \sim n/3$ as $n \to \infty$.

8.11. *Oscar system with house limit $M = 3$ units.* In Example 8.1.4 on p. 286 we evaluated the probability $Q(1, 1)$ of achieving the one-unit goal using the Oscar system (in terms of p and $q := 1 - p$), assuming a house limit of $M = 2$ units. To appreciate how complicated such calculations can be, extend this result to the case $M = 3$.

8.12. *Oscar system with a limited bankroll.* We saw that there is a recursive algorithm for evaluating the probability of achieving the one-unit goal using the Oscar system (in terms of p and $q := 1 - p$) *without ever making a bet that exceeds a house limit of M units.* Show that the same is true with the italicized phrase replaced by *with an initial bankroll of M units*, assuming that one must abort the system if it calls for a bet that exceeds what is left of one's bankroll.

8.13. *Second moment of the duration of the Oscar system when $p = \frac{1}{2}$.* If $p = \frac{1}{2}$, we know from Problem 4.14 on p. 154 that the duration N of the Oscar system satisfies $\mathrm{E}[N] < \infty$. Use Problem 3.9 on p. 112 to show that $\mathrm{E}[N^2] = \infty$.

8.14. *Modified d'Alembert system.* Change the rules of the d'Alembert system by requiring that the gambler never bet more than necessary to achieve a cumulative profit of one unit. Revise Tables 8.7 on p. 296 and 8.8 on p. 297, including this modified d'Alembert system instead of the original one. (Thus, all six systems have a win target of one unit, making them more directly comparable.)

Hint: The method used to analyze the Oscar system works here as well.

8.15. *Babbage's version of the d'Alembert system.* The first bet is one unit and the second bet is one unit. After two consecutive losses, the bet size is increased by one unit, and after two consecutive wins, the bet size, if positive, is decreased by one unit. Finally, after two coups comprising a win and a loss in either order, the bet size remains unchanged. Let us refer to the latter scenario as a *split*. The system terminates if the first nonsplit is a pair of wins or, if not, when the number of wins equals the number of losses. Extend Table 8.7 on p. 296 to this system.

Hint: As an intermediate step, extend Example 8.1.5 on p. 289 to the case $P(X = 1) = p > 0$, $P(X = -1) = q > 0$, and $P(X = 0) = r \geq 0$, where $p + q + r = 1$.

8.16. *Markov property of distinct bet sizes in Blundell system.* Consider the Blundell system with unlimited credit and no house limit. Let B_n be the amount bet at the nth coup for each $n \geq 1$ (with $B_n = 0$ if $n > N$, where N is the number of coups needed to achieve the one-unit goal). Let $N_1 = 1$ and define $N_2, N_3, \ldots$ by $N_k = \min\{n > N_{k-1} : B_n \neq B_{N_{k-1}}\}$. These are the times at which the bet size changes. Let $B_k^* = B_{N_k}$ for each $k \geq 1$, where $B_\infty := 0$, and argue that $\{B_k^*,\ k \geq 1\}$ is a Markov chain and determine its transition probabilities. Also, find the probability, starting at 1, of eventual absorption at 0.

8.17. *Condition for finiteness of mean duration of Blundell system.* In the Blundell system with unlimited credit and no house limit, let N be the number of coups needed to achieve the one-unit goal. Just as we used (8.79) on p. 294 to find a necessary and sufficient condition on p for $P(N < \infty) = 1$, use (8.81) on p. 294 to find a necessary and sufficient condition on p for $E[N] < \infty$.

8.18. *Modified Fitzroy system.* Here is a very slight modification of the Fitzroy system, another system for even-money games. The bet size starts at one unit and increases by one unit after each coup, win or lose, with two exceptions: (a) From one unit it increases to three units rather than to two units. (b) The bettor stops betting as soon as he has won one unit per coup played, and he reduces his bet size when necessary to avoid the risk of overshooting that goal.

Give a more formal description of this system in terms of a Markov chain, much as in (4.7) and (4.8) on p. 121. Extend Table 8.7 on p. 296 so as to include the modified Fitzroy system.

8.19. *Cover system.* In this system, as in the Labouchere system (Example 8.1.3) on p. 282, the gambler's bet size at each coup is determined by a list of positive integers kept on his score sheet, which is updated after each coup. Let us denote such a list by $l = (l_1, \ldots, l_j)$, where $j \geq 0$ denotes its length. (A list of length 0 is empty, denoted by $l = \varnothing$.)

The system begins with a one-unit bet. In the cases of a win, no further bets are made. In the case of a loss, the gambler's initial list is $l^1 := (1)$.

Given a list $l = (l_1, \ldots, l_j)$, the gambler's bet size at the next coup is

$$b(l) := \begin{cases} 0 & \text{if } j = 0, \\ l_1 + 1 & \text{if } j \geq 1. \end{cases} \tag{8.126}$$

Following the resolution of this bet, the list is updated as follows: After a win, it is replaced by

$$l[1] := \begin{cases} \varnothing & \text{if } j \in \{0, 1\}, \\ (l_2, \ldots, l_j) & \text{if } j \geq 2, \end{cases} \tag{8.127}$$

that is, the leftmost term is canceled. After a loss, it is replaced by

$$l[-1] := \begin{cases} \varnothing & \text{if } j = 0, \\ (l_1, \ldots, l_j, b(l)) & \text{if } j \geq 1, \end{cases} \tag{8.128}$$

that is, the amount just lost is appended to the list, if nonempty, as a new last term.

The system is terminated after an initial win or when the list becomes empty (i.e., when the number of wins equals the number of losses and is nonzero).

(a) Show that the cover system shares some properties of the d'Alembert system, such as (8.72) on p. 291.

(b) Extend Tables 8.7 on p. 296 and 8.8 on p. 297 so as to include the cover system.

8.20. *Belgian progression.* This system, like the Blundell system (Example 8.1.6 on p. 292), depends on a positive-integer parameter K, but it need not be even. (In the original formulation, $K = 5$.) The gambler's status at any time is completely specified by a list of positive integers on his score sheet, the sum of whose terms indicates how far behind he is at that time, and by his bet size at the next coup, both of which are updated after each coup. The rules are as follows:

The list is initially empty and the first bet is one unit. After a loss the amount just lost is appended to the list as a single term on the right. The bet size remains the same unless the amount just lost appears K times in the new list; in the latter case the bet size is increased by one unit. After a win sufficiently many terms on the left of the list to sum to the amount just won are deleted. If this cannot be done exactly, the smallest number of terms on the left of the list that sum to a total greater than the amount just won are deleted, and the difference is added to the list as a single term on the left. The bet size remains the same unless the amount just won exceeds the sum of the terms in the new list by more than one; in the latter case the bet size is reduced to an amount sufficient to exceed the sum of the terms in the new list by exactly one. Once the gambler achieves a cumulative profit of one unit, he quits.

(a) Show that the system can be described more easily by a three-dimensional Markov chain $\{(Y_n, Z_n, W_n)\}_{n \geq 0}$, where Y_n and Z_n are as in the Oscar system and W_n is the number of times Z_n appears in the gambler's list following the nth coup. Determine the transition probabilities of this Markov chain.

(b) Extend Table 8.7 on p. 296 so as to include the Belgian progression with arbitrary K.

8.21. *House advantage for craps system of Example 8.2.6.* Find the expected total amount bet in the craps system of Example 8.2.6 on p. 304 (keeping in mind that no free odds bets are made on come-out rolls), thereby obtaining the house advantage (pushes included) for the system.

8.22. *Ponzer system at craps.* This system is identical to that of Example 8.2.6 on p. 304 with one exception: No come bets are made when there are already two unresolved come bets. (This of course limits the number of free odds bets that can be made.) Modify Example 8.2.6 on p. 304 to take this additional complication into account.

8.4 Notes

The memoirs of Giacomo Casanova, Chevalier de Seingalt [1725–1798], written during 1789–1798, contain the following sentence describing an incident that occurred in 1754 (Casanova 1967, Volume 4, Chapter 7):

> I went there and took all the gold I found, and, determinedly doubling my stakes according to the system known as the martingale, I won three or four times a day during the rest of the Carnival.

Thus, the martingale system dates at least as far back as 1754. The term "martingale" appeared with this meaning in the *Dictionnaire de l'Académie française* in 1762. The earliest such usage cited by the *Oxford English Dictionary* (Simpson and Weiner 1989) is 1815. Arguably, the martingale system appeared even earlier, without being named as such, in the fable *St. Pierre et le jongleur*, which dates back to the early 13th century (MacGillavry 1978).

Among the first mathematicians to use the term were Sylvestre François Lacroix [1765–1843] and Charles Babbage [1791–1871]. Lacroix (1822, p. 124; perhaps also in the first edition of 1816) wrote,[2]

> Gamblers with little capital cannot follow a scheme like this, which is called making the martingale, without hurting or ruining themselves, and they are often forced to give up, losing everything they have laid out. As for the rich, they can make much better use of their capital, [...]

---

[2] Translation from Shafer and Vovk (2001, p. 51).

while Babbage (1821) wrote,

> The first and most simple plan, is that of doubling the stake whenever a loss occurs. This is well known, and has been so frequently practised, as to have acquired a peculiar name; it is technically called the *martingal*; [...]

The etymology of the word "martingale" was investigated by Mansuy (2009) but remains somewhat obscure. Since the 17th century and perhaps earlier, the term has meant a strap designed to keep a horse's head down. It is tempting to conjecture that the gambling interpretation derives from this. Specifically, the martingale system is a device intended to keep a gambler's losses down (though it often has the opposite effect). Support for this conjecture was provided by Richard A. Proctor [1837–1888] (1887, p. 250):

> The following is a martingale (as gamblers call these devices for preventing fortune from rearing against them) which has misled many: [...]

Proctor went on to describe the Labouchere system. It should also be noted that "martingale" has been used as a verb (Persius 1823, p. 165):

> After the red or black, it is quite indifferent which, wins five times, begin to *martingale* on the losing colour.

According to Kilby, Fox, and Lucas (2005, pp. 186–190),

> Some [casino] executives believe that the maximum bet protects the casino from "betting systems." Nothing could be further from the truth.

They go on to illustrate their point by considering a martingale bettor playing even-money roulette ($p = 18/38$) in the absence of a house limit. Such a bettor is constrained only by his bankroll, which in their example is assumed to be $F_0 = 2^{20} - 1 = 1{,}048{,}575$ units (one unit is \$1). They argue that such a player is not really a threat to the casino because he will likely experience 20 consecutive losing coups (for a loss of $2^{20} - 1$ units) long before he can double his initial bankroll one unit at a time. This is certainly true.

But suppose they had considered a reverse-martingale bettor (Problem 8.2), whose initial bankroll could be as little as one unit but whose goal is 20 consecutive winning coups (for a win of $2^{20} - 1$ units). And what if a unit in this case were \$10,000? We believe that most casino executives would be uncomfortable with that level of risk and that the assertion of Kilby, Fox, and Lucas is too categorical.

The Fibonacci system is based on the famous Fibonacci sequence, $1, 1, 2, 3, 5, 8, 13, 21, 34, 55, \ldots$, which first appeared in the West in a problem in *Liber Abaci* (*The Book of Calculations*, 1202) by Leonardo Pisano Fibonacci [1170–1250]. In a translation published on the 800th anniversary of the original (Sigler 2002, pp. 283–284), the problem that motivated it reads as follows:

> A certain man had one pair of rabbits together in a certain enclosed place, and one wishes to know how many are created from the pair in one year when it is the nature of them in a single month to bear another pair, and in the second month those born to bear also. [...]

If f_m denotes the number of pairs after m months, then (8.15) holds because the number of pairs after m months is the number of pairs after $m-1$ months plus the number pairs that are fertile, which is the number of pairs after $m-2$ months. This gives the sequence $2, 3, 5, 8, 13, 21, 34, 55, 89, 144, 233, 377, \ldots$, so the answer to Fibonacci's problem is 377. Equation (8.25) is named for Jacques Binet [1786–1856], although it was known to De Moivre much earlier.

Actually, the Fibonacci sequence dates back to ancient India (Singh 1985). The origin of the Fibonacci betting system is uncertain.

The Labouchere system was popularized by British journalist and Member of Parliament Henry Du Pré Labouchere [1831–1912], who described it as follows (Thorold 1913, pp. 65–66; essentially from *Truth*, Feb. 15, 1877, according to Rouge et Noir 1898, Chapter 2):

> I used at one time to take the waters every year at Homburg, and I invariably paid the expenses of my trip out of my winnings at the gambling tables. It may have been luck, or it may have been system; but I give my system for what it is worth. I used to write the following figures on a piece of paper: 3, 4, 5, 6, 7. My stake was always the top and bottom figures added together. If I won, I scratched out these figures; if I lost, I wrote down the stake at the bottom of the figures, and I went on playing until all the figures on my piece of paper were erased.

Labouchere then explained, not entirely successfully, the basis for his system:

> The basis of the 'system' was this. Before reaching the maximum, I could play a series of even chances for about two hours, and if during these two hours I won one quarter as many times as the bank, plus five, all my figures were erased. During these two hours an even chance would be produced two hundred times. If, therefore, I won fifty-five times, and the bank won one hundred and forty-five times, I was the winner of twenty-five napoleons, florins, or whatever was my unit. Now let any one produce an even chance by tossing up a coin and always crying 'heads,' he will find that he may go on until Doomsday before the 'tails' exceed the 'heads,' or the 'heads' exceed the 'tails,' by ninety five. I found this system in a letter from Condorcet to a friend, which I read in a book that I purchased at a stall on the 'Quai' at Paris. It may have been, as I have said, only luck; but all I can say is, that whenever I played it I invariably won.

To assure completion of the Labouchere system in n coups, assuming an initial list of length j_0, one must win at least $\lceil (n+j_0)/3 \rceil$ of the n coups, or at least 69 of 200 in Labouchere's case ($j_0 = 5$). Of course, this analysis does not take the house maximum into account.

We do not know whether the Labouchere system originated with French mathematician and philosopher Marie Jean Antoine Nicolas de Caritat, Marquis de Condorcet [1743–1794]. Downton (1980b) pointed out that "no probability analysis specific to the system appears to have been made," and this is the reason the Labouchere system is excluded from Table 8.8. The system was described in verse by Player (1925, p. 105) as follows:[3]

> There is a system that is told you in the night
> With weird incantations and a red and flickering light.

---

[3] We have taken the liberty of correcting the punctuation.

It was a system of the Chaldean Priest of old,
And roulette was finished if you would but be bold.
It was perfectly simple and perfectly sound,
And to the everlasting problem an answer had been found.
You wrote down little numbers, say 1, 2, 3 and 4,
And if you lost, well, you simply added more.
And when winning you erased them, and when all have been crossed out,
Then the game is finished, and you have won the bout.
But still, as sometimes happens, the system didn't win.
(The croupiers smile quite broadly as they rake the jetons in.)
And at last the écart[4] gets you, as eventually it must,
And then you go on playing, and eventually get bust.
But from Africa, and Egypt, from London and from Rome,
To Monte Carlo rush the crowd to take easy money home.
They have all been told the secret, they all think it something new,
They score on scraps of paper that a passer may not view.
And when they lose they stand aghast; "The tables can't be fair."
A stranger passed, and sighed and said, "Ye Gods! The Labouchere!"

The Oscar system was discussed at length by Allan N. Wilson [1924–2001] in his book *The Casino Gambler's Guide* (1965, pp. 246–253). He learned of it from an unidentified Southern California dice player presumably named Oscar. Oscar "had used it consistently over a period of several years to win the price of many weekend trips to Las Vegas." The system was first described precisely by Wilson (1962) in the problem section of the *American Mathematical Monthly*. (An imprecise description was given by Player 1925, p. 109.) The problem sought the "exact general expression" for the probability, using the Oscar system, of successfully achieving the one-unit goal before the system calls for a bet exceeding a house limit of M units. To date, no solution has been published in the *Monthly*. The recursive solution sketched in Example 8.1.4 is due to Ethier (2000) and was rejected by *Monthly* problem editors in 1995. The result that $P(N = \infty) > 0$ if $p < \frac{1}{2}$ was first proved by Ethier (1996), but the proof given here is that of Borovkov (1997). For additional information about the Oscar system (including proofs of the assertions about N in Table 8.7), see Ethier (1996), Borovkov (1996, 1997), and Novikov (1996). See also Problem 4.14.

As reported by Wilson (1965, pp. 250–252), a computer simulation of the Oscar system was done in the early 1960s by Julian H. Braun [1929–2000] (see Section 21.5 for more about him). He took $p = 244/495$ and $M = 500$, and obtained estimates of the odds against having to abort the system and the conditional expected loss, based on a sample size of 280,000. His figure for the latter quantity was 13,100, which differs slightly from our 12,678. The distinction is that our gambler stops betting as soon as he has no chance at recovery, whereas Braun's gambler continues until the system calls for a bet of 501 units. For more-extensive simulation results see Johnson (1997), but be aware that Tables 3.9 and 3.10 therein are inaccurate; corrected versions are available.

---

[4] The term "écart" is French for "deviation."

The d'Alembert system is named for Jean le Rond d'Alembert [1717–1783], a French mathematician. In 1744 d'Alembert published *Traité de l'équilibre et du mouvement des fluides*, containing his "Theory of Equilibrium." It seems likely that he was therefore credited with this betting system, which relies on the "Theory of Equilibrium" in probability, the belief (erroneous unless $p = \frac{1}{2}$) that wins and losses must eventually equalize. The earliest description of the system that we have seen, as well as a clever derivation of (8.71), is due to Babbage (1821).

The Blundell system is named for Wilfred Blundell, a British resident of Monte Carlo in the 1880s, about whom we have been unable to find any information. (It is possible that the name was a pseudonym.) The system was described by Victor Bethell in his book *Ten Days at Monte Carlo at the Bank's Expense* (1898, pp. 21–23) and was used by London stockbroker Francis N. Curzon [1865–1941] to pay for his annual vacations to Monte Carlo. According to Russell T. Barnhart [1926–2003], who as an undergraduate (Princeton '52) knew William Feller [1906–1970], footnote 2 on p. 346 of Feller (1968) concerns Curzon:

> A certain man used to visit Monte Carlo year after year and was always successful in recovering the cost of his vacations. He firmly believed in a magic power over chance. Actually his experience is not surprising. [. . .]

The term "conservation of fairness" comes from Dubins and Savage (1976, p. 5), but the concept is much older. Theorem 8.2.3 is from Billingsley (1995, Theorem 7.1) but dates back to Doob (1936), and it debunks the *Monte Carlo fallacy* (so-named by Maxim 1904, pp. 36, 46), which, as noted in Section 1.7, is the belief that a series of failures makes a success more likely; the law of averages is often improperly cited as justification. Theorem 8.2.4 is from Breiman (1968), while Theorem 8.2.5 and Example 8.2.6 are slight reformulations of results of Ethier (1998).

Problem 8.5 is due to Grimmett and Stirzaker (2001, Problem 12.9.15). Problem 8.6 was first addressed by Downton (1982). Problem 8.7 is due to Downton (1980b); see Estafanous and Ethier (2009) for a generalization. Problem 8.8 is from Ethier (2008), although parts (a) and (b) are classical results; in fact, part (a) is the so-called *ballot theorem*, which dates back to the 19th century (see Takács 1997; Addario-Berry and Reed 2008). Problem 8.9 is also from Ethier (2008).

The reverse Labouchere system (Problem 8.10) was introduced by Seton R. Beresford [1868–1928] in his book *The Future at Monte Carlo* (1923). Norman Leigh's *Thirteen Against the Bank* (1976) is the story of how in 1966 a team of 13 English men and women won 800,000 francs from the Casino Municipal in Nice, France, over an eight-day period using the reverse Labouchere system. They played six simultaneous reverse Laboucheres (one on each of the six even-money wagers at the same table) for 12 hours per day starting with initial list $1, 2, 3, 4$ and a five-franc initial bet and restarting whenever the list became empty or the system called for a bet exceeding the house limit of 2,600 francs. It would seem that Leigh's book, although not intended as such, is a work of fiction.

Problem 8.12 was suggested by Ramaswamy G. Krishnan.

The actual Fitzroy system (cf. Problem 8.18) was discussed by Barnhart (1983a, Part 5, Chapter 4) and Schwartz (2006, pp. 315–316). It was precisely defined by Bethell (1901, pp. 141–146) and Maxim (1904, pp. 264–269). Here is the definition. The bet size starts at one unit and increases to $n + 1$ units at the nth coup ($n \geq 2$), with one exception: The bettor stops betting as soon as he has won one unit per coup played, and he reduces his bet size when necessary to avoid the risk of overshooting that goal. James Francis Harry St. Clair-Erskine [1869–1939], Fifth Earl of Rosslyn, and his brother, Alexander Fitzroy St. Clair-Erskine [1870–1914], believed that their betting system, also called the Rosslyn system, was infallible. In 1908 they were challenged to prove it by Hiram S. Maxim [1840–1916], best known as inventor of the machine gun. According to the *New York Times* (September 30, 1908),

> The unique gambling contest between Sir Hiram Maxim and Lord Rosslyn came to an end this afternoon and shows defeat for the system advanced by Rosslyn.
>
> Lord Rosslyn claimed that by his system of play it was possible to win at roulette against the Monte Carlo Bank. Sir Hiram said that this was impossible, and in order to settle the controversy the two men began playing roulette in a Piccadilly club ten days ago, Sir Hiram conducting the game in the same way it is played at Monte Carlo. Lord Rosslyn started to play with $50,000 in "dummy" money and the bank had an equal amount.
>
> Lord Rosslyn's system has been proved to be utterly fallacious. At one period of the contest he was about $16,000 ahead, but for the last three days the bank has been a steady winner, and his Lordship's capital in "dummy" money became exhausted this afternoon.

Our description of the cover system (Problem 8.19) comes from Szabo (1962, p. 101–102), though it dates back at least to Albigny (1902, Chapter 1). See also Player (1925, p. 113) and Barnhart (1978, p. 131–133).

The Belgian progression (Problem 8.20) is believed to be due to Édouard Suau de Varennes, a minor French novelist of the 19th century. Suau de Varennes led the "Contrebanque" assault on the Homburg casino in 1844, which was fictionalized by William Makepeace Thackeray [1811–1863] in his 1850 Christmas book, *The Kickleburys on the Rhine* (see the quote on p. 623). As far as we know, the earliest description of the system appears in Richardson (1929, Chapter 5). A different system of the same name, la montant belge, was described by Albigny (1902, Système no. 8). Other descriptions of the Belgian progression can be found in Roxbury (1959) and Comtat (1988, p. 69) with the latter calling it the Holland progression. For additional information about this system, see Ethier (1999) and Johnson (1997).

The Ponzer system (Problem 8.22) is due to Ponzer (1977).

Chapter 9
Bold Play

I was, I may say, the author of our common good fortune, by putting boldness into our play.

William Makepeace Thackeray, *The Luck of Barry Lyndon*

A gambler with a fixed goal is said to use bold play if at each coup he bets his entire fortune or just enough to achieve his goal in the event of a win, whichever amount is smaller. Section 9.1 proves that bold play is optimal at subfair red-and-black, that is, at games of independent coups in which each coup is won, and paid even money, with probability $p < \frac{1}{2}$ and lost otherwise. Section 9.2 treats the case of a house betting limit at subfair red-and-black, and Section 9.3 generalizes Section 9.1 to the case of a subfair primitive casino, the distinction being that, instead of payoff odds of 1 to 1, the payoff odds are a_0 to 1, with $a_0 > 0$ and $p < 1/(a_0 + 1)$.

9.1 Red-and-Black

Red-and-black[1] is the name of the generic game of independent coups in which a single betting opportunity is available at each coup (of course, the gambler has discretion over how much to bet), namely a bet that pays even money, winning with probability p and losing with probability $q := 1 - p$. To avoid trivialities, we assume throughout that $0 < p < 1$.

We have already seen in Table 7.1 on p. 243 (cf. (7.11) on p. 243) that the probability of doubling one's fortune of 10^m units while making only bets of 10^l units $(0 \le l \le m)$ is

$$\frac{1}{(q/p)^{10^{m-l}} + 1},\tag{9.1}$$

---

[1] Not to be confused with rouge et noir, for which see Chapter 20.

S.N. Ethier, *The Doctrine of Chances*, Probability and its Applications, DOI 10.1007/978-3-540-78783-9_9, © Springer-Verlag Berlin Heidelberg 2010

which is increasing in l when $p < \frac{1}{2}$. This suggests that the optimal strategy at subfair red-and-black is to bet all 10^m units, double-or-nothing, on a single coup. Such a strategy is known as *bold play* and is the principal subject of this chapter.

More precisely, *bold play* consists of betting, at each coup, either one's entire fortune or just enough to achieve one's goal in the event of a win, whichever amount is smaller. It will be convenient, and will involve no loss of generality, to measure money in units of the gambler's goal, or, equivalently, to assume that his goal is 1 unit. We further assume that betting units are infinitely divisible. Then the bold player at red-and-black bets $f \wedge (1 - f)$ when his fortune is $f \in [0, 1]$.

Let X be a random variable satisfying

$$P(X = 1) = p \quad \text{and} \quad P(X = -1) = q. \qquad (9.2)$$

In addition, let $X_1, X_2, \ldots$ be a sequence of i.i.d. random variables with common distribution that of X, with X_n representing the gambler's profit per unit bet at coup n. Given an initial fortune $f \in [0, 1]$, let $F_n(f)$ represent the bold player's fortune after n coups. Then

$$F_n(f) = F_{n-1}(f) + [F_{n-1}(f) \wedge (1 - F_{n-1}(f))]X_n, \qquad n \geq 1, \qquad (9.3)$$

where $F_0(f) := f$.

This has an important consequence. If $f \in (0, 1)$ is rational, then there is a finite set $G \subset [0, 1]$ such that, with probability 1, $F_n(f) \in G$ for all $n \geq 1$. Indeed, if j and k are positive integers such that $f = j/k$, then we may take $G := \{i/k : i = 0, 1, \ldots, k\}$.

Actually, the bold player is ruined or achieves his goal rather quickly. Let $T(f)$ be the coup at which the last positive bet is made, assuming an initial fortune of $f \in [0, 1]$. (By convention, $T(0) = T(1) = 0$.) Then, for $0 < f < 1$,

$$
\begin{aligned}
P(T(f) \geq n + 1) &= P(0 < F_1(f) < 1, \ldots, 0 < F_n(f) < 1) \\
&= E[1_{\{0 < F_1(f) < 1, \ldots, 0 < F_{n-1}(f) < 1\}} \\
&\qquad\qquad \cdot P(0 < F_n(f) < 1 \mid X_1, \ldots, X_{n-1})] \\
&\leq P(0 < F_1(f) < 1, \ldots, 0 < F_{n-1}(f) < 1)(q \vee p) \\
&= P(T(f) \geq n)(q \vee p) \\
&\;\;\vdots \\
&\leq (q \vee p)^n, \qquad\qquad\qquad\qquad\qquad\qquad (9.4)
\end{aligned}
$$

where we have used Theorem 2.2.7 on p. 83, implying by Theorem 1.4.2 on p. 30 that

$$E[T(f)] = \sum_{n=0}^{\infty} P(T(f) \geq n + 1) \leq \frac{1}{1 - (q \vee p)} = \frac{1}{p \wedge q}. \qquad (9.5)$$

(This bound can be improved; see Problem 9.7 on p. 351.)

Our primary interest is in the expected utility of the bold player's final fortune when the utility function is given by the indicator of $[1, \infty)$. This is simply

$$\begin{aligned}
Q(f) &:= P(F_{T(f)}(f) = 1) \\
&= P(F_n(f) = 1 \text{ for some } n \geq 1), \qquad 0 \leq f \leq 1.
\end{aligned} \tag{9.6}$$

Conditioning on X_1, we find that

$$\begin{aligned}
Q(f) &= pP(F_n(f) = 1 \text{ for some } n \geq 1 \mid X_1 = 1) \\
&\quad + qP(F_n(f) = 1 \text{ for some } n \geq 1 \mid X_1 = -1) \\
&= pP(F_n(f) = 1 \text{ for some } n \geq 2 \mid X_1 = 1) \\
&\quad + qP(F_n(f) = 1 \text{ for some } n \geq 2 \mid X_1 = -1) \\
&= pP(F_n(f + [f \wedge (1 - f)]) = 1 \text{ for some } n \geq 1) \\
&\quad + qP(F_n(f - [f \wedge (1 - f)]) = 1 \text{ for some } n \geq 1) \\
&= pQ(f + [f \wedge (1 - f)]) + qQ(f - [f \wedge (1 - f)]),
\end{aligned} \tag{9.7}$$

or, since $Q(0) = 0$ and $Q(1) = 1$ by (9.6),

$$Q(f) = \begin{cases} pQ(2f) & \text{if } 0 \leq f \leq \frac{1}{2}, \\ p + qQ(2f - 1) & \text{if } \frac{1}{2} \leq f \leq 1. \end{cases} \tag{9.8}$$

It would be hard to overemphasize the importance of (9.8) in what follows.

Let us say that $f \in [0, 1]$ is a *dyadic rational* if $f = l/2^N$, where l and N are nonnegative integers with $l \leq 2^N$. We can use (9.8) to evaluate $Q(f)$ for all such f. (Since $0 < p < 1$ and $q := 1 - p$, the equations $Q(0) = 0$ and $Q(1) = 1$, which were used in the derivation of (9.8), also follow from it.) First, (9.8) implies that

$$Q(\tfrac{1}{2}) = p. \tag{9.9}$$

Second, (9.8) and (9.9) imply that

$$Q(\tfrac{1}{4}) = p^2, \qquad Q(\tfrac{3}{4}) = p + qp. \tag{9.10}$$

Clearly, we could continue in this way, evaluating $Q(\tfrac{1}{8})$, $Q(\tfrac{3}{8})$, $Q(\tfrac{5}{8})$, $Q(\tfrac{7}{8})$, and so on.

Every dyadic rational $f \in [0, 1)$ can be written in terms of its terminating binary expansion as

$$f = \frac{l}{2^N} = .u_1 u_2 \cdots u_N := \sum_{k=1}^{N} \frac{u_k}{2^k}, \tag{9.11}$$

where $N \geq 1$, $l \in \{0, \ldots, 2^N - 1\}$, and $u_k = 0$ or 1 for $k = 1, \ldots, N$. It will be useful to define the *order* of the dyadic rational (9.11) to be N if l and 2^N are relatively prime (i.e., if l is odd or, equivalently, if $u_N = 1$). Also, 0 and 1 will be said to have order 0. Notice that the dyadic rationals $f \in [0, 1)$ of the form (9.11) are those of order at most N.

It follows by induction on N that, for all dyadic rationals $f \in [0, 1)$ of order at most N, $Q(f)$ is uniquely determined by (9.8). Better yet, we can now derive an explicit formula for $Q(f)$ for all dyadic rationals $f \in [0, 1)$.

Lemma 9.1.1. *For each $N \geq 1$ and $u_k = 0$ or 1 for $k = 1, \ldots, N$, we have*

$$Q(.u_1 \cdots u_N + 2^{-N}) - Q(.u_1 \cdots u_N) = q^{u_1 + \cdots + u_N} p^{N-(u_1 + \cdots + u_N)}, \quad (9.12)$$

and therefore

$$Q(.u_1 \cdots u_N) = \sum_{k=1}^{N} u_k q^{u_1 + \cdots + u_{k-1}} p^{k-(u_1 + \cdots + u_{k-1})}. \quad (9.13)$$

Alternatively, letting $B(i)$ denote the number of 1s in the binary expansion of the nonnegative integer i, we have

$$Q(l/2^N) = \sum_{j=0}^{N} A_j(l) q^j p^{N-j} \quad (9.14)$$

for all $N \geq 1$ and $l \in \{0, \ldots, 2^N\}$, where $A_j(l) = |\{0 \leq i \leq l - 1 : B(i) = j\}|$ (and $A_j(0) := 0$).

Remark. Of the two formulas (9.13) and (9.14), the former is more convenient for computational purposes, while the latter is needed in Section 9.3.

Proof. We prove (9.12) by induction on N. The case $N = 1$ is immediate from $Q(0) = 0$ and (9.9). Let $N \geq 2$, and assume that (9.12) holds with $N - 1$ in place of N. We show that it holds as stated, by considering two cases.

Case 1. $.u_1 \cdots u_N + 2^{-N} \leq \frac{1}{2}$. Here $u_1 = 0$, so by (9.8),

$$\begin{aligned}
Q(.u_1 \cdots u_N &+ 2^{-N}) - Q(.u_1 \cdots u_N) \\
&= p[Q(.u_2 \cdots u_N + 2^{-(N-1)}) - Q(.u_2 \cdots u_N)] \\
&= pq^{u_2 + \cdots + u_N} p^{N-1-(u_2 + \cdots + u_N)} \\
&= q^{u_1 + \cdots + u_N} p^{N-(u_1 + \cdots + u_N)}.
\end{aligned} \quad (9.15)$$

Case 2. $.u_1 \cdots u_N + 2^{-N} > \frac{1}{2}$. Here $u_1 = 1$, so by (9.8),

$$\begin{aligned}
Q(.u_1 \cdots u_N &+ 2^{-N}) - Q(.u_1 \cdots u_N) \\
&= q[Q(.u_2 \cdots u_N + 2^{-(N-1)}) - Q(.u_2 \cdots u_N)]
\end{aligned}$$

$$= qq^{u_2+\cdots+u_N}p^{N-1-(u_2+\cdots+u_N)}$$
$$= q^{u_1+\cdots+u_N}p^{N-(u_1+\cdots+u_N)}. \tag{9.16}$$

In both cases, the second equality uses the induction hypothesis, and (9.12) is proved.

As for (9.13), the result is clearly true if $.u_1 \cdots u_N = 0$, so we assume that $.u_1 \cdots u_N > 0$. Letting $N_1 = \max\{1 \le k \le N : u_k = 1\}$, we have

$$Q(.u_1 \cdots u_N) = Q(.u_1 \cdots u_{N_1})$$
$$= Q(.u_1 \cdots u_{N_1-1}0 + 2^{-N_1})$$
$$= Q(.u_1 \cdots u_{N_1-1}0) + q^{u_1+\cdots+u_{N_1-1}}p^{N_1-(u_1+\cdots+u_{N_1-1})}$$
$$= Q(.u_1 \cdots u_{N_1-1}) + \sum_{k=N_1}^{N} u_k q^{u_1+\cdots+u_{k-1}}p^{k-(u_1+\cdots+u_{k-1})}$$

$$\vdots$$

$$= \sum_{k=1}^{N} u_k q^{u_1+\cdots+u_{k-1}}p^{k-(u_1+\cdots+u_{k-1})}, \tag{9.17}$$

where the third equality uses (9.12) and the fourth holds because the additional terms are 0.

Finally, as for (9.14),

$$Q(l/2^N) = \sum_{i=0}^{l-1}[Q((i+1)/2^N) - Q(i/2^N)]$$
$$= \sum_{i=0}^{l-1} q^{B(i)}p^{N-B(i)}$$
$$= \sum_{j=0}^{N} \sum_{0 \le i \le l-1 : B(i)=j} q^j p^{N-j}$$
$$= \sum_{j=0}^{N} A_j(l)q^j p^{N-j}, \tag{9.18}$$

where the first equality is based on a telescoping sum and the second uses (9.12). This completes the proof. ♠

We next extend (9.13) to all of $[0,1]$.

Theorem 9.1.2. *The function Q is continuous and increasing on $[0,1]$ and is given by the following formula. If $f \in [0,1]$ has binary expansion*

$$f = .u_1u_2\cdots := \sum_{k=1}^{\infty} \frac{u_k}{2^k}, \tag{9.19}$$

where $u_k = 0$ or 1 for every $k \geq 1$, then

$$Q(f) = Q(.u_1 u_2 \cdots) = \sum_{k=1}^{\infty} u_k q^{u_1 + \cdots + u_{k-1}} p^{k-(u_1 + \cdots + u_{k-1})}. \tag{9.20}$$

In particular, if $p = \frac{1}{2}$, then $Q(f) = f$ for all $f \in [0, 1]$.

Proof. We first prove that Q is nondecreasing. By (9.3),

$$F_n(f) = \begin{cases} [2F_{n-1}(f)] \wedge 1 & \text{if } X_n = 1, \\ 0 \vee [2F_{n-1}(f) - 1] & \text{if } X_n = -1, \end{cases} \tag{9.21}$$

so by induction on n, we see that $F_n(f)$ is $[0, 1]$-valued and nondecreasing in $f \in [0, 1]$ for each $n \geq 1$. Therefore, if $0 \leq f \leq g \leq 1$, then

$$\begin{aligned} Q(f) &= \mathrm{P}(F_n(f) = 1 \text{ for some } n \geq 1) \\ &\leq \mathrm{P}(F_n(g) = 1 \text{ for some } n \geq 1) = Q(g). \end{aligned} \tag{9.22}$$

Now a nondecreasing function is continuous if it has no jump discontinuities, and it is increasing if there are no nonempty open intervals on which it is constant. Recalling (9.12), we see that Q has both properties. Finally, the continuity of Q allows us to deduce (9.20) from Lemma 9.1.1. ♠

It is an easy consequence of the theorem (and $q := 1 - p$) that, if $f \in (0, 1)$ is a rational number, then $Q(f)$ is a rational function of p, that is, it is the ratio of two polynomials in p. The following corollary provides a formula.

Corollary 9.1.3. *Given $f \in (0, 1)$ rational, write $f = c/d$ with c and d relatively prime positive integers. Let m be the largest nonnegative integer such that 2^m divides d, and let n be the smallest positive integer such that the odd integer $d/2^m$ divides $2^n - 1$. (For the existence of n, see Theorem A.1.3 on p. 746.) Then we can write f as*

$$f = \frac{k(2^n - 1) + l}{2^m(2^n - 1)}, \tag{9.23}$$

where $0 \leq k \leq 2^m - 1$ and $0 \leq l \leq 2^n - 2$. If, in addition, r and s are the numbers of 1s in the terminating binary expansions of $k/2^m$ and $l/2^n$, respectively, we have

$$Q(f) = Q(k/2^m) + \frac{q^r p^{m-r} Q(l/2^n)}{1 - q^s p^{n-s}}. \tag{9.24}$$

Remark. (a) By Lemma 9.1.1, $Q(k/2^m)$ and $Q(l/2^n)$ are polynomials in p. Therefore, (9.24) implies that $Q(f)$ is a rational function of p.

(b) We do not actually need the assumption that c and d are relatively prime, but it makes for easier application of the corollary.

(c) The definitions of m and n ensure that the terms in the binary expansion of f repeat in blocks of length n following an initial block of length m. This can be seen by expressing (9.23) as

$$f = \frac{k}{2^m} + \frac{l}{2^{m+n}} + \frac{l}{2^{m+2n}} + \frac{l}{2^{m+3n}} + \cdots . \tag{9.25}$$

Proof. By the definitions of m and n, we can write

$$f = \frac{a}{2^m(2^n - 1)}, \tag{9.26}$$

where a is a positive integer less than $2^m(2^n - 1)$. Divide a by $2^n - 1$, and let k be the quotient and l be the remainder. This gives (9.23). Writing

$$\frac{k}{2^m} = .u_1 \cdots u_m \quad \text{and} \quad \frac{l}{2^n} = .v_1 \cdots v_n, \tag{9.27}$$

where each of $u_1, \ldots, u_m$ and $v_1, \ldots, v_n$ equals 0 or 1, (9.25) implies that

$$f = .u_1 \cdots u_m v_1 \cdots v_n v_1 \cdots v_n v_1 \cdots . \tag{9.28}$$

By (9.20),

$$Q(f) = Q(k/2^m) + q^r p^{m-r} Q(l/2^n) + q^{r+s} p^{m+n-r-s} Q(l/2^n)$$
$$+ q^{r+2s} p^{m+2n-r-2s} Q(l/2^n) + \cdots , \tag{9.29}$$

and (9.24) follows.   ♠

Example 9.1.4. *Evaluation of* $Q(1/10)$. Suppose that the bold player wants to multiply his initial fortune by a factor of 10. His probability of achieving this goal is $Q(1/10)$, which can be evaluated using either (9.8), Theorem 9.1.2, or Corollary 9.1.3.

First, repeated application of (9.8) implies that

$$Q(2/10) = pQ(4/10) = p^2 Q(8/10) = p^2[p + qQ(6/10)]$$
$$= p^3 + qp^2[p + qQ(2/10)] = (1 + q)p^3 + q^2 p^2 Q(2/10), \tag{9.30}$$

so

$$Q(1/10) = pQ(2/10) = \frac{(1 + q)p^4}{1 - q^2 p^2}. \tag{9.31}$$

Second, since $1/10$ has binary expansion

$$\frac{1}{10} = .0001100110011 \cdots , \tag{9.32}$$

the theorem tells us that

$$Q(1/10) = p^4 + qp^4 + q^2p^6 + q^3p^6 + q^4p^8 + q^5p^8 + \cdots$$
$$= (1+q)p^4(1 + q^2p^2 + q^4p^4 + \cdots)$$
$$= \frac{(1+q)p^4}{1 - q^2p^2}. \tag{9.33}$$

Third, since

$$\frac{1}{10} = \frac{3}{2(2^4 - 1)}, \tag{9.34}$$

the corollary applies with $m = 1$, $n = 4$, $k = 0$, and $l = 3$. Now $0/2 = .0$ and $3/2^4 = .0011$, so $r = 0$ and $s = 2$, and therefore

$$Q(1/10) = Q(0/2) + \frac{pQ(3/2^4)}{1 - q^2p^2} = \frac{p(p^3 + qp^3)}{1 - q^2p^2} = \frac{(1+q)p^4}{1 - q^2p^2}. \tag{9.35}$$

When evaluating the function Q numerically for a specific p, the series representation (9.20) is probably simplest, as the next example suggests. ♠

Example 9.1.5. *Numerical evaluation of $Q(1/2,000)$ for $p = 949/1,925$.* To evaluate $Q(1/2,000)$, we could apply Corollary 9.1.3, but the binary expansion of $1/2,000$ repeats in blocks of length $n = 54$ after an initial block of length $m = 4$. If our concern is not with $Q(1/2,000)$ as a rational function of p, but merely with $Q(1/2,000)$ for a specific value of p, such as $p = 949/1,925$ (from craps), then the series representation (9.20) leads to a simple algorithm.

In general, how many terms should be evaluated for a specified degree of accuracy? If we truncate the series (9.20) by discarding all terms beyond the Nth one, the error can be bounded as follows:

$$0 \leq Q(.u_1 u_2 \cdots) - \sum_{k=1}^{N} u_k q^{u_1 + \cdots + u_{k-1}} p^{k - (u_1 + \cdots + u_{k-1})}$$
$$= q^{u_1 + \cdots + u_N} p^{N+1 - (u_1 + \cdots + u_N)}$$
$$\cdot \sum_{k=N+1}^{\infty} u_k q^{u_{N+1} + \cdots + u_{k-1}} p^{k - N - 1 - (u_{N+1} + \cdots + u_{k-1})}$$
$$\leq q^{u_1 + \cdots + u_N} p^{N+1 - (u_1 + \cdots + u_N)} \sum_{k=N+1}^{\infty} (q \vee p)^{k - N - 1}$$
$$= \frac{p}{p \wedge q} q^{u_1 + \cdots + u_N} p^{N - (u_1 + \cdots + u_N)}. \tag{9.36}$$

The latter expression is bounded by q^N if $p \leq \frac{1}{2}$, which allows us to determine the minimal N that will ensure accuracy to within a specified degree, regardless of the value of f. For example, with $p = 949/1,925$, $N = 41$ terms ensure accuracy to within 10^{-12}, regardless of f. If we also know f, then (9.36) allows greater precision. For example, if $f = 1/2,000$, the first 41 terms in the binary expansion of f have 14 1s, so the bound in (9.36) becomes $q^{14}p^{27} < 4 \cdot 10^{-13}$, a slight improvement. Regardless, $Q(1/2,000) \approx 0.000427532425$, rounded to 12 decimal places. ♠

We now turn to the problem of establishing the main result of this section, namely that bold play is optimal at subfair red-and-black. The key step in the proof is part (a) of the following lemma.

Lemma 9.1.6. (a) *Assume that* $p \leq \frac{1}{2}$, *and extend* Q *to* $[0, \infty)$ *by defining* $Q(f) := 1$ *if* $f > 1$. *Then, for every* $f \in [0, 1]$ *and* b *satisfying* $0 \leq b \leq f$,

$$Q(f) \geq pQ(f + b) + qQ(f - b). \tag{9.37}$$

(b) *Assume that* $p < \frac{1}{2}$. *If* $0 < f - b < \frac{1}{2} < f + b < 1$, *then strict inequality holds in part* (a).

Remark. Part (a) says that the bold player at subfair (or fair) red-and-black cannot improve his chances by betting nonboldly at the first coup and boldly thereafter.

Proof. (a) By the monotonicity of Q, it is enough to prove (9.37) for $0 \leq b \leq f \wedge (1 - f)$. By the continuity of Q, it is enough to prove the following statement for each $N \geq 0$: (9.37) holds for all dyadic rationals f and b of order at most N, where $0 \leq b \leq f \wedge (1 - f)$. We proceed by induction. The case $N = 1$ is trivial. Let $N \geq 2$, and assume that the statement holds with $N - 1$ in place of N. We show that it holds as stated. Let f and b be dyadic rationals of order at most N with $0 \leq b \leq f \wedge (1 - f)$, and consider four cases.

Case 1. $f + b \leq \frac{1}{2}$. By (9.8),

$$\begin{aligned}
Q(f) &- pQ(f + b) - qQ(f - b) \\
&= p[Q(2f) - pQ(2f + 2b) - qQ(2f - 2b)]. \tag{9.38}
\end{aligned}$$

Case 2. $f - b \geq \frac{1}{2}$. By (9.8),

$$\begin{aligned}
Q(f) &- pQ(f + b) - qQ(f - b) \\
&= q[Q(2f - 1) - pQ(2f - 1 + 2b) - qQ(2f - 1 - 2b)]. \tag{9.39}
\end{aligned}$$

Case 3. $f - b \leq f \leq \frac{1}{2} \leq f + b$. By (9.8),

$$\begin{aligned}
Q(f) &- pQ(f + b) - qQ(f - b) \\
&= pQ(2f) - p[p + qQ(2f - 1 + 2b)] - qpQ(2f - 2b) \\
&= p[Q(2f) - p - qQ(2f - 1 + 2b) - qQ(2f - 2b)] \\
&= pq[Q(4f - 1) - Q(2f - 1 + 2b) - Q(2f - 2b)] \\
&= q[Q(2f - \tfrac{1}{2}) - pQ(2f - 1 + 2b) - pQ(2f - 2b)] \tag{9.40} \\
&\geq q[Q(2f - \tfrac{1}{2}) - pQ(2f - \tfrac{1}{2} + |2b - \tfrac{1}{2}|) - qQ(2f - \tfrac{1}{2} - |2b - \tfrac{1}{2}|)].
\end{aligned}$$

Here the third equality uses $2f \geq f + b \geq \frac{1}{2}$, the fourth equality uses $2f - \frac{1}{2} \leq 2(\frac{1}{2}) - \frac{1}{2} = \frac{1}{2}$, and the inequality uses $p \leq q$.

Case 4. $f - b \leq \frac{1}{2} \leq f \leq f + b$. Again by (9.8),

$$
\begin{aligned}
Q(f) &- pQ(f + b) - qQ(f - b) \\
&= p + qQ(2f - 1) - p[p + qQ(2f - 1 + 2b)] - qpQ(2f - 2b) \\
&= q[p + Q(2f - 1) - pQ(2f - 1 + 2b) - pQ(2f - 2b)] \\
&= qp[1 + Q(4f - 2) - Q(2f - 1 + 2b) - Q(2f - 2b)] \\
&= p[q + Q(2f - \tfrac{1}{2}) - p - qQ(2f - 1 + 2b) - qQ(2f - 2b)] \qquad (9.41) \\
&\geq p[Q(2f - \tfrac{1}{2}) - pQ(2f - \tfrac{1}{2} + |2b - \tfrac{1}{2}|) - qQ(2f - \tfrac{1}{2} - |2b - \tfrac{1}{2}|)].
\end{aligned}
$$

Here the third equality uses $2f - 1 \leq f - b \leq \frac{1}{2}$, the fourth equality uses $2f - \frac{1}{2} \geq 2(\frac{1}{2}) - \frac{1}{2} = \frac{1}{2}$, and the inequality again uses $p \leq q$.

The induction hypothesis applies in all four cases, telling us that the right sides of (9.38)–(9.41) are nonnegative, hence so too are the left sides.

(b) If $0 < f - b < \frac{1}{2} < f + b < 1$, then either case 3 or case 4 of part (a) applies. Further, inequalities (9.40) and (9.41) hold without the restriction to dyadic rationals. The inequality in (9.40) is strict if both $Q(2f - 1 + 2b) > 0$ and $Q(2f - 2b) > 0$, and these inequalities hold by the assumptions on f and b. The inequality in (9.41) is strict if both $Q(2f - 1 + 2b) < 1$ and $Q(2f - 2b) < 1$, and these inequalities also hold by the assumptions on f and b. Since the right sides of both inequalities are nonnegative by part (a), the proof is complete.   ♠

We are now ready to establish the optimality of bold play at subfair red-and-black. First, let us consider an arbitrary alternative strategy. As in Chapter 8, such a strategy consists of a sequence of bet sizes, $B_1, B_2, \ldots$, with B_n representing the amount bet at coup n and depending on only the results of the previous coups, $X_1, \ldots, X_{n-1}$. (Here $X_1, X_2, \ldots$ are i.i.d. with common distribution given by (9.2).) More precisely,

$$
B_1 = b_1 \geq 0, \qquad B_n = b_n(X_1, \ldots, X_{n-1}) \geq 0, \quad n \geq 2, \qquad (9.42)
$$

where b_1 is a constant and b_n is a nonrandom function of $n - 1$ variables for each $n \geq 2$. The gambler's fortune F_n after n coups satisfies

$$
F_n = F_{n-1} + B_n X_n, \qquad n \geq 1, \qquad (9.43)
$$

F_0 being his initial fortune (a nonnegative constant). The only restriction is that he cannot bet more than he has, that is,

$$
B_n \leq F_{n-1}, \qquad n \geq 1. \qquad (9.44)
$$

Finally, we define

$$
T := \min\{n \geq 0 : F_n \in \{0\} \cup [1, \infty)\}, \qquad (9.45)
$$

and we assume that $P(T < \infty) = 1$.

Recall the notation (9.3).

Theorem 9.1.7. *Assume that* $p \leq \frac{1}{2}$. *If* $F_0 = f \in [0,1]$, *then*

$$P(F_T \geq 1) \leq Q(f) := P(F_{T(f)}(f) = 1). \tag{9.46}$$

Remark. The theorem says that bold play is optimal at subfair (or fair) red-and-black. Here optimality means that no strategy achieves the gambler's goal with a higher probability than does bold play.

Proof. As in the statement of Lemma 9.1.6, we extend Q to $[0, \infty)$ by defining $Q(f) := 1$ if $f > 1$. Observe that

$$\begin{aligned}
E[Q(F_n) \mid X_1, \ldots, X_{n-1}] &= E[Q(F_{n-1} + B_n X_n) \mid X_1, \ldots, X_{n-1}] \\
&= pQ(F_{n-1} + B_n) + qQ(F_{n-1} - B_n) \\
&\leq Q(F_{n-1}), \tag{9.47}
\end{aligned}$$

where the second equality uses Theorem 2.2.4 on p. 81 and the inequality uses Lemma 9.1.6(a) if $F_{n-1} \leq 1$ and is obvious otherwise. Consequently, $\{Q(F_n)\}_{n \geq 0}$ is a supermartingale with respect to $\{X_n\}_{n \geq 1}$, and the optional stopping theorem (Theorem 3.2.2 on p. 100) implies that

$$P(F_T \geq 1) = E[Q(F_T)] \leq Q(F_0) = Q(f), \tag{9.48}$$

as required. ♠

We conclude this section by addressing the question of whether bold play is uniquely optimal at subfair red-and-black. We now require $p < \frac{1}{2}$.

Let us consider the strategy (9.42)–(9.45). Because Q is increasing on $[0,1]$, it is suboptimal to overbet bold play, so we replace (9.44) by

$$B_n \leq F_{n-1} \wedge (1 - F_{n-1}), \qquad n \geq 1. \tag{9.49}$$

Further, we can assume without loss of generality that $B_n = 0$ if $n \geq T+1$ (if not, we replace B_n by $B_n 1_{\{n \leq T\}}$). Then $F_n \to F_T$ a.s. as $n \to \infty$, so

$$Q(F_0) \geq E[Q(F_1)] \geq E[Q(F_2)] \geq \cdots \geq E[Q(F_n)] \to E[Q(F_T)]. \tag{9.50}$$

Therefore, a necessary and sufficient condition for equality in (9.46) is equality throughout (9.50), or, equivalently, equality with probability 1 in (9.47) for each $n \geq 1$.

Let us say that a bet $b \in [0, f \wedge (1-f)]$ is *conserving* at $f \in [0,1]$ if

$$Q(f) = pQ(f + b) + qQ(f - b). \tag{9.51}$$

Trivially, the zero bet is conserving at $f \in [0,1]$, and by (9.8), the bold bet

$$b_0(f) := f \wedge (1 - f) \qquad (9.52)$$

is also conserving at $f \in [0, 1]$. Are there any others?

Given a closed interval $[\alpha, \beta] \subset [0, 1]$, let us define *bold play in* $[\alpha, \beta]$ to be the strategy that bets $(f - \alpha) \wedge (\beta - f)$ at $f \in [\alpha, \beta]$ and stops betting as soon as α or β is reached. Bold play in $[0, 1]$ will be referred to simply as *bold play*. Consider the following strategy.

If $f \in [0, \frac{1}{2}]$, use bold play in $[0, \frac{1}{2}]$; if the goal of $\frac{1}{2}$ is achieved, use bold play from $\frac{1}{2}$. The probability of achieving the goal of 1 in this way is

$$(1 - Q(2f))Q(0) + Q(2f)Q(\tfrac{1}{2}) = pQ(2f) = Q(f). \qquad (9.53)$$

If $f \in [\frac{1}{2}, 1]$, use bold play in $[\frac{1}{2}, 1]$; if the goal of 1 fails to be achieved, use bold play from $\frac{1}{2}$. The probability of achieving the goal of 1 in this way is

$$(1 - Q(2f - 1))Q(\tfrac{1}{2}) + Q(2f - 1)Q(1) = p + qQ(2f - 1) = Q(f). \qquad (9.54)$$

This suggests that the bet

$$b_1(f) := \begin{cases} \frac{1}{2}b_0(2f) & \text{if } 0 \le f \le \frac{1}{2}, \\ \frac{1}{2}b_0(2f - 1) & \text{if } \frac{1}{2} \le f \le 1, \end{cases} \qquad (9.55)$$

is conserving at f. In fact we can iterate this procedure, inductively defining

$$b_N(f) := \begin{cases} \frac{1}{2}b_{N-1}(2f) & \text{if } 0 \le f \le \frac{1}{2}, \\ \frac{1}{2}b_{N-1}(2f - 1) & \text{if } \frac{1}{2} \le f \le 1, \end{cases} \qquad (9.56)$$

for each $N \ge 1$. One can check that $b_N(f)$ is the distance between f and the closest dyadic rational to f of order at most N.

Theorem 9.1.8. *Assume that $p < \frac{1}{2}$. Then equality holds in (9.46), where $F_0 = f \in [0, 1]$, if and only if B_n is conserving at F_{n-1} with probability 1 for each $n \ge 1$. Furthermore, $b \in [0, f \wedge (1 - f)]$ is conserving at $f \in [0, 1]$ if and only if either $b = 0$ or $b = b_N(f)$ for some $N \ge 0$.*

Proof. The first assertion is a consequence of the preceding discussion.

For the second assertion, first we show, by induction on N, that $b_N(f)$ is conserving at f for each $f \in [0, 1]$. We have already noted that the statement is true for $N = 0$. Assume that $N \ge 1$ and $b_{N-1}(f)$ is conserving at f for each $f \in [0, 1]$. Given $f \in [0, 1]$, if $0 \le f \le \frac{1}{2}$, then

$$pQ(f + b_N(f)) + qQ(f - b_N(f))$$
$$= pQ(\tfrac{1}{2}(2f + b_{N-1}(2f))) + qQ(\tfrac{1}{2}(2f - b_{N-1}(2f)))$$
$$= p[pQ(2f + b_{N-1}(2f))] + q[pQ(2f - b_{N-1}(2f))]$$
$$= p[pQ(2f + b_{N-1}(2f)) + qQ(2f - b_{N-1}(2f))]$$

$$= pQ(2f)$$
$$= Q(f), \tag{9.57}$$

where the second equality uses $\frac{1}{2}[2f \pm b_{N-1}(2f)] \in [0, \frac{1}{2}]$; if $\frac{1}{2} \leq f \leq 1$, then

$$
\begin{aligned}
pQ(f &+ b_N(f)) + qQ(f - b_N(f)) \\
&= pQ(\tfrac{1}{2}(2f + b_{N-1}(2f-1))) + qQ(\tfrac{1}{2}(2f - b_{N-1}(2f-1))) \\
&= p[p + qQ(2f - 1 + b_{N-1}(2f-1))] \\
&\quad + q[p + qQ(2f - 1 - b_{N-1}(2f-1))] \\
&= p + q[pQ(2f - 1 + b_{N-1}(2f-1)) + qQ(2f - 1 - b_{N-1}(2f-1))] \\
&= p + qQ(2f - 1) \\
&= Q(f), \tag{9.58}
\end{aligned}
$$

where the second equality uses $2f - 1 \pm b_{N-1}(2f-1)] \in [0,1]$, hence $\frac{1}{2}[2f \pm b_{N-1}(2f-1)] \in [\frac{1}{2}, 1]$. In either case $b_N(f)$ is conserving at f.

Turning to the converse, suppose, to get a contradiction, that there exists a pair (f, b) with $f \in [0, 1]$ and $b \in [0, b_0(f)]$ such that (a) b is conserving at f, (b) b differs from 0 and from $b_N(f)$ for each $N \geq 0$, and (c) b is more than half as large as the supremum of all such b. It cannot be the case that $f + b \leq \frac{1}{2}$, for if it were, then

$$
\begin{aligned}
Q(2f) &= p^{-1}Q(f) \\
&= p^{-1}[pQ(f + b) + qQ(f - b)] \\
&= p^{-1}p[pQ(2f + 2b)] + p^{-1}q[pQ(2f - 2b)] \\
&= pQ(2f + 2b) + qQ(2f - 2b), \tag{9.59}
\end{aligned}
$$

so $2b$ would be conserving at $2f$, and we would have a contradiction to our choice of b. (Note that $2b = b_N(2f)$ would imply $b = b_{N+1}(f)$.) Similarly, it cannot be the case that $f - b \geq \frac{1}{2}$, for if it were, then

$$
\begin{aligned}
Q(2f - 1) &= q^{-1}[Q(f) - p] \\
&= q^{-1}[pQ(f + b) + qQ(f - b) - p] \\
&= q^{-1}p[p + qQ(2f - 1 + 2b)] + q^{-1}q[p + qQ(2f - 1 - 2b)] - q^{-1}p \\
&= pQ(2f - 1 + 2b) + qQ(2f - 1 - 2b), \tag{9.60}
\end{aligned}
$$

so $2b$ would be conserving at $2f - 1$, and we would again have a contradiction to our choice of b. (Note that $2b = b_N(2f - 1)$ would imply $b = b_{N+1}(f)$.)

The only other possibility is that $0 < f - b < \frac{1}{2} < f + b < 1$, but in this case b cannot be conserving for f by Lemma 9.1.6(b). This contradiction completes the proof. ♠

Example 9.1.9. *Comparison of two optimal strategies at* $f = \frac{1}{3}$. We could use Theorem 9.1.2 or Corollary 9.1.3 to evaluate $Q(\frac{1}{3})$, but (9.8) is arguably simpler: The system

$$Q(\tfrac{1}{3}) = pQ(\tfrac{2}{3}),$$
$$Q(\tfrac{2}{3}) = p + qQ(\tfrac{1}{3}),$$

(9.61)

implies that $Q(\tfrac{1}{3}) = p^2/(1 - qp)$ and $Q(\tfrac{2}{3}) = p/(1 - qp)$.

Similarly, if $E(f)$ denotes the expected number of positive bets made by the bold player with initial fortune $f \in [0, 1]$, then $E(\tfrac{1}{2}) = 1$ and

$$E(\tfrac{1}{3}) = 1 + pE(\tfrac{2}{3}),$$
$$E(\tfrac{2}{3}) = 1 + qE(\tfrac{1}{3}).$$

(9.62)

It follows that $E(\tfrac{1}{3}) = (1 + p)/(1 - qp)$ and $E(\tfrac{2}{3}) = (1 + q)/(1 - qp)$.

Note that, for the bold player with initial fortune $f = \tfrac{1}{3}$, the only positive bets he makes are of size $\tfrac{1}{3}$, so the expected number of positive bets made is $E(\tfrac{1}{3})$, while the expected total amount bet is $\tfrac{1}{3}E(\tfrac{1}{3})$.

On the other hand, for the player who uses the optimal nonbold strategy with initial fortune $f = \tfrac{1}{3}$ that begins with bold play in $[0, \tfrac{1}{2}]$, the expected number of positive bets made is

$$E(\tfrac{2}{3}) + Q(\tfrac{2}{3})E(\tfrac{1}{2}) = \frac{1 + q}{1 - qp} + \frac{p}{1 - qp} = \frac{2}{1 - qp} > E(\tfrac{1}{3}),$$

(9.63)

while the expected total amount bet is

$$\tfrac{1}{6}E(\tfrac{2}{3}) + \tfrac{1}{2}Q(\tfrac{2}{3})E(\tfrac{1}{2}) = \tfrac{1}{3}E(\tfrac{1}{3}).$$

(9.64)

Thus, the nonbold strategy takes longer than bold play on average but the amount bet is the same on average. If we think of the house advantage as taking a percentage of each unit bet, it is the expected total amount bet, not the expected number of positive bets, that we seek to minimize.   ♠

9.2 Red-and-Black with a House Limit

We continue with our study of red-and-black, a game of independent coups in which a single betting opportunity is available at each coup, namely a bet that pays even money, winning with probability p and losing with probability $q := 1 - p$. We assume throughout that $0 < p < 1$. We also continue to measure money in units of the gambler's goal, or, equivalently, to assume that his goal is 1 unit, and that units are infinitely divisible.

In this section we assume a house-imposed maximum betting limit L, which we require to be the reciprocal of a positive integer M. This simply amounts to the assumption that the gambler's goal (namely, 1) is a positive-integer multiple of the house limit. Hereafter, $L := M^{-1}$.

Here *bold play* consists of betting, at each coup, either one's entire fortune, just enough to achieve one's goal in the event of a win, or the house limit, whichever amount is smaller. Therefore, the bold player at red-and-black with house limit L bets $f \wedge (1 - f) \wedge L$ when his fortune is $f \in [0, 1]$. The cases $M = 1$ and $M = 2$ are indistinguishable from unlimited bold play, so we can assume that $M \geq 3$.

Let $X_1, X_2, \ldots$ be a sequence of i.i.d. random variables with common distribution given by (9.2) on p. 318, with X_n representing the gambler's profit per unit bet at coup n. Given an initial fortune $f \in [0, 1]$, let $F_n(f)$ represent the bold player's fortune after n coups. Then

$$F_n(f) = F_{n-1}(f) + [F_{n-1}(f) \wedge (1 - F_{n-1}(f)) \wedge L]X_n \qquad (9.65)$$

for each $n \geq 1$, where $F_0(f) := f$.

This has an important consequence. If $f \in (0, 1)$ is rational, then there is a finite set $G \subset [0, 1]$ such that, with probability 1, $F_n(f) \in G$ for all $n \geq 1$. Indeed, if j and k are positive integers such that $f = j/k$, and if m is the least common multiple of k and M, then we may take $G := \{i/m : i = 0, 1, \ldots, m\}$.

Our primary interest is in the expected utility of the bold player's final fortune when the utility function is given by the indicator of $[1, \infty)$. This is simply

$$P(f) := \mathrm{P}(F_n(f) = 1 \text{ for some } n \geq 1), \qquad 0 \leq f \leq 1. \qquad (9.66)$$

Conditioning on X_1, we find as in (9.7) on p. 319 that

$$P(f) = pP(f + [f \wedge (1 - f) \wedge L]) + qP(f - [f \wedge (1 - f) \wedge L]), \qquad (9.67)$$

or, since $P(0) = 0$ and $P(1) = 1$,

$$P(f) = \begin{cases} pP(2f) & \text{if } f \in [0, L] \cap C, \\ pP(f + L) + qP(f - L) & \text{if } f \in [L, 1 - L] \cap C, \\ p + qP(2f - 1) & \text{if } f \in [1 - L, 1] \cap C, \end{cases} \qquad (9.68)$$

with $C = [0, 1]$. (The reason for C will be seen shortly.)

We introduce, for use in this section only, the following compact notation for the gambler's ruin formula (see Theorem 7.1.1 on p. 242), stated in terms of the probability of achieving the goal: For $j = 0, 1, \ldots, M$,

$$u(j) := \begin{cases} ((q/p)^j - 1)/((q/p)^M - 1) & \text{if } p \neq \frac{1}{2}, \\ j/M & \text{if } p = \frac{1}{2}. \end{cases} \qquad (9.69)$$

Of course $u(0) = 0$ and $u(M) = 1$. We further introduce, also for use in this section only, notation for the fractional part of a nonnegative real number x, namely $\{x\} := x - \lfloor x \rfloor$.

We begin with a technical lemma, in which C is the set of dyadic rational multiples of L in $[0, 1]$ (i.e., $f \in C$ if $f = (l/2^N)L$ for some $N \geq 1$ and $l \in \{0, 1, \ldots, 2^N M\}$), whereas D is the set of dyadic rationals in $[0, 1]$.

Lemma 9.2.1. *The function P^* on $[0, 1]$, defined in terms of the function Q^* on $[0, 1]$ by*

$$P^*(f) = (1 - Q^*(\{Mf\}))u(\lfloor Mf \rfloor) + Q^*(\{Mf\})u(\lfloor Mf \rfloor + 1), \qquad (9.70)$$

satisfies

$$P^*(f) = \begin{cases} pP^*(2f) & \text{if } f \in [0, L] \cap C, \\ pP^*(f + L) + qP^*(f - L) & \text{if } f \in [L, 1 - L] \cap C, \\ p + qP^*(2f - 1) & \text{if } f \in [1 - L, 1] \cap C, \end{cases} \qquad (9.71)$$

if and only if Q^* satisfies

$$Q^*(g) = \begin{cases} pQ^*(2g) & \text{if } g \in [0, \tfrac{1}{2}] \cap D, \\ p + qQ^*(2g - 1) & \text{if } g \in [\tfrac{1}{2}, 1] \cap D. \end{cases} \qquad (9.72)$$

Remark. The lemma will allow us to prove that P of (9.66) and Q of (9.6) on p. 319 are related in the same way that P^* and Q^* are related in (9.70).

Proof. Given $f \in [0, 1]$, define $g := \{Mf\}$. Notice that $f \in C$ if and only if $g \in D$. There are six cases to consider, the totality of which will complete the proof.

Case 1. $0 \leq f < \frac{1}{2}L$. Here $g = Mf \in [0, \frac{1}{2})$, so $\{M(2f)\} = 2g$ and

$$P^*(f) - pP^*(2f) = [Q^*(g) - pQ^*(2g)]u(1). \qquad (9.73)$$

Case 2. $\frac{1}{2}L \leq f < L$. Here $g = Mf \in [\frac{1}{2}, 1)$, so $\{M(2f)\} = 2g - 1$ and

$$\begin{aligned} P^*(f) - pP^*(2f) &= Q^*(g)u(1) - p[(1 - Q^*(2g - 1))u(1) + Q^*(2g - 1)u(2)] \\ &= [Q^*(g) - p - qQ^*(2g - 1)]u(1) \end{aligned} \qquad (9.74)$$

since $u(1) = pu(2)$, hence $u(2) - u(1) = (q/p)u(1)$.

Case 3. $L \leq f < 1 - L$. Let $k := \lfloor Mf \rfloor \in \{1, \ldots, M - 2\}$. Since $\{M(f - L)\} = \{M(f + L)\} = g$, we have

$$\begin{aligned} P^*(f) &- pP^*(f + L) - qP^*(f - L) \\ &= (1 - Q^*(g))u(k) + Q^*(g)u(k + 1) \\ &\qquad - p[(1 - Q^*(g))u(k + 1) + Q^*(g)u(k + 2)] \\ &\qquad - q[(1 - Q^*(g))u(k - 1) + Q^*(g)u(k)] \end{aligned}$$

$$= (1 - Q^*(g))[u(k) - pu(k+1) - qu(k-1)]$$
$$+ Q^*(g)[u(k+1) - pu(k+2) - qu(k)]$$
$$= 0. \tag{9.75}$$

Case 4. $1 - L \leq f < 1 - \frac{1}{2}L$. Here $g = Mf - (M-1) \in [0, \frac{1}{2})$, so $\{M(2f-1)\} = 2g$ and

$$P^*(f) - p - qP^*(2f-1)$$
$$= (1 - Q^*(g))u(M-1) + Q^*(g)$$
$$\qquad - p - q[(1 - Q^*(2g))u(M-2) + Q^*(2g)u(M-1)]$$
$$= [Q^*(g) - pQ^*(2g)][1 - u(M-1)] \tag{9.76}$$

since $u(M-1) = p + qu(M-2)$, hence $u(M-1) - u(M-2) = (p/q)[1 - u(M-1)]$.

Case 5. $1 - \frac{1}{2}L \leq f < 1$. Here $g = Mf - (M-1) \in [\frac{1}{2}, 1)$, so $\{M(2f-1)\} = 2g - 1$ and

$$P^*(f) - p - qP^*(2f-1)$$
$$= (1 - Q^*(g))u(M-1) + Q^*(g)$$
$$\qquad - p - q[(1 - Q^*(2g-1))u(M-1) + Q^*(2g-1)]$$
$$= [Q^*(g) - p - qQ^*(2g-1)][1 - u(M-1)]. \tag{9.77}$$

Case 6. $f = 1$. This case is trivial. ♠

Theorem 9.2.2. *The function P of (9.66) is continuous and increasing on $[0,1]$ and is given by*

$$P(f) = (1 - Q(\{Mf\}))u(\lfloor Mf \rfloor) + Q(\{Mf\})u(\lfloor Mf \rfloor + 1), \tag{9.78}$$

where Q is as in (9.6) on p. 319. In particular, if $p = \frac{1}{2}$, then $P(f) = f$ for all $f \in [0,1]$.

Remark. Notice that (9.78) implies that there is an alternative to bold play having the same expected utility. Specifically, if $f \in ((k-1)L, kL)$ for some $k \in \{1, \ldots, M\}$, the player begins with bold play in the interval $[(k-1)L, kL]$. Once he has reached either of the endpoints, he bets L at each coup until he is ruined or the goal is achieved.

Proof. First, we need a simple fact about certain tridiagonal systems of linear equations: If

$$A(j) = pA(j+1) + qA(j-1), \qquad j = 1, \ldots n-1, \tag{9.79}$$

then

$$A(j) = A(0)\left(1 - \frac{(q/p)^j - 1}{(q/p)^n - 1}\right) + A(n)\frac{(q/p)^j - 1}{(q/p)^n - 1} \tag{9.80}$$

for $j = 1, \ldots, n - 1$. Thus, each of the interior terms $A(1), \ldots, A(n-1)$ is uniquely determined by the two boundary terms $A(0)$ and $A(n)$. See Problem 7.3 on p. 267.

We use this fact to show that (9.68), with C being the dyadic rational multiples of L in $[0, 1]$, uniquely determines $P(f)$ for all $f \in [0, 1]$ that are dyadic rational multiples of L. More precisely, we prove the following statement for each $N \geq 0$: $P(f)$ *is uniquely determined for all* $f \in \{(k/2^N)L : k = 0, 1, \ldots, 2^N M\}$. We proceed by induction on N. The case $N = 0$ can be written as

$$P(kL) = pP((k+1)L) + qP((k-1)L), \qquad k = 1, \ldots, M-1, \qquad (9.81)$$

so, by virtue of (9.80), $P(L), P(2L), \ldots, P((M-1)L)$ are uniquely determined by $P(0) = 0$ and $P(1) = 1$.

Let $N \geq 1$ and assume that the statement holds with $N - 1$ in place of N. If $f := (k/2^N)L$ belongs to $[0, L] \cup [1 - L, 1]$, then (9.68) expresses $P(f)$ in terms of $P(2f)$ or $P(2f - 1)$, which are uniquely determined by virtue of the induction hypothesis. If, on the other hand, $f := (k/2^N)L$ belongs to $(L, 1-L)$, we can assume that k is odd, for otherwise the induction hypothesis would apply. Let $f_0 := \{k/2^N\}L$ and $f_j := f_0 + jL$ for $j = 1, \ldots, M-1$. Then

$$P(f_j) = pP(f_{j+1}) + qP(f_{j-1}), \qquad j = 1, \ldots, M-2, \qquad (9.82)$$

$f_0 \in (0, L)$, $f_{M-1} \in (1 - L, 1)$, and $f = f_j$ for some $j \in \{1, \ldots, M-2\}$. It follows from (9.80) that $P(f)$ is uniquely determined in terms of $P(f_0)$ and $P(f_{M-1})$, which we have already seen are uniquely determined. This completes the induction step.

Let us define $P^*(f)$ by (9.70) with the role of Q^* played by Q of (9.6) on p. 319. We need to show that $P(f) = P^*(f)$ for all $f \in [0, 1]$. Since Q satisfies (9.8) on p. 319, Lemma 9.2.1 shows that P^* satisfies (9.71) with C being the set of dyadic rational multiples of L in $[0, 1]$. The uniqueness assertion just proved implies that $P(f) = P^*(f)$ for all $f \in [0, 1]$ that are dyadic rational multiples of L.

By (9.65),

$$F_n(f) = \begin{cases} [2F_{n-1}(f)] \wedge 1 \wedge [F_{n-1}(f) + L] & \text{if } X_n = 1, \\ 0 \vee [2F_{n-1}(f) - 1] \vee [F_{n-1}(f) - L] & \text{if } X_n = -1, \end{cases} \qquad (9.83)$$

for each $n \geq 1$, and just as in (9.22) on p. 322 we conclude that $P(f)$ is nondecreasing in $f \in [0, 1]$. On the other hand, the facts that Q is continuous and increasing on $[0, 1]$ imply that $P^*(f)$ is continuous and increasing on $[(k-1)L, kL]$ for $k = 1, \ldots, M$. (Left continuity at the right endpoint follows from the continuity of Q at 1.) Thus, P^* is continuous and increasing on $[0, 1]$.

To summarize, P and P^* are nondecreasing on $[0, 1]$, P^* is also continuous and increasing, and $P(f) = P^*(f)$ for all f in a dense subset of $[0, 1]$. It follows that $P(f) = P^*(f)$ for all $f \in [0, 1]$, proving (9.78).

Finally, in the case $p = \frac{1}{2}$, $Q(f) \equiv f$ and $u(j) \equiv j/M$, so (9.78) yields

$$P(f) = (1 - \{Mf\})\frac{\lfloor Mf \rfloor}{M} + \{Mf\}\frac{\lfloor Mf \rfloor + 1}{M} = f, \qquad (9.84)$$

and the proof is complete. ♠

We now turn to the problem of establishing the main result of this section, namely that bold play is optimal at subfair red-and-black with a house limit, provided the player's goal is an integer multiple of the house limit. The key step in the proof is the following lemma.

Lemma 9.2.3. *Assume that $p \leq \frac{1}{2}$, and extend P to $[0, \infty)$ by defining $P(f) := 1$ if $f > 1$. Then, for all $f \in [0, 1]$ and b satisfying $0 \leq b \leq f \wedge L$,*

$$P(f) \geq pP(f + b) + qP(f - b). \qquad (9.85)$$

Remark. The lemma says that the bold player at subfair (or fair) red-and-black with a house limit of $L := M^{-1}$ units cannot improve his chances by betting nonboldly at the first coup and boldly thereafter.

Proof. If (9.85) were to fail for some pair (f, b) with $0 \leq b \leq f \wedge L$ and $f + b > 1$, then it would also fail for the pair $(f, 1 - f)$, so we can assume that $0 \leq b \leq f \wedge (1 - f) \wedge L$.

Define $u(j)$ by (9.69) for $j = 0, 1, \ldots, M$. Fix $f \in [0, 1]$ and define

$$\begin{aligned} k := \lfloor Mf \rfloor, \qquad k_1 := \lfloor M(f + b) \rfloor, \qquad k_0 := \lfloor M(f - b) \rfloor, \\ g := \{Mf\}, \qquad g_1 := \{M(f + b)\}, \qquad g_0 := \{M(f - b)\}, \end{aligned} \qquad (9.86)$$

$c := Mb \leq 1$, and $V := u(k + 1) - u(k) > 0$. Then

$$\begin{aligned} P(f) &- pP(f + b) - qP(f - b) \\ &= (1 - Q(g))u(k) + Q(g)u(k + 1) \\ &\quad - p[(1 - Q(g_1))u(k_1) + Q(g_1)u(k_1 + 1)] \\ &\quad - q[(1 - Q(g_0))u(k_0) + Q(g_0)u(k_0 + 1)] \\ &= [Q(g) - p(\delta_{k_1, k+1} - \delta_{k_0, k-1}) \\ &\quad - p(q/p)^{k_1 - k}Q(g_1) - q(p/q)^{k - k_0}Q(g_0)]V. \qquad (9.87) \end{aligned}$$

Case 1. $k_1 = k$, $k_0 = k$. Here $g_1 = g + c \in [0, 1)$ and $g_0 = g - c \in [0, 1)$, so $0 \leq c \leq g \wedge (1 - g)$ and

$$\begin{aligned} P(f) &- pP(f + b) - qP(f - b) \\ &= [Q(g) - pQ(g + c) - qQ(g - c)]V. \qquad (9.88) \end{aligned}$$

Case 2. $k_1 = k+1$, $k_0 = k$. Here $g_1 = g+c-1 \in [0,1)$ and $g_0 = g-c \in [0,1)$, so $1-g \leq c \leq g$ or, equivalently, $|c - \frac{1}{2}| \leq g - \frac{1}{2}$, and

$$
\begin{aligned}
P(f) &- pP(f+b) - qP(f-b) \\
&= [Q(g) - p - qQ(g+c-1) - qQ(g-c)]V \\
&= q[Q(2g-1) - Q(g+c-1) - Q(g-c)]V \\
&= (q/p)[Q(g - \tfrac{1}{2}) - pQ(g+c-1) - pQ(g-c)]V \qquad (9.89) \\
&\geq (q/p)[Q(g - \tfrac{1}{2}) - pQ(g - \tfrac{1}{2} + |c - \tfrac{1}{2}|) - qQ(g - \tfrac{1}{2} - |c - \tfrac{1}{2}|)]V,
\end{aligned}
$$

where the second and third equalities use (9.8) on p. 319, and the inequality uses $p \leq q$.

Case 3. $k_1 = k$, $k_0 = k-1$. Here $g_1 = g+c \in [0,1)$ and $g_0 = g-c+1 \in [0,1)$, so $g < c < 1-g$ or, equivalently, $|c - \frac{1}{2}| < \frac{1}{2} - g$, and

$$
\begin{aligned}
P(f) &- pP(f+b) - qP(f-b) \\
&= [Q(g) + p - pQ(g+c) - pQ(g-c+1)]V \\
&= p[Q(2g) + 1 - Q(g+c) - Q(g-c+1)]V \\
&= (p/q)[Q(g + \tfrac{1}{2}) - p + q - qQ(g+c) - qQ(g-c+1)]V \qquad (9.90) \\
&\geq (p/q)[Q(g + \tfrac{1}{2}) - pQ(g + \tfrac{1}{2} + |c - \tfrac{1}{2}|) - qQ(g + \tfrac{1}{2} - |c - \tfrac{1}{2}|)]V,
\end{aligned}
$$

where the second and third equalities use (9.8) on p. 319, and the inequality uses $p \leq q$.

Case 4. $k_1 = k+1$, $k_0 = k-1$. Here $g_1 = g+c-1 \in [0,1)$ and $g_0 = g-c+1 \in [0,1)$, so $g \vee (1-g) \leq c \leq 1$ or, equivalently, $0 \leq 1-c \leq g \wedge (1-g)$, and

$$
\begin{aligned}
P(f) &- pP(f+b) - qP(f-b) \\
&= [Q(g) - qQ(g+c-1) - pQ(g-c+1)]V \\
&= [Q(g) - pQ(g+(1-c)) - qQ(g-(1-c))]V. \qquad (9.91)
\end{aligned}
$$

In all four cases, the final expression is nonnegative by Lemma 9.1.6(a) on p. 325, so the proof is complete. ♠

We are now ready to establish the optimality of bold play at subfair red-and-black with a house limit of the form $L := M^{-1}$ for some integer $M \geq 3$. The argument is virtually identical to the one used to prove Theorem 9.1.7 on p. 327.

First, let us consider an arbitrary alternative strategy. As in Section 9.1, such a strategy consists of a sequence of bet sizes, $B_1, B_2, \ldots$, with B_n representing the amount bet at coup n and depending on only the results of the previous coups, $X_1, \ldots, X_{n-1}$. (Here $X_1, X_2, \ldots$ are i.i.d. with common distribution given by (9.2) on p. 318.) More precisely, (9.42) on p. 326 holds, where b_1 is a constant and b_n is a nonrandom function of $n-1$ variables for each $n \geq 2$. The gambler's fortune F_n after n coups satisfies (9.43) on p. 326,

F_0 being his initial fortune (a nonnegative constant). The only restrictions are that he cannot bet more than he has and he cannot bet more than L units, that is,

$$B_n \leq F_{n-1} \wedge L, \qquad n \geq 1. \tag{9.92}$$

Finally, we define T by (9.45) on p. 326, and we assume that $P(T < \infty) = 1$. Recall the notation (9.65).

Theorem 9.2.4. *Assume that $p \leq \frac{1}{2}$. If $F_0 = f \in [0, 1]$, then*

$$P(F_T \geq 1) \leq P(f) := P(F_{T(f)}(f) = 1), \tag{9.93}$$

where $T(f)$ is the coup at which the last positive bet by the bold player is made, assuming an initial fortune of $f \in [0, 1]$.

Remark. The theorem says that bold play is optimal at subfair (or fair) red-and-black with a house limit, provided the gambler's goal is a positive-integer multiple of the house limit. Here optimality means that no strategy achieves the gambler's goal with a higher probability than does bold play.

Proof. As in the statement of Lemma 9.2.3, we extend P to $[0, \infty)$ by defining $P(f) := 1$ if $f > 1$. Observe that

$$E[P(F_n) \mid X_1, \ldots, X_{n-1}] = E[P(F_{n-1} + B_n X_n) \mid X_1, \ldots, X_{n-1}]$$
$$= pP(F_{n-1} + B_n) + qP(F_{n-1} - B_n)$$
$$\leq P(F_{n-1}), \tag{9.94}$$

where the second equality uses Theorem 2.2.4 on p. 81 and the inequality uses Lemma 9.2.3 if $F_{n-1} \leq 1$ and is obvious otherwise. Consequently, $\{P(F_n)\}_{n \geq 0}$ is a supermartingale with respect to $\{X_n\}_{n \geq 1}$, and the optional stopping theorem (Theorem 3.2.2 on p. 100) implies that

$$P(F_T \geq 1) = E[P(F_T)] \leq P(F_0) = P(f), \tag{9.95}$$

as required.   ♠

Example 9.2.5. *Bold play at don't pass with a house limit.* Consider a gambler who wants to turn one unit into 50,000 units making only don't-pass bets in a casino with a 2,000-unit house limit at the craps table ($p = 949/1{,}925$). Since 50,000 is an integer multiple of 2,000, the results of this section apply.

The bold player bets as follows. He bets everything at each coup until he is ruined or has 2,048 units. Assuming the latter, he is limited to betting 2,000 units at the next coup. He continues to bet 2,000 units at each coup until he reaches either 48 units or 48,048 units. In the former case he starts over betting everything at each coup until he is ruined or has $48 \cdot 64 = 3{,}072$ units, and in the other case he reduces his bet size to 1,952 units, hoping to achieve 50,000 units exactly. If he fails he is back to 46,096 units and bet sizes of 2,000 units. Finding the probability that he achieves the 50,000-unit goal directly, with all these contingencies, would be complicated.

Fortunately, we have Theorem 9.2.2, which corresponds to an optimal strategy that is nonbold. Here the player bets boldly until he is ruined or has 2,000 units. Specifically, he bets everything at each coup until he is ruined or has 1,024 units. Assuming the latter, he bets only 976 units at the next coup, hoping to achieve 2,000 units exactly. Assuming he achieves this intermediate goal, he bets 2,000 units at each coup until he is ruined or has 50,000 units. By (9.78), his probability of achieving the 50,000-unit goal is

$$P(1/50{,}000) = Q(1/2{,}000)\, \frac{(q/p) - 1}{(q/p)^{25} - 1}, \tag{9.96}$$

where $p = 949/1{,}925$. Referring to Example 9.1.5 on p. 324 for $Q(1/2{,}000)$, we find that the odds against success, for either the bold player or the optimal nonbold player, are 83,564.1 to 1.

Here is a third strategy. The player bets everything at each coup until he is ruined or has 2,048 units. Assuming the latter, he puts 48 of these units away, and bets 2,000 units at each coup until he is ruined or has 50,000 units. Of course, if he is ruined he can start anew with the leftover 48 units. As one might expect, this strategy is suboptimal, because the player has violated a fundamental principle of bold play: He has risked overshooting his goal, effectively attempting to achieve a 50,048-unit goal. His odds against success are 83,666.3 to 1.

It has been suggested that one could improve the latter strategy by betting 63 units instead of 64 after the sixth consecutive win. But this strategy is also suboptimal. (A bet of 61 units in this situation would retain optimality, however.) Even if the 63-unit bet were followed by optimal play, the player's odds against success would be 83,582.7 to 1, slightly worse than with bold play. ♠

9.3 Primitive Casinos

A *primitive casino* offers the generic game of independent coups in which a single betting opportunity is available at each coup, namely a bet that pays a_0 to 1, winning with probability p and losing with probability $q := 1 - p$. To avoid trivialities, we assume throughout that $0 < p < 1$ and $a_0 > 0$. Actually, it will be more convenient to parametrize the problem by p and $p_0 := 1/(a_0 + 1)$, so that $0 < p_0 < 1$. We let $q_0 := 1 - p_0$ and observe that $a_0 = q_0/p_0$. When the player bets b units, his fortune increases by $(q_0/p_0)b$ units with probability p and decreases by b units with probability q.

As in Sections 9.1 and 9.2 we will be concerned with bold play in such a casino. By definition, *bold play* consists of betting, at each coup, either one's entire fortune or just enough to achieve one's goal in the event of a win, whichever amount is smaller. It will be convenient, and will involve no loss of generality, to measure money in units of the gambler's goal, or, equivalently, to assume that his goal is 1 unit. We further assume that units are infinitely divisible. Then the bold player bets $f \wedge [(p_0/q_0)(1 - f)]$ when his fortune is $f \in [0, 1]$.

Let X be a random variable satisfying

$$P(X = q_0/p_0) = p \quad \text{and} \quad P(X = -1) = q. \tag{9.97}$$

In addition, let $X_1, X_2, \ldots$ be a sequence of i.i.d. random variables with common distribution that of X, with X_n representing the gambler's profit per unit bet at coup n. Given an initial fortune $f \in [0, 1]$, let $F_n(f)$ represent the bold player's fortune after n coups. Then

$$F_n(f) = F_{n-1}(f) + \{F_{n-1}(f) \wedge [(p_0/q_0)(1 - F_{n-1}(f))]\} X_n \tag{9.98}$$

for each $n \geq 1$, where $F_0(f) := f$.

Unlike in the two preceding sections, it is not the case here that $\{F_n(f)\}_{n \geq 0}$ is confined to a finite set with probability 1 when $f \in (0, 1)$ is rational, even if p_0 is rational (Problem 9.13 on p. 352).

Our primary interest is in the expected utility of bold player's final fortune when the utility function is given by the indicator of $[1, \infty)$. This is simply

$$S(f) := P(F_n(f) = 1 \text{ for some } n \geq 1). \tag{9.99}$$

When we want to emphasize the parameters p and p_0, we will write S_{p,p_0} for S. Conditioning on X_1, we find that

$$\begin{aligned} S(f) = {} & pS(f + [(q_0/p_0)f] \wedge (1 - f)) \\ & + qS(f - f \wedge [(p_0/q_0)(1 - f)]), \end{aligned} \tag{9.100}$$

or, since $S(0) = 0$ and $S(1) = 1$ by (9.99),

$$S(f) = \begin{cases} pS(f/p_0) & \text{if } 0 \leq f \leq p_0, \\ p + qS((f - p_0)/q_0) & \text{if } p_0 \leq f \leq 1. \end{cases} \tag{9.101}$$

Theorem 9.3.1. *The function S is continuous and increasing on $[0, 1]$ and is given by*

$$S(f) = S_{p,p_0}(f) = Q_p(Q_{p_0}^{-1}(f)), \tag{9.102}$$

where Q_p is the function Q of Section 9.1. In particular, if $p = p_0$, then $S(f) = f$ for all $f \in [0, 1]$.

Proof. We begin by showing that (9.101) has at most one bounded solution. Indeed, if S_1 and S_2 are both solutions, then their difference $\Delta S := S_1 - S_2$

satisfies

$$\Delta S(f) = \begin{cases} p\Delta S(f/p_0) & \text{if } 0 \le f \le p_0, \\ q\Delta S((f - p_0)/q_0) & \text{if } p_0 \le f \le 1. \end{cases} \tag{9.103}$$

Suppose there is an $f_0 \in [0,1]$ such that $|\Delta S(f_0)| > 0$. Defining $f_1, f_2, \ldots$ recursively by

$$f_n := \begin{cases} f_{n-1}/p_0 & \text{if } 0 \le f_{n-1} < p_0, \\ (f_{n-1} - p_0)/q_0 & \text{if } p_0 \le f_{n-1} \le 1, \end{cases} \tag{9.104}$$

we find that

$$|\Delta S(f_n)| \ge \frac{|\Delta S(f_{n-1})|}{q \vee p} \ge \cdots \ge \frac{|\Delta S(f_0)|}{(q \vee p)^n} \tag{9.105}$$

for each $n \ge 1$, so S_1 and S_2 cannot both be bounded.

Let us define $S^*(f) := Q_p(Q_{p_0}^{-1}(f))$ for all $f \in [0,1]$. By Theorem 9.1.2 on p. 321, Q is continuous and increasing on $[0,1]$ and $Q(0) = 0$ and $Q(1) = 1$. Therefore, the same can be said about Q^{-1}. It follows that S^* is continuous and increasing on $[0,1]$ and $S^*(0) = 0$ and $S^*(1) = 1$. If we could show that S^* satisfies (9.101), then S and S^* would both be bounded solutions of (9.101), and the first part of the proof would imply that $S(f) = S^*(f)$ for all $f \in [0,1]$.

Given $f \in [0, p_0]$, define $g := Q_{p_0}^{-1}(f) \in [0, \frac{1}{2}]$. Then $f = Q_{p_0}(g) = p_0 Q_{p_0}(2g)$, so $2Q_{p_0}^{-1}(f) = 2g = Q_{p_0}^{-1}(f/p_0)$ and

$$\begin{aligned} S^*(f) := Q_p(Q_{p_0}^{-1}(f)) &= pQ_p(2Q_{p_0}^{-1}(f)) \\ &= pQ_p(Q_{p_0}^{-1}(f/p_0)) = pS^*(f/p_0). \end{aligned} \tag{9.106}$$

Given $f \in [p_0, 1]$, define $g := Q_{p_0}^{-1}(f) \in [\frac{1}{2}, 1]$. Then $f = Q_{p_0}(g) = p_0 + q_0 Q_{p_0}(2g - 1)$, so $2Q_{p_0}^{-1}(f) - 1 = 2g - 1 = Q_{p_0}^{-1}((f - p_0)/q_0)$ and

$$\begin{aligned} S^*(f) := Q_p(Q_{p_0}^{-1}(f)) &= p + qQ_p(2Q_{p_0}^{-1}(f) - 1) \\ &= p + qQ_p(Q_{p_0}^{-1}((f - p_0)/q_0)) = p + qS^*((f - p_0)/q_0). \end{aligned} \tag{9.107}$$

This completes the proof.   ♠

Example 9.3.2. *Numerical evaluation of $S_{1/38,1/36}(\frac{1}{2})$.* What is the probability that a roulette player will double his fortune betting boldly on a single number? Assume a 38-number roulette wheel. The single-number bet wins with probability $p = 1/38$ and pays 35 to 1, so $p_0 = 1/36$. Thus, we need to evaluate $S_{1/38,1/36}(\frac{1}{2})$.

More generally, to evaluate $S_{p,p_0}(f)$ for some $f \in (0,1)$, we apply Theorems 9.3.1 and 9.1.2 on p. 321, obtaining the following formula. If

$$f = \sum_{k=1}^{\infty} u_k q_0^{u_1 + \cdots + u_{k-1}} p_0^{k-(u_1 + \cdots + u_{k-1})}, \tag{9.108}$$

where $u_k = 0$ or 1 for every $k \geq 1$, then

$$S_{p,p_0}(f) = \sum_{k=1}^{\infty} u_k q^{u_1 + \cdots + u_{k-1}} p^{k-(u_1 + \cdots + u_{k-1})}. \tag{9.109}$$

If we truncate (9.109) at N terms, we find as in (9.36) on p. 324 that

$$0 \leq S_{p,p_0}(f) - \sum_{k=1}^{N} u_k q^{u_1 + \cdots + u_{k-1}} p^{k-(u_1 + \cdots + u_{k-1})}$$

$$\leq \frac{p}{p \wedge q} q^{u_1 + \cdots + u_N} p^{N-(u_1 + \cdots + u_N)}. \tag{9.110}$$

To apply this to a specific f, we need an algorithm for determining the sequence $u_1, u_2, \ldots$. Given $f \in (0,1)$, let $f_0 := f$, and define recursively, for each $n \geq 1$,

$$(u_n, f_n) := \begin{cases} (0, f_{n-1}/p_0) & \text{if } 0 \leq f_{n-1} < p_0, \\ (1, (f_{n-1} - p_0)/q_0) & \text{if } p_0 \leq f_{n-1} \leq 1. \end{cases} \tag{9.111}$$

The resulting sequence $u_1, u_2, \ldots$ will satisfy (9.108) (see Problem 9.14 on p. 352).

With $p = 1/38$, $p_0 = 1/36$, and $f = \frac{1}{2}$, we find that the first 100 terms of the sequence $u_1, u_2, \ldots$ are all 1s except for terms 25, 59, 72, and 74. The error bound in (9.110) becomes $q^{96} p^4 < 4 \cdot 10^{-8}$. We conclude that $S_{1/38,1/36}(\frac{1}{2}) \approx 0.480890$. It may be of interest to note that this figure is greater than $Q_{18/38}(\frac{1}{2}) = 18/38 \approx 0.473684$. Thus, the bold player who attempts to double his fortune by betting on a single number has a better chance of success than the bold player who bets double-or-nothing on an even-money proposition, despite the fact that the latter player has the advantage of learning his fate in just one coup. See Example 9.3.8 for a stronger conclusion and Problem 9.16 on p. 352 for the definitive conclusion on this matter. ♠

Our next main goal in this section is to prove that bold play is optimal in a subfair primitive casino. The proof is more difficult than that of the special case treated in Section 9.1. Therefore we break up the proof, doing most of the work in proving two preliminary results.

Lemma 9.3.3. *Assume that $p \leq p_0$. Then the function S of Theorem 9.3.1 satisfies the two inequalities*

$$S(g + h) \geq S(g) + S(h), \qquad g, h \in [0,1], \ g + h \leq 1, \tag{9.112}$$

and

$$S(g + h - 1) \geq S(g) + S(h) - 1, \qquad g, h \in [0, 1], \ g + h \geq 1. \qquad (9.113)$$

Remark. A function S satisfying (9.112) is said to be *superadditive*. If we temporarily define $S^*(f) := 1 - S(1 - f)$ for $0 \leq f \leq 1$, then (9.113) is equivalent to saying that S^* is *subadditive*, that is, $-S^*$ is superadditive. Unfortunately, these inequalities do not have a simple intuitive interpretation.

Proof. We will reformulate several times what remains to be proved. Each reformulation is set in italics as a signpost for the reader.

Given $g, h \in [0, 1]$ with $g + h \leq 1$, let $a = Q_{p_0}^{-1}(g + h)$, $b = Q_{p_0}^{-1}(g)$, and $c = Q_{p_0}^{-1}(h)$, and define the function D on $(0, 1)$ by

$$D(p) := Q_p(a) - Q_p(b) - Q_p(c). \qquad (9.114)$$

Noting that $D(p_0) = g + h - g - h = 0$, we see that for (9.112) it will suffice to show that $D(p) \geq 0$ whenever $p \leq p_0$, or that D is nonnegative to the left of any zero of D. (We say that p is a *zero* of D if $D(p) = 0$.)

Given $g, h \in [0, 1]$ with $g + h \geq 1$, let $a = Q_{p_0}^{-1}(g + h - 1)$, $b = Q_{p_0}^{-1}(g)$, and $c = Q_{p_0}^{-1}(h)$, and again define D on $(0, 1)$ by (9.114). Noting that $D(p_0) + 1 = 0$, we see that for (9.113) it will suffice to show that $D(p) + 1 \geq 0$ whenever $p \leq p_0$, or that $D + 1$ is nonnegative to the left of any zero of $D + 1$. We can recast this in a more attractive form by using the identity

$$Q_p(a) = 1 - Q_q(1 - a); \qquad (9.115)$$

see Problem 9.2 on p. 350. Defining the function $\bar{D}$ on $(0, 1)$ by

$$\bar{D}(q) := Q_q(1 - a) - Q_q(1 - b) - Q_q(1 - c), \qquad (9.116)$$

we observe that $D(p) + 1 = -\bar{D}(q)$. Noting that $\bar{D}(q_0) = 0$, we see that for (9.113) it will suffice to show that $\bar{D}(q) \leq 0$ whenever $q \geq q_0$, or that $\bar{D}$ is nonpositive to the right of any zero of $\bar{D}$.

Now $\bar{D}$ is defined in the same way that D is, except for the choice of the parameters a, b, and c, so *it will suffice to show that, for all $a, b, c \in [0, 1]$, D is nonnegative to the left and nonpositive to the right of any zero of D.*

In fact, we can restrict our attention to the case in which a, b, and c are dyadic rationals. Indeed, writing $D(p; a, b, c)$ for $D(p)$ when we want to emphasize the dependence of D on its parameters, suppose there exist $a, b, c \in [0, 1]$ and $0 < p < p_0 < 1$ such that $D(p; a, b, c) < 0 = D(p_0; a, b, c)$. By the continuity of Q_p, we can choose dyadic rationals $a_1, b_1, c_1 \in [0, 1]$ such that $a_1 \geq a$, $b_1 \leq b$, $c_1 \leq c$, and $D(p; a_1, b_1, c_1) < 0$. Then, by the fact that Q_{p_0} is increasing on $[0, 1]$,

$$\begin{aligned} D(p_0; a_1, b_1, c_1) &= Q_{p_0}(a_1) - Q_{p_0}(b_1) - Q_{p_0}(c_1) \\ &\geq Q_{p_0}(a) - Q_{p_0}(b) - Q_{p_0}(c) \\ &= D(p_0; a, b, c) \end{aligned}$$

$$= 0, \tag{9.117}$$

so, by the continuity of D on $(0,1)$ (see Problem 9.12 on p. 352), there exists $p_1 \in (p, p_0]$ such that $D(p_1; a_1, b_1, c_1) = 0$. Similarly, if there exist $a, b, c \in [0, 1]$ and $0 < p_0 < p < 1$ such that $D(p_0; a, b, c) = 0 < D(p; a, b, c)$, then there exist dyadic rationals $a_1, b_1, c_1 \in [0, 1]$ and $p_1 \in [p_0, p)$ such that $D(p_1; a_1, b_1, c_1) = 0 < D(p; a_1, b_1, c_1)$. Consequently, *it will suffice to show that, for all dyadic rationals $a, b, c \in [0, 1]$, D is nonnegative to the left and nonpositive to the right of any zero of D.*

Let $N \geq 1$ and $k, l, m \in \{0, 1, \ldots, 2^N\}$. By Lemma 9.1.1 on p. 320,

$$D(p) = D(p; k/2^N, l/2^N, m/2^N) = Q_p(k/2^N) - Q_p(l/2^N) - Q_p(m/2^N)$$

$$= \sum_{j=0}^{N} [A_j(k) - A_j(l) - A_j(m)] q^j p^{N-j}. \tag{9.118}$$

(Recall that $A_j(k) = |\{0 \leq i \leq k - 1 : B(i) = j\}|$, where $B(i)$ denotes the number of 1s in the binary expansion of i.) Defining the sequence of coefficients

$$C_j(k, l, m) := A_j(k) - A_j(l) - A_j(m), \tag{9.119}$$

and the polynomial D^* on $(0, \infty)$ by

$$D^*(\rho) = D^*(\rho; k, l, m) := \sum_{j \geq 0} C_j(k, l, m) \rho^j, \tag{9.120}$$

we see that $D^*(q/p) = p^{-N} D(p)$ for all $p \in (0, 1)$. (Notice that D^* does not depend on N.) Clearly, *it will suffice to show that, for all $k, l, m \geq 0$, D^* is nonpositive to the left and nonnegative to the right of any zero of D^*.* (The reversal of inequalities occurs because p appears in the denominator of the argument of D^*.)

By virtue of Descartes's rule of signs (Theorem A.1.4 on p. 746), *it will suffice to show that, for all $k, l, m \geq 0$, C is Cartesian,* that is, the sequence of coefficients $C_0(k, l, m), C_1(k, l, m), \ldots$ (of which only finitely many terms are nonzero) satisfies the following condition:

$$C_i(k, l, m) < 0 < C_j(k, l, m) \qquad \text{implies} \qquad i < j. \tag{9.121}$$

We prove this by induction on k, l, m. More precisely, we show by induction on n that C is Cartesian if $k, l, m \in \{0, 1, \ldots, 2^n + 1\}$.

The case $n = 0$, which amounts to showing that C is Cartesian in each of the 27 cases $k, l, m \in \{0, 1, 2\}$, is easy. Note that $C_j(k, l, m) \leq 0$ for $j = 0, 1, \ldots$ if $k \leq l \vee m$. This leaves only five cases to check, namely $(k, l, m) = (1, 0, 0), (2, 0, 0), (2, 0, 1), (2, 1, 0),$ and $(2, 1, 1)$. Next note that $C_j(k, l, m)$ is either nonnegative for $j = 0, 1, \ldots$ or nonpositive for $j = 0, 1, \ldots$ if $l = 0$ or $m = 0$. This leaves only one case to check, namely $(k, l, m) = (2, 1, 1)$. But

$C_0(2,1,1) = A_0(2) - 2A_0(1) = -1$, $C_1(2,1,1) = A_1(2) - 2A_1(1) = 1$, and $C_j(2,1,1) = 0$ for each $j \geq 2$. Thus, C is Cartesian in this case as well.

For the inductive step, *it will suffice to show that, given $k, l, m \geq 0$, if C is Cartesian for all eight choices of $k + \kappa$, $l + \lambda$, $m + \mu$, where $\kappa, \lambda, \mu \in \{0, 1\}$, then C is Cartesian for all eight choices of $2k + \kappa$, $2l + \lambda$, $2m + \mu$, where $\kappa, \lambda, \mu \in \{0, 1\}$.* For this we will need some properties of the functions $A_j(k)$. First,

$$A_j(2k) = A_j(k) + A_{j-1}(k),$$
$$A_j(2k+1) = A_j(k+1) + A_{j-1}(k), \tag{9.122}$$

where $A_{-1}(k) := 0$, which can be seen by writing

$$\{0, 1, \ldots, 2k - 1\} = \{0, 2, \ldots, 2(k-1)\} \cup \{1, 3, \ldots, 2(k-1)+1\},$$
$$\{0, 1, \ldots, 2k\} = \{0, 2, \ldots, 2k\} \cup \{1, 3, \ldots, 2(k-1)+1\}. \tag{9.123}$$

A second property, which follows from the definitions, is

$$A_j(k+1) = A_j(k) + \delta_{j,B(k)}. \tag{9.124}$$

We conclude that

$$
\begin{aligned}
C_j&(2k+\kappa, 2l+\lambda, 2m+\mu) \\
&= A_j(2k+\kappa) - A_j(2l+\lambda) - A_j(2m+\mu) \\
&= A_j(k+\kappa) - A_j(l+\lambda) - A_j(m+\mu) + [A_{j-1}(k) - A_{j-1}(l) - A_{j-1}(m)] \\
&= C_j(k+\kappa, l+\lambda, m+\mu) + C_{j-1}(k,l,m) \\
&= C_j(k,l,m) + [\kappa\delta_{j,B(k)} - \lambda\delta_{j,B(l)} - \mu\delta_{j,B(m)}] + C_{j-1}(k,l,m), \quad (9.125)
\end{aligned}
$$

where the second equality uses (9.122) and the fourth uses (9.124). If we denote by C^* the sequence of coefficients given by $C_j^* := C_{j-1}$ with $C_{-1} := 0$ and by Δ the sequence of coefficients with Δ_j equal to the term within brackets on the right side of (9.125), then *it will suffice to show that, given $k, l, m \geq 0$, if $C + \Delta$ is Cartesian for all eight choices of $\kappa, \lambda, \mu \in \{0, 1\}$, then $C + \Delta + C^*$ is Cartesian for all eight choices of $\kappa, \lambda, \mu \in \{0, 1\}$.*

Fix $k, l, m \geq 0$, and assume that $C + \Delta$ is Cartesian for all eight choices of $\kappa, \lambda, \mu \in \{0, 1\}$. Now let $\kappa, \lambda, \mu \in \{0, 1\}$ be arbitrary. We claim that for every pair $i < j$, none of the following pairs of inequalities holds:

$$C_i > 0 > C_j + \Delta_j, \tag{9.126}$$
$$C_i + \Delta_i > 0 > C_j, \tag{9.127}$$
$$C_i > 0 > C_j^*, \tag{9.128}$$
$$C_i^* > 0 > C_j, \tag{9.129}$$
$$C_i > 0 > C_j + \Delta_j + C_j^*, \tag{9.130}$$
$$C_i + \Delta_i + C_i^* > 0 > C_j, \tag{9.131}$$

$$C_i + \Delta_i + C_i^* > 0 > C_j + \Delta_j + C_j^*. \tag{9.132}$$

We prove the seven assertions successively.

First, suppose there exist $i < j$ such that (9.126) holds. Since C is Cartesian, $C_j \geq 0$, hence $\Delta_j < 0$. From the definition of Δ it follows that $j = B(l)$ or $j = B(m)$ (or both). If we define Δ^* to be Δ with $\kappa = 0$, $\lambda = \delta_{j,B(l)}$, and $\mu = \delta_{j,B(m)}$, we have $C_i + \Delta_i^* = C_i > 0 > C_j + \Delta_j \geq C_j + \Delta_j^*$, contradicting the assumption the $C + \Delta^*$ is Cartesian.

Second, suppose there exist $i < j$ such that (9.127) holds. Since C is Cartesian, $C_i \leq 0$, hence $\Delta_i > 0$. From the definition of Δ it follows that $i = B(k)$. If we define Δ^* to be Δ with $\kappa = 1$ and $\lambda = \mu = 0$, we have $C_i + \Delta_i^* = C_i + \Delta_i > 0 > C_j = C_j + \Delta_j^*$, contradicting the assumption the $C + \Delta^*$ is Cartesian.

Third, since C is Cartesian, there cannot exist $i < j$ such that $C_i > 0 > C_{j-1}$, which is equivalent to (9.128).

Fourth, since C is Cartesian, there cannot exist $i < j$ such that $C_{i-1} > 0 > C_j$, which is equivalent to (9.129).

Fifth, suppose there exist $i < j$ such that (9.130) holds. Then either (9.126) or (9.128) holds, a contradiction.

Sixth, suppose there exist $i < j$ such that (9.131) holds. Then either (9.127) or (9.129) holds, a contradiction.

Seventh and finally, suppose there exist $i < j$ such that (9.132) holds. By (9.130) with i replaced by $i - 1$, it cannot be the case that $C_{i-1} > 0$, so $C_i^* = C_{i-1} \leq 0$. It follows from (9.132) that

$$C_i + \Delta_i > 0. \tag{9.133}$$

By (9.131) with j replaced by $j - 1$, it cannot be the case that $C_{j-1} < 0$. (If $j - 1 = i$, this follows not from (9.131) but from the fact that $\Delta_i \leq 1$.) Therefore, $C_j^* = C_{j-1} \geq 0$. It follows from (9.132) that

$$C_j + \Delta_j < 0. \tag{9.134}$$

But (9.133) and (9.134) contradict the assumption that $C + \Delta$ is Cartesian. We conclude from (9.132) that $C + \Delta + C^*$ is Cartesian, as required. ♠

Lemma 9.3.4. *Assume that $p \leq p_0$, and extend S of Theorem 9.3.1 to $[0, \infty)$ by defining $S(f) := 1$ if $f > 1$. Then S satisfies*

$$S(f) \geq pS(f + (q_0/p_0)b) + qS(f - b) \tag{9.135}$$

whenever $0 \leq b \leq f \leq 1$.

Proof. Suppose that (9.135) fails for some pair (f, b) satisfying $0 \leq b \leq f \leq 1$. If $f + (q_0/p_0)b > 1$, then (9.135) fails with the pair $(f, (p_0/q_0)(1 - f))$ by the monotonicity of S, so we can assume that $0 \leq b \leq f \wedge [(p_0/q_0)(1 - f)]$. Of course, $b > 0$. We will now derive a contradiction.

First we claim that (9.135) fails for some pair (f, b) satisfying

$$0 < b \le f \quad \text{and} \quad f - b < p_0 < f + (q_0/p_0)b \le 1. \tag{9.136}$$

Indeed, let us choose a pair (f, b) satisfying

$$0 < b \le f \quad \text{and} \quad f + (q_0/p_0)b \le 1 \tag{9.137}$$

for which (9.135) fails and b is nearly as large as possible. Then the inequality $f + (q_0/p_0)b \le p_0$ cannot hold, for if it did, (9.101) would imply that

$$S(f/p_0) - pS((f + (q_0/p_0)b)/p_0) - qS((f - b)/p_0)$$
$$= p^{-1}[S(f) - pS(f + (q_0/p_0)b) - qS(f - b)] < 0. \tag{9.138}$$

Therefore, (9.135) would fail for the pair $(f/p_0, b/p_0)$. Moreover, (9.137) would hold with $(f/p_0, b/p_0)$ in place of (f, b) by virtue of $f + (q_0/p_0)b \le p_0$. This would contradict the near maximality of b. Similarly, the inequality $f - b \ge p_0$ cannot hold, for if it did, the second half of (9.101) would imply that

$$S((f - p_0)/q_0) - pS((f + (q_0/p_0)b - p_0)/q_0) - qS((f - b - p_0)/q_0)$$
$$= q^{-1}[S(f) - pS(f + (q_0/p_0)b) - qS(f - b)] < 0. \tag{9.139}$$

Therefore, (9.135) would fail for the pair $((f - p_0)/q_0, b/q_0)$. Moreover, (9.137) would hold with $((f - p_0)/q_0, b/q_0)$ in place of (f, b) by virtue of $f - b \ge p_0$ and (9.137). This again would contradict the near maximality of b.

Now, given a pair (f, b) satisfying (9.136) but for which (9.135) fails, define

$$g := \frac{f + (q_0/p_0)b - p_0}{q_0} \in [0, 1], \qquad h := \frac{f - b}{p_0} \in [0, 1]. \tag{9.140}$$

Then

$$pS(f + (q_0/p_0)b) = p\left(p + qS\left(\frac{f + (q_0/p_0)b - p_0}{q_0}\right)\right) = p^2 + qpS(g) \tag{9.141}$$

and

$$qS(f - b) = qpS\left(\frac{f - b}{p_0}\right) = qpS(h). \tag{9.142}$$

We consider two cases, depending on the value of

$$g + h = \frac{(f/p_0) - p_0}{q_0}. \tag{9.143}$$

Case 1. $g + h \le 1$. This is equivalent to $f \le p_0$. On the other hand, $g + h \ge 0$ is equivalent to $f/p_0 \ge p_0$. Therefore,

$$S(f) = pS(f/p_0) = p\left[p + qS\left(\frac{(f/p_0) - p_0}{q_0}\right)\right] = p^2 + qpS(g + h). \quad (9.144)$$

From (9.142), (9.144), and (9.141) we find that

$$S(f) - pS(f + (q_0/p_0)b) - qS(f - b)$$
$$= qp[S(g + h) - S(g) - S(h)], \quad (9.145)$$

and this is nonnegative by Lemma 9.3.3.

Case 2. $g + h \geq 1$. This is equivalent to $f \geq p_0$. On the other hand, $g + h \leq 2$ is equivalent to $(f - p_0)/q_0 \leq p_0$. Therefore,

$$S(f) = p + qS((f - p_0)/q_0) = p + q\left[pS\left(\frac{f - p_0}{q_0 p_0}\right)\right] = p + qpS(g + h - 1). \quad (9.146)$$

From (9.146), (9.141), and (9.142), we find that

$$S(f) - pS(f + (q_0/p_0)b) - qS(f - b)$$
$$= qp[S(g + h - 1) - S(g) - S(h) + 1], \quad (9.147)$$

and this is nonnegative by Lemma 9.3.3.

In either case, the assumption that (9.135) fails is contradicted, proving the lemma. ♠

We are now ready to establish the optimality of bold play in a subfair primitive casino. The argument very similar to the ones used to prove Theorems 9.1.7 on p. 327 and 9.2.4 on p. 337.

First, let us consider an arbitrary alternative strategy. As in Section 9.1, such a strategy consists of a sequence of bet sizes, $B_1, B_2, \ldots$, with B_n representing the amount bet at coup n and depending on only the results of the previous coups, $X_1, \ldots, X_{n-1}$. (Recall that $X_1, X_2, \ldots$ are i.i.d. with common distribution given by (9.97).) More precisely, (9.42) on p. 326 holds, where b_1 is a constant and b_n is a nonrandom function of $n - 1$ variables for each $n \geq 2$. The gambler's fortune F_n after n coups satisfies (9.43) on p. 326, F_0 being his initial fortune (a nonnegative constant). The only restriction is that he cannot bet more than he has, that is, (9.44) on p. 326 holds. Finally, we define T by (9.45) on p. 326, and we assume that $P(T < \infty) = 1$.

Recall the notation (9.98).

Theorem 9.3.5. *Assume that $p \leq p_0$. If $F_0 = f \in [0, 1]$, then*

$$P(F_T \geq 1) \leq S(f) := P(F_{T(f)}(f) = 1), \quad (9.148)$$

where $T(f)$ is the coup at which the last positive bet by the bold player is made, assuming an initial fortune of $f \in [0, 1]$.

Remark. The theorem says that bold play is optimal in a subfair (or fair) primitive casino. Here optimality means that no strategy achieves the gambler's goal with a higher probability than does bold play.

Proof. As in the statement of Lemma 9.3.4, we extend S to $[0, \infty)$ by defining $S(f) := 1$ if $f > 1$. Observe that

$$
\begin{aligned}
\mathrm{E}[S(F_n) \mid X_1, \ldots, X_{n-1}] &= \mathrm{E}[S(F_{n-1} + B_n X_n) \mid X_1, \ldots, X_{n-1}] \\
&= pS(F_{n-1} + (q_0/p_0)B_n) + qS(F_{n-1} - B_n) \\
&\leq S(F_{n-1}), \tag{9.149}
\end{aligned}
$$

where the second equality uses Theorem 2.2.4 on p. 81 and the inequality uses Lemma 9.3.4 if $F_{n-1} \leq 1$ and is obvious otherwise. Consequently, $\{S(F_n)\}_{n \geq 0}$ is a supermartingale with respect to $\{X_n\}_{n \geq 1}$, and the optional stopping theorem (Theorem 3.2.2 on p. 100) implies that

$$
\mathrm{P}(F_T \geq 1) = \mathrm{E}[S(F_T)] \leq S(F_0) = S(f), \tag{9.150}
$$

as required.   ♠

As in the case of red-and-black, bold play is not uniquely optimal in a subfair (or fair) primitive casino. There is an analogue of Theorem 9.1.8 on p. 328 here, but its formulation and proof are left to the interested reader (see Problem 9.15 on p. 352).

We pause to record the following corollary to Theorem 9.3.5.

Corollary 9.3.6. *Let $p \leq p_0$. If $g \in [0, 1]$ and $0 \leq h_0 \leq h_1 \leq 1$, then*

$$
S(gh_1 + (1-g)h_0) \geq S(g)S(h_1) + (1 - S(g))S(h_0). \tag{9.151}
$$

In particular,
$$
S(gh) \geq S(g)S(h), \qquad g, h \in [0, 1], \tag{9.152}
$$

and

$$
S(g + (1-g)h) \geq S(g) + (1 - S(g))S(h), \qquad g, h \in [0, 1]. \tag{9.153}
$$

Proof. The first inequality holds because, with initial fortune $gh_1 + (1-g)h_0 \in [h_0, h_1]$, bold play in $[h_0, h_1]$ followed by bold play cannot outperform bold play. The second and third inequalities are the special cases of the first one with $(h_0, h_1) := (0, h)$ and $(h_0, h_1) := (h, 1)$.   ♠

We conclude this section by comparing two subfair primitive casinos.

Theorem 9.3.7. *Let $p < p_0$ and $p^* < p_0^*$. If*

$$
S_{p^*, p_0^*}(p_0) \geq S_{p, p_0}(p_0), \tag{9.154}
$$

then

$$S_{p^*,p_0^*}(f) \geq S_{p,p_0}(f) \tag{9.155}$$

for all $f \in [0,1]$.

Remark. The theorem says that bold play in a primitive casino with parameters $p^* < p_0^*$ performs at least as well as bold play in a primitive casino with parameters $p < p_0$, regardless of the initial fortune, provided it does so with initial fortune p_0. The result is intuitively reasonable, because p_0 is the unique nontrivial initial fortune in the (p, p_0)-casino for which bold play is most "efficient," in the sense that bold play requires just one coup. Note that $S_{p,p_0}(p_0) = p$.

Proof. Consider those $f \in [0,1)$ of the form

$$f = \sum_{k=1}^{N} u_k q_0^{u_1 + \cdots + u_{k-1}} p_0^{k - (u_1 + \cdots + u_{k-1})}, \tag{9.156}$$

where $u_k = 0$ or 1 for $k = 1, \ldots, N$. We will prove (9.155) for all f of this form by induction on N. The case $N = 1$ includes $f = 0$, for which (9.155) is obvious, and $f = p_0$, which is (9.154). We let $N \geq 2$ and assume that (9.155) holds for all f of the form (9.156) with N replaced by $N - 1$.

If $0 \leq f \leq p_0$, then

$$S_{p^*,p_0^*}(f) \geq S_{p^*,p_0^*}(p_0) S_{p^*,p_0^*}(f/p_0) \geq S_{p,p_0}(p_0) S_{p,p_0}(f/p_0)$$
$$= p S_{p,p_0}(f/p_0) = S_{p,p_0}(f), \tag{9.157}$$

where the first inequality uses Corollary 9.3.6 and the second uses (9.154) and the induction hypothesis. If $p_0 \leq f < 1$, then

$$S_{p^*,p_0^*}(f) = S_{p^*,p_0^*}(p_0 + q_0(f - p_0)/q_0)$$
$$\geq S_{p^*,p_0^*}(p_0) + [1 - S_{p^*,p_0^*}(p_0)] S_{p^*,p_0^*}((f - p_0)/q_0)$$
$$\geq S_{p,p_0}(p_0) + [1 - S_{p,p_0}(p_0)] S_{p^*,p_0^*}((f - p_0)/q_0)$$
$$\geq p + q S_{p,p_0}((f - p_0)/q_0) = S_{p,p_0}(f), \tag{9.158}$$

where the first inequality uses Corollary 9.3.6, the second uses (9.154), and the third uses the induction hypothesis. ♠

Example 9.3.8. *Majorized wagers in roulette and craps.* Let us say that a wager corresponding to a primitive casino with parameters $p^* < p_0^*$ *majorizes* a wager corresponding to a primitive casino with parameters $p < p_0$, if the conclusion of Theorem 9.3.7 holds.

In 38-number roulette, the single-number bet majorizes all others. Indeed, for the single-number bet, $p^* = 1/38$ and $p_0^* = 1/36$, while for the k-number bet ($k = 1, 2, 3, 4, 6, 12, 18$), $p = k/38$ and $p_0 = k/36$. By Lemma 9.3.3,

$$S_{p^*,p_0^*}(p_0) = S_{p^*,p_0^*}(kp_0^*) \geq k S_{p^*,p_0^*}(p_0^*) = kp^* = p = S_{p,p_0}(p_0), \tag{9.159}$$

so Theorem 9.3.7 applies. It does not immediately follow from this that the strategy of betting boldly on a single number at each coup is optimal at roulette. Nevertheless, this conclusion is correct, and the reader is referred to Problem 9.16 on p. 352 for an outline of the proof.

In craps, the don't-pass bet has the lowest house advantage of all bets corresponding to primitive casinos, and it majorizes all but two of them. (See Chapter 15 for a description of these bets.) We applied Theorem 9.3.7 to the one-roll bets on $2, 3, 7, 11, 12$, the four hardway bets, the big-six and big-eight bets, the six place bets, the six buy bets, the six lay bets, and the pass-line bet. Except for the place bets on six and eight ($p = 5/11$, $p_0 = 6/13$), condition (9.154) is satisfied.   ♠

9.4 Problems

9.1. *Bold play at red-and-black with pushes.* Red-and-black with pushes is the generic game of independent coups in which a single betting opportunity is available at each coup, namely a bet that pays even money, winning with probability p, losing with probability q, and pushing with probability r, where $p, q, r > 0$ and $p + q + r = 1$. Find an expression for the probability that the bold player achieves his goal of 1 unit starting with initial fortune $f \in [0, 1]$.

9.2. *Duality of Q.* Write Q_p for the function Q of Section 9.1 to emphasize its dependence on p. Show that $Q_p(f) = 1 - Q_{1-p}(1 - f)$ for all $p \in (0, 1)$ and $f \in [0, 1]$.

9.3. *Subfair vs. fair red-and-black.* Assume that $p < \frac{1}{2}$, and, with $Q(f)$ as in Section 9.1, show that $Q(f) < f$ whenever $0 < f < 1$.

9.4. *Graph of the bold player's expected loss at subfair red-and-black.* Fix $p = 949/1,925$ (from craps), and use computer graphics to print a graph of the function L defined on $(0, 1)$ by $L(f) := f - Q(f)$. Evaluate the function at sufficiently many points to make the graph appear continuous, and choose the scale of the vertical axis to best advantage.

9.5. *Q as a cumulative distribution function.* Let Z be a random variable satisfying $\mathrm{P}(Z = 0) = p$ and $\mathrm{P}(Z = 1) = 1 - p$, and let $Z_1, Z_2, \ldots$ be a sequence of i.i.d. random variables with common distribution that of Z. Define the random variable W by

$$W := \sum_{n=1}^{\infty} \frac{Z_n}{2^n}. \tag{9.160}$$

Show that the function Q of Section 9.1 satisfies $Q(f) = \mathrm{P}(W \le f)$ for all $f \in [0, 1]$. In particular, W is not a discrete random variable.

Hint: The functional equation (9.8) on p. 319 has a unique nondecreasing solution.

9.6. *Optimality of every reasonable strategy at fair red-and-black.* We know that bold play is optimal at fair red-and-black. Show that every reasonable strategy is optimal, in the sense that equality holds in (9.46) on p. 327. By "reasonable" we mean that, in addition to the conditions assumed in Theorem 9.1.7 on p. 327, the player never risks overshooting his goal.

9.7. *Expected number of coups for the bold player at red-and-black.* Define $T(f)$ for all $f \in [0, 1]$ as in (9.4) on p. 318, and let $E(f) := \mathrm{E}[T(f)]$ for all $f \in [0, 1]$.

(a) Show that $E(0) = E(1) = 0$ and

$$E(f) = \begin{cases} 1 + pE(2f) & \text{if } f \in (0, \frac{1}{2}], \\ 1 + qE(2f - 1) & \text{if } f \in [\frac{1}{2}, 1). \end{cases} \tag{9.161}$$

(b) Use part (a) to find a formula for $E(f)$ when f is a dyadic rational in $(0, 1)$.

(c) Extend the formula in part (b) to all $f \in (0, 1)$.

(d) In the case $p = \frac{1}{2}$, show that the formula for $E(f)$ simplifies considerably. Notice that the dyadic rationals in $[0, 1]$ are precisely the points of discontinuity of E.

(e) Without assuming that $p = \frac{1}{2}$, show that the dyadic rationals in $[0, 1]$ are precisely the points of discontinuity of E.

9.8. *Expected total amount bet by the bold player at red-and-black.* Define $B(f)$ to be the expected total amount bet by the bold player at red-and-black with initial fortune $f \in [0, 1]$.

(a) Show that

$$B(f) = \begin{cases} f + pB(2f) & \text{if } f \in [0, \frac{1}{2}], \\ 1 - f + qB(2f - 1) & \text{if } f \in [\frac{1}{2}, 1]. \end{cases} \tag{9.162}$$

(b) Use part (a) to find a formula for $B(f)$ when f is a dyadic rational in $[0, 1)$.

(c) Extend the formula in part (b) to all $f \in [0, 1]$, and show that B is continuous on $[0, 1]$.

(d) Find another formula for B, assuming that $p \neq \frac{1}{2}$, using Theorem 8.2.1 on p. 298.

9.9. *Bold play at red-and-black-and-green with partager.* Red-and-black-and-green with partager is the generic game of independent coups in which a single betting opportunity is available at each coup, namely a bet that wins the amount bet with probability $p \in (0, \frac{1}{2})$, loses the amount bet with probability p, and loses half the amount bet ("partager" is French for "to share") with probability $r := 1 - 2p$. Find an expression for the probability that the bold player achieves his goal of 1 unit starting with initial fortune $f = \frac{1}{2}$, and show that it is not a rational function of p. Evaluate it numerically for $p = 18/37$.

Hint: For each $n \geq 0$, let p_n be the probability of achieving the goal starting with 2^{-n} units. Choose $u > 0$ so that $p/u + ru = 1$ and $0 < ru < \frac{1}{2}$, and define $q_n := p_n u^{-n}$ for each $n \geq 0$. Show that, for large n, the quantities $q_1, q_2, \ldots, q_n$ satisfy a tridiagonal system to which Problem 7.3 on p. 267 applies.

9.10. *A house limit of 2/5.* In subfair red-and-black with a house limit of $L = 1/3$, Theorem 9.2.2 on p. 333 says that it is optimal to use bold play in $[0, 1/3]$, $[1/3, 2/3]$, and $[2/3, 1]$, followed by bold play from $1/3$ or $2/3$ if necessary. Show that, if $0 < p < \frac{1}{2}$, the analogous statement is false with a house limit of $L = 2/5$. More precisely, show that the strategy of using bold play in $[0, 2/5]$, $[2/5, 3/5]$, and $[3/5, 1]$, followed by bold play from $2/5$ or $3/5$ if necessary, is suboptimal for some initial fortune f.

Hint: It suffices to consider $f \in \{i/10 : i = 0, 1, \ldots, 10\}$.

9.11. *Nonoptimality of bold play with house limit 2/5.*

(a) Let the function P be as in (9.68) on p. 331 with $L := 2/5$. If m is a positive integer and $N := 2^m \cdot 5$, show that the probabilities $P(1/N), P(2/N), \ldots, P((N-1)/N)$ satisfy a system of $N-1$ linear equations that can be solved recursively; in particular, the solution is unique.

(b) Assuming a house limit of $2/5$, show that, for each $p \in (0, \frac{1}{2})$, the bettor with initial fortune $f = 19/40 - 1/640$ is better off to stake $s = 2/5 - 1/640$ at the first coup and then play boldly than he is to play boldly ($s = 2/5$) from the outset.

9.12. *Continuity of $Q_p(f)$ as a function of p.* Denote by Q_p the function Q of Section 9.1. Fix $f \in (0, 1)$, and show that $Q_p(f)$ is continuous in $p \in (0, 1)$. (In fact, it is real analytic.)

9.13. *Finiteness of the Markov chain $\{F_n(f)\}_{n \geq 0}$ for specified f.* In red-and-black, we saw that, if f is rational, then the Markov chain $\{F_n(f)\}_{n \geq 0}$ that models the bold player's sequence of fortunes is confined to a finite state space. Show that this is not necessarily true in a primitive casino. Is it ever true in a primitive non-red-and-black casino?

9.14. *Evaluation of $Q_{p_0}^{-1}(f)$.* Let $p_0 \in (0, 1)$ and put $q_0 := 1 - p_0$. Given $f \in (0, 1)$, let $f_0 := f$, and define recursively, for each $n \geq 1$,

$$(u_n, f_n) := \begin{cases} (0, f_{n-1}/p_0) & \text{if } 0 \leq f_{n-1} < p_0, \\ (1, (f_{n-1} - p_0)/q_0) & \text{if } p_0 \leq f_{n-1} \leq 1. \end{cases} \tag{9.163}$$

Show that the resulting sequence $u_1, u_2, \ldots$ satisfies (9.108) on p. 341.

9.15. *Other optimal strategies in a subfair primitive casino.* Assume that $p < p_0$, and find and prove the analogue of Theorem 9.1.8 on p. 328 in the setting of Section 9.3.

9.16. *Optimal strategy at roulette.* Consider a generalized roulette wheel with d numbers, and suppose that a bet on a single number pays $d_0 - 1$ to 1. Here

d and d_0 are integers with $1 < d_0 < d$. (Typically, d is 37 or 38, and d_0 is 36.)
Put $p := 1/d$ and $p_0 := 1/d_0$, and let $S := S_{p,p_0}$ be as in Section 9.3. Show
that the strategy of betting boldly on a single number is optimal at roulette
by supplying the details of the proof outlined below.

(a) For each $N \geq 1$, show that

$$S(f/(1 - Nf)) \geq S(f)/(1 - NS(f)), \qquad 0 \leq f \leq (N+1)^{-1}. \qquad (9.164)$$

(b) Given $f \in [0,1]$, let $\boldsymbol{b} = (b_1, \ldots, b_d)$ satisfy $b_1 \geq b_2 \geq \cdots \geq b_d \geq 0$,
$b_1 + \cdots + b_d \leq f$, and $f + d_0 b_j - (b_1 + \cdots + b_d) \leq 1$ for $j = 1, \ldots, d$. If $\boldsymbol{b}$ has $N+1$
nonzero components, where $N \in \{0, 1, \ldots, d-1\}$, define $\boldsymbol{b}^* = (b_1^*, \ldots, b_d^*)$ with
at most N nonzero components and satisfying the same conditions as $\boldsymbol{b}$ by
$b_j^* := b_j - \alpha_0 b_{N+1}$ for $j = 1, \ldots, N$ and $b_j^* := 0$ for $j = N + 1, \ldots, d$, where
$\alpha_0 = 1$ if $N \geq d_0$ and $\alpha_0 = (d_0 - N)^{-1}$ if $N < d_0$. Show that

$$\frac{1}{d} \sum_{j=1}^{d} S(f + d_0 b_j^* - (b_1^* + \cdots + b_d^*))$$

$$\geq \frac{1}{d} \sum_{j=1}^{d} S(f + d_0 b_j - (b_1 + \cdots + b_d)). \qquad (9.165)$$

(c) Extend S to $[0, \infty)$ by defining $S(f) := 1$ for all $f > 1$. Show that

$$S(f) \geq \frac{1}{d} \sum_{j=1}^{d} S(f + d_0 b_j - (b_1 + \cdots + b_d)) \qquad (9.166)$$

whenever $f \in [0,1]$, $b_1 \geq 0, \ldots, b_d \geq 0$, and $b_1 + \cdots + b_d \leq f$.

(d) Define $\boldsymbol{e}_1, \ldots, \boldsymbol{e}_d \in \mathbf{R}^d$ by $\boldsymbol{e}_1 := (1, 0, \ldots, 0)$, $\boldsymbol{e}_2 := (0, 1, 0, \ldots, 0)$, and
so on, and put $\mathbf{1} := (1, 1, \ldots, 1) \in \mathbf{R}^d$. Let $\boldsymbol{X}$ be a random variable satisfying
$P(\boldsymbol{X} = d_0 \boldsymbol{e}_j - \mathbf{1}) = p$ for $j = 1, \ldots, d$, and let $\boldsymbol{X}_1, \boldsymbol{X}_2, \ldots$ be i.i.d. with
common distribution that of $\boldsymbol{X}$. A roulette strategy consists of a sequence
$\boldsymbol{B}_1, \boldsymbol{B}_2, \ldots$, with $\boldsymbol{B}_1 = \boldsymbol{b}_1 \geq \boldsymbol{0}$ and $\boldsymbol{B}_n = \boldsymbol{b}_n(\boldsymbol{X}_1, \ldots, \boldsymbol{X}_{n-1}) \geq \boldsymbol{0}$ for each
$n \geq 2$. The gambler's fortune F_n after n coups satisfies $F_n = F_{n-1} + \boldsymbol{B}_n \cdot \boldsymbol{X}_n$
for each $n \geq 1$, F_0 being his initial fortune (a nonnegative constant). The
only restriction is that he cannot bet more than he has, that is, $\boldsymbol{B}_n \cdot \mathbf{1} \leq$
F_{n-1} for each $n \geq 1$. Finally, define T by (9.45) on p. 326, and assume that
$P(T < \infty) = 1$. Show that $\{S(F_n)\}_{n \geq 0}$ is a supermartingale with respect to
$\{\boldsymbol{X}_n\}_{n \geq 1}$, and therefore, if $F_0 = f \in [0,1]$, that $P(F_T \geq 1) \leq S(f)$.

Hint: (a) For $N = 1$, use (9.112) on p. 341 and (9.152) on p. 348 to obtain
$S(f/(1 - f)) \geq S(f) + S(f^2/(1 - f)) \geq S(f) + S(f)S(f/(1 - f))$. Proceed
by induction using

$$f^* := \frac{f}{1 - (N-1)f} \quad \text{implies} \quad \frac{f^*}{1 - f^*} = \frac{f}{1 - Nf}. \qquad (9.167)$$

(b) Use (9.151) on p. 348 and part (a) with $f = p_0$.

9.17. *Optimality of timid play at discrete superfair red-and-black.* Consider a game that pays even money, with the player winning with probability $p \in (\frac{1}{2}, 1)$ and losing otherwise. Assume that successive coups are independent and that only nonnegative-integer bet sizes are accepted. Let L and W be positive integers, and show that the probability that the player wins W units before losing L units is maximized by *timid play*, in which the player bets only one unit at each coup until he has lost L units or won W units.
Hint: Define $M := L + W$ and

$$Q(i) := \frac{1 - (q/p)^i}{1 - (q/p)^M}, \qquad i = 0, 1, \ldots, M. \tag{9.168}$$

By the gambler's ruin formula (Theorem 7.1.1 on p. 242), $Q(L)$ is the probability that the timid player, starting with L units, will reach his M-unit goal before being ruined. Show that it suffices to prove that

$$Q(i) \geq pQ(i+j) + qQ(i-j), \quad i = 0, 1, \ldots, M; \; j = 0, 1, \ldots, i \wedge (M - i). \tag{9.169}$$

Prove this inequality by a convexity argument.

9.18. *A unifying lemma.* Notice that the proofs of Theorems 9.1.7 on p. 327, 9.2.4 on p. 337, and 9.3.5 on p. 347 are similar. Find and prove a lemma that makes each of these theorems an almost immediate consequence of the results that precede it.

9.19. *Vardi's casino.* In Section 9.3 we considered the primitive casino with fixed parameters p and p_0 (p is the probability of a win, which pays $(1-p_0)/p_0$ to 1). Fix $c \in (0, 1)$ and consider the casino in which the set of permitted wagers at $f \in (0, 1)$ is the union over all $p_0 \in (0, 1)$ of the set of permitted wagers at f in the primitive casino with parameters $(1 - c)p_0$ and p_0. Regardless of which p_0 the player chooses at f, the house advantage of every permitted nonzero wager is c.

(a) One possible interpretation of bold play in this casino is to choose p_0 in such a way that, after one coup, either the gambler is ruined or he has achieved his goal of one unit. Show that the probability $S(f)$ of achieving the goal, starting with initial fortune f, is given by $S(f) = (1-c)f$ for $0 \leq f < 1$.

(b) Show that, regardless of the strategy used, the probability $S(f)$ of achieving the goal, starting with initial fortune f, must satisfy $S(f) \leq \overline{S}(f) := 1 - (1 - f)^{1-c}$ for $0 \leq f \leq 1$.

(c) Given $\alpha \in (0, 1)$, consider the strategy that adopts the wager of part (a) if $f \leq \alpha$ (bet f with $p_0 = f$), but bets $[\alpha/(1 - \alpha)](1 - f)$ with $p_0 = \alpha$ if $f \geq \alpha$. Show that the probability $S_\alpha(f)$ of achieving the goal, starting with initial fortune f, is given by

$$S_\alpha(f) = 1 - [1 - (1 - c)\alpha]^n \left(c + \frac{(1 - c)(1 - f)}{(1 - \alpha)^n} \right) \tag{9.170}$$

for $1 - (1 - \alpha)^n \leq f < 1 - (1 - \alpha)^{n+1}$ and $n \geq 0$.

(d) With $\overline{S}(f)$ as in part (b) and $S_\alpha(f)$ as in part (c), show that

$$\lim_{\alpha \to 0+} S_\alpha(f) = \overline{S}(f), \qquad 0 \leq f < 1. \tag{9.171}$$

In other words, the upper bound in part (b) is best possible.

Hint: (b) Show that $\overline{S}(f)$ satisfies the casino inequality

$$(1 - c)p_0\overline{S}(f + (q_0/p_0)b) + [1 - (1 - c)p_0]\overline{S}(f - b) \leq \overline{S}(f) \tag{9.172}$$

for $0 \leq f \leq 1$, $0 < p_0 < 1$, $q_0 := 1 - p_0$, and $0 \leq b \leq f \wedge (p_0/q_0)(1 - f)$.

9.5 Notes

The primary source for this chapter is the classic monograph by Lester E. Dubins [1920–2010] and Leonard J. Savage [1917–1971] titled *How to Gamble If You Must: Inequalities for Stochastic Processes* (1965, 1976). See also the more recent monograph by Ashok P. Maitra [1938–2008] and William D. Sudderth [1940–] titled *Discrete Gambling and Stochastic Games* (1996) for additional material on the same subject.

Julian Lowell Coolidge [1873–1954] (1909) stated the following result, which he later termed the fundamental theorem of games of chance (Coolidge 1925, p. 55):

> The Player's best chance of winning a stated sum at an unfavourable game is to stake the sum which will bring that return in one turn. If that be not allowed, he should stake at each turn the largest amount that the Banker will accept.

Dubins and Savage (1976, p. 4) pointed out that the proof given by Coolidge is incorrect, as is the implied uniqueness of bold play. Further, the claimed optimality of bold play in the case of a house limit is not always correct.

In fact the belief that betting boldly in a subfair game is optimal is perhaps not as widely held as one might imagine. In his first successful novel, *La peau de chagrin* (*The Wild Ass's Skin*, 1831) the French writer Honoré de Balzac [1799–1850] used these words to describe a young man who bet boldly and lost:[2]

> "He's no gambler," the croupier responded. "Otherwise he would have divided his money into three lots to stand a better chance."

Lemmas 9.1.1 and 9.1.6 as well as Theorems 9.1.2, 9.1.7, and 9.1.8 are from Dubins and Savage (1976, Chapter 5). Section 1 benefitted from the expository treatments of Billingsley (1995, Chapter 1, Section 7) and Siegrist (2008).

---

[2] Translation from Kavanagh (2005, p. 138).

Lemmas 9.2.1 and 9.2.3 as well as Theorems 9.2.2 and 9.2.4 are due to J. Ernest Wilkins, Jr. [1923–] (1972). His assumption that the player's goal is an integer multiple of the house limit cannot be omitted, as shown by Heath, Pruitt, and Sudderth (1972), Schweinsberg (2005), and Chen, Chung, Hsiau, and Yao (2008). It was noted in the Heath, Pruitt, and Sudderth reference that Theorem 9.2.4 was proved by Aryeh Dvoretzky [1916–2008] in 1963 but was not published. Example 9.2.5 was motivated by a discussion in the gambling literature: Stupak (1984) and Griffin (1991, pp. 161–162). The two suboptimal strategies in the example were proposed by Griffin, apparently without knowledge of Wilkins's results. Griffin (1991, p. 162) later calculated the optimal probability.

Lemmas 9.3.3 and 9.3.4, Theorems 9.3.1, 9.3.5, and 9.3.7, Corollary 9.3.6, and (9.159) are due to Dubins and Savage (1976, Chapter 6).

Problem 9.2 is from Dubins and Savage (1976, p. 86). Problem 9.5 is as well (p. 85), but it is credited there to W. Forrest Stinespring. Problem 9.7 comes from Pendergrass and Siegrist (2001); see also Siegrist (2008). Problem 9.11(b) is a result of Heath, Pruitt, and Sudderth (1972). Problem 9.12 is from Dubins and Savage (1976, p. 103), as is Problem 9.15 (p. 110). Problem 9.16 is due to Smith (1967), but our outline follows Dubins's (1968) simplification of Smith's argument. See Dubins (1972) for a more general result. Problem 9.17 is from Maitra and Sudderth (1996, p. 32), who credited Ross (1974) with an alternative derivation; see Thorp (1969) for yet another argument. Vardi's casino was named for its proposer, Yehuda Vardi [1946–2005], and Problem 9.19 is from Shepp [1936–] (2006). A slightly weaker conclusion (in which p_0 can depend on one's initial fortune f but cannot vary from coup to coup) was obtained by Dubins and Savage (1976, p. 182); a preliminary version of that result appeared in an unpublished paper of Savage (1957) titled "The casino that takes a percentage and what you can do about it."

Chapter 10
Optimal Proportional Play

Play ought not to be an affair of calculation, but of inspiration. I have calculated infallibly; and what has been the effect?

William Makepeace Thackeray, *The Newcomes*

This chapter is concerned with a betting system for superfair games, known as optimal proportional play or the Kelly system, that maximizes the long-term geometric rate of growth of the gambler's fortune. Section 10.1 considers the case in which a single superfair betting opportunity is available at each coup. Section 10.2 assumes that multiple betting opportunities, at least one of which is superfair, are available at each coup. Finally, Section 10.3 establishes several optimality properties of the Kelly system.

10.1 A Single Wager

Consider a game of chance that is superfair. Let X represent the gambler's profit (positive, negative, or zero) per unit bet. We assume that X is a $[-1, \infty)$-valued discrete random variable with finite expectation and $E[X] > 0$. To avoid technicalities, we further assume that X takes on only finitely many values. It is typically the case that

$$P(X = -1) > 0, \qquad (10.1)$$

that is, the gambler loses the amount of his bet with positive probability, but we do not require this.

Now let $X_1, X_2, \ldots$ be i.i.d. with common distribution that of X. Here X_n represents the gambler's profit per unit bet at coup n. A betting system consists of a sequence of bet sizes, $B_1, B_2, \ldots$, with B_n representing the

S.N. Ethier, *The Doctrine of Chances*, Probability and its Applications,
DOI 10.1007/978-3-540-78783-9_10, © Springer-Verlag Berlin Heidelberg 2010

amount bet at coup n and depending only on the results of the previous coups, $X_1, \ldots, X_{n-1}$. More precisely,

$$B_1 = b_1 \geq 0, \qquad B_n = b_n(X_1, \ldots, X_{n-1}) \geq 0, \quad n \geq 2, \qquad (10.2)$$

where b_1 is a constant and b_n is a nonrandom function of $n - 1$ variables for each $n \geq 2$. The gambler's fortune F_n after n coups satisfies $F_n = F_{n-1} + B_n X_n$ for each $n \geq 1$, and therefore

$$F_n = F_0 + \sum_{l=1}^{n} B_l X_l, \qquad n \geq 1, \qquad (10.3)$$

F_0 being his initial fortune (a positive constant). We assume that the gambler cannot bet more than he has, that is,

$$B_n \leq F_{n-1}, \qquad n \geq 1. \qquad (10.4)$$

Since the game is superfair (i.e., $E[X] > 0$), the gambler can maximize his expected profit by making the largest bet possible at each coup, namely,

$$B_n = F_{n-1}, \qquad n \geq 1. \qquad (10.5)$$

But then $F_n = F_{n-1}(1 + X_n)$ for each $n \geq 1$, and therefore

$$F_n = F_0 \prod_{l=1}^{n} (1 + X_l), \qquad n \geq 1. \qquad (10.6)$$

In particular, if at any of the first n coups the gambler loses the amount of his bet, he is ruined. It follows that, if (10.1) holds, $P(F_n = 0) = 1 - [1 - P(X = -1)]^n \to 1$ as $n \to \infty$. Thus, maximizing expected profit may result in an unsatisfactory betting system.

A better approach is for the gambler to bet a fixed proportion $f \in [0, 1]$ of his current fortune at each coup, that is,

$$B_n = f F_{n-1}, \qquad n \geq 1. \qquad (10.7)$$

In this case $F_n = F_{n-1}(1 + f X_n)$ for each $n \geq 1$, and therefore

$$F_n = F_0 \prod_{l=1}^{n} (1 + f X_l), \qquad n \geq 1. \qquad (10.8)$$

To determine the optimal choice of the betting proportion f, we define

$$r_n(f) := n^{-1} \ln(F_n / F_0) \qquad (10.9)$$

and

$$\mu(f) := \mathrm{E}[\ln(1 + fX)] \tag{10.10}$$

for all $f \in [0, 1]$ with $f \neq 1$ if (10.1) holds. Noting that $F_n = F_0 \exp\{r_n(f)n\}$, we can interpret $r_n(f)$ as the average geometric rate of growth of the gambler's fortune over the first n coups. By the strong law of large numbers (Theorem 1.5.5 on p. 43),

$$\lim_{n \to \infty} r_n(f) = \lim_{n \to \infty} \frac{1}{n} \sum_{l=1}^{n} \ln(1 + fX_l) = \mu(f) \quad \text{a.s.,} \tag{10.11}$$

so $\mu(f)$ can be interpreted as the long-term geometric rate of growth of the gambler's fortune.

The choice f^* of f that maximizes $\mu(f)$ results in a betting system known as *optimal proportional play* or the *Kelly system*. That there is a unique such f^* is a consequence of the following lemma.

Lemma 10.1.1. *Assume that X is a $[-1, \infty)$-valued random variable that takes on only finitely many values, and that $\mathrm{E}[X] > 0$. Consider four cases:*
(a) *(10.1) holds.*
(b) *(10.1) fails, $\mathrm{E}[X/(1 + X)] < 0$, and $\mathrm{E}[\ln(1 + X)] < 0$.*
(c) *(10.1) fails, $\mathrm{E}[X/(1 + X)] < 0$, and $\mathrm{E}[\ln(1 + X)] \geq 0$.*
(d) *(10.1) fails and $\mathrm{E}[X/(1 + X)] \geq 0$.*
The function $f \mapsto \mu(f)$ is defined for all $f \in [0, 1)$ in case (a) and for all $f \in [0, 1]$ in cases (b), (c), and (d). In all cases, it is strictly concave and there exists a unique $f^* \in [0, 1]$ such that $\mu(f)$ achieves a positive maximum at $f = f^*$. In cases (a), (b), and (c), $0 < f^* < 1$. In case (d), $f^* = 1$. In cases (a) and (b), there exists a unique $f_0 \in (f^*, 1)$ such that $\mu(f_0) = 0$.

Proof. By writing expectations as sums, it is clear that $\mu(f)$ is twice continuously differentiable (in fact, infinitely differentiable), with

$$\mu'(f) = \mathrm{E}\left[\frac{X}{1 + fX}\right] \quad \text{and} \quad \mu''(f) = -\mathrm{E}\left[\frac{X^2}{(1 + fX)^2}\right]. \tag{10.12}$$

The fact that $\mu''(f) < 0$ for all $f \in [0, 1)$ implies the strict concavity of $\mu(f)$ (Lemma 1.4.20 on p. 37). The facts that $\mu'(0) = \mathrm{E}[X] > 0$ and either $\mu'(1-) = -\infty$ (case (a)) or $\mu'(1) < 0$ (cases (b) and (c)) imply the existence of $f^* \in (0, 1)$ such that $\mu'(f^*) = 0$. Since $\mu''(f^*) < 0$, $\mu(f)$ has a local maximum at f^*, and its strict concavity implies that $\mu(f)$ has a global maximum there. In all cases, since $\mu(0) = 0$ and $\mu'(0) > 0$, we must have $\mu(f) > 0$ for all sufficiently small $f > 0$, and therefore $\mu(f^*) > 0$. In case (d), $\mu'(f)$ is decreasing in f and $\mu'(1) \geq 0$, so $\mu'(f) > 0$ for all $f \in [0, 1)$, that is, $\mu(f)$ is increasing in f. The facts that $\mu(f^*) > 0$ and either $\mu(1-) = -\infty$ (case (a)) or $\mu(1) < 0$ (case (b)) imply the existence of $f_0 \in (f^*, 1)$ such that $\mu(f_0) = 0$. The uniqueness of f^* and f_0 follows from the strict concavity of $\mu(f)$. ♠

Case (a) of Lemma 10.1.1 is of primary interest. In this case, $\mu(f) > 0$ if $0 < f < f_0$ and $\mu(f) < 0$ if $f_0 < f < 1$. See Example 10.1.3 below. For examples of the other cases of the lemma, see Problem 10.17 on p. 387.

It is often difficult to determine f^* and f_0 of the lemma explicitly. However, simple approximations are available. If $M \geq 1$, $\mathrm{P}(|X| \leq M) = 1$, and $0 < f^* < 1/M$, then f^* satisfies

$$0 = \mathrm{E}\left[\frac{X}{1 + f^* X}\right] = \mathrm{E}[X(1 - f^* X + (f^* X)^2 - \cdots)]$$
$$= m_1 - m_2 f^* + m_3 (f^*)^2 - \cdots, \tag{10.13}$$

where m_k denotes the kth moment of X. If, as is often the case, f^* is small, higher-order terms in (10.13) can be safely ignored. This yields the approximation

$$f^* \approx \frac{m_1}{m_2}. \tag{10.14}$$

If $M \geq 1$, $\mathrm{P}(|X| \leq M) = 1$, and $0 < f_0 < 1/M$, then f_0 satisfies

$$0 = \mathrm{E}[\ln(1 + f_0 X)] = \mathrm{E}\left[f_0 X - \frac{1}{2}(f_0 X)^2 + \frac{1}{3}(f_0 X)^3 - \cdots\right]$$
$$= m_1 f_0 - \frac{1}{2} m_2 (f_0)^2 + \frac{1}{3} m_3 (f_0)^3 - \cdots. \tag{10.15}$$

If f_0 is small, then

$$f_0 \approx \frac{2 m_1}{m_2}, \tag{10.16}$$

and it follows that $f_0 \approx 2 f^*$.

We can now state the main conclusions of this section. It will be convenient to denote F_n in (10.8) by $F_n(f)$ to emphasize its dependence on f. We will also need, in addition to $\mu(f)$ of (10.10),

$$\sigma^2(f) := \mathrm{Var}[\ln(1 + f X)]. \tag{10.17}$$

Theorem 10.1.2. *Under the above assumptions (namely, X is as in Lemma 10.1.1, and $X_1, X_2, \ldots$ are i.i.d. as X), the following conclusions hold. Let $f \in [0, 1]$ with $f \neq 1$ if (10.1) holds.*
   (a) $\lim_{n \to \infty} (F_n(f)/F_0)^{1/n} = \exp\{\mu(f)\}$ *a.s.*
   (b) *If $\mu(f) > 0$, then $\lim_{n \to \infty} F_n(f) = \infty$ a.s.*
   (c) *If $\mu(f) < 0$, then $\lim_{n \to \infty} F_n(f) = 0$ a.s.*
   (d) *If $\mu(f) = 0$ and $f \neq 0$, then*

$$\limsup_{n \to \infty} F_n(f) = \infty \ \text{a.s.} \quad \text{and} \quad \liminf_{n \to \infty} F_n(f) = 0 \ \text{a.s.} \tag{10.18}$$

   (e) *If $f \neq f^*$, then $\lim_{n \to \infty} F_n(f^*)/F_n(f) = \infty$ a.s.*
   (f) *If $\sigma(f) > 0$, then*

$$\frac{\sqrt{n}}{\sigma(f)}\left(\frac{1}{n}\ln\left(\frac{F_n(f)}{F_0}\right) - \mu(f)\right) \xrightarrow{d} N(0,1). \tag{10.19}$$

Remark. Part (a) is the exponentiated form of the strong law of large numbers (10.11). In this case the weak law of large numbers is perhaps more informative: For every $\varepsilon > 0$,

$$\lim_{n\to\infty} \mathrm{P}(F_n(f) \in [F_0 e^{(\mu(f)-\varepsilon)n}, F_0 e^{(\mu(f)+\varepsilon)n}]) = 1. \tag{10.20}$$

Part (b) (resp., (c)) says that, if $\mu(f) > 0$ (resp., $\mu(f) < 0$), the gambler's fortune approaches infinity (resp., zero) as $n \to \infty$. Part (d) says that, if $\mu(f) = 0$, the gambler's fortune oscillates between zero and infinity. Part (e) says that the Kelly system outperforms any other proportional betting system with a constant betting proportion, when applied to the same sequence of wins and losses. Denoting by Z_n the left side of (10.19), we can write

$$F_n(f) = F_0 e^{\mu(f)n + \sigma(f)\sqrt{n}Z_n} \tag{10.21}$$

with $Z_n \to N(0,1)$ in distribution. Thus, part (f) can be regarded as a refinement of (10.20).

Proof. Part (a) is equivalent to (10.11), and parts (b), (c), and (e) follow from part (a). Part (d) is a consequence of Theorem 3.3.4 on p. 109 applied to the martingale $M_n := \ln(F_n(f)/F_0)$, and part (f) is immediate from the central limit theorem (Theorem 1.5.15 on p. 48). ♠

Part (f) of the theorem can be used as follows. Defining

$$L_n(f,\alpha) := F_0 \exp\{\mu(f)n - z_{1-\alpha}\sigma(f)\sqrt{n}\} \tag{10.22}$$

and

$$U_n(f,\alpha) := F_0 \exp\{\mu(f)n + z_{1-\alpha}\sigma(f)\sqrt{n}\}, \tag{10.23}$$

where $z_{1-\alpha}$ is the $1-\alpha$ quantile of the standard-normal distribution, we have

$$\lim_{n\to\infty} \mathrm{P}(F_n(f) \in [L_n(f,\alpha/2), U_n(f,\alpha/2)]) = 1 - \alpha. \tag{10.24}$$

In particular, when n is large, $F_n(f)$ will belong to the interval $[L_n(f,\alpha/2), U_n(f,\alpha/2)]$ with probability approximately $1 - \alpha$. We use the future tense here because such an interval is often called a $100(1-\alpha)\%$ *prediction interval* for $F_n(f)$. (The term "confidence interval" is best reserved for situations in which one is estimating an unknown parameter.) An alternative form is

$$\lim_{n\to\infty} \mathrm{P}(F_n(f) \in [L_n(f,\alpha), \infty)) = 1 - \alpha. \tag{10.25}$$

Example 10.1.3. *Wagers with three possible outcomes: win, loss, or push.* Let the probabilities of a win, loss, and push be respectively p, q, and r, where $p > 0$, $q > 0$, $r \geq 0$, and $p + q + r = 1$. Suppose a win pays a to 1, where $a > 0$. Then $P(X = a) = p$, $P(X = -1) = q$, and $P(X = 0) = r$. Assume $E[X] > 0$, that is, $ap - q > 0$. Then

$$\mu(f) = p\ln(1 + af) + q\ln(1 - f) \tag{10.26}$$

and

$$\sigma^2(f) = p(1 - p)\{\ln(1 + af)\}^2 + q(1 - q)\{\ln(1 - f)\}^2$$
$$-2pq\ln(1 + af)\ln(1 - f), \tag{10.27}$$

the latter simplifying when $r = 0$ to

$$\sigma^2(f) = pq\left[\ln\left(\frac{1 + af}{1 - f}\right)\right]^2. \tag{10.28}$$

Since

$$\mu'(f) = \frac{ap - q - a(p + q)f}{(1 + af)(1 - f)}, \tag{10.29}$$

it follows that

$$f^* = \frac{ap - q}{a(p + q)} = \frac{E[X \mid X \neq 0]}{a}. \tag{10.30}$$

However, even in this simplest of examples, we cannot find an explicit expression for f_0 of Lemma 10.1.1. Therefore we consider a numerical example.

Take $a = 1$ and $p = 0.51 = 1 - q$. Then $f^* = 0.02$ and, using Newton's method, $f_0 \approx 0.0399893$. Also, $\mu(f^*) \approx 0.000200013$ and $\sigma(f^*) \approx 0.0199987$. The 95% prediction intervals from (10.24) and (10.25) are illustrated in Table 10.1. For example, the Kelly bettor's fortune after 1,000 coups will, with probability approximately 0.95, belong to the interval $[0.3536, 4.219]$, assuming initial fortune $F_0 = 1$. As can be seen from the table, users of the Kelly system can expect a considerable amount of volatility. ♠

Implicit in our discussion so far (and throughout the remainder of this chapter) is the assumption that capital is infinitely divisible, so that any bet that is called for can be made exactly, without rounding. One consequence of this slightly unrealistic assumption is that

$$P(F_n(f) > 0 \text{ for all } n \geq 0) = 1, \qquad 0 \leq f < 1, \tag{10.31}$$

that is, ruin is impossible if $f < 1$. We have also implicitly assumed no betting limits and an adversary with unlimited wealth or credit. These are of course idealizations of reality.

Table 10.1 95% prediction intervals for the Kelly bettor's fortune after n coups, in units of initial fortune. Here $P(X = 1) = 0.51$ and $P(X = -1) = 0.49$, and we apply (10.24) and (10.25) with $F_0 = 1$, $f = f^* = 0.02$, and $\alpha = 0.05$. Numbers are rounded to three significant digits.

n	$[L_n(f^*, 0.025), U_n(f^*, 0.025)]$	$[L_n(f^*, 0.05), \infty)$
100	$[0.689, 1.51]$	$[0.734, \infty)$
1,000	$[0.354, 4.22]$	$[0.432, \infty)$
10,000	$[0.147, 373.]$	$[0.275, \infty)$
100,000	$[0.201 \cdot 10^4, 0.118 \cdot 10^{15}]$	$[0.147 \cdot 10^5, \infty)$

10.2 Simultaneous Wagers

In this section we assume that there are d betting opportunities available at each coup, at least one of which is superfair. (Typically, $d \geq 2$, but we do not exclude the possibility that $d = 1$.) Let X_i represent the gambler's profit per unit bet on the ith betting opportunity. We assume for $i = 1, \ldots, d$ that X_i is a $[-1, \infty)$-valued random variable that takes on only finitely many values, and that

$$\max_{1 \leq i \leq d} E[X_i] > 0. \tag{10.32}$$

The random variables $X_1, \ldots, X_d$ are jointly distributed but typically not independent. Put

$$\boldsymbol{X} := (X_1, \ldots, X_d). \tag{10.33}$$

Now let $\boldsymbol{X}_1, \boldsymbol{X}_2, \ldots$ be i.i.d. with common distribution that of $\boldsymbol{X}$, and consider the betting system $\boldsymbol{B}_1, \boldsymbol{B}_2, \ldots$. Here

$$\boldsymbol{X}_n := (X_{n,1}, \ldots, X_{n,d}), \qquad \boldsymbol{B}_n := (B_{n,1}, \ldots, B_{n,d}), \tag{10.34}$$

with $X_{n,i}$ representing the gambler's profit per unit bet on the ith betting opportunity at the nth coup, and $B_{n,i}$ representing the amount bet on the ith betting opportunity at the nth coup. Of course, $\boldsymbol{B}_n$ can depend only on the results of the previous coups, $\boldsymbol{X}_1, \ldots, \boldsymbol{X}_{n-1}$. More precisely,

$$\boldsymbol{B}_1 = \boldsymbol{b}_1 \geq \boldsymbol{0}, \qquad \boldsymbol{B}_n = \boldsymbol{b}_n(\boldsymbol{X}_1, \ldots, \boldsymbol{X}_{n-1}) \geq \boldsymbol{0}, \quad n \geq 2, \tag{10.35}$$

where $\boldsymbol{b}_1$ is a constant vector and $\boldsymbol{b}_n$ is a nonrandom vector function of $n-1$ variables for each $n \geq 2$. (Vector inequalities hold componentwise.) The gambler's fortune F_n after n coups satisfies $F_n = F_{n-1} + \boldsymbol{B}_n \cdot \boldsymbol{X}_n$ for each $n \geq 1$, and therefore

$$F_n = F_0 + \sum_{l=1}^{n} \boldsymbol{B}_l \cdot \boldsymbol{X}_l, \qquad n \geq 1, \tag{10.36}$$

F_0 being his initial fortune (a positive constant). Here $\boldsymbol{b} \cdot \boldsymbol{x} := \sum_{i=1}^{d} b_i x_i$ is the *inner product*. We assume that the gambler cannot bet more than he has, that is,

$$B_{n,1} + \cdots + B_{n,d} \leq F_{n-1}, \qquad n \geq 1. \tag{10.37}$$

Because one or more of the available wagers is advantageous, expectation can be maximized by betting everything on the most favorable betting opportunity, but as we saw in the preceding section, this may result in an unsatisfactory betting system. Let us define

$$\Delta := \left\{ \boldsymbol{f} = (f_1, \ldots, f_d) : f_1 \geq 0, \ldots, f_d \geq 0, \sum_{i=1}^{d} f_i \leq 1 \right\}. \tag{10.38}$$

A better approach is for the gambler to choose $\boldsymbol{f} \in \Delta$ and bet the fixed proportion f_i of his current fortune on the ith betting opportunity for $i = 1, \ldots, d$ at each coup, that is,

$$\boldsymbol{B}_n = \boldsymbol{f} F_{n-1}, \qquad n \geq 1. \tag{10.39}$$

In this case $F_n = F_{n-1}(1 + \boldsymbol{f} \cdot \boldsymbol{X}_n)$ for each $n \geq 1$, and therefore

$$F_n = F_0 \prod_{l=1}^{n} (1 + \boldsymbol{f} \cdot \boldsymbol{X}_l), \qquad n \geq 1. \tag{10.40}$$

To determine the optimal choice of the vector of betting proportions $\boldsymbol{f}$, we define

$$r_n(\boldsymbol{f}) := n^{-1} \ln(F_n / F_0) \tag{10.41}$$

and

$$\mu(\boldsymbol{f}) := \mathrm{E}[\ln(1 + \boldsymbol{f} \cdot \boldsymbol{X})] \tag{10.42}$$

on the subset

$$\Delta_0 := \{ \boldsymbol{f} \in \Delta : \mathrm{P}(\boldsymbol{f} \cdot \boldsymbol{X} = -1) = 0 \}. \tag{10.43}$$

Noting that $F_n = F_0 \exp\{r_n(\boldsymbol{f})n\}$, we can interpret $r_n(\boldsymbol{f})$ as the average geometric rate of growth of the gambler's fortune over the first n coups. By the strong law of large numbers (Theorem 1.5.5 on p. 43),

$$\lim_{n \to \infty} r_n(\boldsymbol{f}) = \lim_{n \to \infty} \frac{1}{n} \sum_{l=1}^{n} \ln(1 + \boldsymbol{f} \cdot \boldsymbol{X}_l) = \mu(\boldsymbol{f}) \quad \text{a.s.,} \tag{10.44}$$

so $\mu(\boldsymbol{f})$ can be interpreted as the long-term geometric rate of growth of the gambler's fortune.

Any choice $\boldsymbol{f}^*$ of $\boldsymbol{f}$ that maximizes $\mu(\boldsymbol{f})$ results in a betting system known as *optimal proportional play* or the *Kelly system*. That there is such an $\boldsymbol{f}^*$ (though not necessarily a unique one) is a consequence of the following lemma.

Lemma 10.2.1. *Under the above assumptions on $\boldsymbol{X}$ (namely, X_i is a $[-1, \infty)$-valued random variable that takes on only finitely many values for $i = 1, \ldots, d$, and (10.32) holds), the function $\boldsymbol{f} \mapsto \mu(\boldsymbol{f})$, defined for all $\boldsymbol{f} \in \Delta_0$, is concave and achieves a positive maximum. If the maximum is achieved at both $\boldsymbol{f}_0^*$ and $\boldsymbol{f}_1^*$, then $\mathrm{P}(\boldsymbol{f}_0^* \cdot \boldsymbol{X} = \boldsymbol{f}_1^* \cdot \boldsymbol{X}) = 1$.*

Remark. In the special case $d = 1$, the conclusions of this lemma are weaker than those of Lemma 10.1.1 on p. 359.

Proof. The function $h(u) := \ln(1 + u)$ is strictly concave and continuous from $(-1, \infty)$ to $(-\infty, \infty)$, so $\mu(\boldsymbol{f}) = \mathrm{E}[h(\boldsymbol{f} \cdot \boldsymbol{X})]$ is continuous on Δ_0 and therefore on every compact set of the form $\{\boldsymbol{f} \in \Delta_0 : \mu(\boldsymbol{f}) \geq -C\}$. It therefore achieves its maximum. It is concave because, if $\boldsymbol{f}_0, \boldsymbol{f}_1 \in \Delta_0$ and $0 < \lambda < 1$, then, by the concavity of h,

$$\begin{aligned}
\mu(\lambda \boldsymbol{f}_0 + (1 - \lambda)\boldsymbol{f}_1) &= \mathrm{E}[h(\lambda \boldsymbol{f}_0 \cdot \boldsymbol{X} + (1 - \lambda)\boldsymbol{f}_1 \cdot \boldsymbol{X})] \\
&\geq \mathrm{E}[\lambda h(\boldsymbol{f}_0 \cdot \boldsymbol{X}) + (1 - \lambda)h(\boldsymbol{f}_1 \cdot \boldsymbol{X})] \\
&= \lambda \mathrm{E}[h(\boldsymbol{f}_0 \cdot \boldsymbol{X})] + (1 - \lambda)\mathrm{E}[h(\boldsymbol{f}_1 \cdot \boldsymbol{X})] \\
&= \lambda \mu(\boldsymbol{f}_0) + (1 - \lambda)\mu(\boldsymbol{f}_1),
\end{aligned} \tag{10.45}$$

and equality holds if and only if $\mathrm{P}(\boldsymbol{f}_0 \cdot \boldsymbol{X} = \boldsymbol{f}_1 \cdot \boldsymbol{X}) = 1$. This implies the final assertion. By (10.32) and Lemma 10.1.1 on p. 359, $\mu(\boldsymbol{f}) > 0$ for some $\boldsymbol{f} \in \Delta_0$ with only one positive component, and therefore $\mu(\boldsymbol{f}^*) > 0$. ♠

We can now state the main conclusions of this section. It will be convenient to denote F_n in (10.40) by $F_n(\boldsymbol{f})$ to emphasize its dependence on $\boldsymbol{f}$. We will also need, in addition to $\mu(\boldsymbol{f})$ of (10.42),

$$\sigma^2(\boldsymbol{f}) := \mathrm{Var}[\ln(1 + \boldsymbol{f} \cdot \boldsymbol{X})]. \tag{10.46}$$

Theorem 10.2.2. *Under the above assumptions (namely, $\boldsymbol{X}$ is as in Lemma 10.2.1, and $\boldsymbol{X}_1, \boldsymbol{X}_2, \ldots$ are i.i.d. as $\boldsymbol{X}$), the following conclusions hold. Let $\boldsymbol{f} \in \Delta_0$.*
 (a) $\lim_{n \to \infty}(F_n(\boldsymbol{f})/F_0)^{1/n} = \exp\{\mu(\boldsymbol{f})\}$ a.s.
 (b) If $\mu(\boldsymbol{f}) > 0$, then $\lim_{n \to \infty} F_n(\boldsymbol{f}) = \infty$ a.s.
 (c) If $\mu(\boldsymbol{f}) < 0$, then $\lim_{n \to \infty} F_n(\boldsymbol{f}) = 0$ a.s.
 (d) If $\mu(\boldsymbol{f}) = 0$ and $\sigma(\boldsymbol{f}) > 0$, then

$$\limsup_{n \to \infty} F_n(\boldsymbol{f}) = \infty \text{ a.s.} \quad and \quad \liminf_{n \to \infty} F_n(\boldsymbol{f}) = 0 \text{ a.s.} \tag{10.47}$$

 (e) If $\mu(\boldsymbol{f}) < \mu(\boldsymbol{f}^*)$, then $\lim_{n \to \infty} F_n(\boldsymbol{f}^*)/F_n(\boldsymbol{f}) = \infty$ a.s.
 (f) If $\sigma(\boldsymbol{f}) > 0$, then

$$\frac{\sqrt{n}}{\sigma(\boldsymbol{f})}\left(\frac{1}{n}\ln\left(\frac{F_n(\boldsymbol{f})}{F_0}\right) - \mu(\boldsymbol{f})\right) \xrightarrow{d} N(0,1). \qquad (10.48)$$

Remark. These are the analogues of (a)–(f) of Theorem 10.1.2 on p. 360. See the remark following the statement of that theorem for discussion. In part (d), it is not sufficient to assume that $\mu(\boldsymbol{f}) = 0$ and $\boldsymbol{f} \neq \boldsymbol{0}$. In part (e), it is not sufficient to assume that $\boldsymbol{f} \neq \boldsymbol{f}^*$.

Proof. The proof is analogous to that of Theorem 10.1.2 on p. 360.   ♠

As in Section 10.1, we can use part (f) of the theorem to derive a $100(1 - \alpha)\%$ *prediction interval* for $F_n(\boldsymbol{f})$. Defining

$$L_n(\boldsymbol{f}, \alpha) := F_0 \exp\{\mu(\boldsymbol{f})n - z_{1-\alpha}\sigma(\boldsymbol{f})\sqrt{n}\} \qquad (10.49)$$

and

$$U_n(\boldsymbol{f}, \alpha) := F_0 \exp\{\mu(\boldsymbol{f})n + z_{1-\alpha}\sigma(\boldsymbol{f})\sqrt{n}\}, \qquad (10.50)$$

we have

$$\lim_{n\to\infty} P(F_n(\boldsymbol{f}) \in [L_n(\boldsymbol{f}, \alpha/2), U_n(\boldsymbol{f}, \alpha/2)]) = 1 - \alpha \qquad (10.51)$$

and

$$\lim_{n\to\infty} P(F_n(\boldsymbol{f}) \in [L_n(\boldsymbol{f}, \alpha), \infty)) = 1 - \alpha. \qquad (10.52)$$

Example 10.2.3. $d \geq 2$ *simultaneous betting opportunities with one and only one winner.* Consider a game of chance with $d \geq 2$ possible outcomes, which will be denoted by $1, 2, \ldots, d$. Denote their respective probabilities by $p_1, p_2, \ldots, p_d$, where $p_1 > 0$, $p_2 > 0$, $\ldots$, $p_d > 0$, and $\sum_{i=1}^{d} p_i = 1$. Suppose that, for $i = 1, \ldots, d$, a bet on outcome i is available and pays a_i to 1, where $a_i > 0$, and that

$$\max_{1\leq i\leq d}\{(a_i + 1)p_i - 1\} > 0, \qquad (10.53)$$

that is, (10.32) holds. How should the Kelly bettor allocate his betting capital on these d betting opportunities?

Define $\boldsymbol{e}_1, \ldots, \boldsymbol{e}_d \in \Delta$ by $\boldsymbol{e}_1 := (1, 0, \ldots, 0)$, $\boldsymbol{e}_2 := (0, 1, 0, \ldots, 0)$, and so on, and put $\boldsymbol{1} := (1, 1, \ldots, 1) \in \mathbf{R}^d$. The distribution of $\boldsymbol{X}$ is specified by

$$P(\boldsymbol{X} = (a_i + 1)\boldsymbol{e}_i - \boldsymbol{1}) = p_i, \qquad i = 1, \ldots, d, \qquad (10.54)$$

and the problem is to find the unique $\boldsymbol{f}^* = (f_1^*, \ldots, f_d^*) \in \Delta_0$ that maximizes

$$\mu(\boldsymbol{f}) := \mathrm{E}[\ln(1 + \boldsymbol{f} \cdot \boldsymbol{X})] = \sum_{i=1}^{d} p_i \ln\left(1 + (a_i + 1)f_i - \sum_{j=1}^{d} f_j\right). \qquad (10.55)$$

Here Δ_0 is the set of all $\boldsymbol{f} \in \Delta$ except those for which $\sum_{j=1}^{d} f_j = 1$ and $f_i = 0$ for some $i \in \{1, 2, \ldots, d\}$.

Actually, $\boldsymbol{f}^*$ need not be unique. However, a necessary and sufficient condition for uniqueness is

$$\sum_{i=1}^{d}(a_i + 1)^{-1} \neq 1. \tag{10.56}$$

To confirm the sufficiency, suppose that the maximum is achieved at $\boldsymbol{f}_0^*$ and at $\boldsymbol{f}_1^*$. Then, by Lemma 10.2.1,

$$\mathrm{P}(\boldsymbol{f}_0^* \cdot \boldsymbol{X} = \boldsymbol{f}_1^* \cdot \boldsymbol{X}) = 1. \tag{10.57}$$

Since $p_i > 0$ for $i = 1, \ldots, d$, (10.57) and (10.54) imply that

$$(a_i + 1)f_{0,i}^* - \sum_{j=1}^{d} f_{0,j}^* = (a_i + 1)f_{1,i}^* - \sum_{j=1}^{d} f_{1,j}^*, \qquad i = 1, \ldots, d. \tag{10.58}$$

Multiply both sides of (10.58) by $(a_i + 1)^{-1}$ and sum over i to get

$$\left(1 - \sum_{i=1}^{d}(a_i + 1)^{-1}\right)\sum_{j=1}^{d} f_{0,j}^* = \left(1 - \sum_{i=1}^{d}(a_i + 1)^{-1}\right)\sum_{j=1}^{d} f_{1,j}^*. \tag{10.59}$$

By virtue of (10.56), this yields $\sum_{j=1}^{d} f_{0,j}^* = \sum_{j=1}^{d} f_{1,j}^*$, and it follows from (10.58) that $\boldsymbol{f}_0^* = \boldsymbol{f}_1^*$.

For the necessity of (10.56), see Problem 10.20 on p. 387.

We assume (10.56) hereafter. Letting w denote the proportion of betting capital withheld by the gambler, we can write (10.55) as

$$\mu(w, \boldsymbol{f}) := \sum_{i=1}^{d} p_i \ln(w + (a_i + 1)f_i), \tag{10.60}$$

and treat the problem as a constrained maximization problem to which the Karush–Kuhn–Tucker theorem (Theorem A.2.9 on p. 749) can be applied. The inequality constraints are $w \geq 0, f_1 \geq 0, \ldots, f_d \geq 0$, and the equality constraint is $1 - w - f_1 - \cdots - f_d = 0$. Let $(w^*, \boldsymbol{f}^*)$ denote a pair $(w, \boldsymbol{f}) \in [0, 1] \times \Delta_0$ that maximizes $\mu(w, \boldsymbol{f})$ subject to the constraints. By the theorem, there exist $\kappa_0 \geq 0, \kappa_1 \geq 0, \ldots, \kappa_d \geq 0$ and λ real such that, at $(w^*, \boldsymbol{f}^*)$,

$$\frac{\partial \mu}{\partial w} + \kappa_0 - \lambda = 0 \quad \text{and} \quad \kappa_0 w^* = 0, \tag{10.61}$$

$$\frac{\partial \mu}{\partial f_j} + \kappa_j - \lambda = 0 \quad \text{and} \quad \kappa_j f_j^* = 0, \qquad j = 1, \ldots, d. \tag{10.62}$$

We define $I := \{j \in \{1, \ldots, d\} : f_j^* > 0\}$ and $I^c := \{j \in \{1, \ldots, d\} : f_j^* = 0\}$, and we consider two cases.

Case 1. $w^* > 0$. First we want to observe that it must be the case that

$$\sum_{i=1}^{d} (a_i + 1)^{-1} > 1. \tag{10.63}$$

For if the opposite inequality $(<)$ held, then with the withheld capital we could bet on outcome i in proportion to $(a_i + 1)^{-1}$ for $i = 1, \ldots, d$ and be assured of a positive profit solely from these additional bets, contradicting the optimality of $(w^*, \boldsymbol{f}^*)$.

Now, at $(w^*, \boldsymbol{f}^*)$,

$$\frac{\partial \mu}{\partial w} - \lambda = 0 \quad \text{or} \quad \sum_{i=1}^{d} \frac{p_i}{w^* + (a_i + 1) f_i^*} = \lambda, \tag{10.64}$$

$$\frac{\partial \mu}{\partial f_j} - \lambda = 0 \quad \text{or} \quad \frac{(a_j + 1) p_j}{w^* + (a_j + 1) f_j^*} = \lambda, \quad j \in I, \tag{10.65}$$

$$\frac{\partial \mu}{\partial f_j} - \lambda \leq 0 \quad \text{or} \quad \frac{(a_j + 1) p_j}{w^*} \leq \lambda, \quad j \in I^c, \tag{10.66}$$

and $w^* + \sum_{i=1}^{d} f_i^* = 1$. Let us define $q := \sum_{j \in I} p_j$ and $b := \sum_{j \in I} (a_j + 1)^{-1}$. Then rewriting (10.65) as $p_j = \lambda (a_j + 1)^{-1} w^* + \lambda f_j^*$ and summing over $j \in I$, we conclude that

$$q = \lambda b w^* + \lambda (1 - w^*). \tag{10.67}$$

Next, multiplying (10.64) by w^* and substituting (10.65), we obtain

$$\lambda w^* = \sum_{j \in I} \frac{p_j w^*}{w^* + (a_j + 1) f_j^*} + \sum_{j \in I^c} \frac{p_j w^*}{w^*}$$

$$= \sum_{j \in I} \frac{\lambda w^*}{a_j + 1} + 1 - \sum_{j \in I} p_j = \lambda b w^* + 1 - q, \tag{10.68}$$

or

$$q = \lambda b w^* + 1 - \lambda w^*. \tag{10.69}$$

Comparing (10.67) and (10.69), we find that $\lambda = 1$ and therefore (10.69) is equivalent to $w^*(1 - b) = 1 - q$. Since $w^* > 0$, we must have $q < 1$ and therefore $b < 1$, for otherwise $q = 1$ and therefore $I = \{1, \ldots, d\}$ and $b = 1$, contradicting (10.63). Consequently, using (10.65),

$$w^* = \frac{1 - q}{1 - b}, \quad f_j^* = \begin{cases} p_j - (a_j + 1)^{-1} w^* & \text{if } j \in I, \\ 0 & \text{if } j \in I^c, \end{cases} \tag{10.70}$$

and $I^c \neq \varnothing$. This will be the solution to the problem once we identify the set I.

For this we use (10.65) and (10.66), which imply that

$$(a_j + 1)p_j = w^* + (a_j + 1)f_j^* > w^*, \qquad j \in I,$$
$$(a_j + 1)p_j \le w^*, \qquad j \in I^c. \tag{10.71}$$

It will be convenient to relabel the outcomes in such a way that

$$(a_1 + 1)p_1 \ge (a_2 + 1)p_2 \ge \cdots \ge (a_d + 1)p_d, \tag{10.72}$$

the advantage being that then I will have the form $\{1, 2, \ldots, k\}$. To determine the optimal k, we introduce the notation

$$q_k := \sum_{j=1}^{k} p_j, \qquad b_k := \sum_{j=1}^{k} (a_j + 1)^{-1}, \qquad w_k := \frac{1 - q_k}{1 - b_k}, \tag{10.73}$$

where empty sums are 0. In view of (10.71), it suffices to show that there is a unique $k \in \{1, \ldots, d-1\}$ such that $b_k < 1$ and

$$(a_j + 1)p_j > w_k, \qquad j = 1, \ldots, k,$$
$$(a_j + 1)p_j \le w_k, \qquad j = k+1, \ldots, d, \tag{10.74}$$

or, equivalently, $b_k < 1$ and $(a_{k+1} + 1)p_{k+1} \le w_k < (a_k + 1)p_k$.

Let us define, for $k = 1, \ldots, d$,

$$r_k := q_k + (1 - b_k)(a_k + 1)p_k = q_{k-1} + (1 - b_{k-1})(a_k + 1)p_k. \tag{10.75}$$

With this notation we need only show that there exists a unique $k \in \{1, \ldots, d-1\}$ such that $b_k < 1$ and $r_{k+1} \le 1 < r_k$. First, the uniqueness is easy because, by (10.72), $r_{k+1} \le r_k$ for all $k \in \{1, \ldots, d-1\}$ for which $b_k < 1$. For existence we define $k_0 = \max\{k \in \{1, \ldots, d-1\} : r_k > 1\}$. Since $r_1 = (a_1 + 1)p_1 > 1$ by (10.53) and (10.72), and $r_d \le 1$ by (10.75) and (10.63), we must have $r_{k_0+1} \le 1 < r_{k_0}$. Finally, (10.75) shows that $b_{k_0} < 1$, for otherwise $r_{k_0} > 1$ would fail.

Case 2. $w^* = 0$. Here $I = \{1, 2, \ldots, d\}$, for otherwise ruin would occur with positive probability. From (10.61) and (10.62), we have

$$\frac{\partial \mu}{\partial w} - \lambda \le 0 \quad \text{or} \quad \sum_{i=1}^{d} \frac{p_i}{(a_i + 1)f_i} \le \lambda, \tag{10.76}$$

$$\frac{\partial \mu}{\partial f_j} - \lambda = 0 \quad \text{or} \quad \frac{p_j}{f_j^*} = \lambda, \qquad j = 1, \ldots, d, \tag{10.77}$$

and $\sum_{i=1}^{d} f_i^* = 1$. Rewriting (10.77) as $p_j = \lambda f_j^*$ and summing over j gives $\lambda = 1$, and therefore $f_j^* = p_j$ for $j = 1, \ldots, d$. Thus, (10.76) and (10.56) imply that

$$\sum_{i=1}^{d} (a_i + 1)^{-1} < 1. \tag{10.78}$$

In the notation of (10.73) and (10.75), $b_d < 1$, $w_d = 0$, and $r_d > 1$.

To summarize our solution, we assume (10.53), (10.56), and (10.72) and conclude that there is a unique optimal $\boldsymbol{f}^*$ that can be specified (in either case 1 or case 2) by first defining

$$k_0 := \max\{k \in \{1,\ldots,d\} : q_{k-1} + (1 - b_{k-1})(a_k + 1)p_k > 1\}, \qquad (10.79)$$

and then letting

$$f_j^* = \begin{cases} p_j - (a_j + 1)^{-1}w_{k_0} & \text{if } j = 1,\ldots,k_0, \\ 0 & \text{if } j = k_0 + 1,\ldots,d. \end{cases} \qquad (10.80)$$

Here we are using the notation (10.73).

It is worth stating explicitly that, assuming (10.53) and (10.56), case 1 $(w^* > 0)$ implies that $f_i^* = 0$ for some $i \in \{1,\ldots,d\}$ and (10.63) holds, while case 2 $(w^* = 0)$ implies that $f_i^* > 0$ for all $i \in \{1,\ldots,d\}$ and (10.78) holds. For the case in which (10.56) fails, see Problem 10.20 on p. 387.

It is curious that (still assuming (10.53) and (10.56)) the necessary and sufficient condition to bet everything is (10.78), which does not depend on the probabilities $p_1,\ldots,p_d$. On the other hand, the optimal betting proportions in this case are $f_i^* = p_i$ $(i = 1,\ldots,d)$, which do not depend on the payoff odds $a_1,\ldots,a_d$.

For future reference, we note that

$$\mu(\boldsymbol{f}^*) = \sum_{i=1}^{k_0} p_i \ln((a_i + 1)p_i) + (1 - q_{k_0})\ln(w_{k_0}) \qquad (10.81)$$

and

$$\sigma^2(\boldsymbol{f}^*) = \sum_{i=1}^{k_0} p_i[\ln((a_i + 1)p_i)]^2 + (1 - q_{k_0})[\ln(w_{k_0})]^2 - \mu(\boldsymbol{f}^*)^2, \qquad (10.82)$$

where $0 \ln 0 := 0$.   ♠

Example 10.2.4. *Biased roulette with known probabilities.* Consider a biased roulette wheel with $z = 1$ or $z = 2$ zeros. This is the special case of the preceding example in which $d = 36 + z$ and $a_i = 35$ for $i = 1, 2, \ldots, 36 + z$. We relabel the numbers so that

$$p_1 \geq p_2 \geq \cdots \geq p_{36+z}, \qquad (10.83)$$

and we assume that $36p_1 > 1$. Then (10.53), (10.63), and (10.72) hold, and the unique optimal $\boldsymbol{f}^*$ is specified in terms of

$$k_0 := \max\{k \in \{1,\ldots,35\} : p_1 + \cdots + p_{k-1} + (36 - (k - 1))p_k > 1\} \quad (10.84)$$

and

$$f_j^* = \begin{cases} p_j - (36 - k_0)^{-1}(1 - p_1 - \cdots - p_{k_0}) & \text{if } j = 1, \ldots, k_0, \\ 0 & \text{if } j = k_0 + 1, \ldots, 36 + z. \end{cases}$$

$$(10.85)$$

In particular, the Kelly bettor bets on number k if and only if

$$p_1 + \cdots + p_{k-1} + (36 - (k - 1))p_k \qquad (10.86)$$

is greater than 1, a condition that is implied by but does not imply (unless $k = 1$) that a bet on number k is superfair, that is, $36p_k > 1$.

Turning to a specific example, consider two biased 38-number roulette wheels.

Wheel 1 has probabilities $p_1 = 1/18$ and $p_2 = p_3 = \cdots = p_{38} < 1/39$. Number 1 is favorable, while number 2–38 are unfavorable. The Kelly bettor bets a proportion $f_1^* = 1/35$ on number 1 and nothing on the unfavorable numbers. Here

$$\mathrm{E}[\boldsymbol{f}^* \cdot \boldsymbol{X}] \approx 0.0285714, \quad \mu(\boldsymbol{f}^*) \approx 0.0111311, \quad \sigma(\boldsymbol{f}^*) \approx 0.165413. \quad (10.87)$$

Wheel 2 has probabilities $p_k = 1/(24 + k)$ for $k = 1, 2, \ldots, 16$ and $p_{17} = p_{18} = \cdots = p_{38} < 1/44$, so (10.83) holds. Numbers 1–11 are favorable, number 12 is neutral, and numbers 13–38 are unfavorable. Nevertheless, the Kelly bettor bets on each of the first 16 numbers! His total proportion bet is 0.104653, and

$$\mathrm{E}[\boldsymbol{f}^* \cdot \boldsymbol{X}] \approx 0.0256994, \quad \mu(\boldsymbol{f}^*) \approx 0.0120570, \quad \sigma(\boldsymbol{f}^*) \approx 0.159398, \quad (10.88)$$

where we have used (10.81) and (10.82). The details are given in Table 10.2.

If offered a choice between the two wheels, the astute Kelly bettor would prefer wheel 2, despite the fact that his expected profit per coup would be smaller. The point is that $\mu(\boldsymbol{f}^*)$ would be larger and $\sigma(\boldsymbol{f}^*)$ would be smaller, so he would realize a higher long-term geometric rate of growth of his fortune with lower volatility. The greater diversification offered by wheel 2 is responsible for these desirable properties.   ♠

10.3 Optimality Properties

In this section we discuss three optimality properties of the Kelly system. Although, as we have seen, the Kelly system does not maximize the gambler's mean fortune, it does maximize his median fortune. It also minimizes the expected time required to reach a goal. Actually, both of these results are valid only in an asymptotic sense, which we make precise in this section. Finally, we

Table 10.2 The Kelly system applied to two biased 38-number roulette wheels. Wheel 1 has one favorable number, and only one bet is made. Wheel 2 has 11 favorable numbers, and 16 bets are made. The criterion (10.86) must be greater than 1 for a bet to be made on number k.

wheel 1

no. k	exact p_k	approx. p_k	expectation $36p_k - 1$	criterion (10.86)	Kelly propor. f_k^*
1	1/18	.055 556	1.000 000	2.000 000	.028 571
2–38	(equal)	.025 526	−.081 081	.948 949	.000 000
totals	1	1.000 000			.028 571

wheel 2

no. k	exact p_k	approx. p_k	expectation $36p_k - 1$	criterion (10.86)	Kelly propor. f_k^*
1	1/25	.040 000	.440 000	1.440 000	.015 129
2	1/26	.038 462	.384 615	1.386 154	.013 591
3	1/27	.037 037	.333 333	1.337 721	.012 166
4	1/28	.035 714	.285 714	1.294 070	.010 844
5	1/29	.034 483	.241 379	1.254 661	.009 612
6	1/30	.033 333	.200 000	1.219 029	.008 463
7	1/31	.032 258	.161 290	1.186 771	.007 387
8	1/32	.031 250	.125 000	1.157 537	.006 379
9	1/33	.030 303	.090 909	1.131 022	.005 432
10	1/34	.029 412	.058 824	1.106 958	.004 541
11	1/35	.028 571	.028 571	1.085 109	.003 701
12	1/36	.027 778	.000 000	1.065 268	.002 907
13	1/37	.027 027	−.027 027	1.047 250	.002 156
14	1/38	.026 316	−.052 632	1.030 891	.001 445
15	1/39	.025 641	−.076 923	1.016 046	.000 770
16	1/40	.025 000	−.100 000	1.002 585	.000 129
17–38	(equal)	.022 610	−.186 048	.954 780	.000 000
totals	1	1.000 000			.104 653

consider a refinement of the optimality property given in Theorems 10.1.2(e) on p. 360 and 10.2.2(e) on p. 365.

Although the median-maximizing property of the Kelly system is valid in considerable generality, we will limit our discussion to a special case in which stronger results are known. We assume that $P(X = a) = p$ and $P(X = -1) =$

$q := 1 - p$, where $a > 0$ and $0 < p < 1$. In other words, the game pays odds of a to 1 on the occurrence of an event of probability p. We require that $E[X] > 0$, that is, $ap - q > 0$. Let $X_1, X_2, \ldots$ be a sequence of i.i.d. random variables with common distribution that of X. We define $F_n(f)$ by (10.8) on p. 358 and let f^* be the Kelly betting proportion, that is, the $f \in (0,1)$ that maximizes $\mu(f) := E[\ln(1 + fX)]$.

Before stating the relevant theorem, we recall the definition of the *median* of a random variable X. We define $\mathrm{median}(X) := m$ if

$$P(X \le m) \ge \frac{1}{2} \quad \text{and} \quad P(X \ge m) \ge \frac{1}{2}. \tag{10.89}$$

Typically, there is a unique m satisfying (10.89), but in general, the set of all m for which (10.89) holds is a closed interval $[m_{\min}, m_{\max}]$. In such cases, we adopt the convention that $\mathrm{median}(X) := m_{\max}$. In other words, we define the median maximally if it would otherwise not be uniquely defined.

We will need the fact (Theorem A.2.12 on p. 750) that the mean and median of the binomial(n, p) distribution differ by less than $\ln 2 \approx 0.693147$. Thus,

$$\gamma_{n,p} := \mathrm{median}(\mathrm{binomial}(n, p)) - np \tag{10.90}$$

satisfies $|\gamma_{n,p}| < \ln 2$ for all $n \ge 1$ and $0 < p < 1$. (As a function of p, $\mathrm{median}(\mathrm{binomial}(n, p))$ is a nondecreasing step function with n discontinuities, and by virtue of our convention it assumes values in $\{0, 1, \ldots, n\}$ and is right continuous.)

Theorem 10.3.1. *Under the above assumptions and for each $n \ge 1$, the choice $\tilde{f}$ of f that maximizes the median of $F_n(f)$ is given uniquely by*

$$\tilde{f} = f^* + \frac{a+1}{a} \frac{\gamma_{n,p}}{n}. \tag{10.91}$$

Remark. In particular, $\tilde{f}$ and f^* are asymptotically equivalent. Moreover, if the binomial mean np is an integer, then, because the binomial median is always an integer and they differ by less than 1, it must be the case that $\gamma_{n,p} = 0$ and therefore $\tilde{f} = f^*$.

Proof. Fix $n \ge 1$ and let B_n be the number of wins in the first n coups (i.e., $B_n := |\{1 \le i \le n : X_i = a\}|$). Then B_n is binomial(n, p), and

$$F_n(f) = F_0(1 + af)^{B_n}(1 - f)^{n - B_n} = F_0 \left(\frac{1 + af}{1 - f} \right)^{B_n} (1 - f)^n, \tag{10.92}$$

hence

$$E[\ln(F_n(f))] = \ln(F_0) + np \ln(1 + af) + nq \ln(1 - f). \tag{10.93}$$

Now by properties of the median,

$$\ln(\text{median}(F_n(f))) = \text{median} \ln(F_n(f))$$

$$= \text{median}\left(\ln(F_0) + B_n \ln\left(\frac{1+af}{1-f}\right) + n\ln(1-f) \right)$$

$$= \ln(F_0) + \text{median}(B_n) \ln\left(\frac{1+af}{1-f}\right) + n\ln(1-f)$$

$$= \ln(F_0) + (np + \gamma_{n,p}) \ln\left(\frac{1+af}{1-f}\right) + n\ln(1-f)$$

$$= \ln(F_0) + (np + \gamma_{n,p}) \ln(1+af) + (nq - \gamma_{n,p})\ln(1-f)$$

$$= \mathrm{E}[\ln(F_n^\circ(f))], \tag{10.94}$$

where $F_n^\circ(f)$ is $F_n(f)$ with p replaced by

$$p_n := p + \frac{\gamma_{n,p}}{n} = \frac{\text{median}(B_n)}{n}. \tag{10.95}$$

The last step in (10.94) uses (10.93). Since $\mathrm{E}[\ln(F_n(f))] = \ln(F_0) + n\mu(f)$ is uniquely maximized by the Kelly proportion

$$f^* = \frac{(a+1)p - 1}{a} = \frac{\mathrm{E}[X]}{a}, \tag{10.96}$$

it follows from (10.94) that $\text{median}(F_n(f))$ is uniquely maximized by

$$\tilde{f} = \frac{(a+1)p_n - 1}{a} = f^* + \frac{a+1}{a}\frac{\gamma_{n,p}}{n}, \tag{10.97}$$

as required.   ♠

Example 10.3.2. *Kelly proportion vs. median-maximizing proportion.* Define $p_{n,0} := 0$ and $p_{n,n+1} := 1$, and for $k = 1,\ldots,n$, define $p_{n,k}$ to be the unique $p \in (0,1)$ such that $\mathrm{P}(\text{binomial}(n,p) \le k-1) = \frac{1}{2}$. Then, for $k = 0,1,\ldots,n$,

$$\text{median}(\text{binomial}(n,p)) = k \quad \text{if} \quad p_{n,k} \le p < p_{n,k+1}. \tag{10.98}$$

The numbers $p_{n,1},\ldots,p_{n,n}$ can be evaluated numerically. We have done this in the case $n = 100$, and results are summarized in Table 10.3. We find, for example, that if $p = 0.505$, then $f^* = 0.01$ and $\tilde{f} = 0.02$. On the other hand, if $p = 0.51$, then $f^* = \tilde{f} = 0.02$. This is in the nature of approximating a continuous function by a step function.

This suggests that a more useful way to state (10.97) might be to divide it by (10.96), obtaining

$$\frac{\tilde{f}}{f^*} = 1 + \frac{a+1}{\mathrm{E}[X]}\frac{\gamma_{n,p}}{n}, \tag{10.99}$$

Table 10.3 Comparing the Kelly proportion f^* with the median maximizing proportion $\tilde{f}$ when $n = 100$. Here $P(X = 1) = p = 1 - P(X = -1)$, so $f^* = (2p - 1)^+$ and $\tilde{f} = (2\,\text{median}(\text{binomial}(100, p))/100 - 1)^+$. Column $(*)$ contains median(binomial$(100, p)$). Results are rounded to six decimal places.

range of values of p	$(*)$	range of values of f^*	$\tilde{f}$
$[.495\ 017, .504\ 983)$	50	$[.000\ 000, .009\ 967)$	0.00
$[.504\ 983, .514\ 950)$	51	$[.009\ 967, .029\ 900)$	0.02
$[.514\ 950, .524\ 917)$	52	$[.029\ 900, .049\ 833)$	0.04
$[.524\ 917, .534\ 883)$	53	$[.049\ 833, .069\ 766)$	0.06
$[.534\ 883, .544\ 850)$	54	$[.069\ 766, .089\ 699)$	0.08
$[.544\ 850, .554\ 816)$	55	$[.089\ 699, .109\ 633)$	0.10

which shows that, the smaller $E[X]$ is, the larger n must be to ensure a relative error within specified bounds. ♠

Now let us turn to the minimum expected time required to reach a goal property. We assume the conditions and notation of Theorem 10.2.2 on p. 365. Given $f \in \Delta_0$ with $\mu(f) > 0$, we define $M(f)$ to be the maximum value taken on with positive probability by the random variable $\ln(1 + f \cdot X)$, and we define for each win target $W > 1$ the stopping time

$$N_W(f) := \min\{n \geq 1 : F_n(f)/F_0 \geq W\}. \tag{10.100}$$

Theorem 10.3.3. *For all $f \in \Delta_0$ with $\mu(f) > 0$, we have*

$$\frac{\ln(W)}{\mu(f)} \leq E[N_W(f)] \leq \frac{\ln(W) + M(f)}{\mu(f)}. \tag{10.101}$$

In particular,

$$\lim_{W \to \infty} \frac{E[N_W(f)]}{\ln(W)} = \frac{1}{\mu(f)}, \tag{10.102}$$

which is minimized at $f = f^*$.

Remark. Result (10.102) implies that the vector of Kelly betting proportions asymptotically minimizes the expected time required to reach a goal. The case $d = 1$ is included.

Proof. This follows from Corollary 7.3.9 on p. 266, after noting that

$$F_n(\boldsymbol{f})/F_0 \geq W \quad \text{if and only if} \quad \sum_{l=1}^{n} \ln(1 + \boldsymbol{f} \cdot \boldsymbol{X}_l) \geq \ln(W). \qquad (10.103)$$

With probability 1, the maximum overshoot does not exceed $M(\boldsymbol{f})$.   ♠

Because of the limit in (10.102), one may wonder whether the conclusion applies when W is only moderately large. The next example suggests that it may, approximately.

Example 10.3.4. *Expected number of coups required for the proportional bettor to reach a goal.* Here we investigate whether the asymptotic expected time minimization property of the Kelly system can be said to hold approximately for fixed goals such as $W = 2$. For this we need a method of evaluating $\mathrm{E}[N_W(\boldsymbol{f})]$. Let us specialize to the case $d = 1$, $\mathrm{P}(X = a) = p$, and $\mathrm{P}(X = -1) = q := 1 - p$, where $ap - q > 0$. Let $X_1, X_2, \ldots$ be a sequence of i.i.d. random variables with common distribution that of X. As in the proof of Theorem 10.3.1, we define $B_n := |\{1 \leq i \leq n : X_i = a\}|$. Then the following inequalities are equivalent:

$$F_n(f)/F_0 \geq W, \qquad (10.104)$$

$$\sum_{l=1}^{n} \ln(1 + fX_l) \geq \ln(W), \qquad (10.105)$$

$$B_n \ln(1 + af) + (n - B_n)\ln(1 - f) \geq \ln(W), \qquad (10.106)$$

$$B_n \geq \lceil \alpha + \beta n \rceil, \qquad (10.107)$$

where $\alpha := \ln(W)/[\ln(1 + af) - \ln(1 - f)]$ and $\beta := -\ln(1 - f)/[\ln(1 + af) - \ln(1 - f)]$. Thus, $N_W(f) = \min\{n \geq 1 : B_n \geq \lceil \alpha + \beta n \rceil\}$.

We can evaluate the expectation $\mathrm{E}[N_W(f) \wedge m]$ using a Markov chain argument. Indeed, $Y_n := (n, B_n)$ defines a Markov chain in the countable state space $S := \{(n, k) : n \geq 0, \ 0 \leq k \leq n\}$ with one-step transition probabilities

$$P((n, k), (n + 1, k + 1)) := p, \quad P((n, k), (n + 1, k)) := q, \qquad (10.108)$$

and $\mathrm{E}[N_W(f) \wedge m]$ is the mean exit time of $\{Y_l\}_{l \geq 0}$ from the finite set $S_0 := \{(n, k) : 0 \leq n < m, \ 0 \leq k < \lceil \alpha + \beta n \rceil\}$ starting at $Y_0 = (0, 0)$. Letting

$$E(n, k) := \mathrm{E}[\min\{l \geq 0 : Y_l \notin S_0\} \mid Y_0 = (n, k)], \qquad (10.109)$$

we have

$$E(n, k) = \begin{cases} 1 + pE(n + 1, k + 1) + qE(n + 1, k) & \text{if } (n, k) \in S_0. \\ 0 & \text{if } (n, k) \in S - S_0. \end{cases}$$
$$(10.110)$$

Start with $E(m, k) = 0$ for $0 \leq k \leq m$, and successively evaluate $E(n, \cdot)$ for $n = m - 1, m - 2, \ldots, 0$. Finally, $E(0, 0) = \mathrm{E}[N_W(f) \wedge m]$. By choosing m sufficiently large, we can determine $\mathrm{E}[N_W(f)]$ to any desired degree of accuracy.

In Table 10.4, we take $a = 1$, $p = 0.55$, and $W = 2$. As a function of f, $\mathrm{E}[N_2(f)]$ is not even continuous, so it is difficult to say where it is minimized. However, among f-values that are integer multiples of 0.001, the minimum occurs at $f = 0.096$, which is not too far from $f^* = 0.100$. ♠

We turn finally to the optimal growth properties of the Kelly bettor's fortune. We have already seen in Theorem 10.2.2(e) on p. 365 that

$$\lim_{n \to \infty} F_n(\boldsymbol{f}^*)/F_n(\boldsymbol{f}) = \infty \text{ a.s.} \tag{10.111}$$

whenever $\mu(\boldsymbol{f}) < \mu(\boldsymbol{f}^*)$. In fact, Theorem 10.2.2(a) on p. 365 implies that

$$\lim_{n \to \infty} (F_n(\boldsymbol{f}^*)/F_n(\boldsymbol{f}))^{1/n} = \exp\{\mu(\boldsymbol{f}^*) - \mu(\boldsymbol{f})\} > 1 \text{ a.s.} \tag{10.112}$$

But this compares the Kelly system with only proportional betting systems with *constant* betting proportions. Can we obtain similar results with varying (perhaps random) betting proportions? Note that whenever (10.37) on p. 364 holds, $\boldsymbol{B}_n = \boldsymbol{f} F_{n-1}$ for a random $\boldsymbol{f} \in \Delta$, so a proportional betting system with varying and random betting proportions is rather general.

Let $\boldsymbol{f}_n^\circ$ be the vector of betting proportions at the nth coup, and assume that

$$\boldsymbol{f}_1^\circ = \boldsymbol{f}_1, \qquad \boldsymbol{f}_n^\circ = \boldsymbol{f}_n(\boldsymbol{X}_1, \ldots, \boldsymbol{X}_{n-1}), \quad n \geq 2, \tag{10.113}$$

where $\boldsymbol{f}_1$ is a constant vector in Δ_0, and $\boldsymbol{f}_n$ is a nonrandom Δ_0-valued function of $n - 1$ variables. We define the Kelly bettor's fortune after n coups as well as the generic proportional bettor's fortune after n coups as

$$F_n^* := F_n(\boldsymbol{f}^*) = F_0 \prod_{l=1}^{n}(1 + \boldsymbol{f}^* \cdot \boldsymbol{X}_l), \qquad F_n^\circ := F_0 \prod_{l=1}^{n}(1 + \boldsymbol{f}_l^\circ \cdot \boldsymbol{X}_l). \tag{10.114}$$

It is no longer necessarily the case that $\lim_{n \to \infty} F_n^*/F_n^\circ = \infty$ a.s. if the sequence $(\boldsymbol{f}_n^\circ)$ differs from $\boldsymbol{f}^*$, though this does hold if it is sufficiently different. In fact, it is not even necessarily true that $\lim_{n \to \infty} F_n^*/F_n^\circ \geq 1$ a.s. Indeed, take $d = 1$ and consider the system that uses betting proportion $f^\circ \neq f^*$ at the first coup and betting proportion f^* thereafter. Then

$$\lim_{n \to \infty} \frac{F_n^*}{F_n^\circ} = \frac{1 + f^* X_1}{1 + f^\circ X_1}, \tag{10.115}$$

which is less than 1 if $f^\circ < f^*$ and $X_1 < 0$, or if $f^\circ > f^*$ and $X_1 > 0$.

Theorem 10.3.5. *Assume the hypotheses of Theorem 10.2.2 on p. 365 and (10.113).*

Table 10.4 Expected number of coups required for the proportional bettor to double his fortune, as a function of his betting proportion f. Here $P(X = 1) = 0.55 = 1 - P(X = -1)$. The bounds are from (10.101). Results are rounded to six significant digits.

f	lower bound	exact	upper bound
0.01	729.605	734.704	740.079
0.02	385.033	390.331	396.033
0.03	271.748	277.289	283.337
0.04	216.507	222.383	228.758
0.05	184.711	190.855	197.712
0.06	164.879	171.491	178.740
0.07	152.158	158.934	167.010
0.08	144.200	152.119	160.211
0.09	139.806	147.852	157.188
0.10	138.398	147.248	157.428
0.11	139.810	149.067	160.859
0.12	144.233	155.549	167.814
0.13	152.281	164.162	179.131
0.14	165.222	179.574	196.455
0.15	185.554	202.493	222.968
0.087	140.793	148.495	157.738
0.088	140.434	148.113	157.522
0.089	140.105	147.923	157.338
0.090	139.806	147.852	157.188
0.091	139.536	147.895	157.069
0.092	139.296	147.612	156.983
0.093	139.084	147.181	156.928
0.094	138.902	147.045	156.905
0.095	138.747	147.326	156.914
0.096	138.621	146.699	156.954
0.097	138.524	146.777	157.025
0.098	138.454	146.828	157.128
0.099	138.412	146.893	157.262
0.100	138.398	147.248	157.428
0.101	138.412	147.187	157.625
0.102	138.454	147.129	157.854
0.103	138.524	147.778	158.115
0.104	138.622	147.610	158.409
0.105	138.748	148.226	158.734
0.106	138.903	148.509	159.092

(a) *Then* $\{F_n^\circ/F_n^*\}_{n\geq 0}$ *is a supermartingale with respect to* $\{\boldsymbol{X}_n\}_{n\geq 1}$*. Consequently,*

$$\lim_{n\to\infty} \frac{F_n^\circ}{F_n^*} \text{ exists in } [0,\infty) \text{ a.s., and } \mathrm{E}\left[\lim_{n\to\infty} \frac{F_n^\circ}{F_n^*}\right] \leq 1. \qquad (10.116)$$

(b) *Assume, for* $i = 1,\ldots,d$*, that* $f_{n,i}^\circ = 0$ *if* $f_i^* = 0$*, and assume that* $\sum_{1\leq j\leq d} f_{n,j}^\circ = 1$ *if* $\sum_{1\leq j\leq d} f_j^* = 1$*. Then* $\{F_n^\circ/F_n^*\}_{n\geq 0}$ *is a martingale with respect to* $\{\boldsymbol{X}_n\}_{n\geq 1}$*. In particular,*

$$\mathrm{E}\left[\frac{F_n^\circ}{F_n^*}\right] = 1, \qquad n \geq 1. \qquad (10.117)$$

Remark. To interpret the meaning of part (a), we could define the (random) *asymptotic efficiency* of the $(\boldsymbol{f}_n^\circ)$ strategy to be $\lim_{n\to\infty} F_n^\circ/F_n^*$. A strategy that asymptotically outperforms Kelly would have an asymptotic efficiency greater than 1; a strategy that is asymptotically outperformed by Kelly would have an asymptotic efficiency less than 1. Part (a) tells us that no strategy can have an *expected* asymptotic efficiency greater than 1. This can be regarded as an optimality property of the Kelly system.

However, the conclusion of part (b) raises questions about the concept of expected asymptotic efficiency. Under its mild assumptions (namely, the $(\boldsymbol{f}_n^\circ)$ bettor avoids betting opportunities that Kelly avoids and bets all if Kelly does), any strategy $(\boldsymbol{f}_n^\circ)$ that satisfies $\boldsymbol{f}_n^\circ = \boldsymbol{f}^*$ for all $n \geq N + 1$, no matter how bad the strategy is at the first N coups and no matter how large N is, has expected asymptotic efficiency 1.

Proof. Define $H_i := \mathrm{E}[X_i/(1 + \boldsymbol{f}^* \cdot \boldsymbol{X})]$ for $i = 1,\ldots,d$ and $\boldsymbol{H} := (H_1,\ldots,H_d)$. We claim that

$$\boldsymbol{f}_n^\circ \cdot \boldsymbol{H} \leq \boldsymbol{f}^* \cdot \boldsymbol{H}, \qquad (10.118)$$

with equality holding under the assumptions of part (b).

To see this, let $I := \{i \in \{1,\ldots,d\} : f_i^* > 0\}$ and $I^c := \{1,\ldots,d\} - I$. Define $\boldsymbol{e}_1,\ldots,\boldsymbol{e}_d \in \Delta$ by $\boldsymbol{e}_1 := (1,0,\ldots,0)$, $\boldsymbol{e}_2 := (0,1,0,\ldots,0)$, and so on. For $i,j = 1,\ldots,d$, define the functions g_i and g_{ij} by $g_i(t) := \mu(\boldsymbol{f}^* + t\boldsymbol{e}_i)$ and $g_{ij}(t) := \mu(\boldsymbol{f}^* + t\boldsymbol{e}_i - t\boldsymbol{e}_j)$, with domains chosen maximally in $\mathbf{R}$.

If $i \in I$ and $\sum_{1\leq j\leq d} f_j^* < 1$, then $H_i = g_i'(0) = 0$. If $i \in I$ and $\sum_{1\leq j\leq d} f_j^* = 1$, then $H_i = g_i'(0) := \lim_{t\to 0-} t^{-1}[g_i(t) - g_i(0)] \geq 0$. If $i,j \in I$ are distinct, then $H_i - H_j = g_{ij}'(0) = 0$. If $i \in I$ and $j \in I^c$, then $H_i - H_j = g_{ij}'(0) \geq 0$. Let $H \geq 0$ denote the common value of H_i for $i \in I$, and note that $H_j \leq H$ if $j \in I^c$ and that $H = 0$ if $\sum_{1\leq j\leq d} f_j^* < 1$. We conclude that

$$\boldsymbol{f}_n^\circ \cdot \boldsymbol{H} \leq \boldsymbol{f}_n^\circ \cdot (H,H,\ldots,H) \leq H = \boldsymbol{f}^* \cdot \boldsymbol{H}, \qquad (10.119)$$

proving (10.118).

It follows that

$$
\mathrm{E}\left[\frac{1 + \boldsymbol{f}_n^\circ \cdot \boldsymbol{X}_n}{1 + \boldsymbol{f}^* \cdot \boldsymbol{X}_n} \,\middle|\, \boldsymbol{X}_1, \ldots \boldsymbol{X}_{n-1}\right]
$$

$$
= \mathrm{E}\left[\frac{1}{1 + \boldsymbol{f}^* \cdot \boldsymbol{X}_n} \,\middle|\, \boldsymbol{X}_1, \ldots \boldsymbol{X}_{n-1}\right] + \boldsymbol{f}_n^\circ \cdot \mathrm{E}\left[\frac{\boldsymbol{X}_n}{1 + \boldsymbol{f}^* \cdot \boldsymbol{X}_n} \,\middle|\, \boldsymbol{X}_1, \ldots \boldsymbol{X}_{n-1}\right]
$$

$$
= \mathrm{E}\left[\frac{1}{1 + \boldsymbol{f}^* \cdot \boldsymbol{X}_n}\right] + \boldsymbol{f}_n^\circ \cdot \mathrm{E}\left[\frac{\boldsymbol{X}_n}{1 + \boldsymbol{f}^* \cdot \boldsymbol{X}_n}\right]
$$

$$
\leq \mathrm{E}\left[\frac{1}{1 + \boldsymbol{f}^* \cdot \boldsymbol{X}_n}\right] + \boldsymbol{f}^* \cdot \mathrm{E}\left[\frac{\boldsymbol{X}_n}{1 + \boldsymbol{f}^* \cdot \boldsymbol{X}_n}\right]
$$

$$
= \mathrm{E}\left[\frac{1 + \boldsymbol{f}^* \cdot \boldsymbol{X}_n}{1 + \boldsymbol{f}^* \cdot \boldsymbol{X}_n}\right]
$$

$$
= 1, \tag{10.120}
$$

where the first step uses linearity of conditional expectation and Theorem 2.2.7 on p. 83, the second uses independence, the third uses (10.118), the fourth again uses linearity of expectation, and last step is obvious. We conclude that

$$
\mathrm{E}\left[\frac{F_n^\circ}{F_n^*} \,\middle|\, \boldsymbol{X}_1, \ldots, \boldsymbol{X}_{n-1}\right] = \frac{F_{n-1}^\circ}{F_{n-1}^*} \mathrm{E}\left[\frac{1 + \boldsymbol{f}_n^\circ \cdot \boldsymbol{X}_n}{1 + \boldsymbol{f}^* \cdot \boldsymbol{X}_n} \,\middle|\, \boldsymbol{X}_1, \ldots \boldsymbol{X}_{n-1}\right]
$$

$$
\leq \frac{F_{n-1}^\circ}{F_{n-1}^*}, \qquad n \geq 1. \tag{10.121}
$$

The remaining conclusions of part (a) follow from the martingale convergence theorem (Theorem 3.3.2 on p. 107).

The two additional assumptions of part (b) ensure that the two inequalities in (10.118) become equalities, and therefore the same is true of the inequalities in (10.120) and (10.121).   ♠

Theorem 10.3.6. *Under the assumptions of Theorem 10.3.5, $\{F_n^*/F_n^\circ\}_{n \geq 0}$ is a submartingale with respect to $\{\boldsymbol{X}_n\}_{n \geq 1}$. In particular, for each $n \geq 1$,*

$$
\mathrm{E}\left[\frac{F_n^*}{F_n^\circ}\right] \geq 1, \tag{10.122}
$$

with equality if only if $\mathrm{P}(\boldsymbol{f}_l^\circ \cdot \boldsymbol{X}_l = \boldsymbol{f}^* \cdot \boldsymbol{X}_l) = 1$ *for $l = 1, \ldots, n$. In addition,*

$$
\lim_{n \to \infty} \frac{F_n^*}{F_n^\circ} \text{ exists in } (0, \infty] \text{ a.s., and } \mathrm{E}\left[\lim_{n \to \infty} \frac{F_n^*}{F_n^\circ}\right] \geq 1. \tag{10.123}
$$

Remark. The fact that strict inequality holds in (10.122) unless $F_n^\circ = F_n^*$ tells us that, if the definition of asymptotic efficiency as discussed in the remark following the statement of Theorem 10.3.5 were based on any system other than the Kelly system, the conclusion that no strategy has expected asymptotic efficiency greater than 1 would fail. This is another optimality property of the Kelly system.

Proof. First, we observe that

$$
\begin{aligned}
&\mathrm{E}\left[\frac{1 + \boldsymbol{f}^* \cdot \boldsymbol{X}_n}{1 + \boldsymbol{f}_n^\circ \cdot \boldsymbol{X}_n} \,\middle|\, \boldsymbol{X}_1 = \boldsymbol{x}_1, \dots \boldsymbol{X}_{n-1} = \boldsymbol{x}_{n-1}\right] \\
&= \mathrm{E}\left[\frac{1 + \boldsymbol{f}^* \cdot \boldsymbol{X}_n}{1 + \boldsymbol{f}_n(\boldsymbol{x}_1, \dots, \boldsymbol{x}_{n-1}) \cdot \boldsymbol{X}_n} \,\middle|\, \boldsymbol{X}_1 = \boldsymbol{x}_1, \dots \boldsymbol{X}_{n-1} = \boldsymbol{x}_{n-1}\right] \\
&= \mathrm{E}\left[\frac{1 + \boldsymbol{f}^* \cdot \boldsymbol{X}_n}{1 + \boldsymbol{f}_n(\boldsymbol{x}_1, \dots, \boldsymbol{x}_{n-1}) \cdot \boldsymbol{X}_n}\right] \\
&\geq \exp\{\mathrm{E}[\ln(1 + \boldsymbol{f}^* \cdot \boldsymbol{X}_n) - \ln(1 + \boldsymbol{f}_n(\boldsymbol{x}_1, \dots, \boldsymbol{x}_{n-1}) \cdot \boldsymbol{X}_n)] \\
&= \exp\{\mu(\boldsymbol{f}^*) - \mu(\boldsymbol{f}_n(\boldsymbol{x}_1, \dots, \boldsymbol{x}_{n-1}))\} \\
&\geq 1, \quad\quad\quad\quad\quad\quad\quad\quad\quad\quad\quad\quad\quad\quad\quad\quad\quad (10.124)
\end{aligned}
$$

where the first step uses (10.113), the second uses independence, the third uses Jensen's inequality, and the fourth and fifth use the definition and properties of $\mu(\boldsymbol{f})$. Because the log function is strictly concave, the inequalities are strict unless $\mathrm{P}(\boldsymbol{f}_n^\circ \cdot \boldsymbol{X}_n = \boldsymbol{f}^* \cdot \boldsymbol{X}_n) = 1$. We conclude that

$$
\begin{aligned}
\mathrm{E}\left[\frac{F_n^*}{F_n^\circ} \,\middle|\, \boldsymbol{X}_1, \dots, \boldsymbol{X}_{n-1}\right] &= \frac{F_{n-1}^*}{F_{n-1}^\circ} \mathrm{E}\left[\frac{1 + \boldsymbol{f}^* \cdot \boldsymbol{X}_n}{1 + \boldsymbol{f}_n^\circ \cdot \boldsymbol{X}_n} \,\middle|\, \boldsymbol{X}_1, \dots \boldsymbol{X}_{n-1}\right] \\
&\geq \frac{F_{n-1}^*}{F_{n-1}^\circ}, \quad\quad n \geq 1. \quad\quad\quad (10.125)
\end{aligned}
$$

The last two conclusions of the theorem follow respectively from Theorem 10.3.5 and Jensen's inequality (if $Y > 0$, then $\mathrm{E}[Y] \geq 1/\mathrm{E}[1/Y]$, where $1/0 := \infty$). ♠

Theorem 10.3.7. *Under the assumptions of Theorem 10.3.5, with probability 1,*

$$
\lim_{n \to \infty} \frac{F_n^*}{F_n^\circ} = \infty \text{ if and only if } \sum_{n=1}^{\infty} [\mu(\boldsymbol{f}^*) - \mu(\boldsymbol{f}_n^\circ)] = \infty. \quad (10.126)
$$

Remark. The result says that the Kelly system outperforms any other system that differs significantly from it. This is a substantial generalization of Theorem 10.2.2(e) on p. 365.

Proof. Fix $\varepsilon > 0$. First let $\boldsymbol{x}_1, \dots, \boldsymbol{x}_k$ be the distinct values that $\boldsymbol{X}$ takes on with positive probability, put $p_i := \mathrm{P}(\boldsymbol{X} = \boldsymbol{x}_i) > 0$ for $i = 1, \dots, k$, and let $p := \min_{1 \leq i \leq k} p_i > 0$. Then, if g is a function such that $\mathrm{E}[g(\boldsymbol{X})] \geq \varepsilon > 0$, it must be the case that $\mathrm{P}(g(\boldsymbol{X}) \geq \varepsilon) \geq p$. Indeed, if not, then $\mathrm{P}(g(\boldsymbol{X}) \geq \varepsilon) = 0$ and hence $\mathrm{E}[g(\boldsymbol{X})] < \varepsilon$.

It follows that $\mu(\boldsymbol{f}^*) - \mu(\boldsymbol{f}_n^\circ) \geq \varepsilon$ implies

$$P(\ln(1 + \boldsymbol{f}^* \cdot \boldsymbol{X}_n) - \ln(1 + \boldsymbol{f}_n^\circ \cdot \boldsymbol{X}_n) \geq \varepsilon \mid \boldsymbol{X}_1, \ldots \boldsymbol{X}_{n-1}) \geq p. \qquad (10.127)$$

Let us define the event $A := \{\mu(\boldsymbol{f}^*) - \mu(\boldsymbol{f}_n^\circ) \geq \varepsilon \text{ i.o.}\}$. Then, a.s. on A,

$$\sum_{n=1}^{\infty} P(\ln(1 + \boldsymbol{f}^* \cdot \boldsymbol{X}_n) - \ln(1 + \boldsymbol{f}_n^\circ \cdot \boldsymbol{X}_n) \geq \varepsilon \mid \boldsymbol{X}_1, \ldots \boldsymbol{X}_{n-1}) = \infty, \quad (10.128)$$

which by the conditional Borel–Cantelli lemma (Theorem 3.3.5 on p. 110) is equivalent to

$$\ln(1 + \boldsymbol{f}^* \cdot \boldsymbol{X}_n) - \ln(1 + \boldsymbol{f}_n^\circ \cdot \boldsymbol{X}_n) \geq \varepsilon \text{ i.o.} \qquad (10.129)$$

Now

$$\ln\left(\frac{F_n^*}{F_n^\circ}\right) = \sum_{l=1}^{n}[\ln(1 + \boldsymbol{f}^* \cdot \boldsymbol{X}_l) - \ln(1 + \boldsymbol{f}_l^\circ \cdot \boldsymbol{X}_l)] \qquad (10.130)$$

and $\lim_{n \to \infty} \ln(F_n^*/F_n^\circ)$ exists in $(-\infty, \infty]$ a.s. by Theorem 10.3.6. Because the terms of a convergent series must tend to zero, it follows from (10.129) that this limit is infinite a.s. on A. We conclude that both conditions in (10.126) hold a.s. on A.

To deal with A^c, let us define a new strategy $(\boldsymbol{f}_n^\varepsilon)$ by

$$\boldsymbol{f}_n^\varepsilon := \begin{cases} \boldsymbol{f}_n^\circ & \text{if } \mu(\boldsymbol{f}^*) - \mu(\boldsymbol{f}_n^\circ) < \varepsilon, \\ \boldsymbol{f}^* & \text{otherwise.} \end{cases} \qquad (10.131)$$

We then let $F_n^\varepsilon := F_0 \prod_{l=1}^{n}(1 + \boldsymbol{f}_l^\varepsilon \cdot \boldsymbol{X}_l)$ for each $n \geq 1$, and define the martingale

$$M_n := \ln\left(\frac{F_n^*}{F_n^\varepsilon}\right) - \sum_{l=1}^{n}[\mu(\boldsymbol{f}^*) - \mu(\boldsymbol{f}_l^\varepsilon)], \qquad n \geq 0, \qquad (10.132)$$

with respect to $\{\boldsymbol{X}_n\}_{n \geq 1}$. We claim that the increments

$$M_n - M_{n-1} = \ln(1 + \boldsymbol{f}^* \cdot \boldsymbol{X}_n) - \ln(1 + \boldsymbol{f}_n^\varepsilon \cdot \boldsymbol{X}_n) - [\mu(\boldsymbol{f}^*) - \mu(\boldsymbol{f}_n^\varepsilon)] \quad (10.133)$$

are uniformly bounded. To see this, define

$$a := \max_{1 \leq i \leq k} \ln(1 + \boldsymbol{f}^* \cdot \boldsymbol{x}_i), \qquad (10.134)$$

$$b := \min_{1 \leq i \leq k} \ln(1 + \boldsymbol{f}^* \cdot \boldsymbol{x}_i), \qquad (10.135)$$

$$c := \max_{1 \leq i \leq k} \max_{\boldsymbol{f} \in \Delta} \ln(1 + \boldsymbol{f} \cdot \boldsymbol{x}_i), \qquad (10.136)$$

$$D_n := \ln(1 + \boldsymbol{f}_n^\varepsilon \cdot \boldsymbol{X}_n), \qquad (10.137)$$

so that $c \geq a \geq b$ and $c \geq 0$. Then

$$b - c - \varepsilon \leq M_n - M_{n-1} \leq a - D_n \quad \text{a.s.} \tag{10.138}$$

Also, $\mathrm{E}[\ln(1 + \boldsymbol{f}^* \cdot \boldsymbol{X}_n) - D_n \mid \boldsymbol{X}_1, \ldots, \boldsymbol{X}_{n-1}] = \mu(\boldsymbol{f}^*) - \mu(\boldsymbol{f}_n^\varepsilon) < \varepsilon$, and therefore $\mathrm{E}[D_n \mid \boldsymbol{X}_1, \ldots, \boldsymbol{X}_{n-1}] > b - \varepsilon$. By an argument similar to the one used in the first paragraph of this proof, $\mathrm{P}(D_n \geq (b - c - \varepsilon)/p \mid \boldsymbol{X}_1, \ldots, \boldsymbol{X}_{n-1}) = 1$. Indeed, if not, then $\mathrm{P}(D_n < (b - c - \varepsilon)/p \mid \boldsymbol{X}_1, \ldots, \boldsymbol{X}_{n-1}) \geq p$, hence

$$\mathrm{E}[D_n \mid \boldsymbol{X}_1, \ldots, \boldsymbol{X}_{n-1}] < [(b - c - \varepsilon)/p]p + c(1 - p) \leq b - \varepsilon, \tag{10.139}$$

a contradiction. It follows from (10.138) that

$$b - c - \varepsilon \leq M_n - M_{n-1} \leq a - (b - c - \varepsilon)/p \quad \text{a.s.} \tag{10.140}$$

We can now apply Theorem 3.3.4 on p. 109 to conclude that, with probability 1, either (a) $\lim_{n\to\infty} M_n$ exists in $\mathbf{R}$ or (b) $\limsup_{n\to\infty} M_n = \infty$ and $\liminf_{n\to\infty} M_n = -\infty$. Since we know that $\lim_{n\to\infty} \ln(F_n^*/F_n^\varepsilon)$ exists in $(-\infty, \infty]$ a.s. by Theorem 10.3.6, the first conclusion of case (b) implies $\lim_{n\to\infty} \ln(F_n^*/F_n^\varepsilon) = \infty$ a.s., while the second conclusion of case (b) implies $\sum_{n=1}^{\infty} [\mu(\boldsymbol{f}^*) - \mu(\boldsymbol{f}_n^\varepsilon)] = \infty$. On the other hand, case (a) tells us that, with probability 1, $\lim_{n\to\infty} \ln(F_n^*/F_n^\varepsilon) = \infty$ if and only if $\sum_{n=1}^{\infty} [\mu(\boldsymbol{f}^*) - \mu(\boldsymbol{f}_n^\varepsilon)] = \infty$. We conclude that (10.126) holds with $(\boldsymbol{f}_n^\circ)$ replaced by $(\boldsymbol{f}_n^\varepsilon)$. On A^c this is equivalent to (10.126). ♠

Example 10.3.8. *Illustration of Theorem 10.3.7.* First consider, in the setting of Lemma 10.1.1(a), (b), or (c) on p. 359, the nonrandom betting proportions $\boldsymbol{f}_n^\circ := \boldsymbol{f}^* + \alpha_n$, where α_n is chosen so that $\boldsymbol{f}_n^\circ \in [0, 1)$ for each $n \geq 1$. Then, because $\mu'(\boldsymbol{f}^*) = 0$ and $\mu''(\boldsymbol{f}^*) < 0$,

$$\lim_{n\to\infty} \frac{F_n^*}{F_n^\circ} = \infty \quad \text{a.s. if and only if} \quad \sum_{n=1}^{\infty} \alpha_n^2 = \infty. \tag{10.141}$$

For example, if $\boldsymbol{f}_n^\circ := (\boldsymbol{f}^* - 1/\sqrt{n})^+$ for each $n \geq 1$, then both conditions in (10.141) hold.

Next, consider a gambler who at each coup either bets according to Kelly or does not bet at all. That is, $\boldsymbol{f}_n^\circ = \boldsymbol{f}^*$ or $\boldsymbol{f}_n^\circ = \mathbf{0}$ for each $n \geq 1$. The decision to bet or not can depend on the results of previous coups. Then

$$\lim_{n\to\infty} \frac{F_n^*}{F_n^\circ} = \infty \quad \text{a.s. if and only if} \quad \boldsymbol{f}_n^\circ = \mathbf{0} \text{ infinitely often a.s.} \tag{10.142}$$

For example, just to illustrate a point, suppose $\boldsymbol{f}_n^\circ = \mathbf{0}$ if n is a perfect square, and $\boldsymbol{f}_n^\circ = \boldsymbol{f}^*$ otherwise. Then (10.142) holds. However, $F_n^*/F_n^\circ = \prod_{l=1}^{\lfloor \sqrt{n} \rfloor} (1 + \boldsymbol{f}^* \cdot \boldsymbol{X}_{l^2})$ for each $n \geq 1$, so

$$\lim_{n\to\infty} \left(\frac{F_n^*}{F_n^\circ} \right)^{1/n} = 1 \quad \text{a.s.} \tag{10.143}$$

Compare this example with Theorem 10.2.2(e) on p. 365, which is proved via (10.112). For the proof of the theorem it suffices to consider the asymptotic behavior of the nth root of the ratio. But for the example above, that approach is too crude.   ♠

10.4 Problems

10.1. *Double-Kelly in an even-money game.* Assuming $P(X = 1) = p \in (\frac{1}{2}, 1)$ and $P(X = -1) = 1 - p$, use (10.15) on p. 360 to show that $f_0 < 2f^*$, hence $\mu(2f^*) < 0$.

10.2. *Relaxing the assumption that $X \geq -1$.* Show that the assumption made in Section 10.1 that $X \geq -1$ can be relaxed to $X \geq -K$, where $K > 1$. Reformulate Lemma 10.1.1 on p. 359 and Theorem 10.1.2 on p. 360 in the more general setting.

Hint: Restrict f to $[0, 1/K]$.

10.3. *Comparing betting proportions on the same sequence of coups vs. independent sequences.* Let $X_1, X_2, \ldots$ and $X_1^\circ, X_2^\circ, \ldots$ be independent and i.i.d. as X in Lemma 10.1.1 on p. 359. For each $n \geq 1$ and $f \in [0, 1]$, define $F_n(f)$ in terms of $X_1, \ldots, X_n$ and $F_n^\circ(f)$ in terms of $X_1^\circ, \ldots, X_n^\circ$ as in (10.8) on p. 358. Let $f_1, f_2 \in [0, 1]$, $f_1 \neq f_2$, and $f_1, f_2 \neq 1$ if (10.1) on p. 357 holds.

(a) Find the mean and variance of $\ln(F_n(f_1)/F_n(f_2))$, and state the corresponding central limit theorem.

(b) Find the mean and variance of $\ln(F_n(f_1)/F_n^\circ(f_2))$, and state the corresponding central limit theorem.

10.4. *Evaluating the error in the approximation (10.14).* Under the assumptions of Example 10.1.3 on p. 361 and the notation of (10.14) on p. 360, show that

$$f^* - \frac{m_1}{m_2} = \frac{a(a-1)}{E[X^2 \mid X \neq 0]}(f^*)^2. \tag{10.144}$$

10.5. *Kelly and large deviations.* What does Cramér's theorem on large deviations (Theorem 6.3.4 on p. 228) tell us about the behavior of $P(F_n(f) \leq F_0)$ as $n \to \infty$? Assume the conditions of Theorem 10.1.2(b) on p. 360.

10.6. *Nonconvexity of $\sigma^2(f)$.* Let $X \geq -1$ assume finitely many values, and define $\sigma^2(f) := \mathrm{Var}[\ln(1 + fX)]$ for $0 < f < 1$.

(a) Show that $\sigma^2(f)$ is nondecreasing in f.

(b) Show by example that $\sigma^2(f)$ need not be a convex function of f (although it *is* convex in the special case of (10.28)).

Hint: (a) Use Problem 1.38 on p. 60.

10.7. *Extension of Table 10.1 to other betting proportions.* Extend Table 10.1 on p. 363, replacing f^* by $(i/3)f^*$, $i = 1, 2, 3, 4, 5$. Noting that $\sigma(f)$ is increasing in f, it is clear that there is no reason to choose $f > f^*$ (slower growth and higher volatility), whereas $f < f^*$ may be worth considering (slower growth but lower volatility).

10.8. *Wagers with four possible outcomes, including loss and push.* This problem extends Example 10.1.3 on p. 361. Assume that $P(X = a_1) = p_1$, $P(X = a_2) = p_2$, $P(X = -1) = q$, and $P(X = 0) = r$, where $p_1 > 0$, $p_2 > 0$, $q > 0$, $r \geq 0$, and $p_1 + p_2 + q + r = 1$. Assume also that a_1 and a_2 are distinct and positive, and that $E[X] > 0$, that is, $a_1 p_1 + a_2 p_2 - q > 0$. Find the unique Kelly betting proportion $f^* \in (0,1)$.

10.9. *Kelly and video poker.* Video poker machines, although sometimes offering a favorable bet, require constant bet sizes and therefore are not really suited to the Kelly system. However, it would not be technologically difficult to design a machine that accepts user-specified bet sizes. With this in mind, find the Kelly betting proportion $f^* \in (0,1)$, assuming optimal play, for the Deuces Wild machine described in Table 7.5 on p. 265. (Use Newton's method to solve the required equation numerically.) Evaluate as well the approximation (10.14) on p. 360. Are the assumptions underlying the derivation of that approximation satisfied?

10.10. *Return on investment under proportional betting.* Under the assumptions of Theorem 10.1.2(b) on p. 360 (in particular, $\mu(f) > 0$), define F_n as in (10.8) on p. 358, and define the proportional bettor's return on investment after n coups by

$$R_n := \frac{F_n - F_0}{f \sum_{l=0}^{n-1} F_l}. \tag{10.145}$$

(a) By reversing the order of $X_1, \ldots, X_n$, show that

$$R_n^\circ := \frac{1 - (F_0/F_n)}{f \sum_{l=1}^{n} (F_0/F_l)} \tag{10.146}$$

has the same distribution as R_n.

(b) Using the root test for convergence of infinite series, show that

$$R := \frac{1}{f \sum_{l=1}^{\infty} (F_0/F_l)} \tag{10.147}$$

defines a random variable (though not necessarily a discrete one) and that $\lim_{n \to \infty} R_n^\circ = R$ a.s. We call R the proportional bettor's *asymptotic return on investment*.

(c) Define R' in terms of $X_2, X_3, \ldots$ exactly as R is defined in terms of $X_1, X_2, \ldots$, and show that

$$R = (1 + f X_1)\left(f + \frac{1}{R'}\right)^{-1}. \tag{10.148}$$

(d) Noting that R' is independent of X_1 and distributed as R in (10.147), apply Jensen's inequality to show that $E[R] < E[X_1]$. Thus, the expected asymptotic return on investment of the proportional bettor is smaller than that of the constant bettor.

10.11. *Proportional play at a fair game.* Given positive integers m and n, let X have distribution $P(X = n/m) = m/(m + n) = 1 - P(X = -1)$, so that $E[X] = 0$. It has been claimed that proportional play is a losing strategy for the following reason. On average, in $m + n$ coups, the bettor will win m times and lose n times. Hence, with betting proportion $f = 1/n$, the bettor's fortune will, on average, be multiplied by

$$\left(1 + \frac{n}{m}\frac{1}{n}\right)^m \left(1 - \frac{1}{n}\right)^n < e\frac{1}{e} = 1. \tag{10.149}$$

Find the flaw in this argument, yet show that the claim is nevertheless correct.

10.12. *Proportional play at subfair games.* Given an initial fortune of one unit, which of the following three proportional betting scenarios maximizes the probability of ever reaching or exceeding a target of 20 units? In each case the betting proportion f and wager are specified.

(a) $f = 1/10$; payoff odds are 9 to 1; true odds are 10 to 1.
(b) $f = 2/5$; payoff odds are 1 to 1; true odds are 10 to 9.
(c) $f = 1/10$; payoff odds are 1 to 1; true odds are 251 to 244.

These wagers were motivated by (a) hard eight at craps, (b) red at roulette, and (c) pass line at craps.

10.13. *Independent even-money wagers.* Let $p_1, p_2 \in (0, 1)$, and assume that $P(X_1 = 1) = p_1$, $P(X_1 = -1) = q_1 := 1 - p_1$, $P(X_2 = 1) = p_2$, $P(X_2 = -1) = q_2 := 1 - p_2$, and X_1 and X_2 are independent. Assume that $p_1 - q_1 > 0$ and $p_2 - q_2 > 0$, and argue that the Kelly proportions f_1^* and f_2^* are uniquely determined. Evaluate them, showing that $f_1^* < p_1 - q_1$ and $f_2^* < p_2 - q_2$. Thus, independent wagers at the same coup call for smaller betting proportions than independent wagers at successive coups.

Hint: For the inequalities, express f_1^* and f_2^* in terms of $\mu_1 := p_1 - q_1$ and $\mu_2 := p_2 - q_2$.

10.14. $d \geq 2$ *i.i.d. even-money wagers.* Assume that $P(X = 1) = p \in (\frac{1}{2}, 1)$ and $P(X = -1) = q := 1 - p$, and let $\boldsymbol{X} = (X_1, \ldots, X_d)$, where $X_1, \ldots, X_d$ are i.i.d. as X. Argue that the Kelly proportions $f_1^*, \ldots, f_d^*$ are uniquely determined and equal. Use the approximation (10.14) on p. 360 to show that

$$f_i^* \approx \frac{p - q}{1 + (d - 1)(p - q)^2}, \qquad i = 1, \ldots, d. \tag{10.150}$$

Show that the approximation is exact for $d = 1$ and $d = 2$. Determine its accuracy for $p = 0.54$ and $d = 3, 4, \ldots, 10$.

10.15. *Correlated even-money wagers.* Let $\frac{1}{2} < p < 1$, $q := 1 - p$, and $-q^2 \le c \le pq$. Assume that $P(\boldsymbol{X} = (1, 1)) = p^2 + c$, $P(\boldsymbol{X} = (1, -1)) = P(\boldsymbol{X} = (-1, 1)) = pq - c$, and $P(\boldsymbol{X} = (-1, -1)) = q^2 + c$, so that X_1 and X_2 have the same marginal distribution $P(X_1 = 1) = P(X_2 = 1) = p$ and $P(X_1 = -1) = P(X_2 = -1) = q$, but have $\mathrm{Cov}(X_1, X_2) = 4c$. Argue that the Kelly proportions f_1^* and f_2^* are uniquely determined (unless $c = pq$) and that therefore $f_1^* = f_2^*$. Evaluate f_1^* and show that it is a decreasing function of c.

10.16. *Two betting opportunities with at most one winner.* Assume that $P(\boldsymbol{X} = (a_1, -1)) = p_1 > 0$, $P(\boldsymbol{X} = (-1, a_2)) = p_2 > 0$, and $P(\boldsymbol{X} = (-1, -1)) = p_3 > 0$, where $a_1 > 0$, $a_2 > 0$, and $p_1 + p_2 + p_3 = 1$, and assume that $\max\{(a_1 + 1)p_1 - 1, (a_2 + 1)p_2 - 1\} > 0$, that is, (10.53) on p. 366 holds. Find the Kelly proportions f_1^* and f_2^* by using the results of Example 10.2.3 on p. 366 with $d = 3$.

Hint: Choose a_3 positive but small enough that $f_3^* = 0$.

10.17. *Examples of Lemma 10.1.1.* Assume that $P((X_1, X_2) = (a_1, a_2)) = p_1 > 0$, $P((X_1, X_2) = (a_1, -1)) = p_2 > 0$, and $P((X_1, X_2) = (-1, a_2)) = p_3 > 0$, where $a_1 > 0$, $a_2 > 0$, and $p_1 + p_2 + p_3 = 1$. Consider the wager $X := X_1 + X_2$, and give examples of Lemma 10.1.1(b), (c), and (d) on p. 359 by choosing p_1, p_2, p_3, a_1, and a_2 appropriately.

10.18. *Simplest case of nonuniqeness of Kelly proportions.* Let $P(X_1 = 1) = p \in (\frac{1}{2}, 1)$, $P(X_1 = -1) = 1 - p$, and $P(X_2 = -X_1) = 1$, and put $\boldsymbol{X} = (X_1, X_2)$. Noting that

$$\mu(\boldsymbol{f}) := \mathrm{E}[\ln(1 + \boldsymbol{f} \cdot \boldsymbol{X})] = \mathrm{E}[\ln(1 + (f_1 - f_2)X_1)], \qquad (10.151)$$

find all $\boldsymbol{f}^* = (f_1^*, f_2^*) \in \Delta_0$ that maximize $\mu(\boldsymbol{f})$.

10.19. *Favorable and unfavorable bets on opposite sides of a fair coin.* Assume that $P(\boldsymbol{X} = (a_1, -1)) = P(\boldsymbol{X} = (-1, a_2)) = \frac{1}{2}$, where $a_1 > 1 > a_2 > 0$. Show that the Kelly bettor makes one bet if $a_1 a_2 < 1$, two bets if $a_1 a_2 > 1$, and one or two bets if $a_1 a_2 = 1$. (In the latter case, note that $P(X_2 = -a_2 X_1) = 1$, so the method of Problem 10.18 applies.) Determine the Kelly betting proportions in each case.

10.20. *Nonuniqueness of $\boldsymbol{f}^*$ in Example 10.2.3.* Suppose (10.56) on p. 367 fails, that is, $\sum_{i=1}^d (a_i + 1)^{-1} = 1$. Assume also (10.53) on p. 366 and (10.72) on p. 369.

(a) Use ideas from Example 10.2.3 on p. 366 to show that there is a unique $\boldsymbol{f}^*$ with $f_d = 0$ that maximizes $\mu(\boldsymbol{f})$ over all $\boldsymbol{f} \in \Delta_0$.

(b) However, if we allow $f_d > 0$, show that there is a one-parameter family of $\boldsymbol{f}^*$ that maximizes $\mu(\boldsymbol{f})$. More specifically, let $k_0 := \max\{k \in \{1, \ldots, d - 1\} : (a_k + 1)p_k > (a_d + 1)p_d\}$ and define $\boldsymbol{f}^*$ by (10.80) on p. 370. Show that $\boldsymbol{f}_c^*$, defined for $0 \le c \le w_{k_0}$ by

$$f_{c,j}^* := f_j^* + (a_j + 1)^{-1}c, \tag{10.152}$$

maximizes $\mu(\boldsymbol{f})$.

10.21. *Biased roulette with no zeros.*

(a) Using the results of Problem 10.20, extend Example 10.2.4 on p. 370 to the case of roulette wheels with no zeros. Notice that there is an optimal Kelly strategy that wagers on every number except the least frequent one(s).

(b) Therefore, instead of the two wheels described in the example, consider the wheel with frequencies

$$p_1 = p_2 = \cdots = p_{34} = \frac{1}{35}, \quad p_{35} = \frac{3}{5}\frac{1}{35}, \quad p_{36} = \frac{2}{5}\frac{1}{35}. \tag{10.153}$$

Confirm directly that every strategy that wagers only on numbers 1–34 is suboptimal.

10.22. *Biased roulette in which it is optimal to bet on 34 unfavorable numbers.* Example 10.2.4 on p. 370 provides an example of a biased wheel for which it is optimal to bet on 4 unfavorable numbers. Create another example (with 37 or 38 numbers) for which it is optimal to bet on 34 unfavorable numbers (this is the maximum number possible) together with one favorable one. Compare this optimal strategy with the strategy of betting on only the favorable number (optimally, of course).

10.23. *Betting more on unfavorable wagers.* Find an example involving two simultaneous wagers, one favorable and one unfavorable, in which the Kelly bettor bets more on the unfavorable wager than on the favorable one.

10.24. *A corollary of Theorem 10.3.5(a).* Under the assumptions of Theorem 10.3.5(a) on p. 377, show that $\limsup_{n\to\infty} n^{-1}\ln(F_n^\circ/F_n^*) \le 0$ a.s. Show further that this is equivalent to $P(F_n^\circ \ge e^{n\varepsilon}F_n^*$ i.o.$) = 0$, that is, no strategy can exceed Kelly infinitely often by a factor that grows exponentially fast.

10.25. *Outperforming Kelly with high probability.* Consider a game with $P(X = 1) = p \in (\frac{1}{2}, 1)$ and $P(X = -1) = 1 - p$. Let $X_1, X_2, \ldots$ be i.i.d. with common distribution that of X, and define F_n^* and F_n° for each $n \ge 1$ as in (10.114) on p. 377. We noted in (10.115) on p. 377 the existence of a strategy for which $P(\limsup_{n\to\infty} F_n^*/F_n^\circ < 1) = p > \frac{1}{2}$. Given $\varepsilon > 0$, find a strategy for which $P(\limsup_{n\to\infty} F_n^*/F_n^\circ < 1) \ge 1 - \varepsilon$.

10.5 Notes

The Kelly system was introduced by John L. Kelly, Jr. [1923–1965] in a paper titled "A new interpretation of information rate" (1956). (The prepublication title was "Information theory and gambling"; Poundstone 2005, p. 76.) Kelly considered a gambler betting at even money on a baseball game, the winner of which is already known and is transmitted to him over a noisy binary communication channel. With probability $p \in (\frac{1}{2}, 1)$ he receives the correct signal, and with probability $q := 1 - p$ he receives the incorrect one. Letting X denote the gambler's profit per unit bet, we have $P(X = 1) = p$ and $P(X = -1) = q := 1 - p$. Assuming he uses the information received and bets according to the Kelly system, the long-term geometric rate of growth of his capital is

$$\mu(f^*) = p\ln(1 + f^*) + q\ln(1 - f^*) = \ln 2 + p\ln p + q\ln q, \qquad (10.154)$$

using $f^* = p - q$, and this can be recognized as the rate of transmission of the communication channel as defined by Claude E. Shannon [1916–2001] (1948).

Kelly's idea was anticipated to some extent by Daniel Bernoulli [1700–1782] (1738), who introduced the concept of utility. As we have noted, he used logarithmic utility to resolve the St. Petersburg paradox (see Example 1.4.5 and Problem 1.34). He also suggested maximizing the geometric mean of wealth when choosing between risky ventures. This geometric-mean criterion was introduced into the economics literature by John Burr Williams [1900–1989] (1936) and, more explicitly, by Henry Allen Latané [1907–1984] (1959), but it was controversial, with Paul A. Samuelson [1915–2009] (1971, 1979) and other economists arguing against it. See Poundstone (2005, Part 4) for a thorough discussion of the controversy.

Proportional betting also predates Kelly. William Allen Whitworth [1840–1905] (1901, Proposition 68) argued that the proportional bettor at a fair game loses in the long run, although his argument was flawed. See Problem 10.11.

There is a considerable amount of literature on the Kelly system, largely because of its importance in economics and finance. Survey articles include those of Thorp (1969, 2000b) and Christensen (2005). If one bets according to the Kelly system, one is said to be using the *Kelly criterion*, a term introduced by Thorp and Walden (1966).

Lemma 10.1.1 is related to Theorem 10.5.2 of Dubins and Savage (1965), in which other results concerning the Kelly system can be found. Our formulation of Theorems 10.1.2 and 10.2.2 was influenced by Thorp (2000b). Lemma 10.2.1 was proved by Finkelstein and Whitley (1981) under less restrictive assumptions on $\boldsymbol{X}$, though the concavity had previously been established by Thorp (1975). The uniqueness condition in the lemma was known to Breiman (1961).

Example 10.2.3 is due to Kelly (1956). We have followed Kelly's original argument, filling in details where necessary. Breiman (1961) obtained case 2 of the example as his Proposition 2 but neglected to mention the needed assumption (10.78) ($\sum o_i^{-1} < 1$ in his notation). Algoet and Cover (1988) derived the same result using an information-theoretic approach and apparently assumed (10.78) implicitly ($\sum \beta^j < 1$ in their notation). Klotz (2000) gave a derivation that is also rather sparse. See Enns and Tomkins (1993) for a different approach to this example.

Theorem 10.3.1 is from Ethier (2004). The idea that the Kelly system maximizes median fortune had previously been pointed out by Ethier (1988) in the context of a geometric Brownian motion model of proportional betting and also by Maslov and Zhang (1998), albeit with a dubious derivation. Leib (2000) questioned this criterion, asserting that

> [M]aximizing median final bankroll is a concept with a face only a mathematician could love.

Theorem 10.3.3 is a poor-man's version of a theorem of Leo Breiman [1928–2005] (1961). Breiman proved, under the notation of Theorem 10.3.3 and the assumptions of Theorem 10.3.7 (together with the mild assumption that $\ln(1 + \boldsymbol{f}^* \cdot \boldsymbol{X})$ is nonlattice), that

$$\lim_{W \to \infty} \{\mathrm{E}[N_W^\circ] - \mathrm{E}[N_W^*]\} = \frac{1}{\mu(\boldsymbol{f}^*)} \sum_{n=1}^{\infty} \{\mu(\boldsymbol{f}^*) - \mathrm{E}[\mu(\boldsymbol{f}_n^\circ)]\}, \qquad (10.155)$$

where $N_W^* := N_W(\boldsymbol{f}^*)$ and $N_W^\circ := \min\{n \geq 1 : F_n^\circ/F_0 \geq W\}$, the main point being that the limit is nonnegative. Theorem 10.3.3 includes the special case of this in which $(\boldsymbol{f}_n^\circ)$ is constant and nonrandom.

Theorems 10.3.5(a) and 10.3.7 are also due to Breiman (1960, 1961), though our proof of the former follows Finkelstein and Whitley (1981). In the case of the latter, we give Breiman's (1961) proof. Theorems 10.3.5(b) and 10.3.6 are due to Finkelstein and Whitley (1981). Algoet and Cover (1988) showed how to replace the underlying sequence of i.i.d. random variables in Theorem 10.3.5(a) by a stationary ergodic one.

Another justification of the Kelly system, due to Bell and Cover (1980), is that it achieves competitive short-term optimality in a game-theoretic sense. But because this property requires continuous random variables and infinite games, it is outside the scope of our considerations.

Problems 10.3, 10.13, and 10.15 are from Thorp (2000b). Problem 10.8 was motivated by Friedman (1982), who considered the special case in which $X = \frac{1}{2}(X_1 + X_2)$, where X_1 and X_2 are i.i.d. with common distribution $\mathrm{P}(X_1 = a) = p = 1 - \mathrm{P}(X_1 = -1)$ and $\mathrm{E}[X_1] > 0$. Friedman in turn was motivated by sports betting, in which case $a = 10/11$, explaining why we allow a_1 or a_2 to be negative. Problem 10.10 is from Ethier and Tavaré (1983). The conclusion of part (d) had been conjectured by Wong (1982). See Griffin (1984) for discussion. Problem 10.11 was motivated by Whitworth (1901, Proposition 68), as noted above. Problem 10.12 is due to Lorden (1980), though the answer he gave (namely, (a)) is incorrect because of overshoot. Problem 10.14 is due to Griffin (1991). Problem 10.16 was motivated by Friedman (1982), who proposed using method (a) but did not provide an explicit formula. Problem 10.24 is from Algoet and Cover (1988). Problem 10.25 is due to Thorp (1993).

Chapter 11
Card Theory

[W]ho shuffles the cards, and brings trumps, honour, virtue, and prosperity back again? You call it chance; ay, and so it is chance, [...]

William Makepeace Thackeray, *Catherine: A Story*

This chapter is concerned with aspects of card games that are common to a number of games. Section 11.1 treats shuffling; in particular, we justify the well-known result that seven riffle shuffles are both necessary and sufficient to adequately mix a deck of 52 distinct cards. Section 11.2 deals with dealing and the concept of exchangeability, and Section 11.3 establishes the so-called fundamental theorem of card counting.

11.1 Shuffling

Consider a deck of N distinct cards, which for convenience we will assume are labeled $1, 2, \ldots, N$. For specificity, we also label the positions of the cards in the deck as follows: With the cards face down, the top card in the deck is in position 1, the second card is in position 2, and so on. In particular, when the cards are dealt, the card in position 1 is dealt first.

We define a *permutation* π of $(1, 2, \ldots, N)$ to be a one-to-one mapping $\pi : \{1, 2, \ldots, N\} \mapsto \{1, 2, \ldots, N\}$. The *symmetric group* S_N is the group of all permutations of $(1, 2, \ldots, N)$, with group operation given by composition of mappings, that is, $\pi_1 \pi_2 := \pi_2 \circ \pi_1$, where $(\pi_2 \circ \pi_1)(i) := \pi_2(\pi_1(i))$. The number of permutations in S_N is of course $N!$.

Think of $\pi(i) = j$ as meaning that the card in position i is moved to position j under the permutation. If the cards are in natural order $(1, 2, \ldots, N)$ to begin with, then the equivalent statement $\pi^{-1}(j) = i$ says that position j contains card i after the permutation is applied. Thus, a convenient representation for a permutation $\pi \in S_N$ is

S.N. Ethier, *The Doctrine of Chances*, Probability and its Applications,
DOI 10.1007/978-3-540-78783-9_11, © Springer-Verlag Berlin Heidelberg 2010

$$(\pi^{-1}(1), \dots, \pi^{-1}(N)), \tag{11.1}$$

which gives the order of the deck after permutation π is applied, assuming the deck was in natural order $(1, 2, \dots, N)$ originally. Thus, a *shuffle* is a permutation of $(1, 2, \dots, N)$, which is typically random but need not be.

Example 11.1.1. *Perfect riffle shuffle.* Assume $N = 52$. A perfect riffle shuffle of the deck, originally in natural order $(1, 2, \dots, 52)$, is accomplished in two steps: First, the deck is split into two 26-card packets, $(1, 2, \dots, 26)$ and $(27, 28, \dots, 52)$. Then the two packets are perfectly interlaced with card 1 remaining on top. The new deck order is

$$(1, 27, 2, 28, \dots, 26, 52). \tag{11.2}$$

Notice that the resulting permutation $\pi \in S_{52}$ satisfies $\pi^{-1}(2) = 27$, or equivalently $\pi(27) = 2$. Thus, the card originally in position 27, namely card 27, is moved to position 2 by the shuffle.

A second perfect riffle shuffle, applied to the deck (11.2), changes the order of the deck to

$$(1, 14, 27, 40, 2, 15, 28, 41, \dots, 13, 26, 39, 52). \tag{11.3}$$

This is $\pi^2 := \pi \circ \pi \in S_{52}$. Perhaps surprisingly, eight successive perfect riffle shuffles return the deck to its original order, that is, $\pi^8 = $ identity $\in S_{52}$ (Problem 11.1 on p. 421). ♠

Of course, actual riffle shuffles are imperfect. In general, a riffle shuffle is a two-step procedure. First, we break the deck into two packets, the first containing the cards in positions $1, \dots, n$ and the second containing the cards in positions $n+1, \dots, N$, where the size n of the first packet satisfies $0 \le n \le N$. Second, we interlace the two packets, preserving the order of the cards within each packet, to form the shuffled deck. There are $\binom{N}{n}$ ways to do this. One possible result of this interlacing is necessarily the identity permutation, but the remaining $\binom{N}{n} - 1$ permutations are n-dependent. Therefore, the number of permutations in S_N resulting from riffle shuffles is exactly

$$1 + \sum_{n=0}^{N} \left[\binom{N}{n} - 1 \right] = 2^N - N, \tag{11.4}$$

which is less than $N!$ if $N \ge 3$.

Moreover, actual riffle shuffles are random. We adopt a precise mathematical model for a random riffle shuffle, which we will call a *binomial riffle shuffle*: The size n of the first packet is assumed to have the binomial$(N, \frac{1}{2})$ distribution, and the $\binom{N}{n}$ possible interlacings of the two packets are assumed equally likely. An alternative, but equivalent (see Problem 11.3 on p. 421), formulation of the interlacing assumption is the rule that cards drop one-at-a-time from the first or second packets with probabilities proportional to the sizes of the packets. A binomial riffle shuffle results in a random permutation of the deck, that is, an S_N-valued random variable. Table 11.1 illustrates the definition in the case $N = 4$.

Table 11.1 The binomial riffle shuffle of a deck of size $N = 4$, initially in natural order. Notice that there are 12 distinct permutations, consistent with (11.4).

break	probab.	equally likely card orders after shuffle
|1234	1/16	1234
1|234	4/16	1234, 2134, 2314, 2341
12|34	6/16	1234, 1324, 1342, 3124, 3142, 3412
123|4	4/16	1234, 1243, 1423, 4123
1234|	1/16	1234

It will be convenient to generalize the (nonrandom) riffle shuffle to what will be called a (nonrandom) *a-shuffle*, with $a \geq 2$ an integer. Like a riffle shuffle, an a-shuffle is a two-step procedure. First, we break the deck into a packets of sizes $n_1 \geq 0, \dots, n_a \geq 0$, where $n_1 + \cdots + n_a = N$. Specifically, packet 1 contains the cards in positions $1, \dots, n_1$, packet 2 contains the cards in positions $n_1 + 1, \dots, n_1 + n_2$, and so on. Second, we interlace the a packets, preserving the order of the cards within each packet, to form the shuffled deck. There are $\binom{N}{n_1, \dots, n_a}$ ways to do this. The number of permutations in S_N resulting from a-shuffles is at most a^N. Of course, a 2-shuffle is just a riffle shuffle.

Analogous to the binomial riffle shuffle is what we will call a *multinomial a-shuffle*. The vector of packet sizes $(n_1, \dots, n_a)$ is assumed to have the multinomial$(N, a^{-1}, \dots, a^{-1})$ distribution, and the $\binom{N}{n_1, \dots, n_a}$ possible interlacings of the a packets are assumed equally likely. An alternative, but equivalent (see Problem 11.3 on p. 421), formulation of the interlacing assumption is the rule that cards drop one-at-a-time from packet 1 or packet 2 or ... or packet a with probabilities proportional to the sizes of the packets.

Notice that a multinomial 2-shuffle is simply a binomial riffle shuffle. Essential to the analysis is the following remarkable lemma.

Lemma 11.1.2. *The random permutation resulting from a multinomial a-shuffle followed by an independent multinomial b-shuffle is distributed as the random permutation resulting from a single multinomial ab-shuffle.*

Remark. In particular, the random permutation resulting from m successive independent binomial riffle shuffles (or multinomial 2-shuffles) is distributed as the random permutation resulting from a single multinomial 2^m-shuffle.

Proof. As we have seen, an a-shuffle is performed in two steps: First, we break the deck into a packets of sizes $n_1 \geq 0, \ldots, n_a \geq 0$, where $n_1 + \cdots + n_a = N$, and second, we interlace the a packets, preserving order within each packet, to form the shuffled deck. Let us call this pair of operations an *a-break-and-interlace*.

We observe that there is a one-to-one correspondence between a-break-and-interlaces of an N-card deck and N-digit base-a numbers

$$A = \langle a_1 a_2 \cdots a_N \rangle := \sum_{l=1}^{N} a_l a^{N-l}, \tag{11.5}$$

with $a_1, a_2, \ldots, a_N \in \{0, 1, \ldots, a-1\}$. (The notation $\langle a_1 a_2 \cdots a_N \rangle$ is temporarily adopted to distinguish the number in (11.5) from the product $a_1 a_2 \cdots a_N$. The base a is suppressed but implicit in the notation.) Indeed, the numbers of 0s, 1s, $\ldots$, $(a-1)$s in A tell us the packet sizes, and the positions of the 0s, 1s, $\ldots$, $(a-1)$s in A describe the interlacing of the packets.

For example, if $A = \langle 021011 \rangle$ and $a = 3$, then A determines a 3-shuffle of the deck $(1, 2, 3, 4, 5, 6)$ as follows: A has two 0s, three 1s, and one 2, so the first two cards, 1 and 2, comprise packet 1, the next three cards, 3, 4, and 5, comprise packet 2, and the last card, 6, comprises packet 3. In the 3-shuffled deck, the cards of the first packet appear in the positions of the 0s of A, the cards of the second packet appear in the positions of the 1s of A, and so on. Thus, the shuffled deck has the form $(1, 6, 3, 2, 4, 5)$.

Conversely, given the 3-shuffled deck $(1, 6, 3, 2, 4, 5)$ and the packet sizes 2, 3, and 1, A will have 0s at the positions of cards 1 and 2, the first packet, 1s at the positions of cards 3, 4, and 5, the second packet, and a 2 at the position of card 6, the third packet. Thus, $A = \langle 021011 \rangle$.

Given an N-digit base-a number A, let us denote by π_A the permutation in S_N resulting from the a-break-and-interlace corresponding to A. Now suppose that A is random with the discrete uniform distribution over $\{0, 1, \ldots, a^N - 1\}$ (i.e., $P(A = \langle a_1 a_2 \cdots a_N \rangle) = a^{-N}$ for all $a_1, a_2, \ldots, a_N \in \{0, 1, \ldots, a-1\}$). Then we claim that the random permutation π_A is distributed as the random permutation resulting from a multinomial a-shuffle. To see this, it suffices to check that if A has the discrete uniform distribution over $\{0, 1, \ldots, a^N - 1\}$, then its base-$a$ digits $a_1, a_2, \ldots, a_N$ are independent and have the discrete uniform distribution over $\{0, 1, \ldots, a-1\}$. It follows that, if n_j is the number of $(j-1)$s among $a_1, \ldots, a_N$ for $j = 1, \ldots, a$, then $(n_1, \ldots, n_a)$ is multinomial$(N, a^{-1}, \ldots, a^{-1})$, and that, given the values of $n_1, \ldots, n_a$, each arrangement of n_1 0s, n_2 1s, $\ldots$, n_a $(a-1)$s among $a_1, \ldots, a_N$ is equally likely. This establishes the claim.

Again given an N-digit base-a number A, let us also define what we will call the *inverse a-shuffle* corresponding to A: For $l = 1, \ldots, N$ label the card in position l with the lth digit of A, namely a_l. Those cards labeled with a $j - 1$ are placed in packet j, retaining original order. Then the packets are stacked in order $(1, 2, \ldots, a)$ to form the inverse a-shuffled deck. Let the resulting permutation be denoted by $\bar{\pi}_A$, and observe that $\bar{\pi}_A \circ \pi_A$ is the identity permutation, so $\bar{\pi}_A$ is simply π_A^{-1}.

Next, let $A := \langle a_1 a_2 \cdots a_N \rangle$ be an N-digit base-a number, and let $B := \langle b_1 b_2 \cdots b_N \rangle$ be an N-digit base b number. We claim that

$$\pi_A^{-1} \circ \pi_B^{-1} = \pi_C^{-1}, \tag{11.6}$$

where $C := \langle c_1 c_2 \cdots c_N \rangle$ is the N-digit base-ab number given by

$$c_l := a_{\pi_B^{-1}(l)} b + b_l, \qquad l = 1, \ldots, N. \tag{11.7}$$

It suffices to show that both sides of (11.6) produce the same result when applied to a deck in natural order. For the inverse b-shuffle π_B^{-1}, imagine writing $b_1, \ldots, b_N$ on cards $1, \ldots, N$, respectively. The order of the cards after π_B^{-1} is applied is $(\pi_B(1), \ldots, \pi_B(N))$. Now applying the inverse a-shuffle π_A^{-1}, imagine writing $a_1, \ldots, a_N$ on cards $\pi_B(1), \ldots, \pi_B(N)$, respectively, writing the digit of A to the left of the digit of B already written. We write a_k on card $\pi_B(k)$, hence $a_{\pi_B^{-1}(l)}$ on card l, together with b_l. Now if we define $C := \langle c_1 c_2 \cdots c_N \rangle$ as above, the inverse ab-shuffle π_C^{-1} is realized by labeling the cards exactly as they were labeled by $\pi_A^{-1} \circ \pi_B^{-1}$, proving (11.6).

Finally, observe that if A and B are independent with discrete uniform distributions, then C has the discrete uniform distribution. Indeed, the N digits of C are independent with discrete uniform distributions on $\{0, 1, \ldots, ab-1\}$. To see this, let U_a and U_b be independent with discrete uniform distributions on $\{0, 1, \ldots, a-1\}$ and $\{0, 1, \ldots, b-1\}$, respectively; then $U_a b + U_b$ has the discrete uniform distribution on $\{0, 1, \ldots ab-1\}$. It follows from (11.6) that

$$\pi_B \circ \pi_A = (\pi_A^{-1} \circ \pi_B^{-1})^{-1} \overset{d}{=} (\pi_C^{-1})^{-1} = \pi_C, \tag{11.8}$$

which implies the stated conclusion. ♠

Next, given an arbitrary permutation $\pi \in S_N$, we want to determine the probability that a multinomial a-shuffle results in π. For this we need the notion of a rising sequence, which we illustrate by example before defining it. The permutation (11.2) has two rising sequences, $1, 2, \ldots, 26$ and $27, 28, \ldots, 52$. The permutation (11.3) has four rising sequences, $1, 2, \ldots, 13$ and $14, 15, \ldots, 26$ and $27, 28, \ldots, 39$ and $40, 41, \ldots, 52$. In general, a *rising sequence* in the permutation π represented by (11.1) is a maximal set of consecutive integers $j, j+1, \ldots, k$ from $\{1, 2, \ldots, N\}$ such that $\pi(j) < \pi(j+1) < \cdots < \pi(k)$. Here maximal means (a) $j = 1$ or $\pi(j-1) > \pi(j)$ and (b) $k = N$ or $\pi(k+1) < \pi(k)$.

Lemma 11.1.3. *Given a permutation $\pi \in S_N$ with exactly r rising sequences, the probability that π is the result of a multinomial a-shuffle is*

$$\binom{N+a-r}{N} a^{-N}, \tag{11.9}$$

where the binomial coefficient is 0 *if* $r > a$.

Proof. Suppose that $\pi \in S_N$ consists of exactly r rising sequences, namely

$$
\begin{aligned}
&1, 2, \ldots, k_1, \\
&k_1 + 1, \ldots, k_2, \\
&\qquad\vdots \\
&k_{r-1} + 1, \ldots, N.
\end{aligned}
\tag{11.10}
$$

Numbers appear within each row in the same order that they appear in (11.1), but $k_1 + 1$ is to the left of k_1 there, $\ldots$, $k_{r-1} + 1$ is to the left of k_{r-1} there.

A multinomial a-shuffle produces at most a rising sequences, so for a multinomial a-shuffle to produce this π, we must have $a \geq r$.

If $a = r$, then the only way to produce this π is to break the deck into packets of sizes

$$
n_1 := k_1, \ n_2 := k_2 - k_1, \ \ldots, \ n_a := N - k_{a-1}
\tag{11.11}
$$

and for the interlacing to be the unique one that results in π. The probability is

$$
\binom{N}{n_1, \ldots, n_a}\left(\frac{1}{a}\right)^N \cdot \binom{N}{n_1, \ldots, n_a}^{-1} = a^{-N}.
\tag{11.12}
$$

If $a > r$, there are a number of ways to produce π. We must break the deck into a packets, then interlace the packets to get the r rising sequences of π. Clearly, a break must occur between each of the $r - 1$ pairs of successive rising sequences, and the remaining $a - r \ [= (a - 1) - (r - 1)]$ breaks can occur arbitrarily. There are N cards and $a - r$ remaining breaks, hence

$$
\binom{N + a - r}{N}
\tag{11.13}
$$

ways of choosing these remaining breaks (Theorem 1.1.6 on p. 5). The probability of a particular a-shuffle with packet sizes $n_1, \ldots, n_a$ is (11.12), so the probability that π results from an a-shuffle is

$$
\binom{N + a - r}{N} a^{-N},
\tag{11.14}
$$

as required.   ♠

Next we need to count the number of permutations $\pi \in S_N$ with exactly r rising sequences. For this we introduce the *Eulerian numbers* $\left\langle {N \atop r} \right\rangle$. See Table 11.2. These numbers are generated by the recursion

$$
\left\langle {N \atop r} \right\rangle := (N - r + 1)\left\langle {N - 1 \atop r - 1} \right\rangle + r\left\langle {N - 1 \atop r} \right\rangle
\tag{11.15}
$$

for $r = 1, \ldots, N$, where $\left\langle {1 \atop 1} \right\rangle := 1$ and $\left\langle {N-1 \atop 0} \right\rangle = \left\langle {N-1 \atop N} \right\rangle := 0$. This should be reminiscent of (1.7) on p. 4. It will be a consequence of the next lemma that $\sum_{r=1}^{N} \left\langle {N \atop r} \right\rangle = N!$.

Table 11.2 The first six rows of the Eulerian triangle. The Nth row contains the Eulerian numbers $\left\langle {N \atop 1} \right\rangle, \ldots, \left\langle {N \atop N} \right\rangle$. Compare with Table 1.1 on p. 5.

					1					
				1		1				
			1		4		1			
		1		11		11		1		
	1		26		66		26		1	
1		57		302		302		57		1

Lemma 11.1.4. *The number of permutations $\pi \in S_N$ with exactly r rising sequences is $\left\langle {N \atop r} \right\rangle$.*

Proof. We claim that the number of rising sequences in π equals the number of increasing segments of π^{-1}.

For example, consider the case $N = 6$, with π and π^{-1} given respectively by

$$(3, 1, 4, 6, 2, 5) \qquad \text{and} \qquad (2, 5, 1, 3, 6, 4). \qquad (11.16)$$

Then π above has rising sequences $(1, 2)$, $(3, 4, 5)$; and (6), while π^{-1} has increasing segments $(2, 5)$, $(1, 3, 6)$, and (4).

To justify the claim, recall that π and π^{-1} are represented respectively by

$$(\pi^{-1}(1), \ldots, \pi^{-1}(N)) \qquad \text{and} \qquad (\pi(1), \ldots, \pi(N)), \qquad (11.17)$$

and note that the rising sequence in π containing $\pi^{-1}(i)$ terminates with that term if and only if $\pi^{-1}(i) = N$ or $\pi^{-1}(i) + 1$ appears to the left of $\pi^{-1}(i)$ if and only if $\pi^{-1}(i) = N$ or $\pi(\pi^{-1}(i) + 1) < \pi(\pi^{-1}(i)) = i$ if and only if $i = \pi(N)$ or the increasing segment of π^{-1} containing i terminates with that term. It follows that the number of permutations in S_N with r rising sequences is equal to the number of permutations in S_N with r increasing segments.

Let $A(N, r)$ denote the number of permutations in S_N with r increasing segments. Then

$$A(N, r) = (N - r + 1)A(N - 1, r - 1) + rA(N - 1, r) \qquad (11.18)$$

for $r = 1, \ldots, N$. Indeed, every permutation in S_N can be obtained from one in S_{N-1} by inserting an N in any of N positions. It will have r increasing segments if either (a) the permutation in S_{N-1} has r increasing segments and N is inserted after the final term of any of the r increasing segments, or (b) the permutation in S_{N-1} has $r-1$ increasing segments and N is inserted anywhere except after the final term of any of the $r-1$ increasing segments.

Clearly $A(1,1) = 1$ and $A(N-1,0) = A(N-1,N) = 0$. Comparing (11.18) with (11.15), it follows by induction on N that

$$A(N,r) = \left\langle \begin{matrix} N \\ r \end{matrix} \right\rangle \tag{11.19}$$

for $r = 1, \ldots, N$, and the proof is complete.   ♠

Let R_m denote the distribution of the random permutation of $(1, 2, \ldots, N)$ resulting from m successive independent binomial riffle shuffles. Let U denote the discrete uniform distribution on S_N, that is, $U(\pi) = 1/N!$ for all $\pi \in S_N$. We would like to determine how large m must be for R_m to be close to U in some sense.

For two probability distributions P_1 and P_2 on S_N, we define the *total-variation distance* between them to be

$$\|P_1 - P_2\| := \max_{B \subset S_N} |P_1(B) - P_2(B)|. \tag{11.20}$$

An equivalent formulation is

$$\|P_1 - P_2\| := \frac{1}{2} \sum_{\pi \in S_N} |P_1(\{\pi\}) - P_2(\{\pi\})|. \tag{11.21}$$

Noting that $|P_1(B) - P_2(B)| = \frac{1}{2}[|P_1(B) - P_2(B)| + |P_1(B^c) - P_2(B^c)|]$, the triangle inequality shows that the right side of (11.20) is less than or equal to the right side of (11.21). Taking $B = \{\pi \in S_N : P_1(\{\pi\}) \geq P_2(\{\pi\})\}$ shows that equality holds.

While (11.20) is conceptually simpler, (11.21) is easier to work with. In any case, the total-variation distance cannot exceed 1.

We are finally in a position to state and prove the main result of this section.

Theorem 11.1.5. *Given a deck of N distinct cards, the total-variation distance between the distribution of the random permutation resulting from m successive independent binomial riffle shuffles and the discrete uniform distribution on S_N is given by*

$$\|R_m - U\| = \frac{1}{2} \sum_{r=1}^{N} \left\langle \begin{matrix} N \\ r \end{matrix} \right\rangle \left| \binom{N + 2^m - r}{N} 2^{-mN} - \frac{1}{N!} \right|, \tag{11.22}$$

where the coefficients $\left\langle \begin{matrix} N \\ r \end{matrix} \right\rangle$ are the Eulerian numbers, and the binomial coefficient $\binom{N+2^m-r}{N}$ is 0 if $r > 2^m$.

Proof. The work has already been done in proving the lemmas. By definition,

$$\|R_m - U\| = \frac{1}{2} \sum_{\pi \in S_N} |R_m(\{\pi\}) - U(\{\pi\})|. \tag{11.23}$$

By Lemmas 11.1.2 and 11.1.3, the term $|R_m(\{\pi\}) - U(\{\pi\})|$ in (11.23) is

$$\left| \binom{N + 2^m - r}{N} 2^{-mN} - \frac{1}{N!} \right| \tag{11.24}$$

if π has r rising sequences, and by Lemma 11.1.4 there are $\left\langle {N \atop r} \right\rangle$ such terms in (11.23). Therefore, (11.22) holds.   ♠

Of particular interest is a deck of $N = 52$ distinct cards. In Table 11.3 we give numerical values of (11.22) in this case. The results are surprising. Four, or even five, shuffles clearly do not suffice to achieve adequate randomness. It has been argued that seven shuffles are both necessary and sufficient. If, in view of Table 11.3, this conclusion seems a bit arbitrary, we quote the *New York Times* (see the chapter notes) for justification:

> The cards do get more and more randomly mixed if a person keeps on shuffling more than seven times, but seven shuffles is a transition point, the first time that randomness is close. Additional shuffles do not appreciably alter things.

Table 11.3 The total-variation distance between the distribution of the random permutation resulting from m successive independent binomial riffle shuffles of a deck of 52 distinct cards and the discrete uniform distribution on S_{52}.

m	$\|R_m - U\|$	m	$\|R_m - U\|$
1	1.000 000 000	6	.613 549 597
2	1.000 000 000	7	.334 060 999
3	1.000 000 000	8	.167 158 642
4	.999 999 533	9	.085 420 193
5	.923 732 929	10	.042 945 549

Throughout this section we have assumed that the N cards are distinct. In many games this is not the case. For example, in single-deck twenty-one, suits play no role and there is no distinction between tens, jacks, queens, and kings. Therefore the 52-card deck has 10 distinct cards, four each of the denominations ace, $2, 3, \ldots, 9$ together with 16 ten-valued cards. Clearly, the total-variation distance to uniformity after m shuffles will be smaller with such a deck than with a deck of 52 distinct cards, and therefore fewer shuffles may suffice. This issue has been the subject of recent research; see the chapter notes.

11.2 Dealing

Again we consider a deck of N distinct cards labeled $1, 2, \ldots, N$. We also label the positions of the cards in the deck as before: With the cards face down, the top card in the deck is in position 1, the second card is in position 2, and so on. In particular, when the cards are dealt, the card in position 1 is dealt first.

In this section we assume that the deck is *well shuffled*. More precisely, if Π denotes the random permutation of $(1, 2, \ldots, N)$ that describes the deck ordering, we assume Π to have the discrete uniform distribution. Recall that $\Pi(i) = j$ means that the card in position i is moved to position j by the permutation. If the cards are in natural order $(1, 2, \ldots, N)$ initially, then

$$(\Pi^{-1}(1), \ldots, \Pi^{-1}(N)) \tag{11.25}$$

is the order of the cards in the deck after Π is applied. For example, card j is in position 1 after the shuffle if $\Pi^{-1}(1) = j$, or equivalently if $\Pi(j) = 1$, that is, if the card in position j initially, namely card j, is moved to position 1 by the shuffle.

Our first theorem tells us that a well-shuffled deck is *exchangeable*, that is, if the cards are permuted in an arbitrary nonrandom way (or even in a random way under certain conditions), the deck will remain well shuffled. This has important implications for card dealing.

Theorem 11.2.1. *Let Π be an S_N-valued random variable with the discrete uniform distribution.*

(a) If $\pi_1, \pi_2 \in S_N$, then $\pi_1 \circ \Pi$ and $\Pi \circ \pi_2$ have the discrete uniform distribution.

(b) If Π_1 (resp., Π_2) is an S_N-valued random variable independent of Π, then $\Pi_1 \circ \Pi$ (resp., $\Pi \circ \Pi_2$) has the discrete uniform distribution.

(c) If $\lambda : S_N \mapsto S_N$ is one-to-one and nonrandom, then $\lambda(\Pi)$ has the discrete uniform distribution.

(d) Π^{-1} has the discrete uniform distribution.

Proof. (a) $P(\pi_1 \circ \Pi = \pi) = P(\Pi = \pi_1^{-1} \circ \pi) = 1/N!$ for all $\pi \in S_N$. The other case is similar.

(b) This follows from part (a) by conditioning on Π_1 (resp., Π_2).

(c) $P(\lambda(\Pi) = \pi) = P(\Pi = \lambda^{-1}(\pi)) = 1/N!$ for all $\pi \in S_N$.

(d) This is a special case of part (c).   ♠

Actually, the notion of exchangeability is more general. If $X_1, X_2, \ldots, X_N$ are jointly distributed discrete random variables, we say that they are *ex-*

changeable if the joint distribution of $X_{\pi(1)}, X_{\pi(2)}, \ldots, X_{\pi(N)}$ coincides with that of $X_1, X_2, \ldots, X_N$ for all permutations $\pi \in S_N$. For example, if Π is as in Theorem 11.2.1 and $X_j = \Pi^{-1}(j)$ for $j = 1, 2, \ldots, N$, then $X_1, X_2, \ldots, X_N$ are exchangeable.

Example 11.2.2. *Implications of the exchangeability of a well-shuffled deck.* We give four examples to illustrate the theorem.

(a) In some card games the first card is *burned*. This has nothing to do with combustion, but rather means that the first card is removed from play without revealing its identity. What is the probability that the second card in the deck (which is actually the first card in play) is an ace? One might be tempted to condition on whether the first (burned) card is an ace. Thus, with $D = \{$first card is an ace$\}$ and $E = \{$second card is an ace$\}$, we have

$$\mathrm{P}(E) = \mathrm{P}(D)\mathrm{P}(E \mid D) + \mathrm{P}(D^c)\mathrm{P}(E \mid D^c) = \frac{1}{13}\frac{3}{51} + \frac{12}{13}\frac{4}{51} = \frac{1}{13}. \quad (11.26)$$

Theorem 11.2.1(a) allows us to reach the same conclusion without such calculations. To see this, consider the permutation $\pi_1 \in S_{52}$ that transfers the top card to the bottom of the deck. (Specifically, $\pi_1(1) = 52$ and $\pi_1(i) = i - 1$ for $i = 2, \ldots, 52$.) If the ordering of the well shuffled deck is described by Π before the transfer, then it is described by $\pi_1 \circ \Pi$ afterwards. Indeed, the ordering of the cards changes from $(\Pi^{-1}(1), \ldots, \Pi^{-1}(52))$ to

$$
\begin{aligned}
(\Pi^{-1}&(2), \ldots, \Pi^{-1}(52), \Pi^{-1}(1)) \\
&= (\Pi^{-1}(\pi_1^{-1}(1)), \ldots, \Pi^{-1}(\pi_1^{-1}(52))) \\
&= ((\pi_1 \circ \Pi)^{-1}(1), \ldots, (\pi_1 \circ \Pi)^{-1}(52)). \quad (11.27)
\end{aligned}
$$

Thus, the theorem tells us that the deck remains well shuffled after transferring the top card to the bottom of the deck, and so the probability $\mathrm{P}(E)$ in (11.26) is just the probability that the top card of the new deck is an ace, which is of course $1/13$.

(b) Here is a more complicated example. Suppose a deck of N distinct cards is (partially) dealt to m players with each player receiving l cards ($ml \leq N$). Let us label the players as player 1, $\ldots$, player m, starting with the player to the dealer's left and moving in a clockwise direction. The usual procedure is to distribute the first m cards to the m players in order, the first card to player 1, the second to player 2, and so on. This round of dealing is repeated an additional $l - 1$ times until each player has l cards. The remaining $N - ml$ cards may be used later, depending on the game. Call this method of dealing method 1.

Let us consider an alternative method of dealing, which we call method 2. The first l cards are dealt to player 1, the next l cards are dealt to player 2, and so on. With m players, this accounts for ml cards, and the remaining $N - ml$ cards may be used later, depending on the game.

Assuming a well-shuffled deck, we claim that the two methods produce (probabilistically) equivalent results. Let $H_1(i)$ $(1 \leq i \leq m)$ be the set of positions in the deck of the cards dealt to player i using method 1. Define $H_2(i)$ $(1 \leq i \leq m)$ similarly, using method 2. Clearly,

$$
\begin{aligned}
H_1(i) &= \{i, m+i, 2m+i, \ldots, (l-1)m+i\}, \\
H_2(i) &= \{(i-1)l+1, (i-1)l+2, \ldots, il\}.
\end{aligned} \tag{11.28}
$$

Choose $\pi_1 \in S_N$ satisfying $\pi_1(H_1(i)) = H_2(i)$ for $i = 1, \ldots, m$. Then

$$
\begin{aligned}
\{(\pi_1 \circ \Pi)^{-1}(k) : k \in H_2(i)\} &= \{\Pi^{-1}(\pi_1^{-1}(k)) : k \in H_2(i)\} \\
&= \{\Pi^{-1}(j) : \pi_1(j) \in H_2(i)\} \\
&= \{\Pi^{-1}(j) : j \in H_1(i)\}
\end{aligned} \tag{11.29}
$$

for $i = 1, \ldots, m$. This says that the cards dealt to the m players using method 1 and deck ordering Π coincide with the cards dealt to the m players using method 2 and deck ordering $\pi_1 \circ \Pi$. Since both Π and $\pi_1 \circ \Pi$ correspond to well-shuffled decks by Theorem 11.2.1(a), the induced probability distributions coincide.

(c) Next, we consider the effect of cutting the deck. If the N distinct cards are in natural order $(1, 2, \ldots, N)$, a *uniform cut* chooses an integer n at random from $\{0, 1, \ldots, N-1\}$ and interchanges the packet consisting of the top n cards with the packet consisting of the bottom $N-n$ cards. Note that Π_1, the random permutation describing a uniform cut, may be assumed independent of Π, the random permutation describing the ordering of the deck. Of course, $\Pi_1 \circ \Pi$ is the random permutation describing the ordering of the deck after a uniform cut. Theorem 11.2.1(b) says that a uniform cut of a well-shuffled deck leaves the deck well shuffled.

(d) Finally, we consider the game of trente et quarante. (See Chapter 20 for more information.) Trente et quarante is played with six standard 52-card decks mixed together, resulting in a six-deck shoe comprising 312 cards. Suits do not matter except insofar as they determine color: Diamonds and hearts are red, whereas clubs and spades are black. Aces have value one, court cards have value 10, and every other card has value equal to its nominal value. Two rows of cards are dealt. In the first row, called Black, cards are dealt until the total value is 31 or greater. In the second row, called Red, cards are again dealt until the total value is 31 or greater. The winning row is the one with the lower total.

We claim that the joint distribution of the Black total and the Red total is symmetric. This will imply, for example, that the probability that the Red total is less than the Black total equals the probability that the Black total is less than the Red total.

Let K and L denote the numbers of cards in the Black row and the Red row, respectively, and let B and R denote the Black and Red totals. These random variables are nonrandom functions of $\Pi \in S_{312}$, the random permutation that determines the order of the cards in the deck. Define $\Pi_1 \in S_{312}$ to be the random permutation that interchanges the cards in positions $1, \ldots, K$ with those in positions $K+1, \ldots, K+L$, keeping in mind that K and L depend on Π. Notice that $\Pi_1 = \lambda(\Pi)$ for a nonrandom λ and that $\lambda \circ \lambda$ is the identity mapping, hence λ is one-to-one. Theorem 11.2.1(c) therefore applies, and we conclude that Π_1 also has the discrete uniform distribution. Moreover, if the nonrandom function g satisfies $(B, R) = g(\Pi)$, then $(R, B) = g(\Pi_1)$ and the claim follows.

In fact, a slight modification of this argument shows that, if the first n cards have been seen and if sufficiently many cards remain to complete both rows with probability 1, then the conditional joint distribution of the Black total and the Red total, given the cards already seen, is symmetric. This implies that card counting is useless to bettors on the Red and Black rows. ♠

Example 11.2.3. *Casino War.* Casino War is a casino version of a popular children's game. It is played with d standard 52-card decks mixed together. (Usually, $d = 6$.) Denominations are ranked in the usual way (2, 3, 4, 5, 6, 7, 8, 9, 10, J, Q, K, A) and suits do not matter. After the player makes a bet, the dealer deals a card to the player and one to himself, both face up. There are three possibilities:

If the player's denomination outranks the dealer's, the player wins and is paid even money.

If the dealer's denomination outranks the player's, the player loses his bet.

If the two denominations are equal, the player has the option of forfeiting one half of his bet or going to war. To go to war, the player must first double his bet. Then the dealer deals a card to the player and one to himself, both face up. (Before each of these last two cards is dealt, one or more cards are burned, i.e., discarded face down.) There are two possibilities:

If the player's denomination outranks or equals the dealer's, the player wins and is paid 1 to 2.

If the dealer's denomination outranks the player's, the player loses his (doubled) bet.

We relabel the 13 denominations as 2–14 and denote the $52d$ cards by random variables $X_1, X_2, \ldots, X_{52d}$. Specifically, $(X_1, X_2, \ldots, X_{52d})$ assumes each of the $(52d)!/[(4d)!]^{13}$ permutations of $4d$ 2s, $4d$ 3s, ..., $4d$ 14s with equal probability. We regard X_1, X_2, X_3, X_4 as the first four cards seen rather than the first four cards in the deck. (If there are two or more players competing against the dealer, we ignore the cards dealt to the other players.)

We begin by evaluating the relevant probabilities, assuming in the case of a tie that the player goes to war. (See Problem 11.17 on p. 423 for the other case.) First, the probability of a tie is

$$\mathrm{P}(X_1 = X_2) = 13\frac{\binom{4d}{2}\binom{48d}{0}}{\binom{52d}{2}} = \frac{4d - 1}{52d - 1}, \tag{11.30}$$

so by symmetry,

$$P(X_1 < X_2) = P(X_1 > X_2) = \frac{1}{2}(1 - P(X_1 = X_2)) = \frac{24d}{52d - 1}. \quad (11.31)$$

Second, the probability of a tie followed by a tie at war is

$$\begin{aligned}
P(X_1 = X_2,\ X_3 = X_4) &= P(X_1 = X_2)P(X_3 = X_4 \mid X_1 = X_2) \\
&= \frac{4d - 1}{52d - 1}\ \frac{\binom{4d-2}{2}\binom{48d}{0} + 12\binom{4d}{2}\binom{48d-2}{0}}{\binom{52d-2}{2}} \\
&= \frac{(4d - 1)_3 + 48d(4d - 1)^2}{(52d - 1)_3}, \quad (11.32)
\end{aligned}$$

so by symmetry,

$$\begin{aligned}
P(X_1 = X_2,\ X_3 < X_4) &= P(X_1 = X_2,\ X_3 > X_4) \quad (11.33) \\
&= \frac{1}{2}(P(X_1 = X_2) - P(X_1 = X_2,\ X_3 = X_4)).
\end{aligned}$$

Let Y denote the player's profit, in units, from a one-unit initial bet, assuming that he goes to war in the case of a tie. By the rules of the game,

$$Y = \begin{cases} 1 & \text{on } \{X_1 > X_2\} \cup \{X_1 = X_2,\ X_3 \geq X_4\}, \\ -1 & \text{on } \{X_1 < X_2\}, \\ -2 & \text{on } \{X_1 = X_2,\ X_3 < X_4\}, \end{cases} \quad (11.34)$$

and this allows us to evaluate $E[Y]$ directly. However, we can simplify the algebra slightly by noticing that the game is nearly symmetric. In fact, the only two departures from symmetry occur in a war: The player wins tied wars, but when the dealer wins a war, he wins double. We conclude that

$$\begin{aligned}
E[Y] &= P(X_1 > X_2) + P(X_1 = X_2,\ X_3 \geq X_4) \\
&\quad - P(X_1 < X_2) - 2P(X_1 = X_2,\ X_3 < X_4) \\
&= P(X_1 = X_2,\ X_3 = X_4) - P(X_1 = X_2,\ X_3 < X_4) \\
&= \frac{3}{2}P(X_1 = X_2,\ X_3 = X_4) - \frac{1}{2}P(X_1 = X_2) \\
&= \frac{3}{2}\frac{(4d - 1)_3 + 48d(4d - 1)^2}{(52d - 1)_3} - \frac{1}{2}\frac{4d - 1}{52d - 1} \\
&= -\frac{2(4d - 1)(520d^2 - 14d - 3)}{(52d - 1)_3}. \quad (11.35)
\end{aligned}$$

When $d = 6$ this becomes $-857{,}118/(311)_3 \approx -0.028771$. Notice that the game is advantageous to the casino (not surprisingly), despite the fact that the player wins more often than does the dealer.   ♠

This is a good place to describe several generalizations of the hypergeometric distribution. In each case we assume a deck of N cards. The deck contains one or more categories of cards (e.g., suits) and is randomly dealt into one or more hands.

The simplest case assumes that category 1 consists of m of the N cards and hand 1 consists of n of the N cards. Let X be the number of cards from category 1 dealt into hand 1. Then X has the hypergeometric distribution, that is,

$$P(X = k) = \frac{\binom{m}{k}\binom{N-m}{n-k}}{\binom{N}{n}}, \tag{11.36}$$

where $(n - (N - m))^+ \leq k \leq n \wedge m$.

Next assume $c \geq 2$ categories of sizes $m_1 \geq 1, \ldots, m_c \geq 1$. We may as well assume that the categories exhaust the deck, so that $m_1 + \cdots + m_c = N$. Again we assume a single hand of n of the N cards. Let X_i be the number of cards from category i dealt into hand 1, and put $\boldsymbol{X} = (X_1, \ldots, X_c)$. Then $\boldsymbol{X}$ has the multivariate hypergeometric distribution, that is,

$$P(\boldsymbol{X} = \boldsymbol{k}) = \frac{\binom{m_1}{k_1} \cdots \binom{m_c}{k_c}}{\binom{N}{n}}, \tag{11.37}$$

where $0 \leq k_i \leq m_i$ for $i = 1, \ldots, c$ and $k_1 + \cdots + k_c = n$.

Next we return to the assumption that category 1 consists of m of the N cards, but now we assume $d \geq 2$ hands of sizes $n_1 \geq 1, \ldots, n_d \geq 1$. We may as well assume that the hands exhaust the deck, so that $n_1 + \cdots + n_d = N$. Let X_j be the number of cards from category 1 dealt into hand j, and put $\boldsymbol{X} = (X_1, \ldots, X_d)$. Then

$$P(\boldsymbol{X} = \boldsymbol{k}) = \frac{\binom{m}{k_1, \ldots, k_d}\binom{N-m}{n_1 - k_1, \ldots, n_d - k_d}}{\binom{N}{n_1, \ldots, n_d}}, \tag{11.38}$$

where $0 \leq k_j \leq n_j$ for $j = 1, \ldots, d$ and $k_1 + \cdots + k_d = m$.

Finally, we assume $c \geq 2$ categories of sizes $m_1 \geq 1, \ldots, m_c \geq 1$ with $m_1 + \cdots + m_c = N$. We also assume $d \geq 2$ hands of sizes $n_1 \geq 1, \ldots, n_d \geq 1$ with $n_1 + \cdots + n_d = N$. Let X_{ij} be the number of cards from category i dealt into hand j, and let $\boldsymbol{X}$ be the resulting $(c \times d)$-matrix-valued random variable. Then

$$P(\boldsymbol{X} = \boldsymbol{k}) = \frac{\binom{m_1}{k_{11}, \ldots, k_{1d}} \cdots \binom{m_c}{k_{c1}, \ldots, k_{cd}}}{\binom{N}{n_1, \ldots, n_d}}, \tag{11.39}$$

where $\boldsymbol{k}$ is a $c \times d$ matrix of nonnegative integers with row sums $\sum_j k_{ij} = m_i$ for $i = 1, \ldots, c$ and column sums $\sum_i k_{ij} = n_j$ for $j = 1, \ldots, d$.

Notice that (11.39) contains (11.36)–(11.38) as special cases. More precisely, (11.36) corresponds to $c = d = 2$, (11.37) to $c \geq d = 2$, and (11.38) to $2 = c \leq d$.

We now state a so-called *duality theorem* that uses the full generality of (11.39).

Theorem 11.2.4. *Consider two decks of N cards each.*

Deck 1 consists of $c \geq 2$ categories of sizes $m_1, \ldots, m_c$, where $m_1, \ldots, m_c$ are positive integers that sum to N. It is randomly dealt into $d \geq 2$ hands of sizes $n_1, \ldots, n_d$, where $n_1, \ldots, n_d$ are positive integers that sum to N. Let $\boldsymbol{X}$ be the $(c \times d)$-matrix-valued random variable whose (i, j)th entry equals the number of cards from category i dealt into hand j.

Deck 2 has its category sizes and hand sizes reversed. Specifically, it consists of d categories of sizes $n_1, \ldots, n_d$ and is randomly dealt into c hands of sizes $m_1, \ldots, m_c$. Let $\boldsymbol{Y}$ be the $(d \times c)$-matrix-valued random variable whose (i, j)th entry equals the number of cards from category i dealt into hand j.

Then $\boldsymbol{X}$ is distributed as $\boldsymbol{Y}^\mathsf{T}$.

Remark. (a) The example that originally motivated this theorem came not from gambling but from contract bridge. Take $N = 52$, $c = d = 4$, and $m_1 = \cdots = m_c = n_1 = \cdots = n_d = 13$. Let the categories in decks 1 and 2 be suits, and let the hands in decks 1 and 2 be bridge hands. Since row 1 of $\boldsymbol{X}$ is distributed as column 1 of $\boldsymbol{Y}$, we conclude that the joint distribution of the numbers of representatives of a particular suit in the four players' hands coincides with the joint distribution of the numbers of representatives of the four suits in a particular player's hand. This implies, for example, that the probability that no player has more than four cards of a particular suit coincides with the probability that a particular player has more than four cards in no suit.

(b) The special case of the theorem in which $c = d = 2$ is nothing but the property of the hypergeometric distribution identified in Problem 1.21 on p. 56 and Example 2.2.6 on p. 82.

Proof. We provide two proofs. The first is easier but the second may be more revealing.

For the first proof, let $\boldsymbol{k}$ be a $c \times d$ matrix of nonnegative integers with row sums $\sum_j k_{ij} = m_i$ for $i = 1, \ldots, c$ and column sums $\sum_i k_{ij} = n_j$ for $j = 1, \ldots, d$. Then

$$
\begin{aligned}
\mathrm{P}(\boldsymbol{X} = \boldsymbol{k}) &= \frac{\binom{m_1}{k_{11}, \ldots, k_{1d}} \cdots \binom{m_c}{k_{c1}, \ldots, k_{cd}}}{\binom{N}{n_1, \ldots, n_d}} \\
&= \frac{\binom{n_1}{k_{11}, \ldots, k_{c1}} \cdots \binom{n_d}{k_{1d}, \ldots, k_{cd}}}{\binom{N}{m_1, \ldots, m_c}} = \mathrm{P}(\boldsymbol{Y}^\mathsf{T} = \boldsymbol{k}), \qquad (11.40)
\end{aligned}
$$

where the second equality is immediate upon expressing all multinomial coefficients in terms of factorials. The first and third equalities use (11.39).

For the second proof, let us assume that the cards in both decks are numbered $1, 2, \ldots, N$. We define $C_1(i)$ to be the set of cards in category i in deck

1 $(i = 1, \ldots, c)$, and $H_1(j)$ to be the set of positions (of the random permutation that defines the ordering of the deck) of hand j in deck 1 $(j = 1, \ldots, d)$; see Example 11.2.2(b). We define $C_2(j)$ $(j = 1, \ldots, d)$ and $H_2(i)$ $(i = 1, \ldots, c)$ analogously. We first fix two permutations $\pi_1, \pi_2 \in S_N$ such that

$$\pi_1(C_1(i)) = H_2(i) \quad \text{and} \quad \pi_2(C_2(j)) = H_1(j) \tag{11.41}$$

for $i = 1, \ldots, c$ and $j = 1, \ldots, d$. That π_1 and π_2 exist (but not uniquely) is a consequence of the assumption that $|C_1(i)| = |H_2(i)| = m_i$ for $i = 1, \ldots, c$ and $|C_2(j)| = |H_1(j)| = n_j$ for $j = 1, \ldots, d$.

Let Π denote the random permutation that determines the order of the cards in deck 1. Of course Π has a discrete uniform distribution, that is, all $N!$ deck orderings are equally likely. We now define a new random permutation $\bar{\Pi}$ by

$$\bar{\Pi} = \pi_1 \circ \Pi^{-1} \circ \pi_2. \tag{11.42}$$

By Theorem 11.2.1, $\bar{\Pi}$ also has a discrete uniform distribution. We let $\bar{\Pi}$ determine the order of the cards in deck 2. The reason for defining $\bar{\Pi}$ in this way is that, if the two decks are dealt simultaneously, every time a card from category i is dealt into hand j of deck 1, a card from category j will be dealt into hand i of deck 2. The result is that $\boldsymbol{X} = \boldsymbol{Y}^\mathsf{T}$. To see this in another way,

$$\begin{aligned}
Y_{ji} &= |\bar{\Pi}(C_2(j)) \cap H_2(i)| \\
&= |\pi_1(\Pi^{-1}(\pi_2(C_2(j)))) \cap H_2(i)| \\
&= |\pi_1(\Pi^{-1}(H_1(j))) \cap \pi_1(C_1(i))| \\
&= |\pi_1(\Pi^{-1}(H_1(j) \cap \Pi(C_1(i))))| \\
&= |H_1(j) \cap \Pi(C_1(i))| \\
&= X_{ij}, \qquad 1 \le i \le c, \quad 1 \le j \le d. \tag{11.43}
\end{aligned}$$

In summary, the random ordering of deck 1 defines a different random ordering of deck 2 such that $\boldsymbol{X} = \boldsymbol{Y}^\mathsf{T}$. And because both orderings have the discrete uniform distribution, the conclusion of the theorem follows.

This is an example of a *coupling*, in which two random variables are shown to be equal in distribution by constructing them on the same sample space in such a way that they are equal with probability 1.   ♠

We have included Theorem 11.2.4 more for its elegance than for its applicability. Nevertheless, here is an example.

Example 11.2.5. *Application of duality theorem to poker.* Consider two poker games. Game 1 is five-card draw and there are four players. We observe the game immediately after the deal. Game 2 is five-card stud and there are five players. We observe the game immediately after the fourth card has been dealt to each remaining player. In game 1 there are six categories: aces, kings, queens, jacks, tens, and all others. In game 2 there are five categories: AKQJT of clubs, AKQJT of diamonds, AKQJT of hearts, AKQJT of spades, and all others. Theorem 11.2.4 applies with $c = 6$, $(m_1, \ldots, m_6) = (4, 4, 4, 4, 4, 32)$, $d = 5$, and $(n_1, \ldots, n_5) = (5, 5, 5, 5, 32)$. We conclude, for example, that the probability that at least one of the four players in game 1 is dealt four-of-a-kind with tens or better coincides with the probability that at least one of the five players in game 2 has a four-card royal flush (i.e., four of the five cards needed for a royal flush). In the notation of Theorem 11.2.4,

$$
P\left(\max_{1 \le i \le 5} \max_{1 \le j \le 4} X_{ij} = 4 \right) = P\left(\max_{1 \le i \le 4} \max_{1 \le j \le 5} Y_{ij} = 4 \right). \tag{11.44}
$$

This is a remarkable result. To convince the reader of this we write out both probabilities, which we denote by P_1 and P_2, using the inclusion-exclusion law (Theorem 1.1.18 on p. 13):

$$
P_1 = \binom{4}{1} \frac{(5)_1 \binom{4}{4} \binom{48}{1}}{\binom{52}{5}} - \binom{4}{2} \frac{(5)_2 \binom{4}{4}^2 \binom{44}{1,1,42}}{\binom{52}{5,5,42}}
$$

$$
+ \binom{4}{3} \frac{(5)_3 \binom{4}{4}^3 \binom{40}{1,1,1,37}}{\binom{52}{5,5,5,37}} - \binom{4}{4} \frac{(5)_4 \binom{4}{4}^4 \binom{36}{1,1,1,1,32}}{\binom{52}{5,5,5,5,32}}, \tag{11.45}
$$

$$
P_2 = \binom{5}{1} \frac{(4)_1 \binom{5}{4} \binom{47}{0}}{\binom{52}{4}} - \binom{5}{2} \frac{(4)_2 \binom{5}{4}^2 \binom{42}{0,0,42}}{\binom{52}{4,4,44}}
$$

$$
+ \binom{5}{3} \frac{(4)_3 \binom{5}{4}^3 \binom{37}{0,0,0,37}}{\binom{52}{4,4,4,40}} - \binom{5}{4} \frac{(4)_4 \binom{5}{4}^4 \binom{32}{0,0,0,0,32}}{\binom{52}{4,4,4,4,36}}. \tag{11.46}
$$

One can check directly that $P_1 = P_2$ (in fact, the two expressions are equal term by term), but the theorem implies the same conclusion without calculations. Incidentally, $P_1 = P_2 \approx 0.000369321575$.

We have made one implicit assumption, which may be unreasonable. We have assumed that in game 2 no player with the potential of a four-card royal has previously folded. ♠

11.3 Card Counting

In this section we prove that the variance of the player's conditional expected profit in a card game, given knowledge of the cards previously dealt, increases with the depletion of the deck. This result has been called the fundamental theorem of card counting. The theorem has application to baccarat, trente et quarante, and twenty-one.

As usual we consider a deck of N distinct cards, labeled $1, 2, \ldots, N$. We also label the positions of the cards in the deck in the usual way. For example, the first card dealt is the card in position 1. Let Π be an S_N-valued random variable with the discrete uniform distribution (i.e., all $N!$ possible values are equally likely). Recall that $\Pi(i) = j$ means that the card in position i is moved to position j by the permutation Π. If the cards are in natural order $(1, 2, \ldots, N)$ initially, their order after Π is applied is $(\Pi^{-1}(1), \ldots, \Pi^{-1}(N))$. We define $X_j := \Pi^{-1}(j)$ for $j = 1, \ldots, N$, so that X_j is the label of the card in position j. By Theorem 11.2.1(a) on p. 400 $X_1, \ldots, X_N$ is an *exchangeable sequence*. This means that $(X_{\pi(1)}, \ldots, X_{\pi(N)})$ has the same distribution as $(X_1, \ldots, X_N)$ for every permutation $\pi \in S_N$.

We begin with an example in which the variance of the player's conditional expected profit can be calculated directly.

Example 11.3.1. *Simplest example of a fixed strategy.* Consider the following simple game. Assume that N is even and that the player is allowed to bet, at even money, that the next card dealt is odd. If the first n cards have been seen ($0 \le n \le N - 1$) and the player bets on the next card, his profit per unit bet is

$$Y_n := 2 \cdot 1_{\{X_{n+1} \text{ is odd}\}} - 1, \tag{11.47}$$

so his conditional expected profit per unit bet is

$$
\begin{aligned}
Z_n &:= \mathrm{E}[Y_n \mid X_1, \ldots, X_n] \\
&= \frac{2}{N-n}\left(\frac{N}{2} - \sum_{i=1}^{n} 1_{\{X_i \text{ is odd}\}} \right) - 1,
\end{aligned}
\tag{11.48}
$$

being twice the proportion of odd cards in the unseen deck less 1. Notice that we can rewrite this as

$$
\begin{aligned}
Z_n &= \frac{n}{N-n} - \frac{2}{N-n} \sum_{i=1}^{n} 1_{\{X_i \text{ is odd}\}} \\
&= \frac{1}{N-n} \sum_{i=1}^{n} (1 - 2 \cdot 1_{\{X_i \text{ is odd}\}}) \\
&= \frac{1}{N-n} \sum_{i=1}^{n} (-1)^{X_i}.
\end{aligned}
\tag{11.49}
$$

The latter formula has practical implications. Suppose the player assigns to each odd card seen the point value -1 and to each even card seen the point value 1. The *running count* is the sum of these point values over all cards seen and is adjusted each time a new card is seen. The *true count* is the running count divided by the number of unseen cards. Equation (11.49) says that the true count provides the player with his exact expected profit per unit bet on the next card. This information can be used to select a suitable bet size.

Note that $\mathrm{E}[Z_n] = \mathrm{E}[Y_n] = 0$ for $0 \le n \le N - 1$. More importantly, using the fact that

$$\mathrm{Cov}\left(1_{\{X_i \text{ is odd}\}}, 1_{\{X_j \text{ is odd}\}}\right)$$

$$= P(X_i \text{ and } X_j \text{ are odd}) - P(X_i \text{ is odd})P(X_j \text{ is odd})$$

$$= \frac{\frac{1}{2}N(\frac{1}{2}N - 1)}{N(N - 1)} - \frac{1}{4}$$

$$= -\frac{1}{4(N - 1)} \tag{11.50}$$

if $i \neq j$, we calculate that

$$\text{Var}(Z_n) = \frac{4}{(N - n)^2} \text{Var}\left(\sum_{i=1}^{n} 1_{\{X_i \text{ is odd}\}} \right)$$

$$= \frac{4}{(N - n)^2} \left(\sum_{i=1}^{n} \text{Var}(1_{\{X_i \text{ is odd}\}}) \right.$$

$$\left. + \sum_{i \neq j} \text{Cov}(1_{\{X_i \text{ is odd}\}}, 1_{\{X_j \text{ is odd}\}}) \right)$$

$$= \frac{4}{(N - n)^2} \left(\frac{n}{4} - \frac{n(n - 1)}{4(N - 1)} \right)$$

$$= \frac{n}{(N - n)(N - 1)}, \tag{11.51}$$

which increases from 0 to 1 as n increases from 0 to $N - 1$. The conclusions that $E[Z_n]$ is constant in n while $\text{Var}(Z_n)$ is increasing in n are typical of games with a fixed strategy, as we will see in Theorem 11.3.4 below.

To illustrate how a central limit theorem could be used in this context, suppose we want to find the probability that the player has an advantage greater than β. Using (11.51) and a normal approximation with a continuity correction (based on Theorem 1.5.14 on p. 48), we find that, if n is even, then $\sum_{i=1}^{n}(-1)^{X_i}$ is also even, so

$$P(Z_n > \beta) = P\left(\frac{1}{2} \sum_{i=1}^{n} (-1)^{X_i} > \left\lfloor \frac{N - n}{2} \beta \right\rfloor + \frac{1}{2} \right) \tag{11.52}$$

$$\approx 1 - \Phi\left(\frac{2}{N - n} \left\{ \left\lfloor \frac{N - n}{2} \beta \right\rfloor + \frac{1}{2} \right\} \sqrt{\frac{(N - n)(N - 1)}{n}} \right),$$

where Φ is the standard-normal distribution function.

For example, assuming that one fourth of the cards in a 312-card deck have been seen (i.e., $N = 312$, $n = 78$), the probability that the player has an advantage greater than 1.25 percent (i.e., $\beta = 0.0125$) is approximately $1 - \Phi(0.391603) \approx 0.347676$. Here the exact probability can be calculated as

$$\sum_{k=0}^{37} \frac{\binom{156}{k}\binom{156}{78-k}}{\binom{312}{78}} \approx 0.347522, \tag{11.53}$$

showing that the normal approximation is quite good.   ♠

However, when the player's strategy is variable, depending on the cards seen, the situation becomes a bit more complicated, as shown by the next example.

Example 11.3.2. *Simplest example of a variable strategy.* We continue to assume that N is even, but let us now suppose that the player is allowed to make either of two even-money bets, one that the next card dealt is odd, and the other that the next card dealt is even. An obvious optimal strategy is to bet on odd if the number of odd cards seen is less than or equal to the number of even cards seen (or equivalently, if the true count, defined in Example 11.3.1, is nonnegative), and to bet on even otherwise. If the first n cards have been seen ($0 \le n \le N - 1$) and the player employs this strategy to bet on the next card, his conditional expected profit per unit bet is

$$Z_n := \frac{2}{N-n}\left[\frac{N}{2} - \min\left(\sum_{i=1}^{n} 1_{\{X_i \text{ is odd}\}}, \; n - \sum_{i=1}^{n} 1_{\{X_i \text{ is odd}\}}\right)\right] - 1 \quad (11.54)$$

instead of (11.48), this being twice the proportion of odd cards or of even cards in the unseen deck, whichever is greater, less 1. As before, we can rewrite this as

$$\begin{aligned}
Z_n &= \frac{n}{N-n} - \frac{2}{N-n}\min\left(\sum_{i=1}^{n} 1_{\{X_i \text{ is odd}\}}, \; n - \sum_{i=1}^{n} 1_{\{X_i \text{ is odd}\}}\right) \\
&= \frac{1}{N-n}\left|\sum_{i=1}^{n}(1 - 2 \cdot 1_{\{X_i \text{ is odd}\}})\right| \\
&= \frac{1}{N-n}\left|\sum_{i=1}^{n}(-1)^{X_i}\right|.
\end{aligned} \quad (11.55)$$

The interpretation is as in Example 11.3.1: The player bets on odd if the true count is nonnegative and on even otherwise. In either case the absolute value of the true count provides the player with his exact expected profit per unit bet on the next card.

It follows from (11.54) that

$$\mathrm{E}[Z_n] = \frac{2}{N-n}\left(\frac{N}{2} - \sum_{k=0}^{n}\min(k, n-k)\frac{\binom{N/2}{k}\binom{N/2}{n-k}}{\binom{N}{n}}\right) - 1. \quad (11.56)$$

Although it is not clear how to write this in closed form, it is clear that $\mathrm{E}[Z_n]$ is no longer constant in n. One way to see this is to note that $Z_0 = 0$, $Z_1 = 1/(N - 1)$, and $Z_{N-1} = 1$. These, and only these, three cases have $\mathrm{Var}(Z_n) = 0$. This also shows that $\mathrm{Var}(Z_n)$, which could also be expressed in a way similar to (11.56), is no longer increasing (or even nondecreasing) in

n. Thus, the conclusions that hold for a fixed strategy may fail for a variable strategy.

Note that the analogue of (11.52) is

$$P(Z_n > \beta) \approx 2\left[1 - \Phi\left(\frac{2}{N-n}\left\{\left\lfloor \frac{N-n}{2} \right\rfloor \beta \right\rfloor + \frac{1}{2}\right\}\sqrt{\frac{(N-n)(N-1)}{n}}\right)\right]$$
$$\tag{11.57}$$

for n even. In the special case $N = 312$, $n = 78$, and $\beta = 0.0125$, this becomes $2(1 - \Phi(0.391603)) \approx 0.695352$. The exact probability is 0.695044. ♠

Example 2 is rather special in that it permits betting on opposite sides of the same proposition. Here is a generalization that is perhaps more typical.

Example 11.3.3. *A more general example of a variable strategy.* Fix a positive integer K, assume that N is divisible by $2K$, and define $A := \{1, 3, \ldots, 2K-1\}$ (the odd positive integers less than $2K$). Let B be a subset of $\{0, 1, 2, \ldots, 2K-1\}$ of cardinality $|B| = K$ and with $B \neq A$, and define $L := |A \cap B|$, so that $0 \leq L \leq K - 1$.

Let us now suppose that the player is allowed to make either of two even-money bets, one that the next card dealt is odd (or, equivalently, is congruent mod $2K$ to an element of A) and the other that the next card dealt is congruent mod $2K$ to an element of B. If the first n cards have been seen ($0 \leq n \leq N - 1$) and the player employs an obvious optimal strategy to bet on the next card (namely, bet on odd unless the other bet is more favorable), his conditional expected profit per unit bet is $Z_n := \max(Z_n^A, Z_n^B)$, where

$$Z_n^A := \frac{1}{N-n}\sum_{i=1}^{n}(1 - 2 \cdot 1_{\{X_i(\bmod\ 2K)\in A\}}) \tag{11.58}$$

and

$$Z_n^B := \frac{1}{N-n}\sum_{i=1}^{n}(1 - 2 \cdot 1_{\{X_i(\bmod\ 2K)\in B\}}). \tag{11.59}$$

Observe that

$$\mathrm{Cov}(1_{\{X_i(\bmod\ 2K)\in A\}}, 1_{\{X_i(\bmod\ 2K)\in B\}}) = \frac{L}{2K} - \frac{1}{4} = \frac{2(L/K) - 1}{4}, \tag{11.60}$$

while if $i \neq j$,

$$\mathrm{Cov}(1_{\{X_i(\bmod\ 2K)\in A\}}, 1_{\{X_j(\bmod\ 2K)\in B\}})$$
$$= \frac{(\frac{1}{2}N)[(L/K)(\frac{1}{2}N - 1) + (1 - L/K)(\frac{1}{2}N)]}{N(N-1)} - \frac{1}{4}$$
$$= -\frac{2(L/K) - 1}{4(N-1)}. \tag{11.61}$$

It follows that

$$\text{Cov}(Z_n^A, Z_n^B) = \frac{4}{(N-n)^2}\left\{\sum_{i=1}^{n}\text{Cov}(1_{\{X_i(\text{mod } 2K)\in A\}}, 1_{\{X_i(\text{mod } 2K)\in B\}})\right.$$

$$\left. + \sum_{i\neq j}\text{Cov}(1_{\{X_i(\text{mod } 2K)\in A\}}, 1_{\{X_j(\text{mod } 2K)\in B\}})\right\}$$

$$= \frac{n}{(N-n)(N-1)}\left(2\frac{L}{K}-1\right), \tag{11.62}$$

and since Z_n^A and Z_n^B are distributed as Z_n of Example 11.3.1, (11.51) implies that

$$\rho := \text{Corr}(Z_n^A, Z_n^B) = 2\frac{L}{K}-1. \tag{11.63}$$

Example 11.3.2 is the special case $K=1$ and $L=0$, in which case $Z_n^B = -Z_n^A$ and $Z_n = |Z_n^A|$. If $L\neq 0$, the situation becomes significantly more complicated. Here the card counter must keep two counts (or what is called a *two-parameter count*) to track his conditional expected profit per unit bet.

For example, suppose $K=2$ and $L=1$. Then $\rho=0$. Therefore the analogue of (11.52) and (11.57) is[1]

$$\text{P}(Z_n > \beta) \approx 1 - \left[\Phi\left(\frac{2}{N-n}\left\{\left\lfloor\frac{N-n}{2}\beta\right\rfloor+\frac{1}{2}\right\}\sqrt{\frac{(N-n)(N-1)}{n}}\right)\right]^2 \tag{11.64}$$

for n even. In the special case $N=312$, $n=78$, and $\beta=0.0125$, this becomes $1-\Phi(0.391603)^2 \approx 0.574473$. The exact probability can be calculated as

$$\sum_{i,j,k\geq 0:\ i+j+k\leq 78,\ i+\min(j,k)\leq 37}\frac{\binom{78}{i}\binom{78}{j}\binom{78}{k}\binom{78}{78-i-j-k}}{\binom{312}{78}} \approx 0.574307, \tag{11.65}$$

which again shows that the normal approximation is more than adequate. ♠

Now that we have a fair understanding of what happens in the simplest of games, we turn to the general situation. Consider a game that requires up to m cards to complete a coup. (The deck is reshuffled if fewer than m cards remain.) Let us assume also that the player employs a fixed strategy, one that does not vary with the cards seen. We let $X_1,\ldots,X_N$ be as defined at the beginning of this section. If the first n cards have been seen ($0\leq n\leq N-m$) and the player bets on the next coup, his profit per unit bet has the form

$$Y_n := f(X_{n+1},\ldots,X_{n+m}) \tag{11.66}$$

for a suitable nonrandom function f, thereby generalizing (11.47), so his conditional expected profit per unit bet is

---

[1] Because Z_n^A and Z_n^B are uncorrelated and can be shown to be asymptotically bivariate normal (a concept we have not discussed), they are asymptotically independent.

$$Z_n := \mathrm{E}[Y_n \mid X_1, \ldots, X_n] = \mathrm{E}[f(X_{n+1}, \ldots, X_{n+m}) \mid X_1, \ldots, X_n]. \quad (11.67)$$

We can now state the *fundamental theorem of card counting*.

Theorem 11.3.4. *Under the assumptions of the preceding paragraph, the sequence* $\{Z_n\}_{n=0,\ldots,N-m}$ *is a martingale with respect to* $\{X_n\}_{n=1,\ldots,N-m}$. *In particular,*

$$\mathrm{E}[Z_0] = \cdots = \mathrm{E}[Z_{N-m}] \quad (11.68)$$

and

$$0 = \mathrm{Var}(Z_0) \leq \cdots \leq \mathrm{Var}(Z_{N-m}). \quad (11.69)$$

Let I *be an interval such that* $\mathrm{P}(Z_0 \in I, \ldots, Z_{N-m} \in I) = 1$. *If* $\varphi : I \mapsto \mathbf{R}$ *is convex, then* $\{\varphi(Z_n), \ n = 0, \ldots, N - m\}$ *is a submartingale with respect to* $\{X_n\}_{n=1,\ldots,N-m}$. *In particular,*

$$\mathrm{E}[\varphi(Z_0)] \leq \cdots \leq \mathrm{E}[\varphi(Z_{N-m})]. \quad (11.70)$$

Finally, for each inequality in (11.69), the inequality is strict unless both sides of that inequality are 0. If φ *is strictly convex, then, for each inequality in (11.70), the inequality is strict unless both sides of the corresponding inequality in (11.69) are 0.*

Remark. (a) In Example 11.3.1 conclusions (11.68) and (11.69) were verified directly. In more complicated examples this might not be possible, but in any case it is unnecessary, thanks to the theorem.

(b) It should be noted that the martingale $\{Z_n\}_{n=0,\ldots,N-m}$ differs from the usual stochastic model of a fair game. For example, the player who bets $B_{n+1} := 1_{\{Z_n>0\}}$ at coup $n + 1$ can enjoy a considerable advantage over the house.

Proof. First, the martingale property is a consequence of

$$Z_n = \mathrm{E}[Y_{N-m} \mid X_1, \ldots, X_n], \qquad n = 0, \ldots, N - m, \quad (11.71)$$

which holds by virtue of the exchangeability of $X_1, \ldots, X_N$. From this follow (11.68), the submartingale property of $\{\varphi(Z_n)\}_{n=0,\ldots,N-m}$ (Problem 3.1 on p. 111), and (11.70). Taking $\varphi(u) := u^2$ in (11.70) and using (11.68), we deduce (11.69).

Now let us assume that φ is strictly convex. Fix $n \in \{0, \ldots, N - m - 1\}$ and suppose that $\mathrm{E}[\varphi(Z_n)] = \mathrm{E}[\varphi(Z_{n+1})]$. Then

$$\mathrm{E}[\varphi(\mathrm{E}[Z_{n+1} \mid X_1, \ldots, X_n])] = \mathrm{E}[\mathrm{E}[\varphi(Z_{n+1}) \mid X_1, \ldots, X_n]], \quad (11.72)$$

so by the condition for equality in the conditional form of Jensen's inequality (Theorem 2.2.3 on p. 81), the conditional distribution of Z_{n+1} given $X_1, \ldots, X_n$ is degenerate. By the definition of conditional expectation, there exists a nonrandom function h_{n+1} such that $Z_{n+1} = h_{n+1}(X_1, \ldots, X_{n+1})$.

Further, h_{n+1} is a symmetric function of its variables. Since the conditional distribution of $h_{n+1}(X_1, \ldots, X_{n+1})$ given $X_1, \ldots, X_n$ is degenerate, the symmetry of h_{n+1} implies that Z_{n+1} is constant and hence its variance is 0. The stated conclusions follow.   ♠

It must be kept in mind that the main purpose of card counting is to identify favorable situations and vary bet size accordingly. Therefore, a statistic simpler than Z_n may suffice for this purpose. Let us define

$$e(j) := \mathrm{E}[Y_1 \mid X_1 = j] - \mathrm{E}[Y_0] \qquad (11.73)$$

for $j = 1, \ldots, N$. These numbers are the so-called *effects of removal*. They arise from the following hypothesis. Let D be the set of cards $\{1, 2, \ldots, N\}$, and assume that to each card $j \in D$ there is associated a number $c(j)$ such that, with U denoting the set of unseen cards, the player's expected profit E_U per unit bet on the next coup is given by

$$E_U = \frac{1}{|U|} \sum_{j \in U} c(j). \qquad (11.74)$$

Letting $\mu := E_D$ be the full-deck expectation, we find that

$$\begin{aligned}
c(j) &= \sum_{i \in D} c(i) - \sum_{i \in D - \{j\}} c(i) \\
&= N\mu - (N-1)E_{D-\{j\}} \\
&= \mu - (N-1)(E_{D-\{j\}} - \mu) \\
&= \mu - (N-1)e(j), \qquad j = 1, \ldots, N. \qquad (11.75)
\end{aligned}$$

Since $\sum_{j=1}^{N} e(j) = 0$, it follows that, when $X_1, \ldots, X_n$ have been seen, the player's conditional expected profit per unit bet on the next coup is

$$\begin{aligned}
\widehat{Z}_n &:= \frac{1}{N-n} \sum_{j=n+1}^{N} \{\mu - (N-1)e(X_j)\} \\
&= \mu + \frac{1}{N-n} \sum_{j=1}^{n} (N-1)e(X_j). \qquad (11.76)
\end{aligned}$$

Notice that this generalizes (11.49), and the interpretation is similar.

However, we emphasize that the derivation of (11.76) is based on hypothesis (11.74), which is really only an approximation. (Example 11.3.1 is unusual in that the approximation is exact in that case.) Our next theorem provides a statistical justification for this approximation.

Theorem 11.3.5. *The quantities $\mu - (N-1)e(j)$ in (11.75) are the least squares estimators of the parameters $c(j)$ in the linear model*

$$E_U = \frac{1}{|U|} \sum_{j \in U} c(j) + \varepsilon_U, \qquad U \subset D, \ |U| = N - n, \tag{11.77}$$

for fixed n, $1 \leq n \leq N - m$.

Remark. The approximation (11.74) can be made exact by including error terms ε_U as in (11.77). The theorem says that the choice of $(c(1), \ldots, c(N))$ that provides the best fit, in the sense of minimizing the sum of squares

$$\sum_{U \subset D: \ |U| = N-n} \left(E_U - \frac{1}{|U|} \sum_{i \in U} c(i) \right)^2, \tag{11.78}$$

is $c(j) := \mu - (N - 1)e(j)$ for $j = 1, \ldots, N$.

Proof. It will be convenient to use the standard notation of linear statistical models, namely $\boldsymbol{Y} = \boldsymbol{X\beta} + \boldsymbol{\epsilon}$. We number the subsets $U \subset D$ with $|U| = N - n$, and if U is the ith one, we let $Y_i = E_U$, $\epsilon_i = \varepsilon_U$, and $X_{ij} = 1_U(j)$ and $\beta_j = c(j)/|U|$ for $j = 1, \ldots, N$.

We want to minimize

$$\sum_i \left(Y_i - \sum_{j=1}^N X_{ij} \beta_j \right)^2 \tag{11.79}$$

over all $\boldsymbol{\beta}$. Setting the partial derivative with respect to β_k equal to 0 for each k, we get the so-called *normal equations*

$$\sum_i \left(Y_i - \sum_{j=1}^N X_{ij} \beta_j \right) X_{ik} = 0 \tag{11.80}$$

or

$$\sum_i Y_i X_{ik} = \left(\sum_i X_{ik}^2 \right) \beta_k + \sum_{j: j \neq k} \left(\sum_i X_{ij} X_{ik} \right) \beta_j \tag{11.81}$$

for $k = 1, \ldots, N$. Now

$$\sum_i X_{ik}^2 = \binom{N-1}{N-n-1}, \quad \sum_i X_{ij} X_{ik} = \binom{N-2}{N-n-2}, \quad j \neq k, \tag{11.82}$$

and

$$\mu = \frac{\sum_i Y_i X_{ik} + \binom{N-1}{N-n}(\mu + e(k))}{\binom{N}{N-n}}. \tag{11.83}$$

The latter equation follows from the fact that if we average E_U over all subsets $U \subset D$ with $|U| = N - n$, considering separately those U for which $k \in U$ (the first term in the numerator) and those U for which $k \notin U$ (the second term in the numerator), we get μ. We also require here that $N - n \geq m$, so that the next coup can be completed. Now (11.83) implies that

$$\sum_i Y_i X_{ik} = \binom{N}{N-n}\mu - \binom{N-1}{N-n}(\mu + e(k))$$

$$= \binom{N-1}{N-n-1}\mu - \binom{N-1}{N-n}e(k), \qquad (11.84)$$

which when substituted together with (11.82) back into (11.81) yields

$$\binom{N-1}{N-n-1}\mu - \binom{N-1}{N-n}e(k) = \binom{N-1}{N-n-1}\beta_k + \binom{N-2}{N-n-2}\sum_{j:j\neq k}\beta_j.$$
$$(11.85)$$

Letting $\beta := \beta_1 + \cdots + \beta_N$, we can simplify this algebraically to

$$\mu - \frac{n}{N-n}e(k) = \beta_k + \frac{N-n-1}{N-1}(\beta - \beta_k). \qquad (11.86)$$

Summing this equation over $k = 1, \ldots, N$, we find that

$$N\mu = \beta + \frac{N-n-1}{N-1}(N\beta - \beta) = (N-n)\beta, \qquad (11.87)$$

and therefore that $\beta = N\mu/(N-n)$. Substituting back into (11.86) and solving for β_k, we conclude that

$$\beta_k = \frac{\mu - (N-1)e(k)}{N-n}, \qquad (11.88)$$

as required. ♠

It may be helpful to consider an example in which we can evaluate both Z_n and its approximation $\widehat{Z}_n$.

Example 11.3.6. *A game with player expectation quadratic in the running count.* Assume that N is even and that the player is allowed to bet, with payoff odds of 3 to 1, that both of the next two cards dealt are odd. If the first n cards have been seen ($0 \leq n \leq N-2$) and the player bets on the next two cards, his profit per unit bet is

$$Y_n := 4 \cdot 1_{\{X_{n+1} \text{ and } X_{n+2} \text{ are odd}\}} - 1, \qquad (11.89)$$

so his conditional expected profit per unit bet is

$$Z_n := \mathrm{E}[Y_n \mid X_1, \ldots, X_n] = \frac{4}{(N-n)_2}\left(\frac{N}{2} - \sum_{i=1}^n 1_{\{X_i \text{ is odd}\}}\right)_2 - 1, \quad (11.90)$$

where $(k)_2 := k(k-1)$. We know that $\mathrm{E}[Z_n]$ is constant in n and equal to $\mu = \mathrm{E}[Z_0] = -1/(N-1)$. We can check from (11.90) that

$$Z_1 = \begin{cases} -3/(N-1) & \text{if } X_1 \text{ is odd,} \\ 1/(N-1) & \text{if } X_1 \text{ is even,} \end{cases} \tag{11.91}$$

and it follows that $(N-1)e(j) = 2(-1)^j$ for $j = 1, \ldots, N$. We conclude that

$$\widehat{Z}_n = -\frac{1}{N-1} + \frac{2}{N-n} \sum_{j=1}^{n} (-1)^{X_j}. \tag{11.92}$$

The approximation is quite good when $N - n$ is large. ♠

The central limit theorem for samples from a finite population (Theorem A.2.13 on p. 750) applies to (11.76). Letting $\sigma_e^2 = \mathrm{Var}(e(X_1))$, we have

$$\frac{\widehat{Z}_n - \mu}{\sigma_e \sqrt{n(N-1)/(N-n)}} \xrightarrow{d} N(0,1). \tag{11.93}$$

Of course, in practice, the quantities $E_i := (N-1)e(i)$ of (11.76) are replaced by integers J_i that are highly correlated with the given numbers and for which $J_1 + J_2 + \cdots + J_N = 0$. Such a card-counting system is said to be *balanced*. Writing (11.76) as

$$\widehat{Z}_n = \mu + \frac{1}{N-n} \sum_{j=1}^{n} E_{X_j}, \tag{11.94}$$

we can approximate $\widehat{Z}_n$ by

$$Z_n^* := \mu + \gamma \left(\frac{1}{N-n} \sum_{j=1}^{n} J_{X_j} \right), \tag{11.95}$$

where the coefficient γ is given by

$$\gamma = \sum_{i=1}^{N} E_i J_i \bigg/ \sum_{i=1}^{N} J_i^2 = \rho \frac{\sigma_E}{\sigma_J}; \tag{11.96}$$

here ρ is the correlation between E_{X_1} and J_{X_1}, and σ_E and σ_J are the standard deviations of E_{X_1} and J_{X_1}. The parameter γ is called a *regression coefficient*, and it can be found by the method of least squares, specifically by choosing γ to minimize the sum of squares

$$\sum_{i=1}^{N} (E_i - \gamma J_i)^2. \tag{11.97}$$

As in Example 11.3.1, the quantity within parentheses in (11.95) is called the *true count*. It must be adjusted by the constants γ and μ to estimate the player's advantage. A *level-k card-counting system* uses integers whose maximum absolute value is k. Level-one systems are the most popular.

The central limit theorem for samples from a finite population (Theorem A.2.13 on p. 750) applies to (11.95) as well. Indeed, we have

$$\frac{Z_n^* - \mu}{\gamma \sigma_J} \sqrt{\frac{(N-n)(N-1)}{n}} \xrightarrow{d} N(0,1). \tag{11.98}$$

Since J_{X_1} is integer valued, we can improve any normal approximation with a continuity correction.

This allows us, for example, to approximate the probability that a superfair bet is available after n cards have been seen. Even more useful would be to approximate the expected profit available per unit bet after n cards have been seen. Assuming that the player bets one unit if the bet is superfair and nothing if it is subfair, we need to approximate $E[(Z_n^*)^+]$. Noting that Z_n^* is asymptotically normal with mean μ and variance

$$\sigma_n^2 := \frac{\gamma^2 \sigma_J^2 n}{(N-n)(N-1)}, \tag{11.99}$$

and denoting the left side of (11.98) by ζ_n, we obtain the approximation

$$\begin{aligned}
E[(Z_n^*)^+] &= E[(\mu + \sigma_n \zeta_n)^+] \\
&\approx \frac{1}{\sqrt{2\pi}} \int_{-\infty}^{\infty} (\mu + \sigma_n \zeta)^+ e^{-\zeta^2/2}\, d\zeta \\
&= \frac{1}{\sqrt{2\pi}} \int_{-\mu/\sigma_n}^{\infty} (\mu + \sigma_n \zeta)\, e^{-\zeta^2/2}\, d\zeta \\
&= \mu(1 - \Phi(-\mu/\sigma_n)) + \sigma_n \phi(-\mu/\sigma_n) \\
&= \mu\Phi(\mu/\sigma_n) + \sigma_n \phi(\mu/\sigma_n), \tag{11.100}
\end{aligned}$$

where

$$\phi(z) := \Phi'(z) = (2\pi)^{-1/2} e^{-z^2/2} \tag{11.101}$$

is the standard-normal density function. It must be acknowledged that the above approximation is not fully justified, but it will prove useful nonetheless. (Notice that the right side of (11.100) is always positive, even if μ is negative, by virtue of Problem 1.54 on p. 63.)

We turn finally to the general case of a variable strategy. Here we assume that the player has a number of strategies available and therefore a number of choices of f in (11.66). With m as defined before Theorem 11.3.4, let us denote by S the set of such functions f. We assume that S is finite. If the first n cards have been observed ($0 \le n \le N - m$) and the player bets on the next coup, we assume that he chooses the optimal $f \in S$. It will depend on $X_1, \ldots, X_n$, so we denote it by $f_{X_1, \ldots, X_n}$. Here optimality means that

$$\mathrm{E}[f_{X_1,\ldots,X_n}(X_{n+1},\ldots,X_{n+m}) \mid X_1,\ldots,X_n]$$
$$\geq \mathrm{E}[g_{X_1,\ldots,X_n}(X_{n+1},\ldots,X_{n+m}) \mid X_1,\ldots,X_n] \qquad (11.102)$$

for every possible choice $g_{X_1,\ldots,X_n} \in S$. The player's profit per unit bet is

$$Y_n := f_{X_1,\ldots,X_n}(X_{n+1},\ldots,X_{n+m}), \qquad (11.103)$$

so his conditional expected profit per unit bet is

$$Z_n := \mathrm{E}[Y_n \mid X_1,\ldots,X_n] = \mathrm{E}[f_{X_1,\ldots,X_n}(X_{n+1},\ldots,X_{n+m}) \mid X_1,\ldots,X_n].$$
$$(11.104)$$

Here is the weaker form of the *fundamental theorem* in the case of a variable strategy.

Theorem 11.3.7. *Under the above assumptions, $\{Z_n\}_{n=0,\ldots,N-m}$ is a submartingale with respect to $\{X_n\}_{n=1,\ldots,N-m}$. In particular,*

$$\mathrm{E}[Z_0] \leq \cdots \leq \mathrm{E}[Z_{N-m}]. \qquad (11.105)$$

More generally, let I be an interval such that $\mathrm{P}(Z_0 \in I, \ldots, Z_{N-m} \in I) = 1$. If $\varphi : I \mapsto \mathbf{R}$ is convex and nondecreasing, then $\{\varphi(Z_n)\}_{n=0,\ldots,N-m}$ is a submartingale with respect to $\{X_n\}_{n=1,\ldots,N-m}$. In particular,

$$\mathrm{E}[\varphi(Z_0)] \leq \cdots \leq \mathrm{E}[\varphi(Z_{N-m})]. \qquad (11.106)$$

Finally, if φ is strictly convex and increasing, then, for $n = 0, \ldots N-m-1$, $\mathrm{E}[\varphi(Z_n)] = \mathrm{E}[\varphi(Z_{n+1})]$ if and only if Z_n and Z_{n+1} are constant and equal.

Remark. The theorem tells us, for example, that $\mathrm{E}[Z_n^2]$ is increasing in n in Example 11.3.2. (Take $I = [0,1]$ and $\varphi(u) := u^2$.) Recall from Example 11.3.2 that (11.68) and (11.69) may fail here.

Proof. The submartingale property of $\{Z_n\}_{n=0,\ldots,N-m}$ follows by noting that, for $n = 0, \ldots, N-m-1$,

$$\mathrm{E}[Z_{n+1} \mid X_1,\ldots,X_n]$$
$$= \mathrm{E}[\mathrm{E}[f_{X_1,\ldots,X_{n+1}}(X_{n+2},\ldots,X_{n+m+1}) \mid X_1,\ldots,X_{n+1}] \mid X_1,\ldots,X_n]$$
$$\geq \mathrm{E}[\mathrm{E}[f_{X_1,\ldots,X_n}(X_{n+2},\ldots,X_{n+m+1}) \mid X_1,\ldots,X_{n+1}] \mid X_1,\ldots,X_n]$$
$$= \mathrm{E}[f_{X_1,\ldots,X_n}(X_{n+2},\ldots,X_{n+m+1}) \mid X_1,\ldots,X_n]$$
$$= \mathrm{E}[f_{X_1,\ldots,X_n}(X_{n+1},\ldots,X_{n+m}) \mid X_1,\ldots,X_n]$$
$$= Z_n, \qquad (11.107)$$

where the inequality uses (11.102) and the fact that $f_{X_1,\ldots,X_n}$ is of the form $g_{X_1,\ldots,X_{n+1}}$, and the next-to-last equality uses the exchangeability of $X_1,\ldots,X_N$. From this follow (11.105), the submartingale property of $\{\varphi(Z_n)\}_{n=0,\ldots,N-m}$ (Problem 3.1 on p. 111), and (11.106).

Assume next that φ is strictly convex and increasing. Fix $n \in \{0, \ldots, N - m - 1\}$ and suppose that $\mathrm{E}[\varphi(Z_n)] = \mathrm{E}[\varphi(Z_{n+1})]$. Then, by (11.107) and the conditional form of Jensen's inequality (Theorem 2.2.3 on p. 81),

$$\mathrm{E}[\varphi(Z_n)] \leq \mathrm{E}[\varphi(\mathrm{E}[Z_{n+1} \mid X_1, \ldots, X_n])]$$
$$\leq \mathrm{E}[\mathrm{E}[\varphi(Z_{n+1}) \mid X_1, \ldots, X_n]] = \mathrm{E}[\varphi(Z_{n+1})], \quad (11.108)$$

hence the inequalities are equalities. The argument in the proof of Theorem 11.3.4 applies to the second equality in (11.108), resulting in $\mathrm{Var}(Z_{n+1}) = 0$. Moreover, the first equality in (11.108), together with the increasing property of φ and (11.107), tells us that $\mathrm{Var}(Z_n) = 0$. The stated conclusion follows. ♠

11.4 Problems

11.1. *Period of a perfect riffle shuffle.* Consider a deck of N distinct cards, where N is even. Show that the number of successive perfect riffle shuffles (see Example 11.1.1 on p. 392) needed to return the deck to its original order is $\min\{n \geq 1 : 2^n \equiv 1 \pmod{N-1}\}$. (Theorem A.1.3 on p. 746 gives an upper bound on this number.) For example, $\min\{n \geq 1 : 2^n \equiv 1 \pmod{51}\} = 8$, so eight successive perfect riffle shuffles are needed to return the 52-card deck to its original order.

Hint: Label cards and positions by $0, 1, \ldots, N-1$ rather than by $1, 2, \ldots, N$. To what position does the card in position i move under a single perfect riffle shuffle? ... under n perfect riffle shuffles?

11.2. *A multinomial 3-shuffle of a deck of four cards.* Create the analogue of Table 11.1 on p. 393 for a multinomial 3-shuffle of a deck of size $N = 4$, initially in natural order. Confirm the conclusion of Lemma 11.1.3 on p. 395 in this case.

11.3. *Equivalence of interlacing assumptions.* In the definition of a multinomial a-shuffle of an N-card deck there are two ways to state the interlacing assumption, assuming packet sizes $n_1, \ldots, n_a$: (a) The $\binom{N}{n_1, \ldots, n_a}$ possible interlacings of the a packets are equally likely. (b) Cards drop one-at-a-time from packet 1 or packet 2 or ... or packet a with probabilities proportional to the sizes of the packets. Show that these two assumptions are equivalent.

11.4. *Alternative way to perform a multinomial a-shuffle.* Show that a multinomial a-shuffle of an N-card deck can be performed as follows. First, break the deck into a packets of sizes $n_1 \geq 0, \ldots, n_a \geq 0$, where $n_1 + \cdots + n_a = N$, with the vector of packet sizes $(n_1, \ldots, n_a)$ having the multinomial$(N, a^{-1}, \ldots, a^{-1})$ distribution. Then riffle shuffle packets 1 and 2, with all $\binom{n_1+n_2}{n_1}$ interlacings assumed equally likely. Next, riffle shuffle the newly created packet with packet 3, with all $\binom{n_1+n_2+n_3}{n_1+n_2}$ interlacings assumed equally likely, and so on.

This shows, for example, that one does not need three hands to perform a multinomial 3-shuffle.

11.5. *Probability that the top card remains on top.* Find the probability that the card in position 1 remains in position 1 following a binomial riffle shuffle or, more generally, a multinomial a-shuffle.

11.6. *Distance to uniformity of a deck with one card seen.* Suppose that a deck of N distinct cards is well shuffled but that the card on the bottom is seen to be card i. Find the distance to uniformity of this compromised deck.

11.7. *Distance to uniformity for a single binomial riffle shuffle.* Derive formulas for the Eulerian numbers $\left\langle {N \atop 1} \right\rangle$ and $\left\langle {N \atop 2} \right\rangle$ using the recursion (11.15) on p. 396, and find the exact distance to uniformity for a single binomial riffle shuffle of a deck of N distinct cards. Evaluate this distance numerically for $N = 52$, more accurately than in Table 11.3 on p. 399.

11.8. *Color and inverse at trente et quarante.* Recall Example 11.2.2(d) on p. 401 concerning trente et quarante. Let C be the event that the color of the row with the lower total coincides with the color of the first card dealt to Black, and let I be the event that these two colors are different. (Note that $(C \cup I)^c$ is the event that the two rows have the same total.) Show that $P(C) = P(I)$, assuming a well-shuffled six-deck shoe.

11.9. *A geometric-like distribution for sampling without replacement.* Suppose that a well-shuffled deck of N cards contains m cards of type 1.

(a) Find the distribution of the number of cards that must be dealt to obtain the first card of type 1.

(b) Show that the mean of this distribution is $(N + 1)/(m + 1)$.

11.10. *Exchangeability of distances between cards of a specified type.* Suppose that a well-shuffled deck of N cards contains m cards of type 1. Let X_1 be the number of cards that must be dealt to obtain the first card of type 1, let X_2 be the number of *additional* cards that must be dealt to obtain the second card of type 1, and so on. This defines $X_1, X_2, \ldots, X_m$. Finally, let X_{m+1} be 1 plus the number of cards that remain in the deck immediately after the mth card of type 1 has been dealt. Show that the random variables $X_1, X_2, \ldots, X_m, X_{m+1}$ are exchangeable. Use this fact to rederive the mean in Problem 11.9.

11.11. *Application of duality theorem for card dealing.* Call the first 13 cards of a 52-card deck the first quarter of the deck, the second 13 cards the second quarter, and so on. Show that the probability that each quarter of the deck contains an ace coincides with the probability that the first four cards of a poker hand consist of four different suits. Do this first without calculation, using Theorem 11.2.4 on p. 406, and second by direct calculation.

11.12. *Probability of an adjacent ace and jack.* Find the probability that, in a well-shuffled standard 52-card deck, there exist at least one ace and at least one jack that are adjacent (in either order).

11.13. *Betting on odd, once and only once.* Recall the game of Example 11.3.1 on p. 409, and suppose that the player is required to bet one unit, once and only once, that the next card dealt will be odd. The bet pays even money. In other words, before each card is dealt, the player must decide whether to bet on that card (if he has not yet bet), based on the cards dealt up to that point. Is there a strategy that gives the player an advantage? (If the cards are red or black instead of odd or even, this game is called "say red.")

What if the player is allowed to choose his bet size?

11.14. *Moments of the true count in odd-or-even game.* Evaluate the mean, variance, and second moment of Z_n ($0 \leq n \leq N - 1$) in Example 11.3.2 on p. 411 for $N = 52$. Confirm that $\mathrm{E}[Z_n]$ is nondecreasing in n and $\mathrm{E}[Z_n^2]$ is increasing in n.

11.15. *Expected full-deck profit in odd-or-even game.* Under the assumptions and notation of Example 11.3.2 on p. 411, show that

$$\mathrm{E}[Z_0 + Z_1 + \cdots + Z_{N-1}] = \sum_{n=1}^{N/2} \frac{\binom{N/2}{n}\binom{N/2}{n}}{\binom{N}{2n}} = \frac{2^N}{\binom{N}{N/2}} - 1. \qquad (11.109)$$

Hint: Let O_n and E_n be the numbers of odd and even cards among the first n, and consider $S_n := O_n - E_n$, $n = 0, 1, \ldots, N - 1$. If $S_n = 0$ (and hence $|S_{n+1}| = |S_n| + 1$), then the player will win or lose at coup $n + 1$ with probability $\frac{1}{2}$ each. If $S_n \neq 0$ and $|S_{n+1}| = |S_n| + 1$, then the player will lose at coup $n + 1$. If $|S_{n+1}| = |S_n| - 1$, then the player will win at coup $n + 1$.

11.16. *A game for which effects of removal are zero.* Consider a deck of N distinct cards, which are labeled $1, 2, \ldots, N$. Assume that N is even and that the player is allowed to bet, at even money, that the next two cards have the same parity (i.e., are both odd or both even). If the first n cards have been seen ($0 \leq n \leq N - 2$) and the player bets on the next two cards, find a simple expression for Z_n, his conditional expected profit per unit bet. Evaluate $\mathrm{E}[Z_n]$. Show that the effects of removal are 0. Show that nevertheless there is a viable card-counting system.

11.17. *Casino War calculations.*

(a) Find the player's expected profit, assuming a one-unit initial bet, at Casino War (Example 11.2.3 on p. 403), assuming that he forfeits one half of his stake in the case of a tie, instead of going to war. Which strategy is superior?

(b) Find the house advantage at Casino War (with respect to total bet, pushes included and excluded), under the strategy of Example 11.2.3 on p. 403.

(c) The official rules of the game differ slightly from those of Example 11.2.3 on p. 403. In the case of a tied war, the player is paid even money on his (doubled) bet, not just 1 to 2. How does this affect the house advantage?

11.18. *Card counting at Casino War.*

(a) Show that the effects of removal for the basic bet in six-deck Casino War (Example 11.2.3 on p. 403) are all equal to 0. Are there any deck compositions for which the basic bet has positive expectation?

(b) Analyze the tie bet, which pays 10 to 1 for an initial tie, from the card-counter's perspective.

11.5 Notes

The result that seven riffle shuffles are necessary and sufficient to adequately randomize a 52-card deck is due to Dave Bayer [1955–] and Persi Diaconis [1945–] (1992). Specifically, they proved Theorem 11.1.5 and Lemmas 11.1.2 and 11.1.3. More generally, Bayer and Diaconis showed that, for a deck of N distinct cards, if $m_N := \frac{3}{2} \log_2 N + \theta$, then

$$\|R_{m_N} - U\| = 1 - 2\,\Phi\left(-\frac{2^{-\theta}}{4\sqrt{3}}\right) + O(N^{-1/4}) \qquad (11.110)$$

as $N \to \infty$. Here $O(N^{-1/4})$ is an error term that is bounded in absolute value by $C\,N^{-1/4}$ for all N, where C is an unspecified constant. For example, $m_{52} \approx 7$ if $\theta \approx -1.55066$, and with this θ, we find that $1 - 2\,\Phi(-2^{-\theta}/(4\sqrt{3})) \approx 0.328$, which is not far from the exact value for $m = 7$ in Table 11.3.

Earlier work by David Aldous [1952–] and Diaconis (1986) had shown that

$$\|R_m - U\| \leq 1 - \prod_{j=1}^{N-1}\left(1 - \frac{j}{2^m}\right), \qquad (11.111)$$

suggesting that 11 or 12 riffle shuffles suffice when $N = 52$. This bound is often referred to as the *birthday bound* because it is the probability that, among N individuals, there are at least two with the same birthday, assuming only that birthdays are assigned independently and at random from a set of 2^m possible days (in this context, 2^m is usually replaced by 365).

The present treatment benefitted from the expository articles by Mann (1995) and Lawler and Coyle (1999, Lectures 5 and 6). The shuffling model we call a binomial riffle shuffle is due to Edgar N. Gilbert (1955), Claude E. Shannon [1916–2001], and Jim Reeds (unpublished, 1981) and is also known as the GSR model. There are several proofs available for Lemma 11.1.2; see Bayer and Diaconis (1992). Our proof is from Mann (1995), where it is attributed to John Finn. Lemma 11.1.4 was undoubtedly known to Leonhard Euler [1707–1783]. Rising sequences were first exploited by magician Charles O. Williams (1912), according to Bayer and Diaconis (1992). The *New York Times* quote is from an article by Kolata (1990). For work on the problem of determining how many shuffles are needed when the cards are not distinct (as in twenty-one), see Conger and Viswanath (2006a,b) and Assaf, Diaconis, and Soundararajan (2009a,b). Further developments in the study of riffle shuffles were discussed by Diaconis (2003). Hannum (2000) has done a statistical study of actual shuffles by professional dealers.

Example 11.2.2(d) is due to Thorp and Walden (1973). Casino War is a proprietary game owned by Shuffle Master that debuted in 1996 (Stratton 2004a). Theorem 11.2.4 (duality theorem) and its proof via coupling are due to Jeffrey S. Rosenthal [1967–] (1995); the theorem generalizes an earlier result of Sobel and Frankowski (1994).

The basic ideas of card counting go back to Edward O. Thorp [1932–] (1962b). The fundamental theorem of card counting (Theorem 11.3.4) is due to Thorp and William E. Walden [1929–] (1973). Our formulation and proof as well as Example 11.3.3 and Theorem 11.3.7 are from Ethier and Levin (2005), though there are precedents in the gambling literature, such as M'hall's (1996) "true-count theorem." Example 11.3.1 was suggested by Thorp and Walden (1973), and Example 11.3.2 was studied by Knopfmacher and Prodinger (2001), among others. Theorem 11.3.5 is due to Griffin (1976, 1999). As Griffin (1999, p. 36) put it,

> But here we appeal to the method of least squares not to estimate what is assumed to be linear, but to best approximate what is almost certainly not quite so.

Problem 11.1 (period of a perfect shuffle) was said by Diaconis, Graham, and Kantor (1983) to be "well known," citing a 1939 number theory textbook. What we have called a perfect shuffle is actually one of two types of perfect shuffles. In an "out shuffle" the top card remains on top. In an "in shuffle" the top card moves to second position. Problems 11.3 and 11.4 are from Bayer and Diaconis (1992), Problems 11.5 and 11.7 are from Lawler and Coyle (1999, p. 91), and Problem 11.6 is from Aldous and Diaconis (1986). Problem 11.8 is due to Thorp and Walden (1973). Problem 11.10 is from Ross (2006, p. 309). Problem 11.12 is from Grosjean (2009, p. 608, 611, 612), where it is credited to N. B. Winkless and William Chen. Problem 11.13 concerns the game of "say red" due to Connelly (1974) and generalized by Kallenberg (2005, Chapter 5). Problem 11.15 is due to Jostein Lillestol [1943–] (Székely 2003), though the second equality in (11.109) is a result of Helmut Prodinger (2005, personal communication). The "official rules" of Casino War (Problem 11.17) are those found on Shuffle Master's website, but the rules used in Example 11.2.3 are the more common ones. Card counting for the tie bet at Casino War (Problem 11.18) has been studied by Grosjean (2009, Chapter 69).

Part II
Applications

Chapter 12
Slot Machines

What a fierce excitement of doubt, hope, and pleasure! What tremendous hazards
of loss or gain! What were all the games of chance he had ever played compared to
this one?

William Makepeace Thackeray, *Vanity Fair*

The three-reel slot machine, invented in San Francisco in 1898, underwent
substantial evolution over the course of the 20th century. For example, while
classical slots were purely mechanical or electro-mechanical, modern ones are
electronic and controlled by microprocessors with random-number generators. Slot machines are the only casino games for which evaluating the house
advantage requires information not generally available to the gambling public. In this chapter we consider three slot machines, two classical ones and a
modern one, for which the required information is available. In Section 12.1
we evaluate the expected payout for each machine. In Section 12.2 we address
volatility and ruin issues.

12.1 Expected Payout

We begin by quoting an academic-sounding definition of a *slot machine* that
was published in 1978 (see the chapter notes):

> The slot-machine is essentially a cabinet, housing 3 or more narrow cylindrical
> drums, commonly called reels, which are marked with symbols. Vertically disposed
> on a common axis, the reels are caused to revolve freely, when player activates ma-
> chine and pulls a lever-like handle affixed in the side of the cabinet. Awards, which
> are paid automatically, are based on the horizontal alignment of symbols, when the
> spinning reels come to a position of inertial rest.

Unstated but implicit in the definition was the requirement that the player
insert one or more coins into a slot (hence the name "slot machine") in order
to activate the machine.

S.N. Ethier, *The Doctrine of Chances*, Probability and its Applications,
DOI 10.1007/978-3-540-78783-9_12, © Springer-Verlag Berlin Heidelberg 2010

With only minor changes, the definition also applies to modern slot machines, even though they work rather differently. The main distinction is that almost all slots manufactured since about 1985 are electronic and controlled by a microprocessor. The microprocessor typically contains an *EPROM* (erasable programmable read-only memory) on which is programmed a random-number generator that operates continually.[1] The moment the player pulls the handle or presses the button to activate the machine, the most recently generated random numbers (one for each reel) determine the outcome almost instantly. The machine's reels (if it has physical reels, not just video representations of them) are controlled by what are called *stepper motors*, which ensure that the reels stop in a position consistent with the predetermined result. Despite all this, the underlying mathematics of the modern slot machine is not much different from that of the classical machine, the primary distinction being that the stopping positions of the reels need not be equally likely.

There are a few other inconsequential differences as well. Paper currency is preferred to coins, with the player receiving a certain number of credits for each bill inserted. In fact cashless slots (referred to by the acronym *TITO*— ticket-in/ticket-out) are becoming the norm. Also, most players today prefer to activate the machine by pressing a button rather than by pulling a handle because it causes less fatigue. The result of these two developments is that slot machines frequently no longer have coin slots, and "one-armed bandits" frequently no longer have arms.

To evaluate the house advantage of a machine requires information not generally available to the gambling public. In this section we consider three specific slot machines, two classical ones and a modern one, for which the required information is available. This will allow us to understand the underlying mathematical principles.

We already evaluated the expected payout of the three-reel, 10-stop, Fey "Liberty Bell" mechanical slot machine from 1899 in Example 1.4.1 on p. 28. Our first example in this chapter is, in principle, no more difficult.

We begin with a three-reel, 20-stop, Mills mechanical slot machine from 1949. The pay table is listed in Table 12.1. The major innovation here is the jackpot, in fact four of them. The reel-strip labels and symbol-inventory table are provided in Table 12.3. Notice that one stop on each reel has two symbols. If either symbol results in a positive payout, that is the one that applies. The payout analysis is shown in Table 12.2. There are $(20)^3 = 8,000$ equally likely outcomes. (It is the number of stops, not the number of symbols, that is relevant.)

Letting R denote the number of coins paid out from a one-coin bet, we find that

---

[1] We understand that EPROMs are gradually becoming obsolete because of their limited memory capacities.

Table 12.1 Pay table from a three-reel, 20-stop, Mills mechanical slot machine (1949).

reel 1	reel 2	reel 3	payout
7	7	7	200
bar	bar	bar	150
melon	melon	melon	150
melon	melon	bar	150
bell	bell	bell	18
bell	bell	bar	18
plum	plum	plum	14
plum	plum	bar	14
orange	orange	orange	10
orange	orange	bar	10
cherry	cherry	anything	5
cherry	not cherry	anything	2

Table 12.2 Payout analysis from the Mills slot machine of Table 12.1.

reel 1	reel 2	reel 3	payout	number of ways		product
7	7	7	200	$1 \cdot 1 \cdot 1 =$	1	200
bar	bar	bar	150	$1 \cdot 1 \cdot 1 =$	1	150
melon	melon	melon	150	$1 \cdot 1 \cdot 1 =$	1	150
melon	melon	bar	150	$1 \cdot 1 \cdot 1 =$	1	150
bell	bell	bell	18	$1 \cdot 5 \cdot 8 =$	40	720
bell	bell	bar	18	$1 \cdot 5 \cdot 1 =$	5	90
plum	plum	plum	14	$5 \cdot 3 \cdot 3 =$	45	630
plum	plum	bar	14	$5 \cdot 3 \cdot 1 =$	15	210
orange	orange	orange	10	$5 \cdot 4 \cdot 2 =$	40	400
orange	orange	bar	10	$5 \cdot 4 \cdot 1 =$	20	200
cherry	cherry	anything	5	$3 \cdot 6 \cdot 20 =$	360	1,800
cherry	not cherry	anything	2	$3 \cdot 14 \cdot 20 =$	840	1,680
total					1,369	6,380

$$\mathrm{E}[R] = 200\left(\frac{1}{8,000}\right) + 150\left(\frac{1+1+1}{8,000}\right) + 18\left(\frac{40+5}{8,000}\right) + 14\left(\frac{45+15}{8,000}\right)$$
$$+ 10\left(\frac{40+20}{8,000}\right) + 5\left(\frac{360}{8,000}\right) + 2\left(\frac{840}{8,000}\right)$$

Table 12.3 Reel-strip labels and symbol-inventory table from the Mills slot machine of Table 12.1. Notice that stop 1 of each reel has two symbols. If either symbol results in a positive payout, that is the symbol that applies.

stop no.	reel 1	reel 2	reel 3
1	lemon/7	orange/7	bar/7
2	bell	cherry	lemon
3	orange	bell	bell
4	cherry	plum	orange
5	lemon	bell	bell
6	orange	cherry	lemon
7	lemon	orange	bell
8	plum	bell	lemon
9	orange	melon	plum
10	plum	plum	bell
11	cherry	bell	orange
12	lemon	cherry	lemon
13	bar	bar	bell
14	plum	orange	melon
15	orange	cherry	bell
16	melon	bell	plum
17	plum	orange	lemon
18	cherry	cherry	bell
19	plum	plum	plum
20	orange	cherry	bell

symbol	reel 1	reel 2	reel 3
7	1	1	1
melon	1	1	1
bar	1	1	1
bell	1	5	8
plum	5	3	3
orange	5	4	2
cherry	3	6	0
lemon	4	0	5
total	21	21	21

$$= \frac{6{,}380}{8{,}000} = 0.7975, \tag{12.1}$$

concluding that the house advantage is $H_0 = 1 - E[R] = 0.2025$. The probability of a positive payout is, by Table 12.2,

$$P(R > 0) = \frac{1,369}{8,000} = 0.171125. \tag{12.2}$$

Our next machine is a three-reel, 22-stop, five-payline, Bally electro-mechanical slot machine from 1969. In addition to the electro-mechanical design (in which the payouts are determined by electrical circuits while the spinning of the reels is purely mechanical), an important innovation here is the use of five paylines. The machine permits the player to bet from one to five coins, and the number of coins played determines the number of paylines to which the pay table applies. What do we mean by multiple paylines? Most three-reel slot machines display not only the three symbols on the payline but also the three symbols above the payline and the three symbols below the payline, displaying what is effectively a 3×3 matrix of symbols. The first payline is the traditional one, namely row 2 of the matrix. The second payline is row 1 of the matrix, and the third is row 3. The fourth payline is main diagonal of the matrix (the $(1,1), (2,2), (3,3)$ entries), and the fifth payline is the diagonal perpendicular to the main diagonal (the $(3,1), (2,2), (1,3)$ entries). As an incentive to the player to bet the maximum number of coins, the fifth payline has a substantially higher jackpot. See Table 12.4.

The reel-strip labels and symbol-inventory table are provided in Table 12.6, and the payout analysis is shown in Table 12.5. There are $(22)^3 = 10{,}648$ equally likely outcomes.

For $k = 1, 2, 3, 4, 5$, let R_k denote the number of coins paid out from a one-coin bet on the kth payline. Notice that R_1, R_2, R_3, R_4 are identically distributed because the first four paylines have the same pay table. R_5 has a slightly different distribution because of the higher jackpot for the fifth payline. However, the five random variables are certainly not independent. Of course, the lack of independence causes no problems in evaluating the expected total payout.

Indeed,

$$\begin{aligned}
E[R_1] &= E[R_2] = E[R_3] = E[R_4] \\
&= 200\left(\frac{1}{10{,}648}\right) + 100\left(\frac{9}{10{,}648}\right) + 18\left(\frac{100+10}{10{,}648}\right) + 14\left(\frac{50+10}{10{,}648}\right) \\
&\quad + 10\left(\frac{60+12}{10{,}648}\right) + 5\left(\frac{264}{10{,}648}\right) + 2\left(\frac{1{,}188}{10{,}648}\right) \\
&= \frac{8{,}336}{10{,}648} \approx 0.782870,
\end{aligned} \tag{12.3}$$

while

$$E[R_5] = E[R_1] + (3{,}000 - 200)\frac{1}{10{,}648} = \frac{11{,}136}{10{,}648} \approx 1.045830. \tag{12.4}$$

It is interesting that $E[R_5] > 1$.

Table 12.4 Pay table for a three-reel, 22-stop, five-payline, Bally electro-mechanical slot machine (1969).

reel 1	reel 2	reel 3	payout per payline	
			paylines 1–4	payline 5
7	7	7	200	3,000
bar	bar	bar	100	100
bell	bell	bell	18	18
bell	bell	bar	18	18
plum	plum	plum	14	14
plum	plum	bar	14	14
orange	orange	orange	10	10
orange	orange	bar	10	10
cherry	cherry	anything	5	5
cherry	not cherry	anything	2	2

Table 12.5 Payout analysis for the Bally slot machine of Table 12.4. The payout applies to each of the first four paylines.

reel 1	reel 2	reel 3	payout	number of ways		product
7	7	7	200	$1 \cdot 1 \cdot 1 =$	1	200
bar	bar	bar	100	$3 \cdot 3 \cdot 1 =$	9	900
bell	bell	bell	18	$1 \cdot 10 \cdot 10 =$	100	1,800
bell	bell	bar	18	$1 \cdot 10 \cdot 1 =$	10	180
plum	plum	plum	14	$10 \cdot 1 \cdot 5 =$	50	700
plum	plum	bar	14	$10 \cdot 1 \cdot 1 =$	10	140
orange	orange	orange	10	$4 \cdot 3 \cdot 5 =$	60	600
orange	orange	bar	10	$4 \cdot 3 \cdot 1 =$	12	120
cherry	cherry	anything	5	$3 \cdot 4 \cdot 22 =$	264	1,320
cherry	not cherry	anything	2	$3 \cdot 18 \cdot 22 =$	1,188	2,376
total					1,704	8,336

We denote the payout *per coin bet* for an n-coin bet by

$$\overline{R}_n := \frac{1}{n} \sum_{k=1}^{n} R_k, \qquad n = 1, 2, 3, 4, 5. \tag{12.5}$$

It follows from (12.3) and (12.4) that

Table 12.6 Reel-strip labels and symbol-inventory table from the Bally slot machine of Table 12.4.

stop no.	reel 1	reel 2	reel 3
1	7	7	7
2	bell	plum	plum
3	plum	bell	bell
4	orange	cherry	orange
5	plum	bell	bell
6	cherry	orange	plum
7	plum	bell	bell
8	bar	bar	orange
9	plum	bell	bell
10	orange	cherry	plum
11	plum	bell	bell
12	cherry	orange	bar
13	plum	bell	bell
14	bar	bar	orange
15	plum	bell	bell
16	orange	cherry	plum
17	plum	bell	bell
18	cherry	orange	orange
19	plum	bell	bell
20	bar	bar	plum
21	plum	bell	bell
22	orange	cherry	orange

symbol	reel 1	reel 2	reel 3
7	1	1	1
bar	3	3	1
bell	1	10	10
plum	10	1	5
orange	4	3	5
cherry	3	4	0
total	22	22	22

$$\mathrm{E}[\overline{R}_1] = \mathrm{E}[\overline{R}_2] = \mathrm{E}[\overline{R}_3] = \mathrm{E}[\overline{R}_4] = \frac{8{,}336}{10{,}648} \approx 0.782870, \tag{12.6}$$

while

$$\mathrm{E}[\overline{R}_5] = \frac{8{,}896}{10{,}648} \approx 0.835462. \tag{12.7}$$

In particular, the house advantage for a bet of four or fewer coins is $H_0 = 1 - E[\overline{R}_n] \approx 0.217130$ ($n = 1, 2, 3, 4$). For a five-coin bet it is $H_0 = 1 - E[\overline{R}_5] \approx 0.164538$.

The probability of a positive payout from a one-coin bet is, by Table 12.5,

$$P(R_1 > 0) = \frac{1{,}704}{10{,}648} \approx 0.160030, \tag{12.8}$$

while the probability of a positive payout from a multiple-coin bet requires additional analysis. We will return to this point in Section 12.2.

We finally consider the three-reel, 22-stop, five-payline, Bally "In the Money" electronic slot machine from 2000, which can be regarded as a modern version of the example just discussed. This machine has several innovations. As mentioned earlier, like almost all slots manufactured since about 1985, it is controlled by a microprocessor with an EPROM on which is programmed a random-number generator that determines where the reels stop.

Modern single-payline slot machines (ignoring video slots) typically have three reels with each reel having 22 stops, of which 11 have labels and 11 are blank. With a classical slot machine, this would mean only $(22)^3 = 10{,}648$ equally likely outcomes, limiting the size of possible jackpots. But with a modern machine, the 22 stops are no longer equally likely, and larger jackpots are possible. This is the approach taken by the Bally machine. The pay table is shown in Table 12.7, and the bonus feature, another innovation, is described in Table 12.8. It can happen that more than one payline wins a bonus, and if that happens, the random bonus amount is multiplied by the number of paylines to which it applies.

This machine, like most single-payline machines, has 22 stops on each reel, and the labels are shown in Table 12.9. This information, however, even with Tables 12.7 and 12.8, is insufficient. What we also need are the weights of each of the stops on each reel, or equivalently the probabilities with which each stop appears on the payline. In the Bally example, some stops are weighted up to seven times as much as others, and these weights are indicated in Table 12.9 as well. These multiply weighted stops are counted according to their weights in the symbol-inventory table, with the result that each reel can be regarded as having 144 equally likely *virtual stops*, rather than 22 unequally likely *physical stops*.

The payout analysis for this machine is not unlike those of the previously discussed classical machines. But there is one additional complication. In the pay tables for the older machines, the various categories were designed in such a way that they were mutually exclusive. In the pay table for the newer machine, the 37 categories are not mutually exclusive, and when more than one payout applies to a particular combination of reel labels on a particular payline, it is the higher payout that takes precedence. For example, the outcome 1B-1B-1B satisfies the conditions for two of the payouts, namely 1B-1B-1B and XB-XB-XB, which for a one-coin bet pays 10 coins and 5 coins, respectively. Thus, the machine pays out 10 coins in this case (not

Table 12.7 Pay table for the three-reel, 22-stop, five-payline, Bally "In the Money" electronic slot machine (2000). The payout shown applies to the jth payline. AM is MU or MB. A7 is R7 or G7. XB is 1B, 2B, or 3B.

reel 1	reel 2	reel 3	payout	reel 1	reel 2	reel 3	payout
R7	R7	R7	$1{,}000j$	5X	1B	1B	50
G7	G7	G7	100	1B	5X	1B	50
5X	5X	5X	1,000	1B	1B	5X	50
3B	5X	5X	500	AM	AM	AM	bonus
5X	3B	5X	500	3B	3B	3B	20
5X	5X	3B	500	A7	A7	A7	20
2B	5X	5X	375	2B	2B	2B	15
5X	2B	5X	375	1B	1B	1B	10
5X	5X	2B	375	5X	XB	XB	25
1B	5X	5X	250	XB	5X	XB	25
5X	1B	5X	250	XB	XB	5X	25
5X	5X	1B	250	XB	XB	XB	5
5X	3B	3B	100	CH	CH	CH	10
3B	5X	3B	100	CH	CH	·	5
3B	3B	5X	100	CH	·	CH	5
5X	2B	2B	75	·	CH	CH	5
2B	5X	2B	75	CH	·	·	2
2B	2B	5X	75	·	CH	·	2
				·	·	CH	2

Table 12.8 Bonus-feature payouts for the Bally slot machine of Table 12.7. The probabilities are conditional on AM-AM-AM appearing on the payline.

pay	prob.	prod.	pay	prob.	prod.	pay	prob.	prod.
200	0.0020	0.4000	85	0.0040	0.3400	45	0.0100	0.4500
150	0.0030	0.4500	80	0.0040	0.3200	40	0.0200	0.8000
140	0.0040	0.5600	75	0.0045	0.3375	35	0.0300	1.0500
130	0.0030	0.3900	70	0.0030	0.2100	30	0.0402	1.2060
120	0.0015	0.1800	65	0.0030	0.1950	25	0.0500	1.2500
110	0.0040	0.4400	60	0.0030	0.1800	20	0.1000	2.0000
100	0.0045	0.4500	55	0.0020	0.1100	15	0.1500	2.2500
90	0.0040	0.3600	50	0.0050	0.2500	10	0.3000	3.0000
						5	0.2453	1.2265
					total		1.0000	18.4050

Table 12.9 Reel-strip labels and symbol-inventory table for the Bally slot machine of Table 12.7.

stop no.	reel 1		reel 2		reel 3	
	symbol	weight	symbol	weight	symbol	weight
1	R7	2	R7	2	R7	2
2	—	2	—	2	—	2
3	5X	2	2B	2	1B	4
4	—	2	—	8	—	8
5	CH	2	1B	12	3B	8
6	—	2	MB	12	—	8
7	2B	12	MU	12	G7	8
8	—	14	MB	12	—	8
9	2B	4	2B	12	1B	6
10	—	2	—	10	—	11
11	3B	2	G7	10	2B	12
12	—	12	—	10	MB	12
13	1B	6	1B	6	MU	12
14	—	6	—	6	MB	12
15	G7	6	3B	6	1B	12
16	—	6	—	6	—	3
17	1B	12	2B	6	CH	3
18	MB	12	—	2	—	3
19	MU	12	CH	2	5X	3
20	MB	12	—	2	—	3
21	3B	12	5X	2	2B	2
22	—	2	—	2	—	2
total		144		144		144

symbol	symbol name	reel 1	reel 2	reel 3
R7	red seven	2	2	2
G7	gold seven	6	10	8
5X	five times bar	2	2	3
MU	In the Money feature	12	12	12
MB	In the Money special blank	24	24	24
3B	triple bar	14	6	8
2B	double bar	16	20	14
1B	single bar	18	18	22
CH	cherry	2	2	3
—	blank	48	48	48
total		144	144	144

Table 12.10 Payout analysis for the Bally slot machine of Table 12.7. The payout shown applies to the jth payline ($j = 1, 2, 3, 4, 5$).

r 1	r 2	r 3	payout	number of ways		product
R7	R7	R7	1,000j	$2 \cdot 2 \cdot 2 =$	8	8,000j
G7	G7	G7	100	$6 \cdot 10 \cdot 8 =$	480	48,000
5X	5X	5X	1,000	$2 \cdot 2 \cdot 3 =$	12	12,000
3B	5X	5X	500	$14 \cdot 2 \cdot 3 =$	84	42,000
5X	3B	5X	500	$2 \cdot 6 \cdot 3 =$	36	18,000
5X	5X	3B	500	$2 \cdot 2 \cdot 8 =$	32	16,000
2B	5X	5X	375	$16 \cdot 2 \cdot 3 =$	96	36,000
5X	2B	5X	375	$2 \cdot 20 \cdot 3 =$	120	45,000
5X	5X	2B	375	$2 \cdot 2 \cdot 14 =$	56	21,000
1B	5X	5X	250	$18 \cdot 2 \cdot 3 =$	108	27,000
5X	1B	5X	250	$2 \cdot 18 \cdot 3 =$	108	27,000
5X	5X	1B	250	$2 \cdot 2 \cdot 22 =$	88	22,000
5X	3B	3B	100	$2 \cdot 6 \cdot 8 =$	96	9,600
3B	5X	3B	100	$14 \cdot 2 \cdot 8 =$	224	22,400
3B	3B	5X	100	$14 \cdot 6 \cdot 3 =$	252	25,200
5X	2B	2B	75	$2 \cdot 20 \cdot 14 =$	560	42,000
2B	5X	2B	75	$16 \cdot 2 \cdot 14 =$	448	33,600
2B	2B	5X	75	$16 \cdot 20 \cdot 3 =$	960	72,000
5X	1B	1B	50	$2 \cdot 18 \cdot 22 =$	792	39,600
1B	5X	1B	50	$18 \cdot 2 \cdot 22 =$	792	39,600
1B	1B	5X	50	$18 \cdot 18 \cdot 3 =$	972	48,600
AM	AM	AM	18.405	$36 \cdot 36 \cdot 36 =$	46,656	858,703.68
3B	3B	3B	20	$14 \cdot 6 \cdot 8 =$	672	13,440
A7	A7	A7	20	$8 \cdot 12 \cdot 10 - 488 =$	472	9,440
2B	2B	2B	15	$16 \cdot 20 \cdot 14 =$	4,480	67,200
1B	1B	1B	10	$18 \cdot 18 \cdot 22 =$	7,128	71,280
5X	XB	XB	25	$2 \cdot 44 \cdot 44 - 1,448 =$	2,424	60,600
XB	5X	XB	25	$2 \cdot 44 \cdot 44 - 1,464 =$	2,760	69,000
5X	XB	XB	25	$48 \cdot 44 \cdot 3 - 2,184 =$	4,152	103,800
XB	XB	XB	5	$48 \cdot 44 \cdot 44 - 12,280 =$	80,648	403,240
CH	CH	CH	10	$2 \cdot 2 \cdot 3 =$	12	120
CH	CH	·	5	$2 \cdot 2 \cdot 141 =$	564	2,820
CH	·	CH	5	$2 \cdot 142 \cdot 3 =$	852	4,260
·	CH	CH	5	$142 \cdot 2 \cdot 3 =$	852	4,260
CH	·	·	2	$2 \cdot 142 \cdot 141 =$	40,044	80,088
·	CH	·	2	$142 \cdot 2 \cdot 141 =$	40,044	80,088
·	·	CH	2	$142 \cdot 142 \cdot 3 =$	60,492	120,984
total					298,578	see text

15). In computing the number of ways each payout can occur, we must take this into account, subtracting any combinations already accounted for with a higher payout. For example, for the case in which each reel shows any bar (XB) on the payline, it is clear that there are $48 \cdot 44 \cdot 44 = 92{,}928$ ways in which this can occur, but $12{,}280 = 7{,}128 + 4{,}480 + 672$ of them have already been accounted for in categories with higher payouts. Thus, the number of ways for this category is $92{,}928 - 12{,}280 = 80{,}648$. Table 12.10 completes the analysis. Since we have used the virtual stops of the symbol inventory table in doing these calculations, we must count all possible outcomes in the same way, so there are $(144)^3 = 2{,}985{,}984$ equally likely outcomes.

For $k = 1, 2, 3, 4, 5$, let R_k be the number of coins paid out from a one-coin bet on the kth payline. Then

$$
\begin{aligned}
\mathrm{E}[R_k] &= 1{,}000k\left(\frac{8}{(144)^3}\right) + 1{,}000\left(\frac{12}{(144)^3}\right) + 500\left(\frac{84 + 36 + 32}{(144)^3}\right) + \cdots \\
&\quad + 18.405\left(\frac{46{,}656}{(144)^3}\right) + \cdots + 2\left(\frac{40{,}044 + 40{,}044 + 60{,}492}{(144)^3}\right) \\
&= \frac{2{,}595{,}923.68 + 8{,}000k}{2{,}985{,}984} \\
&\approx \begin{cases}
0.872049 & \text{if } k = 1, \\
0.874728 & \text{if } k = 2, \\
0.877407 & \text{if } k = 3, \\
0.880086 & \text{if } k = 4, \\
0.882766 & \text{if } k = 5.
\end{cases}
\end{aligned}
\tag{12.9}
$$

With $\overline{R}_n$ denoting the payout *per coin bet* for an n coin bet, it follows easily that

$$
\mathrm{E}[\overline{R}_n] = \frac{2{,}595{,}923.68 + 4{,}000(n + 1)}{2{,}985{,}984} \approx \begin{cases}
0.872049 & \text{if } n = 1, \\
0.873388 & \text{if } n = 2, \\
0.874728 & \text{if } n = 3, \\
0.876068 & \text{if } n = 4, \\
0.877407 & \text{if } n = 5,
\end{cases}
\tag{12.10}
$$

and the house advantage for a bet of n coins is

$$
H_0 = 1 - \mathrm{E}[\overline{R}_n] \approx \begin{cases}
0.127951 & \text{if } n = 1, \\
0.126612 & \text{if } n = 2, \\
0.125272 & \text{if } n = 3, \\
0.123932 & \text{if } n = 4, \\
0.122593 & \text{if } n = 5.
\end{cases}
\tag{12.11}
$$

The probability of a positive payout from a one-coin bet is, by Table 12.10,

$$P(R_1 > 0) = \frac{298{,}578}{2{,}985{,}984} \approx 0.099994, \tag{12.12}$$

while the probability of a positive payout from a multiple-coin bet requires additional analysis. We will return to this point in Section 12.2.

12.2 Volatility and Ruin

While the expected payout of a slot machine is perhaps its most important characteristic, it does not tell the whole story. Also important is its volatility, as measured by the variance of the payout. Ruin probabilities are of interest as well. In this section we address these issues. The first step is to evaluate the payout distribution for each of the machines considered in Section 12.1.

For the Mills mechanical slot machine of Table 12.1 on p. 431, Table 12.2 on p. 431 contains the needed information. It suffices to consider the union of the three events that pay 150 coins, and similarly for the two that pay 18, the two that pay 14, and the two that pay 10. The results are shown in Table 12.11.

Table 12.11 Payout distribution of the Mills slot machine of Table 12.1 on p. 431.

payout	probability
200	1/8,000
150	3/8,000
18	45/8,000
14	60/8,000
10	60/8,000
5	360/8,000
2	840/8,000
0	6,631/8,000

Letting R denote the number of coins paid out from a one-coin bet, we evaluate the variance of R as

$$\text{Var}(R)$$
$$= E[R^2] - (E[R])^2$$

$$= (200)^2 \left(\frac{1}{8{,}000} \right) + (150)^2 \left(\frac{3}{8{,}000} \right) + (18)^2 \left(\frac{45}{8{,}000} \right) + (14)^2 \left(\frac{60}{8{,}000} \right)$$
$$+ (10)^2 \left(\frac{60}{8{,}000} \right) + (5)^2 \left(\frac{360}{8{,}000} \right) + (2)^2 \left(\frac{840}{8{,}000} \right) - \left(\frac{6{,}380}{8{,}000} \right)^2$$
$$= \frac{1{,}176{,}895{,}600}{(8{,}000)^2} \approx 18.388994. \tag{12.13}$$

Due to the substantial jackpots, the variance is much larger than for the Fey "Liberty Bell" slot machine of Examples 1.4.1 on p. 28 and 1.4.7 on p. 31.

For the Bally electro-mechanical slot machine of Table 12.4 on p. 434, the payout analysis (Table 12.5 on p. 434) is insufficient to determine the payout distributions. In fact, in this example the reel-strip labels (Table 12.6 on p. 435), and not just the symbol-inventory table (also Table 12.6), are needed.

Let us code the six symbols, cherry, orange, plum, bell, bar, and 7, by the six numbers, 1, 2, 3, 4, 5, and 7. Then we can summarize the pay table in Table 12.4 on p. 434 by defining the functions p and p^* from $\{1,2,3,4,5,7\}^3$ into $\mathbf{Z}_+$ as follows: $p(7,7,7) = 200$, $p^*(7,7,7) = 3{,}000$, $p(5,5,5) = p^*(5,5,5) = 100$, $p(4,4,4) = p^*(4,4,4) = p(4,4,5) = p^*(4,4,5) = 18$, and so on. We can also summarize Table 12.6 on p. 435 by defining the function $s : \{1,2,\ldots,22\} \times \{1,2,3\} \mapsto \{1,2,3,4,5,7\}$ for which $s(i,j)$ is the symbol at stop i of reel j. For example, $s(1,1) = s(1,2) = s(1,3) = 7$, $s(2,1) = 4$, $s(2,2) = 3$, $s(2,3) = 3$, and so on. It will be convenient to extend the function s to $\{0,1,2,\ldots,23\} \times \{1,2,3\}$ by defining $s(0,j) := s(22,j)$ and $s(23,j) := s(1,j)$ for $j = 1,2,3$.

Now let X_1, X_2, and X_3 be independent discrete uniform random variables on $\{1,2,\ldots,22\}$, representing the stop numbers of reels 1, 2, and 3 that appear on payline 1 when the reels come to rest. The payout random variables R_1, R_2, R_3, R_4, R_5 defined in Section 12.1 can now be expressed more precisely as

$$R_1 = p(s(X_1,1), s(X_2,2), s(X_3,3)), \tag{12.14}$$
$$R_2 = p(s(X_1 - 1,1), s(X_2 - 1,2), s(X_3 - 1,3)), \tag{12.15}$$
$$R_3 = p(s(X_1 + 1,1), s(X_2 + 1,2), s(X_3 + 1,3)), \tag{12.16}$$
$$R_4 = p(s(X_1 - 1,1), s(X_2,2), s(X_3 + 1,3)), \tag{12.17}$$
$$R_5 = p^*(s(X_1 + 1,1), s(X_2,2), s(X_3 - 1,3)). \tag{12.18}$$

Then, for example, the payout distribution for the player who bets five coins is given by

$$\mathrm{P}(R_1 + R_2 + R_3 + R_4 + R_5 = m)$$
$$= (22)^{-3} |\{(i,j,k) \in \{1,2,\ldots,22\}^3 :$$
$$p(s(i,1), s(j,2), s(k,3))$$

$$+ p(s(i-1,1), s(j-1,2), s(k-1,3))$$
$$+ p(s(i+1,1), s(j+1,2), s(k+1,3))$$
$$+ p(s(i-1,1), s(j,2), s(k+1,3))$$
$$+ p^*(s(i+1,1), s(j,2), s(k-1,3)) = m\}|. \quad (12.19)$$

We evaluate these probabilities by cycling through all $(22)^3 = 10{,}648$ possible outcomes and counting the numbers of ways (see Problem 12.4 on p. 452 for an alternative approach). Naturally, a computer eases the burden. In Table 12.12 we display the payout distribution for the player who bets one, two, three, four, or five coins.

Table 12.12 Payout distributions of the Bally slot machine of Table 12.4 on p. 434 as a function of the number of coins played.

	number of coins played				
	1	2	3	4	5
payout	frequency				
0	8,944	7,276	5,671	5,497	5,284
2	1,188	2,340	3,429	2,196	1,119
4	0	0	0	897	1,686
5	264	528	792	528	264
7	0	0	0	528	1,056
10	72	144	162	174	240
12	0	0	54	108	0
14	60	84	108	150	288
16	0	36	72	72	66
18	110	220	330	452	574
28	0	0	0	6	18
30	0	0	0	0	3
100	9	18	18	27	27
102	0	0	9	0	0
104	0	0	0	9	18
200	1	2	3	4	4
3,000	0	0	0	0	1
total	10,648	10,648	10,648	10,648	10,648

It is clear the probability of a positive payout increases with the number of coins played. The actual values are

$$P(\overline{R}_1 > 0) = \frac{1{,}704}{10{,}648} \approx 0.160030, \quad (12.20)$$

$$P(\overline{R}_2 > 0) = \frac{3{,}372}{10{,}648} \approx 0.316679, \tag{12.21}$$

$$P(\overline{R}_3 > 0) = \frac{4{,}977}{10{,}648} \approx 0.467412, \tag{12.22}$$

$$P(\overline{R}_4 > 0) = \frac{5{,}151}{10{,}648} \approx 0.483753, \tag{12.23}$$

$$P(\overline{R}_5 > 0) = \frac{5{,}364}{10{,}648} \approx 0.503757. \tag{12.24}$$

We hasten to add that a positive payout need not be a profitable payout.
Table 12.12 also yields the variances

$$\mathrm{Var}(\overline{R}_1) = \frac{2{,}017{,}008{,}000}{(10{,}648)^2} \approx 17.789819, \tag{12.25}$$

$$\mathrm{Var}(\overline{R}_2) = \frac{3{,}916{,}504{,}576}{2^2(10{,}648)^2} \approx 8.635800, \tag{12.26}$$

$$\mathrm{Var}(\overline{R}_3) = \frac{5{,}738{,}355{,}840}{3^2(10{,}648)^2} \approx 5.623528, \tag{12.27}$$

$$\mathrm{Var}(\overline{R}_4) = \frac{7{,}635{,}381{,}888}{4^2(10{,}648)^2} \approx 4.208959, \tag{12.28}$$

$$\mathrm{Var}(\overline{R}_5) = \frac{104{,}657{,}160{,}767}{5^2(10{,}648)^2} \approx 36.922649. \tag{12.29}$$

Notice the effect of the large jackpot for a fifth-payline 7-7-7.

For the Bally "In the Money" electronic slot machine of Table 12.7 on
p. 437, Tables 12.7–12.9 contain the needed information. Here we code the
ten symbols —, CH, 1B, 2B, 3B, MB, MU, 5X, G7, R7 (see Table 12.9 on
p. 438) as $0, 1, 2, 3, 4, 5, 6, 7, 8, 9$, respectively. Then we can summarize
the pay table in Table 12.7 by defining the function $p : \{0, 1, 2, \ldots, 9\}^3 \mapsto \mathbf{Z}_+$
as follows: $p(9, 9, 9) = 0$, $p(8, 8, 8) = 100$, $p(7, 7, 7) = 1{,}000$, $p(4, 7, 7) = p(7, 4, 7) = p(7, 7, 4) = 500$, and so on, including $p(i, j, k) = 0$ if $i, j, k \in \{5, 6\}$, so that $p(i, j, k)$ represents the payout for a given payline when (i, j, k)
appears on that payline, exclusive of any R7-R7-R7 payout and any bonus-
feature payout. Therefore

$$p_n(i, j, k) = p(i, j, k) + 1{,}000 n\, \delta_{i,9}\, \delta_{j,9}\, \delta_{k,9} \tag{12.30}$$

represents the payout for the nth payline when (i, j, k) appears on that payline
($n = 1, 2, 3, 4, 5$), exclusive of any bonus-feature payout. We can summarize
the bonus-feature payouts in Table 12.8 on p. 437 by defining the functions
$b : \{1, 2, 3, \ldots, 25\} \mapsto \mathbf{N}$ by $b(1) = 5$, $b(2) = 10$, $\ldots$, $b(24) = 150$, and
$b(25) = 200$ and $c : \{1, 2, 3, \ldots, 25\} \mapsto [0, 1]$ by $c(1) = 0.2453$, $c(2) = 0.3000$,
$\ldots$, $c(24) = 0.0030$, and $c(25) = 0.0020$. (Here b stands for bonus and
c stands for conditional probability.) Finally, we can also summarize Ta-
ble 12.9 on p. 438 by defining the function $s : \{1, 2, \ldots, 22\} \times \{1, 2, 3\} \mapsto$

$\{0, 1, 2, \ldots, 9\}$ for which $s(i, j)$ is the symbol at stop i of reel j and the function $w : \{1, 2, \ldots, 22\} \times \{1, 2, 3\} \mapsto \mathbf{N}$ for which $w(i, j)$ is the weight of stop i of reel j. For example, $s(1, 1) = s(1, 2) = s(1, 3) = 9$ and $w(1, 1) = w(1, 2) = w(1, 3) = 2$, $s(2, 1) = s(2, 2) = s(2, 3) = 0$ and $w(2, 1) = w(2, 2) = w(2, 3) = 2$, and so on. It will be convenient to extend the function s to $\{0, 1, 2, \ldots, 23\} \times \{1, 2, 3\}$ by defining $s(0, j) := s(22, j)$ and $s(23, j) := s(1, j)$ for $j = 1, 2, 3$.

Now let X_1, X_2, and X_3 be independent discrete random variables with values in $\{1, 2, \ldots, 22\}$ and with joint distribution

$$P((X_1, X_2, X_3) = (i, j, k)) = \frac{w(i, 1)}{144} \frac{w(j, 2)}{144} \frac{w(k, 3)}{144}, \tag{12.31}$$

representing the stop numbers of reels 1, 2, and 3 that appear on payline 1 when the reels come to rest. Let Y be a discrete random variable independent of (X_1, X_2, X_3) with distribution

$$P(Y = b(m)) = c(m), \qquad m = 1, 2, \ldots, 25. \tag{12.32}$$

The payout random variables R_1, R_2, R_3, R_4, R_5 defined in Section 12.1 can now be expressed more precisely as

$$R_1 = p_1(s(X_1, 1), s(X_2, 2), s(X_3, 3)), \tag{12.33}$$
$$R_2 = p_2(s(X_1 - 1, 1), s(X_2 - 1, 2), s(X_3 - 1, 3)), \tag{12.34}$$
$$R_3 = p_3(s(X_1 + 1, 1), s(X_2 + 1, 2), s(X_3 + 1, 3)), \tag{12.35}$$
$$R_4 = p_4(s(X_1 - 1, 1), s(X_2, 2), s(X_3 + 1, 3)), \tag{12.36}$$
$$R_5 = p_5(s(X_1 + 1, 1), s(X_2, 2), s(X_3 - 1, 3)), \tag{12.37}$$

except that we have excluded bonus-feature payouts. To include them, we define

$$M_1 := 1_{\{5,6\}^3}(s(X_1, 1), s(X_2, 2), s(X_3, 3)), \tag{12.38}$$
$$M_2 := 1_{\{5,6\}^3}(s(X_1 - 1, 1), s(X_2 - 1, 2), s(X_3 - 1, 3)), \tag{12.39}$$
$$M_3 := 1_{\{5,6\}^3}(s(X_1 + 1, 1), s(X_2 + 1, 2), s(X_3 + 1, 3)), \tag{12.40}$$
$$M_4 := 1_{\{5,6\}^3}(s(X_1 - 1, 1), s(X_2, 2), s(X_3 + 1, 3)), \tag{12.41}$$
$$M_5 := 1_{\{5,6\}^3}(s(X_1 + 1, 1), s(X_2, 2), s(X_3 - 1, 3)), \tag{12.42}$$

with M_n being the indicator of the event that the nth payline qualifies for a bonus-feature payout.

Letting $S_n := R_1 + \cdots + R_n$ and $N_n := M_1 + \cdots + M_n$ for $n = 1, 2, 3, 4, 5$, we find that the payout distribution for the player who bets n coins is given by

$$P(S_n + N_n Y = m), \qquad m = 0, 1, 2, \ldots; \; n = 1, 2, 3, 4, 5. \tag{12.43}$$

To evaluate this for $n = 5$, for example, we observe that

$$P(S_5 + N_5 Y = m) = (144)^{-3} \sum_{(i,j,k,l) \in A(m)} w(i,1)w(j,2)w(k,3)c(l), \quad (12.44)$$

where

$$
\begin{aligned}
A(m) := \{&(i,j,k,l) \in \{1,2,\ldots,22\}^3 \times \{1,2,\ldots,25\} : \\
&p_1(s(i,1), s(j,2), s(k,3)) + p_2(s(i-1,1), s(j-1,2), s(k-1,3)) \\
&+ p_3(s(i+1,1), s(j+1,2), s(k+1,3)) \\
&+ p_4(s(i-1,1), s(j,2), s(k+1,3)) \\
&+ p_5(s(i+1,1), s(j,2), s(k-1,3)) \\
&+ [1_{\{5,6\}^3}(s(i,1), s(j,2), s(k,3)) \\
&\quad + 1_{\{5,6\}^3}(s(i-1,1), s(j-1,2), s(k-1,3)) \\
&\quad + 1_{\{5,6\}^3}(s(i+1,1), s(j+1,2), s(k+1,3)) \\
&\quad + 1_{\{5,6\}^3}(s(i-1,1), s(j,2), s(k+1,3)) \quad\quad (12.45) \\
&\quad + 1_{\{5,6\}^3}(s(i+1,1), s(j,2), s(k-1,3))]b(l) = m\}.
\end{aligned}
$$

We evaluate these probabilities by cycling through all $(22)^3(25) = 266{,}200$ possible outcomes and summing the weights. Naturally, a computer is essential. In Table 12.13 we display the payout distribution for the player who bets five coins.

To check that it makes sense, consider how a payout of exactly 1,000 can occur. First, if R7-R7-R7 appears on the first payline, Tables 12.7 on p. 437 and 12.9 on p. 438 show that the total payout is 1,000, and this occurs with frequency $2 \cdot 2 \cdot 2 = 8$. If R7-R7-R7 appears on any other payline, the total payout is greater than 1,000. Second, if 5X-5X-5X appears on the first payline or the 5th payline, again the total payout is 1,000, and each occurs with frequency $2 \cdot 2 \cdot 3 = 12$. If 5X-5X-5X appears on any of the other three paylines, the total payout is greater than 1,000. It is easy to check that payouts such as $500 + 500$ or $500 + 250 + 250$ are impossible. Finally, all five paylines will qualify for the bonus feature if MU-MU-MU appears on the first payline, hence with frequency $12 \cdot 12 \cdot 12\,(0.002) = 3.456$ the total payout is $5 \cdot 200 = 1{,}000$. The resulting frequency of payouts of 1,000 is $8 + 12 + 12 + 3.456 = 35.456$, confirming one of the entries in Table 12.13.

Using Table 12.13, we find that

$$P(\overline{R}_5 > 0) = 1 - \frac{1{,}977{,}032}{2{,}985{,}984} \approx 0.337896 \quad (12.46)$$

and

$$\mathrm{Var}(\overline{R}_5) = \frac{4{,}602{,}952{,}484{,}026{,}140}{5^2(2{,}985{,}984)^2} \approx 20.650070. \quad (12.47)$$

We leave the evaluations of $P(\overline{R}_n > 0)$ and $\mathrm{Var}(\overline{R}_n)$ for $n = 1, 2, 3, 4$ to the reader (Problem 12.10 on p. 453).

Table 12.13 Payout distribution for a five-coin bet on the Bally slot machine of Table 12.7 on p. 437. Results are exact (no rounding).

pay	frequency	pay	frequency	pay	freq.	pay	freq.
0	1,977,032.0000	60	3,566.5920	195	41.4720	384	120.0000
2	160,464.0000	65	418.8672	200	387.0720	389	12.0000
4	149,760.0000	70	1,516.7360	205	22.4640	390	41.4720
5	334,661.6768	75	6,140.8000	210	41.4720	400	91.5840
6	43,308.0000	77	488.0000	220	145.1520	405	6.9120
7	5,320.0000	79	3,040.0000	225	93.3120	420	55.2960
8	48.0000	80	1,795.7760	240	117.5040	425	6.9120
9	20,184.0000	82	112.0000	245	5.1840	440	13.8240
10	90,102.1248	84	120.0000	250	640.6400	450	132.3840
11	2,760.0000	85	442.3680	252	88.0000	480	5.1840
12	336.0000	90	2,114.5728	254	500.0000	500	383.7760
14	2,268.0000	95	48.3840	255	55.2960	502	32.0000
15	38,033.9840	100	4,178.7520	257	108.0000	504	220.0000
17	472.0000	102	224.0000	259	120.0000	507	84.0000
18	24.0000	104	696.0000	260	114.0480	509	36.0000
19	1,120.0000	105	1,179.5520	265	10.3680	520	10.3680
20	29,267.7568	110	908.5440	270	55.2960	550	6.9120
25	32,599.0784	115	20.7360	275	75.4560	560	13.8240
27	2,832.0000	120	685.8752	280	148.6080	575	12.0000
29	13,500.0000	125	101.9520	285	13.8240	600	40.6080
30	16,116.2112	130	373.2480	300	186.6240	650	5.1840
32	228.0000	135	158.9760	305	10.3680	700	6.9120
34	300.0000	140	566.7840	320	13.8240	750	5.1840
35	3,353.0112	145	24.1920	325	5.1840	800	6.9120
39	12.0000	150	755.6736	330	55.2960	1,000	35.4560
40	6,765.6000	155	25.9200	340	13.8240	1,004	12.0000
45	3,538.8800	160	207.3600	350	5.1840	1,007	12.0000
50	9,818.2400	165	41.4720	360	34.5600	1,009	12.0000
52	792.0000	170	138.2400	375	551.7760	2,025	8.0000
54	3,744.0000	175	65.6640	377	56.0000	3,025	8.0000
55	1,017.7920	180	214.2720	379	448.0000	4,000	8.0000
59	108.0000	185	13.8240	382	96.0000	5,000	8.0000
total							2,985,984.0000

Turning next to the general situation, let T denote the number of coins paid out from a one-coin bet, or the number of coins paid out *per coin bet* from a multiple-coin bet. We put

$$\mu := \mathrm{E}[T] \quad \text{and} \quad \sigma^2 := \mathrm{Var}(T). \tag{12.48}$$

We also let $T_1, T_2, \ldots$ be a sequence of i.i.d. random variables with common distribution that of T, representing the results of a sequence of coups at a specific machine. Then

$$\overline{T}_n := \frac{T_1 + T_2 + \cdots + T_n}{n} \tag{12.49}$$

represents the proportion paid out after n coups. By the central limit theorem (Theorem 1.5.15 on p. 48),

$$\lim_{n\to\infty} \mathrm{P}\left(\mu - \frac{z_{1-\alpha/2}\sigma}{\sqrt{n}} \le \overline{T}_n \le \mu + \frac{z_{1-\alpha/2}\sigma}{\sqrt{n}}\right) = 1 - \alpha, \tag{12.50}$$

where $z_{1-\alpha/2} := \Phi^{-1}(1 - \alpha/2)$ is a standard-normal quantile. We conclude that

$$\left[\mu - \frac{z_{1-\alpha/2}\sigma}{\sqrt{n}}, \mu + \frac{z_{1-\alpha/2}\sigma}{\sqrt{n}}\right] \tag{12.51}$$

is an approximate $100(1-\alpha)\%$ *prediction interval* for $\overline{T}_n$. In other words, the observed proportion of coins paid out, namely $\overline{T}_n$, will belong to this interval with approximate probability $1-\alpha$ for large n. The product $z_{1-\alpha/2}\sigma$ is known in the casino industry as the *volatility index*. Frequently one takes $\alpha = 0.1$, in which case $z_{1-\alpha/2} \approx 1.644853$. Table 12.14 provides some illustrations of this idea.

The Fey "Liberty Bell" has a slightly smaller expected payout than the other machines, but a significantly smaller standard deviation, due to its lack of a jackpot. Notice the effect this has on the approximate prediction intervals in Table 12.14.

While there is no simple gambler's ruin formula in the context of slot machines, it is straightforward to adapt the iterative method of Section 7.2 to these examples. For example, consider the Fey "Liberty Bell" of Example 1.4.1 on p. 28. If we want to evaluate the probability that the player achieves a profit of W or more coins before he achieves a loss of L coins, it suffices to define successively for $n = 0, 1, 2, \ldots,$

$$u_{n+1}^-(i) = \{736u_n^-(i-1) + 200u_n^-(i+1) + 50u_n^-(i+3) + 8u_n^-(i+7)$$
$$+ 3u_n^-(i+11) + u_n^-(i+15) + 2u_n^-(i+19)\}/1{,}000 \tag{12.52}$$

and

$$u_{n+1}^+(i) = \{736u_n^+(i-1) + 200u_n^+(i+1) + 50u_n^+(i+3) + 8u_n^+(i+7)$$

Table 12.14 Approximate 90% prediction intervals for proportion of coins paid out for the three slot machines of this chapter (and the one of Section 1.4) and various numbers of coups n.

machine	coins	μ	σ	n	lower	upper
Fey 1898	1	.756 000	1.740 248	10^2	.469 755	1.042 245
				10^3	.665 481	.846 519
				10^4	.727 375	.784 625
				10^5	.746 948	.765 052
Mills 1949	1	.797 500	4.288 239	10^3	.574 448	1.020 552
				10^4	.726 965	.868 035
				10^5	.775 195	.819 805
Bally 1969	5	.835 462	6.076 401	10^3	.519 399	1.151 525
				10^4	.735 514	.935 410
				10^5	.803 856	.867 068
Bally 2000	5	.877 407	4.544 235	10^3	.641 039	1.113 775
				10^4	.802 661	.952 153
				10^5	.853 770	.901 044

$$+ 3u_n^+(i+11) + u_n^+(i+15) + 2u_n^+(i+19)\}/1{,}000, \quad (12.53)$$

starting with

$$u_0^-(i) := \begin{cases} 0 & \text{if } i = -L, \ldots, W-1 \\ 1 & \text{if } i = W, \ldots, W+18 \end{cases} \quad (12.54)$$

and

$$u_0^+(i) := \begin{cases} 0 & \text{if } i = -L \\ 1 & \text{if } i = -L+1, \ldots, W+18. \end{cases} \quad (12.55)$$

By Theorem 7.2.4 on p. 256, the sequence $\{u_n^-(0)\}$ is nondecreasing, the sequence $\{u_n^+(0)\}$ is nonincreasing, and both sequences converge to the probability in question. It remains only to iterate sufficiently often to achieve the desired degree of accuracy.

To compare the behavior of different machines, possibly with different bet sizes, we define the amount bet per coup to be one unit, and we measure L and W in the preceding paragraph in terms of units. Table 12.15 reports our results.

It might seem surprising that the probability of doubling an initial stake of W units before losing it is not decreasing in W, as it is in a subfair even-money game (see (7.11) on p. 243). The reason is related to the fact that one can overshoot the goal. To see this more clearly, consider an extremely simple

Table 12.15 The probability of doubling (or more) an initial stake of W units before losing it, at the three slot machines of this chapter (and the one of Section 1.4). One unit is the number of coins bet per coup. Results are rounded to six significant digits.

W	Fey 1898 (bet one coin)	Mills 1949 (bet one coin)	Bally 1969 (bet five coins)	Bally 2000 (bet five coins)
20	.069 555 1	.143 583	.084 626 9	.220 612
50	.003 036 64	.093 406 1	.023 901 5	.222 111
100	.000 013 957 6	.130 120	.026 219 3	.155 603
200		.047 215 4	.037 177 8	.066 777 6
500		.001 562 77	.028 442 2	

slot machine that, for a one-coin bet, pays out either a jackpot of J coins or nothing at all. Let p be the probability of winning the jackpot at a given coup. Then, if W is a positive integer no larger than $(J+1)/2$, the probability of doubling an initial stake of W coins before losing it is just the probability of winning the jackpot within the first W coups, that is, $1 - (1 - p)^W$, and this is clearly increasing in W $[\leq (J + 1)/2]$.

Finally, we would like to point out another application of the house advantage associated with a slot machine. Suppose that a player has L coins, and bets one coin per coup. At each coup the machine pays out a nonnegative integer number of coins with expected payout less than 1. Thus, if R is the number of coins paid out from a one-coin bet, then $X := R - 1$ is the player's profit and $E[X] < 0$. Suppose further that the player continues to bet as long as he has at least one coin remaining. Let N denote the number of coups played. How long will his L units last on average, that is, what is $E[N]$? The answer, by Corollary 7.3.9 on p. 266 (in which $L = L^*$), is

$$E[N] = \frac{L}{H_0}, \tag{12.56}$$

where $H_0 := |E[X]|$. (The result is exact because there is no goal to overshoot.) Thus, the smaller the house advantage of a machine (or, equivalently, the greater the expected payout), the longer one's initial capital can be expected to last. See Problem 12.11 on p. 453 for a generalization.

12.3 Problems

12.1. *A three-reel, 25-stop, Pace mechanical slot machine.* To describe this machine in condensed form, we code the seven symbols on the reels by integers: cherry = 1, orange = 2, plum = 3, bell = 4, bar = 5, bonus = 6, and super = 7. The pay table can then be described by the function $p : \{1, 2, 3, 4, 5, 6, 7\}^3 \mapsto \mathbf{Z}_+$ given by $p(7, 7, 7) = 1,000$, $p(6, 6, 6) = 300$, $p(5, 5, 5) = 150$, $p(4, 4, 4) = p(4, 4, 5) = 18$, $p(3, 3, 3) = p(3, 3, 5) = 14$, $p(2, 2, 2) = p(2, 2, 5) = p(1, 1, 1) = 10$, $p(1, 1, 1') = 5$, and $p(1, 1', 1') = 3$; otherwise $p = 0$. Here $1'$ denotes any symbol except 1. The three reel strips can be described as follows:

reel 1 : 7, 2, 3, 4, 5, 1, 3, 5, 2, 3, 2, 6, 3, 1, 5, 3, 2, 4, 6, 3, 2, 5, 1, 3, 2.

reel 2 : 7, 2, 1, 3, 4, 5, 4, 1, 2, 4, 5, 4, 1, 2, 4, 6, 4, 1, 3, 4, 5, 4, 1, 2, 4.

reel 3 : 7, 2, 1, 3, 2, 3, 1, 2, 6, 3, 1, 2, 4, 3, 1, 2, 5, 3, 1, 2, 4, 3, 1, 2, 3.

Find the payout distribution and its mean and variance.

12.2. *A four-reel, 20-stop, Pace mechanical slot machine.* To describe this machine in condensed form, we code the six symbols on the reels by integers: cherry = 1, orange = 2, plum = 3, bell = 4, bar = 5, and special = 6. The pay table can then be described by the function $p : \{1, 2, 3, 4, 5, 6\}^4 \mapsto \mathbf{Z}_+$ given by $p(6, 6, 6, 6) = 5,000$, $p(6, 6, 6, 6') = p(5, 5, 5, 5') = p(4, 4, 4, 4) = 150$, $p(3, 3, 3, 3) = p(2, 2, 2, 2) = p(1, 1, 1, 1) = 20$, $p(4, 4, 4, 4') = 18$, $p(3, 3, 3, 3') = 14$, $p(2, 2, 2, 2') = p(1, 1, 1, 1') = 10$, $p(1, 1, 1', 1') = 5$, and $p(1, 1', 1', 1') = 3$; otherwise $p = 0$. Here j' denotes any symbol except j. The four reel strips can be described as follows:

reel 1 : 6, 2, 1, 3, 5, 2, 3, 4, 5, 2, 4, 1, 5, 2, 3, 4, 5, 2, 3, 2.

reel 2 : 6, 4, 1, 3, 4, 1, 4, 5, 1, 2, 3, 4, 1, 5, 4, 1, 2, 4, 1, 4.

reel 3 : 6, 2, 1, 3, 2, 3, 2, 3, 1, 3, 5, 2, 4, 3, 1, 3, 2, 3, 2, 3.

reel 4 : 6, 1, 3, 4, 3, 1, 3, 2, 3, 1, 2, 4, 3, 1, 3, 2, 3, 1, 3, 4.

Find the payout distribution and its mean and variance.

12.3. *Formulas for expected payout and variance of payout.* Many of the classical one-coin slot machines had a similar payout structure, allowing one to derive a formula for the expected payout. We assume that there are r reels, each with n stops, and that there are m symbols, denoted by $1, \ldots, m$, exactly one of which appears at each stop on each reel. We further assume that each stop (of each reel) is equally likely to appear on the payline, of which there is only one, and that the reels behave independently. Let the function $s : \{1, \ldots, n\} \times \{1, \ldots, r\} \mapsto \{1, \ldots, m\}$ describe the reel-strip labels, so that $s(i, j)$ is the symbol at stop i on reel j. Let the function $p : \{1, \ldots, m\}^r \mapsto \mathbf{Z}_+$ describe the pay table, so that $p(k_1, \ldots, k_r)$ is the number of coins paid out when symbols $k_1, \ldots, k_r$ appear on the payline in that order.

(a) Derive formulas for the expected payout and the variance of the payout in terms of r, n, m, s, and p.

(b) Let the function $f : \{1, \ldots, m\} \times \{1, \ldots, r\}$ describe the symbol-inventory table, so that $f(i, k)$ is the frequency with which symbol i appears on reel k. Show that the first formula can be expressed in terms of r, n, m, f, and p. (In other words, knowledge of s is unnecessary when f is known.)

12.4. *Extended symbol-inventory table for machines with multiple paylines.* Derive the extended symbol-inventory table for the Bally electro-mechanical slot machine of Table 12.4 on p. 434. Instead of counting the number of times each symbol appears on each reel, this table counts the number of times each ordered triple of symbols appears on each reel. For example, bell/orange/bell appears 0 times on reel 1, 3 times on reel 2, and 4 times on reel 3. Use this table to rederive Table 12.12 on p. 443.

12.5. *Covariances between paylines.* Find the covariances of the payouts between each pair of paylines for the Bally electro-mechanical slot machine of Table 12.4 on p. 434. Use these results to find the variance of the payout per unit bet from a five-coin bet. (We have already evaluated this variance at (12.47) on p. 446. This gives an alternative derivation that avoids having to find the payout distribution.)

12.6. *A three-reel, 22-stop, two-coin, IGT electronic slot machine.* The pay table for this machine is shown in Table 12.16. To describe the reel strips, we code the five symbols as blank $= 0$, 1B $= 1$, 5B $= 2$, 7B $= 3$, and JW $= 4$. Then the three reel strips can be described as follows:

reel 1 : 4, *0*, *1*, 0, 2, *0*, *1*, 0, 2, 0, 1, *0*, 3, *0*, *1*, 0, 2, 0, *1*, 0, 2, *0*.

reel 2 : 4, *0*, 1, *0*, *2*, 0, 1, *0*, 3, *0*, 1, 0, 2, *0*, *1*, 0, 1, *0*, 2, 0, 1, *0*.

reel 3 : 4, *0*, 1, *0*, 2, 0, 1, *0*, 3, *0*, 1, *0*, 2, *0*, 1, 0, 1, *0*, 2, *0*, 1, *0*.

Here ordinary numbers have weight 1 and italicized numbers have weight 2. Find the mean and variance of the payout from a one-coin bet and the payout per unit bet from a two-coin bet.

12.7. *Three-reel, 22-stop, three-coin, IGT "Red, White, and Blue" electronic slot machine.* This machine has pay table shown in Table 12.17. To describe the reel strips, we code the seven symbols as blank $= 0$, 1 bar $= 1$, 2 bar $= 2$, 3 bar $= 3$, and blue 7 $= 4$, white 7 $= 5$, and red 7 $= 6$. Then the physical reels 1, 2, and 3 are identical:

reels 1–3 : 2, 0, 3, 0, 5, 0, 1, 0, 4, 0, 2, 0, 3, 0, 6, 0, 3, 0, 2, 0, 1, 0.

The virtual reels, on the other hand, are different:

reel 1 weights : 3, 2, 2, 3, 6, 3, 3, 3, 6, 3, 2, 2, 1, 5, 1, 5, 3, 2, 2, 2, 3, 2.

reel 2 weights : 2, 2, 2, 3, 1, 3, 4, 3, 7, 3, 2, 2, 2, 5, 3, 5, 3, 2, 2, 2, 4, 2.

reel 3 weights : 3, 2, 1, 3, 7, 3, 5, 3, 1, 3, 3, 2, 1, 5, 1, 5, 3, 2, 3, 2, 4, 2.

Find the mean and variance of the payout from a one-coin bet and the payouts per unit bet from a two-coin and three-coin bet.

Table 12.16 Pay table from a three-reel, 22-stop, two-coin, IGT electronic slot machine. See Problem 12.6.

reel 1	reel 2	reel 3	one-coin payout	two-coin payout
JW	JW	JW	400	1,000
7B or JW	7B or JW	7B or JW	200	500
5B or JW	5B or JW	5B or JW	50	125
1B or JW	1B or JW	1B or JW	10	25
not blank	not blank	not blank	5	10
JW	JW	not JW	5	10
JW	not JW	JW	5	10
not JW	JW	JW	5	10
JW	not JW	not JW	2	4
not JW	JW	not JW	2	4
not JW	not JW	JW	2	4

12.8. *Effect of physical vs. virtual stops* . Find the expected payout per unit bet for one-, two-, and three-coin bets on the IGT machine of Problem 12.7 under the assumption that, for each reel, each of the 22 physical stops is equally weighted.

12.9. *Further analysis of the IGT "Red, White, and Blue" slot machine.* For the IGT machine of Problem 12.7, assume a three-coin bet and extend Tables 12.14 and 12.15 on p. 450 to include this example.

12.10. *Further analysis of the Bally "In the Money" slot machine.* For the Bally electronic slot machine of Table 12.7 on p. 437, evaluate $P(\overline{R}_n > 0)$ and $\text{Var}(\overline{R}_n)$ for $n = 1, 2, 3, 4$. This was done in the text for $n = 5$ only.

12.11. *Expected number of coups, assuming a multiple-coin machine.* Generalize (12.56) on p. 450 to allow for multiple-coin play, using Corollary 7.3.9 on p. 266.

12.12. *A Markovian slot machine: The Mills "Futurity."* This machine, dating back to 1936, has two unusual features, one readily apparent and the other less so. The readily apparent feature is that, if the player loses 10 times in a row, the 10 lost coins are returned. At the top of the machine is a pointer that advances by 1 after each loss, and resets at 0 after a win or after 10 consecutive losses. The less apparent feature is that there are 20 symbols on each of the three reels but only the ones in even-numbered positions can appear on the payline if the machine is in mode E, while only the ones in odd-numbered positions can appear on the payline if the machine is in mode O. The mode is nonrandom and is determined by a cam that rotates through

Table 12.17 Pay table from the three-reel, 22-stop, two-coin, IGT "Red, White, and Blue" electronic slot machine. (1 bar is red, 2 bar is white, and 3 bar is blue.) See Problem 12.7.

reel 1	reel 2	reel 3	one-coin payout	two-coin payout	three-coin payout
red 7	white 7	blue 7	2,400	4,800	10,000
red 7	red 7	red 7	1,199	2,400	5,000
white 7	white 7	white 7	200	400	600
blue 7	blue 7	blue 7	150	300	450
any 7	any 7	any 7	80	160	240
1 bar	2 bar	3 bar	50	100	150
3 bar	3 bar	3 bar	40	80	120
2 bar	2 bar	2 bar	25	50	75
any red	any white	any blue	20	40	60
1 bar	1 bar	1 bar	10	20	30
any bar	any bar	any bar	5	10	15
any red	any red	any red	2	4	6
any white	any white	any white	2	4	6
any blue	any blue	any blue	2	4	6
blank	blank	blank	1	2	3

10 positions, advancing one position with each coup and resulting in a specific mode pattern of length 10, EEEEEOEEEO, which is repeated ad infinitum.

To describe the machine in detail, we code the six symbols as lemon = 0, cherry = 1, orange = 2, plum = 3, and bell = 4, and bar = 5. The pay table can then be described by the function $p : \{0, 1, 2, 3, 4, 5\}^3 \mapsto \mathbf{Z}_+$ given by $p(5, 5, 5) = 150$, $p(4, 4, 4) = p(4, 4, 5) = 18$, $p(3, 3, 3) = p(3, 3, 5) = 14$, $p(2, 2, 2) = p(2, 2, 5) = 10$, $p(1, 1, 4) = p(1, 1, 0) = 5$, and $p(1, 1, 5) = p(1, 1, 3) = p(1, 1, 2) = 3$; otherwise $p = 0$. The three reel strips can be described as follows, in which we have italicized the odd-numbered stops for convenience:

$$\text{reel } 1 : 1, 5, 1, 2, 1, 5, 1, 5, 1, 3, 1, 2, 5, 1, 4, 3, 1, 5, 1, 2.$$
$$\text{reel } 2 : 1, 4, 1, 3, 1, 4, 1, 2, 1, 4, 1, 4, 1, 2, 1, 2, 4, 1, 5, 4.$$
$$\text{reel } 3 : 3, 4, 2, 0, 3, 4, 2, 0, 4, 0, 2, 3, 2, 4, 2, 4, 5, 2, 3, 5.$$

Of course, the reels operate independently, and the 10 possible positions at which each reel can stop (given the mode) are equally likely.

(a) A Markov chain $\{(X_n, Y_n)\}_{n \geq 0}$ with state space $S := \{0, 1, \ldots, 9\} \times \{0, 1, \ldots, 9\}$ controls the behavior of the "Futurity." It is in state (i, j) if the cam position is i and the pointer position is j. Find the transition probabilities and stationary distribution of this Markov chain (cf. Problem 4.17 on p. 156). Use this information to find the expected payout for the machine.

(b) Under what conditions on the state of the Markov chain does the strategy "play until, and only until, a payout occurs" offer a positive expectation?

12.13. *Slot machine design.* Design a modern slot machine (i.e., specify the pay table together with the physical reel-strip labels and the weights of the stops on each reel), taking into account the following factors. First, it must have a theme that will appeal to potential players. Second, it must offer sufficient complexity to keep players from becoming bored. Third, payouts must be potentially large and sufficiently frequent to attract and retain players. Fourth, it must offer adequate expected profit to the casino.

12.14. *An early 20th-century poker machine.* This is a five-reel, 11-stop, mechanical slot machine. The pay table can be described as follows. Royal flush = 100; straight flush = 50; four of a kind = 30; full house = 20; flush = 10; straight = 8; three of a kind = 5; two pair = 3; pair of 10s or better = 1. The five reel strips are as follows:

$$\text{reel } 1: A\diamondsuit, 2\clubsuit, 9\diamondsuit, 10\heartsuit, K\spadesuit, 4\diamondsuit, 8\spadesuit, Q\clubsuit, 5\heartsuit, 7\clubsuit, 3\spadesuit.$$
$$\text{reel } 2: 2\diamondsuit, 3\clubsuit, J\heartsuit, 8\heartsuit, 10\diamondsuit, 6\clubsuit, 5\diamondsuit, J\clubsuit, K\clubsuit, 5\spadesuit, 9\heartsuit.$$
$$\text{reel } 3: 3\diamondsuit, A\heartsuit, 8\diamondsuit, 4\clubsuit, 2\spadesuit, Q\spadesuit, 7\diamondsuit, 9\spadesuit, 6\heartsuit, Q\heartsuit, 2\spadesuit.$$
$$\text{reel } 4: 6\diamondsuit, 10\spadesuit, 2\heartsuit, A\clubsuit, 7\heartsuit, 9\clubsuit, 6\diamondsuit, J\diamondsuit, 10\clubsuit, 4\heartsuit, A\spadesuit.$$
$$\text{reel } 5: 6\spadesuit, Q\diamondsuit, 4\spadesuit, K\heartsuit, 5\clubsuit, 3\heartsuit, K\diamondsuit, 7\spadesuit, 8\clubsuit, J\spadesuit, 3\heartsuit.$$

Notice that each of the 52 cards appears at least once (with $2\spadesuit$, $3\heartsuit$, and $6\diamondsuit$ each appearing twice). Find the payout distribution and its mean and variance.

12.4 Notes

For the history of slot machines, see the superb coffee-table books by Fey (2002) and Bueschel (1995). Here we confine ourselves to the highlights. Our source is Fey (2002) except where otherwise noted.

Coin-operated gambling devices, or what were called "nickel-in-the-slot machines," date back to the late 1880s. The most popular types in the last decade of the 19th century were machines with a single reel having up to 100 symbols, as well as poker machines, called dropcard machines, with ten cards on each of five drums. While the single-reel machines had an automatic payout feature, the poker machines did not, and an attendant was required to pay winning hands. In 1898 Charles August Fey [1862–1944] of San Francisco invented a card machine with three reels instead of five that paid awards automatically. This was the first automatic-pay three-reel slot machine. Only one of these machines survives today (serial number 5), rescued by Fey himself

from the 1906 San Francisco earthquake and fire. In 1899 Fey introduced his
"Liberty Bell" slot machine,[2] only a minor improvement of its predecessor
but one which had a huge impact on the slot-machine industry. California
Historical Marker No. 937 reads as follows:

> The first slot machines were manufactured by the inventor Charles Fey just west
> of this site at 406 Market from 1896 till the factory was destroyed in the 1906
> earthquake and fire. Fey, a Bavarian immigrant, dubbed his invention the "Liberty
> Bell" in honor of the famous symbol of freedom. Ultimately, the slot machine became
> the most famous gambling device of all time.

Courts ruled that slot machines were not patentable, lacking the element of
utility. Fey's production was limited; the highest known "Liberty Bell" serial
number is 93. He did not sell or lease his machines, but instead placed them
in saloons and cigar stores around San Francisco on a profit-sharing basis.
In 1905 one of his "Liberty Bell" slots was stolen from a saloon and turned
up at the Chicago factory of Mills Novelty Company. The company was run
by Herbert S. Mills [1872–1929], who was to become the Henry T. Ford of
slot machines. The Mills "Liberty Bell" slot machine debuted around 1906,
and some 30,000 of these machines and their successors were manufactured
by World War I.[3] The Mills machine used the same mechanism as the Fey
machine but added several improvements, including 20 stops per reel instead
of 10 and enlarging the payout window to show three symbols per reel instead
of one, with the intention of revealing a near miss.

Mills Novelty Company of Chicago remained the largest manufacturer of
slot machines for some 60 years. Other major manufacturers included Watling
Manufacturing Company (named for Thomas W. B. Watling), O. D. Jennings
and Company (named for Ode D. Jennings [1874–1953]), and Pace Manufacturing Company (named for Edwin Walton Pace [1879–1954]), all of Chicago,
and Caille Brothers Company (named for Adolph and Arthur Caille) of Detroit. Watling, Jennings, Pace, and Caille also manufactured what came to
be called bell slots.

As for the legality of slot machines, consider the case of San Francisco. In
1893 the Board of Supervisors required that the machines be licensed and
taxed. Still, various reformers and do-gooders wanted to ban them. As the
Morning Call wrote on February 2, 1896 (as quoted by Fey 2002, p. 14),

> Who hasn't dropped a nickel in the slot of those seductive little machines to be
> found almost any place where men most do congregate? ... They are located in
> nearly all the saloons and cigar-stores in the City, eating up nickels as fast as the
> "tiger" can eat up dollars.

---

[2] This date is uncertain. We follow Marshall Fey [1928–] (2002, p. 41), grandson of Charles
Fey. However, Bueschel (1995, p. 54) has argued that the year of the "Liberty Bell"'s
introduction was more likely 1905.

[3] The manner in which the Fey machine was transferred to the Mills factory is in dispute.
We follow Fey (2002, p. 85). However, Bueschel (1995, p. 56) has claimed that Mills Novelty
Company paid Fey for his design by giving him the first 50 machines produced. This claim
was based on interviews with Bert E. Mills, the last surviving brother of Herbert S. Mills,
as well as with a former advertising manager of Mills Novelty.

(See Section 18.4 for the significance of the tiger reference.) In 1897 a Superior Court judge ruled that slot machines were legal, but in 1898 the Board of Supervisors passed an ordinance prohibiting the machines from paying out in money. Machines were then modified to pay out in trade checks, which could be exchanged for merchandise (e.g., cigars or drinks). In 1909 the Board of Supervisors banned slot machines entirely. Nevada outlawed the machines in 1910, and the rest of California did so in 1911.

Slot machine manufacturers reacted by turning their gambling machines into vending machines. In 1910 Mills introduced the "Liberty Bell" Gum Fruit machine, which dispensed a package of gum with each nickel played. Any additional awards were described as "profit sharing." The symbols on this machine were simply the flavors of the gum: spearmint, lemon, orange, and plum. The traditional bell symbol was also included, and the only other symbol was the label from a package of gum. In a subsequent Mills model, the spearmint symbol was replaced by a cherry cluster, and these reel symbols, with the bar replacing the Bell-Fruit-Gum label, were the standard slot machine reel symbols for the next 70 years. Even the interpretations of the symbols remained largely unchanged: The lemon indicated no payout (hence the term "lemon" for a defective product) and the cherry indicated a small payout (hence the term *cherry dribbler* for a machine with frequent small payouts; see Rivlin 2004).

During Prohibition slot machine sales increased as slots found homes in the speakeasies of the 1920s. Gum was replaced by mints (longer lasting), but the most important innovation of that era was the introduction of the jackpot in 1928, which immediately became an essential feature of every machine. The original jackpots were what are called "drop jackpots," in which a transparent container on the front of the machine held a large number of coins. When Nevada legalized gambling in 1931, regulations required that the jackpot amount have a specified value, and this put an end to the drop jackpots (Jones 1978, pp. 4–5). The 1930s, often called the golden age of slots, produced some of the most beautiful (and collectible) designs. It was also a time of reform, led by crusading politicians such as New York's Mayor Fiorello La Guardia [1882–1947] and California's Attorney General Earl Warren [1891–1974], who cracked down on illegal slots. Production of slot machines came to a standstill during World War II but resumed in 1945. In 1951 the Johnson Act (named after Senator Edwin C. Johnson [1884–1970] of Colorado), prohibiting the interstate shipment of "gambling devices" except to those states where they were legal, came into effect, and it devastated the slot machine industry. In particular, it put Watling out of business. (Caille had already ceased production before the war.) Of the major manufacturers, only Mills, Jennings, and Pace remained.

In 1963 Illinois legalized the manufacture of slot machines, and Bally Manufacturing Corporation of Chicago entered the market with a revolutionary design, the *electro-mechanical slot machine*, which made the existing purely mechanical machines obsolete. The two primary advantages of the electro-mechanical design were (a) awards were paid from a "bottomless" hopper capable of automatic payouts of up to 500 coins, in contrast to the 20-coin slide of the mechanical models, and (b) more than 50 different payout combinations could be sensed by the electric circuitry, in contrast to the much more limited "finger" pay design of the mechanical slots. (The reel-spinning mechanism, and therefore the determination of the outcome, was still purely mechanical.) The new design allowed Bally to offer multi-coin multipliers (payout is multiplied by the number of coins played, perhaps with a bonus for maximum-coin play), multi-line play (up to five paylines with as many paid as the number of coins played), right-to-left and left-to-right pay (for example, both cherry-cherry-anything and anything-cherry-cherry qualify for a payout), linked progressives (a progressively growing jackpot is paid to the first of several machines to hit the winning combination), and so on. Bally completely dominated the market for two decades. In particular, Pace went out of business in the 1960s, and Mills and Jennings survived, after a merger, until the mid-1980s.

In 1980 Bally upgraded its electro-mechanical design to an electronic one. But the most important advance of the 1980s came in 1985 with the introduction of the stepper-motor-driven reels, which with the use of microprocessors and random-number generators allowed much larger jackpots on three-reel machines than had been previously possible. Stepper motors were first manufactured by JPM in Wales, UK. In 1984 Inge S. Telnaes of Reno patented an "Electronic gaming device utilizing a random number generator for selecting the reel stop positions" (U.S. patent 4,448,419). The patent abstract reads as follows:

> A gaming machine of the type utilizing rotating reels which carry on the periphery a plurality of indicia, a brake to stop the reels at a selected position and a random number generator for selecting the reel stopping position. Numbers are assigned to the reel stopping positions and entered into the random number generator with each number being entered one or more times to control the payout odds of each particular stopping position being selected thereby enabling any odds to be set without changing the physical characteristics of the machine.

For example, the IGT machine of Problem 12.6 would have the numbers 0–31 assigned to the 22 stops of each reel, two numbers to each of the 10 italicized stops and one number to each of the other 12 stops. Such a machine can be described as having 22 physical stops and 32 virtual stops.

By this time Bally had a major new competitor, International Game Technology (IGT), which has since come to dominate the market. In 1986 IGT initiated "Megabucks," which Crevelt and Crevelt (1988, p. 19) called the *pièce de résistance* for slot players. This is a statewide bank of linked progressive dollar slot machines offering multi-million dollar jackpots designed to compete with state lotteries. With each reel having 368 virtual stops, the probability of hitting the jackpot is $1/(368)^3$, or one chance in 49,836,032 (Robison 2009).

Today, slot machines have "themes" and "bonus rounds," and the lemons, cherries, oranges, plums, and bells, which date back to 1910, are all but obsolete. A sense of the modern slot can be gleaned from the title of a *New York Times Magazine* article (Rivlin 2004):

> The chrome-shiny, lights-flashing, wheel-spinning, touch-screened, Drew-Carey-wise-cracking, video-playing, 'sound events'-packed, pulse-quickening bandit: How the slot machine was remade, and how it's remaking America.

Not discussed in this chapter are the increasingly popular five-reel video slot machines. Instead of physical reels they have video representations of such, and most of them have multiple paylines. Like modern slots with physical reels, results are determined by a random-number generator on an EPROM, but unlike the modern single-payline machines with physical reels, the various stops are equally weighted. The reason is that the number of stops per reel is not limited to 22 or so as it is with physical reels and can be as large as necessary to allow attractive jackpots.

The mathematical details of a machine (including pay table, reel-strip labels, etc.), are described in a document produced by the manufacturer called a *par sheet*. When a manufacturer offers a machine for sale (or lease), it gives the purchaser (the casino) a choice of several different EPROMs, depending on the desired expected payout (ranging perhaps from 0.85 to 0.98), and each has its own par sheet. Michael Shackleford [1965–] (2002) received complete sets of par sheets for five types of five-reel video slot machines from an industry insider, and used this information to rank some 67 Las Vegas casinos according to the average of their expected payouts for the five types of machines. It was sufficient to play a machine of the specified type long enough that the 3×5 matrix of visible symbols yielded results that uniquely determined the corresponding par sheet. Only the results of the study were published, not the par sheets themselves.

The definition of a slot machine quoted at the beginning of the chapter is due to Jones (1978, p. 1).

The data needed for the tables of this chapter come from different sources. Tables 12.1 and 12.3 are from Mead's (1983, p. 128) *Handbook of Slot Machine Reel Strips*, which contains data for more than 200 mechanical machines. This particular Mills model belonged to their High Top line, produced from 1947 to 1961 (Fey 2002, p. 141); a virtually identical machine was said by Mead (2005, p. 229) to be a 1949 model. The Bally machine of Table 12.4, which is in the author's personal collection, is a 1969 model (Fey 2002, p. 207), and Table 12.6 was obtained by opening the machine. The data for the Bally "In the Money" machine (Tables 12.7–12.10) come from a Bally par sheet provided to us by George Stamos of Bally Gaming, though Table 12.13 was evaluated incorrectly on the par sheet. A par sheet for a slot machine of the same name appears in Kilby, Fox, and Lucas (2005, pp. 121–123), but that is a rather different machine.

The data for Problems 12.1 and 12.2 are from Mead (1983, pp. 148, 136). The four-reel Pace machine dates back to the 1950s (probably 1956), as pointed out by Fey (2002, p. 195). Problem 12.3 was motivated by Mead (1983, Section 2). The data for the IGT machine of Problem 12.6 are from Kilby and Fox (1998, pp. 114–115). The data for Problem 12.7 ("Red, White, and Blue") is in the public domain, having been published on Michael Shackleford's web site (`www.wizardofodds.com`). For Problem 12.12, see Geddes (1980), Geddes and Saul (1980), and Ethier and Lee (2010). Other applications of Markov chains to slot machines can be found in Oses (2008).

Chapter 13
Roulette

Above the Babel tongues and the clang of the music, as you listen in the great saloon, you hear from a neighbouring room a certain sharp ringing clatter, and a hard clear voice cries out, "Zero rouge," or "Trente-cinq noir. Impair et passe." And then there is a pause of a couple of minutes, and then the voice says, "Faites le jeu, Messieurs. Le jeu est fait, rien ne va plus"—and the sharp ringing clatter recommences. You know what that room is? That is Hades.

William Makepeace Thackeray, *The Kickleburys on the Rhine*

Roulette, the quintessential casino game for system players, is of French origin and dates back to about 1796. In Section 13.1 we discuss the many wagers available at roulette, observing that the game is unbeatable when the wheel is unbiased. Section 13.2 is concerned with biased wheels, and how to identify and exploit them. In particular, we describe a Bayesian strategy that is optimal in a certain sense.

13.1 Unbiased Wheels

We assume that the reader knows what a roulette wheel is. For our purposes, a *roulette wheel* is a device designed to produce a discrete uniform random variable U on $\{0, 1, 2, \ldots, 36\}$ or on $\{0, 00, 1, 2, \ldots, 36\}$. To treat both 37- and 38-number wheels simultaneously, we assume z zeros, $z \in \{1, 2\}$. The player can bet on certain subsets of numbers. If he bets on a permitted subset of m numbers and U belongs to that subset, he is paid

$$\frac{36}{m} - 1 \quad \text{to} \quad 1. \tag{13.1}$$

(There is one exception to this rule, to be explained in the next paragraph.) Therefore, denoting by X his profit from a one-unit bet on such a subset, the

S.N. Ethier, *The Doctrine of Chances*, Probability and its Applications, DOI 10.1007/978-3-540-78783-9_13, © Springer-Verlag Berlin Heidelberg 2010

house advantage associated with this wager is

$$\mathrm{H_0} = \mathrm{H} = -\left[\left(\frac{36}{m} - 1\right)\frac{m}{36+z} + (-1)\left(1 - \frac{m}{36+z}\right)\right] = \frac{z}{36+z}, \quad (13.2)$$

and the variance is

$$\mathrm{Var}(X) = \left(\frac{36}{m} - 1\right)^2 \frac{m}{36+z} + (-1)^2\left(1 - \frac{m}{36+z}\right) - \left(-\frac{z}{36+z}\right)^2$$

$$= \left(\frac{36}{36+z}\right)^2\left(\frac{36+z}{m} - 1\right). \quad (13.3)$$

A single coup at roulette is often called a *spin*.

To define the collection of subsets on which bets are permitted, we need to introduce the roulette *layout*. The numbers from 1 to 36 are arranged on the layout in the form of a 12×3 matrix with entries in lexicographical order, that is, the entry in row i and column j is $3(i-1) + j$. There is also a row 0, containing only 0 if $z = 1$ and both 0 and 00 if $z = 2$. One can bet on (a) a single number, (b) two adjacent numbers (this includes 0-1, 0-2, and 0-3 if $z = 1$; 0-1, 0-2, 00-2, 00-3, and 0-00 if $z = 2$), (c) three numbers comprising a row (also 0-1-2 and 0-2-3 if $z = 1$; 0-1-2, 0-00-2, and 00-2-3 if $z = 2$), (d) four numbers comprising a 2×2 submatrix (also 0-1-2-3 if $z = 1$), (e) the five numbers 0-00-1-2-3 if $z = 2$ (this bet pays only 6 to 1, the sole exception to (13.1)), (f) six numbers comprising two adjacent rows, (g) 12 numbers (the first, second, and third dozens [1–12, 13–24, and 25–36], and the first, second, and third columns), (h) 18 numbers (low [1–18], high [19–36], odd, even, red, and black), or (i) 24 numbers (two adjacent dozens or two adjacent columns) if $z = 1$. By convention, the sets of even and odd numbers exclude 0 and 00. For the color bets, it is necessary to know that each number has a color associated with it. For 1–10 and 19–28, odd numbers are red and even numbers are black, while for 11–18 and 29–36, the opposite is true.[1] The zeros are typically green.

Thus, the possible values of m in (13.2) and (13.3) are 1, 2, 3, 4, 6, 12, 18, and (if $z = 1$) 24. (5 is also possible if $z = 2$, but then the formulas must be modified.) Numerical values of the house advantage and standard deviation are given in Table 13.1. Notice that there are 161 distinct wagers at roulette (disregarding the amount wagered, and assuming that only one of the three methods of handling the even-money bets is available) in both cases, $z = 1$ and $z = 2$.

---

[1] An arguably simpler way to remember this rule is as follows: Define the function $f : \{1, 2, \ldots, 36\} \mapsto \{1, 2, \ldots, 11\}$ by letting $f(k)$ be the sum of the digits of k. Then, with two exceptions, k is red if $f(f(k))$ is odd, and k is black if $f(f(k))$ is even. The two exceptions are 10 and 28, which are black. The reason this works can be seen by considering the numbers in blocks of nine: 1–9, 10–18, 19–27, and 28–36. In the first and third blocks, odd numbers are red and even numbers are black. In the second and fourth blocks, the opposite is true except for the numbers 10 and 28.

To complete our description of the game, we need to address one more issue. The six 18-number even-money bets, which are also known as the *simple chances*, are sometimes treated more favorably by the casino. A casino offering the *partager* ("to share" in French) rule takes only half of the player's even-money bet when a zero appears. If $z = 2$, this rule is sometimes called the *surrender* rule.

Some casinos offer *en prison* ("in prison" in French). Let us describe the classical form of this rule. The casino moves the player's even-money bet into prison 1 when a zero appears, to await the result of the next coup. If the next coup results in a win for this bet, the bet is released from prison 1 and returned to the player. If it results in a loss, the player's bet is lost to the casino. If it results in another zero, the player's bet is moved into prison 2. More generally, suppose that the player's even-money bet is in prison $n \geq 1$. If the next coup results in a win for this bet, the bet is moved back to prison $n - 1$, where prison 0 is freedom. If it results in a loss, the player's bet is lost to the casino. If it results in another zero, the player's bet is moved into prison $n + 1$.

Let N be the number of coups required to release a bet from prison 1, with $N = \infty$ if the bet is lost and therefore never released. Letting

$$p = q = \frac{18}{36 + z} \quad \text{and} \quad r = \frac{z}{36 + z}, \tag{13.4}$$

we find that $P(N = 2n) = 0$ for all $n \geq 1$ and

$$P(N = 1) = p, \tag{13.5}$$
$$P(N = 3) = rp^2, \tag{13.6}$$
$$P(N = 5) = 2r^2 p^3, \tag{13.7}$$
$$P(N = 7) = 5r^3 p^4, \tag{13.8}$$

and so on. In fact,

$$P(N = 2n + 1) = \frac{1}{n + 1} \binom{2n}{n} r^n p^{n+1}, \qquad n \geq 0, \tag{13.9}$$

and therefore

$$P := P(N < \infty) = \sum_{n=0}^{\infty} \frac{1}{n + 1} \binom{2n}{n} r^n p^{n+1} = \frac{1 - \sqrt{1 - 4rp}}{2r}. \tag{13.10}$$

This is the probability that a bet in prison 1 is eventually released. The verification of (13.9) and (13.10) is left to the reader (Problem 13.1 on p. 474). See Problem 9.9 on p. 351 for a different approach to an equivalent problem.

The house advantage for an even-money bet with partager is $H_0 = \frac{1}{2} z/(36 + z)$, while the variance is

Table 13.1 The various permitted bets at roulette.

37-number wheel

number of numbers	payoff odds	number of permitted bets	house advantage	standard deviation
1	35 to 1	37	.027 027	5.837 838
2	17 to 1	60	.027 027	4.070 238
3	11 to 1	14	.027 027	3.275 515
4	8 to 1	23	.027 027	2.794 652
6	5 to 1	11	.027 027	2.211 597
12	2 to 1	6	.027 027	1.404 366
18	1 to 1	6	.027 027	.999 635
18[1]	1 to 1	6	.013 514	.989 721
18[2]	1 to 1	6	.013 701[3]	.993 220[3]
24	1 to 2	4	.027 027	.716 089
total		161[4]		

[1]with partager   [2]with en prison   [3]pushes included
[4]counting only one of the 18-number rows

38-number wheel

number of numbers	payoff odds	number of permitted bets	house advantage	standard deviation
1	35 to 1	38	.052 632	5.762 617
2	17 to 1	62	.052 632	4.019 344
3	11 to 1	15	.052 632	3.235 879
4	8 to 1	22	.052 632	2.762 030
5	6 to 1	1	.078 947	2.366 227
6	5 to 1	11	.052 632	2.187 854
12	2 to 1	6	.052 632	1.394 489
18	1 to 1	6	.052 632	.998 614
18[1]	1 to 1	6	.026 316	.979 711
18[2]	1 to 1	6	.027 046[3]	.986 754[3]
total		161[4]		

[1]with partager   [2]with en prison   [3]pushes included
[4]counting only one of the 18-number rows

$$\text{Var}(X) = 1 - \frac{3z}{4(36+z)} - \frac{z^2}{4(36+z)^2}. \tag{13.11}$$

The house advantage for an even-money bet with en prison is $H_0 = z(1 - P)/(36 + z)$ (pushes included) or $H = z(1 - P)/(36 + z(1 - P))$ (pushes excluded), while the variance (pushes included) is

$$\text{Var}(X) = 1 - \frac{zP}{36+z} - \left(\frac{z(1-P)}{36+z}\right)^2. \tag{13.12}$$

Numerical values are provided in Table 13.1. If a casino offers the option of choosing partager or en prison, the former is slightly more favorable, but the system player may prefer the latter, so as not to have to deal with fractional units.

The typical roulette player is not satisfied with making just one bet at each spin of the wheel, but prefers to make numerous bets of various types and sizes. With at least 161 bets available at roulette, this is easy to do.

But in a certain sense there are not really that many bets available. Excluding the even-money bets with partager or en prison (and the unattractive 5-number bet), there are in fact only $36 + z$ distinct bets available, namely the $36 + z$ single-number bets. The point is that the one-unit bet on m numbers ($m = 2, 3, 4, 6, 12, 18, 24$), which pays $36/m - 1$ to 1, is precisely equivalent to m single-number bets, each for $1/m$ units. To verify this, it will help to introduce some notation. Given a permitted subset $A \subset \{0, 1, 2, \ldots, 36\}$ (or $A \subset \{0, 00, 1, 2, \ldots, 36\}$) with $|A| \in \{1, 2, 3, 4, 6, 12, 18, 24\}$, we define X_A to be the player's profit from a one-unit bet on the subset A. More precisely,

$$X_A = \begin{cases} 36/|A| - 1 & \text{if } U \in A, \\ -1 & \text{if } U \notin A. \end{cases} \tag{13.13}$$

Then we claim that

$$X_A = \frac{1}{|A|} \sum_{j \in A} X_{\{j\}}. \tag{13.14}$$

The proof is straightforward. Indeed, if $U \in A$, then

$$X_A = \frac{36}{|A|} - 1 = \frac{(35)(1) + (-1)(|A| - 1)}{|A|} = \frac{1}{|A|} \sum_{j \in A} X_{\{j\}}, \tag{13.15}$$

while if $U \notin A$, then

$$X_A = -1 = \frac{(-1)|A|}{|A|} = \frac{1}{|A|} \sum_{j \in A} X_{\{j\}}. \tag{13.16}$$

We hasten to add that $1/m$ betting units may not be a legitimate bet size, either because it is too small or because it is not an integer multiple of the smallest unit of currency accepted. This, however, does not prevent us from using this identity to good effect.

First, we need two preliminary calculations. For $i \neq j$, we have

$$\text{Cov}(X_{\{i\}}, X_{\{j\}})$$
$$= \frac{(35)(-1)(1) + (-1)(35)(1) + (-1)^2(34+z)}{36+z} - \left(-\frac{z}{36+z}\right)^2$$
$$= \frac{-36+z}{36+z} - \left(\frac{z}{36+z}\right)^2 = -\left(\frac{36}{36+z}\right)^2, \qquad (13.17)$$

and

$$\text{Var}(X_{\{i\}}) = \frac{(35)^2(1) + (-1)^2(35+z)}{36+z} - \left(-\frac{z}{36+z}\right)^2$$
$$= \frac{35 \cdot 36 + z}{36+z} - \left(\frac{z}{36+z}\right)^2 = \left(\frac{36}{36+z}\right)^2(35+z). \qquad (13.18)$$

These results allow us to conclude from (13.14) that

$$\text{Cov}(X_A, X_B) = \frac{1}{|A|\,|B|} \text{Cov}\left(\sum_{i \in A} X_{\{i\}}, \sum_{j \in B} X_{\{j\}}\right)$$
$$= \frac{1}{|A|\,|B|} \sum_{i \in A} \sum_{j \in B} \text{Cov}(X_{\{i\}}, X_{\{j\}})$$
$$= \frac{1}{|A|\,|B|} \big[|A \cap B| \text{Var}(X_{\{1\}})$$
$$\qquad\qquad + (|A|\,|B| - |A \cap B|)\text{Cov}(X_{\{1\}}, X_{\{2\}})\big]$$
$$= \left(\frac{36}{36+z}\right)^2 \left[(36+z)\frac{|A \cap B|}{|A|\,|B|} - 1\right], \qquad (13.19)$$

and therefore that

$$\text{Corr}(X_A, X_B) = \frac{(36+z)|A \cap B| - |A|\,|B|}{\sqrt{|A|(36+z-|A|)|B|(36+z-|B|)}}. \qquad (13.20)$$

For example, consider the betting system that calls for a one-unit bet on black and a one-unit bet on the third column. This was claimed by its author to yield a positive expected profit owing to a "flaw" in the roulette layout, namely that the third column contains eight red numbers and only four black ones. Of course, we know this claim is nonsense without any further calculations, because the sum of two negative-expectation random variables cannot have positive expectation. Let A be the set of black numbers, and let B be the set of numbers in the third column. Then, using (13.19),

$$\text{Cov}(X_A, X_B) = \left(\frac{36}{36+z}\right)^2 \left(\frac{36+z}{54} - 1\right) < 0, \qquad (13.21)$$

so we could construct a hedge (Example 1.4.19 on p. 37) from these two wagers, but it would have little to recommend it.

Let us next formulate a result to the effect that (unbiased) roulette is an unbeatable game. Let $U_1, U_2, \ldots$ be a sequence of discrete uniform random variables on $\{0, 1, 2, \ldots, 36\}$ or on $\{0, 00, 1, 2, \ldots, 36\}$. We know that at each coup the roulette player has $d \geq 161$ betting opportunities, all subfair. Let $X_{n,i}$ be the player's profit per unit bet on the ith betting opportunity at the nth coup, and let $B_{n,i}$ be the amount bet on the ith betting opportunity at the nth coup. Put

$$\boldsymbol{X}_n := (X_{n,1}, \ldots, X_{n,d}), \qquad \boldsymbol{B}_n := (B_{n,1}, \ldots, B_{n,d}). \qquad (13.22)$$

We assume that

$$\boldsymbol{B}_1 = \boldsymbol{b}_1 \geq \boldsymbol{0}, \qquad \boldsymbol{B}_n = \boldsymbol{b}_n(U_1, \ldots, U_{n-1}) \geq \boldsymbol{0}, \quad n \geq 2, \qquad (13.23)$$

where $\boldsymbol{b}_1$ is a constant and $\boldsymbol{b}_n$ is a nonrandom function of $n-1$ variables for each $n \geq 2$. (Vector inequalities hold componentwise.) The player's fortune F_n after n coups is given by

$$F_n = F_0 + \sum_{l=1}^{n} \boldsymbol{B}_l \cdot \boldsymbol{X}_l, \qquad (13.24)$$

F_0 being his initial fortune (a positive constant).

Assume bounded bet sizes, that is, there exists a constant $\boldsymbol{C}$ such that $\boldsymbol{0} \leq \boldsymbol{B}_n \leq \boldsymbol{C}$ for all $n \geq 1$. Let N be a stopping time with $\text{P}(N < \infty) = 1$. Then Theorem 8.2.5 on p. 303 yields two conclusions:

(a) If (13.24) satisfies $F_n \geq 0$ for all $n \geq 0$, then

$$\text{E}[F_N] \leq F_0. \qquad (13.25)$$

(b) Assume $\text{E}[N] < \infty$. Then (13.25) holds; moreover,

$$\text{E}[F_N - F_0] = \text{E}\left[\sum_{l=1}^{N} \boldsymbol{B}_l \cdot \boldsymbol{X}_l\right] = \text{E}\left[\sum_{l=1}^{N} \boldsymbol{B}_l\right] \cdot \text{E}[\boldsymbol{X}_1]. \qquad (13.26)$$

Every component of $\text{E}[\boldsymbol{X}_1]$ being negative (see Table 13.1), (13.26) is a stronger conclusion than (13.25). In effect, (13.26) says that, regardless of the system used, one can expect to lose the sum, over all i, of the expected amount bet on the ith betting opportunity multiplied by the corresponding house advantage. (The even-money bets with en prison are the only ones in which pushes can occur and the only ones that may require more than one coup to resolve. Here the house advantage H_0 with pushes included is the correct one to use.) If one avoids bets with partager or en prison as well as the 5-number bet, then all bets have the same house advantage, thereby simplifying the conclusion: Regardless of the system used, one can expect to lose the expected total amount bet multiplied by the common house advantage. Of course, the gambler's actual profit $F_N - F_0$ may turn out to be positive, but its expectation can never be.

Let us consider the gambler's ruin formula in the context of roulette. Let k be a positive integer, and assume that the random variable X satisfies

$$P(X = k) = p, \quad P(X = -1) = q, \quad P(X = 0) = r, \tag{13.27}$$

where $p, q, r \geq 0$ and $p + q + r = 1$. This is general enough to include all one-unit wagers at roulette with the exception of the even-money bets with partager and the 24-number bets. Now let $X_1, X_2, \ldots$ be a sequence of i.i.d. random variables with common distribution that of X, representing the results of independent repetitions of the original wager. Then

$$S_n := X_1 + X_2 + \cdots + X_n \tag{13.28}$$

represents the gambler's cumulative profit after n such wagers; of course, $S_0 = 0$. Given positive integers W and L, we suppose that the gambler stops betting as soon as he wins W or more betting units or loses L betting units. Consequently, his last bet occurs at coup

$$N(-L, W) := \min\{n \geq 0 : S_n = -L \ \text{ or } \ S_n \geq W\}. \tag{13.29}$$

(It is possible to overshoot the upper boundary but not the lower one.)

By Example 7.2.2 on p. 253, we know that

$$P(S_{N(-L,W)} \geq W) = 1 - \left(\frac{q}{p+q}\right)^L \frac{a(W)}{a(L+W)}, \tag{13.30}$$

where

$$a(M) := \sum_{j=0}^{\lfloor (M-1)/(k+1) \rfloor} (-1)^j \binom{M - jk - 1}{j} \left(\frac{pq^k}{(p+q)^{k+1}}\right)^j. \tag{13.31}$$

We can apply this to the bet on m numbers by taking

$$k := \left\lfloor \frac{36}{m} \right\rfloor - 1, \quad p := \frac{m}{36 + z}, \quad q := 1 - p, \quad r := 0, \tag{13.32}$$

where $m \in \{1, 2, 3, 4, 5, 6, 12, 18\}$. Actually, (13.31) is numerically unstable, containing some very large terms when M is large, and so high-precision arithmetic may be needed. But for $m = 12$ we can apply Example 7.2.3 on p. 255, and for $m = 18$ we can apply Theorem 7.1.1 on p. 242. For even-money bets with en prison, we can apply Theorem 7.1.1 once again with

$$p := \frac{18}{36+z}, \quad q := p + \frac{z(1-P)}{36+z}, \quad r := \frac{zP}{36+z}, \tag{13.33}$$

where P is as in (13.10). For even-money bets with partager, we can apply the
iteration method of Section 7.2, although this is time-consuming for large M.
A better approach is suggested in Problem 13.5 on p. 475. For 24-number bets
we can apply the result of Problem 7.10 on p. 268. Results are summarized
in Table 13.2.

Table 13.2 is quite revealing. The primary implication is that the roulette
player attempting to achieve a specified goal with constant betting should
make only single-number bets. Of course, if his goal is so small that he is
likely to overshoot the goal significantly, this advice may not apply (see the
$W = 10$ column in the table).

In fact the same conclusion holds for the bold player. It can be shown
that betting boldly on a single number is optimal at roulette. The proof is
outlined in Problem 9.16 on p. 352.

13.2 Biased Wheels

So far, we have assumed that the roulette wheel achieves perfect randomness,
that is, we have assumed that each of the $36 + z$ numbers is equally likely.
There is an abundance of anecdotal evidence that this is not always the case
(see the chapter notes). The question of how to detect and exploit biased
roulette wheels is an interesting one that we attempt to address in this section.

Suppose we have "clocked" a roulette wheel for n coups, and the most
frequent number has occurred a proportion $1/25$ of the time. Is it a biased
wheel? Although we can never be absolutely certain, we can nevertheless
quantify our degree of confidence. It will of course depend on n.

To make this precise, we consider a generalized roulette wheel with $r \geq 36$
numbers, namely $1, 2, \ldots, r$. Assuming that it is unbiased, we can model the
numbers of 1s, 2s, $\ldots$, rs in n coups by letting $(X_1, X_2, \ldots, X_r)$ have the
multinomial distribution with sample size n and with r cells, each with probability $1/r$, a distribution we have denoted by multinomial$(n, 1/r, \ldots, 1/r)$.

First we need a preliminary inequality. We claim that

$$P(X_1 \geq l, \ X_2 \geq l) \leq P(X_1 \geq l)P(X_2 \geq l) \tag{13.34}$$

for $l = 0, 1, \ldots, n$. This says that $1_{\{X_1 \geq l\}}$ and $1_{\{X_2 \geq l\}}$ are negatively correlated, which is plausible because X_1 and X_2 are negatively correlated. A
proof is sketched in Problem 13.14 on p. 476.

It follows from the inclusion-exclusion inequalities (Theorem 1.1.19 on
p. 13) that

Table 13.2 The probability that a roulette player will double his fortune of W betting units before losing it, if at each coup he bets one unit on a set of m numbers. (Notice that he may overshoot his goal and, in the case of the 18-number bet with partager and the 24-number bet, he may lose an extra half-unit.) Results are rounded to six significant digits.

37-number wheel

m	$W = 10$	$W = 100$	$W = 1{,}000$
1	.172 094	.432 285	.170 727
2	.300 474	.405 719	.038 237 8
3	.386 143	.368 830	.006 883 87
4	.424 473	.328 931	.001 082 94
6	.439 998	.247 713	.000 018 367 1
12	.423 364	.061 513 5	$.159\ 799 \cdot 10^{-11}$
18	.368 031	.004 466 28	$.330\ 297 \cdot 10^{-23}$
18^1	.430 181	.059 250 0	$.104\ 009 \cdot 10^{-11}$
18^2	.431 007	.058 556 6	$.866\ 593 \cdot 10^{-12}$
24	.256 765	.000 022 092 7	$.277\ 070 \cdot 10^{-46}$

[1]with partager   [2]with en prison

38-number wheel

m	$W = 10$	$W = 100$	$W = 1{,}000$
1	.166 086	.393 647	.043 229 8
2	.290 033	.331 735	.001 775 85
3	.372 032	.264 168	.000 057 692 7
4	.406 437	.201 192	.000 001 516 51
5	.391 816	.060 597 6	$.192\ 855 \cdot 10^{-11}$
6	.412 602	.102 633	$.513\ 773 \cdot 10^{-9}$
12	.360 404	.004 818 54	$.831\ 206 \cdot 10^{-23}$
18	.258 533	.000 026 560 7	$.174\ 787 \cdot 10^{-45}$
18^1	.363 572	.004 106 54	$.158\ 718 \cdot 10^{-23}$
18^2	.363 091	.003 612 45	$.392\ 408 \cdot 10^{-24}$

[1]with partager   [2]with en prison

$$r\mathrm{P}(X_1 \geq l) - \binom{r}{2}(\mathrm{P}(X_1 \geq l))^2$$

$$\leq r\mathrm{P}(X_1 \geq l) - \binom{r}{2}\mathrm{P}(X_1 \geq l,\ X_2 \geq l)$$

$$\leq \mathrm{P}\left(\max_{1 \leq i \leq r} X_i \geq l \right) \leq r\mathrm{P}(X_1 \geq l). \tag{13.35}$$

Now X_1 is binomial$(n, 1/r)$, so we can apply De Moivre's limit theorem (Theorem 1.5.13 on p. 48) to conclude from the upper bound in (13.35) that

$$\limsup_{n \to \infty} \mathrm{P}\left(\max_{1 \leq i \leq r} X_i \geq l_n^+ \right) \leq \alpha \tag{13.36}$$

if

$$l_n^+ := \frac{n}{r} + \Phi^{-1}\left(1 - \frac{\alpha}{r} \right)\sqrt{\frac{n}{r}\left(1 - \frac{1}{r} \right)}. \tag{13.37}$$

Furthermore, by the quadratic formula,

$$r\lambda - \binom{r}{2}\lambda^2 = \alpha \quad \text{implies} \quad \lambda = \frac{1 \pm \sqrt{1 - 2\alpha(1 - 1/r)}}{r - 1}, \tag{13.38}$$

so we can conclude from the lower bound in (13.35) that

$$\liminf_{n \to \infty} \mathrm{P}\left(\max_{1 \leq i \leq r} X_i \geq l_n^- \right) \geq \alpha \tag{13.39}$$

if

$$l_n^- := \frac{n}{r} + \Phi^{-1}\left(1 - \frac{1 - \sqrt{1 - 2\alpha(1 - 1/r)}}{r - 1} \right)\sqrt{\frac{n}{r}\left(1 - \frac{1}{r} \right)}. \tag{13.40}$$

In words, (13.37) and (13.40) are asymptotic upper and lower bounds on the $1 - \alpha$ quantile of the distribution of $\max_{1 \leq i \leq r} X_i$. The standard-normal quantiles appearing in (13.37) and (13.40) are evaluated in Table 13.3 for $r = 36, 37, 38$ and several values of α.

Table 13.3 Standard-normal quantiles needed in (13.37) and (13.40).

α	$\Phi^{-1}\left(1 - \frac{\alpha}{r}\right)$			$\Phi^{-1}\left(1 - \frac{1 - \sqrt{1 - 2\alpha(1 - 1/r)}}{r - 1}\right)$		
	$r = 36$	$r = 37$	$r = 38$	$r = 36$	$r = 37$	$r = 38$
0.01	3.452	3.460	3.467	3.451	3.459	3.466
0.02	3.261	3.269	3.276	3.258	3.266	3.273
0.05	2.991	3.000	3.008	2.984	2.992	3.000
0.1	2.773	2.782	2.790	2.756	2.765	2.773
0.2	2.539	2.549	2.558	2.498	2.508	2.518
0.5	2.200	2.211	2.222	1.981	1.992	2.002

Using $r = 37$ or $r = 38$ gives a test for determining whether the wheel is biased, based on the frequency of the most frequent number. Actually, we are not concerned simply with whether the wheel is biased. The relevant question is whether there is a *favorable number*, that is, one that results in a superfair bet; a necessary and sufficient condition is that its probability p satisfy $36p - 1 > 0$, or $p > 1/36$. To answer this question, we can use the same criterion, but with $r = 36$.

As justification, we note that the multinomial probabilities belong in general to

$$\Delta := \{\boldsymbol{p} \in [0, 1]^r : p_1 + \cdots + p_r = 1\}. \tag{13.41}$$

The case of no favorable numbers corresponds to

$$\Delta_0 := \{\boldsymbol{p} \in \Delta : p_1 \le 1/36, \ldots, p_r \le 1/36\}, \tag{13.42}$$

and it suffices to verify that

$$\max_{\boldsymbol{p} \in \Delta_0} \mathrm{P}_{n,\boldsymbol{p}} \left(\max_{1 \le i \le r} X_i \ge l \right) = \mathrm{P}_{n,(1/36,1/36,\ldots,1/36)} \left(\max_{1 \le i \le 36} X_i \ge l \right), \tag{13.43}$$

where the subscripts on the probabilities denote the multinomial parameters. The proof of (13.43) is sketched in Problem 13.15 on p. 477.

For an example, let us return to the question raised at the beginning of this section. Suppose we have clocked a roulette wheel for n coups, and the most frequent number has occurred a proportion $q > 1/36$ of the time. Given that we want a $100(1 - \alpha)\%$ level of confidence that one or more favorable numbers exist, we let $r = 36$ and find the smallest n for which $l_n^+/n \le q$ (using the more conservative upper bound). The requirement is

$$n \ge 35 \left[\frac{\Phi^{-1}(1 - \alpha/36)}{36q - 1} \right]^2. \tag{13.44}$$

When $q = 1/25$ and $\alpha = 0.05$, this amounts to $n \ge 1{,}618$.

A more subtle question than whether to bet is how to bet. When the probabilities of the various numbers are known (an admittedly unlikely scenario), we have seen in Section 10.2 that one should use the Kelly system. Consider a biased roulette wheel with $r \ge 36$ numbers. We relabel the numbers as $1, 2, \ldots, r$ with

$$p_1 \ge p_2 \ge \cdots \ge p_r, \tag{13.45}$$

and we assume that $p_1 > 1/36$. Then the unique vector of optimal betting proportions $\boldsymbol{f}^*$ is specified in terms of

$$k_0 := \max\{k \in \{1, \ldots, 35\} : p_1 + \cdots + p_{k-1} + (36 - (k - 1))p_k > 1\} \tag{13.46}$$

by

$$f_j^* = \begin{cases} p_j - (36 - k_0)^{-1}(1 - p_1 - \cdots - p_{k_0}) & \text{if } j = 1, \ldots, k_0, \\ 0 & \text{if } j = k_0 + 1, \ldots, r. \end{cases} \tag{13.47}$$

In particular, the optimal proportional bettor bets on number k if and only if

$$p_1 + \cdots + p_{k-1} + (36 - (k - 1))p_k > 1, \tag{13.48}$$

a condition that is implied by but does not imply (unless $k = 1$) that number k is favorable, that is, $p_k > 1/36$.

Of course, if $p_1 \leq 1/36$, then there are no favorable numbers and no bets are made, that is, $\boldsymbol{f}^* = \boldsymbol{0}$. For numerical examples, refer to Example 10.2.4 on p. 370.

Now label the $r \geq 36$ numbers by $1, 2, \ldots, r$, and suppose (as is more typical) that the probabilities $p_1, p_2, \ldots, p_r$ are unknown. Of course, $p_1 + \cdots + p_r = 1$, but we do not assume (13.45). A very useful approach, known as *Bayes estimation*, requires as a first step that we regard the unknown probabilities $p_1, p_2, \ldots, p_r$ as random variables with a specified distribution known as the *prior distribution*, representing our understanding of the situation before any data have been seen. For this purpose we assume that $(p_1, \ldots, p_r)$ has the discrete Dirichlet distribution introduced in Problem 1.31 on p. 58. (Readers familiar with the continuous Dirichlet distribution may use that here instead.) And since we have no a priori knowledge of the various probabilities, we assume that the parameters $\theta_1, \ldots, \theta_r$ of the discrete Dirichlet distribution are all equal to some $\theta \geq 1$. Thus,

$$\mathrm{P}\left((p_1, \ldots, p_r) = \left(\frac{m_1}{N}, \ldots, \frac{m_r}{N}\right)\right) = C_N(\theta, \ldots, \theta)\left(\frac{m_1}{N}\right)^{\theta-1} \cdots \left(\frac{m_r}{N}\right)^{\theta-1}, \tag{13.49}$$

where $m_1, \ldots, m_r$ are nonnegative integers that sum to N. Here N is some very large number. For example, with $N = 10^9$, we are effectively assuming that each of the probabilities $p_1, \ldots, p_r$ has been rounded to nine decimal places (in such a way that the sum of the probabilities is exactly 1).

After n coups, the observed statistics are

$$N_j(n) := \text{number of times } j \text{ has occurred in } n \text{ coups} \tag{13.50}$$

for $j = 1, \ldots, r$. The conditional distribution of $(p_1, \ldots, p_r)$ given the observations $(N_1(n), \ldots, N_r(n))$ is known as the *posterior distribution*, and by Bayes's law (Theorem 1.2.7 on p. 19) it is given by

$$\mathrm{P}\left((p_1, \ldots, p_r) = \left(\frac{m_1}{N}, \ldots, \frac{m_r}{N}\right) \Big| (N_1(n), \ldots, N_r(n)) = (n_1, \ldots, n_r)\right)$$

$$= \mathrm{P}\left((p_1, \ldots, p_r) = \left(\frac{m_1}{N}, \ldots, \frac{m_r}{N}\right)\right)$$

$$\cdot \mathrm{P}\left((N_1(n), \ldots, N_r(n)) = (n_1, \ldots, n_r) \,\Big|\, (p_1, \ldots, p_r) = \left(\frac{m_1}{N}, \ldots, \frac{m_r}{N}\right)\right)$$

$$\cdot [\mathrm{P}((N_1(n), \ldots, N_r(n)) = (n_1, \ldots, n_r))]^{-1}$$

$$= C_N(\theta, \ldots, \theta) \left(\frac{m_1}{N}\right)^{\theta-1} \cdots \left(\frac{m_r}{N}\right)^{\theta-1} \binom{n}{n_1, \ldots, n_r} \left(\frac{m_1}{N}\right)^{n_1} \cdots \left(\frac{m_r}{N}\right)^{n_r}$$

$$\cdot [\mathrm{P}((N_1(n), \ldots, N_r(n)) = (n_1, \ldots, n_r))]^{-1}$$

$$= C_N(\theta + n_1, \ldots, \theta + n_r) \left(\frac{m_1}{N}\right)^{\theta+n_1-1} \cdots \left(\frac{m_r}{N}\right)^{\theta+n_r-1}. \tag{13.51}$$

The last step may require additional explanation. The last two expressions, as functions of $(m_1, \ldots, m_r)$, are proportional and are both probability distributions, hence they are equal. Thus, the posterior distribution is, like the prior distribution, a discrete Dirichlet distribution, but it has new parameters, namely $\theta + n_1, \ldots, \theta + n_r$.

In Bayes estimation, the mean vector of the posterior distribution is often taken as the best estimate of the parameter vector. By Problem 1.31 on p. 58, the mean vector of the discrete Dirichlet distribution (13.51) is approximately

$$\left(\frac{\theta + n_1}{r\theta + n}, \ldots, \frac{\theta + n_r}{r\theta + n}\right) = \frac{r\theta}{r\theta + n} \left(\frac{1}{r}, \ldots, \frac{1}{r}\right) + \frac{n}{r\theta + n} \left(\frac{n_1}{n}, \ldots, \frac{n_r}{n}\right). \tag{13.52}$$

This identity shows that the Bayes estimate of the parameter vector is a weighted average of the prior estimate (all r numbers equally likely) and the empirical estimate based on the n observations.

We can now substitute these estimates into the optimal-bet-size formulas (13.46) and (13.47) (after relabeling to achieve (13.45)), obtaining a Bayes strategy that has been shown to be asymptotically optimal in a certain sense. The choice of θ is critical. If θ is chosen too large, it will be a long time before any bias is detectable. If θ is chosen too small, the estimates will be too sensitive to random fluctuations.

13.3 Problems

13.1. *Catalan numbers and the distribution of prison-release time.*

(a) Given $n \geq 1$, let C_n be the number of permutations of n 0s and n 1s such that no initial segment has more 1s than 0s. Show that

$$C_n = \frac{1}{n+1} \binom{2n}{n}, \qquad n \geq 1. \tag{13.53}$$

These are the *Catalan numbers.*

(b) Show that the Catalan numbers have generating function

$$\sum_{n=0}^{\infty} C_n x^n = \frac{1 - \sqrt{1 - 4x}}{2x}, \qquad |x| < 1/4. \qquad (13.54)$$

(c) Use (a) and (b) to prove (13.9) and (13.10) on p. 463.

Hint: (a) Use the result of Problem 3.4 on p. 111.

13.2. *Irrationality of en prison house advantage.* Prove that the house advantage of the even-money bet at roulette with en prison (and one or two zeros) is an irrational number. Do any other casino wagers have this property?

13.3. *Independence of roulette bets.* As in Section 13.1, let X_A be the player's profit from a one-unit bet on the subset A. Assuming a 37- or 38-number wheel, are there any two permitted subsets A and B for which X_A and X_B are independent? Answer the same question for the 36-number wheel (no zeros). Ignore partager and en prison.

13.4. *Correlations between outside bets.* Let X_i denote the profit from a single unit wagered on the ith outside bet, $i = 1, 2, \ldots, 12$ (ordered as follows: low, high, odd, even, red, black; first, second, third dozen; first, second, third column). For each $1 \leq i < j \leq 12$, evaluate $\text{Corr}(X_i, X_i)$ numerically. Consider 38-, 37-, and 36-number wheels, and ignore partager and en prison. (Cf. Problem 13.3.)

13.5. *Gambler's ruin formula for partager.* Using the method of Example 7.2.3 on p. 255, find an explicit gambler's ruin formula for the even-money bet at roulette with partager. More generally, consider the game of red-and-black-and-green with partager (Problem 9.9 on p. 351). Specialize to 37- and 38-number wheels, and use the two resulting formulas to confirm the six relevant entries in Table 13.2 on p. 470.

Hint: Use Theorem A.1.5.

13.6. *Expected number of distinct numbers in n coups.* Find an expression for the expected number of distinct roulette numbers to appear in n coups, without finding the distribution. Assume a 37-number wheel, and find the smallest n for which the expectation exceed 36.

Hint: Use indicators.

13.7. *Distribution of number of distinct numbers in n coups.* Find an explicit expression (i.e., not recursive) for the distribution of the number of distinct roulette numbers to appear in n coups. Assume a 37-number wheel, and determine the most likely number of distinct numbers in 37 coups.

Hint: Use Example 1.4.24 on p. 39.

13.8. *Expected number of coups for all numbers to appear at least once.* Find an expression for the expected number of coups needed for all roulette numbers to appear at least once, assuming a 37-number wheel. Do the same for the standard deviation.

13.9. *A system with a high probability of winning.* Assume a 37-number wheel with partager. Consider the system that bets 72 units on 19–36, 48 units on the first dozen (1–12), 12 units on the row comprising 16–18, 8 units on the adjacent numbers 14 and 15, and 3 units on zero. Find the distribution of the player's profit, showing in particular that it is positive with probability 36/37. Does that make this an infallible system?

13.10. *A system for the player whose lucky number is 17.* Assume a 38-number wheel without partager or en prison. Consider the system that bets one unit on each of the permitted subsets of $\{0, 00, 1, 2, \ldots, 36\}$ containing the number 17. Find the distribution of the player's profit. Use this to find the ratio of the player's expected profit to his total amount bet. Is the result as expected?

13.11. *Insurance at roulette.* Twenty-one and trente et quarante have insurance bets, and in roulette one can construct such a bet as follows. Assume a 37-number wheel with partager. Given a one-unit even-money bet, what fraction of a unit should be bet on zero, in order to pay off the claim (loss of the even-money bet) in the event of a zero occurring? How does this "insurance" affect the house advantage and the standard deviation?

13.12. *Variations of en prison.* Some casinos may not offer en prison according to the classical rule we have specified. Specifically, let us say that a casino offers *en prison of the nth degree* $(n \geq 1)$ if, whenever the player's bet would be moved into prison $n + 1$ under the classical en prison rule, it is instead lost to the casino. Let P_n denote the probability of escape from prison 1, assuming en prison of the nth degree.

(a) Evaluate P_n and the resulting house advantage (pushes included) for $n = 1, 2, 3, 4$.

(b) Argue that P_n is increasing in n, and find its limit P.

Hint: With $p = 18/(36 + z)$ and $r = z/(36 + z)$ as in Section 13.1, show that $P_n = p + r P_{n-1} P_n$ for each $n \geq 1$, where $P_0 := 0$.

13.13. *Boule.* Boule is a simple roulette-like game played in Europe. There are nine numbers, 1–9, which are equally likely to occur. One can bet on a single number, paying 7 to 1, or on a subset of four of the nine numbers, paying even money. The six permitted subsets are low $(1, 2, 3, 4)$, high $(6, 7, 8, 9)$, odd $(1, 3, 7, 9)$, even $(2, 4, 6, 8)$, black $(1, 3, 6, 8)$, and red $(2, 4, 7, 9)$. The number 5, which is yellow, plays the role of roulette's zero(s). There is no partager or en prison. Evaluate the analogues of Tables 13.1 on p. 464 and 13.2 on p. 470 for the game of boule.

13.14. *Multinomial correlation inequality.* Let the random vector (S_n, T_n, U_n) be multinomial(n, p, q, r), where n is a positive integer, $p > 0$, $q > 0$, $r > 0$, and $p + q + r = 1$. Prove by induction on n that

$$\mathrm{P}(S_n \geq s,\ T_n \geq t) \leq \mathrm{P}(S_n \geq s)\mathrm{P}(T_n \geq t) \tag{13.55}$$

for $s, t = 0, 1, \ldots, n$.

Hint: Show that, if $s, t \geq 1$, then

$$P(S_{n+1} \geq s,\ T_{n+1} \geq t) = pP(S_n \geq s - 1,\ T_n \geq t) + qP(S_n \geq s,\ T_n \geq t - 1)$$
$$+ rP(S_n \geq s,\ T_n \geq t) \tag{13.56}$$

and

$$0 \leq P(S_n \geq s - 1,\ T_n \geq t - 1) - P(S_n \geq s - 1,\ T_n \geq t)$$
$$- P(S_n \geq s,\ T_n \geq t - 1) + P(S_n \geq s,\ T_n \geq t). \tag{13.57}$$

13.15. *An inequality concerning the maximum cell frequency of a multinomial.* Prove (13.43) on p. 472 in three steps:

(a) If X is binomial(n, p), show that $f(p) := P(\max(X, n - X) \geq l)$ is nondecreasing in $p \in [\frac{1}{2}, 1)$ for $l = \lfloor n/2 \rfloor + 1, \ldots, n$.

(b) If $\boldsymbol{X}$ is an r-dimensional multinomial$(n, \boldsymbol{p} - t\boldsymbol{e}_j + t\boldsymbol{e}_k)$, where $j, k \in \{1, 2, \ldots, r\}$ are distinct and $0 < p_j \leq p_k < 1$, show that

$$g(t) := P\left(\max_{1 \leq i \leq r} X_i \geq l \right) \tag{13.58}$$

is nondecreasing in $t \in [0, p_j]$ for $l = 1, 2, \ldots, n$. Here $\boldsymbol{e}_1, \ldots, \boldsymbol{e}_r$ are defined by $\boldsymbol{e}_1 := (1, 0, \ldots, 0)$, $\boldsymbol{e}_2 := (0, 1, 0, \ldots, 0)$, and so on.

(c) Show that (13.43) on p. 472 follows from the conclusion of part (b).

Hint: (b) Use Problem 2.13 on p. 90 and part (a).

13.16. *Number of coups needed to conclude that a favorable number exists.* We saw that, if the most frequent number in n coups has occurred a proportion $1/25$ of the time, we can conclude with 95% confidence that one or more favorable numbers exist if $n \geq 1{,}618$. Create a table that generalizes this example. Let the confidence level be $100(1 - \alpha)\%$, where $\alpha = 0.01, 0.02, 0.05, 0.1, 0.2, 0.5$. Let the observed proportion of occurrences of the most frequent number be $1/k$, where $k = 19, 20 \ldots, 35$.

13.4 Notes

The definitive work on the history of roulette is a paper by Barnhart (1988), the principal results of which we summarize below.

In much the same form as we know it today, roulette first appeared at the Palais Royal in Paris in about 1796, with both 0 and 00. Some authors have credited roulette to Blaise Pascal [1623–1662], but this seems highly unlikely because Pascal died more than a century before roulette made its appearance. Others have credited Antoine de Sartine [1729–1801], the Lieutenant General of Police of Paris from 1759 to 1774. As Barnhart (1988) noted, this too is unlikely, for it was Sartine who was responsible for outlawing what was then called roulette in 1759. But there is little doubt that the inventor was French. As Davis (1956, p. 54) put it,

> Roulette, as the name indicates, is an invention of the French—by instigation of the devil.[2]

In any case, like many inventions, roulette relied heavily on several of its predecessors.

The first horizontal-wheel game, called *roly poly* or *roulet*, appeared in London around 1720. Shortly thereafter another wheel game called *ace of hearts* was introduced. In 1739 the House of Commons outlawed ace of hearts (as well as faro, bassette, and hazard). To circumvent this law, a gambling house proprietor named Cook introduced the game of *EO* (even-odd), which was subsequently outlawed in 1745. In EO the wheel had 40 compartments, 19 labeled E, 19 labeled O, one labeled Ē, and one labeled Ō. A bet on E paid even money if the ball landed in an E compartment, was lost if it landed in O or Ō, and was pushed if it landed in Ē. A bet on O was treated analogously. (See Ashton 1968, p. 201.) Roly poly was very similar, with the colors black and white in place of the letters E and O, and likely also 40 compartments. On the other hand, ace of hearts had 25 or 31 compartments labeled by playing cards, and allowed betting on a single card, paying 23 to 1 or 28 to 1. At some point, roulet was exported to Paris, where it was called "roulette." It was described by Huyn (1788) as essentially unchanged from the English version (still an even-money game only). In 1791 the game was referred to as red and black roulette, the colors evidently having been changed from black and white.

The Italian game of *hoca* was introduced to the court of Louis XIV by Cardinal Mazarin [1602–1661] in about 1648. In hoca the numbers 1–30 were arranged in the form of a 10×3 matrix, and players could bet on a single number, paying 28 to 1, two adjacent numbers, paying 13 to 1, and so on. The croupier would then draw a ball at random from a leather sack and announce the winning number. The game was repeatedly outlawed in Paris but continued to be played at Versailles. Around the end of the 17th century, another Italian game, which the French called *biribi*, arrived in Paris. In the casino version of biribi, the numbers 1–70 were arranged in nine columns, each with eight numbers except for the middle column, which had six numbers reserved for the banker. Players could bet on a single number, paying 63 to 1, two adjacent numbers, paying 31 to 1, and so on. Also available were even-money bets on red, black, odd, even, low, and high. The method of drawing the winning number was as in hoca. By the 1780s there was a street version of biribi, which had to be portable and therefore smaller. In this version the numbers 1–36 were arranged in the form of a 6×6 matrix, and one could

---

[2] Those who believe that roulette is the work of the devil may find some support for their position by noticing that the sum of the numbers on a roulette wheel is $1 + 2 + \cdots + 36 = \binom{37}{2} = 666$ (see *Revelation* 13:18).

bet on a single number, paying 31 to 1, two adjacent numbers, paying 15 to 1, and so on. Even-money bets on odd, even, low, and high (but not red or black) were available.

It seems likely that from knowledge of these earlier games some ingenious Frenchman, whose name has not been recorded, created roulette as it was played at the Palais Royal in about 1796. In effect, he took the hoca layout with the even-money bets of biribi and with the 36 numbers of the street version of biribi, to which he added the 0 and 00, and he labeled the compartments of the existing red and black roulette wheel with these 38 numbers.

Barnhart's (1988) paper contains an extensive list of errors made by historians of gambling in attempting to understand the origin of roulette. This resulted in what he called a "scholarly disaster," in which facts became accepted once they had been misstated often enough. The reader is referred to that paper for further details.

Originally, 0 was red and 00 was black. (See the Thackeray quote on p. 461.) Then, just as in EO, a bet on red, for example, was pushed when a 0 appeared and lost when a 00 appeared. For the same purpose, according to Foster (1953b, p. 500), 0 was regarded as odd and low, and 00 was regarded as even and high. According to Corti (1934, p. 198), the single-zero roulette wheel made its debut at Monte Carlo in 1864, thanks to proprietor François Blanc [1806–1877]. But the same author also wrote (p. 47) that the single-zero wheel had already appeared in Homburg by around 1842, introduced by the Blanc brothers to attract gamblers from Baden-Baden, Wiesbaden, and Ems. A likely explanation for the confusion was supplied by Quinn (1892, p. 118): The single-zero wheel was introduced at Homburg, and the en prison rule was introduced at Monte Carlo. Some 19th-century roulette wheels in the United States had a third zero, represented by an eagle. Today, the 38-number wheel is standard in the United States (with surrender in Atlantic City), while the 37-number wheel is standard in Europe (with partager in England, and either partager or en prison, at the player's discretion, at Monte Carlo). As Richardson (1929, p. 9) remarked,

> At some casinos the bank claims two zeros and "gets away with it," but it is a piece of unblushing effrontery to ask the public to submit to such an exorbitant tax on the turnover.

It has been stated by several authors (e.g., Epstein 1977, p. 110; Sifakis 1990, p. 256) that roulette is the oldest casino game still in operation. As we will see, this is incorrect. Trente et quarante is certainly older and is still in operation. (Keno is also older, but it did not become a casino game until the 20th century. Faro too is older, but it is no longer in operation.)

The rule we have stated for en prison is the classical one, as is confirmed by Gall (1883, p. 69), Silberer (1910, Chapter 19), and Boll (1951, p. 47–48). Modern casinos may use a slightly different rule in the case of repeated zeros. Our calculation of P in (13.10) is slightly different from that of Davies and Ross (1979). Sagan (1981) and Dieter and Ahrens (1984) found essentially the same result.

In our view, the en prison rule should be as follows: The casino moves the player's even-money bet into prison when a zero appears, to await the result of the next nonzero coup. If the next nonzero coup results in a win for this bet, the bet is released from prison and returned to the player. If it results in a loss, the player's bet is lost to the casino.

The rule of footnote 1 for remembering which numbers are red and which are black is due to Boll (1951, p. 13). The covariance formula derived in (13.19) has been extended to allow for partager by Hannum (2007).

The order of the numbers on the roulette wheel has not yet been mentioned because it is irrelevant. (In fact, arranging the numbers in natural order would make sense because then a near miss would be more apparent.) Nevertheless, the order is standardized, and we provide it here for the sake of completeness. In clockwise order, for the 37-number wheel,

$$0, 32, 15, 19, 4, 21, 2, 25, 17, 34, 6, 27, 13, 36, 11, 30, 8, 23, 10,$$
$$5, 24, 16, 33, 1, 20, 14, 31, 9, 22, 18, 29, 7, 28, 12, 35, 3, 26, \tag{13.59}$$

and for the 38-number wheel,

$$0, 28, 9, 26, 30, 11, 7, 20, 32, 17, 5, 22, 34, 15, 3, 24, 36, 13, 1,$$
$$00, 27, 10, 25, 29, 12, 8, 19, 31, 18, 6, 21, 33, 16, 4, 23, 35, 14, 2. \tag{13.60}$$

In both cases, no two numbers of the same color are adjacent.

Roulette has played a role in several important literary works. Perhaps the best-known example is Fyodor Dostoevsky's [1821–1881] *The Gambler* (1867), a classic study of compulsive gambling. The game of choice of the narrator, Alexey Ivanovitch, is roulette (Dostoevsky 1967, pp. 179–180):[3]

> I know for certain that I am not mean; I believe that I am not even a spendthrift—and yet with what a tremor, with what a thrill at my heart, I hear the croupier's cry: *trente-et-un, rouge, impair et passe*, or: *quatre, noir, pair et manque!* With what avidity I look at the gambling table on which louis d'or, friedrichs d'or and thalers lie scattered: on the piles of gold when they are scattered from the croupier's shovel like glowing embers, or at the piles of silver a yard high that lie round the wheel. Even on my way to the gambling hall, as soon as I hear, two rooms away, the clink of the scattered money I almost go into convulsions.
>
> Oh! that evening, when I took my seventy gulden to the gambling table, was remarkable too. I began with ten gulden, staking them again on *passe*. I have a prejudice in favour of *passe*. I lost. I had sixty gulden left in silver money; I thought a little and chose *zéro*. I began staking five gulden at a time on *zéro*; at the third turn the wheel stopped at *zéro*; I almost died of joy when I received one hundred and seventy-five gulden; I had not been so delighted when I won a hundred thousand gulden. I immediately staked a hundred on *rouge*—it won; the two hundred on *rouge*—it won; the whole of the four hundred on *noir*—it won; the whole eight hundred on *manque*—it won; altogether with what I had before it made one thousand seven hundred gulden and that—in less than five minutes! Yes, at moments like that one forgets all one's former failures! Why, I had gained this by risking more than life itself, I dared to risk it, and—there I was again, a man among men.

---

[3] Translation by Constance Garnett.

In Ian Fleming's [1908–1964] *Casino Royale* (1953, Chapter 7), James Bond

> decided to play one of his favorite gambits and back two—in this case the first two—dozens, each with the maximum—one hundred thousand francs.

It was pointed out by Spanier (1980, pp. 20–21) that this was not a knowledgeable play. Had Bond bet 150,000 francs on 1–18 and 50,000 francs on 19–24, he would have done no worse than he did and he might have done better, by virtue of the en prison rule. The same observation was made, although of course without reference to Fleming or Bond, by Silberer (1910, Chapter 18).

A not entirely unrelated matter is the existence of the 24-number bet. It is certainly permitted on the 37-number layout. See, for example, Boll (1936, p. 190; 1951, p. 22), Downton and Holder (1972), and Spanier (1980, pp. 12–13). On the other hand, most U.S. gambling authors, such as Scarne (1974), fail to mention this bet. More significantly, according to Lemmel (1964, p. 85) and Friedman (1996, p. 147), placing a bet on the boundary between the second and third dozens of the 38-number layout is interpreted, oddly enough, as a bet on the two numbers 0 and 00. (The intention was to make it easier for players sitting near the third dozen to make this bet.) Thus, the 24-number bet does not exist on the 38-number layout.

The system described below (13.20) first appeared in 1959 in *Bohemia*, a Cuban monthly magazine, according to Scarne (1961, p. 373). Its author analyzed the system for each of the four possible outcomes (black or not, third column or not), and slipped up, perhaps intentionally, on one of the four calculations, allowing him to claim infallibility.

Karl Pearson [1857–1936], one of the founders of modern statistics, studied color runs at Monte Carlo roulette in 1892 (Pearson 1894, 1897), obtaining his data from *Le Monaco*, a weekly Paris publication that listed the previous week's numbers. His results were striking:

> Monte Carlo roulette, if judged by returns which are published apparently with the sanction of the *Société*, is, if the laws of chance rule, from the standpoint of exact science the most prodigious miracle of the nineteenth century. Yet even the supernatural would be discredited by fortnightly recurrences; we are forced to accept as alternative that the random spinning of a roulette manufactured and daily readjusted with extraordinary care is not obedient to the laws of chance, but is chaotic in its manifestations! [...] Clearly, since the Casino does not serve the valuable end of a huge laboratory for the preparation of probability statistics, it has no *scientific raison d'être*. Men of science cannot have their most refined theories disregarded in this shameless manner! The French Government must be urged by the hierarchy of science to close the gaming-saloons; it would be, of course, a graceful act to hand over the remaining resources of the Casino to the Académie des Sciences for the endowment of a laboratory of orthodox probability; [...]

As noted by Stigler (1986, p. 329), the statistical tests performed for the article were naive, but the work led to Pearson's (1900) chi-squared goodness-of-fit test for the multinomial distribution.

The first author to discuss biased roulette in some detail was Wilson (1965, Chapters 1–4). A more thorough treatment was provided by Barnhart (1992). Also useful are Billings and Fredrickson (1996), Zender (2006, Chapter 7), and Grosjean (2009, Chapter 75). A number of wheel clockers over the years have won considerable sums. These include Joseph Jaggers [1830–1892] (Monte Carlo, 1873), Helmut Berlin (Mar del Plata, 1951), Richard Jarecki (San Remo, 1969, 1971), Pierre Basieux (Bad Wiessee, 1981), and Billy Walters (Atlantic City, 1986). Their stories were recounted by Barnhart (1992, Chapters 4–10).

Basieux (2001) (resp., Barnhart 1992, p. 188) used (13.37) with $r = 37$ and $\alpha = 0.05$ (resp., $\alpha = 0.20$) as a betting criterion. As noted in the text, the choice $r = 36$ is more appropriate. (See Grosjean 2009, Chapter 75, if there is any doubt.) Equation (13.35) is due to Kozelka (1956), while (13.40) is due to Fuchs and Kenett (1980). A more recent study by Pascual and Wilslef (1996) is mathematically flawed: Their Appendix B contains several egregious errors.

The Bayes strategy of Section 13.2 was proposed by Thomas G. Kurtz in 1981 (see Ethier 1982b). It was subsequently shown by Klotz (2000) to be asymptotically optimal. Klotz also carried out computer simulations of the Bayes strategy assuming $p_1 = 1/30$ and $p_2 = \cdots = p_{38} = 29/1,110$, and found promising results for $\theta = 200$ and $\theta = 500$. See Billings and Fredrickson (1996, Chapter 7) for further information.

The fact that the strategy requires a computer, which would not likely be welcome in a casino, need not be a concern. Indeed, updating the Bayes strategy every 100 coups, say, should prove adequate. In any case, Nevada enacted its device law in 1985 (Forte 2004, p. 41), which makes it a felony to use, at a licensed gaming establishment, any device to assist in "analyzing the probability of occurrence of an event relating to" a game.

The Catalan numbers (Problem 13.1) are named for Eugène Catalan [1814–1894]. Problem 13.6 comes from Boll (1951, p. 44). Problem 13.7 was motivated by Billings and Fredrickson (1996, p. 73). Problem 13.9 is due to Mahl (1977). Boule (Problem 13.13) is covered at great length—more than it deserves—in Boll (1936, Chapters 10 and 11). Problem 13.12 is due to Sagan (1981). Instead of using Problem 13.14 to prove (13.34), we could use the FKG inequality, named for Fortuin, Kasteleyn, and Ginibre (1971). See Liggett (1985, Corollary 2.12) for a statement of this result.

Of course, the challenge issued years ago by Wilson (1965, p. 40) remains open today.

> The challenge also remains for all young adventurers, the minimum requirement being lots of time for the search. As long as man makes roulette wheels, there will be imperfections, [...]. But let these researchers do their homework well, and gird strongly for the battle. The whirling ladies respect only those who come well armed with knowledge!

Chapter 14
Keno

[B]ut think how mysterious and often unaccountable it is—that lottery of life which gives to this man the purple and fine linen, and sends to the other rags for garments and dogs for comforters.

William Makepeace Thackeray, *Vanity Fair*

Keno is an ancient lottery-type game of Chinese origin that offers high payoffs, albeit with small probabilities and large house advantages. In Section 14.1 we analyze the basic wager, the so-called m-spot ticket. In Section 14.2 we consider the case of multiple wagers made simultaneously. These are known as way tickets and are what make keno more interesting than lotteries.

14.1 The m-Spot Ticket

The basic wager at *keno* is the *m-spot ticket*, which can be described as follows. The keno player chooses m numbers ($1 \leq m \leq 15$) from the set $\{1, 2, \ldots, 80\}$, marks them on his keno ticket, and pays an entry fee. The casino then randomly draws 20 numbers from the set $\{1, 2, \ldots, 80\}$ without replacement. If the player's m numbers and the casino's 20 numbers have k numbers in common, where $0 \leq k \leq m$, the player receives a payoff depending on k, m, and the size of the entry fee, according to a specified payoff schedule to be explained below. This describes one coup, which at keno is called a game or a *race*.

Before getting to the payoffs, let us first consider the probabilities at keno. Although the casino's 20 numbers are drawn sequentially, their order has no relevance, and consequently there are $\binom{80}{20}$ equally likely outcomes. The number of these that result in exactly k numbers in common is $\binom{m}{k}\binom{80-m}{20-k}$, so

S.N. Ethier, *The Doctrine of Chances*, Probability and its Applications,
DOI 10.1007/978-3-540-78783-9_14, © Springer-Verlag Berlin Heidelberg 2010

$$P_m(k) := \mathrm{P}(\text{catch } k \text{ of } m) = \frac{\binom{m}{k}\binom{80-m}{20-k}}{\binom{80}{20}}. \tag{14.1}$$

(The word "catch" is used instead of "match" for historical reasons; see the chapter notes.) This is of course the hypergeometric distribution, introduced in Section 1.3. By Problem 1.21 on p. 56 or Example 2.2.6 on p. 82, (14.1) is equivalent to the formula we get by interchanging m and 20, that is,

$$P_m(k) := \mathrm{P}(\text{catch } k \text{ of } m) = \frac{\binom{20}{k}\binom{80-20}{m-k}}{\binom{80}{m}}. \tag{14.2}$$

Of course, (14.2) would follow directly if the player chose his m numbers at random without replacement (and independently of the casino's selection), but this is contrary to the way in which the game is usually played. Several authors, preferring (14.2) to (14.1) because of its smaller denominator (and therefore smaller numerator), have tried to justify (14.2) directly, without reference to (14.1), usually with dubious results (see the chapter notes).

Nevertheless it is possible to establish (14.2) without reference to (14.1). The argument can be formulated as follows.

As noted, the 20 numbers selected by the casino are drawn sequentially (without replacement). We can imagine that all 80 numbers are drawn sequentially, with only the first 20 "counting" as the casino's numbers. This leads to a random permutation of $(1, 2, \ldots, 80)$, one that has a discrete uniform distribution (i.e., all 80! orderings are equally likely). Let us say that the first number drawn is in position 1 of the random permutation, the second number drawn is in position 2, and so on. The event in question, namely that the player's m numbers and the casino's 20 numbers have k numbers in common ("catch k of m"), is simply the event that k of the positions 1 through 20 of the random permutation are occupied by k of the player's numbers (and therefore $m - k$ of the positions 21 through 80 are occupied by the remaining $m - k$ of the player's numbers). The number of ways in which this can occur (without regard to which k of the player's numbers occupy these positions or in what order) is $\binom{20}{k}\binom{80-20}{m-k}$. And of course there are $\binom{80}{m}$ ways in which m of the positions 1 through 80 of the random permutation can be occupied by the player's m numbers (without regard to order). Moreover, these $\binom{80}{m}$ ways are equally likely by virtue of the random permutation's discrete uniform distribution. Therefore, (14.2) holds.

Let

$$N_m(k) := \binom{20}{k}\binom{60}{m-k} \tag{14.3}$$

be the numerator in (14.2). Then, for $m = 1, 2, \ldots, 15$, $N_m(m) = \binom{20}{m}$ and

$$N_m(k-1) = \frac{k}{21-k}\frac{60-m+k}{m-k+1}N_m(k), \quad k = m, m-1, \ldots, 1. \tag{14.4}$$

This permits easy evaluation of the keno probabilities (14.2) using

$$P_m(k) = \frac{N_m(k)}{\binom{80}{m}}, \qquad k = 0, 1, \ldots, m. \tag{14.5}$$

In Table 14.1 we list the results for the popular 8-spot ticket and the traditional 10-spot ticket. The reason that the 8-spot ticket is popular is that it offers the best chance (approximately one chance in 230,115) of a five-digit payoff. No comparable payoff is available on a 7-spot ticket. A comparable payoff on a 9-spot ticket is only one-sixth as likely.

It is important to recognize that we have made no assumptions on how the player selects his m numbers. He may do so at random (and some casinos offer the option of a computer-generated random selection), or he may use a deterministic algorithm. The probabilities are the same either way. For example, the 8-spot ticket on which the numbers

$$1, 2, 3, 4, 5, 6, 7, 8 \tag{14.6}$$

are marked has the same distribution of catches (given by (14.2) with $m = 8$) as the 8-spot ticket on which the numbers

$$4, 14, 23, 34, 42, 50, 59, 72 \tag{14.7}$$

are marked. It does not matter that the latter set of numbers may appear more random.[1] A knowledgeable player should have no preference for the ticket with (14.7) marked over the one with (14.6) marked. (However, as we will see, the term "knowledgeable player" is something of an oxymoron.)

Next we introduce the payoffs, concerning which there are two complications. First, the payoffs have not been standardized, so they typically vary from one casino to the next. Second, and more importantly, while each casino has a minimum betting limit, there are no maximum betting limits at keno. Instead, each casino limits its liability by announcing a *maximum aggregate payout* per race. To see how this works, it will be convenient to measure money in units of the minimum betting limit. For $k = 0, 1, \ldots, m$, let $a_m(k)$ represent the number of units paid out to the player when his m-spot ticket, purchased for one unit, catches k of his m numbers. These payoffs are nonnegative rational numbers, usually integers. Now let M denote the maximum aggregate payout in units, and assume that $\max_{0 \leq k \leq m} a_m(k) \leq M$. Then, for $k = 0, 1, \ldots, m$, the number of units paid out to the player when his m-spot ticket, purchased for $b \geq 1$ units, catches k of his m numbers is $(a_m(k)b) \wedge M$ units. In particular, no matter how much he bets, the maximum possible amount paid out to the player is M units.

---

[1] In fact, it is equally deterministic, being the first eight numbered subway stops by the northbound A-train in New York City.

Table 14.1 Distribution of the number of catches by 8- and 10-spot keno tickets.

8 spots

catch	number of ways	reciprocal of probability
0	2,558,620,845	11.329 360 192 9
1	7,724,138,400	3.752 850 563 89
2	9,512,133,400	3.047 427 525 56
3	6,226,123,680	4.655 792 052 95
4	2,362,591,575	12.269 381 410 1
5	530,546,880	54.637 089 091 9
6	68,605,200	422.526 822 311
7	4,651,200	6,232.270 629 08
8	125,970	230,114.607 843
sum	28,987,537,150	

10 spots

catch	number of ways	reciprocal of probability
0	75,394,027,566	21.838 495 213 4
1	295,662,853,200	5.568 816 279 42
2	486,137,960,550	3.386 882 415 55
3	440,275,888,800	3.739 682 667 17
4	242,559,401,700	6.787 995 429 49
5	84,675,282,048	19.444 778 574 1
6	18,900,732,600	87.112 608 011 8
7	2,652,734,400	620.677 332 084
8	222,966,900	7,384.468 771 46
9	10,077,600	163,381.371 569
10	184,756	8,911,711.176 47
sum	1,646,492,110,120	

If we define the random variable $R_m(b)$ to be the number of units paid out to the player *per unit bet* from an m-spot ticket purchased for $b \geq 1$ units, then

$$\mathrm{E}[R_m(b)] = \sum_{k=0}^{m} [a_m(k) \wedge (M/b)] P_m(k) \qquad (14.8)$$

and

$$\text{Var}(R_m(b)) = \sum_{k=0}^{m} [a_m(k) \wedge (M/b)]^2 P_m(k) - (\text{E}[R_m(b)])^2. \qquad (14.9)$$

The player's expected profit per unit bet is $\text{E}[R_m(b)] - 1$ units, so the house advantage (pushes included) is $\text{H}_0 = 1 - \text{E}[R_m(b)]$.

Observe that (14.8) and (14.9) do not depend on b as long as

$$1 \leq b \leq \frac{M}{\max_{0 \leq k \leq m} a_m(k)}. \qquad (14.10)$$

Increasing b beyond this threshold will decrease (14.8) and therefore increase the house advantage, though perhaps only slightly.

To make this more concrete, Table 14.2 displays the minimum bet size, the maximum aggregate payout, the payoffs, the house advantage, and the standard deviation (assuming (14.10)) of the number of units paid out *per unit bet* for various 8- and 10-spot tickets available on the Las Vegas Strip. While the house advantage figures are high relative to those of other casino wagers, so too are the standard deviations.

We have not yet explained the reason for the word "aggregate" in the term "maximum aggregate payout." The reason is that the maximum aggregate payout is the limit of the casino's liability for the aggregate payout to all participants in a specific keno race. If the aggregate payout called for exceeds the specified maximum aggregate payout, each winning player receives a prorated share of the maximum aggregate payout. In practice this rule is not rigorously enforced. Typically, all small and moderate payouts are paid in full and only the largest ones are subject to *proration*.

How likely is it that proration will be needed? We can estimate the probability by making some (admittedly unrealistic) simplifying assumptions. Let us assume that there are n players, each betting one unit on an 8-spot ticket and each selecting his 8 numbers at random (and independently of the other players). We further assume that the payoffs and maximum aggregate payout are those at Harrah's et al. (Table 14.2). Finally, to simplify matters, let us suppose that all catches of 7 of 8 or fewer are paid in full, and only catches of 8 of 8 are prorated. The number N of catches of 8 of 8 is of course a binomial$(n, P_8(8))$ random variable, and proration is needed if and only if $N \geq 4$. This occurs with probability

$$P(N \geq 4) = P((N)_4 \geq (4)_4) \leq \frac{\text{E}[(N)_4]}{4!} = \binom{n}{4} p^4 \leq \frac{(np)^4}{24}, \qquad (14.11)$$

where $p := P_8(8)$ and we have used Markov's inequality (Lemma 1.5.1 on p. 42) and the fourth factorial moment of the binomial distribution. The upper bound is small if n is much smaller than $1/p \approx 230{,}115$, as it typically is. In short, the probability that proration will be needed due to multiple winners is extremely small and can be safely ignored.

Table 14.2 Statistics for 8- and 10-spot keno games on the Las Vegas Strip, January 2008. MAP is maximum aggregate payout, HA is house advantage, and SD is standard deviation. (Only eight of the 23 casinos surveyed offered live keno.)

8 spots

casino	min bet (=: 1 unit)	MAP (in units)	units paid out per unit bet* for catching						HA(%)	SD
			0–3	4	5	6	7	8		
Excalibur	$1	100,000	0	0	9	90	1,500	20,000	29.468	46.037
Harrah's et al.**	$2	100,000	0	0	7	80	1,500	30,000	31.149	65.480
Imperial Palace	$1	200,000	0	1	5	80	1,480	25,000	29.153	55.522
Sahara	$1	100,000	0	0	9	90	1,500	25,000	27.295	55.652
Treasure Island	$1	100,000	0	0	8	90	1,500	20,000	31.298	46.034

10 spots

casino	min bet (=: 1 unit)	MAP (in units)	units paid out per unit bet* for catching							HA(%)	SD
			0–4	5	6	7	8	9	10		
Excalibur	$2	50,000	0	2	20	130	1,000	4,000	25,000	29.540	18.303
Harrah's et al.**	$2	100,000	0	0	23	150	1,000	5,000	40,000	32.379	22.580
Imperial Palace	$1	200,000	0	1	25	125	1,000	4,000	30,000	29.693	19.138
Sahara	$2	50,000	0	2	20	130	1,000	4,000	25,000	29.540	18.303
Treasure Island	$1	100,000	0	1	22	132	960	3,800	25,000	32.729	17.792

*regardless of bet size, number of units paid out cannot exceed maximum aggregate payout
**includes Bally's, Caesars Palace, Flamingo, and Harrah's

14.2 Way Tickets

If a player wants to bet on multiple m-spot tickets simultaneously (possibly with several values of m), he can simply create a *way ticket*. (Some authors distinguish between way tickets and combination tickets—we do not.) We begin with the necessary definitions. Given a subset $A \subset \{1, 2, \ldots, 80\}$ satisfying $1 \leq |A| \leq 15$, let us define the random variable R_A to be the number of units paid out to the player from an $|A|$-spot ticket on which the elements of A are marked, after purchasing the ticket for the minimum price of one unit.

The most general way ticket depends on three things, which we now describe. First, let $A_1, A_2, \ldots, A_r$ be mutually exclusive nonempty subsets of $\{1, 2, \ldots, 80\}$, where $r \geq 1$. These are the groups of numbers that will be combined in various ways to create the way ticket. (Here the word "group" is used in the nontechnical sense—no algebraic structure is intended.) Second, let $\mathscr{I}$ be a nonempty collection of nonempty subsets of $\{1, 2, \ldots, r\}$ with the property that

$$I \in \mathscr{I} \quad \text{implies} \quad \left| \bigcup_{i \in I} A_i \right| \leq 15. \tag{14.12}$$

$\mathscr{I}$ is the set of ways in which the groups $A_1, A_2, \ldots, A_r$ will be combined. Third, let $b : \mathscr{I} \mapsto (0, \infty)$ be the function that determines how much will be bet on each way. Typically, the minimum allowable bet per way on a way ticket is considerably less than the minimum allowable bet on a single m-spot ticket, so $b < 1$ is possible.

The payout from the way ticket specified by the groups $A_1, A_2, \ldots, A_r$, the ways $\mathscr{I}$, and the bet-size function b is simply

$$\left(\sum_{I \in \mathscr{I}} b(I) R_{\cup_{i \in I} A_i} \right) \wedge M \tag{14.13}$$

units, where M is the maximum aggregate payout.

Here is an important class of examples. Let r, s, and t be positive integers with $rs \leq 80$, $st \leq 15$, and $t \leq r$. Assume that each of the r (mutually exclusive) groups $A_1, A_2, \ldots, A_r$ has s elements, and let $\mathscr{I}$ be the collection of subsets $I \subset \{1, 2, \ldots, r\}$ with $|I| = t$, so that each way comprises t of the r groups and therefore has st spots. With b constant, the ticket with payout (14.13) could be referred to as an (r, s, t) ticket, but it is more conventional to call it an $\binom{r}{t}$-way st-spot ticket. Examples include $(r, s, t) = (20, 4, 2)$ (a 190-way 8-spot ticket) and $(r, s, t) = (10, 2, 5)$ (a 252-way 10-spot ticket).

Of course, there is no requirement that the groups $A_1, A_2, \ldots, A_r$ be of the same size, or that the subsets in $\mathscr{I}$ be of the same size, or that the bets on each way be of the same size. Some players take pride in creating complex way tickets.

Were it not for the maximum aggregate payout, the house advantage of a way ticket would be simply the weighted average of the house advantages of the individual ways, with weights proportional to the amounts bet (see (6.91) and below on p. 222). Moreover, the variance of the payout from a way ticket would be computable in terms of the variances and covariances of the individual ways and pairs of ways. Let us see how this works.

First, we need a formula for the covariance between the payout from an arbitrary one-unit m-spot ticket and the payout from an arbitrary one-unit n-spot ticket. Let $A \subset \{1, 2, \ldots, 80\}$ satisfy $|A| = m$, let $B \subset \{1, 2, \ldots, 80\}$ satisfy $|B| = n$, and define $l = |A \cap B|$. Then

$$\begin{aligned}
\mathrm{Cov}(R_A, R_B) \\
&= \mathrm{E}[R_A R_B] - \mathrm{E}[R_A]\mathrm{E}[R_B] \\
&= \sum_{i,j,k \geq 0:\, i+j+k \leq 20} a_m(i+k) a_n(j+k) \frac{\binom{m-l}{i}\binom{n-l}{j}\binom{l}{k}\binom{80-m-n+l}{20-i-j-k}}{\binom{80}{20}} \\
&\quad - \sum_{i=0}^{m} a_m(i) \frac{\binom{20}{i}\binom{60}{m-i}}{\binom{80}{m}} \sum_{j=0}^{n} a_n(j) \frac{\binom{20}{j}\binom{60}{n-j}}{\binom{80}{n}},
\end{aligned} \tag{14.14}$$

ignoring the maximum aggregate payout, where we use the convention that a binomial coefficient is 0 if its lower entry exceeds its upper entry. As intuition might suggest, the covariance depends on how many numbers A and B have in common. We illustrate this numerically in Table 14.3.

Table 14.3 The covariance and correlation between the payouts from two one-unit 8-spot tickets, assuming the payoffs are those at Harrah's et al. (Table 14.2 on p. 488) and ignoring the maximum aggregate payout.

numbers in common	covariance	correlation
0	−.414 519 631	−.000 096 677
1	−.211 337 028	−.000 049 289
2	.632 412 239	.000 147 495
3	4.036 864 755	.000 941 501
4	17.560 136 620	.004 095 479
5	70.960 375 988	.016 549 798
6	281.403 606 618	.065 630 609
7	1,106.171 823 008	.257 987 919
8	4,287.688 467 527	1.000 000 000

To treat the general way ticket with payout (14.13) would be extremely complicated, so, for the remainder of this section, we limit consideration to the special class of way tickets defined above, which depends on positive integers r, s, and t with $rs \leq 80$, $st \leq 15$, and $t \leq r$. Specifically, there are r (mutually exclusive) groups $A_1, A_2, \ldots, A_r$, each with s elements, and $\mathscr{I}$ is the collection of subsets $I \subset \{1, 2, \ldots, r\}$ with $|I| = t$. For simplicity, we assume that $b \equiv 1$, so that the entry fee is $\binom{r}{t}$ units. Then, ignoring the maximum aggregate payout, the payout from the resulting $\binom{r}{t}$-way st-spot ticket has expectation

$$\mathrm{E}\left[\sum_{I \subset \{1,2,\ldots,r\}: \ |I|=t} R_{\cup_{i \in I} A_i}\right] = \binom{r}{t} \mathrm{E}[R_{\cup_{i=1}^t A_i}] \tag{14.15}$$

and variance

$$\mathrm{Var}\left(\sum_{I \subset \{1,2,\ldots,r\}: \ |I|=t} R_{\cup_{i \in I} A_i}\right)$$

$$= \sum_{I \subset \{1,2,\ldots,r\}: \ |I|=t} \mathrm{Var}(R_{\cup_{i \in I} A_i}) + \sum_{I \neq J: \ |I|=|J|=t} \sum \mathrm{Cov}(R_{\cup_{i \in I} A_i}, R_{\cup_{j \in J} A_j})$$

$$= \binom{r}{t} \mathrm{Var}(R_{\cup_{i=1}^t A_i}) \tag{14.16}$$

$$+ \binom{r}{t} \sum_{k=(2t-r)\vee 0}^{t-1} \binom{t}{k}\binom{r-t}{t-k} \mathrm{Cov}(R_{\cup_{i=1}^t A_i}, R_{\cup_{j=t+1-k}^{2t-k} A_j}).$$

Let us specialize to the case of the 190-way 8-spot ticket (i.e., $(r, s, t) = (20, 4, 2)$). Then $A_1, A_2, \ldots, A_{20}$ partition $\{1, 2, \ldots, 80\}$ into 20 groups, each with 4 elements. Assuming the payoffs to be those at Harrah's et al. (Table 14.2 on p. 488), the number of units paid out to the player is

$$R := \sum_{1 \leq i < j \leq 20} R_{A_i \cup A_j} \tag{14.17}$$

units, and, from (14.15) and (14.16), the mean and standard deviation of the payout per unit bet are

$$\mathrm{E}\left[\frac{R}{190}\right] \approx 0.688507756 \quad \text{and} \quad \mathrm{SD}\left(\frac{R}{190}\right) \approx 5.055706459, \tag{14.18}$$

ignoring the aggregate maximum payout. Notice that the covariance terms in (14.16) make a positive contribution to the variance.

How does the maximum aggregate payout affect these results? To answer this question, we would need to find the distribution, not just the mean and the standard deviation, of the payout per unit bet from the 190-way 8-spot ticket. This is easier than one might expect. We therefore return to the general case (i.e., (r, s, t) arbitrary).

Let $\mathscr{C}$ be the random set of numbers drawn by the casino so that $\mathscr{C} \subset \{1, 2, \ldots, 80\}$ and $|\mathscr{C}| = 20$. Define random variables $N_0, N_1, \ldots, N_s$ by

$$N_j := |\{1 \le i \le r : |A_i \cap \mathscr{C}| = j\}|, \qquad j = 0, 1, \ldots, s, \qquad (14.19)$$

so that N_j is the number of groups with exactly j catches. Then we find that $(N_0, N_1, \ldots, N_s)$ has distribution

$$P((N_0, N_1, \ldots, N_s) = (n_0, n_1, \ldots, n_s))$$

$$= \binom{r}{n_0, n_1, \ldots, n_s} \frac{\binom{s}{0}^{n_0} \binom{s}{1}^{n_1} \cdots \binom{s}{s}^{n_s} \binom{80-rs}{20-n_1-2n_2-\cdots-sn_s}}{\binom{80}{20}}, \qquad (14.20)$$

where $n_0, n_1, \ldots, n_s \ge 0$, $n_0 + n_1 + \cdots + n_s = r$, and $(rs - 60)^+ \le n_1 + 2n_2 + \cdots + sn_s \le 20$. The first factor comes from Theorem 1.1.6 on p. 5, and the second factor is the multivariate hypergeometric distribution (Section 1.3).

Once we know $(N_0, N_1, \ldots, N_s)$, we can calculate the amount paid out to the player. More precisely, suppose that $(N_0, N_1, \ldots, N_s) = (n_0, n_1, \ldots, n_s)$. Recall that the player's $\binom{r}{t}$ st-spot tickets are indexed by subsets $I \subset \{1, 2, \ldots, r\}$ with $|I| = t$. The number of these subsets I satisfying

$$k_j = |I \cap \{1 \le i \le r : |A_i \cap \mathscr{C}| = j\}|, \qquad j = 0, 1, \ldots, s, \qquad (14.21)$$

(i.e., k_j is the number of groups on ticket I with j catches) is

$$\binom{n_0}{k_0} \binom{n_1}{k_1} \cdots \binom{n_s}{k_s}, \qquad (14.22)$$

provided $0 \le k_j \le n_j$ for $j = 0, 1, \ldots, s$ and $k_0 + k_1 + \cdots + k_s = t$. (As a check, notice that the sum of the products (14.22) is $\binom{r}{t}$, by virtue of the multivariate hypergeometric distribution.) We conclude that the number of units paid out to the player is

$$R := \sum_{k_1=0}^{N_1 \wedge t} \sum_{k_2=0}^{N_2 \wedge (t-k_1)} \cdots \sum_{k_s=0}^{N_s \wedge (t-k_1-\cdots-k_{s-1})} \binom{N_0}{t - k_1 - \cdots - k_s}$$

$$\cdot \binom{N_1}{k_1} \binom{N_2}{k_2} \cdots \binom{N_s}{k_s} a_{st}(k_1 + 2k_2 + \cdots + sk_s), \qquad (14.23)$$

again ignoring the maximum aggregate payout. From this and (14.20) we can find the distribution of $R \wedge M$, which is the payout to the player when the maximum aggregate payout is taken into account.

Returning to the 190-way 8-spot ticket, from Table 14.2 on p. 488, $M = 100{,}000$ units. We find that

$$E\left[\frac{R \wedge M}{190}\right] \approx 0.688507389 \quad \text{and} \quad SD\left(\frac{R \wedge M}{190}\right) \approx 5.055652886. \qquad (14.24)$$

Thus, the expectation of payout is reduced only in the seventh significant digit by the maximum aggregate payout, and the standard deviation is reduced only in the fifth significant digit. However, to say that the way-ticket player should never be concerned with the maximum aggregate payout would be overstating the conclusion (see Problem 14.12 on p. 495).

14.3 Problems

14.1. *Payout from betting on every 8-spot ticket.* Suppose a keno player bets $\binom{80}{8} = 28{,}987{,}537{,}150$ units, one on each 8-spot ticket. Assume that the payoffs are those at Harrahs et al. (Table 14.2 on p. 488), but that the casino has agreed to suspend the maximum aggregate payout rule. What is the expected payout to the player? What is the variance?

14.2. *Relationship between bet size, maximum aggregate payout, and house advantage.* Consider the 8-spot ticket, and assume that the payoffs are those at Harrah's et al. (Table 14.2 on p. 488). Compute the house advantage for various bet sizes (e.g., 1, 2, 5, 10, 20, 50, 100 units) and various values of the maximum aggregate payout (e.g., 50, 100, 250, 500, 1,000, 2,500, 5,000, in thousands of units).

14.3. *Probability of no two consecutive numbers.* Find the probability that the casino's 20 numbers, drawn at random without replacement from the set $\{1, 2, \ldots, 80\}$, contain no two consecutive numbers. (Note that 10 and 11, for example, are consecutive, even though they are not adjacent on the keno ticket.)

14.4. *Probability of a number in every row and every column.* The keno ticket is effectively an 8×10 matrix whose (i, j)th entry is $10(i - 1) + j$. Find the probability that the casino's 20 numbers, drawn at random without replacement from the set $\{1, 2, \ldots, 80\}$, include at least one member of each row and at least one member of each column of the keno ticket.

14.5. *Edge ticket.* The *edge* or border ticket is in effect a 32-spot ticket with the numbers on the edge of the keno ticket marked. (Specifically, numbers congruent to 0 or 1 (mod 10) and numbers differing from 1 or 80 by at most 9 are marked.) Assume that the payoffs are as in Table 14.4. Evaluate the house advantage as well as the standard deviation of the number of units paid out.

14.6. *Top-bottom or left-right ticket.* The *top-bottom* ticket is in effect a 40-spot ticket with the numbers 1–40 marked. The *left-right* ticket is in effect a 40-spot ticket with the numbers congruent to 1, 2, 3, 4, or 5 (mod 10) marked. Assume that the payoffs are as in Table 14.5. Evaluate the house advantage as well as the standard deviation of the number of units paid out.

14.7. *Catch-all tickets.*

Table 14.4 Edge ticket payoffs (in units) at the Sahara (Jan. 2008). Minimum bet is $5 (one unit). Maximum aggregate payout is 20,000 units.

catch	payoff	catch	payoff	catch	payoff
0	5,000	7	0	14	40
1	600	8	0	15	200
2	80	9	0	16	2,000
3	8	10	0	17	10,000
4	2	11	1	18	12,000
5	1	12	2	19	15,000
6	0	13	8	20	20,000

Table 14.5 Top-bottom (or left-right) ticket payoffs (in units) at the Excalibur, December 2007. Bet is $5 (one unit). Maximum aggregate payout is 20,000 units.

catch	payoff	catch	payoff	catch	payoff
0 or 20	20,000	3 or 17	120	6 or 14	2
1 or 19	10,000	4 or 16	25	7 or 13	1
2 or 18	600	5 or 15	6	8 or 12	4/5
				9, 10, or 11	0

(a) For each m, $1 \leq m \leq 15$, find the probability of catching all m spots marked on an m-spot ticket.

(b) For $m = 4, 5, 6$, Treasure Island offers a one-unit m-spot catch-all ticket that pays out $a(m)$ units for catching all m spots and pays out nothing otherwise (Jan. 2008). Specifically, $a(4) = 225$, $a(5) = 1,100$, and $a(6) = 5,500$. Find the house advantage and standard deviation in each of the three cases.

14.8. *Defining 8-spot tickets in terms of the 10-spot ticket.* Given the 10-spot payoffs $a_{10}(k)$ in Table 14.2 on p. 488, use (14.26) on p. 499 to derive the corresponding 8-spot payoffs $a_8(k)$. Compare these theoretical payoffs with the actual ones in the same table.

14.9. *Defining m-spot tickets in terms of the 10-spot ticket.* Given arbitrary 10-spot payoffs $a_{10}(k)$, define m-spot payoffs by (14.25) on p. 499 if $11 \leq m \leq 15$ and by (14.26) on p. 499 if $1 \leq m \leq 9$. Ignoring the maximum aggregate payout, show that all 15 tickets have the same expected payout.

14.10. *252-way 10-spot ticket.* Analyze the 252-way 10-spot ticket (i.e., $(r, s, t) = (10, 2, 5)$), just as we did for the 190-way 8-spot ticket. Assume that the payoffs and maximum aggregate payout are those at Harrah's et al. (Table 14.2 on p. 488).

(a) First, find the mean and standard deviation of the payout, disregarding the maximum aggregate payout.

(b) Find the payoff distribution. Specifically, for each triple (n_0, n_1, n_2) of nonnegative integers with $n_0 + n_1 + n_2 = 10$, of which there are $\binom{10+3-1}{3-1} = 66$, find the payout and the probability when n_i groups catch i numbers $(i = 0, 1, 2)$. Use this to find the mean and standard deviation of the payout, taking the maximum aggregate payout into account.

14.11. *Probability of catching 10 of 10 on a 252-way 10-spot ticket.* Find the probability of catching 10 of 10 on at least one way of a 252-way 10-spot ticket. Compare this with the probability of catching 10 of 10 on at least one of 252 independent 10-spot tickets.

14.12. *91,390-way 8-spot ticket.* The $\binom{20}{2}$-way 8-spot ticket is popular, but the $\binom{40}{4}$-way 8-spot ticket (i.e., $(r, s, t) = (40, 2, 4)$) is not. Analyze this ticket, finding in particular the expected payout, first ignoring the maximum aggregate payout and then taking it into account. Explain why this ticket deserves to be unpopular. Assume that the payoffs and maximum aggregate payout are those at Harrah's et al. (Table 14.2 on p. 488).

14.13. *3,160-way 2-spot ticket.* The $\binom{80}{2}$-way 2-spot ticket (i.e., $(r, s, t) = (80, 1, 2)$) is another ticket that should be avoided. Explain why, assuming that a 2-spot ticket pays 12 for 1 for two catches and nothing for one catch or none.

14.14. *A keno system.* Consider the player who bets one unit on an 8-spot ticket at each race until, and only until, he catches 8 of 8. Find the mean and standard deviation of his loss. Assume that the payoffs are those at Harrah's et al. (Table 14.2 on p. 488).

Hint: Use Wald's identity (Corollary 3.2.5 on p. 103).

14.15. *Keno vs. craps.* Consider a gambler with an initial fortune of one unit and a goal of 30,000 units. He has two possible strategies. He can play craps, betting boldly on don't pass ($p = 949/1{,}925$), but restricted by a house betting limit of 500 units. Alternatively, he can buy an 8-spot keno ticket whose payoffs are those at Harrah's et al. (Table 14.2 on p. 488) and, with the money won, if any, bet boldly on don't pass as before. Which of the two strategies has the greater chance of success? Find the odds against success in both cases.

14.16. *Four-color keno.* In this keno variant, the house draws all 80 numbers, in four groups of 20 each. Instead of having just 20 numbers lit up on the keno display board, all 80 are lit, with 20 red, 20 green, 20 blue, and 20 yellow. A player can specify which color he wants to play, or he can play all four colors, assured that each of his numbers will be drawn and will appear in one of the four colors. Consider a player who plays the same 8-spot ticket on each of the four colors, paying a one-unit entry fee per color. Assuming the payoffs are those at Harrah's et al. (Table 14.2 on p. 488), evaluate the expectation and standard deviation of the number of units paid out. Compare these results with those for the player who plays a one-unit 8-spot ticket on four independent races.

14.17. *Bingo.* Bingo has some superficial similarities with keno. A bingo card is a 5×5 matrix. The $(3,3)$ entry is marked "free," and each of the 24 remaining entries is a number from $\{1, 2, \ldots, 75\}$. More precisely, column j contains distinct numbers from $\{15(j-1)+1, 15(j-1)+2, \ldots, 15j\}$ $(j = 1, 2, 3, 4, 5)$, not necessarily ordered. Thus, there are $(15)_5(15)_5(15)_4(15)_5(15)_5$ distinct bingo cards possible. The player purchases one or more such preprinted cards. Just as in keno, the bingo operator then successively draws numbers from $\{1, 2, \ldots, 75\}$ at random without replacement, with the player marking them on his bingo cards, until some player wins the game by filling in a prespecified pattern. Unlike at keno, the order in which the balls are drawn is important.

(a) One bingo game with the potential for large payoffs is the coverall game. Here the player must cover all 24 numbers on his bingo card to win. One casino offers a \$100,000 grand prize if this is done with 47 or fewer numbers drawn, \$20,000 with 48, \$5,000 with 49. Given a bingo card, find the distribution of the number of numbers that need to be drawn to cover the card. Determine the median. What is the probability of winning the grand prize?

(b) Here is a more complicated issue. Given a bingo card, find the distribution of the number of numbers that need to be drawn to cover any one of the five rows, any one of the five columns, or either main diagonal, whichever of the 12 events occurs first. The free space is always assumed covered. Determine the median.

14.4 Notes

McClure (1983) gave a very thorough treatment of the history of keno. Our summary is based on his work, except where otherwise noted, and we have not tried to independently verify his findings.

Originally, the game had 120 characters rather than 80. It was introduced by Cheung Leung of the Han Dynasty (205–187 B.C.) to raise money for the army. By perhaps the sixth century A.D., it had evolved into a game with 80 characters, of which 10 were selected by the players and 20 by the "house," corresponding to today's 10-spot ticket. According to Cowles (2003, p. 14), the game is even older:

> Legend tells us that keno came about as a result of the "Canon of Change," written in 1130 B.C. by Chou Kung, son of the founder of the Kung Dynasty, and was used to finance construction of the Great Wall, one of the wonders of the world.

The 80 characters were not the numbers 1–80 but rather the first 80 characters of a poem known as "The Thousand Character Classic," written by Chou Hsing-szu between 507 and 521 A.D. There are several legends about its composition, which can be summarized as follows (Chou 1963, p. 3):[2]

> In a history of the Liang dynasty, it is stated that the emperor commanded his minister Wang Hsi-chih to write out a thousand characters, and give them to Chou Hsing-szu, that he might form them into an ode, or rhythmical composition. This he did and presented it to His Majesty, who pronounced it excellent and rewarded him with rich presents. In another work it is said that Wu-ti [464–529], the last emperor of the Liang family, commanded all the princes and high officers of his court to write on some subject which they might choose for themselves; he then ordered one of his ministers to select from their writings 1000 characters, and copy them on a thousand separate slips of paper. These being thrown together in confusion, the emperor summoned Chou Hsing-szu, and asked him if he was able to form them into an ode. He immediately undertook the task, and completed it in a single night; but such was the labor of his mind that all the hair of his head turned white! Another story is that this task was imposed on Chou Hsing-szu as a punishment for some crime he had committed.

The poem was written in the classical style, extremely condensed and cryptic without punctuation. For example, the first eight characters may be literally translated, "Heaven earth, dark yellow. Space time, vast barren." Expanded slightly, this becomes, "Heaven is dark, the earth is yellow. The universe is vast and barren." (See Chou 1963, pp. 38–39.)

The number of characters correctly guessed later became known as the number "caught." This arose from an unusual way of thinking about the game (McClure 1983, p. 8):

> The Chinese envisioned eighty mountains with twenty loose monkeys, and calculated how many monkeys might be caught by ten soldiers searching one mountain each.

By the mid-19th century, with the influx of Chinese immigrants, the game had come to the United States. It was called *pák kòp piú*, which means "white pigeon ticket." This arose from the fact that, in China where lotteries were illegal, carrier pigeons were used to carry tickets and payoffs between the lottery office and its patrons. Americans referred to the game as simply *Chinese lottery*, at least until the 1930s. Gradually, Caucasians began operating the game, and by 1934 there were four Caucasian-run games operating in Butte, Montana. In these games numbers replaced Chinese characters on the keno tickets.

Gambling had been legalized in Nevada in 1931, but lotteries were still illegal. Lotteries were also illegal in Montana. Thus, Chinese lottery could not be legally played. So around 1935 the name of the game was changed to *race horse keno*. Keno was the name of what is now called bingo (see Problem 14.17), and during the 1920s and 1930s the game was adopted by social organizations for the purpose of fundraising (see Quinn 1912, pp. 167–169, for a description of keno). To avoid the gambling connotations of the

---

[2] Editorial remarks by Francis W. Parr.

word "keno," these organizations began calling it beano or bingo. It was to differentiate Chinese lottery from the old keno, which was still remembered, that the compound adjective "race horse" was added. Names of 80 race horses were included on the keno ticket, one for each number.

The term "keno," incidentally, is an adaptation of the French word "quine," meaning a set of five winning numbers in a lottery, according to the *Oxford English Dictionary* (Simpson and Weiner 1989). More precisely, a quine was effectively a five-spot ticket in the French Lottery (Laurent 1893, Section 21); see Problem 1.37. 19th-century spellings include keeno, quino, kino, and keno, the latter as early as 1871.

Race horse keno came to Reno in 1936, once it had been established that it was a "banking game" played with a "mechanical device," and not a lottery. By 1960 it was no longer necessary to differentiate race horse keno from what used to be called keno, so the race horse names were dropped, and the game became known simply as keno. According to Mechigian (1972, p. 11), the change was due to a 1951 law that taxed off-track betting. In any case, individual coups at keno are still often referred to as races.

In the game of pák kòp piú as played in Philadelphia in 1880, the method of drawing the 20 house characters was described by Culin (1891, pp. 9–10) as follows:

> Eighty pieces of white paper have been provided, upon which have been written or printed the eighty characters of the tickets, one on each, a box of hand stamps for the purpose, forming part of the equipment of most lotteries. The manager carefully rolls the eighty pieces of paper into as many pellets, so that they cannot be distinguished, one from another, and places them in a large tin pan. He mixes them thoroughly, and then, one at a time, counts twenty of the pellets into a white china bowl, distinguished by a white paper label marked "one." He then counts twenty more into another bowl labeled "two," and, in turn, places the remainder, in the same way, into two other bowls marked "three" and "four."
>
> One of the players, who is paid a small gratuity, is now asked to select one of the bowls, and the one he designates is declared to contain the winning numbers. These the manager carefully unrolls, one at a time, at once pasting them on a board in the back part of his office.

Various minor improvements in this scheme were made until the early 1930s, when keno operators began to draw numbered ping-pong balls from a wire cage. A blower device was introduced in 1946 to further ensure randomness. Similar devices are still used today. Some of today's casinos take a more high-tech approach, using random number generators to select the numbers. (In fact, this is mandated in Atlantic City.) According to Cowles (2003, p. 178), they would be more widely used in Nevada, were it not for the fact that "Keno players just don't like them." Perhaps there is concern about having the same computer both record the players' numbers and select the casino's numbers. This brings to mind a remark by comedian Bill Cosby, as quoted by Cowles (1996, p. 230):

> To play keno, all you have to do is mark the numbers you want to win on a piece of paper, and give the paper and your money to The Man.

Now remember, you're telling him which numbers you want him to pick, and if he picks them, he's going to have to pay you a whole lot of money.

Tell me, do you *really* think he's going to pick your numbers?

For centuries the 10-spot ticket was the only ticket available at what was to become keno, but by the late 19th century way tickets began to appear, and this led to m-spot tickets for various values of m. To explain how this works, some additional terminology is helpful. Recall that a way ticket is specified by its groups and its ways. The unique element of any singleton group is called a *king*, and a group that is used in every way is called a *field*.

First, if $11 \leq m \leq 15$, we can regard an m-spot ticket as an $\binom{m}{10}$-way 10-spot ticket. Indeed, think of the m spots marked as kings, any 10 of which combine to form a way. This allows us to define

$$a_m(k) := \sum_{j=(10-m+k)^+}^{k \wedge 10} \frac{\binom{k}{j}\binom{m-k}{10-j}}{\binom{m}{10}} a_{10}(j). \qquad (14.25)$$

To see this more clearly, imagine that the m-spot ticket has k catches, and we know exactly what they are. Now randomly choose one of the $\binom{m}{10}$-way 10-spot tickets. The probability that the 10-spot ticket has j catches is the coefficient of $a_{10}(j)$ in (14.25).

Second, if $1 \leq m \leq 9$, we can regard an m-spot ticket as an $\binom{80-m}{10-m}$-way 10-spot ticket. Indeed, think of the m marked spots as a field and the $80-m$ unmarked spots as kings, any $10-m$ of which combine with the field to form a way. This allows us to define

$$a_m(k) := \sum_{j=0}^{10-m} \frac{\binom{20-k}{j}\binom{60-m+k}{10-m-j}}{\binom{80-m}{10-m}} a_{10}(k+j). \qquad (14.26)$$

As before, imagine that the m-spot ticket has k catches, and we know exactly what they are. Now randomly choose one of the $\binom{80-m}{10-m}$-way 10-spot tickets. The probability that the 10-spot ticket has $k+j$ catches is the coefficient of $a_{10}(k+j)$ in (14.26).

Thus, all m-spot tickets can be defined in terms of the 10-spot ticket. Although this was done originally, modern payoff schedules do not usually satisfy (14.25) or (14.26) (see Problem 14.8).

Keno played a role in David Kranes's contemporary novel *Keno Runner: A Romance* (1989, Chapter 10). When Janice Stewart was asked why she had chosen to become a keno runner[3] in Las Vegas, she replied, "I bring the numbers." Elaborating, she continued,

I bring the one-spots and the three-spots. And all the four- and five-way combinations. I carry 63s! And 39s! And 4s! And 16s, 17s, and 18s—all in my hand! I carry 52s! I carry 80s! I carry 3s and 2s and 1s! I take them all! Old women in

---

[3] A *keno runner* is a casino employee "who carries tickets, wagers and wins between players and the game" (McClure 1983, p. 217).

wheelchairs. Truckers. Marketing executives. Mad handkerchief-chewing women in shawls. Junkies. Nuns. Murderers. I find them all! [. . .] And I am a keno runner in the City of Death—running, moving, moving back. And I love always what I do. And I wouldn't change it!

There is little doubt that today keno is in decline, just as faro was a century ago. As we noted in Table 14.2, live keno was available at eight casinos on the Las Vegas Strip in January 2008. It was not offered at 15 others.

Who was the first to evaluate the keno probabilities? Characteristically, Scarne (1961, p. 437) claimed priority. While he did correctly calculate the probabilities for the 10-spot ticket, he was almost certainly not the first to do so. We quote Culin (1891, p. 10) once again:

> The price which should be charged for more than ten numbers, with the prizes to be paid, and the methods of calculating the company's chances, and what its profits should be, are contained in a book known as the *pák kòp piú t'ò*, of which several editions are current among the gamblers in American cities. One in general use, entitled "*Shang ts'oi tsit king*" or "A quick way to get rich," may be purchased in the Chinese shops.

Unfortunately, we have not been able to locate a copy of this publication.

We mentioned in Section 14.1 that there are a number of authors who have attempted to derive (14.2) directly under the assumptions that led to (14.1). A good example of this is Packel (2006, p. 73):

> To see how products of combinations arise, consider the case of marking exactly 6 winning numbers [on a 10-spot ticket]. Assume, for the purposes of our reasoning process, that the 20 random numbers have been determined but remain unknown to the player. In how many ways can exactly 6 out of 10 marked numbers appear among the 20 numbers drawn? (In actuality, the player marks his ticket before the numbers are drawn, but our reinterpretation will not affect the results.)

He answered that "there is a total of $\binom{20}{6}\binom{60}{4}$ equally likely ways to have exactly 6 of the 10 marked numbers drawn." This is misleading because it obscures the underlying assumption (in the reinterpretation) that the player's 10 marked numbers are chosen at random. Incidentally, Theorem 1.1.1 requires that *all* outcomes in the sample space be equally likely, not just those corresponding to the event in question. Thus, Packel's derivation is flawed, but it can be fixed using the ideas in the paragraph above (14.3).

Problem 14.3 was motivated by Drakakis (2005). Problem 14.4 is due to Alspach (2001a). A solution to most of Problem 14.10 can be found in Mechigian (1972, pp. 105–106). Problem 14.13 was suggested by Claussen (1982, pp. 54, 112). Problem 14.15 was motivated by an article of Stupak (1984) and subsequent replies by Griffin (1991, p. 161) and Joel Friedman (unpublished). Four-color keno (Problem 14.16) does not currently exist, to the best of our knowledge. The game was suggested by a remark of McClure (1983, p. 24). We do not know who first solved Problem 14.17(a), but Scarne (1961, p. 196) claimed priority. Problem 14.17(b) is due to Agard and Shackleford (2002).

Chapter 15
Craps

"The bones?" cries Caroline, with a bewildered look.

"The dice, my dear! 'Seven's the main' was my ruin. Seven's the main and eleven's the nick to seven. That used to be the little game!"

William Makepeace Thackeray, *The Adventures of Philip*

Craps is a 19th-century American simplification of hazard, a dice game thought to be of Arab origin that dates back to the Crusades. Mathematically, the most interesting feature of craps is that the principal bets require a random number of rolls for resolution. Section 15.1 examines these wagers, the so-called line bets, together with certain associated fair side bets called free odds bets. In Section 15.2 we study the shooter's hand, that is, the sequence of rolls from the initial come-out roll to the seven out.

15.1 Line Bets and Free Odds

The game of *craps* is played by rolling a pair of dice repeatedly. A single roll of the dice can be modeled by (U, V), where U and V are independent discrete uniform random variables on $\{1, 2, 3, 4, 5, 6\}$. Actually, because the dice are typically indistinguishable, the observer sees only $U_0 := \min(U, V)$ and $V_0 := \max(U, V)$. Let $(U_1, V_1), (U_2, V_2), \ldots$ be a sequence of i.i.d. random variables with common distribution that of (U_0, V_0), and define

$$T_n := U_n + V_n, \qquad n \geq 1. \tag{15.1}$$

Letting

$$\mathscr{S} := \{2, 3, 4, \ldots, 12\}, \tag{15.2}$$

we conclude that $T_1, T_2, \ldots$ is a sequence of i.i.d. $\mathscr{S}$-valued random variables with common distribution that of $T_0 := U_0 + V_0 = U + V$, namely,

S.N. Ethier, *The Doctrine of Chances*, Probability and its Applications, DOI 10.1007/978-3-540-78783-9_15, © Springer-Verlag Berlin Heidelberg 2010

$$\pi_j := \mathrm{P}(T_0 = j) = \frac{(j-1) \wedge (13-j)}{36} = \frac{6 - |j-7|}{36}, \qquad j \in \mathscr{S}. \qquad (15.3)$$

(See Example 1.1.9 on p. 6.)

To describe the so-called *line bets*, it will be convenient to define the set of *point numbers* (also called *box numbers*) to be

$$\mathscr{P} := \{4, 5, 6, 8, 9, 10\}. \qquad (15.4)$$

A *pass-line bet* or *come bet* made prior to the nth roll of the dice is resolved following roll N_n, where

$$N_n := \begin{cases} n & \text{if } T_n \in \mathscr{S} - \mathscr{P}, \\ \min\{k \ge n+1 : T_k = T_n \text{ or } T_k = 7\} & \text{if } T_n \in \mathscr{P}. \end{cases} \qquad (15.5)$$

The bet is won if $T_n \in \{7, 11\}$ (a *natural* is rolled) or if both $T_n \in \mathscr{P}$ (a *point* is established) and $T_{N_n} = T_n$ (the point is *made*). It is lost if $T_n \in \{2, 3, 12\}$ (a *craps number* is rolled) or if both $T_n \in \mathscr{P}$ (a point is established) and $T_{N_n} = 7$ (the point is *missed*). A win pays even money.

If this description seems a bit terse, the reader should refer back to Examples 1.2.2 on p. 17 and 1.2.8 on p. 19.

A *free odds bet* associated with a pass-line bet or a come bet becomes available when a point is established; it is also referred to as *taking the odds*. This is in effect a side bet that the original bet will be won, and it pays fair odds. More precisely, the payoff odds are 2 to 1 if the point is 4 or 10, 3 to 2 if the point is 5 or 9, and 6 to 5 if the point is 6 or 8. Because it is a fair bet, its size is limited: The gambler in an *m-times odds* casino is permitted to bet no more on free odds than m times the amount of his pass-line bet or come bet. Usually, m is 1, 2, or 3, in which case we say that the casino offers single, double, or triple odds. (One could define m_4-m_5-m_6-times free odds so as to generalize Examples 1.5.7 on p. 44 and 6.1.7 on p. 205, which defined 3-4-5-times free odds. See Problem 15.6 on p. 514.)

A *don't-pass bet* or *don't-come bet* made prior to the nth roll of the dice is also resolved following roll N_n. The bet is won if $T_n \in \{2, 3\}$ (a *nonbarred* craps number is rolled) or if both $T_n \in \mathscr{P}$ (a point is established) and $T_{N_n} = 7$ (the point is missed). It is lost if $T_n \in \{7, 11\}$ (a natural is rolled) or if both $T_n \in \mathscr{P}$ (a point is established) and $T_{N_n} = T_n$ (the point is made). Finally, the bet is pushed is $T_n = 12$ (the *barred* craps number is rolled). The bet can be withdrawn following a push. Again, a win pays even money.

A *free odds bet* associated with a don't-pass bet or a don't-come bet becomes available when a point is established; it is also referred to as *laying the odds*. This is in effect a side bet that the original bet will be won, and it pays fair odds. More precisely, the payoff odds are 1 to 2 if the point is 4 or 10, 2 to 3 if the point is 5 or 9, and 5 to 6 if the point is 6 or 8. Because it is a fair bet, its size is limited: The gambler in an m-times odds casino is permitted to bet no more on free odds than the amount just sufficient to win m times the amount of his don't-pass bet or don't come bet.

The distinction between a pass-line bet and a come bet, and between a don't-pass bet and a don't-come bet, can be described as follows. If we recursively define

$$C_1 := 1, \qquad C_n := N_{C_{n-1}} + 1, \quad n \geq 2, \tag{15.6}$$

then $C_1, C_2, \ldots$ represents the sequence of *come-out rolls*. Every roll that is not a come-out roll is a *point roll*. Pass-line bets and don't-pass bets are made prior to come-out rolls, whereas come bets and don't-come bets are made prior to point rolls. We can also define at this time

$$D_n := C_{n+1} - C_n, \qquad n \geq 1, \tag{15.7}$$

so that D_n represents the random number of rolls required for the nth pass-line decision. Of course, $D_1, D_2, \ldots$ is a sequence of i.i.d. random variables.

There are many other bets available at craps, but they are all less favorable to the player than those just described.

The probabilities of winning, losing, and pushing a don't-pass bet (or a don't-come bet) are respectively

$$p_0 := \pi_2 + \pi_3 + \sum_{j \in \mathscr{P}} \pi_j \frac{\pi_7}{\pi_j + \pi_7} = \frac{949}{1{,}980}, \tag{15.8}$$

$$q_0 := \pi_7 + \pi_{11} + \sum_{j \in \mathscr{P}} \pi_j \frac{\pi_j}{\pi_j + \pi_7} = \frac{976}{1{,}980} = \frac{244}{495}, \tag{15.9}$$

$$r_0 := \pi_{12} = \frac{55}{1{,}980} = \frac{1}{36}, \tag{15.10}$$

so the house advantage is $H_0 = q_0 - p_0 = 27/1{,}980 = 3/220$ or $H = (q_0 - p_0)/(1 - r_0) = 27/1{,}925$. When m-times odds are taken into account, the house advantage is

$$H_0 = \frac{3}{220} \frac{1}{1+m} \qquad \text{or} \qquad H = \frac{27}{1{,}925} \frac{1}{1+(36/35)m}. \tag{15.11}$$

The probabilities of winning and losing a pass-line bet (or a come bet) are respectively q_0 and $p_0 + r_0$, so the house advantage is $H_0 = H = p_0 + r_0 - q_0 = 28/1{,}980 = 7/495$. When m-times odds are taken into account, the house advantage is

$$H_0 = H = \frac{7}{495} \frac{1}{1+(2/3)m}. \tag{15.12}$$

Numerical values of these house advantages are shown in Table 15.1. Notice that a free odds bet reduces the overall house advantage because, while the player's expected loss remains the same, his expected amount of action increases from 1 to $1+(2/3)m$ in the case of a pass-line bet, with an even greater increase in the case of a don't-pass bet. Does this suggest that the pass-line bettor should always take free odds? Typically, yes. See Example 6.1.7 on p. 205 for elaboration. Of course, one could argue that his expected loss could be reduced to nothing by betting nothing. But then whatever benefit the gambler accrues by betting (for example, entertainment value) would also be reduced to nothing.

Table 15.1 House advantages (in percentage terms) for pass-line and don't-pass bets, with and without free odds. Results are identical for come and don't-come bets. Recall that pushes are included in H_0 and excluded in H.

wager	H_0	H
pass line	1.414 141	1.414 141
pass line, single odds	.848 485	.848 485
pass line, double odds	.606 061	.606 061
pass line, triple odds	.471 380	.471 380
pass line, 3-4-5-times odds[1]	.374 332	.374 332
pass line, 5-times odds	.326 340	.326 340
pass line, 10-times odds	.184 453	.184 453
pass line, 20-times odds	.098 661	.098 661
don't pass	1.363 636	1.402 597
don't pass, single odds	.681 818	.691 421
don't pass, double odds	.454 545	.458 794
don't pass, triple odds	.340 909	.343 293
don't pass, 3-4-5-times odds[1]	.272 727	.274 251
don't pass, 5-times odds	.227 273	.228 330
don't pass, 10-times odds	.123 967	.124 281
don't pass, 20-times odds	.064 935	.065 020

[1]see Examples 1.5.7 on p. 44 and 6.1.7 on p. 205

The expected number of rolls required for a pass-line decision is (cf. Example 2.1.4 on p. 78)

$$E[D_1] = 1 + \sum_{j \in \mathscr{P}} \pi_j \frac{1}{\pi_j + \pi_7} = \frac{557}{165} = 3.3\overline{75}. \qquad (15.13)$$

More generally, the distribution of D_1 is given by

$$P(D_1 = n) = \begin{cases} 1 - \sum_{j \in \mathscr{P}} \pi_j & \text{if } n = 1, \\ \sum_{j \in \mathscr{P}} \pi_j (1 - \pi_j - \pi_7)^{n-2}(\pi_j + \pi_7) & \text{if } n \geq 2, \end{cases} \qquad (15.14)$$

so that the conditional distribution of $D_1 - 1$, given that $D_1 \geq 2$, is a mixture of three geometric distributions. It follows that the probability generating function of D_1 has the form

$$h(z) := \mathrm{E}[z^{D_1}] = \left(1 - \sum_{j \in \mathscr{P}} \pi_j\right) z + \sum_{j \in \mathscr{P}} \frac{\pi_j(\pi_j + \pi_7)z^2}{1 - (1 - \pi_j - \pi_7)z}. \qquad (15.15)$$

From either (15.14) or (15.15) we can find the variance of D_1. Alternatively, we can condition on the result of the come-out roll to obtain

$$\mathrm{Var}(D_1) = 1 + \sum_{j \in \mathscr{P}} \pi_j \left[\frac{1}{\pi_j + \pi_7} + \frac{2}{(\pi_j + \pi_7)^2}\right] - (\mathrm{E}[D_1])^2$$

$$= \frac{245{,}672}{27{,}225} \approx 9.023765, \qquad (15.16)$$

where we have used the fact that if X is geometric(p), then $\mathrm{E}[(1 + X)^2] = 1 + p^{-1} + 2p^{-2}$. See Example 2.2.14 on p. 87 for yet another method.

15.2 The Shooter's Hand

A shooter rolls the dice until he *sevens out*, that is, until he misses a point. When this happens, all bets are resolved and a new shooter comes out. The sequence of rolls by a shooter, from the initial come out to the seven out, is known as the *shooter's hand*. With N denoting the number of pass-line decisions in the shooter's hand, the length L of the shooter's hand (i.e., the number of rolls) is

$$L := D_1 + D_2 + \cdots + D_N. \qquad (15.17)$$

Noting that N is geometrically distributed with parameter

$$q := \sum_{j \in \mathscr{P}} \pi_j \frac{\pi_7}{\pi_j + \pi_7} = \frac{196}{495} = 0.39\overline{5}, \qquad (15.18)$$

the expected number of pass-line decisions in the shooter's hand is

$$\mathrm{E}[N] = \frac{1}{q} = \frac{495}{196} \approx 2.525510, \qquad (15.19)$$

so the expected length of the shooter's hand follows easily from Wald's identity (Corollary 3.2.5 on p. 103)—see Example 3.2.6 on p. 103:

$$\mathrm{E}[L] = \mathrm{E}[N]\mathrm{E}[D_1] = \frac{1}{q}\mathrm{E}[D_1] = \frac{1{,}671}{196} \approx 8.525510. \qquad (15.20)$$

An alternative derivation of (15.20) can be obtained by conditioning on the result of the first come-out roll:

$$E[L] = \left(1 - \sum_{j \in \mathscr{P}} \pi_j\right)(1 + E[L])$$

$$+ \sum_{j \in \mathscr{P}} \pi_j \left(1 + \frac{1}{\pi_j + \pi_7} + \frac{\pi_j}{\pi_j + \pi_7} E[L]\right). \qquad (15.21)$$

But from (15.13) on p. 504 and (15.18), we recognize this as

$$E[L] = E[D_1] + (1 - q)E[L], \qquad (15.22)$$

which can also be seen directly, and (15.22) reduces to (15.20). Of course, one must show *a priori* that $E[L] < \infty$. See Problem 15.10 on p. 514 for still another approach.

More generally, one can obtain the distribution of L recursively: If $t(n) := P(L \geq n)$ for each $n \geq 1$, then $t(1) = t(2) = 1$ and

$$t(n) = \left(1 - \sum_{j \in \mathscr{P}} \pi_j\right)t(n - 1) + \sum_{j \in \mathscr{P}} \pi_j(1 - \pi_j - \pi_7)^{n-2}$$

$$+ \sum_{j \in \mathscr{P}} \pi_j \sum_{l=2}^{n-1} (1 - \pi_j - \pi_7)^{l-2} \pi_j \, t(n - l) \qquad (15.23)$$

for each $n \geq 3$. The interpretation of (15.23) should be clear: For the event that the shooter sevens out in no fewer than n rolls to occur, consider the result of the initial come-out roll. If a natural or a craps number occurs, then, beginning with the next roll, the shooter must seven out in no fewer than $n - 1$ rolls. If a point number occurs, there are two possibilities. Either the point is still unresolved after $n - 2$ additional rolls, or it is made at roll l for some $l \in \{2, 3, \ldots, n-1\}$ and the shooter subsequently sevens out in no fewer than $n - l$ rolls. Numerical values of $P(L \geq n)$ are displayed in Table 15.2 for $n = 2, 3, \ldots, 100$.

We could also get the probabilities $t(n)$ from a Markov chain; see Example 4.1.6 on p. 125. A recursion similar to (15.23) holds for $s(n) := P(L = n)$ for each $n \geq 2$ (Problem 15.12 on p. 515), but it is $P(L \geq n)$ that is of primary interest. It was recently discovered that the distribution of L is a linear combination of four geometric distributions and that there is a closed-form expression for $t(n)$. See the chapter notes for the details.

Let us define the sequence of i.i.d. random vectors

$$\boldsymbol{X}_1 := (T_1, \ldots, T_{D_1}), \quad \boldsymbol{X}_2 := (T_{D_1+1}, \ldots, T_{D_1+D_2}), \quad \cdots \quad ; \qquad (15.24)$$

this is a sequence of random vectors whose dimensions are also random. Notice that $(\boldsymbol{X}_1, \ldots, \boldsymbol{X}_N)$ is precisely the shooter's hand. Our principal aim

Table 15.2 The distribution of the length L of the shooter's hand. Results are rounded to six significant digits.

n	$P(L \geq n)$	n	$P(L \geq n)$	n	$P(L \geq n)$
2	1.000 000	35	.007 921 21	68	.000 060 037 3
3	.888 889	36	.006 831 87	69	.000 051 780 6
4	.772 119	37	.005 892 34	70	.000 044 659 4
5	.667 353	38	.005 082 01	71	.000 038 517 6
6	.576 129	39	.004 383 11	72	.000 033 220 4
7	.497 210	40	.003 780 33	73	.000 028 651 7
8	.429 044	41	.003 260 44	74	.000 024 711 4
9	.370 191	42	.002 812 05	75	.000 021 312 9
10	.319 391	43	.002 425 32	76	.000 018 381 8
11	.275 547	44	.002 091 78	77	.000 015 853 8
12	.237 710	45	.001 804 11	78	.000 013 673 5
13	.205 062	46	.001 556 00	79	.000 011 793 1
14	.176 892	47	.001 342 01	80	.000 010 171 2
15	.152 588	48	.001 157 45	81	.000 008 772 39
16	.131 620	49	.000 998 270	82	.000 007 565 96
17	.113 531	50	.000 860 982	83	.000 006 525 44
18	.097 926 2	51	.000 742 575	84	.000 005 628 02
19	.084 465 4	52	.000 640 451	85	.000 004 854 02
20	.072 854 0	53	.000 552 373	86	.000 004 186 46
21	.062 838 2	54	.000 476 407	87	.000 003 610 72
22	.054 198 9	55	.000 410 889	88	.000 003 114 15
23	.046 747 1	56	.000 354 381	89	.000 002 685 87
24	.040 319 5	57	.000 305 644	90	.000 002 316 49
25	.034 775 5	58	.000 263 610	91	.000 001 997 91
26	.029 993 7	59	.000 227 357	92	.000 001 723 15
27	.025 869 4	60	.000 196 089	93	.000 001 486 17
28	.022 312 1	61	.000 169 122	94	.000 001 281 78
29	.019 243 9	62	.000 145 863	95	.000 001 105 50
30	.016 597 5	63	.000 125 803	96	.000 000 953 468
31	.014 315 1	64	.000 108 502	97	.000 000 822 341
32	.012 346 5	65	.000 093 580 1	98	.000 000 709 248
33	.010 648 6	66	.000 080 710 4	99	.000 000 611 708
34	.009 184 24	67	.000 069 610 6	100	.000 000 527 582

is to provide an alternative description of this random vector by conditioning on N. More precisely, we will effectively use Bayes's law to find the conditional distribution of $(\boldsymbol{X}_1, \ldots, \boldsymbol{X}_N)$, given the event that $N = n$.

Let us denote by A the set of all sequences of dice rolls that result in a single pass-line decision that does not culminate in a seven out. Thus, A contains (2), (3), (7), (11), (12), and all sequences of the form

$$(j_1, j_2, \ldots, j_{k-1}, j_1), \qquad j_1 \in \mathscr{P}, \quad j_2, \ldots, j_{k-1} \in \mathscr{S} - \{j_1, 7\}, \quad k \geq 2.$$
$$(15.25)$$

Similarly, let us denote by B the set of all sequences of dice rolls that result in a single pass-line decision culminating in a seven out. Thus, B contains all sequences of the form

$$(j_1, j_2, \ldots, j_{k-1}, 7), \qquad j_1 \in \mathscr{P}, \quad j_2, \ldots, j_{k-1} \in \mathscr{S} - \{j_1, 7\}, \quad k \geq 2.$$
$$(15.26)$$

In terms of (15.24) we can formally define N by

$$N := \min\{n \geq 1 : \boldsymbol{X}_n \in B\}. \tag{15.27}$$

Fix $n \geq 1$. We claim that the conditional distribution of $(\boldsymbol{X}_1, \ldots, \boldsymbol{X}_N)$, given that $N = n$, is equal to the distribution of $(\boldsymbol{Y}_1, \ldots, \boldsymbol{Y}_{n-1}, \boldsymbol{Z}_n)$, where $\boldsymbol{Y}_1, \ldots, \boldsymbol{Y}_{n-1}$ are i.i.d. with common distribution equal to the conditional distribution of $\boldsymbol{X}_1$ given $\boldsymbol{X}_1 \in A$, and $\boldsymbol{Z}_n$ is independent of $\boldsymbol{Y}_1, \ldots, \boldsymbol{Y}_{n-1}$ with distribution equal to the conditional distribution of $\boldsymbol{X}_1$ given $\boldsymbol{X}_1 \in B$.

Indeed, if $\boldsymbol{j}_1, \ldots, \boldsymbol{j}_{n-1} \in A$ and $\boldsymbol{j}_n \in B$, then

$$\begin{aligned}
P((\boldsymbol{X}_1, \ldots, \boldsymbol{X}_n) &= (\boldsymbol{j}_1, \ldots, \boldsymbol{j}_n) \mid N = n) \\
&= P((\boldsymbol{X}_1, \ldots, \boldsymbol{X}_n) = (\boldsymbol{j}_1, \ldots, \boldsymbol{j}_n)) / P(N = n) \\
&= P(\boldsymbol{X}_1 = \boldsymbol{j}_1) \cdots P(\boldsymbol{X}_{n-1} = \boldsymbol{j}_{n-1}) P(\boldsymbol{X}_n = \boldsymbol{j}_n) / P(N = n) \\
&= \frac{P(\boldsymbol{X}_1 = \boldsymbol{j}_1)}{P(\boldsymbol{X}_1 \in A)} \cdots \frac{P(\boldsymbol{X}_{n-1} = \boldsymbol{j}_{n-1})}{P(\boldsymbol{X}_{n-1} \in A)} \frac{P(\boldsymbol{X}_n = \boldsymbol{j}_n)}{P(\boldsymbol{X}_n \in B)} \\
&= P(\boldsymbol{X}_1 = \boldsymbol{j}_1 \mid \boldsymbol{X}_1 \in A) \cdots P(\boldsymbol{X}_1 = \boldsymbol{j}_{n-1} \mid \boldsymbol{X}_1 \in A) \\
&\qquad\qquad\qquad\qquad P(\boldsymbol{X}_1 = \boldsymbol{j}_n \mid \boldsymbol{X}_1 \in B) \\
&= P(\boldsymbol{Y}_1 = \boldsymbol{j}_1, \ldots, \boldsymbol{Y}_{n-1} = \boldsymbol{j}_{n-1}, \boldsymbol{Z}_n = \boldsymbol{j}_n), \tag{15.28}
\end{aligned}$$

and the conclusion follows.

Moreover, we can evaluate the conditional distributions of $\boldsymbol{X}_1$ given $\boldsymbol{X}_1 \in A$ and given $\boldsymbol{X}_1 \in B$. By Bayes's law and (15.18),

$$P(T_1 = j \mid \boldsymbol{X}_1 \in A) = \pi_j^A \quad \text{and} \quad P(T_1 = j \mid \boldsymbol{X}_1 \in B) = \pi_j^B, \tag{15.29}$$

where

$$\pi_j^A := \begin{cases} \pi_j / (1-q) & \text{if } j \in \mathscr{S} - \mathscr{P}, \\ \pi_j^2 / [(\pi_j + \pi_7)(1-q)] & \text{if } j \in \mathscr{P}, \end{cases} \tag{15.30}$$

and

$$\pi_j^B := \begin{cases} 0 & \text{if } j \in \mathscr{S} - \mathscr{P}, \\ \pi_j \pi_7 / [(\pi_j + \pi_7)q] & \text{if } j \in \mathscr{P}. \end{cases} \tag{15.31}$$

We now claim that the conditional distribution of $\boldsymbol{X}_1$ given $\boldsymbol{X}_1 \in A$ can be constructed as follows. On the come-out roll, biased dice with distribution (π_j^A) are rolled. Given that a point is established, fair dice are then rolled until the point is repeated or 7 appears. Regardless, the last roll is counted as the point.

We claim also that the conditional distribution of $\boldsymbol{X}_1$ given $\boldsymbol{X}_1 \in B$ can be constructed as follows. On the come-out roll, biased dice with distribution (π_j^B) are rolled to establish a point. Then fair dice are rolled until the point is repeated or 7 appears. Regardless, the last roll is counted as 7.

To see this, let T_1^A and T_1^B be random variables independent of $T_2, T_3, \ldots$ and with distributions (π_j^A) and (π_j^B) We continue to assume that $T_2, T_3, \ldots$ are i.i.d. with common distribution that of T_0 of (15.3) on p. 502. Let $N_A := \min\{l \geq 2 : T_l \in \{T_1^A, 7\}\}$ if $T_1^A \in \mathscr{P}$ and $N_B := \min\{l \geq 2 : T_l \in \{T_1^B, 7\}\}$. Our claims, stated more precisely, are that the conditional distribution of $\boldsymbol{X}_1$ given $\boldsymbol{X}_1 \in A$ is equal to the distribution of

$$\boldsymbol{X}_1^A := \begin{cases} (T_1^A) & \text{if } T_1^A \in \mathscr{S} - \mathscr{P}, \\ (T_1^A, T_2, \ldots, T_{N_A - 1}, T_1^A) & \text{if } T_1^A \in \mathscr{P}, \end{cases} \tag{15.32}$$

and that the conditional distribution of $\boldsymbol{X}_1$ given $\boldsymbol{X}_1 \in B$ is equal to the distribution of

$$\boldsymbol{X}_1^B := (T_1^B, T_2, \ldots, T_{N_B - 1}, 7). \tag{15.33}$$

Indeed, if $\boldsymbol{j} = j_1 \in \mathscr{S} - \mathscr{P}$, then

$$\begin{aligned} \mathrm{P}(\boldsymbol{X}_1 = \boldsymbol{j} \mid \boldsymbol{X}_1 \in A) &= \frac{\mathrm{P}(\boldsymbol{X}_1 = \boldsymbol{j})}{\mathrm{P}(\boldsymbol{X}_1 \in A)} = \frac{\mathrm{P}(T_1 = j_1)}{\mathrm{P}(\boldsymbol{X}_1 \in A)} = \frac{\pi_{j_1}}{1 - q} \\ &= \mathrm{P}(T_1^A = j_1) = \mathrm{P}(\boldsymbol{X}_1^A = \boldsymbol{j}); \end{aligned} \tag{15.34}$$

if $k \geq 2$ and $\boldsymbol{j} = (j_1, j_2, \ldots, j_{k-1}, j_1) \in A$, then

$$\begin{aligned} \mathrm{P}(\boldsymbol{X}_1 = \boldsymbol{j} \mid \boldsymbol{X}_1 \in A) &= \frac{\mathrm{P}(\boldsymbol{X}_1 = \boldsymbol{j})}{\mathrm{P}(\boldsymbol{X}_1 \in A)} = \frac{\pi_{j_1} \pi_{j_2} \cdots \pi_{j_{k-1}} \pi_{j_1}}{1 - q} \\ &= \pi_{j_1}^A \pi_{j_2} \cdots \pi_{j_{k-1}} (\pi_{j_1} + \pi_7) \\ &= \mathrm{P}(\boldsymbol{X}_1^A = \boldsymbol{j}); \end{aligned} \tag{15.35}$$

and if $k \geq 2$ and $\boldsymbol{j} = (j_1, j_2, \ldots, j_{k-1}, 7) \in B$, then

$$\begin{aligned} \mathrm{P}(\boldsymbol{X}_1 = \boldsymbol{j} \mid \boldsymbol{X}_1 \in B) &= \frac{\mathrm{P}(\boldsymbol{X}_1 = \boldsymbol{j})}{\mathrm{P}(\boldsymbol{X}_1 \in B)} = \frac{\pi_{j_1} \pi_{j_2} \cdots \pi_{j_{k-1}} \pi_7}{q} \\ &= \pi_{j_1}^B \pi_{j_2} \cdots \pi_{j_{k-1}} (\pi_{j_1} + \pi_7) \\ &= \mathrm{P}(\boldsymbol{X}_1^B = \boldsymbol{j}). \end{aligned} \tag{15.36}$$

It should be noted that independence holds between the n pass-line decisions but not within them. Within each pass-line decision the rolls are conditionally independent, given the result of the come-out roll.

To illustrate these results, we give a derivation of (15.20) without the use of Wald's identity. Conditioning on N, we obtain

$$
\begin{aligned}
\mathrm{E}[L] &= \sum_{n=1}^{\infty} \mathrm{P}(N = n)\mathrm{E}[L \mid N = n] \\
&= \sum_{n=1}^{\infty} \mathrm{P}(N = n)\mathrm{E}[D_1 + \cdots + D_{n-1} + D_n \mid N = n] \\
&= \sum_{n=1}^{\infty} \mathrm{P}(N = n)\{(n-1)\mathrm{E}[D_1 \mid \boldsymbol{X}_1 \in A] + \mathrm{E}[D_1 \mid \boldsymbol{X}_1 \in B]\} \\
&= (\mathrm{E}[N] - 1)\mathrm{E}[D_1 \mid \boldsymbol{X}_1 \in A] + \mathrm{E}[D_1 \mid \boldsymbol{X}_1 \in B] \\
&= \frac{(1-q)\mathrm{E}[D_1 \mid \boldsymbol{X}_1 \in A] + q\mathrm{E}[D_1 \mid \boldsymbol{X}_1 \in B]}{q} \\
&= \frac{\mathrm{E}[D_1]}{q} = \frac{1{,}671}{196} \approx 8.525510,
\end{aligned}
\tag{15.37}
$$

as required. It is interesting to evaluate, using (15.30) and (15.31),

$$
\mathrm{E}[D_1 \mid \boldsymbol{X}_1 \in A] = 1 + \sum_{j \in \mathscr{P}} \pi_j^A \frac{1}{\pi_j + \pi_7} = \frac{42{,}457}{16{,}445} \approx 2.581757
\tag{15.38}
$$

and

$$
\mathrm{E}[D_1 \mid \boldsymbol{X}_1 \in B] = 1 + \sum_{j \in \mathscr{P}} \pi_j^B \frac{1}{\pi_j + \pi_7} = \frac{1{,}766}{385} \approx 4.587013.
\tag{15.39}
$$

Similarly, we can determine the probability generating function of L. First, the conditional probability generating functions of D_1, given $\boldsymbol{X}_1 \in A$ and given $\boldsymbol{X}_1 \in B$, are

$$
h_A(z) := \mathrm{E}[z^{D_1} \mid \boldsymbol{X}_1 \in A] = \left(1 - \sum_{j \in \mathscr{P}} \pi_j^A\right)z + \sum_{j \in \mathscr{P}} \pi_j^A \frac{(\pi_j + \pi_7)z^2}{1 - (1 - \pi_j - \pi_7)z}
\tag{15.40}
$$

and

$$
h_B(z) := \mathrm{E}[z^{D_1} \mid \boldsymbol{X}_1 \in B] = \sum_{j \in \mathscr{P}} \pi_j^B \frac{(\pi_j + \pi_7)z^2}{1 - (1 - \pi_j - \pi_7)z}.
\tag{15.41}
$$

Therefore, the probability generating function of L is

$$f(z) := \mathrm{E}[z^L] = \mathrm{E}[\mathrm{E}[z^L \mid N]] = \sum_{n=1}^{\infty} \mathrm{P}(N = n)\mathrm{E}[z^{D_1 + \cdots + D_{n-1} + D_n} \mid N = n]$$

$$= \sum_{n=1}^{\infty} (1 - q)^{n-1} q\, h_A(z)^{n-1} h_B(z) = \frac{q\, h_B(z)}{1 - (1 - q)h_A(z)}. \tag{15.42}$$

After some calculations, this leads to

$$\mathrm{Var}(L)$$
$$= f''(1) + \mathrm{E}[L] - (\mathrm{E}[L])^2$$
$$= \frac{(1 - q)h_A''(1) + q h_B''(1)}{q} + 2\left(\frac{1}{q} - 1\right)\frac{(1 - q)h_A'(1) + q h_B'(1)}{q}\, h_A'(1)$$
$$\qquad + \mathrm{E}[L] - (\mathrm{E}[L])^2$$
$$= \frac{\mathrm{E}[D_1(D_1 - 1)]}{q} + 2\left(\frac{1}{q} - 1\right)\frac{\mathrm{E}[D_1]}{q}\,\mathrm{E}[D_1 \mid \boldsymbol{X}_1 \in A] + \frac{\mathrm{E}[D_1]}{q} - \left(\frac{\mathrm{E}[D_1]}{q}\right)^2$$
$$= \frac{q\,\mathrm{E}[D_1^2] + 2(1 - q)\mathrm{E}[D_1]\,\mathrm{E}[D_1 \mid \boldsymbol{X}_1 \in A] - (\mathrm{E}[D_1])^2}{q^2}$$
$$= \frac{\dfrac{196}{495}\left[\dfrac{245{,}672}{27{,}225} + \left(\dfrac{557}{165}\right)^2\right] + 2\left(1 - \dfrac{196}{495}\right)\dfrac{557}{165}\dfrac{42{,}457}{16{,}445} - \left(\dfrac{557}{165}\right)^2}{(196/495)^2}$$
$$= \frac{1{,}768{,}701}{38{,}416} \approx 46.040738. \tag{15.43}$$

Here is another application of the Bayesian approach. A recent addition to craps is the *Fire Bet*. This is a bet that at least four *distinct* point numbers will be made during the shooter's hand. At one Las Vegas casino, it pays 25 for 1 for four distinct point numbers made, 250 for 1 for five, and 1,000 for 1 for six. To analyze this bet, we need to find the distribution of the number of distinct points made during the shooter's hand.

First we need a preliminary result. Consider a sequence of n independent trials, each with $k + 1$ possible outcomes, $1, 2, \ldots, k + 1$, with probabilities $p_1 > 0$, $p_2 > 0$, $\ldots$, $p_{k+1} > 0$ satisfying $p_1 + p_2 + \cdots + p_{k+1} = 1$. We want to find the distribution of the number of events $1, 2, \ldots, k$ that occur at least once in the n trials. Let A_i be the event that outcome i does not occur during the n trials, and define M to be the number of the events $A_1^c, A_2^c, \ldots, A_k^c$ that occur; also, define N to be the number of the events $A_1, A_2, \ldots, A_k$ that occur. Noting that $M = k - N$ and using Example 1.4.24 on p. 39, we find that

$$\mathrm{P}(M = j) = \mathrm{P}(N = k - j) = \sum_{l=k-j}^{k} (-1)^{l-k+j}\binom{l}{k-j} S_l, \tag{15.44}$$

where $S_0 := 1$, $S_1 := \sum_{i=1}^{k} \mathrm{P}(A_i)$, $S_2 := \sum\sum_{1 \le i < j \le k} \mathrm{P}(A_i \cap A_j)$, and so on. If $I \subset \{1, 2, \ldots, k\}$, then

$$\mathrm{P}\left(\bigcap_{i \in I} A_i\right) = \left(1 - \sum_{i \in I} p_i\right)^n, \qquad (15.45)$$

hence

$$\mathrm{P}(M = j) = \sum_{l=k-j}^{k} (-1)^{l-k+j} \binom{l}{k-j} \sum_{I \subset \{1,2,\ldots,k\}:|I|=l} \left(1 - \sum_{i \in I} p_i\right)^n. \quad (15.46)$$

Now, returning to the original question, we condition on the number of pass-line decisions in the hand. If there are n, then we apply (15.46) with n replaced by $n-1$, $k = 6$, and $p_i = \pi_i^A$ for $i \in \mathscr{P}$. The key observation is that, at any of the first $n - 1$ pass-line decisions, a point is made if and only if that point is established on the come-out roll. We then multiply the resulting probability by $(1 - q)^{n-1}q$ and sum over $n \ge 1$. We conclude that, with M being the number of distinct points made during the shooter's hand,

$$\mathrm{P}(M = j) = \sum_{l=6-j}^{6} (-1)^{l+j} \binom{l}{6-j} q \sum_{I \subset \mathscr{P}:|I|=l} \left(1 - (1-q)\left(1 - \sum_{i \in I} \pi_i^A\right)\right)^{-1},$$
$$(15.47)$$

We evaluate this formula numerically in Table 15.3 and conclude that the house advantage for the Fire Bet with the stated payoffs is about 0.207628. The odds against four or more distinct points being made are about 93.335 to 1. The odds against all six being made are about 6,155.318 to 1.

Table 15.3 Distribution of the number of distinct points made during the shooter's hand, rounded to 12 decimal places.

number	probability
0	.593 939 393 939
1	.260 750 492 004
2	.101 275 355 549
3	.033 434 212 179
4	.008 798 178 440
5	.001 639 933 139
6	.000 162 434 749

15.3 Problems

15.1. *Hardway, place, buy, and lay bets.*

(a) There are four *hardway bets*. An even dice total is said to be *hard* if both dice show the same number. Otherwise the total is said to be *easy*. For $j = 2, 3, 4, 5$, a *hard-$2j$ wager* is a bet that a hard $2j$ will appear before the first of an easy $2j$ and 7. Payoff odds are 7 to 1 on the hard 4 and hard 10, 9 to 1 on the hard 6 and hard 8. Find the house advantages.

(b) There are six *place bets* and six *buy bets*. A place bet or buy bet on the number $j \in \mathscr{P}$ is a bet that j will appear before 7 appears. The six place bets (previously described in Example 6.1.5 on p. 203) pay short odds: 9 to 5 for 4 or 10, 7 to 5 for 5 or 9, and 7 to 6 for 6 or 8. The six buy bets pay fair odds but are charged a 5% commission. Thus, the payoff odds are effectively 39 to 21 for 4 or 10, 29 to 21 for 5 or 9, and 23 to 21 for 6 and 8. Occasionally, the 5% commission is charged only on winning bets. In this case the payoff odds are effectively 39 to 20 for 4 or 10, 29 to 20 for 5 or 9, and 23 to 20 for 6 or 8. Compare the house advantages of the 18 bets described.

(c) There are six *place bets to lose* (though these are outdated and now rarely offered) and six *lay bets*. A place bet to lose or lay bet on the number $j \in \mathscr{P}$ is a bet that 7 will appear before j appears. The six place bets to lose pay short odds: 5 to 11 for 4 or 10, 5 to 8 for 5 or 9, and 4 to 5 for 6 or 8. The six lay bets pay fair odds but are charged a 5% commission. Thus, the payoff odds are effectively 19 to 41 for 4 or 10, 19 to 31 for 5 or 9, and 19 to 25 for 6 and 8. Compare the house advantages of the 12 bets described.

15.2. *Put bets.* The pass-line bet is an example of a *contract bet*, that is, a bet that cannot be withdrawn until it is resolved. The reason is clear: If a pass-line bet could be withdrawn once a point is established, the player would enjoy a significant advantage over the house because $\pi_7 + \pi_{11} = 2(\pi_2 + \pi_3 + \pi_{12})$.

On the other hand, players often *are* allowed to put a bet on the pass line *after* a point has been established. Such bets are known as *put bets*. The only motivation for making such bets is to take advantage of the free odds bets. Show that, for each $j \in \mathscr{P}$, there is an odds factor m_j such that a put bet with m-times odds on the point j has house advantage less than or equal to the house advantage of a place bet on j if $m \geq m_j$.

15.3. *Setter's advantage at hazard.* Find the advantage of the setter at hazard for each choice of the main (English rules) as well as when the main is chosen randomly (French rules). See Section 15.4 for details.

15.4. *Variance of D_1 via its probability generating function.* Find the variance of D_1 using only its probability generating function (15.15) on p. 505.

15.5. *Some conditional expectations.* Let X denote the player's profit from a one-unit pass-line bet, and let D denote the number of rolls required by that pass-line decision. Find the joint distribution of X and D. Use this to find the conditional expectations $E[X \mid D]$ and $E[D \mid X]$. Show, in particular, that $\lim_{n\to\infty} E[X \mid D = n]$ exists, and evaluate it.

15.6. *m_4-m_5-m_6-times free odds.* Define m_4-m_5-m_6-times free odds for both pass-line and don't-pass bets so as to generalize 3-4-5-times free odds of Example 6.1.7 on p. 205 as well as the m-times free odds of this chapter, corresponding to $m_4 = m_5 = m_6 = m$. Find formulas that generalize (15.11) and (15.12) on p. 503 and include all of the results of Table 15.1 on p. 504 as special cases.

15.7. *Correlation between successive line bets.* Consider the player who bets one unit on the pass line at the come-out roll, and then one unit on come if the second roll is a point roll, or on the pass line if the second roll is a come-out roll. Let X_1 (resp., X_2) be the player's profit from the first (resp., second) bet. Find the correlation between X_1 and X_2. Before doing the necessary calculation, make an educated guess as to whether the answer is positive or negative (or zero).

15.8. *Asymptotic variance for pass-come system.* Suppose a player bets one unit at each roll of the dice, making a pass line bet at each come-out roll and a come bet at each point roll. Let X_n be his profit from the bet made at the nth roll. Show that

$$\lim_{n\to\infty} \frac{1}{n}\mathrm{Var}(X_1 + X_2 + \cdots + X_n) \tag{15.48}$$

exists, and evaluate it.

Hint: $X_1, X_2, \ldots$ is a *stationary* sequence (i.e., for each $k \geq 1$, the joint distribution of $X_{n+1}, X_{n+2} \ldots, X_{n+k}$ does not depend on n) but not an independent sequence, as Problem 15.7 shows.

15.9. *A positive (conditional) expectation.* Suppose a player bets as described in Problem 15.8. Consider the player's conditional expected profit from the $(n+1)$th bet, given that he (eventually) loses the first n bets. Show that this tends to a positive limit, and evaluate it. What is the smallest n for which it is positive?

15.10. *Expected length of the shooter's hand by conditioning.* Let E be the expected number of rolls in the shooter's hand, and let E_j be the conditional expected number of *additional* rolls in the shooter's hand, given that the point $j \in \mathscr{P}$ has been established. Argue that

$$E = 1 + \left(1 - \sum_{j \in \mathscr{P}} \pi_j\right)E + \sum_{j \in \mathscr{P}} \pi_j E_j \tag{15.49}$$

and

$$E_k = 1 + \pi_k E + (1 - \pi_k - \pi_7)E_k \quad \text{and} \quad E_k = E_{14-k}, \qquad k \in \mathcal{P}, \tag{15.50}$$

to obtain a system of linear equations for E, E_4, E_5, and E_6. Solve the system.

15.11. *Expected length of the shooter's hand vs. expected time until the first 7.* Let L be the length of the shooter's hand, let S be the number of rolls needed to achieve the first 7, and let N be the number of pass-line decisions in the shooter's hand. Find a direct probabilistic argument showing that

$$E[L] = E[S] + E[N]. \tag{15.51}$$

(We can *confirm* that (15.51) is true using (15.19) and (15.20) on p. 505. The problem is to *explain* why (15.51) is true.)

15.12. *Probability generating function of the length of the shooter's hand via recursion.* Let L be the length of the shooter's hand, and derive a recursive formula for $s(n) := P(L = n)$ $(n \geq 2)$ similar to (15.23) on p. 506. Use this recursion to give an alternative derivation of the probability generating function (15.42) on p. 511.

15.13. *Distribution of the number of sevens in the shooter's hand.* Determine the distribution of the number of sevens in the shooter's hand. Find the mean and variance as well.

15.14. *Expected number of rolls until a winning 11; a losing 12.* Find the expected number of rolls required to achieve an 11 on a come-out roll; a 12 on a come-out roll.

15.15. *Number of winning points of each type in the shooter's hand.* For $j \in \mathscr{P}$ let W_j be the number of times point j is established on a come-out roll in the shooter's hand and subsequently made.
 (a) Find the mean of W_j for each $j \in \mathscr{P}$.
 (b) Find the distribution of W_j for each $j \in \mathscr{P}$.
 (c) Find the joint distribution of $W_4, W_5, W_6, W_8, W_9, W_{10}$.

15.16. *Decomposition of the shooter's hand.* In the shooter's hand let W_7 and W_{11} be the numbers of winning 7s and 11s (i.e., 7s and 11s on come-out rolls), let L_2, L_3, and L_{12} be the numbers of losing 2s, 3s, and 12s, let E_j $(j \in \mathscr{P})$ be the number of times point j is established, let W_j $(j \in \mathscr{P})$ be the number of winning js (i.e., the number of times point j is made), let $L_7 = 1$ be the number of losing 7s, and let I_j $(j \in \mathscr{P})$ be the number of inconclusive rolls when j is established as the point (i.e., inconclusive for pass-line bets). Find the expectations of these 24 random variables (as fractions, not decimals), and check that they sum to $E[L]$, the expected length of the shooter's hand.

15.17. *Lack-of-memory property of the shooter's hand.* Let $\boldsymbol{X}_1, \boldsymbol{X}_2, \ldots$ be as in (15.24) on p. 506, and define N by (15.27) on p. 508. Show that the conditional distribution of $(\boldsymbol{X}_n, \ldots, \boldsymbol{X}_N)$, given that $N \geq n$, coincides with the unconditional distribution of $(\boldsymbol{X}_1, \ldots, \boldsymbol{X}_N)$.

15.18. *Five-count system.* It has been suggested that some shooters have the ability to control the dice (see the chapter notes). This has led to a craps system called the *five-count system*. The idea is to avoid betting on shooters who cannot control the dice. Specifically, one waits to bet on a new shooter until he has qualified by achieving the five-count. A new shooter achieves the five-count by rolling a point number (called the one-count), then rolling three arbitrary numbers (the two-, three-, and four-counts), and finally rolling a point number again (the five-count). Of course he must do all this before sevening out.

(a) Assuming an unskilled shooter, what is the probability that he will achieve the five-count?

(b) If one bets only on shooters *after* they have achieved the five-count, what proportion of rolls does one bet on, assuming all shooters are unskilled?

15.19. *Crapless Craps.* In *Crapless Craps*, 7 is a natural and all other numbers are point numbers. Thus, the player can win the pass-line bet either by rolling 7 on the come-out roll, or by rolling any other number and then repeating that number before 7 appears. The bet pays even money. Find the house advantage, the expected duration of a pass-line decision, and the expected length of the shooter's hand.

15.20. *No-Crap Craps.* Crapless Craps (Problem 15.19) has two serious drawbacks: There is a large house advantage on the pass-line bet (with the result that the game is regarded as a carnival game and generates little interest), and there is no don't-pass bet. *No-Crap Craps* was created to overcome these problems. Here 7 is a natural, and there are eight point numbers: 2-or-3, 4, 5, 6, 8, 9, 10, and 11-or-12. Otherwise the rules for the pass-line bet are unchanged. For example, if a 12 appears on the come-out roll, the shooter's point is 11-or-12. To win the pass-line bet, he must roll either 11 or 12 before 7 appears. The bet pays even money. As in standard craps, the don't-pass bet is the reverse of pass-line bet (that is, the criteria for winning and losing are reversed, and the bet pays even money), with one exception: If the shooter's point is 11-or-12 and he sevens out, the don't-pass bettor is paid 1 to 2 instead of 1 to 1. Find the house advantage of the pass-line bet and the don't-pass bet, the expected duration of a pass-line decision, and the expected length of the shooter's hand.

15.21. *Bold play at craps.* Consider the problem of doubling one's initial fortune at craps before losing it. Notice that bold play on don't pass succeeds with probability 949/1,925. Assuming the availability of 3-4-5-times odds, show that this success probability can be improved.

15.22. *Ponzer system.* One of the most widely used systems at craps is the so-called *Ponzer system*, which works as follows. The player makes a one-unit pass-line bet or come bet at each roll of the dice, with this exception: Whenever there are two unresolved come bets, no bet is made at the next roll. In addition, a free odds bet is made whenever possible.

The system is more complicated than it first appears. Find a suitable finite Markov chain to describe the state of the system. Use it to evaluate the gambler's expected loss during the shooter's hand when playing the Ponzer system.

15.4 Notes

This chapter is an expanded version of Ethier (2007b).

Craps is a simplification of the dice game *hazard*, so we consider that game first. Several authors (e.g., Asbury 1938, p. 40) have cited William of Tyre [c. 1130–1186] as stating that hazard was invented by English Crusaders during the siege of an Arab castle called Hazart or Asart in 1125. As the *Oxford English Dictionary* (Simpson and Weiner 1989) noted, "The true Arab name of the castle appears to have been Ain Zarba." However, Kendall (1956) wrote (with the *OED* concurring),

> There can be little doubt that the word [hazard] derives from the Arabic *al zhar*, meaning a die.

Would English Crusaders have given a game of their own invention a name based on an Arabic word? It seems more likely that hazard already existed in Palestine at the time of the siege, and is therefore of Arab origin.

One of literature's first references to hazard occurs in Jehan Bodel's [c. 1165–1210] play *Le jeu de Saint Nicolas* (c. 1200). In fact there is sufficient detail in that work and in other sources that scholars have been able to specify the rules of the 12th-century game almost completely. (See Kavanagh 2005, pp. 44–45, for a summary.) This version of the game is played with three dice, and only the sum of the three numbers obtained is relevant. Numbers 3–6 are called *low hazards* and numbers 15–18 are called *high hazards*. Together, the low and high hazards are called *hazards*. We will refer to the player throwing the dice as the *caster* to conform with modern terminology. First, the caster rolls the dice, winning if he rolls a hazard and otherwise establishing the *main*. (Again, this is modern terminology.) In the latter case, the caster rolls again, losing if he rolls a hazard, and otherwise establishing the *chance*. If the chance is established, the caster continues to roll the dice, until either he wins by rolling a high hazard or the chance, or he loses by rolling a low hazard or the main. The only ambiguity in this description is what happens when the chance equals the main. We believe that this should result in a draw, for then the game is nearly symmetric, the only departure from symmetry being that the caster wins in the case of a hypothetical "double hazard."

At some point, presumably during the 13th or 14th centuries, hazard evolved into a game played with two dice instead of three.

Hazard played a role in "The Pardoner's Tale," one of Geoffrey Chaucer's [c. 1343–1400] *The Canterbury Tales* (see Chaucer 1980, pp. 448–449), a sermon that preaches against the sins of lechery, gluttony, sloth, false oaths, and gambling:[1]

> 'By goddes precious herte,' and 'By his nayles,'
> And 'By the blood of Crist that is in Hayles,
> Seuene is my chance and thyn is cynk and treye',

---

[1] Edited by N. F. Blake.

'By goddes armes if thow falsly pleye,
This dagger shal thurghout thyn herte go'.
This frut cometh of the bicche bones two:
Forsweryng, ire, falsnesse, homycide.

With modern spelling the third line would read, "Seven is my chance and thine is cinq and trey." If the last three words are interpreted as "five plus three," there is a direct relationship to the modern game of hazard, as will become clear in the next paragraph.

There are two versions of two-dice hazard, English and French. Let us describe the English version first. The two players are called the *caster* and the *setter*. First, the caster calls out a number, known as the *main*, equal to 5, 6, 7, 8, or 9, at his discretion. Then the caster rolls the dice. If he rolls a number that *nicks* the main, the caster wins. By definition, every number nicks itself, 5 and 9 are nicked by only themselves, 6 and 8 are nicked also by 12, and 7 is nicked also by 11. (One can also use "nick" as a noun. See the Thackeray quote on page 501.) The caster loses if he rolls *crabs*, that is, any of the numbers 2, 3, 11, or 12 that does not nick the main. Any other number, of which there are six, becomes the *chance*. In that case, the caster continues to roll the dice until the chance or the main appears. If the chance appears first, the caster wins the amount of his bet from the setter. If the main appears first, the caster loses the amount of his bet to the setter. See Table 15.4 for a summary of these rules.

Table 15.4 The structure of hazard.

main	nicks	crabs	chances
5	5	2, 3, 11, 12	4, 6, 7, 8, 9, 10
6	6, 12	2, 3, 11	4, 5, 7, 8, 9, 10
7	7, 11	2, 3, 12	4, 5, 6, 8, 9, 10
8	8, 12	2, 3, 11	4, 5, 6, 7, 9, 10
9	9	2, 3, 11, 12	4, 5, 6, 7, 8, 10

In the French version of hazard, instead of calling the main, the caster rolls the dice until he obtains a number between 5 and 9 inclusive. This becomes the main, and the game proceeds as in the English version. Another distinction is that the players bet against each other in the English game, while they bet against the house in the French game.

For the earliest description of these rules, specifically the French version, see Cotton [1630–1687] (1674, Chapter 34). Cotton remarked,

Certainly *Hazzard* is the most bewitching Game that is plaid on the Dice; for when a man begins to play he knows not when to leave off; and having once accustom'd himself to play at Hazzard he hardly ever after minds any thing else.

Montmort (1708, pp. 113–114; 1713, pp. 177–179) also described the French version. Huyn (1788, pp. 8–22) discussed the English version at some length and called it *krabs*. But the most extensive mathematical analysis of hazard was published by George Lambert (1812) under the title *The Game of Hazard Investigated*. To give a sense of Lambert's treatise, which Bohn (1850, pp. 353–376) reproduced,[2] we quote the following typical excerpt (p. 362):

> That the Caster does not throw in twice successively when 6 is the main, is $\frac{6961}{14256}$ multiplied by $\frac{6961}{14256}$ that is $\frac{48455521}{203233536}$ or $3l.$ $3s.$ $10\frac{1}{2}d.$ to $1l.$[3]
> That the Caster does not throw in three times successively is $\frac{48455521}{203233536}$ multiplied by $\frac{6961}{14256}$ that is $\frac{337298881681}{2897297289216}$ or $7l.$ $11s.$ $9\frac{1}{2}d.$ to $1l.$

Bohn (1850, p. 352) remarked that

> [I]t is the faculty of reducing the effect of the fractions that draws the line,—the mighty demarcation between the professor and the pupil of Hazard.

Lambert was a professor of hazard.

Rouse (1814, pp. 133–134), in a book titled *The Doctrine of Chances*, gave perhaps the clearest explanation of the English game. Van-Tenac (1847, pp. 30–36) cited both, calling the French version "hasard" and the English version "krabs."

It is clear how craps was derived from the English version of hazard. The caster simply calls the main of 7 every time. Since this choice gives the setter his smallest advantage, it is likely that the simplification evolved in the same way as the drawing rules in chemin de fer.

There are two theories on how craps got its name. According to Tinker (1933, pp. 2–3), craps was first brought to America by Bernard Xavier Philippe de Marigny de Mandeville [1785–1868]. Actually, it was hazard that the young Marigny learned about in London and brought back to his Creole friends in New Orleans shortly after the turn of the century. It achieved instant popularity, and Americans called it "Johnny Crapaud's" game, after a British term for the French, originating from the latters' supposed predilection for eating frog legs. Eventually, "crapaud's" was shortened to "craps." This theory was supported by Chase (1997, p. 81).

However, there is a more likely explanation for the origin of the term "craps": It is simply a corruption of "krabs," the French name for English hazard. "Krabs" is actually the French spelling of the English "crabs," which originally meant two aces, as in crab eyes (equivalent to today's snake eyes). Later, it came to mean any of the numbers 2, 3, 11, and 12 that do not nick the main.

---

[2] Oddly, Bohn referred to the author as George Lowbut, but Lambert is the correct surname.

[3] To *throw in* is to win, and to *throw out* is to lose. Here he has calculated the probability that the caster *does* throw in twice successively when 6 is the main, and he has converted to odds, expressing them in terms of pounds, shillings, and pence using $1l. = 20s.$ and $1s. = 12d.$ The calculations are correct.

In any case, Marigny, who inherited large land holdings from his father, was forced to sell much of the land to pay his gambling debts, including parcels on both sides of a newly opened extension of Burgundy Street into the Faubourg Marigny (said to be the first suburb of New Orleans), which in 1808 he christened "Rue de Craps" (Chase 1997, p. 85). According to Tinker (1933, pp. 3–4),

> Craps Street it remained for over fifty years, until a church was built in the neighborhood and the good parishioners were too horrified to walk on Sundays through a thoroughfare with such an unregenerate name, so they petitioned the city fathers [...]

and on November 20, 1852, Craps Street became Burgundy Street. Yet locals continued to call it Craps Street throughout the 19th century, just as New Yorkers still refer to Avenue of the Americas (so named in 1945) by its original name, Sixth Avenue. The church Tinker cited could not have been the Craps Street German Methodist Episcopal Church of the South, which was founded in 1847 (Deiler 1983, pp. 49–51), because it was not until 1899 that its name was changed to Burgundy Street Methodist Episcopal Church.

Who is responsible for this simplification of hazard, and when did it occur? Scarne (1980, p. 18) stated that craps is of American Black origin, but Barnhart (1991) disputed this assertion, arguing that 19th-century Blacks did not have the education necessary to create the game. We find the latter claim dubious: How much education is needed to recognize that 7 is the caster's best choice for the main? Anyone who keeps records will, with sufficient patience, discover this fact empirically.

Unfortunately, it is difficult to pin down when and where craps was introduced. Jonathan Harrington Green [1813–1887] (1843, p. 88) said of craps,

> This is a game lately introduced into New Orleans, and is fully equal to faro in its vile deception and ruinous effects.

But we do not know whether the game he was speaking of was, in today's terminology, hazard or craps. John Philip Quinn [1846–1916] (1892, p. 277), referring to an incident that occurred on a Mississippi steamboat at some time between 1873 and 1881 (see p. 42), wrote,

> I went out on deck, and my attention was arrested by hearing a negro crying in a stentorian voice, "come 7 or 11," then another man calling out, "chill'en cryin' fo' bread." This was followed by the sound of something rolling on the floor. My curiosity was aroused, and I went below to learn what was going on. Here I first saw the game of "craps" and my introduction to it cost me precisely $15.

Quinn went on to describe the rules of the game, which were exactly those of today's pass-line bet. Thus, craps was introduced before 1881, and certainly after 1800, because the game Marigny brought back from England, and named a street after, was, in today's terminology, hazard.

Craps may have played a role in one of the most famous disasters in American history. On September 28, 1944, Northwestern University's Medill

School of Journalism announced a gift of $35,000 from the estate of Chicago importer Louis M. Cohn [1853–1942]. The last paragraph of the press release contained the following admission (DeBartolo 1997):

> Mr. Cohn had an interesting connection with the origin of the Great Chicago Fire [October 8–10, 1871]. He steadfastly maintained that the traditional story of the cause of the fire—Mrs. O'Leary's cow that kicked over a lantern—was untrue. He asserted that he and Mrs. O'Leary's son, in the company of several other boys, were shooting dice in the hayloft ... by the light of a lantern, when one of the boys accidentally overturned the lantern, thus setting the barn afire. Mr. Cohn never denied that when the other boys fled, he stopped long enough to scoop up the money.

See DeBartolo (1997, 1998) for research that tends to corroborate this story.

The free odds bet in craps is a carryover from hazard. As Steinmetz (1870, Chapter 10) noted, quoting from the *Pall Mall Gazette* of Sept. 3, 1869,

> For example, suppose the caster "sets"—that is, places on the table—a stake of £10, and it is covered by an equal amount, and he then calls 7 as his main and throws 5; the groom-porter at once calls aloud, "5 to 7"—that means, 5 is the number to win and 7 the number to lose, and the player continues throwing until the event is determined by the turning up of either the main or the chance. During this time, however, a most important feature in the game comes into operation—the laying and taking of the odds caused by the relative proportions of the main and the chance. These, as has been said, are calculated with mathematical nicety, are proclaimed by the groom-porter, and are never varied. In the above instance, as the caster stands to win with 5 and to lose with 7, the odds are declared to be 3 to 2 against him, inasmuch as there are three ways of throwing 7, and only two of throwing 5. As soon as the odds are declared, the caster may increase his stake by any sum he wishes, and the other players may cover it by putting down (in this instance) two-thirds of the amount, the *masse*, or entire sum, to await the turning up of either main or chance.

The don't-pass (and don't-come) bet was introduced by John H. Winn in 1907 (Scarne 1980, p. 20). Initially, there was a 5 percent commission. Later the barred number became standard. Winn also introduced what was known as the Philadelphia layout, the predecessor of today's craps layout, and was more responsible than anyone for making craps a casino game.

With the decline of faro, craps had become the most popular gambling game in the United States, and it remained so through two world wars. It even came to Monte Carlo in 1948 (Chafetz 1960, p. 421). Ultimately, the popularity of craps was eclipsed in the U.S. by that of twenty-one and slot machines.

In recent years craps has attracted the interest of "advantage players." Books have been written and seminars offered on the topic of dice control. By setting the dice and then throwing them carefully, it is claimed that one can reduce the probability of a 7 and thereby obtain an advantage over the house. Authors who have endorsed this point of view include Frank Scoblete [1947–] (1991, 1993, 2000, 2003), Sharpshooter (2002), Stanford Wong [1943–] (2005), Scoblete and Dominator (2005), and Gambino (2006). On the other hand, Grosjean (2009, Chapter 78) argued that dice control is not a viable advantage-play technique, while Zender (2006, Chapter 8; 2008, Chapter 33) suggested that the question is still open. Incidentally, it was established by Cabot and Hannum (2005) that dice control is within the rules of the game and is therefore not illegal, even though such shooters are attempting to alter the game's probabilities.

As for the mathematics of craps, John Arbuthnot [1667–1735] (1692, p. 90), Pierre Rémond de Montmort [1678–1719] (1713, p. 178), and Abraham De Moivre [1667–1754] (1718, p. 156) knew that the probability of winning at hazard when the main is 7 is $244/495 = 0.49\overline{2}$. Some two centuries later, the identical problem of finding the probability of winning (a pass-line bet) at craps was posed in the problem section of the *American Mathematical Monthly*, where it was solved incorrectly by Zerr (1903); he obtained $722/1{,}485 \approx 0.486195$. When it was posed again by Wilson (1905), the *Monthly* problem editor remarked,

> A correct solution of this problem is given by Dr. Zerr, in Vol. X, No. 3, p. 81. Mr. J. E. Sanders sent in a different solution, his answer being $\frac{244}{495}$, in favor of the player.

(Mr. Sanders was correct.) The same problem was posed once again in the *Monthly* by the respected mathematician Warren Weaver [1894–1978] (1919). He stated the relevant probability as 0.49847, which is also incorrect. Decades later, Weaver (1963, p. 329) redeemed himself.

Bancroft H. Brown [1894–1974] (1919) was the first to evaluate the expected number of rolls in a pass-line decision, and Smith (1968) found the distribution and the variance as well. Boll (1936, p. 115) had previously evaluated the distribution numerically, albeit rather inaccurately. Brown (1919) also found the expected number of pass-line decisions in the shooter's hand, and the present author used these results in 1985 to find the expected length of the shooter's hand (Griffin 1991, Chapter 16). Stewart (1986) found the median, and finally Braun (1987) numerically evaluated the distribution without revealing his algorithm. The variance in (15.43) is from Ethier (2007b), but a similar calculation in a more general context can be found in Zhang and Lai (2004).

The longest recorded hand in craps history is 154 rolls, having occurred to Patricia DeMauro on May 23, 2009, at the Borgata Hotel Casino & Spa in Atlantic City, over the course of four hours and 18 minutes. The probability that a random hand lasts at least this long is, from (15.23),

$$t(154) \approx 0.178882426 \cdot 10^{-9}, \tag{15.52}$$

which amounts to one chance in 5.59 billion, approximately.

The record was previously held by Stanley Fujitake [1923–2000] (118 rolls, May 28, 1989, California Hotel and Casino, Las Vegas). One might ask how reliable these numbers (154 and 118) are. In Mr. Fujitake's case, casino personnel replayed the surveillance videotape to confirm the number of rolls and the duration of time (three hours and six minutes). See Akane (2008).

There is also a report (Scoblete 2007, Part 4) that Fujitake's record was broken earlier by a gentleman known only as the Captain [1923–] (148 rolls, July 2005, Atlantic City, two hours and 18 minutes), although the documentation is sparse, and the credibility of this story has been questioned by Grosjean (2009, p. 480).

The DeMauro incident motivated Ethier and Hoppe (2010) to find a closed-form expression for $t(n) := P(L \geq n)$:

$$t(n) = c_1 e_1^{n-1} + c_2 e_2^{n-1} + c_3 e_3^{n-1} + c_4 e_4^{n-1}, \qquad n \geq 1, \qquad (15.53)$$

where

$$e_1 := e(1,1), \quad e_2 := e(1,-1), \quad e_3 := e(-1,1), \quad e_4 := e(-1,-1), \quad (15.54)$$

$$e(u,v) := \frac{5}{8} + \frac{u}{72}\sqrt{\frac{349+\alpha}{3}} + \frac{v}{72}\sqrt{\frac{698-\alpha}{3} - 2{,}136u\sqrt{\frac{3}{349+\alpha}}}, \quad (15.55)$$

$$\alpha := 2\sqrt{9{,}829}\,\cos\left[\frac{1}{3}\cos^{-1}\left(-\frac{710{,}369}{9{,}829\sqrt{9{,}829}}\right)\right], \qquad (15.56)$$

$$c_1 := f(e_1), \quad c_2 := f(e_2), \quad c_3 := f(e_3), \quad c_4 := f(e_4), \qquad (15.57)$$

$$f(w) := \frac{1{,}975 - 16{,}346w + 33{,}828w^2 - 20{,}736w^3}{3w^2(6{,}107 - 34{,}356w + 58{,}320w^2 - 31{,}104w^3)}. \qquad (15.58)$$

We did not say that the expression would be simple! But it yields the unexpected result that the distribution of L is a linear combination of four geometric distributions.

The Fire Bet, discussed at the end of Section 15.2, was patented by Perry B. Stasi in 2003 (U.S. patent 6,655,689). His patent application contains the relevant probabilities.

> An exact probability analysis was performed to derive the probability distribution of all points made. The accuracy of the analysis was verified by a one-billion roll computer simulation. In that a shooter could, in theory, have an infinitely long streak of making points without sevening out, approximations were made. However, the error is estimated to be within an insignificant 0.000002%.

Table 15.3 shows that Stasi's analysis is indeed accurate to the stated degree of precision. See Grosjean (2009, pp. 472, 476) for a recursive solution. The payoff odds vary from one casino to the next. Other existing payoff odds, cited by Shackleford (2005a, p. 39), are 10 to 1, 200 to 1, and 2,000 to 1.

Shackleford (2009) addressed Problem 15.2. Problem 15.3 was solved by Montmort and De Moivre early in the 18th century, as suggested above. $E[D \mid X]$ from Problem 15.5 was found by Ross (2006, p. 370–371). Problem 15.10 and 15.14 are due to Wong (2005, pp. 165–167), while Problem 15.11 was motivated by an observation of Wong (2005, p. 167). The mean of the distribution of Problem 15.13 was found by Wong (2005, p. 167). Problems 15.15 and 15.16 are also results of Wong (2005, pp. 167–168), who used a different approach. The five-count system (Problem 15.18) was attributed by Scoblete (1991, Chapter 3) to the Captain.

Crapless Craps (Problem 15.19) is a registered trademark of Bob Stupak. The game was offered by Stupak at his Vegas World in 1981 and has more recently been offered at the Stratosphere in Las Vegas as well as in other locations under the names Never Ever Craps and Craps No More. Writing during World War II, Scarne and Rawson (1945, p. 354) described the game as a "recent innovation." In fact, it may be even older, because Sham (1930) called it "everything a point" and correctly evaluated the house advantage. No-Crap Craps (Problem 15.20) was patented by the present author in 2006 (U.S. patent 7,134,660). It was approved for use in Nevada by the Nevada Gaming Control Board in 2005 but has not yet appeared in casinos.

Problem 15.21 is due to Ethier (1987). It was motivated by a widely reported incident that occurred in 1980 in which an unidentified gambler bet $777,000 double or nothing on don't pass at Binion's Horseshoe Club in downtown Las Vegas, and won (*Time*, October 6, 1980). The gambler was later identified as William Lee Bergstrom [1951–1985] of Austin, Texas. See Drummond (1985) for the complete story, including that of Bergstrom's 1985 suicide.

Ponzer (1977) is responsible for the Ponzer system (Problem 15.22).

Chapter 16
House-Banked Poker

But that there poker ain't fair, you know.

William Makepeace Thackeray, *The Wolves and the Lamb*

Since 1990 a number of new casino games have been introduced, and here we consider two of the most successful such games, Let It Ride and Three Card Poker (Sections 16.1 and 16.2, respectively). Each is house banked and poker based, and each offers a greater advantage to the house than do the traditional games of craps, faro, baccarat, trente et quarante, and twenty-one. The analysis of each game is straightforward in principle but exhausting in practice.

16.1 Let It Ride

Let It Ride is a proprietary house-banked poker-based game that is widely available at U.S. casinos. It is played with a standard 52-card deck, which is reshuffled between coups, typically with an automatic shuffling machine. There are two betting opportunities, the *basic bet* and the *bonus side bet*. We will describe the more interesting basic bet, leaving the bonus side bet to Problem 16.1 on p. 539.

Each player makes a three-unit bet and receives three cards face down. Two community cards are then dealt face down to the dealer. Each player, after examining his hand, has the option of either reducing his bet size by one unit or letting it ride. The dealer then reveals the first community card, which is each player's fourth card. Each player then has the option of reducing his bet size by one unit (regardless of his previous decision) or letting it ride. The dealer than reveals the second community card, which is each player's fifth and final card. If a player has a pair of tens or better, he is paid according

S.N. Ethier, *The Doctrine of Chances*, Probability and its Applications,
DOI 10.1007/978-3-540-78783-9_16, © Springer-Verlag Berlin Heidelberg 2010

to the value of his hand. A royal flush pays 1,000 to 1, a straight flush pays 200 to 1, four of a kind pays 50 to 1, a full house pays 11 to 1, a flush pays 8 to 1, a straight pays 5 to 1, three of a kind pays 3 to 1, two pair pays 2 to 1, and a high pair (tens or better) pays 1 to 1. See Table 16.1. Note that these odds are paid on whatever remains of the player's initial three-unit bet (one, two, or three units). If the player has a pair of 9s or worse, he loses whatever remains of his initial three-unit bet. There is no possibility of a push.

Table 16.1   Analysis of the first bet at Let It Ride (no cards seen).

five-card hand	number of ways	profit	product
royal flush	4	1,000	4,000
straight flush	36	200	7,200
four of a kind	624	50	31,200
full house	3,744	11	41,184
flush	5,108	8	40,864
straight	10,200	5	51,000
three of a kind	54,912	3	164,736
two pair	123,552	2	247,104
high pair*	422,400	1	422,400
other	1,978,380	−1	−1,978,380
sum	2,598,960		−968,692

*tens or better

Observe that, although the player begins by placing three units on the betting surface, at that point he is placing only one unit at risk and is making what we will call the *first bet*. After seeing his three-card hand, the player has the option of reclaiming one of his three units. If he lets it ride, he is putting a second unit at risk and is making what we will call the *second bet*. After the first community card is revealed, the player has the option, on the basis of his four-card hand, of reclaiming one of his two or three remaining units. If he lets it ride, he is putting another unit at risk and is making what we will call the *third bet*. Each of the three bets is a one-unit bet.[1]

Our plan in this section is to analyze each of these three bets separately. We will determine under what conditions the second and third bets should be made, using the traditional criterion that the conditional expected profit be nonnegative. (See Example 6.1.10 on p. 211 for a different criterion.) The question of whether the first bet should be made is of course tantamount to

---

[1] Our terminology departs slightly from casino practice, in which what we have called the first, second, and third bets are placed in circles labeled $, 1, and 2, respectively.

asking whether one should play Let It Ride, and on this point we do not offer any advice.

Table 16.1 is essentially self-explanatory, with the $\binom{52}{5} = 2{,}598{,}960$ poker hands categorized much as in Example 1.1.10 on p. 6. The only new category is that of high pair, and the number of such hands is

$$\binom{5}{1}\binom{12}{3}\binom{4}{2}\binom{4}{1}^3 = 422{,}400. \tag{16.1}$$

We conclude from Table 16.1 that the player's profit X_1 from his (one-unit) first bet has expected value

$$\mathrm{E}[X_1] = \frac{-968{,}692}{\binom{52}{5}} = -\frac{242{,}173}{649{,}740} \approx -0.372723. \tag{16.2}$$

We caution the reader to withhold judgment until we have analyzed the second and third bets, both of which help to mitigate the effect of this rather unfavorable first bet.

Tables 16.2 and 16.3 summarize the analysis of the player's second bet. Let X_2 denote the player's profit from his (one-unit) second bet, assuming he chooses to make this bet, and let Y_3 denote the player's initial three-card hand. Table 16.2 tells us that

$$\mathrm{E}[X_2] = \frac{-9{,}686{,}920}{\binom{52}{3}\binom{49}{2}} = -\frac{242{,}173}{649{,}740} \approx -0.372723 \tag{16.3}$$

and

Table 16.2 Analysis of the second bet at Let It Ride (three cards seen). Column (a) contains the number of ways, column (b) contains the player's conditional expected profit (from a one-unit bet), and column (c) is column (b) multiplied by $\binom{49}{2}$.

three-card hand	(a)	(b)	(c)	(a) × (c)
three of a kind	52	5.408 163	6,360	330,720
high pair	1,440	1.436 224	1,689	2,432,160
low pair	2,304	−.060 374	−71	−163,584
no pair, suited	1,144	(finer breakdown needed—		
no pair, unsuited	17,160	see Table 16.3)		
sum	22,100			−9,686,920
restricted sum*	1,608			2,836,136

*sum over rows with entry in column (c) nonnegative

$$\mathrm{E}[X_2\, 1_{\{\mathrm{E}[X_2|Y_3]\ge 0\}}] = \frac{2{,}836{,}136}{\binom{52}{3}\binom{49}{2}} = \frac{354{,}517}{3{,}248{,}700} \approx 0.109126. \tag{16.4}$$

Equation (16.3) tells us that the player who makes the second bet regardless of his cards has the same expected profit on the second bet as he does on the first, a fact that is of course obvious. Equation (16.4) tells us that the player who makes the second bet only when his conditional expected profit is nonnegative has a significant positive expected profit on the second bet.

We will later return to the question of how Tables 16.2 and 16.3 were computed, but let us next consider the third bet. Tables 16.4 and 16.5 summarize the analysis of the player's third bet. Let X_3 denote the player's profit from his (one-unit) third bet, assuming he chooses to make this bet, and let Y_4 denote the player's initial four-card hand. Table 16.4 tells us that

$$\mathrm{E}[X_3] = \frac{-4{,}843{,}460}{\binom{52}{4}\binom{48}{1}} = -\frac{242{,}173}{649{,}740} \approx -0.372723 \tag{16.5}$$

and

$$\mathrm{E}[X_3\, 1_{\{\mathrm{E}[X_3|Y_4]\ge 0\}}] = \frac{2{,}969{,}836}{\binom{52}{4}\binom{48}{1}} = \frac{742{,}459}{3{,}248{,}700} \approx 0.228540. \tag{16.6}$$

Equation (16.5) tells us that the player who makes the third bet regardless of his cards has the same expected profit on the third bet as he does on the first, a fact that is again obvious. Equation (16.6) tells us that the player who makes the third bet only when his conditional expected profit is nonnegative has a significant positive expected profit on the third bet, in fact more than twice that of the second bet.

By (16.2), (16.4), and (16.6), the player's profit X at Let It Ride, under optimal play, has expected value

$$\begin{aligned}
\mathrm{E}[X] &= \mathrm{E}[X_1 + X_2\, 1_{\{\mathrm{E}[X_2|Y_3]\ge 0\}} + X_3\, 1_{\{\mathrm{E}[X_3|Y_4]\ge 0\}}] \\
&= -\frac{242{,}173}{649{,}740} + \frac{354{,}517}{3{,}248{,}700} + \frac{742{,}459}{3{,}248{,}700} \\
&= -\frac{37{,}963}{1{,}082{,}900} \approx -0.035057.
\end{aligned} \tag{16.7}$$

By Tables 16.2 and 16.4 his total amount bet B has expected value

$$\begin{aligned}
\mathrm{E}[B] &= \mathrm{E}[1 + 1_{\{\mathrm{E}[X_2|Y_3]\ge 0\}} + 1_{\{\mathrm{E}[X_3|Y_4]\ge 0\}}] \\
&= 1 + \frac{1{,}608}{\binom{52}{3}} + \frac{43{,}133}{\binom{52}{4}} = \frac{333{,}556}{270{,}725} \approx 1.232084.
\end{aligned} \tag{16.8}$$

We conclude that the house advantage at Let It Ride, under optimal play, is

Table 16.3 Analysis of the second bet at Let It Ride (three cards seen), continued. Column (a) contains the number of ways, column (b) contains the player's conditional expected profit (from a one-unit bet), and column (c) is column (b) multiplied by $\binom{49}{2}$. Hands with three distinct denominations are described by (s, h), where s is the number of straights—disregarding suits—that can be made from the hand, and h is the number of high cards (A, K, Q, J, T) in the hand.

(s, h)	(a)	(b)	(c)	(a) $\times$ (c)
		no pair, suited		
$(0, 0)$	112	$-.505\ 102$	-594	$-66{,}528$
$(0, 1)$	496	$-.311\ 224$	-366	$-181{,}536$
$(0, 2)$	280	$-.117\ 347$	-138	$-38{,}640$
$(1, 0)$	48	$-.265\ 306$	-312	$-14{,}976$
$(1, 1)$	48	$-.071\ 429$	-84	$-4{,}032$
$(1, 2)$	24	$.122\ 449$	144	$3{,}456$
$(1, 3)$	24	$.996\ 599$	$1{,}172$	$28{,}128$
$(2, 0)$	44	$-.025\ 510$	-30	$-1{,}320$
$(2, 1)$	12	$.168\ 367$	198	$2{,}376$
$(2, 2)$	12	$.362\ 245$	426	$5{,}112$
$(2, 3)$	12	$1.236\ 395$	$1{,}454$	$17{,}448$
$(3, 0)$	20	$.214\ 286$	252	$5{,}040$
$(3, 1)$	4	$.408\ 163$	480	$1{,}920$
$(3, 2)$	4	$.602\ 041$	708	$2{,}832$
$(3, 3)$	4	$1.476\ 190$	$1{,}736$	$6{,}944$
		no pair, unsuited		
$(0, 0)$	1,680	$-.849\ 490$	-999	$-1{,}678{,}320$
$(0, 1)$	7,440	$-.655\ 612$	-771	$-5{,}736{,}240$
$(0, 2)$	4,200	$-.461\ 735$	-543	$-2{,}280{,}600$
$(1, 0)$	720	$-.767\ 857$	-903	$-650{,}160$
$(1, 1)$	720	$-.573\ 980$	-675	$-486{,}000$
$(1, 2)$	360	$-.380\ 102$	-447	$-160{,}920$
$(1, 3)$	360	$-.186\ 224$	-219	$-78{,}840$
$(2, 0)$	660	$-.686\ 224$	-807	$-532{,}620$
$(2, 1)$	180	$-.492\ 347$	-579	$-104{,}220$
$(2, 2)$	180	$-.298\ 469$	-351	$-63{,}180$
$(2, 3)$	180	$-.104\ 592$	-123	$-22{,}140$
$(3, 0)$	300	$-.604\ 592$	-711	$-213{,}300$
$(3, 1)$	60	$-.410\ 714$	-483	$-28{,}980$
$(3, 2)$	60	$-.216\ 837$	-255	$-15{,}300$
$(3, 3)$	60	$-.023\ 959$	-27	$-1{,}620$

$$H_0(B,X) = H(B,X) := \frac{-E[X]}{E[B]} = \frac{37{,}963}{1{,}334{,}224} \approx 0.028453. \qquad (16.9)$$

By optimal play we mean making the second and third bets if and only if one's conditional expected profit is nonnegative (see Example 6.1.10 on p. 211 for an alternative formulation). Tables 16.2–16.5 give precise criteria, and we summarize these criteria in Table 16.6. In most published accounts the strategies for the second and third bets are called, for obvious reasons, the *three-card strategy* and the *four-card strategy*. We adopt this terminology.

Table 16.4 Analysis of the third bet at Let It Ride (four cards seen). Column (a) contains the number of ways, column (b) contains the player's conditional expected profit (from a one-unit bet), and column (c) is column (b) multiplied by $\binom{48}{1}$.

four-card hand	(a)	(b)	(c)	(a) × (c)
four of a kind	13	50.000 000	2,400	31,200
three of a kind	2,496	4.479 167	215	536,640
two pair	2,808	2.750 000	132	370,656
high pair	31,680	1.208 333	58	1,837,440
low pair	50,688	−.458 333	−22	−1,115,136
no pair, suited	2,860	(finer breakdown needed—		
no pair, unsuited	180,180	see Table 16.5)		
sum	270,725			−4,843,460
restricted sum*	43,133			2,969,836

*sum over rows with entry in column (c) nonnegative

In the case of a no-pair unsuited hand with $(s,h) = (1,4)$ or $(s,h) = (2,0)$, the optimal four-card strategy is arguable, owing to the fact that the player's conditional expected profit is 0. We have recommended letting it ride in such cases for the same reason that we recommend the free-odds bet at craps: It reduces the house advantage shown in (16.9). (For further details on this point, see Example 6.1.7 on p. 205.) Those who prefer to avoid such bets would follow the strategy of Table 16.7

The last issue we take up is the question of how Tables 16.2–16.5 were computed. Let us first consider column (c), the player's expected profit. In each case there is a formula. For Table 16.3 the relevant formulas are

Table 16.5 Analysis of the third bet at Let It Ride (four cards seen), continued. Column (a) contains the number of ways, column (b) contains the player's conditional expected profit (from a one-unit bet), and column (c) is column (b) multiplied by $\binom{48}{1}$. Hands with four distinct denominations are described by (s, h), where s is the number of straights—disregarding suits—that can be made from the hand, and h is the number of high cards (A, K, Q, J, T) in the hand.

(s, h)	(a)	(b)	(c)	(a) × (c)
		no pair, suited		
$(0, 0)$	212	.687 500	33	6,996
$(0, 1)$	1,084	.812 500	39	42,276
$(0, 2)$	1,100	.937 500	45	49,500
$(0, 3)$	300	1.062 500	51	15,300
$(1, 0)$	48	5.062 500	243	11,664
$(1, 1)$	32	5.187 500	249	7,968
$(1, 2)$	16	5.312 500	255	4,080
$(1, 3)$	16	5.437 500	261	4,176
$(1, 4)$	16	22.229 167	1,067	17,072
$(2, 0)$	20	9.437 500	453	9,060
$(2, 1)$	4	9.562 500	459	1,836
$(2, 2)$	4	9.687 500	465	1,860
$(2, 3)$	4	9.812 500	471	1,884
$(2, 4)$	4	26.604 167	1,277	5,108
		no pair, unsuited		
$(0, 0)$	13,356	−1.000 000	−48	−641,088
$(0, 1)$	68,292	−.875 000	−42	−2,868,264
$(0, 2)$	69,300	−.750 000	−36	−2,494,800
$(0, 3)$	18,900	−.625 000	−30	−567,000
$(1, 0)$	3,024	−.500 000	−24	−72,576
$(1, 1)$	2,016	−.375 000	−18	−36,288
$(1, 2)$	1,008	−.250 000	−12	−12,096
$(1, 3)$	1,008	−.125 000	−6	−6,048
$(1, 4)$	1,008	.000 000	0	0
$(2, 0)$	1,260	.000 000	0	0
$(2, 1)$	252	.125 000	6	1,512
$(2, 2)$	252	.250 000	12	3,024
$(2, 3)$	252	.375 000	18	4,536
$(2, 4)$	252	.500 000	24	6,048

$\mathrm{E}[X_2 \mid Y_3$ a no-pair, suited (s, h) hand$]$

$$= \binom{49}{2}^{-1} \{1{,}001\delta_{h,3} + 201(s - \delta_{h,3}) + 9(45 - s) + 6(15s) + 4(9) + 3(27)$$
$$+ 2[(3h)(40) + (5 - h)(6)]\} - 1 \qquad (16.10)$$

and

$\mathrm{E}[X_2 \mid Y_3$ a no-pair, unsuited (s, h) hand$]$ $\qquad (16.11)$

$$= \binom{49}{2}^{-1} \{6(16s) + 4(9) + 3(27) + 2[(3h)(40) + (5 - h)(6)]\} - 1,$$

and the three remaining cases in Table 16.2 are straightforward. For Table 16.5 the relevant formulas are

$\mathrm{E}[X_3 \mid Y_4$ a no-pair, suited (s, h) hand$]$ $\qquad (16.12)$

$$= \binom{48}{1}^{-1} [1{,}001\delta_{h,4} + 201(s - \delta_{h,4}) + 9(9 - s) + 6(3s) + 2(3h)] - 1$$

and

$\mathrm{E}[X_3 \mid Y_4$ a no-pair, unsuited (s, h) hand$]$

$$= \binom{48}{1}^{-1} [6(4s) + 2(3h)] - 1, \qquad (16.13)$$

and the five remaining cases in Table 16.4 are straightforward.

Table 16.6 Optimal strategy at Let It Ride. Second and third bets are made only with nonnegative conditional expected profit. Here s is the number of straights—disregarding suits—that can be made from the hand, and h is the number of high cards (A, K, Q, J, T) in the hand.

Three-card strategy: Let it ride if holding ...
• a made payoff
• a suited hand with $s + h \geq 3$

Four-card strategy: Let it ride if holding ...
• a made payoff
• a suited hand
• a no-pair unsuited hand with $s = 2$ or $h = 4$

Next we consider column (a) of Tables 16.2–16.5, the number of ways each type of hand can occur. The eight cases in Tables 16.2 and 16.4 are

Table 16.7 Alternative optimal strategy at Let It Ride. Second and third bets are made only with *positive* conditional expected profit. Here s is the number of straights—disregarding suits—that can be made from the hand, and h is the number of high cards (A, K, Q, J, T) in the hand.

Three-card strategy: Let it ride if holding ...
• a made payoff • a suited hand with $s + h \geq 3$
Four-card strategy: Let it ride if holding ...
• a made payoff • a suited hand • a no-pair unsuited hand with $s = 2$ and $h \geq 1$

straightforward using the methods of Example 1.1.10 on p. 6. Column (a) of Table 16.3 is slightly simpler than column (a) of Table 16.5, so let us consider the latter. First, for each four distinct denominations, there are four suited four-card hands and

$$\binom{4}{1}^4 - \binom{4}{1} = 252 \tag{16.14}$$

unsuited ones, so it suffices to consider the $\binom{13}{4} = 715$ four-card hands without regard to suits, and determine how many of them have specified values of (s, h). Of the 715 hands, $\binom{5}{0}\binom{8}{4} = 70$ have $h = 0$. Of these, five hands have $s = 2$, namely 5-4-3-2, 6-5-4-3, 7-6-5-4, 8-7-6-5, and 9-8-7-6. Twelve have $s = 1$, namely 6–2, 7–3, 8–4, 9–5, each with one (interior) gap. The remaining 53 subsets have $s = 0$. Of the 715 hands, $\binom{5}{1}\binom{8}{3} = 280$ have $h = 1$. Of these, only one, T-9-8-7, has $s = 2$. Eight have $s = 1$, namely 4-3-2-A, 5-3-2-A, 5-4-2-A, 5-4-3-A, T-8-7-6, T-9-7-6, T-9-8-6, and J-9-8-7. The remaining 271 subsets have $s = 0$. Of the 715 hands, $\binom{5}{2}\binom{8}{2} = 280$ have $h = 2$. Of these, only one, J-T-9-8, has $s = 2$. Four have $s = 1$, namely J-T-8-7, J-T-9-7, Q-T-9-8, and Q-J-9-8. The remaining 275 hands have $s = 0$. Of the 715 hands, $\binom{5}{3}\binom{8}{1} = 80$ have $h = 3$. Of these, only one, Q-J-T-9, has $s = 2$. Four have $s = 1$, namely Q-J-T-8, K-J-T-9, K-Q-T-9, and K-Q-J-9. The remaining 75 hands have $s = 0$. Finally, of the 715 hands, $\binom{5}{4}\binom{8}{0} = 5$ have $h = 4$. Of these, only one, K-Q-J-T, has $s = 2$. The remaining four have $s = 1$. This confirms column (a) of Table 16.5. The analogous argument for column (a) of Table 16.3 is left to the reader.

16.2 Three Card Poker

Three Card Poker is a proprietary house-banked poker-based game that has
become quite popular in recent years, perhaps because of the ease with which
the optimal strategy can be mastered. It is played with a standard 52-card
deck, which is reshuffled between coups, typically with an automatic shuffling machine. There are two betting opportunities, the *ante-play wager* and
the *pair-plus wager*. We will describe the more interesting ante-play wager,
leaving the pair-plus wager to Problem 16.6 on p. 540.

Each player makes an *ante wager* and receives three cards face down. The
dealer also receives three cards face down. Each player, after examining his
hand, must decide whether to fold or play. If he folds, he loses his ante wager.
If he plays, he must make an additional wager, called a *play wager*, equal in
size to the ante wager. The dealer then reveals his hand to determine whether
it *qualifies* with a rank of queen high or better. If the dealer's hand fails to
qualify, the player's ante wager is paid even money and his play wager is
pushed. If the dealer's hand qualifies, it is compared with the player's hand.
If the player's hand outranks the dealer's hand, both of the player's wagers
are paid even money. If the dealer's hand outranks the player's hand, both
of the player's wagers are lost. If the player's hand and the dealer's hand are
equal in rank, both of the player's wagers are pushed.

The ranking of hands in Three Card Poker is slightly different from that
in traditional poker. First, the categories two pair, full house, and four of
a kind do not occur. Second, the relative ranks of the categories three of
a kind, straight, and flush are reversed. The reason for this reversal can
be seen from Table 16.8. Hands in different categories are ranked as the
categories are ranked, namely (from best to worst) straight flush, three of
a kind, straight, flush, one pair, and no pair. Hands in the same categories
are ranked with the help of the following rules. Straight flushes are ranked
according to the highest denomination (which is 3 in the case of 3-2-A). Three-
of-a-kind hands are ranked according to the common denomination. Straights
are ranked according to the highest denomination (which is 3 in the case of 3-
2-A). Flushes are ranked first according to the highest denomination, second
according to the middle denomination, and third according to the lowest
denomination. One-pair hands are ranked first according to the denomination
of the pair and second according to the denomination of the third card. No-
pair hands are ranked first according to the highest denomination, second
according to the middle denomination, and third according to the lowest
denomination.

There is one remaining detail called the *ante bonus*. Regardless of whether
the player plays, regardless of whether the dealer qualifies, and regardless
of the outcome of the ante-play wager, if the player's hand is a straight or
better, he receives a bonus payoff on his ante wager. This bonus amounts to
1 to 1 for a straight, 4 to 1 for three of a kind, and 5 to 1 for a straight flush.

Table 16.8 Ranking of hands in Three Card Poker. For each expression that is the product of two factors, the first is the number of ways to choose the hand's denominations, and the second is the number of ways to choose the suits for the chosen denominations.

rank	number of ways	
straight flush	$\binom{12}{1}\binom{4}{1}$	48
three of a kind	$\binom{13}{1}\binom{4}{3}$	52
straight	$\binom{12}{1}\left[\binom{4}{1}^3 - \binom{4}{1}\right]$	720
flush	$\left[\binom{13}{3} - \binom{12}{1}\right]\binom{4}{1}$	1,096
one pair	$\binom{13}{1,1,11}\left[\binom{4}{2}\binom{4}{1}\right]$	3,744
A-, K-, or Q-high	$\left\{\binom{13}{3} - \binom{12}{1} - \left[\binom{10}{3} - \binom{8}{1}\right]\right\}\left[\binom{4}{1}^3 - \binom{4}{1}\right]$	9,720
J-high or worse	$\left[\binom{10}{3} - \binom{8}{1}\right]\left[\binom{4}{1}^3 - \binom{4}{1}\right]$	6,720
sum	$\binom{52}{3}$	22,100

The primary issue in Three Card Poker is to determine, given the player's three-card hand, whether the player should fold or play. Let us consider an example. Suppose that the player has Q-6-4 with three distinct suits, and assume a one-unit ante wager. If he folds, he loses one unit, for an expected profit of -1. If he plays, we must analyze the various possibilities for the dealer's hand, which can be regarded as having been dealt from the residual 49-card deck. (If there are other players in the game, their hands are unknown and can be disregarded.) This analysis is carried out in Table 16.9. Denoting by X the player's profit, we note that $X = 1$ if the dealer fails to qualify, $X = 2$ if the dealer qualifies with a hand worse than unsuited Q-6-4, $X = 0$ if the dealer qualifies with an unsuited Q-6-4, and $X = -2$ if the dealer qualifies with a hand better than unsuited Q-6-4. Thus, the player's conditional expected profit is

$$
\begin{aligned}
\mathrm{E}[X \mid \text{Q-6-4 with three distinct suits}] \\
= (1)\frac{5{,}758}{18{,}424} + (2)\frac{305}{18{,}424} + (0)\frac{26}{18{,}424} \\
+ (-2)\frac{39 + 43 + 589 + 907 + 3{,}114 + 6{,}189 + 1{,}454}{18{,}424} \\
= -\frac{18{,}302}{18{,}424} \approx -0.993378. \tag{16.15}
\end{aligned}
$$

This is a slight improvement over an expected profit of -1, so we conclude that the player should play this hand.

Incidentally, Table 16.9 provides more information than is really needed for (16.15) because the conditional probability that $X = -2$ is $1 - (5{,}758 + 305 + 26)/18{,}424$ by complementation. But an advantage of the full table is that we can check it for accuracy by confirming that the entries sum to $\binom{49}{3}$.

We have considered a player hand of Q-6-4 with three distinct suits. Thus, we have treated $(4)_3 = 24$ of the $\binom{52}{3} = 22{,}100$ player hands simultaneously. Let us call two hands *equivalent* if they have the same three denominations and if the corresponding denominations have the same suits after a permutation of $(\clubsuit, \diamondsuit, \heartsuit, \spadesuit)$. Thus, for example, the equivalence class containing Q$\clubsuit$-6$\diamondsuit$-4$\heartsuit$ has 24 hands, while the one containing Q$\clubsuit$-6$\clubsuit$-4$\diamondsuit$ has 12 hands and the one containing Q$\clubsuit$-6$\clubsuit$-4$\clubsuit$ has four hands.

There are 1,755 equivalence classes, as can be seen in Table 16.10. We have carried out the rather tedious calculation analogous to Table 16.9 for each of the other 1,754 equivalence classes of player hands, a very abbreviated summary of which is presented in Table 16.11. We find that the optimal player strategy is to fold with unsuited Q-6-3 or worse, and to play with unsuited Q-6-4 or better. As in Section 16.1, we are using the traditional criterion of accepting the conditional wager if the conditional expected profit for doing so is nonnegative. See Example 6.1.9 on p. 209 for an alternative approach.

All that remains is to find the house advantage when the player employs the optimal strategy. First, the number of player hands that should be folded, that is, the number of player hands of unsuited Q-6-3 or worse, is

$$\binom{8}{1}\left[\binom{4}{1}^3 - \binom{4}{1}\right] + \left[\binom{10}{3} - \binom{8}{1}\right]\left[\binom{4}{1}^3 - \binom{4}{1}\right] = 7{,}200, \quad (16.16)$$

because, disregarding suits, there are eight queen-high hands of Q-6-3 or worse (namely, Q-6-3, Q-6-2, Q-5-4, Q-5-3, Q-5-2, Q-4-3, Q-4-2, and Q-3-2). Therefore, the player folds with probability $7{,}200/22{,}100 = 72/221 \approx 0.325792$. Since he bets one unit if he folds and two units if he plays, his bet size B has expected value

$$E[B] = \frac{72}{221} + 2\left(1 - \frac{72}{221}\right) = \frac{370}{221} \approx 1.674208. \quad (16.17)$$

Finally, with the help of a computer, we can show that

$$E[X] = \cdots + (-1)\frac{24}{22{,}100} + (-1)\frac{12}{22{,}100} + (-1)\frac{12}{22{,}100} + (-1)\frac{12}{22{,}100}$$
$$+ \left(\frac{-18{,}302}{18{,}424}\right)\frac{24}{22{,}100} + \left(\frac{-18{,}312}{18{,}424}\right)\frac{12}{22{,}100} + \left(\frac{-18{,}316}{18{,}424}\right)\frac{12}{22{,}100}$$
$$+ \left(\frac{-18{,}325}{18{,}424}\right)\frac{12}{22{,}100} + \cdots + (1)\frac{720}{22{,}100} + (4)\frac{52}{22{,}100} + (5)\frac{48}{22{,}100}$$

Table 16.9 Classification of dealer hands in Three Card Poker when the player has Q-6-4 with three distinct suits.

rank	number of ways	
straight flush	$\binom{4}{1}\binom{4}{1} + \binom{7}{1}\binom{3}{1} + \binom{1}{1}\binom{2}{1}$	39
three of a kind	$\binom{10}{1}\binom{4}{3} + \binom{3}{1}\binom{3}{3}$	43
straight	$\binom{4}{1}\left[\binom{4}{1}^3 - \binom{4}{1}\right] + \binom{7}{1}\left[\binom{4}{1}^2\binom{3}{1} - \binom{3}{1}\right]$ $+ \binom{1}{1}\left[\binom{4}{1}\binom{3}{1}^2 - \binom{2}{1}\right]$	589
flush	$\left[\binom{10}{3}\binom{3}{0} - \binom{4}{1}\right]\binom{4}{1} + \left[\binom{10}{2}\binom{3}{1} - \binom{7}{1}\right]\binom{3}{1}$ $+ \left[\binom{10}{1}\binom{3}{2} - \binom{1}{1}\right]\binom{2}{1} + \binom{10}{0}\binom{3}{3}\binom{1}{1}$	907
one pair	$\binom{10}{1}\binom{9}{1}\binom{4}{2}\binom{4}{1} + \binom{10}{1}\binom{3}{1}\binom{4}{2}\binom{3}{1}$ $+ \binom{3}{1}\binom{10}{1}\binom{3}{2}\binom{4}{1} + \binom{3}{1}\binom{2}{1}\binom{3}{2}\binom{3}{1}$	3,114
A-high or K-high	$\left\{\binom{10}{3}\binom{3}{0} - \binom{4}{1} - \left[\binom{8}{3}\binom{3}{0} - \binom{3}{1}\right]\right\}\left[\binom{4}{1}^3 - \binom{4}{1}\right]$ $+ \left\{\binom{10}{2}\binom{3}{1} - \binom{7}{1} - \left[\binom{8}{2}\binom{3}{1} - \binom{5}{1}\right]\right\}\left[\binom{4}{1}^2\binom{3}{1} - \binom{3}{1}\right]$ $+ \left\{\binom{10}{1}\binom{3}{2} - \binom{1}{1} - \left[\binom{8}{1}\binom{3}{2} - \binom{1}{1}\right]\right\}\left[\binom{4}{1}\binom{3}{1}^2 - \binom{2}{1}\right]$	6,189
Q-high		
> Q-6-4	$\left[\binom{8}{2}\binom{2}{0} - \binom{3}{2}\binom{2}{0} - \binom{2}{2}\binom{2}{0}\right]\left[\binom{4}{1}^2\binom{3}{1} - \binom{3}{1}\right]$ $+ \left[\binom{5}{1}\binom{2}{1} + \binom{1}{1}\binom{1}{1}\right]\left[\binom{4}{1}\binom{3}{1}^2 - \binom{2}{1}\right]$	1,454
= Q-6-4	$\binom{3}{1}^3 - \binom{1}{1}$	26
< Q-6-4	$\binom{3}{1}\left[\binom{4}{1}^2\binom{3}{1} - \binom{3}{1}\right] + \binom{5}{1}\left[\binom{4}{1}\binom{3}{1}^2 - \binom{2}{1}\right]$	305
J-high or worse	$\left[\binom{8}{3}\binom{2}{0} - \binom{3}{1}\right]\left[\binom{4}{1}^3 - \binom{4}{1}\right]$ $+ \left[\binom{8}{2}\binom{2}{1} - \binom{4}{1}\right]\left[\binom{4}{1}^2\binom{3}{1} - \binom{3}{1}\right]$ $+ \left[\binom{8}{1}\binom{2}{2} - \binom{1}{1}\right]\left[\binom{4}{1}\binom{3}{1}^2 - \binom{2}{1}\right]$	5,758
sum	$\binom{49}{3}$	18,424

$$= \frac{-35{,}253{,}012}{(18{,}424)(22{,}100)} + \frac{1{,}168}{22{,}100} = -\frac{686{,}689}{20{,}358{,}520} \approx -0.033730. \qquad (16.18)$$

Here the first eight listed terms correspond to the first eight of the 12 rows of Table 16.11, the ellipses signify the 1,747 missing terms, and the last three terms correspond to the ante bonus (see Table 16.8). Note that the weights 24 and 12 come from Table 16.10.

Table 16.10 Number of equivalence classes of player hands in Three Card Poker. (a) is the number of ways of choosing the denominations, (b) is the number of ways of choosing the suits for each set of denominations, (c) is the number of distinct suits in each hand, (d) is the number of equivalence classes corresponding to the entry in column (c) for each set of denominations, and (e) is the size of each such equivalence class.

rank	(a)	(b)	(a) × (b)	(c)	(d)	(e)	(a) × (d)
straight flush	12	4	48	1	1	4	12
three of a kind	13	4	52	3	1	4	13
straight	12	60	720	3	1	24	12
				2	3	12	36
flush	274	4	1,096	1	1	4	274
one pair	156	24	3,744	3	1	12	156
				2	1	12	156
ace high or worse	274	60	16,440	3	1	24	274
				2	3	12	822
sum			22,100				1,755

The bottom line is a house advantage of

$$\mathrm{H}_0(B, X) := \frac{-\mathrm{E}[X]}{\mathrm{E}[B]} = \frac{686{,}689}{34{,}084{,}400} \approx 0.020147 \qquad (16.19)$$

if pushes are included. In fact pushes should be included. Indeed, after a push in Three Card Poker, one's bet is double what it was initially, so the player is unlikely to regard a push as merely a temporary delay, as he might if the bet were unchanged.

Table 16.11 Player's conditional expected profit E in Three Card Poker (from a one-unit ante wager and a one-unit play wager) when playing with unsuited Q-6-4, Q-6-3, or Q-6-2.

player hand	E	$\binom{49}{3}E$
Q-6-4, three distinct suits	$-.993\ 378$	$-18{,}302$
Q-6-4, two distinct suits, 4 is of odd suit	$-.993\ 921$	$-18{,}312$
Q-6-4, two distinct suits, 6 is of odd suit	$-.994\ 138$	$-18{,}316$
Q-6-4, two distinct suits, Q is of odd suit	$-.994\ 627$	$-18{,}325$
Q-6-3, three distinct suits	$-1.002\ 551$	$-18{,}471$
Q-6-3, two distinct suits, 3 is of odd suit	$-1.002\ 877$	$-18{,}477$
Q-6-3, two distinct suits, 6 is of odd suit	$-1.003\ 311$	$-18{,}485$
Q-6-3, two distinct suits, Q is of odd suit	$-1.003\ 962$	$-18{,}497$
Q-6-2, three distinct suits	$-1.012\ 375$	$-18{,}652$
Q-6-2, two distinct suits, 2 is of odd suit	$-1.012\ 484$	$-18{,}654$
Q-6-2, two distinct suits, 6 is of odd suit	$-1.013\ 135$	$-18{,}666$
Q-6-2, two distinct suits, Q is of odd suit	$-1.013\ 786$	$-18{,}678$

16.3 Problems

16.1. *Let It Ride: Bonus side bet.* The bonus side bet at Let It Ride is a bet that the player's five-card hand will rank three of a kind or better. It pays 20,000 for 1 for a royal flush, 2,000 for 1 for a nonroyal straight flush, 400 for 1 for four of a kind, 200 for 1 for a full house, 50 for 1 for a flush, 25 for 1 for a straight, and 5 for 1 for three of a kind. (Actually, these numbers vary from one casino to the next. The stated pay schedule was offered at the MGM Grand in December 2007.) Find the probability of a win as well as the house advantage associated with this wager.

16.2. *Let It Ride: Joint probabilities.*

(a) Let A_2 (resp., A_3) be the event that the basic bettor at Let It Ride has nonnegative conditional expectation at his second (resp., third) bet, or, to put it another way, at his first (resp., second) opportunity to let it ride. Tables 16.2 on p. 527 and 16.4 on p. 530 yield

$$P(A_2) = \frac{1{,}608}{\binom{52}{3}} \quad \text{and} \quad P(A_3) = \frac{43{,}133}{\binom{52}{4}}. \tag{16.20}$$

Find $P(A_2 \cap A_3)$, and use this to evaluate the distribution of the total amount bet when using the strategy of Table 16.6 on p. 532. Confirm that results are consistent with (16.8) on p. 528.

(b) How do things change when "nonnegative" is replaced by "positive"?

(c) How do things change when "nonnegative conditional expectation" is replaced by "conditional expectation per unit bet no less than $-5/196$"? See Example 6.1.10 on p. 211 for the significance of this.

16.3. *Let It Ride: A cautious strategy.* Suppose that the basic bettor at Let It Ride adopts the cautious strategy of reclaiming his bet at each opportunity unless he has a made payoff. Find the house advantage associated with this strategy, and compare it with the house advantage under optimal play.

16.4. *Let It Ride: Correlations between bets.* In terms of the notation of Section 16.1, define

$$Z_1 := X_1, \quad Z_2 := X_2 \, 1_{\{E[X_2|Y_3] \geq 0\}}, \quad Z_3 := X_3 \, 1_{\{E[X_3|Y_4] \geq 0\}}. \quad (16.21)$$

Find the correlations between Z_1 and Z_2, between Z_1 and Z_3, and between Z_2 and Z_3. Find the variance of $Z_1 + Z_2 + Z_3$ without first finding its distribution.

16.5. *Let It Ride: Distribution of profit.* Find the distribution of the basic bettor's profit at Let It Ride (assuming the usual three-unit initial bet and optimal play). Use this to find the variance.

16.6. *Three Card Poker: Pair-plus wager.* The pair-plus wager at Three Card Poker is a bet that the player will be dealt a pair or better. It pays 40 to 1 for a straight flush, 30 to 1 for three of a kind, 6 to 1 for a straight, 4 to 1 for a flush, and 1 to 1 for a pair. Find the probability of a win as well as the house advantage associated with this wager.

16.7. *Three Card Poker: Cost of mimicking the dealer.* Suppose that an ante-play bettor at Three Card Poker adopts the strategy of playing with a queen high or better, effectively mimicking the dealer's mandatory strategy. Find the resulting house advantage (including pushes), and compare it with the house advantage under optimal play.

16.8. *Three Card Poker: Bluffing vs. folding.* An ante-play bettor at Three Card Poker is said to "bluff" if he plays with jack high or worse. Such a player is effectively betting that the dealer will not qualify.

(a) Without taking the player's cards into account, find the probability that the dealer qualifies. Use this to show, again without taking the player's cards into account, that the player's loss in expectation by bluffing instead of folding is $1{,}940/\binom{52}{3} \approx 0.087783$ (as a proportion of the ante).

(b) Now take the player's cards into account and argue that the player's real loss in expectation by bluffing instead of folding is *at least* $2{,}446/\binom{49}{3} \approx 0.132762$, more than 50 percent larger. Explain the discrepancy.

16.9. *Three Card Poker: Player hand with smallest positive conditional expected profit.* Find the player hand for which the conditional expected profit from the ante-play wager at Three Card Poker is minimal positive. Be sure to take suits into account.

16.10. *Three Card Poker: Nonmonotonicity of player conditional expectation in poker ranks.* Evaluate the player's conditional expected profit from the ante-play wager at Three Card Poker (assuming a one-unit ante), given his three cards, for each of the best 25 (equivalence classes of) hands, the 12 straight flushes and 13 three of a kinds, and notice the nonmonotonicity of these conditional expectations. Explain why this was to be expected. Explain, in particular, why three jacks is better than three queens, and why a jack-high straight flush is better than an ace-high straight flush.

16.11. *Three Card Poker: Probability of a push and evaluation of the house advantage.* Use combinatorics to find the probability of a push for the ante-play wager at Three Card Poker, assuming optimal play. (Notice that every hand, except three of a kind, can be tied, but that tied straight flushes and tied straights do not result in pushes because of the ante bonus.) Use this to find the house advantage H (pushes excluded) associated with this wager, even though, as we have noted, H_0 (pushes included) is more appropriate for this game.

16.12. *Three Card Poker: Distribution of profit.* Find the distribution of the ante-play bettor's profit at Three Card Poker (assuming a one-unit ante and optimal play). Use this to find the variance.

16.13. *Caribbean Stud Poker. Caribbean Stud Poker* is similar to Three Card Poker. It is played with a standard 52-card deck, which is reshuffled between coups. We describe only the *ante-call wager*. Each player makes an *ante wager* and receives five cards face down. The dealer also receives five cards, four face down and one face up. Each player, after examining his hand and the dealer's upcard, must decide whether to fold or call. If he folds, he loses his ante wager. If he calls, he must make an additional wager, known as a *call wager*, equal in size to exactly twice the ante wager. The dealer then reveals his hand to determine whether it *qualifies* with a rank of ace-king or better. If the dealer's hand fails to qualify, the player's ante wager is paid even money and his call wager is pushed. If the dealer's hand qualifies, it is compared with the player's hand. If the player's hand outranks the dealer's hand, the player's ante wager is paid even money and his call wager is paid at the rate shown in Table 16.12. If the dealer's hand outranks the player's hand, both of the player's wagers are lost. If the player's hand and the dealer's hand are equal in rank, both wagers are pushed.

The ranking of hands in Caribbean Stud Poker is exactly as in traditional poker. If two hands belong to the same category in Table 16.12, they are ranked as in Problem 1.6 on p. 53.

(a) Suppose that the player has A♣-K♣-Q♢-7♢-2♡ and the dealer's upcard is 6♣. Should the player fold or call?

(b) In Let It Ride there were 33 equivalence classes of three-card hands and 33 equivalence classes of four-card hands. In Three Card Poker there were 1,755 equivalence classes of hands. Define a suitable equivalence relation in Caribbean Stud Poker and determine the number of equivalence classes.

Table 16.12 Payoff odds in Caribbean Stud Poker.

rank	ways	payoff
royal flush	4	100 to 1
straight flush	36	50 to 1
four of a kind	624	20 to 1
full house	3,744	7 to 1
flush	5,108	5 to 1
straight	10,200	4 to 1
three of a kind	54,912	3 to 1
two pair	123,552	2 to 1
one pair	1,098,240	1 to 1
ace-king high	167,280	1 to 1
other	1,135,260	
total	2,598,960	

(c) Find the optimal strategy at Caribbean Stud Poker, that is, the strategy that calls if and only if the conditional expected profit is nonnegative. Evaluate the associated house advantage (pushes included).

16.14. *Caribbean Stud Poker: Alternative optimal strategy.* Apply Theorem 6.1.8 on p. 207 to find the optimal strategy in Caribbean Stud Poker in the sense of minimizing house advantage (pushes included, and with respect to expected total bet).

16.15. *Caribbean Stud Poker: Optimal strategy without seeing dealer's upcard.* If none of the dealer's cards is exposed (or if the player simply ignores the dealer's upcard), the player's optimal strategy is much simpler. Find it and determine the resulting house advantage.

16.4 Notes

There are literally dozens of proprietary games vying for the attention of gamblers. Most of them are ultimately unsuccessful. Let It Ride and Three Card Poker are two of the more successful such games, which is why we have chosen to analyze them here. An early draft of this chapter included a section devoted to Caribbean Stud Poker, but we came to realize that the game is too similar to Three Card Poker to warrant such treatment; also, CSP has declined in popularity in recent years. For analyses of some of the other proprietary games, we refer the reader to Grosjean (2009) and Shackleford (2009).

According to Stratton (2004b), Let It Ride was invented by Shuffle Master founder John G. Breeding as a marketing tool for his automatic shuffling machines, and it debuted in 1993. It was patented in 1994 (U.S. patent 5,288,081), and "Let It Ride" is a registered trademark of Shuffle Master, Inc. The first analysis of the game, as far as we know, was done by Brooks (1994). The most complete mathematical study of Let It Ride is due to Ko (1997). Ko's approach is different from ours, but the results are consistent. One minor point on which we would take issue is Ko's recommendation to avoid making bets that have zero conditional expected profit, for the purpose of reducing the fluctuations in the player's bankroll. In subfair games, variance is good. See the discussion motivating (6.119). Grosjean (2009, Chapter 59) treated the three bets separately, much as in our treatment. A recent analysis of Let It Ride by Stelzer (2006, pp. 125–134) found a suboptimal strategy.

Again according to Stratton (2004b), Three Card Poker was invented by Derek Webb and introduced in 1996. It was patented in 1997 (U.S. patent 5,685,774), and "Three Card Poker" is a registered trademark. (That is why we have omitted the hyphen in the compound adjective "three-card.") The U.S. rights are now owned by Shuffle Master, and the U.K. rights are owned by Webb's Prime Table Games. The first published mathematical study of Three Card Poker is due to Ko (2001). Ko's analysis is consistent with ours except in one minor respect. In place of our four conditional player expectations given the player hand of unsuited Q-6-4 (see Table 16.11) Ko evaluated a single conditional player expectation given the player hand of unsuited Q-6-4, in effect assuming that the player makes no use of the suit information other than to confirm that he does not have a flush. It is conceivable *a priori* that the optimal strategy is suit-dependent, but it turns out not to be, so Ko's argument leads to the correct solution. Three Card Poker has been analyzed by Grosjean (2009, Chapter 60) and Frome (2003) as well.

From Stratton (2004b) once again, we find that Caribbean Stud Poker (Problem 16.13) dates back to 1990. Since the appearance of that article, the game was acquired by Shuffle Master. The first mathematical study of Caribbean Stud is due to Frome and Frome (1996), who found a good approximation to the optimal strategy. Then Griffin and Gwynn (2000; announced in 1994) found a nearly optimal strategy, and it was left to Ko (1998) to obtain the exact optimal strategy. Because the optimal strategy is so complicated, several authors have proposed nearly optimal strategies that cost the player little. Griffin and Gwynn (2000) considered three strategy levels, A, B, and C. In Strategy A the player does not take the dealer's upcard into account (see Problem 16.15); in Strategy B the player takes the denomination of the dealer's upcard into account; and in Strategy C the player takes both the denomination and the suit of the dealer's upcard into account. Ko (1998) also proposed three strategy levels, his Level Three Strategy being the optimal one.

In each of these games, players are not allowed, at most casinos, to share information about their hands with other players. Grosjean (2007b) has argued that this restriction is completely unnecessary.

We have ordered the two games studied in this chapter according to computational difficulty, which in turn is related to the number of ways that a game can be played. A one-player game of Let It Ride can be played in

$$\binom{52}{3,1,1,47} = \frac{(52)_5}{6} = 51{,}979{,}200 \qquad (16.22)$$

ways, while a one-player game of Three Card Poker can be played in

$$\binom{52}{3,3,46} = \frac{(52)_6}{36} = 407{,}170{,}400 \qquad (16.23)$$

ways. By way of comparison, a one-player game of Caribbean Stud Poker (Problem 16.13) can be played in

$$\binom{52}{5,1,4,42} = \frac{(52)_{10}}{2{,}880} = 19{,}993{,}230{,}517{,}200 \qquad (16.24)$$

ways.

Several of the problems, such as Problem 16.1, are straightforward and can be found in various sources. Problem 16.2(b) was done by Ko (1997, p. 24) and Kilby and Fox (1998). Problems 16.6 and 16.7 are from Ko (2001, pp. 6–7, 17), and Problem 16.8(a) is from Hannum and Cabot (2005, p. 153). Ko (2001, p. 19) has results related to Problem 16.12. Problem 16.13 is from Ko (1998). Problem 16.15 is due to Griffin and Gwynn (2000).

Chapter 17
Video Poker

Video meliora, deteriora sequor, as we said at college.

William Makepeace Thackeray, *The Virginians*

Video poker is an electronic form of five-card draw poker that dates back to the late 1970s. The player is dealt five cards and is allowed to replace any number of them by an equal number of cards drawn from the unseen deck. The rank of the resulting hand (and the bet size) determines the amount paid out to the player. We focus on two specific video poker games, Jacks or Better (Section 17.1), the version closest to five-card draw, and Deuces Wild (Section 17.2), a version with the additional complication of four wild cards. With optimal play, the Deuces Wild player has a slight advantage over the house, while the Jacks or Better player has a slight disadvantage.

17.1 Jacks or Better

Video poker is played on a machine that resembles a slot machine with a video monitor. Typically, the player inserts from one to five coins (currency and tickets are also accepted) to place a bet. He then receives five cards face up on the screen, with each of the $\binom{52}{5} = 2{,}598{,}960$ possible hands equally likely. (The order of the five cards is irrelevant.) For each card, the player must then decide whether to hold or discard that card. Thus, there are $2^5 = 32$ ways to play the hand. If he discards k cards, he is dealt k new cards, with each of the $\binom{47}{k}$ possibilities equally likely. The player then receives his payout, which depends on the amount he bet and the rank of his final hand.

The payout schedule for full-pay *Jacks or Better* is shown in Table 17.1. Unlike in Example 1.1.10 on p. 6, here we distinguish between a royal flush and a (nonroyal) straight flush. We use the term "full-pay" to emphasize the

S.N. Ethier, *The Doctrine of Chances*, Probability and its Applications,   545
DOI 10.1007/978-3-540-78783-9_17, © Springer-Verlag Berlin Heidelberg 2010

fact that there are similar machines with less favorable payout schedules. It should also be mentioned that, typically, to qualify for the 800 for 1 payout on a royal flush, the player must bet five coins. We assume that the player does this, and we define the total value of these five coins to be one unit.

Table 17.1 The full-pay Jacks or Better payoff odds, assuming a maximum-coin bet, and the pre-draw frequencies.

rank	payoff odds	number of ways
royal flush	800 for 1	4
straight flush	50 for 1	36
four of a kind	25 for 1	624
full house	9 for 1	3,744
flush	6 for 1	5,108
straight	4 for 1	10,200
three of a kind	3 for 1	54,912
two pair	2 for 1	123,552
pair of jacks or better	1 for 1	337,920
other	0 for 1	2,062,860
total		2,598,960

The primary issue in Jacks or Better is to determine, given the player's initial five-card hand, which cards should be held. Let us consider an example. Suppose that the player is dealt A♣-Q♢-J♣-T♣-9♣ (in any order, of course). There are several plausible strategies.

If the player holds the four-card one-gap straight A♣-Q♢-J♣-T♣, his payout R from a one-unit bet has conditional expected value

$$E[R] = 4\left(\frac{4}{47}\right) + 1\left(\frac{9}{47}\right) + 0\left(\frac{34}{47}\right) = \frac{25}{47} \approx 0.531915. \qquad (17.1)$$

For convenience we do not write out the conditioning event, namely "player is dealt A♣-Q♢-J♣-T♣-9♣ and holds A♣-Q♢-J♣-T♣."

If the player holds the four-card open-ended straight Q♢-J♣-T♣-9♣,

$$E[R] = 4\left(\frac{8}{47}\right) + 1\left(\frac{6}{47}\right) + 0\left(\frac{33}{47}\right) = \frac{38}{47} \approx 0.808511. \qquad (17.2)$$

If the player holds the four-card flush A♣-J♣-T♣-9♣,

$$E[R] = 6\left(\frac{9}{47}\right) + 1\left(\frac{6}{47}\right) + 0\left(\frac{32}{47}\right) = \frac{60}{47} \approx 1.276596. \qquad (17.3)$$

If the player holds the three-card open-ended straight flush J♣-T♣-9♣,

$$E[R] = 50 \frac{3\binom{2}{2}\binom{45}{0}}{\binom{47}{2}} + 6 \frac{\left[\binom{9}{2} - 3\binom{2}{2}\binom{7}{0}\right]\binom{38}{0}}{\binom{47}{2}}$$

$$+ 4 \frac{2\binom{4}{1}\binom{3}{1}\binom{40}{0} + \binom{4}{1}^2\binom{39}{0} - 3\binom{2}{2}\binom{45}{0}}{\binom{47}{2}}$$

$$+ 3 \frac{3\binom{3}{2}\binom{44}{0}}{\binom{47}{2}} + 2 \frac{3\binom{3}{1}^2\binom{41}{0}}{\binom{47}{2}} + 1 \frac{\binom{3}{1}\binom{6}{0}\binom{38}{1} + 2\binom{3}{2}\binom{44}{0} + \binom{4}{2}\binom{43}{0}}{\binom{47}{2}}$$

$$= \frac{703}{1{,}081} \approx 0.650324, \tag{17.4}$$

where we have omitted the term corresponding to no payout because it does not contribute.

Finally, if the player holds the three-card two-gap royal flush A♣-J♣-T♣,

$$E[R] = 800 \frac{\binom{2}{2}\binom{45}{0}}{\binom{47}{2}} + 6 \frac{\left[\binom{9}{2} - \binom{2}{2}\binom{7}{0}\right]\binom{38}{0}}{\binom{47}{2}} + 4 \frac{\binom{4}{1}\binom{3}{1}\binom{40}{0} - \binom{2}{2}\binom{45}{0}}{\binom{47}{2}}$$

$$+ 3 \frac{3\binom{3}{2}\binom{44}{0}}{\binom{47}{2}} + 2 \frac{3\binom{3}{1}^2\binom{41}{0}}{\binom{47}{2}} + 1 \frac{2\binom{3}{1}\binom{6}{0}\binom{38}{1} + \binom{4}{2}\binom{43}{0} + \binom{3}{2}\binom{44}{0}}{\binom{47}{2}}$$

$$= \frac{1{,}372}{1{,}081} \approx 1.269195. \tag{17.5}$$

While there do not seem to be any other reasonable strategies, one should not rely too heavily on one's intuition in these situations, so we have computed the conditional expectations in the other 27 cases, listing all 32 results in Table 17.2.

The five suggested strategies are indeed the five best. In particular, we conclude that holding A♣-J♣-T♣-9♣ is the optimal strategy, because it maximizes the player's conditional expected payout. This suggests that drawing to a four-card flush is marginally better than drawing to a three-card two-gap royal flush. However, one has to be careful when making such generalizations.

Let us consider a closely related example to illustrate this point. Suppose that the player is dealt A♣-J♣-T♣-9♣-6♢ (in any order). (The Q♢ in the preceding example has been replaced by the 6♢.) The analogues of the two best strategies in the preceding example are the following.

If the player holds the four-card flush A♣-J♣-T♣-9♣, his payout R from a one-unit bet has conditional expected value (17.3).

If the player holds the three-card two-gap royal flush A♣-J♣-T♣, (17.5) is replaced by

$$E[R] = 800 \frac{\binom{2}{2}\binom{45}{0}}{\binom{47}{2}} + 6 \frac{\left[\binom{9}{2} - \binom{2}{2}\binom{7}{0}\right]\binom{38}{0}}{\binom{47}{2}} + 4 \frac{\binom{4}{1}^2\binom{39}{0} - \binom{2}{2}\binom{45}{0}}{\binom{47}{2}}$$

Table 17.2 The 32 possible values of the player's conditional expected payout at Jacks or Better when he is dealt A♣-Q♦-J♣-T♣-9♣.

cards held					$k :=$ no. of cards drawn	conditional expected payout E	$\binom{47}{k}E$
A♣	Q♦	J♣	T♣	9♣	0	.000 000	0
A♣	Q♦	J♣	T♣		1	.531 915	25
A♣	Q♦	J♣		9♣	1	.191 489	9
A♣	Q♦		T♣	9♣	1	.127 660	6
A♣		J♣	T♣	9♣	1	1.276 596	60
	Q♦	J♣	T♣	9♣	1	.808 511	38
A♣	Q♦	J♣			2	.441 258	477
A♣	Q♦		T♣		2	.338 575	366
A♣	Q♦			9♣	2	.294 172	318
A♣		J♣	T♣		2	1.269 195	1,372
A♣		J♣		9♣	2	.493 987	534
A♣			T♣	9♣	2	.391 304	423
	Q♦	J♣	T♣		2	.427 382	462
	Q♦	J♣		9♣	2	.382 979	414
	Q♦		T♣	9♣	2	.280 296	303
		J♣	T♣	9♣	2	.650 324	703
A♣	Q♦				3	.464 693	7,535
A♣		J♣			3	.495 776	8,039
A♣			T♣		3	.361 640	5,864
A♣				9♣	3	.352 760	5,720
	Q♦	J♣			3	.482 455	7,823
	Q♦		T♣		3	.348 319	5,648
	Q♦			9♣	3	.339 439	5,504
		J♣	T♣		3	.391 243	6,344
		J♣		9♣	3	.382 362	6,200
			T♣	9♣	3	.266 482	4,321
A♣					4	.434 614	77,520
	Q♦				4	.450 784	80,404
		J♣			4	.436 722	77,896
			T♣		4	.272 279	48,565
				9♣	4	.275 822	49,197
		(none)			5	.308 626	473,414

$$+ 3 \frac{3\binom{3}{2}\binom{44}{0}}{\binom{47}{2}} + 2 \frac{3\binom{3}{1}\binom{3}{1}\binom{41}{0}}{\binom{47}{2}} + 1 \frac{2\binom{3}{1}\binom{6}{0}\binom{38}{1} + 2\binom{4}{2}\binom{43}{0}}{\binom{47}{2}}$$

$$= \frac{1{,}391}{1{,}081} \approx 1.286772. \tag{17.6}$$

None of the other 30 strategies is as good as these two. We conclude that the optimal strategy is to hold A♣-J♣-T♣.

Notice that the two best strategies (hold the four-card flush; hold the three-card two-gap royal flush) are ordered differently in the two examples. The reason is easy to see. The player's expected payout when holding the four-card flush is the same in the two examples. However, his expected payout when holding the three-card two-gap royal flush is smaller when he discards the Q◇ (first example) than when he discards the 6◇ (second example). Clearly, the absence of the Q◇ from the residual 47-card deck reduces the chance that the player will make a straight. (It also reduces the chance that he will make a pair of jacks or better.) Thus, the Q◇ is called a straight-penalty card.

More generally, when a particular discard reduces the probability that the player will make a straight or a flush (over what it would have been otherwise), that discard is called a *straight-* or a *flush-penalty card*. Another example may help to clarify this concept.

Suppose that the player is dealt A♣-Q◇-J♣-T♣-9♡ (in any order). (The 9♣ in the first example has been replaced by the 9♡.) The player can no longer hold a four-card flush, but the strategy of holding the three-card two-gap royal flush A♣-J♣-T♣ is still viable. In this case the player's payout R from a one-unit bet has conditional expected value

$$\mathrm{E}[R] = 800 \frac{\binom{2}{2}\binom{45}{0}}{\binom{47}{2}} + 6 \frac{\left[\binom{10}{2} - \binom{2}{2}\binom{8}{0}\right]\binom{37}{0}}{\binom{47}{2}} + 4 \frac{\binom{3}{1}\binom{4}{1}\binom{40}{0} - \binom{2}{2}\binom{45}{0}}{\binom{47}{2}}$$

$$+ 3 \frac{3\binom{3}{2}\binom{44}{0}}{\binom{47}{2}} + 2 \frac{3\binom{3}{1}^2\binom{41}{0}}{\binom{47}{2}} + 1 \frac{2\binom{3}{1}\binom{6}{0}\binom{38}{1} + \binom{4}{2}\binom{43}{0} + \binom{3}{2}\binom{44}{0}}{\binom{47}{2}}$$

$$= \frac{1{,}426}{1{,}081} \approx 1.319149. \tag{17.7}$$

Here the player's expected payout when holding the three-card two-gap royal flush is smaller when he discards the 9♣ (first example) than when he discards the 9♡ (third example). Thus, the 9♣ in the first example is a flush-penalty card.

It is now clear what must be done to determine the optimal strategy at Jacks or Better. We simply create a table analogous to Table 17.2 for each of the player's $\binom{52}{5} = 2{,}598{,}960$ possible initial hands, determine for each such hand which of the $2^5 = 32$ ways to play it maximizes his conditional expected payout, summarize the resulting player strategy, and average the conditional expectations thus obtained to evaluate the overall expected payout. Of course the variance of the payout and more generally its distribution would also be of interest.

Actually, we can reduce the amount of work required by nearly a factor of 20 by taking equivalence of initial hands into account, just as we did in Section 16.2. Let us call two initial hands *equivalent* if they have the same five denominations and if the corresponding denominations have the same suits after a permutation of $(\clubsuit,\diamondsuit,\heartsuit,\spadesuit)$. This is an equivalence relation in the sense of Theorem A.1.1. Thus, for example, the equivalence class containing A$\clubsuit$-A$\diamondsuit$-A$\heartsuit$-K$\clubsuit$-Q$\diamondsuit$ has $\binom{4}{3}\binom{3}{1}\binom{2}{1} = 24$ hands, the one containing A$\clubsuit$-A$\diamondsuit$-A$\heartsuit$-K$\clubsuit$-Q$\spadesuit$ has $\binom{4}{3}\binom{3}{1} = 12$ hands, and the one containing A$\clubsuit$-A$\diamondsuit$-A$\heartsuit$-K$\spadesuit$-Q$\spadesuit$ has $\binom{4}{3} = 4$ hands.

How many equivalence classes are there of a given size associated with a particular set of denominations? This question can be answered on a case-by-case basis, so we consider just one such case. Consider a hand with five distinct denominations m_1, m_2, m_3, m_4, m_5 with $14 \geq m_1 > m_2 > m_3 > m_4 > m_5 \geq 2$. (Here 14, 13, 12, 11 correspond to A, K, Q, J. There are $\binom{13}{5} = 1{,}287$ ways to choose the denominations.) We number the suits of denominations m_1, m_2, m_3, m_4, m_5 by $n_1, n_2, n_3, n_4, n_5 \in \{1, 2, 3, 4\}$. Since we are concerned only with equivalence classes, we choose n_1, n_2, n_3, n_4, n_5 successively, using the smallest available integer for each suit that does not appear in a higher denomination. Thus,

$$n_1 = 1$$
$$n_2 \leq n_1 + 1$$
$$n_3 \leq \max(n_1, n_2) + 1 \qquad\qquad (17.8)$$
$$n_4 \leq \max(n_1, n_2, n_3) + 1$$
$$n_5 \leq \max(n_1, n_2, n_3, n_4) + 1.$$

It is easy to see that there is a one-to-one correspondence between the set of such $(n_1, n_2, n_3, n_4, n_5)$ and the set of equivalence classes of hands with denominations $(m_1, m_2, m_3, m_4, m_5)$. By direct enumeration (rather than by combinatorial analysis) we find that there are 51 equivalence classes. See Table 17.3, which also includes the other types of hands.

Table 17.3 shows that there are exactly

$$\binom{13}{5}51 + \binom{13}{1,3,9}20 + \binom{13}{2,1,10}8 + \binom{13}{1,2,10}5 + \binom{13}{1,1,11}3 = 134{,}459$$
$$(17.9)$$

equivalence classes. As a check, we compute the total number of hands by summing the sizes of the equivalence classes:

$$1{,}287(1\cdot 4 + 15\cdot 12 + 35\cdot 24) + 2{,}860(8\cdot 12 + 12\cdot 24) \qquad (17.10)$$
$$+ 858(1\cdot 4 + 7\cdot 12 + 5\cdot 24) + 156(1\cdot 4 + 2\cdot 12) = 2{,}598{,}960.$$

Table 17.3 List of equivalence classes of initial player hands in Jacks or Better, together with the size of each equivalence class. The hand A♣-Q♢-J♣-T♣-9♣, for example, belongs to the equivalence class A-Q-J-T-9 $(1,2,1,1,1)$, as do 11 other hands with the same denominations.

---

five distinct denominations (a,b,c,d,e): $\binom{13}{5} = 1{,}287$ ways
(includes hands ranked no pair, straight, flush, straight flush, royal flush)

$(1,1,1,1,1)$	4	$(1,1,2,3,3)$	24	$(1,2,2,1,2)$	12	$(1,2,3,2,1)$	24
$(1,1,1,1,2)$	12	$(1,1,2,3,4)$	24	$(1,2,2,1,3)$	24	$(1,2,3,2,2)$	24
$(1,1,1,2,1)$	12	$(1,2,1,1,1)$	12	$(1,2,2,2,1)$	12	$(1,2,3,2,3)$	24
$(1,1,1,2,2)$	12	$(1,2,1,1,2)$	12	$(1,2,2,2,2)$	12	$(1,2,3,2,4)$	24
$(1,1,1,2,3)$	24	$(1,2,1,1,3)$	24	$(1,2,2,2,3)$	24	$(1,2,3,3,1)$	24
$(1,1,2,1,1)$	12	$(1,2,1,2,1)$	12	$(1,2,2,3,1)$	24	$(1,2,3,3,2)$	24
$(1,1,2,1,2)$	12	$(1,2,1,2,2)$	12	$(1,2,2,3,2)$	24	$(1,2,3,3,3)$	24
$(1,1,2,1,3)$	24	$(1,2,1,2,3)$	24	$(1,2,2,3,3)$	24	$(1,2,3,3,4)$	24
$(1,1,2,2,1)$	12	$(1,2,1,3,1)$	24	$(1,2,2,3,4)$	24	$(1,2,3,4,1)$	24
$(1,1,2,2,2)$	12	$(1,2,1,3,2)$	24	$(1,2,3,1,1)$	24	$(1,2,3,4,2)$	24
$(1,1,2,2,3)$	24	$(1,2,1,3,3)$	24	$(1,2,3,1,2)$	24	$(1,2,3,4,3)$	24
$(1,1,2,3,1)$	24	$(1,2,1,3,4)$	24	$(1,2,3,1,3)$	24	$(1,2,3,4,4)$	24
$(1,1,2,3,2)$	24	$(1,2,2,1,1)$	12	$(1,2,3,1,4)$	24		

---

one pair (a,a,b,c,d): $\binom{13}{1,3,9} = 2{,}860$ ways

$(1,2,1,1,1)$	12	$(1,2,1,2,3)$	24	$(1,2,3,1,1)$	24	$(1,2,3,3,3)$	12
$(1,2,1,1,2)$	12	$(1,2,1,3,1)$	24	$(1,2,3,1,2)$	24	$(1,2,3,3,4)$	12
$(1,2,1,1,3)$	24	$(1,2,1,3,2)$	24	$(1,2,3,1,3)$	24	$(1,2,3,4,1)$	24
$(1,2,1,2,1)$	12	$(1,2,1,3,3)$	24	$(1,2,3,1,4)$	24	$(1,2,3,4,3)$	12
$(1,2,1,2,2)$	12	$(1,2,1,3,4)$	24	$(1,2,3,3,1)$	24	$(1,2,3,4,4)$	12

---

two pair (a,a,b,b,c): $\binom{13}{2,1,10} = 858$ ways

$(1,2,1,2,1)$	12	$(1,2,1,3,1)$	24	$(1,2,1,3,3)$	24	$(1,2,3,4,1)$	12
$(1,2,1,2,3)$	12	$(1,2,1,3,2)$	24	$(1,2,1,3,4)$	24	$(1,2,3,4,3)$	12

---

three of a kind (a,a,a,b,c): $\binom{13}{1,2,10} = 858$ ways

$(1,2,3,1,1)$	12	$(1,2,3,1,4)$	12	$(1,2,3,4,4)$	4
$(1,2,3,1,2)$	24	$(1,2,3,4,1)$	12		

---

full house (a,a,a,b,b): $\binom{13}{1,1,11} = 156$ ways

$(1,2,3,1,2)$	12	$(1,2,3,1,4)$	12

---

four of a kind (a,a,a,a,b): $\binom{13}{1,1,11} = 156$ ways

$(1,2,3,4,1)$	4

Needless to say, a computer is a necessity for this kind of problem. Our program methodically cycles through each of the 134,459 equivalence classes. For each one it computes the 32 conditional expectations and determines which is largest and if it is uniquely the largest. It stores this information in a file as it proceeds. Finally, it computes the payout distribution under the optimal strategy, first for each equivalence class and then for the game as a whole.

Let us now consider the issue of uniqueness. In one obvious case the optimal strategy is nonunique: If a player is dealt four of a kind, he may hold or discard the card of the odd denomination with no effect. This accounts for 156 equivalence classes or 624 hands. There is only one other situation for which uniqueness fails. With K-Q-J-T-T it is optimal to discard one of the tens—it does not matter which one—unless three or more of the cards are of the same suit. Of the 20 equivalence classes for this set of denominations, 3 of size 12 and 9 of size 24 have nonunique optimal strategies. This accounts for another 12 equivalence classes or 252 hands. However, it is important to note that the payout distribution is unaffected by the choice of optimal strategy in each of these cases of nonuniqueness.

Thus, the payout distribution for Jacks or Better played optimally is uniquely determined, and we display it in Table 17.4. Here it is worth giving exact results. A common denominator (not necessarily the least one) is

$$\text{l.c.m.}\left\{\binom{52}{5}\binom{47}{k} : k = 0, 1, 2, 3, 4, 5\right\} = \binom{52}{5}\binom{47}{5}5, \qquad (17.11)$$

where l.c.m. stands for least common multiple. In fact, the least common denominator is this number divided by 12.

It follows that the mean payout under optimal play is

$$\frac{1,653,526,326,983}{1,661,102,543,100} \approx 0.995439043695, \qquad (17.12)$$

while the variance of the payout is 19.514676427086. Thus, the house has a slight advantage (less than half of one percent) over the optimal Jacks or Better player.

There remains an important issue that has not yet been addressed. What exactly is the optimal strategy at Jacks or Better? Our computer program provides one possible answer: specific instructions for each of the 134,459 equivalence classes. However, what we need is something simpler, a strategy that can actually be memorized. The usual approach is to construct a so-called *hand-rank table*. Each of the various types of holdings is ranked according to its conditional expectation (which varies slightly with the cards discarded). One then simply finds the highest-ranked hand in the table that is applicable to the hand in question. Only relatively recently has a hand-rank table been found for Jacks or Better that reproduces the optimal strategy precisely, and that table is presented as Table 17.5.

Table 17.4 The distribution of the payout R from a one-unit bet on Jacks or Better. Assumes maximum-coin bet and optimal drawing strategy.

result	R	probability	probability $\times \binom{52}{5}\binom{47}{5}5/12$
royal flush	800	.000 024 758 268	41,126,022
straight flush	50	.000 109 309 090	181,573,608
four of a kind	25	.002 362 545 686	3,924,430,647
full house	9	.011 512 207 336	19,122,956,883
flush	6	.011 014 510 968	18,296,232,180
straight	4	.011 229 367 241	18,653,130,482
three of a kind	3	.074 448 698 571	123,666,922,527
two pair	2	.129 278 902 480	214,745,513,679
high pair (jacks or better)	1	.214 585 031 126	356,447,740,914
other	0	.545 434 669 233	906,022,916,158
sum		1.000 000 000 000	1,661,102,543,100

Let us consider the example, A♣-Q◇-J♣-T♣-9♣, discussed near the beginning of this section. Recall that we identified the five most promising strategies: (a) hold the four-card one-gap straight, (b) hold the four-card open-ended straight, (c) hold the four-card flush, (d) hold the three-card open-ended straight flush, and (e) hold the three-card two-gap royal flush.

(a) is not in the table, (b) is ranked 12th, (c) is ranked seventh, (d) is ranked 13th ($s = 3$ and $h = 1$), and (e) is ranked eighth. Holding the four-card flush ranks highest and is therefore the correct strategy, as we have already seen.

Finally, we reconsider the closely related example A♣-J♣-T♣-9♣-6◇. The two best strategies are (c) and (e) from the preceding example.

(c) is now ranked ninth, while (e) is still ranked eighth. Holding the three-card two-gap royal flush ranks highest and is therefore the correct strategy, as we have already seen. This emphasizes the fact the our hand-rank table is sensitive enough to take penalty cards into account.

17.2 Deuces Wild

As with Jacks or Better, the *Deuces Wild* video-poker player receives five cards face up on the screen, with each of the $\binom{52}{5} = 2{,}598{,}960$ possible hands equally likely. For each card, he must then decide whether to hold or discard that card. Thus, there are $2^5 = 32$ ways to play the hand. If he discards

Table 17.5 The optimal strategy at Jacks or Better. Choose the applicable strategy ranked highest in the table, holding only the cards listed in that strategy. If none applies, draw five new cards. Abbreviations: RF = royal flush, SF = straight flush, 4K = four of a kind, FH = full house, F = flush, S = straight, 3K = three of a kind, 2P = two pair, HP = high pair (jacks or better), LP = low pair. n-RF, n-SF, n-F, and n-S refer to n-card hands that have the potential to become RF, SF, F, and S, respectively. 3-3K, 4-2P, 2-HP, and 2-LP have a slightly different meaning: For example, 3-3K is a 3-card three of a kind, i.e., the potential is already realized. A, K, Q, J, T denote ace, king, queen, jack, and ten. H denotes any high card (A, K, Q, J). If two or more cards are italicized, that indicates that they are of the same suit. s is the number of straights, disregarding suits, that can be made from the hand, and h denotes the number of high cards in the hand. fp, sp, and 9sp denote flush penalty, straight penalty, and 9 straight penalty.

rank	description	rank	description
1	5-RF, 5-SF, 5-4K, or 5-FH	17	2-RF: AH or KH
2	3-3K	18	3-SF: $s + h = 2$, no sp
3	4-2P or 4-RF	19	4-S: AHHT or KQJ9
4	5-F or 5-S	20	3-SF: $s + h = 2$
5	4-SF	21	3-S: KQJ
6	2-HP	22	2-S: QJ
7	4-F: $AHTx$ + K, Q, J, or T	23	2-S: KJ if JT fp
8	3-RF	24	2-RF: JT
9	4-F	25	2-S: KH
10	4-S: KQJT	26	2-S: AQ if QT fp
11	2-LP	27	2-RF: QT
12	4-S: 5432–QJT9	28	2-S: AH
13	3-SF: $s + h \geq 3$	29	1-RF: K if KT fp and 9sp
14	4-S: AKQJ if QJ fp or 9p	30	2-RF: KT
15	2-RF: QJ	31	1-RF: A, K, Q, or J
16	4-S: AKQJ	32	3-SF: $s + h = 1$

k cards, he is dealt k new cards, with each of the $\binom{47}{k}$ possibilities equally likely. The player then receives his payout, which depends on the amount he bet and the rank of his final hand.

In Deuces Wild, as the name suggests, the four deuces (i.e., twos) are wild cards, and this affects the payout schedule. A *wild card* is a card that can play the role of any one of the 52 cards, even one that already appears in the hand. For example, A♣-A◇-A♡-A♠-2♣ counts as five of a kind. When wild cards are present in a hand, there may be more than one way to interpret the hand. The interpretation with the highest payout is the one that applies.

For example, A♣-2♣-2◇-2♡-2♠ counts as four deuces, not five of a kind, because four deuces pays more than five of a kind. Similarly, A♣-A◇-K♣-2♣-2◇ counts as four of a kind, not a full house.

The payout schedule for full-pay Deuces Wild is shown in Table 17.6. A *wild royal flush* is a royal flush with at least one wild card. A *natural royal flush* is a royal flush with no wild cards. We again use the term "full-pay" to emphasize the fact that there are similar machines with less favorable payout schedules. It should also be mentioned that, typically, to qualify for the 800 for 1 payout on a natural royal flush, the player must bet five coins. We assume that the player does this, and we define the total value of these five coins to be one unit.

Table 17.6 The full-pay Deuces Wild payoff odds and pre-draw frequencies.

rank	payoff odds	number of ways
natural royal flush	800 for 1	4
four deuces	200 for 1	48
wild royal flush	25 for 1	480
five of a kind	15 for 1	624
straight flush	9 for 1	2,068
four of a kind	5 for 1	31,552
full house	3 for 1	12,672
flush	2 for 1	14,472
straight	2 for 1	62,232
three of a kind	1 for 1	355,080
other	0 for 1	2,119,728
total		2,598,960

In Tables 17.7 and 17.8 we show how the pre-draw frequencies in Table 17.6 were evaluated. The key is to first classify each poker hand according to the number of deuces it contains (4, 3, 2, 1, or 0). Then it is relatively straightforward to count the numbers of hands of each type within each of these five classes. For example, to describe a straight with two deuces, we first specify the two deuces $\left[\binom{4}{2} \text{ ways}\right]$. Then we specify the lowest non-deuce denomination in the straight. If it is an ace, then one of the deuces must be used as a deuce and the hand must contain two of the three denominations 3, 4, and 5 $\left[\binom{1}{1}\binom{3}{2} \text{ ways}\right]$. If it is 3–T, then the hand must contain two of the next four denominations $\left[\binom{8}{1}\binom{4}{2} \text{ ways}\right]$. If it is a jack, the hand must contain two of the denomination Q, K, A $\left[\binom{1}{1}\binom{3}{2} \text{ ways}\right]$. If it is a queen, the hand must contain both of the denominations K and A $\left[\binom{1}{1}\binom{2}{2} \text{ ways}\right]$. Finally, we must specify the suits of the three non-deuces $\left[\binom{4}{1}^3 - \binom{4}{1}\text{ ways}\right]$, which cannot be the same or we would have a straight flush or a wild royal flush.

The primary issue in Deuces Wild is to determine, given the player's initial five-card hand, which cards should be held. Let us consider an example. Suppose that the player is dealt K♣-Q◇-J◇-T◇-8◇ (in any order, of course). There are three plausible strategies.

If the player holds the four-card open-ended straight K♣-Q◇-J◇-T◇, his payout R from a one-unit bet has conditional expected value

$$\text{E}[R] = 2\left(\frac{12}{47}\right) + 0\left(\frac{35}{47}\right) = \frac{24}{47} \approx 0.510638. \tag{17.13}$$

For convenience we do not write out the conditioning event, namely "player is dealt K♣-Q◇-J◇-T◇-8◇ and holds K♣-Q◇-J◇-T◇."

If the player holds the four-card one-gap straight flush Q◇-J◇-T◇-8◇, his payout R from a one-unit bet has conditional expected value

$$\text{E}[R] = 9\left(\frac{5}{47}\right) + 2\left(\frac{7}{47}\right) + 2\left(\frac{3}{47}\right) + 0\left(\frac{32}{47}\right) = \frac{65}{47} \approx 1.382979. \tag{17.14}$$

If the player holds the three-card royal flush Q◇-J◇-T◇, his payout R on a one-unit bet has conditional expected value

$$\text{E}[R] = 800 \frac{\binom{2}{2}\binom{45}{0}}{\binom{47}{2}} + 25 \frac{\binom{2}{2}\binom{43}{0} + \binom{4}{1}\binom{2}{1}\binom{41}{0}}{\binom{47}{2}} + 9 \frac{\binom{4}{1}\binom{1}{1}\binom{42}{0} + \binom{4}{0}\binom{2}{2}\binom{41}{0}}{\binom{47}{2}}$$

$$+ 2 \frac{\binom{4}{1}\binom{5}{1}\binom{38}{0} + \binom{4}{0}\left[\binom{8}{2} - 2\binom{2}{2}\binom{6}{0}\right]\binom{35}{0}}{\binom{47}{2}}$$

$$+ 2 \frac{\binom{4}{1}\binom{11}{1}\binom{32}{0} + \binom{4}{0}\left[3\binom{4}{1}\binom{3}{1}\binom{36}{0} - 2\binom{2}{2}\binom{41}{0}\right]}{\binom{47}{2}}$$

$$+ 1 \frac{\binom{4}{1}\binom{9}{1}\binom{34}{0} + \binom{4}{0}3\binom{3}{2}\binom{40}{0}}{\binom{47}{2}}$$

$$= \frac{1,488}{1,081} \approx 1.376503. \tag{17.15}$$

While there do not seem to be any other reasonable strategies, one should not rely too heavily on one's intuition in these situations, so we have computed the conditional expectations in the other 29 cases, listing all 32 results in Table 17.9.

The three suggested strategies are indeed the three best. In particular, we conclude that holding Q◇-J◇-T◇-8◇ is the optimal strategy, because it maximizes the player's conditional expected payout. This suggests that drawing to a four-card one-gap straight flush is marginally better than drawing to a three-card royal flush. However, one has to be careful when making such generalizations.

Table 17.7 The full-pay Deuces Wild pre-draw frequencies.

deuces	rank	number of ways	
4	four deuces	$\binom{4}{4}\binom{48}{1}$	48
3	wild royal flush	$\binom{4}{3}\binom{5}{2}\binom{4}{1}$	160
3	five of a kind	$\binom{4}{3}\binom{12}{1}\binom{4}{2}$	288
3	straight flush	$\binom{4}{3}\left[\binom{1}{1}\binom{3}{1}+\binom{7}{1}\binom{4}{1}\right]\binom{4}{1}$	496
3	four of a kind	$\binom{4}{3}\binom{48}{2}-\text{subtotal}$	3,568
2	wild royal flush	$\binom{4}{2}\binom{5}{3}\binom{4}{1}$	240
2	five of a kind	$\binom{4}{2}\binom{12}{1}\binom{4}{3}$	288
2	straight flush	$\binom{4}{2}\left[\binom{1}{1}\binom{3}{2}+\binom{7}{1}\binom{4}{2}\right]\binom{4}{1}$	1,080
2	four of a kind	$\binom{4}{2}\binom{12}{1}\binom{11}{1}\binom{4}{2}\binom{4}{1}$	19,008
2	flush	$\binom{4}{2}\left[\binom{12}{3}-\binom{5}{3}-\binom{1}{1}\binom{3}{2}-\binom{7}{1}\binom{4}{2}\right]\binom{4}{1}$	3,960
2	straight	$\binom{4}{2}\left[\binom{1}{1}\binom{3}{2}+\binom{8}{1}\binom{4}{2}+\binom{1}{1}\binom{3}{2}\right.$ $\left.+\binom{1}{1}\binom{2}{2}\right]\left[\binom{4}{1}^{3}-\binom{4}{1}\right]$	19,800
2	three of a kind	$\binom{4}{2}\binom{48}{3}-\text{subtotal}$	59,400
1	wild royal flush	$\binom{4}{1}\binom{5}{4}\binom{4}{1}$	80
1	five of a kind	$\binom{4}{1}\binom{12}{1}\binom{4}{4}$	48
1	straight flush	$\binom{4}{1}\left[\binom{1}{1}\binom{4}{4}+\binom{7}{1}\binom{4}{3}\right]\binom{4}{1}$	464
1	four of a kind	$\binom{4}{1}\binom{12}{1}\binom{11}{1}\binom{4}{3}\binom{4}{1}$	8,448
1	full house	$\binom{4}{1}\binom{12}{2}\binom{4}{2}^{2}$	9,504
1	flush	$\binom{4}{1}\left[\binom{12}{4}-\binom{5}{4}-\binom{1}{1}\binom{4}{4}-\binom{7}{1}\binom{4}{3}\right]\binom{4}{1}$	7,376
1	straight	$\binom{4}{1}\left[\binom{1}{1}\binom{3}{3}+\binom{8}{1}\binom{4}{3}\right.$ $\left.+\binom{1}{1}\binom{3}{3}\right]\left[\binom{4}{1}^{4}-\binom{4}{1}\right]$	34,272
1	three of a kind	$\binom{4}{1}\binom{12}{1}\binom{11}{2}\binom{4}{2}\binom{4}{1}^{2}$	253,440
1	one pair*	$\binom{4}{1}\binom{48}{4}-\text{subtotal}$	464,688

continued in Table 17.8

*no payout

Let us consider a closely related example to illustrate this point. Suppose that the player is dealt Q◊-J◊-T◊-8◊-7♣ (in any order). (The K♣ in the preceding example has been replaced by the 7♣.) The analogues of the two best strategies in the preceding example are the following.

If the player holds the four-card one-gap straight flush Q◊-J◊-T◊-8◊, his payout R from a one-unit bet has conditional expected value (17.14).

Table 17.8 Continuation of Table 17.7: The full-pay Deuces Wild pre-draw frequencies.

deuces	rank	number of ways	
0	natural royal flush	$\binom{4}{0}\binom{5}{5}\binom{4}{1}$	4
0	straight flush	$\binom{4}{0}\binom{7}{1}\binom{4}{4}\binom{4}{1}$	28
0	four of a kind	$\binom{4}{0}\binom{12}{1}\binom{11}{1}\binom{4}{4}\binom{4}{1}$	528
0	full house	$\binom{4}{0}\binom{12}{1}\binom{11}{1}\binom{4}{3}\binom{4}{2}$	3,168
0	flush	$\binom{4}{0}\left[\binom{12}{5}-\binom{5}{5}-\binom{7}{1}\binom{4}{4}\right]\binom{4}{1}$	3,136
0	straight	$\binom{4}{0}\binom{8}{1}\left[\binom{4}{1}^{5}-\binom{4}{1}\right]$	8,160
0	three of a kind	$\binom{4}{0}\binom{12}{1}\binom{11}{2}\binom{4}{3}\binom{4}{1}^{2}$	42,240
0	no payout	$\binom{4}{0}\binom{48}{5}-$ subtotal	1,655,040
total (Tables 17.7 and 17.8)			2,598,960

If the player holds the three-card royal flush Q◊-J◊-T◊, his payout R from a one-unit bet has conditional expected value

$$
\begin{aligned}
E[R] &= 800\,\frac{\binom{2}{2}\binom{45}{0}}{\binom{47}{2}} + 25\,\frac{\binom{4}{2}\binom{43}{0}+\binom{4}{1}\binom{2}{1}\binom{41}{0}}{\binom{47}{2}} + 9\,\frac{\binom{4}{1}\binom{1}{1}\binom{42}{0}+\binom{4}{0}\binom{2}{2}\binom{41}{0}}{\binom{47}{2}}\\[1mm]
&\quad + 2\,\frac{\binom{4}{1}\binom{5}{1}\binom{38}{0}+\binom{4}{0}\left[\binom{8}{2}-2\binom{2}{2}\binom{6}{0}\right]\binom{35}{0}}{\binom{47}{2}}\\[1mm]
&\quad + 2\,\frac{\binom{4}{1}\binom{12}{1}\binom{31}{0}+\binom{4}{0}\left[2\binom{4}{1}^{2}\binom{35}{0}+\binom{4}{1}\binom{3}{1}\binom{36}{0}-2\binom{2}{2}\binom{41}{0}\right]}{\binom{47}{2}}\\[1mm]
&\quad + 1\,\frac{\binom{4}{1}\binom{9}{1}\binom{34}{0}+\binom{4}{0}3\binom{3}{2}\binom{40}{0}}{\binom{47}{2}}\\[1mm]
&= \frac{1{,}512}{1{,}081} \approx 1.398705. \tag{17.16}
\end{aligned}
$$

Table 17.9 The 32 possible values of the player's conditional expected payout at Deuces Wild when he is dealt K♣-Q◇-J◇-T◇-8◇.

cards held					$k :=$ no. of cards drawn	conditional expected payout E	$\binom{47}{k}E$
K♣	Q◇	J◇	T◇	8◇	0	.000 000	0
K♣	Q◇	J◇	T◇		1	.510 638	24
K♣	Q◇	J◇		8◇	1	.000 000	0
K♣	Q◇		T◇	8◇	1	.000 000	0
K♣		J◇	T◇	8◇	1	.000 000	0
	Q◇	J◇	T◇	8◇	1	1.382 979	65
K♣	Q◇	J◇			2	.178 538	193
K♣	Q◇		T◇		2	.178 538	193
K♣	Q◇			8◇	2	.047 179	51
K♣		J◇	T◇		2	.178 538	193
K♣		J◇		8◇	2	.047 179	51
K♣			T◇	8◇	2	.047 179	51
	Q◇	J◇	T◇		2	1.376 503	1,488
	Q◇	J◇		8◇	2	.295 097	319
	Q◇		T◇	8◇	2	.295 097	319
		J◇	T◇	8◇	2	.377 428	408
K♣	Q◇				3	.169 164	2,743
K♣		J◇			3	.169 164	2,743
K♣			T◇		3	.169 164	2,743
K♣				8◇	3	.126 981	2,059
	Q◇	J◇			3	.238 606	3,869
	Q◇		T◇		3	.238 606	3,869
	Q◇			8◇	3	.178 292	2,891
		J◇	T◇		3	.263 275	4,269
		J◇		8◇	3	.202 960	3,291
			T◇	8◇	3	.233 117	3,780
K♣					4	.250 397	44,662
	Q◇				4	.236 633	42,207
		J◇			4	.245 805	43,843
			T◇		4	.257 220	45,879
				8◇	4	.323 817	48,117
(none)					5	.323 817	496,716

None of the other 30 strategies is as good as these two. We conclude that the optimal strategy is to hold Q$\diamond$-J$\diamond$-T$\diamond$.

Notice that the two best strategies (hold the four-card one-gap straight flush; hold the three-card royal flush) are ordered differently in the two examples. The reason is easy to see. The player's expected payout when holding the four-card one-gap straight flush is the same in the two examples. However, his expected payout when holding the three-card royal flush is smaller when he discards the K$\clubsuit$ (first example) than when he discards the 7$\clubsuit$ (second example). Clearly, the absence of the K$\clubsuit$ from the residual 47-card deck reduces the chance that the player will make a straight. Thus, the K$\clubsuit$ is called a straight-penalty card.

More generally, when a particular discard reduces the probability that the player will make a straight or a flush (over what it would have been otherwise), that discard is called a *straight-* or a *flush-penalty card*. There are other types of penalty cards as well, as the following example indicates.

Suppose that the player is dealt five of a kind with three deuces. Should he hold the pat five of a kind or discard the nondeuce pair to draw for the fourth deuce? The only question is whether the latter strategy provides a conditional expected payout from a one-unit bet of more than 15 units, which is the guaranteed payout for the pat five of a kind. Surprisingly, the answer depends on the denomination of the nondeuce pair. For example, if the nondeuce pair has denomination 9, the player's payout R has conditional expected value

$$E[R] = 5 + 195 \frac{\binom{1}{1}\binom{46}{1}}{\binom{47}{2}} + 20 \frac{\binom{5}{2}\binom{4}{1}}{\binom{47}{2}} + 10 \frac{11\binom{4}{2}\binom{43}{0} + \binom{2}{2}\binom{45}{0}}{\binom{47}{2}}$$
$$+ 4 \frac{2\binom{4}{1}\binom{4}{1} + 4[\binom{2}{1}\binom{4}{1} + \binom{2}{1}\binom{3}{1}] + \binom{2}{1}\binom{4}{1} + \binom{4}{1}\binom{3}{1}}{\binom{47}{2}}$$
$$= \frac{16{,}277}{1{,}081} \approx 15.057354. \tag{17.17}$$

For a second example, if the nondeuce pair has denomination ten, the player's payout R has conditional expected value

$$E[R] = 5 + 195 \frac{\binom{1}{1}\binom{46}{1}}{\binom{47}{2}} + 20 \frac{\binom{1}{0}\binom{4}{2}\binom{4}{1} + \binom{1}{1}\binom{4}{1}\binom{2}{1}}{\binom{47}{2}} + 10 \frac{11\binom{4}{2}\binom{43}{0} + \binom{2}{2}\binom{45}{0}}{\binom{47}{2}}$$
$$+ 4 \frac{3\binom{4}{1}\binom{4}{1} + 4[\binom{2}{1}\binom{4}{1} + \binom{2}{1}\binom{3}{1}] + \binom{4}{1}\binom{3}{1}}{\binom{47}{2}}$$
$$= \frac{16{,}149}{1{,}081} \approx 14.938945. \tag{17.18}$$

In evaluating both conditional expectations, we noted that the player is assured a 5-unit payout, so we added to this guaranteed amount the additional contributions from hands better than four of a kind. (This allowed us to avoid counting the number of four-of-a-kind hands.)

Thus, we find that the optimal strategy with a three-deuce hand of five 9s is to hold only the deuces, while the optimal strategy with a three-deuce hand of five tens is to hold all five cards. In the case of the three-deuce hand of five tens, the two tens, if discarded, can be regarded as wild-royal-flush-penalty cards.

It is now clear what must be done to determine the optimal strategy at Deuces Wild. We simply create a table analogous to Table 17.9 for each of the player's $\binom{52}{5} = 2{,}598{,}960$ possible initial hands, determine for each such hand which of the $2^5 = 32$ ways to play it maximizes his conditional expected payout, summarize the resulting player strategy, and average the conditional expectations thus obtained to evaluate the overall expected payout.

Actually, we can reduce the amount of work required by more than a factor of 20 by taking equivalence of initial hands into account, much as we did in Section 17.1. Let us call two initial hands *equivalent* if they have the same five denominations and if the corresponding nondeuce denominations have the same suits after a permutation of $(\clubsuit,\diamondsuit,\heartsuit,\spadesuit)$. The reason for the word "nondeuce" in the definition is that the suits of deuces do not matter. This is an equivalence relation in the sense of Theorem A.1.1. Thus, for example, the equivalence class containing T$\clubsuit$-T$\diamondsuit$-2$\clubsuit$-2$\diamondsuit$-2$\heartsuit$ has $\binom{4}{2}\binom{4}{3} = 24$ hands, including T$\clubsuit$-T$\diamondsuit$-2$\diamondsuit$-2$\heartsuit$-2$\spadesuit$. (These two hands are not equivalent under the equivalence relation of Section 17.1.)

Table 17.10 shows that there are exactly

$$\binom{12}{5}51 + \binom{12}{1,3,8}20 + \binom{12}{2,1,9}8 + \binom{12}{1,2,9}5 + \binom{12}{1,1,10}3$$
$$+ \binom{12}{4}15 + \binom{12}{1,2,9}6 + \binom{12}{2}3 + \binom{12}{1,1,10}2 + \binom{12}{1}1 \quad (17.19)$$
$$+ \binom{12}{3}5 + \binom{12}{1,1,10}2 + \binom{12}{1}1 + \binom{12}{2}2 + \binom{12}{1}2 = 102{,}359$$

equivalence classes. As a check, we compute the total number of hands by summing the sizes of the equivalence classes:

$$792(1 \cdot 4 + 15 \cdot 12 + 35 \cdot 24) + 1{,}980(8 \cdot 12 + 12 \cdot 24)$$
$$+ 660(1 \cdot 4 + 7 \cdot 12 + 5 \cdot 24) + 132(1 \cdot 4 + 2 \cdot 12)$$
$$+ 495(1 \cdot 16 + 7 \cdot 48 + 7 \cdot 96) + 660(4 \cdot 48 + 2 \cdot 96)$$
$$+ 66(2 \cdot 24 + 1 \cdot 96) + 132(1 \cdot 16 + 1 \cdot 48) + 12(1 \cdot 4)$$
$$+ 220(1 \cdot 24 + 3 \cdot 72 + 1 \cdot 144) + 132(2 \cdot 72) + 12(1 \cdot 24)$$
$$+ 66(1 \cdot 16 + 1 \cdot 48) + 12(1 \cdot 4 + 1 \cdot 24) = 2{,}598{,}960. \quad (17.20)$$

Again, a computer is a necessity for this kind of problem. Our program methodically cycles through each of the 102,359 equivalence classes. For each

Table 17.10 List of equivalence classes of initial player hands in Deuces Wild, together with the size of each equivalence class. a, b, c, d represent distinct nondeuce denominations.

no deuces: see Table 17.3[1] on p. 551			
$(2, a, b, c, d)$: $\binom{12}{4} = 495$ ways			
$(*, 1, 1, 1, 1)$ 16	$(*, 1, 1, 2, 3)$ 96	$(*, 1, 2, 2, 1)$ 48	$(*, 1, 2, 3, 2)$ 96
$(*, 1, 1, 1, 2)$ 48	$(*, 1, 2, 1, 1)$ 48	$(*, 1, 2, 2, 2)$ 48	$(*, 1, 2, 3, 3)$ 96
$(*, 1, 1, 2, 1)$ 48	$(*, 1, 2, 1, 2)$ 48	$(*, 1, 2, 2, 3)$ 96	$(*, 1, 2, 3, 4)$ 96
$(*, 1, 1, 2, 2)$ 48	$(*, 1, 2, 1, 3)$ 96	$(*, 1, 2, 3, 1)$ 96	
$(2, a, a, b, c)$: $\binom{12}{1,2,10} = 660$ ways			
$(*, 1, 2, 1, 1)$ 48	$(*, 1, 2, 1, 3)$ 96	$(*, 1, 2, 3, 3)$ 48	
$(*, 1, 2, 1, 2)$ 48	$(*, 1, 2, 3, 1)$ 96	$(*, 1, 2, 3, 4)$ 48	
$(2, a, a, b, b)$: $\binom{12}{2} = 66$ ways			
$(*, 1, 2, 1, 2)$ 24	$(*, 1, 2, 1, 3)$ 96	$(*, 1, 2, 3, 4)$ 24	
$(2, a, a, a, b)$: $\binom{12}{1,1,10} = 132$ ways			
$(*, 1, 2, 3, 1)$ 48	$(*, 1, 2, 3, 4)$ 16		
$(2, a, a, a, a)$: $\binom{12}{1} = 12$ ways			
$(*, 1, 2, 3, 4)$ 4			
$(2, 2, a, b, c)$: $\binom{12}{3} = 220$ ways			
$(*, *, 1, 1, 1)$ 24	$(*, *, 1, 2, 1)$ 72	$(*, *, 1, 2, 3)$ 144	
$(*, *, 1, 1, 2)$ 72	$(*, *, 1, 2, 2)$ 72		
$(2, 2, a, a, b)$: $\binom{12}{1,1,10} = 132$ ways			
$(*, *, 1, 2, 1)$ 72	$(*, *, 1, 2, 3)$ 72		
$(2, 2, a, a, a)$: $\binom{12}{1} = 12$ ways			
$(*, *, 1, 2, 3)$ 24			
$(2, 2, 2, a, b)$: $\binom{12}{2} = 66$ ways			
$(*, *, *, 1, 1)$ 16	$(*, *, *, 1, 2)$ 48		
$(2, 2, 2, a, a)$: $\binom{12}{1} = 12$ ways			
$(*, *, *, 1, 2)$ 24			
$(2, 2, 2, 2, a)$: $\binom{12}{1} = 12$ ways			
$(*, *, *, *, 1)$ 4			

[1]except replace 13 by 12 in the multinomial coefficients, and adjust the corresponding partitions of 13

one it computes the 32 conditional expectations and determines which is largest and if it is uniquely the largest. It stores this information in a file as it proceeds. Finally, it computes the payout distribution under the optimal strategy, first for each equivalence class and then for the game as a whole.

Let us now consider the issue of uniqueness, which is more complicated than with Jacks or Better. There are some obvious cases of nonuniqueness: If a player is dealt four deuces, he may hold or discard the nonwild card with no effect. If he is dealt two pair without deuces, it is usually optimal to discard one of the pairs—it does not matter which one. Finally, it is frequently the case that a hand with two four-card straights has two optimal one-card discards. For example, with A♣-K♣-Q♢-J-♡-9♣, we can discard the ace or the 9 with no effect. In all of these examples, the payout distribution is unaffected by the choice of optimal strategy. But that is not true in general.

Consider the hand A♣-K♢-Q♢-J♡-9♢. As in the preceding example, there are two optimal one-card discards (the ace and the 9), but here there is a third optimal strategy, namely to hold the three-card straight flush. To confirm that all three strategies are optimal, it suffices to evaluate the expected payout in each case. If we hold either of the four-card straights, it is

$$E[R] = 2\left(\frac{8}{47}\right) + 0\left(\frac{39}{47}\right) = \frac{16}{47} \approx 0.340426. \qquad (17.21)$$

If we hold the three-card straight flush, it is

$$E[R] = 9\,\frac{\binom{6}{2}}{\binom{47}{2}} + 2\,\frac{\binom{4}{0}\left[\binom{9}{2}-1\right] + \binom{4}{1}\binom{7}{1} + \binom{4}{0}\left[\binom{4}{1}\binom{3}{1}-1\right] + \binom{4}{1}\binom{5}{1}}{\binom{47}{2}}$$

$$+ 1\,\frac{\binom{4}{0}\binom{3}{1}\binom{3}{2} + \binom{4}{1}\binom{3}{1}\binom{3}{1}}{\binom{47}{2}} = \frac{368}{1{,}081} = \frac{16}{47} \approx 0.340426. \quad (17.22)$$

The four terms in the numerator of the second fraction correspond to (a) flushes without deuces, (b) flushes with one deuce, (c) straights without deuces, and (d) straights with one deuce.

This is an example of an equivalence class for which the optimal strategy is not only nonunique but is what we will call *essentially nonunique*, in that the choice of strategy affects the payout distribution. There are 572 equivalence classes with this property, 286 of size 12 and 286 of size 24. Notice that the variance corresponding to (17.21) is smaller than that corresponding to (17.22). The optimal player who discards only one card in each such situation is using the *minimum-variance optimal strategy*, whereas the optimal player who discards two cards in each such situation is using the *maximum-variance optimal strategy*.

Thus, the payout distribution for Deuces Wild played according to the minimum-variance optimal strategy is uniquely determined, and we display it in Table 17.11. The same is true of the payout distribution for Deuces Wild played according to the maximum-variance optimal strategy, but we do not display it.

Here it is worth giving exact results. A common denominator (not the least one) is (17.11) on p. 552. The least common denominator is this number divided by 12.

It follows that the mean payout under optimal play is

$$\frac{1{,}673{,}759{,}500{,}036}{1{,}661{,}102{,}543{,}100} = \frac{32{,}187{,}682{,}693}{31{,}944{,}279{,}675} \approx 1.007619612039. \tag{17.23}$$

We have discovered something remarkable: Deuces Wild is a rare example of a casino game that offers positive expected profit to the knowledgeable player (about 3/4 of one percent). The variance of the payout, under the minimum-variance optimal strategy, is 25.834618052354.

Table 17.11 The distribution of the payout R from a one-unit bet on the video poker game Deuces Wild. Assumes maximum-coin bet and the minimum-variance optimal drawing strategy.

result	R	probability	probability $\times \binom{52}{5}\binom{47}{5}5/12$
natural royal flush	800	.000 022 083 864	36,683,563
four deuces	200	.000 203 703 199	338,371,902
wild royal flush	25	.001 795 843 261	2,983,079,808
five of a kind	15	.003 201 603 965	5,318,192,488
straight flush	9	.004 119 878 191	6,843,540,140
four of a kind	5	.064 938 165 916	107,868,952,548
full house	3	.021 229 137 790	35,263,774,770
flush *or* straight	2	.073 145 116 685	121,501,539,340
three of a kind	1	.284 544 359 823	472,657,359,726
other	0	.546 800 107 307	908,291,048,815
total		1.000 000 000 000	1,661,102,543,100

There remains an important issue that has not yet been addressed. What exactly is the optimal strategy at Deuces Wild? Our computer program provides one possible answer: specific instructions for each of the 102,359 equivalence classes. However, what we need is something simpler, a strategy that can actually be memorized. The usual approach is to construct a so-called *hand-rank table*. Each of the various types of holdings is ranked according to its conditional expectation (which varies slightly with the cards discarded). One then simply finds the highest-ranked hand in the table that is applicable to the hand in question. Only relatively recently has a hand-rank table been

Table 17.12 The (minimum-variance) optimal strategy for Deuces Wild. Count the number of deuces and choose the corresponding strategy ranked highest in the table, holding only the cards listed in that strategy. If none applies, hold only the deuces. Abbreviations: RF = royal flush, 4D = four deuces, 5K = five of a kind, SF = straight flush, 4K = four of a kind, FH = full house, F = flush, S = straight, 3K = three of a kind, 2P = two pair, 1P = one pair. n-RF, n-SF, n-F, and n-S refer to n-card hands that have the potential to become RF, SF, F, and S, respectively. 4-4D, 3-3K, 4-2P, and 2-1P have a slightly different meaning: For example, 3-3K is a 3-card three of a kind, i.e., the potential is already realized. A, K, Q, J, T denote ace, king, queen, jack, and ten. If two or more cards are italicized, that indicates that they are of the same suit. s is the number of straights, disregarding suits, that can be made from the hand, *excluding 2-low straights*. fp, sp, 8p, etc. denote flush penalty, straight penalty, 8 straight penalty, etc. Finally, unp. S pot. stands for unpenalized straight potential.

rank	description	rank	description
four deuces		**no deuces**	
1	5-4D	1	5-RF
		2	4-RF
three deuces		3	5-SF
1	5-RF	4	4-4K
2	5-5K: 222AA–222TT	5	5-FH, 5-F, or 5-S
		6	3-3K
two deuces		7	4-SF: $s = 2$
1	5-RF, 5-5K, or 5-SF	8	3-RF: *QJT*, no Kp
2	4-RF or 4-4K	9	4-SF: $s = 1$
3	4-SF: $s = 4$	10	3-RF
		11	2-1P
one deuce		12	4-F or 4-S: $s = 2$
1	5-RF, 5-5K, 5-SF, or 5-FH	13	3-SF: $s \geq 2$
2	4-RF or 4-4K	14	3-SF: *JT7*, Ap+Kp
3	4-SF: $s = 3$		or Qp but not Qp+8p
4	5-F or 5-S	15	2-RF: *JT*
5	3-3K	16	2-RF: *QJ* or *QT*, no sp
6	4-SF: $s \leq 2$	17	3-SF: $s = 1$, except A-low, no sp
7	3-RF: Q-high or J-high	18	2-RF: *QJ* or *QT*, no fp, unp. S pot.
8	3-RF: K-high	19	2-RF: *QT* if *QT*876
9	3-SF: $s = 4$	20	4-S: $s = 1$, except A-low
10	3-RF: A-high, no fp or sp*	21	3-SF: $s = 1$, except A-low
		22	2-RF: *QJ* or *QT*
		23	2-RF: *KQ, KJ,* or *KT*, no fp or sp*

*there are exceptions (see Table 17.17)

found for Deuces Wild that reproduces the optimal strategy precisely, and that table is presented as Table 17.12.

Let us consider the example, K♣-Q◇-J◇-T◇-8◇, discussed near the beginning of this section. Recall that we identified the three most promising strategies: (a) hold the four-card open-ended straight, (b) hold the four-card one-gap straight flush, and (c) hold the three-card royal flush.

Under no deuces, (a) is ranked 12th, (b) is ranked ninth, and (c) is ranked tenth. Holding the four-card one-gap straight flush ranks highest and is therefore the correct strategy, as we have already seen.

Finally, we reconsider the closely related example Q◇-J◇-T◇-8◇-7♣. The two best strategies are (b) and (c) from the preceding example.

(b) is still ranked ninth, but (c) is now ranked eighth. Holding the three-card royal flush ranks highest and is therefore the correct strategy, as we have already seen. This emphasizes the fact the our hand-rank table is sensitive enough to take penalty cards into account.

17.3 Problems

17.1. *A Jacks or Better strategy decision.* In Jacks or Better it is optimal to hold K♣-Q♣-J♣-T♣-9♣ for a guaranteed payout of 50 units rather than to discard the 9 and draw for the royal flush. How large would the payout on a royal flush have to be, all other things being equal, to make it optimal to discard the 9?

17.2. *A Deuces Wild strategy decision.* In Deuces Wild it is optimal to hold A♣-K♣-Q♣-J♣-2◇ for a guaranteed payout of 25 units rather than to discard the deuce and draw for the natural royal flush. How large would the payout on a natural royal flush have to be, all other things being equal, to make it optimal to discard the deuce?

17.3. *Jacks or Better practice hands.* For each initial hand listed in Table 17.13, (a) guess the optimal strategy, (b) use Table 17.5 on p. 554 to determine the optimal strategy, and (c) compute the conditional expectations of all promising strategies, using either combinatorics or a computer program.

17.4. *Deuces Wild practice hands.* For each initial hand listed in Table 17.14, (a) guess the optimal strategy, (b) use Table 17.12 on p. 565 to determine the optimal strategy, and (c) compute the conditional expectations of all promising strategies, using either combinatorics or a computer program.

17.5. *Five of a kind with three deuces.* In Deuces Wild we showed in (17.17) and (17.18) on p. 560 that it is optimal to hold all five cards if dealt three deuces and two tens but not if dealt three deuces and two 9s.

(a) Extend this analysis to all three-deuce five-of-a-kind hands, of which there are 12 equivalence classes. Use combinatorial analysis.

Table 17.13 Some Jacks or Better practice hands.

1.	A♦-Q♦-J♦-T♦-6♦	11.	J♥-A♣-9♦-K♣-Q♥
2.	K♣-Q♦-J♦-T♦-9♦	12.	J♥-A♣-8♥-K♣-Q♥
3.	Q♣-Q♦-J♣-9♣-8♣	13.	A♥-Q♦-T♦-9♣-8♦
4.	K♦-Q♦-J♦-T♦-9♦	14.	A♥-J♣-Q♠-T♠-8♠
5.	A♣-Q♣-J♣-9♣-7♦	15.	7♥-A♥-Q♥-K♠-T♥
6.	K♣-J♣-T♣-9♣-9♦	16.	K♥-T♥-3♣-9♣-5♥
7.	9♦-8♣-7♣-6♥-6♦	17.	T♥-K♥-8♥-9♦-J♠
8.	K♠-Q♣-J♣-8♣-7♣	18.	A♥-K♥-T♥-5♥-K♠
9.	A♦-K♣-Q♥-J♣-6♠	19.	T♣-K♦-6♣-9♦-J♣
10.	K♦-Q♠-J♦-7♠-5♠	20.	9♠-J♣-3♥-4♥-7♥

Table 17.14 Some Deuces Wild practice hands.

1.	K♣-Q♣-J♣-T♣-9♣	11.	7♣-K♦-J♣-A♦-T♣
2.	K♣-Q♣-J♣-9♣-9♦	12.	8♥-Q♠-J♥-T♠-7♥
3.	K♦-8♦-5♦-3♦-3♥	13.	3♠-2♥-K♠-7♦-A♠
4.	9♠-8♠-2♣-2♦-2♥	14.	9♦-T♥-5♦-Q♥-8♦
5.	T♥-9♥-3♦-2♥-2♠	15.	6♦-K♣-7♠-Q♣-8♦
6.	A♣-K♣-K♦-T♣-9♣	16.	7♦-2♣-A♥-6♦-J♥
7.	7♣-6♥-5♦-2♦-2♠	17.	T♥-Q♥-7♥-A♣-K♦
8.	A♥-A♠-K♥-K♠-Q♥	18.	T♥-8♠-7♥-Q♣-J♥
9.	K♣-Q♦-J♦-T♦-9♦	19.	J♠-8♠-A♥-Q♠-T♠
10.	A♠-J♠-T♠-9♠-2♥	20.	5♥-A♣-3♥-4♣-6♥

(b) In the cases of part (a) where it is optimal to draw, evaluate the variance of the payout. In these cases, departing from optimal play reduces the expected payout very slightly but reduces the variance of the payout dramatically.

17.6. *Variance of the payout under maximum-variance optimal strategy.* In Deuces Wild we found that the variance of the payout is approximately 25.834618052 under the minimum-variance optimal strategy. Find the corresponding figure for the maximum-variance optimal strategy.

17.7. *Distribution of the number of cards held.*

(a) Assuming the optimal strategy at Jacks or Better (and standing pat with four of a kind), find the joint distribution of the payout and the number of cards held. How many of the 60 joint probabilities are 0? Find the marginal distribution of the number of cards held.

(b) Do the same for Deuces Wild, assuming the minimum-variance optimal strategy (and standing pat with four deuces).

17.8. *Deuces Wild conditional probabilities.*

(a) Find the conditional probabilities of the 10 payouts in Deuces Wild (assuming the minimum-variance optimal strategy), given the number of deuces in the hand before the draw. The unconditional probabilities are displayed in Table 17.11 on p. 564. How many of these 50 conditional probabilities are 0?

(b) Find the conditional expected payouts in Deuces Wild, given the number of deuces in the hand before the draw.

17.9. *A dubious notion of optimality.*

(a) In Jacks or Better, suppose the goal is to maximize the probability of obtaining a royal flush. In other words, instead of choosing the strategy that maximizes the conditional expected payout, we choose one that maximizes the conditional probability of a royal flush. (Alternatively, we could adjust the payoff odds so that only a royal flush has a positive payout.) What is this maximum probability? Use combinatorial analysis.

(b) Is the answer the same for Deuces Wild and a natural royal flush?

17.10. *n-play video poker.* Fix an integer $n \geq 1$. In n-play video poker, the player bets n units. He then receives five cards face up on the screen, with each of the $\binom{52}{5}$ possible hands equally likely. For each card, the player must then decide whether to hold or discard that card. Thus, there are 2^5 ways to play the hand. If he discards k cards, then the following occurs n times with the results conditionally independent: He is dealt k new cards, with each of the $\binom{47}{k}$ possibilities equally likely. The player then receives his payout, which depends on the payout schedule and assumes one unit bet on each of the n hands.

(a) Argue that the optimal strategy does not depend on n.

(b) Show that the variance of the *payout per unit bet* is decreasing in n. More precisely, let $X_1, \ldots, X_n$ be the payouts from the n plays and let Y denote the initial hand. Then $X_1, \ldots, X_n$ are conditionally i.i.d. given Y. Use the conditioning law for variances (Theorem 2.2.13 on p. 87) to show that

$$\mathrm{Var}\left(\frac{X_1 + \cdots + X_n}{n}\right) = \mathrm{Var}(X_1) - \left(1 - \frac{1}{n}\right)\mathrm{E}[\mathrm{Var}(X_1|Y)]. \qquad (17.24)$$

17.11. *Probability of a royal at n-play Jacks or Better vs. at n independent games of Jacks or Better.* Fix an integer $n \geq 2$. Assuming the optimal strategy, which is more likely, at least one royal flush at one game of n-play Jacks or Better (Problem 17.10) or at least one royal flush at n *independent* games of (1-play) Jacks or Better?

17.12. *Jacks or Better: A mathematical simplification.* In Jacks or Better, assume for simplicity that when k cards are discarded, the k new cards are chosen at random without replacement from the $47 + k$ cards not held rather than from the 47 cards not seen. (To put it another way, we assume that the discards are shuffled back into the deck before their replacements are chosen.)

(a) Determine the optimal strategy under this simplification.

(b) Determine the expected payout using this simplified strategy in the real game.

Hint: (a) Let H denote the set of cards held. Evaluate the expected payout E_H for each H, starting with $H = \varnothing$, then $|H| = 1$ (52 cases), then $|H| = 2$ (1,326 cases), and so on. Keep track only of those E_H for which $E_H \geq E_G$ whenever $G \subset H$.

17.13. *Double Bonus video poker.* Double Bonus Poker is played with a standard 52-card deck and there are no wild cards. The payout schedule is shown in Table 17.15. Find the expected payout under optimal play.

Table 17.15 Double Bonus Poker payoff odds and pre-draw frequencies.

rank	payoff odds	number of ways
royal flush	800 for 1	4
straight flush	50 for 1	36
four aces	160 for 1	48
four 2s, 3s, or 4s	80 for 1	144
four of a kind (others)	50 for 1	432
full house	10 for 1	3,744
flush	7 for 1	5,108
straight	5 for 1	10,200
three of a kind	3 for 1	54,912
two pair	1 for 1	123,552
pair of jacks or better	1 for 1	337,920
other	0 for 1	2,062,860
total		2,598,960

17.14. *Joker Wild video poker.* Joker Wild adds a joker, which acts as a wild card, to the standard 52-card deck. The payout schedule is shown in Table 17.16.

(a) Confirm the pre-draw frequencies shown in the table.

(b) Define an equivalence relation on the set of all five-card hands, analogous to the one in Section 17.2, and determine the number of equivalence classes.

(c) Find the expected payout under optimal play.

Table 17.16 Joker Wild payoff odds and pre-draw frequencies.

rank	payoff odds	number of ways without a joker	number of ways with a joker
royal flush (natural)	800 for 1	4	0
five of a kind	200 for 1	0	13
royal flush (joker)	100 for 1	0	20
straight flush	50 for 1	36	144
four of a kind	20 for 1	624	2,496
full house	7 for 1	3,744	2,808
flush	5 for 1	5,108	2,696
straight	3 for 1	10,200	10,332
three of a kind	2 for 1	54,912	82,368
two pair	1 for 1	123,552	0
pair of aces or kings	1 for 1	168,960	93,996
other	0 for 1	2,231,820	75,852
total		2,598,960	270,725

17.4 Notes

Video poker became firmly established with the introduction by SIRCOMA (Si Redd's [1911–2003] Coin Machines, which evolved into International Game Technology, or IGT) of "Draw Poker" in 1979. See Fey (2002, p. 217) for a picture of this machine, whose pay table was identical to that for full-pay Jacks or Better, except that the 1 for 1 payout on a pair of jacks or better was absent. As explained by Paymar (2004, p. 10), the manufacturer had not yet figured out how to evaluate the game's expected payback, so they initially played it safe. However, there was little interest by the gambling public in such an unfair game, so the computer chip was modified and a placard was attached to the machine saying "BET RETURNED ON A PAIR OF JACKS OR BETTER." This was the origin of today's full-pay Jacks or Better. However, there were several predecessors, so the year of video poker's debut is arguable.

Bally Manufacturing Corporation introduced a video poker machine in 1976, according to Weber and Scruggs (1992), though it was not mentioned by Jones (1978). Dale Electronics introduced an electronic machine called "Poker Matic" in 1970, but it did not have a video monitor, so strictly speaking it was not video poker. But if the video aspect is inessential to the nature of the game, then so too is the electronic aspect, and we have to go back to 1901, when Charles August Fey [1862–1944] introduced the first poker machine with a draw feature (Fey 2002, p. 76). There were 10 cards on each of five reels, so only $(10)^5 = 100,000$ hands, not 2,598,960, were possible. It is this issue, we believe, not the lack of a video monitor, that disqualifies the Fey machine as the first video poker machine.

Deuces Wild came later and remains one of the more popular forms of video poker. Today there are dozens of variations on the original game. Unfortunately, according to Bob Dancer (personal communication, 2009), full-pay Deuces Wild "is down to perhaps 100 machines in Las Vegas—and five years ago there were more than 1,000."

Because of the complications of video poker, optimal strategies were not immediately forthcoming. Early attempts used computer simulation rather than exact computation. Some of the earliest authors include Wong (1988), Frome (1989), and Weber and Scruggs (1992).

For a group-theoretic approach to counting the 134,459 equivalence classes of pre-draw hands at Jacks or Better, see Alspach (2007).

The Deuces Wild pre-draw frequencies of Tables 17.7 and 17.8 are well known. However, Goren's *Hoyle* (1961, p. 110) has them wrong, giving for example 4,072 ways to be dealt a nonroyal straight flush instead of the correct 2,068. Russell (1983, pp. 6, 83) and Percy (1988, p. 16) have the same errors.

Exactly optimal strategies are available from only a few sources. Our Tables 17.5 and 17.12 are adapted from Dancer and Daily (2004, Chapter 6; 2003, Chapter 6). The asterisks in Table 17.12 hide a number of exceptions, for which the reader is referred to Table 17.17 below. Paymar (2004) has made the case that the mathematically optimal strategy is less efficient in practice than a simplified version of it that he called "precision play."

Marshall (2006) published the complete list of the 134,459 equivalence classes of hands in Deuces Wild. (He regarded deuces as having suits, unlike our approach.) This required some 357 pages, seven columns per page, 55 hands per column. For each equivalence class he provided the (more precisely, an) optimal play. For example, the equivalence class containing A♣-K♦-Q♦-J♡-9♦ is listed as KQ9·A·J with the cards to be held underlined. It appears that his strategy is the minimum-variance one.

As for exactly optimal strategies for Jacks or Better, we quote the following summary of the situation as of the mid-1990s (Gordon 2000):

> Paymar (1994, p. 9) claims that the 99.54 figure he reported is an absolute maximum. That claim is surprising since Frome's 99.6 had previously been published and Paymar was aware of it. No basis for that claim is given other than unidentified "independent analyses." Such upper bounds are difficult to establish in much simpler analyses and are impossible for a complicated analysis like this. Frome's counter-example is sufficient to render the claim invalid.

See (17.12) for the exact figure. Frome's number was simply inaccurate. Gordon (2000) went on to claim a 99.75 percent expected payout, which is even less accurate.

The exact optimal strategy for Jacks or Better was said by Paymar (1998, p. 45) to have been published, though he did not say where or by whom. As far as we know, the first exact evaluation of the payout distribution for Deuces Wild was obtained by Jensen (2001), and his numbers were used by Ethier and Khoshnevisan (2002). However, they differ slightly from those of Table 17.11 because he found the maximum-variance optimal strategy, apparently unaware of the nonuniqueness issue.

Most of the examples in Problems 17.3 and 17.4 are from Dancer and Daily (2004, 2003), as is Problem 17.5 (2003, Appendix A). Double Bonus video poker (Problem 17.13) and Joker Wild video poker (Problem 17.14) have been studied by the same authors.

Table 17.17 Exceptions to Table 17.12. From Dancer and Daily (2003, Appendices B and C); two minor mistakes have been corrected.

Exceptions to one-deuce rule 10: 3-RF: A-high, no fp or sp

---

hold 3-RF, even though it is penalized:

3-RF: AJT7 when T7 are unsuited with each other and with AJ
3-RF: $AJ\,T$7 when J7 are unsuited with each other and with AT

hold only the deuce, even though 3-RF is unpenalized:

3-RF: AK and 93, *83*, 73, 64, 63, 54, 53, or 43
3-RF: AQ and 63, *54*, 53, or 43
3-RF: AJ and *53* or 43
3-RF: AT and *43*

Exceptions to no-deuce rule 23: 2-RF: KQ, KJ, or KT, no fp or sp

---

hold 2-RF, even though it is penalized:

2-RF: KQ9 and 87 or 76
2-RF: KQ9 and 86 when nonKQ cards are of different suits or *96*
2-RF: KQ9 and 85 or 75 when nonKQ cards are of different suits
2-RF: AKJ and 87 when nonKJ cards are of different suits
2-RF: KJ9 and 86, 85, 76, or 75
2-RF: KJ9 and 74 or 65 when nonKJ cards are of different suits
2-RF: AKT and 87 when nonKT cards are not all of the same suit
2-RF: AKT and 86 or 76 when nonKT cards are of different suits
2-RF: KT9 and 85, 75, or 65
2-RF: KT9 and 84 or 74 when nonKT cards are not all of the same suit
2-RF: KT9 and 73 or 64 when nonKT cards are of different suits

hold nothing, even though 2-RF is unpenalized:

2-RF: KQ and 743 when *74* or *43*
2-RF: KQ and 653 when two of the nonKQ cards are suited
2-RF: KQ and 643 or 543
2-RF: KJ and 643 when *64* or *43*
2-RF: KJ and 543

Chapter 18
Faro

All these luxuries becoming my station could not, of course, be purchased without credit and money, to procure which, as our patrimony had been wasted by our ancestors, and we were above the vulgarity and slow returns and doubtful chances of trade, my uncle kept a faro bank.

William Makepeace Thackeray, *The Luck of Barry Lyndon*

Faro is a card game that originated in 17th-century France and reached its height of popularity in the 19th-century American West. Unfortunately, it is now extinct, having been last dealt in Nevada in the 1980s. Nevertheless, faro has several interesting features that make it worthy of our attention. First, it is the only casino game in which card counting is performed by the house for the benefit of the players. Second, there is no single house advantage associated with the principal bet at faro, the denomination bet. Third, faro has been studied by Montmort, three of the Bernoullis, De Moivre, Euler, and Thorp. In Section 18.1 we evaluate the house advantage(s) of the denomination bet. In Section 18.2 we analyze a complete deal comprising 25 turns.

18.1 The Denomination Bet

Faro is played with a standard 52-card deck. Suits are irrelevant, and only the 13 denominations, A, 2–10, J, Q, K, matter. It will be convenient to regard these denominations as 1, 2–10, 11, 12, 13, respectively. The shuffled deck is placed face-up in an open dealing box, exposing the top card, which is called the *soda*. Bets are placed prior to the first hand, or *turn*, and are then resolved as follows. The soda is removed, exposing the *losing card*. Then the losing card is removed, exposing the *winning card*. Bets are settled, and new bets are placed (or old ones are allowed to remain) prior to the second turn,

S.N. Ethier, *The Doctrine of Chances*, Probability and its Applications, DOI 10.1007/978-3-540-78783-9_18, © Springer-Verlag Berlin Heidelberg 2010

which is resolved in the same way, by exposing the next two cards. There are 25 turns in a deal, accounting for all but the last card, which is called the *hock*. An unusual feature of faro is that a casino employee keeps track of the cards played on an abacus-like device called the *case*, and this information is available for all to see.

Let us denote the 52 cards by random variables $X_1, X_2, \ldots, X_{52}$. Specifically, $(X_1, X_2, \ldots, X_{52})$ assumes each of the $52!/(4!)^{13}$ permutations of

$$(1, 1, 1, 1, 2, 2, 2, 2, \ldots, 13, 13, 13, 13) \tag{18.1}$$

with equal probability. Then X_1 is the soda, X_{2n} and X_{2n+1} are the losing and winning cards of the nth turn ($n = 1, 2, \ldots, 25$), and X_{52} is the hock.

The principal bet in faro is the *denomination bet*. A one-unit bet on denomination i to win at the nth turn provides the bettor with a profit of

$$Y_n^i := \begin{cases} 1 & \text{if } X_{2n} \neq i \text{ and } X_{2n+1} = i, \\ -1 & \text{if } X_{2n} = i \text{ and } X_{2n+1} \neq i, \\ -\frac{1}{2} & \text{if } X_{2n} = i \text{ and } X_{2n+1} = i, \\ 0 & \text{otherwise.} \end{cases} \tag{18.2}$$

In words, the bet is won (and paid even money) if the winning card is of denomination i and the losing card is not. The bet is lost if the losing card is of denomination i and the winning card is not. Half the bet is lost if both losing and winning cards are of denomination i. This is called a *split*, and it is the sole source of the casino's advantage. Finally, if neither losing nor winning card is of denomination i, the result is a push. Bets can be withdrawn or left in place following a push.

Similarly, a one-unit bet on denomination i to lose at the nth turn provides the bettor with a profit of

$$\bar{Y}_n^i := \begin{cases} 1 & \text{if } X_{2n} = i \text{ and } X_{2n+1} \neq i, \\ -1 & \text{if } X_{2n} \neq i \text{ and } X_{2n+1} = i, \\ -\frac{1}{2} & \text{if } X_{2n} = i \text{ and } X_{2n+1} = i, \\ 0 & \text{otherwise.} \end{cases} \tag{18.3}$$

There are no separate spaces on the layout for denomination bets to lose, so these are signified by *coppering* one's bet, that is, by placing a small six-sided black chip on top of it.

The house advantage of the denomination bet to win at the nth turn can be easily evaluated. (The house advantage of the denomination bet to lose at the nth turn is identical, so we can disregard it.) It is simply the conditional expectation of $-Y_n^i$, given that $Y_n^i \neq 0$ and given the $2n - 1$ cards seen prior to the nth turn. This is the analogue of (6.3) on p. 200, which in the present setting has the form

$$H(Y_n^i \mid X_1, \ldots, X_{2n-1}) := \frac{-E[Y_n^i \mid X_1, \ldots, X_{2n-1}]}{P(Y_n^i \neq 0 \mid X_1, \ldots, X_{2n-1})}. \tag{18.4}$$

Prior to the nth turn, there are

$$m := 52 - (2n - 1) \tag{18.5}$$

unseen cards in the deck. If $l \geq 1$ of these m cards are of denomination i, that is, if

$$l := |\{2n \leq k \leq 52 : X_k = i\}| \geq 1, \tag{18.6}$$

then, using an exchangeability argument to show that the probabilities of winning and losing are equal, we have

$$E[Y_n^i \mid X_1, \ldots, X_{2n-1}] = -\frac{1}{2} P(X_{2n} = X_{2n+1} = i \mid X_1, \ldots, X_{2n-1})$$

$$= -\frac{1}{2} \frac{(l)_2}{(m)_2} \tag{18.7}$$

and

$$P(Y_n^i \neq 0 \mid X_1, \ldots, X_{2n-1}) = 1 - P(X_{2n} \neq i, \ X_{2n+1} \neq i \mid X_1, \ldots, X_{2n-1})$$

$$= 1 - \frac{(m - l)_2}{(m)_2}. \tag{18.8}$$

It follows that

$$H(Y_n^i \mid X_1, \ldots, X_{2n-1}) = \frac{1}{2} \frac{(l)_2}{[(m)_2 - (m - l)_2]}$$

$$= \frac{l - 1}{2(2m - l - 1)}, \tag{18.9}$$

where m and l are given by (18.5) and (18.6). It is easy to check from (18.9) that the house advantage increases, if $l \geq 2$, with each turn at which the bet remains unresolved, and that it increases at a specified turn with the number of unseen cards of denomination i.

There are 25 possible values for m, corresponding to the 25 turns in a complete deal, and four possible values for l, with $l = 0$ excluded as irrelevant. Further, the pairs $(m, l) = (51, 1), (51, 2), (3, 4)$ cannot occur, so there are 97 instances (!) that must be considered in evaluating the house advantage of the denomination bet to win. When $l = 1$, a split is impossible and the house advantage is zero; such a bet is called a *case bet*. Excluding the 24 such instances as trivial, the remaining 73 instances are evaluated in Table 18.1.

But that is not all. There is another, equally valid (but with a slightly different interpretation), definition of the house advantage. Instead of considering the bettor's conditional expected loss at the nth turn, given that the bet is resolved at the nth turn and given the $2n - 1$ cards already seen, we could consider the bettor's conditional expected loss, assuming that the bet remains in place until a decision occurs. More precisely, we define

$$N_n^i := \min\{k \geq n : Y_k^i \neq 0\} \tag{18.10}$$

and

$$H'(Y_n^i \mid X_1, \ldots, X_{2n-1}) := -E[Y_{N_n^i}^i \mid X_1, \ldots, X_{2n-1}]. \tag{18.11}$$

Again, we define m and l by (18.5) and (18.6). Clearly, (18.11) is 0 if $l = 1$, so we assume that $l \geq 2$. Then, letting

$$t := \left\lfloor \frac{m-l}{2} \right\rfloor + 1 \tag{18.12}$$

be the number of turns needed to guarantee a decision, we have

$$H'(Y_n^i \mid X_1, \ldots, X_{2n-1})$$
$$= \frac{1}{2} \sum_{k=n}^{n+t-1} P(X_j \neq i \text{ if } 2n \leq j < 2k, \ X_{2k} = X_{2k+1} = i \mid X_1, \ldots, X_{2n-1})$$
$$= \frac{1}{2} \frac{\binom{2}{2}\binom{m-2}{l-2} + \binom{2}{0}\binom{2}{2}\binom{m-4}{l-2} + \binom{4}{0}\binom{2}{2}\binom{m-6}{l-2} + \cdots + \binom{2(t-1)}{0}\binom{2}{2}\binom{m-2t}{l-2}}{\binom{m}{l}}$$
$$= \frac{1}{2} \frac{\binom{m-2}{l-2} + \binom{m-4}{l-2} + \binom{m-6}{l-2} + \cdots + \binom{m-2t}{l-2}}{\binom{m}{l}}$$
$$= \begin{cases} 1/(2m) & \text{if } l = 2, \\ 3(m-1)/[4m(m-2)] & \text{if } l = 3, \\ (2m-1)/[2m(m-2)] & \text{if } l = 4, \end{cases} \tag{18.13}$$

where the second equality can be understood by choosing the l positions of the denomination-i cards among the m positions that remain, much as in the paragraph above (14.3) on p. 484; and the last step uses $t = (m-1)/2$ if $l = 2$ or $l = 3$, and $t = (m-3)/2$ if $l = 4$ (see Problem 18.4 on p. 586). Again, it is easy to check from (18.13) that the house advantage increases with the number of the turn at which the bet is first made (assuming that the number of unseen cards of denomination i remains unchanged), and increases at a specified turn with the number of unseen cards of denomination i.

In Table 18.1 we list the values of (18.4) and (18.11) for all meaningful values of n, m, and l (except $l = 1$, in which case the house advantage is zero).

What is the distinction between the two versions of the house advantage? The latter version (18.11) effectively assumes that the player leaves his bet in place until it is resolved, while the former version (18.4) does not. The latter is a weighted average of the present and future values of the former. Thus, we believe that (18.4) is the more useful statistic.

Table 18.1 House advantages for the denomination bet at faro. The four blank entries correspond to cases that cannot occur.

turn	unseen cards	house advantage (18.4)			house advantage (18.11)		
		unseen cards of specified denomination					
		4	3	2	4	3	2
1	51	.015 464	.010 204		.020 202	.015 006	
2	49	.016 129	.010 638	.005 263	.021 052	.015 632	.010 204
3	47	.016 854	.011 111	.005 495	.021 977	.016 312	.010 638
4	45	.017 647	.011 628	.005 747	.022 987	.017 054	.011 111
5	43	.018 519	.012 195	.006 024	.024 094	.017 867	.011 628
6	41	.019 481	.012 821	.006 329	.025 314	.018 762	.012 195
7	39	.020 548	.013 514	.006 667	.026 662	.019 751	.012 821
8	37	.021 739	.014 286	.007 042	.028 163	.020 849	.013 514
9	35	.023 077	.015 152	.007 463	.029 841	.022 078	.014 286
10	33	.024 590	.016 129	.007 937	.031 733	.023 460	.015 152
11	31	.026 316	.017 241	.008 475	.033 879	.025 028	.016 129
12	29	.028 302	.018 519	.009 091	.036 335	.026 820	.017 241
13	27	.030 612	.020 000	.009 804	.039 174	.028 889	.018 519
14	25	.033 333	.021 739	.010 638	.042 490	.031 304	.020 000
15	23	.036 585	.023 810	.011 628	.046 414	.034 161	.021 739
16	21	.040 541	.026 316	.012 821	.051 128	.037 594	.023 810
17	19	.045 455	.029 412	.014 286	.056 889	.041 796	.026 316
18	17	.051 724	.033 333	.016 129	.064 076	.047 059	.029 412
19	15	.060 000	.038 462	.018 519	.073 260	.053 846	.033 333
20	13	.071 429	.045 455	.021 739	.085 315	.062 937	.038 462
21	11	.088 235	.055 556	.026 316	.101 515	.075 758	.045 455
22	9	.115 385	.071 429	.033 333	.123 016	.095 238	.055 556
23	7	.166 667	.100 000	.045 455	.185 714	.128 571	.071 429
24	5	.300 000	.166 667	.071 429	.300 000	.200 000	.100 000
25	3		.500 000	.166 667		.500 000	.166 667

The action reaches a climax at the 25th and final turn. The denominations of the last three cards are known, but their order is not. If the three denominations are distinct, a bet on a denomination to win or to lose is of course a fair bet since a split is impossible. If two of the last three cards are of the same denomination, a split of that denomination occurs with probability 1/3, whereas a split of the odd denomination is impossible. Finally, in the rare situation (probability 1/425) in which the last three cards are of the same denomination, a split is certain, so a denomination bet is inadvisable. (In fact, according to one author, denomination bets to win or to lose are replaced by bets on the card of the odd color. Because this rule is rather obscure, and because it complicates (18.13), we disregard it here; but see Problem 18.7 on p. 588.) However, in each instance there is another bet available, known as *calling the turn*, which we do not consider.

18.2 From Soda to Hock

Recall that a faro deal begins with the soda, ends with the hock, and has 25 turns in between. We begin by examining the random variable S defined to be the number of splits in these 25 turns. More precisely, define the events $A_1, A_2, \ldots, A_{25}$ by letting A_n be the event that a split occurs at the nth turn. Then

$$S := \sum_{n=1}^{25} 1_{A_n}. \tag{18.14}$$

It follows from exchangeability that $P(A_1) = \cdots = P(A_{25})$, hence

$$E[S] = \sum_{n=1}^{25} P(A_n) = 25\,P(A_1) = 25 \cdot \frac{13(4)_2}{(52)_2} = \frac{25}{17} \approx 1.470588. \tag{18.15}$$

Therefore, splits occur fewer than three times in two deals on average.

A similar argument, using $P(A_m \cap A_n) = P(A_1 \cap A_2)$ for $1 \le m < n \le 25$, yields the variance:

$$
\begin{aligned}
\mathrm{Var}(S) &= \sum_{n=1}^{25} \mathrm{Var}(1_{A_n}) + 2 \sum\sum_{1 \le m < n \le 25} \mathrm{Cov}(1_{A_m}, 1_{A_n}) \\
&= 25\,P(A_1)(1 - P(A_1)) + 2\binom{25}{2}[P(A_1 \cap A_2) - P(A_1)^2] \\
&= 25 \cdot \frac{13(4)_2}{(52)_2}\left(1 - \frac{13(4)_2}{(52)_2}\right) \\
&\quad + 25 \cdot 24\left[\frac{13(4)_2}{(52)_2} \cdot \frac{12(4)_2}{(50)_2} + \frac{13(4)_4}{(52)_4} - \left(\frac{13(4)_2}{(52)_2}\right)^2\right] \\
&= \frac{19{,}984}{14{,}161} \approx 1.411200.
\end{aligned}
\tag{18.16}
$$

Note that the covariance terms are positive, albeit quite small.

More generally, we can find the distribution of S with the help of Example 1.4.24 on p. 39. For $n = 1, 2, \ldots, 25$ and $k = 0, 1, \ldots, \lfloor n/2 \rfloor$, let $P_n(k)$ be the probability of n splits in the first n turns, with the splits at turns $2, 4, \ldots, 2k$, and at only those turns, exhausting a denomination. More explicitly,

$$P_n(0) = P(X_2 = X_3, \ X_4 = X_5, \ \ldots, \ X_{2n} = X_{2n+1},$$
$$\text{with } X_2, X_4, \ldots, X_{2n} \text{ distinct}),$$
$$P_n(1) = P(X_2 = X_3 = X_4 = X_5, \ X_6 = X_7, \ \ldots, \ X_{2n} = X_{2n+1},$$
$$\text{with } X_6, X_8 \ldots, X_{2n} \text{ distinct}), \tag{18.17}$$
$$P_n(2) = P(X_2 = X_3 = X_4 = X_5, \ X_6 = X_7 = X_8 = X_9,$$
$$X_{10} = X_{11}, \ \ldots, \ X_{2n} = X_{2n+1},$$
$$\text{with } X_{10}, X_{12} \ldots, X_{2n} \text{ distinct}),$$

and so on. Then

$$P_n(0) = \frac{13(4)_2}{(52)_2} \cdot \frac{12(4)_2}{(50)_2} \cdots \frac{[13-(n-1)](4)_2}{(52-2(n-1))_2},$$
$$P_n(1) = \frac{13(4)_4}{(52)_4} \cdot \frac{12(4)_2}{(48)_2} \cdot \frac{11(4)_2}{(46)_2} \cdots \frac{[13-(n-2)](4)_2}{(52-2(n-1))_2}, \tag{18.18}$$
$$P_n(2) = \frac{13(4)_4}{(52)_4} \cdot \frac{12(4)_4}{(48)_4} \cdot \frac{11(4)_2}{(44)_2} \cdot \frac{10(4)_2}{(42)_2} \cdots \frac{[13-(n-3)](4)_2}{(52-2(n-1))_2},$$

and so on. The relevance of these probabilities is that we can express $P(A_1)$, $P(A_1 \cap A_2)$, $P(A_1 \cap A_2 \cap A_3)$, $\ldots$, in terms of them. Indeed,

$$P(A_1) = P_1(0),$$
$$P(A_1 \cap A_2) = P_2(0) + \binom{2}{2} P_2(1),$$
$$P(A_1 \cap A_2 \cap A_3) = P_3(0) + \binom{3}{2} P_3(1), \tag{18.19}$$
$$P(A_1 \cap A_2 \cap A_3 \cap A_4) = P_4(0) + \binom{4}{2} P_4(1) + \frac{1}{2!}\binom{4}{2} P_4(2),$$

and, more generally,

$$P(A_1 \cap A_2 \cap \cdots \cap A_n) = P_n(0) + \binom{n}{2} P_n(1) + \frac{1}{2!}\binom{n}{2,2,n-4} P_n(2)$$
$$+ \frac{1}{3!}\binom{n}{2,2,2,n-6} P_n(3) + \cdots \tag{18.20}$$
$$+ \frac{1}{\lfloor n/2 \rfloor!}\binom{n}{2,\ldots,2,n-2\lfloor n/2 \rfloor} P_n(\lfloor n/2 \rfloor)$$

for $1 \le n \le 25$.

Now

$$P(S = j) = \sum_{n=j}^{25} (-1)^{n-j} \binom{n}{j}\binom{25}{n} P_n, \qquad 0 \le j \le 25, \tag{18.21}$$

where $P_0 = 1$ and, for $1 \le n \le 25$, $P_n := \mathrm{P}(A_1 \cap A_2 \cap \cdots \cap A_n)$. Table 18.2 contains an evaluation of the distribution of S. For example, the probability of no splits in a faro deal is slightly greater than $2/9$.

Table 18.2 The distribution of the number of splits in a complete deal consisting of 25 turns. Results are rounded to six significant digits.

splits	probability	splits	probability
0	.223 067	13	$.517\,032 \cdot 10^{-9}$
1	.341 279	14	$.313\,939 \cdot 10^{-10}$
2	.254 469	15	$.166\,286 \cdot 10^{-11}$
3	.123 019	16	$.772\,484 \cdot 10^{-13}$
4	.043 268 7	17	$.316\,619 \cdot 10^{-14}$
5	.011 777 0	18	$.115\,196 \cdot 10^{-15}$
6	.002 575 79	19	$.374\,150 \cdot 10^{-17}$
7	.000 463 977	20	$.109\,001 \cdot 10^{-18}$
8	.000 069 994 1	21	$.285\,786 \cdot 10^{-20}$
9	.000 008 945 62	22	$.674\,315 \cdot 10^{-22}$
10	.000 000 976 347	23	$.143\,799 \cdot 10^{-23}$
11	.000 000 091 506 5	24	$.256\,784 \cdot 10^{-25}$
12	.000 000 007 394 69	25	$.534\,967 \cdot 10^{-27}$

Is there a betting strategy that is "optimal" in some sense? Let us assume that the player makes a one-unit denomination bet to win at every turn. Clearly he should bet on a denomination (not necessarily unique) that has the smallest positive number of representatives in the unseen deck. For example, he should bet on the soda at the first turn, and at later turns he should make a case bet if at least one such bet is available. This defines a system and there are a number of questions we could ask about it. For example, what is the player's expected loss over the course of the deal?

We can answer such questions by introducing a Markov chain $\{\boldsymbol{D}_n\}_{n \ge 0}$ (cf. Example 4.1.5 on p. 124) in the state space

$$\Sigma := \{\boldsymbol{d} = (d_0, d_1, d_2, d_3, d_4) \in \mathbf{Z}_+^5 : d_0 + d_1 + d_2 + d_3 + d_4 = 13\}, \quad (18.22)$$

which has $\binom{13+5-1}{5-1} = 2{,}380$ states (Theorem 1.1.8 on p. 6). Specifically, $\boldsymbol{D}_n = \boldsymbol{d}$ if, after n cards have been dealt $(n = 0, 1, \ldots, 52)$, the unseen deck of size $52 - n$ has d_i denominations with i representatives $(i = 0, 1, 2, 3, 4)$. In particular, the initial state is $\boldsymbol{D}_0 = (0, 0, 0, 0, 13)$, and the one-step transition matrix $\boldsymbol{P}$ is given by

$$P(\boldsymbol{d}, \boldsymbol{d} - \boldsymbol{e}_j + \boldsymbol{e}_{j-1}) := \frac{j d_j}{\mu(\boldsymbol{d})}, \qquad j = 1, 2, 3, 4, \qquad (18.23)$$

for all $d \in \Sigma - \{(13, 0, 0, 0, 0)\}$, where $e_0 := (1, 0, 0, 0, 0)$, $e_1 := (0, 1, 0, 0, 0)$, and so on, and the function $\mu : \Sigma \mapsto \{0, 1, \ldots, 52\}$, defined by

$$\mu(d) := d_1 + 2d_2 + 3d_3 + 4d_4, \qquad (18.24)$$

is the number of cards remaining. Finally, state $(13, 0, 0, 0, 0)$ is absorbing. Let us define

$$\Sigma_n := \{d \in \Sigma : \mu(d) = 52 - n\}, \qquad n = 0, 1, \ldots, 52. \qquad (18.25)$$

These sets partition Σ. Notice that $P(D_n \in \Sigma_{n \wedge 52}) = 1$ for all $n \geq 0$.

It is straightforward to recursively compute the distribution of D_n for $n = 1, 2, \ldots, 52$. Indeed,

$$P(D_n = d) = \sum_{0 \leq j \leq 3:\, d_j \geq 1} \frac{(j+1)(d_{j+1}+1)}{\mu(d)+1} P(D_{n-1} = d - e_j + e_{j+1})$$

$$(18.26)$$

for all $d \in \Sigma_n$ and $n = 1, 2, \ldots, 52$. For example,

$$P(D_0 = (0, 0, 0, 0, 13)) = 1, \qquad (18.27)$$

$$P(D_1 = (0, 0, 0, 1, 12)) = 1, \qquad (18.28)$$

$$P(D_2 = d) = \begin{cases} 3/51 & \text{if } d = (0, 0, 1, 0, 12), \\ 48/51 & \text{if } d = (0, 0, 0, 2, 11), \end{cases} \qquad (18.29)$$

and

$$P(D_3 = d) = \begin{cases} (3)_2/(51)_2 & \text{if } d = (0, 1, 0, 0, 12), \\ (3 \cdot 48 + 48 \cdot 6)/(51)_2 & \text{if } d = (0, 0, 1, 1, 11), \\ (48 \cdot 44)/(51)_2 & \text{if } d = (0, 0, 0, 3, 10). \end{cases} \qquad (18.30)$$

Of course, a computer is needed to complete the list.

Of interest in faro is the induced Markov chain

$$\{D_{2n-1}\}_{n=1,2,\ldots,25} \qquad (18.31)$$

corresponding to the unseen-deck compositions prior to each of the 25 turns. Its initial distribution is (18.28) and its one-step transition matrix is P^2, the two-step transition matrix for the original chain. We define the function $m : \Sigma - \{(13, 0, 0, 0, 0)\} \mapsto \{1, 2, 3, 4\}$ by

$$m(d) := \min\{j \geq 1 : d_j \geq 1\}. \qquad (18.32)$$

(See Table 18.3.) Recall that we are assuming that at each turn the player bets one unit on a denomination (not necessarily unique) that has the smallest positive number of representatives in the unseen deck. At the nth turn, this

Table 18.3 An incomplete list of the elements of $\cup_{n=1}^{25} \Sigma_{2n-1}$, and the function m.

turn n	list of all $d \in \Sigma_{2n-1}$	$m(d)$	turn n	list of all $d \in \Sigma_{2n-1}$	$m(d)$
1	$(0,0,0,1,12)$	3			
2	$(0,1,0,0,12)$	1	24	$(8,5,0,0,0)$	1
	$(0,0,1,1,11)$	2		$(9,3,1,0,0)$	1
	$(0,0,0,3,10)$	3		$(10,1,2,0,0)$	1
				$(10,2,0,1,0)$	1
3	$(1,0,0,1,11)$	3		$(11,0,1,1,0)$	2
	$(0,1,1,0,11)$	1		$(11,1,0,0,1)$	1
	$(0,1,0,2,10)$	1			
	$(0,0,2,1,10)$	2	25	$(10,3,0,0,0)$	1
	$(0,0,1,3,9)$	2		$(11,1,1,0,0)$	1
	$(0,0,0,5,8)$	3		$(12,0,0,1,0)$	3

number is $m(\boldsymbol{D}_{2n-1})$. Notice that, if $m(\boldsymbol{d}) = 4$, then $\mu(\boldsymbol{d})$ is a multiple of 4, hence even, but $\mu(\boldsymbol{D}_{2n-1}) = 53 - 2n$ is odd. Therefore $m(\boldsymbol{D}_{2n-1})$ is never equal to 4. We also define

$$N_j := \sum_{n=1}^{25} 1_{\{m(\boldsymbol{D}_{2n-1})=j\}}, \qquad j = 1, 2, 3, \tag{18.33}$$

so that $N_1 + N_2 + N_3 = 25$. In words, N_1 is the number of turns at which a case bet is available (hence made), N_2 is the number of turns at which a bet is made on a denomination with two representatives unseen, and N_3 is the number of turns (including the first turn) at which a bet is made on a denomination with three representatives unseen. Using

$$\mathrm{E}[N_j] = \sum_{n=1}^{25} \mathrm{P}(m(\boldsymbol{D}_{2n-1}) = j)$$

$$= \sum_{n=1}^{25} \sum_{\boldsymbol{d} \in \Sigma_{2n-1}: \, m(\boldsymbol{d})=j} \mathrm{P}(\boldsymbol{D}_{2n-1} = \boldsymbol{d}), \qquad j = 1, 2, 3, \tag{18.34}$$

computations show that

$$\mathrm{E}[N_1] \approx 18.164238, \quad \mathrm{E}[N_2] \approx 4.222628, \quad \mathrm{E}[N_3] \approx 2.613134. \tag{18.35}$$

It is perhaps surprising that the expected number of turns at which a case bet is available is as large as 18.

We can now evaluate the player's expected loss over the course of the 25 turns, assuming the betting strategy described above. Let L_n be the player's loss at the nth turn, so that $L := L_1 + L_2 + \cdots + L_{25}$ is his total loss. Then

$$
\begin{aligned}
\mathrm{E}[L] &= \sum_{n=1}^{25} \mathrm{E}[L_n] \\
&= \sum_{n=1}^{25} \sum_{\boldsymbol{d} \in \Sigma_{2n-1}} \mathrm{P}(\boldsymbol{D}_{2n-1} = \boldsymbol{d}) \, \mathrm{E}[L_n \mid \boldsymbol{D}_{2n-1} = \boldsymbol{d}] \\
&= \sum_{n=1}^{25} \sum_{\boldsymbol{d} \in \Sigma_{2n-1}} \mathrm{P}(\boldsymbol{D}_{2n-1} = \boldsymbol{d}) \, \frac{1}{2} \frac{(m(\boldsymbol{d}))_2}{(53 - 2n)_2} \\
&\approx 0.007281.
\end{aligned}
\tag{18.36}
$$

Thus, the player's expected loss *per deal* is less than 3/4 of 1 percent of the amount bet *per turn*.

We can find the house advantage for this betting system by dividing the expected loss by the expected total amount of action. Let B_n be the indicator of the event that the bet at the nth turn is resolved at that turn, so that $B := B_1 + B_2 + \cdots + B_{25}$ is the total amount of action. Then

$$
\begin{aligned}
\mathrm{E}[B] &= \sum_{n=1}^{25} \mathrm{E}[B_n] \\
&= \sum_{n=1}^{25} \sum_{\boldsymbol{d} \in \Sigma_{2n-1}} \mathrm{P}(\boldsymbol{D}_{2n-1} = \boldsymbol{d}) \, \mathrm{E}[B_n \mid \boldsymbol{D}_{2n-1} = \boldsymbol{d}] \\
&= \sum_{n=1}^{25} \sum_{\boldsymbol{d} \in \Sigma_{2n-1}} \mathrm{P}(\boldsymbol{D}_{2n-1} = \boldsymbol{d}) \left(1 - \frac{(53 - 2n - m(\boldsymbol{d}))_2}{(53 - 2n)_2} \right) \\
&\approx 3.631584.
\end{aligned}
\tag{18.37}
$$

Of the approximately 3.631584 resolved bets expected, about 2.905763 are from case bets on average. We conclude finally that the house advantage for the player adopting our proposed system is $\mathrm{H} := \mathrm{E}[L]/\mathrm{E}[B] \approx 0.002005$, just over 1/5 of 1 percent.

Finally, the last topic we take up is a bit more complicated than what we have done until now. Specifically, we evaluate the distribution of N_1, the number of turns at which a case bet is available. Unfortunately, there is no formula available such as (18.21). Instead, we need to examine the sample paths of the Markov chain $\{\boldsymbol{D}_{2n-1}, \, n = 1, 2, \ldots, 25\}$. Since there are at most $\binom{4}{1} + \binom{4}{2} = 10$ possible one-step transitions from each state, there are at most $3 \cdot 6 \cdot 10^{20} \cdot 6 \cdot 3 \cdot 1 = 3.24 \cdot 10^{22}$ sample paths. This is a crude upper bound, but it suggests that even the actual number is beyond the capability of computers. However, by using a trick, we will see that it suffices to examine only the first 13 turns, $\{\boldsymbol{D}_{2n-1},\ n = 1, 2, \ldots, 13\}$, in which case the bound on the number of sample paths is replaced by the square root of the previous bound, $1.8 \cdot 10^{11}$, which is within the capability of computers. The trick is based on the observation that turns 14–25 are in a certain sense probabilistically identical to turns 2–13. This is formalized in Problem 18.13 on p. 589, which implies that

$$(\boldsymbol{D}_1, \boldsymbol{D}_3, \ldots, \boldsymbol{D}_{51}) \stackrel{d}{=} (\widehat{\boldsymbol{D}}_{51}, \widehat{\boldsymbol{D}}_{49}, \ldots, \widehat{\boldsymbol{D}}_1), \tag{18.38}$$

where $\widehat{\boldsymbol{d}} := (d_4, d_3, d_2, d_1, d_0)$ if $\boldsymbol{d} = (d_0, d_1, d_2, d_3, d_4)$.

We define $h_j(\boldsymbol{d}) := 1_{\{d_j \geq 1\}}$ for $j = 1$ and $j = 3$ and

$$M_j := \sum_{n=2}^{13} h_j(\boldsymbol{D}_{2n-1}) + \sum_{n=14}^{25} h_j(\boldsymbol{D}_{2n-1}) =: M_j^* + M_j^{**}. \tag{18.39}$$

Then $N_1 = M_1 = M_1^* + M_1^{**}$, so it will suffice to find the joint distribution of M_1^* and M_1^{**}. Now

$$
\begin{aligned}
&\mathrm{P}(M_1^* = m,\ M_1^{**} = n) \\
&= \sum_{\boldsymbol{d} \in \Sigma_{25}} \sum_{\boldsymbol{d}' \in \Sigma_{27}} \mathrm{P}(M_1^* = m,\ \boldsymbol{D}_{25} = \boldsymbol{d},\ M_1^{**} = n,\ \boldsymbol{D}_{27} = \boldsymbol{d}') \\
&= \sum_{\boldsymbol{d} \in \Sigma_{25}} \sum_{\boldsymbol{d}' \in \Sigma_{27}} \mathrm{P}(M_1^* = m,\ \boldsymbol{D}_{25} = \boldsymbol{d})\mathrm{P}(\boldsymbol{D}_{27} = \boldsymbol{d}' \mid M_1^* = m,\ \boldsymbol{D}_{25} = \boldsymbol{d}) \\
&\qquad\qquad\qquad \cdot \mathrm{P}(M_1^{**} = n \mid M_1^* = m,\ \boldsymbol{D}_{25} = \boldsymbol{d},\ \boldsymbol{D}_{27} = \boldsymbol{d}') \\
&= \sum_{\boldsymbol{d} \in \Sigma_{25}} \sum_{\boldsymbol{d}' \in \Sigma_{27}} \frac{\mathrm{P}(M_1^* = m,\ \boldsymbol{D}_{25} = \boldsymbol{d})P^2(\boldsymbol{d}, \boldsymbol{d}')\mathrm{P}(M_1^{**} = n,\ \boldsymbol{D}_{27} = \boldsymbol{d}')}{\mathrm{P}(\boldsymbol{D}_{27} = \boldsymbol{d}')} \\
&= \sum_{\boldsymbol{d} \in \Sigma_{25}} \sum_{\boldsymbol{d}' \in \Sigma_{27}} F_m(\boldsymbol{d})P^2(\boldsymbol{d}, \boldsymbol{d}')G_n(\widehat{\boldsymbol{d}'})/\mathrm{P}(\boldsymbol{D}_{27} = \boldsymbol{d}') \\
&= \sum_{\boldsymbol{d}, \boldsymbol{d}' \in \Sigma_{25}} F_m(\boldsymbol{d})P^2(\boldsymbol{d}, \widehat{\boldsymbol{d}'}) \frac{G_n(\boldsymbol{d}')}{G_0(\boldsymbol{d}') + G_1(\boldsymbol{d}') + \cdots + G_{12}(\boldsymbol{d}')}, \tag{18.40}
\end{aligned}
$$

where $F_m(\boldsymbol{d}) := \mathrm{P}(M_1^* = m,\ \boldsymbol{D}_{25} = \boldsymbol{d})$ and $G_m(\boldsymbol{d}) := \mathrm{P}(M_3^* = m,\ \boldsymbol{D}_{25} = \boldsymbol{d})$. To elaborate on the next-to-last equality in (18.40),

$$
\begin{aligned}
\mathrm{P}(M_1^{**} = n,\ \boldsymbol{D}_{27} = \boldsymbol{d}') &= \mathrm{P}\left(\sum_{n=14}^{25} h_1(\boldsymbol{D}_{2n-1}) = n,\ \boldsymbol{D}_{27} = \boldsymbol{d}' \right) \\
&= \mathrm{P}\left(\sum_{n=2}^{13} h_1(\widehat{\boldsymbol{D}}_{2n-1}) = n,\ \widehat{\boldsymbol{D}}_{25} = \boldsymbol{d}' \right)
\end{aligned}
$$

$$= \mathrm{P}\left(\sum_{n=2}^{13} h_3(\boldsymbol{D}_{2n-1}) = n, \ \boldsymbol{D}_{25} = \widehat{\boldsymbol{d}'} \right)$$

$$= \mathrm{P}(M_3^* = n, \ \boldsymbol{D}_{25} = \widehat{\boldsymbol{d}'})$$

$$= G_n(\widehat{\boldsymbol{d}'}). \tag{18.41}$$

It remains to evaluate $F_m(\boldsymbol{d})$ and $G_m(\boldsymbol{d})$ for $m = 0, 1, \ldots, 12$ and $\boldsymbol{d} \in \Sigma_{25}$ (note that $|\Sigma_{25}| = 102$), which can be done be examining each sample path of $\{\boldsymbol{D}_{2n-1}, \ n = 1, 2, \ldots, 13\}$. Results are summarized in Table 18.4. We find, for example, that the median of N_1 is 18, as is the mode (the most likely value). We have seen that the mean is about 18.164238, and the variance can now be calculated to be about 4.438598.

Table 18.4 The distribution of the number of turns at which a case bet is available. Results are rounded to six significant digits.

n	$\mathrm{P}(N_1 = n)$	n	$\mathrm{P}(N_1 = n)$
0	.000 000 000 000 000 019 255 5	12	.001 105 43
1	.000 000 000 000 000 719 080	13	.007 167 09
2	.000 000 000 000 014 749 5	14	.027 503 9
3	.000 000 000 000 224 302	15	.067 974 7
4	.000 000 000 002 893 57	16	.119 980
5	.000 000 000 034 505 7	17	.163 023
6	.000 000 000 405 185	18	.179 199
7	.000 000 004 890 33	19	.164 259
8	.000 000 061 628 3	20	.127 207
9	.000 000 797 887	21	.082 583 1
10	.000 010 133 8	22	.042 914 5
11	.000 117 236	23	.015 271 8
		24	.001 683 18

There have been a number of explanations put forward for faro's demise, including changing tastes, the anti-gambling laws of the late 19th and early 20th centuries, and the widely acknowledged fact that many faro games were dishonest. But the results of this section, especially (18.36), yield another explanation: Faro is simply too close to being a fair game, with the result that it is not sufficiently profitable for the casino.

18.3 Problems

18.1. *High-card bet.* The player may bet on high card to win (resp., lose), which is an even-money bet that the higher of the two cards at the next turn is the winning (resp., losing) card. (Recall that A, 2–10, J, Q, K are regarded as 1, 2–10, 11, 12, 13.) In the event of a split, the player loses half his bet.[1]

(a) Find a formula for the house advantage (pushes excluded) of the high-card bet. It will depend on $(l_1, l_2, \ldots, l_{13})$, where l_i is the number of cards remaining of denomination i.

(b) Evaluate the formula in the case of a high-card bet made at the first turn. Does it depend on the soda?

(c) Show that the house advantage lies in the interval $[0, \frac{1}{2}]$ and that both endpoints are achieved.

18.2. *Odd and even bets.* A bet on odd (resp., even) pays even money if the winning card is odd (resp., even) and the losing card is not. The bet is lost if the losing card is even (resp., odd) and the winning card is not. (Recall that A, 2–10, J, Q, K are regarded as 1, 2–10, 11, 12, 13. Note that there are more odd cards than even ones initially.) If both winning and losing cards are even or both are odd, and if their denominations are different, the result is a push. In the event of a split, the player loses half his bet.

(a) Find a formula for the house advantage (pushes excluded) of the bet on even. It will depend on $(l_1, l_2, \ldots, l_{13})$, where l_i is the number of cards remaining of denomination i.

(b) Evaluate the formula in the case of a bet on even made at the first turn when the soda is even; … when the soda is odd.

(c) Show that the house advantage lies in the interval $[0, \frac{1}{2}]$ and that both endpoints are achieved.

18.3. *Commission on winning case bets.* Some casinos charged a 5 percent commission on winning case bets, that is, such bets are paid 19 to 20. Extend Table 18.1 on p. 577 to include case bets under this assumption, obtaining the analogues of both (18.4) on p. 575 and (18.11) on p. 576. (In the case of the latter, include pushes; a push occurs if the specified case card is the hock.)

18.4. *House advantage algebra.* Verify the last equality in (18.13) on p. 576.

18.5. *House advantage at 18th-century faro.*

(a) Rederive (18.13) on p. 576 under the rules of faro prevalent in the 18th century (see Section 18.4). More precisely, let $m := 54 - 2n$ be the number of cards that remain at the nth turn, and show that

$$\mathrm{H}'(Y_n^i \mid X_1, \ldots, X_{2n-2})$$

---

[1] In some casinos a split was a push for the high-card bettor, in which case the bet was fair.

$$= \begin{cases} 1/m & \text{if } l = 1, \\ (m+2)/[2m(m-1)] & \text{if } l = 2, \\ 3/[4(m-1)] & \text{if } l = 3, \\ (2m-5)/[2(m-1)(m-3)] & \text{if } l = 4, \end{cases} \tag{18.42}$$

where the notation is analogous to that in (18.13) on p. 576. (As in Problem 18.3, include pushes; a push occurs if $l = 1$ and the case card is the last card.)

(b) Suppose that 12 cards remain, including all four aces. Find the house advantage of a bet on the ace denomination. Do the same if eight cards remain, including all four aces. These scenarios occurred in a game dealt by Casanova (see the chapter notes).

18.6. *Formulas of Montmort, Nicolaus Bernoulli, and Euler.*

(a) Define, for positive integers n and nonnegative integers r,

$$A(n,r) := \sum_{j=0}^{\lfloor (n-1)/2 \rfloor} \binom{n+r-1-2j}{r}. \tag{18.43}$$

Using Problem 1.3 on p. 52 in the first equation and grouping adjacent terms together in the second, show that

$$\sum_{j=1}^{n} \binom{n+r-j}{r} = \binom{n+r}{r+1}, \tag{18.44}$$

$$\sum_{j=1}^{n} (-1)^{j-1} \binom{n+r-j}{r} = A(n, r-1), \tag{18.45}$$

and therefore that

$$A(n,r) = \frac{1}{2} \left\{ \binom{n+r}{r+1} + A(n, r-1) \right\} \tag{18.46}$$

and

$$A(n,r) = \frac{1}{2} \left\{ \binom{n+r+1}{r+1} - A(n+1, r-1) \right\}. \tag{18.47}$$

Use these identities to show respectively that

$$A(n,r) = \sum_{j=1}^{r} 2^{-j} \binom{n+r+1-j}{r+2-j}$$
$$+ 2^{-(r+1)} \{ n + \tfrac{1}{2}[1 - (-1)^n] \} \tag{18.48}$$

and

$$A(n,r) = \sum_{j=1}^{r}(-1)^{j-1}2^{-j}\binom{n+r+1}{r+2-j}$$

$$+ (-1)^r 2^{-(r+1)}\{n+r+\tfrac{1}{2}[1-(-1)^{n+r}]\}. \qquad (18.49)$$

(b) Starting with (18.53) on p. 594, use (18.48) to prove Montmort's formula (18.55) on p. 594.

(c) Starting with (18.53) on p. 594, use (18.49) to prove Nicolaus Bernoulli's formula (18.56) on p. 594.

(d) Assume that m is even and deduce Euler's formula (18.57) on p. 594 from Nicolaus Bernoulli's formula (18.56) on p. 594, using the partial-fraction expansion

$$\frac{1}{(x+1)(x+2)\cdots(x+j)} = \frac{A_1}{x+1} + \frac{A_2}{x+2} + \cdots + \frac{A_j}{x+j}. \qquad (18.50)$$

18.7. *Rule variation for the last turn.* According to one author, in the rare case in which the last three cards in the deck are of the same denomination, denomination bets (to win or to lose) at the 25th turn are replaced by bets on the card of the odd color. How does this rule affect (18.13) on p. 576? (Include pushes; a push occurs if the odd color is the hock.) Specifically, how does it affect the house advantage (18.13) of a bet on the soda at the first turn?

18.8. *Faro and the fundamental theorem of card counting.*

(a) Use the fundamental theorem of card counting (Theorem 11.3.4 on p. 414) to show that the player's conditional expected profit from a one-unit bet on denomination i at turn n has expectation constant in n and variance increasing in n. Establish the same conclusions directly.

(b) Now consider the player who bets according to the strategy suggested in Section 18.2, that is, he bets on a denomination (not necessarily unique) that has the smallest positive number of representatives in the unseen deck. Evaluate the player's expected profit from a one-unit bet at turn n $(n = 1, 2, \ldots, 25)$. Is this sequence nondecreasing? If not, explain why Theorem 11.3.7 on p. 420 does not apply.

18.9. *Two simultaneous denomination bets to win.* Consider two simultaneous denomination bets to win, on denominations $i \neq j$, at the nth turn. Noting that only one of these bets can be lost at a single turn, the pair of bets is ordinarily regarded as a single bet, and therefore the player need only stake enough for one bet. Find the house advantage

$$H(Y_n^i + Y_n^j \mid X_1, \ldots, X_{2n-1}) := \frac{-E[Y_n^i + Y_n^j \mid X_1, \ldots, X_{2n-1}]}{P(Y_n^i + Y_n^j \neq 0 \mid X_1, \ldots, X_{2n-1})}. \qquad (18.51)$$

Assume that there are m unseen cards, with m given by (18.5) on p. 575, and that k and l of the m unseen cards are in denominations i and j, respectively.

When $k = l$, show that the result is greater than (18.4) on p. 575 if $l \geq 2$. Evaluate the two numbers for bets at the first turn ($m = 51$) on nonsoda denominations ($k = l = 4$).

18.10. *Two simultaneous denomination bets, one to win and the other to lose.* Consider two simultaneous denomination bets, one to win and the other to lose, on denominations $i \neq j$, at the nth turn. Unlike in Problem 18.9, both bets can be lost on a single turn (when this happens, the bettor is said to be *whipsawed*). Here it would be improper to regard this wager as a single wager, even though in practice this was sometimes done (when the bettor was whipsawed, he would owe money to the house). Assume that there are m unseen cards, with m given by (18.5) on p. 575, and that k and l of the m unseen cards are in denominations i and j, respectively. Evaluate the house advantage of this composite wager. When $k = l$, show that the result is less than (18.4) on p. 575 if $l \geq 2$. Evaluate the two numbers for bets at the first turn ($m = 51$) on nonsoda denominations ($k = l = 4$).

18.11. *Covariance between two denomination bets.*

(a) Fix $i \neq j$ and find the conditional covariance between Y_n^i and Y_n^j, given $X_1, \ldots, X_{2n-1}$, for all possible values of n, k, and l as in Problem 18.9. Are these covariances ever positive?

(b) Do the same for Y_n^i and $\bar{Y}_n^j$ as in Problem 18.10. Are these covariances ever negative?

18.12. *Probability that every turn results in a split.* Find a combinatorial expression for the probability that a split occurs at each of the 25 turns of a faro deal. (One possible approach is via (18.21) on p. 579, but a more direct argument may be simpler.) Using scientific notation, evaluate to six significant digits.

18.13. *A reversibility-like property.* For each $\boldsymbol{d} = (d_0, d_1, d_2, d_3, d_4) \in \Sigma$, where Σ is as in (18.22) on p. 580, define $\widehat{\boldsymbol{d}} := (d_4, d_3, d_2, d_1, d_0) \in \Sigma$. For the Markov chain $\{\boldsymbol{D}_n\}_{n \geq 0}$ in Σ with initial state $(0, 0, 0, 0, 13)$, transition probabilities (18.23) on p. 580, and absorbing state $(13, 0, 0, 0, 0)$, show that

$$(\boldsymbol{D}_0, \boldsymbol{D}_1, \ldots, \boldsymbol{D}_{52}) \overset{d}{=} (\widehat{\boldsymbol{D}}_{52}, \widehat{\boldsymbol{D}}_{51}, \ldots, \widehat{\boldsymbol{D}}_0). \qquad (18.52)$$

18.14. *Expected number of case bets at the nth turn.* For $n = 1, 2, \ldots, 25$, use the Markov chain of Section 18.2 to find the expected number of case bets available at the nth turn.

18.4 Notes

This chapter is an expanded version of Ethier (2007a).

One of the most thorough treatments of the history of faro was given by Asbury (1938, Chapter 1), and we summarize the principal facts.

Faro originated in 17th-century France, where it was called *pharaon* and was studied by Montmort (1708, pp. 4–29; 1713, pp. 77–104). It is undoubtedly a descendant of the very similar Venetian game of *bassette* (for which there are several spellings), a game studied by Jacob Bernoulli [1654–1705] (1713; 2006, Part 3, Problem 21) and even earlier by Joseph Sauveur [1653–1716] (1679). Bassette was in turn a descendant of the German game of *lansquenet* dating back to 1400. The name pharaon is said to be due to the fact that the backs of early French cards bore a picture of an Egyptian king.

Under the name faro, a corruption of pharaon, the game became popular in England until it was prohibited by an Act of Parliament in 1738. It was returned to favor around 1785. In the meantime, faro had made its way to the United States, probably through New Orleans. From there it spread north and east, though it was not until the American Revolution that it reached the large eastern cities. According to Asbury (1938, p. 6),

> The dissemination of Faro was quickened by the Louisiana Purchase [1803], and within a decade after that historic event it had become the most widely played game in the United States, and had gained a foothold from which it was not dislodged for more than a hundred years.

During the Civil War there were no less than 163 gambling establishments in Washington, D.C. According to Milton (1930),

> Thousands of dollars changed hands in a single night. The patrons were drawn "largely from members of Congress." The great men of the country could be seen at the principal Faro banks on the nights when no official reception was scheduled.

Faro was especially popular in the cattle-trading and mining towns of the 19th-century American West. According to DeArment (1982, p. 28), the California mining town of Columbia was said to have had 143 faro banks at the height of the gold rush. Among the more colorful characters who dealt faro in Dodge City, Kansas, and Tombstone, Arizona Territory, were Wyatt Earp [1848–1929], Doc Holliday [1851–1887], and Bat Masterson [1853–1921] (see DeArment 1982, p. 102). Dealing faro was considered a respectable profession. Nevertheless, when Masterson ran for town marshall in Trinidad, Colorado, in 1883, the editor of the local newspaper wrote (Sasuly 1982, p. 68),

> There are now two bankers running for city officer—one on each ticket—Mr. Taylor of the Las Animas County Bank, and Mr. Masterson of the bank of Fair O. Both have large numbers of depositors—the one of time depositors and the other receives his deposits for keeps.

Masterson lost the election decisively.

Alice Ivers [1851–1930], later to become known as "Poker Alice," was born in England but was raised in the mining town of Lake City, Colorado, where she learned how to use a gun and how to gamble. She also got a formal education in Virginia at an academy that catered to daughters of elite families. While still a young woman, she sat down at Jack "Lucky" Hardesty's faro table at the Gold Dust Gambling House in Lake City. He said, "I'm not gonna play you, ma'am. Faro is a man's game." Alice responded,

Reckon we'll sit here till you do, cause I'm not leaving this table until you deal me in.

Several hours later, Alice had won the money and the game was closed. "Beginner's luck," declared Hardesty.

"Beginner's luck, hell!" Alice retorted. "I specialized in mathematics while I was at the academy so I had the odds calculated."

See Kelly (1995, pp. 21–25).

Faro banks were often advertised with the likeness of a tiger, and consequently, playing faro became known as "bucking the tiger." Reformed professional gambler and anti-gambling crusader John Philip Quinn [1846–1916] (1892, p. 192) suggested the reason for this metaphor:

Faro has been happily likened to the "tiger," which, crafty, treacherous, cruel and relentless, hides under cover waiting, with impatient eagerness, for the moment when it may bury its velvet covered claws within the vitals of its unsuspecting victim and slake its fiery, unquenchable thirst with his life blood.

Faro had a well-deserved reputation as a dishonest game. In his book on cheating methods, Maskelyne (1894, p. 184) wrote,

Faro may almost be said to occupy in America the position of a national game. The methods of cheating used in connection with it are so numerous and so ingenious that it becomes really necessary to devote an entire chapter specially to them.

According to an earlier reformed gambler and anti-gambling crusader, Jonathan Harrington Green [1813–1887] (1843, p. 174),

[A] man, it will clearly appear, would act more rationally and correctly, to burn his money than to bet it on faro. In both cases he would lose his money, but in the former it would be lost without the sin of gambling being committed, his time wasted and his reputation injured; [...]

Even professionals were susceptible to cheating, as related by Davis (1956, p. 82):

There's a story in this connection about a Mississippi steamboat dealer who went broke bucking the tiger in a river town dive. A friend expressed astonishment at such stupidity and asked if the gambler didn't know that the town game was crooked. The faro dealer shrugged his shoulders and said, "Sure, I knew the game was crooked. But what could I do? It was the only game in town."

As Maurer (1943) put it,

Honest faro is probably the fairest of banking games, since there is a very small percentage in favor of the bank; however, honest faro has not been played in this country [the U.S.] for many, many years.

By the early 20th century faro was in decline, as gambling became illegal throughout the United States. By the time gambling was legalized in Nevada in 1931, other games such as craps had become more popular. Quoting Maurer (1943) once again,

So, although faro is not dead, it has withdrawn from popular patronage to such an extent that many persons who consider themselves gamblers have never seen it played.

By 1961 there were only about seven faro banks left in Nevada (Scarne 1961, p. 214). The last faro game was dealt at the Reno Ramada in the 1980s.

In Example 11.1.1 we discussed perfect riffle shuffles. These are sometimes called *faro shuffles*. The reason for this term is easy to understand. In faro the dealer places losing cards in the *losing pile* and winning cards in the *winning pile*. If the soda begins the winning pile and the hock finishes the losing pile, a perfect riffle shuffle (with the soda on top) of the two resulting packets will result in a deck in which every split from the preceding deal remains a split in the next deal (in fact, the two deals will be identical). As Maskelyne (1894, p. 204) put it,

> Thus it is evident that those cards which have been placed, with malice aforethought, in corresponding positions in the two piles, will come together in a shuffle of this kind, and form splits.

Faro was mentioned some 75 times in Giacomo Casanova's [1725–1798] autobiography (Casanova 1894). An example is provided at the end of this section.

Faro plays a major role is Aleksandr Pushkin's [1799–1837] classic 1834 short story "The Queen of Spades" (see Pushkin 1983, pp. 232–233):[2]

> Hermann picked a card and placed it on the table, covering it with a stack of bank notes. It was like a duel. A profound silence reigned over the gathering.
>   Chekalinskii started dealing with trembling hands. On his right showed a queen, on his left an ace.
>   "The ace has won!" said Hermann and turned his card face up.
>   "Your lady has been murdered," said Chekalinskii affably.
>   Hermann shuddered: indeed, instead of an ace, the queen of spades lay before him. He could not believe his eyes; he could not fathom how he could possibly have pulled the wrong card.

Notice that in the faro game of Pushkin's Russia, the player chose a card from his own deck (not at random) and placed it face down with his wager on top, concealing from the dealer the denomination on which he was betting, perhaps to reduce the likelihood of cheating. In the story, Hermann had fully intended to bet on the ace.

Faro also appeared in Leo Tolstoy's [1828–1910] *War and Peace* of 1869 (Tolstoy 1889, Volume 2, Part 1, Chapter 13):[3]

> He could not bring himself to believe that blind chance, by throwing the seven of hearts to the right rather than to the left, might deprive him of all this just comprehended and just appreciated happiness, and plunge him into the abyss of a wretchedness never before experienced, and of which he had no adequate idea.

---

[2] Translation by Paul Debreczeny.

[3] Translation by Nathan Haskell Dole.

Originally, faro was dealt from a face-down deck and there were 26 turns. In 1825 a Cincinnati watchmaker named Graves invented the *open dealing box* (Asbury 1938, p. 10), which was an immediate success, with the result that the game was dealt from a face-up deck and there were only 25 turns. As we have noted, the first card is called the soda (a corruption of "zodiac," according to Asbury 1938, p. 16) and the last card is called the hock. The latter was a remnant of the early days of faro, when the last card was called the "hockelty" card and was not a winning card (Asbury 1938, p. 15). With the alternative spelling "hocly," this term was said to mean "a certainty" in an 1842 British Hoyle; see H. 1842. It was pointed out by Asbury (1938, p. 16), Lemmel (1964, p. 96), and DeArment (1982, p. 2) that the phrase "from soda to hock," which we have adopted as the title of Section 18.2, was a popular 19th-century slang expression meaning "from beginning to end" or, as we might say today, "from A to Z."

Later in the 19th century an inventor applied for a patent on a faro dealing box and was rejected. He appealed to the Patent Office examiners-in-chief, who responded on May 7, 1878, as follows (Bueschel 1995, p. 16):

> The instrument before us is said by the examiner to be a "faro deal box." There is no essential need or demand for such an instrument for any legitimate harmless game. Its manifest principle [sic] use is therefore what the examiner alleges. It is true as urged the instrument can be used in connection with innocent games but the probable application of it being essentially pernicious and contrary to public policy and morals and the best interests of society, this seems to be a proper case to exercise discretion as to whether the invention is sufficiently useful and important to warrant the grant of a patent. At best there is no essential utility of purpose. The most that can be said is that it facilitates and affords a convenient means of dealing cards which are used in an idle game having no intrinsic commercial value.

Let us describe more fully the rules of 18th-century faro (Montmort 1713, p. 77; De Moivre 1756, p. 77.) The game was identical to modern faro with the following exceptions. First, as we said, the game was dealt from a face-down deck and there were 26 turns. Second, there must have been some restriction on withdrawing a bet prior to resolution. Though not explicitly stated by Montmort or De Moivre, this is implicit in the next two rules. (However, according to Bohn 1850, p. 335, a bet could be withdrawn if 10 or more cards remained to be dealt. This was confirmed by H. 1842.) Third, when betting on a denomination with only two unseen representatives, a split at the last turn cost the player his entire bet. And fourth, when betting on a case card, a win for the player at the last turn was counted as a push. Montmort (1713, p. 97), De Moivre (1756, Problem 14), and Euler (1764) each correctly derived the analogue of (18.13) for this version of the game. But, as noted by Todhunter (1865, Article 158), Montmort's (1713, p. 105) table of exact house advantages contains some inaccuracies. (Specifically, the first 17 entries in the second column and the first four entries in the fourth column are inaccurate, albeit only slightly in most cases.)

Pierre Rémond de Montmort [1678–1719] (1713, p. 90), with the help of Johann Bernoulli [1667–1748], derived (18.42) by starting with the alternative formula

$$\frac{1}{2} \sum_{k=1}^{t} \frac{(l)_2 (m-l)_{2k-2}}{(m)_{2k}}, \tag{18.53}$$

which can be rewritten using $(m)_{a+b} = (m)_a (m-a)_b = (m)_b (m-b)_a$ as

$$\frac{1}{2} \frac{(l)_2}{(m)_l} \sum_{k=1}^{t} (m-2j)_{l-2}. \tag{18.54}$$

Here t is the number of turns needed to guarantee resolution of the bet. Using a combinatorial identity (see Todhunter 1865, Article 156, and Problem 18.6), this was written by Montmort (1713, p. 98) as

$$\frac{1}{2} \sum_{j=1}^{l-1} 2^{-j} \frac{(l)_j}{(m)_j} + 2^{-l} \frac{l!}{(m)_l} \frac{1+(-1)^{m-l}}{2}, \tag{18.55}$$

and by Nicolaus Bernoulli [1687–1759] (see Montmort 1713, p. 99) as

$$\frac{1}{2} \sum_{j=1}^{l-1} (-1)^{j-1} 2^{-j} \frac{(l)_j}{(m-l+j)_j} + (-1)^{l-1} 2^{-l} \frac{l!}{(m)_l} \frac{1-(-1)^m}{2}. \tag{18.56}$$

We have expressed the last two formulas in such a way that they apply with either m even (18th-century rules) or m odd (modern rules), at least if $l \geq 3$. Assuming that m is even, Leonhard Euler [1707–1783] (1764) discovered that (18.53) can be written as

$$\frac{l}{2^l} \sum_{k=1}^{\lfloor l/2 \rfloor} \frac{(l-1)_{2k-1}}{(2k-1)!(m-2k+1)}. \tag{18.57}$$

Euler did not supply a proof, but Todhunter (1865, pp. 244–245) showed that (18.57) follows from (18.56) and a partial-fractions expansion. Alternative formulas were found by Nicolaus Bernoulli and Montmort (1713, p. 304) and by Thorp (1973, 1976). Daniel Bernoulli [1700–1782] also contributed to the subject (Hald 1990, p. 303).

The first correct analysis of the modern game was given by Edward O. Thorp [1932–] (1973, 1976), who also pointed out the flaws in earlier 20th-century studies of faro. Results (18.9) and (18.13) are due to him. The best source for the modern rules is Lemmel (1964, Chapter 9).

Result (18.15) was known, at least empirically, in the 18th century: Huyn (1788, p. 39) noted that one can expect three splits (then called "doublets") in two deals. The exact expected number of splits per deal in the 18th-century game was $26/17 \approx 1.529412$.

The strategy analyzed in Section 18.2 was motivated by an observation of Thorp (1973):

[T]he player who wishes to maximize his (conditional) expectation per bet that is resolved should, if he can alter his bet after each turn, limit his bets to ranks with a minimum (positive) number of cards remaining.

This is not unlike the strategy recommended by Nicholas (Nick the Greek) Dandolos [1883–1966] (Thackrey 1968, p. 132):

Bet the cases. Bet the High-Low. Don't call the turn unless you feel lucky. Very lucky. It's even odds ... remember that! Even odds. You'll never get a better break in this world!

Faro was Nick's favorite game.

Problems 18.1 and 18.2 were discussed by Thorp (1976). According to Riddle (1963, p. 135) and Thackrey (1968, p. 129), a split was a push for the high-card bettor. Lemmel (1964, p. 97) was ambiguous on this point, while under the rules in effect at the Reno Ramada in the 1980s half the bet was lost in the case of a split. Problem 18.3 was motivated by the Reno Ramada rules. Problem 18.5(a) is due to Montmort (1713, p. 97), De Moivre (1756, Problem 14), and Euler (1764), while Problem 18.5(b) was motivated by an incident in Casanova's (1894, Volume 3, Chapter 20) autobiography, in which he was a faro dealer:[4]

While I was shuffling a fresh pack of cards, the youngest of them drew out of his pocket-book a paper which he spewed to his two companions. It was a bill of exchange. "Will you stake the value of this bill on a card, without knowing its value?" said he.

"Yes," I replied, "if you will tell me upon whom it is drawn, and provided that it does not exceed the value of the bank."

After a rapid glance at the pile of gold before me, he said, "The bill is not for so large a sum as your bank, and it is payable at sight by Zappata, of Turin."

I agreed, he cut, and put his money on an ace, the two friends going half shares. I drew and drew and drew, but no ace appeared. I had only a dozen cards left.

"Sir," said I, calmly to the punter, "you can draw back if you like."

"No, go on."

Four cards more, and still no ace; I had only eight cards left.

"My lord," said I, "it's two to one that I do not hold the ace, I repeat you can draw back."

"No, no, you are too generous, go on."

I continued dealing, and won; I put the bill of exchange in my pocket without looking at it.

The rule variation cited in Problem 18.7 was described by Lemmel (1964, p. 107).

---

[4] Translation by Arthur Machen.

Chapter 19
Baccarat

After dinner, you may be sure that cards were not wanting, and that the company who played did not play for love merely.

William Makepeace Thackeray, *The Luck of Barry Lyndon*

Baccarat is a 20th-century Argentinian descendant of the elegant card games chemin de fer (Example 5.1.4 on p. 167) and baccara en banque, which are of French origin and date back to the 19th century. Unlike its ancestors, baccarat offers no discretionary strategy decisions. This makes the game easier to analyze but less interesting mathematically than chemin de fer or baccara en banque. In Section 19.1 we evaluate the probabilities and house advantages associated with the player, banker, and tie wagers. In Section 19.2 we explore the potential for card counting at baccarat.

19.1 Player vs. Banker

Baccarat, also known as *punto banco*, is played with an eight-deck shoe comprising 416 cards. (Occasionally, only six decks are used; we will ignore this case.) Denominations A, 2–9, 10, J, Q, K have values 1, 2–9, 0, 0, 0, 0, respectively. The value of a hand, consisting of two or three cards, is the sum of the values of the cards, modulo 10. In other words, only the final digit of the sum is used to evaluate a hand. For example, $5 + 7 \equiv 2 \pmod{10}$ and $5 + 7 + 9 \equiv 1 \pmod{10}$.

House-banked bets are available on player, banker, and tie. We think of player and banker not as player 1 and player 2 as in chemin de fer, but rather as the two principal betting opportunities at baccarat, much like the pass line and don't pass at craps. Two cards are dealt to player and two to banker. The object of the game is to have the higher-valued hand (closer to 9) at the end of play. A two-card hand of value 8 or 9 is a *natural*. If either hand

S.N. Ethier, *The Doctrine of Chances*, Probability and its Applications,
DOI 10.1007/978-3-540-78783-9_19, © Springer-Verlag Berlin Heidelberg 2010

Table 19.1 Mandatory draw/stand rules at baccarat if neither player nor banker has a natural. X_3 denotes player's third card ($= \varnothing$ if player stands). Notice the 13 departures from symmetric play.

player's two-card total	player's rules	banker's two-card total	banker's rules
0–5	draw	0–2	draw
6, 7	stand	3	draw unless $X_3 = 8$
		4	draw unless $X_3 \in \{0, 1, 8, 9\}$
		5	draw unless $X_3 \in \{0, 1, 2, 3, 8, 9\}$
		6	stand unless $X_3 \in \{6, 7\}$
		7	stand

is a natural, the game is over and the higher-valued hand wins. Hands of equal value result in tie. If neither hand is a natural, player must draw a third card if the value of his two-card hand is 5 or less, and must stand if that value is 6 or 7. The third card is dealt face up. Next, banker draws a third card or stands according to the value of his two-card hand and the third card, if any, dealt to player. Table 19.1 shows that there are 13 departures from symmetric play, which were chosen to give banker an advantage. This completes the game, and the higher-valued hand wins. Again, hands of equal value result in a tie. The *player bet* pays even money if player wins, is lost if banker wins, and is pushed in the case of a tie. The *banker bet* pays 19 to 20 if banker wins (in other words, there is a 5 percent commission charged on a winning banker bet), is lost if player wins, and is pushed in the case of a tie. Finally, there is also a *tie bet*, which pays 8 to 1 for a tie and is otherwise lost.

Let us suppose that the baccarat shoe comprises d standard 52-card decks, so that the number of cards in the shoe is $N := 52d$. The case $d = 8$ (or $N = 416$) is of primary interest. We define $n_0 := 16d$ to be the number of cards of value 0 in the shoe, and $n_j := 4d$ to be the number of cards of value j in the shoe for $j = 1, 2, \ldots, 9$. A coup in baccarat requires 4, 5, or 6 cards, randomly sampled without replacement. The probabilities of such 4-, 5-, and 6-card sequences can be evaluated as follows, keeping in mind that the order of the cards within each initial two-card hand is irrelevant. With $i_1 \leq i_2$ (resp., $j_1 \leq j_2$) corresponding to player's (resp., banker's) first two cards in either order, and with k and l corresponding to the fifth and sixth cards, respectively, if necessary, the relevant probabilities are

$$P_4(i_1, i_2, j_1, j_2)$$

$$:= \frac{n_{i_1}(n_{i_2} - \delta_{i_1 i_2})(2 - \delta_{i_1 i_2})}{N(N-1)} \tag{19.1}$$

$$\cdot \frac{(n_{j_1} - \delta_{i_1 j_1} - \delta_{i_2 j_1})(n_{j_2} - \delta_{i_1 j_2} - \delta_{i_2 j_2} - \delta_{j_1 j_2})(2 - \delta_{j_1 j_2})}{(N-2)(N-3)},$$

$$P_5(i_1, i_2, j_1, j_2, k) := P_4(i_1, i_2, j_1, j_2) \frac{n_k - \delta_{i_1 k} - \delta_{i_2 k} - \delta_{j_1 k} - \delta_{j_2 k}}{N-4}, \tag{19.2}$$

and

$$P_6(i_1, i_2, j_1, j_2, k, l) := P_5(i_1, i_2, j_1, j_2, k)$$
$$\cdot \frac{n_l - \delta_{i_1 l} - \delta_{i_2 l} - \delta_{j_1 l} - \delta_{j_2 l} - \delta_{kl}}{N-5}. \tag{19.3}$$

Because in studying chemin de fer in Examples 5.1.4 on p. 167 and 5.2.7 on p. 183 we assumed sampling with replacement to simplify the analysis, we would like to treat that case here for purposes of comparison, along with the more realistic case of sampling without replacement. We can get the relevant probabilities by letting $d \to \infty$ in (19.1)–(19.3), or we can deduce them directly. We find, using the same notation,

$$P_4(i_1, i_2, j_1, j_2) := \frac{(1 + 3\delta_{i_1 0})(1 + 3\delta_{i_2 0})(2 - \delta_{i_1 i_2})}{(13)^2}$$
$$\cdot \frac{(1 + 3\delta_{j_1 0})(1 + 3\delta_{j_2 0})(2 - \delta_{j_1 j_2})}{(13)^2}, \tag{19.4}$$

$$P_5(i_1, i_2, j_1, j_2, k) := P_4(i_1, i_2, j_1, j_2) \frac{1 + 3\delta_{k0}}{13}, \tag{19.5}$$

and

$$P_6(i_1, i_2, j_1, j_2, k, l) := P_5(i_1, i_2, j_1, j_2, k) \frac{1 + 3\delta_{l0}}{13}. \tag{19.6}$$

Next we want to find the joint distribution of the values of player's and banker's hands at the conclusion of play. For this purpose we introduce several random variables. Let X denote the value of player's two-card hand and let Y denote the value of banker's two-card hand. On the event $\{X \le 7, Y \le 7\}$, let X_3 denote the value of player's third card if he draws, and let $X_3 := \varnothing$ if he stands. Similarly, let Y_3 denote the value of banker's third card if he draws, and let $Y_3 := \varnothing$ if he stands. (X_3 and Y_3 can be defined arbitrarily on the complement of $\{X \le 7, Y \le 7\}$ if it seems desirable to do so.)

It will be convenient to define $M : \{0, 1, \ldots\} \mapsto \{0, 1, \ldots, 9\}$ by $M(i) = j$ if $i \equiv j \pmod{10}$, to let $D \subset \{0, 1, \ldots, 7\} \times \{0, 1, \ldots, 9, \varnothing\}$ be the set of pairs (j, k) such that $(Y, X_3) = (j, k)$ results in a banker "draw" (see Table 19.1), and to let $S \subset \{0, 1, \ldots, 7\} \times \{0, 1, \ldots, 9, \varnothing\}$ be the set of pairs (j, k) such that $(Y, X_3) = (j, k)$ results in a banker "stand." Note that D and S are complementary subsets of $\{0, 1, \ldots, 7\} \times \{0, 1, \ldots, 9, \varnothing\}$.

Case 1. Neither player nor banker draws a third card. This occurs if $X \in \{8, 9\}$, or if $Y \in \{8, 9\}$, or if both $X \in \{6, 7\}$ and $Y \in \{6, 7\}$. In this case the probability that player's hand has value i and banker's hand has value j is

$$P_{00}(i, j) := \mathrm{P}(X = i,\ Y = j) \tag{19.7}$$
$$= \sum\sum_{0 \leq i_1 \leq i_2 \leq 9: M(i_1+i_2)=i}\ \sum\sum_{0 \leq j_1 \leq j_2 \leq 9: M(j_1+j_2)=j} P_4(i_1, i_2, j_1, j_2)$$

if $i \in \{8, 9\}$, or if $j \in \{8, 9\}$, or if both $i \in \{6, 7\}$ and $j \in \{6, 7\}$; otherwise $P_{00}(i, j) = 0$.

Case 2. Only banker draws a third card. This occurs if $X \in \{6, 7\}$ and $Y \leq 5$. In this case the probability that player's hand has value i and banker's hand has value j' is

$$P_{01}(i, j') := \mathrm{P}(X = i,\ Y \leq 5,\ M(Y + Y_3) = j')$$
$$= \sum\sum_{0 \leq i_1 \leq i_2 \leq 9: M(i_1+i_2)=i}\ \sum\sum_{0 \leq j_1 \leq j_2 \leq 9: M(j_1+j_2) \leq 5} \tag{19.8}$$
$$\cdot \sum_{0 \leq l \leq 9: M(j_1+j_2+l)=j'} P_5(i_1, i_2, j_1, j_2, l)$$

if $i \in \{6, 7\}$ and $j' \in \{0, 1, \ldots, 9\}$; otherwise $P_{01}(i, j') := 0$.

Case 3. Only player draws a third card. This occurs if $X \leq 5$, $Y \leq 7$, and $(Y, X_3) \in S$. In this case the probability that player's hand has value i' and banker's hand has value j is

$$P_{10}(i', j) = \mathrm{P}(X \leq 5,\ Y = j,\ (j, X_3) \in S,\ M(X + X_3) = i')$$
$$= \sum\sum_{0 \leq i_1 \leq i_2 \leq 9: M(i_1+i_2) \leq 5}\ \sum\sum_{0 \leq j_1 \leq j_2 \leq 9: M(j_1+j_2)=j} \tag{19.9}$$
$$\cdot \sum_{0 \leq k \leq 9: (j,k) \in S,\ M(i_1+i_2+k)=i'} P_5(i_1, i_2, j_1, j_2, k)$$

if $i' \in \{0, 1, \ldots, 9\}$ and $j \in \{0, 1, \ldots, 7\}$; otherwise $P_{10}(i', j) := 0$.

Case 4. Both player and banker draw a third card. This occurs if $X \leq 5$, $Y \leq 7$, and $(Y, X_3) \in D$. In this case the probability that player's hand has value i' and banker's hand has value j' is

$$P_{11}(i', j') := \mathrm{P}(X \leq 5,\ Y \leq 7,\ (Y, X_3) \in D,$$
$$M(X + X_3) = i',\ M(Y + Y_3) = j')$$
$$= \sum\sum_{0 \leq i_1 \leq i_2 \leq 9: M(i_1+i_2) \leq 5}\ \sum\sum_{0 \leq j_1 \leq j_2 \leq 9: M(j_1+j_2) \leq 7}$$

$$\sum_{0\le k\le 9:(M(j_1+j_2),k)\in D,\ M(i_1+i_2+k)=i'} \qquad (19.10)$$

$$\cdot \sum_{0\le l\le 9:M(j_1+j_2+l)=j'} P_6(i_1,i_2,j_1,j_2,k,l)$$

if $i' \in \{0,1,\ldots,9\}$ and $j' \in \{0,1,\ldots,9\}$; otherwise $P_{11}(i',j') := 0$.

Having computed the probabilities in (19.7)–(19.10), we can evaluate the joint distribution of the values of player's and banker's hands at the conclusion of play via

$$P(i,j) := P_{00}(i,j) + P_{01}(i,j) + P_{10}(i,j) + P_{11}(i,j), \qquad (19.11)$$

where $i,j \in \{0,1,\ldots,9\}$. These probabilities are evaluated numerically in Tables 19.2 (8-deck shoe) and 19.3 (infinite-deck shoe).

Then the probabilities of a win, loss, and push for player (or, equivalently, loss, win, and push for banker) are respectively

$$p := \sum_{i=1}^{9}\sum_{j=0}^{i-1} P(i,j), \quad q := \sum_{i=0}^{8}\sum_{j=i+1}^{9} P(i,j), \quad r := \sum_{i=0}^{9} P(i,i). \qquad (19.12)$$

These probabilities are evaluated numerically in Table 19.4. They immediately lead to the house advantage of the player bet, namely $q - p$, and that of the banker bet, namely $p - (19/20)q$. These figures are divided by $1 - r$ if pushes are to be excluded. Finally, the house advantage of the tie bet is $1 - 9r$.

Thus, the house advantage for the player bet is about 1.235 percent in the eight-deck game if pushes are included. We might call this figure the *banker's advantage* in baccarat. (We emphasize that we are now considering only the even-money player bet.) What is the source of the banker's advantage? Clearly, if banker were to follow the same strategy as player (namely, draw to 5 or less, stand on 6 or 7), the game would be completely symmetric, in that player and banker could exchange roles with no effect. In particular, the banker's advantage in that scenario would be zero. Consequently, the banker's advantage in baccarat is entirely due to banker's 13 departures from symmetric play (see Table 19.1). Let us see how much each one of these departures contributes.

They include, for example, the requirement that banker stand with 3 when player draws an 8. We refer to this departure as $(3,8)$. The set of all 13 such departures is

$$A := \{(3,8),(4,0),(4,1),(4,8),(4,9),$$
$$(5,0),(5,1),(5,2),(5,3),(5,8),(5,9),(6,6),(6,7)\}. \qquad (19.13)$$

Therefore, conditioning on (Y,X_3) when $X \le 5$ and $Y \le 7$ (so that neither player nor banker has a natural, and player draws a third card),

Table 19.2 Joint distribution of player's final total i and banker's final total j, assuming sampling without replacement from an 8-deck shoe. Probabilities are rounded to six decimal places. Row B contains the column sums, and is therefore the distribution of the value of banker's hand. Row P contains the row sums, and is therefore the distribution of the value of player's hand.

				j		
i	0	1	2	3	4	5
0	.005 798	.004 860	.004 844	.005 393	.009 210	.009 237
1	.004 929	.004 101	.004 095	.004 642	.008 121	.008 154
2	.004 844	.004 024	.004 003	.004 555	.007 720	.008 582
3	.004 751	.003 928	.003 914	.004 452	.007 632	.009 025
4	.004 930	.004 108	.004 094	.004 106	.007 261	.008 359
5	.005 047	.004 227	.004 212	.004 222	.006 867	.007 939
6	.011 478	.009 797	.009 777	.009 803	.010 307	.011 389
7	.011 674	.009 992	.009 961	.009 995	.009 970	.011 055
8^3	.003 702	.003 140	.003 130	.003 968	.003 956	.004 496
9^3	.003 683	.003 120	.003 110	.003 657	.004 472	.004 483
8^2	.013 942	.008 974	.008 939	.008 974	.008 940	.008 974
9^2	.013 997	.009 009	.008 974	.009 009	.008 974	.009 009
B	.088 775	.069 281	.069 054	.072 776	.093 429	.100 704
P	.093 989	.074 527	.074 317	.074 527	.074 309	.074 511

				j		
i	6	7	8^3	9^3	8^2	9^2
0	.009 731	.010 820	.003 079	.003 079	.013 942	.013 997
1	.008 124	.009 208	.002 584	.002 584	.008 974	.009 009
2	.008 556	.009 112	.002 504	.002 504	.008 939	.008 974
3	.008 999	.009 034	.002 404	.002 404	.008 974	.009 009
4	.009 163	.009 198	.002 587	.002 587	.008 940	.008 974
5	.009 290	.009 325	.002 696	.002 701	.008 974	.009 009
6	.019 240	.020 183	.006 663	.006 675	.008 940	.008 974
7	.018 988	.020 350	.006 842	.006 856	.008 974	.009 009
8^3	.005 540	.006 613	.002 079	.002 085		
9^3	.005 530	.006 604	.002 062	.002 062		
8^2	.008 940	.008 974			.008 900	.008 975
9^2	.008 974	.009 009			.008 975	.008 971
B	.121 076	.128 431	.033 501	.033 538	.094 532	.094 903
P	.133 227	.133 666	.038 707	.038 784	.094 532	.094 903

^{3}three cards ^{2}two cards (natural)

Table 19.3 Joint distribution of player's final total i and banker's final total j, assuming sampling with replacement. Entries must be divided by $(13)^6$. Row B contains the column sums, and therefore gives the distribution of the value of banker's hand. Row P contains the row sums, and therefore gives the distribution of the value of player's hand.

			j			
i	0	1	2	3	4	5
0	28,432	23,572	23,572	26,132	44,692	44,692
1	23,924	19,793	19,793	22,353	39,153	39,153
2	23,556	19,425	19,425	21,985	37,345	41,345
3	23,044	18,913	18,913	21,473	36,833	43,393
4	23,956	19,825	19,825	19,825	35,185	40,305
5	24,468	20,337	20,337	20,337	33,137	38,257
6	55,716	47,265	47,265	47,265	49,825	54,945
7	56,484	48,033	48,033	48,033	48,033	53,153
8^3	17,956	15,121	15,121	19,121	19,121	21,681
9^3	17,844	15,009	15,009	17,569	21,569	21,569
8^2	67,600	43,264	43,264	43,264	43,264	43,264
9^2	67,600	43,264	43,264	43,264	43,264	43,264
B	430,580	333,821	333,821	350,621	451,421	485,021
P	455,780	359,021	359,021	359,021	359,021	359,021

			j			
i	6	7	8^3	9^3	8^2	9^2
0	47,252	52,372	14,932	14,932	67,600	67,600
1	39,153	44,273	55,713	55,713	43,264	43,264
2	41,345	43,905	55,345	55,345	43,264	43,264
3	43,393	43,393	54,833	54,833	43,264	43,264
4	44,305	44,305	55,745	55,745	43,264	43,264
5	44,817	44,817	56,257	56,257	43,264	43,264
6	93,345	97,345	75,505	75,505	43,264	43,264
7	91,553	98,113	76,273	76,273	43,264	43,264
8^3	26,801	31,921	10,081	10,081		
9^3	26,689	31,809	9,969	9,969		
8^2	43,264	43,264			43,264	43,264
9^2	43,264	43,264			43,264	43,264
B	585,181	618,781	161,805	161,805	456,976	456,976
P	643,981	643,981	187,005	187,005	456,976	456,976

^{3}three cards ^{2}two cards (natural)

banker's advantage

$\qquad$ = E[banker's gain if banker follows banker's strategy]

$\qquad\qquad$ − E[banker's gain if banker follows player's strategy]

$\qquad$ $= \displaystyle\sum_{(j,k)\in A} P(X \le 5,\ Y \le 7,\ (Y, X_3) = (j, k))$

$\qquad\qquad$ $\cdot \big\{$E[banker's gain if banker follows banker's strategy $|$

$\qquad\qquad\qquad$ $X \le 5,\ Y \le 7,\ (Y, X_3) = (j, k)]$

$\qquad\qquad\qquad$ − E[banker's gain if banker follows player's strategy $|$

$\qquad\qquad\qquad\qquad$ $X \le 5,\ Y \le 7,\ (Y, X_3) = (j, k)]\big\},$ $\qquad$ (19.14)

where we have used the fact that the difference in conditional expectations is 0 if $(j, k) \notin A$.

Now we need to evaluate the individual summands in (19.14). We let G_{stand} and G_{draw} represent banker's gain when he stands and draws. Then, for each $(j, k) \in A$ with $j \le 5$, the difference in conditional expectations in (19.14) is equal to

$$\big\{\mathrm{E}[G_{\text{stand}} \mid X \le 5,\ Y \le 7,\ (Y, X_3) = (j, k)]$$
$$- \mathrm{E}[G_{\text{draw}} \mid X \le 5,\ Y \le 7,\ (Y, X_3) = (j, k)]\big\}, \qquad (19.15)$$

whereas for $(j, k) = (6, 6)$ and $(j, k) = (6, 7)$, it is equal to the negative of (19.15). The reader may recall Table 5.3 from which these conditional expectations can be determined. However, that table assumed sampling with replacement, and here we want to treat both that case and the slightly more complicated case of sampling without replacement.

Let $E_{\text{stand}}(i_1, i_2, j_1, j_2, k)$ and $E_{\text{draw}}(i_1, i_2, j_1, j_2, k)$ represent the conditional expectation of G_{stand} and G_{draw}, respectively, given that $i_1 \le i_2$ are player's first two cards in either order and $M(i_1 + i_2) \le 5$, $j_1 \le j_2$ are banker's first two cards in either order and $M(j_1 + j_2) \le 7$, and k is player's third card. Then the (j, k)th summand in (19.14) equals

$$\sum_{0 \le i_1 \le i_2 \le 9:M(i_1+i_2)\le 5} \ \sum_{0 \le j_1 \le j_2 \le 9:M(j_1+j_2)=j} P_5(i_1, i_2, j_1, j_2, k) \qquad (19.16)$$
$$\cdot [E_{\text{stand}}(i_1, i_2, j_1, j_2, k) - E_{\text{draw}}(i_1, i_2, j_1, j_2, k)]$$

if $j \le 5$, and equals the negative of this quantity if $j = 6$. Furthermore,

$$E_{\text{stand}} = \operatorname{sgn}(M(j_1 + j_2) - M(i_1 + i_2 + k)) \qquad (19.17)$$

and

$$E_{\text{draw}} = \sum_{l=0}^{9} \frac{(n_l - \delta_{i_1 l} - \delta_{i_2 l} - \delta_{j_1 l} - \delta_{j_2 l} - \delta_{kl})}{N - 5}$$

Table 19.4 Some probabilities in baccarat, and the house advantage. Assumes sampling without replacement in the eight-deck case, and with replacement in the infinite-deck case. Results are rounded to nine decimal places.

	8 decks	∞ decks	∞ decks
event		probability	
player wins	.446 246 609	.446 146 512	$2{,}153{,}464/(13)^6$
banker wins	.458 597 423	.458 427 918	$2{,}212{,}744/(13)^6$
player and banker tie	.095 155 968	.095 425 570	$460{,}601/(13)^6$
wager		house advantage	
player, pushes included	.012 350 813	.012 281 406	$4{,}560/(13)^5$
player, pushes excluded	.013 649 660	.013 576 999	$(*)$
banker, pushes included	.010 579 058	.010 639 990	$256{,}786/(5(13)^6)$
banker, pushes excluded	.011 691 582	.011 762 426	$(**)$
tie	.143 596 288	.141 169 870	$681{,}400/(13)^6$

$(*)$ $59{,}280/((13)^6 - 460{,}601)$ $(**)$ $256{,}786/(5((13)^6 - 460{,}601))$

$$\cdot \operatorname{sgn}(M(j_1 + j_2 + l) - M(i_1 + i_2 + k)). \qquad (19.18)$$

Thus, we have all the ingredients we need to evaluate the 13 summands in (19.14), and we display the results in Table 19.5. The table suggests that the inventors of baccarat made two mistakes!

19.2 Card Counting

By redefining N and $n_0, n_1, \ldots, n_9$ in (19.1)–(19.3) on pp. 598–599, we can calculate the expected gain from a one-unit bet on player, banker, or tie with an arbitrary shoe composition. This leads immediately to the effects of removal for the three wagers. We list them in Table 19.6.

Table 19.7 gives balanced level-two card-counting systems for the three wagers. For each wager, the correlation with the effects of removal is at least 0.959. We see that the player may use a one-parameter counting system to track his advantage at both the player bet and the banker bet. The situation is analogous to that of Example 11.3.2 on p. 411. The tie bet requires a different counting system.

We next use a normal approximation to approximate the probability that the player's approximate advantage exceeds a certain level as a function of the number of unseen cards. This is complicated by the fact that the player has

Table 19.5 Contributions to the banker's advantage in baccarat from banker's 13 departures from symmetric play. Value of banker's two-card hand is j, and player's third card is k. Entries in the last column should be multiplied by $16/(13)^6$.

	8 decks	∞ decks	∞ decks
(j,k)	contribution to banker's advantage		
$(3,8)$	.000 010 099	.000 016 574	5
$(4,0)$	.001 411 537	.001 392 224	420
$(4,1)$	$-$.000 008 423	$-$.000 016 574	-5
$(4,8)$	.000 545 305	.000 546 945	165
$(4,9)$	.000 712 335	.000 712 686	215
$(5,0)$	.005 653 481	.005 635 193	1,700
$(5,1)$	.001 049 581	.001 044 168	315
$(5,2)$	.000 684 376	.000 679 538	205
$(5,3)$	.000 332 006	.000 314 908	95
$(5,8)$	.000 543 629	.000 546 945	165
$(5,9)$	.001 244 519	.001 243 057	375
$(6,6)$	.000 019 738	.000 016 574	5
$(6,7)$	.000 152 629	.000 149 167	45
total	.012 350 813	.012 281 406	3,705
$(6,\varnothing)$	.001 416 641	.001 378 965	416
total[1]	.013 775 877	.013 676 945	4,126

[1]including $(6,\varnothing)$ but excluding $(4,1)$

two primary bets to choose from, player and banker, as well as the secondary bet, tie.

Let Z_n^P, Z_n^B, and Z_n^T be the analogues of Z_n^* of (11.95) on p. 418 for the player, banker, and tie bets, and similarly let P_i, B_i, and T_i correspond to J_i; μ_P, μ_B, and μ_T to μ; γ_P, γ_B, and γ_T to γ; and σ_P, σ_B, and σ_T to σ. Then $P_i = -B_i$, $\gamma_P > 0$, and $\gamma_B > 0$, so

$$\mathrm{P}(\max(Z_n^P, Z_n^B) > \beta) = \mathrm{P}(Z_n^P > \beta) + \mathrm{P}(Z_n^B > \beta) \qquad (19.19)$$

if $\beta > \max(\mu_P, \mu_B)$. Assuming this, we have from (11.98) on p. 419 that

$$\mathrm{P}(Z_n^P > \beta)$$
$$= \mathrm{P}\left(\mu_P + \frac{\gamma_P}{N-n}\sum_{j=1}^{n} P_{X_j} > \beta\right)$$

Table 19.6 Effects of removal, multiplied by 415, for the player bet, the banker bet, and the tie bet in eight-deck baccarat. Results are exact, except for rounding.

card value	player bet	banker bet	tie bet
0	−.007 377	.007 784	.212 844
1	−.018 591	.018 276	.053 639
2	−.022 516	.021 677	−.099 284
3	−.027 887	.026 943	−.088 836
4	−.049 601	.048 023	−.121 329
5	.034 885	−.034 318	−.109 726
6	.046 829	−.046 995	−.481 205
7	.033 917	−.034 327	−.452 934
8	.022 121	−.020 814	.271 516
9	.010 351	−.009 601	.176 782

Table 19.7 Balanced level-two card-counting systems for the player bet, banker bet, and tie bet in baccarat.

card value	player bet	banker bet	tie bet
0	0	0	1
1	−1	1	0
2	−1	1	0
3	−1	1	0
4	−2	2	−1
5	1	−1	−1
6	2	−2	−2
7	1	−1	−2
8	1	−1	1
9	0	0	1
correlation with Table 19.6	.959 118	.966 844	.968 208

$$= \mathrm{P}\bigg(\sum_{j=1}^{n} P_{X_j} > \Big\lfloor (N-n)\frac{\beta - \mu_P}{\gamma_P} \Big\rfloor + \frac{1}{2} \bigg) \tag{19.20}$$

$$\approx 1 - \Phi\bigg(\frac{1}{(N-n)\sigma_P} \Big\{ \Big\lfloor (N-n)\frac{\beta - \mu_P}{\gamma_P} \Big\rfloor + \frac{1}{2} \Big\} \sqrt{\frac{(N-n)(N-1)}{n}} \bigg),$$

and from (11.100) on p. 419 that

$$\mathrm{E}[(Z_n^P)^+] \approx \mu_P \Phi\left(\frac{\mu_P}{\sigma_{P,n}}\right) + \sigma_{P,n}\phi\left(\frac{\mu_P}{\sigma_{P,n}}\right), \tag{19.21}$$

where

$$\sigma_{P,n}^2 := \frac{\gamma_P^2 \sigma_P^2 n}{(N-n)(N-1)}. \tag{19.22}$$

Exactly analogous formulas hold for Z_n^B and Z_n^T with the P subscript in μ_P, γ_P, σ_P, and $\sigma_{P,n}$ changed to B or T where appropriate.

Here $N = 416$ and, from Table 19.4 on p. 605, $\mu_P \approx -0.012351$, $\mu_B \approx -0.010579$, and $\mu_T \approx -0.143596$. From Tables 19.6 and 19.7 we have $\gamma_P \approx 0.025198$, $\gamma_B \approx 0.024742$, and $\gamma_T \approx 0.212438$. Furthermore, $\sigma_P^2 = \sigma_B^2 = 14/13$ and $\sigma_T^2 = 16/13$. With n cards seen, the estimated probability that a one-unit player bet or a one-unit banker bet has an expected value of at least β when n cards remain unseen is given by (19.20) and evaluated for several choices of n and β in Table 19.8. We hasten to add that these numbers are based on three approximations, namely (11.74) on p. 415, (11.95) on p. 418, and (19.20). Because of the small sample sizes, the results are not very reliable, but they do suggest that card counting can overcome the house advantage only deep in the shoe, and even then only occasionally.

It might also be of interest to estimate the probability that our proposed counting system suggests that the player bet is more favorable (or less unfavorable) than the banker bet. This is simply

$$P(Z_n^P > Z_n^B)$$

$$= P\left(\mu_P + \frac{\gamma_P}{N-n}\sum_{j=1}^n P_{X_j} > \mu_B + \frac{\gamma_B}{N-n}\sum_{j=1}^n B_{X_j}\right)$$

$$= P\left(\sum_{j=1}^n P_{X_j} > \left\lfloor (N-n)\frac{\mu_B - \mu_P}{\gamma_P + \gamma_B}\right\rfloor + \frac{1}{2}\right) \tag{19.23}$$

$$\approx 1 - \Phi\left(\frac{1}{(N-n)\sigma_P}\left\{\left\lfloor (N-n)\frac{\mu_B - \mu_P}{\gamma_P + \gamma_B}\right\rfloor + \frac{1}{2}\right\}\sqrt{\frac{(N-n)(N-1)}{n}}\right).$$

Numerical values are evaluated in Table 19.9.

19.3 Problems

19.1. *Analysis of one of banker's departures from symmetric play.* Give as simple an explanation as possible as to why it is optimal for banker to stand with a two-card total of 3 when player's third card is 8. (For simplicity, assume an infinite deck.) This may be contrary to intuition because six cards (namely, 1–6) increase banker's total while only three (namely, 7–9) decrease it.

Table 19.8 Estimating the probability that a one-unit baccarat bet has expected profit of at least β units as a function of the number of cards that remain unseen. The last column approximates the card counter's estimated expected profit when he bets one unit in situations perceived as favorable and nothing otherwise.

cards left	β				$E[(Z^*)^+]$
	$-.01$	$.00$	$.01$	$.02$	
		player bet			
6	.422	.161	.015	.001	.000 641
13	.342	.039	.001	.000	.000 122
26	.313	.007	.000	.000	.000 010
52	.260	.000	.000	.000	.000 000
		banker bet			
6	.422	.161	.037	.001	.000 843
13	.446	.068	.002	.000	.000 202
26	.461	.012	.000	.000	.000 026
52	.415	.001	.000	.000	.000 001
		tie bet			
6	.098	.048	.048	.048	.002 793
13	.016	.016	.008	.004	.000 290
26	.001	.001	.000	.000	.000 008
52	.000	.000	.000	.000	.000 000

19.2. *Expected values of player's and banker's hands.*

(a) Using Table 19.2 on p. 602 for the joint distribution (19.11) on p. 601 of the values of player's and banker's hands at the conclusion of play, find the marginal means and variances. Show that, perhaps surprisingly, player's hand has a higher expected value than banker's.

(b) Find the correlation between the values of player's and banker's hands at the conclusion of play.

19.3. *Conditional expectation given player's first two cards.*

(a) Given player's first two cards, and given that neither player nor banker has a natural, evaluate the conditional expectation from a one-unit player bet. (There are 44 cases to consider.)

(b) It has been suggested that casinos offer a *surrender* option at baccarat. This would mean that, assuming no naturals have been dealt, bettors on

Table 19.9 Estimating the probability that the player bet is more favorable than the banker bet, as perceived by the counting system.

cards left	probab.	cards left	probab.	cards left	probab.
13	.500	104	.331	260	.190
26	.423	156	.279	312	.115
52	.388	208	.254	364	.032

player can surrender half their bets for the privilege of exiting from the coup. The decision would be made after exposing player's first two cards but before exposing banker's first two cards (while confirming that banker does not have a natural). Show that it is never correct to surrender. In fact, show that, perhaps surprisingly, even if the cost of surrendering were only a quarter of the amount bet, it would still never be correct to surrender.

19.4. *Distribution of the number of cards required for a coup.*

(a) Assuming an eight-deck shoe, evaluate the probabilities of the following events: both player and banker stand; player stands and banker draws; player draws and banker stands; both player and banker draw; player draws; banker draws; four cards required for coup; five cards required; six cards required. Also, find the mean number of cards required for the coup. Results should be accurate to nine decimal places.

(b) Repeat part (a) for an infinite-deck shoe, except that results should be exact. (A common denominator is $(13)^6$.)

(c) Assume an infinite-deck shoe, and assume that no coup is begun once 411 or more cards have been seen (as would be the case for an eight-deck shoe). Use Wald's identity to derive upper and lower bounds on the mean number of coups. What does the renewal theorem (Theorem 4.4.1 on p. 146) tell us about the probability that the 411th card begins a new coup? What does it tell us about the probability that 416 cards are used? Can we be more precise about the mean number of coups? What is the mean number of cards required for the last coup (the one that uses the 411th card)?

19.5. *Exact house advantage.* The eight-deck house advantage figures in Table 19.4 on p. 605 are rounded to nine decimal places, and are therefore only approximations (albeit very good ones). Find the exact figures as rational numbers. Note, for example, in the case of the player bet with pushes included, a possible denominator is

$$\binom{416}{2,2,1,1,410} = \frac{(416)_6}{4} = 1{,}249{,}599{,}568{,}875{,}840. \tag{19.24}$$

19.6. *House advantage as a function of the number of decks.* Although we emphasized the case of $d = 8$ decks, the formulas given apply more generally. In fact, as long as d is a positive-integer multiple of $\frac{1}{4}$, baccarat can be played with d decks. Does the house advantage for the player bet (pushes included and excluded) appear to be a monotone function of d? Answer the same question for the banker bet.

19.7. *Eliminating the 5 percent commission and EZ Baccarat.* Instead of charging a 5 percent commission on a winning banker bet, which is a nuisance and slows down the game, various alternatives have been suggested that allow an even-money payoff, yet give the house a suitable advantage.

(a) In *EZ Baccarat*, the casino bars a winning banker three-card 7. In other words, if banker's hand comprises three cards and has value 7 at the conclusion of play while the player's hand comprises two or three cards and has value 6 or less, then the banker bet is declared a push. Show that the house advantage of the modified banker bet (pushes included) is equivalent to the house advantage of the standard banker bet (pushes included) with a commission of about 4.914 percent on winning bets.

(b) Suppose that the casino bars a banker 8 or 9 vs. a player 2. In other words, if the value of banker's hand is 8 or 9 (natural or not) at the conclusion of play and that of player's hand is 2, then the banker bet is declared a push. Show that the house advantage of the modified banker bet (pushes included) is equivalent to the house advantage of the standard banker bet (pushes included) with a commission of about 4.998 percent on winning bets.

(c) Suppose that the casino declares a loss for banker on all natural pushes. In other words, if banker's two-card total is 8 or 9 and player's is the same, then the banker bet is lost (and the player bet is pushed). Find the house advantage of this modified banker bet (pushes included and excluded).

(d) Suppose that the casino declares a loss for banker on all pushes of 8 or 9, natural or not. Find the house advantage of this modified banker bet (pushes included and excluded).

19.8. *Bahama Baccarat.* Bahama Baccarat is a simplification of standard baccarat. First, banker's draw/stand rules are the same as those of player. Specifically, banker draws to 0–5 and stands on 6 and 7, just as player does. Second, there is no commission on winning banker bets. Third, to give the house an edge, winning hands with value 2 are barred, that is, they result in a push, for either player or banker. Notice that the player bet and the banker bet are identically distributed.

(a) Find the house advantage.

(b) Argue that the card counter never has a positive-expectation bet.

19.9. *Six-card residual subsets.* Assume, unrealistically, that the eight-deck shoe is dealt to the bottom, with any incomplete coups void. Suppose that six cards remain, so that exactly one coup remains.

(a) For the player bet, the banker bet, and the tie bet, determine the probability that the bet is superfair as well as the conditional expectation of the advantage (pushes included), given that it is superfair.

(b) Determine for each bet which subsets achieve the maximal expectation.

19.10. *Accuracy of card counting.* Estimate the card counter's expected profit from a one-unit bet on player if 32 cards valued 0, 8 cards each valued 1, 2, 3, and 4, and 24 cards each valued 5, 6, 7, 8, and 9, have been played from the 416-card shoe, leaving 232 cards unseen. Do this first using the level-two counting system in Table 19.7 on p. 607, second using the more accurate effects of removal in Table 19.6 on p. 607, and third by exact calculation.

19.11. *Tie bet at 9:1.* Consider the tie bet, a bet that the next coup will result in a tie. Typically, this bet pays 8 to 1, but assume here that it pays 9 to 1 instead. (This still gives the house an approximately 4.844 percent advantage.) Evaluate the effects of removal for this bet, multiplied by 415, and analyze its card-counting potential.

19.12. *Bets on natural 8 or natural 9.* Although no longer available, there used to be bets on player natural 8, player natural 9, banker natural 8, and banker natural 9. Each of the four bets paid 9 to 1. For specificity, consider the bet on player natural 9.

(a) Find the effects of removal and a suitable point count. Estimate $E[(Z_n^*)^+]$ using (11.100) on p. 419, for $n := 416 - m$, where $m = 6, 13, 26, 52$.

(b) Compute the expectations of part (a) exactly.

Hint: (b) Suppose m cards remain. The number m_9 of 9s among them has a hypergeometric distribution. Further, given m_9, we can determine the conditional probability of a player natural 9 and therefore the conditional expectation of a one-unit bet on that event.

19.13. *Additivity of effects of removals.* It has been claimed that

> the removal of a six, seven, eight and nine from the deck produced a far greater reduction in the casino's advantage on player-hand bets than would be thought possible by adding up the effects of removing each card separately.

Show that this statement is incorrect by calculating the exact house advantage (pushes included) as well as the house advantage estimated using the effects of removal. Assume an eight-deck shoe.

19.14. *Tamburin–Rahm count.* Evaluate the vector $(h_0, h_1, \ldots, h_9)$ of (19.25) on p. 621, and show that it has high correlation with the Tamburin–Rahm level-3 count $(-1, 0, 1, 2, 3, 1, 0, 0, -1, -2)$. Can the correlation be increased without increasing the level?

19.15. *Correlation between successive coups.* Consider two successive baccarat coups, and let X_1^P (resp., X_1^B) denote the profit from a one-unit bet on player (resp., banker) at the first coup, and let X_2^P (resp., X_2^B) have a similar interpretation for the second coup. Find the correlations between X_1^P

and X_2^P; between X_1^P and X_2^B; between X_1^B and X_2^P; and between X_1^B and X_2^B.

19.16. *Banker's strategy at baccarat en banque.* See the chapter notes for details of this game, and assume an infinite deck. It appears that Table 19.10 on p. 615 was derived under the dubious assumption that both players adopt the $(\frac{1}{2}, \frac{1}{2})$ mixed strategy. Under this assumption derive a more accurate version of Table 19.10. Normalize by requiring the $(5, 4)$ entry to be $\frac{1}{2}$.

19.4 Notes

The early history of baccarat was summarized by Epstein (1977, p. 193):

> Introduced into France from Italy during the reign of Charles VIII (*ca.* 1490), the game was apparently devised by a gambler named Felix Falguiere who based it on the old Etruscan ritualism of the "Nine Gods."[1] According to legend, twenty-six centuries ago in "The Temple of Golden Hair" in Etruscan Rome, the "Nine Gods" prayed standing on their toes to a golden-tressed virgin who cast a *novem dare* (nine-sided die) at their feet. If her throw was 8 or 9, she was crowned a priestess. If she threw a 6 or 7, she was disqualified from further religious office and her vestal status summarily transmuted. And if her cast was 5 or under, she walked gracefully into the sea. Baccarat was designed with similar partitions (albeit less dramatic payoffs) of the numbers, modulo 10.

Unfortunately, this charming story is of dubious validity. As Parlett (1991, pp. 81–82) put it,

> Unsupported protestations of mythic antiquity notwithstanding, Baccara ('Baccarat' in British and Nevadan casinos) does not grace the realms of recorded history before the nineteenth century, when it became firmly entrenched in French casinos.

This is consistent with Foster (1953b, p. 485), who stated that baccarat

> originated in the south of France, and came into vogue during the latter part of the reign of Louis Philippe.

(His reign lasted from 1830 to 1848.) If baccarat had existed as early as the 17th century, it would likely have been mentioned by Montmort (1713) or Huyn (1788), or in 18th-century French literature. But there is no such mention.

The earliest citation by *The Oxford English Dictionary* (Simpson and Weiner 1989) is 1866. The earliest citation by *Trésor de la langue française* (Imbs 1974) is 1851. This source states that "the points of 10, 20, 30 are named *baccara*," but that the origin of the word is unknown. (It is often said that baccara means "zero" in Italian, but this is incorrect.) Finally, the earliest citation by *Grande dizionario della lingua italiana* (Battaglia 1961) is

---

[1] Macaulay, *Horatius at the Bridge*: "Lars Porsena of Clusium/By the Nine Gods he swore/...". (Epstein's footnote.)

1855, and it is acknowledged there that the word is of French origin, albeit of unknown etymology. Many authors have stated that the game is of Italian origin, but in light of the above, this seems doubtful. Of course, games evolve. Should a game that is likely ancestral to baccarat be regarded as an early version of the game? We do not believe so, unless the game is fundamentally similar.

At least two older games share some of baccarat's features, and may therefore be ancestral to it. One is *macao*, which shares card values with baccarat as well as the value of the best possible hand (nine). According to *Grande dizionario della lingua italiana* (Battaglia 1961), macao is known as Italian-style baccara, and this may explain why baccarat is often thought to be of Italian origin. However, only one card is dealt initially and modulo-10 arithmetic is not used. Naturals include 7, 8, and 9, with bonuses paid for 8 and 9. See Foster (1953b, p. 484) or Sifakis (1990, p. 191) for details. Another possible ancestor is the Korean domino game *kōl ye-se* (or card game *ye-se*), which shares with baccarat the best possible hand (nine) and modulo-10 arithmetic. See Culin (1895, p. 107) and Musante (1985) for further details. Finally, there is *ronfle*, which dates back to the 16th century and was said by Billard (1883, p. 192) to be an ancestor of baccarat. This conclusion, however, was apparently based on only the fact that ronfle is what Parlett (1991, Chapter 7) has called a face-count game, and is therefore unpersuasive.

There are three versions of baccarat. First, there is *chemin de fer*, introduced in Example 5.1.4. Second, there is *baccara en banque* or *baccara à deux tableaux*. And finally, there is *punto banco* or *baccarat*, the subject of this chapter. The first two versions date back to the 19th century, while the third is more recent and originated in Argentina (Renzoni and Knapp 1973, p. 11). It came to Nevada in November, 1959, thanks to Francis "Tommy" Renzoni (Renzoni and Knapp 1973, p. 69). (See Stuart 2007, p. 60, for the story of Renzoni's suicide.) The tie bet was an innovation of Jess Marcum (Johnston 1992, p. 5). All three versions of baccarat are currently played in Monte Carlo.

Let us describe baccara en banque. Like chemin de fer, it is played with six decks. (In the 19th century, three decks were typically used; see Hoffmann 1891.) Instead of dealing two two-card hands, called player and banker, the dealer deals three two-card hands, which may be called player 1, player 2, and banker. Bets can be placed on player 1, on player 2, or on both in equal proportions. Bets on banker are not allowed. The result is essentially two dependent games of chemin de fer, player 1 vs. banker and player 2 vs. banker. Player 1 announces his decision as to whether to draw or stand, and then player 2 does the same. (Recall that only when a player has a two-card total of 5 does he have any discretion.) Only then are the third cards dealt face up.[2] The banker's decision as to whether to draw or stand will depend on

---

[2] These are the current rules as described by Barnhart (1980). In what Downton and Lockwood (1976) called the "traditional form of the game," player 1's third card (if any) is revealed before player 2 decides whether to take a third card.

player 1's third card (if any), player 2's third card (if any), the total amount of money bet on player 1, and the total amount of money bet on player 2.

The banker's drawing strategy at baccara en banque is ordinarily not revealed to the public. When Barnhart (1990) politely asked the gaming inspector in Monte Carlo's ornate Salle Privée why this strategy was kept secret from the gamblers, the gentleman

> first stared thoughtfully through his dark-tinted glasses at the gilded two-story-high ceiling and then replied with equal politeness: "Because, Monsieur, it does not concern them."

Nevertheless, one possible banker strategy was described by Barnhart (1978, p. 217). Denoting the (j, k) entry in Table 19.10 by $c(j, k)$, banker always draws to 0–2 and always stands on 7. If his two-card total is $j \in \{3, 4, 5, 6\}$, if k_1 is player 1's third card, if k_2 is player 2's third card, if B_1 units are bet on player 1, and if B_2 units are bet on player 2, banker evaluates $c(j, k_1)B_1$ and $c(j, k_2)B_2$. Whichever is larger, he stands if the corresponding entry in Table 19.10 is italic and draws otherwise. (See Boll (1944, p. 67) for a similar but less accurate table.) The casino employee or syndicate member playing the role of banker memorizes the tables and all calculations are done mentally. See Problem 19.16.

Table 19.10 Banker's secret strategy at baccara en banque. See text.

j	0	1	2	3	4	k 5	6	7	8	9	$\varnothing$
3	11	21	31	49	53	38	23	7	*3*	1	50
4	*15*	*4*	6	16	33	38	23	7	*9*	*18*	50
5	*33*	*30*	*20*	*10*	$\frac{1}{2}$	18	23	7	*9*	*24*	37
6	*39*	*49*	*46*	*35*	*25*	*15*	3	7	*9*	*24*	*8*

Aside from the differences in rules between the three variations of baccarat, there are differences in practices. Baccarat (punto banco) is banked by the casino, but the players are given the opportunity to deal the cards. (Recall that there are no strategy decisions to be made.) Baccara en banque is typically banked by a syndicate that shares its profits with the casino. Chemin de fer is not a banked game, and each player has the opportunity of becoming the banker as the shoe moves around the table *counter-clockwise*. When the banker wins, he retains the shoe and the casino collects a 5 percent commission. In both chemin de fer and baccara en banque, the betting limits are set by the banker.

There is also a game called *mini-baccarat*, which is mathematically identical to baccarat (perhaps with only six decks) but is played at a smaller table and at a faster pace, with less formality and (usually) lower stakes, and only the dealer handles the cards.

It is curious that baccara en banque requires perhaps more skill on the part of the banker than any other casino game, whereas baccarat (punto banco) requires absolutely no skill on the part of the players. As Hughes (1976) put it,

> To say that the actual play of Baccara is simple is an understatement. Most children's games are infinitely more complicated, and it is doubtful if Baccara played without stakes could hold the attention of any but the most backward child.

A similar but more charitable view was expressed by Dawson (1950, p. 148):

> Although from a bare description it may sound stupid and uninteresting, it needs but a single experiment in actual play to prove its charm and fascination—that is of course from a gambling point of view.

Baccarat played a central role in an 1890–1891 scandal involving the Prince of Wales [1841–1910], who was later to become King Edward VII upon the death of his mother, Queen Victoria. Briefly, Lieutenant-Colonel Sir William Gordon-Cumming [1848–1930], a personal friend of the Prince, was accused of cheating at a private game of baccarat in which the Prince was banker. It was alleged that Gordon-Cumming repeatedly added to his stake after learning that he had won. The story leaked out, and Gordon-Cumming, to salvage his reputation, brought a civil action against his accusers for slander, which he lost. See Havers, Grayson, and Shankland (1988).

One of literature's most dramatic games of baccarat[3] occurred in Ian Fleming's [1908–1964] *Casino Royale* (1953, Chapters 10–13). At the climax of the story, 32 million francs were at stake on a single coup. The two combatants were British agent James Bond ("007") and Soviet agent Le Chiffre ("The Number").[4]

> The two cards slithered towards him across the green sea.
> Like an octopus under a rock, Le Chiffre watched him from the other side of the table.
> Bond reached out a steady right hand and drew the cards towards him. Would it be the lift of the heart which a nine brings, or an eight brings?
> He fanned the two cards under the curtain of his hand. The muscles of his jaw rippled as he clenched his teeth. His whole body stiffened in a reflex of self-defence.
> He had two queens, two red queens.
> They looked roguishly back at him from the shadows. They were the worst. They were nothing. Zero. Baccarat.
> 'A card,' said Bond fighting to keep hopelessness out of his voice. He felt Le Chiffre's eyes boring into his brain.
> The banker slowly turned his own two cards face up.

---

[3] It was actually what might be called "baccara à un tableau," as there were not enough players for baccara à deux tableaux (Fleming 1953, Chapter 9). Thus, the game was equivalent to chemin de fer with a fixed banker and no constraints on banker's strategy.

[4] From Fleming (2002, Chapter 13, pp. 86–87). Used with permission of Penguin Group USA and Penguin Group UK.

He had a count of three—a king and a black three.

Bond softly exhaled a cloud of tobacco smoke. He still had a chance. Now he was really faced with the moment of truth. Le Chiffre slapped the shoe, slipped out a card, Bond's fate, and slowly turned it face up.

It was a nine, a wonderful nine of hearts, the card known in gipsy magic as 'a whisper of love, a whisper of hate', the card that meant almost certain victory for Bond.

The croupier slipped it delicately across. To Le Chiffre it meant nothing. Bond might have had a one, in which case he now had ten points, or nothing, or baccarat, as it is called. Or he might have had a two, three, four, or even five. In which case, with the nine, his maximum count would be four.

Holding a three and giving nine is one of the moot situations at the game. The odds are so nearly divided between to draw or not to draw. Bond let the banker sweat it out. Since his nine could only be equalled by the banker drawing a six, he would normally have shown his count if it had been a friendly game.

Bond's cards lay on the table before him, the two impersonal pale pink-patterned backs and the faced nine of hearts. To Le Chiffre the nine might be telling the truth or many variations of lies.

The whole secret lay in the reverse of the two pink backs where the pair of queens kissed the green cloth.

The sweat was running down either side of the banker's beaky nose. His thick tongue came out slyly and licked a drop out of the corner of his red gash of a mouth. He looked at Bond's cards, and then at his own, and then back at Bond's.

Then his whole body shrugged and he slipped out a card for himself from the lisping shoe.

He faced it. The table craned. It was a wonderful card, a five.

'*Huit à la banque,*' said the croupier.

As Bond sat silent, Le Chiffre suddenly grinned wolfishly. He must have won.

The croupier's spatula reached almost apologetically across the table. There was not a man at the table who did not believe Bond was defeated.

The spatula flicked the two pink cards over on their backs. The gay red queens smiled up at the lights.

'*Et le neuf.*'

A great gasp went up round the table, [. . .]

The most celebrated baccarat player in history was Nicolas Zographos [1886–1953], one of the four initial members of the Greek Syndicate, the other three being Eli Eliopulo, Zaret Couyoumdjian, and Athanase Vagliano. They banked the baccara en banque games at Deauville, Cannes, and Monte Carlo during the period between the World Wars. It was Zographos who in 1922 instituted the policy of "tout va" (no betting limits). He had a phenomenal memory, and could remember all cards played during a baccarat shoe (as many as 312) and use this information to advantage.

Zographos was born in Athens, the son of a professor of political economy, and he showed mathematical aptitude at an early age. He went to Munich to study engineering. According to Graves (1963, p. 12),

> Years later, when he took the bank at baccarat, his father was informed that his son was a banker. To a dedicated professor, this had only one meaning and nobody ever disillusioned him.

For the full story of the Greek Syndicate, see Graves (1963).

The first study of baccarat, and indeed the first published reference to the game that we have seen, was by Charles Van-Tenac (1847, pp. 84–97). He assumed, without explanation, that player draws to 4 or less and stands with 5 or more. He then analyzed banker's strategy for each player third card. For example, he argued that, if player draws a card and receives 0, then banker should stand on 3. For if player had 0, 1, or 2, banker would win; if he had 3, there would be a tie; and if he had 4, banker would lose. But Van-Tenac did not compute the resulting expectation or compare it to the expectation when banker draws. Thus, his recommendations were rather inaccurate.

The first accurate analysis of baccarat, as far as we know, was carried out by French mining engineer and actuary Émile Dormoy [1829–1891] (1872, 1873). The second citation refers to the book, *Théorie mathématique du jeu de baccarat*, while the first refers to a two-part journal article with the same title. This work contains tables of expectation differences corresponding to Tables 5.3 and 5.4 (see pp. 39 and 49 of his book), except that only banker totals of 3–6 are considered and results are rounded (sometimes inaccurately) to two decimal places. Of Dormoy's 88 recommendations, only two are in error, and both errors can be attributed to rounding: When player draws to 4 or less, banker has 5, and player's third card is 4, Dormoy claimed it is indifferent whether banker draws or stands, and this is nearly correct. In fact, the corresponding entries in Table 5.4 are 300/1,157 (player's expected profit when banker draws) and 299/1,157 (player's expected profit when banker stands), so banker stands. When player draws to 5 or less, banker has 6, and player's third card is 6, Dormoy recommended that banker stand. The corresponding entries in Table 5.3 are 203/1,365 and 208/1,365, so here banker draws.

The next contribution was Badoureau's [1853–1923] (1881) "Étude mathématique sur le jeu de baccarat." According to Deloche and Oguer (2007b), Badoureau was famous in France for having written a supplementary chapter to Jules Verne's *Topsy-Turvy* intended to answer objections from scientists. In his paper Badoureau derived the distribution of banker's final total as in Table 19.3, but his results are inaccurate.

The next study of baccarat was that of Ludovic Billard (1883), who was said to have been an expert baccarat player. In fact, in a preface by Georges de Lamazière, Billard is called "The Baccara's King!" Billard's treatise is titled *Bréviaire du baccara expérimental*, and he was aware of the work of his predecessors, Dormoy and Badoureau. Billard acknowledged using the experimental method (simulation) rather than mathematics. His recommendations coincide with Dormoy's exactly (including the two errors), raising the question of whether he actually did the experiments. His results were reproduced by Hoffmann (1891), who introduced a third error. Finally, there is Laun (c. 1891), whose strategy tables (p. 35) are correct (with one exception) but incomplete, as he neglected to give banker's strategy when player stands.

A contribution to baccarat was made by Joseph Bertrand [1822–1900] (1888, pp. 38–42), who asked whether it is advantageous for player or banker to draw on 5. His analysis was later described by Émile Borel [1871–1956] (1953) as "extremely incomplete." An *American Mathematical Monthly* problem posed by Holmes (1907) raised the similar question of whether player should stand on 5, but the published solution is of little interest because the proposer neglected to mention naturals or the addition-modulo-10 feature of baccarat.

Writing before the concept of a mixed strategy was widely known, Beresford (1926, p. 71) remarked,

> As to the percentage in favor of the banker at Baccara or Chemin de Fer, it is not of course in the power of any man to make an acceptable mathematical calculation. The unknown quantity of the draw, or otherwise, with the point of five must in itself prohibit any definite estimate.

Le Myre (1935) and Boll (1936) both gave thorough treatments of baccarat, including the first use of a mixed strategy, specifically the player strategy that stands on 5 and draws to 5 with probability $\frac{1}{2}$ each. Also, these authors finally give the exactly correct banker strategy for each of player's pure strategies. In addition, one finds in Le Myre (1935) several references to other works on baccarat that we have not yet mentioned. These include Bertezène (1897), Renaudet (1922), Lafrogne (1927), and Lafaye and Krauss (1927). As noted by Deloche and Oguer (2007b), these works did not break new ground.

Although Borel had the tools to solve baccarat in the 1920s, they were not yet sufficiently well known in the 1930s to have come to the attention of Le Myre or Boll. Boll (1944) later provided an even more extensive analysis but it was still lacking a game-theoretic perspective.

It wasn't until 1957 that John G. Kemeny [1926–1992] and J. Laurie Snell [1925–] (1957) were first to find the optimal mixed strategies in baccarat chemin de fer. Independently of Kemeny and Snell, Kendall and Murchland (1964) derived most of Example 5.1.4 (such as the equivalent of Tables 5.3 and 5.4), but made an approximation that led to an incorrect solution: They eliminated from consideration all banker pure strategies except the two that are optimal against player's two pure strategies. Also independently of Kemeny and Snell, Foster (1964) found the correct optimal mixed strategies but rounded his results to a single significant digit. He assumed that player's and banker's two-card hands are each sampled without replacement, but *independently*, and are independent of any subsequent cards. One result of this hybrid assumption is that $(5, 4)$ belongs to T_1 rather than T_3, in the notation of Lemma 5.1.3.

Wilson (1965, Chapter 13) gave what remains to this day the best treatment of baccarat/punto banco intended for the general reader, including perhaps the first correct evaluation of the house advantages of the player and banker bets (sampling with replacement, pushes included). Scarne (1961, pp. xviii, 427) claimed priority, but his figures are slightly inaccurate. Wilson also included a version of our Table 19.5. He pointed out, in particular, the single negative entry in Table 19.5, corresponding to $(j, k) = (4, 1)$, and suggested it was due to a miscalculation long ago. He might have also noted the case $(j, k) = (6, \varnothing)$, which should be one of banker's departures from symmetric play but is not. Downton and Lockwood (1975) offered the more charitable explanation that perhaps these two departures from optimality were intentional, a partial attempt to equalize the house advantages of the player and banker bets.

Downton and Lockwood (1975) regarded chemin de fer as a 2×2^{484} matrix game, effectively allowing banker, but not player, to take the composition of his two-card hand into account.

Each of these works assumed sampling with replacement in whole or in part. The first analysis of baccarat assuming sampling without replacement was carried out by Thorp and Walden (1966), who obtained for eight decks results consistent with our Table 19.4.

The most thorough analysis of baccara en banque to date was provided by Downton and Lockwood (1976). They assumed the "traditional form of the game" (see footnote 2 on p. 614) and considered two kinds of optimal strategies. In the first, which they called the "co-operative optimum strategy," players 1 and 2 jointly choose their strategies to "minimize the banker's maximum expected gain per unit total stake." In the second, which they called the "optimum equilibrium strategy," each player individually chooses his strategy to minimize banker's maximum expected gain per unit stake from that player. Downton and Lockwood suggested that the cooperative approach is of greatest interest to the banker, while the equilibrium approach is academic. Other works on baccara en banque include Kendall and Murchland (1964), F. G. Foster (see the discussion of Kendall and Murchland 1964), Downton and Holder (1972), Barnhart (1980), and Judah and Ziemba (1983).

Card counting at baccarat was first addressed by Thorp and Walden (1966), who found the greatest opportunities in the natural eight and natural nine bets (Problem 19.12), which no longer exist. (See Scarne 1961, p. 419, for a photograph of a baccarat layout exhibiting these bets.) The level-two point count system of Table 19.7 coincides with that of Thorp (1984, Table 3-2). Downton and Lockwood (1975) addressed card counting at chemin de fer based on counting cards valued 0, cards valued 5–8, and all other cards. Sklansky (1982) analyzed certain six-card residual subsets, and Joel Friedman (unpublished) subsequently analyzed all of them. A clear and elementary explanation of why card counting is ineffective at baccarat was given by Vancura (1996, pp. 165–172). Griffin (1999, pp. 216–223) provided another explanation based on computer simulations. He considered the correlation between the estimate of favorability provided by card counting and the actual favorability, and he showed that this correlation decays as the shoe is depleted. As Griffin (1999, p. 227) put it,

> Traditional card counting systems are futile because the only worthwhile wagers occur near the end of the shoe when correlations begin to disintegrate to the degree that the capacity to distinguish the favorable subsets is lost.

The most thorough study of card counting at baccarat was carried out by Grosjean (2007a; 2009, Chapters 84, 85).

Tamburin and Rahm (1983) published what they called "the first effective card counting systems" for baccarat. Their level-three system has correlations 0.431 and 0.414 with the effects of removal for the player bet and the banker bet, respectively. Their level-two system has correlations 0.374 and 0.357. This suggests that their systems are rather less effective than the system of Table 19.7 (correlations 0.959 and 0.967), which is itself only marginally effective. Tamburin and Rahm's systems have high correlation not with the effects of removal but rather with a decomposition $(h_0, h_1, \ldots, h_9)$ of the expectation of the player bet, namely

$$- 0.012281406 \approx \mathrm{E}[G] = \sum_{i=0}^{9}(1 + 3\delta_{i0})h_i, \qquad (19.25)$$

where we assume an infinite-deck shoe and

$$(1 + 3\delta_{i0})h_i := \mathrm{E}[G\, 1_{\{X_3=i,Y_3=\varnothing\}}] + \mathrm{E}[G\, 1_{\{X_3=\varnothing,Y_3=i\}}] \qquad (19.26)$$
$$+ (1/2)(\mathrm{E}[G\, 1_{\{X_3=i,Y_3\neq\varnothing\}}] + \mathrm{E}[G\, 1_{\{X_3\neq\varnothing,Y_3=i\}}])$$

(see Problem 19.14). Although the criterion seems plausible, there is no mathematical basis for it, as far as we know.

Problem 19.1 was discussed by Wilson (1965, pp. 201–202). Problem 19.3 was motivated by May (1998, p. 131), Problem 19.4 was motivated by Barnhart (1980), and Problem 19.7 was motivated by Hannum (1998), who considered a variety of alternatives to the 5 percent commission on winning banker bets. See Grosjean (2009, p. 502) for a similar analysis. EZ Baccarat is marketed by DEQ Systems Corp. Parts (c) and (d) were motivated by a suggestion of May (1998, pp. 129–130). Bahama Baccarat (Problem 19.8), also known as Quick-Draw Baccarat, was patented by Richard and Kurt Lofink in 1994 (U.S. patent 5,362,064); in fact, their patent is broad enough to apply to EZ Baccarat as well. Problem 19.9 was suggested by Sklansky (1982) and subsequent work by Joel Friedman (unpublished) and Griffin (1999, pp. 216–217). Problem 19.11 was done by Griffin (1999, p. 220), and Problem 19.12 was treated at length by Thorp and Walden (1966). The quote in Problem 19.13 is from Tamburin (1983).

Chapter 20
Trente et Quarante

There came, at a time when the chief Lenoir was at Paris, and the reins of government were in the hands of his younger brother, a company of adventurers from Belgium, with a capital of three hundred thousand francs, and an infallible system for playing *rouge et noir*, and they boldly challenged the bank of Lenoir, and sat down before his croupiers, and defied him.

William Makepeace Thackeray, *The Kickleburys on the Rhine*

Trente et quarante ("thirty and forty" in French) is an elegant card game that dates at least as far back as the 17th century and is still played in Monte Carlo. It has been studied by Poisson, De Morgan, Bertrand, and Thorp, among others. In Section 20.1 we describe the four betting opportunities and evaluate the associated probabilities and house advantages. In Section 20.2 we analyze the card-counting potential of trente et quarante.

20.1 Red, Black, Color, Inverse

Trente et quarante, also known as *rouge et noir* ("red and black" in French), is played with a six-deck shoe comprising 312 cards. Suits do not matter except insofar as they determine color: Diamonds and hearts are red, whereas clubs and spades are black. Aces have value one, court cards have value 10, and every other card has value equal to its nominal value. Two rows of cards are dealt. In the first row, called Black, cards are dealt until the total value is 31 or greater. In the second row, called Red, the process is repeated. Thus, each row has associated with it a total between 31 and 40 inclusive.

Four even-money bets are available, called red, black, color, and inverse. (To avoid confusion, we use upper case when referring to the row, e.g., Red, and lower case when referring to the wager, e.g., red.) Let us define the *winning row* (Red or Black) to be the one with the smaller total if the two

S.N. Ethier, *The Doctrine of Chances*, Probability and its Applications, DOI 10.1007/978-3-540-78783-9_20, © Springer-Verlag Berlin Heidelberg 2010

totals are different; of course, the other row is the *losing row*. A bet on *red* wins (resp., loses) if Red is the winning (resp., losing) row. A bet on *black* wins (resp., loses) if Black is the winning (resp., losing) row. A bet on *color* wins (resp., loses) if the color of the first card dealt to Black is the same as the color of the winning (resp., losing) row. A bet on *inverse* wins (resp., loses) if the color of the first card dealt to Black is different from the color of the winning (resp., losing row). For all four bets, a push occurs if the Red and Black totals are equal and greater than 31. If the Red and Black totals are both equal to 31, half the amount of the bet is lost; this is called a *refait* and is the source of the casino's advantage.

For all four bets, if the Red and Black totals are both equal to 31, there is typically an *en prison* option similar to that in roulette. We will not consider this option in what follows (but see Problem 20.8 on p. 636).

Associated with each of the four even-money bets is an *insurance bet* for 1 percent of the original bet. It pays off the loss in the case of a refait though the bet itself is retained by the casino (just as an insurance company retains the premium when it pays off a claim). The insurance bet is lost if the original bet is won or lost, and it is pushed if the original bet is pushed. A drawback to taking insurance is that it restricts one's bets to 100 times the smallest unit of currency accepted, and integer multiples thereof.

We define a *trente-et-quarante sequence* to be a finite sequence $a_1, \ldots, a_K$ of positive integers, none of which exceeds 10, and at most 24 of which are equal to 1, such that

$$a_1 + \cdots + a_{K-1} \leq 30 \quad \text{and} \quad a_1 + \cdots + a_K \geq 31. \tag{20.1}$$

(In particular, at most 16 of the terms are equal to 2, at most 11 are equal to 3, and so on.) Clearly, if $a_1, \ldots, a_K$ is such a sequence, then its length K satisfies $4 \leq K \leq 28$. (24 1s and four 2s give $K = 28$.) The number of trente-et-quarante sequences can be evaluated by noting that, given any such sequence, each permutation of the terms that fixes the last term results in another trente-et-quarante sequence. So we let $p_{10}(k)$ be the set of partitions of the positive integer k with no part greater than 10. Such a partition can be described as $(k_1, \ldots, k_{10})$, with $k_i \geq 0$ being the multiplicity of part i. In particular, $\sum_{i=1}^{10} i k_i = k$. It follows that the number of trente-et-quarante sequences is

$$\sum_{k=21}^{30} \sum_{(k_1, \ldots, k_{10}) \in p_{10}(k): \, k_1 \leq 24} \binom{k_1 + \cdots + k_{10}}{k_1, \ldots, k_{10}}$$
$$\cdot (10 - (31 - k) + 1 - \delta_{k,30} \delta_{k_1,24}). \tag{20.2}$$

Here k represents the penultimate total, that is, the sum of all terms except the last one. The multinomial coefficient is the number of ways of ordering the terms (except for the last one), and the last factor is the number of possible final terms, which will necessarily increase the total of k to 31 or more. (The term $\delta_{k,30}\,\delta_{k_1,24}$ is needed because, if all 24 1s have been used and the penultimate total is 30, the final term cannot be a 1.) To evaluate (20.2), one must cycle through the elements of $p_{10}(k)$ for $k = 21,\ldots,30$. To illustrate how this is done, we provide in Table 20.1 an incomplete list of the elements of $p_{10}(30)$, ordered by putting $(k_{10}, k_9, \ldots, k_1)$ in lexicographical order. This leads to a straightforward algorithm, but it requires a computer because the double sum contains 18,096 terms. The conclusion is that there are 9,569,387,893 trente-et-quarante sequences.

Table 20.1 A (necessarily abbreviated) list of the elements of $p_{10}(30)$, the partitions of 30 with no part exceeding 10. The last column counts the numbers of 10s, 9s, …, 1s.

1	$10 + 10 + 10$	$(3, 0, 0, 0, 0, 0, 0, 0, 0, 0)$
2	$10 + 10 + 9 + 1$	$(2, 1, 0, 0, 0, 0, 0, 0, 0, 1)$
3	$10 + 10 + 8 + 2$	$(2, 0, 1, 0, 0, 0, 0, 0, 1, 0)$
4	$10 + 10 + 8 + 1 + 1$	$(2, 0, 1, 0, 0, 0, 0, 0, 0, 2)$
5	$10 + 10 + 7 + 3$	$(2, 0, 0, 1, 0, 0, 0, 1, 0, 0)$
$\vdots$		
42	$10 + 10 + 1 + 1 + \cdots + 1$	$(2, 0, 0, 0, 0, 0, 0, 0, 0, 10)$
43	$10 + 9 + 9 + 2$	$(1, 2, 0, 0, 0, 0, 0, 0, 1, 0)$
$\vdots$		
530	$10 + 1 + 1 + \cdots + 1$	$(1, 0, 0, 0, 0, 0, 0, 0, 0, 20)$
531	$9 + 9 + 9 + 3$	$(0, 3, 0, 0, 0, 0, 0, 1, 0, 0)$
$\vdots$		
3,590	$1 + 1 + \cdots + 1$	$(0, 0, 0, 0, 0, 0, 0, 0, 0, 30)$

Similar reasoning gives the probabilities of the ten trente-et-quarante totals, assuming a full six-deck shoe. Let the initial counts of the ten card values be $n_1 = \cdots = n_9 = 24$ and $n_{10} = 96$, with $N := n_1 + \cdots + n_{10} = 312$. Then, for $i = 31,\ldots,40$, the total i occurs with probability

$$P(i) := \sum_{k=i-10}^{30} \sum_{(k_1,\ldots,k_{10})\in p_{10}(k)} \binom{k_1 + \cdots + k_{10}}{k_1, \ldots, k_{10}}$$
$$\cdot \frac{(n_1)_{k_1} \cdots (n_{10})_{k_{10}} (n_{i-k} - k_{i-k})}{(N)_{k_1 + \cdots + k_{10} + 1}}. \quad (20.3)$$

The condition $k_1 \leq 24$ can be omitted here, because $(n_1)_{k_1} = 0$ if $k_1 > n_1 = 24$. These numbers too are easily computed (the double sum for $P(31)$

contains 18,115 terms, of which 19 are 0), and we summarize the results in Table 20.2.

Table 20.2 The probabilities of the ten trente-et-quarante totals. Assumes sampling without replacement from the full six-deck shoe. Results are rounded to nine decimal places.

total	probability	total	probability
31	.148 057 777	36	.094 992 448
32	.137 826 224	37	.083 858 996
33	.127 576 652	38	.072 302 455
34	.116 865 052	39	.060 800 856
35	.106 151 668	40	.051 567 873

The distribution of the length of a trente-et-quarante sequence can be found by a slight modification of (20.3). Indeed, a trente-et-quarante sequence has length K with probability

$$
Q(K) := \sum_{i=31}^{40} \sum_{k=i-10}^{30} \sum_{\substack{(k_1,\ldots,k_{10}) \in p_{10}(k):\ k_1+\cdots+k_{10}+1=K}}
$$
$$
\cdot \binom{k_1 + \cdots + k_{10}}{k_1, \ldots, k_{10}} \frac{(n_1)_{k_1} \cdots (n_{10})_{k_{10}} (n_{i-k} - k_{i-k})}{(N)_{k_1+\cdots+k_{10}+1}}. \quad (20.4)
$$

Numerical results are summarized in Table 20.3. Both (20.3) and (20.4) simplify considerably if one assumes sampling with replacement (see Problems 20.3 on p. 634 and 20.5 on p. 635).

A *coup* at trente et quarante involves an ordered pair of trente-et-quarante sequences $a_1, \ldots, a_K$ and $b_1, \ldots, b_L$, with at most 24 of the $K+L$ terms equal to 1 and at most 24 of them equal to 2. Clearly, if $a_1, \ldots, a_K$ and $b_1, \ldots, b_L$ is such a pair, then $8 \leq K+L \leq 44$. (24 1s, four 2s, and 16 2s give $K+L = 44$.)

Let us now find the joint distribution of the Black total and the Red total. Because sampling is without replacement, we cannot assume independence, though doing so gives a reasonable first approximation. Arguing as in (20.3), for $i, j = 31, \ldots, 40$, the totals i for Black and j for Red occur with probability

$$
P(i, j) := \sum_{k=i-10}^{30} \sum_{l=j-10}^{30} \sum_{(k_1,\ldots,k_{10}) \in p_{10}(k)} \sum_{(l_1,\ldots,l_{10}) \in p_{10}(l)}
$$
$$
\cdot \binom{k_1 + \cdots + k_{10}}{k_1, \ldots, k_{10}} \binom{l_1 + \cdots + l_{10}}{l_1, \ldots, l_{10}} \frac{(n_1)_{k_1+l_1} \cdots (n_{10})_{k_{10}+l_{10}}}{(N)_{k_1+\cdots+k_{10}+l_1+\cdots+l_{10}}}
$$

Table 20.3 The distribution of the length of a trente-et-quarante sequence. Assumes sampling without replacement from the full six-deck shoe. Results are rounded to six significant digits.

length	probability	length	probability
4	.260 817	17	$.434\,982 \cdot 10^{-10}$
5	.367 050	18	$.177\,920 \cdot 10^{-11}$
6	.239 624	19	$.578\,250 \cdot 10^{-13}$
7	.096 878 8	20	$.145\,805 \cdot 10^{-14}$
8	.028 043 6	21	$.276\,595 \cdot 10^{-16}$
9	.006 268 16	22	$.379\,031 \cdot 10^{-18}$
10	.001 127 78	23	$.354\,942 \cdot 10^{-20}$
11	.000 167 254	24	$.209\,899 \cdot 10^{-22}$
12	.000 020 728 8	25	$.695\,856 \cdot 10^{-25}$
13	.000 002 162 19	26	$.106\,452 \cdot 10^{-27}$
14	.000 000 190 061	27	$.524\,841 \cdot 10^{-31}$
15	.000 000 014 023 5	28	$.319\,192 \cdot 10^{-35}$
16	.000 000 000 861 242		

$$\cdot \frac{(n_{i-k} - k_{i-k} - l_{i-k})(n_{j-l} - k_{j-l} - l_{j-l} - \delta_{i-k,j-l})}{(N - k_1 - \cdots - k_{10} - l_1 - \cdots - l_{10})_2}. \quad (20.5)$$

It is clear from (20.5) (or see Example 11.2.2(d) on p. 401) that the joint distribution is symmetric; of course, both marginals are given by (20.3). The evaluation of (20.5) requires a fair amount of computing power, inasmuch as the quadruple sum for $P(31, 31)$ contains $(18,115)^2 = 328,153,225$ terms. The most important conclusions from the joint distribution are recorded in Table 20.4. The joint distribution itself is evaluated in Table 20.5.

Finally, the joint distribution of the lengths of the two trente-et-quarante sequences comprising a coup can be found by a slight modification of (20.5). Indeed, a trente-et-quarante coup has a Black sequence of length K and a Red sequence of length L with probability

$$Q(K, L) := \sum_{i=31}^{40} \sum_{k=i-10}^{30} \sum_{(k_1,\dots,k_{10}) \in p_{10}(k): \, k_1+\cdots+k_{10}+1=K}$$

$$\cdot \sum_{j=31}^{40} \sum_{l=j-10}^{30} \sum_{(l_1,\dots,l_{10}) \in p_{10}(l): \, l_1+\cdots+l_{10}+1=L}$$

$$\cdot \binom{k_1+\cdots+k_{10}}{k_1,\dots,k_{10}} \binom{l_1+\cdots+l_{10}}{l_1,\dots,l_{10}} \frac{(n_1)_{k_1+l_1} \cdots (n_{10})_{k_{10}+l_{10}}}{(N)_{k_1+\cdots+k_{10}+l_1+\cdots+l_{10}}}$$

Table 20.4 Some probabilities in trente et quarante, and the house advantage. Assumes sampling without replacement from the full six-deck shoe. Results are rounded to nine decimal places.

event	probability
Red total < Black total	$p := .445\,200\,543$
Red total > Black total	$p := .445\,200\,543$
Red total = Black total ≥ 32	$q := .087\,707\,543$
Red total = Black total $= 31$	$r := .021\,891\,370$

red, black, color, or inverse	house advantage
without insurance, pushes included	$\frac{1}{2}r = .010\,945\,685$
without insurance, pushes excluded	$\frac{1}{2}r/(1-q) = .011\,998\,000$
with insurance, pushes included	$0.01(1-q)/1.01 = .009\,032\,599$
with insurance, pushes excluded	$0.01/1.01 = .009\,900\,990$

$$\cdot \frac{(n_{i-k} - k_{i-k} - l_{i-k})(n_{j-l} - k_{j-l} - l_{j-l} - \delta_{i-k,j-l})}{(N - k_1 - \cdots - k_{10} - l_1 - \cdots - l_{10})_2}. \quad (20.6)$$

We assumed in Section 11.3 that a coup is not begun unless there are enough cards to complete it, but here we add the phrase "with high probability." Regarding K and L as random variables, one can use (20.6) to evaluate the distribution of $K + L$, and this is displayed in Table 20.6. It follows that $P(K + L > 20) < 0.000000606$, so we can safely take $m = 20$ in Section 11.3.

20.2 Card Counting

We can now address the card-counting potential of trente et quarante. Let D be the set of 312 cards (the full six-deck shoe), and let U be an arbitrary subset of D. Let P_U denote conditional probability given that U is the set of unseen cards, with all possible permutations of the cards of U equally likely. Let Y_1 denote the next card to be dealt, which is the first card dealt to Black. Let R (resp., B, C, I) be the event that a bet on red (resp., black, color, inverse) wins, let T be the event that Red and Black tie (at any number, 31 to 40 inclusive), and let T_{31} be the event of a refait. An exchangeability argument (see Example 11.2.2(d) on p. 401) shows that $P_U(R) = P_U(B)$, regardless of U, whereas it is not necessarily true that $P_U(C) = P_U(I)$ (see Problem 20.15 on p. 637); nevertheless, the latter equality *is* true if each card value has equal numbers of red and black representatives in U. More generally, conditioning on Y_1 we find that

Table 20.5 The probability $P(i, j)$ that the Black total is i and the Red total is j. (The distribution is symmetric, i.e., $P(i, j) = P(j, i)$ for $i, j = 31, \ldots, 40$, so it suffices to consider $i \leq j$.) Assumes sampling without replacement from the full six-deck shoe. Results are rounded to nine decimal places.

i	j	$P(i, j)$	i	j	$P(i, j)$	i	j	$P(i, j)$
31	31	.021 891 370	32	40	.007 120 173	35	37	.008 901 928
31	32	.020 388 372	33	33	.016 262 244	35	38	.007 677 199
31	33	.018 879 150	33	34	.014 904 199	35	39	.006 457 497
31	34	.017 300 060	33	35	.013 542 499	35	40	.005 478 069
31	35	.015 719 668	33	36	.012 122 659	36	36	.009 013 711
31	36	.014 071 677	33	37	.010 705 187	36	37	.007 961 839
31	37	.012 426 430	33	38	.009 232 389	36	38	.006 866 552
31	38	.010 716 875	33	39	.007 765 682	36	39	.005 775 434
31	39	.009 014 446	33	40	.006 589 062	36	40	.004 898 978
31	40	.007 649 729	34	34	.013 647 235	37	37	.007 020 594
32	32	.018 976 056	34	35	.012 402 551	37	38	.006 056 911
32	33	.017 573 581	34	36	.011 102 267	37	39	.005 094 553
32	34	.016 103 731	34	37	.009 804 028	37	40	.004 320 694
32	35	.014 632 450	34	38	.008 455 206	38	38	.005 213 743
32	36	.013 098 550	34	39	.007 111 866	38	39	.004 387 532
32	37	.011 566 833	34	40	.006 033 910	38	40	.003 720 409
32	38	.009 975 639	35	35	.011 259 026	39	39	.003 680 545
32	39	.008 390 838	35	36	.010 080 781	39	40	.003 122 462
						40	40	.002 634 388

$$\mathrm{P}_U(C) = \sum_{i=1}^{10} \mathrm{P}_U(Y_1 = \text{red } i)\mathrm{P}_U(R \mid Y_1 = \text{red } i)$$

$$+ \sum_{i=1}^{10} \mathrm{P}_U(Y_1 = \text{black } i)\mathrm{P}_U(B \mid Y_1 = \text{black } i) \quad (20.7)$$

and

$$\mathrm{P}_U(I) = \sum_{i=1}^{10} \mathrm{P}_U(Y_1 = \text{red } i)\mathrm{P}_U(B \mid Y_1 = \text{red } i)$$

$$+ \sum_{i=1}^{10} \mathrm{P}_U(Y_1 = \text{black } i)\mathrm{P}_U(R \mid Y_1 = \text{black } i), \quad (20.8)$$

and therefore that

Table 20.6 The distribution of the number of cards used in a coup of trente et quarante. Assumes sampling without replacement from the full six-deck shoe. Results are rounded to six significant digits.

cards	probability	cards	probability	cards	probability
8	.066 652 7	20	.366 181 $\cdot$ 10^{-5}	32	.118 516 $\cdot$ 10^{-18}
9	.190 628	21	.528 615 $\cdot$ 10^{-6}	33	.363 236 $\cdot$ 10^{-20}
10	.260 886	22	.680 566 $\cdot$ 10^{-7}	34	.925 477 $\cdot$ 10^{-22}
11	.228 217	23	.782 519 $\cdot$ 10^{-8}	35	.192 299 $\cdot$ 10^{-23}
12	.143 936	24	.803 693 $\cdot$ 10^{-9}	36	.318 183 $\cdot$ 10^{-25}
13	.070 038 9	25	.736 607 $\cdot$ 10^{-10}	37	.406 746 $\cdot$ 10^{-27}
14	.027 471 1	26	.601 205 $\cdot$ 10^{-11}	38	.386 110 $\cdot$ 10^{-29}
15	.008 952 12	27	.435 555 $\cdot$ 10^{-12}	39	.257 889 $\cdot$ 10^{-31}
16	.002 476 83	28	.278 846 $\cdot$ 10^{-13}	40	.112 187 $\cdot$ 10^{-33}
17	.000 591 079	29	.156 852 $\cdot$ 10^{-14}	41	.282 234 $\cdot$ 10^{-36}
18	.000 123 091	30	.769 647 $\cdot$ 10^{-16}	42	.335 008 $\cdot$ 10^{-39}
19	.000 022 560 4	31	.326 541 $\cdot$ 10^{-17}	43	.123 827 $\cdot$ 10^{-42}
				44	.418 326 $\cdot$ 10^{-47}

$$\mathrm{P}_U(C) - \mathrm{P}_U(I) = \sum_{i=1}^{10}[\mathrm{P}_U(Y_1 = \text{red } i) - \mathrm{P}_U(Y_1 = \text{black } i)]$$
$$\cdot [\mathrm{P}_U(R \mid Y_1 = i) - \mathrm{P}_U(B \mid Y_1 = i)]. \quad (20.9)$$

Here we are using a simple property of conditional probabilities: If B_1 and B_2 are mutually exclusive and have positive probabilities and $\mathrm{P}(A \mid B_1) = \mathrm{P}(A \mid B_2)$, then both conditional probabilities are equal to $\mathrm{P}(A \mid B_1 \cup B_2)$. In particular, since $\mathrm{P}_U(R \mid Y_1 = \text{red } i) = \mathrm{P}_U(R \mid Y_1 = \text{black } i)$, both conditional probabilities are equal to $\mathrm{P}_U(R \mid Y_1 = i)$; of course, the same is true with B in place of R.

We conclude that, for the color bet the effects of red removals are

$$\mathrm{P}_{D-\{\text{red } j\}}(C) - \mathrm{P}_{D-\{\text{red } j\}}(I) - \frac{1}{2}\mathrm{P}_{D-\{\text{red } j\}}(T_{31})$$
$$- \left(\mathrm{P}_D(C) - \mathrm{P}_D(I) - \frac{1}{2}\mathrm{P}_D(T_{31})\right)$$
$$= -\frac{1}{311}[\mathrm{P}_{D-\{j\}}(R \mid Y_1 = j) - \mathrm{P}_{D-\{j\}}(B \mid Y_1 = j)]$$
$$- \frac{1}{2}[\mathrm{P}_{D-\{j\}}(T_{31}) - \mathrm{P}_D(T_{31})], \quad (20.10)$$

while the effects of black removals are

$$P_{D-\{\text{black } j\}}(C) - P_{D-\{\text{black } j\}}(I) - \frac{1}{2}P_{D-\{\text{black } j\}}(T_{31})$$

$$- \left(P_D(C) - P_D(I) - \frac{1}{2}P_D(T_{31}) \right)$$

$$= \frac{1}{311}[P_{D-\{j\}}(R \mid Y_1 = j) - P_{D-\{j\}}(B \mid Y_1 = j)]$$

$$- \frac{1}{2}[P_{D-\{j\}}(T_{31}) - P_D(T_{31})]. \tag{20.11}$$

Here we are using that fact that, for certain events, the color of a removal does not affect the probability:

$$P_{D-\{\text{red } j\}}(A \mid Y_1 = j) = P_{D-\{j\}}(A \mid Y_1 = j) \quad \text{for} \quad A = B, R,$$
$$P_{D-\{\text{red } j\}}(T_{31}) = P_{D-\{j\}}(T_{31}); \tag{20.12}$$

similar results hold with black j in place of red j.

For the inverse bet, the results are analogous, but with C and I interchanged, so the sign of the term with coefficient $\pm 1/311$ is changed in (20.10) and (20.11). Even simpler, the effects of red removals and black removals are interchanged.

For the color bet with insurance, the term $-\frac{1}{2}[P_{D-\{j\}}(T_{31}) - P_D(T_{31})]$ in (20.10) and (20.11) is replaced by $(0.01)[P_{D-\{j\}}(T) - P_D(T)]$. The same is true of the inverse bet with insurance. (Here "per unit bet" in the definition of Z_n of (11.67) on p. 414 means "per unit bet on color or inverse only.")

Evaluation of these quantities requires easy modifications of the program used for Table 20.3 on p. 627. To be more explicit about this, the probabilities $P_{D-\{j\}}(T_{31})$ in (20.10) and (20.11) as well as $P_{D-\{j\}}(T)$ in the preceding paragraph can be obtained from the joint distribution (20.5) on p. 626 but with n_j and N both reduced by 1. To evaluate $P_{D-\{i_0\}}(R \mid Y_1 = i_0)$, however, requires a bit more thought. Here we are assuming that a card with value i_0 has already been removed from the deck, *and* the first card dealt to black has value i_0. Thus, n_{i_0} and N are both reduced by 2, and the Black row totals i_0 before any cards are dealt. We can use (20.5) on p. 626 once again, but with $p_{10}(k)$ replaced by $p_{10}(k - i_0)$, and with n_{i_0} and N both reduced by 2. Results are summarized in Table 20.7.

Table 20.8 gives what we believe to be the best balanced level-one counting system in two cases: (a) the player bets on color (or inverse), never with insurance, and (b) the player bets on color (or inverse), always with insurance. (For simplicity, we do not consider the case in which the player sometimes takes insurance and other times does not.) In the case without insurance, the correlation with the effects of removal is about 0.924; in the case with insurance, it is about 0.974.

The conclusions are perhaps unexpected. The player who always takes insurance may use a one-parameter counting system to track his advantage at both the color bet and the inverse bet. The situation is analogous to

Table 20.7 Effects of removal, multiplied by 311, for the color bet in trente et quarante. For the inverse bet, effects for red cards and black cards are interchanged. Results are rounded to six decimal places.

	without insurance		with insurance	
card value	red card	black card	red card	black card
1	.051 988	.006 896	.022 942	−.022 151
2	.044 022	−.006 993	.025 593	−.025 422
3	.040 616	−.011 962	.026 236	−.026 342
4	.035 853	−.015 102	.025 329	−.025 626
5	.028 053	−.016 000	.021 816	−.022 237
6	.016 415	−.012 616	.014 292	−.014 739
7	.003 811	−.008 802	.006 105	−.006 508
8	−.010 233	−.002 965	−.003 766	.003 503
9	−.027 035	.004 986	−.016 032	.015 988
10	−.045 133	.014 899	−.029 889	.030 143

that of Example 11.3.2 on p. 411. On the other hand, the player who never takes insurance must use a two-parameter counting system (one parameter for the red cards, the other for the black cards) to track his advantage at both the color bet and the inverse bet. The situation is analogous to that of Example 11.3.3 on p. 412, especially (11.64) on p. 413.

We next use a normal approximation to approximate the probability that the player's approximate advantage exceeds a certain level as a function of the number of unseen cards. This is complicated by the fact that the player has two bets to choose from, color and inverse.

We begin with the case in which the player bets on color or inverse, and always takes insurance. Let Z_n^C and Z_n^I be the analogues of Z_n^* of (11.95) on p. 418 for the color and inverse bets, and similarly let C_i and I_i correspond to J_i, and μ_C and μ_I to μ, γ_C and γ_I to γ, σ_C and σ_I to σ. Then $I_i = -C_i$, $\mu_C = \mu_I < 0$, $\gamma_C = \gamma_I > 0$, and $\sigma_C^2 = \sigma_I^2$, so

$$P(\max(Z_n^C, Z_n^I) > \beta)$$

$$= P\left(\max\left\{ \mu_C + \frac{\gamma_C}{N-n} \sum_{j=1}^{n} C_{X_j}, \ \mu_C - \frac{\gamma_C}{N-n} \sum_{j=1}^{n} C_{X_j} \right\} > \beta \right)$$

$$= P\left(\left| \sum_{j=1}^{n} C_{X_j} \right| > \left\lfloor (N-n) \frac{\beta - \mu_C}{\gamma_C} \right\rfloor + \frac{1}{2} \right) \tag{20.13}$$

$$\approx 2\left[1 - \Phi\left(\frac{1}{(N-n)\sigma_C} \left\{ \left\lfloor (N-n) \frac{\beta - \mu_C}{\gamma_C} \right\rfloor + \frac{1}{2} \right\} \sqrt{\frac{(N-n)(N-1)}{n}} \right) \right].$$

Table 20.8 The best balanced level-one card-counting systems for the color bet in trente et quarante. For the inverse bet, point values for red cards and black cards are interchanged.

card value	without insurance red card	without insurance black card	with insurance red card	with insurance black card
1	1	0	1	−1
2	1	0	1	−1
3	1	0	1	−1
4	1	0	1	−1
5	1	0	1	−1
6	0	0	1	−1
7	0	0	0	0
8	0	0	0	0
9	−1	0	−1	1
10	−1	0	−1	1
correlation with Table 20.7	.923 874		.974 264	

Here $N = 312$, $\mu_C = -0.01(1 - q) \approx -0.009123$ (cf. Table 20.4 on p. 628), $\gamma_C \approx 0.024767$, and $\sigma_C^2 = 11/13$. Numerical values are provided in Table 20.9. We hasten to add that these numbers are based on three approximations, namely (11.74) on p. 415, (11.95) on p. 418, and (20.13).

We turn to the case in which the player bets on color or inverse, and never takes insurance. Using the same notation as before (Z_n^C, Z_n^I, C_i, I_i, μ_C, μ_I, γ_C, γ_I, σ_C, σ_I), we have $\mu_C = \mu_I < 0$, $\gamma_C = \gamma_I > 0$, $\sigma_C^2 = \sigma_I^2$, and $\mathrm{Corr}(C_{X_1}, I_{X_1}) = 0$, so

$$P(\max(Z_n^C, Z_n^I) > \beta)$$

$$= P\left(\max\left\{ \mu_C + \frac{\gamma_C}{N - n} \sum_{j=1}^{n} C_{X_j}, \ \mu_C + \frac{\gamma_C}{N - n} \sum_{j=1}^{n} I_{X_j} \right\} > \beta \right)$$

$$= 1 - P\left(\sum_{j=1}^{n} C_{X_j} \le \left\lfloor (N - n)\frac{\beta - \mu_C}{\gamma_C} \right\rfloor + \frac{1}{2}, \right.$$

$$\left. \sum_{j=1}^{n} I_{X_j} \le \left\lfloor (N - n)\frac{\beta - \mu_C}{\gamma_C} \right\rfloor + \frac{1}{2} \right) \qquad (20.14)$$

$$\approx 1 - \left[\Phi\left(\frac{1}{(N - n)\sigma_C} \left\{ \left\lfloor (N - n)\frac{\beta - \mu_C}{\gamma_C} \right\rfloor + \frac{1}{2} \right\} \sqrt{\frac{(N - n)(N - 1)}{n}} \right) \right]^2$$

as in (11.64) on p. 413. Here $N = 312$, $\mu_C \approx -0.010946$ (from Table 20.4 on p. 628), $\gamma_C \approx 0.040810$, and $\sigma_C^2 = 5/13$. Again, numerical values are provided in Table 20.9).

We conclude that card counting can overcome the house advantage only deep in the shoe, and even then only occasionally.

Table 20.9 Estimate of the probability that the maximum of the expected values of a one-unit color bet and a one-unit inverse bet is at least β when n cards remain unseen.

	without insurance				with insurance			
n	β				β			
	$-.005$	$.000$	$.005$	$.010$	$-.005$	$.000$	$.005$	$.010$
13	.433	.107	.012	.003	.442	.166	.021	.001
26	.233	.032	.001	.000	.317	.035	.001	.000
52	.066	.001	.000	.000	.161	.001	.000	.000
104	.003	.000	.000	.000	.023	.000	.000	.000

20.3 Problems

20.1. *Number of trente-et-quarante sequences with a specified total.* In (20.2) on p. 624 we counted the number of trente-et-quarante sequences. For $m = 31, 32, \ldots, 40$, find the number of trente-et-quarante sequences that sum to m. (Notice that these numbers are *not* proportional to the corresponding probabilities.)

20.2. *Number of coups per shoe.*

(a) Assuming that the six-deck shoe is dealt out completely with any incomplete coup declared void, show that the maximum number of possible coups is 32 and the minimum number of possible coups is 25.

(b) Now assume an infinite-deck shoe, and suppose cards are dealt until the total number of points dealt exceeds 2,040 with the last coup declared void. Estimate the expected number of coups, getting upper and lower bounds from Wald's identity (Corollary 3.2.5 on p. 103). Can the renewal theorem (Theorem 4.4.1 on p. 146) be used to give a better estimate? (Cf. Problem 19.4 on p. 610.)

20.3. *Probabilities of the ten trente-et-quarante totals, assuming sampling with replacement.* When sampling with replacement, the definition of a trente-et-quarante sequence is as in (20.1) on p. 624, except that the condition that at most 24 of the terms are equal to 1 is dropped. Let P_n be the probability that a trente-et-quarante sequence has total n, $31 \le n \le 40$, assuming sampling with replacement (or, equivalently, an infinite-deck shoe).

(a) Show that these probabilities can be generated by the recursion

$$P_n = \frac{1}{13}(P_{(n-1)\wedge 30} + \cdots + P_{n-9}) + \frac{4}{13}P_{n-10} \qquad (20.15)$$

for $n = 1, 2, \ldots, 40$, where $P_n := 0$ if $n < 0$ and $P_0 := 1$. (The subscripts within parentheses in (20.15) are decreasing from left to right, so there are $9 \wedge (40 - n)$ terms.) Evaluate P_n to nine decimal places for $31 \le n \le 40$ and compare the results with those of Table 20.2 on p. 626.

(b) Evaluate the mean of this distribution.

20.4. *Expected length of a trente-et-quarante sequence, assuming sampling with replacement.* When sampling with replacement, the definition of a trente-et-quarante sequence is as in Problem 20.3.

(a) Show that the expected length of a trente-et-quarante sequence is E_{31}, where

$$E_n := 1 + \frac{1}{13}(E_{n-1} + \cdots + E_{n-9}) + \frac{4}{13}E_{n-10} \qquad (20.16)$$

for $n = 1, 2, \ldots, 31$ and $E_n := 0$ if $n \le 0$. Evaluate E_{31} to nine decimal places.

(b) Show that the expectation in part (a) can be derived from Problem 20.3(b) using Wald's identity (Corollary 3.2.5 on p. 103).

20.5. *Joint distribution of the total and length of a trente-et-quarante sequence, assuming sampling with replacement.* When sampling with replacement, the definition of a trente-et-quarante sequence is as in Problem 20.3. Let $P_{n,l}$ be the probability that a trente-et-quarante sequence has total n and length l, $31 \le n \le 40$, $4 \le l \le 31$, assuming sampling with replacement from the six-deck shoe. Show that these probabilities can be generated by the recursion

$$P_{n,l} = \frac{1}{13}(P_{(n-1)\wedge 30,l-1} + \cdots + P_{n-9,l-1}) + \frac{4}{13}P_{n-10,l-1} \qquad (20.17)$$

for $n = 1, 2, \ldots, 40$ and $l = 1, 2, \ldots, 31$, where $P_{n,l} := 0$ if $n < 0$ and $l \ge 0$ or if $n = 0$ and $l \ge 1$, and $P_{0,0} := 1$. (The first subscripts within parentheses in (20.17) are decreasing from left to right, so there are $9 \wedge (40 - n)$ terms.) Evaluate $P_{n,l}$ to six significant digits for $31 \le n \le 40$ and $4 \le l \le 31$, and similarly evaluate the marginal distributions. Compare the marginal results with those of Tables 20.2 on p. 626 and 20.3 on p. 627.

20.6. *Probabilities of the ten trente-et-quarante totals, assuming a 40-card deck.* Consider the 40-card deck obtained from the standard 52-card deck by discarding the eights, nines, and tens. In this case a trente-et-quarante sequence is defined as in (20.1) on p. 624, except that at most four of the terms are equal to j for $1 \le j \le 7$, and none of the terms is equal to 8 or 9.

Modify (20.3) on p. 625 to find the probabilities of the ten trente-et-quarante totals, and evaluate to nine decimal places. See the chapter notes for the significance of this problem.

20.7. *Poisson's formula for $P(31)$.* Prove Poisson's formula (20.20) on p. 639 by showing that (20.20) is equivalent to

$$P(31) = \sum_{j_1,j_2,\ldots,j_{10} \geq 0: \; j_1+2j_2+\cdots+10j_{10}=31} \cdots \sum \frac{\binom{24}{j_1} \cdots \binom{24}{j_9}\binom{96}{j_{10}}}{\binom{312}{j_1+\cdots+j_9+j_{10}}}, \qquad (20.18)$$

where $\binom{24}{j_1} := 0$ if $j_1 > 24$.

20.8. *House advantage with en prison, assuming independent coups.* Evaluate the house advantage (pushes included and excluded) of red, black, color, and inverse, assuming that the player chooses the classical *en prison* option. The rules are the same as in single-zero roulette (see Section 13.1). To make this problem tractable, assume that each coup is dealt from a complete six-deck shoe, independent of previous coups.

20.9. *Correlation between Red total and Black total.* Assuming sampling without replacement from the full six-deck shoe, evaluate to nine decimal places the correlation between the Red total and the Black total.

20.10. *Joint distribution of the total and length of a trente-et-quarante sequence.* Simultaneously generalize (20.3) on p. 625 and (20.4) on p. 626 by finding the joint distribution of the total and the length of a trente-et-quarante sequence. Evaluate to six significant digits, checking that the marginals are consistent with Tables 20.2 on p. 626 and 20.3 on p. 627. The total ranges from 31 to 40 and the length ranges from 4 to 28. Of the $10 \cdot 25$ entries, how many are zero? Find the correlation.

20.11. *Exact probability that a trente-et-quarante sequence (or coup) has maximum possible length.*
   (a) Find a combinatorial expression for the probability that a trente-et-quarante sequence has exactly 28 terms (the maximum possible number). Compare with Table 20.3 on p. 627.
   (b) Find another such expression for the probability that a trente-et-quarante coup requires exactly 44 cards (the maximum possible number). Compare with Table 20.6 on p. 630.

20.12. *Advisability of insurance bet.* It has been argued that since an insurance claim pays only half of the insurer's bet, its real rate is two percent, not one percent, and therefore insurance bets are inadvisable. Does this argument have any merit?

20.13. *Bets on odd and even.* It has been said that there is sometimes an even-money bet that the winning total is odd (and similarly for even), with ties handled in the same way that they are for bets on red or black. Show that this is unlikely by evaluating the resulting house advantages (without insurance, pushes included).

20.14. *Correlation between red and color bets.* Let X_1 be the profit from a one-unit bet on red, and let X_2 be the profit from a one-unit bet on color. Find the correlation between X_1 and X_2, assuming a full six-deck shoe.

20.15. *An eight-card residual subset.* Suppose that the remaining cards comprise one black five and seven red tens, but that their order is unknown.

(a) Evaluate the expected profit from a one-unit bet on color.

(b) Confirm the result using (20.9) on p. 630.

(c) Find an example with higher expectation.

20.16. *Accuracy of card counting.* Estimate the player's advantage on the color bet, without and with insurance, if six red aces, six red 2s, six red 3s, six red 4s, and six red 5s have been exposed in the six-deck shoe, leaving 282 cards unseen. Do this in three ways.

(a) Use the balanced level-one counting system in Table 20.8 on p. 633.

(b) Use the more accurate effects of removal in Table 20.7 on p. 632.

(c) Use exact calculation.

20.17. *Quarter-deck trente et quarante.* In quarter-deck trente et quarante, only 13 cards, one from each denomination, are used. (We ignore the color and inverse bets, so suits are irrelevant.) Notice that a complete coup is assured since the 13 card values sum to 85.

(a) Find the probabilities of the ten trente et quarante totals, and compare them with those of Table 20.2 on p. 626.

(b) Find the probabilities analogous to p, q, and r of Table 20.4 on p. 628.

20.18. *End play at trente et quarante.* Assume that the six-deck shoe is dealt to the bottom, with any incomplete coup declared void.

(a) Suppose that exactly three aces and six 10s remain. It has been argued that a bet on red, insured against a refait, offers positive expectation. (See the chapter notes for the reasoning that led to this conclusion.) Notice that there are $\binom{9}{3} = 84$ equally likely outcomes. By examining each of them, verify that the argument is flawed.

(b) Recall Example 11.2.2(d) on p. 401, where it was argued that the probability that red wins is equal to the probability that black wins, regardless of the composition of the set of unseen cards, *provided only that sufficiently many cards remain to complete both rows with probability 1.* Prove that the result is true without the italicized proviso.

20.4 Notes

The primary source for this chapter is Ethier and Levin (2005).

The history of trente et quarante is obscure, and as Parlett (1991, p. 78) noted,

> Its age is uncertain, partly because there seems no way of proving its identity with such fifteenth/sixteenth century game-names as 'a la terza, a la quarta', and partly because later reports often confuse it with Trente-et-Un.

But it certainly dates at least as far back as the 17th century, having been mentioned in Molière's [1622–1673] 1668 play *The Miser* (Molière 1913, Act 2, Scene 6):[1]

> Moreover, she has a horrible aversion to play, which is not common in women nowadays; and I know of one in our neighbourhood who has lost twenty thousand francs this year at trente-et-quarante.

In French novelist Edmond About's [1828–1885] 1858 novel *Trente et Quarante*, we find the following remarks (About 1899, Chapter 9):[2]

> Meo made acquaintance with trente et quarante under the eye of his mentors. This game, the easiest of all, astonished him by its simplicity. It seemed to him astounding that 70,000 francs could be lost or won in the course of a few deals, because the banker had dealt thirty-seven points to red and thirty-eight to black.

The first published calculation of the probabilities of the ten possible trente-et-quarante totals is due to Mr. D. M. (1739),[3] assuming a nonstandard deck composition, namely a 40-card deck obtained from the standard 52-card deck by removing the eights, nines, and tens. Isaac Todhunter [1820–1884] (1865, Article 358) described Mr. D. M.'s effort contemptuously:

> The problem is solved by examining all the cases which can occur, and counting up the number of ways. The operation is most laborious, and the work is perhaps the most conspicuous example of misdirected industry which the literature of Games of Chance can furnish.

More interesting than the question of whether Mr. D. M.'s industry was misdirected is the question of whether it was correct. The answer is that it was correct in principle but contained errors. Although he assumed sampling without replacement from the 40-card deck and correctly recognized that between four and 14 cards, inclusive, would be needed to complete a row, he effectively dealt only nine cards for the purpose of his calculations. Comparing

---

[1] Translation by Curtis Hidden Page.

[2] Translation by Lord Newton.

[3] Ethier and Levin (2005) credited D. M. Florence, misreading Todhunter's (1865, Article 358) remark that

> We have next to notice a work entitled *Calcul du Jeu appellé par les François le trente-et-quarante, et que l'on nomme á Florence le trente-et-un.* ... *Par Mr D. M.* Florence, 1739.

A careful reading reveals that Florence is not the author's last name but rather the city of publication! It is conceivable, albeit unlikely, that Mr. D. M. was Mr. De Moivre.

his numbers to what he should have obtained under this simplification, each of his 10 results (summarized on his p. 87) has relative error of less than one percent, which is quite respectable for 18th-century work of this complexity.

P. N. Huyn (1788, pp. 28–29) proposed a solution, assuming sampling with replacement from a standard deck, but it is quite inaccurate. Noting that for a total of 40 the last card must have value 10 (4 denominations); for a total of 39 the last card must have value 9 or 10 (5 denominations); ... for a total of 31 the last card must have value 1, 2, ..., or 10 (13 denominations), Huyn concluded that the probability of a total of i is

$$P(i) = \frac{44 - i}{85}, \qquad 31 \le i \le 40, \qquad (20.19)$$

since $4 + 5 + \cdots + 13 = 85$. The argument was sufficiently plausible that several subsequent authors (Grégoire 1853, pp. 37–38; Dormoy 1873, Section 105; Trumps c. 1875, p. 448; Gall 1883, p. 96; Silberer 1910, pp. 72–73; Scrutator 1924, pp. 84–85; and Scarne 1974, p. 518) adopted it as their own. For good measure, Scarne (1974, p. xx), who called himself the "world's foremost gambling authority," added that he was the first to evaluate these probabilities. Incidentally, there is a sense in which (20.19) is correct; see Example 4.4.4.

Siméon-Denis Poisson [1781–1840] (1825) not only pointed out the error in Huyn's work, but he found two correct expressions for the probabilities in question, assuming sampling without replacement from the six-deck shoe. For example, he showed that

$$P(31) = \text{coefficient of } t^{31} \text{ in} \qquad\qquad (20.20)$$
$$313 \int_0^1 (1 - y + yt)^{24} \cdots (1 - y + yt^9)^{24}(1 - y + yt^{10})^{96} \, dy.$$

But he was able to evaluate the probabilities only in an asymptotic case corresponding to sampling with replacement. While acknowledging the "analytical virtuosity" of Poisson's work and certifying its "ingenuity," Bernard Bru [1942–] (2005) used the terms "anodyne," "prosaic," in "bad taste", and "profane" to describe the specific problem that Poisson addressed. Recalling Todhunter's remark about Mr. D. M.'s treatise, we surmise that research on such applied topics as trente et quarante is disdained by the historians of probability.

Independently, Augustus De Morgan [1806–1871] (1838, Appendix 1, p. vii) evaluated the ten probabilities assuming sampling with replacement. Joseph Bertrand [1822–1900] (1888, pp. 35–38) and Marcel Boll [1886–1958] (1936, Chapter 14) treated the same case in their analyses. De Morgan's argument was simpler than Poisson's (and Bertrand followed De Morgan). See Problem 20.3. Of course, this approach does not work when sampling without replacement. To justify the assumption of sampling with replacement, De Morgan argued as follows:

The only ways in which 31, for example, could be obtained from an unlimited number of packs, and which could not equally well be obtained from six packs, are those in which more than 24 aces occur. Now the probability that out of 31 cards drawn at hazard, 24 or more shall be aces, is altogether beneath consideration. It is less than one out of a million of million of millions.

Indeed, it is about $0.283 \cdot 10^{-20}$. Nevertheless, the argument is flawed, for the same reasoning would imply that, if trente et quarante were played with eight decks instead of six, there would be no inaccuracy in assuming an unlimited number of decks.

The next significant contribution was made by Ludwig Oettinger [1797–1869] (1867), who, among other things, evaluated $P(31)$ accurately to eight decimal places under the correct assumptions (sampling without replacement from the six-deck shoe), a remarkable achievement for its era. Poisson's (1825) estimate was accurate to only five decimal places.

Thorp and Walden (1973) also addressed the problem under the correct assumptions but only approximated the probabilities in question by limiting consideration to at most eight cards. Their estimate of $P(31)$ was accurate to only four decimal places. In particular, the 1867 "by hand" analysis was substantially more accurate than the 1973 computer analysis, perhaps surprisingly.

To summarize the history, we list in Table 20.10 the various estimates of the probability of a refait (i.e., $P(31, 31)$) that have been published over the past 225 years. We do not include Mr. D. M. in the table because he assumed a 40-card deck and did not consider the probability of a refait.

Table 20.10 Evaluating the probability of a refait at trente et quarante.

author (year)	published result
Huyn (1788)	$(13/85)^2 \approx 0.023\,391$
Poisson (1825)	$\frac{313}{312}(0.148062)^2 - 0.000026 \approx 0.021\,967$
De Morgan (1838)	$(0.1481)^2 \approx 0.021\,9$
Oettinger (1867)	$(0.1480577)(0.1480575) \approx 0.021\,921\,05$
Bertrand (1888)	$(0.148218)^2 \approx 0.021\,968\,6$
Boll (1936)	$(0.148061)^2 \approx 0.021\,922\,0$
Thorp and Walden (1973)	$(0.148123)^2 \approx 0.021\,940$
Ethier and Levin (2005)	$0.021\,891\,370$

The distribution of the length of a trente-et-quarante sequence was first computed by Boll (1936, p. 200), assuming sampling with replacement, but unfortunately his figures are significantly in error. The source of the error can be inferred from Boll (1945, Figures 2 and 3). Thorp and Walden (1973) correctly evaluated the sequence-length distribution for sequences of length up to eight, assuming sampling without replacement.

Edward O. Thorp [1932–] and William E. Walden [1929–] (1973) observed that bets on color or inverse can have positive expectation, whereas bets on red or black cannot. It seems likely that Martin Gall [1830–1905] was aware of this 90 years earlier, based on these remarks (Gall 1883, pp. 91, 233–234):

> We shall show in what follows that from the mathematical point of view the production of chances on the petit tableau does not present an absolute equality. A flaw exists, but what advantage can be gained from it? This we shall see later. [...]
>
> The all-in wager would not be unreasonable if, at any given moment, one had the odds in one's favor. For example, it would suffice for the player to count the cards as they are dealt for him to be able to know the approximate composition of the remainder of the shoe. By playing on the petit tableau (inverse and color), he will have, although rarely, a few more positive-expectation situations on the last coup of the shoe. But, first, it is very difficult to keep track of the rapidly dealt cards, and, second, the opportunities to make the end-play are extremely rare.

The result is certainly not self-evident, and authors such as Bethell (1901, p. 57) have in fact suggested the contrary:

> If too many small cards have already appeared, and big ones are due, it ought to be good to back the Red, because a high point may be expected in the top line.

The error was repeated by Billiken (1924, p. 197) and Player (1925, p. 41).

May (2004) discovered a counting system whose point values coincide with our with-insurance one:

> The best one-level system for counting these two bets [color and inverse] counts red A–6 and black 9–K as +1, with black A–6 and red 9–K as −1. When the count is above 23, the color bet is favorable. When it is below −23, inverse is favorable.

No further details were provided.

Our results are consistent with the findings of Thorp and Walden (1973): No card-counting system at trente et quarante can yield a "practically important player advantage."

Nevertheless, our analysis provides a better understanding of trente et quarante. When the directors of the Monte Carlo Casino were advised by General Pierre Polovtsoff [1872–1965] (1937, p. 189), President of the International Sporting Club, that an Italian gang was exploiting a weakness in the game, they responded,

> Impossible! Trente-et-quarante has been played here for eighty years, and it is inconceivable that anyone can have discovered anything about it that we do not already know.

See Barnhart (1983b) for further detail on the leader of the Italian gang, Enrico Sassi, as well as discussion of British bookmaker Joseph Owers, said by Emery (1926) to have broken the bank 38 times in one season at Monte Carlo and referred to by Beresford (1926) as the "Trente et Quarante King." What was their system? Barnhart did not say exactly, but suggested that it involved end play.

Billedivoire (1929, Chapter 2) claimed that bets on red are advantageous in certain end-play situations. At that time cards were dealt to the bottom of the shoe with any incomplete coup declared void. (Since then, it has been traditional to insert a cut card between the fifth and sixth cards from the bottom. If the cut card is dealt during a coup, that coup is declared void.) In a very simple special case, Billedivoire's idea is the following. The sum of the card values in the deck is 2,040. Suppose exactly 63 of these "points" remain (e.g., three aces and six 10s). There are several possible outcomes: If Black gets 33 or more points, then Red will get at most 30 points and the coup will be void. If Black gets 32 points, then Red will get 31 points and win. If Black gets 31 points, then Red will get 32 points and lose, so the game is symmetric ... with one exception: If the last card is an ace and Black gets 31 points, then a refait will occur. Therefore, by insuring against a refait, a bet on red should offer positive expectation.

Although appearing plausible at first glance, this argument is flawed and Billedivoire's (1929) "system" is without merit (see Problem 20.18).

Problem 20.1 was suggested by Patrick W. Feitt (though not included in his senior project, Feitt 2008). Problem 20.2(a) was pointed out by Poisson (1825). Problem 20.3 is essentially from De Morgan (1838, Appendix 1). Problem 20.4(b) is from Boll (1945, p. 23). Boll (1945, Figures 2 and 3) gave an incorrect solution to Problem 20.5. Problem 20.6 was motivated by the treatise of Mr. D. M. (1739). The argument in Problem 20.12 comes from Silberer (1910, Chapter 20). Problem 20.13 was motivated by Kavanagh (2005, p. 136). The example in Problem 20.15 is from Thorp and Walden (1973). Problem 20.18(b) was established by Thorp (2008, personal communication).

Chapter 21
Twenty-One

"You double us all round? I will take a card upon each of my two. Thank you, that will do—a ten—now, upon the other, a queen,—two natural vingt-et-uns, and as you doubled us you owe me so and so."

William Makepeace Thackeray, *The Virginians*

Twenty-one, or blackjack, first appeared in 18th-century France and is today's most popular casino table game. Its current level of popularity dates back only to the 1960s, when Thorp published *Beat the Dealer: A Winning Strategy for the Game of Twenty-One*. In Section 21.1 we specify the set of rules assumed, and we note the lack of symmetry of the game. In Section 21.2 we derive what is called (composition-dependent) basic strategy, that is, the optimal strategy that does not employ card counting. In Section 21.3 we discuss card counting at twenty-one with emphasis on the Hi-Lo system.

21.1 Rules

Twenty-one, which we will usually refer to by its more commonly used name, *blackjack*, is played with a standard 52-card deck, or with two such decks mixed together, or with a four-, six-, or eight-deck shoe. The specific rules vary slightly from one casino to the next. Therefore, we assume throughout this chapter the set of rules that was at one time standard on the Las Vegas Strip but is no longer available. The reason for studying this outdated set of rules is that we regard it as the benchmark, against which other sets of rules can be measured. For the reader already familiar with blackjack, we summarize these rules in Table 21.1.

For the reader unfamiliar with the game, let us start at the beginning. In blackjack each player competes against the dealer. We assume that the game is played with a single deck. Aces have value either 1 or 11, court cards (J, Q,

S.N. Ethier, *The Doctrine of Chances*, Probability and its Applications,
DOI 10.1007/978-3-540-78783-9_21, © Springer-Verlag Berlin Heidelberg 2010

Table 21.1 Blackjack rules assumed throughout this chapter. (In parentheses are the acronyms commonly used in the blackjack literature.)

- single deck (1D)
- dealer stands on soft 17 (S17)
- double down on any first two cards (DOA)
- no double down after splits (NDAS)
- split aces once, receiving only one card per ace (SPA1)
- split non-ace pairs up to three times [up to four hands] (SPL3)
- untied player natural pays 3 to 2 (3:2)
- untied dealer natural wins original bets only (OBO)
- no surrender (NS)

K) have value 10, and 2s, 3s, ..., 10s assume their nominal values. Suits do not play a role. A hand comprising two or more cards has value equal to the total of the values of the cards. The total is called *soft* if the hand contains an ace valued as 11, otherwise it is called *hard*.

After making a bet, each player receives two cards, usually face down (but it does not actually matter), and the dealer receives two cards, one face down (the *downcard* or *hole card*) and one face up (the *upcard*). If the player has a two-card total of 21 (a *natural* or *blackjack*) and the dealer does not, the player wins and is paid 3 to 2. If the dealer has a natural and the player does not, the player loses his bet. If both player and dealer have naturals, a push is declared. If the dealer's upcard is a 10-valued card, he checks his downcard to determine whether he has a natural before proceeding. If the dealer's upcard is an ace, each player is given the opportunity to make an *insurance bet* for up to half the size of his original bet. This is a side bet that the player wins if and only if the dealer has a natural. A winning insurance bet pays 2 to 1. The dealer checks his downcard and collects losing insurance bets before proceeding. Of course, if he has a natural, the hand is over.

If the dealer and at least one player fail to have naturals, play continues. Starting with the player to the dealer's left and moving clockwise, each player completes his hand as follows. He must decide whether to *stand* (take no more cards) or to *hit* (take an additional card). If he chooses the latter and his new total does not exceed 21, he must make the same decision again and continue to do so until he either stands or *busts* (his total exceeds 21). If the player busts, his bet is lost, even if the dealer subsequently busts. The player has one or two other options after seeing his first two cards. He may *double down*, that is, double his bet and take one, and only one, additional card. If he has two cards of the same value, he may *split* his pair, that is, make an additional bet equal to his initial one and play two hands, with each of his first two cards being the initial card for one of the two hands and each of his two bets applying to one of the two hands. A two-card 21 after a split is not regarded as a natural and is therefore not entitled to a 3-to-2 payoff; in addition, it pushes a dealer 21 comprising three or more cards. Doubling after splitting is not allowed, and split aces receive one, and only one, card on each ace. One can split non-ace pairs up to three times (up to four hands). Two 10-valued cards are considered a pair, even if they are of different denominations.

As we have already assumed, the dealer checks for a natural when his upcard is an ace or a 10-valued card. This is sometimes stated by saying that an untied dealer natural wins original bets only—additional bets due to doubling and splitting, if they could be made, would be pushed.

After each player has stood, busted, doubled down, or split aces, the dealer acts according to a set of mandatory rules. The dealer draws to hands of 16 or less and stands on hands of 17 or more, including soft totals.

If the dealer busts, all remaining players are paid even money. If the dealer stands, his total is compared to that of each remaining player. If the player's total (which does not exceed 21) exceeds the dealer's total, the player is paid even money. If the dealer's total (which does not exceed 21) exceeds the player's total, the player loses his bet. If the player's total and the dealer's total are equal (and do not exceed 21), the hand is declared a push.

We regard these rules as the traditional ones. The most common departures from them include the following variations: Two, four, six, or eight decks may be used instead of one; the dealer may be required to hit soft 17; doubling down may be permitted only on initial totals of 10 and 11 or of 9, 10, and 11; doubling down may be permitted after splitting; an untied player natural may pay only 6 to 5; resplitting of aces may be permitted; resplitting of pairs may not be permitted; the dealer showing an ace or 10-valued card may not check for a natural until after the players have completed their hands; a surrender option may be available to the player (see Problem 21.22 on p. 675); and so on. We will not discuss these variations except (in some cases) in the problems.

In roulette, craps, faro, baccarat, and trente et quarante, the player can bet on either side of the basic even-money propositions. But in blackjack, because the player's strategy is discretionary, casinos cannot allow betting on the dealer. Blackjack is also less symmetric than these other games, with the player having several options not available to the dealer, and the dealer having the considerable advantage of acting last and in particular winning in the case in which both player and dealer bust.

Suppose that the player does not take advantage of these options and instead decides to *mimic the dealer*. In other words, suppose the player adopts the dealer's mandatory strategy as his own. What is the house advantage (pushes included)? It turns out that this problem has a rather complicated solution, but the same is true of many blackjack problems, so this is as good a place as any to begin.

If the player mimics the dealer, there are only two departures from symmetry. If both player and dealer bust, the dealer wins. On the other hand, if the player is dealt a natural and the dealer is not, the player is paid 3 to 2, not just 1 to 1. Thus, the house advantage we seek can be expressed as

$$H_0 = \text{P(both player and dealer bust)}$$
$$- \frac{1}{2}\, \text{P(player, but not dealer, has natural)}. \qquad (21.1)$$

In Section 20.1 we defined a trente-et-quarante sequence. Analogously, let us define a *blackjack-dealer sequence* to be a finite sequence $a_1, \ldots, a_k$ of positive integers, none of which exceeds 10, and at most four of which are equal to 1, at most four of which are equal to 2, and so on, such that k is the smallest integer $j \geq 2$ for which

$$a_1 + \cdots + a_j \geq 17 \qquad (21.2)$$

or

$$1 \in \{a_1, \ldots, a_j\} \quad \text{and} \quad 7 \leq a_1 + \cdots + a_j \leq 11. \qquad (21.3)$$

Observe that (21.2) signifies a hard total of $a_1 + \cdots + a_j$ and that, since 1s play the role of aces, (21.3) signifies a soft total of $a_1 + \cdots + a_j + 10$. Clearly, the order of the terms is crucial: $8, 8, 10$ is a blackjack-dealer sequence but $8, 10, 8$ and $10, 8, 8$ are not. In general, if $a_1, \ldots, a_k$ is a blackjack-dealer sequence, then its length k satisfies $2 \leq k \leq 10$.

In Example 1.1.11 on p. 8 we observed that there are 48,532 blackjack-dealer sequences. To evaluate the first probability in (21.1), we could consider all ordered pairs of such sequences, but that would be inefficient. Another useful observation from Example 1.1.11 is that the probability of a blackjack-dealer sequence does not depend on the order of the terms. For example, of the $\binom{7}{1,3,2,1} = 420$ permutations of $1, 2, 2, 2, 3, 3, 4$, only 156 are blackjack-dealer sequences, but each of those 156 has the same probability, namely $(4)_1 (4)_3 (4)_2 (4)_1 / (52)_7$.

Given an arbitrary blackjack-dealer sequence $a_1, \ldots, a_k$, we define the *unordered blackjack-dealer sequence* $\boldsymbol{k} = (k_1, k_2, \ldots, k_{10})$ by

$$k_j := |\{1 \leq i \leq k : a_i = j\}|, \qquad j = 1, 2, \ldots, 10, \qquad (21.4)$$

that is, k_1 is the number of aces, k_2 is the number of 2s, $\ldots$, k_{10} is the number of 10-valued cards. It will be convenient to let K denote the set of unordered blackjack-dealer sequences. In Table 21.2 we list the elements of K in lexicographical order. Note that $|K| = 2{,}741$. We define two positive-integer-valued functions on K, namely T and M, as follows: We let $T(\boldsymbol{k})$ be the *total* of the sequence $\boldsymbol{k}$, that is,

$$T(\boldsymbol{k}) := \begin{cases} \sum jk_j & \text{if } 17 \leq \sum jk_j \leq 26, \\ \sum jk_j + 10 & \text{if } k_1 \geq 1 \text{ and } 7 \leq \sum jk_j \leq 11, \end{cases} \qquad (21.5)$$

where the first case signifies a hard total and the second case a soft total. We let $M(\boldsymbol{k})$ be the number of blackjack-dealer sequences $a_1, \ldots, a_k$ (among the 48,532) for which $\boldsymbol{k}$ satisfies (21.4). M stands for *multiplicity*. For example, as we saw in the preceding paragraph, with $\boldsymbol{k} := (1, 3, 2, 1, 0, 0, 0, 0, 0, 0)$, we have $T(\boldsymbol{k}) = 17$ and $M(\boldsymbol{k}) = 156$.

Now we can evaluate (21.1):

$$
\begin{aligned}
\mathrm{H}_0 &= \sum_{\boldsymbol{k}, \boldsymbol{l} \in K:\, T(\boldsymbol{k}) > 21,\, T(\boldsymbol{l}) > 21} M(\boldsymbol{k}) M(\boldsymbol{l}) \frac{(4)_{k_1 + l_1} \cdots (4)_{k_9 + l_9} (16)_{k_{10} + l_{10}}}{(52)_{k_1 + \cdots + k_{10} + l_1 + \cdots + l_{10}}} \\
&\quad - \frac{1}{2} \frac{\binom{4}{1}\binom{16}{1}\binom{32}{0}}{\binom{52}{2}} \left(1 - \frac{\binom{3}{1}\binom{15}{1}\binom{32}{0}}{\binom{50}{2}} \right) \\
&\approx 0.080091780 - 0.023246222 \\
&\approx 0.056845559, \quad\quad\quad\quad\quad\quad\quad\quad\quad\quad\quad\quad\quad\quad\quad\quad (21.6)
\end{aligned}
$$

where $(4)_r := 0$ if $r \geq 5$. Of course, in practice, if each player busts, the dealer does not complete his hand. However, for the purpose of this calculation, we can assume, without loss of generality, that the dealer always completes his hand.

Few players would actually adopt the mimic-the-dealer strategy because it ignores a very important piece of information: the dealer's upcard. In Example 1.1.11 on p. 8 we evaluated the distribution of the dealer's final total. In Table 21.3 we evaluate the *conditional* distribution of the dealer's final total, given the dealer's upcard, and we see that the dependence of this distribution on the upcard is significant.

For example, with an upcard of 5, the dealer has more than twice as great a chance of busting as he does with a 10-valued upcard; with an upcard of 9, he has more than twice as great a chance of busting as he does with an ace. The derivation of Table 21.3 is essentially as in Example 1.1.11 on p. 8, except that the initial term in each blackjack-dealer sequence is regarded as the dealer's upcard. The numbers of sequences for the various upcards and final totals are listed in Table 21.4. These numbers were found by direct enumeration.

Blackjack is a rather asymmetric game. The dealer has the advantage of acting last, thereby winning in the case of a hypothetical (or actual) double bust. The player, in addition to the advantage of the 3-to-2 payoff for an untied natural, has complete discretion over his strategy, allowing him to stand with stiff totals (hard 12–16) and hit soft totals of 17 or more, based on seeing the dealer's upcard. In addition, the player can double down and split pairs. If all of these player advantages are exploited in an optimal way, blackjack is very close to an even game (without card counting). In fact, under the rules assumed here, the player has a slight advantage, which may help to explain why this set of rules is no longer available. In the next section we show how to play optimally by determining the so-called (composition-dependent) basic strategy.

Table 21.2 A (necessarily abbreviated) list of the 2,741 *unordered* blackjack-dealer sequences, in lexicographical order. (Rules: See Table 21.1.)

seq. no.	unordered sequence	total	number of ordered seqs.	example of ordered sequence
1	00000 00002	20	1	10, 10
2	00000 00011	19	2	10, 9
3	00000 00020	18	1	9, 9
4	00000 00101	18	2	10, 8
5	00000 00110	17	2	9, 8
6	00000 00201	26	1	8, 8, 10
7	00000 00210	25	1	8, 8, 9
8	00000 00300	24	1	8, 8, 8
$\vdots$				
805	04210 00000	18	105	4, 3, 3, 2, 2, 2, 2
806	04300 00000	17	35	3, 3, 3, 2, 2, 2, 2
807	10000 00001	21	2	10, 1
808	10000 00010	20	2	9, 1
809	10000 00100	19	2	8, 1
810	10000 00200	17	1	8, 8, 1
811	10000 01000	18	2	7, 1
812	10000 01010	17	2	9, 7, 1
$\vdots$				
1,424	13200 00001	23	10	3, 3, 2, 2, 2, 1, 10
1,425	13200 00010	22	10	3, 3, 2, 2, 2, 1, 9
1,426	13200 00100	21	10	3, 3, 2, 2, 2, 1, 8
1,427	13200 01000	20	10	3, 3, 2, 2, 2, 1, 7
1,428	13200 10000	19	58	3, 3, 2, 2, 2, 1, 6
1,429	13201 00000	18	103	3, 3, 2, 2, 2, 1, 5
1,430	13210 00000	17	156	3, 3, 2, 2, 2, 1, 4
1,431	13300 00001	26	44	3, 3, 3, 2, 2, 2, 1, 10
$\vdots$				
2,734	44100 00001	25	5	3, 2, 2, 2, 2, 1, 1, 1, 1, 10
2,735	44100 00010	24	5	3, 2, 2, 2, 2, 1, 1, 1, 1, 9
2,736	44100 00100	23	5	3, 2, 2, 2, 2, 1, 1, 1, 1, 8
2,737	44100 01000	22	5	3, 2, 2, 2, 2, 1, 1, 1, 1, 7
2,738	44100 10000	21	5	3, 2, 2, 2, 2, 1, 1, 1, 1, 6
2,739	44101 00000	20	5	3, 2, 2, 2, 2, 1, 1, 1, 1, 5
2,740	44110 00000	19	10	3, 2, 2, 2, 2, 1, 1, 1, 1, 4
2,741	44200 00000	18	15	3, 2, 2, 2, 2, 1, 1, 1, 1, 3
total			48,532	

Table 21.3 Conditional distribution of dealer's final total, given dealer's upcard. (Rules: See Table 21.1.)

up-card	dealer's final total						
	17	18	19	20	21^3	21^2	bust
2	.138 976	.131 762	.131 815	.123 948	.120 526		.352 973
3	.130 313	.130 946	.123 761	.123 345	.116 047		.375 588
4	.130 973	.114 163	.120 679	.116 286	.115 096		.402 803
5	.119 687	.123 483	.116 909	.104 694	.106 321		.428 905
6	.166 948	.106 454	.107 192	.100 705	.097 878		.420 823
7	.372 345	.138 583	.077 334	.078 897	.072 987		.259 854
8	.130 857	.362 989	.129 445	.068 290	.069 791		.238 627
9	.121 886	.103 921	.357 391	.122 250	.061 109		.233 442
T	.114 418	.112 879	.114 662	.328 879	.036 466	.078 431	.214 264
A	.126 128	.131 003	.129 486	.131 553	.051 565	.313 725	.116 540

[3]three or more cards [2]two cards (natural)

Table 21.4 Number of (ordered) blackjack-dealer sequences, according to dealer's upcard and dealer's final total. (Rules: See Table 21.1.)

up-card	dealer's final total							sum
	17	18	19	20	21^3	21^2	bust	
2	1,727	1,707	1,849	1,852	1,844		7,411	16,390
3	1,129	1,140	1,153	1,177	1,172		4,738	10,509
4	688	692	709	707	705		2,858	6,359
5	424	427	435	434	432		1,752	3,904
6	245	248	251	250	249		1,012	2,255
7	154	155	157	157	156		635	1,414
8	93	94	95	94	94		382	852
9	62	63	63	63	62		253	566
T	31	32	32	32	31	1	129	288
A	581	685	689	689	688	1	2,662	5,995
sum	5,134	5,243	5,433	5,455	5,433	2	21,832	48,532

[3]three or more cards [2]two cards (natural)

21.2 Basic Strategy

Roughly speaking, *basic strategy* is the optimal (expectation-maximizing) strategy for the player who does not count cards. Since cards from earlier rounds or from other players in the current round are ignored, we can assume that the hand is dealt from the full deck. Basic strategy tells us how the hand should be played (i.e., stand, hit, double, or split). We allow the player to make full use of the individual cards in his hand (not just their hard or soft total) and the dealer's upcard. In other words, we are concerned with *composition-dependent basic strategy*. The only ambiguity in this formulation is the phrase "the individual cards in his hand" when the player splits. Our convention is that the player makes use only of the cards in the hand he is currently playing. (Keeping track of the cards in hands split from the current one is too much like card counting.) With this understanding and with the rules specified, basic strategy is unambiguously defined. We will see that it is also uniquely determined, unlike in Deuces Wild video poker, for example.

Since there are $\binom{10}{1} + \binom{10}{2} = 55$ initial two-card player hands and 10 dealer upcards, basic strategy involves 550 strategy decisions, excluding *multicard hands* (hands with three or more cards). When we include multicard hands, the number of strategy decisions increases to 19,620, as we will see.[1] In this section we show how composition-dependent basic strategy is derived.

First, we introduce our notation. Let $\boldsymbol{X} = \{X_1, X_2\}$ denote the player's two-card initial hand. We write $\{X_1, X_2\}$ to suggest an *unordered* pair, i.e., $\{X_1, X_2\} = \{X_2, X_1\}$. Let Y denote the player's third card, if any, and let U denote the dealer's upcard, D his downcard, and S his final total. Finally, let G_{std} and G_{hit} denote the player's profit from standing and hitting, assuming an initial one-unit bet. The random variables X_1, X_2, U, D, and Y assume values in $\{1, 2, 3, \ldots, 10\}$ (1 corresponds to an ace, denoted by A, and 10 corresponds to a 10-valued card, denoted by T), while S assumes values in $\{17, 18, \ldots, 26\}$ and G_{std} and G_{hit} assume values in $\{-1, 0, 1\}$, at least if we exclude for now player naturals as well as doubling and splitting.

We begin with one of the simpler decisions, a player $\{6, \mathrm{T}\}$ vs. a dealer upcard of 9. To decide whether to stand on or hit the hard 16, we compute the player's conditional expected profit both ways and determine which is larger. First, if the player stands, he wins if and only if the dealer busts (a push is impossible), so

$$
\begin{aligned}
\mathrm{E}[G_{\mathrm{std}} \mid \boldsymbol{X} = \{6, 10\},\, U = 9] &= \mathrm{P}(S > 21 \mid \boldsymbol{X} = \{6, 10\},\, U = 9) \\
&\quad - \mathrm{P}(S \le 21 \mid \boldsymbol{X} = \{6, 10\},\, U = 9) \\
&= 2\,\mathrm{P}(S > 21 \mid \boldsymbol{X} = \{6, 10\},\, U = 9) - 1 \\
&\approx -0.539232. \tag{21.7}
\end{aligned}
$$

---

[1] Actually, this is only an upper bound because it includes some decisions, such as $\{\mathrm{A}, \mathrm{A}, \mathrm{T}\}$ vs. any dealer upcard, that never occur to the basic strategist.

For the last step we used the conditional distribution of S when the player stands, given $\boldsymbol{X} = \{6, 10\}$ and $U = 9$, as shown in the "none" row of Table 21.5. This is computed from the 566 blackjack-dealer sequences starting with 9, with probabilities assigned to each sequence as though the cards (except for the initial 9) were dealt without replacement from a 49-card deck lacking one 6, one T, and one 9.

Next, if the player hits, we assume that he hits only once. The result is

$$\mathrm{E}[G_{\mathrm{hit}} \mid \boldsymbol{X} = \{6, 10\}, \ U = 9]$$
$$= \sum_{k=1}^{10} \mathrm{P}(Y = k \mid \boldsymbol{X} = \{6, 10\}, \ U = 9)$$
$$\cdot \mathrm{E}[G_{\mathrm{hit}} \mid \boldsymbol{X} = \{6, 10\}, \ U = 9, \ Y = k], \qquad (21.8)$$

and, for $k = 1, 2, 3, 4, 5$,

$$\mathrm{E}[G_{\mathrm{hit}} \mid \boldsymbol{X} = \{6, 10\}, \ U = 9, \ Y = k] \qquad (21.9)$$
$$= \mathrm{P}(S < 16 + k \text{ or } S > 21 \mid \boldsymbol{X} = \{6, 10\}, \ U = 9, \ Y = k)$$
$$- \mathrm{P}(16 + k < S \leq 21 \mid \boldsymbol{X} = \{6, 10\}, \ U = 9, \ Y = k),$$

while, for $k = 6, 7, 8, 9, 10$,

$$\mathrm{E}[G_{\mathrm{hit}} \mid \boldsymbol{X} = \{6, 10\}, \ U = 9, \ Y = k] = -1. \qquad (21.10)$$

Putting this all together, we find that

$$\mathrm{E}[G_{\mathrm{hit}} \mid \boldsymbol{X} = \{6, 10\}, \ U = 9] \approx -0.479306. \qquad (21.11)$$

For the last step we used, for $k = 1, 2, 3, 4, 5$, the conditional distribution of S, given $\boldsymbol{X} = \{6, 10\}$, $U = 9$, and $Y = k$, as shown in Table 21.5. This is computed from the 566 blackjack-dealer sequences starting with 9, with probabilities assigned to each sequence as though the cards (except for the initial 9) were dealt without replacement from a 48-card deck lacking one 6, one T, one 9, and one k. With (21.11) being greater than (21.7), we conclude that it is optimal to hit $\{6, \mathrm{T}\}$ vs. 9.

We turn to a slightly more complicated decision, a player $\{6, \mathrm{T}\}$ vs. a dealer upcard of T. To decide whether to stand on or hit the hard 16, we compute the player's conditional expected profit both ways and determine which is larger. However, here the conditioning event must include the event that the dealer fails to have a natural. First, as in (21.7),

$$\mathrm{E}[G_{\mathrm{std}} \mid \boldsymbol{X} = \{6, 10\}, \ U = 10, \ D \neq 1]$$
$$= \mathrm{P}(S > 21 \mid \boldsymbol{X} = \{6, 10\}, \ U = 10, \ D \neq 1)$$
$$- \mathrm{P}(S \leq 21 \mid \boldsymbol{X} = \{6, 10\}, \ U = 10, \ D \neq 1)$$
$$= 2\,\mathrm{P}(S > 21 \mid \boldsymbol{X} = \{6, 10\}, \ U = 10, \ D \neq 1) - 1$$

Table 21.5 Table of conditional probabilities and expectations for $\{6, T\}$ vs. 9. (Rules: See Table 21.1 on p. 644.)

player's 3rd card	dealer's final total					
	17	18	19	20	21	bust
none	.125 606	.106 802	.350 040	.125 524	.061 644	.230 384
A	.125 909	.109 342	.357 567	.107 580	.063 585	.236 019
2	.124 803	.104 541	.355 309	.126 506	.056 468	.232 372
3	.124 383	.105 982	.353 186	.126 536	.062 066	.227 848
4	.125 697	.105 434	.354 581	.124 279	.061 940	.228 069
5	.125 214	.105 998	.354 865	.125 592	.059 792	.228 539

player's 3rd card	player's total	conditional probability	cond'l expected profit	product
A	17	4/49	−.402 054	−.032 821
2	18	4/49	−.181 108	−.014 784
3	19	4/49	.269 610	.022 009
4	20	4/49	.751 842	.061 375
5	21	4/49	.940 208	.076 752
$6, 7, 8, 9, T$	bust	29/49	−1.000 000	−.591 837
sum		1		−.479 306

$$\approx -0.542952. \tag{21.12}$$

For the last step we used the conditional distribution of S when the player stands, given $\boldsymbol{X} = \{6, 10\}$, $U = 10$, and $D \neq 1$ as shown in the "none" row of Table 21.6. This is computed from the 287 blackjack-dealer sequences starting with 10, excluding $(10, 1)$, with probabilities assigned to each sequence as though the cards (except for the initial 10) were dealt without replacement from a 49-card deck lacking one 6 and two Ts. The probabilities must be normalized because of the excluded $(10, 1)$, so we divide by $45/49$.

Next, as in (21.8),

$$E[G_{\text{hit}} \mid \boldsymbol{X} = \{6, 10\}, \ U = 10, \ D \neq 1]$$

$$= \sum_{k=1}^{10} P(Y = k \mid \boldsymbol{X} = \{6, 10\}, \ U = 10, \ D \neq 1) \tag{21.13}$$

$$\cdot E[G_{\text{hit}} \mid \boldsymbol{X} = \{6, 10\}, \ U = 10, \ D \neq 1, \ Y = k],$$

and, for $k = 1, 2, 3, 4, 5$,

$$E[G_{\text{hit}} \mid \boldsymbol{X} = \{6, 10\}, \ U = 10, \ D \neq 1, \ Y = k] \tag{21.14}$$

$$= \mathrm{P}(S < 16 + k \text{ or } S > 21 \mid \boldsymbol{X} = \{6, 10\}, \ U = 10, \ D \neq 1, \ Y = k)$$
$$- \mathrm{P}(16 + k < S \leq 21 \mid \boldsymbol{X} = \{6, 10\}, \ U = 10, \ D \neq 1, \ Y = k),$$

while, for $k = 6, 7, 8, 9, 10$,

$$\mathrm{E}[G_{\mathrm{hit}} \mid \boldsymbol{X} = \{6, 10\}, \ U = 10, \ D \neq 1, \ Y = k] = -1. \tag{21.15}$$

We can evaluate the conditional probability in (21.13) using Bayes's law (Theorem 1.2.7 on p. 19; see Example 1.2.9 on p. 19 for discussion):

$$\mathrm{P}(Y = k \mid \boldsymbol{X} = \{6, 10\}, \ U = 10, \ D \neq 1)$$
$$= \mathrm{P}(Y = k \mid \boldsymbol{X} = \{6, 10\}, \ U = 10)$$
$$\cdot [1 - \mathrm{P}(D = 1 \mid \boldsymbol{X} = \{6, 10\}, \ U = 10, \ Y = k)]$$
$$\cdot [1 - \mathrm{P}(D = 1 \mid \boldsymbol{X} = \{6, 10\}, \ U = 10)]^{-1}. \tag{21.16}$$

As for (21.14), it uses the conditional distribution of S, given $\boldsymbol{X} = \{6, 10\}$, $U = 10$, $D \neq 1$, and $Y = k$ ($k = 1, 2, 3, 4, 5$), as shown in Table 21.6. This is computed from the 287 blackjack-dealer sequences starting with 10, excluding $(10, 1)$, with probabilities assigned to each sequence as though the cards (except for the initial 10) were dealt without replacement from a 48-card deck lacking one 6, two Ts, and one k. The probabilities must be normalized because of the excluded $(10, 1)$, so we divide by $45/48$ if $k = 1$ and by $44/48$ otherwise. Putting this all together, we find that

$$\mathrm{E}[G_{\mathrm{hit}} \mid \boldsymbol{X} = \{6, 10\}, \ U = 10, \ D \neq 1] \approx -0.506929. \tag{21.17}$$

With (21.17) being greater than (21.12), we conclude that it is optimal to hit $\{6, \mathrm{T}\}$ vs. T.

In both of these calculations, we assumed that it is optimal for the player to hit a $\{6, \mathrm{T}\}$ at most once, an assumption that can be verified with a separate calculation. With sufficient patience, we could continue with this analysis, but instead we start anew with the observation that the entire derivation is best done recursively. We will not need the preceding calculations except as motivation for what follows.

Our notation will be much as before, except that instead of denoting the player's hand by $\boldsymbol{X} = \{X_1, X_2, \ldots\}$, we write $\boldsymbol{X} = \boldsymbol{l} = (l_1, l_2, \ldots, l_9, l_{10})$ if the player's hand consists of l_1 aces, l_2 2s, $\ldots$, l_9 9s, and l_{10} 10-valued cards. The number of cards in his hand is

$$|\boldsymbol{l}| := \sum_{j=1}^{10} l_j, \tag{21.18}$$

and his current total is a slight generalization of (21.5) on p. 646:

Table 21.6 Table of conditional probabilities and expectations for $\{6, \text{T}\}$ vs. T, conditioned also on the dealer's not having a natural. (Rules: See Table 21.1 on p. 644.)

player's 3rd card	dealer's final total					
	17	18	19	20	21	bust
none	.129 767	.126 102	.128 062	.348 427	.039 118	.228 524
A	.128 032	.126 240	.128 320	.348 602	.039 371	.229 435
2	.128 325	.125 376	.128 859	.354 286	.037 930	.225 224
3	.128 925	.124 870	.127 746	.354 581	.038 324	.225 554
4	.129 197	.126 278	.127 128	.353 348	.038 592	.225 458
5	.129 749	.125 712	.127 625	.353 693	.037 304	.225 917

player's 3rd card	player's total	conditional probability*	cond'l expected profit	product
A	17	4/48	−.413 098	−.034 425
2	18	(44/48)(4/45)	−.167 526	−.013 650
3	19	(44/48)(4/45)	.086 444	.007 044
4	20	(44/48)(4/45)	.569 468	.046 401
5	21	(44/48)(4/45)	.962 696	.078 442
6, 7, 8, 9, T	bust	(44/48)(29/45)	−1.000 000	−.590 741
sum		1		−.506 929

*see (21.16) or Example 1.2.9 on p. 19

$$
T(\boldsymbol{l}) := \begin{cases} \sum j l_j + 10 & \text{if } l_1 \geq 1 \text{ and } \sum j l_j \leq 11, \\ \sum j l_j & \text{otherwise,} \end{cases} \tag{21.19}
$$

with the two cases corresponding to soft and hard totals. Again, let Y denote the player's next card, if any, and let U denote the dealer's upcard, D his downcard, and S his final total. Finally, let G_{std}, G_{hit}, G_{dbl}, and G_{spl} denote the player's profit from standing, hitting, doubling, and splitting, assuming an initial one-unit bet.

Next it will be convenient to introduce some new notation. First we define the events on which we will condition:

$$
A(\boldsymbol{l}, u) := \begin{cases} \{\boldsymbol{X} = \boldsymbol{l}, \ U = 1, \ D \neq 10\} & \text{if } u = 1, \\ \{\boldsymbol{X} = \boldsymbol{l}, \ U = u\} & \text{if } u = 2, 3, \ldots, 9, \\ \{\boldsymbol{X} = \boldsymbol{l}, \ U = 10, \ D \neq 1\} & \text{if } u = 10. \end{cases} \tag{21.20}
$$

We then define the conditional expectations associated with each player hand, dealer upcard, and strategy:

$$E_{\mathrm{std}}(\boldsymbol{l}, u) := \mathrm{E}[G_{\mathrm{std}} \mid A(\boldsymbol{l}, u)], \tag{21.21}$$

$$E_{\mathrm{hit}}(\boldsymbol{l}, u) := \mathrm{E}[G_{\mathrm{hit}} \mid A(\boldsymbol{l}, u)], \tag{21.22}$$

$$E_{\mathrm{dbl}}(\boldsymbol{l}, u) := \mathrm{E}[G_{\mathrm{dbl}} \mid A(\boldsymbol{l}, u)] \quad (|\boldsymbol{l}| = 2), \tag{21.23}$$

$$E_{\mathrm{spl}}(\boldsymbol{l}, u) := \mathrm{E}[G_{\mathrm{spl}} \mid A(\boldsymbol{l}, u)] \quad (\boldsymbol{l} = 2\boldsymbol{e}_i, \ i = 1, \ldots, 10), \tag{21.24}$$

where $\boldsymbol{e}_1 = (1, 0, 0, \ldots, 0)$, $\boldsymbol{e}_2 = (0, 1, 0, 0, \ldots, 0)$, and so on. Basic strategy split calculations are rather challenging, so we postpone them for the time being. Therefore, we define the maximal conditional expectation for each player hand and dealer upcard as follows:

$$E_{\max}(\boldsymbol{l}, u) := \begin{cases} \max\{E_{\mathrm{std}}(\boldsymbol{l}, u), E_{\mathrm{hit}}(\boldsymbol{l}, u), E_{\mathrm{dbl}}(\boldsymbol{l}, u)\} & \text{if } |\boldsymbol{l}| = 2, \\ \max\{E_{\mathrm{std}}(\boldsymbol{l}, u), E_{\mathrm{hit}}(\boldsymbol{l}, u)\} & \text{if } |\boldsymbol{l}| \geq 3. \end{cases} \tag{21.25}$$

For future reference, we also define

$$E_{\max}^*(\boldsymbol{l}, u) := \max\{E_{\mathrm{std}}(\boldsymbol{l}, u), E_{\mathrm{hit}}(\boldsymbol{l}, u)\}, \qquad |\boldsymbol{l}| \geq 2, \tag{21.26}$$

for situations in which doubling is not permitted.

It will be necessary to specify more precisely the set of player hands and dealer upcards we will consider. We denote the set of all unbusted player hands of two or more cards by

$$\mathscr{L} := \{\boldsymbol{l} : |\boldsymbol{l}| \geq 2, \ T(\boldsymbol{l}) \leq 21\} \tag{21.27}$$

and the set of all pairs of such hands and dealer upcards by

$$\mathscr{M} := \{(\boldsymbol{l}, u) \in \mathscr{L} \times \{1, 2, \ldots, 10\} : l_u \leq 3\}. \tag{21.28}$$

We exclude the possibility that $l_u = 4$ because, if the player's hand has four cards of the same value (necessarily 1, 2, 3, 4, or 5), it is impossible for the dealer's upcard to have that value as well. The cardinality of $\mathscr{L}$ is the sum over $2 \leq n \leq 21$ of the number of partitions of the integer n into two or more parts with no part greater than 10 and no part having multiplicity greater than 4. By direct enumeration, $|\mathscr{L}| = 2{,}008$. Similarly, we can verify that $|\mathscr{M}| = (2{,}008)(10) - 460 = 19{,}620$.

The basic relations connecting the conditional expectations defined above include, for all $(\boldsymbol{l}, u) \in \mathscr{M}$,

$$E_{\mathrm{std}}(\boldsymbol{l}, u) = \mathrm{P}(S < T(\boldsymbol{l}) \text{ or } S > 21 \mid A(\boldsymbol{l}, u)) \\ - \mathrm{P}(T(\boldsymbol{l}) < S \leq 21 \mid A(\boldsymbol{l}, u)), \tag{21.29}$$

$$E_{\mathrm{hit}}(\boldsymbol{l}, u) = \sum_{1 \leq k \leq 10: \ (\boldsymbol{l}+\boldsymbol{e}_k, u) \in \mathscr{M}} p(k \mid \boldsymbol{l}, u) E_{\max}(\boldsymbol{l} + \boldsymbol{e}_k, u) \\ + \sum_{1 \leq k \leq 10: \ \boldsymbol{l}+\boldsymbol{e}_k \notin \mathscr{L}} p(k \mid \boldsymbol{l}, u)(-1), \tag{21.30}$$

$$E_{\mathrm{dbl}}(\boldsymbol{l}, u) = 2 \sum_{1 \le k \le 10:\, (\boldsymbol{l}+\boldsymbol{e}_k, u) \in \mathcal{M}} p(k \mid \boldsymbol{l}, u) E_{\mathrm{std}}(\boldsymbol{l}+\boldsymbol{e}_k, u)$$

$$+ 2 \sum_{1 \le k \le 10:\, \boldsymbol{l}+\boldsymbol{e}_k \notin \mathcal{L}} p(k \mid \boldsymbol{l}, u)(-1) \qquad (|\boldsymbol{l}| = 2), \quad (21.31)$$

where

$$p(k \mid \boldsymbol{l}, u) := \mathrm{P}(Y = k \mid A(\boldsymbol{l}, u)). \tag{21.32}$$

The probabilities $p(k \mid \boldsymbol{l}, u)$ are derived from Bayes's law (Theorem 1.2.7 on p. 19) for $u = 1, 10$, as we saw in (21.16). Specifically, let

$$n_1 = \cdots = n_9 := 4 \quad \text{and} \quad n_{10} := 16 \tag{21.33}$$

be the numbers of cards of each value in the full deck and $N := n_1 + \cdots + n_{10} = 52$ be the total number of cards. Then

$$p(k \mid \boldsymbol{l}, 1) = \frac{n_k - l_k - \delta_{1,k}}{N - |\boldsymbol{l}| - 1 - (n_{10} - l_{10})} \left(1 - \frac{n_{10} - l_{10} - \delta_{10,k}}{N - |\boldsymbol{l}| - 2}\right), \quad (21.34)$$

$$p(k \mid \boldsymbol{l}, u) = \frac{n_k - l_k - \delta_{u,k}}{N - |\boldsymbol{l}| - 1}, \qquad u = 2, 3, \ldots, 9, \tag{21.35}$$

$$p(k \mid \boldsymbol{l}, 10) = \frac{n_k - l_k - \delta_{10,k}}{N - |\boldsymbol{l}| - 1 - (n_1 - l_1)} \left(1 - \frac{n_1 - l_1 - \delta_{1,k}}{N - |\boldsymbol{l}| - 2}\right). \tag{21.36}$$

The quantities (21.29) are computed directly, while those in (21.30) and (21.31) are obtained recursively. They are recursive in the number of player cards $|\boldsymbol{l}|$ and also in the player's hard total

$$T_{\mathrm{hard}}(\boldsymbol{l}) := \sum_{j=1}^{10} j l_j. \tag{21.37}$$

We adopt the latter approach because then the initial conditions for the recursion have the simple form

$$E_{\mathrm{hit}}(\boldsymbol{l}, u) = -1, \qquad T_{\mathrm{hard}}(\boldsymbol{l}) = 21, \quad u = 1, 2, \ldots, 10. \tag{21.38}$$

There is one exception to (21.29), namely that an untied player natural is paid 3 to 2:

$$E_{\mathrm{std}}(\boldsymbol{e}_1 + \boldsymbol{e}_{10}, u) = 1.5, \qquad u = 1, 2, \ldots, 10. \tag{21.39}$$

We are now in a position to let the computer take over. We begin by computing $E_{\mathrm{std}}(\boldsymbol{l}, u)$ for all $(\boldsymbol{l}, u) \in \mathcal{M}$. The number of blackjack-dealer sequences that must be analyzed for each such $\boldsymbol{l}$ is at most 48,532 (see Table 21.4 on p. 649). Then we go back and compute $E_{\max}(\boldsymbol{l}, u)$ for $T_{\mathrm{hard}}(\boldsymbol{l}) = 21, 20, \ldots, 2$ (in that order) and all u using (21.25), (21.30), (21.31), (21.33)–(21.36), (21.38), and (21.39). We notice along the way which strategy achieves the maximum in (21.25), and this gives us our composition-dependent basic strategy, except for splits.

We can also evaluate the player's overall expectation E using the optimal strategy thus derived. It is simply a matter of conditioning on the player's initial two-card hand and the dealer's upcard. Now the 550 events $A(\boldsymbol{l}, u)$ for $|\boldsymbol{l}| = 2$ and $u = 1, 2, \ldots, 10$ do not partition the sample space, but if we include the 110 events

$$B(\boldsymbol{e}_i + \boldsymbol{e}_j, 1) := \{\boldsymbol{X} = \boldsymbol{e}_i + \boldsymbol{e}_j, \ U = 1, \ D = 10\} \qquad (21.40)$$
$$B(\boldsymbol{e}_i + \boldsymbol{e}_j, 10) := \{\boldsymbol{X} = \boldsymbol{e}_i + \boldsymbol{e}_j, \ U = 10, \ D = 1\} \qquad (21.41)$$

as well, where $1 \le i \le j \le 10$, then we do have a partition, and conditioning (Theorem 2.1.3 on p. 78) gives the desired result, namely

$$
\begin{aligned}
E = &\sum_{u=1}^{10} \sum_{1 \le i \le j \le 10} \mathrm{P}(A(\boldsymbol{e}_i + \boldsymbol{e}_j, u)) E_{\max}(\boldsymbol{e}_i + \boldsymbol{e}_j, u) \\
&+ \sum_{1 \le i \le j \le 10: \ (i,j) \ne (1,10)} \mathrm{P}(B(\boldsymbol{e}_i + \boldsymbol{e}_j, 1))(-1) \\
&+ \sum_{1 \le i \le j \le 10: \ (i,j) \ne (1,10)} \mathrm{P}(B(\boldsymbol{e}_i + \boldsymbol{e}_j, 10))(-1) \\
&+ \mathrm{P}(B(\boldsymbol{e}_1 + \boldsymbol{e}_{10}, 1))(0) + \mathrm{P}(B(\boldsymbol{e}_1 + \boldsymbol{e}_{10}, 10))(0) \\
= &\sum_{u=1}^{10} \sum_{1 \le i \le j \le 10} \mathrm{P}(A(\boldsymbol{e}_i + \boldsymbol{e}_j, u)) E_{\max}(\boldsymbol{e}_i + \boldsymbol{e}_j, u) \\
&- \frac{\binom{n_1}{1}\binom{n_{10}}{1}}{\binom{N}{2}} \left(1 - \frac{\binom{n_1-1}{1}\binom{n_{10}-1}{1}}{\binom{N-2}{2}}\right).
\end{aligned}
\qquad (21.42)
$$

The second equality uses the fact that the union of the events $B(\boldsymbol{e}_i + \boldsymbol{e}_j, u)$ $(1 \le i \le j \le 10, \ (i,j) \ne (1,10), \ u \in \{1,10\})$ is the event that the dealer has a natural and the player does not. Finally, we observe that, for $1 \le i < j \le 10$ or $1 \le i \le 10$,

$$\mathrm{P}(A(\boldsymbol{e}_i + \boldsymbol{e}_j, u)) = \frac{\binom{n_i}{1}\binom{n_j}{1}}{\binom{N}{2}} \frac{n_u - \delta_{u,i} - \delta_{u,j}}{N-2}, \qquad (21.43)$$

$$\mathrm{P}(A(2\boldsymbol{e}_i, u)) = \frac{\binom{n_i}{2}}{\binom{N}{2}} \frac{n_u - 2\delta_{u,i}}{N-2}, \qquad (21.44)$$

for $u = 2, 3, \ldots, 9$, and

$$\mathrm{P}(A(\boldsymbol{e}_i + \boldsymbol{e}_j, 1)) = \frac{\binom{n_i}{1}\binom{n_j}{1}}{\binom{N}{2}} \frac{n_1 - \delta_{1,i} - \delta_{1,j}}{N-2}$$

$$\cdot \left(1 - \frac{n_{10} - \delta_{10,i} - \delta_{10,j}}{N - 3}\right), \qquad (21.45)$$

$$\mathrm{P}(A(2e_i, 1)) = \frac{\binom{n_i}{2}}{\binom{N}{2}} \frac{n_1 - 2\delta_{1,i}}{N - 2} \left(1 - \frac{n_{10} - 2\delta_{10,i}}{N - 3}\right), \qquad (21.46)$$

$$\mathrm{P}(A(e_i + e_j, 10)) = \frac{\binom{n_i}{1}\binom{n_j}{1}}{\binom{N}{2}} \frac{n_{10} - \delta_{10,i} - \delta_{10,j}}{N - 2}$$
$$\cdot \left(1 - \frac{n_1 - \delta_{1,i} - \delta_{1,j}}{N - 3}\right), \qquad (21.47)$$

$$\mathrm{P}(A(2e_i, 10)) = \frac{\binom{n_i}{2}}{\binom{N}{2}} \frac{n_{10} - 2\delta_{10,i}}{N - 2} \left(1 - \frac{n_1 - 2\delta_{1,i}}{N - 3}\right), \qquad (21.48)$$

and the derivation is complete. We find that $E \approx -0.003334915963$, but this is a preliminary result because it does not allow splits. It remains to determine composition-dependent basic strategy with splits.

All we need to do is evaluate $E_{\mathrm{spl}}(2e_i, u)$ for $i = 1, \ldots, 10$ and $u = 1, \ldots, 10$, and then redefine

$$E_{\max}(2e_i, u) := \max\{E_{\mathrm{std}}(2e_i, u), E_{\mathrm{hit}}(2e_i, u),$$
$$E_{\mathrm{dbl}}(2e_i, u), E_{\mathrm{spl}}(2e_i, u)\}. \qquad (21.49)$$

We can then reevaluate E as in (21.42) to obtain the overall expectation for the composition-dependent basic strategist.

Before we can do this, we must extend our notation slightly. We define $E_{\mathrm{std}}(l, u \mid m)$ analogously to $E_{\mathrm{std}}(l, u)$, but with $m = (m_1, m_2, \ldots, m_{10})$ indicating that the initial deck is depleted by removing m_1 aces, m_2 2s, $\ldots$, m_{10} 10-valued cards (in addition to the cards in the player's hand and the dealer's upcard). Thus, $E_{\mathrm{std}}(l, u) = E_{\mathrm{std}}(l, u \mid 0)$. Also, we let Y_1 denote the player's next card, Y_2 the card that follows Y_1, and so on. We can then generalize (21.32) to

$$p(i \mid l, u) := \mathrm{P}(Y_1 = i \mid A(l, u)), \qquad (21.50)$$
$$p(i, j \mid l, u) := \mathrm{P}(Y_1 = i, \ Y_2 = j \mid A(l, u)), \qquad (21.51)$$
$$p(i, j, k \mid l, u) := \mathrm{P}(Y_1 = i, \ Y_2 = j, \ Y_3 = k \mid A(l, u)), \qquad (21.52)$$

and so on.

The easiest case is splitting a pair of aces because each ace receives only one card. The result is that

$$E_{\mathrm{spl}}(2e_1, u) = \sum_{j=1}^{10} \sum_{k=1}^{10} p(j, k \mid 2e_1, u)[E_{\mathrm{std}}(e_1 + e_j, u \mid e_1 + e_k)$$
$$+ E_{\mathrm{std}}(e_1 + e_k, u \mid e_1 + e_j)]$$

$$= 2 \sum_{j=1}^{10} \sum_{k=1}^{10} p(j, k \mid 2\boldsymbol{e}_1, u) E_{\mathrm{std}}(\boldsymbol{e}_1 + \boldsymbol{e}_j, u \mid \boldsymbol{e}_1 + \boldsymbol{e}_k)$$

$$= 2 \sum_{j=1}^{10} p(j \mid 2\boldsymbol{e}_1, u) E_{\mathrm{std}}(\boldsymbol{e}_1 + \boldsymbol{e}_j, u \mid \boldsymbol{e}_1), \qquad (21.53)$$

where the first equality follows by conditioning on the two cards dealt to the two aces, the second uses symmetry (and $p(j, k \mid 2\boldsymbol{e}_1, u) = p(k, j \mid 2\boldsymbol{e}_1, u)$), and the third uses conditioning once again (and $p(j, k \mid 2\boldsymbol{e}_1, u) = p(j \mid 2\boldsymbol{e}_1, u) p(k \mid 2\boldsymbol{e}_1 + \boldsymbol{e}_j, u)$).

We finally turn to the more difficult case of splitting a pair of non-aces. First, we define $E^*(\boldsymbol{l}, u \mid \boldsymbol{m})$ to be the player's expectation using the same strategy used to achieve $E^*_{\max}(\boldsymbol{l}, u)$ of (21.26) (since doubling after splitting is not permitted), but with $\boldsymbol{m} = (m_1, m_2, \ldots, m_{10})$ indicating that the initial deck is depleted by removing m_1 aces, m_2 2s, $\ldots$, m_{10} 10-valued cards (in addition to the cards in the player's hand and the dealer's upcard). The values of $\boldsymbol{m}$ that are of interest will be specified shortly. These expectations are computed recursively, much as in (21.29)–(21.31), except that a depleted deck is used instead of the full deck and (21.39) is omitted because a two-card 21 after a split does not count as a natural. (Of course, we use basic strategy optimized for the full deck, not the depleted deck.) More precisely, (21.29)–(21.31) are replaced by

$$E^*(\boldsymbol{l}, u \mid \boldsymbol{m}) = E_{\mathrm{std}}(\boldsymbol{l}, u \mid \boldsymbol{m}) \quad \text{if} \quad E_{\mathrm{std}}(\boldsymbol{l}, u) > E_{\mathrm{hit}}(\boldsymbol{l}, u), \qquad (21.54)$$

$$E^*(\boldsymbol{l}, u \mid \boldsymbol{m}) = \sum_{1 \leq k \leq 10: \ (\boldsymbol{l} + \boldsymbol{e}_k, u) \in \mathcal{M}} p(k \mid \boldsymbol{l} + \boldsymbol{m}, u) E^*(\boldsymbol{l} + \boldsymbol{e}_k, u \mid \boldsymbol{m})$$

$$+ \sum_{1 \leq k \leq 10: \ \boldsymbol{l} + \boldsymbol{e}_k \notin \mathcal{L}} p(k \mid \boldsymbol{l} + \boldsymbol{m}, u)(-1) \qquad (21.55)$$

otherwise. Of course, we need $l_i + m_i + \delta_{u,i} \leq n_i$ for $i = 1, 2, \ldots, 10$ to define $E^*(\boldsymbol{l}, u \mid \boldsymbol{m})$. In addition, for the kth term in the first sum in (21.55) to be present we need $l_k + m_k + \delta_{u,k} \leq n_k - 1$. Finally, we must generalize $p(k \mid \boldsymbol{l}, u)$ beyond the case in which $(\boldsymbol{l}, u) \in \mathcal{M}$, but the formulas (21.34)–(21.36) are unchanged, with the proviso that the numerators in those formulas cannot be allowed to become negative (or the denominators nonpositive).

What happens when a player splits a non-ace pair? In practice, the cards are exposed and separated, he places his original bet by one of the cards and places an equal bet by the other card, and he plays out the two hands one by one. If one of his second cards has the same value as the original two (called a *paircard*), he has a new pair and splits again. (Remember, the information available to the player at the second split is assumed exactly the same as that available at the first split, so the strategy will be the same.) A third split is also possible but a fourth is prohibited by the rules (Table 21.1 on p. 644).

Actually, in a single-deck game, even if there were no limit on the number of splits permitted, only 10-valued cards could be split more than three times.

It will turn out to be more convenient (and it is mathematically equivalent) to assume that all second cards are dealt before any third cards are offered. There are eight ways in which this can happen, and they are listed in Table 21.7.

Table 21.7 List of ways in which a non-ace pair can be split. P denotes a paircard, N denotes a non-paircard, and X denotes a card of either type. (PP) denotes the initial hand. (Rules: See Table 21.1 on p. 644.)

case	card sequence	no. of hands	contribution to $E_{\mathrm{spl}}(2e_i, u)$
1	(PP)NN	2	$2S_1 - 2S_2$
2	(PP)PNNN	3	$3S_2 - 6S_3 + 3S_4$
3	(PP)NPNN	3	$3S_2 - 6S_3 + 3S_4$
4	(PP)PPXXXX	4	$4S_3 + 4T_3$
5	(PP)NPPXXX	4	$4S_3 - 3S_4 + 3T_3 - 3T_4$
6	(PP)PNPXXX	4	$4S_3 - 3S_4 + 3T_3 - 3T_4$
7	(PP)PNNPXX	4	$4S_3 - 6S_4 + 2S_5 + 2T_3 - 4T_4 + 2T_5$
8	(PP)NPNPXX	4	$4S_3 - 6S_4 + 2S_5 + 2T_3 - 4T_4 + 2T_5$

Let us begin by considering case 1, in which both initial paircards receive non-paircards in the split. Denoting the paircard by i, the contribution to $E_{\mathrm{spl}}(2e_i, u)$ from this case is

$$\sum_{j:j\neq i}\sum_{k:k\neq i} p(j,k \mid 2e_i, u)[E^*(e_i + e_j, u \mid e_i + e_k) + E^*(e_i + e_k, u \mid e_i + e_j)]$$

$$= 2\sum_{j:j\neq i}\sum_{k:k\neq i} p(j,k \mid 2e_i, u)E^*(e_i + e_j, u \mid e_i + e_k)$$

$$= 2\sum_{j:j\neq i} p(j \mid 2e_i, u)[E^*(e_i + e_j, u \mid e_i)$$

$$\hspace{4cm} - p(i \mid 2e_i + e_j, u)E^*(e_i + e_j, u \mid 2e_i)]$$

$$= 2S_1 - 2S_2, \tag{21.56}$$

where

$$S_1 := \sum_{j:j\neq i} p(j \mid 2e_i, u)E^*(e_i + e_j, u \mid e_i), \tag{21.57}$$

$$S_2 := \sum_{j:j\neq i} p(i,j \mid 2e_i, u)E^*(e_i + e_j, u \mid 2e_i), \tag{21.58}$$

$$S_3 := \sum_{j:j\neq i} p(i,i,j \mid 2e_i, u) E^*(e_i + e_j, u \mid 3e_i), \qquad (21.59)$$

and so on.

Next we consider case 4, in which both initial paircards receive a paircard in the initial split, spawning two more splits. The maximum number of splits having been reached, each of the four second cards is arbitrary. Here the contribution to $E_{\mathrm{spl}}(2e_i, u)$ is

$$\sum_j \sum_k \sum_l \sum_m p(i,i,j,k,l,m \mid 2e_i, u)$$

$$\cdot [E^*(e_i + e_j, u \mid 3e_i + e_k + e_l + e_m)$$
$$+ E^*(e_i + e_k, u \mid 3e_i + e_j + e_l + e_m)$$
$$+ E^*(e_i + e_l, u \mid 3e_i + e_j + e_k + e_m)$$
$$+ E^*(e_i + e_m, u \mid 3e_i + e_j + e_k + e_l)]$$

$$= 4 \sum_j \sum_k \sum_l \sum_m p(i,i,j,k,l,m \mid 2e_i, u)$$

$$\cdot E^*(e_i + e_j, u \mid 3e_i + e_k + e_l + e_m)$$

$$= 4 \sum_j p(i,i,j \mid 2e_i, u) E^*(e_i + e_j, u \mid 3e_i)$$

$$= 4S_3 + 4T_3, \qquad (21.60)$$

where S_3 is as above and

$$T_3 := p(i,i,i \mid 2e_i, u) E^*(2e_i, u \mid 3e_i), \qquad (21.61)$$
$$T_4 := p(i,i,i,i \mid 2e_i, u) E^*(2e_i, u \mid 4e_i), \qquad (21.62)$$

and so on.

We leave the remaining formulas in Table 21.7 to the reader (Problem 21.16 on p. 674). The bottom line is that

$$E_{\mathrm{spl}}(2e_i, u) = 2S_1 + 4S_2 + 8S_3 - 12S_4 + 4S_5 + 14T_3 - 14T_4 + 4T_5. \quad (21.63)$$

Notice that S_n is 0 unless there are $n+1$ cards of value i and T_n is 0 unless there are $n+2$ cards of value i. Thus,

$$E_{\mathrm{spl}}(2e_i, u) = 2S_1 + 4S_2 + 8S_3 \qquad (21.64)$$

unless $i = 1$ or $i = 10$. In summary, (21.53) applies when $i = 1$, (21.64) applies when $i = 2, \ldots, 9$, and (21.63) applies when $i = 10$.

The final result of all this analysis and computation is that, taking splits into account, the player's overall expectation E is given by

$$E \approx 0.000412515770. \qquad (21.65)$$

Remarkably, this number is positive. The strategy that achieves this positive expectation, composition-dependent basic strategy, is specified in Table 21.8. (See Example 2.2.1 on p. 80 for discussion of the insurance decision.)

Table 21.8 Algorithm for playing a blackjack hand according to composition-dependent basic strategy. This is best described in terms of total-dependent basic strategy with composition-dependent exceptions. (Rules: See Table 21.1 on p. 644.)

1. *Insurance and dealer natural.* Is dealer's upcard an ace or a 10-valued card? If not, go to Step 2.

   - If dealer's upcard is an ace, do not take insurance.

   If dealer has a natural, stop. Otherwise go to Step 2.

2. *Doubling.* Does hand consist of the initial two cards only (but not after a split)? If not, go to Step 3.

   - *Hard totals.* Always double 11. Double 10 vs. 2–9. Double 9 vs. 2–6. Double 8, except for {2, 6}, vs. 5 and 6.
   - *Soft totals.* Double 13–16 vs. 4–6, 17 vs. 2–6, 18 vs. 3–6, and 19 vs. 6.

   If hand is doubled, stop. Otherwise go to Step 3.

3. *Splitting.* Does hand consist of a pair (possibly after a split)? If not, go to Step 4.

   - Always split {A, A} and {8, 8}. Never split {4, 4}, {5, 5}, or {T, T}. Split {2, 2} vs. 3–7, {3, 3} vs. 4–7, {6, 6} vs. 2–6, {7, 7} vs. 2–7, and {9, 9} vs. 2–9 except 7.

   If aces are split, stop. If any other pair is split, apply this algorithm, beginning with Step 3, to each hand. Otherwise go to Step 4.

4. *Hitting and standing.*

   - *Hard totals.* Always stand on 17 or higher. Hit stiffs (12–16) vs. high cards (7, 8, 9, T, A). Stand on stiffs vs. low cards (2, 3, 4, 5, 6), except hit 12 vs. 2 and 3. Always hit 11 or lower. Two-card exceptions: Stand on {7, 7} vs. T, hit {3, T} vs. 2, stand on {5, 7} and {4, 8} vs. 3, and hit {2, T} vs. 4 and 6. For multicard exceptions, see Step 5.
   - *Soft totals.* Hit 17 or less, and stand on 18 or more, except hit 18 vs. 9 and T. For multicard exceptions, see Step 5.

   After standing or busting, stop. After hitting without busting, repeat Step 4.

5. *Multicard exceptions.* If hand consists of three or more cards, the following list of exceptions takes precedence over Step 4.[2]

---

[2] Braces enclosing hands and commas separating cards are omitted here to save space.

- hard 16 vs. T. Stand if at most one card is 6 or more (with aces counting as 1). Stand also on 277, A78, and AA77. Exceptions: Hit 222T, 22336, AA22T, A22236, and AA22226.
- hard 16 vs. 9. Stand if all cards are 5 or less (with aces counting as 1) or if hand contains five or more cards. Stand also on 556, 448, 358, 2446, 2338, 2248, A456, A348, A258, A249, and AA59. Exceptions: Hit 22336, 22237, A3336, A2337, A2247, AA347, AA266, AA257, AA22T, AAA67, AAA3T, A22236, A22227, AA2237, AAA337, and AAAA66.
- hard 16 vs. 8. Stand if all cards are 5 or less (with aces counting as 1). Stand also on AAAA336 and AAAA2226.
- hard 16 vs. 7. Stand if all cards are 5 or less (with aces counting as 1) or if hand contains six or more cards. Stand also on AAA3T. Exceptions: Hit AAA229, AAAA66, and AAAA39.
- hard 15 vs. T. Stand on 555, 456, 366, A455, A356, A266, AA355, AAA66, and AAAA56.
- hard 15 vs. 7. Stand on AAAA2333.
- hard 13 vs. 2. Hit A2T.
- hard 12 vs. 3. Stand. Exceptions: Hit 336, 237, A29, 2334, 2235, 2226, A335, A236, A227, 22233, 22224, A2333, A2234, A2225, and A22223.
- hard 12 vs. 2. Stand on 22224, A22223, and AAAA26.
- soft 18 vs. A. Hit AA6.
- soft 18 vs. T. Stand on A223, AA33, AA222, and AAA23.

There are several ways to simplify this strategy at relatively little cost, such as by using a total-dependent strategy for multicard hands. See Section 21.5 for further discussion.

One problem we have not addressed in this section is how to find the distribution of profit from a hand played according to composition-dependent basic strategy. The possible values that the profit random variable can assume are 0, ± 1, 1.5, ± 2, ± 3, and ± 4. The reader who would like a challenge is encouraged to attempt this. (Caution: This problem is more difficult than any of the problems in Section 21.4.)

21.3 Card Counting

When the unseen deck is rich in aces and 10-valued cards, the blackjack player has a significant advantage. There are several reasons for this: (a) a surplus of aces and 10-valued cards makes naturals, which are more valuable to the player than to the dealer, more likely; (b) a surplus of 10-valued cards helps the player when, for example, he doubles down on 11 or splits aces; (c) a surplus of 10-valued cards hurts when hitting stiffs, which is mandatory for the dealer but not for the player. The primary purpose of card counting is to identify these rich decks and exploit them by increasing one's bet size.

In the preceding section we evaluated E, the expected profit from an initial one-unit blackjack wager, assuming composition-dependent basic strategy, when the deck is full. Let us denote by $E(\boldsymbol{m})$, where $\boldsymbol{m} = (m_1, m_2, \ldots, m_{10})$, the expected profit from an initial one-unit blackjack wager, assuming composition-dependent basic strategy, when the full deck is depleted by m_1 aces, m_2 2s, $\ldots$, m_{10} 10-valued cards. Thus, $E(\boldsymbol{0}) = E$. Of course we would anticipate that basic strategy (optimized for the full deck) is suboptimal for every deck other than the full deck.

We assume familiarity with Section 11.3. With substantial computing effort, we can evaluate the *effects of removal* on the expected profit from an initial one-unit blackjack wager, assuming composition-dependent basic strategy:

$$\text{EoR}(i) := E(\boldsymbol{e}_i) - E(\boldsymbol{0}), \qquad i = 1, 2, \ldots, 10. \qquad (21.66)$$

Here $\boldsymbol{e}_1 := (1, 0, \ldots, 0)$, $\boldsymbol{e}_2 := (0, 1, 0, \ldots, 0)$, and so on. The numbers (21.66), multiplied by 100, are listed in Table 21.9. Notice that

$$\text{EoR}(1) + \cdots + \text{EoR}(9) + 4\,\text{EoR}(10) = 0. \qquad (21.67)$$

Table 21.9 Effects of removal, multiplied by 100, for the basic bet in single-deck blackjack, assuming composition-dependent basic strategy. Results are rounded to six decimal places. (Rules: See Table 21.1 on p. 644.) Also shown are the point values for the Hi-Lo, Knock-Out, Hi-Opt I, Noir, Zen, Halves, and Ultimate card-counting systems.

card value i	$100\,\text{EoR}(i)$	seven card-counting systems						
		HL	KO	HO	N	Z	H*	U
1	$-$.595 250	-1	-1	0	1	-1	-2	-9
2	.382 530	1	1	0	1	1	1	5
3	.436 234	1	1	1	1	1	2	6
4	.553 671	1	1	1	1	2	2	8
5	.702 373	1	1	1	1	2	3	11
6	.414 736	1	1	1	1	2	2	6
7	.284 252	0	1	0	1	1	1	4
8	.002 350	0	0	0	1	0	0	0
9	$-$.168 973	0	0	0	1	0	-1	-3
10	$-$.502 981	-1	-1	-1	-2	-2	-2	-7
correlation		.966	.972	.873	.723	.960	.991	.998

*each point value is multiplied by 2 to avoid half-integers

Recall that, in a *balanced* card-counting system, the sum of the point values over the entire deck is 0. For the system $(J_1, \ldots, J_{10})$, this means that

$$J_1 + \cdots + J_9 + 4J_{10} = 0. \tag{21.68}$$

Table 21.9 lists five balanced card-counting systems, namely the Hi-Lo, Hi-Opt I, Zen, Halves, and Ultimate systems. Listed as well are two unbalanced systems, the Knock-Out and the Noir systems. In each case we indicate the correlation with the effects of removal.

We next use a normal approximation to approximate the probability that the player's estimated advantage, based on the Hi-Lo card-counting system, exceeds a certain level as a function of the number of cards seen. Let Z_n^* be the player's estimated expected profit on the next hand based on the true count after n cards have been seen. Then, by (11.94) on p. 418,

$$Z_n^* := \mu + \frac{\gamma}{52}\left(\frac{52}{52 - n}\sum_{j=1}^{n} J_{X_j}\right) = \mu + \frac{\gamma}{52}\,\mathrm{TC}_n, \tag{21.69}$$

where $X_1, X_2, \ldots, X_{52}$ is the sequence of card values in the order in which they are exposed, $(J_1, \ldots, J_{10}) := (-1, 1, 1, 1, 1, 1, 0, 0, 0, -1)$ is the vector of Hi-Lo point values, $\mu = E$ is the full-deck expectation,

$$\gamma := \sum_{i=1}^{10} w_i E_i J_i \Big/ \sum_{i=1}^{10} w_i J_i^2 \approx 0.259932 \tag{21.70}$$

is the regression coefficient of (11.96) on p. 418, where $w_1 = \cdots = w_9 := 4/52$, $w_{10} := 16/52$, and $E_i := 51\,\mathrm{EoR}(i)$, and finally TC_n is the true count, the quantity within parentheses in (21.69). (It is customary in blackjack to define the true count to be the running count divided by the number of *decks* that remain, not the number of *cards* that remain, which explains the factor of 52.)

Then, with $\sigma^2 := \sum_{i=1}^{10} w_i J_i^2 = 10/13$,

$$\begin{aligned}
\mathrm{P}(Z_n^* > \beta) &= \mathrm{P}\left(\mu + \frac{\gamma}{52 - n}\sum_{j=1}^{n} J_{X_j} > \beta\right) \\
&= \mathrm{P}\left(\sum_{j=1}^{n} J_{X_j} > \left\lfloor(52 - n)\frac{\beta - \mu}{\gamma}\right\rfloor + \frac{1}{2}\right) \tag{21.71} \\
&\approx 1 - \varPhi\left(\frac{1}{\sigma}\left\{\left\lfloor(52 - n)\frac{\beta - \mu}{\gamma}\right\rfloor + \frac{1}{2}\right\}\sqrt{\frac{51}{n(52 - n)}}\right),
\end{aligned}$$

where $\varPhi$ is the standard-normal cumulative distribution function (see (1.162) on p. 42).

We evaluate this function for several values of n and β in Table 21.10. We hasten to add that these numbers are based on three approximations, namely (11.74) on p. 415, (11.95) on p. 418, and (21.71).

Table 21.10 Estimating the probability that a one-unit blackjack bet, assuming composition-dependent basic strategy, has expected profit of at least β units when n cards have been seen. The last column approximates the card counter's estimated expected profit as a function of n when he bets one unit in situations perceived as favorable and nothing otherwise.

n	β				$E[(Z_n^*)^+]$
	.000	.010	.020	.030	
5	.605	.213	.032	.002	.004 363
10	.579	.276	.082	.037	.006 423
15	.569	.302	.194	.060	.008 317
20	.564	.315	.211	.130	.010 276
25	.562	.438	.217	.136	.012 462
30	.563	.437	.317	.214	.015 079
35	.566	.434	.308	.308	.018 481
40	.574	.426	.426	.289	.023 458

Perhaps a better sense of blackjack's earning potential can be found by applying (11.100) on p. 419, which approximates the player's estimated expected profit when he bets one unit in situations perceived as favorable and nothing otherwise, as a function of the number n of cards seen:

$$E[(Z_n^*)^+] \approx \mu\Phi(\mu/\sigma_n) + \sigma_n\phi(\mu/\sigma_n), \qquad (21.72)$$

where $\sigma_n^2 := \gamma^2\sigma^2 n/[51(52-n)]$ and ϕ is the standard-normal density function (11.101) on p. 419. The expectation (21.72) is tabulated for the Hi-Lo system in Table 21.10. Notice its monotonicity in n, which is to be expected in view of the fundamental theorem of card counting (Theorem 11.3.4 on p. 414, especially (11.70) on p. 414).

We conclude that card counting can turn blackjack—an essentially fair game for the basic strategist—into a profitable one.

Although bet variation is the primary reason for card counting, a secondary reason is strategy variation. That is, the count may suggest that a departure from basic strategy is called for. Let us try to understand this concept better in the context of the simplest of all decisions, that of whether to take insurance. We continue to assume that the player is using the Hi-Lo system.

We wish to evaluate the player's expected profit per unit bet on insurance when $n \geq 3$ cards have been seen and the running count is r. Of course, three of those cards are the two cards in the player's hand and the dealer's upcard, an ace, and by exchangeability (Section 11.2) we may assume that those were the first three cards seen. It turns out that all that matters about the player's two cards is whether they are low (2, 3, 4, 5, or 6, denoted by L), neutral (7, 8, or 9, denoted by N), 10-valued cards (T), or aces (A). Let us denote the $\binom{4}{1} + \binom{4}{2} = 10$ expectations for each appropriate n and r by $E_{A,A}(n,r)$, $E_{A,L}(n,r)$, ..., $E_{T,T}(n,r)$.

To evaluate these expectations, we need to introduce some random variables. For $n \geq 3$ define L_n, N_n, H_n, T_n, and A_n to be the number of low cards, neutral cards, high cards, 10-valued cards, and aces among the first n seen ($H_n := T_n + A_n$), and let R_n be the running Hi-Lo count based on the first n cards seen ($R_n := L_n - H_n$). With I denoting the player's profit per unit bet on insurance (we ignore the player's basic bet), we have, for $3 \leq n \leq 51$ and $-[(n - 3 - R_3) \wedge (52 - n)] \leq r \leq (n - 3 + R_3) \wedge (52 - n)$,

$$
\begin{aligned}
E[I \mid R_n = r, \; &L_3, N_3, T_3, A_3] \\
&= \sum_k P(L_n = r + k, \; H_n = k \mid R_n = r, \; L_3, N_3, T_3, A_3) E(n, r, k, T_3, A_3) \\
&= \frac{\sum_k \binom{20-L_3}{r+k-L_3}\binom{12-N_3}{n-r-2k-N_3}\binom{20-H_3}{k-H_3} E(n, r, k, T_3, A_3)}{\sum_k \binom{20-L_3}{r+k-L_3}\binom{12-N_3}{n-r-2k-N_3}\binom{20-H_3}{k-H_3}},
\end{aligned}
\tag{21.73}
$$

where

$$
\begin{aligned}
E(n, r, k, T_3, A_3) &= E[I \mid L_n = r + k, \; H_n = k, \; L_3, N_3, T_3, A_3] \tag{21.74} \\
&= 3\left(\frac{16 - T_3 - (k - H_3)(16 - T_3)/(20 - H_3)}{49 - (n - 3)}\right) - 1
\end{aligned}
$$

and the sums range over all integers k satisfying $L_3 \leq r + k \leq 20$, $N_3 \leq n - r - 2k \leq 12$, and $H_3 \leq k \leq 20$. The conditional probabilities are given by the multivariate hypergeometric distribution (whose denominator $\binom{52-3}{n-3}$ is canceled from (21.73)), while the expression within the large parentheses in (21.74) is the conditional probability p that the dealer's downcard is a 10-valued card. Since a winning insurance bet pays 2 to 1, the expected profit per unit bet on insurance is $2p + (-1)(1 - p) = 3p - 1$. The formula for p is an immediate consequence of Example 2.2.11 on p. 86 with the roles of the balls labeled 1 played by the low and neutral cards, the balls labeled 2 by the tens, and the balls labeled 3 by the aces; and the role of n is played by $n - 3$.

Consider $E_{T,T}(n, r)$, for example. This is just (21.73) with $L_3 = N_3 = 0$, $T_3 = 2$, and $A_3 = 1$. We find that there are 1,173 pairs (n, r) that can occur, and evaluating each of the resulting expectations, we observe that, with two exceptions, the insurance bet has nonnegative expectation if and only if the true count $52R_n/(52-n)$ is greater than or equal to 3. To explain this in more detail, if we sort the true counts in the 1,173 cases in ascending order, the last one (ignoring the two exceptions) with a negative insurance expectation is at $(n, r) = (17, 2)$, and the first (again ignoring the two exceptions) with a positive insurance expectation is at $(n, r) = (18, 2)$. Thus, any number in the interval $(104/35, 104/34]$ would do, so we round up the left endpoint $2.971428\cdots$ to 2.972 and round down the right endpoint $3.058823\cdots$ to 3.058.

Thus, 3 is what is called an *index* for taking insurance (and thereby departing from basic strategy) with $\{T, T\}$ vs. A. The two exceptions we mentioned are $(n, r) = (9, 3)$, in which case insurance has expectation approximately -0.023256 in spite of a true count of about 3.628, and $(n, r) = (50, 0)$, in which case insurance has expectation approximately 0.034483 in spite of a true count of 0. The latter case would rarely occur in practice (the deck is typically reshuffled before 51 cards are dealt) but is included for completeness.

In Table 21.11 we summarize the optimal insurance strategy for the Hi-Lo card counter. In cases where the insurance bet has expectation 0, we assume that the player makes the bet because, combined with his basic blackjack bet, it may help to reduce variance. For example, if the player has a natural, an insurance bet assures "even money" and thereby reduces his variance to zero.

We conclude this section by briefly addressing a more difficult issue: How does the count affect basic strategy decisions other than insurance? For example, how does it affect the player's decision to hit $\{6, T\}$ vs. 9? Intuitively, a positive count might suggest a surplus of 10-valued cards in the unseen deck, which might make it prudent to stand. Conversely, a negative count might suggest a surplus of cards of value 2–5, which might make it prudent to hit.

Notice that the Hi-Lo system is not ideal for this decision. Observed sixes contribute to a positive count and therefore to a decision to stand, but a shortage of sixes in the unseen deck is beneficial when hitting hard 16. Nevertheless, we must work with the count we have chosen.

In much the same way that the pre-deal true count can be used to estimate the expected profit from the next hand using composition-dependent basic strategy and consequently influence the choice of bet size, the true count can also be used to estimate the expectation gain by departing from basic strategy in specific strategic situations such as $\{6, T\}$ vs. 9.

It is a rather easy computation to evaluate the *effects of removal* on the expectation gain by departing from basic strategy with $\{6, T\}$ vs. 9:

$$
\begin{aligned}
\text{EoR}_{\{6,T\},9}(i) :=\ & E_{\text{std}}(e_6 + e_{10}, 9 \mid e_i) - E_{\text{hit}}(e_6 + e_{10}, 9 \mid e_i) \\
& - \{E_{\text{std}}(e_6 + e_{10}, 9 \mid 0) - E_{\text{hit}}(e_6 + e_{10}, 9 \mid 0)\} \\
\approx\ & E_{\text{std}}(e_6 + e_{10}, 9 \mid e_i) - E_{\text{hit}}(e_6 + e_{10}, 9 \mid e_i) \\
& - (-0.059925), \qquad i = 1, 2, \ldots, 10, \qquad (21.75)
\end{aligned}
$$

Table 21.11 Hi-Lo insurance indices. The insurance bet has nonnegative expectation if and only if the true Hi-Lo count is greater than or equal to the insurance index (any number belonging to the interval indicated), with some exceptions. Here L, N, T, and A denote a low card (2, 3, 4, 5, or 6), a neutral card (7, 8, or 9), a 10-valued card, and an ace, while n is the number of cards seen, including the two in the player's hand and the dealer's ace, and r is the running Hi-Lo count based on these n cards. (Rules: See Table 21.1 on p. 644.)

player hand	insurance index interval	complete list of exceptions (n, r) [n = number of cards seen, r = running count]
$\{A, A\}$	$[-2.476, -2.419]$	$(4, -3), (6, -2), (10, -2), (49, 0), (51, 0)$
$\{A, L\}$	$[-1.061, 0.000]$	$(4, -1)$ and $(n, 0)$,
		$n = 40, 41, 42, 43, 44, 45, 46, 47, 49, 51$
$\{A, N\}$	$[-1.405, -1.369]$	$(4, -2), (6, -1), (47, 0), (49, 0), (51, 0)$
$\{A, T\}$	$(0.000, 1.155]$	$(7, 0), (48, 0), (50, 0)$
$\{L, L\}$	$[1.858, 1.925]$	$(4, 1), (50, 0)$
$\{L, N\}$	$(0.000, 1.083]$	$(4, 0), (48, 0), (50, 0)$
$\{L, T\}$	$[2.364, 2.418]$	$(7, 2), (50, 0)$
$\{N, N\}$	$[-1.061, 0.000]$	$(4, -1)$ and $(n, 0)$,
		$n = 6, 40, 41, 42, 43, 44, 45, 47, 49, 51$
$\{N, T\}$	$(1.300, 1.333]$	$(7, 1), (50, 0)$
$\{T, T\}$	$[2.972, 3.058]$	$(9, 3), (50, 0)$

where the numerical value comes from (21.7) on p. 650 and (21.11) on p. 651. These numbers, multiplied by 100, are listed in Table 21.12.

Table 21.12 Effects of removal, multiplied by 100, on the expectation gain by departing from basic strategy with $\{6, T\}$ vs. 9. Results are rounded to six decimal places. (Rules: See Table 21.1 on p. 644.)

card value i	$100 \, \mathrm{EoR}_{\{6,T\},9}(i)$	card value i	$100 \, \mathrm{EoR}_{\{6,T\},9}(i)$
1	.777 268	6	$-1.715\ 214$
2	.829 028	7	$-2.020\ 314$
3	1.231 397	8	0.190 056
4	2.205 477	9	$-0.488\ 656$
5	2.630 734	10	$-1.117\ 532$

The player's estimated expectation is, by analogy with (21.69),

$$Z_n^* := \mu + \frac{\gamma}{49}\left(\frac{49}{49-n}\sum_{j=1}^{n} J_{X_j}\right) = \mu + \frac{\gamma}{49}\,\mathrm{TC}_n^*, \qquad (21.76)$$

where $X_1, X_2, \ldots, X_{49}$ is the sequence of card values *in the 49-card deck* in the order in which they are exposed, $(J_1, \ldots, J_{10})$ is as before, $\mu \approx -0.059925$ is the 49-card-deck expectation gain by departing from basic strategy,

$$\gamma := \sum_{i=1}^{10} w_i E_i J_i \Big/ \sum_{i=1}^{10} w_i J_i^2 \approx 0.455934 \qquad (21.77)$$

is the regression coefficient (11.96) on p. 418, where $w_i := (4 - \delta_{6,i} - \delta_{9,i})/49$ for $i = 1, \ldots, 9$, $w_{10} := 15/49$, and $E_i = 48\,\mathrm{EoR}_{\{6,\mathrm{T}\},9}(i)$, and finally TC_n^* is the adjusted true count, the quantity within parentheses in (21.76). (In particular, the 6, T, and 9 are not included in this count, nor do they count among the cards seen. This is why we say *adjusted* and add an asterisk to TC_n.) It follows that $Z_n^* \geq 0$ if and only if $\mathrm{TC}_n^* \geq -49\mu/\gamma \approx 6.440$, so this is our *index* for departing from basic strategy with $\{6, \mathrm{T}\}$ vs. 9.

Notice that this criterion is really just an approximation based on a linearity assumption such as (11.74) on p. 415. An exact analysis is feasible and would be similar in principle to the one we carried out above for the insurance decision, but it would be much more complicated because the wager's resolution depends on as many as eight of the unseen cards, not just one. We will mercifully withhold the details of such a computation. But see Problem 21.27 on p. 676.

21.4 Problems

21.1. *Number of blackjack hands of a specified size.* We observed in Section 21.2 that there are 2,008 distinct (single-deck) blackjack hands of two or more cards with a hard total of 21 or less (including hands, such as $\{\mathrm{A}, \mathrm{A}, \mathrm{T}\}$, that do not occur to the basic strategist).

(a) Refine that result by finding, for $n = 2, 3, \ldots, 11$, the number of distinct (unbusted) blackjack hands of size n. (For example, there are 55 two-card hands and only one hand of size 11: four aces, four 2s, and three 3s.)

(b) Refine that result further by finding, for $m = 2, 3, \ldots, 21$ and $n = 2, 3, \ldots, 11$, the number of distinct blackjack hands with hard total m and size n.

21.2. *Number of blackjack subsets of a specified size.* Problem 1.1 on p. 52 determined the number of distinguishable subsets of the 52-card blackjack deck. Refine that result by finding, for $n = 0, 1, \ldots, 52$, the number of distinguishable subsets of size n.

21.3. *Player's conditional expectation, given player's abstract final total and dealer's upcard.* Evaluate the player's conditional expectation (per unit bet), given the player's abstract final total and the dealer's upcard. By "abstract" total we mean that the composition of that total is unspecified, so the dealer "selects" his downcard and completes his hand from the 51-card deck obtained by removing his upcard. Allow for player abstract final totals of stiff (12–16), 17, 18, 19, 20, and nonnatural 21. For dealer upcards of A and T, condition on the dealer's not having a natural.

21.4. *6-to-5 payoff for naturals.* Some casinos pay untied player naturals only 6 to 5 instead of the standard 3 to 2. In a single-deck game, what does this cost the basic strategist in terms of expectation? Answer the same question for a d-deck game, $d = 2, 4, 6, 8$.

21.5. *Distribution of the dealer's final total in infinite-deck case.* Under the rules of Table 21.1 on p. 644 except for the single-deck requirement, derive a recursive algorithm for the conditional distribution of the dealer's final total, given his upcard, assuming infinitely many decks or, more precisely, sampling with replacement. In other words, derive the analogue of Table 21.3 on p. 649.

Hint: Define $r(i) := 1/13$ for $i = 1, \ldots, 9$ and $r(10) := 4/13$. Think of the dealer's total as a Markov chain in the state space consisting of states $1, 2, \ldots, 26$ (totals that include no aces) and $1^*, 2^*, \ldots, 26^*$ (hard totals that include one or more aces). States $17, 18, \ldots, 26$ are absorbing, as are states $7^*, 8^*, 9^*, 10^*, 11^*$ and $17^*, 18^*, \ldots, 26^*$. Starting from nonabsorbing states, transitions include $i \mapsto (i + 1)^*$ with probability $r(1)$, $i \mapsto i + j$ with probability $r(j)$ ($j = 2, 3, \ldots, 10$), and $i^* \mapsto (i + j)^*$ with probability $r(j)$ ($j = 1, 2, \ldots, 10$).

Define $P_n(t)$ (resp., $P_n^*(t)$) to be the probability that, after n steps of the Markov chain, it is in state t (resp., t^*). Argue that, for $n \geq 1$ and $t = 1, 2, \ldots, 26$,

$$P_n(t) = P_{n-1}(t)1_{\{t \geq 17\}} + \sum_{i=2\vee(t-16)}^{10\wedge(t-1)} r(i)P_{n-1}(t - i), \tag{21.78}$$

$$P_n^*(t) = P_{n-1}^*(t)1_{\{7 \leq t \leq 11 \text{ or } t \geq 17\}} + r(1)P_{n-1}(t - 1)1_{\{t \leq 17\}} \tag{21.79}$$
$$+ \sum_{i=1\vee(t-6)}^{10\wedge(t-1)} r(i)P_{n-1}^*(t - i) + \sum_{i=1\vee(t-16)}^{10\wedge(t-12)} r(i)P_{n-1}^*(t - i).$$

Let the initial state correspond to the dealer's upcard. The Markov chain then absorbs in 11 or fewer steps. After adjusting for naturals, the modified Table 21.3 should follow.

21.6. *Basic strategy under the assumptions of Baldwin et al.* Modify the rules of Table 21.1 on p. 644 in three respects: Assume infinitely many decks (sampling with replacement), no resplitting of pairs (in the notation of the table, SPL1), and doubling after splits (DAS). Derive (total-dependent) basic strategy. Find the player's overall expectation, accurate to 12 decimal places. Notice that this problem is substantially simpler than its single-deck (sampling without replacement) counterpart.

Hint: Start with Problem 21.5. Use it to compute player standing expectations for each total 4–21 vs. each dealer upcard, $2, 3, \ldots, T, A$. Remember to condition on the dealer's not having a natural. Then recursively compute player hitting expectations for hard totals $21, 20, \ldots, 11$, soft totals $21, 20, \ldots, 12$, and hard totals $10, 9, \ldots, 4$, in that order, vs. each dealer upcard. Then compute player doubling expectations for hard totals 4–21 and soft totals 12–21, vs. each dealer upcard. Finally, compute player splitting expectations for each of the 10 pairs vs. each dealer upcard, remembering to allow for doubling after splits (except with aces) and no resplits. With these computations, basic strategy can be determined, and, after allowing for player and dealer naturals, the player's overall expectation can be evaluated.

21.7. *Analysis of A-8 vs. 6.* Derive the correct basic strategy decision for $\{A, 8\}$ vs. 6, much as was done in Section 21.2 with $\{6, T\}$ vs. 9. But now consider doubling as well as standing and hitting. Assume, as can be verified, that if the player hits his soft 19, it is correct for him to stand with the resulting total, whatever it may be.

21.8. *Analysis of T-T vs. 6.* Derive the correct basic strategy decision for $\{T, T\}$ vs. 6, much as was done in Section 21.2 with $\{6, T\}$ vs. 9. Specifically, compare standing and splitting. (It is clear that hitting and doubling are unwise.) To keep the computations manageable, assume SPL1 (no resplitting) in place of SPL3 (up to three splits). Assume, as can be verified, that if the player splits his tens, it is correct for him to stand with the resulting two-card totals, whatever they may be.

21.9. *Composition-dependent basic strategy practice hands.* Assume the rules of Table 21.1 on p. 644. For each player hand and dealer upcard listed in Table 21.13, (a) state (if known) or guess the correct composition-dependent basic strategy play, (b) use Table 21.8 on p. 662 to determine the correct composition-dependent basic strategy play, and (c) write a computer program to evaluate the conditional expectations for standing and for hitting (and for doubling and for splitting when applicable). Caution: These examples include some of the closest calls in basic strategy. Which decision is *the* closest?

21.10. *Worst basic strategy error.* Find the worst basic strategy error, in terms of loss of expectation (assuming the rules of Table 21.1 on p. 644 and optimal subsequent play, if any), an error so egregious that it has likely never been made.

21.11. *Basic strategy conditional expectations.* Evaluate the conditional expectations for all two-card player hands, dealer upcards, and player strategies, assuming that any subsequent decisions are made in accordance with

Table 21.13 Some composition-dependent basic strategy practice hands. (Rules: See Table 21.1 on p. 644.)

1. $\{7,7\}$ vs. T	11. $\{2,2,2,4,6\}$ vs. 9
2. $\{4,9\}$ vs. 2	12. $\{A,2,3,T\}$ vs. T
3. $\{6,6\}$ vs. 7	13. $\{A,A,2,2,2,8\}$ vs. 7
4. $\{3,9\}$ vs. 3	14. $\{A,A,A,A,2,2,2,6\}$ vs. T
5. $\{2,T\}$ vs. 6	15. $\{A,A,2,5,6\}$ vs. T
6. $\{4,4\}$ vs. 5	16. $\{A,2,T\}$ vs. 2
7. $\{2,6\}$ vs. 5	17. $\{A,2,3,6\}$ vs. 3
8. $\{A,8\}$ vs. 6	18. $\{A,A,A,A,2,6\}$ vs. 2
9. $\{A,6\}$ vs. 2	19. $\{A,A,6\}$ vs. A
10. $\{A,2\}$ vs. 4	20. $\{A,A,A,A,2,2\}$ vs. T

composition-dependent basic strategy. That is, evaluate the 1,750 conditional expectations $E_{\mathrm{std}}(e_i+e_j,u)$, $E_{\mathrm{hit}}(e_i+e_j,u)$, $E_{\mathrm{dbl}}(e_i+e_j,u)$, and $E_{\mathrm{spl}}(2e_i,u)$ of Section 21.2 for $1 \le i \le j \le 10$ and $u = 1,\dots,10$.

21.12. *Composition-dependent basic strategy expectations for hard 16 vs. each of 7, 8, 9, T.* For each of the 145 composition-dependent hard 16s (excluding the pair $\{8,8\}$) vs. an upcard of 7, 8, 9, or T, find the standing and hitting expectations. For which hand is the (absolute) difference in expectations the smallest? ... the largest?

21.13. *Total- and size-dependent basic strategy from composition-dependent basic strategy.* Use the results of Problem 21.12 to determine the size-dependent basic strategy for hard 16 vs. each of $7,8,9,\mathrm{T}$. In other words, suppose that the player is allowed to take the number of cards in the hand into account (and of course the fact that the total is hard 16) but not the composition of the hand. For example, it is optimal to hit a two-card hard 16 vs. each of $7,8,9,\mathrm{T}$, whereas it is optimal to stand on an eight- or nine-card hard 16 vs. each of $7,8,9,\mathrm{T}$, because in such cases there are no composition-dependent exceptions. Now treat hands with $3,4,5,6,7$ cards, but notice that probabilities and expectations must be taken into account.

Hint: How does one evaluate the probabilities? Consider a three-card hard 16 vs. T. Assume for simplicity that the probabilities of the 15 three-card hard 16s are proportional to the relevant multivariate hypergeometric probabilities for a 51-card deck. For example, $\{A,5,T\}$ would have probability proportional to

$$\frac{\binom{4}{1}^2 \binom{4}{0}^7 \binom{15}{1}}{\binom{51}{3}}. \tag{21.80}$$

(Of course, this is only an approximation because it does not take into account that, for example, $\{A,5,T\}$ can be dealt in only four of the six possible orders:

5 cannot be the third card because no one would hit a natural. Readers who are uncomfortable with this approximation are encouraged to do it exactly.)

21.14. *Comparing two hard 16s vs. 9.* Consider the two decisions $\{A, 2, 2, 2, 9\}$ vs. 9 and $\{A, 3, 3, 3, 6\}$ vs. 9. Composition-dependent basic strategy calls for standing on $\{A, 2, 2, 2, 9\}$ and hitting $\{A, 3, 3, 3, 6\}$ in spite of the fact, when hitting hard 16, one would prefer a shortage of 2s to a shortage of 3s. Explain this apparent paradox by evaluating the distribution of the dealer's final total in both cases when the player stands and when the player draws an $A, 2, 3, 4, 5$ (as in Table 21.5 on p. 652).

21.15. *Distribution of the basic strategist's bet size.* Assume a one-unit initial bet on a hand played according to composition-dependent basic strategy. If the player doubles down, the amount bet is two units. If he splits, it is two, three, or four units, depending on the number of splits. Find the distribution of the amount bet. Use this to find the mean amount bet and therefore the house advantage (pushes included) of the basic blackjack bet.[3]

21.16. *Pair-splitting formulas.* Verify the formulas in Table 21.7 on p. 660 using the methods of (21.56) on p. 660 and (21.60) on p. 661.

21.17. *Invariance property of basic strategy expectation.* Consider a single composition-dependent basic strategist under the present rules (Table 21.1 on p. 644), starting with a full well-shuffled deck. Find the largest positive integer m such that, with probability 1, at least m rounds can be dealt before exhausting the deck. Show that the player has the same expectation on each of those m rounds.

Hint: Cf. Example 11.2.2(d) on p. 402.

21.18. *Quarter-deck blackjack.* In quarter-deck blackjack, only 13 cards, one from each denomination, are used. Notice that a complete (one player vs. dealer) hand is assured if the splitting of 10-valued cards is prohibited (no other splitting opportunities occur). Otherwise the rules are as in Table 21.1 on p. 644. Find the quarter-deck composition-dependent basic strategy and the player's expected profit from a one-unit initial bet.

21.19. *Simplified twenty-one: Seven.* Consider the following version of blackjack. The deck comprises two aces, two deuces, and four treys, with aces having value either 1 or 4, and deuces and treys having values 2 and 3, respectively. The target total is 7 instead of 21, and ace-trey is a natural. The dealer stands on 6 or 7, including soft totals, and otherwise hits. The player can stand, hit, double, or split, but there is no resplitting and insurance is not available. In the notation of Table 21.1 on p. 644, we assume S6, DOA, NDAS, SPA1, SPL1, 3:2, OBO, and NS.

Evaluate the composition-dependent basic strategy and the player's expectation from a one-unit initial bet. Notice that there are 14 unbusted player hands in this game compared to 2,008 in blackjack. The point of this problem

---

[3] The formulation of house advantage as in Section 6.1 is not usually applied to blackjack.

is that the game is simple enough that the calculations can be done by hand (with sufficient patience), though it might nevertheless be easier and more reliable to use a computer.

21.20. *A particular residual subset.* Suppose that only six cards remain, namely 2, 4, 6, 7, 9, T. Notice that a complete (one player vs. dealer) hand is assured. Find the player's expectation (a) under composition-dependent basic strategy and (b) under the precisely optimal strategy for this subset.

21.21. *Insuring a natural.* Let p be the proportion of 10-valued cards in the unseen deck. We know that the insurance wager has positive expectation if and only if $p > 1/3$. What is the optimal insurance bet for the Kelly bettor who has been dealt a natural and who has bet a proportion f of his bankroll on the hand? Recall that the insurance bet can be up to half the size of the original bet.

21.22. *Basic strategy with late surrender* How is composition-dependent basic strategy affected by the late surrender option? Under this rule, the player may *surrender* any two-card initial hand, that is, forfeit the right to play the hand at the cost of half of his bet. The adjective "late" refers to the provision that, if the dealer's upcard is an ace or a 10-valued card, he must check for a natural before allowing the player to surrender.

21.23. *Effect of DAS on basic strategy.*

(a) How must the formulas of Section 21.2 be modified if doubling down is permitted after splitting?

(b) How does this affect the composition-dependent basic strategy and the expectation (21.65) on p. 661?

21.24. *Perfect insurance betting.* Consider the unbalanced Noir count of Table 21.9 on p. 664 initialized at $-4d$, assuming a d-deck shoe, where d is a positive integer. Show that the running count can be used to make perfect insurance decisions. That is, the running count tells us whether an insurance bet has positive expectation. Note that we do not need to keep track of the number of cards seen.

21.25. *Strategy variation for 6-T vs. T.* In Section 21.3 we analyzed strategy variation for $\{6, T\}$ vs. 9 when using the Hi-Lo system. Repeat this analysis for $\{6, T\}$ vs. T, remembering to condition on the dealer's not having a natural. Do the effects of removal sum to 0 when summed over the 49 remaining cards? If not, what are the appropriate weights?

21.26. *Approximating strategy variation.* In Section 21.3 we analyzed strategy variation for $\{6, T\}$ vs. 9 when using the Hi-Lo system. Repeat this analysis for an "abstract" hard 16 vs. an "abstract" 9, that is, a hard 16 vs. 9 for which the unseen deck is the full 52-card deck. More precisely, first find the effects of removal on the expectation gain by varying basic strategy. Then find the analogues of (21.76) and (21.77) on p. 670 as well as the critical index for departing from basic strategy in this situation. (This is the standard approach, unlike that of Section 21.3.)

21.27. *An exact index for 6-T vs. 9.* In Section 21.3 we found an approximate index for varying composition-dependent basic strategy with $\{6, T\}$ vs. 9 when using the Hi-Lo system. Can this be done exactly, as we did for the insurance decision, including a list of all exceptions?

Hint: In the insurance case, we decomposed the event $\{R_n = r\}$ as the union over k of the events $\{L_n = r + k, \; H_n = k\}$. Here we need a finer partition of $\{R_n = r\}$, one for which each event allows complete specification of the unseen deck.

21.5 Notes

It is widely agreed (e.g., Parlett 1991, p. 80) that *vingt-et-un* ("twenty-one" in French) appeared in France in about the middle of the 18th century. Presumably, if it had existed by the turn of the 18th century, Montmort and De Moivre would have studied it. Vingt-et-un was the favorite card game of Napoleon Bonaparte [1769–1821].

Possible ancestors of vingt-et-un include the 15th-century game of *trente-et-un* ("thirty-one"), also known as *bone-ace*, the 16th-century French game of *quinze* ("fifteen"), and the 17th-century Italian game of *sette e mezzo* ("seven and a half"). All three games shared several features of vingt-et-un. Bone-ace was the first game to count aces as 1 or 11. In quinze a 52-card deck was used, with aces counting as 1, court cards as 10, and so on. In sette e mezzo a 40-card deck (8s, 9s, and 10s removed) was used and court cards counted as $\frac{1}{2}$. In all three games each player and the dealer made hit-or-stand decisions, busting with totals greater than the target total implied by the name of the game, although in quinze (unlike in vingt-et-un and sette e mezzo) a busted player tied a busted dealer. In William Shakespeare's [1564–1616] *The Taming of the Shrew* one finds the phrase "two-and-thirty, a pip out."[4] (Act 1, Scene 2). Sources for this paragraph were Snyder (2006, Chapter 1) and Parlett (1991, pp. 54, 80–81).

In 1854, Frenchwoman Eleanore Dumont [c. 1830–1879] (a.k.a. Simone Jules), then in her mid-20s, opened the Vingt-et-Un, a gambling house in Nevada City, California. Some years later (accounts differ as to the year) she acquired the nickname "Madame Mustache" when "a growth of dark hair, previously absent or virtually invisible, appeared on her upper lip" (Drago 1969, p. 87). Over the span of 25 years she became legendary, drifting from one mining boomtown to another—in California, Oregon, Washington, Idaho, Montana, and Nevada—always dealing her game of choice, twenty-one. It was that game that finally bankrupted her in Bodie, California, in 1879, resulting in her suicide at about age 49. See DeArment (1982, Chapter 10) for the full story.

---

[4] A pip out was referred to as a burst in the 19th century and is now called a bust.

According to Scarne (1961, pp. 315–316), the name "blackjack" derives from a practice, initiated in 1912 in Evansville, Indiana, of paying a 10-to-1 bonus for a natural comprising A♠-J♠ or A♠-J♣, that is, the ace of spades and a black jack. In 1917, the phrase "Black Jack pays 3 to 2" first appeared on twenty-one tables, and the name "blackjack" has been associated with the game ever since. We do not have independent confirmation of these claims. We looked at casino-supply catalogues from H. C. Evans & Co. and found that blackjack tables were sold in 1923 but not in 1918. On the 1923 tables there were no markings.

Doubling and splitting date back to 18th-century vingt-et-un (see the Thackeray quote on page 643; *The Virginians* is a historical novel set before and during the American Revolution). At that time the rules were different in several respects. Specifically, a player would bet only after seeing his first card; the dealer had the option, after seeing his first card, of doubling all bets; both of the dealer's first two cards were dealt face down; the dealer won ties, except for a natural (two-card 21) vs. natural tie; an untied 21 (natural or not) paid 2 to 1 for either player or dealer; splitting (and resplitting) was allowed for both player and dealer, and a natural after a split paid 2 to 1; the dealer's strategy was discretionary; finally, with the dealer enjoying a considerable advantage, the role of dealer was passed around among the players, usually when one of the players obtained a natural. See Baxter-Wray [Peel, Walter H.] (1891, pp. 61–66) for details.

By the time of Scarne (1949, p. 154), insurance was standard. However, it was not mentioned by MacDougall (1944), perhaps because it was a new development at that time. A 1940 gambling-supply catalogue (*Practical Equipment* by B. C. Wells & Co. of Detroit) shows two blackjack layouts, both with the phrases "Black Jack pays 3 to 2" and "Dealer must stand on 17 and must draw to 16." The layout with only these words was the standard one, whereas the layout with the additional phrase, "Insurance pays 2 to 1," was called the "Johnson style" layout. Evidently, blackjack insurance was introduced by Johnson, but we have no further information about this individual.

In Asbury (1938) blackjack merited only a few sentences in a chapter titled "Small Fry." In Lewis (1953, p. 77) it was said to be fourth in popularity in Nevada behind slots, craps, and roulette. As late as 1956, blackjack was discussed by Davis (1956) in a chapter titled "Minor Games." (Major games, those with their own chapters, included roulette, faro, craps, poker, and slot machines.) This probably understates the popularity of blackjack (and overstates that of faro) in the 1950s. According to Scarne (1961, p. 308), among casino table games blackjack was second only to craps in popularity in the United States.

Wilson (1965, pp. 90–99) described some of the early research on basic strategy in the 1950s at various computer installations around the United States. These included unnamed researchers at the Atomic Energy Commission in Los Alamos, Richard A. Epstein [1927–] at Ramo-Woolridge, Robert Bamford at the Jet Propulsion Laboratory, and Allan N. Wilson [1924–2001] at General Dynamics Astronautics. The first published version of basic strategy at blackjack, which appeared in the *Journal of the American Statistical Association*, is due to Roger R. Baldwin [1929–], Wilbert E. Cantey [1931–2008], Herbert Maisel [1930–], and James P. McDermott [1930–] (1956) of the Aberdeen Proving Ground in Maryland. They called it the "optimum strategy"; the term "basic strategy" was later coined by Thorp (1962b, p. 4). Baldwin, Cantey, Maisel, and McDermott assumed sampling with replacement (often referred to as assuming infinitely many decks), no resplitting of pairs (SPL1 in the notation of Table 21.1), and doubling after splits (DAS), and obtained a player expectation of −0.0062, working only with Monroe and Marchant desktop calculators. Cantey later claimed that, because of an arithmetic error, their figure should have been −0.0032 (Thorp 1962b, p. 15). Actually, the original figure was closer to the correct one under their set of assumptions: −0.005703880. See Problem 21.6.

To give a sense of how dramatically basic strategy changed the conventional wisdom on how to play blackjack, we quote the splitting recommendations of MacDougall (1944, p. 87):

> Here is the strategy followed by most good players.
>
> Always split a pair of aces. Almost always split deuces and threes. Seldom split fours and fives. Practically every time it is a good policy to split sixes and seven. Eights are optional, depending on circumstances. Never split nines.

Notice that MacDougall's strategy, in addition to being vague, does not take the dealer's upcard explicitly into account. See Table 21.8 for the correct strategy.

Edward O. Thorp [1932–] completed his Ph.D. in mathematics (thesis title: "Compact linear operators in normed spaces") in 1958 at UCLA and began a Moore Instructorship at MIT. That same year he noticed the paper by Baldwin, Cantey, Maisel, and McDermott (1956), and soon recognized the potential for a winning strategy at blackjack. With the help of an IBM 704 computer at MIT and the then recently developed FORTRAN programming language, he found his first card-counting system, based on counting fives, and announced it to a standing-room-only crowd at the January 1961 meeting of the American Mathematical Society (Thorp 1960). His talk was titled "Fortune's formula: The game of blackjack." That same month Thorp (1961) published a research paper on the subject in the prestigious *Proceedings of the National Academy of Sciences*, emphasizing his strategy based on fives but also mentioning a strategy based on tens. The paper, titled "A favorable strategy for twenty-one," was refereed by National Academy member Claude E. Shannon [1916–2001]. This title was thought to be more sedate and respectable than the originally proposed title, "A winning strategy for blackjack" (Thorp 2005). In November 1962 Thorp published *Beat the Dealer: A Winning Strategy for the Game of Twenty-One* (Thorp 1962b), which supplied complete details of the ideas sketched in the *PNAS* article.

See Poundstone's (2005) *Fortune's Formula* for the full Thorp story and related material. Other sources include Thorp (1962a,b), Barnhart (2000), and a chapter titled "The Day of the Lamb" in Spanier (1994, Chapter 1). The title was motivated by a remark of a casino executive when asked in a television interview whether his customers ever win (Thorp 1962a):

> When a lamb goes to the slaughter, the lamb might kill the butcher. But we always bet on the butcher.[5]

Was there any precedent for Thorp's discovery? Let us examine the historical record. The idea of keeping track of the cards to gain an advantage is very old. Girolamo Cardano [1501–1576] wrote (Ore 1953, p. 220),[6]

> Since here we exercise judgment in an unknown matter, it follows that the memory of those cards which we have deposited or covered or left should be of some importance, and in certain games it is of the greatest importance, [...].

MacDougall (1944, p. 86) stressed the importance of sitting at third base[7] and therefore having "the advantage of knowing what cards others have drawn." Warren Nelson [1913–1994], who dealt blackjack in Reno in the postwar 1940s, said in reference to that era (Nelson, Adams, King, and Nelson 1994, p. 90),

> If you had too many aces or face cards left in your deck after several go-arounds, you were expected to shuffle up and not give the player the best of it.

This statement comes from an oral history recorded in the 1990s.

John Scarne[8] [1903–1985] wrote in his autobiography (1966, Chapter 10) that he had the ability, as early as 1947, to beat blackjack by card counting, and after having demonstrated his ability that year, he was "the first honest person to be barred from gambling in Nevada casinos." We note that this account postdates Thorp's (1962b) book. Only two years after his alleged barring, Scarne (1949, p. 164) wrote,

> It is impossible for a player to overcome the banker's percentage at Black Jack in the long run, [...].

Scarne's Complete Guide to Gambling (1961) was undoubtedly written without knowledge of Thorp's work. Here (p. 330) he stated that

> Black Jack, as presently dealt in Nevada, is the only casino game that can be beaten by strategy.

He went on to discuss what he called "card casing," but held back details of the system. Elsewhere (p. 310) he quoted a 1958 article from the *Havana Post*

---

[5] This metaphor was also the basis for the title of the book *Always Bet on the Butcher* by Nelson, Adams, King, and Nelson (1994).

[6] Translation by Sydney Henry Gould.

[7] This refers to the seat-position of the last player to act. The blackjack-baseball analogy is not ideal: Blackjack players act in clockwise order but ballplayers run the bases counterclockwise.

[8] Pronounced "Skarney."

about two winning blackjack players who "memorize the number of face cards, including aces, which have fallen." In *Scarne's Guide to Casino Gambling* (1978, p. 101), the author claimed (for the first time) that the "Gambler's Ten-Card Count Strategy" had been used early in the 20th century and that he had written a pamphlet titled "Beware of the Ten-Card Count Black Jack Strategy," which the Army distributed to millions of GIs during World War II. Scarne (1978, pp. 101–102) even quoted four paragraphs from the pamphlet. Snyder (1992) looked into the mystery of why none of these pamphlets has survived, and later concluded (Snyder 1997, p. 189) that they never existed.

It should be noted that Scarne (1966, 1974, 1978) was one of the few experts who questioned Thorp's ability to beat blackjack by card counting. In effect, he was saying that only he, Scarne, had that skill. It was an unfortunate way for the author of some 28 books, a few of which are still unsurpassed in terms of their comprehensiveness, to close out his career. See Barnhart (1997) for a personal account of the life of John Scarne, in which it is alleged that almost all of Scarne's books an gambling were ghostwritten by his one-time coauthor, Clayton Rawson.

The first honest person to be barred from playing blackjack in Las Vegas casinos was actually Jess Marcum [1919–1992] in the early 1950s. Marcum was a Rand Corporation physicist who allegedly derived basic strategy analytically, with paper and pencil, around 1949. He also derived a counting system that gave him an advantage that he estimated at 3%. This is unusually high for card counting (see Table 21.10). The explanation, from Thorp (personal communication, 2008), is that Marcum was engaging in *endplay*. In the 1950s, casinos dealt to the bottom of the (single) deck, which offered great opportunities to players who could estimate the expectation of the last hand before the shuffle. After being barred from playing in Las Vegas, Marcum moved to Reno, then to other casino jurisdictions, including Havana, where he was one of the two players referred to in the *Havana Post* article cited above (the other being New York bookie Manny Kimmel). He ended his playing career in 1962 when Thorp's book was published, and he never revealed his precise methods. See Schaffer (2005) for further details.[9]

Other early counters included such colorful characters as "Greasy John" and Benjamin F. "System Smitty" Smith (Thorp 1962b, Chapter 12). A Las Vegas blackjack dealer was quoted in 1953 as follows (Wechsberg 1953):

> The most dangerous opponents in blackjack are the ones who "case the deck." You recognize them by their tense look, the forehead wrinkled in concentration. They memorize each card that is played out until only a few cards remain in my hand. Suppose that many of the low cards have been drawn after a few games and I'm still holding a relatively high number of face cards, tens and aces. Since the rule is that the dealer must hit (draw up to) sixteen, and stand (draw no more cards) on seventeen, the deck-caser knows there is a better-than-even chance that the dealer

---

[9] Marcum's final triumph occurred in 1990 when he did some consulting for Donald Trump concerning a *whale* (high roller) at the Trump Plaza. See Johnston (1992, Chapters 1 and 24).

will go bust. So the deck-caser will stand on an unusually low hand. It takes real concentration and memory. What a shame those guys don't use their brains in a more productive line of work!

It could be argued that the first published card-counting system is due to Baldwin, Cantey, Maisel, and McDermott (1957)—Chapter 10 is titled "Using the exposed cards to improve your chances." However, they emphasized strategy variation rather than bet variation, with the result that their system was largely ineffective. A modestly effective but unpublished card-counting system was tested by Wilson (1965, pp. 103–116) in 1959.

Thus, it seems that the concept of card counting was not entirely unknown prior to the publication of Thorp's book (1962b). Yet it must be emphasized that none of these precedents was known to Thorp at the time he did his research, which was completed prior to his January 1961 announcement— only the *JASA* paper of Baldwin, Cantey, Maisel, and McDermott (1956) had come to his attention. In the Foreword to the 2008 reprint of Baldwin, Cantey, Maisel, and McDermott (1957), Thorp graciously wrote,

> To paraphrase Isaac Newton, "If I have seen farther than others, it is because I stood on the shoulders of four giants."

Las Vegas casinos reacted to *Beat the Dealer*, which reached the *New York Times* bestseller list in 1964, by changing the rules in two respects: They prohibited splitting aces and limited doubling down to totals of 11. Business fell off and within weeks the original rules were restored. However, over a period of decades, multiple-deck games have gradually replaced the single-deck game.

Thorp (1962b) emphasized the ten-count strategy (equivalent to the balanced point count $(4, 4, 4, 4, 4, 4, 4, 4, -9)$) but also introduced what he called the "Ultimate" point-count strategy (see Table 21.9), which shows that he understood the importance of high correlation between the point count and the effects of removal, more than a decade before Griffin (1976) introduced the betting correlation (see below). Harvey Dubner announced the "Hi-Lo" system in 1963,[10] and this simpler approach was explored in the second edition of *Beat the Dealer* (Thorp 1966a). Charles Einstein [1928–2007] (1968) introduced a system that was later named the Hi-Opt I system (see Table 21.9), and Lawrence Revere (1973) created several of his own systems. Soon there were dozens of card-counting systems vying for the attention of blackjack players. The computing for Thorp's (1966a) revised edition and Epstein (1967, 1977, Chapter 7) was done by IBM programmer Julian H. Braun [1929– 2000], and his programs were used to develop Revere's (1973) systems and

---

[10] This had already been discussed by Thorp and Shannon from November 1960 to June 1961. Thorp called it the Complete Count and preferred it to the Ten Count, but when Manny Kimmel offered to bankroll a test of Thorp's system in Nevada, he insisted on using the Ten Count due to his endplay experience with Jess Marcum and others, and therefore it was the Ten Count that was fully developed in the first edition of *Beat the Dealer* (Thorp, personal communication 2008.)

Lance Humble's Hi-Opt systems (Humble 1976, Humble and Cooper 1980). He used both simulation and exact computation to get results that were for many years state of the art. See his article (Braun 1975) for a description of some of his blackjack programs, including Thorp's original FORTRAN II "arbitrary subset" program and Braun's revision of it.

Peter A. Griffin [1937–1998] (1976; 1999, Chapter 4), in perhaps his most important contribution to blackjack, showed how to evaluate card-counting systems. He defined the *efficiency* E of a system to be "the ratio of the profit accruing from using the system to the total gain possible from perfect knowledge and interpretation of the unplayed set of cards." He used (11.100) to estimate both numerator and denominator, obtaining

$$E = \frac{\mu\Phi(\mu/\sigma_n) + \sigma_n\phi(\mu/\sigma_n)}{\mu\Phi(\mu/\sigma_n^*) + \sigma_n^*\phi(\mu/\sigma_n^*)}, \tag{21.81}$$

where μ is the expectation at the next round and

$$\sigma_n := \rho\sigma_e\sqrt{\frac{n}{51(52-n)}} \quad \text{and} \quad \sigma_n^* := \sigma_e\sqrt{\frac{n}{51(52-n)}}. \tag{21.82}$$

Here ρ is the correlation between the count system and the effects of removal, and σ_e^2 is the weighted average of the squares of the effects of removal. Notice that (21.81) reduces to ρ when $\mu = 0$.

This formula permits the evaluation of both the *betting efficiency* and the *playing efficiency* of a system. For the former we use μ as in (21.65) and the effects of removal as in Table 21.9. Since $\mu \approx 0$, we find that the betting efficiency is approximately ρ, which is also called the *betting correlation*. For the playing efficiency we consider each of a variety of playing decisions, such as $\{6, T\}$ vs. 9 using μ as in (21.76) and the effects of removal as in Table 21.12. Taking a weighted average over 70 playing decisions (hard 10–16 vs. all dealer upcards) and taking $n = 32$, Griffin tabulated the playing efficiencies for a variety of systems.

Most card-counting systems have an extensive list (well over 100) of indices for departing from basic strategy based on the count. (We analyzed just one of these, for $\{6, T\}$ vs. 9, in Section 21.3.) Don Schlesinger [1946–] (1986; 2005, Chapter 5) observed that a player can capture much of the potential expected gain with just 18 of these indices, which he termed the *Illustrious 18*. His study assumed a four-deck shoe, 75% penetration, and bet sizes that are roughly proportional to the true Hi-Lo count. Preliminary work along these lines was done by Griffin in 1979 and 1986 (Griffin 1999, pp. 30, 229). Other attempts to simplify the process of card counting have included the use of unbalanced counts, introduced by Jacques Noir (1968) in a book titled *Casino Holiday* (his "One-Two Count" is the Noir count in Table 21.9). According to Braun (1980, p. 148), Jacques Noir is a pseudonym for a professor at the University of California, Berkeley, but we have been unable to confirm this and determine the author's true identity. Other unbalanced counts, which do not require converting the running count to a true count, include Snyder's (1983) Red Seven and Vancura and Fuchs's (1998) Knock-Out count (see Table 21.9).

Despite the fact that card counting is perfectly legal and ethical, skilled players may legally be barred from playing blackjack in most jurisdictions, including Nevada. In addition, today's rules are generally less favorable than the classic ones we have assumed in this chapter. Shackleford (2005a, p. 20) defined what he called the "Las Vegas Benchmark Rules," which differ from those of Table 21.1 in the following respects:

- six decks (6D)
- dealer hits soft 17 (H17)
- double down after splits (DAS)
- late surrender (LS) [see Problem 21.22]

The first is significantly disadvantageous to the player, the second slightly disadvantageous, and the last two slightly advantageous. Today, hand-held games (single- and double-deck) are frequently offered with only a 6-to-5 payoff for an untied player natural, which makes the game nearly unbeatable. See Zender (2008, Chapter 6) and Problem 21.4.

Let us very briefly survey the blackjack literature. Certainly the most important treatment of blackjack from the mathematical perspective is Griffin's (1999) *The Theory of Blackjack*, first published in 1979. Thorp's (1962b, 1966a) *Beat the Dealer* and the blackjack chapter in Epstein's (1967, 1977, 2009) *The Theory of Gambling and Statistical Logic* are also important in this regard. Among recent works, Schlesinger's (2005) *Blackjack Attack*, first published in 1997, James Grosjean's (2000, 2009) *Beyond Counting*, and N. Richard Werthamer's [1935–] (2009) *Risk and Reward* have significant mathematical content. As for players' guides, Revere's (1973) *Playing Blackjack as a Business*, first published in 1969, was influential for many years. Stanford Wong's [1943–] (1994) *Professional Blackjack*, which emphasizes the Hi-Lo and Halves (Table 21.9) systems and was first published in 1975, and Arnold Snyder's [1948–] (1983, 2005) *Blackbelt in Blackjack*, which emphasizes the Red Seven, Hi-Lo, and Zen (Table 21.9) systems, have stood the test of time. For the history of blackjack we recommend Snyder's (2006, Chapters 1–6) *The Big Book of Blackjack*. For popular accounts of successful team play, see Uston and Rapoport's (1977) *The Big Player* and Mezrich's (2002) *Bringing Down the House*. The 2008 film *21* was loosely based on the Mezrich book, which itself can be described as fictionalized nonfiction.

There are several definitions of basic strategy in the blackjack literature. Everyone agrees that basic strategy is the optimal strategy for the first round of play, but differences in what information the player may use exist. The most complex basic strategy is Griffin's (1999, p. 12):

> It is even conceivable, if not probable, that nobody, experts included, knows precisely what the basic strategy is, if we pursue the definition to include instructions on how to play the second and subsequent cards of a split depending on what cards were

used on the earlier parts. For example, suppose we split eights against the dealer's ten, busting the first hand $(8,7,7)$ and reaching $(8,2,2,2)$ on the second. Quickly now, do we hit or do we stand with the 14?

Schlesinger (2005, p. 389) argued persuasively against this formulation:

> So how many cards in split hands am I allowed to reckon before "basic" turns hopelessly complicated? Again, the purists say, All of them. So, if I'm allowed splits to four hands, and, after playing the first three hands, there are, say, 13 cards lying on the table, and I am now playing my fourth hand, am I supposed to include knowledge of all those previous cards in the play of the current hand while still invoking "basic" strategy? Gimme a break! Permit me to claim how utterly ridiculous that concept is to me.

He went on to define what he called a "practical, workable" definition of basic strategy: composition-dependent basic strategy for two-card hands and total-dependent basic strategy for multicard hands and split hands.

The definition of basic strategy that we have adopted in this chapter is in some respects a compromise between these two approaches. We have emphasized composition-dependent basic strategy for the original hand and for split hands (while ignoring all cards in hands split from the hand currently being played) rather than total-dependent basic strategy because, although it is harder to remember, it is easier to evaluate! Moreover, total-dependent basic strategy is analogous, it seems to us, to a video poker strategy that ignores penalty cards; in other words, it is a strategy that sacrifices optimality for simplicity.

It should be noted that, in Table 21.8, several cases that do not occur to the basic strategist were omitted. These include

- hard 12 vs. 4. Hit AAT.
- hard 12 vs. 3. Hit 228, AAT, AA334, AA2233, AA2224, AAA333, AAA2223, AAAA233, and AAAA2222.
- hard 12 vs. 2. Stand on AAAA8 and AAAA2222.

For example, $\{2,2,8\}$ vs. 3 does not occur because $\{2,2\}$ calls for splitting and $\{2,8\}$ calls for doubling. Also, $\{A,A,A,A,2,2,2,2\}$ does not occur because both seven-card subsets call for standing.

The idea of recursively computing composition-dependent basic strategy is due to Allison R. Manson [1939–2002], Anthony James Barr [1940–], and James H. Goodnight [1943–] (1975). (In 1976, Barr, Goodnight, and two others founded the SAS Institute, Inc.) For the four-deck game, Manson, Barr, and Goodnight obtained exact results except for pair splitting, for which they used computer simulation. Griffin (1999, p. 172) explained their algorithm very succinctly; it is the one we used in Section 21.2.

> The process begins with the formation of all possible player totals of 21, such as $(T,T,A), (T,9,2), (T,9,A,A), \ldots, (2,2,2,A,A,\ldots,A)$. For each of these player hands, the dealer's exact probabilities are figured for the up card being considered and from this the player's standing expectation is computed, stored, and indexed

for retrieval. (The indexing and retrieval mechanism for all possible hands is one of the more difficult aspects of the computer program.)

Next, all possible player totals of 20 are formed, starting with (T, T). Now (playing devil's advocate at this stage) one calculates the player's expectation from hitting the total of 20 by referring to the standing expectations already catalogued for the hands of 21 which might be reached if an ace were drawn. Then this hitting expectation is compared to the player's standing expectation with the currently possessed total of 20, which is calculated as in the previous paragraph.

In this manner the computer cycles downward through the player's totals until finally the exactly correct strategy and expectation is available for any possible player hand. The procedure is not, of course, restricted to four deck analysis; applied to any prespecified set of cards it will yield the absolutely correct *composition* dependent strategy and associated expectation, without any preliminary guesswork as to what *totals* the player should stand with.

Much of the early research on blackjack was based on computer simulation, which is still widely used today. Exceptions included Baldwin, Cantey, Maisel, and McDermott (1956) and Thorp (1962b), who used exact computations in their work. Thorp used simulations only to confirm findings already computed. In the blackjack literature, exact computation is usually referred to as "combinatorial analysis," a term that has a slightly different meaning in the mathematical literature.

The reader may wonder how reliable our computations are. All of the numerical results in Sections 1 and 2 (and 3 with some exceptions) have been obtained before, so it was merely a matter of confirming that our numbers are consistent with earlier work. For example, our mimic-the-dealer calculation matches one done by Cacarulo at bjmath.com in 2003. Several authors have approximated this number by assuming independence between the player's hand and the dealer's, as though they were dealt from separate decks. Thorp (1962b, p. 224) obtained 0.0573 in place of (21.6) in this way, which can be compared with Wilson's (1965, p. 47) 0.045 and Griffin's (1999, p. 18) 0.055. The correct figure under this assumption, from Table 1.5, is approximately (cf. (21.6))

$$H_0 \approx (0.283585403)^2 - \frac{1}{2} \frac{\binom{4}{1}\binom{16}{1}\binom{32}{0}}{\binom{52}{2}} \left(1 - \frac{\binom{4}{1}\binom{16}{1}\binom{32}{0}}{\binom{52}{2}}\right)$$
$$\approx 0.080420681 - 0.022967953$$
$$= 0.057452728. \tag{21.83}$$

Scarne (1961, p. 328) essentially assumed sampling with replacement, an even less realistic assumption, and got 0.059, a result he claimed priority for even though he published it five years after Baldwin, Cantey, Maisel, and McDermott's (1956) 0.056. For the record, the correct figure under this assumption is approximately

$$H_0 \approx (0.281592847)^2 - \frac{1}{2} 2 \frac{1}{13} \frac{4}{13} \left(1 - 2 \frac{1}{13} \frac{4}{13}\right)$$

$$\approx 0.079294532 - 0.022548230$$
$$= 0.056746302, \tag{21.84}$$

as noted by Bewersdorff (2005, p. 124).

Our basic strategy numbers (excluding splits) are consistent with those posted by Cacarulo at `bjmath.com`. The basic strategy split calculation was first done by Cacarulo and Marc Estafanous at `bjmath.com` in 2002–2003 with help from Steve Jacobs. Our final number (21.65) is rounded from a double-precision computation that agrees with that of Estafanous to 15 decimal places, despite different algorithms and different programming languages. It is also consistent with Cacarulo's number.

Table 21.3 was first computed by Thorp (1962b, p. 203). Table 21.4 is an elaboration of Griffin (1999, p. 158), who gave the rightmost column only. Our basic strategy example $\{6, T\}$ vs. 9 (especially Table 21.5) is essentially from Griffin (1999, p. 14). The eight cases in Table 21.7 were first pointed out by Steve Jacobs at `bjmath.com` in 2002, and the expectation formulas were first obtained (in different but equivalent forms) by Marc Estafanous, Cacarulo, and Eric Farmer, also at `bjmath.com` in 2003.

The effects of removal of Table 21.9 are not available elsewhere (owing to our use of composition-dependent basic strategy) but at least they sum to 0, which is partial confirmation of their accuracy. The insurance indices of Table 21.11 were first obtained by Marc Estafanous and Cacarulo at `bjmath.com` in 2004, though they averaged expectations over different depths and running counts with the same true count rather than considering each depth and running count in isolation as we have done. Our approach is apparently similar to that reported by Griffin (1998) in a 1970 letter to Thorp, but the numbers differ significantly in some cases. Our remark that insurance may help to reduce variance can be stated more precisely: Insurance reduces variance for good and mediocre hands—hard totals of 8–11 and 18–21 (Griffin 1988) as well as all soft totals (Grosjean 2009, Chapter 5; Canjar 2007a).

Our analysis of basic strategy departures for $\{6, T\}$ vs. 9 differs slightly from that of the blackjack literature (cf. Griffin 1999, pp. 26–27). The standard approach is to assume that the player has an "abstract" hard 16, and the full-deck expectation gain μ from hitting is found by dealing from a 51-card deck (the full deck less the dealer's 9). The effects of removal on the expectation gain from hitting in this situation are derived using a full 52-card deck (including the dealer's 9), numbers first tabulated by Griffin in 1979 (in the first edition of Griffin 1999) and subsequently refined by Zenfighter (Schlesinger 2005, Appendix D). See Griffin (1999, Appendix A to Chapter 6) for an explanation of why the full deck was used for these tables. The idea of estimating strategy indices using effects of removal is due to Snyder (1982).

Problem 21.2 is due to Griffin (1999, p. 159). Problem 21.3 was motivated by Chambliss and Roginski (1990, p. 73), who chose not to condition on the dealer's not having a natural. Problems 21.5 and 21.6 are due to

Baldwin, Cantey, Maisel, and McDermott (1956), although there are several ways to accomplish this (see, for example, Bewersdorff 2005, Chapter 17; Werthamer 2009, Chapter 7, Section A). Problem 21.8 was suggested by Marc Estafanous. The closest decision in basic strategy (Problem 21.9) was noted by Griffin (1999, p. 20). Problem 21.10 was posed by Sklansky (2001, p. 166), who instead found "the worst mistake you will normally see," a slightly different criterion than that of the problem. Problem 21.11 has been done by several authors, albeit perhaps with slightly different definitions of basic strategy. The most complete published computation along these lines is due to Cacarulo and is found in Schlesinger (2005, Appendix A). A partial solution of Problem 21.12 can be found in Epstein (2009, Appendix, Table D) and a complete one in Milyavskaya (2005). Problem 21.14 was suggested by an observation of Bob Fisher at bj21.com in 2007. Griffin (1999, p. 21) is responsible for Problem 21.17; see also Thorp (2000a). Problems 21.18, 21.20, and 21.21 are from Griffin (1999, pp. 177, 23, and 236, resp.); for the latter, see also Grosjean (2009, pp. 86–87) and Canjar (2007a). Problem 21.19 was motivated by Epstein's (2009, p. 284) *grayjack* (one ace, two each of 2–5, and four 6s; target total is 13), a game still too complicated for hand calculation. Problem 21.24 comes from Noir (1968); see also Grosjean (2009, Chapter 14).

Chapter 22
Poker

[A]nd yet, forsooth, a gallant man who sets him down before the baize and challenges all comers, his money against theirs, his fortune against theirs, is proscribed by your modern moral world.

William Makepeace Thackeray, *The Luck of Barry Lyndon*

Poker is a card game of 19th-century American origin with Persian and European ancestors. It differs from the games treated in previous chapters in that the players compete against each other rather than against the house. In Section 22.1 we describe the rules of the game as well as the concept of pot odds. In Section 22.2 we consider a simplified model of the poker endgame that is amenable to game-theoretic analysis. In Section 22.3 we study the game of Texas hold'em, currently the most widely played form of poker and therefore the subject of much of the recent poker literature.

22.1 Rules and Pot Odds

Poker, unlike the games treated in previous chapters, is not a house-banked game, that is, the players compete against each other rather than against the house. *Poker* is actually a generic name for a variety of games, such as five-card draw, seven-card stud, and Texas hold'em. Although the details of these games differ, they have a number of elements in common.

Most poker games are played with a standard 52-card deck; occasionally, a 53rd card, a *joker*, is added. Over the course of a game, players make bets by adding money to a common *pot*. At the game's conclusion, the winner takes the pot. The object of poker is to win money.

Each game consists of several betting rounds. Within each round players act in clockwise order. If no bet has yet been made in a round, a player may either *check* (decline to bet but remain active in the game) or bet. Once

S.N. Ethier, *The Doctrine of Chances*, Probability and its Applications,
DOI 10.1007/978-3-540-78783-9_22, © Springer-Verlag Berlin Heidelberg 2010

a bet has been made in a round, each subsequent player must either *fold*
(muck his cards, forfeit his chance to win the pot, and become inactive in
the game), *call* (match the amount bet by the most recent player to bet or
raise), or *raise* (increase the size of the bet required to remain active in the
game). A player may check and raise in the same round, provided another
player bets; this practice, called *check-raising* or *sandbagging*, is considered
perfectly acceptable (see Table 22.8 on p. 739 for an example). A betting
round is completed when all active players have checked or when all other
active players have called the most recent bet or raise. At the conclusion
of each round, all active players will have contributed the same amount of
money to the pot. If only one active player remains, he wins the pot and the
game is over. If two or more players remain active after the final round of
betting, a *showdown* occurs. In most games the highest five-card poker hand
wins the pot, where poker hands are ranked as explained in Example 1.1.10
on p. 6 and Problem 1.6 on p. 53. Ties are possible, in which case the pot is
split evenly among the tying players.

Most poker games have some form of forced betting before the deal to
create a pot worth contesting. In some games each player is required to bet
an equal amount. This is called an *ante*. In other games one or more players
(often to the left of the dealer) are required to make *blind bets*, and the
amounts of the blinds are specified. To make the game equitable, the role of
the dealer moves one player to the left after each hand. In the casino setting,
there is a professional dealer, so the position of the nominal dealer is indicated
by the *button*, which moves one player to the left after each hand. (In some
games this is unnecessary because the player in the dealer's position does not
have a significant advantage.) In the first betting round, even if all players
call the so-called big blind) or fold, the big blind is allowed to raise.

Some games may have a forced bet after the deal. For example, in a game
with exposed cards in each hand, the weakest exposed hand may be required
to initiate the betting with a *bring-in*, a small bet of a specified size.

There are three common types of betting limits: structured limit, pot limit,
and no limit. In a *structured-limit* game, the size of a bet or raise is specified,
though it is typically larger in later rounds. For example, in a 2-4 game with
four betting rounds, bets and raising increments are 2 units in the first two
rounds and 4 units in the last two rounds. In a *pot-limit* game, a player
may raise by an amount equal to the size of the pot before the raise (but
including his call). In a *no-limit* game, there are no betting limits aside from
those imposed by the table-stakes rule described below.

Each player begins each game with an amount of money of his choosing
(usually in poker chips) in front of him. He may neither remove money from
the table nor add money to it during the play of the hand, a rule known as
table stakes. When a player has insufficient funds to call a bet or a raise,
he may fold or declare himself *all in*. In the latter case, each active player
who wishes to call or raise contributes the same amount to the main pot as
he does, and a side pot is created, for which the all-in player is ineligible,

containing all bets beyond those in the main pot. The key point is that no player can be forced out of a hand by having an insufficient stake. Poker is highly equitable in this regard. Incidentally, the option to go *all in* is not limited to players with insufficient funds. Rather, it is a term for betting one's entire stack of chips; it may be a raise, a call, or a partial call.

In the casino setting, the professional dealer deducts a small amount from each pot to pay for his services. This is called the *rake*.

These are the principal rules of poker, though there are some subtle points that we have not addressed. We now provide more detail on the most common forms of poker. Some, such as five-card draw and five-card stud, are no longer played much but are of historical interest.

In *five-card draw* each player receives five cards face down. A first betting round is initiated by the player to the dealer's left. Usually, a pair of jacks or better is required to open the betting. Beginning with the player to the dealer's left, each player discards none to five of his cards and receives the same number as replacements. A second betting round follows, initiated by the player who opened. If at least two players remain after the second betting round, there is a showdown with the highest poker hand winning. Often a joker is added to the deck. The joker is not a wild card but rather a *bug*; that is, it may be used to fill a straight, flush, or straight flush; otherwise it is an ace. Five aces beats a straight flush. A full table usually comprises eight players.

In *five-card stud* each player receives one card face down and one card face up. A betting round is initiated with the low card bringing it in. (Ties are broken using suits, with ♣, ♢, ♡, ♠ ranking lowest to highest for this purpose only.) Three more cards are dealt face up to each active player, with a betting round after each, initiated by the player with the highest exposed hand. (In the case of a tie, the eldest hand bets first.) If at least two players remain after the fourth betting round, there is a showdown with the highest poker hand winning. Up to 10 can play.

In *seven-card stud* each player receives two cards face down and one card face up. A betting round is initiated with the low card bringing it in. (Ties are handled as in five-card stud.) Three more cards are dealt face up and one more face down to each active player, with a betting round after each, initiated by the player with the highest exposed hand. If at least two players remain after the fifth betting round, there is a showdown with the highest five-card poker hand (chosen from the player's seven cards) winning. A full table usually comprises eight players.

In *Texas hold'em* the player to the left of the dealer is the small blind and the player to his left is the big blind. Each player receives two *hole cards* face down. A betting round is initiated by the player to the left of the big blind (said to be *under the gun*). Then three community cards are dealt face up (the *flop*), then a fourth one (the *turn*), and finally a fifth one (the *river*). The five community cards are called the *board*. Betting rounds, initiated by the first active player to the dealer's left, follow the flop, the turn, and the

river. If at least two players remain after the fourth round of betting, there is
a showdown with the highest five-card poker hand (chosen from the player's
two hole cards and the five community cards) winning. Although as many
as 22 players can play Texas hold'em (allowing for the five-card board and
the traditional three burn cards), a full table usually comprises nine or 10
players.

Omaha hold'em (or just Omaha) is similar to Texas hold'em, except four
hole cards are dealt face down to each player. In the showdown the highest
hand that comprises two of the four hole cards and three of the five community cards wins.

There are several poker games in which the lowest poker hand wins. *Lowball
draw* is similar to five-card draw, except that the lowest poker hand wins,
ignoring straights, flushes, and straight flushes, and aces are regarded as low.
Thus, 5-4-3-2-A, called a *wheel*, is the best possible hand. Again, a 53-card
deck is usually used, except now the joker is the lowest card not already in
the hand that contains it. *Razz* is similar to seven-card stud, except that the
lowest five-card poker hand wins, as in lowball draw. Again, a wheel is the
best possible hand. In a variation of lowball draw, *deuce-to-seven lowball* or
Kansas City lowball, aces are high and straights, flushes, and straight flushes
are not ignored. Here the best possible hand is 7-5-4-3-2 unsuited. A joker is
not used in razz or in deuce-to-seven lowball.

In *high-low* games, especially seven-card stud and Omaha, the high hand
and the low hand split the pot. Often, the low hand must be eight-high or
better in order to qualify. If no hand qualifies for low, the high hand scoops
the pot. A player can have both high hand and low hand, not necessarily
with the same five cards.

HORSE is an event in such poker tournaments as the World Series of
Poker. It comprises five versions of poker played alternately: Texas hold'em,
Omaha high-low eight or better, razz, seven-card stud, and seven-card stud
high-low eight or better. It is considered the ultimate test of poker skill.

One of the most important concepts in poker is that of pot odds. Suppose,
with all cards out, the pot contains a units and a player is contemplating
calling a bet of b units. Then we say that the pot is offering him odds of a to
b, that is, the *pot odds* are a to b, or a/b to 1. Suppose also that the player
can estimate his probability of winning the hand as p, which is to say that
the odds against winning are $1 - p$ to p, or $(1 - p)/p$ to 1. Then the player's
expected profit from making the call is $(1 - p)(-b) + pa$, and this is positive
if and only if

$$p > \frac{b}{a + b} \qquad \text{or} \qquad \frac{a}{b} > \frac{1 - p}{p}, \tag{22.1}$$

that is, *the pot odds exceed the true odds*. In calculating the expectation above,
it is important to recognize that money in the pot no longer belongs to the
player who contributed it. Instead, it belongs to the pot until the hand is
completed. In economic terms, it is a sunk cost.

With all cards out, the probability p depends on the ability of the player to read his opponents, which in turn requires judgment and experience. But the concept of pot odds applies in other situations as well, in which the probability p can be evaluated in the usual way.

For example, suppose, following the turn in a game of Texas hold'em, a player holds a *flush draw*, that is, a four-card flush. He has reason to believe that he will have the best hand if and only if he makes his flush on the river. Since 46 cards remain unseen, of which 9 are of the flush suit, his odds against winning are 37 to 9. If the pot contains a units and the player is contemplating calling a bet of b units, he should call if and only if $a/b > 37/9 = 4.\overline{1}$.

Now we consider the same situation *before* the turn. The player now has two cards to come. The probability of making his flush is

$$\frac{9}{47} + \left(1 - \frac{9}{47}\right)\frac{9}{46} = \frac{378}{1{,}081}, \tag{22.2}$$

which corresponds to odds of 703 to 378 against. If the pot contains a units and the player is contemplating calling a bet of b units, it no longer makes sense to compare a/b to $703/378$, the point being that the player may have to bet following the turn to find out if he will make his flush on the river. If only one opponent remains, and if the amount the player expects to have to call following the turn is c units, then his effective odds are $a + c$ to $b + c$. He should call the b-unit bet if and only if $(a+c)/(b+c) > 703/378 \approx 1.860$. Pot odds that take into account more than one card to come are called *effective odds*.

Let us return to the first example, in which the pot contains a units and the player is contemplating calling a bet of b units following the turn. Suppose that, if the player makes his flush on the river and bets c units, he expects his opponent to call. This will add c units to the pot on the final betting round (the player's bet of c units is not counted here because it is not yet a sunk cost), with the result that the pot odds increase to $a + c$ to b. These odds are known as *implied odds* because they take future bets by the player's opponents into account. It is often the case that a call that is not justified by the pot odds *is* justified by the implied odds.

A good example of the use of implied odds occurred on the final hand of the 1980 World Series of Poker main event (no-limit Texas hold'em). Doyle Brunson, who had $232,500 in front of him, was *heads up* (i.e., only two players remained) against Stu Ungar, who had $497,500. Brunson was dealt A♡-7♠ and Ungar was dealt 5♠-4♠. After the first betting round, the pot contained $30,000. The flop came A-7-2, giving Brunson two pair and Ungar an inside-straight draw. Ungar checked and Brunson bet $17,000, a bet intended to keep Ungar around. Ungar's odds against making his straight on the turn were 43 to 4, or 10.75 to 1, while the pot was offering only 47 to 17, or about 2.765 to 1. But if he were to make his straight and Brunson had a big hand, he might be able to win Brunson's entire $232,500 plus his own

$15,000 contribution to the pot. This made his implied odds 247.5 to 17, or about 14.559 to 1, so he called. Remarkably, the turn card was a 3, giving Ungar his straight, so he bet $40,000. Brunson raised all-in (betting all of his remaining $200,500) and Ungar called. Brunson failed to catch an ace or a 7 on the river, and Ungar was world champion.

There are also situations in which future bets by the player can make a call less desirable than the pot odds might suggest. For example, suppose a player has a pair of aces in a game of Texas hold'em and three suited non-aces of a suit other than those of the aces appear on the flop. If the pot contains a units and the player is contemplating calling a bet of b units, the pot odds of a to b may be misleading. Indeed, if the player's opponent fails to make a flush, he won't contribute much more to the pot, whereas if he has a flush or makes a flush, the player with a pair of aces has little chance to win the hand and it will cost him to see it to the end. If he expects to have to call two bets of c units to make it to the showdown, his *reverse implied odds* are a to $b + 2c$ (or $a + c$ to $b + 2c$) because his opponent will likely fold if he lacks the cards to win the hand.

22.2 Poker Models and Game Theory

Poker is too complicated to analyze directly using game theory, but there is a long tradition of studying simplified poker models thoroughly in the hope of capturing the essence of the game. We begin with a model of the endgame in poker.

Model 1. Basic endgame. The rules for this game are very simple. Two players, 1 and 2, ante a units each, where $a > 0$ is specified. Player 1 then draws a card that gives him a winning hand with probability P, where $0 < P < 1$, and a losing hand otherwise. Both players know the value of P but only player 1 knows whether he has a winning hand. It is assumed that play begins with a check by player 2. Player 1 may then check or bet b units, where $b > 0$ is specified. If he checks, there is a showdown. If he bets, player 2 may fold or call. If player 2 folds, player 1 wins the pot. If player 2 calls, there is a showdown.

Situations of this type arise in real poker. For example, consider the last round of betting in a heads-up game of five-card stud with structured-limit betting. Player 1 has 8-7-6-5 unsuited and a hidden hole card, and player 2 has A-A-3-2 and a hidden hole card. Player 1 will win with a 9 or a 4 in the hole, regardless of player 2's hole card, and will lose otherwise. Player 2 acts first and it is correct for him to check. If he does so, we are essentially in the situation of basic endgame, with a equal to half the pot size, b equal to player 1's permissible bet, and P equal to the conditional probability that player 1 has a winning hand, given the history of the game.

Actually, basic endgame is not a perfect model for this situation. First, P is a very complicated posterior probability based on the three previous rounds of betting (it is not simply equal to $8/44$), so it is highly unlikely that both players would evaluate it in the same way. Second, player 2's hole card gives him additional information, unknown to player 1, which will influence his estimate of P. Third, in real poker player 2 is not restricted to folding or calling—he may raise player 1's bet. Nevertheless, analysis of the model should provide useful information.

To formulate basic endgame as a matrix game, we need to identify the players' strategies. Player 1 has four pure strategies, namely check or bet with losing or winning hands (four combinations), and player 2 has two pure strategies, namely fold or call if player 1 bets. Thus, the payoff matrix for this game, with rows labeled by player 1's pure strategies, columns labeled by player 2's pure strategies, and entries representing the expected profit of player 1, has the form

$$
\begin{array}{c}
\\
\\
\text{check if loser, check if winner} \\
\text{check if loser, bet if winner} \\
\text{bet if loser, check if winner} \\
\text{bet if loser, bet if winner}
\end{array}
\begin{array}{c}
\text{fold if} \\
\text{player 1 bets} \\
\left(\begin{array}{c} 2aP \\ 2aP \\ 2a \\ 2a \end{array}\right.
\end{array}
\begin{array}{c}
\text{call if} \\
\text{player 1 bets} \\
\begin{array}{c} 2aP \\ (2a+b)P \\ 2aP+(-b)(1-P) \\ (2a+b)P+(-b)(1-P) \end{array}\left.\begin{array}{c}\\\\\\\end{array}\right)
\end{array}.
$$
(22.3)

Notice that money already in the pot is regarded as a sunk cost, so player 1's profit, if he wins, will include the $2a$ units in the pot and the call, if any, made by player 2. Player 1's profit, if he loses, is minus the amount that he bets, excluding his ante.

Observe next that the first and third rows are dominated (but not strictly) by the second and fourth rows, respectively, a result of the fact that, if player 1 has a winning hand, he is assured of at least as much profit by betting as he is by checking. Therefore we eliminate the first and third rows at the risk of not obtaining the complete set of solutions of the game. The new payoff matrix is

$$
\begin{array}{c}
\\
\\
\text{check if loser, bet if winner} \\
\text{bet if loser, bet if winner}
\end{array}
\begin{array}{c}
\text{fold if} \\
\text{player 1 bets} \\
\left(\begin{array}{c} 2aP \\ 2a \end{array}\right.
\end{array}
\begin{array}{c}
\text{call if} \\
\text{player 1 bets} \\
\begin{array}{c} (2a+b)P \\ (2a+b)P+(-b)(1-P) \end{array}\left.\begin{array}{c}\\\\\end{array}\right)
\end{array}.
$$
(22.4)

Labeling the rows and columns 1 and 2 and denoting the entries of this payoff matrix temporarily by $a_{11}, a_{12}, a_{21}, a_{22}$, we note that $a_{11} < a_{12}$, $a_{11} < a_{21}$, and $a_{12} > a_{22}$. If $a_{21} \leq a_{22}$, then $(2,1)$ is a saddle point, whereas if $a_{21} > a_{22}$, then $\min(a_{12}, a_{21}) > \max(a_{11}, a_{22})$, which is the second condition in (5.71) on p. 182. Therefore we arrive at the following conclusions, which depend on

$$P_1 := \frac{2a + b}{2a + 2b}. \tag{22.5}$$

If $2a \leq (2a + b)P + (-b)(1 - P)$, that is, if $P \geq P_1$, then the payoff matrix has a saddle point, it is optimal for player 1 to always bet, and it is optimal for player 2 to fold when player 1 bets. The value of the game (to player 1) is $v = 2a$.

If $2a > (2a + b)P + (-b)(1 - P)$, that is, if $P < P_1$, then both players have optimal mixed strategies that use both of their pure strategies. Recalling Example 5.2.6 on p. 182, it is optimal for player 1 to always bet if he has a winning hand and, if he has a losing hand, to bet (i.e., to bluff) with probability

$$p^* := \frac{a_{12} - a_{11}}{(a_{12} - a_{11}) + (a_{21} - a_{22})} = \frac{b}{2a + b} \frac{P}{1 - P}. \tag{22.6}$$

It is optimal for player 2, if player 1 bets, to call with probability

$$q^* := \frac{a_{21} - a_{11}}{(a_{12} - a_{11}) + (a_{21} - a_{22})} = \frac{2a}{2a + b}. \tag{22.7}$$

The value of the game (to player 1) is

$$v := \frac{a_{12}a_{21} - a_{11}a_{22}}{(a_{12} - a_{11}) + (a_{21} - a_{22})} = 2a \frac{P}{P_1}. \tag{22.8}$$

Taking both cases into account, the value of the game is

$$v = 2a \left(\frac{P}{P_1} \wedge 1 \right), \tag{22.9}$$

which is a continuous function of P. Observe also that p^*, defined by (22.6), is less than 1 if and only if $P < P_1$, so if $p^* \geq 1$, then player 1 should always bet. On the other hand, q^* is always less than 1, so player 2 needs to confirm that $P < P_1$ before applying his optimal mixed strategy.

We observe that player 1's bluffing probability, p^*, decreases as the pot size $2a$ increases and increases as the bet size b increases. On the other hand, player 2's calling probability, q^*, increases as the pot size increases and decreases as the bet size increases, just as the pot odds do. Finally, the value of the game to player 1 is an increasing function (if $P < P_1$) of both the pot size and the bet size.

We can see directly that player 1's optimal mixed strategy is better than either of his pure strategies. (We assume that $P < P_1$, so that both players have optimal mixed strategies.) Indeed, if player 1 *never* bluffs with a losing hand (i.e., he uses row 1 of (22.4)), then player 2 will always fold, and player 1's expected win is

$$2aP < 2a \frac{P}{P_1} = v. \tag{22.10}$$

If player 1 *always* bluffs with a losing hand (i.e., he uses row 2 of (22.4)), then player 2 will always call, and player 1's expected win is

$$(2a + b)P + (-b)(1 - P) = (2a + b)\frac{P}{P_1} - b < 2a\frac{P}{P_1} = v. \qquad (22.11)$$

Thus, bluffing is an essential part of the arsenal of a skillful poker player.

Similarly, player 2's optimal mixed strategy is better than either of his pure strategies, again assuming that $P < P_1$. Indeed, if player 2 *never* calls a potential bluff by player 1 (i.e., he uses column 1 in (22.4)), then player 1 will always bluff with a losing hand, resulting in an expected win (for player 1) of

$$2a > 2a\frac{P}{P_1} = v. \qquad (22.12)$$

If player 2 *always* calls a potential bluff by player 1 (i.e., he uses column 2 in (22.4)), then player 1 will never bluff with a losing hand, resulting in an expected win (for player 1) of

$$(2a + b)P > 2a\frac{P}{P_1} = v \qquad (22.13)$$

since $P_1 > 2a/(2a + b)$.

Notice also that, if $P < P_1$ and player 2 were to play strictly according to pot odds, he would be indifferent to folding or calling player 1's bet. Indeed, if player 1 bets, the pot is offering player 2 odds of $2a + b$ to b, while his odds against winning the hand are P' to $1 - P'$, where P' is the posterior probability that player 1 wins, given that he bets. By Bayes's law (Theorem 1.2.7 on p. 19) and (22.6), we have

$$P' = \frac{P \cdot 1}{P \cdot 1 + (1 - P)p^*} = P_1. \qquad (22.14)$$

But

$$\frac{2a + b}{b} = \frac{P_1}{1 - P_1} = \frac{P'}{1 - P'}, \qquad (22.15)$$

that is, the pots odds coincide with the true odds. The conclusion is consistent with the indifference principle mentioned in Section 5.2.

We turn next to a poker model with two betting rounds in which it is assumed that cards dealt between rounds do not affect the outcome.

Model 2. Basic endgame with two rounds of betting. Two players, 1 and 2, ante a units each, where $a > 0$ is specified. Player 1 then draws a card that gives him a winning hand with probability P, where $0 < P < 1$, and a losing hand otherwise. Both players know the value of P but only player 1 knows whether he has a winning hand. It is assumed that play begins with a check by player 2. Player 1 may then check or bet b_1 units, where $b_1 > 0$ is specified. If he checks, there is a showdown. If he bets, player 2 may fold or call. If player 2 folds, player 1 wins the pot. If player 2 calls, the game enters round 2.

If the game enters round 2, it is assumed that play begins with a check by player 2. Player 1 may then check or bet b_2 units, where $b_2 > 0$ is specified. If he checks, there is a showdown. If he bets, player 2 may fold or call. If player 2 folds, player 1 wins the pot. If player 2 calls, there is a showdown.

As we saw in Model 1, player 1 should never check with a winning hand, so we assume that he always bets with such a hand. Then we have a 3×3 matrix game. Player 1's pure strategy i is to always bet with a winning hand and to bet exactly i times ($0 \le i \le 2$) with a losing hand if he has the opportunity to do so, and player 2's pure strategy j is to call exactly j times ($0 \le j \le 2$) if he has the opportunity to do so. Thus, the payoff matrix for this game has the form

$$
\begin{array}{c}
\quad\quad\quad\text{fold/---}\quad\quad\text{call/fold}\quad\quad\quad\quad\quad\text{call/call} \\
\begin{array}{c}
\text{check/---} \\
\text{bet/check} \\
\text{bet/bet}
\end{array}
\begin{pmatrix}
2aP & (2a+b_1)P & (2a+b_1+b_2)P \\
2a & 2aP + b_1(2P-1) & (2a+b_2)P + b_1(2P-1) \\
2a & 2a+b_1 & 2aP + (b_1+b_2)(2P-1)
\end{pmatrix}.
\end{array}
\tag{22.16}
$$

Rows are labeled by player 1's pure strategies, described by his first/second-round decisions if he has a losing hand. Columns are labeled by player 2's pure strategies, described by his first/second-round decisions if player 1 bets. Entries represent the expected profit of player 1.

The solution of this game can be described as follows. (We postpone the proof because this model is a special case of the following one.) Let

$$
P_2 := \frac{2a+b_1}{2a+2b_1} \frac{2a+2b_1+b_2}{2a+2b_1+2b_2}.
\tag{22.17}
$$

If $P \ge P_2$, then player 2's optimal mixed strategy is $\boldsymbol{q}^* = (1,0,0)$, that is, player 2 should fold on the first round if player 1 bets, and player 1's optimal mixed strategy is $\boldsymbol{p}^* = (0, 1 - p_2^*, p_2^*)$, where

$$
p_2^* := \left(\frac{b_2}{2a+2b_1+b_2} \frac{P}{1-P} \right) \wedge 1,
\tag{22.18}
$$

that is, player 1 should always bet on round 1 and should bet on round 2 with probability p_2^*. The value of the game is $v = 2a$.

If $P < P_2$, then player 2's optimal mixed strategy is $\boldsymbol{q}^* = (r_1, (1 - r_1)r_2, (1-r_1)(1-r_2))$, where $r_1 := b_1/(2a+b_1)$ and $r_2 := b_2/(2a+2b_1+b_2)$, that is, player 2 should fold in round 1 with probability r_1 if player 1 bets, and fold in round 2 with probability r_2 if player 1 bets. Notice that player 2 can achieve the same result by applying the optimal strategy from basic endgame to each of his two decisions. Player 1's optimal mixed strategy is $\boldsymbol{p}^* = (p_0^*, p_1^*, p_2^*)$, where

$$p_2^* = r_2 \frac{P}{1-P}, \tag{22.19}$$

$$p_1^* + p_2^* = \frac{1-P_2}{P_2} \frac{P}{1-P}, \tag{22.20}$$

and $p_0^* + p_1^* + p_2^* = 1$, that is, player 1 should always bet with a winning hand but, with a losing hand, he should bet in round 1 with probability $p_1^* + p_2^*$ and should bet in round 2 with probability $p_2^*/(p_1^* + p_2^*)$, which does not depend on P. The value of the game is $v = 2aP/P_2$.

Taking both cases into account, the value of the game is

$$v = 2a\left(\frac{P}{P_2} \wedge 1\right), \tag{22.21}$$

which is a continuous function of P. Observe also that $p_1^* + p_2^* < 1$ if and only if $P < P_2$, so if $p_1^* + p_2^*$, defined by (22.20), is greater than or equal to 1, then player 1 should always bet. On the other hand, r_1 is always less than 1, so player 2 needs to confirm that $P < P_2$ before applying his optimal mixed strategy.

We conclude by generalizing Models 1 and 2 to the case of n betting rounds.

Model n. Basic endgame with $n \geq 1$ rounds of betting. This model is an extension of Models 1 and 2. Two players, 1 and 2, ante a units each, where $a > 0$ is specified. Player 1 then draws a card that gives him a winning hand with probability P, where $0 < P < 1$, and a losing hand otherwise. Both players know the value of P but only player 1 knows whether he has a winning hand. It is assumed that play begins with a check by player 2. Player 1 may then check or bet b_1 units, where $b_1 > 0$ is specified. If he checks, there is a showdown. If he bets, player 2 may fold or call. If player 2 folds, player 1 wins the pot. If player 2 calls, the game enters round 2 unless $n = 1$, in which case there is a showdown.

If the game enters round k, where $2 \leq k \leq n$, it is assumed that play begins with a check by player 2. Player 1 may then check or bet b_k units, where $b_k > 0$ is specified. If he checks, there is a showdown. If he bets, player 2 may fold or call. If player 2 folds, player 1 wins the pot. If player 2 calls, the game enters round $k+1$ unless $n = k$, in which case there is a showdown.

As we saw in Model 1, player 1 should never check with a winning hand, so we assume that he always bets with such a hand. Then we have an $(n + 1) \times (n + 1)$ matrix game. Player 1's pure strategy i is to always bet with a winning hand and to bet exactly i times ($0 \leq i \leq n$) with a losing hand if he has the opportunity to do so, and player 2's pure strategy j is to call exactly j times ($0 \leq j \leq n$) if he has the opportunity to do so. If we let

$$s_j := a + b_1 + \cdots + b_j \tag{22.22}$$

denote each player's contribution to the pot after the completion of round j, then the payoff matrix $\boldsymbol{A} = (a_{ij})$ has the form

$$a_{ij} = \begin{cases} (a + s_j)P + (a - s_i)(1 - P) & \text{if } 0 \le i \le j \le n, \\ (a + s_j)P + (a + s_j)(1 - P) & \text{if } 0 \le j < i \le n, \end{cases}$$

$$= \begin{cases} a + s_j P - s_i(1 - P) & \text{if } 0 \le i \le j \le n, \\ a + s_j & \text{if } 0 \le j < i \le n. \end{cases} \tag{22.23}$$

To see this, observe that, if player 1 chooses strategy i and player 2 chooses strategy j, there are two cases. If player 1 has a winning hand (probability P), he bets at every opportunity and will win $a + s_j$ units when player 2 folds ($j < n$) or at the final showdown ($j = n$). If player 1 has a losing hand (probability $1 - P$), he will lose $-a + s_i$ units (or win $a - s_i$ units) if he is called every time he bets ($i \le j$) and will win $a + s_j$ units if player 2 folds before player 1 checks ($j < i$).

Let us denote by $\boldsymbol{p}^* = (p_0, p_1, \ldots, p_n)$ and $\boldsymbol{q}^* = (q_0, q_1, \ldots, q_n)$ optimal mixed strategies for players 1 and 2, respectively. We tentatively assume that $p_i^* > 0$ for $i = 0, 1, \ldots, n$ and $q_j^* > 0$ for $j = 0, 1, \ldots, n$, so that we can use Corollary 5.2.4(a) on p. 179 to derive formulas for $\boldsymbol{p}^*$ and $\boldsymbol{q}^*$. We will later confirm that this hypothesis is correct under a certain condition on P. By the lemma,

$$v_i := \left[\boldsymbol{A}(\boldsymbol{q}^*)^{\mathsf{T}} \right]_i = \sum_{j=0}^{n} a_{ij} q_j^* = a + \sum_{j=0}^{i-1} s_j q_j^* + \sum_{j=i}^{n} s_j q_j^* P - s_i \sum_{j=i}^{n} q_j^* (1 - P)$$

$$= a + \sum_{j=0}^{n} s_j q_j^* P + \sum_{j=0}^{i-1} s_j q_j^* (1 - P) - s_i \sum_{j=i}^{n} q_j^* (1 - P) \tag{22.24}$$

does not depend on i. Therefore, for $k = 1, \ldots, n$,

$$0 = v_k - v_{k-1} = \left(s_{k-1} q_{k-1}^* - s_k \sum_{j=k}^{n} q_j^* + s_{k-1} \sum_{j=k-1}^{n} q_j^* \right)(1 - P)$$

$$= \left[(s_k + s_{k-1}) q_{k-1}^* - (s_k - s_{k-1}) \sum_{j=k-1}^{n} q_j^* \right](1 - P). \tag{22.25}$$

This implies that, for $k = 1, \ldots, n$,

$$\frac{q_{k-1}^*}{\sum_{j=k-1}^{n} q_j^*} = \frac{s_k - s_{k-1}}{s_k + s_{k-1}} = \frac{b_k}{2(a + b_1 + \cdots + b_{k-1}) + b_k} =: r_k. \tag{22.26}$$

Consequently, $q_0^* = r_1$, $q_1^* = (1 - q_0^*) r_2 = (1 - r_1) r_2$, and, more generally,

$$q_k^* = \left[\prod_{i=1}^{k}(1 - r_i)\right]r_{k+1}, \qquad k = 0, 1, \ldots, n - 1, \qquad (22.27)$$

where empty products are 1. We conclude that $q_n^* = \prod_{i=1}^{n}(1 - r_i)$.

Now by (22.24),

$$v_0 = a + \sum_{j=0}^{n} s_j q_j^* P - s_0(1 - P) \qquad (22.28)$$

and

$$v_n = a + \sum_{j=0}^{n} s_j q_j^* P + \sum_{j=0}^{n-1} s_j q_j^*(1 - P) - s_n q_n^*(1 - P)$$

$$= a + \sum_{j=0}^{n} s_j q_j^* - 2 s_n q_n^*(1 - P), \qquad (22.29)$$

so since $v_0 = v_n$, we have

$$v_0 = \frac{v_0 - v_n P}{1 - P} = 2 s_n q_n^* P = 2 s_n \left[\prod_{i=1}^{n}(1 - r_i)\right] P. \qquad (22.30)$$

But

$$s_n \prod_{i=1}^{n}(1 - r_i) = s_n \prod_{i=1}^{n} \frac{2 s_{i-1}}{s_i + s_{i-1}} = s_0 \prod_{i=1}^{n} \frac{2 s_i}{s_i + s_{i-1}} = a \prod_{i=1}^{n}(1 + r_i), \quad (22.31)$$

so, letting

$$P_n := \prod_{i=1}^{n}(1 + r_i)^{-1}, \qquad (22.32)$$

we have

$$v_0 = 2a \frac{P}{P_n}. \qquad (22.33)$$

Next we derive $\boldsymbol{p}^*$. By 5.2.4(a) on p. 179,

$$w_j := [\boldsymbol{p}^* \boldsymbol{A}]_j = \sum_{i=0}^{n} p_i^* a_{ij} = a + \sum_{i=0}^{j} p_i^*[s_j P - s_i(1 - P)] + \sum_{i=j+1}^{n} p_i^* s_j$$

$$= a - \sum_{i=0}^{j} p_i^* s_i(1 - P) + \left(1 - \sum_{i=j+1}^{n} p_i^*\right) s_j P + \sum_{i=j+1}^{n} p_i^* s_j$$

$$= a + s_j P + \left[\sum_{i=j+1}^{n} p_i^* s_j - \sum_{i=0}^{j} p_i^* s_i\right](1 - P) \qquad (22.34)$$

does not depend on j. Therefore, for $k = 1, \ldots, n$,

$$0 = w_k - w_{k-1} \tag{22.35}$$

$$= (s_k - s_{k-1})P + \left[\sum_{i=k+1}^{n} p_i^*(s_k - s_{k-1}) - p_k^*(s_k + s_{k-1}) \right](1 - P),$$

so, dividing by $(s_k + s_{k-1})(1 - P)$, we find that

$$p_k^* = r_k \frac{P}{1 - P} + \sum_{i=k+1}^{n} p_i^* r_k \tag{22.36}$$

for $k = 1, \ldots, n$. This implies that $p_n^* = r_n[P/(1 - P)]$, and by backward induction that

$$\sum_{i=k+1}^{n} p_i^* = \left[\prod_{i=k+1}^{n} (1 + r_i) - 1 \right] \frac{P}{1 - P} \tag{22.37}$$

for $k = 0, \ldots, n - 1$. Indeed, by (22.36) and (22.37), we have

$$\sum_{i=k}^{n} p_i^* = p_k^* + \sum_{i=k+1}^{n} p_i^*$$

$$= \left[r_k + \left(\prod_{i=k+1}^{n} (1 + r_i) - 1 \right) r_k + \prod_{i=k+1}^{n} (1 + r_i) - 1 \right] \frac{P}{1 - P}$$

$$= \left[\prod_{i=k}^{n} (1 + r_i) - 1 \right] \frac{P}{1 - P}. \tag{22.38}$$

But to have $p_0^* > 0$, we must have

$$\sum_{i=1}^{n} p_i^* = \left[\prod_{i=1}^{n} (1 + r_i) - 1 \right] \frac{P}{1 - P} = \frac{1 - P_n}{P_n} \frac{P}{1 - P} < 1 \tag{22.39}$$

or, equivalently, $P < P_n$. In this case, by (22.34) and (22.39),

$$w_0 = a + s_0 P + \left[\sum_{i=1}^{n} p_i^* s_0 - p_0^* s_0 \right](1 - P) = 2aP + \left[2a \sum_{i=1}^{n} p_i^* \right](1 - P)$$

$$= 2aP + \left[2a \frac{1 - P_n}{P_n} \frac{P}{1 - P} \right](1 - P)$$

$$= 2a \frac{P}{P_n}. \tag{22.40}$$

But this equals v_0 from (22.33), so $\boldsymbol{p}^*$ and $\boldsymbol{q}^*$ are optimal strategies and the game value is v_0, provided $P < P_n$. (More precisely, we have verified (5.44) and (5.46) on p. 177, with equalities instead of inequalities, for $v = v_0$.)

We continue to assume that $P < P_n$. As in Model 2, player 2's optimal strategy can be described more easily. If player 1 bets at the kth betting round, then it is optimal for player 2 to fold with probability r_k, $1 \le k \le n$. This is again just repeated application of basic endgame. Player 1's optimal strategy also has a simpler description. It is optimal for player 1 to always bet with a winning hand but, with a losing hand, he bets at round 1 with probability (22.39), and he bets at round k ($2 \le k \le n$) with probability

$$\frac{\sum_{i=k}^{n} p_i^*}{\sum_{i=k-1}^{n} p_i^*} = \frac{\prod_{i=k}^{n}(1 + r_i) - 1}{\prod_{i=k-1}^{n}(1 + r_i) - 1} \tag{22.41}$$

(see (22.38)), which does not depend on P. In other words, P influences the decision to bluff at the first round but does not affect that decision at subsequent rounds.

Thus, we have solved the game assuming $P < P_n$. We claim that if $P \ge P_n$, then it is optimal for player 1 to bet at round 1, that is, $p_0^* = 0$, and furthermore the value of the game is $2a$. Since column 0 of $\boldsymbol{A}$ consists of $(2aP, 2a, \ldots, 2a)^{\mathsf{T}}$, player 2 can fold at round 1 ($q_0^* = 1$) to ensure that player 1's profit is at most $2a$. So all we need to show is that there is a mixed strategy $\boldsymbol{p}^* = (0, p_1^*, \ldots, p_n^*)$ for player 1 that guarantees an expected profit of at least $2a$, regardless of what player 2 does.

We therefore need only find $\boldsymbol{p}^* = (0, p_1^*, \ldots, p_n^*)$ so that

$$w_j := [\boldsymbol{p}^* \boldsymbol{A}]_j = \sum_{i=1}^{n} p_i^* a_{ij} \ge 2a, \qquad j = 0, 1, \ldots, n. \tag{22.42}$$

We replace the requirement (22.42) by a slightly stronger one:

$$w_0 = 2a, \qquad w_k - w_{k-1} \ge 0, \quad k = 1, \ldots, n. \tag{22.43}$$

The first of these is automatic. By (22.34), $w_0 = a + s_0 P + [1 \cdot s_0 - 0 \cdot s_0](1 - P) = 2a$. The second is achieved if the right side of (22.35) is nonnegative, or if the equal sign in (22.36) is replaced by $\le$, or if (22.37) is replaced by

$$\sum_{i=k+1}^{n} p_i^* = \left(\left[\prod_{i=k+1}^{n} (1 + r_i) - 1 \right] \frac{P}{1 - P} \right) \wedge 1 \tag{22.44}$$

for $k = 0, 1, \ldots, n - 1$. These equations determine $\boldsymbol{p}^*$. The assumption that $P \ge P_n$ ensures that $p_1^* + \cdots + p_n^* = 1$.

Taking both cases into account, the value of the game is

$$v = 2a\left(\frac{P}{P_n} \wedge 1 \right), \tag{22.45}$$

which is a continuous function of P. Observe also that $p_1^* + \cdots + p_n^*$, defined by (22.39), is less than 1 if and only if $P < P_n$, so if $p_1^* + \cdots + p_n^* \ge 1$, then player 1 should always bet. On the other hand r_1 is always less than 1, so player 2 needs to confirm that $P < P_n$ before applying his optimal mixed strategy.

22.3 Texas Hold'em

The rules of Texas hold'em were explained in Section 22.1. We begin with some of the easier probability calculations relevant to the game.

There are $\binom{52}{2} = 1{,}326$ possible initial hands but, for example, A♠-K♡ is for all practical purposes equivalent to A◊-K♣. More generally, let us call two initial hands *equivalent* if they have the same two denominations and if the corresponding denominations have the same suits after a permutation of (♣,◊,♡,♠). Then there are 169 equivalence classes of initial hands, namely $\binom{13}{1} = 13$ *pocket pairs* (i.e., paired hole cards), $\binom{13}{2} = 78$ (unpaired) suited hands, and $\binom{13}{2} = 78$ unpaired unsuited hands. A hand belonging to a specific pocket-pair equivalence class (such as A-A) is dealt with probability

$$\frac{\binom{4}{2}\binom{48}{0}}{\binom{52}{2}} = \frac{6}{1{,}326} = \frac{1}{221} \approx 0.00452; \qquad (22.46)$$

a hand belonging to a specific suited-hand equivalence class (such as A-K suited) is dealt with probability

$$\frac{\binom{4}{1}\binom{1}{1}\binom{3}{0}\binom{44}{0}}{\binom{52}{2}} = \frac{4}{1{,}326} = \frac{2/3}{221} \approx 0.00302; \qquad (22.47)$$

and a hand belonging to a specific unpaired-unsuited-hand equivalence class (such as A-K unsuited) is dealt with probability

$$\frac{\binom{4}{1}\binom{1}{0}\binom{3}{1}\binom{44}{0}}{\binom{52}{2}} = \frac{12}{1{,}326} = \frac{2}{221} \approx 0.00905. \qquad (22.48)$$

Later, we will consider unordered pairs of initial hands, or what might be called *heads-up matchups* before the flop. For now, however, we continue with simpler matters.

A card that a player needs to assure a winning hand is called an *out*. For example, if a player has hole cards A♠-K♠ and two more spades appear on the flop, the nine unseen spades are considered outs. Any one of them on the turn or the river will result in a *nut* (unbeatable) flush, provided there are no possibilities of better hands (full house, four of a kind, or straight flush).

Therefore, with one or two cards to come, it is often possible to count the outs and determine the player's (approximate) odds of winning the hand. Specifically, if a player has n outs after the flop, the probability that he makes his hand on the turn is $n/47$. If he has n outs after the turn, the probability that he makes his hand on the river is $n/46$. Finally, if a player has n outs after the flop, the probability that he makes his hand on the turn *or* the river is

$$\frac{n}{47} + \left(1 - \frac{n}{47}\right)\frac{n}{46}. \tag{22.49}$$

See Table 22.1, in which these probabilities have been converted to odds.

Table 22.1 Odds to 1 against making a hand in Texas hold'em with one or two cards to come. Here it is assumed that the player needs three of a kind or better to win.

no. of outs	current hand	turn	river	turn or river
20		1.350	1.300	.481
19		1.474	1.421	.538
18		1.611	1.556	.601
17	open-ended straight-flush draw and a pair	1.765	1.706	.673
16		1.937	1.875	.755
15	straight draw (8 outs) and flush draw	2.133	2.067	.848
14	inside straight-flush draw and a pair	2.357	2.286	.955
13		2.615	2.538	1.079
12	inside straight draw and flush draw	2.917	2.833	1.224
11	flush draw and a pair	3.273	3.182	1.397
10	open-ended straight draw and a pair	3.700	3.600	1.605
9	flush draw	4.222	4.111	1.860
8	open-ended straight draw or double inside straight draw*	4.875	4.750	2.179
7		5.714	5.571	2.591
6	inside straight draw and a pair	6.833	6.667	3.142
5		8.400	8.200	3.914
4	inside straight draw or two pair	10.750	10.500	5.073
3		14.667	14.333	7.007
2	one pair	22.500	22.000	10.879
1		46.000	45.000	22.500

*for example, 8-6-5-4-2

While on the subject of draw probabilities at Texas hold'em, let us use the (univariate or multivariate) hypergeometric distribution to evaluate a few more.

If a player has unpaired hole cards, then the probability that the flop pairs neither of them is

$$\frac{\binom{3}{0}\binom{3}{0}\binom{44}{3}}{\binom{50}{3}} = \frac{13,244}{19,600} = \frac{6,622}{9,800} \approx 0.676; \qquad (22.50)$$

the probability that the flop pairs exactly one of them exactly once is

$$2\frac{\binom{3}{1}\binom{3}{0}\binom{44}{2}}{\binom{50}{3}} = \frac{5,676}{19,600} = \frac{2,838}{9,800} \approx 0.290; \qquad (22.51)$$

the probability that the flop pairs both exactly once each (resulting in two pair) is

$$\frac{\binom{3}{1}\binom{3}{1}\binom{44}{1}}{\binom{50}{3}} = \frac{396}{19,600} = \frac{198}{9,800} \approx 0.0202; \qquad (22.52)$$

the probability that the flop pairs one of the hole cards twice but does not pair the other (resulting in a *set*, i.e., three of a kind) is

$$2\frac{\binom{3}{2}\binom{3}{0}\binom{44}{1}}{\binom{50}{3}} = \frac{264}{19,600} = \frac{132}{9,800} \approx 0.0135; \qquad (22.53)$$

the probability that the flop pairs one of the hole cards twice and the other once (resulting in a full house) is

$$2\frac{\binom{3}{2}\binom{3}{1}\binom{44}{0}}{\binom{50}{3}} = \frac{18}{19,600} = \frac{9}{9,800} \approx 0.000918; \qquad (22.54)$$

and the probability that the flop pairs one of hole cards three times (resulting in four of a kind) is

$$2\frac{\binom{3}{3}\binom{3}{0}\binom{44}{0}}{\binom{50}{3}} = \frac{2}{19,600} = \frac{1}{9,800} \approx 0.000102. \qquad (22.55)$$

Notice that these six probabilities sum to 1.

If a player has a pocket pair, then the probability of flopping no card of the same denomination is

$$\frac{\binom{2}{0}\binom{48}{3}}{\binom{50}{3}} = \frac{17,296}{19,600} = \frac{1,081}{1,225} \approx 0.882; \qquad (22.56)$$

the probability of flopping exactly one card of the same denomination is

$$\frac{\binom{2}{1}\binom{48}{2}}{\binom{50}{3}} = \frac{2,256}{19,600} = \frac{141}{1,225} \approx 0.115; \qquad (22.57)$$

and the probability of flopping two cards of the same denomination (resulting in four of a kind) is

$$\frac{\binom{2}{2}\binom{48}{1}}{\binom{50}{3}} = \frac{48}{19,600} = \frac{3}{1,225} \approx 0.00245. \tag{22.58}$$

Notice that these three probabilities sum to 1.

If a player has suited hole cards, then the probability of flopping no cards of the same suit is

$$\frac{\binom{11}{0}\binom{39}{3}}{\binom{50}{3}} = \frac{9,139}{19,600} \approx 0.466; \tag{22.59}$$

the probability of flopping exactly one card of the same suit is

$$\frac{\binom{11}{1}\binom{39}{2}}{\binom{50}{3}} = \frac{8,151}{19,600} \approx 0.416; \tag{22.60}$$

the probability of flopping exactly two cards of the same suit (resulting in a flush draw) is

$$\frac{\binom{11}{2}\binom{39}{1}}{\binom{50}{3}} = \frac{2,145}{19,600} \approx 0.109; \tag{22.61}$$

and the probability of flopping three cards of the same suit (resulting in a flush) is

$$\frac{\binom{11}{3}\binom{39}{0}}{\binom{50}{3}} = \frac{165}{19,600} \approx 0.00842. \tag{22.62}$$

Notice that these four probabilities sum to 1.

We leave the cases of unsuited and suited *connectors* (adjacent hole cards) as problems (Problems 22.15 and 22.16 on p. 728). We could also calculate the probabilities of making these hands by the turn or by the river (Problem 22.17 on p. 728).

The preceding calculations may give the misleading impression that Texas hold'em probabilities are rather straightforward. To dispel that notion, we consider the following problem. Suppose, in a heads-up game, one player has hole cards A♠-K♠ and the other has hole cards 8♡-8♢. It will be convenient to refer to the two players by their hole cards (rather than as player 1 and player 2). If neither player folds before the showdown, what is the probability that A♠-K♠ wins? What is the probability that 8♡-8♢ wins? What is the probability of a tie?

Questions of this nature are important enough that commercial software exists for answering them. However, let us work out the answers by hand using combinatorial analysis to better appreciate the subtleties of Texas hold'em. For those not interested in the derivation, the answers may be found at (22.86)–(22.88).

Since four cards have been seen, 48 remain unseen, and therefore there are

$$\binom{48}{5} = 1{,}712{,}304 \tag{22.63}$$

ways to select the five cards needed to complete the hand, their order being irrelevant to the probabilities of interest. We treat six cases, depending on the number of cards r (among the five on the board) that are other than aces, kings, and 8s.

Case 0. $r = 0$. Here the board consists entirely of aces, kings, and 8s. There are $\binom{8}{5}\binom{40}{0} = 56$ ways altogether. 8♡-8◇ wins if he draws two more 8s while A♠-K♠ fails to draw three more aces or three more kings; A♠-K♠ wins otherwise. Thus, the number of ways for 8♡-8◇ to win is

$$\binom{3}{1}\binom{3}{2}\binom{2}{2}\binom{40}{0} + \binom{3}{2}\binom{3}{1}\binom{2}{2}\binom{40}{0} = 18. \tag{22.64}$$

Notice that the four binomial coefficients in each term correspond to the numbers of ways to choose the As, Ks, 8s, and other cards, respectively. It follows that the number of ways for A♠-K♠ to win is 38.

Case 1. $r = 1$. There are $\binom{8}{4}\binom{40}{1} = 2{,}800$ ways altogether. 8♡-8◇ wins if he draws two more 8s; A♠-K♠ wins otherwise. Thus, the number of ways for 8♡-8◇ to win is

$$\binom{6}{2}\binom{2}{2}\binom{40}{1} = 600. \tag{22.65}$$

It follows that the number of ways for A♠-K♠ to win is 2,200.

Case 2. $r = 2$. There are $\binom{8}{3}\binom{40}{2} = 43{,}680$ ways altogether. Here there is a possibility of a tie. A tie occurs if and only if the board shows all clubs (including, necessarily, A♣-K♣-8♣), and the number of ways is

$$\binom{3}{3}\binom{5}{0}\binom{10}{2}\binom{30}{0} = 45. \tag{22.66}$$

Now 8♡-8◇ wins if he draws two more 8s, or if he draws just one more 8 with three exceptions. One of the exceptions is the tie just noted, the second is the case in which the 8 is a spade, the board does not include a pair of aces or of kings, and the two non-A-K-8 cards are spades as well (for then A♠-K♠'s flush wins), and the third is the case in which the board includes a pair of aces or of kings as well as a pair of some non-A-K-8 denomination (for then A♠-K♠'s full house beats 8♡-8◇'s full house). Thus, the number of ways for 8♡-8◇ to win is

$$\binom{6}{1}\binom{2}{2}\binom{40}{2} + \binom{6}{2}\binom{2}{1}\binom{40}{2} - \binom{3}{3}\binom{5}{0}\binom{10}{2}\binom{30}{0}$$
$$- \binom{3}{1}\binom{3}{1}\binom{1}{1}\binom{1}{0}\binom{10}{2}\binom{30}{0} - 2\binom{3}{2}\binom{3}{0}\binom{2}{1}10\binom{4}{2}\binom{36}{0}$$
$$= 26{,}910. \tag{22.67}$$

It follows that the number of ways for A♠-K♠ to win is 16,725.

Case 3. $r = 3$. There are $\binom{8}{2}\binom{40}{3} = 276{,}640$ ways altogether. Again there is a possibility of a tie. We consider that case first. A tie occurs when the highest five-card hand both for A♠-K♠ and for 8♡-8♢ comprises the five community cards. With $r = 3$ there are two of the remaining three aces, three kings, and two 8s among them. If they are any A-K, a tie occurs if the board shows any ace-high straight or royal flush other than one with Q♠-J♠-T♠ or one with exactly four hearts or exactly four diamonds. If they are A♡-K♡ (or A♢-K♢), a tie also occurs if the board shows a flush with each card 9 or above (there are three such flushes). If they are A♣-K♣, A♣-8♣, or K♣-8♣, a tie also occurs if the board shows all clubs, but we exclude the royal flush already counted. Therefore, the number of ways for a tie to occur is

$$
\binom{3}{1}^2\binom{2}{0}\left[\binom{4}{1}^3\binom{28}{0} - \binom{3}{3}\binom{37}{0}\right] - 2\binom{2}{2}\binom{6}{0}3\binom{3}{1}\binom{2}{2}\binom{35}{0}
$$

$$
- 8\binom{2}{2}\binom{6}{0}\binom{3}{3}\binom{37}{0} + 2\binom{2}{2}\binom{6}{0}3\binom{3}{3}\binom{37}{0}
$$

$$
+ \binom{2}{2}\binom{6}{0}\left[\binom{10}{3}\binom{30}{0} - \binom{3}{3}\binom{37}{0}\right] + 2\binom{2}{2}\binom{6}{0}\binom{10}{3}\binom{30}{0}
$$

$$
= 906. \tag{22.68}
$$

Continuing with case 3, 8♡-8♢ wins with two more 8s unless A♠-K♠ gets a royal flush, and 8♡-8♢ wins with only one more 8 unless A♠-K♠ gets a straight, flush, or royal flush with some exceptions explained below. Finally, 8♡-8♢ wins with no more 8s if he gets a flush. Thus, the number of ways for 8♡-8♢ to win is

$$
\binom{6}{0}\binom{2}{2}\left[\binom{40}{3} - \binom{3}{3}\binom{37}{0}\right] + \binom{6}{1}\binom{1}{1}\binom{1}{0}\left[\binom{40}{3}\right.
$$

$$
- \left\{\binom{4}{1}^3\binom{28}{0} + \binom{10}{3}\binom{30}{0} - \binom{3}{3}\binom{37}{0}\right\} - 4\binom{4}{3}\binom{36}{0}\right]
$$

$$
+ \binom{4}{1}\binom{1}{1}\binom{3}{0}\binom{3}{3}\binom{37}{0} - \binom{2}{1}\binom{1}{1}\binom{5}{0}\left[\binom{10}{3}\binom{30}{0} - \binom{3}{3}\binom{37}{0}\right]
$$

$$
+ \binom{6}{1}\binom{1}{0}\binom{1}{1}\left[\binom{40}{3} - \left\{\binom{4}{1}^3\binom{28}{0} + \binom{10}{3}\binom{30}{0}\right.\right.
$$

$$
+ \binom{10}{2}\binom{6}{0}\binom{24}{1} - \binom{3}{3}\binom{37}{0} - 3\binom{2}{2}\binom{3}{1}\binom{35}{0}\right\} - 4\binom{4}{3}\binom{36}{0}\right]
$$

$$
+ \binom{4}{1}\binom{1}{1}\binom{3}{0}\binom{3}{3}\binom{37}{0} + 2\binom{2}{2}\binom{6}{0}\left[\binom{10}{2}\binom{30}{1} + \binom{10}{3}\binom{30}{0}\right.
$$

$$
- \binom{4}{3}\binom{36}{0}\right] + \binom{2}{1}\binom{2}{1}\binom{4}{0}2\binom{10}{3}\binom{30}{0} + \binom{4}{1}\binom{2}{1}\binom{2}{0}\binom{10}{3}\binom{30}{0}
$$

$$
= 124{,}247. \tag{22.69}
$$

Regarding terms enclosed within brackets as comprising a single term, term 1 corresponds to the case in which the board shows two 8s. Term 2 corresponds to the case in which the 8♣ but not the 8♠ appears on the board. 8♡-8♢ wins unless A♠-K♠ gets a straight, flush, or straight flush, or unless three queens, jacks, tens, or 9s appear on the board. Two other exceptions are covered by the next two terms. Term 3 corresponds to the subcase of the latter case in which 8♡-8♢'s flush beats A♠-K♠'s straight, while term 4 corresponds to another subcase in which the board shows all clubs, which is a tie, not a win for 8♡-8♢. Term 5 corresponds to the case in which the 8♠ but not the 8♣ appears on the board. This is similar to term 2 except that A♠-K♠ will also get a flush if two of the three non-A-K-8 cards on the board are spades and no pair appears (two terms are subtracted within braces to avoid double counting). Term 6 corresponds to the subcase of the latter case in which the board shows the ace or king of hearts or diamonds. The last three terms involve the case of no 8s on the board. They correspond respectively to the cases in which the aces and kings on the board include two hearts or two diamonds, one heart and one diamond, and one heart or diamond and one club. Here 8♡-8♢ can win only with a flush, but note that a heart or diamond flush on the board comprising cards of 9 or higher is a tie. It follows that the number of ways for A♠-K♠ to win is 151,487.

Case 4. $r = 4$. There are $\binom{8}{1}\binom{40}{1} = 731{,}120$ ways altogether. This is the most complicated case, so we break it down further.

Subcase 4a. $r = 4$ and the board includes the A♣. There are $\binom{1}{1}\binom{7}{0}\binom{40}{4} = 91{,}390$ ways altogether. A tie occurs if the board shows four of a kind or all clubs or an ace-low straight except one with three or four spades or four hearts or four diamonds. Thus, the number of ways for a tie to occur is

$$\binom{1}{1}\binom{7}{0}\left[10\binom{4}{4}\binom{36}{0} + \binom{10}{4}\binom{30}{0} + \binom{4}{1}^4\binom{24}{0} - 4\binom{3}{3}\binom{3}{1}\binom{34}{0}\right.$$
$$\left. - 4\binom{4}{4}\binom{36}{0}\right] = 460. \tag{22.70}$$

Here 8♡-8♢ wins if the board shows four hearts or four diamonds or if 8♡-8♢ has a straight that uses one of his 8s but not a queen-high straight or a straight with three or four spades or four clubs. Therefore, the number of ways for 8♡-8♢ to win is

$$\binom{1}{1}\binom{7}{0}\left[2\binom{10}{4}\binom{30}{0} + 4\left\{\binom{4}{1}^4\binom{24}{0} - 4\binom{3}{3}\binom{3}{1}\binom{34}{0} - 2\binom{4}{4}\binom{36}{0}\right\}\right.$$
$$\left. - 2\cdot4\binom{4}{4}\binom{36}{0}\right] = 1{,}380. \tag{22.71}$$

It follows that the number of ways for A♠-K♠ to win is 89,550.

Subcase 4b. $r = 4$ and the board includes the A♡ or A♢. There are $\binom{2}{1}\binom{6}{0}\binom{40}{4} = 182{,}780$ ways altogether. A tie occurs if the board shows four of a kind or an ace-low straight or straight flush except one with three or four spades or four hearts or four diamonds. Thus, the number of ways for a tie to occur is

$$\binom{2}{1}\binom{6}{0}\left[10\binom{4}{4}\binom{36}{0} + \binom{4}{1}^4\binom{24}{0} - 2\cdot4\binom{3}{3}\binom{3}{1}\binom{34}{0} - 2\binom{4}{4}\binom{36}{0}\right]$$
$$= 480. \tag{22.72}$$

Here 8♡-8◇ wins under the same conditions as in subcase 4a, but also if 8♡-8◇ has a straight that uses one of his 8s and four clubs, or if the non-A-K-8 cards include only three cards of the ace suit. There is one other change as well. If the non-A-K-8 cards consist of Q-J-T-9 of the ace suit, then 8♡-8◇ ties instead of winning. Therefore, the number of ways for 8♡-8◇ to win is

$$\binom{2}{1}\binom{6}{0}\left[2\binom{10}{4}\binom{30}{0} + 4\left\{\binom{4}{1}^4\binom{24}{0} - 4\binom{3}{3}\binom{3}{1}\binom{34}{0} - \binom{4}{4}\binom{36}{0}\right\}\right.$$
$$\left. - 2\cdot4\binom{4}{4}\binom{36}{0} - 4\cdot4\binom{3}{3}\binom{3}{1}\binom{34}{0} + \binom{10}{3}\binom{30}{1} - \binom{4}{4}\binom{36}{0}\right]$$
$$= 9,870. \tag{22.73}$$

It follows that the number of ways for A♠-K♠ to win is 172,430.

Subcase 4c. $r = 4$ and the board includes the K♣. There are $\binom{1}{1}\binom{7}{0}\binom{40}{4} = 91,390$ ways altogether. A tie occurs if the board shows all clubs (four of a kind is a win for A♠-K♠), and the number of ways is

$$\binom{1}{1}\binom{7}{0}\binom{10}{4}\binom{30}{0} = 210. \tag{22.74}$$

The number of ways for 8♡-8◇ to win is 1,380, as in subcase 4a. It follows that the number of ways for A♠-K♠ to win is 89,800.

Subcase 4d. $r = 4$ and the board includes the K♡ or K◇. There are $\binom{2}{1}\binom{6}{0}\binom{40}{4} = 182,780$ ways altogether. A tie occurs if the board shows a king-high straight flush, and the number of ways is

$$\binom{2}{1}\binom{6}{0}\binom{4}{4}\binom{36}{0} = 2. \tag{22.75}$$

The number of ways for 8♡-8◇ to win is 9,870, as in subcase 4b. It follows that the number of ways for A♠-K♠ to win is 172,908.

Subcase 4e. $r = 4$ and the board includes the 8♠. There are $\binom{1}{1}\binom{7}{0}\binom{40}{4} = 91,390$ ways altogether. A tie occurs if the board shows a straight or straight flush except a queen-high one and except one with three or four spades or four hearts or four diamonds. Thus, the number of ways for a tie to occur is

$$\binom{1}{1}\binom{7}{0}4\left[\binom{4}{1}^4\binom{24}{0} - 6\binom{2}{2}\binom{3}{1}^2\binom{32}{0} - 4\binom{3}{3}\binom{3}{1}\binom{34}{0}\right]$$

$$-2\binom{4}{4}\binom{36}{0}\Big] = 752. \tag{22.76}$$

Here A♠-K♠ wins if the board shows four of a kind or three, four, or five spades with no pairs or trips on the board (an exception is a straight flush using the 8♠, jack-high or lower), or if A♠-K♠ has an ace-high or ace-low straight with no pairs or trips on the board. The condition of no pairs or trips on the board can be omitted if A♠-K♠ has a royal flush. If four hearts or four diamonds appear on the board, it is a win for 8♡-8◊. Therefore, the number of ways for A♠-K♠ to win is

$$\binom{1}{1}\binom{7}{0}\Big[10\binom{4}{4}\binom{36}{0} + \binom{10}{2}\Big\{\binom{6}{0}\binom{24}{2} - 8\binom{3}{2}\binom{27}{0}\Big\}$$
$$+ \binom{10}{3}\binom{9}{0}\binom{21}{1} + \binom{3}{3}\binom{9}{1}\binom{28}{0} + \Big\{\binom{10}{4} - 4\binom{4}{4}\binom{6}{0}\Big\}\binom{30}{0}$$
$$+ \binom{4}{1}^3\binom{28}{1} + \binom{4}{1}^4\binom{24}{0} - 3\binom{1}{1}\binom{7}{1}\binom{3}{1}^2\binom{26}{0}$$
$$- 3\binom{2}{2}\binom{3}{1}\binom{21}{1}\binom{14}{0} - 3\binom{2}{2}\binom{3}{1}\binom{7}{1}\binom{28}{0} - \binom{3}{3}\binom{21}{1}\binom{16}{0}$$
$$- \binom{3}{3}\binom{7}{1}\binom{30}{0} - 6\binom{2}{2}\binom{3}{1}^2\binom{32}{0} - 4\binom{3}{3}\binom{3}{1}\binom{34}{0}$$
$$- \binom{4}{4}\binom{36}{0} - 2\binom{3}{3}\binom{7}{1}\binom{30}{0} - 2\binom{4}{4}\binom{36}{0}\Big] = 15{,}581. \tag{22.77}$$

Of the last 10 subtracted terms, the first five correspond to ace-high straights for A♠-K♠ that have been counted twice, the next three to ace-low straights that have been counted twice, and the last two to cases in which 8♡-8◊'s flush beats A♠-K♠'s straight. It follows that the number of ways for 8♡-8◊ to win is 75,057.

Subcase 4f. $r = 4$ and the board includes the 8♣. There are $\binom{1}{1}\binom{7}{0}\binom{40}{4} = 91{,}390$ ways altogether. A tie occurs as in subcase 4e, and in addition if the board consists of all clubs. Thus, the number of ways for a tie to occur is

$$\binom{1}{1}\binom{7}{0}\Big[\binom{10}{4}\binom{30}{0} + 4\Big\{\binom{4}{1}^4\binom{24}{0} - 4\binom{3}{3}\binom{3}{1}\binom{34}{0} - \binom{4}{4}\binom{36}{0}$$
$$- 3\binom{4}{4}\binom{36}{0}\Big\}\Big] = 1{,}170. \tag{22.78}$$

Here A♠-K♠ wins under the same conditions as in subcase 4e, except that "three, four, or five" should be replaced by "three or four." Therefore, the number of ways for A♠-K♠ to win is

$$\binom{1}{1}\binom{7}{0}\Big[10\binom{4}{4}\binom{36}{0} + \binom{10}{3}\binom{9}{0}\binom{21}{1} + \binom{3}{3}\binom{9}{1}\binom{28}{0}$$

$$+ \binom{10}{4}\binom{30}{0} + \binom{4}{1}^{3}\binom{28}{1} + \binom{4}{1}^{4}\binom{24}{0} - 3\binom{2}{2}\binom{3}{1}\binom{7}{1}\binom{28}{0}$$

$$- \binom{3}{3}\binom{21}{1}\binom{16}{0} - \binom{3}{3}\binom{7}{1}\binom{30}{0} - 4\binom{3}{3}\binom{3}{1}\binom{34}{0}$$

$$- \binom{4}{4}\binom{36}{0} - 2\binom{3}{3}\binom{7}{1}\binom{30}{0} - 2\binom{4}{4}\binom{36}{0}$$

$$- \binom{3}{3}\binom{7}{1}\binom{30}{0} - \binom{4}{4}\binom{36}{0}\Bigg] = 4{,}669, \tag{22.79}$$

where the last two terms correspond to club flushes on the board. It follows that the number of ways for 8♡-8♢ to win is 85,551.

Combining the results of subcases 4a–4f, we find that the number of ways for A♠-K♠ to win is 544,938, the number of ways for 8♡-8♢ to win is 183,108, and the number of ways for a tie to occur is 3,074.

Case 5. $r = 5$. There are $\binom{8}{0}\binom{40}{5} = 658{,}008$ ways altogether. Again, a tie occurs when the highest five-card hand both for A♠-K♠ and for 8♡-8♢ comprises the five community cards. With $r = 5$, there are no aces, kings, or 8s among them. A tie occurs if the board shows a spade straight flush except for a queen-high one (7-high and 6-high are the only ones that exclude aces, kings, and 8s), a heart or diamond straight flush except for a 7-high one (hence 6-high only), a full house with a pair of 9s or higher, a club flush or straight flush, or a straight, excluding 7-high straights (hence 6-high only), excluding straights with three or four spades, four hearts, or four diamonds, and of course excluding the four 6-high straight flushes that were already counted. Thus, the number of ways for a tie to occur is

$$\binom{8}{0}\Bigg[2\binom{5}{5}\binom{35}{0} + 2\binom{5}{5}\binom{35}{0} + 4\cdot 9\binom{4}{2}\binom{4}{3}\binom{32}{0} + \binom{10}{5}\binom{30}{0}$$

$$+ \binom{4}{1}^{5}\binom{20}{0} - 10\binom{3}{3}\binom{3}{1}^{2}\binom{31}{0} - 3\cdot 5\binom{4}{4}\binom{3}{1}\binom{33}{0}$$

$$- 4\binom{5}{5}\binom{35}{0}\Bigg] = 2{,}005. \tag{22.80}$$

Continuing with case 5, the number of ways for A♠-K♠ to win with a royal flush or an ace-low straight flush is

$$\binom{8}{0}\Bigg[\binom{3}{3}\binom{37}{2} + \binom{4}{4}\binom{1}{0}\binom{35}{1}\Bigg] = 701. \tag{22.81}$$

The number of ways for A♠-K♠ to win with four of a kind is

$$\binom{8}{0}10\binom{4}{4}\binom{36}{1} = 360. \tag{22.82}$$

The number of ways for A♠-K♠ to win with a flush is

$$\binom{8}{0}\left[\left\{\binom{10}{3}-\binom{3}{3}\binom{7}{0}\right\}\left\{\binom{30}{2}-3\binom{3}{2}\binom{27}{0}\right\}+\left\{\binom{10}{4}-\binom{3}{3}\binom{7}{1}\right.\right.$$
$$\left.-\binom{4}{4}\binom{6}{0}\right\}\binom{30}{1}+\left\{\binom{10}{5}-\binom{3}{3}\binom{7}{2}-\binom{4}{4}\binom{1}{0}\binom{5}{1}\right.$$
$$\left.\left.-2\binom{5}{5}\binom{5}{0}\right\}\binom{30}{0}\right]=56{,}978, \tag{22.83}$$

because three, four, or five spades on the board suffice, provided trips do not appear and straight flushes are excluded.

The number of ways for A♠-K♠ to win with an ace-high or ace-low straight is

$$\binom{8}{0}\left[\binom{4}{1}^3\binom{28}{2}-\left\{\binom{3}{3}\binom{9}{0}\binom{28}{2}+3\binom{2}{2}\binom{3}{1}\binom{7}{0}\left(\binom{7}{1}\binom{21}{1}\right.\right.\right.$$
$$\left.+\binom{7}{2}\binom{21}{0}\right)+3\binom{1}{1}\binom{3}{1}^2\binom{5}{0}\binom{7}{2}\binom{21}{0}+2\binom{3}{3}\binom{9}{0}\binom{7}{1}\binom{21}{1}$$
$$+2\cdot3\binom{2}{2}\binom{3}{1}\binom{7}{0}\binom{7}{2}\binom{21}{0}+3\binom{3}{3}\binom{9}{0}\binom{7}{2}\binom{21}{0}\right\}$$
$$+3\binom{4}{2}\binom{4}{1}^2\binom{28}{1}-3\left\{\binom{3}{3}\binom{3}{1}\binom{6}{0}\binom{28}{1}\right.$$
$$+\binom{2}{2}\binom{3}{2}\binom{7}{0}\binom{7}{1}\binom{21}{0}+2\binom{2}{2}\binom{3}{1}^2\binom{4}{0}\binom{7}{1}\binom{21}{0}$$
$$+2\binom{3}{3}\binom{3}{1}\binom{6}{0}\binom{7}{1}\binom{21}{0}\right\}+3\binom{4}{2}^2\binom{4}{1}\binom{28}{0}$$
$$-3\binom{3}{3}\binom{3}{1}^2\binom{3}{0}\binom{28}{0}+\binom{4}{1}^4\binom{4}{0}\binom{20}{1}$$
$$-\left\{\binom{4}{4}\binom{12}{0}\binom{4}{0}\binom{20}{1}+4\binom{3}{3}\binom{3}{1}\binom{10}{0}\binom{4}{0}\binom{20}{1}\right.$$
$$+6\binom{2}{2}\binom{3}{1}^2\binom{4}{0}\binom{5}{1}\binom{15}{0}+2\binom{4}{4}\binom{12}{0}\binom{4}{0}\binom{5}{0}\binom{15}{1}$$
$$+2\cdot4\binom{3}{3}\binom{3}{1}\binom{10}{0}\binom{4}{0}\binom{5}{1}\binom{15}{0}+3\binom{4}{4}\binom{12}{0}\binom{4}{0}\binom{5}{1}\binom{15}{0}\right\}$$
$$+4\binom{4}{2}\binom{4}{1}^3\binom{24}{0}-4\left\{\binom{4}{4}\binom{3}{1}\binom{9}{0}\binom{24}{0}+3\binom{3}{3}\binom{3}{1}^2\binom{7}{0}\binom{24}{0}\right.$$
$$+\binom{3}{3}\binom{3}{2}\binom{10}{0}\binom{24}{0}+2\binom{4}{4}\binom{3}{1}\binom{9}{0}\binom{24}{0}\right\}\right]=34{,}455. \tag{22.84}$$

This requires explanation. The first three positive terms count hands with Q-J-T, resulting in an ace-high straight, and the next two positive terms count hands with 5-4-3-2, resulting in an ace-low straight. But some of these hands qualify as flushes or straight flushes, while others give 8♡-8◇ a full house or a heart or diamond flush or give the board a heart, diamond, or club flush, and these counts must be subtracted off. To make it easier to follow, the subtracted terms appear right after the terms they are subtracted from.

Finally, the only other way in which A♠-K♠ can win is with two pair, both of which are 9 or above, and the number of ways is

$$\binom{8}{0}\left[6\left\{\binom{4}{2}^2\binom{32}{1}-\binom{2}{2}\binom{3}{1}^2\binom{8}{1}\binom{24}{0}\right\}-3\binom{4}{2}^2\binom{4}{1}\binom{28}{0}\right.$$
$$\left.+3\binom{3}{3}\binom{3}{1}^2\binom{31}{0}\right]=6{,}075. \tag{22.85}$$

Note that we have subtracted hands that result in a flush or a straight, adding back once those that were subtracted twice.

In summary, the number of ways for A♠-K♠ to win in case 5 is 98,569. It follows that the number of ways for 8♡-8◇ to win is 557,434.

Summing the results for cases 0–5, we find that

$$P(\text{A♠-K♠ wins}) = \frac{813{,}957}{1{,}712{,}304} \approx 0.475358, \tag{22.86}$$

$$P(\text{8♡-8◇ wins}) = \frac{892{,}317}{1{,}712{,}304} \approx 0.521121, \tag{22.87}$$

$$P(\text{a tie occurs}) = \frac{6{,}030}{1{,}712{,}304} \approx 0.003522. \tag{22.88}$$

The advantage of this combinatorial approach, compared to computer enumeration, is that it gives a much better idea of the wide variety of scenarios that can arise. The disadvantages are that it is much more time-consuming and prone to error. (The author used his own computer-enumeration program[1] to confirm this calculation, so it would be inaccurate to say that the calculation was done entirely "by hand.")

We have just evaluated the probabilities for *one* of the

$$\frac{1}{2}\binom{52}{2,2,48}=812{,}175 \tag{22.89}$$

unordered pairs of initial hands in Texas hold'em, or what we will call heads-up matchups between initial hands. (The factor of $\frac{1}{2}$ accounts for the fact that A♠-K♠ vs. 8♡-8◇ is the same as 8♡-8◇ vs. A♠-K♠.) But this number does not take equivalence into account. Let us call an unordered pair of initial hands *equivalent* if they have the same four denominations and if the corresponding denominations have the same suits after a permutation of

---

[1] The program first creates a 48-card deck—the full deck less the four hole cards—and then deals out all 1,712,304 possible boards. For each one, it evaluates the two resulting seven-card hands, finding the highest five-card hand in each and comparing them.

($\clubsuit,\diamondsuit,\heartsuit,\spadesuit$). Then, as we will see, there are only 47,008 equivalence classes of heads-up matchups between initial hands.

On the other hand, there are

$$\binom{2 + 169 - 1}{2} = 14{,}365 \tag{22.90}$$

heads-up matchups between equivalence classes of initial hands (Theorem 1.1.8 on p. 6). Notice that there are two equivalence relations here, one on initial hands that results in 169 equivalence classes, and the other on unordered pairs of initial hands that results in 47,008 equivalence classes. Table 22.2 may help to clarify the terminology.

Table 22.2 Terminology involving initial hands at Texas hold'em.

term	number
initial hands	1,326
example: A$\spadesuit$-K$\spadesuit$	
equivalence classes of initial hands	169
example: A-K suited	
heads-up matchups between initial hands	812,175
example: A$\spadesuit$-K$\spadesuit$ vs. 8$\heartsuit$-8$\diamondsuit$	
equivalence classes of heads-up matchups between initial hands	47,008
example: A-K suited vs. pocket 8s with no suit matches	
heads-up matchups between equivalence classes of initial hands	14,365
example: A-K suited vs. pocket 8s	

To further elaborate, one of the heads-up matchups between equivalence classes is A-K suited vs. pocket 8s. The evaluation of the probabilities for this confrontation requires the observation that two equivalence classes of heads-up matchups contribute. One can be represented by A$\spadesuit$-K$\spadesuit$ vs. 8$\heartsuit$-8$\diamondsuit$, for which the results were obtained in (22.86)–(22.88), while the other can be represented by A$\spadesuit$-K$\spadesuit$ vs. 8$\spadesuit$-8$\heartsuit$. The distinction is no suit matches between the hands in the first case and one in the second case. In the case of A$\spadesuit$-K$\spadesuit$ vs. 8$\spadesuit$-8$\heartsuit$, we use our computer program to get the answers quickly. The results are

$$P(\text{A}\spadesuit\text{-K}\spadesuit \text{ wins}) = \frac{806{,}758}{1{,}712{,}304} \approx 0.471153, \tag{22.91}$$

$$P(\text{8}\spadesuit\text{-8}\heartsuit \text{ wins}) = \frac{897{,}979}{1{,}712{,}304} \approx 0.524427, \tag{22.92}$$

$$P(\text{a tie occurs}) = \frac{7{,}567}{1{,}712{,}304} \approx 0.004419. \tag{22.93}$$

For A-K suited vs. pocket 8s, it suffices to take a weighted average of these two cases. But what are the weights? By permuting suits if necessary, we can assume that the suit of our suited A-K is spades. Then, of the $\binom{4}{2} = 6$ equally likely ways in which distinct suits can be assigned to the pocket 8s, three of them include no spades, while three others include one spade. Thus, we can weight the two cases, (22.86)–(22.88) and (22.91)–(22.93), equally. We conclude that

$$P(\text{A-K suited wins}) = \frac{1,620,715}{3,424,608} \approx 0.473256, \qquad (22.94)$$

$$P(\text{pocket 8s wins}) = \frac{1,790,296}{3,424,608} \approx 0.522774, \qquad (22.95)$$

$$P(\text{a tie occurs}) = \frac{13,597}{3,424,608} \approx 0.003970. \qquad (22.96)$$

Table 22.3 includes the necessary information to evaluate the probabilities for each of the 14,365 heads-up matchups between equivalence classes of initial hands. To explain this table in more detail, for each of the three types of equivalence classes of initial hands (pocket pair, suited, and unpaired unsuited), we have listed all possible opposing equivalence classes of initial hands. There are 28 subtypes of heads-up matchups of equivalence classes of initial hands. For each subtype we have counted the number of ways to choose the denominations in the second hand (column (a)), assuming that the denominations in the first hand have been specified, as well as the number of ways to choose the suits for each choice of denominations (column (b)). If, for example, we denote by $(a)_1 \cdot (b)_1$ the inner product of column (a) and column (b) for the type 1 entries only, we find that

$$\binom{13}{1}(a)_1 \cdot (b)_1 + \binom{13}{2}[(a)_2 \cdot (b)_2 + (a)_3 \cdot (b)_3] = 1,624,350, \qquad (22.97)$$

which is twice (22.89), as one would expect. The point is that A-K suited vs. 8-8 and 8-8 vs. A-K suited both appear in Table 22.3 (rows 2c and 1e). We have also counted the number of equivalence classes for each choice of denominations (column (c)). Here we find

$$\binom{13}{1}[(a)_1 \cdot (c)_1 + 1] + \binom{13}{2}[(a)_2 \cdot (c)_2 + 1 + (a)_3 \cdot (c)_3 + 2] = 94,016, \qquad (22.98)$$

which is twice the number of equivalence classes of unordered pairs of initial hands. The three extra terms, 1, 1, and 2, arise from the fact that (a, a) vs. (a, a), (a, b)s vs. (a, b)s, and (a, b)u vs. (a, b)u each appear only once in the table, so an adjustment must be made before dividing by 2. Thus, 47,008 is the minimum number of computations of the form (22.86)–(22.88) that must be performed to analyze each of the 14,365 possible heads-up matchups between equivalence classes of initial hands. (In fact it may be simpler, albeit more time-consuming, to work with equivalence classes of *ordered* pairs of initial hands.)

Once we have analyzed each of these heads-up matchups between equivalence classes of initial hands, we can use this information to answer questions of the following sort: If a player has A-K suited before the flop, what are his probabilities of winning, losing, and tying in heads-up play against a *random* hand, assuming neither player folds before showdown? These probabilities will be weighted averages of the probabilities against specific opponents, and the weights are shown in Table 22.3.

The results of this computation are that, when A-K suited competes heads up against a random hand,

$$P(\text{A-K suited wins}) = \frac{1,389,004,215}{2,097,572,400} \approx 0.662196, \qquad (22.99)$$

$$P(\text{random hand wins}) = \frac{673,957,209}{2,097,572,400} \approx 0.321303, \qquad (22.100)$$

$$P(\text{a tie occurs}) = \frac{343,610,976}{2,097,572,400} \approx 0.016500. \qquad (22.101)$$

Here the denominators are $\binom{50}{2,5,43} = 2,097,572,400$. A useful way to summarize these numbers in a single statistic is in terms of expected profit per unit bet:

$$\begin{aligned}
&\text{expected profit for A-K suited heads-up against a random hand} \\
&= (1)P(\text{A-K suited wins}) + (-1)P(\text{random hand wins}) \\
&= \frac{715,047,006}{2,097,572,400} \approx 0.340893. \qquad (22.102)
\end{aligned}$$

Although it requires some time, we can repeat this computation for each of the 169 equivalence classes of initial hands and then rank the equivalence classes by expectation. We have done this in Table 22.4, though to save space and because the numbers are readily available from other sources, we have not listed the expectations. Notice, for example, that A-K suited ranks ahead of 7-7, though in a heads-up matchup between these two hands, the latter wins with probability greater than $\frac{1}{2}$.

This ranking applies only to heads-up play, and even in that setting some authors have raised concerns about its applicability. We briefly discuss this issue in the chapter notes.

Next, we would like to extend this ranking of equivalence classes of initial hands to the case of $n \geq 2$ opponents, where $n = 9$ would be of particular interest. One approach would be, given an initial hand, to deal random hands to all n opponents, assume that none of the $n + 1$ players folds before the

Table 22.3 Ordered pairs of equivalence classes of initial hands in Texas hold'em. s stands for suited, and u stands for unsuited. Column (a) shows the number of ways to choose the denominations in the second hand, assuming that the denominations in the first hand have been specified. Column (b) shows the number of ways to choose the suits in both hands for each choice of denominations. Column (c) shows the number of equivalence classes for each choice of denominations. In the last three columns are the sizes of the various equivalence classes arranged according to the number of suit matches between the two hands. Throughout, a, b, c, d are distinct denominations. For 1e and 1f, $b > c$; for 2 and 3, $a > b$; for 2j, 2k, 3j, and 3k, $c > d$.

type	hands	(a)	(b)	(c)	no. of suits in both hands		
					0	1	2
1a	(a,a) vs. (a,a)	1	6	1	6		
1b	(a,a) vs. (b,b)	12	36	3	6	24	6
1c	(a,a) vs. (a,b)s	12	12	1	12		
1d	(a,a) vs. (a,b)u	12	36	2	12	24	
1e	(a,a) vs. (b,c)s	66	24	2	12	12	
1f	(a,a) vs. (b,c)u	66	72	4	12	24, 24	12
2a	(a,b)s vs. (a,a)	1	12	1	12		
2b	(a,b)s vs. (b,b)	1	12	1	12		
2c	(a,b)s vs. (c,c)	11	24	2	12	12	
2d	(a,b)s vs. (a,b)s	1	12	1	12		
2e	(a,b)s vs. (a,b)u	1	24	1	24		
2f	(a,b)s vs. (a,c)s	11	12	1	12		
2g	(a,b)s vs. (a,c)u	11	36	2	24	12	
2h	(a,b)s vs. (b,c)s	11	12	1	12		
2i	(a,b)s vs. (b,c)u	11	36	2	24	12	
2j	(a,b)s vs. (c,d)s	55	16	2	12	4	
2k	(a,b)s vs. (c,d)u	55	48	3	24	12, 12	
3a	(a,b)u vs. (a,a)	1	36	2	12	24	
3b	(a,b)u vs. (b,b)	1	36	2	12	24	
3c	(a,b)u vs. (c,c)	11	72	4	12	24, 24	12
3d	(a,b)u vs. (a,b)s	1	24	1	24		
3e	(a,b)u vs. (a,b)u	1	84	4	24	24, 24	12
3f	(a,b)u vs. (a,c)s	11	36	2	24	12	
3g	(a,b)u vs. (a,c)u	11	108	5	24	24, 24, 24	12
3h	(a,b)u vs. (b,c)s	11	36	2	24	12	
3i	(a,b)u vs. (b,c)u	11	108	5	24	24, 24, 24	12
3j	(a,b)u vs. (c,d)s	55	48	3	24	12, 12	
3k	(a,b)u vs. (c,d)u	55	144	7	24	24, 24, 24, 24	12, 12

Table 22.4 Ranking of equivalence classes of Texas hold'em initial hands by expectation versus a random hand in heads-up play to the showdown. s stands for suited and u stands for unsuited.

1.	AA	35.	A4s	69.	K5u	103.	96s	137.	85u
2.	KK	36.	A7u	70.	J9u	104.	J2s	138.	64s
3.	QQ	37.	K8s	71.	K2s	105.	Q2u	139.	83s
4.	JJ	38.	A3s	72.	Q5s	106.	T5s	140.	94u
5.	TT	39.	QJu	73.	T8s	107.	J5u	141.	75u
6.	99	40.	K9u	74.	K4u	108.	T4s	142.	82s
7.	88	41.	A5u	75.	J7s	109.	97u	143.	73s
8.	AKs	42.	A6u	76.	Q4s	110.	86s	144.	93u
9.	77	43.	Q9s	77.	Q7u	111.	J4u	145.	65u
10.	AQs	44.	K7s	78.	T9u	112.	T6u	146.	53s
11.	AJs	45.	JTs	79.	J8u	113.	95s	147.	63s
12.	AKu	46.	A2s	80.	K3u	114.	T3s	148.	84u
13.	ATs	47.	QTu	81.	Q6u	115.	76s	149.	92u
14.	AQu	48.	44	82.	Q3s	116.	J3u	150.	43s
15.	AJu	49.	A4u	83.	98s	117.	87u	151.	74u
16.	KQs	50.	K6s	84.	T7s	118.	T2s	152.	72s
17.	66	51.	K8u	85.	J6s	119.	85s	153.	54u
18.	A9s	52.	Q8s	86.	K2u	120.	96u	154.	64u
19.	ATu	53.	A3u	87.	22	121.	J2u	155.	52s
20.	KJs	54.	K5s	88.	Q2s	122.	T5u	156.	62s
21.	A8s	55.	J9s	89.	Q5u	123.	94s	157.	83u
22.	KTs	56.	Q9u	90.	J5s	124.	75s	158.	42s
23.	KQu	57.	JTu	91.	T8u	125.	T4u	159.	82u
24.	A7s	58.	K7u	92.	J7u	126.	93s	160.	73u
25.	A9u	59.	A2u	93.	Q4u	127.	86u	161.	53u
26.	KJu	60.	K4s	94.	97s	128.	65s	162.	63u
27.	55	61.	Q7s	95.	J4s	129.	84s	163.	32s
28.	QJs	62.	K6u	96.	T6s	130.	95u	164.	43u
29.	K9s	63.	K3s	97.	J3s	131.	T3u	165.	72u
30.	A5s	64.	T9s	98.	Q3u	132.	92s	166.	52u
31.	A6s	65.	J8s	99.	98u	133.	76u	167.	62u
32.	A8u	66.	33	100.	87s	134.	74s	168.	42u
33.	KTu	67.	Q6s	101.	T7u	135.	T2u	169.	32u
34.	QTs	68.	Q8u	102.	J6u	136.	54s		

showdown, and then evaluate the given hand's expectation (splitting the pot in the case of ties). The number of games that would have to be analyzed is

$$\binom{50}{2, 2, \ldots, 2, 5, 50 - (2n + 5)}, \qquad (22.103)$$

which is about $1.9 \cdot 10^{12}$ for $n = 2$, $1.6 \cdot 10^{15}$ for $n = 3$, and $2.4 \cdot 10^{31}$ for $n = 9$. (The order of the n opponents is irrelevant, so we may be able to divide this number by $n!$.) And we would have to do this for each of the 169 equivalence classes of initial hands. Not surprisingly, an exact analysis has not been done, even for $n = 2$ (a three-player game), though simulation studies have been.

We propose an alternative approach. We assume that the specified hand goes to the showdown heads up against the best of his n opponents. This associates an expectation with the specified hand, and the resulting expectations allow us to rank the 169 equivalence classes of initial hands in a game with $n+1$ players. But what do we mean by the "best" of the specified hand's opponents? We assume that the 169 equivalence classes of initial hands in an n-player game have already been ranked. In other words, we are describing a recursive procedure.

There is still the computational issue corresponding to (22.103). To get around this, we propose an approximation. Instead of sampling the n hands without replacement from a 50-card deck, we select them with replacement (independently) from the 50-card deck. Although this approximation will introduce some inaccuracy, we do not believe that it will be significant. Still, our results for $n = 9$ opponents remain incomplete, and are therefore not reported here. See the chapter notes for further discussion.

We conclude this section with two examples of how this kind of information can be useful. First, however, we want to describe how Bayes's law (Theorem 1.2.7 on p. 19) can be used to make decisions in Texas hold'em. For simplicity, we assume that the game has reached the point of heads-up play. Therefore, the only unknown, from player 1's perspective, is player 2's initial hand. There are $\binom{50}{2} = 1{,}225$ possibilities and, without additional information, they are equally likely. However, it may be sufficient to consider the 169 equivalence classes of initial hands, or perhaps even a cruder classification (e.g., one that does not distinguish between suited and unsuited hands). We let $H_1, H_2, \ldots, H_n$ be events describing player 2's initial hand, and we assume that they partition the sample space. Their probabilities, calculated under the assumption that the $\binom{50}{2}$ possible hands are equally likely, are the *prior probabilities*. Let A be the event describing the actions of the player's opponent. We assume that the player can estimate the conditional probabilities $P(A \mid H_i)$ for $i = 1, 2, \ldots, n$. Then, by Bayes's law, the *posterior probabilities* are

$$P(H_j \mid A) = \frac{P(H_j)P(A \mid H_j)}{\sum_{i=1}^{n} P(H_i)P(A \mid H_i)}, \qquad j = 1, 2, \ldots, n. \qquad (22.104)$$

It is often assumed (perhaps unrealistically) that player 1 can "put" player 2 on a range of hands, that is, there is an index set I such that

$$P(A \mid H_i) = \begin{cases} 1 & \text{if } i \in I, \\ 0 & \text{if } i \notin I, \end{cases} \tag{22.105}$$

then the posterior probabilities are proportional to the relevant prior probabilities:

$$P(H_j \mid A) = \frac{P(H_j)}{\sum_{i \in I} P(H_i)}, \qquad j \in I. \tag{22.106}$$

In our first example, we consider a 10-player game of no-limit Texas hold'em. The small blind is one unit and the big blind is two units. The players are numbered 1–10 clockwise from the small blind to the button, and we view the game from the perspective of player 10, who holds Q♠-Q♡. Player 3 calls the big blind and player 4 raises to 10 units. Players 5–9 call and player 10 raises to 80 units. Players 1–3 fold and player 4 re-raises all-in to 150 units. Players 5–9 fold, and it is up to player 10. What should player 10 do now?

The key is whether player 10 can read player 4. Player 4 raised from early position and re-raised player 10's substantial raise. He must have a very strong hand, particularly if he is known to be a cautious player. If player 10 can put player 4 on A-A (the best possible equivalence class), he can calculate as in (22.94)–(22.96) that

$$P(\text{Q-Q wins}) = \frac{312{,}251}{1{,}712{,}304} \approx 0.182357, \tag{22.107}$$

$$P(\text{A-A wins}) = \frac{1{,}392{,}614}{1{,}712{,}304} \approx 0.813298, \tag{22.108}$$

$$P(\text{a tie occurs}) = \frac{7{,}439}{1{,}712{,}304} \approx 0.004344. \tag{22.109}$$

Here the decision is straightforward. Ignoring the rake, the pot is offering player 10 odds of 285 to 70, or about 4.071 to 1. On the other hand, his odds against winning the pot are (cf. Problem 22.1 on p. 725)

$$\left(\frac{312{,}251}{1{,}712{,}304} + \frac{1}{2} \frac{7{,}439}{1{,}712{,}304} \right)^{-1} - 1 \approx 4.419 \tag{22.110}$$

to 1. Here the correct decision is to fold.

Suppose instead that player 10 can put player 4 on A-A, K-K, or Q-Q (the three best possible equivalence classes). Then he can calculate that, in a Q-Q vs. K-K matchup

$$P(\text{Q-Q wins}) = \frac{305{,}351}{1{,}712{,}304} \approx 0.178328, \tag{22.111}$$

$$\mathrm{P}(\text{K-K wins}) = \frac{1{,}398{,}993}{1{,}712{,}304} \approx 0.817024, \tag{22.112}$$

$$\mathrm{P}(\text{a tie occurs}) = \frac{7{,}960}{1{,}712{,}304} \approx 0.004649, \tag{22.113}$$

whereas in a Q$\spadesuit$-Q$\heartsuit$ vs. Q$\diamondsuit$-Q$\clubsuit$ matchup, $\mathrm{P}(\text{Q}\spadesuit\text{-Q}\heartsuit \text{ wins}) + \frac{1}{2}\mathrm{P}(\text{a tie occurs}) = \frac{1}{2}$ by symmetry. Now the probability p that player 10 wins is a weighted average over the three possible opponents, with A-A and K-K weighted 6/13 each and Q-Q weighted 1/13. We get

$$P = \frac{6}{13}\left(\frac{312{,}251}{1{,}712{,}304} + \frac{1}{2}\frac{7{,}439}{1{,}712{,}304}\right) + \frac{6}{13}\left(\frac{305{,}351}{1{,}712{,}304} + \frac{1}{2}\frac{7{,}960}{1{,}712{,}304}\right)$$
$$+ \frac{1}{13}\frac{1}{2} = \frac{1{,}535{,}987}{7{,}419{,}984} \approx 0.207007, \tag{22.114}$$

with the result that the odds are $(1 - P)/P \approx 3.831$ to 1 against. Here it is correct to call.

For our second example, we consider an eight-player game of no-limit Texas hold'em. Again, the small blind is one unit and the big blind is two units. The players are numbered 1–8 clockwise from the small blind to the button, and we view the game from the perspective of player 5, who holds 8$\heartsuit$-8$\clubsuit$. Player 1 is known to be *loose* (he plays a lot of hands) and *aggressive* (he bets and raises more than he checks and calls). Table 22.5 shows the betting history up to the point at which player 5 must decide whether to call an all-in bet by player 1 at the turn.

Table 22.5 A game of no-limit Texas hold'em. Player 5 holds 8$\heartsuit$-8$\clubsuit$. Player 1 is known to be loose and aggressive. What should player 5 do? (Bracketed numbers reflect effective bet and pot sizes, owing to the fact that player 5 starts with a shorter stack than does player 1.)

	betting history								pot
player	1sb	2bb	3	4	5	6	7	8b	
chip stack	500	145	300	360	440	100	250	160	
pre-flop	1	2	6	6	6	6	6	f	
	5	f							38
flop: T$\spadesuit$-8$\spadesuit$-4$\diamondsuit$	0		20	20	20	f	f		
	80		60	f	60				298
turn: J$\spadesuit$	414		f		?				712
	[354]								[652]

sb = small blind, bb = big blind, b = button, f = fold, 0 = check

The goal is to infer what player 1 is holding, based on his known style of play and his play in the present game. Of course, we cannot say with certainty what player 1's cards are, but we can narrow down the possibilities. What follows is one player's analysis. Others might argue differently.

Pre-flop, player 1 called a raise from the small blind after four others had called. A loose-aggressive player might do this with just about anything, except that he would have re-raised with the premium hands A-A, K-K, Q-Q, J-J, and A-K. So we rule out those possibilities.

After the flop, player 1 check-raised with three players remaining, and this suggests a high flush draw or a flush/straight combination draw. He would likely have just called with a low flush draw or a straight draw.

After the turn, his all-in bet suggests he has made his flush, in which case his hole cards consist of two spades, at least one of which is 9 or higher. Another possibility is that he has T-T or 4-4 (we have already ruled out J-J), giving him a set.

A final possibility is that he has two pair, with J-T, T-8, T-4, or 8-4 in the hole (J-8 and J-4 are unlikely, as he would not have check-raised on the flop in those cases). But this is less likely than a set, because if he were beaten with a flush, he would have only four outs on the river, as opposed to 10 outs with T-T or 4-4.

If we put player 1 on a flush or a set, we must count the number of ways each such hand can occur and the number of outs player 5 has against each hand. If we include also the four possible cases of two pair, we will get different results. The numbers are summarized in Table 22.6.

Table 22.6 Distribution of hands assigned to player 1 by player 5 in the game described by Table 22.5. See text.

player 1 hand	number of ways	player 5 outs
straight flush (Q♠-9♠, 9♠-7♠)	2	0
flush without 4♠ (not both 8 or less, not A♠-K♠)	23	10
flush with 4♠ (not both 8 or less)	4	9
T-T	3	1
4-4	3	43
J-T	9	40
T-8	3	42
T-4	9	42
8-4	3	44

Player 5 has put 86 units into the pot, so he has only 354 with which to call. Therefore, player 1's all-in bet is effectively only 354 units and hence the actual pot size is only 652 units. (Again, we ignore the rake.) That is, player 5 is getting pot odds of 652 to 354, or about 1.842 to 1.

What are his odds against winning? We first use (22.106) with the 35 hands specified in Table 22.5 above the line. The probability that player 5 wins is

$$P_0 := \frac{2}{35}\frac{0}{44} + \frac{23}{35}\frac{10}{44} + \frac{4}{35}\frac{9}{44} + \frac{3}{35}\frac{1}{44} + \frac{3}{35}\frac{43}{44} = \frac{199}{770}, \tag{22.115}$$

which amounts to odds of 571 to 199 against, or about 2.869 to 1. Here player 5 should fold.

If, however, we include the four two-pair hands among player 1's possibilities, now there are 59 hands, and (22.115) becomes

$$P_1 := \frac{35}{59}P_0 + \frac{9}{59}\frac{40}{44} + \frac{3}{59}\frac{42}{44} + \frac{9}{59}\frac{42}{44} + \frac{3}{59}\frac{44}{44} = \frac{697}{1{,}298}, \tag{22.116}$$

which amounts to odds of 601 to 697 against, or about 0.862 to 1. In this case, player 5 should call.

Notice that we get different conclusions, depending on a choice that relies heavily on judgment and experience. Such is the nature of poker.

22.4 Problems

22.1. *Pot odds with the possibility of a tie.* We saw in (22.1) on p. 692 that if the pot odds exceed the odds against winning the hand, a call offers positive expectation. Show that, if the probabilities of winning, losing, and tying the hand are $p > 0$, $q > 0$, and $r \geq 0$, where $p + q + r = 1$, and if a tie results in winning half the pot, then the same result holds, except that the odds are those corresponding to the probability $P := p + \frac{1}{2}r$.

22.2. *A decision with complete information.* In a game of Texas hold'em, player 1 holds A♣-T♣ in a heads-up game. The flop comes Q♣-J♣-2♡, and the turn card is the 4♠. Player 2 bets b units, bringing the pot to a units, and then turns over his cards by mistake, thinking player 1 had folded, and reveals Q♠-Q◇. Under what conditions on a and b should player 1 call?

22.3. *A pot-odds calculation.* A player holds T♠-9♠ in Texas hold'em, and the flop comes A♠-6◇-2♣. The pot has a units and he is contemplating calling a bet of b units. Assuming he wins if and only if he makes a *backdoor flush* (i.e., a spade at the turn and another at the river), under what conditions on a and b is a call justified? Assume also a structured-limit game in which the bet at the turn will be $2b$ units.

22.4. *Five-card poker probabilities with a bug.* How are the five-card poker probabilities in Table 1.3 on p. 8 affected by the presence of a bug, that is, a joker that may be used to fill a straight, flush, or straight flush and is otherwise an ace?

22.5. *An unusual event in five-card stud.* In a heads-up game of five-card stud, assuming that neither player folds before the showdown, find the odds against the event that the winning hand is a straight flush and the losing hand is a full house. Such a game occurred in the novel *The Cincinnati Kid*.

22.6. *Four of a kind in Omaha hold'em.* Use combinatorial analysis to find the probability that an Omaha hold'em hand is ranked four of a kind, assuming as usual that it is not folded before the showdown.
    Hint: The answer is *not*

$$\binom{13}{1}\frac{\binom{4}{2,2,0}\binom{48}{2,3,43} + \binom{4}{1,3,0}\binom{48}{3,2,43}}{\binom{52}{4,5,43}}. \tag{22.117}$$

For example, consider A♠-A♡-K♠-K♡/A♢-A♣-K♢-K♣-Q♠, which has four aces *and* four kings, and A♠-A♡-K♣-Q♣/A♢-A♣-J♣-T♣-2♡, which has four aces but is ranked as a straight flush.

22.7. *Lowball draw frequencies.* Find the pre-draw frequencies in lowball draw (a) without a joker, and (b) with a joker. (c) Do the same for deuce-to-seven lowball (without a joker). Events include 5-high, 6-high, ..., K-high, and other in (a) and (b), and 7-high, 8-high, ..., A-high, and other in (c).

22.8. *Razz frequencies.* In razz, players are dealt seven cards with which to make the lowest five-card poker hand. Straights, flushes, and straight flushes do not count, and aces are low. Table 22.7 shows the numbers of each of ten categories. Derive combinatorial formulas for these numbers. (Cf. Problem 1.8 on p. 54.)

Table 22.7 Razz frequencies. Assumes the hand is not folded before the showdown.

rank	no. of ways	rank	no. of ways
5-high	781,824	T-high	23,675,904
6-high	3,151,360	J-high	24,837,120
7-high	7,426,560	Q-high	21,457,920
8-high	13,171,200	K-high	13,939,200
9-high	19,174,400	other	6,169,072
sum			133,784,560

22.9. *Saddle points for basic endgame.* Consider Model n of Section 22.2. Show that, if

$$P \geq Q_n := \frac{2a + b_1 + \cdots + b_n}{2a + 2b_1 + \cdots + 2b_n}, \qquad (22.118)$$

then there exists at least one saddle point. Find all saddle points. Show that these results are consistent with the conclusions established in Section 22.2.

Hint: Verify that $Q_n \geq P_n$, with strict inequality if $n \geq 2$. .

22.10. *Basic endgame with pot-limit betting.* Find formulas for the optimal strategies in the basic endgame Model n when all bets are assumed maximal under a pot-limit betting structure. Take $a = 1$.

22.11. *Basic endgame with three-card deck.* Two players, 1 and 2, ante a units each, where $a > 0$ is specified. Each player is then dealt one card at random (without replacement) from a three-card deck $\{1, 2, 3\}$, and each sees only his own card. It is assumed that play begins with a check by player 2. Player 1 may then check or bet b units, where $b > 0$ is specified. If he checks, there is a showdown. If he bets, player 2 may fold or call. If player 2 folds, player 1 wins the pot. If player 2 calls, there is a showdown. In a showdown, the player with the higher card wins. Formulate the game as an 8×8 matrix game, eliminate dominated pure strategies to reduce it to a 2×2 matrix game, and find a solution of the game.

22.12. *Kuhn's poker model.* Two players, 1 and 2, ante a units each, where $a > 0$ is specified. Each player is then dealt one card at random (without replacement) from a three-card deck $\{1, 2, 3\}$, and each sees only his own card. Player 1 may check or bet b units, where $b > 0$ is specified. If player 1 checks, player 2 may check or bet b units. If player 1 bets, player 2 may fold or call. If player 1 checks and player 2 bets, then player 1 may fold or call. If either player folds, the other wins the pot. If both players check or if either calls a bet, there is a showdown. In a showdown, the player with the higher card wins.

(a) Formulate the game as a 27×64 matrix game, and eliminate dominated pure strategies to reduce it to an 8×4 matrix game.

(b) Find a solution of the game (it is not necessarily unique).

Hint: (a) Some dominations are obvious: Never call with a 1, never fold with a 3. Eliminate such strategies before evaluating the payoff matrix.

(b) Consider five cases, depending on whether b/a belongs to $(0, 1)$, $\{1\}$, $(1, 2)$, $\{2\}$, or $(2, \infty)$; it may be possible to combine some of these cases. Restrict attention to a 3×3 submatrix since the reduced payoff matrix has rank three (its four columns are linearly dependent).

22.13. *Color poker.* Two players, 1 and 2, ante one unit each. Each player is dealt one card at random (without replacement) from a standard 52-card deck, and each sees only his own card. In this game, black cards (♣ and ♠) are ranked above red cards (◇ and ♡), but cards of the same color are equivalent. Initially, player 2 checks. Player 1 may check (forcing a showdown) or bet one unit. If player 1 bets, player 2 may fold (thereby losing), call (forcing a showdown), or raise by betting two units. If player 2 raises, player 1 may fold (thereby losing) or call (forcing a showdown). In a showdown, the high card wins the pot; in the case of a tie, the pot is split. Formulate the game as a 9×9 matrix game and solve the game.

Hint: For $i, j = 1, 2, 3$, strategy (i, j) for either player is the strategy that risks (including the ante) up to i units if one's card is red and up to j units if one's card is black.

22.14. *Fold or raise all-in at one-card poker.* Two players, 1 and 2, have equal stacks of $S = 10$ units. Player 1 posts a small blind of one unit, and player 2 posts a big blind of two units. Each player is dealt one card at random (without replacement) from a standard 52-card deck, and each sees only his own card. Player 1 acts first and may fold or raise all-in. If he raises all-in, player 2 may fold or call. If either player folds, the other wins the pot. If player 1 raises all-in and player 2 calls, there is a showdown, in which case the player with the higher card wins the pot of $2S$ units. (In case of a tie, the pot is split.) Formulate this as a 14×14 matrix game (cf. Example 5.1.2 on p. 162), eliminate dominated pure strategies, and find a solution of the game.

Also, investigate the game for other values of S.

For reasons of symmetry, it is convenient to regard player 1's payoff as his profit over the course of the game. Specifically, if player 1 folds, his payoff is -1; if player 1 raises and player 2 folds, player 1's payoff is 2; if player 1 raises and player 2 calls, player 1's payoff is $\pm S$ or 0. Notice that, if there were no blind bets, the game would be symmetric, and player 1 could assure a profit of 0 by folding regardless of his hand.

22.15. *Probabilities of flopping various hands with unsuited connectors.* Assuming unsuited connectors in Texas hold'em with maximal reach, specifically J-T unsuited, T-9 unsuited, ..., or 5-4 unsuited, find the probability of flopping (a) a straight, (b) an open-ended straight draw, (c) an inside straight draw (excluding the case J-T), and (d) a double inside straight draw.

22.16. *Probabilities of flopping various hands with suited connectors.* Assuming suited connectors in Texas hold'em with maximal reach, specifically J-T suited, T-9 suited, ..., or 5-4 suited, find the probability of flopping (a) a straight, (b) a flush, (c) a straight flush, (d) a straight, flush, or straight flush, (e) an open-ended straight draw (exclude a made flush), (f) an inside straight draw (exclude a made flush), (g) a double inside straight draw (exclude a made flush), and (h) a flush draw (include a made straight), (i) an open-ended straight-flush draw (include a made straight or a made flush), and (j) an inside straight-flush draw (include a made straight or a made flush).

22.17. *Probabilities of various hands by the turn and by the river.* Recalculate the probabilities (22.50)–(22.62) on pp. 706–707 of making various hands on the flop, instead finding the probabilities of making them by the turn; ... by the river. Make sure that the probabilities sum to 1 in both cases.

22.18. *A dramatic game of Texas hold'em.* In heads-up play, one player is dealt A♣-A♢ and the other K♡-K♠. The flop comes K♣-Q♣-J♣. The player with three kings moves all in and the player with a royal flush draw calls. Which player is favored, and by how much?

22.19. *Probability of a board that allows flushes.*

(a) Find the probability, from the perspective of an observer with no knowledge of the players' hole cards, that the five-card board in Texas hold'em has three or more cards of some suit and therefore allows for the possibility that some player has a flush.

(b) Find the probability, from the perspective of a player with suited hole cards, that the five-card board in Texas hold'em has three or more cards of some suit.

(c) Find the probability, from the perspective of a player with unsuited hole cards, that the five-card board in Texas hold'em has three or more cards of some suit.

22.20. *Probabilities of overcards when holding a pocket pair.* Given that a Texas hold'em player is dealt pocket tens, find the conditional probability that the flop has at least one *overcard* (i.e., a card of a higher denomination). Find the conditional probability that the full board has at least one overcard. Repeat for each pocket pair, 2s, 3s, . . . , Ks.

22.21. *Probability of an opposing ace.* Given that a player in a game of Texas hold'em with n opponents $(1 \leq n \leq 9)$ is dealt m aces $(m = 0, 1, 2)$, find the probability that at least one opponent is dealt at least one ace.

22.22. *Distribution of aces.* In a 10-handed game of Texas hold'em, find the probabilities that the four aces are distributed over the 10 two-card hands in each of the following forms: none; 1; 2; 1, 1; 1, 2; 1, 1, 1; 2, 2; 1, 1, 2; 1, 1, 1, 1. The results should sum to 1. Repeat for nine-handed games.

22.23. *Probability of duplicate pocket pairs.* In a 10-handed game of Texas hold'em, find the probability that, for at least one denomination, two players are dealt pocket pairs of that denomination.

Notice that the problem is equivalent to finding the probability that, for at least one denomination, two splits of that denomination occur in the first 10 turns of a faro deal (cf. Section 18.2).

22.24. *Probability of a better ace.* Given that a player in a game of Texas hold'em with n opponents $(1 \leq n \leq 9)$ is dealt A-K, find the probability that at least one opponent is dealt a better ace, that is, one with a higher kicker (including pocket aces). Repeat for A-Q, A-J, . . . , A-2.

22.25. *Combinatorial analysis of initial hands.* In Texas hold'em, suppose that player 1's hole cards are A♠-3♣ and player 2's are A♢-2♢. Assuming that neither player folds before the showdown, use combinatorial analysis to determine the probability that player 1 wins; that player 2 wins; that a tie occurs.

Hint: As in Section 22.3, treat six cases, depending on the number of cards r, among the five on the board, that are other than aces, deuces, and treys.

22.26. *Best and worst initial hands vs. pocket aces.*

(a) Which initial hand (not just which equivalence class of initial hands) in Texas hold'em performs worst in heads-up play against pocket aces?

(b) Which initial hand, other than A-A, performs best in heads-up play against pocket aces?

In each case, find the probabilities of winning, losing, and tying with the specified hand, assuming that neither hand is folded before the showdown.

22.27. *Best and worst initial-hand matchups.*

(a) Find two equivalence classes of initial hands A and B in Texas hold'em that maximize the expectation

$$E_{A,B} := \mathrm{P}(A \text{ beats } B) - \mathrm{P}(B \text{ beats } A), \qquad (22.119)$$

assuming that neither hand is folded before the showdown.

(b) Solve part (a) not for equivalence classes of initial hands, but for initial hands.

(c) Find two equivalence classes of initial hands A and B that minimize the expectation (22.119) over all ordered pairs (A, B) for which the expectation is positive.

(d) Solve part (c) not for equivalence classes of initial hands, but for initial hands.

22.28. *Number of equivalence classes of fair heads-up matchups.* Of the 47,008 equivalence classes of heads-up matchups at Texas hold'em, how many have $E_{A,B} = 0$, in the notation of Problem 22.27?

Hint: There are no examples other than the obvious ones.

22.29. *Nontransitivity of initial hands.* Find three equivalence classes of initial hands, A, B, and C, in Texas hold'em such that, in the notation of Problem 22.27, $E_{A,B} > 0$, $E_{B,C} > 0$, and $E_{C,A} > 0$.

Hint: The classic example is A-K unsuited, J-T suited, and 2-2.

22.30. *Frequency of three nontransitive initial hands.* How many sets of three equivalence classes of initial hands, A, B, and C, in Texas hold'em have the nontransitive property of Problem 22.29 in the sense that either $E_{A,B} > 0$, $E_{B,C} > 0$, and $E_{C,A} > 0$, or $E_{A,B} < 0$, $E_{B,C} < 0$, and $E_{C,A} < 0$. This requires checking all $\binom{169}{3} = 790{,}244$ cases.

22.31. *Sklansky–Chubukov numbers.* Evaluate the Sklansky–Chubukov numbers (defined in the chapter notes and illustrated with K-K) for the equivalence classes of initial Texas hold'em hands A-A, Q-Q, A-K suited, and A-K unsuited.

22.5 Notes

The history of poker has been discussed by Foster (1905, pp. 1–16), Asbury (1938, Chapter 2), Parlett (1991, Chapters 8–9), and Wilson (2008). A definitive work by McManus (2009) has just appeared. Poker most likely originated in early-19th-century New Orleans, but initially it was very similar to the 18th- and 19th-century Persian game of as-nas. Coffin (1955, p. 14) attributed the game to

> French inhabitants [of New Orleans] who had been in the French Service in Persia *circa* 1800–20.

In as-nas each player was dealt five cards from a 20-card deck (four each of lion, king, lady, soldier, and dancing girl). A single betting round followed, with the high hand winning the showdown. The ranks of the hands were listed by Foster (1905, p. 5): "Four of a kind; three of a kind and a pair; three of a kind; two pairs; one pair." It is often said that as-nas is an ancient game but Parlett (1991, p. 112) debunked this notion.

In the original form of poker, which McManus (2009, p. 55) called "old poker," each player was dealt five cards from a 20-card deck (A, K, Q, J, T of each suit). A single betting round followed, with the high hand winning the showdown. Straights, flushes, and straight flushes were not yet recognized. The earliest poker game described in print (Cowell 1844, p. 94) was of this type, occurring in December 1829 on a Mississippi steamboat. Green (1843, p. 95) also described the 20-card game. An earlier reference to poker appeared in 1836 in a publication titled *Dragoon Campaigns to the Rocky Mountains*, said to have been authored by James Hildreth (1836, p. 128) or William L. Gordon-Miller. There poker was said to be "A favorite game of cards at the south and west."

Several other "vying" games, of European origin, are ancestral to poker and perhaps as-nas. These include the 18th-century French game of poque, of which "poker" is likely an American mispronunciation, and its ancestor, the 15th-century German game of poch. Others include the French games of ambigu, bouillotte, and brelan, the English games of brag and post-and-pair, and the Italian game of primiera. See Parlett (1991, Chapter 8) for more detail.

Poker spread north by steamboat on the Mississippi River. By 1837 it was adapted to the standard 52-card deck (Coffin 1955, p. 15). Draw poker was introduced in the 1840s, perhaps by influence of the English game brag, which had this feature. It was immediately seen to be superior to old poker, depending more on skill than chance, and it also expedited the demise of the 20-card deck. A subsequent refinement, jackpots, wherein a pair of jacks or better was required to open, with new antes and a new deal if no one opened, is said to have originated in about 1870 in Toldeo, Ohio, as a solution to the problem of overly cautious players.

The rules of draw poker were first put in writing by the Hon. Robert C. Schenck [1809–1890], an American Ambassador to England, at the request of a hostess at a country house in Somersetshire he visited in 1872. Hands were ranked as follows (Schenck 1880, pp. 13–14): Sequence flush, fours, full, flush, sequence, threes, two pairs, one pair, and none of the above. These differ from the modern ranks (Example 1.1.10) in name only. Jackpots were not included in Schenck's rules.

Flushes were recognized by the 1840s, having been included in an 1845 Hoyle (Anners 1845, p. 261). According to Schneir (1993, p. 13), straights were introduced around 1855, but it was not until about 1880 that they were universally accepted. As Trumps (c. 1875, p. 179) put it,

> Straights are not considered in the game, although they are played in some localities, and it should always be determined whether they are to be admitted at the commencement of the game.

Trumps (p. 180) went on to make the case for straights and straight flushes:

> It is strongly urged by some players, that the strongest hand at Draw Poker should be a *Straight, or Royal Flush*, for the reason that it is more difficult to get than four of a kind, and removes from the game the objectionable feature of a known invincible hand. It is *impossible* to tie four Aces or four Kings and an Ace, but it is *possible* for four Straight Flushes of the same value to be out in the same deal. No gentleman would care to bet on a "sure thing," and we therefore think that the Straight Flush should be adopted when gentlemen play at this game.

Modern poker players have no qualms about betting on a sure thing.

Table stakes with sidepots were firmly established by about 1860, although a decade earlier we find, in Bohn (1851, p. 382),

> Not having funds enough to meet the stake put in the pool, entitles you to see an adversary's hand for such an amount as you have.

Prior to that, especially on Mississippi steamboats, the stakes were often unlimited. Keller (1887, p. 51) explained the meaning of this:

> If two or more men agree to play the unlimited game, it is understood that each player is prepared to call any raise that any other player may make. If any player should make a bet to an amount greater than the sum of money immediately at the command of another player, and this second player should desire to call, the second player may have twenty-four hours in which to procure the money necessary to call the first player. In the meantime the cards are to be sealed up and lodged in hands satisfactory to both the players or all the players.

As Winterblossom (1875, p. 56) put it,

> The truth is, there is no science whatever necessary in the unlimited game; it is purely a question of intimidation. The limited game, on the contrary, is highly scientific.

In old poker it was the custom to let the winner of each pot deal the next hand. As Asbury (1938, pp. 27–28) explained,

The practice of passing the deal to the left after each hand was inaugurated soon after the introduction of the draw, which gave the dealer at least a moral advantage, and led to the picturesque custom of using a buck, which originated on the Western frontier during the late 1860's or early 1870's. The buck could be any object, but was usually a knife, and most Western men in those days carried knives with buckhorn handles, hence the name. As first used, the buck simply marked the deal; it was placed in front of the dealer, and passed along at the conclusion of each pot. In some sections a player who didn't wish to deal was permitted to ante and "pass the buck." In gambling-house Stud, where the deal never changed, the buck was used to indicate which player received the first card.

The buck was later replaced by a silver dollar, which is why "buck" is a slang term for "dollar." U.S. President Harry S. Truman [1884–1972], who was a poker player, had a sign on his desk that read "The buck stops here." Today the role of the buck is played by the button, but it indicates the nominal dealer, not the player who receives the first card.

Around the time of the Civil War, five-card stud poker was introduced. It was originally called "stud-horse poker" for reasons that were explained in an 1890 Sacramento newspaper article summarized by Singsen (1988):

[S]tud-horse poker evolved out of a high-stakes Civil War era draw poker game between a Mississippi riverboat gambler McCool and his wealthy New Orleans counterpart Brady. Whenever the gamblers played, they agreed to turn "low cards" face up to see who would buy the next drink. On one such occasion, after the first card had been dealt, McCool called for the next card to be turned up for another round of drinks. A four was dealt to both players, each made large bets, and McCool called for another card to be turned. Once again, both players received the same card, a five. McCool bet again, Brady called, and a third card was turned up. This time, a six fell to both players. After another betting round, two deuces were dealt face up[, one] to each player. At this point McCool, whose first down card was a six, felt confident with his pair. Brady, though, had a three in the hole and his straight was a sure winner. The players agreed to dispense with the draw and play out the hand. Eventually, all of McCool's cash was on the table, and he couldn't meet Brady's $5,000 raise. McCool's last possession was a magnificent black stallion and he offered to put the horse up to match the final bet. Brady accepted, turned over his three, and McCool "disappeared permanently from the river. ... the first man to lose big money at stud-horse poker. ..."

It seems likely that this is an apocryphal story. Stud has some advantages over draw. For example, there are four betting rounds instead of two, and more skill is said to be required (see Oliver P. Carriere's introduction to the book by Dowling 1970, p. 14). However, draw remained the more popular game throughout the remainder of the 19th century.

An 1885 amendment to California Penal Code section 330 prohibited "stud-horse poker" and several other games of chance, but did not specifically prohibit draw poker. This was interpreted to mean that stud poker, as well as variants such as Texas hold'em, were illegal in California's card rooms, while draw poker and it variants, such as lowball, were legal. Finally, in 1987, a Los Angeles court ruled that all forms of poker were legal. See Singsen (1988) for the full story.

A possible ancestor of stud poker was trijaques ("three jacks"), described and analyzed by Jacob Bernoulli (1713, Part 3, Problem 18). This game was not mentioned by Parlett (1991), perhaps because an English translation of Bernoulli's *Ars Conjectandi* (1713) has only recently been published. A 24-card deck (9–A) was used with 9s and the J♣ wild. Two cards were dealt face down to each player, followed by a betting round. Then two cards were dealt face up to each player, followed by a second betting round. The best hands, in order, were four of a kind, flush, and three of a kind. One pair and two pair were not recognized.

Livingston (1971, p. 72) wrote, "In American poker, the standard big-money game for the past fifty years has been Five-Card Stud." Gradually, the five-card version was superseded by the seven-card version. In 1967 Morehead (1967, p. 25) wrote,

> No game has lost popularity as rapidly as [five-card stud]. Thirty years ago two-thirds of the professional games were five-card stud; today, not one-tenth of the games are.

Five-card stud was dropped from the World Series of Poker in 1975, and five-card draw was dropped in 1983 (McManus 2009, pp. 142, 148).

Other developments in the evolution of poker include the introduction of the joker by 1875, at which time it was called the mistigris, and the introduction of lowball, high-low games, and Texas hold'em around the turn of the 20th century.

According to Johnny Moss (Grotenstein and Reback 2005, p. 8), Texas hold'em was already popular in Dallas in the 1920s. House Concurrent Resolution No. 109 of the Texas State Legislature (May 11, 2007) tells us that it was first played in the early 1900s. The last two paragraphs of the seven-paragraph resolution read as follows:

> WHEREAS, It is said that Texas Hold'em takes a minute to learn and a lifetime to master, and this telling statement underscores the high level of skill necessary to win consistently; a successful hold'em player relies on reason, intuition, and bravado, and these same qualities have served many notable Texans well throughout the proud history of the Lone Star State; now, therefore, be it
>
> RESOLVED, That the 80th Legislature of the State of Texas hereby formally recognize Robstown, Texas, as the birthplace of the poker game Texas Hold'em.

The earliest published reference to Texas hold'em, as far as we know, is Livingston (1968), where it is said that the game was called "hold me darling" or "Tennessee hold me" in 1968, at which time it was only a few years old.

Crandell Addington (Brunson 2005, p. 77) noted that high-stakes no-limit Texas hold'em was introduced to Nevada in the 1960s by a small group of Texans, led by Felton "Corky" McCorquodale, and including Doyle "Texas Dolly" Brunson [1933–], Bryan "Sailor" Roberts [1931–1995], Johnny Moss [1907–1995], Thomas "Amarillo Slim" Preston [1928–], Jack "Treetop" Straus [1930–1988], and Crandell Addington [1938–].

As Moss put it (Alvarez 2002, p. 28), "Hold'em is to stud and draw what chess is to checkers." Brunson (2002, p. 419) termed no-limit hold'em "the Cadillac of Poker games."

Indeed, limit hold'em is "a mechanical game," while the no-limit game "takes in everything" (Brunson 2002, p. 440). Alvarez (2002, p. 36) explained the distinction in more detail, saying that most top professionals "look down on limit poker as an unimaginative, mechanical game."

> Jack Straus described it contemptuously as a "disciplined job," saying, "Anybody who wants to work out the mathematics can be a limit player and chisel out an existence. You just have to condition yourself to sit there and wait." Serious players, he meant, know the odds of filling a straight or a flush or a full house with one or two cards to come. In limit poker, where they also know precisely how much this will cost them and how much money the pot will be offering if they call or raise or are reraised, every move can be reduced to mathematics and probabilities.

Grotenstein and Reback (2005, pp. 19–20) went further, saying,

> While limit poker tends to be a very mathematical game whose correct play is usually dictated by odds, no-limit incorporates a larger palette of skills, including controlled aggression, a deep understanding of human psychology, and a certain elevated class of courage that players refer to simply as "heart." Perhaps Crandell Addington, an accomplished no-limit player from San Antonio, Texas, put it best when he said,
>
> "Limit poker is a science, but no-limit is an art. In limit you're shooting at a target. In no-limit the target comes alive and shoots back at you."

A product of Texas gamblers, Texas hold'em has some colorful terminology. An inside straight draw is either a "gutshot" or a "belly-buster" straight draw, and, as we have already noted, the first player to act is "under the gun." It is possible that "the turn" was borrowed from horseracing, though Livingston (1971, Chapter 8) referred to all of the cards on the board as turn cards because they are turned over by the dealer. The term "the river" has a more interesting origin (McManus 2009, p. 146):

> That the seventh card [in seven-card stud] was dealt facedown gave rise to the nickname "Down-the-River" with its echo of "sold down the river," which we now understand to mean cruelly betrayed.[2] (The phrase originally referred to the fate of troublesome slaves in border states punitively sold to planters in the Deep South, where working conditions were even harsher.) Down-the-River was duly shortened to "the river," which eventually stuck as a synonym for the final community card in hold'em and Omaha, even though that card is always dealt faceup.

How did Texas hold'em become the most popular form of poker played today? There are several contributing factors, not the least of which are the World Series of Poker, Internet poker, and the World Poker Tour.

The idea for the World Series of Poker may have originated in 1949, when Nicholas (Nick the Greek) Dandolos [1883–1966] arrived in Las Vegas looking to play high-stakes heads-up poker. Benny Binion [1904–1989], proprietor of

---

[2] In fact, "Down-the-River," was an alternative name for seven-card stud in the 1930s (Jacoby 1947, p. 69).

Binion's Horseshoe Club, recognized the publicity value and arranged for a suitable opponent in poker legend Johnny Moss. The marathon session, played publicly, lasted from January to May. Games played included five-card stud, five-card draw, seven-card stud, seven-card high-low, and lowball, both ace-to-five and deuce-to-seven, but not Texas hold'em (Gordon and Grotenstein 2004, p. 9). It finally came to an end when Nick the Greek, said to be down at least two million dollars and thoroughly exhausted, softly spoke the famous line, "Mr. Moss, I have to let you go." (Alvarez 2002, pp. 27–32.)[3]

A hand from that match is one of the most famous hands in poker history. The game was five-card stud and both players had about \$250,000 in front of them. After the first two cards were dealt, Moss had a 9 in the hole and an exposed 6; the Greek had an exposed 7. In Johnny Moss's own words, as told to A. Alvarez in 1981, here is what happened (Alvarez 2004, p. 96):

> Low man brings it in. I bet two hunnerd with a six, he raises fifteen hunnerd or two thousand, I call him. The next card comes, I catch a nine, he catches a six. I got two nines then. I make a good bet—five thousand, maybe—an' he plays back at me, twenty-five thousand. I jus' call him. I'm figurin' to take all that money of his, an' I don't wanna scare him none. The next card comes, he catches a trey, I catch a deuce. Ain't nuttin' he got can beat my two nines. I check then to trap him, an' he bets, jus' like I wanted. So I raise him wa-ay up there, an' he calls. I got him in there, all right. There's a hunnerd thousand dollars in that pot—maybe more; I don't know exactly—an' I'm a-winnin' it. On the end I catch a trey, he catches a jack. He's high now with the jack an' he bets fifty thousand. I cain't put him on no jack in the hole, you know. He ain't gonna pay all that money jus' for the chance to outdraw me. I don't care what he catches, he's gotta beat those two nines of mine. So I move in with the rest of my money.

From Moss's perspective, the situation was like this:

$$\text{Moss} \quad (9)\text{-}6\text{-}9\text{-}2\text{-}3 \qquad \text{Greek} \quad (?)\text{-}7\text{-}6\text{-}3\text{-}J. \qquad (22.120)$$

In another of his famous lines, the Greek then said, "Mr. Moss, I think I have a jack in the hole." He called Moss's bet and turned over a jack, winning the half-million-dollar pot. Why did the Greek say he *thought* he had a jack in the hole? Amarillo Slim explained (Preston and Dinkin 2003, p. 139):

> You see, great poker players—and the Greek certainly was one—play their *opponent's* hand, not their own. When the Greek made up his mind on third street that Johnny was on a bluff, he didn't look back at his own hand; he didn't need to.

In 1969 the Texas Gamblers Reunion, organized by Texan Tom Moore and by Vic Vickrey [1926–2009] was held at the Holiday Hotel in Reno. It consisted of a series of high-stakes poker matches among the world's best

---

[3] Accounts differ as to the year. Preston and Dinkin (2003, p. 136) and Grotenstein and Reback (2005, p. 15) put it at 1951. But Binion's Horseshoe Club did not open until August, 1951 (Preston and Dinkin 2003, p. 134), so perhaps the match took place at the Westerner, in which Binion was a partner before he opened the Horseshoe. In any case, there is a curious lack of contemporaneous evidence supporting the occurrence of this widely acclaimed event.

players. After learning that the very successful Reunion would not be repeated, Benny Binion decided to host the World Series of Poker at Binion's Horseshoe Club in 1970, and it has been an annual event ever since. In 1970 Johnny Moss was voted world champion by his fellow players, but ever since 1971 a freezeout-style tournament has been used to crown the world champion. The World Series is actually a series of tournaments, but the main event is the $10,000-buy-in no-limit Texas hold'em tournament. Each player starts with $10,000, and the tournament ends when one player has all the money, which is then apportioned among the top finishers. The tournament grew steadily for the first 30 years or so (7 entries in 1970, 73 in 1980, 194 in 1990, and 512 in 2000) and explosively in recent years (839 entries in 2003, 2,576 in 2004, 5,619 in 2005, and 8,773 in 2006).

What explains the tripling of entries in the World Series of Poker main event in 2004? Certainly, one possible explanation is the fact that the $2.5 million first prize in the 2003 main event was won by an amateur with the appropriate name of Chris Moneymaker [1975–]. He earned his $10,000 entry fee in online satellite tournaments with an investment of only $40. His success inspired many amateur players. His own source of inspiration was the 1998 film *Rounders* (screenwriters David Levien and Brian Koppelman) about the underground world of high-stakes poker. It featured a classic scene in which Mike McDermott (Matt Damon) defeated Teddy KGB (John Malkovich) in a heads-up game of no-limit Texas hold'em.

Another explanation is Internet poker, which has also experienced explosive growth since its debut in the late 1990s. Recently, this growth has slowed in the U.S. due to the UIGEA of 2006, the Unlawful Internet Gambling Enforcement Act.

But the most likely explanation is the *World Poker Tour*. This is a series of poker tournaments held in various casinos around the world, in which the play at the final table is filmed, edited, and broadcast at a later date. Beginning on March 30, 2003, it became a hugely successful program on the Travel Channel. Part of its success was due to its use of "hole-card cameras" that revealed to the television audience each player's hole cards. (See Berman and Karlins 2005, Chapter 5.) The idea is often attributed to *Late Night Poker*, broadcast on British Channel 4 from 1999 to 2002, which used a glass table so that the players' hole cards could be filmed from below (Gordon and Grotenstein 2004, p. 17).[4] Both of these programs featured Texas hold'em, perhaps at least in part because it is an easier game to televise than seven-card stud, for example, what with all of the exposed cards being in one place.

Poker has played important roles in literature and film. Just about everyone's favorite literary treatment of poker is Al Alvarez's [1929–] (2002) *The Biggest Game in Town*, first published in 1983, which chronicled the 1981

---

[4] Actually, the concept is due to Henry Orenstein, who was awarded a patent in 1995 for an optical scanning device that would display to the audience the players' hole cards in a poker tournament (U.S. patent 5,451,054).

World Series of Poker. There are also a number of more recent books in the same genre.

In fiction, Richard Jessup's [1925–1982] novel *The Cincinnati Kid* (1963, pp. 132–141) featured a remarkable game of five-card stud, in which the "Cincinnati Kid" faced Lancey "The Man" Hodges. (In a 1965 film based on the novel, their roles were played by Steve McQueen and Edward G. Robinson.) After three cards had been dealt to each, the situation was like this:

$$\text{The Kid}\quad (Q\heartsuit)\text{-T-T}\qquad \text{Lancey}\quad (?)\text{-}7\heartsuit\text{-}8\heartsuit. \qquad (22.121)$$

The Kid bet \$500 to steal a \$250 pot, and Lancey raised \$300. The Kid re-raised \$2,000 and Lancey called. Fourth cards were a queen for the Kid and the T$\heartsuit$ for Lancey. The Kid bet \$1,000 and Lancey called. Fifth cards were another queen for the Kid and the 9$\heartsuit$ for Lancey. To summarize,

$$\text{Kid}\quad (Q\heartsuit)\text{-T-T-Q-Q}\qquad \text{Lancey}\quad (?)\text{-}7\heartsuit\text{-}8\heartsuit\text{-}T\heartsuit\text{-}9\heartsuit. \qquad (22.122)$$

The Kid went all-in with \$1,420, Lancey raised all-in to \$5,520, offering to accept the Kid's marker for \$4,100. (Evidently, table stakes were not played.) The Kid called, and the pot contained \$14,790 plus the marker. Lancey turned over the J$\heartsuit$ for a straight flush, beating the Kid's full house. The betting has been criticized by Spanier (1990, pp. 26–29) as amateurish. (Specifically, Lancey's \$300 raise, his \$2,000 call, the Kid's \$1,000 bet, and his \$1,420 bet were all found wanting.) See Problem 22.5.

In Section 19.4 we quoted from Ian Fleming's *Casino Royale* (1953) a description of a high-stakes baccarat game between James Bond and Le Chiffre. When *Casino Royale* was made into a 2006 film (screenwriters Neal Purvis, Robert Wade, and Paul Haggis), the Cold War had become the war on terrorism and the baccarat game had become a no-limit Texas hold'em tournament. The buy-in was \$10 million (with a re-buy of \$5 million) and 10 players participated. At the final hand (see Table 22.8), four players remained: Bond (Daniel Craig) was in the small blind (\$500,000), player 2 was in the big blind (\$1 million), Felix Leiter (Jeffrey Wright) was under the gun, and Le Chiffre (Mads Mikkelsen) was on the button. The flop came A$\heartsuit$-8$\spadesuit$-6$\spadesuit$ and the turn card was the 4$\spadesuit$. The pot contained \$24 million and the four players each checked. The river card was the A$\spadesuit$. Bond checked, player 2 went all-in with \$6 million, and Leiter went all-in with \$5 million. Le Chiffre raised to \$12 million, and Bond raised all-in to \$40.5 million. Le Chiffre called, putting himself all-in. Player 2 turned over K$\spadesuit$-Q$\spadesuit$ for the top flush. Leiter turned over 8$\heartsuit$-8$\clubsuit$ for a full house, eights full of aces. Le Chiffre turned over A$\clubsuit$-6$\heartsuit$ for a higher full house, aces full of sixes. Finally, Bond turned over 7$\spadesuit$-5$\spadesuit$ for a straight flush, winning the \$115 million pot.

Poker has even contributed to art. Artists inspired by poker include Thomas Hart Benton [1889–1975], LeRoy Neiman [1921–], and several others (see Flowers and Curtis 2000). But perhaps best known is the series of nine paintings of dogs playing poker by Cassius Marcellus Coolidge [1844–1934],

Table 22.8 Summary of the final hand of the Texas hold'em tournament in the 2006 film *Casino Royale*.

	betting history				pot
player:	James Bond	player 2	F. Leiter	Le Chiffre	
	small blind	big blind	utg	button	
chip stack:	$46.5m	$12m	$11m	$45.5m	
blinds:	$0.5m	$1m			
pre-flop:	(betting not specified)				
flop: A♡-8♠-6♠	(betting not specified; each has bet $6m)				$24m
turn: 4♠	check	check	check	check	$24m
river: A♠	check $40.5m ai [$39.5m]	$6m ai	$5m ai	$12m $27.5m ai	$115m [$114m]
hole cards:	7♠-5♠	K♠-Q♠	8♡-8♣	A♣-6♡	

utg = under the gun, m = million, ai = all in

created in 1903. *A Friend in Need*, for example, shows seven dogs seated around a poker table playing five-card draw (and the bulldog is cheating). In 2005 two of the paintings sold at auction for $590,400. See McManus (2009, Chapter 21) for further discussion.

As noted in Section 1.7, the five-card poker probabilities were published by 1879. Trumps (c. 1875, p. 179) ranked straights below three of a kind, and Foster (1905, p. 13) noted that "almost all the questions submitted to Wilkes's *Spirit of the Times* from 1870 to 1874 were disputes about the proper rank of straights." Blackbridge (1880), who called poker "the most interesting game at cards the human mind has yet devised," provided three derivations. The first one, reprinted from the May 9, 1879, issue of the *Rambler*, is due to William Pole, Fellow of the Royal Society, but it gave erroneous results for the numbers of hands ranked three of a kind and one pair. (In fact, Pole's results were apparently published earlier; see Trumps c. 1875, p. 512.) Specifically, Pole calculated the number of hands ranked three of a kind as

$$\binom{13}{1}\binom{4}{3}\binom{49}{2} - \binom{13}{1}\binom{48}{1} - \binom{13}{1}\binom{4}{3}\binom{12}{1}\binom{4}{2} = 56{,}784. \quad (22.123)$$

Because of the factor $\binom{49}{2}$, the first term counts hands ranked four of a kind *four times each* as well as hands ranked full house once each. Thus, the second term should have read $4\binom{13}{1}\binom{48}{1}$. In a letter to the editor of the *Rambler* dated May 23, 1879, William Hoffman, a Lieutenant in the U.S. Army from Fort Concho, Texas, provided the correct numbers, possibly for the first time, and his letter was also reprinted by Blackbridge (1880). Finally, in a second 1880 printing of Blackbridge, an appendix titled "Poker principles and chance laws" by Richard A. Proctor [1837–1888], a British astronomer, was added. This gave the first correct published derivation (not just the correct numbers), as far as we know. Only five years earlier, Henry T. Winterblossom (1875), a professor of mathematics, did not include these probabilities in his book *The Game of Draw-Poker, Mathematically Illustrated*. Of course it was about the same time (1880) that straights and straight flushes became universally accepted. One might imagine that the Blackbridge book settled the issue, but as late as July 25, 1882, another letter to the editor of the *Rambler*, from George S. Clark, argued that threes (three of a kind) should be ranked above straights (Fisher 1934a, p. 14):

> Dear Sir:
>
> Professor Robert Schenk [Schenck], in his remarks on our great national pastime of Poker arbitrarily places a sequence, or Straight, as the fifth hand in the decreasing scale of relative values of the nine grades of hands. He says: "Many experts rate Threes in relative value above a sequence, but in my opinion a sequence should rank first, as being in itself one of the complete hands."
>
> This is no reason at all, my dear Professor Schenk, for our great national pastime is a game of combinations resulting from chance; and the fewer the number of combinations in the pack the higher the relative value of the hand. Now it is well known by all of us who love this finest of all pastimes that in our pack of 52 cards there are only fifty-two combinations of Threes, and at least a hundred of Straights.
>
> There is no getting over this mathematical fact, which alas seems to have escaped our no-doubt well-meaning Professor, and this is the way the game has always been played since its inception on the Mississippi River many years ago.

Many authors found the approximate conditional probabilities associated with five-card draw, but that is about as far as the mathematics of poker went for many years. Although seven-card stud dates at least as far back as the 1920s (some say the 19th century), the best five- of seven-card poker probabilities (Problem 1.8) are relatively recent (Hill 1996b; Alspach 2000c). A masters thesis by Davies (1959) contained erroneous numbers, attesting to the difficulty of this problem. Books by Wallace (1968, Appendix D, Table 4A) and Dangel (1977, p. 235) also got it wrong. A small part of the problem was found explicitly by Pierre Rémond de Montmort [1678–1719] (1708, p. 99): He evaluated the number of ways to get three pairs (of which only two are counted in poker) in a seven-card hand.

There is a substantial body of literature on the mathematics of poker dating back to the late 1930s. Chapter 5 of Émile Borel's [1871–1956] *Applications aux jeux de hasard* (1938) is devoted to poker, and Chapter 19 of John von Neumann [1903–1957] and Oskar Morgenstern's [1902–1977] *Theory of Games and Economic Behavior* (1944) concerns poker and bluffing. Both chapters introduced poker models that were further extended by a number of authors in the 1950s. See Karlin (1959b, Chapter 9) for a summary and Ferguson and Ferguson (2003) for a recent discussion. These models were continuous models, in contrast to the discrete model of Harold W. Kuhn [1925–] (1950), for example (see Problem 22.11). Poker is too complicated for a complete game-theoretic analysis, which explains why poker models have been studied as proxies for the real game. Ankeny (1981) studied the game of five-card draw using game theory.

Other authors, writing primarily for poker players, have contributed to a better understanding of the game. Important works include Herbert O. Yardley's [1889–1958] (1957) *The Education of a Poker Player*, Frank R. Wallace's (1968) *Poker: A Guaranteed Income for Life*, Doyle Brunson's (2002) *Super System: A Course in Power Poker* (originally published in 1978 under the title *How I Made over $1,000,000 Playing Poker*), David Sklansky's [1947–] (1978, 1994) *The Theory of Poker*, and Bill Chen [1970–] and Jerrod Ankenman's (2006) *The Mathematics of Poker*.

In recent years there has been a flood of books published on Texas hold'em. The first such book, by David Sklansky, appeared in 1976 (Sklansky 1976). More comprehensive treatments include Sklansky and Malmuth (1999) and Sklansky and Miller (2006). Books with a mathematical emphasis include Yao (2005) and Guerrera (2007).

Most of the material in this chapter is due to others. The notions of effective, implied, and reverse implied odds are due to Sklansky (1976, 1978). The analysis of the game between Doyle Brunson and Stu Ungar [1953–1998] is from Sklansky (1994, pp. 54–55). Section 2 is a condensed version of an article by Chris Ferguson [1963–] and Tom Ferguson [1929–] (2007); the former was World Poker Champion in 2000. They acknowledged an unpublished article of Cutler (1976) as contributing to their approach. Our treatment differs from that of the Fergusons in only one minor respect (aside from being less thorough). Specifically, each entry of their payoff matrix is less than the corresponding entry of ours by the amount a, player 1's ante. (We regard his ante as belonging to the pot, whereas they regard it as belonging to player 1, at least until the game is resolved.) This affects the value of the game but not the optimal strategies or their derivation.

As for Section 3 the odds table (Table 22.1) for one or two cards to come can be found in many sources. Most of the flop probabilities in Section 3 were found by Petriv (1996), though a few of his results are erroneous. The probabilities of winning, losing, and tying with given initial hands in heads-up play against random hands were apparently first found in 1995 by a researcher named Jazbo. The fact that there are 47,008 equivalence classes of unordered pairs of initial hands was known to Hill (1991b).

Some have objected to this methodology for ranking initial hands. Malmuth (1992) considered A♠-T♣ vs. 2♡-2♢. Although the pair of deuces wins more often than not in heads-up play between the two hands, he argued that A-T is the better hand. This fact is strongly confirmed by Table 22.4. Malmuth also considered these two hands in an eight-player game. If none of the eight players folds before the showdown, A-T wins more often than the deuces, but he argued that 2-2 is the better hand because it will be folded if it does not flop a set, whereas A-T will usually lose even when it flops a big pair. Malmuth concluded that these computations do not represent Texas hold'em as it is typically played.

> Specifically, your typical opponent does not play two random cards, and he does not automatically go to the end.

Alternative rankings are due to Sklansky (Sklansky and Malmuth 1999, pp. 14–15, or Sklansky 1996, p. 19), Caro (2007, Chapter 4), and Bloch (2007, Table 6.1). Sklansky's rankings of 86 playable hands reflect which hands will win the most money at a full table (8–11 players). According to Gordon and Grotenstein (2004, p. 37),

> There's no denying, however, that this is *the* list by which all others are judged. Every serious poker player has committed these hand rankings to memory.

Caro's and Bloch's rankings include all 169 hands. None of the three authors has described the algorithm for generating his rankings in sufficient detail that the results can be replicated. To give a sense of the disparities between the rankings, consider pocket deuces. This hand ranks 87th in Table 22.4; it was ranked 61st by Sklansky, 124th by Caro, 42nd by Bloch, and 42nd in the Sklansky–Chubukov rankings, described below.

An entirely different and very interesting ranking of initial hands is due to David Sklansky and Victor Chubukov (Sklansky and Miller 2006, pp. 299–310). Associated with each of the 169 equivalence classes of initial hands is a number, the *Sklansky–Chubukov number*, defined as follows. If player 1, the one-unit blind, exposes his cards of the given equivalence class in a heads-up game against player 2, the two-unit blind, how many units would player 1 have to have in his stack of chips before it would be better for him to fold than to raise all-in, assuming an optimal call or fold by player 2?

For example, consider pocket kings. Suppose player 1 has S units left in his stack (after posting his blind). His pocket kings will be called with pocket aces (with 0, 1, or 2 suit matches) or pocket kings only. If player 2 folds, then player 1 wins three units (the blinds). If he calls, he bets $S - 1$ units, and a win for player 1 wins $S + 2$ units (the blinds and player 2's bet), a loss loses S units (player 1's bet), and a tie wins 1 unit ($S + 1$ units from the split pot less player 1's bet). Player 1's expected profit is

$$\left(1 - \frac{7}{1{,}225}\right)(3) + \frac{1}{1{,}225} \frac{317{,}694(S+2) + 1{,}388{,}072(-S) + 6{,}538(1)}{1{,}712{,}304}$$

$$+ \frac{4}{1{,}225} \frac{305{,}177(S+2) + 1{,}399{,}204(-S) + 7{,}923(1)}{1{,}712{,}304}$$

$$+ \frac{1}{1{,}225} \frac{292{,}660(S+2) + 1{,}410{,}336(-S) + 9{,}308(1)}{1{,}712{,}304}$$

$$+ \frac{1}{1,225} \, \frac{37,210(S+2) + 37,210(-S) + 1,637,884(1)}{1,712,304}, \tag{22.124}$$

and setting this equal to 0 and solving for S gives the Sklansky–Chubukov number for pocket kings, namely $S = 953.995465$. This concept was originated by David Sklansky, who has been a leading theoretician of poker for more than 30 years. See Problem 22.31.

The two examples at the end of Section 22.3 are due to Guerrera (2007, pp. 62–73), though the first was modified slightly.

Problem 22.2 is due to Guerrera (2007, p. 38). Problem 22.3 is from Sklansky (1997, pp. 103–106). Problem 22.4 is standard, but see for example Brunson (2002, Appendix, Table 1); some sources (Morehead 1967, p. 217; Percy 1988, p. 17) have it wrong. Problem 22.5 was motivated by Holden (1990, p. 72), who wrote,

> The odds against *any* full house being beaten by *any* straight flush, in a two-handed game, are 45,102,784 to 1; [. . .]

This figure is wrong on two counts. First, it assumes that the hands are dealt independently, and second and more importantly, it overlooks the fact that the event also occurs if the two hands are exchanged, meaning that these odds are too high by nearly a factor of 2. Problem 22.6 was first solved by Michael W. Shackleford (`wizardofodds.com/poker`) by computer enumeration. Problem 22.7 is standard, but see for example Brunson (2002, Appendix, Tables 35, 31, and 39). A more complicated version of Problem 22.8 was solved by Alspach (2005b).

Problem 22.10 was addressed by Ferguson and Ferguson (2007) for Model 2 of Section 22.2. Problem 22.11, which is a simplified version of Kuhn's poker model (Problem 22.12), was analyzed by Chen and Ankenman (2006, Example 13.2), who credited Mike Caro with the idea. Problem 22.12 was introduced and first solved by Kuhn (1950, 2003). See Karlin (1959a, pp. 86–92) and Demange and Ponssard (1994, pp. 202–214) for alternative solutions. Chen and Ankenman (2006, Example 15.1) have also studied this model. Problem 22.13 is from Williams (1954, Example 24), except that he implicitly assumed that cards are dealt with replacement. Problem 22.14 is a discrete version of a model introduced by Chen and Ankenman (2006, Example 12.1).

Most of Problem 22.16 was done by Petriv (1996, Annex H), though he adopted different definitions. For example, his straight draws excluded made flushes *and* flush draws. Problem 22.18 is due to Hill (1989). Problem 22.19 is due to Alspach (1998, 2000a). Problems 22.20, 22.22, 22.23, and 22.24 are also due to Alspach (2001b, 2005a, 2001c, and 2005a, respectively). Problem 22.28 is from Hill (1991b). Problem 22.26(a) was posed by Doyle Brunson in 1980, according to Caro (1994); the latter found an answer by computer simulation, getting the correct denominations but the wrong suits. Part (b) of that problem is due to Hill (1990b).

W. Lawrence Hill [1944–2005] worked on Problem 22.27(d) (and, to a lesser extent, (b)) for several years, 1990–1996, publishing his results in his biweekly column in *Card Player* magazine. He could analyze a particular matchup but apparently lacked the computer speed to do an exhaustive analysis of all 47,008 of them. Thus, he repeatedly updated his list of the closest matchups. He wrote (Hill 1994a),

> Many of these matchups were suggested by sharp readers, but this columnist, himself, proudly claims the credit, and any other attendant glory, for the discovery of the closest matchup of all on the list, 3♠ 3♡ against A♠ 10♠ by 50.007% to 49.993%, meaning any pair of threes against A-10 suited, in suit with one of the threes.

The relevant articles include Hill (1990a, 1991a,b,c, 1992, 1994a,b, 1995, 1996a). We can confirm that the list of the 34 closest matchups in Hill (1996a)—those for which the expectation belongs to $(0, 10^{-3})$—is exactly correct.

Problem 22.29 was suggested by Hill (1993) with A-6 instead of A-K. With A-K it was cited by Chen and Ankenman (2006, p. 130) as a "famous proposition bet." Problem 22.31 is from Sklansky and Miller (2006, pp. 214–221).

Appendix A
Results Cited

With which my friend Pan, heaving a great sigh, as if confessing his inability to look
Infinity in the face, sank back resigned, and swallowed a large bumper of Claret.

William Makepeace Thackeray, *The Snobs of England*

We collect here several results that are cited in the text.

A.1 Algebra and Number Theory

We need the basic idea of equivalence classes in several places. We say that
$\sim$ is an *equivalence relation* on a set A if it is *reflexive* ($a \sim a$), *symmetric*
($a \sim b$ implies $b \sim a$), and *transitive* ($a \sim b$ and $b \sim c$ imply $a \sim c$). Let us
define the *equivalence class* containing $a \in A$ by

$$[a] := \{b \in A : b \sim a\}. \tag{A.1}$$

Theorem A.1.1. *Given an equivalence relation $\sim$ on a set A, every two
equivalence classes $[a]$ and $[b]$ are either equal or mutually exclusive. Consequently, the distinct equivalence classes partition the set A.*

The proof is easy; see Herstein (1964, p. 7).
The next result is needed in the proof of Theorem 4.3.2.

Theorem A.1.2. *Let $J \subset \mathbf{N}$ be closed under addition and satisfy g.c.d.$(J) =
1$. Then there exists a positive integer n_0 such that $n \geq n_0$ implies $n \in J$.*

For a proof see Hoel, Port, and Stone (1972, pp. 79–80).
The *Euler φ-function* is defined for each positive integer n by

$$\varphi(n) := |\{1 \leq k \leq n : \text{g.c.d.}(k, n) = 1\}|. \tag{A.2}$$

S.N. Ethier, *The Doctrine of Chances*, Probability and its Applications,
DOI 10.1007/978-3-540-78783-9, © Springer-Verlag Berlin Heidelberg 2010

For example, $\varphi(12) = |\{1, 5, 7, 11\}| = 4$. The next theorem, due to Leonhard Euler [1707–1783], is needed in Corollary 9.1.3 and Problem 11.1.

Theorem A.1.3. *If m and n are positive integers with g.c.d.$(m, n) = 1$, then $m^{\varphi(n)} \equiv 1 \pmod{n}$.*

For a proof, see Herstein (1964, p. 37).

The next result, known as *Descartes's rule of signs*, is due to René Descartes [1596–1650]. It is needed in the proof of Lemma 9.3.3.

Theorem A.1.4. *The number of positive roots (counting multiplicities) of a polynomial $p(x)$ with real coefficients is equal to the number of sign changes of the nonzero coefficients of $p(x)$ or is less than this number by an even positive integer. The number of negative roots (counting multiplicities) of a polynomial $p(x)$ with real coefficients is equal to the number of sign changes of the nonzero coefficients of $p(-x)$ or is less than this number by an even positive integer.*

For example, the nonzero coefficients of $p(x) := x^3 - x^2 - x + 1$ are $1, -1, -1, 1$, so there are two sign changes, and consequently $p(x)$ has 2 or 0 positive roots. In addition, the nonzero coefficients of $p(-x) = -x^3 - x^2 + x + 1$ are $-1, -1, 1, 1$, so there is one sign change, hence $p(x)$ has one negative root. In fact, $p(x) = (x + 1)(x - 1)^2$, so $p(x)$ has two positive roots and one negative root.

For a proof, see Albert (1943).

We turn next to the cubic formula, which was first published by Girolamo Cardano [1501–1576] (1545, 1968) and is often called *Cardano's formula*. It is needed in Problems 7.12 and 13.5. Given a cubic equation

$$a_3 z^3 + a_2 z^2 + a_1 z + a_0 = 0, \tag{A.3}$$

we can assume without loss of generality that $a_3 = 1$.

Theorem A.1.5. *The three roots z_1, z_2, z_3 of the cubic equation (A.3) with real coefficients and $a_3 = 1$ can be described as follows. Let*

$$p := a_1 - \frac{a_2^2}{3}, \quad q := -a_0 + \frac{a_1 a_2}{3} - \frac{2a_2^3}{27}, \quad D := \frac{p^3}{27} + \frac{q^2}{4}, \tag{A.4}$$

and define

$$P := \sqrt[3]{\tfrac{1}{2}q + \sqrt{D}}, \qquad Q := \sqrt[3]{\tfrac{1}{2}q - \sqrt{D}}. \tag{A.5}$$

Then

$$z_1 = P + Q - \frac{a_2}{3}, \tag{A.6}$$

$$z_2 = \omega P + \omega^2 Q - \frac{a_2}{3}, \tag{A.7}$$

$$z_3 = \omega^2 P + \omega Q - \frac{a_2}{3}, \tag{A.8}$$

where $\omega := e^{2\pi i/3} = -\frac{1}{2} + \frac{1}{2}\sqrt{3}\,i$ *is a cube root of unity, as is* $\omega^2 = e^{4\pi i/3} = \bar{\omega} = -\frac{1}{2} - \frac{1}{2}\sqrt{3}\,i$.

Remark. The definitions (A.5) are slightly ambiguous because of the non-uniqueness of the cube roots. [If (P, Q) is replaced in the definitions of z_1, z_2, and z_3 by $(\omega P, \omega^2 Q)$ or by $(\omega^2 P, \omega Q)$, then z_1, z_2, and z_3 are simply permuted.] If $D > 0$, then P and Q can be taken to be real and distinct, in which case z_1 is real and z_2 and z_3 are complex conjugates; in particular, z_1, z_2, and z_3 are distinct. If $D = 0$, then P and Q can be taken to be real and equal, in which case z_1, z_2, and z_3 are real and $z_2 = z_3$. If $D < 0$, then we can take $Q = \bar{P}$, in which case z_1, z_2, and z_3 are real and distinct. In fact, writing $\frac{1}{2}q + \sqrt{-D}\,i = re^{i\theta}$, we can take $P = r^{1/3}e^{i\theta/3}$, $Q = \bar{P}$, and

$$z_1 = 2r^{1/3}\cos(\theta/3) - \frac{a_2}{3}, \tag{A.9}$$

$$z_2 = 2r^{1/3}\cos((\theta + 2\pi)/3) - \frac{a_2}{3}, \tag{A.10}$$

$$z_3 = 2r^{1/3}\cos((\theta + 4\pi)/3) - \frac{a_2}{3}. \tag{A.11}$$

Note that $r^{1/3} = \sqrt{-p/3}$ and $\theta = \cos^{-1}(\frac{1}{2}q/\sqrt{-p^3/27})$.

For a proof, see Mac Lane and Birkhoff (1988, pp. 437–438).

A.2 Analysis and Probability

We need several results about interchanging limits, sums, integrals, and expectations. They can be stated more elegantly using measure theory because sums and expectations are special cases of abstract Lebesgue integrals, but we want to keep the presentation as elementary as possible. The results are due primarily to Henri Lebesgue [1875–1941]. The reader is referred to Rudin (1987, Chapter 1) for proofs.

We also need to define the expectation of a random variable X that is not necessarily discrete. If $X \geq 0$, then

$$\mathrm{E}[X] := \int_0^\infty \mathrm{P}(X \geq t)\,dt \tag{A.12}$$

works, where the right side is an improper Riemann integral. If X is arbitrary, then we can write $X = X^+ - X^-$ and define $\mathrm{E}[X] := \mathrm{E}[X^+] - \mathrm{E}[X^-]$ as long as both expectations are finite. These definitions are consistent with the ones already given for discrete random variables (see Problem 1.32) and suffice for our purposes.

We begin with the *bounded convergence theorem*, needed in Example 1.5.10.

Theorem A.2.1. *Let $X_1, X_2, \ldots$ be a sequence of discrete random variables that converges a.s. to the random variable X. Assume the existence of a positive constant M such that $|X_n| \leq M$ for all $n \geq 1$. Then $\lim_{n \to \infty} \mathrm{E}[X_n] = \mathrm{E}[X]$.*

The point is that we can interchange the limit and the expectation provided only that the sequence of random variables is uniformly bounded.

Next is *Fatou's lemma*, named for Pierre Fatou [1878–1929] and needed in the proofs of Theorems 3.2.2, 3.3.2, and 4.3.3.

Lemma A.2.2. *Let $X_1, X_2, \ldots$ be a sequence of nonnegative discrete random variables. Then $\mathrm{E}[\liminf_{n \to \infty} X_n] \leq \liminf_{n \to \infty} \mathrm{E}[X_n]$.*

This leads to the *monotone convergence theorem*, needed in the proofs of Corollaries 7.1.6 and 7.1.8 and Theorem 8.2.4.

Theorem A.2.3. *Let $X_1, X_2, \ldots$ be a sequence of nonnegative discrete random variables such that $0 \leq X_1 \leq X_2 \leq \cdots$, and let $X = \lim_{n \to \infty} X_n$. Then $\lim_{n \to \infty} \mathrm{E}[X_n] = \mathrm{E}[X]$.*

This theorem gives the same conclusion as Theorem A.2.1 but with a monotonicity assumption instead of a uniform boundedness assumption. It yields the following corollary, needed in the proofs of Theorems 3.2.2 and 4.2.1.

Corollary A.2.4. *Let $X_1, X_2, \ldots$ be a sequence of nonnegative discrete random variables. Then*

$$\mathrm{E}\left[\sum_{n=1}^{\infty} X_n\right] = \sum_{n=1}^{\infty} \mathrm{E}[X_n]. \tag{A.13}$$

Finally, the *dominated convergence theorem* generalizes the bounded convergence theorem. It is needed in the proofs of Theorems 2.1.3, 3.2.2, 4.3.2, 4.3.3, 4.4.2, 6.1.1, and 6.3.4.

Theorem A.2.5. *Let $X_1, X_2, \ldots$ be a sequence of discrete random variables that converges a.s. to the random variable X. Assume the existence of a random variable Y such that $|X_n| \leq Y$ for all $n \geq 1$ and $\mathrm{E}[Y] < \infty$. Then $\lim_{n \to \infty} \mathrm{E}[X_n] = \mathrm{E}[X]$.*

The dominated convergence theorem yields the following corollary, needed in Problem 3.6.

Corollary A.2.6. *Let $X_1, X_2, \ldots$ be a sequence of discrete random variables such that*

$$\sum_{n=1}^{\infty} \mathrm{E}\big[|X_n|\big] < \infty. \tag{A.14}$$

Then

$$\mathrm{E}\left[\sum_{n=1}^{\infty} X_n\right] = \sum_{n=1}^{\infty} \mathrm{E}[X_n]. \tag{A.15}$$

We also state versions of the monotone and dominated convergence theorems for Riemann integrals, needed in Problems 1.32 and 1.58.

Theorem A.2.7. *Let $f_1, f_2, \ldots$ be a sequence of Riemann integrable functions on an interval I such that $0 \leq f_1 \leq f_2 \leq \cdots$, and put $f := \lim_{n \to \infty} f_n$. Assume that f is Riemann integrable on I. Then $\lim_{n \to \infty} \int_I f_n(x)\,dx = \int_I f(x)\,dx$.*

Theorem A.2.8. *Let $f_1, f_2, \ldots$ be a sequence of Riemann integrable functions on an interval I that converges pointwise to the Riemann integrable function f on I. Assume the existence of a Riemann integrable function g on I such that $|f_n| \leq g$ for all $n \geq 1$ and $\int_I g(x)\,dx < \infty$. Then $\lim_{n \to \infty} \int_I f_n(x)\,dx = \int_I f(x)\,dx$.*

The *Karush–Kuhn–Tucker theorem* extends the method of Lagrange multipliers to optimization problems with inequality constraints as well as equality constraints. It was first proved by William Karush [1917–1997] (1939) in a masters thesis. It was rediscovered by Harold W. Kuhn [1925–] and A. W. Tucker [1905–1995] (1951) and is often called the Kuhn–Tucker theorem. We need it in Example 10.2.3.

Theorem A.2.9. *Let $f : \mathbf{R}^n \mapsto \mathbf{R}$, $g_i : \mathbf{R}^n \mapsto \mathbf{R}$ for $i = 1, \ldots, k$, and $h_j : \mathbf{R}^n \mapsto \mathbf{R}$ for $j = 1, \ldots, l$. Consider the problem of maximizing $f(\boldsymbol{x})$, subject to the constraints*

$$g_i(\boldsymbol{x}) \geq 0, \quad i = 1, \ldots, k, \qquad h_j(\boldsymbol{x}) = 0, \quad j = 1, \ldots, l. \qquad (A.16)$$

Suppose that f, g_i, and h_j are continuously differentiable at $\boldsymbol{x}^*$ for $i = 1, \ldots, k$ and $j = 1, \ldots, l$, and that $\boldsymbol{x}^*$ is a local maximizer of f satisfying the constraints. Assume that $\nabla g_i(\boldsymbol{x}^*)$ ($i \in \{1, \ldots, k\}$ for which $g_i(\boldsymbol{x}^*) = 0$) and $\nabla h_j(\boldsymbol{x}^*)$ ($j = 1, \ldots, l$) are linearly independent.

Then there exist constants (Lagrange multipliers) $\kappa_i \geq 0$ ($i = 1, \ldots, k$) and λ_j real ($j = 1, \ldots, l$) such that

$$\nabla f(\boldsymbol{x}^*) + \sum_{i=1}^{k} \kappa_i \nabla g_i(\boldsymbol{x}^*) + \sum_{j=1}^{l} \lambda_j \nabla h_j(\boldsymbol{x}^*) = \mathbf{0}, \qquad (A.17)$$

(A.16) holds at $\boldsymbol{x} = \boldsymbol{x}^*$, and $\kappa_i g_i(\boldsymbol{x}^*) = 0$ for $i = 1, \ldots, k$.

For a proof, see Fletcher (2000, Theorem 9.1.1). The linear independence condition in the theorem is called a *constraint qualification*.

We need the negative binomial series in the definition of the negative binomial distribution in Section 1.3 and in Example 1.4.23. The proof is a straightforward calculus problem.

Theorem A.2.10. *For all complex z with $|z| < 1$ and every positive integer n,*

$$(1-z)^{-n} = \sum_{k=0}^{\infty} \binom{k+n-1}{n-1} z^k. \tag{A.18}$$

The relationship between infinite products and infinite series is needed in Example 8.1.6.

Theorem A.2.11. *Let $0 < p_n < 1$ for each $n \geq 1$. Then*

$$\prod_{n=1}^{\infty}(1-p_n) > 0 \quad \text{if and only if} \quad \sum_{n=1}^{\infty} p_n < \infty. \tag{A.19}$$

Notice, for example, that neither condition holds when $p_n = 1/(n+1)$ for all $n \geq 1$. For a proof, see Rudin (1987, p. 300).

In Section 10.3 we need the following result about the median of the binomial distribution.

Theorem A.2.12. *Let n be a positive integer and let $0 < p < 1$. Then*

$$\text{median}(\text{binomial}(n,p)) \in (np - \ln 2, np + \ln 2). \tag{A.20}$$

The result is due to Edelman (c. 1979) and Hamza (1995).

The central limit theorem for samples from a finite population is needed in Section 11.3.

Theorem A.2.13. *For each $N \geq 2$, let $(X_{N,1}, X_{N,2}, \ldots, X_{N,N})$ have the discrete uniform distribution over all $N!$ permutations of the N (not necessarily distinct but not all equal) numbers $x_{N,1}, x_{N,2}, \ldots, x_{N,N}$, and define*

$$\mu_N := \mathrm{E}[X_{N,1}] = \frac{1}{N} \sum_{j=1}^{N} x_{N,j} \tag{A.21}$$

and

$$\sigma_N^2 := \mathrm{Var}(X_{N,1}) = \frac{1}{N} \sum_{j=1}^{N} (x_{N,j} - \mu_N)^2. \tag{A.22}$$

Assume that

$$\max_{1 \leq j \leq N} \frac{|x_{N,j} - \mu_N|}{\sqrt{N\sigma_N^2}} \to 0 \tag{A.23}$$

as $N \to \infty$. Then, with $S_{N,n} := X_{N,1} + \cdots + X_{N,n}$, we have

$$\frac{S_{N,n} - n\mu_N}{\sqrt{n\sigma_N^2(N-n)/(N-1)}} \overset{d}{\to} N(0,1), \tag{A.24}$$

provided $n, N \to \infty$ in such a way that $n/N \to \alpha \in (0,1)$.

Notice that this generalizes Theorem 1.5.14. One of the first results of this type was proved by Erdős and Rényi (1959); see Billingsley (1968, pp. 208–212) for a generalization.

Bibliography

Each entry is followed by a braced list of page numbers where the entry is cited. In most cases, authors' names are given as they appeared in print; see the index for full names. Occasionally, modern spellings are used (e.g., Kolmogorov instead of Kolmogoroff).

About, Edmond (1899). *Trente et Quarante*. Edward Arnold, London. English translation by Lord Newton. Originally published in 1858. {**638**}

Addario-Berry, L. and Reed, B. A. (2008). Ballot theorems, old and new. In Győri, Ervin, Katona, Gyula O. H., and Lovász, László (Eds.), *Horizons of Combinatorics*, pp. 9–35. Springer, Berlin. {**315**}

Agard, David B. and Shackleford, Michael W. (2002). A new look at the probabilities in bingo. *College Mathematics Journal* **33** (4) 301–305. {**500**}

Akane, Kimo (2008). The man with the golden arm, Parts I and II. *Around Hawaii* (May 1, June 1). {**522**}

Albert, A. A. (1943). An inductive proof of Descartes' rule of signs. *American Mathematical Monthly* **50** (3) 178–180. {**746**}

Albigny, G. d' (1902). *Les martingales modernes: Nouveau recueil du joueur de roulette et de trente-et-quarante contenant les meilleurs systèmes. Revus, corrigés ou inédits et expérimentés avec explications nombreuses et détaillées de différentes attaques, et conseils pratiques pour assurer aux joueurs les plus grandes chances de gain*. Librairie des Mathurins, Paris. {**115, 316**}

Aldous, David and Diaconis, Persi (1986). Shuffling cards and stopping times. *American Mathematical Monthly* **93** (5) 333–348. {**157, 424, 425**}

Algoet, Paul H. and Cover, Thomas M. (1988). Asymptotic optimality and asymptotic equipartition properties of log-optimum investment. *Annals of Probability* **16** (2) 876–898. {**389, 390**}

Allais, M. (1953). Le comportement de l'homme rationnel devant le risque: Critique des postulats et axiomes de l'école américaine. *Econometrica* **21** (4) 503–546. {**198**}

Alspach, Brian (1998). Flushing boards. http://www.math.sfu.ca/~alspach/comp3.pdf. {**743**}

Alspach, Brian (2000a). Flushing boards II. http://www.math.sfu.ca/~alspach/comp4/. {**743**}

Alspach, Brian (2000b). 5-card poker hands. http://www.math.sfu.ca/~alspach/comp18/. {**66**}

Alspach, Brian (2000c). 7-card poker hands. http://www.math.sfu.ca/~alspach/comp20/. {**72, 740**}

Alspach, Brian (2001a). Two problems from RPG. http://www.math.sfu.ca/~alspach/mag67/. {**500**}

Alspach, Brian (2001b). Overcards. http://www.math.sfu.ca/~alspach/comp34/. {**743**}

S.N. Ethier, *The Doctrine of Chances*, Probability and its Applications,
DOI 10.1007/978-3-540-78783-9, © Springer-Verlag Berlin Heidelberg 2010

Alspach, Brian (2001c). Multiple pocket pairs. http://www.math.sfu.ca/~alspach/comp35/. {**743**}

Alspach, Brian (2005a). Distribution of aces among dealt hands. http://www.math.sfu.ca/~alspach/comp47.pdf. {**743**}

Alspach, Brian (2005b). 7-card stud low hands. http://www.math.sfu.ca/~alspach/comp49.pdf. {**743**}

Alspach, Brian (2007). Enumerating poker hands using group theory. In Ethier, Stewart N. and Eadington, William R. (Eds.), *Optimal Play: Mathematical Studies of Games and Gambling*, pp. 121–129. Institute for the Study of Gambling and Commercial Gaming, University of Nevada, Reno. {**571**}

Alvarez, A. (2002). *The Biggest Game in Town*. Chronicle Books LLC, San Francisco. Originally published in 1983. {**735–737**}

Alvarez, A. (2004). *Bets, Bluffs, and Bad Beats*. Chronicle Books LLC, San Francisco. Originally published in 2001. {**736**}

Anderson, L. R. and Fontenot, R. A. (1980). On the gain ratio as a criterion for betting in casino games. *Journal of the Royal Statistical Society, Series A* **143** (1) 33–40. {**239, 240**}

Ankeny, Nesmith C. (1981). *Poker Strategy: Winning with Game Theory*. Basic Books, Inc., New York. {**741**}

Anners, Henry F. (1845). *Hoyle's Games: Containing the Established Rules and Practice of Whist, Quadrille, Piquet, Quinze, Vingt-Un, Cassino, [...]*. Henry F. Anners, Philadelphia. {**732**}

Arbuthnot, John (1692). *Of the Laws of Chance, or, A Method of Calculation of the Hazards of Game, Plainly Demonstrated, and Applied to Games at Present Most in Use, Which May Be Easily Extended to the Most Intricate Cases of Chance Imaginable*. Benj. Motte, London. {**67, 522**}

Arrow, Kenneth J. (1963). Comment on Duesenberry's "The portfolio approach to the demand for money and other assets." *Review of Economics and Statistics* **45** (1, Part 2, Supplement) 24–27. {**198**}

Asbury, Herbert (1938). *Sucker's Progress: An Informal History of Gambling in America from the Colonies to Canfield*. Dodd, Mead and Company, Inc., New York. Reprinted by Patterson Smith, Montclair, NJ, 1969. {**517, 589, 590, 593, 677, 731, 732**}

Ashton, John (1968). *The History of Gambling in England*. Bert Franklin, New York. Originally published by Duckworth and Co., London, 1898. {**478**}

Assaf, Sami, Diaconis, Persi, and Soundararajan, K. (2009a). Riffle shuffles of a deck with repeated cards. arXiv:0905.4698. {**425**}

Assaf, Sami, Diaconis, Persi, and Soundararajan, K. (2009b). A rule of thumb for riffle shuffling. arXiv:0905.4698. {**425**}

Athreya, K. B. and Ney, P. E. (1972). *Branching Processes*. Springer-Verlag, New York. {**117**}

Babbage, Charles (1821). An examination of some questions connected with games of chance. *Transactions of the Royal Society of Edinburgh* **9** 153–177. Reprinted in Campbell-Kelly, Martin (Ed.), *The Works of Charles Babbage, Vol. 1. Mathematical Papers*, pp. 327–343. New York University Press, New York, 1989. {**312, 315**}

Bachelier, Louis (1912). *Calcul des probabilités, Vol. 1*. Gauthier-Villars, Imprimeur-Libraire, Paris. Reprinted by Éditions Jacques Gabay, Sceaux, 1992. {**94, 116, 117**}

Bachelier, Louis (1914). *Le jeu, la chance et le hasard*. Ernest Flammarion, Paris. Reprinted by Éditions Jacques Gabay, Sceaux, 1993. {**vi**}

Badoureau, [Albert] (1881). Étude sur le jeu de baccarat. *La Revue Scientifique de la France et de l'Étranger* **1** 239–246. {**618**}

Baldwin, Roger R., Cantey, Wilbert E., Maisel, Herbert, and McDermott, James P. (1956). The optimum strategy in blackjack. *Journal of the American Statistical Association* **51** (275) 429–439. Corrigenda **54** (288) 810, 1959. {**678, 681, 685, 687**}

Baldwin, Roger R., Cantey, Wilbert E., Maisel, Herbert, and McDermott, James P. (1957). *Playing Blackjack to Win: A New Strategy for the Game of 21*. M. Barrows and Company, New York. Reprinted by Cardoza Publishing, Las Vegas, 2008. {**681**}

Balzac, Honoré de (1831). *La peau de chagrin, roman philosophique*. Charles Gosselin et Urbain Canel, Paris. {**355**}

Barnhart, Russell T. (1978). *Casino Gambling: Why You Win, Why You Lose*. Brandywine Press, E. P. Dutton, New York. {**316, 615**}

Barnhart, Russell T. (1980). *Banker's Strategy at Baccara Chemin-de-Fer, Baccara-en-Banque, and Nevada Baccarat*. GBC Press, Las Vegas. {**614, 620, 621**}

Barnhart, Russell T. (1983a). *Gamblers of Yesteryear*. GBC Press, Las Vegas. {**316**}

Barnhart, Russell T. (1983b). Can trente-et-quarante be beaten? *Gambling Times* **7** (8, Dec.) 74–77. {**641**}

Barnhart, Russell T. (1988). The invention of roulette. In Eadington, William R. (Ed.), *Gambling Research: Proceedings of the Seventh International Conference on Gambling and Risk Taking, Vol. 1. Public Policy and Commercial Gaming Industries throughout the World*, pp. 295–330. Bureau of Business and Economic Research, University of Nevada, Reno. {**477, 479**}

Barnhart, Russell T. (1990). The banker's 'secret' strategy at baccara-en-banque. *Win Magazine* (Nov./Dec.) 54–57. {**615**}

Barnhart, Russell T. (1991). The invention of craps. Unpublished. {**520**}

Barnhart, Russell T. (1992). *Beating the Wheel*. A Lyle Stuart Book, Carol Publishing Group, New York. {**482**}

Barnhart, Russell T. (1997). The historiography of gambling history: John Scarne and Clayton Rawson, Sr. Unpublished. {**680**}

Barnhart, Russell T. (2000). My blackjack trip in 1962 to Las Vegas and Reno with Professor Edward O. Thorp and Mickey MacDougall. *Blackjack Forum* **20** (1) 9–30. {**678**}

Barstow, Frank (1979). *Beat the Casino*. Carlyle Associates, Santa Monica, CA. {**239**}

Basieux, Pierre (2001). *Roulette: Die Zähmung des Zufalls* (fifth ed.). Printul Verlag, Munich. {**482**}

Battaglia, Salvatore (1961). *Grande dizionario della lingua italiana*. Unione Tipografico-Editrice Torinese, Torino. {**613, 614**}

Baxter-Wray [Peel, Walter H.] (1891). *Round Games with Cards. A Practical Treatise on All the Most Popular Games, with Their Different Variations, and Hints for Their Practice*. Frederick A. Stokes Company, New York. {**677**}

Bayer, Dave and Diaconis, Persi (1992). Trailing the dovetail shuffle to its lair. *Annals of Applied Probability* **2** (2) 294–313. {**424, 425**}

Bayes, Thomas (1764). An essay towards solving a problem in the doctrine of chances. *Philosophical Transactions of the Royal Society of London* **53** 370–418. {**68**}

Bell, Robert M. and Cover, Thomas M. (1980). Competitive optimality of logarithmic investment. *Mathematics of Operations Research* **5** (2) 161–166. {**390**}

Bellhouse, D. R. (2000). *De Vetula*: A medieval manuscript containing probability calculations. *International Statistical Review* **68** (2) 123–136. {**65**}

Bellhouse, David (2007). The problem of Waldegrave. *Electronic Journal for History of Probability and Statistics* **3** (2). {**197**}

Benjamin, A. T. and Goldman, A. J. (2002). Analysis of the *N*-card version of the game le her. *Journal of Optimization Theory and Applications* **114** (3) 695–704. {**198**}

Benjamin, Arthur T. and Goldman, A. J. (1994). Localization of optimal strategies in certain games. *Naval Research Logistics* **41** (5) 669–676. {**198**}

Beresford, S. R. (1923). *The Future at Monte Carlo: A Successful Method Explained. The Only System That Can Beat the Bank*. Palmer, Sutton & Co., London. {**315**}

Beresford, S. R. (1926). *Beresford's Monte-Carlo*. J. Beresford, Publishers, Nice. {**619, 641**}

Berman, Lyle and Karlins, Marvin (2005). *I'm All In! High Stakes, Big Business and the Birth of the World Poker Tour*. Cardoza Publishing, New York. {**737**}

Bernoulli, Daniel (1738). Specimen theoriae novae de mensura sortis. *Commentarii Academiae Scientiarum Imperialis Petropolitanae* **5** 175–192. {**69, 73, 198, 389**}

Bernoulli, Daniel (1954). Exposition of a new theory on the measurement of risk. *Econometrica* **22** (1) 23–36. English translation by L. Sommer of Bernoulli (1738). {**73**}

Bernoulli, Jacob (1690). Quaestiones nonnullae de usuris, cum solutione problematis de sorte alearum. *Acta Eruditorum* (Maji) 219–223. {**66**}

Bernoulli, Jacob (1713). *Ars Conjectandi.* Thurnisius, Basilea. {**v, 65, 68, 70, 73, 117, 271, 590, 734**}

Bernoulli, Jacob (2006). *The Art of Conjecturing, together with Letter to a Friend on Sets in Court Tennis.* Johns Hopkins University Press, Baltimore. English translation by Edith Dudley Sylla. {**65, 68, 73, 117, 271, 590**}

Bernstein, S. N. (1926). Sur l'extension du théorème limite du calcul des probabilités aux sommes de quantités dépendantes. *Mathematische Annalen* **97** (1) 1–59. {**157**}

Bertezène, Alfred (1897). *Mémoire à l'Académie des sciences.* Librairie de la Voix de Paris, Paris. {**619**}

Bertrand, J. (1888). *Calcul des probabilités.* Gauthier-Villars et Fils, Paris. {**71, 618, 639, 640**}

Bethell, Victor (1898). *Ten Days at Monte Carlo at the Bank's Expense.* William Heinemann, London. Author's name listed only as V. B. {**315**}

Bethell, Victor (1901). *Monte Carlo Anecdotes and Systems of Play.* William Heinemann, London. Author's name listed only as V. B. {**316, 641**}

Bewersdorff, Jörg (2005). *Luck, Logic, and White Lies: The Mathematics of Games.* A. K. Peters, Ltd., Wellesley, MA. English translation by David Kramer of *Glück, Logik und Bluff,* 2. Auflage, Friedr. Vieweg & Sohn Verlag, Wiesbaden, 2001. {**686, 687**}

Bienaymé, I. J. (1853). Considérations à l'appui de la découverte de Laplace sur la loi de probabilité dans la méthode des moindres carrés. *Comptes Rendus de l'Académie des Sciences, Paris* **37** 309–324. {**70**}

Billard, Ludovic (1883). *Bréviaire du baccara expérimental.* Chez l'auteur, Paris. {**614, 618**}

Billedivoire (1929). *Jouer et gagner.* Éditions Argo, Paris. {**641, 642**}

Billiken (1924). *The Sealed Book of Roulette and Trente-et-Quarante, Being a Guide to the Tables at Monte Carlo, together with Simple Descriptions of Several Unique Systems. Also, Chapters Dealing with Baccarat, Chemin-de-Fer and la Boule.* John Lane, The Bodley Head, Ltd., London. {**641**}

Billings, Mark and Fredrickson, Brent (1996). *The Biased Wheel Handbook.* Saros Designs Publishing. {**482**}

Billingsley, Patrick (1968). *Convergence of Probability Measures.* John Wiley & Sons, Inc., New York. Second edition, 1999. {**750**}

Billingsley, Patrick (1995). *Probability and Measure* (third ed.). John Wiley & Sons, Inc., New York. First edition, 1979; second edition, 1986. {**157, 315, 355**}

Blackbridge, John (1880). *The Complete Poker-Player. A Practical Guide Book to the American National Game: Containing Mathematical and Experimental Analyses of the Probabilities at Draw Poker.* Dick & Fitzgerald, Publishers, New York. {**739, 740**}

Bloch, Andy (2007). No-limit hold 'em: Play before the flop. In Craig, Michael (Ed.), *The Full Tilt Poker Strategy Guide,* pp. 60–116. Grand Central Publishing, New York. {**742**}

Bohn, Henry G. (1850). *The Hand-Book of Games: Comprising New or Carefully Revised Treatises on Whist, Piquet, Écarté, Lansquenet, Boston, Quadrille, Cribbage, and Other Card Games; Faro, Rouge et Noir, Hazard, Roulette; Backgammon, Draughts; Billiards, Bagatelle, American Bowls; etc., etc.* Henry G. Bohn, London. {**519, 593**}

Bohn, Henry G. (1851). *Bohn's New Hand-Book of Games; Comprising Whist, by Deschapelles, Matthews, Hoyle, Carleton. Draughts, by Sturges & Walker. Billiards, by White & Bohn.* Henry F. Anners, Philadelphia. {**732**}

Boll, Marcel (1936). *La chance et les jeux de hasard: Loterie, boule, roulettes, baccara, 30 & 40, dés, bridge, poker, belote, écarté, piquet, manille.* Librairie Larousse, Paris. Revised edition, 1948, reprinted by Éditions Jacques Gabay, Sceaux, 1992. {**116, 481, 482, 522, 619, 639, 640**}

Boll, Marcel (1944). *Le baccara: Chemin de fer — banque.* Le Triboulet, Monaco. {**615, 619**}

Boll, Marcel (1945). *Le trente et quarante.* Le Triboulet, Monaco. {**640, 642**}

Boll, Marcel (1951). *La roulette* (third ed.). Le Triboulet, Monaco. First edition, 1944. {**479–482**}

Borel, Émile (1909). Les probabilités dénombrables et leur applications arithmétiques. *Rendiconti del Circolo Matematico di Palermo* **27** 247–271. {**70, 71, 73**}

Borel, Émile (1938). *Applications aux jeux de hasard.* Gauthier-Villars, Imprimeur-Éditeur, Paris. Written up by Jean Ville. Reprinted by Éditions Jacques Gabay, Sceaux, 1992. {**740**}

Borel, Émile (1949). Sur une martingale mineure. *Comptes Rendus de l'Académie des Sciences, Paris* **229** 1181–1183. {**117**}

Borel, Émile (1953). On games that involve chance and the skill of the players. *Econometrica* **21** (1) 101–115. English translation by Leonard J. Savage of Borel, Émile (1924). *Théorie des probabilités.* Librairie Scientifique, J. Hermann, Paris, pp. 204–224. {**197, 619**}

Borovkov, K. A. (1996). An estimate for the distribution of a stopping time for a stochastic system. *Siberian Mathematical Journal* **37** (4) 683–689. {**314**}

Borovkov, Konstantin (1997). On random walks with jumps scaled by cumulative sums of random variables. *Statistics & Probability Letters* **35** (4) 409–416. {**71, 157, 314**}

Braun, Julian (1987). To seven out or getting it right. *Gambling Times* **10** (14, Jul./Aug.) 43–54. {**522**}

Braun, Julian H. (1975). The development and analysis of winning strategies for the casino game of blackjack. Unpublished. {**682**}

Braun, Julian H. (1980). *How to Play Winning Blackjack.* Data House Publishing Co., Inc., Chicago. {**682**}

Breiman, L. (1961). Optimal gambling systems for favorable games. In Neyman, Jerzy (Ed.), *Proceedings of the Fourth Berkeley Symposium on Mathematical Statistics and Probability, Vol. 1*, pp. 65–78. University of California Press, Berkeley, CA. {**389, 390**}

Breiman, Leo (1960). Investment policies for expanding businesses optimal in a long-run sense. *Naval Research Logistics Quarterly* **7** (4) 647–651. {**390**}

Breiman, Leo (1968). *Probability.* Addison-Wesley Publishing Co., Inc., Reading, MA. {**315**}

Brisman, Andrew (2004). *Mensa Guide to Casino Gambling: Winning Ways.* Sterling Publishing Co., Inc., New York. {**239**}

Brooks, L. R. (1994). A basic strategy for "Let-It-Ride." Unpublished. {**543**}

Brown, Bancroft H. (1919). Probabilities in the game of "shooting craps." *American Mathematical Monthly* **26** (8) 351–352. {**94, 522**}

Bru, Bernard (2005). Poisson, the probability calculus, and public education. *Electronic Journal for History of Probability and Statistics* **1** (2). English translation by Glenn Shafer of Bru, Bernard (1981). Poisson, le calcul des probabilités et l'instruction publique. In Métivier, Michel, Costabel, Pierre, and Dugac, Pierre (Eds.), *Siméon Denis Poisson et la science de son temps*, pp. 51–94. École Polytechnique, Palaiseau. {**639**}

Brunson, Doyle (2002). *Super System: A Course in Power Poker* (third ed.). Cardoza Publishing, New York. Originally published as *How I Made over $ 1,000,000 Playing Poker*, B & G Publishing, Las Vegas, 1978. {**735, 741, 743**}

Brunson, Doyle (2005). *Super System 2: A Course in Power Poker.* Cardoza Publishing, New York. {**734**}

Bueschel, Richard M. (1995). *Lemons, Cherries and Bell-Fruit-Gum: Illustrated History of Automatic Payout Slot Machines.* Royal Bell Books, Denver. {**455, 456, 593**}

Bunyakovsky, V. (1859). Sur quelques inégalités concernant les intégrales ordinaires et les intégrales aux différences finies. *Mémoires de l'Académie Impériale des Sciences de St.-Pétersbourg, VIIe Série* **1** (9) 1–18. {**70**}

Cabot, Anthony N. and Hannum, Robert C. (2005). Advantage play and commercial casinos. *Mississippi Law Journal* **74** (3) 681–777. {**522**}

Canjar, R. Michael (2007a). Advanced insurance play in 21: Risk aversion and composition dependence. In Ethier, Stewart N. and Eadington, William R. (Eds.), *Optimal Play: Mathematical Studies of Games and Gambling*, pp. 3–34. Institute for the Study of Gambling and Commercial Gaming, University of Nevada, Reno. {**198, 686, 687**}

Canjar, R. Michael (2007b). Gambler's ruin revisited: The effects of skew and large jackpots. In Ethier, Stewart N. and Eadington, William R. (Eds.), *Optimal Play: Mathematical Studies of Games and Gambling*, pp. 439–469. Institute for the Study of Gambling and Commercial Gaming, University of Nevada, Reno. {**274**}

Cantelli, F. P. (1917). Sulla probabilità come limite della frequenza. *Atti Reale Accademia Nazionale Lincei* **26** 39–45. {**70, 71**}

Cardano, Gerolamo (1663). *Liber de Ludo Aleae.* Caroli Sponii, ed., Lyons. In *Opera omnia*, Vol. 1, pp. 262–276. See Ore (1953) for an English translation. {**v, viii, 64**}

Cardano, Girolamo (1539). *Practica arithmeticae generalis et mensurandi singularis.* Publisher unknown, Milan. In *Opera omnia*, Vol. 4. {**92**}

Cardano, Girolamo (1545). *Artis Magnae, sive de Regulis Algebraicis, Lib. unus.* Johann Petrieus, Nuremberg. In *Opera omnia*, Vol. 4. {**746**}

Cardano, Girolamo (1968). *The Great Art, or the Rules of Algebra.* The M.I.T. Press, Cambridge, MA. English translation by T. Richard Witmer. {**746**}

Caro, Mike (1994). Hold'em—What's the worst it gets? *Card Player* **7** (18) 9, 29. {**743**}

Caro, Mike (2007). *Caro's Most Profitable Hold 'em Advice: The Complete Missing Arsenal.* Cardoza Publishing, New York. {**742**}

Casanova, Giacomo (1894). *The Memoirs of Jacques Casanova de Seingalt.* F. A. Brockhaus, London. English translation by Arthur Machen in six volumes. {**592, 595**}

Casanova, Giacomo (1967). *History of My Life.* Harcourt, Brace & World, New York. English translation by Willard R. Trask in 12 volumes. {**311**}

Castleman, Deke (2004). *Whale Hunt in the Desert: The Secret Las Vegas of Superhost Steve Cyr.* Huntington Press Publishing, Las Vegas. {**69, 240**}

Catalan, E. (1837). Solution d'un problème de probabilité relatif au jeu de rencontre. *Journal de Mathématiques Pures et Appliquées, 1re Série* **2** 469–482. {**67, 73**}

Catlin, Donald (2003). Mensa mystery. *Casino City Times* (May 4). http://catlin.casinocitytimes.com/articles/6025.html. {**239**}

Catlin, Donald (2009). At risk or in play? *Casino City Times* (Sept. 6). http://catlin.casinocitytimes.com/article/at-risk-or-in-play?-47958. {**238**}

Catlin, Donald E. (2000). Using overall expected return per dollar risked to determine strategy decisions in gambling games. In Vancura, Olaf, Cornelius, Judy A., and Eadington, William R. (Eds.), *Finding the Edge: Mathematical Analysis of Casino Games*, pp. 367–376. Institute for the Study of Gambling and Commercial Gaming, University of Nevada, Reno. {**238, 239**}

Cauchy, Augustin-Louis (1821). *Cours d'analyse de l'École royale polytechnique. Première partie. Analyse algébrique.* Debure frères, Paris. {**69**}

Chafetz, Henry (1960). *Play the Devil: A History of Gambling in the United States from 1492 to 1955.* Clarkson N. Potter, Inc., New York. {**521**}

Chambliss, Carlson R. and Roginski, Thomas C. (1990). *Fundamentals of Blackjack.* GBC Press, Las Vegas. {**686**}

Chase, John Churchill (1997). *Frenchmen, Desire, Good Children and Other Streets of New Orleans* (third ed.). Touchstone, New York. First edition, Simon & Schuster, New York, 1949. {**519, 520**}

Chaucer, Geoffrey (1980). *The Canterbury Tales*. Edward Arnold, London. Edited from the Hengwrt Manuscript by N. F. Blake. {**517**}

Chaundy, T. W. and Bullard, J. E. (1960). John Smith's problem. *Mathematical Gazette* **44** (350) 253–260. {**73**}

Chebyshev, P. L. (1867). Des valeurs moyennes. *Journal de Mathématiques Pures et Appliquées* **12** 177–184. {**70**}

Chen, Bill and Ankenman, Jerrod (2006). *The Mathematics of Poker*. ConJelCo LLC, Pittsburgh. {**741, 743, 744**}

Chen, May-Ru, Chung, Pei-Shou, Hsiau, Shoou-Ren, and Yao, Yi-Ching (2008). On nonoptimality of bold play for subfair red-and-black with a rational-valued house limit. *Journal of Applied Probability* **45** (4) 1024–1038. {**356**}

Chernoff, Herman (1952). A measure of asymptotic efficiency for tests of a hypothesis based on the sum of observations. *Annals of Mathematical Statistics* **23** (4) 493–507. {**239**}

Chou, Hsing-szu (1963). *Ch'ien Tzu Wen, The Thousand Character Classic: A Chinese Primer*. Frederick Ungar Publishing Co., New York. Edited by Francis W. Parr. {**497**}

Chow, Y. S. and Robbins, Herbert (1961). On sums of independent random variables with infinite moments and "fair" games. *Proceedings of the National Academy of Sciences, USA* **47** (3) 330–335. {**73**}

Christensen, Morten Mosegaard (2005). On the history of the growth optimal portfolio. Preprint, University of Southern Denmark. {**389**}

Claussen, Jim (1982). *Keno Handbook*. GBC Press, Las Vegas. {**500**}

Coffin, George S. (1955). *The Poker Game Complete* (second ed.). Faber and Faber Limited, London. First edition, 1950. {**731**}

Comtat, Jean (1988). *Passe, pair ... et gagne!* Albin Michel, Paris. {**316**}

Conger, Mark and Viswanath, D. (2006a). Riffle shuffles of decks with repeated cards. *Annals of Probability* **34** (2) 804–819. {**425**}

Conger, Mark and Viswanath, D. (2006b). Shuffling cards for blackjack, bridge, and other card games. arXiv:math/0606031. {**425**}

Connelly, Robert (1974). Say red. *Pallbearers Review* **9** 702. {**425**}

Coolidge, J. L. (1909). The gambler's ruin. *Annals of Mathematics* **10** (4) 181–192. {**355**}

Coolidge, Julian Lowell (1925). *An Introduction to Mathematical Probability*. Clarendon Press, Oxford, UK. Reprinted by Dover, New York, 1962. {**355**}

Corti, Count (1934). *The Wizard of Homburg and Monte Carlo*. Thornton Butterworth Ltd., London. {**479**}

Cotton, Charles (1674). *The Compleat Gamester: or, Instructions How to Play at Billiards, Trucks, Bowls, and Chess. Together with All Manner of Usual and Most Gentile Games either on Cards or Dice. To Which Is Added, the Arts and Mysteries of Riding, Racing, Archery, and Cock-Fighting*. Printed by A. M. for R. Cutler, London. Reprinted by Cornmarket Reprints, London, 1972. {**518**}

Cowell, Joe (1844). *Thirty Years Passed among the Players in England and America: Interspersed with Anecdotes and Reminiscences of a Variety of Persons, Directly or Indirectly Connected with the Drama during the Theatrical Life of Joe Cowell, Comedian*. Harper & Brothers, New York. {**731**}

Cowles, David W. (1996). *Complete Guide to Winning Keno*. Cardoza Publishing, New York. {**498**}

Cowles, David W. (2003). *Complete Guide to Winning Keno* (third ed.). Cardoza Publishing, New York. {**238, 496, 498**}

Cramér, H. (1937). On a new limit theorem in the theory of probability. In *Colloquium on the Theory of Probability*. Hermann, Paris. {**239**}

Crevelt, Dwight E. and Crevelt, Louise G. (1988). *Slot Machine Mania*. Gollehon Books, Grand Rapids, MI. {**458**}

Csörgő, Sándor and Simons, Gordon (1996). A strong law of large numbers for trimmed sums, with applications to generalized St. Petersburg games. *Statistics & Probability Letters* **26** (1) 65–73. {**71**}

Culin, Stewart (1891). *The Gambling Games of the Chinese in America*. University of Pennsylvania Press, Philadelphia. {**498, 500**}

Culin, Stewart (1895). *Chinese Games with Dice and Dominoes*. Smithsonian Institution, Washington, DC. {**614**}

Cutler, William H. (1976). End-game poker. Unpublished. {**741**}

Dancer, Bob and Daily, Liam W. (2003). *A Winner's Guide to Full Pay Deuces Wild*. Compton Dancer Consulting Inc., Las Vegas. {**571, 572**}

Dancer, Bob and Daily, Liam W. (2004). *A Winner's Guide to Jacks or Better* (second ed.). Compton Dancer Consulting Inc., Las Vegas. {**571, 572**}

Dangel, Philip N. (1977). *Poker: Double Your Skills, Double Your Profit*. GBC Press, Las Vegas. {**740**}

David, F. N. (1962). *Games, Gods and Gambling: A History of Probability and Statistical Ideas*. Charles Griffin & Co., Ltd., London. Reprinted by Dover Publications, Inc., Mineola, NY, 1998. {**v, 64–66, 68, 72, 92, 94**}

Davidson, R. R. and Johnson, B. R. (1993). Interchanging parameters of the hypergeometric distribution. *Mathematics Magazine* **66** (5) 328–329. {**94**}

Davies, Alan Dunbar (1959). An analysis of five and seven card stud poker. Master's thesis, University of Pennsylvania. {**740**}

Davies, P. and Ross, A. S. C. (1979). Repeated zero at roulette. *Mathematical Gazette* **63** (423) 54–56. {**479**}

Davis, Clyde Brion (1956). *Something for Nothing*. J. B. Lippincott Company, Philadelphia. {**478, 591, 677**}

Dawson, Lawrence H. (1950). *Hoyle's Games Modernized* (20th ed.). Routledge & Kegan Paul, Ltd., London. {**616**}

De Moivre, A. (1712). De mensura sortis, seu, de probabilitate eventuum in ludis a casu fortuito pendentibus. *Philosophical Transactions of the Royal Society of London* **27** (329) 213–264. English translation by Bruce McClintock, *International Statistical Review* **52** (3) 237–262, 1984. {**68–70, 73, 93, 94, 271, 272**}

De Moivre, A. (1718). *The Doctrine of Chances: or, A Method of Calculating the Probability of Events in Play*. W. Pearson, London. {**v, vi, viii, 66, 67, 70, 72, 94, 236, 271, 272, 522**}

De Moivre, A. (1730). *Miscellanea Analytica de Seriebus et Quadraturis*. Tonson and Watts, London. {**71, 94**}

De Moivre, A. (1733). Approximatio ad summam terminorum binomii $\overline{a+b}^n$ in seriem expansi. Printed for private circulation under the name A. D. M. Reprinted in English in *The Doctrine of Chances*, second edition, pp. 235–243; third edition, pp. 243–254. {**71**}

De Moivre, A. (1738). *The Doctrine of Chances: or, A Method of Calculating the Probabilities of Events in Play* (second ed.). H. Woodfall, London. Reprinted by Frank Cass and Company Limited, London, 1967. {**v, 67, 69, 70, 72, 94, 271–273**}

De Moivre, A. (1756). *The Doctrine of Chances: or, A Method of Calculating the Probabilities of Events in Play* (third ed.). A. Millar, London. Reprinted by Chelsea Publishing Co., New York, 1967. {**v, 271, 272, 593, 595**}

De Morgan, Augustus (1838). *An Essay on Probabilities, and on Their Application to Life Contingencies and Insurance Offices*. Longman, Brown, Green & Longman's, London. {**639, 640, 642**}

De Morgan, Augustus (1847). *Formal Logic: or, The Calculus of Inference, Necessary and Probable*. Taylor and Walton, London. {**66**}

DeArment, Robert K. (1982). *Knights of the Green Cloth: The Saga of the Frontier Gamblers*. University of Oklahoma Press, Norman. {**69, 590, 593, 676**}

DeBartolo, Anthony (1997). Who caused the Great Chicago Fire? A possible deathbed confession. *Chicago Tribune* (Oct. 8). {**521**}

DeBartolo, Anthony (1998). Odds improve that a hot game of craps in Mrs. O'Leary's barn touched off Chicago Fire. *Chicago Tribune* (Mar. 3). {**521**}

Deiler, J. Hanno (1983). *A History of the German Churches in Louisiana (1823–1893)*. Center for Louisiana Studies, University of Southwestern Louisiana, Lafayette, LA. English translation by Marie Stella Condon. Originally published in 1894. {**520**}

Deloche, Régis and Oguer, Fabienne (2007a). What game is going on beneath *baccarat*? A game-theoretic analysis of the card game *le her*. In Ethier, Stewart N. and Eadington, William R. (Eds.), *Optimal Play: Mathematical Studies of Games and Gambling*, pp. 175–193. Institute for the Study of Gambling and Commercial Gaming, University of Nevada, Reno. {**198**}

Deloche, Régis and Oguer, Fabienne (2007b). *Baccara* and perfect Bayesian equilibrium. In Ethier, Stewart N. and Eadington, William R. (Eds.), *Optimal Play: Mathematical Studies of Games and Gambling*, pp. 195–210. Institute for the Study of Gambling and Commercial Gaming, University of Nevada, Reno. {**198, 618, 619**}

Demange, Gabrielle and Ponssard, Jean-Pierre (1994). *Théorie des jeux et analyse économique*. Presses Universitaires de France, Paris. {**743**}

Devlin, Keith (2008). *The Unfinished Game: Pascal, Fermat, and the Seventeenth-Century Letter That Made the World Modern*. Basic Books, New York. {**93**}

Diaconis, Persi (2003). Mathematical developments from the analysis of riffle shuffling. In Ivanov, A. A., Liebeck, M. W., and Saxl, J. (Eds.), *Groups, Combinatorics & Geometry: Durham 2001*, pp. 73–97. World Scientific Publishing Co., Inc., River Edge, NJ. {**425**}

Diaconis, Persi, Graham, R. L., and Kantor, William M. (1983). The mathematics of perfect shuffles. *Advances in Applied Mathematics* **4** (2) 175–196. {**425**}

Dieter, U. and Ahrens, J. H. (1984). The prison rule in roulette. *Metrika* **31** (1) 227–231. {**479**}

Dimand, Robert W. and Dimand, Mary Ann (1992). The early history of the theory of strategic games from Waldegrave to Borel. In Weintraub, E. Roy (Ed.), *Toward a History of Game Theory*, pp. 15–27. Duke University Press, Durham, NC. {**197**}

Doeblin, W. (1938). Exposé de la théorie des chaînes simples constantes de Markov à un nombre fini d'états. *Revue Mathematique de l'Union Interbalkanique* **2** 77–105. {**157**}

Doob, J. L. (1936). Note on probability. *Annals of Mathematics* **37** (2) 363–367. {**315**}

Doob, J. L. (1940). Regularity properties of certain families of chance variables. *Transactions of the American Mathematical Society* **47** (3) 455–486. {**115, 117**}

Doob, J. L. (1949). Application of the theory of martingales. In *Le calcul des probabilités et ses applications*, pp. 23–27. Centre National de la Recherche Scientifique, Paris. {**115, 116**}

Doob, J. L. (1953). *Stochastic Processes*. John Wiley & Sons, Inc., New York. {**68, 115, 117**}

Dormoy, Émile (1872). Théorie mathématique des jeux de hasard. *Journal des Actuaires Français* **1** (2) 120–146, 232–257. Also **2** (5) 38–57, 1873. {**618**}

Dormoy, Émile (1873). *Théorie mathématique du jeu de baccarat*. Armand Anger, Libraire-Éditeur, Paris. {**618, 639**}

Dostoevsky, Fyodor (1967). *The Gambler*. The Heritage Press, New York. English translation by Constance Garnett. {**480**}

Dowling, Allen (1970). *The Great American Pastime*. A. S. Barnes and Company, Cranbury, NJ. {**733**}

Downton, F. (1969). A note on betting strategies in compound games. *Journal of the Royal Statistical Society, Series A* **132** (4) 543–547. {**238**}

Downton, F. (1980a). Monotonicity and the gain ratio. *Journal of the Royal Statistical Society, Series A* **143** (1) 41–42. {**240**}

Downton, F. (1980b). A note on Labouchere sequences. *Journal of the Royal Statistical Society, Series A* **143** (3) 363–366. {**313, 315**}

Downton, F. (1982). Rational roulette. *Bulletin of the Australian Mathematical Society* **26** (3) 399–420. {**315**}

Downton, F. and Holder, R. L. (1972). Banker's games and the Gaming Act 1968. *Journal of the Royal Statistical Society, Series A* **135** (3) 336–364. {**198, 239, 481, 620**}

Downton, F. and Lockwood, Carmen (1975). Computer studies of baccarat, I: Chemin-de-fer. *Journal of the Royal Statistical Society, Series A* **138** (2) 228–238. {**620**}

Downton, F. and Lockwood, Carmen (1976). Computer studies of baccarat, II: Baccarat-banque. *Journal of the Royal Statistical Society, Series A* **139** (3) 356–364. {**614, 620**}

Drago, Harry Sinclair (1969). *Notorious Ladies of the Frontier.* Dodd, Mead & Company, Inc., New York. {**676**}

Drakakis, Konstantinos (2005). Distances between the winning numbers in lottery. arXiv:math.CO/0507469. {**500**}

Dresher, Melvin (1951). Games of strategy. *Mathematics Magazine* **25** (2) 93–99. {**198**}

Dresher, Melvin (1961). *Games of Strategy: Theory and Applications.* Prentice–Hall, Inc., Englewood Cliffs, NJ. Reprinted as *The Mathematics of Games of Strategy: Theory and Applications,* Dover Publications, Inc., New York, 1981. {**197**}

Drummond, Bob (1985). 'Phantom' gambler lost ultimate game. *Dallas Times Herald* (Feb. 12) 1, 6. {**524**}

Dubins, Lester E. (1968). A simpler proof of Smith's roulette theorem. *Annals of Mathematical Statistics* **39** (2) 390–393. {**356**}

Dubins, Lester E. (1972). On roulette when the holes are of various sizes. *Israel Journal of Mathematics* **11** (2) 153–158. {**356**}

Dubins, Lester E. and Savage, Leonard J. (1965). *How to Gamble If You Must: Inequalities for Stochastic Processes.* McGraw-Hill Book Co., New York. {**239, 355, 389**}

Dubins, Lester E. and Savage, Leonard J. (1976). *Inequalities for Stochastic Processes (How to Gamble If You Must)* (corrected ed.). Dover Publications, Inc., New York. {**vi, 239, 315, 355, 356**}

Dunbar and B., Jeff (Math Boy) (1999). Risk of ruin for video poker and other skewed-up games. *Blackjack Forum* **19** (3) 21–27. {**274**}

Durrett, Richard (2004). *Probability: Theory and Examples* (third ed.). Duxbury Press, Belmont, CA. First edition, Wadsworth & Brooks/Cole, Pacific Grove, CA, 1991. Second edition, Duxbury Press, Belmont, CA, 1996. {**157**}

Edelman, David (c. 1979). Supremum of mean-median differences for the binomial and Poisson distributions: ln 2. Technical report, Department of Mathematical Statistics, Columbia University. {**750**}

Edwards, A. W. F. (1982). Pascal and the problem of points. *International Statistical Review* **50** (3) 259–266. {**69**}

Edwards, A. W. F. (1983). Pascal's problem: The "gambler's ruin." *International Statistical Review* **51** (1) 73–79. {**271**}

Edwards, A. W. F. (1987). *Pascal's Arithmetical Triangle.* Charles Griffin & Company Ltd., London. {**64, 65, 68, 94**}

Einstein, Charles (1968). *How to Win at Blackjack.* Cornerstone Library, New York. {**681**}

Emery, Steuart M. (1926). Monte Carlo gambler has 13 years of luck. *New York Times* (Mar. 28) Section 9, p. 22. {**641**}

Engel, Arthur (1993). The computer solves the three tower problem. *American Mathematical Monthly* **100** (1) 62–64. {**117**}

Enns, E. G. and Tomkins, D. D. (1993). Optimal betting allocations. *Mathematical Scientist* **18** (1) 37–42. {**390**}

Epstein, Richard A. (1967). *The Theory of Gambling and Statistical Logic.* Academic Press, New York. {**vii, 273, 681, 683**}

Epstein, Richard A. (1977). *The Theory of Gambling and Statistical Logic* (revised ed.). Academic Press, New York. {**vii, 479, 613, 681, 683**}

Epstein, Richard A. (2009). *The Theory of Gambling and Statistical Logic* (second ed.). Elsevier/Academic Press, Amsterdam. {**vii, 66, 238, 683, 687**}

Erdős, P., Feller, W., and Pollard, H. (1949). A property of power series with positive coefficients. *Bulletin of the American Mathematical Society* **55** (2) 201–204. {**157**}

Erdős, Paul and Rényi, Alfréd (1959). On the central limit theorem for samples from a finite population. *Publications of the Mathematical Institute of the Hungarian Academy of Sciences* **4** (1) 49–57. {**750**}

Estafanous, Marc and Ethier, S. N. (2009). The duration of play in games of chance with win-or-lose outcomes and general payoffs. *Mathematical Scientist* **34** (2) 99–106. {**315**}

Ethier, S. N. (1982a). On the definition of the house advantage. In Eadington, William R. (Ed.), *The Gambling Papers: Proceedings of the Fifth National Conference on Gambling and Risk Taking, Vol. 13. Quantitative Analysis of Gambling: Stock Markets and Other Games*, pp. 46–67. Bureau of Business and Economic Research, University of Nevada, Reno. {**239**}

Ethier, S. N. (1982b). Testing for favorable numbers on a roulette wheel. *Journal of the American Statistical Association* **77** (379) 660–665. {**482**}

Ethier, S. N. (1987). Improving on bold play at craps. *Operations Research* **35** (6) 814–819. {**524**}

Ethier, S. N. (1988). The proportional bettor's fortune. In Eadington, William R. (Ed.), *Gambling Research: Proceedings of the Seventh International Conference on Gambling and Risk Taking. Vol. 4. Quantitative Analysis and Gambling*, pp. 375–383. Bureau of Business and Economic Research, University of Nevada, Reno. {**390**}

Ethier, S. N. (1996). A gambling system and a Markov chain. *Annals of Applied Probability* **6** (4) 1248–1259. {**157, 314**}

Ethier, S. N. (1998). An optional stopping theorem for nonadapted martingales. *Statistics & Probability Letters* **39** (3) 283–288. {**117, 315**}

Ethier, S. N. (1999). Thackeray and the Belgian progression. *Mathematical Scientist* **24** (1) 1–23. {**316**}

Ethier, S. N. (2000). Analysis of a gambling system. In Vancura, Olaf, Cornelius, Judy A., and Eadington, William R. (Eds.), *Finding the Edge: Mathematical Analysis of Casino Games*, pp. 3–18. Institute for the Study of Gambling and Commercial Gaming, University of Nevada, Reno. {**314**}

Ethier, S. N. (2004). The Kelly system maximizes median fortune. *Journal of Applied Probability* **41** (4) 1230–1236. {**390**}

Ethier, S. N. (2007a). Faro: From soda to hock. In Ethier, Stewart N. and Eadington, William R. (Eds.), *Optimal Play: Mathematical Studies of Games and Gambling*, pp. 225–235. Institute for the Study of Gambling and Commercial Gaming, University of Nevada, Reno. {**589**}

Ethier, S. N. (2007b). A Bayesian analysis of the shooter's hand at craps. In Ethier, Stewart N. and Eadington, William R. (Eds.), *Optimal Play: Mathematical Studies of Games and Gambling*, pp. 311–322. Institute for the Study of Gambling and Commercial Gaming, University of Nevada, Reno. {**517, 522**}

Ethier, S. N. (2007c). Optimal play in subfair compound games. In Ethier, Stewart N. and Eadington, William R. (Eds.), *Optimal Play: Mathematical Studies of Games and Gambling*, pp. 525–540. Institute for the Study of Gambling and Commercial Gaming, University of Nevada, Reno. {**238**}

Ethier, S. N. (2008). Absorption time distribution for an asymmetric random walk. In Ethier, Stewart N., Feng, Jin, and Stockbridge, Richard H. (Eds.), *Markov Processes and Related Topics: A Festschrift for Thomas G. Kurtz*, pp. 31–40. IMS Collections **4**. Institute of Mathematical Statistics, Beachwood, OH. {**315**}

Ethier, S. N. and Hoppe, Fred M. (2010). A world record in Atlantic City and the length of the shooter's hand at craps. *Mathematical Intelligencer* **32** to appear. arXiv:0906.1545. {**523**}

Ethier, S. N. and Khoshnevisan, Davar (2002). Bounds on gambler's ruin probabilities in terms of moments. *Methodology and Computing in Applied Probability* **4** (1) 55–68. {**274, 571**}

Ethier, S. N. and Lee, Jiyeon (2009). Limit theorems for Parrondo's paradox. *Electronic Journal of Probability* **14** 1827–1862. arXiv:0902.2368. {**158**}

Ethier, S. N. and Lee, Jiyeon (2010). A Markovian slot machine and Parrondo's paradox. *Annals of Applied Probability* **20** to appear. arXiv:0906.0792. {**158, 460**}

Ethier, S. N. and Levin, David A. (2005). On the fundamental theorem of card counting with application to the game of trente et quarante. *Advances in Applied Probability* **37** (1) 90–107. {**425, 637, 638, 640**}

Ethier, S. N. and Tavaré, S. (1983). The proportional bettor's return on investment. *Journal of Applied Probability* **20** (3) 563–573. {**390**}

Euler, Leonhard (1753). Calcul de la probabilité dans le jeu de rencontre (E201). *Mémoires de l'Académie des Sciences de Berlin* **7** 255–270. {**67**}

Euler, Leonhard (1764). Sur l'avantage du banquier au jeu de pharaon (E313). *Mémoires de l'Académie de Sciences de Berlin* **20** 144–164. {**236, 593–595**}

Feitt, Patrick W. (2008). An analysis of standard and casino-style *trente-et-quarante*. Senior project, Allegheny College. {**642**}

Feller, W. (1945). Note on the law of large numbers and "fair" games. *Annals of Mathematical Statistics* **16** (3) 301–304. {**71, 73**}

Feller, William (1968). *An Introduction to Probability Theory and Its Applications, Vol. I* (third ed.). John Wiley & Sons, Inc., New York. First edition, 1950; second edition, 1957. {**66, 68, 71, 73, 94, 117, 157, 273, 315**}

Ferguson, C. and Ferguson, T. S. (2003). On the Borel and von Neumann poker models. In Petrosjan, L. A. and Mazalov, V. V. (Eds.), *Game Theory and Applications, Vol. 9*, pp. 17–32. Nova Science Publishers, New York. {**741**}

Ferguson, Chris and Ferguson, Tom (2007). The endgame in poker. In Ethier, Stewart N. and Eadington, William R. (Eds.), *Optimal Play: Mathematical Studies of Games and Gambling*, pp. 79–106. Institute for the Study of Gambling and Commercial Gaming, University of Nevada, Reno. {**741, 743**}

Ferguson, Thomas S. (2009). *Game Theory.* http://www.math.ucla.edu/~tom/Game_Theory/Contents.html. {**198**}

Fey, Marshall (2002). *Slot Machines: America's Favorite Gaming Device* (sixth ed.). Liberty Belle Books, Reno. {**69, 455, 456, 459, 460, 570**}

Finkelstein, Mark and Thorp, Edward O. (2007). Nontransitive dice. In Ethier, Stewart N. and Eadington, William R. (Eds.), *Optimal Play: Mathematical Studies of Games and Gambling*, pp. 293–310. Institute for the Study of Gambling and Commercial Gaming, University of Nevada, Reno. {**198**}

Finkelstein, Mark and Whitley, Robert (1981). Optimal strategies for repeated games. *Advances in Applied Probability* **13** (2) 415–428. {**117, 389, 390**}

Fisher, George Henry (1934a). *Stud Poker Blue Book.* The Stud Poker Press, Los Angeles. {**740**}

Fisher, R. A. (1918). The correlation between relatives on the supposition of Mendelian inheritance. *Transactions of the Royal Society of Edinburgh* **52** 399–433. {**69**}

Fisher, R. A. (1934b). Randomisation, and an old enigma of card play. *Mathematical Gazette* **18** (231) 294–297. {**198**}

Fleming, Ian (1953). *Casino Royale.* Jonathan Cape Ltd., London. {**481, 616, 738**}

Fleming, Ian (2002). *Casino Royale.* Penguin Group (USA) Inc. and Penguin Books Ltd., New York and London. {**616**}

Fletcher, R. (2000). *Practical Methods of Optimization* (second ed.). John Wiley & Sons, Ltd., Chichester, UK. {**749**}

Flowers, Arthur and Curtis, Anthony (2000). *The Art of Gambling: Through the Ages.* Huntington Press, Las Vegas. {**738**}

Forte, Steve (2004). *Casino Game Protection: A Comprehensive Guide.* SLF Publishing, LLC, Las Vegas. {**482**}

Fortuin, C. M., Kasteleyn, P. W., and Ginibre, J. (1971). Correlation inequalities on some partially ordered sets. *Communications in Mathematical Physics* **22** (2) 89–103. {**73, 482**}

Foster, F. G. (1953a). On the stochastic matrices associated with certain queuing processes. *Annals of Mathematical Statistics* **24** (3) 355–360. {**157**}

Foster, F. G. (1964). A computer technique for game-theoretic problems I: Chemin-de-fer analyzed. *Computer Journal* **7** (2) 124–130. {**619**}

Foster, R. F. (1905). *Practical Poker*. Brentano's, New York. {**731, 739**}

Foster, R. F. (1953b). *Foster's Complete Hoyle: An Encyclopedia of Games* (revised and enlarged ed.). J. B. Lippincott, Philadelphia. First edition, Frederick A. Stokes Company, 1897. {**479, 613, 614**}

Friedman, Bill (1996). *Casino Games* (revised and updated ed.). Golden Press, New York. First edition, 1973. {**481**}

Friedman, Joel (1982). Understanding and applying the Kelly criterion. In Eadington, William R. (Ed.), *The Gambling Papers: Proceedings of the Fifth National Conference on Gambling and Risk Taking, Vol. 10. The Blackjack Papers*, pp. 128–139. Bureau of Business and Economic Research, University of Nevada, Reno. {**390**}

Frome, Elliot A. (2003). *Expert Strategy for Three Card Poker*. Compu-Flyers, Bogota, NJ. {**543**}

Frome, Elliot A. and Frome, Ira D. (1996). *Expert Strategy for Caribbean Stud Poker*. Compu-Flyers, Bogota, NJ. {**543**}

Frome, Lenny (1989). *Expert Video Poker for Las Vegas*. Compu-Flyers, Bogota, NJ. {**571**}

Fuchs, Camil and Kenett, Ron (1980). A test for detecting outlying cells in the multinomial distribution and two-way contingency tables. *Journal of the American Statistical Association* **75** (370) 395–398. {**482**}

Gale, David (1960). *The Theory of Linear Economic Models*. McGraw-Hill Book Company, Inc., New York. {**197**}

Gall, Martin [Arnous de Rivière, Jules] (1883). *La roulette et le trente-et-quarante*. Delarue, Libraire-Éditeur, Paris. {**479, 639, 641**}

Gallian, Joseph A. (1994). *Contemporary Abstract Algebra*. D. C. Heath and Company, Lexington, MA. {**73**}

Galton, Francis (1888). Co-relations and their measurement, chiefly from anthropometric data. *Proceedings of the Royal Society of London* **45** 135–145. {**70**}

Gambarelli, Gianfranco and Owen, Guillermo (2004). The coming of game theory. *Theory and Decision* **56** (1–2) 1–18. {**198**}

Gambino, Tino (2006). *The Mad Professor's Crapshooting Bible*. Pi Yee Press, Las Vegas. {**521**}

Gardner, Martin (1978). Mathematical games. *Scientific American* **238** (2) 19–32. {**73**}

Geddes, Robert N. (1980). The Mills Futurity. *Loose Change* **3** (1) 10–14, 21. {**460**}

Geddes, Robert N. and Saul, David L. (1980). The mathematics of the Mills Futurity slot machine. *Loose Change* **3** (4) 22–27. {**460**}

Gilbert, Edgar N. (1955). Theory of shuffling. Bell Telephone Laboratories technical memorandum. {**424**}

Gilliland, Dennis, Levental, Shlomo, and Xiao, Yimin (2007). A note on absorption probabilities in one-dimensional random walk via complex-valued martingales. *Statistics and Probability Letters* **77** (11) 1098–1105. {**273**}

Golomb, Solomon W., Berlekamp, Elwyn, Cover, Thomas M., Gallager, Robert G., Massey, James L., and Viterbi, Andrew J. (2002). Claude Elwood Shannon (1916–2001). *Notices of the American Mathematical Society* **49** (1) 8–16. {**116**}

Gordon, Edward (2000). An accurate analysis of video poker. In Vancura, Olaf, Cornelius, Judy A., and Eadington, William R. (Eds.), *Finding the Edge: Mathematical Analysis of Casino Games*, pp. 379–392. Institute for the Study of Gambling and Commercial Gaming, University of Nevada, Reno. {**571**}

Gordon, Phil and Grotenstein, Jonathan (2004). *Poker: The Real Deal*. Simon Spotlight Entertainment, New York. {**736, 737, 742**}

Goren, Charles H. (1961). *Goren's Hoyle: Encyclopedia of Games*. Chancellor Hall, Ltd., New York. {**571**}

Graves, Charles (1963). *None But the Rich: The Life and Times of the Greek Syndicate*. Cassell & Company, Ltd., London. {**617**}

Green, J. H. (1843). *An Exposure of the Arts and Miseries of Gambling; Designed Especially as a Warning to the Youthful and Inexperienced, against the Evils of That Odious and Destructive Vice.* U. P. James, Cincinnati. {**520, 591, 731**}

Grégoire, G. (1853). *Traité du trente-quarante contenant des analyses et des faits pratiques du plus haut intérêt suivis d'une collection de plus de 40,000 coups de banque.* Comptoir des Imprimeurs-Unis, Paris. {**639**}

Griffin, Peter (1976). The rate of gain in player expectation for card games characterized by sampling without replacement and an evaluation of card counting systems. In Eadington, William R. (Ed.), *Gambling and Society: Interdisciplinary Studies on the Subject of Gambling*, pp. 429–442. Charles C. Thomas, Springfield, IL. {**425, 681, 682**}

Griffin, Peter (1991). *Extra Stuff: Gambling Ramblings.* Huntington Press, Las Vegas. {**69, 72, 94, 236, 273, 274, 356, 390, 500, 522**}

Griffin, Peter (1998). Letter to E. O. Thorp dated October 16, 1970. *Blackjack Forum* **18** (4) 19–20. {**686**}

Griffin, Peter (2000). A short note on the expected duration of the Australian game "two-up." In Vancura, Olaf, Cornelius, Judy A., and Eadington, William R. (Eds.), *Finding the Edge: Mathematical Analysis of Casino Games*, pp. 393–395. Institute for the Study of Gambling and Commercial Gaming, University of Nevada, Reno. {**73**}

Griffin, Peter and Gwynn, John M., (2000). An analysis of Caribbean Stud Poker. In Vancura, Olaf, Cornelius, Judy A., and Eadington, William R. (Eds.), *Finding the Edge: Mathematical Analysis of Casino Games*, pp. 273–284. Institute for the Study of Gambling and Commercial Gaming, University of Nevada, Reno. {**543, 544**}

Griffin, Peter A. (1984). Different measures of win rate for optimal proportional betting. *Management Science* **30** (12) 1540–1547. {**390**}

Griffin, Peter A. (1988). Insure a "good" hand? (Part 2). *Blackjack Forum* **8** (2) 12–13. {**686**}

Griffin, Peter A. (1999). *The Theory of Blackjack: The Compleat Card Counter's Guide to the Casino Game of 21* (sixth ed.). Huntington Press, Las Vegas. First and second editions, GBC Press, Las Vegas, 1979 and 1981. Third edition, Faculty Publishing, Davis, CA, 1986. Fourth and fifth editions, Huntington Press, Las Vegas, 1988 and 1996. {**66, 68, 71, 274, 425, 620, 621, 682–687**}

Grimmett, Geoffrey R. and Stirzaker, David R. (2001). *One Thousand Exercises in Probability.* Oxford University Press, Oxford, UK. {**66, 73, 117, 157, 315**}

Grosjean, James (2000). *Beyond Counting: Exploiting Casino Games from Blackjack to Video Poker.* RGE Publishing, Oakland, CA. {**683**}

Grosjean, James (2007a). Much ado about *baccara.* In Ethier, Stewart N. and Eadington, William R. (Eds.), *Optimal Play: Mathematical Studies of Games and Gambling*, pp. 143–173. Institute for the Study of Gambling and Commercial Gaming, University of Nevada, Reno. {**621**}

Grosjean, James (2007b). Are casinos paranoid? Can players get an edge by sharing information? In Ethier, Stewart N. and Eadington, William R. (Eds.), *Optimal Play: Mathematical Studies of Games and Gambling*, pp. 211–223. Institute for the Study of Gambling and Commercial Gaming, University of Nevada, Reno. {**544**}

Grosjean, James (2009). *Exhibit CAA. Beyond Counting: Exploiting Casino Games from Blackjack to Video Poker.* South Side Advantage Press, Las Vegas. {**68, 69, 72, 236, 425, 482, 521, 523, 542, 543, 621, 683, 686, 687**}

Grosjean, James and Chen, Frederick (2007). Rebated losses revisited. In Ethier, Stewart N. and Eadington, William R. (Eds.), *Optimal Play: Mathematical Studies of Games and Gambling*, pp. 507–524. Institute for the Study of Gambling and Commercial Gaming, University of Nevada, Reno. {**69**}

Grotenstein, Jonathan and Reback, Storms (2005). *All In: The (Almost) Entirely True Story of the World Series of Poker.* Thomas Dunne Books, an imprint of St. Martin's Press, New York. {**734–736**}

Guerrera, Tony (2007). *Killer Poker by the Numbers: The Mathematical Edge for Winning Play*. Lyle Stuart, Kensington Publishing Corp., New York. {**741, 743**}

Guilbaud, G. Th. (1961). Faut-il jouer au plus fin? (Notes sur l'histoire de la théorie des jeux). In *La décision*, pp. 171–182. Centre National de la Recherche Scientifique, Paris. {**198**}

H., G. (1842). *Hoyle's Games, Improved and Enlarged by New and Practical Treatises, with the Mathematical Analysis of the Chances of the Most Fashionable Games of the Day, Forming an Easy and Scientific Guide to the Gaming Table, and the Most Popular Sports of the Field*. Longman, Brown, & Co., London. {**593**}

Hacking, Ian (1975). *The Emergence of Probability: A Philosophical Study of Early Ideas about Probability, Induction and Statistical Inference*. Cambridge University Press, London. {**64**}

Haigh, John (1999). *Taking Chances: Winning with Probability*. Oxford University Press, Oxford, UK. {**66**}

Hald, A. and Johansen, S. (1983). On de Moivre's recursive formulae for the duration of play. *International Statistical Review* **51** (3) 239–253. {**273**}

Hald, Anders (1990). *History of Probability and Statistics and Their Applications before 1750*. John Wiley & Sons, Inc., New York. {**64, 66, 70, 94, 198, 273, 594**}

Hald, Anders (1998). *A History of Mathematical Statistics from 1750 to 1930*. John Wiley & Sons, Inc., New York. {**64**}

Hamza, Kais (1995). The smallest uniform upper bound on the distance between the mean and the median of the binomial and Poisson distributions. *Statistics & Probability Letters* **23** (1) 21–25. {**750**}

Hannum, Robert (1998). Mathematical analysis of baccarat: Expected bank wager outcomes under "barred" bank totals. Unpublished. {**621**}

Hannum, Robert (2000). Casino card shuffles: How random are they? In Vancura, Olaf, Cornelius, Judy A., and Eadington, William R. (Eds.), *Finding the Edge: Mathematical Analysis of Casino Games*, pp. 33–55. Institute for the Study of Gambling and Commercial Gaming, University of Nevada, Reno. {**425**}

Hannum, Robert C. (2007). The partager rule at roulette: Analysis and case of a million-euro win. In Ethier, Stewart N. and Eadington, William R. (Eds.), *Optimal Play: Mathematical Studies of Games and Gambling*, pp. 325–344. Institute for the Study of Gambling and Commercial Gaming, University of Nevada, Reno. {**480**}

Hannum, Robert C. and Cabot, Anthony N. (2005). *Practical Casino Math* (second ed.). Institute for the Study of Gambling and Commercial Gaming, University of Nevada, Reno. First edition, 2001. {**66, 237, 544**}

Hannum, Robert C. and Kale, Sudhir H. (2004). The mathematics and marketing of dead chip programmes: Finding and keeping the edge. *International Gambling Studies* **4** (1) 33–45. {**240**}

Harmer, Gregory P. and Abbott, Derek (2002). A review of Parrondo's paradox. *Fluctuation and Noise Letters* **2** (2) R71–R107. {**158**}

Hausdorff, Felix (1914). *Grundzüge der Mengenlehre*. Verlag von Veit & Comp., Leipzig. {**70**}

Havers, Michael, Grayson, Edward, and Shankland, Peter (1988). *The Royal Baccarat Scandal* (second ed.). Souvenir Press Ltd., London. {**616**}

Heath, David C., Pruitt, William E., and Sudderth, William D. (1972). Subfair red-and-black with a limit. *Proceedings of the American Mathematical Society* **35** (2) 555–560. {**356**}

Henny, Julian (1973). *Niklaus und Johann Bernoullis Forschungen auf dem Gebiet der Wahrscheinlichkeitsrechnung in ihrem Briefwechsel mit Pierre Rémond de Montmort*. Birkhäuser Verlag, Basel. {**70, 73**}

Herstein, I. N. (1964). *Topics in Algebra*. Blaisdell Publishing Company, Waltham, MA. {**745, 746**}

Herstein, I. N. and Milnor, John (1953). An axiomatic approach to measurable utility. *Econometrica* **21** (2) 291–297. {**198**}

Hildreth, James (1836). *Dragoon Campaigns to the Rocky Mountains; Being a History of the Enlistment, Organization, and First Campaigns of the Regiment of United States Dragoons; Together with Incidents of a Soldier's Life, and Sketches of Scenery and Indian Character.* Wiley & Long, New York. Author listed as "A Dragoon." Some say the author was William L. Gordon-Miller. **{731}**

Hill, W. Lawrence (1989). Hollywood hold'em. *Card Player* **2** (31, Dec. 15) 5, 34. **{743}**

Hill, W. Lawrence (1990a). All-time classic heads-up hold'em match-up. *Card Player* **3** (6, Mar 23) 26–27. **{744}**

Hill, W. Lawrence (1990b). Up against a pair of aces. *Card Player* **3** (10, May 18) 20–21. **{743}**

Hill, W. Lawrence (1991a). Minor classic . . . how should we figure jack-10 offsuit versus a pair of fives? *Card Player* **4** (10, May 17) 32. **{744}**

Hill, W. Lawrence (1991b). Hold'em matches and mismatches. *Card Player* **4** (18, Sept. 6) 30. **{741, 743, 744}**

Hill, W. Lawrence (1991c). Closing in on the closest match-up. *Card Player* **4** (26, Dec. 27) 45. **{744}**

Hill, W. Lawrence (1992). Circling the closest matchup. *Card Player* **5** (23, Nov. 13) 81. **{744}**

Hill, W. Lawrence (1993). Scissors, paper, rock. *Card Player* **6** (8, Apr. 23) 56. **{744}**

Hill, W. Lawrence (1994a). Near-missing the closest matchup. *Card Player* **7** (12, June 17) 36. **{744}**

Hill, W. Lawrence (1994b). More close matchups. *Card Player* **7** (22, Nov. 4) 56. **{744}**

Hill, W. Lawrence (1995). Making the list of close matchups. *Card Player* **8** (7, Apr. 7) 48. **{744}**

Hill, W. Lawrence (1996a). Mining out the lode of close matchups. *Card Player* **9** (4, Feb. 23) 52–53. **{744}**

Hill, W. Lawrence (1996b). Count 'em. *Card Player* **9** (16, Aug. 9) 46–47. **{740}**

Hill, W. Lawrence (1997). The median seven-card hand. *Card Player* **10** (26, Dec. 26) 26, 99. **{72}**

Hoel, Paul G., Port, Sidney C., and Stone, Charles J. (1972). *Introduction to Stochastic Processes.* Houghton Mifflin Company, Boston. **{745}**

Hoffman, Paul (1998). *The Man Who Loved Only Numbers: The Story of Paul Erdős and the Search for Mathematical Truth.* Hyperion Books, New York. **{72}**

Hoffmann, Professor [Lewis, Angelo John] (1891). *Baccarat Fair and Foul, Being an Explanation of the Game, and a Warning against Its Dangers.* George Routledge and Sons, Limited, London. Reprinted by GBC Press, Las Vegas, 1977. **{614, 618}**

Holden, Anthony (1990). *Big Deal: A Year as a Professional Poker Player.* Viking, New York. **{743}**

Holmes, A. H. (1907). Problem 192. *American Mathematical Monthly* **14** (10) 189. **{619}**

Hoppe, Fred M. (2007). Branching processes and the effect of parlaying bets on lottery odds. In Ethier, Stewart N. and Eadington, William R. (Eds.), *Optimal Play: Mathematical Studies of Games and Gambling*, pp. 383–395. Institute for the Study of Gambling and Commercial Gaming, University of Nevada, Reno. **{274}**

Howard, J. V. (1994). A geometrical method of solving certain games. *Naval Research Logistics* **41** (1) 133–136. **{198}**

Hughes, Barrie (1976). *The Educated Gambler: A Guide to Casino Games.* Stanley Paul, London. **{616}**

Humble, Lance (1976). *Blackjack Gold: A New Approach to Winning at "21."* International Gaming Incorporated, Toronto. **{682}**

Humble, Lance and Cooper, Carl (1980). *The World's Greatest Blackjack Book* (revised ed.). Doubleday, New York. **{682}**

Huygens, C. (1657). De ratiociniis in ludo aleae. In van Schooten, Frans (Ed.), *Exercitationum Mathematicarum*, pp. 517–534. Johannis Elsevirii, Leiden. **{v, viii, 68, 69, 72, 271}**

Huyn, P. N. (1788). *La théorie des jeux de hasard, ou analyse du krabs, du passe-dix, de la roulette, du trente & quarante, du pharaon, du biribi & du lotto.* Publisher unknown, Paris. {**478, 519, 594, 613, 639, 640**}

Imbs, Paul (1974). *Trésor de la langue française.* Éditions du Centre National de la Recherche Scientifique, Paris. {**115, 613**}

Ionescu Tulcea, C. (1981). *A Book on Casino Craps, Other Dice Games & Gambling Systems.* Van Nostrand Reinhold Co., New York. {**239**}

Jacoby, Oswald (1947). *Oswald Jacoby on Poker* (revised ed.). Doubleday & Company, Inc., Garden City, NY. {**735**}

Jensen, J. L. W. V. (1906). Sur les fonctions convexes et les inégalités entre les valeurs moyennes. *Acta Mathematica* **30** 175–193. {**70**}

Jensen, Stephen (2001). Optimal drawing strategy for Deuces Wild video poker. University of Utah REU report. {**571**}

Jessup, Richard (1963). *The Cincinnati Kid.* Little, Brown and Company, Boston. {**738**}

Johnson, Craig Alan (1997). Simulation of the Belgian progression and Oscar's system. Master's thesis, University of Utah. {**314, 316**}

Johnson, Norman L., Kotz, Samuel, and Balakrishnan, N. (1997). *Discrete Multivariate Distributions.* John Wiley & Sons, Inc., New York. {**73**}

Johnston, David (1992). *Temples of Chance: How America Inc. Bought Out Murder Inc. to Win Control of the Casino Business.* Doubleday, New York. {**614, 680**}

Jones, Herbert B. (1978). *Slot-Machines: An Historical and Technological Report.* Bally Manufacturing Corporation, Chicago. {**457, 459, 570**}

Jordan, C. (1867). De quelques formules de probabilité. *Comptes Rendus de l'Académie des Sciences, Paris* **65** 993–994. {**70**}

Judah, Sherry and Ziemba, William T. (1983). Three person baccarat. *Operations Research Letters* **2** (4) 187–192. {**620**}

Kallenberg, Olav (2005). *Probabilistic Symmetries and Invariance Principles.* Springer, New York. {**425**}

Karlin, Samuel (1959a). *Mathematical Methods and Theory in Games, Programming, and Economics, Vol. I. Matrix Games, Programming, and Mathematical Economics.* Addison-Wesley Publishing Company, Inc., Reading, MA. {**743**}

Karlin, Samuel (1959b). *Mathematical Methods and Theory in Games, Programming, and Economics, Vol. II. The Theory of Infinite Games.* Addison-Wesley Publishing Company, Inc., Reading, MA. {**741**}

Karush, William (1939). Minima of functions of several variables with inequalities as side conditions. Master's thesis, University of Chicago. {**749**}

Kavanagh, Thomas M. (2005). *Dice, Cards, Wheels: A Different History of French Culture.* University of Pennsylvania Press, Philadelphia. {**355, 517, 642**}

Keller, John W. (1887). *The Game of Draw Poker.* White, Stokes, & Allen, New York. {**732**}

Kelly, Bill (1995). *Gamblers of the Old West: Gambling Men and Women of the 1800s. How They Lived—How They Died.* B & F Enterprises, Las Vegas. {**591**}

Kelly, J. L., Jr. (1956). A new interpretation of information rate. *Bell System Technical Journal* **35** 917–926. {**71, 388, 389**}

Kemeny, John G. and Snell, J. Laurie (1957). Game-theoretic solution of baccarat. *American Mathematical Monthly* **64** (7) 465–469. {**198, 619**}

Kemeny, John G., Snell, J. Laurie, and Knapp, Anthony W. (1966). *Denumerable Markov Chains.* D. Van Nostrand Company, Inc., Princeton, NJ. Second edition, Springer-Verlag, New York, 1976. {**157**}

Kendall, M. G. (1956). Studies in the history of probability and statistics: II. The beginnings of a probability calculus. *Biometrika* **43** (1–2) 1–14. {**517**}

Kendall, M. G. and Murchland, J. D. (1964). Statistical aspects of the legality of gambling. *Journal of the Royal Statistical Society, Series A* **127** (3) 359–391. {**198, 619, 620**}

Khinchin, A. (1929). Sur la loi des grands nombres. *Comptes Rendus de l'Académie des Sciences, Paris* **189** 477–479. {**70**}

Kilby, Jim and Fox, Jim (1998). *Casino Operations Management.* John Wiley & Sons, Inc., New York. {**460, 544**}

Kilby, Jim, Fox, Jim, and Lucas, Anthony F. (2005). *Casino Operations Management* (second ed.). John Wiley & Sons, Inc., Hoboken, NJ. {**237, 312, 459**}

Klotz, Jerome H. (2000). A winning strategy for roulette. In Vancura, Olaf, Cornelius, Judy A., and Eadington, William R. (Eds.), *Finding the Edge: Mathematical Analysis of Casino Games*, pp. 397–411. Institute for the Study of Gambling and Commercial Gaming, University of Nevada, Reno. {**389, 482**}

Knopfmacher, Arnold and Prodinger, Helmut (2001). A simple card guessing game revisited. *Electronic Journal of Combinatorics* **8** (2). {**425**}

Ko, Stanley (1997). *Mastering the Game of Let It Ride* (tenth ed.). Gambology, Las Vegas. First edition, 1995. {**543, 544**}

Ko, Stanley (1998). *Mastering the Game of Caribbean Stud Poker* (seventh ed.). Gambology, Las Vegas. First edition, 1995. {**237, 543, 544**}

Ko, Stanley (2001). *Mastering the Game of Three Card Poker* (third ed.). Gambology, Las Vegas. First edition, 1999. {**543, 544**}

Kolata, Gina (1995). In shuffling cards, 7 is winning number. *New York Times* (Jan. 9) Section C, p. 1. {**425**}

Kolmogorov, A. N. (1930). Sur la loi forte des grands nombres. *Comptes Rendus de l'Académie des Sciences, Paris* **191** 910–912. {**71**}

Kolmogorov, A. N. (1933). *Grundbegriffe der Wahrscheinlichkeitsrechnung.* Springer-Verlag, Berlin. {**66, 94**}

Kolmogorov, A. N. (1936). Zur Theorie der Markoffschen Ketten. *Mathematische Annalen* **112** (1) 155–160. {**157**}

Kolmogorov, A. N. (1937). Markov chains with a denumerable number of possible states. *Bulletin Moskov. Gosudarstvennogo Universiteta Matematika i Mekhanika* **1** (3) 1–16. In Russian. {**157**}

Kozek, Andrzej S. (1995). A rule of thumb (not only) for gamblers. *Stochastic Processes and Their Applications* **55** (1) 169–181. {**239, 273**}

Kozek, Andrzej S. (2002). Solution of a roulette-type ruin problem. A correction to: "A rule of thumb (not only) for gamblers." *Stochastic Processes and Their Applications* **100** (1–2) 301–311. {**273**}

Kozelka, Robert M. (1956). Approximate upper percentage points for extreme values in multinomial sampling. *Annals of Mathematical Statistics* **27** (2) 507–512. {**482**}

Kranes, David (1989). *Keno Runner: A Romance.* University of Utah Press, Salt Lake City. {**499**}

Kuhn, H. W. (1950). A simplified two-person poker. In Kuhn, H. W. and Tucker, A. W. (Eds.), *Contributions to the Theory of Games*, pp. 97–103. Princeton University Press, Princeton, NJ. {**741, 743**}

Kuhn, H. W. and Tucker, A. W. (1951). Nonlinear programming. In Neyman, Jerzy (Ed.), *Proceedings of the Second Berkeley Symposium on Mathematical Statistics and Probability*, pp. 481–492. University of California Press, Berkeley, CA. {**749**}

Kuhn, Harold (1968). James Waldegrave: Excerpt from a letter. In Baumol, William J. and Goldfeld, Stephen M. (Eds.), *Precursors in Mathematical Economics: An Anthology*, pp. 3–9. The London School of Economics and Political Science, London. {**197, 198**}

Kuhn, Harold W. (2003). *Lectures on the Theory of Games.* Princeton University Press, Princeton, NJ. {**197, 743**}

Lacroix, S. F. (1822). *Traité élémentaire du calcul des probabilités* (second ed.). Bachelier, Paris. First edition, 1816; third edition, 1833. {**311**}

Lafaye, P. and Krauss, E. (1927). *Nouvel exposé de la théorie mathématique du jeu de baccara.* Papéteries Nouvelles Imp., Paris. {**619**}

Lafrogne, Amiral (1927). *Calcul de l'avantage du banquier au jeu de baccara.* Gauthier-Villars et Cie, Paris. {**619**}

Lagrange, Joseph-Louis (1777). Recherches sur les suites récurrentes dont les termes varient de plusieurs manières différentes, ou sur l'intégration des équations linéaires aux différences finies et partielles; et sur l'usage de ces équations dans la théorie des hasards. *Nouveaux Mémoires de l'Académie Royale des Sciences et Belles-Lettres de Berlin, 1775* 183–272. {**273**}

Lambert, George (1812). *The Game of Hazard Investigated; or, the Difference between the Caster's and Setter's Expectations Correctly Ascertained and Exemplified in a Clear and Concise Manner: Together with Other Calculations on Events Arising Out of the Game.* George Lambert, Newmarket. Published also by Davis and Dickson, London, 1815. Reprinted in Bohn (1850). {**519**}

Laplace, Pierre-Simon (1774). Mémoire sur la probabilité des causes par les évènemens. *Mémoires de l'Académie Royale des Sciences Présentés par Divers Savans* **6** 621–656. English translation by Stephen M. Stigler, *Statistical Science* **1** (3) 359–378, 1986. {**68**}

Laplace, Pierre-Simon (1820). *Théorie analytique des probabilités* (third ed.). Courcier, Paris. First and second editions, 1812 and 1814. {**67, 71, 73**}

Latané, Henry Allen (1959). Criteria for choice among risky ventures. *Journal of Political Economy* **67** (2) 144–155. {**389**}

Laun [Delauney, Julien Felix] (c. 1891). *Traité théorique et pratique du baccarat.* Delarue, Libraire-Éditeur, Paris. {**618**}

Laurent, H. (1893). *Théorie des jeux de hasard.* Gauthier-Villars et Fils, Imprimeurs-Éditeurs, Paris. {**498**}

Lawler, Gregory F. (2006). *Introduction to Stochastic Processes* (second ed.). Chapman & Hall/CRC, Boca Raton, FL. {**157**}

Lawler, Gregory F. and Coyle, Lester N. (1999). *Lectures on Contemporary Probability.* American Mathematical Society, Providence, RI. {**424, 425**}

Le Myre, Georges (1935). *Le baccara.* Hermann & Cie, Éditeurs, Paris. {**619**}

Leib, John E. (2000). Limitations on Kelly, or the ubiquitous "$n \to \infty$." In Vancura, Olaf, Cornelius, Judy A., and Eadington, William R. (Eds.), *Finding the Edge: Mathematical Analysis of Casino Games*, pp. 233–258. Institute for the Study of Gambling and Commercial Gaming, University of Nevada, Reno. {**390**}

Leigh, Norman (1976). *Thirteen against the Bank.* William Morrow and Company, Inc., New York. {**315**}

Lemmel, Maurice (1964). *Gambling: Nevada Style.* Doubleday & Co., Inc., Garden City, NY. {**237, 481, 593–595**}

Lévy, Paul (1937). *Théorie de l'addition des variables aléatoires.* Gauthier-Villars, Éditeur, Paris. {**73, 115, 117**}

Lewis, Oscar (1953). *Sagebrush Casinos: The Story of Legal Gambling in Nevada.* Doubleday & Company, Inc., Garden City, NY. {**677**}

Li, Shuo-Yen Robert (1980). A martingale approach to the study of occurrence of sequence patterns in repeated experiments. *Annals of Probability* **8** (6) 1171–1176. {**117**}

Liggett, Thomas M. (1985). *Interacting Particle Systems.* Springer-Verlag, New York. {**482**}

Lindeberg, J. W. (1922). Eine neue Herleitung des Exponentialgesetzes in der Wahrscheinlichkeitsrechnung. *Mathematische Zeitschrift* **15** 211–225. {**71**}

Livingston, A. D. (1968). 'Hold me': A wild new poker game ... and how to tame it. *Life* (Aug. 16) 40–42. {**734**}

Livingston, A. D. (1971). *Poker Strategy and Winning Play.* Melvin Powers, Wilshire Book Company, North Hollywood, CA. {**734, 735**}

Llorente, Loreto (2007). A profitable strategy in the *pelota* betting market. In Ethier, Stewart N. and Eadington, William R. (Eds.), *Optimal Play: Mathematical Studies of Games and Gambling*, pp. 399–415. Institute for the Study of Gambling and Commercial Gaming, University of Nevada, Reno. {**94**}

Lorden, Gary A. (1980). Gambling with statistics. *[Caltech] Engineering & Science* **44** (2, Dec.) 6–9. {**390**}

Lyapunov, A. M. (1900). Sur une proposition de la théorie des probabilités. *Bulletin de l'Académie Impériale des Sciences de St.-Pétersbourg, Série 5* **13** (4) 359–386. {**71**}

M., D. (1739). *Calcul du jeu appellé par les françois le trente-et-quarante, et que l'on nomme à florence le trente-et-un. Cet ouvrage fait connoître toutes les manieres dont peuvent etre formez 31, 32, 33, 34, 35, 36, 37, 38, 39, & 40 par les XL. cartes du jeu d'hombre, et etablit la proportion qui est entre chacun de ces dix nombres.* Chez Bernard Paperini, Florence. {**638, 642**}

M., P. A. (1950). Review of *Le calcul des probabilités et ses applications. Journal of the Royal Statistical Society, Series A* **113** (4) 583–584. {**116**}

Mac Lane, Saunders and Birkhoff, Garrett (1988). *Algebra* (third ed.). Chelsea Publishing Co., New York. {**747**}

MacDougall, Mickey (1944). *MacDougall on Dice and Cards: Modern Rules, Odds, Hints and Warnings for Craps, Poker, Gin Rummy and Blackjack.* Coward-McCann, Inc., New York. {**677–679**}

MacGillavry, Kitty (1978). Le jeu de dés dans le fabliau de *St Pierre et le jongleur. Marche Romane* **28** 175–179. {**311**}

Machina, Mark (1982). "Expected utility" analysis without the independence axiom. *Econometrica* **50** (2) 277–323. {**198**}

Mahl, Huey (1977). Roulette dinner. *Casino & Sports* **2** 30. {**482**}

Maistrov, L. E. (1974). *Probability Theory: A Historical Sketch.* Academic Press, New York. English translation by Samuel Kotz. {**v, 64**}

Maitra, Ashok P. and Sudderth, William D. (1996). *Discrete Gambling and Stochastic Games.* Springer-Verlag, New York. {**355, 356**}

Malmuth, Mason (1992). A few simulations. *Card Player* **5** (3) 13. {**741**}

Mann, Brad (1995). How many times should you shuffle a deck of cards? In Snell, J. Laurie (Ed.), *Topics in Contemporary Probability and Its Applications*, pp. 261–289. CRC Press, Boca Raton, FL. {**424**}

Manson, A. R., Barr, A. J., and Goodnight, J. H. (1975). Optimum zero-memory strategy and exact probabilities for 4-deck blackjack. *American Statistician* **29** (2) 84–88. Correction **29** (4) 175. {**684**}

Mansuy, Roger (2009). The origins of the word "martingale." *Electronic Journal for History of Probability and Statistics* **5** (1) 1–10. English translation by Ronald Sverdlove of Mansuy, Roger (2005). Histoire de martingales. *Mathématiques et Sciences Humaines* **169** (1) 105–113. {**312**}

Markov, A. A. (1906). Extension of the law of large numbers to dependent events. *Bulletin de la Société Mathématique de Kazan, Série 2* **15** 135–156. {**157**}

Markov, Andrei A. (1912). *Wahrscheinlichkeitsrechnung.* B. G. Teubner, Leipzig. First published in Russian, 1900. Second edition, 1908. German translation of second edition by Heinrich Liebmann. {**273**}

Marsh, D. C. B. (1967). Solution of Problem E 1865. *American Mathematical Monthly* **74** (9) 1136–1137. {**71**}

Marshall, D. C. B. (2006). *The Poker Code: The Correct Draw for Every Video Poker Hand of Deuces Wild.* Marshall House, Inc. {**571**}

Martin-Löf, Anders (1985). A limit theorem which clarifies the "Petersburg paradox." *Journal of Applied Probability* **22** (3) 634–643. {**71**}

Maskelyne, John Nevil (1894). *Sharps and Flats: A Complete Revelation of the Secrets of Cheating at Games of Chance and Skill.* Longmans, Green, and Co., New York. {**591, 592**}

Maslov, Sergei and Zhang, Yi-Cheng (1998). Optimal investment strategy for risky assets. *International Journal of Theoretical and Applied Finance* **1** (3) 377–387. {**390**}

Maurer, David W. (1943). The argot of the faro bank. *American Speech* **18** (1) 3–11. {**591**}

Maurer, David W. (1950). The argot of the dice gambler. *Annals of the American Academy of Political and Social Science* **269** (May) 114–133. {**70, 239**}

Maxim, Hiram S. (1904). *Monte Carlo Facts and Fallacies.* Grant Richards, London. {**315, 316**}

May, John (1998). *Baccarat for the Clueless: A Beginner's Guide to Playing and Winning.* A Lyle Stuart Book, Carol Publishing Group, Secaucus, NJ. {**621**}

May, John (2004). Trente et quarante. http://ourworld.compuserve.com/homepages/ greenbaize21/trenteet.htm. {**641**}

McClure, Wayne (1983). *Keno Winning Ways* (second ed.). GBC Press, Las Vegas. First edition, 1979. {**496, 497, 499, 500**}

McManus, James (2009). *Cowboys Full: The Story of Poker.* Farrar, Straus and Giroux, New York. {**731, 734, 735, 739**}

Mead, Daniel R. (1983). *Handbook of Slot Machine Reel Strips.* Mead Publishing Corporation, Long Beach, CA. {**459, 460**}

Mead, David N. (2005). *Reel History: A Photographic History of Slot Machines.* Mead Publishing Company, Yucca Valley, CA. {**459**}

Mechigian, John (1972). *Encyclopedia of Keno: A Guide to Successful Gambling with Keno.* Funtime Enterprises, Inc., Fresno, CA. {**498, 500**}

Mertens, Jean-François, Samuel-Cahn, Ester, and Zamir, Shmuel (1978). Necessary and sufficient conditions for recurrence and transience of Markov chains, in terms of inequalities. *Journal of Applied Probability* **15** (4) 848–851. {**157**}

Mezrich, Ben (2002). *Bringing Down the House: The Inside Story of Six MIT Students Who Took Vegas for Millions.* Free Press, New York. {**683**}

M'hall, Abdul Jalib (1996). The true count theorem. rec.gambling.blackjack. {**425**}

Milton, George Fort (1930). *The Age of Hate.* Coward-McCann, Inc., New York. {**590**}

Milyavskaya, Polina (2005). Composition dependent strategy at blackjack: Hit or stand on hard sixteen. Unpublished. {**687**}

Minkowski, H. (1910). *Geometrie der Zahlen.* B. G. Teubner, Leipzig. {**197**}

Molière [Poquelin, Jean-Baptiste] (1913). *L'Avare (The Miser).* G. P. Putnam's Sons, New York. English translation by Curtis Hidden Page. {**638**}

Montmort, Pierre Rémond de (1708). *Essay d'analyse sur les jeux de hazard.* Chez Jacque Quillau, Paris. Published anonymously. {**v, vii, 65–67, 69, 72, 94, 236, 274, 519, 590, 740**}

Montmort, Pierre Rémond de (1713). *Essay d'analyse sur les jeux de hazard* (second ed.). Chez Jacque Quillau, Paris. Published anonymously. Reprinted by Chelsea Publishing Co., New York, 1980. {**v, 66, 68–70, 72, 73, 93, 94, 197, 198, 237, 272, 519, 522, 590, 593–595, 613**}

Morehead, Albert H. (1967). *The Complete Guide to Winning Poker.* A Fireside Book, published by Simon & Schuster, Inc., New York. {**734, 743**}

Morgan, J. P., Chaganty, N. R., Dahiya, R. C., and Doviak, M. J. (1991). Let's make a deal: The player's dilemma. *American Statistician* **45** (4) 284–287. {**72**}

Munford, A. G. and Lewis, S. M. (1981). A note on gambler's ruin against an infinitely rich adversary. *International Journal of Mathematical Education in Science and Technology* **12** (2) 165–168. {**274**}

Musante, Michael J. (1985). The evolution of pai gow, baccarat and blackjack. In Eadington, William R. (Ed.), *The Gambling Studies: Proceedings of the Sixth National Conference on Gambling and Risk Taking, Vol. 4. Investments, Wagers, Games and Gambling*, pp. 1–29. Bureau of Business and Economic Research, University of Nevada, Reno. {**614**}

Nelson, Warren, Adams, Ken, King, R. T., and Nelson, Gail K. (1994). *Always Bet on the Butcher: Warren Nelson and Casino Gaming, 1930s–1980s.* University of Nevada Oral History Program, Reno. {**679**}

Noir, Jacques (1968). *Casino Holiday.* Oxford Street Press, Berkeley, CA. {**682, 687**}

Novikov, A. A. (1996). Martingales, a Tauberian theorem, and strategies for games of chance. *Theory of Probability and Its Applications* **41** (4) 716–729. {**314**}

Oettinger, L. (1867). Ueber einige Probleme der Wahrscheinlichkeitsrechnung, ins Besondere über das Rouge et Noire und den Vortheil der Bank bei diesem Spiele. Ein Beitrag zur Wahrscheinlichkeitsrechnung. *Journal für die reine und angewandte Mathematik* **67** 327–359. {**640**}

O'Hara, John (1988). *A Mug's Game: A History of Gaming and Betting in Australia.* New South Wales University Press, Kensington, NSW, Australia. {**73**}

Ore, Oystein (1953). *Cardano, the Gambling Scholar. With a Translation from the Latin of Cardano's* Book on Games of Chance *by Sydney Henry Gould.* Princeton University Press, Princeton, NJ. Reprinted by Dover Publications, Inc., New York, 1965. {**64, 65, 679, 756**}

Oses, Noelia (2008). Markov chain applications in the slot machine industry. *OR Insight* **21** (1) 9–21. {**460**}

Pacioli, Luca (1494). *Summa de arithmetica, geometria et proportionate.* Paganini, Venice. {**92**}

Packel, Edward W. (2006). *The Mathematics of Games and Gambling* (second ed.). Mathematical Association of America, Washington, DC. First edition, 1981. {**66, 500**}

Parlett, David (1991). *A History of Card Games.* Oxford University Press, Oxford, UK. {**613, 614, 638, 676, 731, 734**}

Pascal, Blaise (1665). *Traité du triangle arithmétique, avec quelques autres petits traités sur la même matière.* Desprez, Paris. {**65, 92**}

Pascual, Michael J. and Wilslef, Dennis (1996). *How to Find and Wager on Biased Roulette Wheels.* Self-published. {**482**}

Paymar, Dan (1994). *Video Poker—Precision Play* (eighth ed.). Enhanceware, Las Vegas. {**571**}

Paymar, Dan (1998). *Video Poker—Optimum Play.* ConJelCo LLC, Pittsburgh. {**571**}

Paymar, Dan (2004). *Video Poker—Optimum Play* (second ed.). ConJelCo LLC, Pittsburgh. {**570, 571**}

Pearson, Karl (1894). Science and Monte Carlo. *Fortnightly Review* **55** 183–193. {**481**}

Pearson, Karl (1897). The scientific aspect of Monte Carlo roulette. In Pearson, Karl (Ed.), *The Chances of Death and Other Studies in Evolution,* Vol. 1, pp. 42–62. Edward Arnold, London. {**481**}

Pearson, Karl (1900). On the criterion that a given system of deviations from the probable in the case of a correlated system of variables is such that it can be reasonably supposed to have arisen from random sampling. *Philosophical Magazine, 5th series* **50** 157–175. Correction in *6th series* **1** 670–671, 1901. {**481**}

Pendergrass, Marcus and Siegrist, Kyle (2001). Generalizations of bold play in red and black. *Stochastic Processes and Their Applications* **92** (1) 163–180. {**356**}

Penney, W. (1969). Problem 95: Penney-ante. *Journal of Recreational Mathematics* **2** (4) 241. {**117**}

Percy, George (1988). *Poker: America's Game.* Self-published. {**571, 743**}

Perlman, Michael D. and Wichura, Michael J. (1975). A note on substitution in conditional distribution. *Annals of Statistics* **3** (5) 1175–1179. {**94**}

Persius, Charles [Dunne, Charles] (1823). *Rouge et Noir. The Academicians of 1823; or the Greeks of the Palais Royal, and the Clubs of St. James's.* Lawler and Quick and Stephen Couchman, London. {**312**}

Petriv, Mike (1996). *Hold'em's Odd(s) Book.* Objective Observer, Toronto. {**741, 743**}

Player, A. T. (1925). *Roulette and Trente-et-Quarante as Played at Monte-Carlo. Some 140 Systems, Progressions, Marches and Attacks* (second ed.). A. R. Thackrah & Cie, Menton. {**313, 314, 316, 641**}

Poe, Edgar Allan (1842–1843). The mystery of Marie Rogêt. *Snowden's Ladies' Companion* (Nov., Dec., and Feb.). {**67**}

Poe, Edgar Allan (1984). *Poetry and Tales.* The Library of America, New York. {**67**}

Poisson, S.-D. (1837). *Recherches sur la probabilité des jugements en matière criminelle et en matière civile, précédées des règles générales du calcul des probabilités.* Bachelier, Paris. Reprinted by Éditions Jacques Gabay, Paris, 2003. {**68, 71, 92**}

Poisson, [Siméon-Denis] (1825). Mémoire sur l'avantage du banquier au jeu de trente et quarante. *Annales de Mathématiques Pures et Appliquées* **16** (6) 173–208. {**236, 237, 639, 640, 642**}

Poisson, [Siméon-Denis] (1835). Recherches sur la probabilité des jugements, principalement en matière criminelle. *Comptes Rendus Hebdomadaires des Séances de l'Académie des Sciences* **1** 473–494. {**70**}

Polovtsoff, General Pierre (1937). *Monte Carlo Casino.* Stanley Paul & Co., Ltd., London. {**641**}

Pólya, Georg (1920). Über den zentralen Grenzwertsatz der Wahrscheinlichkeitsrechnung und das Momentenproblem. *Mathematische Zeitschrift* **8** 171–181. {**71**}

Pólya, Georg (1921). Über eine Aufgabe der Wahrscheinlichkeitsrechnung betreffend die Irrfahrt im Straßennetz. *Mathematische Annalen* **84** (1–2) 149–160. {**157**}

Ponzer, Fred (1977). Ponzer dice system. *Gambling Times* **1** (1, Feb.) 24–25. {**316, 524**}

Poundstone, William (1992). *Prisoner's Dilemma.* Doubleday, New York. {**197**}

Poundstone, William (2005). *Fortune's Formula: The Untold Story of the Scientific Betting System that Beat the Casinos and Wall Street.* Hill and Wang, a division of Farrar, Straus and Giroux, New York. {**388, 389, 678**}

Pratt, John W. (1964). Risk aversion in the small and in the large. *Econometrica* **32** (1–2) 122–136. {**198**}

Preston, Amarillo Slim and Dinkin, Greg (2003). *Amarillo Slim in a World Full of Fat People: The Memoirs of the Greatest Gambler Who Ever Lived.* HarperEntertainment, an imprint of HarperCollins Publishers Inc., New York. {**736**}

Proctor, Richard A. (1887). *Chance and Luck: A Discussion of the Laws of Luck, Coincidences, Wagers, Lotteries, and the Fallacies of Gambling; with Notes on Poker and Martingales* (second ed.). Longmans, Green, and Co., London. {**312**}

Pushkin, Alexander (1983). *Complete Prose Fiction.* Stanford University Press, Stanford, CA. English translation by Paul Debreczeny. {**592**}

Puza, Borek D., Pitt, David G. W., and O'Neill, Terence J. (2005). The Monty Hall three doors problem. *Teaching Statistics* **27** (1) 11–15. {**72**}

Quinn, John Philip (1892). *Fools of Fortune: or Gambling and Gamblers, Comprehending a History of the Vice in Ancient and Modern Times, and in Both Hemispheres; an Exposition of Its Alarming Prevalence and Destructive Effects; with an Unreserved and Exhaustive Disclosure of Such Frauds, Tricks and Devices as Are Practiced by "Professional" Gamblers, "Confidence Men" and "Bunko Steerers."* The Anti-Gambling Association, Chicago. {**479, 520, 591**}

Quinn, John Philip (1912). *Gambling and Gambling Devices: Being a Complete Systematic Educational Exposition Designed to Instruct the Youth of the World to Avoid All Forms of Gambling.* John Philip Quinn, Canton, OH. {**497**}

Raleigh [Strutt, John] (1877). On Mr. Venn's explanation of a gambling paradox. *Mind* **2** (7) 409–410. {**116**}

Renaudet, B. (1922). *Le baccara, règles complètes. Baccara à deux tableaux. Baccara chemin de fer. Le trente et quarante. Le lansquenet. Le vingt-et-un. La banque.* Librairie S. Bornemann, Paris. {**619**}

Renzoni, Tommy and Knapp, Don (1973). *Renzoni on Baccarat and the Secrets of Professional Gambling.* Lyle Stuart, Inc., Secaucus, NJ. {**614**}

Revere, Lawrence (1973). *Playing Blackjack as a Business.* Lyle Stuart, Inc., Secaucus, NJ. First published by Paul Mann Publishing Co., Las Vegas, 1969. {**681, 683**}

Richardson, P. W. (1929). *Systems and Chances.* G. Bell & Sons, Ltd., London. {**316, 479**}

Riddle, Major A. (1963). *The Weekend Gambler's Handbook.* Random House, New York. {**595**}

Rivlin, Gary (2004). The chrome-shiny, lights-flashing, wheel-spinning, touch-screened, Drew-Carey-wisecracking, video-playing, 'sound events'-packed, pulse-quickening bandit: How the slot machine was remade, and how it's remaking America. *New York Times Magazine* (May 9) 42–47, 74. {**457, 459**}

Robison, John (2009). The odds of winning Megabucks. *Casino City Times* (Aug. 8). `http://robison.casinocitytimes.com/article/the-odds-of-winning-megabucks-47597`. {**458**}

Rosenhouse, Jason (2009). *The Monty Hall Problem: The Remarkable Story of Math's Most Contentious Brain Teaser*. Oxford University Press, Oxford, UK. {**72**}

Rosenthal, Jeffrey S. (1995). On duality of probabilities for card dealing. *Proceedings of the American Mathematical Society* **123** (2) 559–561. {**425**}

Ross, Andrew M., Benjamin, Arthur T., and Munson, Michael (2007). Estimating winning probabilities in backgammon races. In Ethier, Stewart N. and Eadington, William R. (Eds.), *Optimal Play: Mathematical Studies of Games and Gambling*, pp. 269–291. Institute for the Study of Gambling and Commercial Gaming, University of Nevada, Reno. {**158**}

Ross, Sheldon (2006). *A First Course in Probability* (seventh ed.). Pearson Prentice Hall, Upper Saddle River, NJ. First edition, 1976. {**94, 425, 523**}

Ross, Sheldon M. (1974). Dynamic programming and gambling models. *Advances in Applied Probability* **6** (3) 593–606. {**356**}

Ross, Sheldon M. (2003). *Probability Models* (eighth ed.). Academic Press, San Diego. First edition, 1972. {**94, 117**}

Rouge et Noir [Heckethorn, Charles William] (1898). *The Gambling World: Anecdotic Memories and Stories of Personal Experience in the Temples of Hazard and Speculation with Some Mysteries and Iniquities of Stock Exchange Affairs*. Hutchinson & Co., London. Reissued by Singing Tree Press, Detroit, 1968. {**313**}

Rouse, William (1814). *The Doctrine of Chances, or the Theory of Gaming, Made Easy to Every Person Acquainted with Common Arithmetic, so as to Enable Them to Calculate the Probabilities of Events in Lotteries, Cards, Horse Racing, Dice, &c. with Tables on Chance, Never before Published, Which from Mere Inspection Will Solve a Great Variety of Questions*. Lackington, Allen & Co., London. {**519**}

Roxbury, L. E. (1959). *Your Chances at Roulette: or, You Can Win—but Don't Bet on It* (second ed.). High-Iron Publishers, Warwick, VA. {**316**}

Rudin, Walter (1987). *Real and Complex Analysis* (third ed.). McGraw-Hill Book Co., New York. {**747, 750**}

Russell, Jere (1983). *Compleat Poker: The Gambler's Edge*. Bonne Chance Enterprises, Panorama City, CA. {**571**}

Sagan, Hans (1981). Markov chains in Monte Carlo. *Mathematics Magazine* **54** (1) 3–10. {**479, 482**}

Samuels, S. M. (1975). The classical ruin problem with equal initial fortunes. *Mathematics Magazine* **48** (5) 286–288. {**274**}

Samuelson, Paul A. (1971). The "fallacy" of maximizing the geometric mean in long sequences of investing or gambling. *Proceedings of the National Academy of Sciences, USA* **68** (10) 2493–2496. {**389**}

Samuelson, Paul A. (1977). St. Petersburg paradoxes: Defanged, dissected, and historically described. *Journal of Economic Literature* **15** (1) 24–55. {**71**}

Samuelson, Paul A. (1979). Why we should not make mean log of wealth big though years to act are long. *Journal of Banking and Finance* **3** (4) 305–307. {**389**}

Sasuly, Richard (1982). *Bookies and Bettors: Two Hundred Years of Gambling*. Holt, Rinehart and Winston, New York. {**590**}

Sauveur, Joseph (1679). Supputation des avantages du banquier dans le jeu de la bassette. *Journal des Sçavans* (Feb. 13) 44–52. {**236, 590**}

Savage, Leonard J. (1957). The casino that takes a percentage and what you can do about it. Technical report P-1132, The Rand Corporation. {**356**}

Scarne, John (1949). *Scarne on Cards*. Crown Publishers, New York. {**66, 677, 679**}

Scarne, John (1961). *Scarne's Complete Guide to Gambling*. Simon and Schuster, New York. {**68, 71, 72, 236, 239, 481, 500, 592, 619, 620, 677, 679, 685**}

Scarne, John (1966). *The Odds Against Me: An Autobiography*. Simon and Schuster, New York. {**679, 680**}

Scarne, John (1974). *Scarne's New Complete Guide to Gambling* (fully revised, expanded, updated ed.). Simon and Schuster, New York. {**72, 239, 481, 639, 680**}

Scarne, John (1978). *Scarne's Guide to Casino Gambling.* Simon and Schuster, New York. {**680**}

Scarne, John (1980). *Scarne on Dice* (eighth revised ed.). Crown Publishers, Inc., New York. {**520, 521**}

Scarne, John and Rawson, Clayton (1945). *Scarne on Dice.* The Military Service Publishing Co., Harrisburg, PA. {**524**}

Schaffer, Allan (2005). Jess Marcum, mathematical genius and blackjack legend and the early days of card counting. *Blackjack Forum* **24** (3). {**680**}

Schenck, Robert C. (1880). *Draw. Rules for Playing Poker.* Privately printed, Brooklyn, NY. {**732**}

Schlesinger, Don (1986). Attacking the shoe. *Blackjack Forum* **6** (3) 5–11. {**682**}

Schlesinger, Don (1999a). Before you play, know the SCORE! Part I. *Blackjack Forum* **19** (2) 7–21. {**239**}

Schlesinger, Don (1999b). SCORE! Part II. *Blackjack Forum* **19** (2) 22–34. {**239**}

Schlesinger, Don (2005). *Blackjack Attack: Playing the Pros' Way* (third ed.). RGE Publishing, Ltd., Las Vegas. First edition, 1997; second edition, 2000. {**239, 682–684, 686, 687**}

Schneider, Ivo (1988). The market place and games of chance in the fifteenth and sixteenth centuries. In Hay, Cynthia (Ed.), *Mathematics from Manuscript to Print, 1300–1600,* pp. 220–235. Clarendon Press, Oxford, UK. {**92**}

Schneir, Leonard (1993). *Gambling Collectibles: A Sure Winner.* Schiffer Publishing Ltd., Atglen, PA. {**732**}

Schwartz, David G. (2006). *Roll the Bones: The History of Gambling.* Gotham Books, New York. {**316**}

Schwarz, H. A. (1885). Über ein die Flächen kleinsten Flächeninhalts betreffendes Problem der Variationsrechnung. *Acta Societatis Scientiarum Fennicae* **15** 315–362. {**69**}

Schweinsberg, Jason (2005). Improving on bold play when the gambler is restricted. *Journal of Applied Probability* **42** (2) 321–333. {**356**}

Scoblete, Frank (1991). *Beat the Craps Out of the Casinos: How to Play Craps and Win!* Bonus Books, Inc., Chicago. New expanded edition, 2005. {**521, 524**}

Scoblete, Frank (1993). *The Captain's Craps Revolution.* Paone Press, Lynbrook, NY. {**239, 521**}

Scoblete, Frank (2000). *Forever Craps: The Five-Step Advantage-Play Method.* Bonus Books, Inc., Chicago. {**521**}

Scoblete, Frank (2003). *The Craps Underground: The Inside Story of How Dice Controllers Are Winning Millions from the Casinos.* Bonus Books, Inc., Chicago. {**521**}

Scoblete, Frank (2007). *The Virgin Kiss and Other Adventures.* Research Services Unlimited, Daphne, AL. {**523**}

Scoblete, Frank and Dominator [LoRiggio, Dominic] (2005). *Golden Touch Dice Control Revolution!* Research Services Unlimited, Daphne, AL. {**521**}

Scrutator (1924). *The Odds at Monte Carlo.* John Murray, London. {**639**}

Selvin, Steve (1975a). A problem in probability. *American Statistician* **29** (1) 67. {**72**}

Selvin, Steve (1975b). On the Monty Hall problem. *American Statistician* **29** (3) 134. {**72**}

Seneta, E. (1998). I. J. Bienaymé [1796–1878]: Criticality, inequality, and internationalization. *International Statistical Review* **66** (3) 291–301. {**70**}

Shackleford, Michael (2002). The slot payout comparison. *Anthony Curtis' Las Vegas Advisor* **19** (5) 8–9. {**459**}

Shackleford, Michael (2005a). *Gambling 102.* Huntington Press, Las Vegas. {**237, 238, 523, 683**}

Shackleford, Michael W. (2005b). Frequently asked questions about craps. http://wizardofodds.com/askthewizard/craps-faq.html. {**237**}

Shackleford, Michael W. (2009). Craps. http://wizardofodds.com/craps/. {**523, 542**}

Shafer, Glenn and Vovk, Vladimir (2001). *Probability and Finance: It's Only a Game!* John Wiley & Sons, Inc., New York. {**311**}

Sham [Shampaign, Charles E.] (1930). *Handbook on Percentages: Containing Rules for the Playing of the Games with Descriptions, Technicalities, Probabilities, Percentages, Instructions, Examples, etc. American Method of Playing with Numerous Tables and Tabulations to Which Is Appended an Elaborate Treatise on the Doctrine of Chances* (revised ed.). Joe Treybal Sporting Goods Co., St. Louis. Reprinted by Gambler's Book Club, Las Vegas, 1976. {**237, 524**}

Shannon, C. E. (1948). A mathematical theory of communication. *Bell System Technical Journal* **27** 379–423, 623–656. {**116, 389**}

Sharpshooter [Pawlicki, Christopher] (2002). *Get the Edge at Craps: How to Control the Dice!* Bonus Books, Chicago. {**521**}

Shepp, Larry (2006). Bold play and the optimal policy for Vardi's casino. In Hsiung, Agnes Chao, Ying, Zhiliang, and Zhang, Cun-Hui (Eds.), *Random Walk, Sequential Analysis and Related Topics: A Festschrift in Honor of Yuan-Shih Chow*, pp. 150–156. World Scientific Publishing Co., Hackensack, NJ. {**356**}

Siegrist, Kyle (2008). How to gamble if you must. *Loci* **1** (July). {**355, 356**}

Sifakis, Carl (1990). *Encyclopedia of Gambling*. Facts on File, New York. {**479, 614**}

Sigler, L. E. (2002). *Fibonacci's Liber Abaci: A Translation into Modern English of Leonardo Pisano's Book of Calculation*. Springer-Verlag, New York. {**312**}

Silberer, Victor (1910). *The Games of Roulette and Trente et Quarante as Played at Monte Carlo. Being a Reprint of the Technical Chapters from the Work "Vom grünen Tisch in Monte Carlo."* Harrison & Sons, London. {**479, 481, 639, 642**}

Sileo, Patrick (1992). The evaluation of blackjack games using a combined expectation and risk measure. In Eadington, William R. and Cornelius, Judy A. (Eds.), *Gambling and Commercial Gaming: Essays in Business, Economics, Philosophy and Science*, pp. 551–563. Institute for the Study of Gambling and Commercial Gaming, University of Nevada, Reno. {**239**}

Simpson, J. A. and Weiner, E. S. C. (1989). *Oxford English Dictionary* (second ed.). Clarendon Press, Oxford, UK. {**236, 311, 498, 517, 613**}

Simpson, Thomas (1740). *The Nature and Laws of Chance. The Whole after a New, General, and Conspicuous Manner, and Illustrated with a Great Variety of Examples*. Edward Cave, London. {**70, 73**}

Singh, Parmanand (1985). The so-called Fibonacci numbers in ancient and medieval India. *Historia Mathematica* **12** (3) 229–244. {**313**}

Singsen, Michael Pierce (1988). Where will the buck stop on California Penal Code Section 330?: Solving the stud-horse poker conundrum. *COMM/ENT, Hastings Communications and Entertainment Law Journal* **11** (1) 95–147. {**733**}

Sklansky, David (1976). *Hold 'em Poker*. GBC Press, Las Vegas. {**741**}

Sklansky, David (1978). *Sklansky on Poker Theory*. GBC Press, Las Vegas. {**741**}

Sklansky, David (1982). Card counting and baccarat. *Gambling Times* **5** (11, Mar.) 84–86. {**620, 621**}

Sklansky, David (1994). *The Theory of Poker* (third ed.). Two Plus Two Publishing, Las Vegas. First edition, 1987. {**741**}

Sklansky, David (1996). *Hold 'em Poker*. Two Plus Two Publishing, Henderson, NV. {**742**}

Sklansky, David (1997). *Fighting Fuzzy Thinking in Poker, Gaming, and Life*. Two Plus Two Publishing, Henderson, NV. {**743**}

Sklansky, David (2001). *Getting the Best of It* (second ed.). Two Plus Two Publishing, Henderson, NV. First edition, 1982. {**72, 687**}

Sklansky, David and Malmuth, Mason (1999). *Hold 'em Poker for Advanced Players* (third ed.). Two Plus Two Publishing, Henderson, NV. First edition, 1988. {**741, 742**}

Sklansky, David and Miller, Ed (2006). *No Limit Hold 'em: Theory and Practice*. Two Plus Two Publishing, Henderson, NV. {**741, 742, 744**}

Slutsky, E. (1925). Über stochastische Asymptoten und Grenzwerte. *Metron* **5** (3) 3–89. {**73**}

Smith, Armand V., Jr. (1968). Some probability problems in the game of "craps." *American Statistician* **22** (3) 29–30. {**94, 522**}

Smith, Gerald John (1967). Optimal strategy at roulette. *Zeitschrift für Wahrscheinlichkeitstheorie und verwandte Gebiete* **8** 91–100. {**356**}

Snell, J. Laurie (1997). A conversation with Joe Doob. *Statistical Science* **12** (4) 301–311. {**68, 115**}

Snell, J. Laurie and Vanderbei, Robert (1995). Three bewitching paradoxes. In Snell, J. Laurie (Ed.), *Topics in Contemporary Probability and Its Applications*, pp. 355–370. CRC Press, Boca Raton, FL. {**72**}

Snyder, Arnold (1982). Algebraic approximation of optimum blackjack strategy (revised). In Eadington, William R. (Ed.), *The Gambling Papers: Proceedings of the Fifth National Conference on Gambling and Risk Taking, Vol. 10. The Blackjack Papers*, pp. 34–58. Bureau of Business and Economic Research, University of Nevada, Reno. {**686**}

Snyder, Arnold (1983). *Blackbelt in Blackjack: Playing 21 as a Martial Art*. RGE Publishing, Berkeley, CA. {**683**}

Snyder, Arnold (1992). The John Scarne riddle. *Card Player* **5** (5) 44–46. {**680**}

Snyder, Arnold (1997). *Blackjack Wisdom*. RGE Publishing, Oakland, CA. {**680**}

Snyder, Arnold (2005). *Blackbelt in Blackjack: Playing 21 as a Martial Art* (third ed.). Cardoza Publishing, New York. {**683**}

Snyder, Arnold (2006). *The Big Book of Blackjack*. Cardoza Publishing, New York. {**676, 683**}

Sobel, Milton and Frankowski, Krzysztof (1994). A duality theorem for solving multiple-player multivariate hypergeometric problems. *Statistics & Probability Letters* **20** (2) 155–162. {**425**}

Sodano, John and Yaspan, Arthur (1978). A neglected probability formula. *Two-Year College Mathematics Journal* **9** (3) 145–147. {**72**}

Spanier, David (1980). *The Pocket Guide to Gambling*. Simon and Schuster, New York. {**481**}

Spanier, David (1990). *Total Poker* (revised ed.). Andre Deutsch Limited, London. First published by Martin Secker & Warburg Limited, London, 1977. {**738**}

Spanier, David (1994). *Inside the Gambler's Mind*. University of Nevada Press, Reno. {**679**}

Steinmetz, Andrew (1870). *The Gaming Table: Its Votaries and Victims, in All Times and Countries, Especially in England and in France, Vol. 2.* Tinsley Brothers, London. Reprinted by Patterson Smith Publishing Corp., Montclair, NJ, 1969. {**72, 521**}

Stelzer, Michael A. (2006). *Mathematical Strategies to Winning Casino Poker*. BookSurge, LLC, North Charleston, SC. {**543**}

Stern, Frederick (1975). Conditional expectation of the duration in the classical ruin problem. *Mathematics Magazine* **48** (4) 200–203. {**272**}

Stewart, Peter (1986). Letter to the editor. *Gambling Times* (Oct.) 6. {**522**}

Stigler, Stephen M. (1983). Who discovered Bayes's theorem? *American Statistician* **37** (4) 290–296. {**68**}

Stigler, Stephen M. (1986). *The History of Statistics: The Measurement of Uncertainty before 1900*. The Belknap Press of Harvard University Press, Cambridge, MA. {**64, 69, 481**}

Stirling, J. (1730). *Methodus Differentialis: sive Tractatus de Summatione et Interpolatione Serierum Infinitarum*. G. Strahan, London. {**71**}

Stirzaker, David (1994). Tower problems and martingales. *Mathematical Scientist* **19** (1) 52–59. {**117**}

Stratton, David (2004a). Turning the tables on slots: Table games not yet ready for extinction. *Gaming Today* **29** (25, June 15–21) 3, 6. {**425**}

Stratton, David (2004b). Turning the tables: New games draw players to the "pits." *Gaming Today* **29** (26, June 22–28) 1, 4, 7. {**543**}

Struyck, Nicolaas (1716). *Calculation of the Chances in Play, by Means of Arithmetic and Algebra, Together with a Treatise on Lotteries and Interest.* Publisher unknown, Amsterdam. {**271**}

Stuart, Lyle (2007). *Lyle Stuart on Baccarat* (third ed.). Lyle Stuart, New York. First edition, 1984; second edition, 1997. {**614**}

Stupak, Bob (1984). The worst of it could be the best of it. *Gambling Times* **7** (11, Mar.) 82. {**356, 500**}

Szabo, Aloïs (1962). *How to Win* (fourth ed.). Wehman Bros., Inc., New York. {**316**}

Székely, Gábor J. (2003). Problem corner. *Chance* **16** (4) 52–53. {**425**}

Takács, Lajos (1967). On the method of inclusion and exclusion. *Journal of the American Statistical Association* **62** (317) 102–113. {**70**}

Takács, Lajos (1969). On the classical ruin problems. *Journal of the American Statistical Association* **64** (327) 889–906. {**273**}

Takács, Lajos (1994). The problem of points. *Mathematical Scientist* **19** (2) 119–139. {**93**}

Takács, Lajos (1997). On the ballot theorems. In Balakrishnan, N. (Ed.), *Advances in Combinatorial Methods and Applications to Probability and Statistics*, pp. 97–114. Birkhäuser, Boston. {**315**}

Takahasi, Koiti and Futatsuya, Masao (1984). On a problem of a probability arising from poker. *Journal of the Japan Statistical Society* **14** (1) 11–18. {**71**}

Tamburin, Henry J. (1983). Baccarat card counting, another opinion. *Experts Blackjack Newsletter* **1** (11) 12, 15. {**621**}

Tamburin, Henry J. and Rahm, Dick (1983). *Winning Baccarat Strategies: The First Effective Card Counting Strategies for the Casino Game of Baccarat.* Research Services Unlimited, Greenboro, NC. {**621**}

Tartaglia, Nicolo Fontana (1556). *Trattato generale di numeri e misure.* Curtio Troiano, Venice. {**92**}

Thackeray, William Makepeace (1989a). *Vanity Fair: A Novel without a Hero.* Garland Publishing, Inc., New York. Edited by Peter L. Shillingsburg. {**241, 429, 483**}

Thackeray, William Makepeace (1989b). *The History of Henry Esmond.* Garland Publishing, Inc., New York. Edited by Edgar F. Harden. {**199**}

Thackeray, William Makepeace (1991a). *Flore et Zéphyr; The Yellowplush Correspondence; The Tremendous Adventures of Major Gahagan.* Garland Publishing, Inc., New York. Edited by Peter L. Shillingsburg. {**159**}

Thackeray, William Makepeace (1991b). *The History of Pendennis.* Garland Publishing, Inc., New York. Edited by Peter L. Shillingsburg. {**v, 3, 119**}

Thackeray, William Makepeace (1996). *The Newcomes: Memoirs of a Most Respectable Family.* The University of Michigan Press, Ann Arbor. Edited by Peter L. Shillingsburg. {**75, 95, 357**}

Thackeray, William Makepeace (1999a). *Catherine: A Story. By Ikey Solomons, Esq. Junior.* The University of Michigan Press, Ann Arbor. Edited by Sheldon F. Goldfarb. {**391**}

Thackeray, William Makepeace (1999b). *The Luck of Barry Lyndon; a Romance of the Last Century. By Fitz-Boodle.* The University of Michigan Press, Ann Arbor. Edited by Edgar F. Harden. {**317, 573, 597, 689**}

Thackeray, William Makepeace (2005). *The Snobs of England and Punch's Prize Novelists.* The University of Michigan Press, Ann Arbor. Edited by Edgar F. Harden. {**745**}

Thackrey, Ted, Jr. (1968). *Gambling Secrets of Nick the Greek.* Rand McNally & Company, Chicago. {**595**}

Thorold, Algar Labouchere (1913). *The Life of Henry Labouchere.* G. P. Putnam's Sons, The Knickerbocker Press, New York. {**313**}

Thorp, E. O. (1960). Fortune's formula: The game of blackjack. *Notices of the American Mathematical Society* **7** (7) 935–936. {**678**}

Thorp, E. O. (1966b). Elementary Problem E 1865. *American Mathematical Monthly* **73** (3) 309. {**71**}

Thorp, Edward (1961). A favorable strategy for twenty-one. *Proceedings of the National Academy of Sciences, USA* **47** (1) 110–112. {**678**}

Thorp, Edward O. (1962a). A prof beats the gamblers. *Atlantic Monthly* **209** (6, June) 41–46. {**678, 679**}

Thorp, Edward O. (1962b). *Beat the Dealer: A Winning Strategy for the Game of Twenty-One.* Random House, New York. {**425, 643, 678–681, 683, 685, 686**}

Thorp, Edward O. (1964). Repeated independent trials and a class of dice problems. *American Mathematical Monthly* **71** (7) 778–781. {**68, 72**}

Thorp, Edward O. (1966a). *Beat the Dealer: A Winning Strategy for the Game of Twenty-One* (revised and simplified ed.). Random House, New York. {**681, 683**}

Thorp, Edward O. (1966c). *Elementary Probability.* John Wiley & Sons, Inc., New York. {**66, 71**}

Thorp, Edward O. (1969). Optimal gambling systems for favorable games. *International Statistical Review* **37** (3) 273–293. {**356, 389**}

Thorp, Edward O. (1973). Nonrandom shuffling with applications to the game of faro. *Journal of the American Statistical Association* **68** (344) 842–847. {**237, 594**}

Thorp, Edward O. (1975). Portfolio choice and the Kelly criterion. In Ziemba, W. T. and Vickson, R. G. (Eds.), *Stochastic Optimization Models in Finance*, pp. 599–619. Academic Press, New York. {**389**}

Thorp, Edward O. (1976). Probabilities and strategies for the game of faro. In Eadington, William R. (Ed.), *Gambling and Society: Interdisciplinary Studies on the Subject of Gambling*, pp. 443–466. Charles C. Thomas, Springfield, IL. {**594, 595**}

Thorp, Edward O. (1984). *The Mathematics of Gambling.* Lyle Stuart, Secaucus, NJ. {**620**}

Thorp, Edward O. (1993). The Kelly system: Fact and fallacy. *Blackjack Forum* **13** (2) 18–23. {**390**}

Thorp, Edward O. (2000a). Does basic strategy have the same expectation for each round? Vancura, Olaf, Cornelius, Judy A., and Eadington, William R. (Eds.), *Finding the Edge: Mathematical Analysis of Casino Games*, pp. 115–132. Institute for the Study of Gambling and Commercial Gaming, University of Nevada, Reno. {**71, 687**}

Thorp, Edward O. (2000b). The Kelly criterion in blackjack, sports betting, and the stock market. In Vancura, Olaf, Cornelius, Judy A., and Eadington, William R. (Eds.), *Finding the Edge: Mathematical Analysis of Casino Games*, pp. 163–213. Institute for the Study of Gambling and Commercial Gaming, University of Nevada, Reno. Corrected version in Zenios, S. A. and Ziemba, W. T. (Eds.), *Handbook of Asset and Liability Management, Vol. 1. Theory and Methodology*, pp. 385–428. North-Holland, Elsevier B.V., Amsterdam, 2006. {**389, 390**}

Thorp, Edward O. (2005). The first wearable computer. *Noesis: The Journal of the Mega Society* **176** 12–19. {**678**}

Thorp, Edward O. and Walden, William E. (1966). A favorable side bet in Nevada baccarat. *Journal of the American Statistical Association* **61** (314) 313–328. {**389, 620, 621**}

Thorp, Edward O. and Walden, William E. (1973). The fundamental theorem of card counting with applications to trente-et-quarante and baccarat. *International Journal of Game Theory* **2** (1) 109–119. {**425, 640–642**}

Tierney, John (1991). Behind Monty Hall's doors: Puzzle, debate and answer? *New York Times* (July 21) Section 1, pp. 1, 20. {**72**}

Tinker, Edward Larocque (1933). *The Palingenesis of Craps.* The Press of the Woolly Whale, New York. {**519, 520**}

Todhunter, I. (1865). *A History of the Mathematical Theory of Probability: From the Time of Pascal to That of Laplace.* Macmillan, Cambridge, UK. Reprinted by Chelsea Publishing Co., New York, 1949, 1965. {**viii, 64, 65, 69, 70, 198, 271–273, 593, 594, 638**}

Tolstoy, Leo (1889). *War and Peace.* Thomas Y. Crowell & Co., New York. English translation by Nathan Haskell Dole. {**592**}

Toti Rigatelli, Laura (1985). Il "problema delle parti" in manoscritti del XIV e XV secolo. In Folkerts, Menso and Lindgren, Uta (Eds.), *Mathemata: Festschrift für Helmuth Gericke*, pp. 229–236. Franz Steiner Verlag, Stuttgart. {**92**}

Trumps (c. 1875). *The American Hoyle; or a Gentleman's Hand-Book of Games; Containing All the Games Played in the United States, with Rules, Descriptions, and Technicalities, Adapted to the American Methods of Playing* (ninth ed.). Dick & Fitzgerald, Publishers, New York. {**639, 732, 739**}

Uspensky, J. V. (1937). *Introduction to Mathematical Probability*. McGraw–Hill Book Company, Inc., New York. {**66, 70, 273**}

Uston, Ken and Rapoport, Roger (1977). *The Big Player: How a Team of Blackjack Players Made a Million Dollars*. Holt, Rinehart and Winston, New York. {**683**}

Van-Tenac, [Charles] (1847). *Album des jeux de hasard et de combinaisons en usage dans les salons et dans les cercles. Règles, lois, conventions et maximes. Recueillies et codifiées d'après les meilleures autorités de l'ancienne et de la nouvelle école. Avec un abrégé et des applications de la théorie des probabilités*. Gustave Havard, Éditeur, Paris. {**237, 519, 618**}

Vancura, Olaf (1996). *Smart Casino Gambling: How to Win More and Lose Less*. Index Publishing Group, Inc., San Diego. {**620**}

Vancura, Olaf and Fuchs, Ken (1998). *Knock-Out Blackjack: The Easiest Card-Counting System Ever Devised*. Huntington Press, Las Vegas. {**683**}

Vanniasegaram, Sithparran (2006). Le her with s suits and d denominations. *Journal of Applied Probability* **43** (1) 1–15. {**198**}

Ville, Jean (1938). Sur la théorie générale des jeux où intervient l'habileté des joueurs. In Borel, Émile (Ed.), *Applications aux jeux de hasard*, pp. 105–113. Gauthier-Villars, Imprimeur-Éditeur, Paris. {**197**}

Ville, Jean (1939). *Étude critique de la notion de collectif*. Gauthier-Villars, Imprimeur-Éditeur, Paris. {**115, 116**}

von Mises, Richard (1928). *Wahrscheinlichkeitsrechnung, Statistic und Wahrheit*. Verlag von Julius Springer, Wien. {**116**}

von Neumann, John (1928). Zur Theorie der Gesellschaftsspiele. *Mathematische Annalen* **100** (1) 295–320. {**197**}

von Neumann, John and Morgenstern, Oskar (1944). *Theory of Games and Economic Behavior*. Princeton University Press, Princeton, NJ. Second edition, 1947; third edition, 1953. {**197, 198, 740**}

vos Savant, Marilyn (1990a). Ask Marilyn. *Parade Magazine* (Sept. 9) 15. {**72**}

vos Savant, Marilyn (1990b). Ask Marilyn. *Parade Magazine* (Dec. 2) 25. {**72**}

vos Savant, Marilyn (1991). Ask Marilyn. *Parade Magazine* (Feb. 17) 12. {**72**}

Wald, A. (1945). Sequential tests of statistical hypotheses. *Annals of Mathematical Statistics* **16** (2) 117–186. {**117**}

Wallace, Frank R. (1968). *Poker: A Guaranteed Income for Life by Using the Advanced Concepts of Poker*. I & O Publishing Company, Wilmington, DE. {**740, 741**}

Weaver, Warren (1919). Problem 2787. *American Mathematical Monthly* **26** (8) 366. {**522**}

Weaver, Warren (1963). *Lady Luck: The Theory of Probability*. Anchor Books, Doubleday & Co., Inc., Garden City, NY. Reprinted by Dover Publications, Inc., New York, 1982. {**522**}

Weber, Glenn and Scruggs, W. Todd (1992). A mathematical and computer analysis of video poker. In Eadington, William R. and Cornelius, Judy A. (Eds.), *Gambling and Commercial Gaming: Essays in Business, Economics, Philosophy and Science*, pp. 625–633. Institute for the Study of Gambling and Commercial Gaming, University of Nevada, Reno. {**570, 571**}

Wechsberg, Joseph (1953). Blackjack Pete. *Colliers* **132** (4, July 28) 34–37. {**680**}

Werthamer, N. Richard (2009). *Risk and Reward: The Science of Casino Blackjack*. Springer, New York. {**683, 687**}

Whitworth, William Allen (1901). *Choice and Chance with 1000 Exercises* (fifth ed.). Deighton Bell and Co., Cambridge, UK. First edition, 1867. {**389, 390**}

Wilkins, J. Ernest, Jr. (1972). The bold strategy in presence of house limit. *Proceedings of the American Mathematical Society* **32** (2) 567–570. {**356**}

Williams, C. O. (1912). A card reading. *Magician Monthly* **8** 67. {**425**}

Williams, David (1991). *Probability with Martingales.* Cambridge University Press, Cambridge, UK. {**117**}

Williams, J. D. (1954). *The Compleat Strategyst: Being a Primer on the Theory of Games of Strategy.* McGraw-Hill Book Co., Inc., New York. {**198, 743**}

Williams, John Burr (1936). Speculation and the carryover. *Quarterly Journal of Economics* **50** (3) 436–455. {**389**}

Wilson, Allan N. (1962). Advanced Problem 5036. *American Mathematical Monthly* **69** (6) 570–571. {**314**}

Wilson, Allan N. (1965). *The Casino Gambler's Guide.* Harper & Row, New York. Enlarged edition, 1970. {**236, 314, 482, 619, 621, 677, 681, 685**}

Wilson, Des (2008). *Ghosts at the Table: Riverboat Gamblers, Texas Rounders, Internet Gamers, and the Living Legends Who Made Poker What It Is Today.* Da Capo Press, New York. {**69, 731**}

Wilson, E. B. (1905). Problem 155. *American Mathematical Monthly* **12** (3) 76. {**522**}

Winterblossom, Henry T. (1875). *The Game of Draw-Poker, Mathematically Illustrated; Being a Complete Treatise on the Game, Giving the Prospective Value of Each Hand before and after the Draw, and the True Method of Discarding and Drawing, with a Thorough Analysis and Insight of the Game as Played at the Present Day by Gentlemen.* Wm. H. Murphy, Printer and Publisher, New York. {**732, 740**}

Wong, Chi Song (1977). A note on the central limit theorem. *American Mathematical Monthly* **84** (6) 472. {**73**}

Wong, Stanford (1982). What proportional betting does to your win rate. In Eadington, William R. (Ed.), *The Gambling Papers: Proceedings of the Fifth National Conference on Gambling and Risk Taking, Vol. 10. The Blackjack Papers,* pp. 59–66. Bureau of Business and Economic Research, University of Nevada, Reno. Also appeared in *Blackjack World* **3** 162–168, 1981. {**390**}

Wong, Stanford (1988). *Professional Video Poker.* Pi Yee Press, New York. Revised edition, 2000. {**571**}

Wong, Stanford (1994). *Professional Blackjack.* Pi Yee Press, Las Vegas. First published in 1975. {**683**}

Wong, Stanford (2005). *Wong on Dice.* Pi Yee Press, Las Vegas. {**71, 237, 521, 524**}

Yao, King (2005). *Weighing the Odds in Hold'em Poker.* Pi Yee Press, Las Vegas. {**741**}

Yardley, Herbert O. (1957). *The Education of a Poker Player: Including Where and How One Learns to Win.* Simon and Schuster, Inc., New York. {**741**}

Zender, Bill (2006). *Advantage Play for the Casino Executive.* Self-published. {**482, 522**}

Zender, Bill (2008). *Casino-ology: The Art of Managing Casino Games.* Huntington Press, Las Vegas. {**522, 683**}

Zerr, G. B. M. (1903). Solution of Problem 129. *American Mathematical Monthly* **10** (3) 81. {**522**}

Zhang, Lingyun and Lai, C. D. (2004). Variance of a randomly stopped sum. *Mathematical Scientist* **29** (1) 42–44. {**522**}

Index

Page numbers of primary entries are shown in **bold**. Names of people are indexed if they are explicitly mentioned prior to the bibliography.

A

a-break-and-interlace 394
a-shuffle 393
Abbott, Derek [1960–] 158
About, Edmond [1828–1885] 638
absorbing state 123
ace of hearts 478
acey-deucey 94
action 200
  alternative formulation 218
  composite wager 220
Adams, Ken [1942–] 679
Addario-Berry, Dana Louigi 315
Addington, Crandell [1938–] 734, 735
Adventures of Philip, The (Thackeray) 501
Agard, David B. 500
age process 146
Ahrens, Joachim H. 479
Akane, Kimo 522
al-Kashi, Jamshid *see* Kashi, Jamshid al-
Albert, A. Adrian [1905–1972] 746
Albigny, G. d' 115, 316
Aldous, David [1952–] 157, 424, 425
Alembert, Jean le Rond d' *see* d'Alembert, Jean le Rond
Alex [fictional] 193
Algoet, Paul H. 389, 390
all in 690, **691**
Allais, Maurice [1911–] 198
Allais's paradox **197**, 198
almost surely 42

Alspach, Brian [1938–] 66, 72, 500, 571, 740, 743
Alvarez, Al [1929–] 735–737
Always Bet on the Butcher (Nelson, Adams, King, & Nelson) 679
"Amarillo Slim" *see* Preston, Thomas
American Mathematical Monthly (periodical) 314, 522, 619
Anderson, Larry R. 239, 240
Andersson, Patrik [1981–] viii
Ankenman, Jerrod 741, 743, 744
Ankeny, Nesmith C. [1927–1993] 741
Anners, Henry F. 732
ante 690
any-craps bet 37, **231**
aperiodicity 137
Applications aux jeux de hasard (Borel) 740
Arbuthnot, John [1667–1735] 67, 522
arithmetic mean 38
arithmetic-geometric mean inequality 38
Arnous de Rivière, Jules *see* Gall, Martin
Arrow, Kenneth J. [1921–] 198
Arrow–Pratt measure of relative risk aversion **192**, 198
Ars Conjectandi (Bernoulli) v, 73, 734
Art de gagner à tous les jeux, L' (Robert-Houdin) 72
Asbury, Herbert [1891–1963] 517, 589, 590, 593, 677, 731, 732
Ashton, John [1834–1911] 478
Assaf, Sami H. 425

I

Y

Z